# Handbook of Humidity Measurement

## Methods, Materials, and Technologies

### Volume 3: Sensing Materials and Technologies

**Handbook of Humidity Measurement: Methods, Materials and Technologies, 3-Volume Set**
(ISBN: 978-1-138-29787-6)

**Handbook of Humidity Measurement, Volume 1: Spectroscopic Methods of Humidity Measurement**
(ISBN: 978-1-138-30021-7)

**Handbook of Humidity Measurement, Volume 2: Electronic and Electrical Humidity Sensors**
(ISBN: 978-1-138-30022-4)

**Handbook of Humidity Measurement, Volume 3: Sensing Materials and Technologies**
(ISBN: 978-1-138-48287-6)

# Handbook of Humidity Measurement

## Methods, Materials, and Technologies

### Volume 3: Sensing Materials and Technologies

Ghenadii Korotcenkov

CRC Press
Taylor & Francis Group
Boca Raton London New York

CRC Press is an imprint of the
Taylor & Francis Group, an **informa** business

CRC Press
Taylor & Francis Group
6000 Broken Sound Parkway NW, Suite 300
Boca Raton, FL 33487-2742

First issued in paperback 2021

© 2020 by Taylor & Francis Group, LLC
CRC Press is an imprint of Taylor & Francis Group, an Informa business

No claim to original U.S. Government works

ISBN 13: 978-1-138-48287-6 (hbk)
ISBN 13: 978-1-03-223909-5 (pbk)

DOI: 10.1201/9781351056502

---

**Library of Congress Cataloging-in-Publication Data**

---

Names: Korotcenkov, G. S. (Gennadii Sergeevich), author.
Title: Handbook of humidity measurement : methods, materials, and
technologies / Ghenadii Korotcenkov.
Other titles: Humidity measurement
Description: Boca Raton : CRC Press, Taylor & Francis Group, 2018-[2020] |
Includes bibliographical references. Contents: volume 1. Spectroscopic
methods of humidity measurement -- volume 2. Electronic and electrical
humidity sensors -- volume 3. Sensing materials and technologies.
Identifiers: LCCN 2017049880| ISBN 9781138300217 (hardback : alk. paper : v.
1) | ISBN 9781138300224 (hardback : alk. paper : v. 2) | ISBN
9781138482876 (hardback : alk. paper : v. 3)
Subjects: LCSH: Hygrometry. | Humidity. | Water vapor, Atmospheric. |
Spectrum analysis.
Classification: LCC QC915 .K825 2018 | DDC 551.57/10287--dc23
LC record available at https://lccn.loc.gov/2017049880

---

Visit the Taylor & Francis Web site at
http://www.taylorandfrancis.com

and the CRC Press Web site at
http://www.crcpress.com

# Contents

## SECTION I  Humidity-Sensitive Materials

## SECTION II  Sensor Technologies and Related Materials

# SECTION III   Calibration and Market of Humidity Sensors

# Preface to the Series

On account of unique water properties, humidity greatly affects living organisms, including humans and materials. The amount of water vapor in the air can affect human comfort, and the efficiency and the safety of many manufacturing processes, including drying of products such as paint, paper, matches, fur, and leather; packaging and storage of different products such as tea, cereal, milk, and bakery items; and manufacturing of food products such as plywood, gum, abrasives, pharmaceutical powder, ceramics, printing materials, and tablets. Moreover, industries discussed above are only a small part of the industries where the humidity should be controlled. In agriculture, the measurement of humidity is important for the plantation protection (dew prevention), the soil-moisture monitoring, and so on. In the medical field, humidity control should be used in respiratory equipment, sterilizers, incubators, pharmaceutical processing, and biological products. Humidity measurements at the Earth's surface are also required for meteorological analysis and forecasting, for climate studies, and for many special applications in hydrology, aeronautical services, and environmental studies, because water vapor is the key agent in both weather and climate.

This means that the determination of humidity is of great importance. Therefore, humidity control becomes imperative in all fields of our activity, from production management to creating a comfortable environment for our living, and for understanding the nature of the changes happening to the climate. Humidity can widely change. This means that there are necessary devices and sensors capable of carrying out humidity measurements in the entire range of possible changes in humidity. It is clear that these sensors and measurement systems must be able to work in a variety of climatic conditions, ensuring the functioning of the control and surveillance systems for a long time.

In the past decade, it was done frequently for the development of new methods for measuring humidity, improvements, and optimization of manufacturing technology of already developed humidity sensors, and for the development of different measuring systems with an increased efficiency. As a result, the field of humidity sensors has broadened and expanded greatly. At present, humidity sensors are being used in medicine, agriculture, industry, environmental control, and other fields (read Volume 1). Humidity sensors are being widely used for the continuous monitoring of humidity in diverse applications, such as the baking and drying of food, cigar storage, civil engineering to detect water ingress in soils or in the concrete of civil structures, medical applications, and many other fields. However, the process of developing new humidity sensors and improving older types of devices used for humidity measurement is still ongoing. New technologies and the toughening of ecological standards require more sensitive instruments with faster response times, better selectivity, and improved stability. It is therefore time to resume the developments carried out during this time and identify ways for further development in this area of research. This is important because currently too many approaches and types of devices that can be used for measuring humidity are proposed. They use different measuring principles, different humidity-sensitive materials, and various configurations of devices, making it difficult to conduct an objective analysis of these devices' capabilities. We hope that data presented in this book with detailed information on various types of humidity sensors that are developed by different teams, accompanied by an analysis of their strengths and weaknesses, will allow one to make the cross-comparison and the selection of suitable sensing methods for specific applications. As a result, all the conditions will be created for the development of devices to ensure accurate, reliable, economically viable, and efficient humidity measurements.

This issue is organized as follows. Taking into account the current trends in the development of instruments for measuring humidity, this publication is divided into three parts. The first volume, *Spectroscopic Methods of Humidity Measurement*, focuses on the review of devices based on optical principles of measurement, such as optical UV and fluorescence hygrometers and optical and fiber-optic sensors of various types. As it was indicated before, atmospheric water plays a key role in the climate. Therefore, various methods for monitoring the atmosphere have been developed in recent years, on the basis of measuring the absorption of the electromagnetic field in different spectral ranges. All these methods, covering the optical (FTIR and Lidar techniques), microwave, and THz ranges, are discussed in this volume, and their analyses of strengths and weaknesses are given. The role of humidity-sensitive materials in optical and fiber-optic sensors is also detailed. This volume also describes the reasons that cause us to control the humidity, features of water and water vapors, and units used for humidity measurement. I am confident that this information will be cognitive and interesting for readers.

The second volume, *Electronic and Electrical Humidity Sensors*, as the name implies, is entirely devoted to the consideration of different types of solid-state devices, the operating principles of which are based on other physical principles. Detailed information is given, including advantages and disadvantages of the capacitive, resistive, gravimetric, hygrometric, field ionization, microwave, solid-state electrochemical, and thermal-conductivity-based humidity sensors, followed by a relevant analysis of the properties of humidity-sensitive materials used for the development of such devices. Humidity sensors based on thin-film and field-effect transistors, heterojunctions, flexible substrates, and integrated humidity sensors are also discussed in this volume. This time is an age of automation and control. Therefore, in addition to interest in the properties of sensors, such as accuracy and long-term drift, there is also interest in durability in different environments, component size, digitization, simple- and quick-system integration, and last but not least, the price.

This means that modern humidity sensors should be able to integrate all these demands into one sensor. The experiment showed that these capabilities should be fully realized in electric and electronic sensors manufactured using semiconductor solid-state technology. Such humidity sensors can be fabricated with low cost and are used for moisture control more conveniently. Great attention in this volume is also paid to the consideration of conventional devices, which were used for the measurement of humidity for several centuries. It is important to note that many of these methods are widely used so far.

The title of the third volume, *Sensing Materials and Technologies*, indicates that this volume is focused on considering the properties of various materials suitable for the development of humidity sensors. Polymers, metal oxides, porous semiconductors (Si, SiC), carbon-based materials (black carbon, carbon nanotubes, and graphene), zeolites, silica, and some others, are included in the list of these materials. Features of fabrication of humidity sensors and related materials used in their manufacturing are also considered. Market forces naturally lead to ever-more specialized and innovative products. Sensors should be smaller in size, cheaper, more robust, and accurate in measurement; they should have better sensitivity and stability. This challenge requires new technological solutions, some of which, such as the integration and miniaturization, are considered in this volume. Specificity of the humidity-sensor calibration and the market of humidity sensors are also an object of analysis.

The content of these books shows that materials play a key role in humidity-sensor functioning and that the range of materials that can be used in the development of humidity sensors is very broad. Each material has its advantages and disadvantages, and therefore the selection of optimal sensing material for humidity sensors is a complicated and multivariate task. However, the number of published books or reviews describing the analysis of all possible materials through their application in the field of humidity sensors is very limited. Therefore, it is very difficult to conduct a comparative analysis of various materials and to choose a humidity-sensing material optimal for particular applications. The content of the present book contributes to the solution of this problem.

In these three volumes, the readers, including scientists, can find a comparative analysis of all materials acceptable for humidity-sensor design and can estimate their real advantages and shortcomings. Moreover, throughout these books, numerous strategies for the fabrication and characterization of humidity-sensitive materials and sensing structures, employed in sensor applications, are described. This means that one can consider the present books as a selective guide to materials for humidity-sensor manufacturing.

Thus, these books provide an up-to-date account of the present status of humidity sensors, from understanding the concepts behind humidity sensors to the practical knowledge necessary to develop and manufacture such sensors. In addition, these books contain a large number of tables with information necessary for humidity-sensor design. The tables alone make these books very helpful and comfortable for the user. Therefore, this issue can be utilized as a reference book for researchers and engineers as well as graduate students who are either entering the field of humidity measurement for the first time, or who are already conducting research in these areas but are willing to extend their knowledge in the field of humidity sensors. In this case, these books will act as an introduction to the world of humidity sensors, which may encourage further study and estimate the role that humidity sensors may play in the future. I hope that these books will also be useful to university students, postdocs, and professors. The structure of these books offers the basis for courses in the field of material sciences, chemical sensors, sensor technologies, chemical engineering, semiconductor devices, electronics, medicine, and environmental monitoring. We believe that practicing engineers, measurement experts, laboratory technicians, and project managers in industries and national laboratories who are interested in looking for solutions to humidity-measurement tasks in industrial and environmental applications, but do not know how to design or to select optimal devices for these applications, will also find useful information in these volumes that help them to make the right choices concerning technical solutions and investments.

**Ghenadii Korotcenkov**

# Preface

In the previous volumes, we examined in detail the specifics of using optical (Volume 1) and solid-state sensors (Volume 2) for measuring the humidity of the air. It was also shown that sensing materials play a key role in the successful implementation of humidity sensors, which, year by year, find wider application in various areas from environmental monitoring to climate control and product storage. Humidity sensors can also be found in various industries such as chemical and petrochemical industries, food and drink processing, semiconductor manufacturing, agriculture, fabrication industries, and power generation, as well as the motor, ship, and aircraft industries, where control of air humidity is necessary.

As follows from previous considerations, humidity sensors can be organized on different principles based on the various humidity-sensing effects. This means that each type of humidity sensor requires a specific humidity-sensing material with its specific set of parameters. However, the multidimensional nature of the interactions between function and composition, preparation method, and end-use conditions of sensing materials often make their rational design for application in specific humidity sensors very challenging. Moreover, the world of humidity-sensing materials is very broad, and practically all well-known materials could be used for humidity-sensor elaboration. Therefore, the selection of optimal sensing material for humidity sensor is a complicated and multivariate task.

However, one should note that the number of published review articles describing the analysis of materials through their application in the field of humidity sensors is very limited. Moreover, most of them are devoted to analysis of one specific sensing material, for example, polymers or metal oxides. Therefore, it is very difficult to conduct a comparative analysis of various humidity-sensing materials and to choose sensing material optimal for concrete application. As for the books published in this field, there are no such books at all. The only exception is the two-volume *Handbook of Gas Sensor Materials* published by Springer in 2013. However, this book analyzes sensing materials for all types of gas sensors. Only one small chapter is devoted to the consideration of sensing materials for humidity sensors. Taking this situation into account, I decided to fill this gap and devote the third volume, *Handbook of Humidity Measurements: Sensing Materials and Technologies*, to a detailed review of materials and technologies that are promising for use in humidity sensors. To solve this problem, Volume 3 is organized as follows.

It begins with a detailed review of materials acceptable for humidity-sensor development (Section I). This section analyzes polymers, metal oxides, black carbon, carbon nanotubes (CNTs), graphene, porous silicon, silica, various semiconductors, zeolite, metal phosphates, phosphorene, metal-organic frameworks (MOFs), supramolecular materials, and biomaterials, including descriptions their fundamental and humidity-sensing properties as well as advantages and limitations for application in humidity sensors. The stability of the characteristics and approaches used to optimize the humidity-sensing characteristics are also discussed in this section. There you can also find a description of the processes used to form porous silicon and $Al_2O_3$.

Then in Section II, the technologies developed for humidity-sensor fabrication are discussed. Here we analyze technologies that are used for synthesis, deposition, and processing of humidity-sensing materials, including micromachining technologies. Substrates, electrodes, protective covers, and filters in humidity sensors, as well as various humidity-sensor platforms such as micromachining membranes and microcantilevers, are also described. Features of the synthesis of 1D materials, nanofibers, mesoporous and hierarchical structures and their use in humidity sensors are also discussed in this section. Here one can find a description of the approaches used for packaging, air cleaning, and storage of humidity sensors.

As for Section III, here we summarize our consideration of humidity sensors and humidity-sensitive materials. This section can be thought of as a guide for humidity-sensor selection and operation, where one can find practical advice for humidity-sensor selection, installation, testing, and use. A comparative analysis of humidity sensors of various types and their advantages and shortcomings are also given in this section. A general description of the calibration process and an analysis of the market of electronic humidity sensors and conventional humidity-measuring devices are also provided.

Analyzing the contents of this book, we can conclude that this volume can justifiably be regarded as an encyclopedia of humidity-sensitive materials and sensor technologies used in the fabrication of humidity sensors, and methods acceptable for their testing.

Despite the fact that this book is devoted to humidity sensors, it can be used as a basis for studying the principles of functioning and development of any gas and vapor sensors.

**Ghenadii Korotcenkov**

# Acknowledgments

My sincere gratitude goes to CRC Press for the opportunity to write this book. I would also like to give acknowledgment to Gwangju Institute of Science and Technology, Gwangju, South Korea, and Moldova State University, Chisinau, Rep. of Moldova, for inviting me and supporting my research in various programs and projects. I would like to thank my wife and the love of my life, Irina Korotcenkov, for always being there for me, inspiring and supporting all my endeavors. She gives me purpose, motivates me to continue my work, and makes my life so much more exciting. Also, I am grateful to my daughters, Anya and Anastasia, for being a part of my life and for encouraging my work. My special thanks go to my friends, colleagues, and coauthors for their support and collaboration over the years. Great thanks to all of you. This would not be possible without you being by my side.

# Author

**Ghenadii Korotcenkov** earned his PhD in material sciences from the Technical University of Moldova, Chisinau, Moldova, in 1976 and his doctor of science degree (doctor habilitate) in physics from the Academy of Science of Moldova in 1990 (Highest Qualification Committee of the USSR, Moscow). He has more than 45 years of experience as a scientific researcher. For a long time, he has been the leader of the gas-sensor group and manager of various national and international scientific and engineering projects carried out in the Laboratory of Micro- and Optoelectronics, Technical University of Moldova. His research has been receiving financial support from international foundations and programs, such as the CRDF, MRDA, ICTP, INTAS, INCO-COPERNICUS, COST, and NATO. From 2007 to 2008, he was an invited scientist at the Korea Institute of Energy Research, Daejeon, South Korea. Then, until the end of 2017, Dr. G. Korotcenkov was a research professor at the School of Materials Science and Engineering at Gwangju Institute of Science and Technology, Gwangju, South Korea. Currently, Dr. G. Korotcenkov is the chief scientific researcher at the Department of Physics and Engineering at the Moldova State University, Chisinau, the Rep. of Moldova.

Specialists from the former Soviet Union know Dr. G. Korotcenkov's research results in the field of study Schottky barriers, MOS structures, native oxides, and photoreceivers on the basis of III–Vs compounds, such as InP, GaP, AlGaAs, and InGaAs. His present scientific interests, starting from 1995, include materials science, focusing on the metal-oxide film deposition and characterization, surface science, and the design of thin-film gas sensors and thermoelectric convertors. These studies were carried out in cooperation with scientific teams from Ioffe Institute (St. Petersburg, Russia), University of Michigan (Ann Arbor, USA), Kiev State University (Kiev, Ukraine), Charles University (Prague, Czech Republic), St. Petersburg State University (St. Petersburg, Russia), Illinois Institute of Technology (Chicago, USA), University of Barcelona (Barcelona, Spain), Moscow State University (Moscow, Russia), University of Brescia (Brescia, Italy), Belarus State University (Minsk, Belarus), and South-Ukrainian University (Odessa, Ukraine).

Dr. G. Korotcenkov is the author or editor of 39 books, including the 11-volume *Chemical Sensors* series published by the Momentum Press (USA), the 15-volume *Chemical Sensors* series published by Harbin Institute of Technology Press (China), three-volume *Porous Silicon: From Formation to Application* published by CRC Press (USA), two-volume *Handbook of Gas Sensor Materials* published by Springer (USA), and the three-volume *Handbook of Humidity Measurement*, which is published by CRC Press (USA). In addition, at present, Dr. G. Korotcenkov is a series' editor of the *Metal Oxides* series, which is being published by Elsevier. Starting from 2017, already 16 volumes have been published within the framework of that series.

Dr. G. Korotcenkov is the author and coauthor of more than 600 scientific publications, including 30 review papers, 38 book chapters, and more than 200 articles published in peer-reviewed scientific journals (h-factor = 42 [Scopus] and h-factor = 50 [Google Scholar citation]). In the majority of publications, he is the first author. Besides, Dr. G. Korotcenkov is a holder of 17 patents. He has presented more than 250 reports at national and international conferences, including 17 invited talks. Dr. G. Korotcenkov was co-organizer of several international conferences. His name and activities have been listed by many biographical publications, including *Who's Who*. His research activities are honored by an Award of the Academy of Sciences of Moldova (2019) and of the Supreme Council of Science and Advanced Technology of the Republic of Moldova (2004) for valuable results obtained in the field of exact and engineering sciences; Prize of the Presidents of the Ukrainian, Belarus, and Moldovan Academies of Sciences for significant scientific achievements obtained during joint development of a scientific problem (2003); and National Youth Prize of the Republic of Moldova in the field of science and technology (1980), among others. Dr. G. Korotcenkov also received a fellowship from the International Research Exchange Board (IREX, United States, 1998), Brain Korea 21 Program (2008–2012), and Brainpool Program (Korea, 2007–2008 and 2015–2017).

# Section I

## Humidity-Sensitive Materials

# 1 Polymers

## 1.1 INTRODUCTION

### 1.1.1 POLYMERS AS SENSING MATERIALS

Polymers form a large class of materials that are promising for a wide range of applications. In particular, a large number of articles and reviews were published on chemical sensors, including humidity sensors, where many polymers were used as sensing materials (Armstrong and Horvai 1990; Bidan 1992; Adhikari and Majumdar 2004; Persaud 2005; Adhikari and Kar 2010). Unlike discrete low-molecular-weight compounds, polymers are macromolecules. With the exception of a few, polymers are organic macromolecules consisting of carbon and hydrogen atoms in a significant percentage with some heteroatoms, such as nitrogen, oxygen, sulfur, phosphorous, and halogens, as minor constituents. In terms of size and molecular weight, in general, polymers are more than a million times larger than small molecular compounds. A polymer molecule is formed by the repeated unit of a large number of reactive small molecules in a regular sequence (see Figure 1.1). The repeated unit in the backbone of the polymer molecule is known as the *mer unit*, and the reactive small molecule from which the polymer is formed is called the *monomer*. The simplest example of a polymer is polyethylene, in which the ethylene moiety is the mer unit.

Properties of polymers, in general, depend on their chemical composition, molecular structure, molecular weight, molecular-weight distribution, and morphology. Morphologically, polymers are quasi-crystalline in nature, having small crystallites dispersed in an amorphous matrix in which one molecule may extend from one amorphous region to a distant amorphous region while passing through several crystallite regions. Both the bulk and the surface of a polymer sample may or may not contain active functional groups. Although primary covalent bonds are prevalent in polymers, secondary bonding influences both their processing and their functional performance. Extensive secondary bonding interaction can be a major cause of insolubility of a polymer in solvents, which may restrict its processability to produce suitable devices, such as gas and humidity sensors.

As it was shown in previous volumes, there are a large number of polymers, which can be used in solid-state humidity sensors as humidity-sensitive material. Some of them are listed in Table 1.1.

As it was shown in Volume 2, polymer-based humidity sensors use different principles of humidity detection, which means that polymers that allow implementation of these devices must have a different set of mechanical, chemical, electrophysical, electrochemical, piezoelectric, and optical properties (read Volume 2 of this series). In this regard, as a class, polymers are unique compared to other materials. Experience has shown that this situation has become possible, since polymers have high tailorability of their molecular structure and composition. As a result, a wide range of properties can be easily obtained with polymers (Adhikari and Kar 2010). Some polymers retain their size and volume when interacting with water vapor, while in other polymers the swelling effect appears (read Chapter 10, Volume 1). The first polymers are mainly used in capacitive sensors (Chapter 10, Volume 2), while the swelling is the main mechanism of response in conductometric humidity sensors based on nanocomposites (Chapter 11, Volume 2), which include nonconductive polymers and conductive nanoparticles.

Polymers can also have both hydrophobic and hydrophilic surface properties, that is, they interact with water vapor in different ways. Some polymers may have insulating properties (see Table 1.2), while the others can exhibit ionic or electronic conduction properties. In general, polymers are electrically insulating in nature due to the nonavailability of free electrons, since the four valence electrons of carbon are fully saturated. However, some organic polymer molecules are semiconducting in nature, by virtue of their extended π-electron conjugation along the backbone chain of the macromolecule. The ability of conducting polymers to conduct electricity depends on the alternating double bond–single bond structure in the polymer backbone, coupled with the formation of some charged centers on the chain by partial oxidation. In other polymers, which are polyelectrolytes, conductivity appears after the introduction of ionizable functional groups (read Chapter 11, Volume 2). The introduction of such extra charges in the polymer by doping (in analogy to inorganic semiconductors) alters the conductivity of such polymers from almost insulators to something approaching a metallic conductor. Due to the more reactive nature of the ionic groups in polyelectrolytes and π-electron conjugation in conducting polymers, these are the important sensing sites for corresponding analytes. Because of its long backbone chain and flexible nature, the polymer can accommodate a large amount of foreign substance in the form of fine particles (as filler) or fibers or even a highly viscous liquid. Thus, the polymer in its solid state contains ample free volume in the bulk. Depending on its affinity toward a foreign agent, the polymer can hold it for quite a long period.

In addition, polymers are easy to synthesize and process to develop a suitable device for sensor application (Chapter 19, Volume 3). Unlike metal and metal-oxide semiconductor materials, there is no need for special clean rooms, high temperatures, or special high-cost technological processes for the

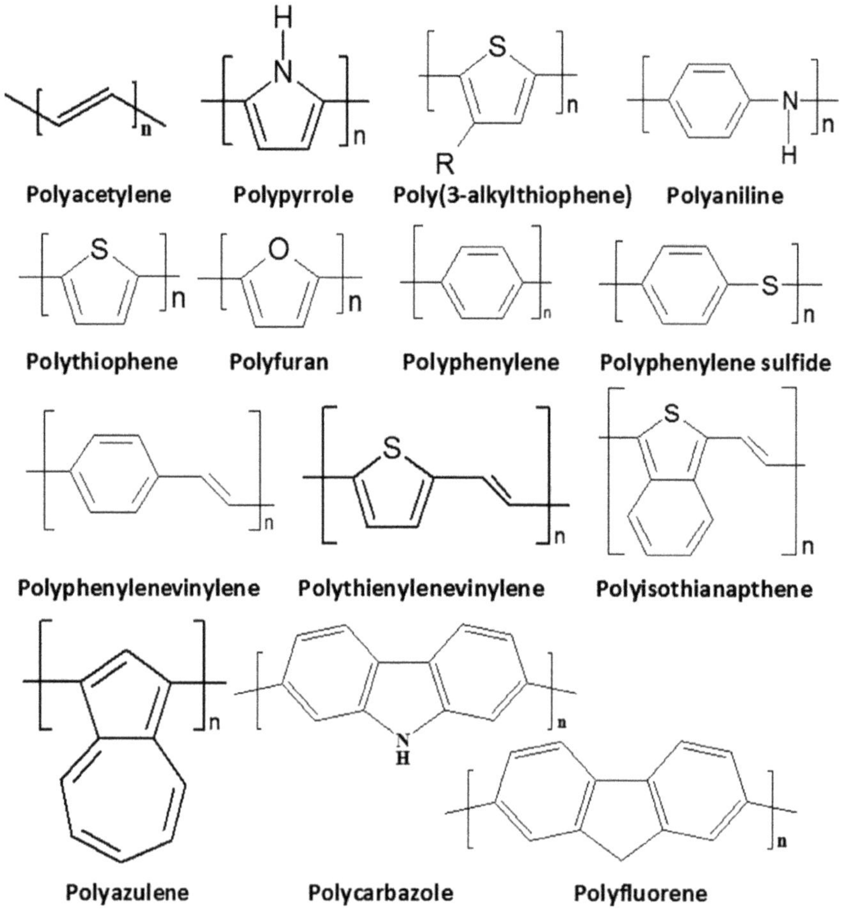

**FIGURE 1.1**   Repeating unit structures of some common conducting polymers.

## TABLE 1.1
## Polymers Used for Humidity Sensor Fabrication

| Type of Humidity Sensor | Polymer Tested as Humidity Sensitive Material |
|---|---|
| Capacitive | Polyimide (PI); cellulose acetate butyrate (CAB); poly(methyl methacrylate) (PMMA); poly(vinyl crotonate); poly(ethyleneterephthalate) (PETT); polyethersulphone (PES); polysulfone (PSF); divinyl siloxane benzocyclobutene (BCB); hexamethyldisilazane (HMDSN); polyphenylacetylene (PPA); polydimethylphosphazene (PDMP); etc. |
| Resistive | |
| *Polyelectrolytes* | Poly(dimethyldiallylammonium chloride); poly(sodium p-styrene sulfonate); poly(propargyl alcohol) doped with sulfuric acid; poly(p-diethynylbenzene-co-propargyl alcohol) doped with iron trichloride; poly(2-acrylamido-2-methylpropane sulfonic acid); 2-amylamido-2-methylpropane sulfonic acid; etc. |
| *Ionic conductive* | PVA: KOH; Nafion: ($H^+$, $Li^+$, and $Na^+$); etc. |
| *Semiconductive* | Polyaniline (PANI); poly(p-diethynylbenzene) (PDEB); poly(propargyl benzoate) (PPBT); p-diethynylbenzene-co-propargyl alcohol; poly(3,4-ethylenedioxythiophene) (PEDOT); etc. |
| *Composites* | PVA/carbon black (CB); poly(4-vinylpyridine) (PVP)/CB; $SiO_2$/Nafion; $BaTiO_3$/polystyrene sulfonic sodium (PSS); $SiO_2$/poly(2-Acrylamido-2-methylpropane sulfonate) (poly(AMPS)); ZnO/sodium polystyrenesulfonate (NaPSS); etc. |
| Mass sensitive | |
| *SAW* | Polyvinyl alcohol (PVA); polyethynylfluorenol (PEFL); cellulose acetate; hexamethyldisiloxane (HMDSO); polyimide; etc. |
| *QCM* | Polyimide; polyvinyl alcohol (PVA); etc. |
| *Cantilever, membrane* | Polyimide; polyvinyl-difluorene (PVDF); etc. |

**TABLE 1.2**

**Properties of Polymeric Insulating Materials Used for Humidity Sensors Fabrication**

| Property | Nylon | PI | PE | PTFE | PVC | PVDF |
|---|---|---|---|---|---|---|
| Density, $\rho_m$ (kg/m³) | 1,150 | 1,300 | 935 | 2,200 | 1,350 | 1,760 |
| Maximum working point, $T_{max}$ (°C) | 100 | — | 71–93 | 260 | 70–74 | 150 |
| Thermal conductivity, $\kappa$ (W/mK) | ~0.26 | 0.15 | ~0.36 | ~0.24 | 0.16 | 0.1 |
| Specific heat capacity, $C_p$ (J/K kg) | ~1,750 | 1,100 | 1,900 | 1,050 | ~1,000 | — |
| Temperature expansivity, $\alpha_l$ ($10^{-5}K^{-1}$) | 28 | — | 14–16 | 10 | 5–18 | 8–14 |
| Dielectric constant, $\varepsilon_r$ | 3.7–5.5 | — | 2.3 | 2 | 3.0–4.0 | 2.9 |
| Bulk resistivity, $\Omega \cdot m$ | $>10^8$ | $~10^{17}$ | $>10^{14}$ | $~10^{16}$ | $>10^{12}$ | $>10^{12}$ |
| Young's modulus, $E_m$ (GPa) | 1–4 | ~3.1 | 0.4–1.3 | 0.4 | 2.9 | 2.1 |
| Tensile strength, $Y_m$ (MPa) | 50–90 | 69–104 | 8–24 | 10–31 | 34–62 | 36–56 |

*Source:* Data extracted from Gardner J.W. et al., Microsensors, *MEMS, and Smart Devices*, John Wiley & Sons, Chichester, UK, 2002. These values are intended only as a guide, and they should be validated against other sources.

Nylon is a thermoplastic silky material based on aliphatic or semi-aromatic polyamides; PE—polythene; PI—polyimide; PTEF—polytetrafluoroethylene, trend name is Teflon; PVC—polyvinyl chloride; PVDF—polyvinylidene fluoride or polyvinylidene difluoride.

manufacture of sensor devices using polymers. The simplicity of creating new functional groups on the polymer backbone, their ability to respond to redox systems, their charge-carrying capability, and their ability to respond to optical stimulation are other advantages of their use in sensor devices.

### 1.1.2 STABILITY OF POLYMERS

In the first and second volumes of the present *Handbook of Humidity Measurements*, we have already discussed the main properties of humidity-sensitive polymers, which allow the implementation of a large number of humidity sensors operating on the basis of various principles. Therefore, in this chapter, we will focus on the stability of their properties. It is stability that largely determines the possibility of using one or another polymer in the development of humidity sensors intended for the market (Korotcenkov 2013). It is known that organic materials are inherently more susceptible to chemical degradation under the influence of oxygen or water than inorganic materials.

In polymers, some percentages of heteroatoms such as nitrogen, oxygen, sulfur, phosphorous, and halogens are present along with a long chain of carbon and hydrogen. The carbon–carbon or carbon–hydrogen bonds are comparatively more stable than the bonds between carbon and the heteroatoms (Adhikari and Kar 2010). Therefore, the polymer that has very good electrical conductivity may not have good stability in the ambient environment, or its stability may decrease when it is in contact with the analyte compound. This is because unsaturated bonds in conducting polymers are often very reactive when exposed to environmental agents, such as oxygen or moisture. Apart from the inherent stability of the polymers, the stability of some foreign chemical compounds, which are used as dopants in conducting polymers, are also important. These dopants may not be very stable within the polymer matrix and may also react with environmental agents.

Stability and degradation of the polymers is also important when polymers can be exposed to chemical environments, especially at high temperature (Allen and Edge 1992; Adhikari and Kar 2010). The polymer must be resistant to degradation or dissolution in organic solvents but must interact with the reactive target gas or vapor, such as water vapor. In addition, the selected polymer must maintain its intrinsic sensing characteristics over a wide temperature range. The polymers must also maintain their properties during the manufacturing process of the sensors themselves. As is known, the sensitive layer in the sensors during their manufacture can be subjected to various influences, including high-temperature treatments. Therefore, information on the stability and degradation of polymers intended for use in chemical sensors such as gas and humidity sensors is important and needs attention.

According to Ehrenstein and Pongratz (2013), the behavior of polymer materials during operation and manufacture of sensors depends on many factors, including the specific thermal properties and the duration and type of the temperature exposure, as well as the loading. Material design and processing are also crucial. Ehrenstein and Pongratz (2013) believe that changing the properties of polymers is associated with processes such as softening, physical aging, and chemical degradation. Softening behavior plays a decisive role mainly under high thermal loads (such as when plastics substrates are welded) and is determined by the glass transition temperature in amorphous polymers and by the melting temperature in semicrystalline plastics. The softening behavior is characterized by the heat-distortion temperature. This is the temperature at which a test specimen is deformed up to a limit value under a certain external load. Heat-distortion temperature also depends on the crystalline structure, orientations, and residual stresses as well as on the thermal-expansion behavior.

In the process of physical aging, a morphological change occurs. Physical aging processes are relaxation processes of residual stresses and orientations, post-crystallization, separation, and agglomeration as well as loss of plasticizer, plasticizer migration, or plasticizer extraction (Ehrenstein and Pongratz 2013). Physical aging processes are always the result of thermodynamically unstable states caused by process-dependent cooling conditions during the manufacture of polymer products. Although physical aging can occur without any external influence, its rate is found to be dependent on several factors, such as temperature, gas environment, and polymer structure. In particular, physical aging processes are accelerated strongly by elevated temperatures. More importantly, it was observed that physical aging occurs faster in a thin polymer film (that is, the rate of aging in a film is inversely proportional to its thickness), which is obvious for both conventional polymers, such as polyimide and polysulfone and high free-volume polymers such as poly [1-(trimethylsilyl)-1-propyne], Teflon AF, Hyflon AD, and various polymers of intrinsic microporosity (Low et al. 2018). It is believed that the accelerated aging of thin films (50–800 nm) is due to increased mobility near the surface, or the diffusion of free volume to the surface of the film, which allows the polymer to achieve a lower free-volume state faster than bulk samples (Low et al. 2018). Post-crystallization takes place in semicrystalline plastics mainly at elevated temperatures. However, this can occur at room temperature as well. The change in physical structure caused by physical aging processes is often accompanied by a change in size, which creates mechanical stresses in the material due to inhibition of elongation or shrinkage. Cracks or fractures are the result of such stresses. Physical aging processes also change properties, such as water absorption and diffusion, and oxygen diffusion (Low et al. 2018). This has consequences for mechanical properties and the progress of chemical-degradation processes. Oxygen diffusion decreases with increasing crystallization and orientation. This process is reversible through melting.

It is known that the fundamental weaknesses of polymer materials are caused by their macromolecular structure and relatively weak bonding forces. Polymers are thus subject to decidedly greater influences by the effects of heat, light, and oxygen, than are metals or inorganic materials such as metal oxides (Ehrenstein and Pongratz 2013). Chemical aging processes in polymer materials cause changes on the molecular scale and lead to the chain cleavage but also to cross-linking and cyclization. This process is not reversible via melting. In chemical aging processes (chemical degradation), the macromolecules are degraded either beginning at the surface of the polymer layer or homogeneously. These and additional external influences, such as chemicals and radiation, initiate chemical aging processes that cause the deterioration of properties and thus shorten the life span of polymer-based sensors.

According to Ehrenstein and Pongratz (2013), the short-term behavior of polymers under thermal loading is determined by their softening behavior and physical aging processes, whereas the long-term behavior is mainly dominated by chemical degradation.

It should be noted that unlike the measure of the magnitude of sensor response, there is no single measure for

stability, and for this reason most of the reports on stability of polymer gas sensors employ different measures and different experimental conditions. A number of studies have been carried out, and they showed that the stability/degradation issue is rather complicated and certainly not yet fully understood, though progress has been made (Allen and Edge 1992; Scurlock et al. 1995; Abdou et al. 1997; Cumston and Jensen 1998; Dam et al. 1999; Katz et al. 2001; Yang et al. 2005; Jørgensen et al. 2008; Ehrenstein and Pongratz 2013). It is thus impossible to directly compare stability reports due to the differences in materials, conditions, data acquisition, etc. For example, Swaidan et al. (2014) suggested that the gas permeability of a high free-volume polymer should be reported about 10–15 days after film conditioning, as there existed a quasi-steady state after a rapid initial drop in the gas permeability and the increase in the gas selectivity. It is assumed that reporting the gas permeability within the first 2 weeks does not truly represent the performance of the polymer film because the permeation data are highly dependent on the time at which the permeation is measured and the conditions of the environment. In particular, aging of a high free-volume polymer occurs rapidly in the first few days/weeks, followed by slower aging over several years. Therefore, in this chapter the analysis of polymer-based sensor's stability will have only general character.

## 1.2 POLYMER DEGRADATION

Polymer degradation is a change in the properties—tensile strength, color, shape, conductivity, etc.—of a polymer or polymer-based product under the influence of one or more environmental factors, such as heat, light, or chemicals. Degradation agents and polymers most susceptible to their influence are given in Table 1.3.

Experiments showed that polymers, containing sites of unsaturation, such as polyisoprene and the polybutadienes, are most susceptible to oxidation by oxygen and ozone. However, most other polymers, including polystyrene, polypropylene, nylons, polyethylenes, and most natural and naturally derived polymers also show some susceptibility to such degradation (Carraher 2008). Most heterochained polymers, including condensation polymers, are susceptible to aqueous-associated acid or base degradation. Some polymer-deterioration reactions occur without loss in molecular weight. These include a wide range of reactions where free radicals (most typical) or ions are formed and a cross-linking or other nonchain session reaction occurs. Cross-linking discourages chain and segmental chain movement. At times, this cross-link is desirable, for example, in permanent press fabric and in elastomeric materials. Often cross-links lead to increased brittleness beyond the desired. Some degradation reactions occur without increasing cross-linking or reducing chain length. Thus, with minor amounts of HCl, water, ester, etc., elimination can occur with vinyl polymers, giving localized sites of double-bond formation. Since such sites are less flexible and more susceptible to further degradation, these reactions are usually considered as unwanted.

## TABLE 1.3
### Degradation Agents and Polymers Most Susceptible to Their Influence

| Degradation Agent | Most Susceptible Polymer Types | Examples |
|---|---|---|
| Acids and bases | Heterochain polymers | Polyesters, polyurethanes |
| Moisture | Heterochain polymers | Polyesters, nylons, polyurethanes |
| High-energy radiation | Aliphatic polymers with quaternary carbons | Polypropylene, LDPE, PMMA, poly(alpha-methylstyrene) |
| Ozone | Unsaturated polymers | Polybutadienes, polyisoprene |
| Organic liquids/vapors | Amorphous polymers | |
| Biodegradation | Heterochain polymers | Polyesters, nylons, polyurethanes |
| Heat | Vinyl polymers | PVC, poly(alpha-methylstyrene) |
| Mechanical (applied stresses) | Polymers below $T_g$ | |

*Source:* Data extracted from Carraher, C.E. Jr., *Polym. Chem.*, CRC Press, Boca Raton, FL, 2008.

Taking into account the mechanism of environment influence on polymer properties, one can select the following degradation reactions taking place in polymers (Herman and Sheldon 1965; Atkins and Batich 1993; White and Turnbull 1994; George and Goss 2001; Ray and Cooney 2018):

1. Thermal degradation
2. Oxidative degradation
   - Photo-oxidation
   - Thermal oxidation
3. Hydrolysis degradation

It should be noted that degradation of polymers also occurs under the influence of radiation exposure (Davenas et al. 2002) and microbial attack (biodegradation) (Bilibin and Zorin 2006; Shah et al. 2008). Mechanical degradation caused by mechanical stress is also possible. It should also be borne in mind that in most cases polymer degradation often is caused by a combination of degradation agents and can involve several chemical and mechanical mechanisms (Figure 1.2). A typical example of this is the simultaneous action of light, oxygen, and other atmospheric agents, or the simultaneous influence of heat, mechanical stresses, and oxygen. At that, the contribution of each of these factors cannot always be established.

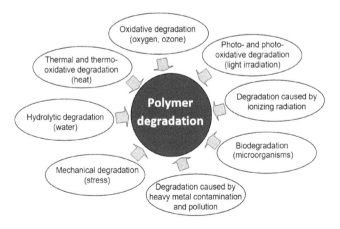

**FIGURE 1.2** The factors affecting the aging of polymer materials.

### 1.2.1 THERMAL DEGRADATION

The way of polymer degradation under the influence of thermal energy in an inert atmosphere is determined, on the one hand, by the chemical structure of the polymer itself, and on the other, by the presence of traces of unstable structures (impurities or additions). Thermal degradation does not occur until the temperature is so high that primary chemical bonds are separated. Table 1.4 summarizes data related to polymer thermal degradation. Degradation of polymers begins typically at temperatures around 150°C–200°C and the rate of degradation increases as the temperature increases (Beyler and Hirschler 1995; Van Krevelen and Nijenhuis 2009; Ray and Cooney 2018). For many polymers, thermal degradation is characterized by the breaking of the weakest bond and is consequently determined by a bond dissociation energy. According to Beyler and Hirschler (1995), there are four general classes of chemical mechanisms important in the thermal decomposition of polymers: (1) random-chain scission, in which chain scissions occur at apparently random locations in the polymer chain; (2) end-chain scission, in which individual monomer units are successively removed at the chain end; (3) chain-stripping, in which atoms or groups not part of the polymer chain (or backbone) are cleaved; and (4) cross-linking, in which bonds are created between polymer chains. It was established that in all cases the thermal degradation usually is a multistep, free-radical reaction with all the general features of such reaction mechanisms: initiation, propagation, branching, and termination steps.

Although thermal degradation of polymers and polymer composites occurs due to the application of heat, there are several factors that can influence this degradation process (Ray and Cooney 2018; Król-Morkisz and Pielichowska 2019). As a result, the nature of thermal degradation of different polymers or their composites is also different. Even within the same type of polymer, such changes can be seen. The main factors that play a decisive role in the degradation process are the macromolecular structure, environmental conditions, and additives.

*Macromolecular structure:* The fundamental reason of the degradation process involves bond scission, which is related to the bond energy. Consequently, all issues

## TABLE 1.4
### Data Related to Thermal Degradation of Polymers

| Polymer | $T_{d,o}$ (K) | $T_{d,1/2}$ (K) |
|---|---|---|
| Poly(vinyl acetate) | — | 542 |
| Poly(vinyl chloride) | 443 | 543 |
| Poly(vinyl alcohol) | 493 | 547 |
| Poly(a-methyl styrene) | — | 559 |
| Poly(propylene oxide) | — | 586 |
| Poly(isoprene) | 543 | 596 |
| Cellulose | 500 | 600 |
| Poly(methyl acrylate) | — | 601 |
| Poly(methyl methacrylate) | 553 | 610 |
| Poly(ethylene oxide) | — | 618 |
| Poly(isobutylene) | — | 621 |
| Poly(m-methyl styrene) | — | 631 |
| Poly(styrene) | 600 | 637 |
| Poly(vinyl cyclohexane) | — | 642 |
| Poly(chloro-trifluoro ethylene) | — | 653 |
| Poly(propylene) | 593 | 660 |
| Poly(vinyl fluoride) | 623 | 663 |
| Poly(ethylene) (branched) | 653 | 677 |
| Poly(butadiene) | 553 | 680 |
| Poly(trifluoro ethylene) | 673 | 685 |
| Poly(methylene) | 660 | 687 |
| Poly(hexamethylene adipamide) | 623 | 693 |
| Poly(benzyl) | — | 703 |
| Poly($\varepsilon$-caproamide) (Nylon 6) | 623 | 703 |
| Poly(p-xylylene) = poly(p-phenylene-ethylene) | — | 715 |
| Poly(ethylene terephthalate) | 653 | 723 |
| Poly(acrylonitril) | 563 | 723 |
| Poly(dian terephthalate) | 673 | ~750 |
| Poly(dian carbonate) | 675 | ~750 |
| Poly(2,6-dimethyl p-phenylene oxide) | 723 | 753 |
| Poly(tetrafluoro ethylene) | — | 782 |
| Poly(p-phenylene terephthalamide) | ~720 | ~800 |
| Poly(m-phenylene 2,5-oxadiazole) | 683 | ~800 |
| Poly (pyromellitide) (Kapton) | 723 | ~840 |
| Poly(p-phenylene) | >900 | >925 |

*Source:* Data extracted from Madorsky, S.L. and Straus, S., *SCI Monograph*, 13, 60–74, 1961; Arnold, C., *J. Polym. Sci.: Macromol. Rev.*, 14, 265–378, 1979; Van Krevelen, D.W. and Nijenhuis, K.Te., *Properties of Polymers: Their Correlation with Chemical Structure; Their Numerical Estimation and Prediction from Additive Group Contributions*, Elsevier, Amsterdam, the Netherlands, 2009.

$T_{d,o}$ is the temperature of initial decomposition, that is, this is the temperature at which the loss of weight during heating is just measurable; $T_{d,1/2}$ is the temperature of half decomposition, that is, this is the temperature at which the loss of weight during pyrolysis reaches 50% of its initial value.

directly related to the binding energies in the macromolecular structure are responsible, namely, elemental constituents of the polymer, the type of covalent or noncovalent bonds, and the degree of unsaturation. For example, the higher thermal stability of polytetrafluoroethylene (PTFE) as compared to polyethylene (PE) can be explained by the difference in bond energies of C-F and C-H (116 and 97 kcal/mol,

respectively). The influences of intermolecular forces due to secondary valence bonds as a result of induction, dipole-dipole interaction, and hydrogen bonding are important for common macromolecules and especially for functional polymers. For example, the presence of hydrogen bonding helps enhance the thermal stability of polyamides. High-resonance energy also contributes to high thermal stability as it happens for aromatic polymers (Ray and Cooney 2018).

***Environmental conditions:*** Experiments have shown that thermal degradation of polymeric materials are highly influenced by the environmental conditions, namely, humidity, temperature, and UV light (read the following sections). A synergistic effect of high temperature and the factors mentioned above can cause rapid degradation of polymer films.

Regarding ***additives***, then the impurities such as catalysts, residues of polymerization catalysts, or additives incorporated in polymeric materials to meet the requirement of particular application (flexibility–plasticizers, strength–fillers, conductivity-ionic liquids, color–dyes, or pigments) and for safe processing can also influence the degradation process (Ray and Cooney 2018). Some additives may affect the aging process and act as stabilizers or prodegradants in polymer materials. For example, inorganic nanoparticles (nanoclays, metal oxides, carbon-based nanostructures, etc.) can improve the thermal stability of polymers (Król-Morkisz and Pielichowska 2019). However, the compatibility between inorganic nanoadditives and polymer matrix in nanocomposites is often quite poor. The nanoparticles are prone to aggregation and, in consequence, the dispersion of nanoadditives in polymer matrix is very difficult (Khrenov et al. 2007). As a result, the influence of nanoadditives on the improvement of composite material thermal stability is limited. The effect of *other additives* on the stability of polymers is discussed in the following sections.

### 1.2.2 Oxidative Degradation

The polymer can also be degraded by chemical changes due to reaction with components in the environment (Van Krevelen and Nijenhuis 2009). The most important of these degrading reagents is oxygen (air). It was established that polymer degradation is almost always faster in the presence of oxygen, due primarily to the auto-accelerating nature of reactions between oxygen and carbon-centered radicals. Hence, most oxidation reactions are of an autocatalytic nature. If the oxidation is induced by light, the phenomenon is called photo-oxidation. If the oxidation is induced by purely thermal factors, the term *thermal oxidation* is used. In photo-oxidation, a radical is formed by absorption of $h\upsilon$ and in thermal oxidation by $\Delta T$, shear, or even by residues of the polymerization catalysts. Under the influence of these factors on the polymer-free radicals are formed, which together with oxygen initiate a chain reaction. The mechanism of polymer degradation in the presence of oxygen is shown in Figure 1.3. It is seen that interactions with oxygen lead to an increase in the concentration

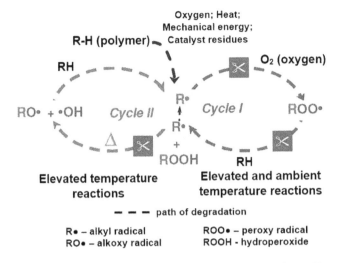

**FIGURE 1.3** Polymer autooxidation cycle. (Idea from Van Beusichem, B. and Ruberto, M.A., Introduction to polymer additives and stabilization. Product quality research institute, (http://pqri.org/wp-content/uploads/2015/08/pdf/Polymer_Additives_PQRI_Poster.pdf; http://www.everspringchem.com.)

**TABLE 1.5**

**Products of Polymer Degradation**

| Product/Function | Examples |
|---|---|
| Active products: | Alcyl radical (R•) (carbon-centered free radical) |
| Intermediates of | Peroxy radical (R-OO•) (oxygen-centered radical) |
| degradation | Alkoxy radical (R-O•) |
| | Hydroperoxide (R-OOH → R-O• + •OH) |
| Inactive products: | Alcohol (R-OH) |
| Influence | Aldehyde (R-CHO) |
| organoleptics | Ketone (R-C=O-R) |
| Modified properties | Polymer |

*Source:* Data extracted from Van Beusichem, B. and Ruberto, M.A., Introduction to polymer additives and stabilization. Product quality research institute (http://pqri.org/wp-content/uploads/2015/08/pdf/Polymer_Additives_PQRI_Poster.pdf), 2015.

of polymer alkyl radicals (R•), and therefore to higher levels of scission and cross-linking products. Possible products of polymer degradation are presented in Table 1.5. We need to say that fragmentation reactions of oxygen-centered radicals (RO•) can yield new species (oxidation products), not found in polymers processed under air-free conditions.

One should note that the presence of ozone in the air, even in very small concentrations, markedly accelerates the aging of polymeric materials. Ozone mainly affects vulcanized rubbers with unsaturation in the main polymer chain and causes cracking in stretched form in rubber. Ozone normally attacks the unsaturation in unsaturated polymers, and this reaction generally occurs in three principal steps (see Figure 1.4) (Cataldo and Angelini 2006; Lee and Coote 2013). The first step is a cycloaddition of ozone to the olefin double bond to form ozone-olefin adduct referred to as the "primary ozonide," which is an unstable species because it contains two very weak O–O bonds. The second step in the ozonolysis mechanism is the decomposition of the primary ozonide to carbonyl compounds and a carbonyl oxide. The carbonyl oxide is considered to be the key intermediate in the C=C bond ozonolysis mechanism. The third step is the fate of the carbonyl oxide, which depends on its source, as well as on its environment. The carbonyl oxide flips over with the nucleophilic oxyanion attacking the carbon atom of the carbonyl group (Anachkov et al. 2000; Ozen et al. 2003).

### 1.2.2.1 Photochemical Oxidation

In photochemical decomposition, activation energy is provided by sunlight (White and Turnbull 1994). The energies required to break single covalent bonds, with rare exceptions, are in the range from 165 to 420 kJ/mol. These energies correspond to radiation of wavelengths from 720 to 280 nm. This means that the radiation in the near ultraviolet region (300–400 nm) is energetic enough to break most single covalent bonds, except for strong bonds such as C–H and O–H. Primary chain breakage or radical formation in the various photochemical processes is often followed by embrittlement due to cross-linking, but secondary reactions, especially in the presence of oxygen, cause further degradation of the polymer. The resulting superoxide or hydrogen peroxide, generated as a result of the oxygen activation by UV illumination, will then aggressively attack any organic substance present, including active polymers. As a result, the properties of polymers, including optical, electrophysical, and mechanical, can deteriorate drastically. Currently, the autocatalytic nature of the reaction is believed to be due to the decomposition of the hydroperoxides (White and Turnbull 1994; Denisov 2000; Al-Malaika 2003):

$$ROOH \xrightarrow{hv} RO\bullet + \bullet OH \quad (1.1)$$

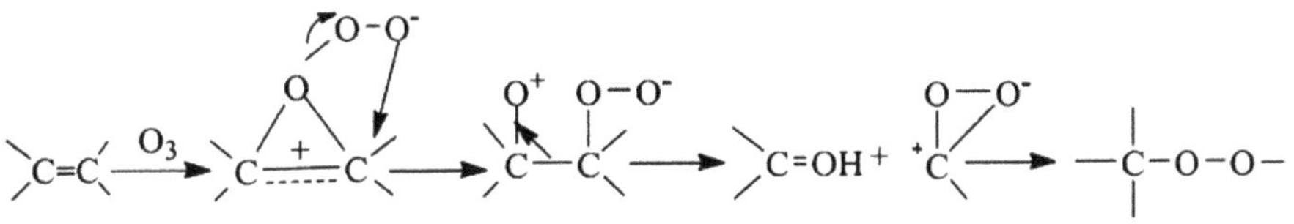

**FIGURE 1.4** Mechanism of ozone interaction with polymers. (Idea from Marzec A., The effect of dyes, pigments and ionic liquids on the properties of elastomer composites. PhD Thesis, Université Claude Bernard—Lyon I, Lyon, France, 2014.)

followed by:

$$RO \bullet + RH \rightarrow ROH + R \bullet \quad (1.2)$$

$$HO \bullet + RH \rightarrow H_2O + R \bullet \quad (1.3)$$

The hydroperoxides also cause secondary reactions in which colored resinous products are formed (via carbonyl compounds).

The experiment showed that some materials are more susceptible to degradation than others, and therefore the challenge is to select polymers with desirable gas-sensing properties that are also resistant to chemical and photochemical degradation. For example, polyphenylene vinylene (PPV)-type polymers are especially prone to photochemical degradation. Scurlock et al. (1995) and Dam et al. (1999) investigated the singlet oxygen photodegradation of oligomers of phenylene vinylenes as model compounds for the degradation of PPVs. They believed that the singlet oxygen reacts with the vinylene groups in PPVs through a 2 + 2 cyclo addition reaction. The intermediate adduct may then break down leading to chain scission. They found that the rate of reaction with oxygen strongly depends on the nature of the substituents. Electron-donating groups increase the rate, while electron-withdrawing substituents slow down the rate of these reactions. A possible scheme of PPV photodegradation is shown in Figure 1.5.

Poly-3-hexylthiophene (P3HT) is significantly more stable, but devices based on this material are also susceptible to chemical degradation (Jørgense et al. 2008). The reaction of P3HT with oxygen has not yet been studied in detail. It is known, however (Abdou et al. 1997), that poly(3-alkylthiophenes) form charge-transfer complexes with oxygen, shown in Figure 1.6. Most pure, organic synthetic polymers [polyethylene, polypropylene, poly(vinyl chloride), polystyrene, etc.] do not absorb at wavelengths longer than 300 nm owing to their ideal structure, and hence should not be affected by sunlight. However, even these polymers, for example PPy, often degrade by exposure to sunlight. It was found that this effect is due to the presence of small amounts of impurities or structural defects, which absorb light and initiate the degradation (Rabek 1995; Fang et al. 2002). Although the exact nature of the impurities or structural defects responsible for the photosensitivity is not exactly known, it is generally accepted that these impurities are various types of carbonyl groups (ketones, aldehydes) and also peroxides. We must also take into account that the effect of sunlight on the rate of oxidation can be exacerbated by the presence of atmospheric pollutants, that under influence of the sunlight can be activated to free-radical species.

This is especially true for nitrogen and sulfur oxides, which are frequently components of industrial atmospheres. For example, the rate of physical deterioration of a polymer such as polyolefins (e.g., polypropylene) in the presence of nitrogen oxides can be increased by almost an order of magnitude by light. This can seriously limit the use of polyolefins in the outdoor environment (Van Krevelen and Nijenhuis 2009).

It was established that the presence of heterogeneous catalysts, usually used for surface functionalizing of gas-sensing materials, accelerates the degradation of polymers as well. It was assumed that heterogeneous catalysts were also supposed to play the same role as oxygen to provide large quantities of radicals. For example, cobalt compounds were reported to accelerate the degradation of polymers and organic compounds, such as polypropylene, low-density polyethylene, and cyclohexane (Perkas et al. 2001; Anipsitakis et al. 2005; Roy et al. 2005) due to their ability to produce radicals by electron transfer in the 3d subshell (Roy et al. 2005). Other transition metals such as Ni and Fe also have the potential to participate in the radical-formation reactions (Osawa 1988; Perkas et al. 2001; Anipsitakis and Dionysiou 2004). However, cobalt compounds show the strongest catalytic effect in certain environments.

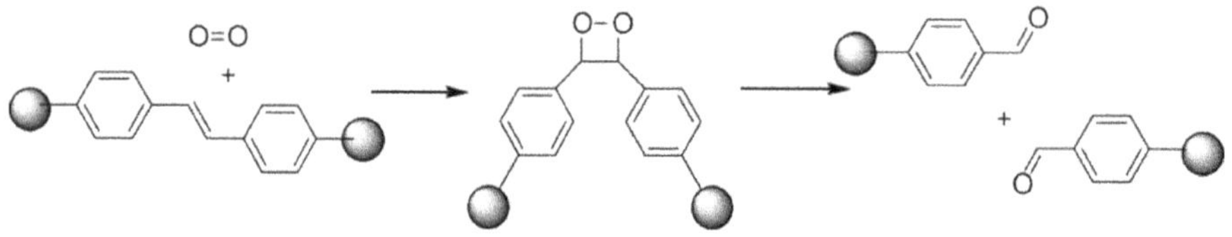

**FIGURE 1.5** Initial reaction of a PPV polymer with singlet oxygen. Singlet oxygen adds to the vinylene bond forming an intermediate dioxetane followed by the chain scission. The aldehyde products shown can react further with oxygen. (Reprinted from *Sol. Energy Mater. Sol. Cells*, 92, Jørgensen, M. et al., Stability/degradation of polymer solar cells, 686–714, Copyright 2008, with permission from Elsevier.)

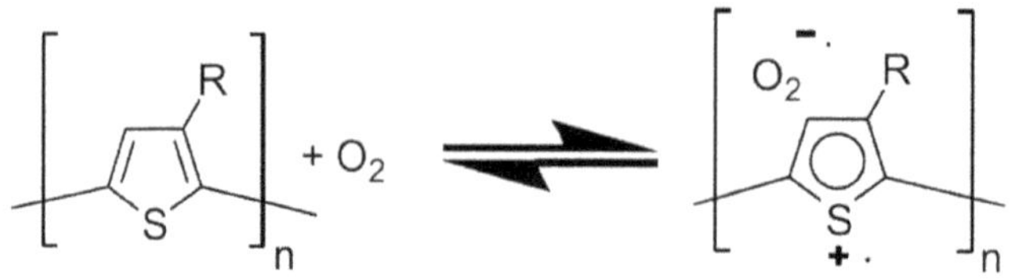

**FIGURE 1.6** Reversible formation of a charge transfer complex between P3AT (A = alkyl) and oxygen. *R* represents an alkyl group. (Reprinted from *Sol. Energy Mater. Sol. Cells*, 92, Jørgensen, M. et al., Stability/degradation of polymer solar cells, 686–714, Copyright 2008, with permission from Elsevier.)

We must note that photochemical degradation strongly depends on the temperature. At normal temperature, polymers usually react so slowly with oxygen that the oxidation only becomes apparent only after a long time. For example, if polystyrene is stored in air in the dark for several years, the UV spectrum does not change significantly. On the other hand, if UV light under similar conditions irradiates the same polymer for 12 days, there appear strong bands in the spectrum (Van Krevelen and Nijenhuis 2009). The same applies to other polymers, such as polyethylene and natural rubber. Therefore, in essence, the problem is not the oxidizability as such, but the synergistic effect of various factors, such as electromagnetic radiation and thermal energy on oxidation.

### 1.2.2.2 Thermal Oxidation

In the absence of light, most polymers are stable for very long periods at ambient temperatures. However, above room temperature many polymers begin to degrade in an atmosphere of air, even without exposure to light. For example, a number of polymers already show a deterioration of the mechanical properties after heating for some days at about 100°C and even at lower temperatures [e.g., polyethylene, polypropylene, poly(oxy methylene), and poly(ethylene sulfide)]. Measurements have shown that the oxidation at 140°C of low-density polyethylene increased exponentially after an induction period of 2 hours. It was concluded that thermal oxidation, such as photo-oxidation, is caused by autoxidation, the only difference being that the formation of radicals from the hydroperoxide is now activated by heating. The primary reaction can be a direct reaction with oxygen (White and Turnbull 1994; Van Krevelen and Nijenhuis 2009).

$$RH + O_2 \rightarrow R \bullet + \bullet OOH \qquad (1.4)$$

Studies have shown that in the case of thermo-oxidative degradation, the presence of C–H bonds or unsaturation in the polymer chain enhances the possibility of the formation of peroxy radicals. Thus, polymers with saturated structure (PE) are more resistant to thermo-oxidative degradation than the unsaturated polymers (Ray and Cooney 2018). Similarly, a polymer with no hydrogen at all or with unreactive methyl and phenyl groups shows resistance to oxidation. The following factors associated with the macromolecular structure also significantly affect the above process: functional group(s), molecular weight and size, molecular weight distribution, degree of branching, cross-linking, crystallinity, and amorphousness. For example, incorporation of carbonyl group in polyolefins makes these polymers susceptible to thermo-oxidative degradation. Thermal stability can be improved by increasing the molecular weight and size of the macromolecules. It was reported that the amorphous regions in the polymer are more resistant to thermal oxidation than crystalline areas due to their high permeability to molecular oxygen. The presence of branching in the polymer reduces the intra- and intermolecular forces and, therefore, affects the crystallinity. Thus, in most cases, this reduces the thermal stability of polymers, since they tend to oxidize more easily than linear structures. Also, the oxidation rate increases with the degree of branching.

It is important to note that additives, used to improve the sensitive properties of polymer films, can also affect thermo-oxidative degradation (Ray and Cooney 2018). For example, traces of transition metals can accelerate thermal oxidative processes of polyolefins by inducing hydroperoxide decomposition. In particular, metals, for example, manganese (Mn), can act as pro-oxidants in polyolefins, making the polymer susceptible for thermo-oxidative degradation.

The experiment showed that among the polymers used in the development of humidity sensors, polyimide (PI) films are characterized by good parameters' stability (Diaham et al. 2012). However, even in this material, at temperatures above 200°C, the effect of long-time aging, which becomes apparent in weight loss and change in chemical and mechanical properties, is found in an oxidizing environment. (Meador et al. 1997; Tandon et al. 2008; Ruggles-Wrenn and Broeckert 2009). However, it has been found that, although thermal degradation in an oxidative environment occurs throughout the material, the oxidative degradation occurs mainly in a thin surface layer, where oxygen diffuses into the material. It was also shown that increasing the number of benzene rings contributes to an increase in the degradation temperature (Sroog et al. 1965). It was established that the degradation temperature can be also affected by the presence of low thermo-stable bonds in the macromolecular structure. For example, even if BPDA-PDA [poly(p-phenylene benzophenonetetracarboximide)] and PMDA-ODA [poly(4,4'-oxydiphenylene pyromellitimide)] (Kapton-type) own the same number of benzene rings (i.e., three in elementary monomer backbone), the absence of the C–O–C ether group in the case of BPDA-PDA [poly(4-4'-oxydiphenylene biphenyltetracarboximide)] allows increasing $T_d$ (defined at 10% wt. loss) of 48°C in nitrogen and 100°C in air in comparison to $T_d$ of PMDA-ODA (see Figure 1.7). Moreover, if the p-phenylene diamine (ODA) is replaced by p-phenylene diamine (PDA) in BPDA-based PIs, $T_d$ increases to 68°C in nitrogen and 105°C in air. Indeed, this is due to the lower thermal stability of the ether bonds inducing earlier degradations than the rest of the structure (Sroog et al. 1965; Tsukiji et al. 1990). It is important that in inert atmosphere PI does not change properties even after aging at 300°C during 1000 h.

It should be noted that thermal oxidation of polymers can also be accompanied by combustion. Although some polymers, such as PVC, do not easily ignite, most organic polymers, such as hydrocarbons, will burn. Some will support combustion, such as polyolefins, styrene-butadiene rubber (SBR), wood, and paper, when lit with a match or some other source of flame. Thermally, simple combustion of polymeric materials gives a complex of compounds that vary according to the particular reaction conditions. In particular, for vinyl polymers thermal degradation in the air (combustion) produces the expected products of water, carbon dioxide (or carbon monoxide if insufficient oxygen is present), and char along with numerous hydrocarbon products (Carraher 2008). Application of heat under controlled conditions can result in true depolymerization, usually occurring

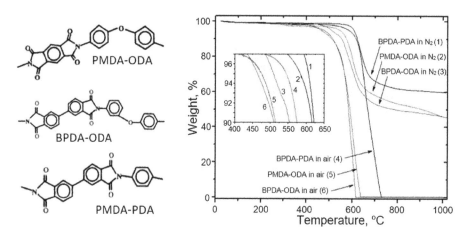

**FIGURE 1.7** Dynamical thermogravimetric analysis (TGA) of polyimide films PMDA-ODA, BPDA-ODA, and BPDA-PDA in air and nitrogen. (Reprinted from Diaham, S. et al., BPDA-PDA polyimide: Synthesis, characterizations, aging and semiconductor device passivation, in: Médard Abadie M.J. (ed.) *High Performance Polymers—Polyimides Based—From Chemistry to Applications*, InTech, Rijeka, Croatia, pp. 15–36. Published by InTech as open access.)

via an unzipping. Such depolymerization can be related to the ceiling temperature of the particular polymer. Polymers such as poly(methylmethacrylate) (PMMA) and poly(alpha-methylstyrene) depolymerize to give large amounts of monomer when heated under the appropriate conditions. Thermal depolymerization usually leads to some charring and the formation of smaller molecules, including water, methanol, and carbon dioxide.

### 1.2.2.3 Hydrolytic Degradation

Hydrolytic degradation plays a role if hydrolysis is the potential key reaction in breaking bonds, as in polyesters and polycarbonates (White and Turnbull 1994; Van Krevelen and Nijenhuis 2009). A water attack can be quick if the temperature is high enough; attack by acids depends on acid strength and temperature. Degradation under the influence of basic substances is highly dependent on the penetration of the agent; ammonia and amines can cause much greater degradation than substances such as caustic soda, which mainly attack the surface. The amorphous regions are attacked first and most quickly, but crystalline regions are not free from attack. It was established that in general, condensation polymers that contain functional groups in the polymer chain, notably polyesters, polyamides, and polyurethanes, are much more subject to hydrolytic and biodegradation than polymers containing a carbon-carbon backbone.

### 1.2.3 BIODEGRADATION

In a broad sense, biodegradation involves destruction occurring under the influence of environmental factors, such as water, biological fluids, and microorganisms, such as bacteria and fungi (Gu et al. 2000). Many synthetic polymers, especially carbochain polymers, that is, those that are built exclusively of carbon atoms in the main chain, are not biodegradable and remain in the environment for many years (Albertsson 1980; Cruz-Pinto et al. 1994; Albertsson et al. 1994; Shah et al. 2008).

On the contrary, biopolymers (natural polymers and their synthetic analogues) are biodegradable, turning into low-molecular-mass compounds. The main mechanism of biodegradation of a high-molecular-weight polymer is the oxidation or hydrolysis by an enzyme to form functional groups that improve its hydrophylicity. Consequently, the main polymer chains degrade, resulting in a polymer with a low molecular weight and feeble mechanical properties, which makes it more accessible for further microbial assimilation (Albertsson et al. 1987; Huang et al. 1990; Albertsson and Karlsson 1990). Examples of synthetic polymers that biodegrade include poly(vinyl alcohol), poly(lactic acid), aliphatic polyesters, polycaprolactone, and polyamides. Several oligomeric structures that biodegrade are known: oligomeric ethylene, styrene, isoprene, butadiene, acrylonitrile, and acrylate. Similar to the previously discussed mechanisms, the physical properties of polymers, such as crystallinity, orientation, and surface area, also affect the rate of degradation (Huang et al. 1992). Features of biodegradation in comparison with other mechanisms are listed in Table 1.6.

### 1.2.4 DEGRADATION CAUSED BY IONIZING RADIATIONS

The degradation of polymer properties under ionizing radiation is widely spread due to the growing use of polymer-based sensors in hard radiation environments encountered in nuclear power plants, space crafts, high-energy particle accelerators, and in disposable medical-device sterilization by electron beams or gamma rays. Polymer degradation also occurs during the photolithography process carried out using e-beams or X-rays.

Polymers are generally classified as predominantly undergoing degradation and cross-linking when exposed to ionizing radiation. Harmful effects caused by high-energy radiation include color development, loss of tensile strength, and increased propensity to brittle failure (Sing and Silverman 1991). For example, Navarro et al. (2018) have shown that irradiation with an electron beam of polycaprolactone (PCL)

**TABLE 1.6**

**Comparison of Various Polymer Degradation Routes**

| Factors (Requirement/Activity) | Photo-Degradation | Thermo-Oxidative Degradation | Biodegradation |
|---|---|---|---|
| Active agent | UV-light or high-energy radiation | Heat and oxygen | Microbial agents |
| Requirement of heat | Not required | Higher than ambient temperature required | Not required |
| Rate of degradation | Initiation is slow. But propagation is fast | Fast | Moderate |
| Other consideration | Environment friendly if high-energy radiation is not used | Environmentally not acceptable | Environment friendly |

*Source:* Reprinted from *Biotechnol. Adv.*, 26, Shah, A.A. et al., Biological degradation of plastics: A comprehensive review, 246–265, Copyright 2008, with permission from Elsevier.

was accompanied by a significant change in melting point, glass transition middle point ($T_g$), and crystallization temperature. Mechanical properties are also severely affected by irradiation. Stress at break and strain at break have continuously decreased with increasing dose until brittle material is obtained, regardless of the initial molecular weight of PCL. Changes in the mechanical, physical, and sensing properties of polymers are the result of the usual competition between chain scission and cross-linking (see Figure 1.8), which depends on the polymer, the atmosphere, the percent crystallinity, additives, dose, dose rate, and the temperature (Atkins and Batich 1993; Davenas et al. 2002). Polymers in which ionizing radiation causes cross-linking often have better mechanical properties. If chain scissioning dominates, then low-molecular-weight fragments, gas evolution (odor), and unsaturated bonds (color) may appear. Free radicals determine initiating centers for scissioning and/or cross-linking, therefore, for the induced modifications within the macromolecular chain. It is important to note that ionizing radiation is less specific than UV or visible radiation in that almost all chemical bonds undergo slow homolytic cleavage to produce radicals and secondary electrons. Many of them are combined or terminated by interaction with a stabilizer.

### 1.2.5 Conducting Polymers Dedoping

Dedoping (undoping) of conducting polymers can also be considered as degradation mechanisms for polymer-based gas sensors. Conducting polymers are used mainly in resistive-type sensors (chemiresistors, thin film transistors (TFT), and field effect transistors (FET)), and therefore a change in the conductivity of the film causes both a shift in the sensor baseline, and a change in the sensor response. As is well known, undoped, conjugated polymers consist of high-resistance semiconductors or insulators. For example, undoped conjugated polymers, such as polythiophenes and polyacetylenes, have an electrical conductivity of around $10^{-10}$–$10^{-8}$ S/cm. Therefore, for use in gas sensors, conducting polymers must be doped. The mechanism of conductivity in conducting polymers was discussed in Chapter 3 (Volume 1).

Figure 1.9 illustrates how strong the temporal change in electroconductivity of as-prepared conducting polymers can be, especially in the first days. Lima and de Andrade (2009) deposited poly(o-methoxyaniline) (POMA) thin films and after annealing at 60°C for 30 min immersed these films in a HCl solution (pH ~0.8) for 60 seconds. After drying with nitrogen, the sensors were kept in vials. The electrical resistance was

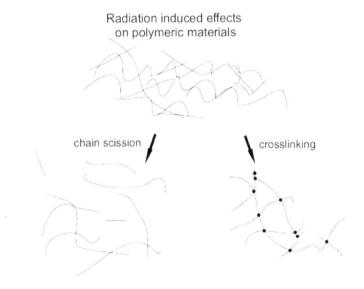

Radiation induced effects on polymeric materials

chain scission

crosslinking

**FIGURE 1.8** Ionizing radiation-induced cross-linking and chain scissioning of polymeric materials. (Idea from Marzec A., The effect of dyes, pigments and ionic liquids on the properties of elastomer composites. PhD Thesis, Université Claude Bernard—Lyon I, France, 2014.)

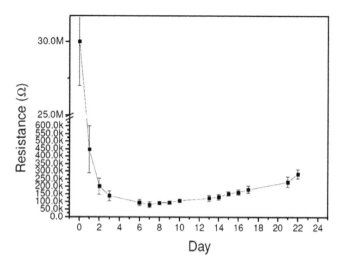

**FIGURE 1.9** Electrical resistance of POMA (3000 rpm) in function of days after the chemical treatment in HCl. (Data extracted from Lima, J.P.H. and de Andrade A.M., Stability study of conducting polymers as gas sensors, in: *Proceedings of 11th International Conference on Advanced Materials*, ICAM 2009, September 20–25, Rio de Janeiro, Brazil, 1566, 2009.)

measured daily. It was seen that a great decrease in the values of electrical resistance took place in the first five days, followed by a slight increase afterward. It should be noted that the dedoping effect is typical for all conducting polymers, including such polymers as PPy, PTh, etc., which are easily dedoped when exposed to air. For example, Jiang et al. (2005) reported that the sensitivity of PPy/PVA composite sensor was only maintained for 2 weeks, while the sensitivity of a pure PPy sensor was maintained over 1 month. According to results obtained by Hosseini et al. (2005), the stabilities of doped polyaniline films with hydrogen halides, hydrogen cyanide, hydrogen sulfide, halogens, and halomethyl compounds ranged from one week to several months.

No doubt, the stability of conductivity in air is highly dependent on dopants. For example, Wang et al. (1991) reported that polypyrroles doped with arylsulfonates exhibit excellent stability in inert atmospheres but are somewhat less stable in the presence of dry or humid air. Polypyrrole samples doped with the tosylate anion were found to be the most stable, while polypyrroles doped with longer sidechain substituted benzenesulfonates exhibited poorer stability. It is believed that polymers with longer sidechains are more flexible and thus more sensitive to the thermal undoping process. Wang et al. (1991) have shown that the tosylate-doped polypyrrole in humid air was approximately 10 times more stable than the $FeCl_3$-doped poly(3-butylthiophene) (P3BT) in dry nitrogen. Budrowski et al. (1990) found that instability of chemically prepared PPy doped with ferric chloride under ambient conditions occurs due to the absorption of water molecules, which leads to a weakening of hydrogen bonding between the polymer matrix and the anion. Li and Wan (1999) studied the stability of doped polyaniline films and found that for 6 months in air at room temperature the conductivity of PANI film doped with $H_2SO_4$, *p*-toluene sulfonic acid (*p*-TSA) and camphor sulfonic acid (CSA) changed little, while the

conductivity of PANI films doped with HCl and $HClO_4$ has decreased by about 40%. At that, it is important to note that polypyrroles, polythiophenes, and polyaniline are considered as polymers with good stability of doped state. The stability of doped polyacetylene, polyphenylene, polyphenylene sulfide (PPS), and PPV is considerably worse.

Studies have shown that the dedoping process is greatly accelerated with increasing temperature. Rannou and Nechtschein (1999) carried out kinetic studies of the conductivity of doped PEDT thin films and found that the half-life of the conductivity in a device operating in air at 100°C is only about 150 h (see Figure 1.10). Wang and Rubner (1990) have shown that the electrical conductivity of $FeCl_3$-doped poly(3-hexylthiophene) in a laboratory environment at $T = 110°C$ is decreased to negligibly small values in approximately 1 h. Among other regularities established by Wang and Rubner (1990) during stability studies of poly(3-butylthiophene), poly(3-hexylthiophene), poly(3-octylthiophene), and poly(3-decylthiophene), it is necessary to note the following: (1) the thermal stability of poly(alkyl thiophene)s doped with $FeCl_3$ by chemical methods is slightly better than those doped electrochemically; (2) the nature of the counterion affects the stability; and (3) the stability decreases when the length of the alkyl chain increases.

The findings of Truong (1992) are also important for understanding the processes responsible for polymer aging. Truong (1992) studied polypyrrole films doped with *p*-toluene sulfonate and established the following: (1) At high aging temperature (120°C and 150°C), the oxygen diffusion into the bulk is predominant, that is, the conductivity decay shows a $t^{1/2}$ dependence. The diffusion coefficient $D$ derived from

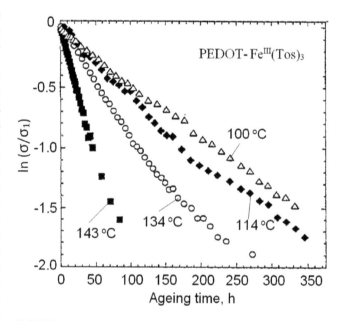

**FIGURE 1.10** Logarithm of the reduced conductivity versus aging time for air-aged conductive PEDOT thin layer laid on polycarbonate substrate. Iron(III) tris-p-toluenesulfonate ($Fe^{III}(Tos)_3$)-doped poly(3,4-ethylenedioxythiophene) (PEDOT-$Fe^{III}(Tos)_3$) thin layer had the thickness equaled 500 nm. (Reprinted from Rannou, P. and Nechtschein, M., *Synth. Met.*, 101, 474–474, Copyright 1999, with permission from Elsevier.)

the experimental results is a material property independent of thickness; and (2) At low aging temperatures (70°C–90°C), conductivity degradation of the thick film (43 μm) apparently exhibits first-order reaction kinetics, which suggest chemical-reaction-controlled degradation. This study has revealed that as the thickness increases and aging temperature decreases, thermal degradation of the PPy films exhibits a transition from a diffusion-controlled mechanism to a reaction-controlled mechanism. Therefore, any extrapolation of data from high temperature to low temperature would be inappropriate.

Additional annealing of doped polymers also accelerates the process of dedoping. Li and Wan (1999) have shown that the temporal change in the conductivity of a doped polyaniline film at room temperature greatly increases after the thermal treatment of the doped polymer at $T = 150°C$. The results of this study are shown in Figure 1.11. We must recognize that this effect limits the use of any thermal treatments during gas-sensor fabrication. It was noted that the conductivity of PANI–CSA and PANI–$p$-TSA are the most stable below 200°C, while the stability of PANI films doped with $H_2SO_4$ and $H_3PO_4$ are much better than the PANI–$HClO_4$

and PANI–HCl after thermal treatment at high temperature (200°C) (Li and Wan 1999).

It is important that processes of degradation and dedoping discussed above are also typical for organic semiconductors. For example, Oester et al. (1993) studied oligothiophene thin films doped by $FeCl_3$ and found that $FeCl_3$ was not stable as a dopant in the presence of water and oxygen. After exposure to the atmosphere for about two months the concentration of Cl in organic semiconductor was beyond the detection limit of analytical devices. For information, oligothiophenes (Fichow 1999) are a promising class of organic semiconductors, and their derivatives have been employed as the active layer in both chemiresistor and thin-film transistor gas sensors (Torsi et al. 2000; Chang et al. 2006). Read also Chapter 20 (Volume 1) and Chapter 10 (Volume 2).

## 1.3  APPROACHES TO POLYMER STABILIZATION

It is necessary to recognize that over the past decades, numerous studies have been carried out aimed at stabilizing the parameters of polymers. Different stabilizers for polymers,

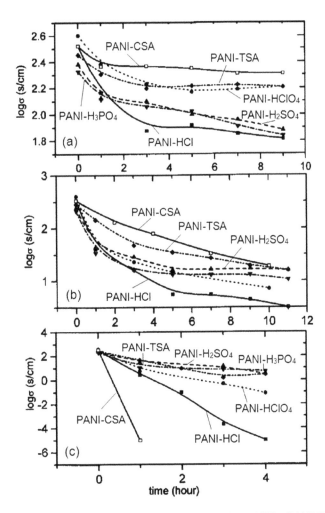

**FIGURE 1.11**  Room-temperature conductivity of doped PANI films (PANI-HCl, PANI-HClO₄, PANI-H₂SO₄, PANI-H₃PO₄, PANI-TSA, and PANI-CSA) treated at different temperatures in air: (a) 100°C; (b) 150°C; and (c) 200°C. Free-standing films of doped PANI were synthesized by the doping–dedoping–redoping method. (From Li, W. and Wan, M.: Stability of polyaniline synthesized by a doping–dedoping–redoping method. *Journal of Applied Polymer Science.* 1999. 71. 615–621. Wiley-VCH Verlag GmbH & Co. KGaA. Reproduced with permission.)

that prevent various effects of chemical degradation, such as oxidation, chain scission, and uncontrolled recombinations and cross-linking reactions, which are caused by photo-oxidation of polymers, have been proposed (Scott 1981; Moss and Zweifel 1989; Al-Malaika 1994; Allen and Edge 1992; George and Goss 2001; Krohnke 2001; Van Krevelen and Nijenhuis 2009; Wilen and Pfaendner 2013; Low et al. 2018). Usually stabilizers are chemical substances that are added to polymers in small amounts (at most 1–2 w%) and are able to capture the emerging free radicals or unstable intermediate products (such as hydroperoxides) during autooxidation and turn them into stable end products. Antioxidants, light stabilizers, antiozonats, UV absorbants, fire retardants, etc., are examples of such stabilizers. Several examples of such stabilizers are listed in Table 1.7.

Experiments have shown that activity and efficiency of these stabilizers depend mainly on the following factors: (1) the internal activity of the stabilizer, which is primarily influenced by the structure of the molecule, including factors such as intramolecular interactions; (2) compatibility/mobility of the stabilizer, which will again be determined by intra- and intermolecular interactions in the molecule usually in the direction opposite to that indicated above; and (3) the volatility of the stabilizer, which will be determined by molecular weight and molecular interaction in the polymer. So, for real applications, stabilizers should correspond to several requirements, such as compatibility with the polymer, nonvolatility, light fastness, heat stability and also, for some applications, the resistance to water and high humidity.

It should be noted that in addition to the use of consumable stabilizers, the addition of fillers can also be used to stabilize polymers (Allen and Edge 1992; George and Goss 2001). For example, for this purpose one can use fillers, such as a carbon in various forms (carbon black, carbon nanotubes (CNTs), graphene) or an inorganic nanopowders, such as $TiO_2$ or silica (Allen et al. 1998). In particular, carbon black is a UV absorber, as well as a hydroperoxide decomposing agent, a quencher of excited states, and radical scavenger (Allen et al. 1998). In addition, carbon nanostructures are a reinforcement in elastomers adding to the mechanical strength and modulus. The rutile and

## TABLE 1.7
### Materials Used as Stabilizers for Polymers

| Material | Function | Mechanism of Stabilization |
|---|---|---|
| Hindered phenols; aromatic amines; benzofuranones; etc. | Antioxidants | Antioxidants are used to terminate the oxidation reactions taking place due to different weathering conditions and reduce the degradation of organic materials. Antioxidants function by interfering with radical reactions that lead to polymer oxidation and, in turn, to degradation. Primary antioxidants are generally radical scavengers or H-donors. Secondary antioxidants are typically hydroperoxide decomposers. |
| Organosulfur compounds; hindered amines; organophosphates; etc. | Thermal stabilizers | Thermal stabilizers are antioxidants protected polymers during thermal treatments. For example, organosulfur compounds are efficient hydroperoxide decomposers, which thermally stabilize the polymers, while hindered amines efficiently scavenge radicals, which are produced by heat. |
| Paraffin waxes; ethylene diurea (EDU); p-phenylenediamines; etc. | Antiozonats | Antiozonants prevent or slow down the degradation of material caused by ozone gas in the air. For example, paraffin waxes form a surface barrier for ozone. |
| Oxanilides (for polyamides); benzophenones (for PVC); benzotriazoles and hydroxy-phenyltriazines (for polycarbonate); $TiO_2$; etc. | UV absorbers | The UV absorbers dissipate the absorbed light energy from UV ray as heat by reversible intra molecular proton transfer. This reduces the absorption of UV ray by polymer matrix and hence reduces the rate of weathering. |
| Hindered amine | Light stabilizers | Hindered amine scavenges radicals which are produced by light. This effect may be explained by the formation of nitroxyl radicals. |
| Nickel compounds (for polyolefins) | Quenchers | A quencher induces harmless dissipation of the energy of photo-excited states. |
| Magnesium or aluminum hydroxide; organobromide compounds; sodium carbonate; etc. | Fire retardant | When heated, hydrooxides dehydrate to form aluminum oxide, releasing water vapor in the process. This reaction absorbs a great deal of heat, cooling the material over which it is coated. Brominated flame retardants have an inhibitory effect on the ignition of combustible organic materials. Sodium carbonate, which releases carbon dioxide when heated, shields the reactants from oxygen. |

*Source:* Data extracted from Krohnke, C., Polymer stabilization, in: Buschow K.H.J., Cahn R.W., Flemings M.C., Ilschner B., Kramer E.J., Mahajan S., Veyssière P. (eds.) *Encyclopedia of Materials: Science and Technology*, Elsevier, Oxford, UK, pp. 7507–7516, 2001; Mark, H.F. (ed.), *Encyclopedia of Polymer Science and Technology*, 3rd edn., Vol. 5, Wiley-Interscience, New York, 2007; Van Krevelen, D.W. and Nijenhuis, K.Te., *Properties of Polymers. Their Correlation with Chemical Structure; Their Numerical Estimation and Prediction from Additive Group Contributions*, 4th ed. Elsevier, Amsterdam, the Netherlands, 2009.

the anatase phase of $TiO_2$ have different stabilization efficiencies in polymers. $TiO_2$ is a UV absorber; however, the relaxation process can lead to hydroxyl radicals or singlet oxygen. These radicals are believed to be formed by electron transfer from the excited $TiO_2$ to water or oxygen, respectively (Allen et al. 1998). Anatase has also been shown to increase the sensitivity of the polymer to certain UV wavelengths (365 nm) that rutile does not, while both pigments can stabilize at short wavelengths (254 nm).

It was also found that polymers with a cross-linked structure have a higher thermal stability since in this case the degradation requires simultaneous breakdown of many bonds in order to reduce the molecular weight. The structural order of polymers can also contribute to thermal stability. It was established that polymers having ordered (tactic) structures cause an increase in thermal and oxidative stability. Head-to-head type structures have lower thermal stability than head-to-tail arrangement. Thus, head-to-head linkage in PMMA enhances thermal degradation of the polymer. Increased intermolecular forces can be achieved by copolymerization and, therefore, it can enhance thermal stability (Ray and Cooney 2018).

There are also other approaches to the stabilization of polymers (Aldiss 1989). For example, it was found that the formation of composites of two conducting polymers, one of which is air stable, improves the stability of polymeric materials. Experiments carried out with pyrrole/polyacetylene and polyaniline/polyacetylene composites have shown that the composites appeared to be more stable than doped polyacetylene and possessed mechanical properties, similar to polyacetylene. Stabilization can also be achieved chemically by copolymerization. In particular, it was found that copolymerization of acetylene with other monomers, such as styrene, isoprene, ethylene, or butadiene was accompanied by an improvement in the stability of the polymer (Aldiss 1989). Crispin et al. (2003) established that poly(3,4-ethylene dioxythiophene) (PEDT or PEDOT) degrades under influence of UV. UV light induces the photo-oxidation of the conjugated PEDT chains, which is accompanied by a reduction in the electrical conductivity. Photo-oxidation leads to shorter conjugation lengths because of sulfone group formation and chain scission accompanied by the addition of carboxyl groups. The photo-oxidation mechanism of PEDT is similar to the case of polythiophene: a $\pi$–$\pi$* transition generated by UV light is followed by energy transfer to oxygen, which consequently reaches its singlet state and reacts with the conjugated chain. The copolymerization of PEDT and poly(styrene sulfonate) (PSS) clearly helps to block this degradation pathway. However, the degradation was not blocked completely. Due to hydroscopicity of PSS, in the presence of ambient humidity, additional effects occur on the PSS portion of the PEDT/PSS polymer blend. As a result, the degradation process leads to the appearance of nitrogen at the surface of PSS exposed to UV light in air. This occurs through the formation of ammonium sulfate salts formed in a complex chain reaction. Pei and Inganas (1992) have shown that copolymers of 3-octylthiophene and 3-methylthiophene (POTMT) doped by $FeCl_3$ are more stable than doped poly(3-octylthiophene) (P3OT) (see Figure 1.12).

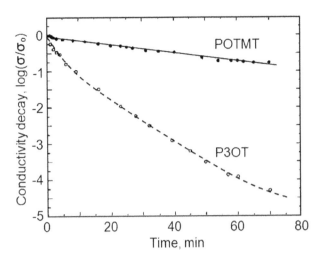

**FIGURE 1.12** Conductivity decay of copolymer POTMT at 110°C in laboratory air compared with that of poly(3-octylthiophene) (P3OT). The samples were doped with $FeCl_3$. (Reprinted from *Synth. Met.*, 45, Pei, Q. and Inganas, O., Poly(3-octylthiophene-co-3-methylthiophene), a processible and stable conducting copolymer, 353–357, Copyright 1992, with permission from Elsevier.)

Surface protection also gives a positive result. It was found that when polyacetylene is doped, the charge transfer complex is stable under vacuum or inert atmosphere. Thus, if the system is protected against moisture and especially against oxygen, the system would be stable as well. A large variety of plastic polymers is available for this purpose. For instance, a 50-µm polyvinylchloride (PVC) film could keep the efficiency of a hetero-junction [doped or undoped $(CH)_x$/conventional semiconductor] almost constant for several months. When a cell with the same characteristics was kept in the air, its properties deteriorated after one or two days (Aldiss 1989).

Magnoni et al. (1996) have shown that rational synthesis of a polymer with a favored topology of the dopant molecule also is a very promising approach to improve stability of doped polymers. This concept, based on the results of simple quantum mechanical calculations (Lopez-Navarrete and Zerbi 1990), supposes that the position in the space of the dopant relative to the polymer molecule is an important factor, determining the chemical stability of these systems. If the dopant can sit near the molecule in an energetically favorable position, the charge-transfer (CT) bond becomes stronger, thus favoring stability. It follows that the path of approach of the donor/acceptor dopant molecule to the acceptor/donor polymer is determined by the free volume that the approaching dopant molecule finds along its reaction coordinate. The steric hindrance of the side group attached to the backbone chain then plays the dominant role in affecting the strength, hence the stability, of the CT bond. In particular, the preferred site of attack is found where the steric hindrance is minimized. Following these principles, regiospecific poly(3,3''-dihexyl-2,2':5',2''-terthiophene) (PDHTT) was synthesized and doped with $FeCl_3$. A method for the synthesis of highly regiospecific polyalkylthiophenes with a large content of head-to-tail linking was developed by McCullough et al. (1993a, 1993b). Experiments carried out by Magnoni et al. (1996) have shown that the regular periodic chemical structure of poly(3,3″-dihexyl-2,2′:5′,2″-terthiophene) opens periodic pockets along the polymer backbone, large and ordered enough to allow the dopant to enter and develop preferential CT interactions with the delocalized $\pi$ electrons. As a result, PDHTT in the doped state showed an extremely great stability in the air even at relatively high temperatures. In poly(3,3″-dihexyl-2,2′:5′,2″-terthiophene) chemically doped with $FeCl_3$ the charge carriers remained unaltered even after one year of standing as films in open air and at room temperature. The experiments carried out at $T = 140°C$ and $T = 170°C$ showed that the dedoping of the samples of PDHTT doped with $FeCl_3$ is certainly much slower than that observed for other materials of the class of polyalkylthiophenes (Wang and Rubner 1990). At $T = 100°C$, the concentration of Cl in PDHTT was decreased in two times during 150 h. For comparison, the concentration of I in iodine doped PDHTT was decreased in two times at $T = 100°C$ during already 15 min. Magnoni et al. (1996) believe that the approach previously mentioned is possibly useful for the synthesis of other stable doped conjugated polymer materials. According to opinion of Magnoni et al. (1996), the best way to achieve these conditions is by regiospecific synthesis, which allows the spacing between side chains to be modulated at will in order to fit the space required by the dopant molecule.

However, it should be noted that the proposed solutions only slow down the process of polymer degradation. Most polymers are susceptible to degradation under natural radiation, sunlight, and high temperatures even in the presence of stabilizers, such as antioxidants. The results presented in Figure 1.13 illustrate this situation. As it is seen, the presence of antioxidant in $I_2$-doped polyacetylene decreases the rate of degradation only.

Moreover, it is necessary to take into account that in many cases the incorporation of stabilizers in a polymer or an additional surface protective layer is accompanied by a strong deterioration of sensing characteristics, since in order to achieve high stability we need to reduce both the reactivity of the polymer and the gas permeability. Therefore, the stabilization of a polymer material intended for use in gas and humidity sensors continues to be an important technical area requiring great industrial and scientific attention. In addition, we must note that there are some fundamental limitations to achieve the required temporal and thermal stability of polymers with high reactivity. At first, the unsaturated bonds in conducting polymers that are involved in gas-sensitive effects, are often very reactive when exposed to environmental agents such as oxygen or moisture. Second, the information presented in Refs. (Razumovskii and Zaikov 1982; Bhuiyan 1984) for nylon-type polymers shows that the melting point of the polymer decreases with increasing chain length. This means that the complication of the polymer will inevitably be accompanied by a decrease in the melting temperature (see Figure 1.14), and, therefore, by a drop in the stability of a polymer. The above confirms once again that the problem of

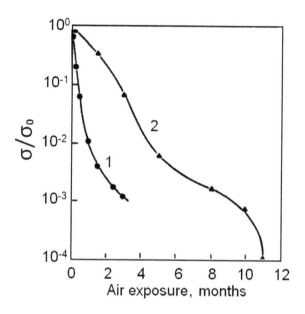

**FIGURE 1.13** Conductivity vs. air exposure time for $I_2$-doped polyacetylene: (1) in the absence of and (2) in the presence of a antioxidizing agent (2-terbutyl,6-methyl phenol). (Data extracted from Aldiss, M., *Inherently Conducting Polymers: Processing, Fabrication, Applications, Limitations.* Noyes Data Corporation, Park Ridge, NJ, 1989.)

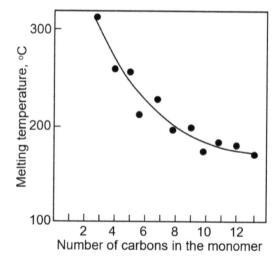

**FIGURE 1.14** Influence of the number of carbon atoms in the monomer on approximate melting temperature of nylon-type polymers. (Data extracted from *Polymer,* 25, Bhuiyan, A.L., Some aspects of the thermal stability action of the structure in aliphatic polyamides and polyacrylamides, 1699–1710, Copyright 1984, With permission from Elsevier.)

stability and reliability of gas sensors is a determining factor for the practical use of any gas-sensing material.

With regard to physical aging, the main approaches to slowing down this process are shown in Figure 1.15. It is seen that there are various approaches to improve the aging resistance of high free-volume polymers that are used usually in gas and humidity sensors. They include treatments such as polymer backbone design, postsynthetic modification, cross-linking, heat treatment, blending of nanomaterials, and

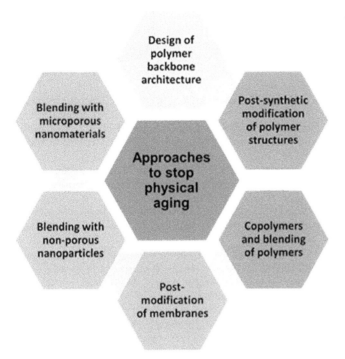

**FIGURE 1.15** Methods to stop physical aging of polymers. (Reprinted with permission from Low, Z.-X. et al., *Chem. Rev.*, 118, 5871–5911, 2018. Copyright 2018 American Chemical Society.)

combinations of the preceding methods (Beyler and Hirschler 1995; Low et al. 2018). These strategies are usually aimed at increasing the stiffness of the polymer or increasing the interaction between polymer chains. It is clear that increasing the crystallinity of the polymer increases the interaction between the polymer chains. Cross-linking also raises the melting temperature and improves the stability of polymers. However, it is necessary to take into account all these previously mentioned approaches greatly reduce the effect of swelling, which is usually used in polymer-based gas and humidity sensors of a conductometric type. At the same time, the preceding approaches can significantly increase the lifetime of the capacitive sensors.

## REFERENCES

Abdou M.S.A., Orfino F.P., Son S., Holdcroft S. (1997) Interaction of oxygen with conjugated polymers: Charge transfer complex formation with poly(3-alkylthiophenes). *J. Am. Chem. Soc.* **119**, 4518–4524.

Adhikari B., Kar P. (2010) Polymers in chemical sensors. In: Korotcenkov G. (ed.) *Chemical Sensors: Fundamentals of Sensing Materials*. Vol. 3: *Polymers and Other Materials*. Momentum Press, New York, pp. 1–76.

Adhikari B., Majumdar S. (2004) Polymers in sensor applications. *Prog. Polym. Sci.* **29**, 699–766.

Albertsson A.C. (1980) The shape of the biodegradation curve for low and high density polyethylenes in prolonged series of experiments. *Eur. Polym. J.* **16**, 623–630.

Albertsson A.C., Andersson S.O., Karlsson S. (1987) The mechanism of biodegradation of polyethylene. *Polym. Degrad. Stab.* **18**, 73–87.

Albertsson A.C., Barenstedt C., Karlsson S. (1994) Abiotic degradation products from enhanced environmentally degradable polyethylene. *Acta Polym.* **45**, 97–103.

Albertsson A.C., Karlsson S. (1990) The influence of biotic and abiotic environments on the degradation of polyethylene. *Prog. Polym. Sci.* **15**, 177–192.

Aldiss M. (1989) *Inherently Conducting Polymers: Processing, Fabrication, Applications, Limitations.* Noyes Data Corporation, Park Ridge, NJ.

Allen N.S., Edge M. (1992) *Fundamentals of Polymer Degradation and Stabilisation.* Chapman and Hall, London, UK.

Allen N.S., Edge M., Corrales T., Childs A., Liauw C.M., Catalina F., et al. (1998) Ageing and stabilisation of filled polymers: An overview. *Polym. Degrad. Stab.* **61**(2), 183–199.

Al-Malaika S. (1994) Some aspects of polymer stabilization. *Int. J. Polym. Mater.* **24**(1–4), 47–58.

Al-Malaika S. (2003) Oxidative degradation and stabilization of polymers. *Intern. Mater. Rev.* **48**, 165–185.

Anachkov M.P., Rakovski S.K., Stefanova R.V. (2000) Ozonolysis of 1,4-cispolyisoprene and 1,4-transpolyisoprene in solution. *Polym. Degrad. Stab.* **67**, 355–363.

Anipsitakis G.P., Dionysiou D.D. (2004) Radical generation by the interaction of transition metals with common oxidants. *Environ. Sci. Technol.* **38**(13), 3705–3712.

Anipsitakis G.P., Stathatos E., Dionysiou D.D. (2005) Heterogeneous activation of oxone using $Co_3O_4$. *J. Phys. Chem. B* **109**(27), 13052–13055.

Armstrong R.D., Horvai G. (1990) Review article: Properties of PVC based membranes used in ion-selective electrodes. *Electrochim. Acta* **35**(1), 1–7.

Arnold C. (1979) Stability of high-temperature polymers. *J. Polym. Sci. Macromol. Rev.* **14**, 265–378.

Atkins T., Batich C. (1993) Environmental stability of polymers. *MRS Bull.* **18**(9), 40–44.

Beyler C.L., Hirschler M.M. (1995) Thermal decomposition of polymers. In: DiNenno P.J. (ed.) *The SFPE Handbook of Fire Protection Engineering*, 2nd ed. NFPA, Quincy, MA, Ch. 1–7, pp. 110–131.

Bhuiyan A.L. (1984) Some aspects of the thermal stability action of the structure in aliphatic polyamides and polyacrylamides. *Polymer* **25**, 1699–1710.

Bidan G. (1992) Electroconducting conjugated polymers: New sensitive matrices to build up chemical or electrochemical sensors: A review. *Sens. Actuators B* **6**, 45–56.

Bilibin A.Yu., Zorin I.M. (2006) Polymer degradation and its role in nature and modern medical technologies. *Russ. Chem. Rev.* **75**(2), 133–145.

Budrowski C., Przytuski J., Kucharski Z., Suwalski J. (1990) Stability of doped polypyrrole studied by Mossbauer spectroscopy. *Synth. Met.* **35**, 151–154.

Carraher C.E. Jr. (2008) *Polymer Chemistry.* CRC Press, Boca Raton, FL.

Cataldo F., Angelini G. (2006) Some aspects of the ozone degradation of poly(vinyl alcohol). *Polym. Deg. Stab.* **91**, 2793–2800.

Chang J.B., Liu V., Subramanian V., Sivula K., Luscombe C., Murphy A., Liu J., and Frechet M.J. (2006) Printable polythiophene gas sensor array for low cost electronic noses. *J. Appl. Phys.* **100**, 014506.

Crispin X., Marciniak S., Osikowicz W., Zotti G., Van der Gon A.W.D., Louwet F., et al. (2003) Conductivity, morphology, interfacial chemistry, and stability of poly(3,4-ethylene dioxythiophene)–poly(styrene sulfonate): A photoelectron spectroscopy study. *J. Polymer Sci. B* **41**, 2561–2583.

Cruz-Pinto J.J.C., Carvalho M.E.S., Ferreira J.F.A. (1994) The kinetics and mechanism of polyethylene photo-oxidation. *Angew Makromol. Chem.* **216**, 113–133.

Cumston B.H., Jensen K.F. (1998) Photooxidative stability of substituted poly(phenylene vinylene) (PPV) and poly(phenylene acetylene) (PPA). *J. Appl. Polymer Sci.* **69**, 2451–2458.

Dam N., Scurlock R.D., Wang B., Ma L., Sundahl M., Ogilby P.R. (1999) Singlet oxygen as a reactive intermediate in the photodegradation of phenylenevinylene oligomers. *Chem. Mater.* **11**, 1302–1305.

Davenas J., Stevenson I., Celette N., Cambon S., Gardette J.L., Rivaton A., Vignoud L. (2002) Stability of polymers under ionising radiation: The many faces of radiation interactions with polymers. *Nuclear Instrum. Methods Phys. Res. B* **191**, 653–661.

Denisov E.T. (2000) Polymer oxidation and antioxidant action. In: Hamid S.H. (ed.) *Handbook of Polymer Degradation*, 2nd ed. Marcel Dekker, New York, Chap. 9.

Diaham S., Locatelli M.-L., Khazaka R. (2012) BPDA-PDA polyimide: Synthesis, characterizations, aging and semiconductor device passivation. In: Médard Abadie M.J. (ed.) *High Performance Polymers—Polyimides Based—From Chemistry to Applications*, InTech, Rijeka, Croatia, pp. 15–36.

Ehrenstein G., Pongratz S. (2013) *Resistance and Stability of Polymers*, Hanser Publishers, Munich, Germany.

Fang Q., Chetwynd D.G., Gardner J.W. (2002) Conducting polymer films by UV-photo processing. *Sens. Actuators A* **99**, 74–77.

Fichow D. (ed.) (1999) *Handbook of Oligo- and Polythiophenes*, Wiley-VCH, Weinheim, Germany.

Gardner J.W., Varadan V.K., Awadelkarim O.A. (2002) *Microsensors, MEMS, and Smart Devices*, John Wiley & Sons, Chichester, UK.

George G.A., Goss B. (2001) Elastomers in oxygen and ozone, degradation of. In: Buschow K.H.J., Cahn R.W., Flemings M.C., Ilschner B., Kramer E.J., Mahajan S., Veyssière P. (eds.) *Encyclopedia of Materials: Science and Technology*. Elsevier, New York, pp. 2466–2469.

Gu J.D., Ford T.E., Mitton D.B., Mitchell R. (2000) Microbial corrosion of metals. In: Revie W. (ed.) *The Uhlig's Corrosion Handbook*. 2nd edn. Wiley, New York, pp. 915–927.

Herman F.M., Sheldon H.A. (1965) Principles of polymer stability. *Polym. Eng. Sci.* **5**(3), 204–207.

Hosseini S.H., Oskooei S.H.A., Entezami A.A. (2005) Toxic gas and vapour detection by polyaniline gas sensors. *Iranian Polymer J.* **14**(4), 333–344.

Huang J., Shetty A.S., Wang M. (1990) Biodegradable plastics: A review. *Adv. Polym. Technol.* **10**, 23–30.

Huang S.J., Roby M.S., Macri C.A., Cameron J.A. (1992) The effects of structure and morphology on the degradation of polymers with multiple groups. In: Vert M., Feijen J., Albertsson A., Scott G., and Chiellini E. (eds.) *Biodegradable Polymers and Plastic*. Royal Society of Chemistry, London, UK, p. 149.

Jiang L., Jun H.-K., Hoh Y.-S., Lim J.-O., Lee D.-D., Huh J.-S. (2005) Sensing characteristics of polypyrrole–poly(vinyl alcohol) methanol sensors prepared by in situ vapor state polymerization. *Sens. Actuators B* **105**, 132–137.

Jørgensen M., Norrman K., Krebs F.C. (2008) Stability/degradation of polymer solar cells. *Sol. Energy Mater. Sol. Cells* **92**, 686–714.

Katz E.A., Faiman D., Tuladhar S.M., Kroon J.M., Wienk M.M., Fromherz T., et al. (2001) Temperature dependence for the photovoltaic device parameters of polymer-fullerene solar cells under operating conditions. *J. Appl. Phys.* **90**, 5343–5350.

Khrenov V., Schwager F., Klapper M., Koch M., Müllen K. (2007) Compatibilization of inorganic particles for polymeric nanocomposites. Optimization of the size and the compatibility of ZnO particles. *Polym. Bull.* **58**(5), 799–807.

Korotcenkov G. (2013) *Handbook of Gas Sensor Materials*, Vol. 2: *New Trends and Technologies*. Springer, New York.

Krohnke C. (2001) Polymer stabilization. In: Buschow K.H.J., Cahn R.W., Flemings M.C., Ilschner B., Kramer E.J., Mahajan S., Veyssière P. (eds.) *Encyclopedia of Materials: Science and Technology*. Elsevier, Oxford, UK, pp. 7507–7516.

Król-Morkisz K., Pielichowska K. (2019) Thermal decomposition of polymer nanocomposites with functionalized nanoparticles. In: Pielichowski K., Majka T.M. (eds.) *Polymer Composites with Functionalized Nanoparticles*. Elsevier, New York, pp. 405–435.

Lee R., Coote M.L. (2013) New insights into 1,2,4-trioxolane stability and the crucial role of ozone in promoting polymer degradation. *Phys. Chem. Phys.* **15**, 16428–16431.

Li W., Wan M. (1999) Stability of polyaniline synthesized by a doping–dedoping–redoping method. *J. Appl. Polymer Sci.* **71**, 615–621.

Lima J.P.H., de Andrade A.M. (2009) Stability study of conducting polymers as gas sensors. In: *Proceedings of 11th International Conference on Advanced Materials*, ICAM 2009, Sept. 20–25, Rio de Janeiro, Brazil, I566.

Lopez-Navarrete J.T., Zerbi G. (1990) On the stability of doped conducting polymers: Electrostatic contributions and sterical effects. *Chem. Phys. Lett.* **175**, 125–129.

Low Z.-X., Budd P.M., McKeown N.B., Patterson D.A. (2018) Gas permeation properties, physical aging, and its mitigation in high free volume glassy polymers. *Chem. Rev.* **118**(12), 5871–5911.

Madorsky S.L., Straus S. (1961) High temperature resistance and thermal degradation of polymers. *SCI Monograph* **13**, 60–74.

Magnoni M.C., Gallazzi M.C., Zerbi G. (1996) Search for conducting doped polymers with great chemical stability: Regiospecific poly(3,3″-dihexyl-2,2′:5′,2″-terthiophene). *Acta Polymer.* **47**, 228–233.

Mark H.F. (ed.) (2007) *Encyclopedia of Polymer Science and Technology*, 3rd edn., Vol. 5. Wiley-Interscience, New York.

Marzec A. (2014) The effect of dyes, pigments and ionic liquids on the properties of elastomer composites. PhD Thesis, Université Claude Bernard—Lyon I, Lyon, France.

McCullough R., Lowe R.D., Jayaraman M., Anderson D.L. (1993a) Design, synthesis, and control of conducting polymer architectures: Structurally homogeneous poly(3-alkylthiophenes). *J. Org. Chem.* **58**, 904–912.

McCullough R., Tristiam-Nagle S., Williams S.P., Lowe R.D., Jayaraman M. (1993b) Self-oriented poly(3-alkylthiophenes): New insights on structure-property relationships in conducting polymers. *J. Am. Chem. Soc.* **115**, 4910–4911.

Meador M.A.B., Johnston C.J., Cavano P.J., Frimer A.A. (1997). Oxidative degradation of Nadic-end-capped polyimides. 2. Evidence for reactions occurring at high temperatures. *Macromolecules.* **30**, 3215–3223.

Moss S., Zweifel H. (1989) Degradation and stabilization of high density polyethylene during multiple extrusions. *Polym. Degrad. Stabil.* **25**(2–4), 217–245.

Navarro R., Burillo G., Adem E., Marcos-Fernández A. (2018) Effect of ionizing radiation on the chemical structure and the physical properties of polycaprolactones of different molecular weight. *Polymers* **10**, 397.

Oester D., Ziegler Ch., Gopel W. (1993) Doping and stability of ultrapure α-oligothiophene thin films. *Synth. Met.* **61**, 147–150.

Osawa Z. (1988) Role of metals and metal-deactivators in polymer degradation. *Polym. Degrad. Stabil.* **20**(3–4), 203–236.

Ozen B.F., Mauer L.J., Floros J.D. (2003) Effects of ozone exposure on the structural, mechanical and barrier properties of select plastic packaging films. *Pack. Technol. Sci.* **15**, 301–311.

Pei Q., Inganas O. (1992) Poly(3-octylthiophene-co-3-methylthiophene), a processible and stable conducting copolymer. *Synth. Met.* **45**, 353–357.

Perkas N., Koltypin Y., Palchik O., Gedanken A., Chandrasekaran S. (2001) Oxidation of cyclohexane with nanostructured amorphous catalysts under mild conditions. *Appl. Catal. A* **209**(1–2), 125–130.

Persaud K.C. (2005) Polymers for chemical sensing. *Mater. Today* **8**, 38–45.

Rabek J.F. (1995) *Polymer Photodegradation: Mechanism and Experimental Methods.* Chapman & Hall, London, UK.

Rannou P., Nechtschein M. (1999) Ageing of poly(3,4-ethylene-dioxythiophene): Kinetics of conductivity decay and lifespan. *Synth. Met.* **101**, 474–474.

Ray S., Cooney R.P. (2018) Thermal degradation of polymer and polymer composites. In: Kutz M. (ed.) *Handbook of Environmental Degradation of Materials.* 2nd edn. Elsevier, New York, pp. 185–206.

Razumovskii S.D., Zaikov G.Y. (1982) Effect of ozone on saturated polymers. *Polym. Sci. USSR* **24**(10), 2805–2325.

Roy P.K., Surekha P., Rajagopal C., Chatterjee S.N., Choudhary V. (2005) Effect of benzil and cobalt stearate on the aging of low density polyethylene films. *Polym. Degrad. Stabil.* **90**(3), 577–585.

Ruggles-Wrenn M.B., Broeckert J.L (2009) Effects of prior aging at 288°C in air and in argon environments on creep response of PMR-15 neat resin. *J. Appl. Polym. Sci.* **111**, 228–236.

Scott G. (1981) Mechanism of polymer stabilization. In: Scott G. (ed.) *Developments in Polymer Stabilization.* 4th edn. Applied Science, London, UK, pp. 276–289.

Scurlock R.D., Wang B., Ogilby P.R., Sheats J.R., Clough R.L. (1995) Singlet oxygen as a reactive intermediate in the photo-degradation of an electroluminescent polymer. *J. Am. Chem. Soc.* **117**, 10194–10202.

Shah A.A., Hasan F., Hameed A., Ahmed S. (2008) Biological degradation of plastics: A comprehensive review. *Biotechnol. Adv.* **26**, 246–265.

Sing A., Silverman J. (eds.) (1991) *Radiation Processing of Polymers*, Hanser Pub, New York.

Sroog C.E., Endrey A.L., Abramo S.V., Berr C.E., Edwards W.M., Olivier K.L. (1965). Aromatic polypyromellitimides from aromatic polyamic acid. *J. Polym. Sci. A Polym. Chem.* **3**, 1373–1390.

Swaidan R., Al-Saeedi M., Ghanem B., Litwiller E., Pinnau I. (2014) Rational design of intrinsically ultramicroporous polyimides containing bridgehead-substituted triptycene for highly selective and permeable gas separation membranes. *Macromolecules* **47**, 5104–5114.

Tandon G.P., Pochiraju K.V., Schoeppner G.A. (2008) Thermo-oxidative behavior of high-temperature PMR-15 resin and composites. *Mater. Sci. Eng. A.* **498**, 150–161.

Torsi L., Dodabalapur A., Sabbatini L., Zambonon P.G. (2000) Multi-parameter gas sensors based on organic thin-film-transistors. *Sens. Actuators B* **67**, 312–316.

Truong V.-T. (1992) Thermal degradation of polypyrrole: effect of temperature and film thickness. *Synth. Met.* **52**, 33–44.

Tsukiji M., Bitoh W., Enomoto J. (1990) Thermal degradation and endurance of polyimide films. In: *Proceedings of the IEEE International Symposium on Electrical Insulation (IEEE)*, 3–6 June, Toronto, ON, pp. 88–91.

Van Beusichem B., Ruberto M.A. (2015) Introduction to polymer additives and stabilization. Product quality research institute (http://pqri.org/wp-content/uploads/2015/08/pdf/Polymer_Additives_PQRI_Poster.pdf).

Van Krevelen D.W., Nijenhuis K.Te. (2009) *Properties of Polymers. Their Correlation with Chemical Structure; Their Numerical Estimation and Prediction from Additive Group Contributions*, 4th ed. Elsevier, Amsterdam, the Netherlands.

Wang Y., Rubner M.F., Buckley J. (1991) Stability studies of electrically conducting polyheterocycles. *Synth. Met.* **41–43**, 1103–1108.

Wang Y., Rubner M.F. (1990) Stability studies of the electrical conductivity of various poly (3-alkylthiophenes). *Synth. Met.* **39**, 153–175.

White J.R., Turnbull A. (1994) Weathering of polymers: mechanisms of degradation and stabilization, testing strategies and modeling, *J. Mater. Sci.* **29**, 584–613.

Wilen C.E., Pfaendner R. (2013) Improving weathering resistance of flame-retarded polymers. *J. Appl. Polym. Sci.* **129**, 925–944.

Yang X., Loos J., Veenstra S.C., Verhees W.J.H., Wienk M.M., Kroon J.M., et al. (2005) Nanoscale morphology of high-performance polymer solar cells. *Nano Lett.* **5**, 579–583.

# 2 Metal Oxide in Humidity Sensors

## 2.1 MAIN ADVANTAGES AND LIMITATIONS OF METAL OXIDES AND THEIR CHARACTERIZATION

At the present time, the materials for commercially developed humidity sensors are mainly organic polymer films and porous ceramics. However, all the available materials show some limitations. As it was shown earlier in Chapter 1, polymer films cannot operate at high temperature and high humidity, and show degradation upon exposure to UV, some solvents, or to electrical shocks. Ceramics, in particular metal oxides, have shown advantages in terms of their mechanical strength, their chemical resistance in most environments, good reproducibility of the electrical properties, and their thermal and physical stability (Nitta 1981, 1988; Fagan and Amarakoon 1993; Pelino et al. 1994; Traversa 1995). In addition, it was established that ceramic materials possess a unique structure consisting of grains, grain boundaries, surfaces, and pores, which makes them suitable for humidity sensors. Metal oxides tested as humidity-sensitive materials are listed in Table 2.1. A great amount of various oxides were tested from binary oxides to multicomponent ones (Blank et al. 2016). It was also shown that metal oxides can be used in the development of all types of humidity sensors, including ionic, electronic, capacitive, and solid-electrolyte type sensors (read Vols. 1 and 2). This means that metal oxides used in humidity sensors can be dielectrics or conductors, have electronic and ionic conductivity, and have semiconductor properties of $n$- and $p$-types.

Metal oxides indicated in Table 2.1 also have a different crystallographic structure. A large group of metal oxides are binary oxides having the structure of corundum ($\alpha$-$Al_2O_3$, $\alpha$-$Fe_2O_3$; $In_2O_3$), wurtzite (ZnO) cubic ($\gamma$-$Fe_2O_3$; $In_2O_3$), rutile ($TiO_2$, $SnO_2$), or tetragonal ($WO_3$) types. Spinel compounds ($AB_2O_4$: $ZnCr_2O_4$; $MgAl_2O_4$; $MgFe_2O_4$; $ZnCr_2O_4$; etc.), perovskites ($AXO_3$: $BaTiO_3$; $LaFeO_3$; $NaTaO_3$; $ZnSnO_3$; $CaZrO_3$; etc.), and solid solutions ($TiO_2$-$SnO_2$; etc.) also form large groups of humidity-sensitive oxides. Metal oxides can also be amorphous ($\gamma$-$Al_2O_3$) or refer to various compounds that do not have a specific name, such as $MnWO_4$, $Bi_2WO_6$, $Na_2Ti_3O_7$. This suggests that the crystallographic structure of metal oxides is not a factor in determining their humidity-sensitive properties.

As for the shortcomings, it was also established that the problems for metal-oxide humidity sensors are mainly related to their need for periodic regeneration by heat cleaning to recover their humidity-sensitive properties; prolonged exposure to humid environments leads to the gradual formation of stable chemisorbed OH- on the surface, causing a progressive drift in the resistance of the ceramic humidity sensor (Traversa 1995). The hydroxyl ions are removed by heating to temperatures higher than 400°C (Morimoto et al. 1969). It is important that contaminants, which act in the same way as chemisorbed water, may also be removed by heating. Therefore, the most advanced metal oxide humidity sensors have a built-in heater for this purpose (Qu and Meyer 1998; Kang and Wise 2000).

## 2.2 MECHANISMS OF HUMIDITY SENSITIVITY

Various humidity-sensing mechanisms and operating principles have been identified for metal oxide ceramics (Yamazoe et al. 1979; Fagan and Amarakoon 1993; Traversa 1995; Chen and Lu 2005; Farahani et al. 2014). It was established that adsorption, diffusion, and condensation are the main processes of humidity sensing. It was also shown that the type of conduction may be ionic or electronic one. Most current humidity sensors are based on porous sintered bodies of ionic-type humidity-sensitive ceramics. Other humidity sensors using different sensing mechanisms are of the solid-electrolyte and semiconductor types (Yagi and Ichikawa 1993), or use the hetero contacts between $p$- and $n$-type semiconducting oxides (Ushio et al. 1993). All these mechanisms have been analyzed previously in Volume 2 (Chapters 10–19) and therefore will not be considered in detail in this chapter. More details on mechanisms of water interaction with metal-oxide surfaces can be found in the published excellent articles and reviews (Morimoto et al. 1969; McCafferty et al. 1971; Yamazoe et al. 1979; Thiel and Madey 1987; Freund 1995; Brown et al. 1999; Henderson 2002; Diebold 2003; Chen and Lu 2005; Mu et al. 2017). We will only give a diagram illustrating the main stages of water adsorption on the surface of metal oxides (see Figure 2.1).

At the first stage of adsorption, a water molecule is chemically adsorbed on an activated site (Figure 2.1a) to

**TABLE 2.1**

**Examples of Metal Oxides Tested as Humidity Sensitive Materials**

| Sensor Type | Metal Oxide Used |
|---|---|
| *Resistive* (electronic) | $ZnO$; $Fe_2O_3$; $WO_3$; $TiO_2$; $SnO_2$; $SnO_2$:Sb; $TiO_2$–$SnO_2$; $Cr_2O_3$–$WO_3$; $ZnO/TiO_2$; Al-doped $ZnO$:$TiO_2$; $TiO_2$–$WO_3$; $ZnMoO_4$:$ZnO$; $SrSnO_3$; $BaTiO_3$; $Sr_{1-x}La_xSnO_3$; etc. |
| *Resistive* (ionic) | $Al_2O_3$; $AlO(OH)$; $Li^+$-doped $Fe_2O_3$; $MgAl_2O_4$; $SnO_2$:$ZrO_2$; $MnO_2$–$Mn_3O_4$; $MgCr_2O_4$-$TiO_2$; $ZnCr_2O_4$-$LiZnVO_4$; $TiO_2$–$Cu_2O$–$Na_2O$; $KTaO_3/TiO_2$; $TiO_2$–$K_2O$–$LiZnVO_4$; $K^+$-doped $SnO_2$–$LiZnVO_4$; $Li^+$-doped $NiMoO_4$–$MoO_3$, $CuMoO_4$–$MoO_3$ and $PbMoO_4$–$MoO_3$; $MgAl_2O_4$–$MgFe_2O_4$; etc. |
| *Capacitive* | $ZnO$; $TiO_2$; $WO_3$; $ZrO_2$; $Al_2O_3$; $Mn_2O_3$; $Fe_3O_4$–Si; $Ta_2O_5$–Si; $ZnO/TiO_2$; $TiO_2$–$WO_3$; Mo–$SiO_2$; $Li^+$-doped $Fe_2O_3$; $LaFeO_3$; $MgAl_2O_4$; $BaTiO_3$; $Bi_2WO_6$; $Ba_{0.5}Sr_{0.5}TiO_3$; $TiO_2$–$K_2O$–$LiZnVO_4$; $K^+$-doped $LaCo_{0.3}Fe_{0.7}O_3$; etc. |
| *Impedance* | $ZnO$; $In_2O_3$; $HfO_2$; $ZrO_2$; p-$CuO$; $ZnO$–$SiO_2$; $ZnO$–$In_2O_3$; Cu–Zn/$CuO$–$ZnO$; $TiO_2$:$ZnO$; $TiO_2$:$ZrO_2$; $TiO_2$–$WO_3$; $Fe_2O_3$/$SiO_2$; Zn/$CuO$–$ZnO$; $K^+$-doped $ZnO$, $SnO_2$, $TiO_2$, $TiO_2$–$SnO_2$; $Li^+$-doped $ZnO$, $ZrO_2$–$TiO_2$; $Mn^{2+}$-doped $ZnO$; $BaTiO_3$; $NaTaO_3$; $ZnSnO_3$; $LaFeO_3$; $ZrTiO_4$; $ZnWO_4$; $Bi_2WO_6$; $Na_2Ti_3O_7$; In-doped $CaZrO_3$; $Na^+$-doped $Al_2O_3$-$SiO_2$-$ZrO_2$; $Bi_{0.5}K_{0.5}TiO_3$; $ZnCr_2O_4$–$K_2CrO_4$; $MgCr_2O_4$–$TiO_2$; $Bi_{0.5}Na_{0.5}TiO_3$–$Bi_{0.5}K_{0.5}TiO_3$; $TiO_2$–$K_2O$–$LiZnVO_4$; etc. |
| *p-n and hetero-junctions* | $NiO/ZnO$; $CuO/ZnO$; $La_2CuO_4/ZnO$; $ZnO/Si$ |
| *Schottky barrier* | $Pd/ZnO$; $Au/ZnO$; $Ag/ZnO$ |
| *QCM* | $ZnO$; $Al_2O_3$; $TiO_2$; $SnO_2$ |
| *SAW* | $ZnO$; $TiO_2$; $LiNbO_3$ |
| *Cantilever, membrane* | $Fe_2O_3$; $Al_2O_3$; $SnO_2$; $ZnO$ |
| *Optical and fiber-optic* | $TiO_2$; $SnO_2$; $MgO$; $Co_3O_4$; $In_2O_3$; $In_2O_3$:Sn; $ZnO$ |

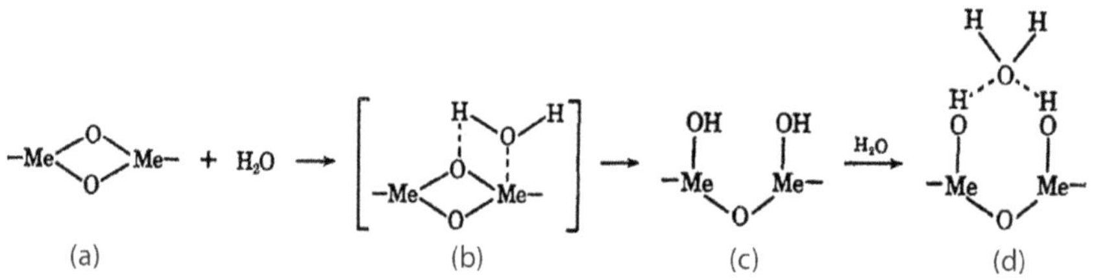

**FIGURE 2.1** Four stages of the adsorption. (Reprinted with permission from Morimoto, T. et al., *J. Phys. Chem.*, 73, 243–248, 1969. Copyright 1969 American Chemical Society.)

form an adsorption complex (Figure 2.1b), which subsequently transfers to surface hydroxyl groups (Figure 2.1c). Then, another water molecule comes to be adsorbed through hydrogen bonding on the two neighboring hydroxyl groups as shown in Figure 2.1d. The top water molecule condensed cannot move freely due to the restriction from the two hydrogen bonding (Figure 2.1d). Thus, this layer or the first physically adsorbed layer is immobile, and there are not hydrogen bonds formed between the water molecules in this layer. Therefore, no proton could be conducted in this stage. As water continues to condense on the surface of the ceramic, an extra layer on the top of the first physically adsorbed layer forms (Figure 2.2). This layer is less ordered than the first physically adsorbed. For example, there may be only one hydrogen bond locally. If more layers condensed, the ordering from the initial surface may gradually disappear, and protons may have more and more freedom to move inside the condensed water through the Grotthuss mechanism. In other words, from the second physisorbed layer, water molecules become mobile and finally almost identical to the bulk liquid water, and the Grotthuss mechanism becomes dominant (Agmon 1995). A brief illustration of the Grotthuss mechanism is shown in Figure 2.3.

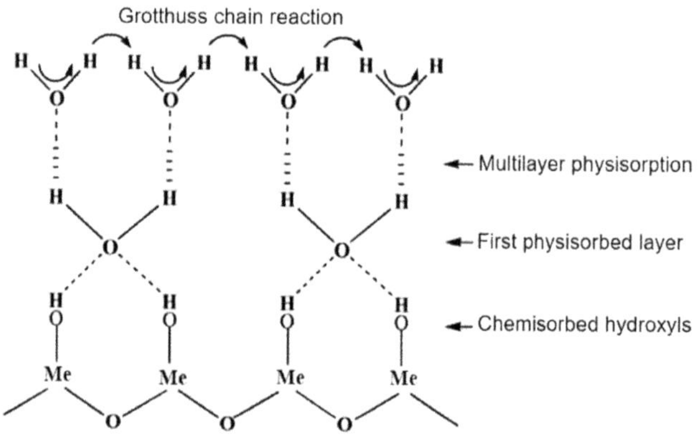

**FIGURE 2.2** Multilayer structure of condensed water. (Idea from McCafferty, E. and Zettlemoyer, A.C., *Faraday Discussions*, 52, 239–254, 1971; Wang Z. et al., *Nanotechnology*, 22, 275502, 2011.)

It is important to note that despite the similarity of the processes occurring on the surface of metal oxides during the adsorption of water, each metal oxide has its own specific interaction with water vapor. For example, on the $Fe_3O_4(111)$ and rutile $TiO_2(110)$ surfaces, water dissociation is exothermic and thermodynamically more preferred than

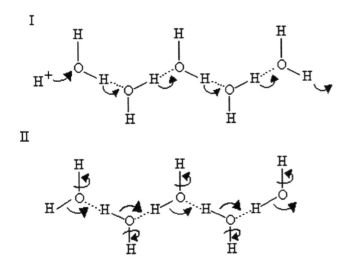

**FIGURE 2.3** Brief illustration of the Grotthuss mechanism. Water is a good conductor of protons because of the H-bonded networks between water molecules. Conduction occurs through a "hop-turn" mechanism, first suggested by Grotthuss, and often referred to as the Grotthuss mechanism. According to this mechanism, in the "hop" part of the mechanism, a proton first hops from the end of the H-bonded chain to an adjacent group (I, right); transfer of H-bond strength then allows it to be replaced by a $H^+$ binding at the other end, to give the structure in II. In the "turn" phase, rotation of the waters as shown in II then restores the starting structure (I). (Idea from De Grotthuss, C.J.T., *Ann. Chim.*, LVIII, 54–74, 1806; Nagle, J.F. and Morowitz, H.J., *Proc. Natl. Acad. Sci. USA*, 75, 298–302, 1978; Cukierman S.I., *Biochimica et Biophysica Acta, Bioenerg.*, 1757, 876–885, 2006.)

molecular adsorption at low coverage (Zhou and Cheng 2012). However, the dissociation processes need to overcome moderate activation barriers. To this end, elevated temperatures are necessary to promote the reaction. At high coverage, the hydrogen-bonding network formed by the preadsorbed water molecules and the dissociated –OH surface species can mediate the dissociation, and coexistence of molecularly and dissociatively adsorbed species is often observed. Surface defects, particularly O vacancies, can facilitate the dissociation. On the $\alpha$-$Al_2O_3(0001)$ surface, water dissociation was found to be both kinetically and thermodynamically facile at both low and high coverage. As a consequence, the surface is normally hydroxylated upon exposure to air.

It should also be borne in mind that in semiconductor metal oxides the adsorption of water in molecular and hydroxyl forms may be accompanied by electronic exchange between the adsorbed molecules and the conduction band. This exchange is accompanied by the band-bending induced by the formation of surface hydroxyls, which leads to an increase in the conductivity of $n$-type ceramics and to a decrease in the conductivity of $p$-type ceramics (Boyle and Jones 1977; Avani and Nanis 1981). This effect has been attributed to the donation of electrons from the chemically adsorbed water molecules to the metal-oxide surface (Boyle and Jones 1977). The magnitude of the band bending and conductivity change depends strongly on the electronic structure of the metal oxide and

concentration of point defects. The differences in the electronic structure lead to different acido-basic properties of the surface sites and, consequently, to the different degree of water dissociation. As a result, it leads to the formation of qualitatively different hydrogen-bond networks, which govern amounts and structure of adsorbed water, as well as proton-transfer and molecular diffusion dynamics. In the process of adsorption, water molecules can also replace the previously adsorbed and ionized oxygen ($O^-$, $O^{2-}$, etc.), which also releases the electrons from the ionized oxygen (Yamazoe et al. 1979; Shimizu et al. 1989). This process also has the "donor effect." The surface geometry also affects the dissociation of water and the thermodynamic stability of adsorbed layers. In this case, different surfaces can stabilize either molecular or dissociative adsorption, depending on distances between surface sites, which may strengthen or weaken hydrogen bonding between neighboring molecules (Vlcek et al. 2012).

Thus, in metal oxides, the conductivity, depending on the state of the surface, can have an electron or proton nature. In the latter case, the charge transport is governed by proton hopping between neighboring molecules in the continuous film of adsorbed water. Proton conductivity is possible only at room temperature ($<100°C$), when moisture can effectively condense on the surface. At higher temperatures, chemisorption of water molecules is only responsible for changes in the electrical conductivity of ceramics. It should be also noted that sensors based purely on water-phase protonic conduction would not be quite sensitive to low humidity, at which the water vapor could rarely form continuous mobile layers on the sensor surface. It was established that the two immobile layers, the chemisorbed and the first physisorbed ones, cannot contribute to proton-conducting activity. However, experiment and simulations have shown that in this case the electron tunneling between donor water sites is possible (Khanna and Nahar 1986; Yeh et al. 1989). The tunneling effect, along with the energy induced by the surface anions, facilitates electrons to the hop along the surface that is covered by the immobile layers and therefore contributes to the conductivity. This mechanism is quite helpful for detecting low humidity levels, at which there is not effective protonic conduction. Nonetheless, the tunneling effect is definitely not the semiconducting mechanism. However, this mechanism was used to explain the humidity-sensing effects in porous $Al_2O_3$ (Khanna and Nahar 1986) and $TiO_2$ (Yeh et al. 1989).

## 2.3 MULTILAYER WATER ADSORPTION AT SOLID SURFACES

Let's look now at some theoretical background that allows multilayer water adsorption phenomena on low-temperature solid surfaces to be analyzed. To do this, we use the description and the results of the calculations presented by Helwig et al. (2009). It is important to note that the phenomena described in the present section take place on the surfaces of all solid-state materials, including metal oxides and various semiconductors and dielectrics.

### 2.3.1 Brunauer-Emmett-Teller (BET) Isotherm

It is known that multilayer adsorption of gases on solid surfaces is described by the BET isotherm. This isotherm was first introduced by Brunauer et al. in 1938. The main assumptions leading to this isotherm are the following (Dorfler 2002):

1. Polymolecular adsorption of spherical particles;
2. Homogeneous surface with constant adsorption centre density;
3. The heat of adsorption of the first adsorbed monomolecular layer drops to the heat of liquefaction for all following layers;
4. All following polymolecular layers act like a liquid; and
5. The number of adsorbate layers can reach infinity.

With these assumptions the relative surface coverage $\Theta(T, P)$ is obtained:

$$\theta(P,T) = \left(\frac{P}{P_0(T) - p}\right) \cdot \left[\frac{c(\mathrm{T})}{1 + \left(\dfrac{P}{P_0(T)}\right) \cdot (c(T) - 1)}\right] \quad (2.1)$$

In this equation, $P$ and $P_0$ represent the equilibrium and saturation vapor pressures of the adsorbate gas at the temperature $T$, and $c(T)$ is the BET constant:

$$c(T) = exp\left(\frac{\varepsilon_1 - \varepsilon_L}{k_B \cdot T}\right) \quad (2.2)$$

In this equation, $\varepsilon_1$ and $\varepsilon_L$ represent the heats of adsorption and liquefaction, respectively. A particularly simple form of the BET isotherm is obtained in case the first layer is adsorbed as tightly as all the others. Then $c(T) = 1$ and the expression for the BET isotherm reduces to Eq. (2.3).

$$\theta(P,T) = \frac{P}{P_0(T) - P} \quad (2.3)$$

It must be borne in mind that there exists a temperature $T_{sat}$ at which the vapor pressure $P$ in the air equals the saturation pressure $P_0(T_{sat})$. At all temperatures $T$ lower than $T_{sat}$, the maximum vapor pressure $P(T)$ is limited by the saturation pressure at that temperature, that is, $P(T) = P_0(T)$. Equation (2.3) then predicts that under conditions of saturated water vapor, that is, at $T \leq T_{sat}$, a solid surface will adsorb infinitely many layers of water and thus form a closed layer of condensed water on the surface. For temperatures higher than $T_{sat}$, the water vapor pressure in the air is supply limited and lower than $P_0(T)$. In this limiting case, $P$ is much smaller than $P_0(T)$, and the BET equation reduces to the familiar Langmuir isotherm (Langmuir 1916):

$$\theta(P,T) = \frac{c(T) \cdot P}{c(T) \cdot P + P_0(T)} \quad (2.4)$$

As $P_0(T)$ increases exponentially with temperature, the surface coverage $\Theta(T, P)$ in this temperature range drops exponentially as the temperature is increased beyond $T_{sat}$.

### 2.3.2 Water Adsorption at Solid Surfaces

Considering the main constituents of the ambient air ($N_2$, $O_2$, $H_2O$, and $CO_2$) as well as the range of analyte gases that have frequently been studied in gas sensing experiments, water vapor is outstanding insofar as it can occur in high concentrations (Figure 2.4) and at the same time exhibits a high boiling point (Figure 2.5a) and a concomitantly high heat of vaporization (Figure 2.5b). Because of these properties, adsorbed water vapor can easily undergo multilayer adsorption discussed before, leading to thin liquid electrolyte layers on solid surfaces.

In order to estimate the water coverage on a metal-oxide surface using the BET equation, an estimate of the BET constant (2) is required. The value of this constant depends on two parameters: the heat of liquefaction $\varepsilon_L$ and the heat of adsorption $\varepsilon_1$. The heat of liquefaction of water is known from the literature [$\varepsilon_L = 0.424$ eV per molecule (Kohlrausch 1986)]. The heat of adsorption describes the adsorption energy of the first monolayer on the adsorbent surface. This latter value can vary strongly with the chemical composition and the morphology of the adsorbing solid. A first estimate for the heat of adsorption of water on MOX surfaces is 0.52 eV (Qi et al. 2000). With $c(T)$ being known, the surface coverage of water on a MOX surface can now be obtained from Eq. (2.1).

Assuming that the water vapor saturation in the air is reached at 30°C, it is revealed from Figure 2.6 that a MOX surface will always be covered by multiple layers of water as long as the substrate temperature remains at room temperature or below ($0 < T < 30$°C). In case the substrate temperature is raised beyond the saturation temperature $T_{sat}$, the water surface coverage will rapidly drop into the range of a single monolayer as $T \sim 70$°C is approached. As a single monolayer of

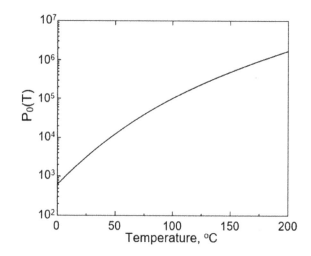

**FIGURE 2.4** Saturation vapor pressure of water as a function of the temperature in the air. (Reprinted from Helwig A. et al., *J. Sensors*, 2009, 620720, 2009. Published by Hindawi Publishing Corporation as open access.)

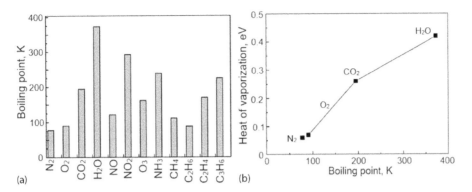

(a)　(b)

**FIGURE 2.5** (a) Boiling points of the main air constituents and of a number of often-studied analyte gas molecules; and (b) heat of vaporization of the main air constituents. (Reprinted from Helwig A. et al., *J. Sensors*, 2009, 620720, 2009 Published by Hindawi Publishing Corporation as open access.)

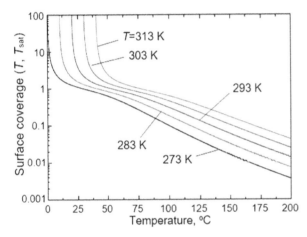

**FIGURE 2.6** Surface coverage with water as a function of the MOX substrate temperature as evaluated from the BET isotherm. The heat of adsorption of the first monolayer was assumed to be $\varepsilon_1 = 0.52$ eV (Qi et al. 2000). (Reprinted from Helwig A. et al., *J. Sensors*, 2009, 620720, 2009 Published by Hindawi Publishing Corporation as open access.)

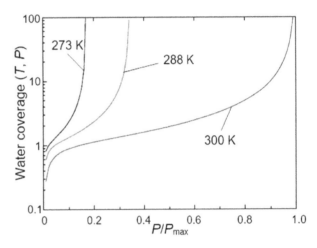

**FIGURE 2.7** Water coverage as a function of vapour pressure and substrate temperature (low-temperature region $T \leq T_{sat}$). $T_{sat} = 27°C$ (300 K). (Reprinted from Helwig A. et al., *J. Sensors*, 2009, 620720, 2009 Published by Hindawi Publishing Corporation as open access.)

water is sufficient to prevent oxygen ionosorption, the normal combustive detection mechanism still will not work (Helwig et al. 2009). This means that in this temperature range the conductivity should be controlled mainly by Grotthuss mechanism described before in Section 2.2. However, it must be borne in mind that even in this case, cross sensitivity is possible. It occurs when gas or vapor molecules are dissolved in the surface water layer and dissociated there. This indirect sensing mechanism is similar to what is observed in pH sensors (Yates et al. 1974).

As it is seen in Figure 2.7, raising the substrate temperature toward 150°C, the water coverage drops to about 0.1 monolayers, which means that the large patches of free semiconductor surface start to occur, providing adsorption sites for oxygen. As a consequence, it can be expected that at temperatures above 150°C, MOX gas sensors can have a conductivity response to combustible gases and water vapors controlled by a chemisorption mechanism.

Figures 2.7 and 2.8 refer to two situations in which the water vapor pressure in the gas phase is varied from zero to the

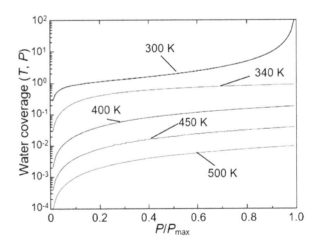

**FIGURE 2.8** Water coverage as a function of vapor pressure and substrate temperature (high-temperature region $T \geq T_{sat}$). $T_{sat} = 27°C$ (300 K). (Reprinted from Helwig A. et al., *J. Sensors*, 2009, 620720, 2009 Published by Hindawi Publishing Corporation as open access.)

saturation vapor pressure, considering that the air temperature is 30°C. The first diagram reveals that with the MOX substrate being kept at room temperature or below, a multilayer water adsorption will take place also at water vapor pressures much lower than the saturation pressure level. A normal conductivity response characteristic for heated sensors (Barsan and Weimar 2001; Korotcenkov and Sysoev 2011; Korotcenkov 2012; Barsan and Schierbaum 2018), therefore, cannot be expected in this lower $H_2O$ vapor pressure range as well. This situation changes when the MOX substrates are heated to 150°C or more. In this latter case, the surface water coverage varies over the entire $H_2O$ pressure range and remains in the submonolayer range. In such situations, an oxygen ionosorption and a normal combustive gas response can occur, and, in addition, single $H_2O$ molecules may directly adsorb on the MOX surface (Zhou and Cheng 2012) and thus give rise to a water-vapor-dependent response signal (Prades et al. 2012; Vlcek et al. 2012)

## 2.4 HOW TO OPTIMIZE SENSOR PERFORMANCE

Earlier in Volume 2 (Chapters 10–19), a detailed analysis was given of all types of humidity sensors developed on the basis of metal oxides. Therefore, in this chapter we consider only the approaches used to optimize metal-oxide humidity sensors.

As a result of the analysis, it was concluded that regardless of the type of humidity sensors, to achieve their optimal parameters, the metal oxides used in these devices should have high-surface reactivity with water and a controlled microstructure (Shimizu et al. 1985). First of all, they should have open porosity, great specific surface area and a large pore volume, as well as they should be highly porous in order to allow water vapor to pass easily through the pores and condensate in the capillary-like pores between the grain surfaces (Shimizu et al. 1985; Zhang et al. 2017; Jeseentharani et al. 2018). Water condensation in pores, besides the electrolytic

conduction occurring in the liquid layer of water condensed within capillary pores, gives the maximum contribution to the change in the dielectric constant of the metal oxide. It is important to note that the specific surface area is the principal microstructure for sensing humidity under low relative humidity (RH) conditions, while the pore volume dominates under high RH conditions. Under low humidity, water adsorption on the sample surface is likely the dominant factor for electrical conduction. Therefore, a higher surface area would provide more sites for water adsorption and produce more charge carriers for electrical conduction. Under high RH, for example, above 80%, as a rule, mesopores are already filled with water and therefore exactly the pore volume determines the total quantities of water condensed in metal-oxide matrix, that is, the magnitude of the change in dielectric constant.

Optimal samples should also have a specific distribution of pore sizes, since water condenses in the pores with different radii as a function of the RH according to the Kelvin equation (Shah et al. 2007). The smaller the value of pore size, the more easily water condensation takes place (De Bore 1958). Best results were shown by the samples with a wide distribution of pore sizes from 2 nm up to radii in which capillary condensation cannot take place at any humidity at their operating temperatures (Shimizu et al. 1985). According to the prediction of the Kelvin equation, water vapor starts to condense at room temperature in mesopores of size 2 nm around 15% RH and continues to around 100 nm under saturated atmosphere. The presence of micropores causes a slow response time, while the absence of micropores is accompanied by a decrease in sensitivity. Compacts with a given microstructure can be produced by controlling the different relatively simple steps of the ceramic production process, such as mixing raw materials, forming the part, and sintering (Gusmano et al. 1991). In particular, the required porosity and surface area is easily produced by controlling the sintering conditions. An example of such process is shown in Figure 2.9. It was also shown that the reduction in the grain size led to more

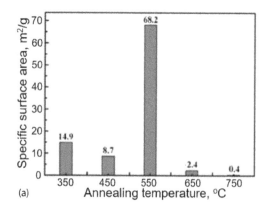
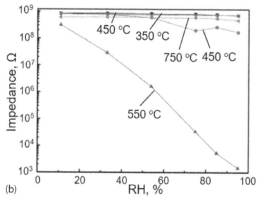

**FIGURE 2.9** Influence of sintering on the surface area and humidity sensitivity of $Bi_{3.25}La_{0.75}Ti_3O_{12}$ (BLT): (a) A specific surface area comparison for BLT powders; and (b) impedance versus RH curves of BLT-350, BLT-450, BLT-550, BLT-650, and BLT-750 at 100 Hz. It is seen that BLT-550 with maximum specific surface area exhibited the best humidity sensitivity. (With kind permission from Springer Science+Business Media: *J. Electron. Mater.*, Effect of annealing temperature on of $Bi_{3.25}La_{0.75}Ti_3O_{12}$ powders for humidity sensing properties, 46, 377–385, Zhang Y., et al. Copyright 2017: Springer.)

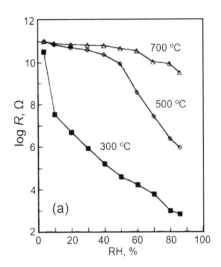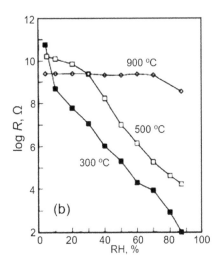

**FIGURE 2.10** The RH dependence of the resistance of (a) K$^+$-doped TiO$_2$ (10%); and (b) iron-oxide films heated to various temperatures, measured at 40°C. (Reprinted from *Sens. Actuators B*, 31, Traversa E. et al., Ceramic thin films by sol-gel processing as novel materials for integrated humidity sensors, 59–70, Copyright 1996, with permission from Elsevier.)

grain boundaries and pores. Usually the materials with grain size of nanometer order have nanoporous structure. That is why the nanograined materials offer higher surface area for water-vapor detection and have more surface active sites and stronger absorption ability in comparison with micrograined materials (Tai et al. 2003; Biju and Jain 2008; Faia et al. 2009; Biswas et al. 2013).

It is important to note that sintering, apparently, is one of the most crucial processes in the production of humidity sensors. As can be seen from the data shown in Figure 2.10, there is always an optimal annealing temperature conducive to achieving maximum sensitivity. Moreover, even relatively small deviations from these temperatures are accompanied by significant changes in the sensor response, which requires careful monitoring of the sensor manufacturing process. It should also be borne in mind that the so-called optimal sintering modes are not universal and depend both on the nature of the material used and the technology of their synthesis.

As for the surface reactivity with water vapor, according to theoretical modeling and experimental studies, it was established that water on the surface of metal oxides really adsorbs both molecularly and dissociatively (Brown et al. 1999; Diebold 2003). At that, water molecules react predominantly with defect sites, such as point defects and terraces. Moreover, it was found that the exposed coordinatively unsaturated metal cations, which act as Lewis acid sites, are also the preferred sites for water adsorption. There, water can strongly interact via its oxygen lone pairs and can be stabilized in its molecular form even at room temperature. The stronger the water–metal interaction is, the more acidic the O–H proton becomes thus leading to higher Brønsted acidity for the adsorbed water molecules and a higher propensity for its dissociation. This means that an improvement in the humidity-sensor performance can be achieved through optimization of the stoichiometry of the metal oxides used. This is confirmed by experiments showing

that the adsorption state of the interfacial water (molecular vs. dissociative) can change dramatically even for the same material as the structure and termination of the surface changes (Mu et al. 2017). To affect the stoichiometry of metal oxides, heat treatments can be used in an appropriate atmosphere or doping with appropriate additives.

It was also shown that variation in chemical composition of ceramic materials permits both performance optimization in sensors exploiting their electrical properties and tailoring to specific requirements (Traversa 1995; Blank et al. 2016). As a rule, this effect is achieved by optimizing the microstructure of humidity-sensitive metal oxides and increasing the concentration of mobile charge carriers (Table 2.2). For example, most TiO$_2$-based sensors have low sensitivity, especially at low humidity levels. The experiment showed that doping with electrolytes or ions, for example, P$_2$O$_5$ (Makita et al. 1997) and alkali ions (Slunecko et al. 1996; Traversa et al. 1996; Jain et al. 1999; Ying et al. 2000) may improve the conductivity of TiO$_2$ and thus considerably enhance their sensitivity. In particular, for K$^+$-doped TiO$_2$ film, it was capable to sense humidity levels lower than 10% (Traversa et al. 1996). Adding SnO$_2$ in TiO$_2$ increases its porosity and thus enhances the sensitivity at high RH range (>70% RH) (Tai and Oh 2002a). In addition, bilayered TiO$_2$/SnO$_2$ (Tai and Oh 2002b), and TiO$_2$/Al-doped ZnO (Tai and Oh 2003) were reported to have less hysteresis than pure TiO$_2$ and Al-doped ZnO.

Of course, the surface reactivity can also significantly change due to the formation of composites and solid solutions of the required composition. For example, Gu et al. (2011) believe that improved performance of TiO$_2$-ZnO humidity sensors in comparison with ZnO ones takes place since TiO$_2$-ZnO materials are more hydrophilic than ZnO due to dissociative adsorption of water at Ti$^{3+}$ defect sites.

Processes occurring in nanocomposites and multicomponent oxides were considered in detail by Korotcenkov and

**TABLE 2.2**

**Additives Improving the Performance of Humidity Sensors**

| Additive | Based Oxide | Effect |
|---|---|---|
| $K^+$ (KCl, $K_2O$), $Li^+$, $Na^+$ | $SnO_2$; $TiO_2$; $Fe_2O_3$; ZnO; $SiO_2$; $ZrO_2$–$TiO_2$; $TiO_2$–$SnO_2$; $SnO_2$–$LiZnVO_4$; $NiMoO_4$–$MoO_3$; $CuMoO_4$–$MoO_3$; $PbMoO_4$–$MoO_3$; $Al_2O_3$-$SiO_2$-$ZrO_2$; $TiO_2$–$K_2O$–$LiZnVO_4$; $SnO_2$–$K_2O$–$LiZnVO_4$; etc. | Direct contribution of $K^+$, $Li^+$, $Na^+$ to the conduction as a mobile carrier |
| La, Sr, Y, Nb, Ga, In, Sn, Mn doping | ZnO; $MgFe_2O_4$; $BaZrO_3$; $SrCeO_3$; $CoAl_2O_4$; $BaAl_2O_4$; $ZnAl_2O_4$; etc. | Microstructure optimization (increase in porosity and surface area). Influence on conductivity via increasing concentration of defects, charge carriers, and adsorption centers |
| Forming nanocomposites and multicomponent metal oxides | $TiO_2$–$SnO_2$; ZnO-$TiO_2$; ZnO–$In_2O_3$; $TiO_2$–$WO_3$; $Zn_2SnO_4$–$LiZnVO_4$; $Mg_{0.9}Sn_{0.1}Fe_2O_4$; $Fe_2O_3$-$SiO_2$; $Al_2O_3$-$SiO_2$-$ZrO_2$; $ZnMoO_4$-ZnO; $Cr_2O_3$–$WO_3$; $NiMoO_4$–$MoO_3$; $MnO_2$–$Mn_3O_4$; etc. | Improvement of the microstructure (decrease in grain and pore sizes). Influence on adsorption properties |
| Carbon nanotubes, graphene oxide (GO) | $TiO_2$; $WO_3$:Fe | Optimization of structure. Specific surface properties of CNTs and GO |

Cho (2017). They have shown that in addition to improving the parameters of the sensors, the complexity of the composition creates certain difficulties that may limit the possibilities of this approach in the development of humidity sensors designed for the market. First, the use of nanocomposites, mixed metal oxides, and multicomponent metal oxides does not provide a considerable improvement in gas-sensing characteristics in comparison with binary metal oxides. Often, improvement in the parameters of nanocomposite sensors is being observed when compared with materials having a very low sensitivity, which is not peculiar to the material with the parameters optimal for use in gas and humidity sensors. This means that in some cases, optimization of technological process and control of the grain size of the binary oxides is more acceptable solution when developing sensors than using nanocomposites.

Second, the complex nature of these materials limits their use for integrated gas and humidity sensors especially fabricated using a thin-film technology. The large number of elements in these metal oxides makes it hard to deposit thin films with good and repeatable stoichiometric ratios. It is also difficult to control the size and morphology of the oxide composites, which, in turn, has influence on their physicochemical properties. Furthermore, the results obtained in one laboratory while developing composite-based gas sensors cannot be correctly reproduced in another one. Technologies are unique in the use of technological equipment, precursors, and their concentrations; they have a specific temperature and time parameters of synthesis, annealing and the subsequent formation of the gas sensitive layer. Therefore, there is always a factor or parameter that cannot be reproduced when transferring technology.

Third, one should take into account that the complication of the composition of the gas sensing matrix is always being accompanied by deterioration of sensor parameters' reproducibility. Too many additional factors, which can affect gas-sensing properties of materials, appear in complex metal oxides and nanocomposites. It was established that the change

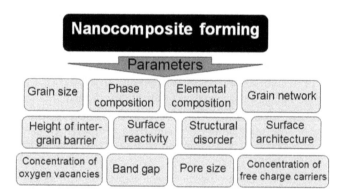

**FIGURE 2.11** Parameters subjected to modification during metal oxide–metal oxide nanocomposite formation. (Reprinted from *Sens. Actuators B*, 244, Korotcenkov, G. and Cho, B.K., Metal oxide composites in conductometric gas sensors: Achievements and challenges, 182–210, Copyright 2017, with permission from Elsevier.)

of additional component concentration is the same powerful factor of influencing the gas-sensing materials' parameters as the change of the deposition temperature (synthesis) or technological route. All main parameters of gas-sensing material (see Figure 2.11) significantly depend on the type and the concentration of additives. This means that for achievement of required reproducibility of sensor parameters, we must provide careful monitoring of a large number of parameters of both material and technological process. At that, fabrication technology used for a nanocomposite preparation must be able to effectively control the grain size, concentration, and dispersion of all phases presented in nanocomposite.

Fourth, for achieving optimal effect, we have to find a specific composition of nanocomposites because as a rule an optimizing effect is being observed at certain concentration of one of the components only. When deviations from the optimal concentration in one direction or another occur, as a rule, sensor performance significantly deteriorates (see Figure 2.12). One of the mechanisms of such a change is shown in Figure 2.13. One can see that at low concentration

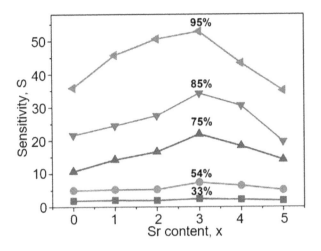

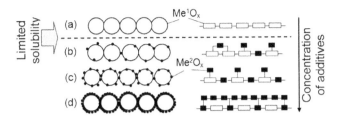

FIGURE 2.13 The change in the film structure and electrical scheme of the film conductivity during a phase modification of metal oxides. (Reprinted from *Sens. Actuators B*, 107, Korotcenkov, G., Gas response control through structural and chemical modification of metal oxides: State of the art and approaches, 209–232, Copyright 2005, with permission from, Elsevier.)

FIGURE 2.12 Humidity sensitivity curves of $Ba_{5-x}Sr_xNb_4O_{15}$ solid solutions at 11%–95% RH. Metal oxides were sintered at 900°C for 4 h. (Reprinted from *Ceram. Intern.*, 44, Ji, G.-J. et al., Molten-salt synthesis of $Ba_{5-x}Sr_xNb_4O_{15}$ solid solutions and their enhanced humidity sensing properties, 477–483, Copyright 2018, with permission from Elsevier.)

of additive, the second phase only modifies surface activity and the surface properties of the base oxide (Figure 2.13b). At somewhat higher concentration, it can contribute to limiting the electroconductivity of the metal-oxide matrix due to the formation of additional interfaces on the way of the current transport (Figure 2.13c). The appearance of a second phase may also help to decrease the grain size and optimize structure of pores and intergrain necks. Figure 2.14 shows how the grain sizes change in composites with a change in

the concentration of components. And at the final stage (Figure 2.13d), at a certain combination of electroconductivity and gas sensitivity for two metal-oxide phases in the gas-sensing matrix, the second oxide phase can produce either full blockage of the contacts between the grains of the base oxide or shunting of the matrix of the base oxide through a more conductive second metal-oxide phase.

Fifth, it is important to know that sometimes in devices, elaborated on the base of nanocomposites, the increase of the sensitivity can be attained at the expense of worsening other exploitation parameters of sensors, for example, stability or selectivity.

As for the technological features of the synthesis and the formation of thin and thick layers of humidity-sensitive metal oxides, they will be discussed in detail in a separate Chapters 20 and 23.

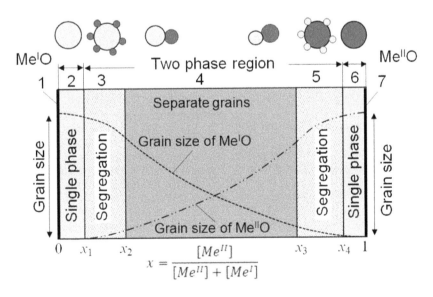

FIGURE 2.14 Scheme of mutual distribution of components in the $M^I O$–$M^{II} O$ nanocomposites. Possible transformation of the crystallite size and the grain structure of the major components of the nanocomposite is also shown in the figure. $x_1$ and $x_4$ correspond to the limit solubility of $Me^{II}$ in $Me^I O$ and $Me^I$ in $Me^{II} O$. (Reprinted from *Sens. Actuators B*, 244, Korotcenkov, G. and Cho, B.K., Metal oxide composites in conductometric gas sensors: Achievements and challenges, 182–210, Copyright 2017, with permission from Elsevier.; Idea from Rumyantseva, M.N. and Gaskov, A.M., *Russ. Chem. Bull. Int. Ed.*, 57, 1106–1125, 2008 and Gas'kov, A. and Rumyantseva, M. Metal-oxide nanocomposites: Synthesis and characterization in relation with gas sensing phenomena, in: Baraton M.I. (Ed.), *Sensors for Environment, Health and Security.* Springer Science + Business Media B.V., Dordrecht, the Netherlands, pp. 3–29, 2009.)

## REFERENCES

Agmon N. (1995) The Grotthuss mechanism. *Chem. Phys. Lett.* **24**, 456–462.

Avani G.N., Nanis L. (1981) Effects of humidity on hydrogen sulfide detection by $SnO_2$ solid state gas sensors. *Sens. Actuators B* **2**, 201–206.

Barsan N., Schierbaum K. (eds.) (2018) *Gas Sensors Based on Conducting Metal Oxides. Basic Understanding, Technology and Applications.* (Elsevier Metal Oxide Series edited by G. Korotcenkov), Elsevier, Cambridge, MA.

Barsan N., Weimar U. (2001) Conduction model of metal oxide gas sensors. *J. Electroceram.* **7**(3) 143–167.

Biju K.P., Jain M.K. (2008) Sol–gel derived $TiO_2$:$ZrO_2$ multilayer thin films for humidity sensing application. *Sens. Actuators B* **128**, 407–413.

Biswas P., Kundu S., Banerji P., Bhunia S. (2013) Super rapid response of humidity sensor based on MOCVD grown ZnO nanotips array. *Sens. Actuators B* **178**, 331–338.

Blank T.A., Eksperiandova L.P., Belikov K.N. (2016) Recent trends of ceramic humidity sensors development: A review. *Sens. Actuators B* **228**, 416–442.

Boyle J.F., Jones K.A. (1977) The effects of CO, water vapor and surface temperature on the conductivity of a $SnO_2$ gas sensor. *J. Electronic Mater.* **6**, 717–733.

Brown G.E. Jr., Henrich V., Casey W., Clark D., Eggleston C., Felmy A. et al. (1999) Metal oxide surfaces and their interactions with aqueous solutions and microbial organisms. *Chem. Rev.* **99**, 77–174.

Brunauer S., Emmett P.H., Teller E. (1938) Adsorption of gases in multimolecular layers. *J. Am. Chem. Soc.* **60**(2), 309–319.

Chen Z., Lu C. (2005) Humidity sensors: A review of materials and mechanisms. *Sens. Lett.* **3**, 274–295.

Cukierman S.I. (2006) Et tu, Grotthuss! and other unfinished stories. *Biochim. Biophys. Acta Bioenerg.* **1757**, 876–885.

De Bore J.H. (1958) *The Shape of Capillaries, the Structure and Properties of Porous Materials.* Butterworth, London, UK, pp. 68–87.

De Grotthuss C.J.T. (1806) Sur la décomposition de l'eau et des corps qu'elle tient en dissolution à l'aide de l'électricité galvanique (On the decomposition of water and of the bodies that it holds in solution by means of galvanic electricity). *Ann. Chim.* **LVIII**, 54–74.

Diebold U. (2003) The surface science of titanium dioxide. *Surf. Sci. Rep.* **48**, 53–229.

Dorfler H.D. (2002) *Grenzflachen und kolloid-Disperse Systeme: Physik und Chemie,* Springer, Berlin, Germany.

Fagan J.G., Amarakoon V.R.W. (1993) Reliability and reproducibility of ceramic sensors: Part III, humidity sensors. *Am. Ceram. Soc. Bull.* **72**, 119–130.

Faia P.M., Ferreira A.J., Furtado C.S. (2009) Establishing and interpreting an electrical circuit representing a $TiO_2$–$WO_3$ series of humidity thick film sensors. *Sens. Actuators B* **140**, 128–133.

Farahani H., Wagiran R., Hamidon M.N. (2014) Humidity sensors principle, mechanism, and fabrication technologies: A comprehensive review. *Sensors* **14**, 7881–7939.

Freund H.-J. (1995) Metal oxide surfaces: Electronic structure and molecular adsorption. *Phys. Stat. Sol. (b)* **192**, 407–440.

Gas'kov A., Rumyantseva M. (2009) Metal oxide nanocomposites: Synthesis and characterization in relation with gas sensing phenomena. In: Baraton M.I. (ed.) *Sensors for Environment, Health and Security.* Springer Science + Business Media B.V., Dordrecht, the Netherlands, pp. 3–29.

Gu L., Zheng K., Zhou Y., Li J., Mo X., Greta Patzke R., Chen G. (2011) Humidity sensors based on $ZnO$/$TiO_2$ core/shell nanorod arrays with enhanced sensitivity. *Sens. Actuators B* **159**, 1–7.

Gusmano O., Montesperelli G., Nunziante P., Traversa E. (1991) Influence of the powder synthesis process on the a.c. impedance response of $MgAl_2O_4$ spinel pellets at different environmental humidities. In: Hirano S.I., Messing G.L., Hausner H. (eds.) *Ceramic Transactions,* Vol. 22, *Ceramic Powder Science IV,* Am. Ceram. Soc., Westerville, OH, pp. 545–551.

Helwig A., Muller G., Sberveglieri G., Eickhoff M. (2009) On the low-temperature response of semiconductor gas sensors. *J. Sensors* **2009**, 620720.

Henderson M.A. (2002) The interaction of water with solid interface: Fundamental aspects revisited. *Surf. Sci. Rep.* **46**, 1–308.

Jain M.K., Bhatnagar M.C., Sharma G.L. (1999) Effect of $Li^+$ doping on $ZrO_2$–$TiO_2$ humidity sensor. *Sens. Actuators B* **55**, 180–185.

Jeseentharani V., Dayalan A., Nagaraja K.S. (2018) Nanocrystalline composites of transition metal molybdate ($Ni_{1-x}Co_xMoO_4$; x=0, 0.3, 0.5, 0.7, 1) synthesized by a co-recipitation method as humidity sensors and their photoluminescence properties. *J. Phys. Chem. Solids* **115**, 75–83.

Ji G.-J., Zhang L.-X., Zhu M.-Y., Li S.-M., Yin J., Zhao L.-X., et al. (2018) Molten-salt synthesis of $Ba_{5-x}Sr_xNb_4O_{15}$ solid solutions and their enhanced humidity sensing properties. *Ceram. Intern.* **44**, 477–483.

Kang U., Wise K.D. (2000) A high-speed capacitive humidity sensor with on-chip thermal reset. *IEEE Trans. Electron. Dev.* **47** (4), 702–710.

Khanna V.K., Nahar R.K. (1986) Carrier-transfer mechanisms and $Al_2O_3$ sensors for low and high humidities. *J. Phys. D: Appl. Phys.* **19**, L141–L145.

Kohlrausch F. (1986) *Praktische Physik 3, Tabellen und Diagramme,* 23 edn. Alberts W.G. (ed.). B.G. Teubner, Stuttgart, Germany.

Korotcenkov G. (2005) Gas response control through structural and chemical modification of metal oxides: State of the art and approaches. *Sens. Actuators B* **107**, 209–232.

Korotcenkov G. (ed.) (2012) *Chemical Sensors: Simulation and Modeling.* Vol. 2: *Conductometric Gas Sensors,* Momentum Press, New York.

Korotcenkov G., Cho B.K. (2017) Metal oxide composites in conductometric gas sensors: Achievements and challenges. *Sens. Actuators B* **244**, 182–210.

Korotcenkov G., Sysoev V. (2011) Conductometric metal oxide gas sensors. In: *Chemical Sensors: Comprehensive Sensor Technologies.* Vol. 4. *Solid State Devices,* Korotcenkov G. (ed.), Momentum Press, New York, pp. 53–186.

Langmuir I. (1916) The constitution and fundamental properties of solids and liquids—part I: Solids. *J. Am. Chem. Soc.* **38**(11), 2221–2295.

Makita K., Nogami M., Abe Y. (1997) Sol–gel synthesis of high-humidity-sensitive amorphous $P_2O_5$–$TiO_2$ films. *J. Mater. Sci. Lett.* **16**, 550–552.

McCafferty E., Zettlemoyer A.C. (1971) Adsorption of water vapour on $\alpha$-$Fe_2O_3$. *Faraday Discuss.* **52**, 239–254.

Morimoto T., Nagao M., Fukuda F. (1969) The relation between the amounts of chemisorbed and physisorbed water on metal oxides. *J. Phys. Chem.* **73**, 243–248.

Mu R., Zhao Z.-J., Dohnálek Z., Gong J. (2017) Structural motifs of water on metal oxide surfaces. *Chem. Soc. Rev.* **46**, 1785–1806.

Nagle J.F., Morowitz H.J. (1978) Molecular mechanisms for proton transport in membranes. *Proc. Natl. Acad. Sci. USA* **75**, 298–302.

Nitta T. (1981) Ceramic humidity sensor. *Ind. Eng. Chem. Prod. Res. Dev.* **20**, 669–674.

Nitta T. (1988) Development and application of ceramic humidity sensors. In: Seiyama T. (ed.) *Chemical Sensor Technology*, Vol. 1, Kodansha, Tokyo/Elsevier, Amsterdam, the Netherlands, pp. 57–78.

Pelino M., Cantalini C., Faccio M. (1994) Principles and applications of ceramic humidity sensors. *Active Passive Elec. Comp.* **16**, 69–87.

Prades J.D., Cirera A., Korotcenkov G., Cho B.K. (2012) Microstructural and surface modeling of $SnO_2$ using DFT calculations. In: Korotcenkov G. (ed.) *Chemical Sensors: Simulation and Modeling.* Vol. 1. *Microstructural Characterization and Modeling of Metal Oxides.* Momentum Press, New York, pp. 81–161.

Qi S., Hay K.J., Rood M.J., Cal M.P. (2000) Equilibrium and heat of adsorption for water vapor and activated carbon. *J. Environ. Eng.* **126**(3), 267–271.

Qu W., Meyer J.U. (1998) A novel thick-film ceramic humidity sensor. *Sens. Actuators B* **40**, 175–182.

Rumyantseva M.N., Gaskov A.M. (2008) Chemical modification of nanocrystalline metal oxides: Effect of the real structure and surface chemistry on the sensor properties. *Russ. Chem. Bull. Int. Ed.* **57**, 1106–1125.

Shah J., Kotnala R.K., Singh B., Kishan H. (2007) Microstructure-dependent humidity sensitivity of porous $MgFe_2O_4$–$CeO_2$ ceramic. *Sens. Actuators B* **128**, 306–311.

Shimizu Y., Arai H., Seiyama T. (1985) Theoretical studies on the impedance-humidity characteristics of ceramic humidity sensors. *Sens. Actuators* **7**, 11–22.

Shimizu Y., Shimabukuro M., Arai H., Seiyama T. (1989) Humidity-sensitive characteristics of $La^{3+}$-doped and undoped $SrSnO_3$. *J. Electrochem. Soc.* **136**, 1206–1210.

Slunecko J., Holc J., Kosec M., Kolar D. (1996) Thin film humidity sensor based on porous titania. *Informacije Midem-J. Microelectronics Electron. Comp. Mater.* **26**, 285–290.

Tai W., Oh J. (2002a) Fabrication and humidity sensing properties of nanostructured $TiO_2$–$SnO_2$ thin films. *Sens. Actuators B* **85**, 154–157.

Tai W., Oh J. (2002b) Preparation and humidity sensing behaviors of nanocrystalline $SnO_2$/$TiO_2$ bilayered films. *Thin Solid Films* **422**, 220–224.

Tai W.-P., Kim J.-G., Oh J.-H. (2003) Humidity sensitive properties of nanostructured Al-doped $ZnO$:$TiO_2$ thin films. *Sens. Actuators B* **96**, 477–481.

Thiel P.A., Madey T.E. (1987) The interaction of water with solid surfaces: Fundamental aspects. *Surf. Sci. Rep.* **7**, 211–385.

Traversa E. (1995) Ceramic sensors for humidity detection: The state-of-the-art and future developments. *Sens. Actuators B* **23**, 135–156.

Traversa E., Gnappi G., Montenero A., Gusmano G. (1996) Ceramic thin films by sol-gel processing as novel materials for integrated humidity sensors. *Sens. Actuators B* **31**, 59–70.

Ushio Y., Miyayama M., Yanagida H. (1993) Fabrication of thin film $CuO$/$ZnO$ hetero-junction and its humidity-sensing properties. *Sens. Actuators B* **12**, 135–139.

Vlcek L., Ganesh P., Bandura A., Mamontov E., Predota M., Cummings P.T., Wesolowski D.J. (2012) Modeling interactions of metal oxide surfaces with water. In: Korotcenkov G. (ed.) *Chemical Sensors: Modeling and Simulation*, Vol. 1. *Microstructural Characterization and Modeling of Metal Oxides*, Momentum Press, New York, NY, pp. 217–262.

Wang Z., Shi L., Wu F., Yuan S., Zhao Y., Zhang M. (2011) The sol–gel template synthesis of porous $TiO_2$ for a high-performance humidity sensor. *Nanotechnology* **22**, 275502.

Yagi H., Ichikawa K. (1993) Humidity sensing characteristics of a limiting current type planar oxygen sensor for high temperatures. *Sens. Actuators B* **13**, 92–95.

Yamazoe N., Fuchigami J., Kishikawa M., Seiyama T. (1979) Interactions of tin oxide surface with $O_2$, $H_2O$ and $H_2$. *Surf. Sci.* **86**, 335–344.

Yates D.E., Levine S., Healy T.W. (1974) Site-binding model of the electrical double layer at the oxide/water interface. *J. Chem. Soc. Faraday Trans.* **70**, 1807–1818.

Yeh Y., Tseng T., Chang D. (1989) Electrical properties of porous titania ceramic humidity sensors. *J. Am. Ceram. Soc.* **72**, 1472–1475.

Ying J., Wan C., He P. (2000) Sol–gel processed $TiO_2$–$K_2O$–$LiZnVO_4$ ceramic thin films as innovative humidity sensors. *Sens. Actuators B* **62**, 165–170.

Zhang Y., He J., Yuan M., Jiang B., Li P., Tong Y., Zheng X. (2017) Effect of annealing temperature on of $Bi_{3.25}La_{0.75}Ti_3O_{12}$ powders for humidity sensing properties. *J. Electron. Mater.* **46**(1), 377–385.

Zhou C., Cheng H. (2012) density functional theory study of water dissociative chemisorption on metal oxide surfaces. In: Korotcenkov G. (ed.) *Chemical Sensors: Modeling and Simulation*, Vol. 1. *Microstructural Characterization and Modeling of Metal Oxides*, pp. 263–292.

# 3 Al₂O₃ as a Humidity-Sensitive Material

## 3.1 Al₂O₃-BASED HUMIDITY SENSORS

A detailed analysis of the properties of $Al_2O_3$ and features of its use in humidity sensors can be found in Chen and Lu (2005). According to this analysis, $Al_2O_3$ is one of the most favorable ceramic humidity-sensing materials. Moreover, porous $Al_2O_3$ was one of the first ceramics to be used in humidity capacitance-type sensors. The first humidity-sensitive $Al_2O_3$ layer formed through anodization on the Al metal surface was reported in 1953 (Ansbacher and Jason 1953). The anodization was carried out in 3% $H_2CrO_3$ at 50 V. The capacitance increased linearly while the resistance decreased exponentially to the relative humidity (see Figure 3.1a). With increasing temperature, the capacitance (Figure 3.1b) and the sensitivity of the sensors decrease (Juhasz and Mizsei 2010).

Today this material is being used in many commercial impedance humidity sensors (read Chapter 28)—on the one hand, because the etching technology is well established. On the other hand, $Al_2O_3$ has proven to be stable at elevated temperature and at high humidity level (Khanna and Nahar 1984; Sberveglieri et al. 1994, 1995; Nahar and Khanna 1998; Mai et al. 2000; Nahar 2000; Varghese and Grimes 2003; Cheng et al. 2011). Such sensors have relatively good properties. It was established that humidity sensors based on the porous anodic aluminum-oxide ($Al_2O_3$) layer with optimal structure exhibit high sensitivity and rapid response, and they are useful in a wide range of humidity detection scenarios (Nahar 2000). They are relatively stable, have small hysteresis, work at wide range of temperature and pressure changes, and they do not require any specific service (Chakraborty et al. 1999a). Some of $Al_2O_3$-based humidity sensors have also very good linear response (Chakraborty et al. 1999b).

As a rule, $Al_2O_3$-based impedance humidity sensors have sandwiched structure, where porous $Al_2O_3$ film is between a top-porous Au electrode and an under-gate Al electrode (Chakraborty et al. 1999a). The typical structure is shown in Figure 3.2. Sometimes, humidity sensors have a more complex structure. Chakraborty et al. (1999b) suggested use of the $SiO_2/Si_3N_4/Ta/Ta_2O_5/Al_2O_3$ structure. The sensor is excited with a low-voltage alternating current at a fixed frequency. It is widely established that the porosity and morphology of the $Al_2O_3$ layer is responsible for unique properties of alumina-based humidity sensors. On a microscopic level, the aluminum oxide appears as a matrix with many parallel pores. When exposed to even small amounts of water vapor, the superstructure enables water molecules to permeate into the matrix where adsorption and microcondensation occur (Khanna and Nahar 1984; Nahar et al. 1984).

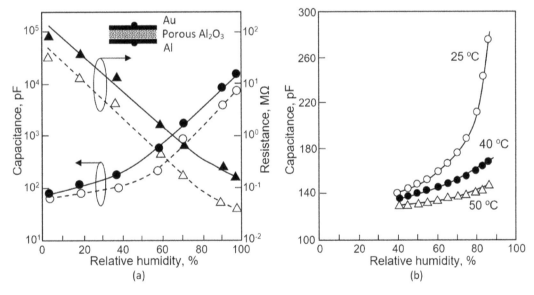

**FIGURE 3.1** (a) Typical experimental C (•, ○) and R (Δ, ▲) characteristics of an $Al_2O_3$ sensor measured at 120 Hz (full curves) and 1 kHz (broken curves). The porous $Al_2O_3$ film was grown in 10% $H_2SO_4$ electrolyte at 10 mA/cm up to a thickness of 10 μm. The film porosity was 16% and the average pore diameter 13 nm; and (b) sensing characteristics of $Al_2O_3$-based humidity sensors at different temperatures. The Al was anodized in 20 wt.% sulfuric acid solution in a single step at 10 V. (a) (Reprinted with permission from Nahar R.K. et al. (1984) On the origin of the humidity-sensitive electrical properties of porous aluminium oxide. *J. Phys. D.,* 1, 2087–2095, 1984. Copyright 1984, Institute of Physics.) (b) (Reprinted from *Procedia Eng.* 5, Juhasz, L. and Mizsei, J., A simple humidity sensor with thin film porous alumina and integrated heating, 5, 701–704, Copyright 2010, with permission from Elsevier.)

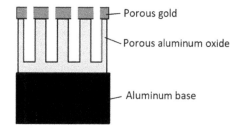

**FIGURE 3.2** The structure of $Al_2O_3$-based humidity sensor.

Mechanisms of these processes were discussed in Chapter 2. Since the dielectric constants of dry gases are significantly lower than for gases containing moisture (about an 80:1 ratio for nitrogen or standard air), each pore acts as a microcapacitor. As the microcapacitors are in a parallel arrangement, the total capacitance is additive. In essence, the sensor acts as a water molecule counter.

The impedance of the sensor relates to the water-vapor pressure by the following relationship:

$$Z^{-1} = A \cdot e^{B \cdot P_W} + C \qquad (3.1)$$

where $Z$ is the impedance; $P_w$ is a partial pressure of water; and $A$, $B$, and $C$ are constants. Each sensor is calibrated at multiple dew/frost points, the partial pressure of water being a function of the dew/frost point temperature ($T_d$). The impedance at each dew/frost point is recorded and entered into a digital look-up table either imbedded in the memory of the sensor module or programmed into an analyzer. The analyzer utilizes a polynomial expansion equation to convert the measured impedance by reference to the look-up table to produce direct readout in dew/frost point temperature. Typical accuracy is $\pm 2^{\circ}C$ $T_d$ from $+60^{\circ}C$ to $-65^{\circ}C$ $T_d$ and $\pm 3^{\circ}C$ $T_d$ from $-66^{\circ}C$ to $-110^{\circ}C$ $T_d$.

For analysis of the impedance of the sensor, the equivalent circuit is usually used as shown in Figure 3.3a (Falk et al. 1992; Gonzalez et al. 1999). However, Chachuiski et al. (2004) believe that the model (Figure 3.3b) (Nitsch et al. 1998), which includes diffusion processes, better describes the results obtained, especially in the low-frequency range.

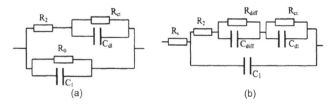

**FIGURE 3.3** (a, b) Electric equivalent circuits for aluminum humidity sensor usually used for analysis of the impedance of the humidity sensors. The $C_i$ capacity and the $R_0$ resistance correspond to impedance of the alumina layer between layer of the aluminum and the external electrode. $R_2$ corresponds to the resistance along walls of pores, $R_{ct}$ and $C_{dl}$ express the resistance and the capacitance of the oxide layer between the bottoms of the canals and upper golden electrode. $R_{diff}$ and $C_{diff}$ correspond to the diffusive processes of adsorbed water in pores and have significant meaning at low frequencies.

As for the mechanism of conductivity of $Al_2O_3$ film in a humid atmosphere, according to Khanna and Nahar (1986), the low-humidity and high-humidity sensitivity of $Al_2O_3$ sensors are controlled by different surface-conduction mechanisms. Khanna and Nahar (1986) believe that the aqueous adsorbed layer can provide both electron migration and ion migration. At low humidities, an electron-tunneling mechanism is responsible for the carrier transport, while at high humidities a proton-transport process is dominated. According to Khanna and Nahar (1986), an electron tunneling takes place between donor water sites. Statz and De Mars (1958) have shown that water introduces donor surface states near the Fermi level. The surface anions also provide donor energy levels and help in this conduction mechanism. The hopping conductivity is expected to be of the form (Madelung 1978).

$$\sigma_W = \sigma_0 \cdot exp\left(-\frac{2r}{\lambda} - \frac{E_a}{k_B T}\right) \qquad (3.2)$$

where $\sigma_0$ is a temperature-independent pre-exponential factor, $r$ is the tunneling distance, $\lambda$ is the localization length of water state, $E_a$ is the activation energy, and $k_B$ is the Boltzmann constant. The mechanism of proton conductivity was described in Chapter 2.

Experimental studies have shown that sensor parameters of $Al_2O_3$-based devices (sensitivity, humidity range, response time, stability, etc.) strongly depend on the thickness and the density of the porous layer, and the size of the pores as well (Figure 3.4). It was found that the minimum detectable humidity decreases as the pore radius decreases (Banerjee and Sengupta 2002; Varghese et al. 2002). This means that the detection limit could be set very low by shrinking the pore size. For example, Chen et al. (1990) when optimizing the pore size, achieved sensitivity as low as 1 ppmv. Chakraborty et al. (1999b) also reported that their $Al_2O_3$-based humidity sensors exhibited sensitivity less than 1 ppmv. It was also established that the diameters and depths of the pores can be controlled by the technological parameters, such as concentration and temperature of the electrolyte and the current density in the anodizing cell. In particular, $Al_2O_3$ exhibits a pore size distribution that depends on the etching parameters, such as current density (Masuda et al. 1993). Chen et al. (2014) have shown that even a magnetic field can be used to improve the parameters of humidity sensors (see Figure 3.5). Chen et al. (2014) believe that the water molecules under a magnetic field preferentially form water clusters in which the dipole moments of the water molecules have a more orderly arrangement.

Cunningham and Williams (1980) and Seiyama et al. (1983) established that the response time of ceramic humidity sensors is generally limited by diffusion. It was shown that in nanoporous $Al_2O_3$ in contrast to molecular oxygen (that moves rapidly in and out of the alumina), the exchange of water vapor is found very slow (Svensson et al. 2010). This means that the response and recovery times are controlled by the size of pores and the thickness of the layer (Sun et al. 1989). However, Chakraborty et al. (1999b) have shown

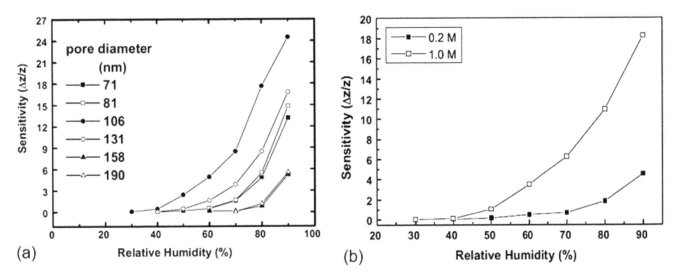

**FIGURE 3.4** The influence of (a) the pore size; and (b) the concentration of electrolyte on the sensitivity of impedance sensor to relative humidity. Alumina templates were prepared in phosphoric acid electrolyte under 185 V anodization voltage. (Reprinted with permission from *Sens. Actuators A*, 174, Almasi Kashi, M. et al., Capacitive humidity sensors based on large diameter porous alumina prepared by high current anodization, 174, 69–74, Copyright 2012 with permission from Elsevier.)

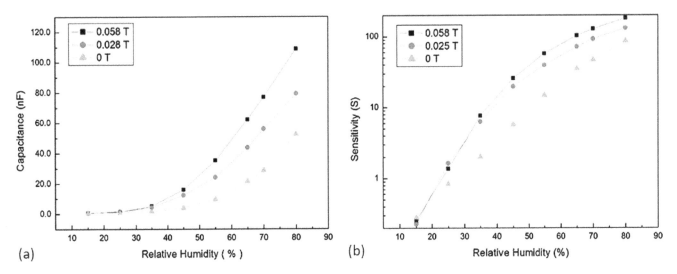

**FIGURE 3.5** (a) Capacitance vs. relative humidity (RH) of porous AAO sensor measured at 1 kHz with various magnetic field strengths (0 ~ 0.058 T); and (b) mean value of the sensitivity of the sensor to humidity. (Reprinted with permission from *Sens. Actuators B*, 2014, Chen S.W. et al., Sensitivity evolution and enhancement mechanism of porous anodic aluminum oxide humidity sensor using magnetic field, 199, 384–388. Copyright 2014: Elsevier.)

that in certain conditions this response can be fast. According to Chakraborty et al. (1999b), at optimal structure of porous layer, the response time can be less than 1 second. In other conditions, the response time can reach several minutes. As a rule, to achieve a fast response, the pore size should be maximal (see Figure 3.6), and the thickness of the sensitive layer should be minimal. The thinner layer means water molecules will travel faster in and out of the sensor pores; therefore, the response will also be faster. The thinner oxide layer of the humidity sensor results also in bigger capacitance changes because capacitance is inversely proportional to the distance of the capacitors' plates from each other (the distance between the aluminum core and the gold film deposited on the oxide layer). Exactly this approach was used in Teledyne Analytical Instruments to develop humidity sensors of 8800 series (www.teledyne-ai.com). From the empirical point of view, it is clear that when developing devices with highly humidity-sensitive properties in a wide range of humidity, it is necessary to use the Al$_2$O$_3$ layer with a large pore volume and a wide dispersion of pore sizes over from nano- to macropores (Traversa 1995). Thus, a controlling of microstructure of the Al$_2$O$_3$ layer, as well as for metal oxides, is important for sensing characteristics, and the film preparation conditions should be carefully chosen.

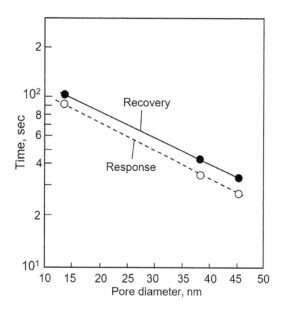

**FIGURE 3.6** Dependence of the response time and recovery time on an average pore size. Porous $Al_2O_3$ films were fabricated using anodization process. (Reprinted from Dickey, E.C. et al., *Sensors*, 2, 91–110, 2002. Published by MDPI as open access.)

It is important to note that in addition to impedance humidity sensors, porous $Al_2O_3$ can also be used in the design of other types of humidity sensors. For example, Boytsova et al. (2015) on the base of porous $Al_2O_3$ developed cantilever arrays. In the humidity range of 10%–20% RH, the sensitivity of porous anodic aluminum oxide cantilevers ($\Delta f/\Delta m$) and the humidity sensitivity were about 56 Hz/pg and about 100 Hz/%, respectively.

## 3.2 FEATURES OF Al ANODIZATION

### 3.2.1 GENERAL CONSIDERATION

Although amorphous AAO (anodic aluminum oxide) was found to be humidity-sensitive in 1950s, but until 1978 the researchers did not discover that it could form regular microstructure (Thompson et al. 1978). In general, the anodizing of aluminum can result in two different types of oxide film: a barrier-type anodic film and a porous oxide film (see Figure 3.7). It was generally accepted that the nature of an electrolyte used for anodizing aluminum is a key factor, which determines the type of oxide grown on the surface. The barrier-type anodic films on aluminum that a strongly adherent, non-porous, and non-conducting can be formed by anodizing in neutral or basic solutions (pH = 5–7) in which the anodic oxide layer is not chemically affected and stays practically insoluble. These films are extremely thin and dielectrically compact. The group of electrolytes used for this barrier-type film formation includes boric acid, ammonium borate, ammonium tartrate, and aqueous phosphate solutions, as well as tetraborate in ethylene glycol, perchloric acid with ethanol, and some organic electrolytes such as citric, malic, succinic, and glycolic acids (Sulka 2008). A characteristic of a barrier-type oxide films is that the thickness of the oxide is

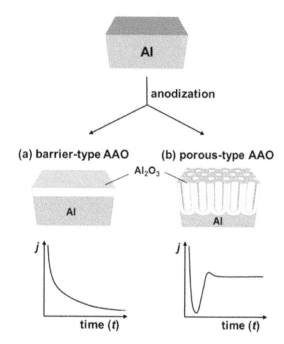

**FIGURE 3.7** Two different types of anodic aluminum oxide (AAO) formed by (a) barrier-type; and (b) porous-type anodizations, along with the respective current ($j$)–time ($t$) transients under potentiostatic conditions. It has been experimentally verified that the thickness of barrier-type film is directly proportional to the applied potential ($U$). On the other hand, current density ($j$) in porous-type anodization under potentiostatic conditions remains almost constant within a certain range of values during the anodization process, due to the constant thickness of the barrier layer at the pore bottom. The thickness of the resulting porous oxide film is linearly proportional to the total amount of charge (i.e., anodization time, $t$) involved in the electrochemical reaction. (Reprinted with permission from Lee, W. and Park, S.-J., *Chem. Rev.*, 114, 7487–7556, Copyright 2014 with permission from American Chemical Society.)

not affected by the electrolyzing time or temperature of the electrolyte but only affected by the applied voltage (about 1.4 nm/V). The maximum thickness of the barrier oxide is restricted by the oxide breakdown at high voltages, typically around 500–700 V.

In contrast, porous aluminum oxides are most commonly grown in strongly acidic electrolytes, such as dilute sulfuric acid, typically 10 wt%. But there are also commercial processes using phosphoric acid, chromic acid, oxalic acid, and mixtures of inorganic and organic acids (see Table 3.1). Typical electrolytes used to produce porous $Al_2O_3$ layer have a pH that is less than 5. These electrolytes slowly dissolve the forming oxide layer. Electrolytes that are composed of less concentrated acids tend to produces oxide coatings that are harder, thicker, less porous, and more wear resistant than those composed of higher concentrated acids.

It was established that by using Al anodization, porous $Al_2O_3$ films 100-μm thick can easily be made. At that, unlike barrier films, a high voltage is not needed to make a thick porous film because of the unique structure of these films. Moreover, these porous structures can be prepared in a simple wet chemistry lab without complicate equipment (Figure 3.8).

**TABLE 3.1**

**Major Acid Components of Typical Electrolyte Types Used to Produce Porous Oxide Layer on an Aluminum Substrate**

| Main Acid Used in Electrolyte | Molecular Formula | Concentration (M) | Pore Size Range (nm) |
|---|---|---|---|
| Chromic | $H_2CrO_4$ | 0.3 to 0.44 | 17 to 100 |
| Sulfuric | $H_2SO_4$ | 0.18 to 2.5 | 12 to 100 |
| Phosphoric | $H_3PO_4$ | 0.04 to 1.1 | 30 to 235 |
| Acetic | $CH_3CO_2H$ | 1 | Not specified |
| Citric | $HO_2CCH_2(OH)(CO_2H)CH_2CO_2H$ | 0.1 to 2 | 90 to 250 |
| Tartaric | $HO_2CCH(OH)CH(OH)CO_2H$ | 0.1 to 3 | Not specified |
| Glycolic | $CH_2(OH)CO_2H$ | 1.3 | 35 |
| Malic | $HO_2CH_2CH(OH)CO_2H$ | 0.15 to 0.3 | Not specified |
| Malonic | $CH_2(CO_2H)_2$ | 0.1 to 5 | Not specified |
| Oxalic | $C_2H_2O_4$ | 0.2 to 0.5 | 20 to 80 |

*Source:* Poinern G.E.J. et al., *Materials*, 4, 487–526, 2011. Published by MDPI as open access.

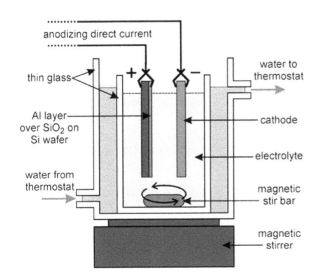

**FIGURE 3.8** Typical configuration of experimental equipment used to produce anodized aluminum oxide. (Reprinted from *Thin Solid Films*, 517, Juhász L. and Mizsei J., Humidity sensor structures with thin film porous alumina for on-chip integration, 6198–6201, Copyright 2009, with permission from Elsevier.)

Typical anodization working ranges and the resulting pore-to-pore distances are 10–25 V and 35–70 nm for sulfuric acid, and 30–60 V and 80–150 nm for oxalic acid, respectively. These conditions are considered "mild anodization," and the AAO growth rate is relatively slow. Recently, the working ranges for oxalic acid has been extended to 120–150 V with the corresponding pore-to-pore distance expanded to 220–300 nm under the "hard anodization" condition (Lee et al. 2006). With a combined oxalic acid anodization followed by phosphoric acid anodization at 185 V, a pore-to-pore distance over 400 nm can be reached. Anodization time can vary from 10 minutes to 10 hours (Poinern et al. 2011).

It is necessary to take into account that during anodization, the surface of aluminum heats up (the reaction involved in the anodic oxidation of aluminum is exothermic), leading to a requirement for a cooling system (Mankotia et al. 2014). Therefore, during the anodization, temperature should be kept lower than room temperature to prevent the formed oxide structure from being dissolved in acidic electrolyte. The temperature should be set between 5°C and 20°C for oxalic acid and between 0°C and 5°C for sulfuric acid. The second reason to keep the temperature low as possible is to avoid a local heating at the bottom of pores during anodization. The local heat causes an inhomogeneous electric field distribution at the bottom, leading to electrical breakdown of the oxide. In fact, bursts and crack of the oxide film are generated if porous film was produced without temperature controlling. The speed of growth of porous alumina is affected by the temperature. The lower the temperature is, the lower is the growth rate.

An idealized sketch of the film structure is shown in Figure 3.9. It is seen that the oxide has a cellular structure with a central pore in each cell. The pores (a pore along with its hexagonal vicinity is called a "cell") are arranged in a regular-packed hexagonal pattern of close-packed cylindrical pores perpendicular to the metal surface with a uniform radius, and the spacing between any two pores (or "interpore distance" or "periodicity") is the same (Lu and Chen 2011). All pores also have uniform depth. The average number of pores on the unit surface area is called the packing density of the pore array. Between the pore bottom and the aluminum metal layer that has not been anodized, there is an alumina film, called the barrier layer. As follows from this model, hexagonal cells should be uniform, but the experiment showed that most anodization conditions produce films with more disorder, with a distribution of cell sizes and pore diameters (Asoh et al. 2001). The typical image of AAO is shown in Figure 3.10.

Cell and pore dimensions depend on bath composition, temperature, and voltage, but the end result is always an extremely high density of the fine pores. The nanopore formation based on the driving force from self-assembly demonstrates beautifully the simplicity and power of the bottom-up approach.

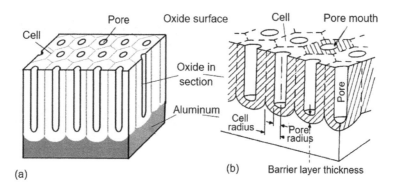

**FIGURE 3.9** (a, b) Idealized structure of anodic porous aluminum oxide represented by various authors: (a) (Reprinted with permission from Asoh, H. et al., *J. Electrochem. Soc.*, 148, B152–B156, Copyright 2001, with permission from Electrochemical Society); and (b) (Reprinted by permission from Macmillan Publishers Ltd., *Nature*, Furneaux, R.C. et al., 1989, Copyright 1989, Nature).

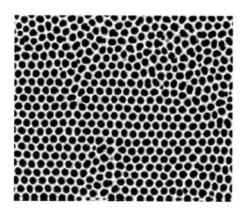

**FIGURE 3.10** Typical image of honeycomb structure of anodic aluminum oxide (AAO) formed in an acidic electrolyte solution. It is seen that AAO consists of highly aligned nanopores "self-assembled" in a hexagonally close-packed pattern). (Reprinted from Asoh, H. et al., *J. Electrochem. Soc.*, 148, B152–B156, Copyright 2001, with permission from Electrochemical Society.)

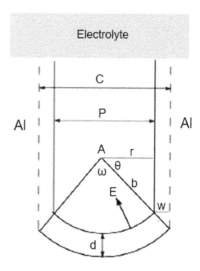

**FIGURE 3.11** Common model of a single pore in AAO: A—the center of curvature, E—electrical field, C—cell diameter, P—pore diameter, $r$—pore radius, $d$—barrier layer thickness, $w$—thickness of cell wall.

The diameters and depths of the pores can be controlled by tuning the anodization conditions. The naturally occurred pore diameter and pore-to-pore distance (the cell diameter) in anodic porous $Al_2O_3$ films range between ~30 and 300 nm and between ~50 nm and submicron, respectively. The pore diameter is typically 1/3 to 1/2 of the cell diameter. For example, the film thickness of 20–50 μm and 40–60 nm diameter cells with 20 nm pores are typical for coatings grown in sulfuric acid.

### 3.2.2 MECHANISM OF PORE FORMATION

The formation mechanism of AAO membranes was first proposed in the early fifties (Keller et al. 1953), and then was expanded significantly in the seventies (O'Sullivan and Wood 1970) and the nineties (Thompson and Wood 1983). This model is based on a field-assisted dissolution and field-assisted growth of the oxide, in which the heterogeneity of $Al_2O_3$ dissolution is caused and maintained by various electric densities on different sites on the film. When describing the process of pore formation, AAO is generally regarded as a membrane, consisting of straight nanopores with a spherical bottom less than a hemisphere

(see Figure 3.11). The contour of the cell base pattern can be described with a center of curvature and a radius of curvature. The center of curvature (A) is continuously moving downward during anodization. The active layer during the growth of nanopore is the bottom barrier layer with thickness ($d$).

There are two active interfaces on the barrier layer. The outer one is associated with anodization of aluminum to alumina, while the inner one is associated with dissolution and deposition of alumina to and from the etching solution. In the early stages of the anodization process, $Al^{3+}$ ions migrate from the metal across the metal/oxide interface into the forming oxide layer (Patermarakis 1998). Meanwhile, $O^{2-}$ ions formed from water at the oxide/electrolyte interface travel into the oxide layer. During this stage, approximately 70% of the $Al^{3+}$ ions and the $O^{2-}$ ions contribute to the formation of the barrier-oxide layer (Palbroda 1995), the remaining $Al^{3+}$ ions are dissolved into the electrolyte. This condition was shown to be the prerequisite for the porous oxide growth, in which the Al–O bonds in the oxide lattice break to release

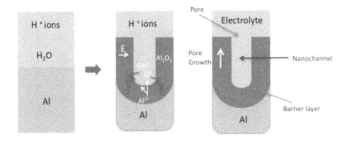

**FIGURE 3.12** Schematic of the pore formation mechanism in an acidic electrolyte. (Reprinted from Poinern, G.E.J. et al., *Materials*, 4, 487–526, 2011. Published by MDPI as open access.)

Al³⁺ ions (Shawaqfeh and Baltus 1999). During the oxide formation, the barrier layer constantly regenerates with further oxide growth and transforms into a semi-spherical oxide layer of constant thickness that forms the pore bottom, as shown in Figure 3.12. This electrochemical process can be described by the following reduction-oxidation (i.e., redox) equations (Thompson and Wood 1983; Despic and Parkhutik 1989; Li et al. 1998b):

1. Al³⁺ ions form at the metal/oxide interface, distribute in the oxide layer near the oxide/metal interface (it is assumed that an oxide layer is present on the surface of Al):

$$Al \rightarrow Al^{3+} + 3e^-  \qquad (3.3)$$

2. Aluminum cations diffuse within the oxide-barrier layer.
3. The electrolysis of water (a water-splitting reaction) occurs at the pore bottom near the electrolyte/oxide interface:

$$2H_2O \rightarrow 2O^{2-} + 4H^+  \qquad (3.4)$$

4. Due to the electric field, the O²⁻ ions migrate within the barrier layer from the electrolyte/oxide interface to the oxide/metal interface, and react with the Al³⁺ ions there, forming Al₂O₃:

$$2Al^{3+} + 3O^{2-} \rightarrow Al_2O_3(s)  \qquad (3.5)$$

5. There is an electric-field-enhanced oxide dissolution at the electrolyte/oxide interface:

$$Al_2O_3(s) + 6H^+(aq) = 2Al^{3+}(aq) + 3H_2O  \qquad (3.6)$$

6. At the electrolyte-cathode interface (cathode), hydrogen evolution takes place:

$$6H^+(aq) + 6e^- \rightarrow 3H_2(g)  \qquad (3.7)$$

The process of porosification is driven by the local electric field (*E*), which is defined by the applied current (*I*) over conductivity (s) times the surface area of the spherical bottom.

At the bottom of a pore, the electric field is converged, and thus a high field density is expected. As a result, the rate of the field-assisted dissolution at the pore bottom is faster than in other regions, and the pore is deepened through the anodization process. During the anodization, if a pore was widened, the field convergence is then reduced, and the field-assisted dissolution at the bottom becomes inhabited, the pore will finally be flattened and disappear. In another case, if a pore is initially narrower than average, the intensive electric field at its bottom will cause fast dissolving, and the diameter will become larger. Under a constant applied current (or potential *E*) and during equilibrium growth, each nanopore reaches an optimized solid angle, and radius of curvature *b* (Wang et al. 2003). Each nanopore slowly moves its position with respect to its closest neighbors in order to even out the mechanical stress among them. This self-assembly process leads to nanopores with uniform pore diameter and arranged in a two-dimensional hexagonally close packed array. This self-assembly process in AAO formation is a slow process and takes hours to reach equilibrium.

### 3.2.3 Parameters of Anodic Porous Alumina and How to Control Them

As it follows from the previous consideration, anodic porous alumina can be characterized by parameters such as pore diameter or radius, interpore distance, and porosity. The experiment has shown that the diameter of the pore is controlled by the field-assisted anodization conditions. A summary of the influence of anodizing parameters on the pore diameter of a nanostructure formed under potentiostatic conditions is presented schematically in Figure 3.13. If the solution composition and temperature remain unchanged through the whole process, the average radius (*r*) of the pores is simply controlled by the applied bias voltage ($U_a$) (see Figure 3.14). It was verified that the average radius is almost proportional to the applied voltage ($r \sim U_a$), and therefore high voltage

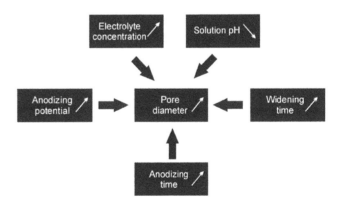

**FIGURE 3.13** The influence of anodization parameters on the pore diameter of nanostructure formed by anodization of aluminum at potentiostatic regime. (From Sulka, G.D.: Highly ordered anodic porous alumina formation by self-organized anodizing, in: Eftekhari A. (ed.), *Nanostructured Materials in Electrochemistry*, 2008. Copyright Wiley-VCH Verlag GmbH & Co. KGaA. Reproduced with permission.)

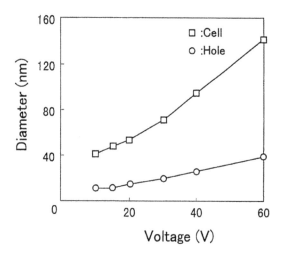

**FIGURE 3.14** Voltage dependence of nanohole diameter and cell diameter, formed by oxalic anodic oxidation. (Reprinted with permission from Shingubara S., et al., Ordered two-dimensional nanowire array formation using self-organized nanoholes of anodically oxidized aluminum, *Jpn. J. Appl. Phys.*, 36, 7791–7795, 1997. Copyright 1997, with permission from the Japan Society of Applied Physics.)

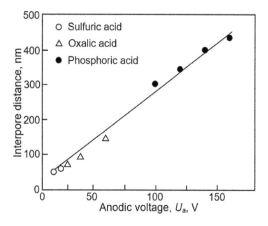

**FIGURE 3.15** Linear relationship between the interpore distance and applied bias voltage. For Al anodization three acids-oxalic, sulfuric, and phosphoric acids—were used. (Reprinted with permission from Li A.P. et al., *J. Appl. Phys.* 84, 6023, Copyright 1998a by the American Institute of Physics.)

causes wide pores. In refined anodization conditions, under which well-organized hexagonal pore arrays can be achieved, the interpore distance was also found to increase linearly with the bias voltage, as shown in Figure 3.15. Since the packing density ($\rho$) is proportional to the reciprocal of the square of the interpore distance ($\rho \sim 1/d^2$) and therefore roughly proportional to $1/U_a^2$, doubling the bias voltage causes the packing density to decrease with a factor of about $\sqrt{2}/2$.

The thickness of the barrier layer ($t_b$) is one of the most important structural parameters of porous AAO (Lee and Park 2014). Like other structural parameters, the barrier-layer thickness ($t_b$) is also dependent on the anodizing potential ($U$) (see Figure 3.16). The potential dependence of the barrier-layer thickness has also been known as "anodizing ratio (AR = $t_b/U$)," the inverse of which corresponds to the electric

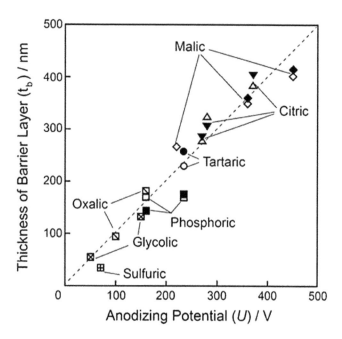

**FIGURE 3.16** Influence of anodizing potential ($U$) on the barrier-layer thickness ($t_b$) for porous AAO formed in different acid electrolytes. (Solid symbols, measured values; open symbols, calculated values from the half-thickness of the pore walls.) (Reprinted with permission from Lee, W. and Park, S.-J., *Chem. Rev.*, 114, 7487–7556, 2014. Copyright 2014 American Chemical Society.; Data extracted from Chu, S.Z. et al., *J. Electrochem. Soc.*, 153, B384–B391, 2006.)

field ($E$) across the barrier layer, and it determines the ionic current density ($j$).

As for the porosity of the layers being formed, then the porosity of the nanostructures formed by the aluminum anodizing depends heavily on the rate of oxide growth, the rate of chemical dissolution of oxide in acidic electrolyte, and anodizing conditions, such as the type of electrolyte, the concentration of electrolyte, time of anodization, the anodizing potential (Figure 3.17), and temperature (Sulka 2008). The most important factor governing the porosity of the structure is the anodizing potential and pH of the solution. One should note that there is a great inconsistency among experimental data on the porosity of nanostructures, with the estimated porosity of anodic porous alumina varying from about 8% to 30%, and even more. An exponential decrease in the porosity with increasing anodization potential was reported for anodizing in sulfuric acid and oxalic acid (Sulka and Parkola 2006). A decrease in the porosity of nanostructures with increasing anodizing potential was observed for constant potential anodizations, conducted in sulfuric, oxalic, phosphoric, and chromic acids (Ono and Masuko 2003; Ono et al. 2004). On the other hand, a slight increase in the porosity is observed with increasing anodizing potential for anodization carried out in sulfuric acid (Chu et al. 2005). As might be expected, the porosity of nanostructures may also be affected by the anodizing time, an extension of which usually results in increasing porosity of the nanostructure formed in phosphoric acid solutions (Nakamura et al. 1992). Increasing the anodizing temperature decreases the

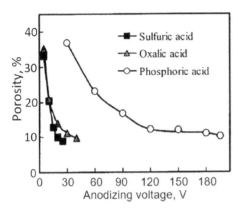

**FIGURE 3.17** Changes in the porosity of anodic Al$_2$O$_3$ films with a voltage formation measured at three electrolytes. Anodizing was carried out at constant voltage. (Reprinted with permission from Ono, S. et al., *J. Electrochem. Soc.*, 151, B473–B478, 2004. Copyright 2004 Electrochemical Society.)

porosity of the nanostructure formed in oxalic acid (Bocchetta et al. 2002); the opposite effect has been observed in sulfuric acid (Sulka and Parkola 2007).

### 3.2.4 Two-Step Anodization

A commonly accepted procedure for preparation of the well-ordered AAO membranes is called a two-step anodization (Masuda and Satoh 1996). It was found that AAO membranes fabricated using a one-step anodization process were far from an idealized model. The lack of a highly ordered pore/cell array results from the random nature of the initial formation mechanism. The concept of the two-step anodization is to first generate aligned nanopores. For this purpose, in a two-step process, after a long first anodization, the top-distorted first alumina layer is removed from the Al foil, leaving a highly ordered concave pattern on the surface of the Al foil. Thus, this process during the first step generates highly ordered indents on the unreacted aluminum surface. Then a second anodization is carried out on this surface-patterned Al foil, resulting in alumina membranes with regular pore arrays. In this case, alumina membranes have a densely packed ideal hexagonal configuration of straight and parallel pore channels from the metal/oxide interface to the oxide/electrolyte interface. Typical parameters of two-step anodization are listed in Table 3.2.

To improve the pore regularity and the surface smoothness of the anodic alumina membranes, some pretreatments

of the Al foils are necessary before the anodization, such as annealing and electropolishing (Lei et al. 2007). The annealing process of Al foils under vacuum conditions can remove the mechanical stresses in the Al foils and increases the grain size. This will facilitate the self-organization process of the pores in the following anodization. The electropolishing process (normally using mixture solutions of perchloric acid and ethanol) of Al foils can result in a smooth Al surface for the anodization. Experimental evidence suggests that the details of the surface topology of the Al foil, for example, its surface roughness, are important for the regularity of the pore structure. However, it should be mentioned that even when using two-step anodization, the pore arrangement of anodic porous alumina is usually far from an ideally packed hexagonal columnar array over a sizable region of, say, millimeter dimensions. The defect-free areas of the pore arrays are typically several square micrometers. The size of the defect-free areas increases with the anodization time. There are approaches that allow an increase of the size of the defect-free areas (Masuda et al. 1997; Lei et al. 2007), but these approaches are not used in the development of humidity sensors.

More information about the Al anodization process can be found in (Shingubara et al. 1999; Lei et al. 2007; Sulka 2008; Lu and Chen 2011; Poinern et al. 2011; Abdel-Karim and El-Raghy 2014; Lee and Park 2014; Santos et al. 2014; Losic and Santos 2015).

## 3.3 LIMITATIONS OF Al$_2$O$_3$-BASED HUMIDITY SENSORS AND HOW RESOLVE THIS PROBLEM

Exploitation of Al$_2$O$_3$-based humidity sensors showed that the primary problem of anodized-based Al$_2$O$_3$ humidity sensors is that when exposed for a long duration in high humidity, significant degradation in the sensitivity and the long-term drift in the capacitance characteristics would be expected (Emmer et al. 1985; Visscher and Kornet 1994; Chen and Lu 2005). The typical drift is around 2°C per year. This is a major drawback for the aluminum-oxide sensor. Even the commercial aluminum-oxide humidity sensors have to be calibrated twice a year to assure their accuracy (Chen and Lu 2005). Nahar (2000) believed that this effect was attributed to the widening of the pores due to diffusion of the adsorbed water. However, in other research it was found that the basic reason for the long-term drift of sensors

**TABLE 3.2**

**Typical Two-Step Anodization Parameters for Common Porous-Forming Electrolytes**

| Electrolyte | Concentration (M) | Potential (V) | 1st Anodization (Hours) | 2nd Anodization (Hours) |
|---|---|---|---|---|
| Sulfuric acid | 0.3 | 24 | 5 | 3 |
| Oxalic acid | 0.3 | 30 | 8 | 5 |
| Oxalic acid | 0.3 | 60 | 5 | 3 |
| Phosphoric acid | 2.5 | 60 | 8 | 5 |

*Source:* Poinern G.E.J. et al., *Materials*, 4, 487–526, 2011. Published by MDPI as open access.

using porous anodic aluminum oxide films is related to the structure transformation of Al$_2$O$_3$ during exploitation.

It was established that anodic aluminum oxides are mainly amorphous oxides, or are amorphous with some forms of crystalline γ-Al$_2$O$_3$ (Sulka 2008). However, it was shown that under certain conditions it is also possible to form the α-Al$_2$O$_3$ (corundum) phase. Studies have shown that γ-Al$_2$O$_3$ and amorphous phases are more sensitive than the α-Al$_2$O$_3$ due to its high porosity. Therefore, many Al$_2$O$_3$-based humidity sensing applications use the γ-phase or amorphous phase Al$_2$O$_3$. However, it was established that the α-Al$_2$O$_3$ is a thermodynamically stable phase, while γ-phase and amorphous Al$_2$O$_3$ suffer from degradation associated with hydration. When the structure exposes to a humid atmosphere, the γ-phase or amorphous Al$_2$O$_3$ changes to γ-Al$_2$O$_3$·H$_2$O (boehmite) (Young 1961). This irreversible phase change causes a volume expansion of aluminum oxide, resulting in the gradual decrease of the surface area and porosity (Alwitt 1971; Emmer et al. 1985; Chen et al. 1991), or a change in dielectric properties of the structure (Khanna 2015). In addition, high sensitive Al$_2$O$_3$-based humidity sensors have significant response times in the wet-to-dry direction.

Many researchers tried to improve the anodic aluminum oxide sensors by optimizing the anodization process (Emmer et al. 1985; Chen et al. 1990; Varghese et al. 2002; Varghese and Grimes 2003), aging the aluminum oxide in boiling water or macerating the films in some ion solutions (Emmer et al. 1985). Mai et al. (2000) proposed using thermal annealing at about 400°C. However, the drift cannot be completely eliminated. This problem seriously hinders the widespread use of aluminum-oxide moisture sensors. Chen and Jin (1992a) proposed other approaches to improving stability. As mentioned above, α-Al$_2$O$_3$ is a very stable phase. In addition, studies showed that the α-alumina similarly to γ-Al$_2$O$_3$ can adsorb the first few water layers even at room temperature (Al-Abadleh and Grassian 2003). This means that α-Al$_2$O$_3$ can be used as a humidity-sensitive material. However, it is very difficult to form α-Al$_2$O$_3$. It is very difficult to obtain α-Al$_2$O$_3$, because the temperature for the phase change from γ-Al$_2$O$_3$ phase to α-Al$_2$O$_3$ phase is at least 900°C (Chen and Jin 1992). Fortunately, there is an electrochemical approach to deposit α-Al$_2$O$_3$ without a high-temperature process. It was established that the anodization technique can be divided into two categories: the low-voltage (<100 V) anodization and anodic-spark deposition (usually >100 V). The low-voltage anodization produces γ-phase or amorphous Al$_2$O$_3$, and the anodic spark deposition results in porous α-Al$_2$O$_3$.

Anodic spark deposition is a unique process for forming certain ceramic coatings. Various coatings can be formed on a wide variety of substrates by this method (Brown et al. 1971; Koshkarian and Kriven 1988). In particular, Brown et al. (1971) deposited α-Al$_2$O$_3$ coatings with thickness of more than 100 μm by anodic spark deposition in aluminate aqueous solutions with 150 V applied to the anode. However, these films were not good candidates for humidity sensors because of very little porosity and large thickness. Much better results were obtained by Tajima et al. (1959). Tajima et al.

(1959) deposited porous α-l$_2$O$_3$ films on aluminum plates with thickness less than 10 μm in the melt of bisulfates (NaHSO$_4$-KHSO$_4$) mixture, and based on these films they fabricated humidity sensors with capability of detecting moisture level as low as 1 ppmv. Due to the tremendous energy dissipated at very large instantaneous current density (~10$^4$ A/cm$^2$), the already deposited Al$_2$O$_3$ barrier film breaks down and electric sparks occur. The extremely high temperature resulting from the electric sparks melt the Al$_2$O$_3$ film locally, resulting in a porous structure (Figure 3.18). Figure 3.18 shows the SEM images of the anodic-spark-deposited α-Al$_2$O$_3$ films (Chen et al. 1991). These α-Al$_2$O$_3$ films exhibited a continuous open-pore structure with the average pore size in the range from 1 to 2 μm, and the porosity was about 30%. However, it was found that the base of the pores was the aluminum substrate, not a barrier film. This causes short-circuit of the sensors when a 0.5 V (AC) voltage is applied to sensors. In order to overcome this difficulty, the α-Al$_2$O$_3$ film is reanodized in diluted sulfuric acid or borax solution for a short time to form a thin barrier layer as shown in the inset in Figure 3.19a. Re-anodization of the porous α-Al$_2$O$_3$ in certain acid solutions at a low voltage was found to be effective in increasing the film resistance so that no short-circuit occurred (Chen et al. 1991). The schematic of the sensor structure and the moisture sensing probe is shown in Figure 3.19b. The fabricated sensors showed high sensitivity for moisture levels from 1000 to 1 ppmv or −20°C to −76°C dew/frost point (D/F PT) (Chen and Jin 1992a). The response was also very fast. From the low to the high moisture level, the response time was about 5 s and from the high to the low moisture level about 20 s. But the most important in obtained results was that the fabricated sensors demonstrated excellent long-term stability; for more than six months, there was no drift in the response of tested sensors (see Figure 3.20) (Chen and Jin 1992; Chen and Lu 2005).

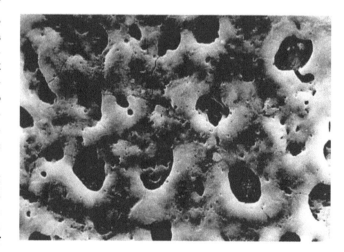

**FIGURE 3.18** Porous Al$_2$O$_3$ by anodic spark deposition. (From Chen, Z. et al.,: Properties of modified anodic-spark-deposited alumina porous ceramic films as humidity sensors., *Journal of the American Ceramic Society*, 1991. 74. 1325–1330. Copyright Wiley-VCH Verlag GmbH & Co. KGaA. Reproduced with permission.)

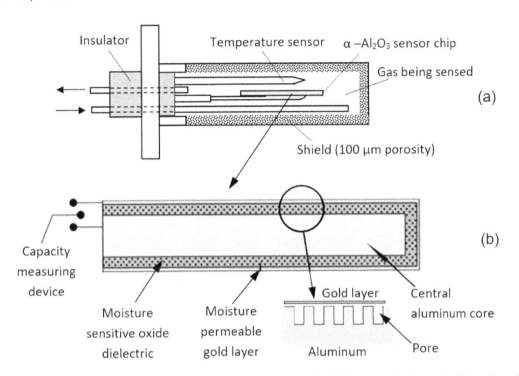

**FIGURE 3.19** (a) Schematic structure of the α-Al$_2$O$_3$ moisture sensor chip (the detailed structure is shown in the inset); and (b) moisture-sensing probe assembly. (Idea from Chen, Z. and Jin, M.-C., An alpha-alumina moisture sensor for relative and absolute humidity measurement, in *Proceedings of the 27th Annual Conference of IEEE Industry Application Society*, Houston, TX, 2, 1668–1675, 1992a; Chen, Z. and Lu, C., *Sens. Lett.*, 3, 274–295, 2005.)

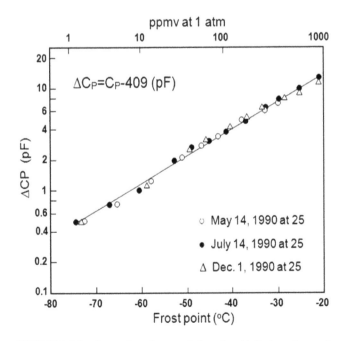

**FIGURE 3.20** Operating characteristics of α-Al$_2$O$_3$-based capacitance humidity sensors during a long-term stability test. (Data extracted from Chen, Z., Jin M.-C., An alpha-alumina moisture sensor for relative and absolute humidity measurement, in *Proceedings of the 27th Annual Conference of IEEE Industry Application Society*, Houston, TX, 2, 1668–1675, 1992a; Chen, Z. and Lu, C., *Sens. Lett.*, 3, 274–295, 2005.)

## 3.4 Al$_2$O$_3$-BASED HUMIDITY SENSORS FABRICATED USING OTHER TECHNOLOGIES

Other methods, such as electron beam evaporation (Shamala et al. 2004), reactive evaporation (Chen et al. 1990), spray pyrolysis (Shamala et al. 2004), and reactive-ion plating (Suzuki et al. 1983) were also utilized to deposit Al$_2$O$_3$ thin films. It was shown that the humidity sensitivity strongly depends on the parameters of films deposition, such as deposition rate and oxygen partial pressure. At optimal parameters of deposition, the sensitivity can be high (see Figure 3.21). Cathodically grown aluminum hydroxide, or hydrated Al$_2$O$_3$, can also be used as a humidity-sensing materials (Mitchell et al. 1998). Using saturated Al$_2$(SO$_4$)$_3$ as the solution and a hydrogen-adsorbing metal (palladium) as the cathode, the aluminum-hydroxide film can be deposited on the palladium. Although this film has good response at high humidity, it is not sensitive to low humidity.

Unfortunately, similar to the films formed at low-voltage anodization, Al$_2$O$_3$ films prepared by vacuum methods at lower substrate temperatures are usually γ-phase or amorphous, which suffer from degradation, as mentioned before. Especially strong degradation takes place during the first days. For example, Suzuki et al. (1983) observed double decrease in capacity during the first 20 days. Then capacity becomes more stable. The decrease in capacity for 100 days was about 2%.

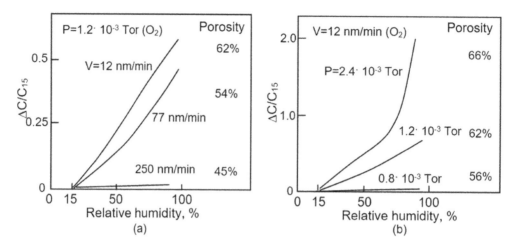

FIGURE 3.21 Influence of (a) deposition rate; and (b) ambient pressure on sensitivity to RH of $Al_2O_3$ films deposited by reactive-ion plating. (Data extracted from Suzuki, K. et al., Alumina thin film humidity sensor controlling of humidity characteristics and aging. in *Proceedings of the 3rd Sensor Symposium*, Tsukuba, Japan, 251–256, 1983.)

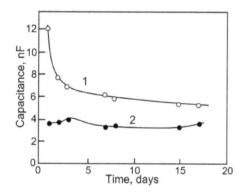

FIGURE 3.22 Variations of capacitance of two humidity sensors fabricated using reactive evaporation with time at 70% RH. Both sensors were fabricated in the same vacuum run at the same time. 1—with special treatment, 2—sensor was treated in 0.1 M $NaH_2PO_4$ solution at 90°C for 10 min. (Reprinted from *Sens. Actuators B*, 2, Chen, Z., et al., Humidity sensors with reactively evaporated $Al_2O_3$ films as porous dielectrics, 167–171 Copyright 1990, with permission from Elsevier.)

Suzuki et al. (1983) also found that by applying the boiling treatment, the aging period of the sensor was considerably shortened. Chen et al. (1990) have shown that appropriate treatment in $NaH_2PO_4$ aqueous solution also contributes to the stabilization of sensor parameters (see Figure 3.22).

An effective method for obtaining porous $\alpha$-$Al_2O_3$ (stable) for humidity sensing is reactive evaporation at elevated substrate temperatures (800°C–1300°C) (Chen and Jin 1992b), in which the metal aluminum is evaporated and oxidized before the oxide particles are deposited on the substrate. Reactively evaporated $Al_2O_3$ films have been reported to be sensitive to moisture levels as low as 1 ppmv (Chen et al. 1990).

Capacitive humidity sensors based on the bulk-sintered $Al_2O_3$ films formed using various technologies were also developed (Basu et al. 2001; Chatterjee et al. 2001; Kumar Mistry et al. 2005). The $Al_2O_3$ film thickness in these sensors

varied in the range of 40–70 μm. It was established that the sensitivity of such sensors depends on the technology used. In the case of mixing, milling, and sintering of the powders, sensors were only sensitive to water-vapor levels higher than 50–100 ppmv (Figure 3.23a). Possibly, low sensitivity is conditioned by the low porosity of the formed layers and big pore size of pores due to a big size of $Al_2O_3$ powders used. In the case of $Al_2O_3$ films prepared using sol-gel technology, the sensitivity was much better. Sensors with nanoporous structure were sensitive even in the range of 3–10 ppmv (Kumar Mistry et al. 2005; Saha et al. 2005). Saha et al. (2005) reported that $\gamma$-$Al_2O_3$ ceramics contained pore size less than 4 nm. The response time of the sensors also depended on the technology of fabrication and can be varied from 3 to 40 s (Chatterjee et al. 2001). It is important that sensors fabricated by sol-gel technology had improved linearity of capacitive change in comparison with sensors developed using standard ceramics technology (Basu et al. 2001; Chatterjee et al. 2001; Kumar Mistry et al. 2005). It was quasi-linear over the range of moisture (0–100 ppmv) (Figure 3.23b).

Impedance humidity sensors based on the lithium-stabilized bulk-sintered Na-$\beta''$-$Al_2O_3$ films were developed by Li et al. (2018). Lithium-stabilized Na-$\beta''$-alumina (LSBA) is known by its high sodium ion conductivity. LSBA powder was synthesized by conventional solid-state reaction. The LSBA-based paste was coated onto an Ag-interdigitated electrode and then was fired at 400°C for 2 h. The sensor showed good sensitivity to RH changes. A good linearity relationship spanning six orders of magnitude was obtained when RH changed from 11.3% to 93.6% (see Figure 3.24). Mechanism of conductivity in metal oxides doped by alkali ions was discussed in Chapter 2 (Volume 2). The response and recovery time at 84.3% RH were 195 s and 68 s, while at lower RH circumstance, the response time and recovery time reduced quickly. The LSBA powder-based sensor also showed good stability in 30 days and a negative temperature effects.

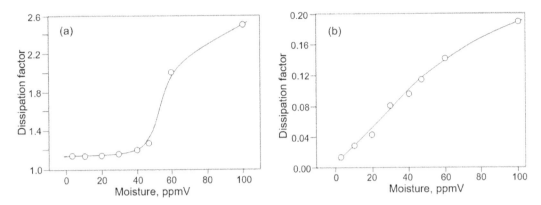

**FIGURE 3.23**  (a) Change in dissipation factor with moisture for sensor prepared by thick-film technology (mixing, milling, and sintering); and (b) change in dissipation factor with moisture of sensor prepared by sol-gel technique. (Reprinted from *Sens. Actuators B*, 109, Saha, D. et al., Dependence of moisture absorption property on sol–gel process of transparent nano-structured γ-Al$_2$O$_3$ ceramics, 363–366. Copyright 2005, with permission from Elsevier.)

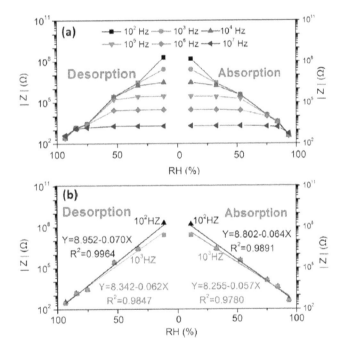

**FIGURE 3.24**  (a) Impedance modulus (| Z |) versus RH curve at different frequency; and (b) linearity fit results of | Z | to RH curve at 100 Hz and 1 kHz for the LSBA powder-based sensor in the absorption and desorption processes, respectively. (Reprinted from *Sens. Actuators B*, 255, Li, H. et al., Highly sensitive humidity sensor based on lithium stabilized Na-β″-alumina: DC and AC analysis, 1445–1454, Copyright 2018, with permission from Elsevier.)

# REFERENCES

Abdel-Karim R., El-Raghy S.M. (2014) Fabrication of nanoporous alumina. In: Ebothe J., Ahmed W. (eds.) *Nanofabrication Using Nanomaterials*, One Central Press, Altrincham, UK, pp. 197–218.

Al-Abadleh H.A., Grassian V.H. (2003) FT-IR study of water adsorption on aluminum oxide surfaces. *Langmuir* **19**, 341–347.

Almasi Kashi M., Ramazani A., Abbasian H., Khayyatian A. (2012) Capacitive humidity sensors based on large diameter porous alumina prepared by high current anodization. *Sens. Actuators A* **174**, 69–74.

Alwitt R.S. (1971) Some physical and dielectric properties of hydrous alumina films. *J. Electrochem. Soc.* **118**, 1730–1733.

Ansbacher F., Jason A.C. (1953) Effects of water vapour on the electrical properties of anodized aluminium. *Nature* **24**, 177–178.

Asoh H., Nishio K., Nakao M., Tamamura T., H. Masuda H. (2001) Conditions for fabrication of ideally ordered anodic porous alumina using pretextured Al. *J. Electrochem. Soc.* **148**, B152–B156.

Banerjee G., Sengupta K. (2002) Pore size optimisation of humidity sensor—A probabilistic approach. *Sens. Actuators B* **86**, 34–41.

Basu S., Chatterjee S., Saha M., Bandyopadhay S., Mistry K., Sengupta K. (2001) Study of electrical characteristics of porous alumina sensors for detection of low moisture in gases. *Sens. Actuators B* **79**, 182–186.

Bocchetta P., Sunseri C., Bottino A., Capannelli G., Chiavarotti G., Piazza S., Di Quarto F. (2002) Asymmetric alumina membranes electrochemically formed in oxalic acid solution. *J. Appl. Electrochem.* **32**, 977–985.

Boytsova O., Klimenko A., Lebedev V., Lukashin A., Eliseev A. (2015) Nanomechanical humidity detection through porous alumina cantilevers. *Beilstein J. Nanotechnol.* **6**, 1332–1337.

Brown S.D., Kuna K.J., Van T.B. (1971) Anodic spark deposition from aqueous solutions of NaAlO$_2$ and Na$_2$SiO$_3$. *J. Am. Ceram. Soc.* **54**, 384–390.

Chachuiski B., Jasinski G., Zajt T., Nowakowski A., Jasinski P. (2004) Properties of humidity sensors with porous Al$_2$O$_3$ as a dielectric layer. *Proc. SPIE* **5505**, 95–100.

Chakraborty S., Hara K., Lai P.T. (1999b) New microhumidity field-effect transistor sensor in ppmv level. *Rev. Sci. Instrum.* **70**, 1565–1567.

Chakraborty S., Nemoto K., Hara K., Lai P.T. (1999a) Moisture sensitive field effect transistors using SiO$_2$/Si$_3$N$_4$/Al$_2$O$_3$ gate structure. *Smart Mater. Struct.* **8**, 274–277.

Chatterjee S., Basu S., Chattopadhyay D., Mistry K., Sengupta K. (2001) Humidity sensor using porous tape cast alumina substrate. *Rev. Sci. Instrum.* **72**, 2792–2795.

Chen S.W., Khor O.K., Liao M.W., Chung C.K. (2014) Sensitivity evolution and enhancement mechanism of porous anodic aluminum oxide humidity sensor using magnetic field. *Sens. Actuators B* **199**, 384–388.

Chen Z., Jin M.-C. (1992a) An alpha-alumina moisture sensor for relative and absolute humidity measurement. In: *Proceedings of the 27th Annual Conference of IEEE Industry Application Society*, Vol. 2. Houston, TX, pp. 1668–1675.

Chen Z., Jin M.-C. (1992b) Effect of high substrate temperatures of crystalline growth of Al$_2$O$_3$ films deposited by reactive evaporation. *J. Mater. Sci. Lett.* **11**, 1023–1025.

Chen Z., Jin M.-C., Zhen C. (1990) Humidity sensors with reactively evaporated Al$_2$O$_3$ films as porous dielectrics. *Sens. Actuators B* **2**, 167–171.

Chen Z., Jin M.-C., Zhen C., Chen G. (1991) Properties of modified anodic-spark-deposited alumina porous ceramic films as humidity sensors. *J. Am. Ceram. Soc.* **74**, 1325–1330.

Chen Z., Lu C. (2005) Humidity sensors: A review of materials and mechanisms. *Sens. Lett.* **3** (4), 274–295.

Cheng B., Tian B., Xie C., Xiao Y., Lei S. (2011) Highly sensitive humidity sensor based on amorphous Al$_2$O$_3$ nanotubes. *J. Mater. Chem.* **21**, 1907–1912.

Chu S.Z., Wada K., Inoue S., Isogai M., Katsuta Y., Yasumori A. (2006) Large-scale fabrication of ordered nanoporous alumina films with arbitrary pore intervals by critical-potential anodization. *J. Electrochem. Soc.* **153**, B384–B391.

Chu S-Z., Wada K., Inoue S., Isogai M., Yasumori A. (2005) Fabrication of ideally ordered nanoporous alumina films and integrated alumina nanotubule arrays by high-field anodization. *Adv. Mater.* **17**, 2115–2119.

Cunningham R.E., Williams R.J.J. (1980) *Diffusion in Gases and Porous Media*. Plenum, New York.

Despic A., Parkhutik V.P. (1989) Electrochemistry of aluminum in aqueous solutions and physics of its anodic oxide. In: Bokris J.O., White R.E., Conwa B.E. (eds.) *Modern Aspects of Electrochemistry*, Vol. 23. Plenum Press, New York, pp. 401–504.

Dickey E.C., Varghese O.K., Ong K.G., Gong D., Paulose M., Grimes C.A. (2002) Room temperature ammonia and humidity sensing using highly ordered nanoporous alumina films. *Sensors* **2**, 91–110.

Emmer I., Hajek Z., Repa P. (1985) Surface adsorption of water vapour on hydrated layers of Al$_2$O$_3$. *Surf. Sci.* **162**, 303–309.

Falk E., Lacquet B.M., Swart P.L. (1992) Determination of equivalent circuit parameters of porous dielectric humidity sensors. *Electron. Lett.* **28**, 166–167.

Furneaux R.C., Rigly W.R., Davidson A.P. (1989) The formation of controlled-porosity membranes from anodically oxidized aluminium. *Nature* **337**, 147–149.

Gonzalez J.A., Lopez V., Bautista A., Otero E., Novoa X.R. (1999) Characterization of porous aluminium oxide films from a.c. impedance measurements. *J. Appl. Electrochem.* **29**, 229–238.

Juhász L., Mizsei J. (2009) Humidity sensor structures with thin film porous alumina for on-chip integration. *Thin Solid Films* **517**, 6198–6201.

Juhasz L., Mizsei J. (2010) A simple humidity sensor with thin film porous alumina and integrated heating. *Procedia Eng.* **5**, 701–704.

Keller F., Hunter M.S., Robinson D.L. (1953) Structural features of oxide coatings on aluminum. *J. Electrochem. Soc.* **100**, 411–419.

Khanna V.K. (2015) A plausible ISFET drift-like aging mechanism for Al$_2$O$_3$ humidity sensor. *Sens. Actuators B* **213**, 351–359.

Khanna V.K., Nahar R.K. (1984) Effect of moisture on the dielectric properties of porous alumina films. *Sens. Actuators* **5**, 187–198.

Khanna V.K., Nahar R.K. (1986) Carrier-transfer mechanisms and Al$_2$O$_3$ sensors for low and high humidities. *J. Phys. D: Appl. Phys.* **19**, L141–L145.

Koshkarian K.A., Kriven W.M. (1988) Investigation of a ceramic-metal interface prepared by anodic spark deposition. *J. Phys. Colloq.* **49**, C5, 213–218.

Kumar Mistry K., Saha D., Sengupta K. (2005) Sol–gel processed Al$_2$O$_3$ thick film template as sensitive capacitive trace moisture sensor. *Sens. Actuators B* **106**, 258–262.

Lee W., Ji R., Gösele U., Nielsch K. (2006) Fast fabrication of long-range ordered porous alumina membranes by hard anodization. *Nature Mater.* **5**, 741–747.

Lee W., Park S.-J. (2014) Porous anodic aluminum oxide: Anodization and templated synthesis of functional nanostructures. *Chem. Rev.* **114**, 7487–7556.

Lei Y., Cai W., Wilde G. (2007) Highly ordered nanostructures with tunable size, shape and properties: A new way to surface nano-patterning using ultra-thin alumina masks. *Prog. Mater. Sci.* **52**, 465–539.

Li A.P., Muller F., Birner A., Nielsch K., Gosele U. (1998a) Hexagonal pore arrays with a 50–420 nm interpore distance formed by self-organization in anodic alumina. *J. Appl. Phys.* **84**, 6023.

Li F., Zhang L., Metzger R.M. (1998b) On the growth of highly ordered pores in anodized aluminum oxide. *Chem. Mater.* **10**, 2470–2480.

Li H., Fan H., Liu Z., Zhang J., Wen Y., Lu J., Jiang X., Chen G. (2018) Highly sensitive humidity sensor based on lithium stabilized Na-β″-alumina: DC and AC analysis. *Sens. Actuators B* **255**, 1445–1454.

Losic D., Santos A. (eds.) (2015) *Nanoporous Alumina. Fabrication, Structure, Properties and Applications*, Springer, New York.

Lu C., Chen Z. (2011) Anodic aluminum oxide-based nanostructures and devices. In: Nalwa H.S. (ed.) *Encyclopedia of Nanoscience and Nanotechnology*, Vol. 11. American Scientific Publisher, pp. 235–259.

Madelung O. (1978) *Introduction to Solid State Theory*. Vol. 2. Springer, Berlin, Germany, p. 448.

Mai L.H., Hoa P.T.M., Binh N.T., Ha N.T.T., An D.K. (2000) Some investigation results of the instability of humidity sensors based on alumina and porous silicon materials. *Sens. Actuators B* **66**, 63–65.

Mankotia D., Sharma Y.C., Sharma S.K. (2014) Review of highly ordered anodic porous alumina membrane development. *Intern. J. Recent Res. Aspects* **1**(2), 171–176.

Masuda H., Nishio K., Baba N. (1993) Fabrication of a one-dimensional microhole array by anodic oxidation of aluminium. *Appl. Phys. Lett.* **63**, 3155–3157.

Masuda H., Satoh M. (1996) Fabrication of gold nanodot array using anodic porous alumina as an evaporation mask. *Jpn. J. Appl. Phys.* **35**, L126–L129.

Masuda H., Yamada H., Satoh M., Asoh H. (1997) Highly ordered nanochannel-array architecture in anodic alumina. *Appl. Phys. Lett.* **71**, 2770–2772.

Mitchell P.J., Mortimer R.J., Wallace A. (1998) Evaluation of a cathodically precipitated aluminium hydroxide film at a hydrogen-sorbing palladium electrode as a humidity sensor, *J. Chem. Soc. Faraday Trans.* **94**, 2423–2428.

Nahar R.K. (2000) Study of the performance degradation of thin film aluminum oxide sensor at high humidity. *Sens. Actuators B* **63**, 49–54.

Nahar R.K., Khanna V.K. (1998) Ionic doping and inversion of the characteristic thin film porous Al$_2$O$_3$ humidity sensor. *Sens. Actuators B* **46**, 35–41.

Nahar R.K., Khanna V.K., Khokle W.S. (1984) On the origin of the humidity-sensitive electrical properties of porous aluminium oxide. *J. Phys. D: Appl. Phys.* **1**(7), 2087–2095.

Nakamura S., Saito M., Huang Li-F., Miyagi M., Wada K. (1992) Infrared optical constants of anodic alumina films with micropore arrays. *Jpn. J. Appl. Phys.* **31**, Part. 1(11), 3589–3593.

Nitsch K., Licznerski B.W., Teterycz H., Golonka L.J., Wisniewski K. (1998) AC equivalent circuits of thick film humidity sensors. *Vacuum* **50**, 131–137.

O'Sullivan J.P., Wood G.C. (1970) The morphology and mechanism of formation of porous anodic films on aluminium. *Proc. Roy. Soc. Lond. A* **317**, 511–543.

Ono S., Masuko, N. (2003) Evaluation of pore diameter of anodic porous films formed on aluminum. *Surf. Coat. Technol.* **169–170**, 139–142.

Ono S., Saito M., Ishiguro M., Asoh H. (2004) Controlling factor of self-ordering of anodic porous alumina. *J. Electrochem. Soc.* **151**(8), B473–B478.

Palbroda, E. (1995) Aluminium porous growth—II. On the rate determining step. *Electrochim. Acta* **40**, 1051–1055.

Patermarakis, G. (1998) Development of a theory for the determination of the composition of the anodising solution inside the pores during the growth of porous anodic Al$_2$O$_3$ films on aluminium by a transport phenomenon analysis. *J. Electroanal. Chem.* **447**, 25–41.

Poinern G.E.J., Ali N., Fawcett D. (2011) Progress in nano-engineered anodic aluminum oxide membrane development. *Materials* **4**, 487–526.

Saha D., Kumar Mistry K., Giri R., Guha A., Sensgupta K. (2005) Dependence of moisture absorption property on sol–gel process of transparent nano-structured γ-Al$_2$O$_3$ ceramics. *Sens. Actuators B* **109**, 363–366.

Santos A., Kumeria T., Losic D. (2014) Nanoporous anodic alumina: A versatile platform for optical biosensors. *Materials* **7**(6), 4297–4320.

Sberveglieri G., Murri R., Pinto N. (1995) Characterisation of porous Al$_2$O$_3$-SiO$_2$/Si sensor for low and medium humidity ranges. *Sens. Actuators B* **23**, 177–180.

Sberveglieri G., Rinchetti G., Groppelli S., Faglia G. (1994) Capacitive humidity sensor with controlled performances, based on porous Al$_2$O$_3$ thin film grown on SiO$_2$-Si substrate. *Sens. Actuators B* **18**, 551–553.

Seiyama T., Yamazoe N., Arai H. (1983) Ceramic humidity sensors. *Sens. Actuators* **4**, 85–96.

Shamala K.S., Murthy L.C.S., Rao K.N. (2004) Studies on optical and dielectric properties of Al$_2$O$_3$ thin films prepared by electron beam evaporation and spray pyrolysis method. *Mater. Sci. Eng. B* **106**, 269–274.

Shawaqfeh, A.T., Baltus, R.E. (1999) Fabrication and characterization of single layer and multi-layer anodic alumina membrane. *J. Membrane Sci.* **157**, 147–158.

Shingubara S., Okino O., Sayama Y., Sakaue H., Takahagi T. (1997) Ordered two-dimensional nanowire array formation using self-organized nanoholes of anodically oxidized aluminum. *Jpn. J. Appl. Phys.* **36**, (12), 7791–7795.

Shingubara S., Okino O., Sayama Y., Sakaue H., Takahagi T. (1999) Two-dimensional nanowire array formation on Si substrate using self-organized nanoholes of anodically oxidized aluminum. *Solid State Electron.* **43**, 1143–1146.

Statz H., De Mars G.A. (1958) Electrical conduction via slow surface states on semiconductors. *Phys. Rev.* **111**, 169–182.

Sulka G.D. (2008) Highly ordered anodic porous alumina formation by self-organized anodizing. In: Eftekhari A. (ed.) *Nanostructured Materials in Electrochemistry*. Wiley-VCH Verlag, Weinheim, Germany, pp. 1–116.

Sulka G.D., Parkola, K. (2006) Anodising potential influence on well-ordered nanostructures formed by anodisation of aluminium in sulphuric acid. *Thin Solid Films* **515**, 338–345.

Sulka G.D., Parkola, K.G. (2007) Temperature influence on well-ordered nanopore structures grown by anodization of aluminium in sulphuric acid. *Electrochim. Acta* **52**, 1880–1888.

Sun H.T., Ming-Tang W., Ping L., Xi Y. (1989) Porosity control of humidity-sensitive ceramics and theoretical model of humidity-sensitive characteristics. *Sens. Actuators* **19**, 61–70.

Suzuki K., Koyama K., Inuzuka T., Nabeta Y-I. (1983) Alumina thin film humidity sensor controlling of humidity characteristics and aging. In: *Proceedings of the 3rd Sensor Symposium*, Tsukuba, Japan, pp. 251–256.

Svensson T., Lewander M., Svanberg S. (2010) Laser absorption spectroscopy of water vapor confined in nanoporous alumina: Wall collision line broadening and gas diffusion dynamics. *Opt. Express* **18**(16), 16460–16473.

Thompson C.E., Wood G.C. (1983) Anodic films on aluminium. In: Scully J.C. (ed.) *Treatise on Materials Science and Technology*. Academic Press, New York, pp. 205–329.

Thompson G.E., Furneaux R.C., Wood G.C., Richardson J.A., Goode J.S. (1978) Nucleation and growth of porous anodic films on aluminium. *Nature* **272**, 433–435.

Traversa E. (1995) Ceramic sensors for humidity detection: the state-of-the-art and future developments. *Sens. Actuators B* **23**, 1335–1356.

Varghese O.K., Gong D., Paulose M., Ong K.G., Grimes C.A., Dickey E.C. (2002) Highly ordered nanoporous alumina films: Effect of pore size and uniformity on sensing performance. *J. Mater. Res.* **17**(5), 1162–1171.

Varghese O.K., Grimes C.A. (2003) Metal oxide nanoarchitectures for environmental sensing. *J. Nanosci. Nanotechnol.* **3**(4), 277–293.

Visscher G.J.W., Kornet J.G. (1994) Long-term tests of capacitive humidity sensors. *Meas. Sci. Technol.* **5**, 1294–1302.

Wang H.H., Han C.Y., Willing G.A., Xiao Z. (2003) Nanowire and nanotube syntheses through self-assembled nanoporous AAO templates. *Mat. Res. Soc. Symp. Proc.* **775**, 107–112.

Young L. (1961) *Anodic Oxide Films*. Academic, New York.

# 4 Carbon-Based Materials

## 4.1 INTRODUCTION

In recent years, research of carbon-based materials such as fullerenes, graphene, amorphous carbon, carbon nanotubes (CNTs), carbon nanofibers (CNFs), and nanostructured carbon films (carbon nanosheets and nanohoneycombs) is a very popular trend. The main advantages of these materials are a large sensing area, high-chemical inertness, large porosity, and a sensitivity to doping, which makes it possible to change their physical, chemical, and electrical properties (see Table 4.1). In addition, studies have shown that properties of these materials exhibit sensitivity to air humidity, and therefore they can be incorporated into sensors as humidity-sensitive materials, using for these purposes different techniques such as physical vapor deposition, thermal carbonization, magnetron sputtering, thin-film deposition, and so on. In particular, it was found that carbon-based materials can be used for developing resistive, capacitive, quartz crystal microbalance (QCM), surface acoustic wave (SAW) and fiber-optic humidity sensors (see Table 4.2).

## 4.2 CARBON BLACK

Carbon black (CB) is one of numerous forms of carbon (see Table 4.1). CB is a material produced by the incomplete combustion of heavy petroleum products such as fluid catalytic cracking (FCC) tar, coal tar, ethylene cracking tar, and a small amount from vegetable oil. CB is a form of amorphous carbon that has a high surface-area-to-volume ratio. However, in spite of this fact, carbon black, due to specific conductivity and mechanical properties, is very rarely used as sensing material in humidity sensors. There are only a few examples of this application. Chen et al. (2010) suggest using CB to design capacitive sensors based on CB-n-Si heterostructures, while Afify et al. (2017) used carbonized bamboo (a variation of carbon black) in the manufacture of resistive humidity sensors. Characteristics of these sensors are given in Figures 4.1 and 4.2.

It is seen that with the relative humidity (RH) changing from 11% to 95%, a capacitive response of over 200% was achieved at 1 kHz, and the capacitive response is highly linear

## TABLE 4.1
### The Properties of Carbon Allotropes

| Parameter | Carbon Allotropes | | | | |
| --- | --- | --- | --- | --- | --- |
| | Graphite | Diamond | Fullerene ($C_{60}$) | Carbon Nanotube | Graphene |
| Hybridized form | $sp^2$ | $sp^3$ | Mainly $sp^2$ | Mainly $sp^2$ | $sp^2$ |
| Dimension | 3-D | 3-D | 0-D | 1-D | 2-D |
| Crystal system | Hexagonal | Octahedral | Tetragonal | Icosahedral | Hexagonal |
| Bond length, nm | 0.142 (C=C) 0.144 (C-C) | 0.154 (C-C) | 0.14 (C=C) 0.146 (C-C) | 0.144 (C=C) | 0.142 (C=C) |
| Experimental specific surface area ($m^2/g$) | ~10–20 | 20–160 | 80–90 | ~1300 | ~1500 |
| Density ($g/cm^3$) | 2.09–2.26 | 3.5–3.53 | 1.72 | 1.2–2.0 | >1 |
| Optical properties | Uniaxial | Isotropic | Non-linear optical response | Structure-dependent properties | 97.7% of optical transmittance |
| Thermal conductivity (W/m·K) | 1500–2000[a], 5–10[c] | 900–2320 | 0.4 | 3500 | 4840–5300 |
| Hardness | High | Ultrahigh | High | High | Highest (single layer) |
| Tenacity | Flexible nonelastic | — | Elastic | Flexible elastic | Flexible elastic |
| Electronic properties | Electrical conductor | Insulator, Semicond. ($E_g$=5.47 eV) | Semicond. ($E_g$=1.9 eV) | Metallic and semiconducting | Semimetal, zero-gap semiconductor |
| Electrical conductivity (S/cm) | Anisotropic, 2–3·$10^{4(a)}$; $6^b$ | | $10^{-10}$ | Structure-dependent | 2000 |

*Source:* Saito, R. et al., *Physical Properties of Carbon Nanotubes*, Imperial College Press, London, UK, 1998; Wu, Z.-S. et al., *Nano Energy*, 1, 107–131, 2012.

[a] a-direction

[b] c-direction.

**TABLE 4.2**

**Types of Humidity Sensors Developed Using Carbon-Based Materials**

| Type of Sensor | Carbon Black | Fullerenes | CNTs | Graphene | Diamond |
|---|---|---|---|---|---|
| Capacitance | Chen et al. (2010) | Li et al. (2018) | Varghese et al. (2001); Snow et al. (2005); Yeow and She (2006); Chen et al. (2009); Liu et al. (2011) | Bi et al. (2013); Ali et al. (2016) | |
| Resistive | Tsubokawa et al. (1997); Okazaki et al. (1999); Afify et al. (2017) | Sberveglieri et al. (1996); Berdinsky et al. (2000) | Jiang et al. (2007); Su and Wang (2007); Lee et al. (2013); Tang et al. (2011) | Ghosh et al. (2009); Borini et al. (2013); Smith et al. (2015); Zhang et al. (2016b) | Wang et al. (2007a); Kromka et al. (2010) |
| QCM | | Radeva et al. (1997); Lin and Shih (2003) | Zhang et al. (2005); Su et al. (2006); Wisitsoraat et al. (2008) | Yao et al. (2011); Yao and Xue (2015) | Yao et al. (2014b); Yao and Xue (2015) |
| SAW | Penza and Anisimkin (1999) | Lin and Shih (2003) | Penza et al. (2004); Sheng et al. (2011) | Chen et al. (2016); Yuan et al. (2016); Le et al. (2018) | Chevallier et al. (2009); Hribšek et al. (2010) |
| Other types | | | Cusano et al. (2006); Chen et al. (2017) | Ghadiry et al. (2016) | Possas-Abreu et al. (2017) |

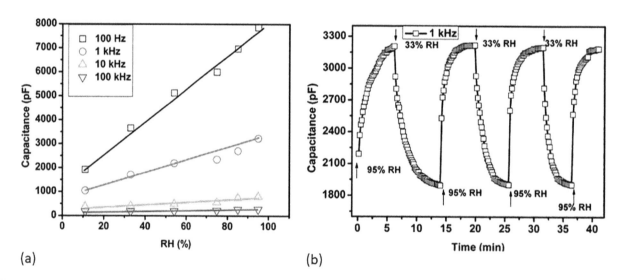

**FIGURE 4.1**   (a) Capacitance–RH curves of the CB/Si junction (prepared at 6.0 Pa) at set frequencies; and (b) experimental capacitance–time curve of CB/Si junction (prepared at 6.0 Pa) measured by carrying out vapor adsorption–desorption cycles between RH = 33% and RH = 95%. (Reprinted from *Sens. Actuators B*, 150, Chen, H.-J. et al., Capacitive humidity sensor based on amorphous carbon film/n-Si heterojunctions, 487–489, Copyright 2010, with permission from Elsevier.)

with RH for the given frequencies. This means that the CB/Si junctions have potential application as humidity gas sensors. The increase in capacitance of the CB/Si junction with increasing RH can be attributed to the increase in the amount of physisorbed water having a dipole moment. As for resistive sensors, it can be concluded that despite the rather wide range in which RH can influence resistance and reasonably fast response and recovery (less than 2 min), the sensor response of resistive sensors is characterized by a large hysteresis. Lower sensitivity of annealed CB (CBA)-based sensors can be

explained based on the results of BET measurements: annealing led to a drastic reduction of open porosity.

### 4.2.1 CB-POLYMER COMPOSITES AND THEIR APPLICATION IN HUMIDITY SENSORS

Another possibility for CB to be incorporated into humidity sensors is associated with the use of composites, where gas-sensitive properties are possessed by another material, while CB plays the role of a filler characterized by high conductivity

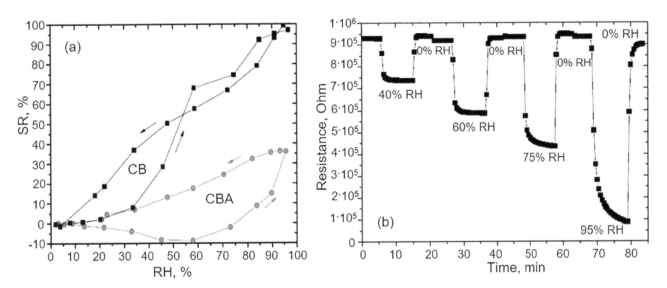

**FIGURE 4.2** (a) Sensor response [S(%) = 100 $(R_o-R_g)/R_o$] toward relative humidity at room temperature [26.3°C and 27.4°C, respectively, for carbonized bamboo (CB) and annealed CB (CBA)]; arrows indicate increasing and decreasing RH values; and (b) the response and recovery time of CB sensor. (Reprinted from *Sens. Actuators B*, 239, Afify, A.S. et al., Elaboration and characterization of novel humidity sensor based on micro-carbonized bamboo particles, 1251–1256, Copyright 2017, with permission from Elsevier.)

and high dispersion. The features of such composites were discussed earlier in Chapter 11 (Volume 2). The key CB properties useful for composite design are excellent dispersion, integrity of the CB structure or network, consistent particle size, specific resistance, structure, and purity. As a rule, CB is used mainly in polymer-based composites. The CB endows electrical conductivity to the films, whereas the different organic polymers, such as poly(ethylene glycol) (PEG) (Okazaki et al. 1999), polyethyleneimine (PEI) (Tsubokawa et al. 1997), polyvinyl alcohol (PVA) (Barkauskas 1997; Li and Chen 2012), poly(4-vinylpyridine) (PVP) (Li et al. 2007; Jiang et al. 2014; Ziegler et al. 2017), amino-functional copolymer [N, N,-dimethyl-1,3-propanediamine] (Matsuguchi et al. 2013), hydroxyethyl cellulose (HEC) (Ma et al. 2015), and poly(1,5-diaminonaphthalene) (DAN) (Shim and Park 2000), are the source of chemical diversity between elements in the sensor array. In addition, polymers function as the insulating phase of the CB composites. The concentration of CB in composites is varied in the range from 2% to 40% (w/w). The conductivity of these materials and their response to compression, or expansion, can be explained using percolation theory (McLachlan et al. 1990). The compression of a composite prepared by mixing conducting and insulating particles leads to increased conductivity and, conversely, expansion leads to decreased conductivity. This effect is especially strong in the composites with compositions around the percolation threshold; an extremely small volume change of the phase due to an extrinsic perturbation brings about the resistivity change of the composite (see Figure 4.3). This means that the swelling of the polymer upon exposure to a water vapor increases the resistance of the film, thereby providing an extraordinarily simple means for monitoring the presence of water vapors. For crystalline polymers, it was also considered

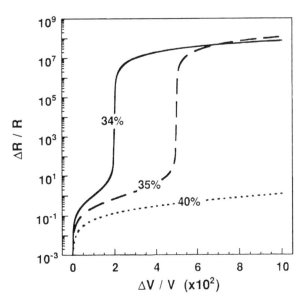

**FIGURE 4.3** Relative differential resistance change, $\Delta R/R$, predicted by percolation theory (see Eq. 6 and associated text) as a function of the relative volume change, $\Delta V/V$, of a carbon black-polymer composite upon swelling. The volume of carbon black is assumed to be unaffected by swelling and the polymer matrix is assumed to have a conductivity 11 orders of magnitude lower than that of carbon black. The three separate lines are for composites with differing initial volume percentages of carbon black, as indicated. The percolation threshold for the system is at *CB content* = 0.33. The total volume change results in a change in the effective carbon black content. When this value drops below the percolation threshold, a sharp increase in response is observed. Of course, the position of this sharp increase depends on the value of the percolation threshold. (Reprinted with permission from Lonergan, M.C. et al., *Chem. Mater.*, 8, 2298–2312, 1996. Copyright 1996 American Chemical Society.)

that the polymer matrix within the composite is dissolved by solvent absorption, and the movement of CB particles in the amorphous regions causes the destruction of conductive networks, which results in the increase in electrical resistance consequently (Chen and Tsubokawa 2000; Dong et al. 2004).

Typical operating characteristics of CB-polymer composite-based humidity sensors are shown in Figures 4.4 and 4.5. It is seen that the composites' resistance increases during water-vapor absorption and returns to the initial value when the vapor desorbs completely. It is necessary to note that the percolation threshold strongly depends on both the parameters of CB used and the technology of composite preparing.

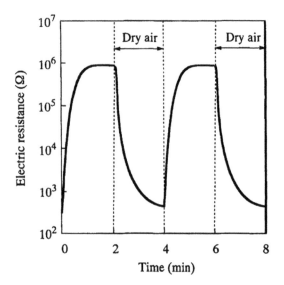

**FIGURE 4.4** Effect of humidity on the electric resistance of composite from PEG-grafted carbon black at 25°C. (Reprinted by permission from Macmillan Publishers Ltd. *Polymer J.*, Okazaki, M. et al., 1999, copyright 1999.)

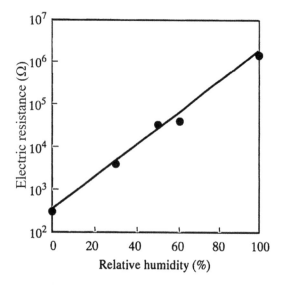

**FIGURE 4.5** Relationship between electric resistance of composite from PEG-1540-grafted carbon black and relative humidity at 25°C. The thickness of the composite on the electrode was about 120 μm. (Reprinted by permission from Macmillan Publishers Ltd. *Polymer J.*, Okazaki, M. et al., 1999, copyright 1999.)

In different articles, the percolation threshold was observed at CB content, which was varied from <3% (Chen et al. 2005a) to 33% (w/w) (Lonergan et al. 1996). This means that reproducibility of sensor parameters designed on the base of CB-polymer composites is not high.

Unfortunately, at present it is impossible to draw a definite conclusion regarding the optimal compositions of CB-based composites for the manufacture of humidity sensors. There is too little information for this. It can only be noted that from published data it is possible to distinguish the results published by Barkauskas (1997) and Shim and Park (2000). Barkauskas (1997) reported that PVA-CB sensors had the response time approximately 45 seconds and detection limit was 0.17%. According to Shim and Park (2000), their poly(1,5-diaminonaphthalene) (DAN)-CB sensors (15:1) showed a straight line over the entire measured range of percent relative humidity (12%–93% RH) without hysteresis. Moreover, the current drift was less than ±1% for the lower humidity.

It was established that many factors can influence the response of electrical resistance of CB-polymer-based composite sensors to water vapors. For example, during the modification of the CB surface by grafting polymerization (Chen and Tsubokawa 2000a, 200b), the response, reproducibility and stability of the composites were closely related to the crystallinity and molecular weight of the polymer matrix (Chen et al. 2002), as well as to the content and dispersivity of CB in the composites (Dong et al. 2003). It was also found that the electrical response of the composites is a function of temperature (Matzger et al. 2000). For example, elevated temperature usually increases the rate of response to saturated vapors. However, at operating temperatures higher than 35°C–50°C the decrease of the sensor response takes place. It was also established that, as a rule, the relationships between the electrical response and the vapor concentration or partial pressure are non-linear (see Figure 4.3). According to Patel et al. (2000), the non-linear relationship was once attributed to the interference from the content of CB, that is, when the composites are near the percolation threshold, the next small addition of analyte causes a disproportionately large increase in electric resistance. This means that far off the percolation threshold the relationships between the electrical response and vapor concentration can be linear (Shim and Park 2000). It should be noted that in spite of the fact that previous results were obtained for sensors in the process of detecting solvent vapors, the conclusions made can be used with full confidence and applied to humidity sensors. The processes occurring during the detection of water vapor and solvents are very similar.

Analysis of obtained results has shown that CB-polymer composite humidity sensors have advantages: they are high sensitive, inexpensive, easily controlled, and robust in many different environments. In addition, they have simple fabrication processing and good compatibility with modern CMOS VLSI technology. As a result, CB-polymer-based sensors are promising for design of various e-nose systems (Ryan et al. 2004; Hossain et al. 2012); they may have diverse responses by choosing the used materials of insulating polymers, additive plasticizers, and conductive CBs and by regulating relative

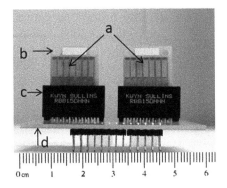

**FIGURE 4.6** A schematic representation of the sensor array connected with the edge connector: (a) sensing electrodes; (b) printed circuit board; (c) edge connector; and (d) base. (With kind permission from Springer Science+Business Media: *Agric. Res.*, Carbon black polymer sensor array for incipient grain spoilage monitoring, 1, 2012, 87–94, Hossain, M.E. et al.)

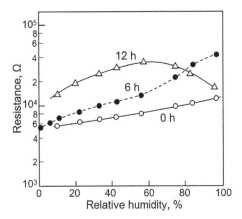

**FIGURE 4.8** The resistance of composites as a function of humidity with different cross-linking times (feed ratio of CB to PVP was 1:9). (Reprinted from *Sens. Actuators B*, 123, Li, Y. et al., Poly(4-vinylpyridine)/carbon black composite as a humidity sensor, 554–559, Copyright 2007, with permission from Elsevier.)

quantities among them. Moreover, these sensors provide the opportunity to fabricate very small size, low-power, and lightweight sensor arrays (Munoz et al. 1999; Matzger et al. 2000; Kim et al. 2005; Xie et al. 2006). One of the examples of such sensor arrays is shown in Figure 4.6. The mixture of polymer and CB was sprayed on interdigitated electrodes and then dried in the room air for 1 day.

However, it is necessary to keep in mind that, like all types of polymer sensors, CB-polymer composite-based humidity sensors are not selective. For example, as is seen in Figure 4.7, the electric resistance of the composite such as PEI-CB in addition to a water-vapor sensitivity, drastically increased in methanol, ethanol, and other organic vapors, which are good solvents for PEI. On the contrary, the electric resistance hardly changed in toluene and hexane, which are nonsolvent of PEI. A similar situation is observed for other composites (Lonergan et al. 1996).

It was also found that polymer-CB sensors are prone to aging during operation (Hossain et al. 2012). Experiments showed that thermal annealing or cross-linking reaction can

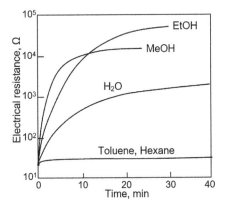

**FIGURE 4.7** Effect of various vapors on the electric resistance of the composite prepared from PEI(C) and PEI(C)-grafted carbon black. (With kind permission from Springer Science+Business Media: *Polym. Bull.*, Responsibility of electric resistance of polyethyleneimine-grafted carbon black against alcohol vapor and humidity, 39, 1997, 217–224, Tsubokawa, N. et al.)

be used to improve stability of the sensitive films based on CB and PVA and PVP (Li et al. 2007; Jiang et al. 2014). In particular, Li et al. (2007) established that the composite of CB and optimized cross-linked PVP can be used because humidity-sensitive materials are able to not only detect low humidity but also work in the whole humidity range (0%–97%RH) with enhanced sensitivity (see Figure 4.8) and good stability due to the network structure. Hossain et al. (2012) have also established that co-polymers with styrene showed a low-sensitivity drop compared to non-styrene co-polymers. Therefore, it was concluded that styrene co-polymer sensors were best suited to build a stable sensor for the sensor array.

There are attempts to use other CB-based composites for gas and humidity-sensor design (Liou and Lin 2007; Llobert et al. 2014; Bhargava et al. 2017). For example, Bhargava et al. (2017) proposed using a CB alum composite. However, such an approach does not give any improvement in operating characteristics in comparison with conventional metal oxide or CB-polymer-based gas and humidity sensors.

Experiment has shown that for preparing polymer-CB composites can be applied methods, such as physical mixing of CB and the polymer matrix (Mather and Thomas 1997), in-situ polymerization in the presence of CB (Dong et al. 2004), and ultrasonic mixing of CB and polymer powders (Ramos et al. 2005). However, it is necessary to note that generally, for most of methods used, it is difficult to attain good dispersion of the CB into the polymer matrix, which affects the percolation limit of the composite. This is due to the high surface energy, small particle size and strong agglomeration tendency of the CB, limited shear force of the mixer, and the high viscosity of the polymer solution. To obtain CB-polymer based composites with high CB dispersibility, the CB is usually modified by coating the surface of CB with an organic compound, such as an oligomer with a terminal active group (Chen and Tsubokawa 2000). However, Chen and Tsubokawa (2000) reported that this process of surface modification is difficult to implement because there are almost no active groups on the surface of conductive CB to be used for such treatment reaction. So, it is impossible to bind the organic compound directly onto the surface by a chemical reaction. Therefore, usually, a two-step modification process is applied. In particular, Chen and Tsubokawa (2000) at the first step introduced carboxyl groups onto the CB surface through the trapping of 4-cyanopentanoic acid radicals, which came from the decomposition of 4,4'-Azobis(4-cyanopentanoic acid) (ACPA). Then as a second step, poly(ethylene-block-ethylene oxide) (PE-b-PEO) was grafted onto the surface by direct condensation between terminal hydroxyl groups of PE-b-PEO and carboxyl groups on the CB surface in the presence of N, N'-Dicyclohexylcarbodiimide (DCC) as a condensing agent. Chen and Tsubokawa (2000) also established that the conductometric response of LDPE/CB composite with PE-b-PEO-grafted CB to various gases is more stable and reproducible than the response of composites with untreated CB. Arshak et al. (2005) found that the treatment of composites with surfactant such as Hypermer PS3 and Hypermer PS4 (Uniquema) also gives the improvement of gas-sensing characteristics. The percolation curves of surfactant-treated composites showed that the resistivity of the composite was increased due to better dispersion of the CB to various gases and also the prevention of the CB from reagglomerating after shear mixing.

## 4.3  FULLERENES

Fullerenes are closed-cage carbon molecules containing pentagonal and hexagonal rings arranged in such a way that they have the formula $C_{20+m}$, with $m$ being an integer number (Dresselhaus et al. 1996; Mauter and Elimelech 2008) (see Figure 4.9). They are the fifth allotropic form of carbon, the others being graphite, diamond, CNT, and graphene (see Table 1.1). Fullerenes comprise a wide range of isomers and homologous series, from the most studied $C_{60}$ or $C_{70}$ to the so-called higher fullerenes, $C_{240}$, $C_{540}$, and $C_{720}$. The first of these compounds were discovered in 1985 through spectrometric measurements on interstellar dust, and their structure was confirmed later in the laboratory (Kroto et al. 1985). Kroto, Smalley, and Curl received the Nobel Prize in 1996 for their work.

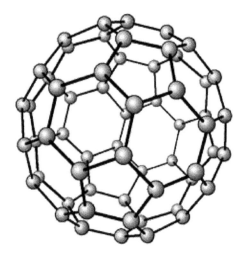

**FIGURE 4.9**  Typical diagram illustrating the structure of fullerene $C_{60}$.

Physicists, chemists, and material scientists or engineers, among others, have found unusual potential in these new spherical carbon structures for use as superconductor materials, sources of new compounds, self-assembling nanostructures, and several optical and electronic devices (Dresselhaus et al. 1996). The reasons why fullerenes and their films are attractive for the creation of devices can be formulated as follows: spherical shape and large size of fullerene molecule provide a substantial volume of empty intermolecular space in a face-centered cubic lattice of solid fullerite. As a consequence, this material is easily intercalated by different impurities changing its properties. Pure fullerene is a dielectric, but it can change its properties from dielectric to superconducting as a result of intercalation. This initial attention led to an increasing number of investigations that showed the special properties of fullerenes, some of which might lead to practical applications (Mauter and Elimelech 2008). Although a wide range of uses has been explored and several applications developed, fullerenes are not, so far, fulfilling their initial spectacular promise (Baena et al. 2002).

It was found that a characteristic feature of fullerene is its affinity to various organic molecules. Therefore, fullerene $C_{60}$ with 60 $\pi$-electrons potentially can be applied as a good adsorbent to adsorb and detect nonpolar and some polar organic molecules (Baena et al. 2002) and inorganic molecules (Synowczyk and Heinze 1993) using SAW (Lin and Shih 2003) and QCM (Chao and Shin 1998; Shih et al. 2001) sensors. In particular, it was established that $NH_3$ adsorption onto the fullerene film reduces the film resistance, resulting in a charge transfer to the electronic system. Sensitivity levels of a few milligrams per liter of $NH_3$ in air were achieved, but there were still some problems, such as the lack of selectivity versus other gas vapors (which were also adsorbed, leading to the same electrical signal), response times of the order of seconds, the influence of the humidity level on the calibration, or instability of the sensor when exposed to air several times.

It was also established that fullerenes similarly to CB can be used in humidity sensors. In particular, Radeva et al. (1997) have shown that mass-sensitive sensors coated by $C_{60}$ film have high sensitivity to humidity, and Ding et al. (2018) have found that the QCM humidity sensors based on fullerene/graphene oxide (GO) nanocomposites showed better performances at quality factor and dynamic response/recovery behavior than the GO-based QCM humidity sensor. Based on the admittance analysis and frequency response analysis of the sensors, Ding et al. (2018) speculated that the introduction of $C_{60}$ molecules into GO reduced the aggregation of the GO sheets and formed some hydrophobic isolation layers between the GO sheets, which could inhibit the water molecules' permeation and maintain the mechanical stiffness of the $C_{60}$/GO film.

The resistance of the $C_{60}$ film is also strongly changed in humid atmosphere (Sberveglieri et al. 1996; Berdinsky et al. 2000). For example, Figure 4.10 shows the dependence of the $C_{60}$ film resistance on the RH of the environment. It is seen

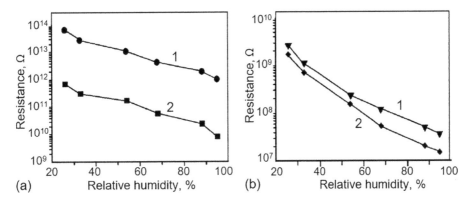

**FIGURE 4.10** The dependencies of the resistance of porous films $C_{60}$ on relative humidity: 1—initial film, 2—the film oxidized in air at (a) 390°C; and (b) 300°C. The synthesis, isolation, and purification of fullerenes were carried out according to the standard technology with graphite evaporation in an electric arc (Lamb and Huffman 1993). The thickness of deposited films was 2.4 mm. (Reprinted with permission from Berdinsky, A.S. et al., *Chem. Sustainable Develop.*, 8, 141–146, 2000. Copyright 2000. Siberian Branch of the RAS.)

that due to lower resistance and better sensitivity, oxidized films are better suitable for use in humidity sensors. In addition, fullerenes can easily form thermally stable films with polymers, such as polystyrene (PS), and possesses useful electronic and photochemical properties, such as a fairly long lifetime for the photo-excited triplet state (~100 μs). However, one should recognize that the unavailability and high cost of fullerenes in comparison with CB have deterred their use in sensorics. An additional limitation is high resistance of the $C_{60}$ film (see Figure 4.10a). However, this limitation is important when designing resistive sensors. For capacitive sensors or QCM sensors, this property of $C_{60}$ is an advantage (Li et al. 2018).

## 4.4 CARBON NANOTUBES

CNTs were first fabricated in 1991 by Iijima (1991). Starting from this time, a great deal of effort has been devoted to the fundamental understanding of their properties and of their use in a wide range of applications, such as electronics, catalysis, filters, and sensors (Schnorr and Swager 2011). It was established that there are two types of CNT morphology (Dresselhaus et al. 1996; Saito et al. 1998; Varghese et al. 2001; Terrones et al. 2004). On the one hand, single-walled carbon nanotubes (SWCNTs) consist of a honeycomb network of carbon atoms and can be visualized as a cylinder rolled from a graphitic sheet. On the other hand, multi-walled carbon nanotubes (MWCNTs) are a coaxial assembly of graphitic cylinders generally separated by a plane space of graphite (Dresselhaus et al. 1996) (see Figure 4.11). Each tubule in MWCNTs has diameter ranged typically from 2 to 25 nm size with a 0.34-nm distance between sheets close to the interlayer spacing in graphite. The diameter and the length of SWCNTs typically vary from 0.5 to 3 nm and from 1 to 100 μm, respectively. CNTs have the tendency to aggregate, usually forming bundles that consist of tens to hundreds of nanotubes in parallel and in contact with each other. This effect is due to strong Van der Waals interactions between nanotubes.

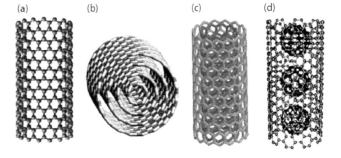

**FIGURE 4.11** Schematic diagrams of (a) a single-wall carbon nanotube (SWNT); (b) a multi-wall carbon nanotube (MWNT); (c) a double-wall carbon nanotube (DWNT); and (d) a peapod nanotube consisting of a SWNT filled with fullerenes (e.g., $C_{60}$). (Reprinted from *Mater. Sci. Eng. C*, 23, Dresselhaus, M.S. et al., Nanowires and nanotubes, 129–140, Copyright 2003, with permission from Elsevier.)

### 4.4.1 SYNTHESIS OF CNTs

The synthesis of CNTs (single or multiwalled) for humidity sensor applications can be performed by different methods, such as arc discharge, laser ablation, pyrolysis, chemical vapor deposition (CVD), physical vapor deposition (PVD), and gas-phase catalytic growth (Bruzzi et al. 2004; Mamalis et al. 2004; Terrones et al. 2004; Kuchibhatla et al. 2007; Zhang and Zhang 2009; Chu et al. 2013; Lekawa-Raus et al. 2015). However, till now these methods do not produce a monodisperse product with controlled physical and chemical properties.

At present, the gas-phase processes are the most common. All these methods have a carbon source, catalyst nanoparticle species, and energy input (Journet et al. 2012). Depending on the carbon source, the synthesis can be classified as follows: when the carbon source is provided in liquid or gaseous form, the generic term of "medium- or low-temperature method" is used. The term "high-temperature method" is used to define processes that involve the sublimation of the

**TABLE 4.3**

**Comparison of CNTs Production Techniques**

| Parameter | Laser Ablation | Arc Discharge | CVD |
|---|---|---|---|
| Process control | Difficult | Difficult | Easy, can be automated |
| Raw materials availability | Difficult | Difficult | Easy, abundantly available |
| Production rate | Low | Low | High |
| Purity of product | High | High | High |
| Energy requirement | High | High | Moderate |
| Product purity | High | High | High |
| Cost | High | High | Low |

*Source:* Ali Rafique and Iqbal (2011). Published by Scientific Researcher as open access.

solid source (graphite) at temperatures greater than 3200°C. The first CNTs were synthesized by this method.

The catalyst species necessary for CNT synthesis are typically transition-metal nanoparticles (diameter less than 10 nm) formed on a support substrate. For this purpose, iron, cobalt, and nickel are the most effective transition metals used. Other metals such as chrome and copper and their alloys are also promising for these purposes (Azam et al. 2015). Metal catalyst nanoparticles are crucial in CNTs growth. At high temperature, these metals have high solubility of carbon and a high diffusion rate, which are very helpful during the CNTs growth process (Azam et al. 2015). Usually the growth temperature is in the range of 700°C–900°C (Homma et al. 2003).

Currently, the synthesis of CNTs is usually realized using laser ablation (Journet et al. 2012; Azam et al. 2015), electric arc discharge (Sinnott and Andrews 2001; Kundrapu et al. 2012), and CVD techniques (Sinnott and Andrews 2001), including plasma-enhanced CVD (PECVD) (Meyyappan 2009; Gautier et al. 2016). A comparison of these techniques is shown in Table 4.3. The CVD and PECVD techniques are the most reported in the literature owing to advantages such as lower processing temperature, possibility of controlling the CNT structure, and high purity, as well as large-quantity production (Chen et al. 2004).

### 4.4.2 FEATURES OF CNTS AND PROSPECTS FOR THEIR APPLICATIONS IN HUMIDITY SENSORS

It was established that these novel materials show extraordinary physical, mechanical, and chemical properties. Actually, CNTs have demonstrated very high carrier mobility in the field-effect transistors, a very high electromigration threshold, a very high thermal conductivity, and exceptional mechanical properties. At that, the electronic structure of SWCNTs can be either metallic or semiconducting, depending on their diameter and chirality or helicity (symmetry of the two-dimensional carbon lattice) (see Figure 4.12) (Dresselhaus et al. 1996, 2003; Odom et al. 1998; Varghese et al. 2001). Semiconducting SWCNTs are *p*-type semiconductors with holes as the main charge carriers. The bandgap of semiconducting SWCNTs is

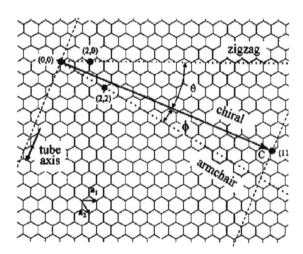
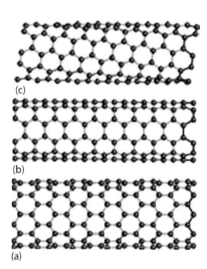

**FIGURE 4.12** The carbon lattice and the ways it can be rolled up to form a (a) zigzag; (b) armchair; or (c) chiral single-walled carbon nanotubes, depicted with its chiral angle. (Reprinted from *Precision Eng.*, 28, Mamalis, A.G. et al., Nanotechnology and nanostructured materials: Trends in carbon nanotubes, 16–30, 2004, Copyright 2004, with permission from Elsevier.)

inversely related to their diameter and corresponds to ~0.8 eV for a tube with a diameter of 1 nm. It is important to note that theoretical calculations indicate that all armchair tubules have metallic electronic properties only (Saito et al. 1992). It is supposed (Valentini et al. 2003) that these diverse electronic properties of CNTs give possibility to develop nanoelectronic devices as metal/semiconductor heterojunctions by combining metallic and semiconducting nanotubes. A possible approach is the modification of different parts of a single nanotube to have different electronic properties using controlled mechanical or chemical processes (e.g., nanotube bending or gas-molecule adsorption).

It is necessary to note that CNTs have the same developed surface as fullerenes, and therefore their applications should lie in the same general area, namely in the field of analytical chemistry, in particular, gas sensing (Mauter and Elimelech 2008). Moreover, CNTs seem to be more suitable for adsorption and detection of gases because small diameter and hollow structure makes them extremely sensitive to changes in their surroundings; all of the atoms on a CNT are exposed to its environment, and the extremely small diameter forces electrical signals traveling along the tube to interact with even tiny defects on or near the tube. As a result, gas adsorption on CNTs, including water-vapor adsorption, is today the focus of intense experimental and theoretical studies (Chen et al. 2001; Zhao et al. 2002; Snow et al. 2005; Kyakuno et al. 2011; Lekawa-Raus et al. 2015).

The results of research carried out in this area have shown that CNTs may really find successful applications in design of room-temperature adsorption/desorption-type gas sensors such as SAW (Penza et al. 2004; Sheng et al. 2011) and QCM (Zhang et al. 2005; Su et al. 2006; Chappanda et al. 2018), where their peculiar structural features could be realized (Varghese et al. 2001; Kuchibhatla et al. 2007; Zhang et al. 2008; Schnorr and Swager 2011). Moreover, it was established that such sensors can be extremely sensitive. For example, research conducted by Penza et al. (2004) showed that at room temperature, CNT-based SAW sensors were up to 3–4 orders

of magnitude more sensitive than existing organic layer-coated SAW sensors. The mass sensitivity of carbon-nanotube sensors can reach zeptograms ($10^{-21}$ g). Therefore, they had a very low limit of detection, and 1 ppm of ethanol or toluene was easily sensed. Several studies have shown that usually SWCNTs-based sensors have better performance compared to MWCNT sensors, while preparation of MWCNT is easier. However, there are studies (Su et al. 2006) where it is reported that MWCNT-based sensors have a higher sensitivity. Typical parameters of QCM humidity sensors are listed in Table 4.4.

Since CNTs have porous structures, and chemisorption and capillary condensation happen in the MWCNT films, humidity sensors of capacitance type can be designed as well (Varghese et al. 2001; Yeow and She 2006; Chen et al. 2009). Capacitance response of MWCNTs-based sensors to RH developed by Chen et al. (2009) is shown in Figure 4.13. A 90% response and recovery time of these sensors operated at room temperatures were 45 and 15 s, correspondingly. As for SWCNT-based capacitance sensors, then Snow et al. (2005) believe that the SWCNT surface generates a large fringing electrical field while applying a gate bias, and this electrical field polarizes the adsorbed molecules on the SWCNT surface. By calculation, they found that SWCNT capacitance was affected by both dielectric polarization and the charge transfer upon exposure to chemical vapor molecules, but the charge contribution was less than 10% (Snow et al. 2006). Therefore, they attribute the capacitance response to a field-induced polarization of surface dipoles. Thus, the SWCNT capacitance can be highly sensitive to a wide group of polar and nonpolar vapors (e.g., chlorobenzene, water, acetone) that give an adsorbate polarized by the fringing electric fields radiated from the surface of SWCNT electrode. The experiment has shown that such SWCNT chemicapacitors can be fast, highly sensitive, and completely reversible.

CNT-based gas sensors have shown good electrical response as well (Pati et al. 2002; Consales et al. 2008; Zhang et al. 2008). It was established that chemiresistors and chemical-field-effect transistors (ChemFETs) are probably the

## TABLE 4.4
## Comparison of Carbon-Based QCM Humidity Sensors

| Sensing Material | RH Range (%) | Sensitivity (Δf/f) per Percent RH | Hysteresis (RH%) | Response | Response/ Recovery Time (s) | References |
|---|---|---|---|---|---|---|
| CNT/Nafion | 0.035–61 | $20 \times 10^{-6}$ | NR | Linear | 5/5 | Chen et al. (2005) |
| Ball milling and hydrogen plasma treated MWCNT | 5–97 | $4.3 \times 10^{-6}$ | NR | Linear | 60 70 | Zhang et al. (2005) |
| MWCNT | 3–82 | $5.1 \times 10^{-6}$ | NR | Non-linear | NR | Wisitsoraat et al. (2008) |
| Graphene oxide film | 6.4–93.5 | $3.1 \times 10^{-6}$ | <5 | Non-linear | 18'/12 | Yao et al. (2011) |
| Poly(diallyldimethylammonium chloride)/SWCNT composite | 20.9–80.2 | $0.25 \times 10^{-6}$ | <8 | Non-linear | NR | Jing et al. (2012) |
| Nanodiamond films | 11.3–97.3 | $5.1 \times 10^{-6}$ | <4 | Non-linear | 25/3 | Yao et al. (2014) |
| Graphene oxide/Nafion composite | 11.3–97.3 | $6.4 \times 10^{-6}$ | <3 | Non-linear | 22/5 | Chen et al. (2016) |
| Graphene oxide/poly(ethyleneimine) | 11.3–97.3 | $2.7 \times 10^{-6}$ | <1 | Non-linear | 42/5 | Yuan et al. (2016) |
| MWCNT-HKUST-1 composite | 5–75 | $25 \times 10^{-6}$ | <5 | Non-linear | 250/265 | Chappanda et al. (2018) |

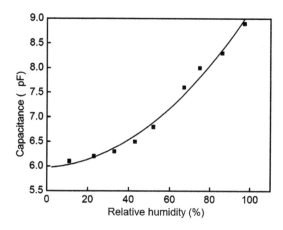

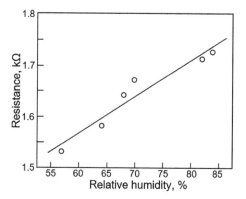

FIGURE 4.13 Capacitance response of MWCNT-based sensors to RH. The area and the thickness of the MWCNT-SiO$_2$ film
were1 × 0.8 cm and about 175 µm, correspondingly. The role of
SiO$_2$ powder was to increase the adhesive properties of the film.
The CNTs were first purified by heating at 400°C. Then they were
functionalized by a H$_2$SO$_4$ and HNH$_3$ mixture (v/v = 3:1) at 100°C
for 2 hours to functionalize the CNTs with –OH and –COOH
groups, which can make the CNTs hydrophilic and thus increase
the sensor sensitivity. (Reprinted from Chen, W.-P. et al., *Sensors*, 9,
7431–7444, 2009. Published by MDPI as open access.)

FIGURE 4.14 Resistance of MWNTs as a function of humidity. (With kind permission from Taylor & Francis: *J. Experiment.
Nanosci.*, Fabrication of humidity sensors by multi-walled carbon
nanotubes 5, 2010, 302–309, Tsai, J.T.H., Lu, C.-C., and Li, J.G.)

most promising types of gas sensors designed on the base of
CNTs (Kong et al. 2000; Bondavalli et al. 2009; Wang and
Yeow 2009; Zhang and Zhang 2009; Hu et al. 2010). Many
studies have shown that although CNs are robust and inert
structures, their electrical properties are extremely sensitive
to the effects of charge transfer due to interaction of CNTs
with analyte (Peng and Cho 2000; Zhao et al. 2001, 2002;
Bauschlicher and Ricca 2004) and chemical doping by various molecules (Peng and Cho 2000; Zhao et al. 2001, 2002;
Bauschlicher and Ricca 2004). For example, Zahab et al. (2000)
have shown that H$_2$O adsorbed at the CNT's surface acts as a
donor, and due to this adsorption, the conductivity type of the
SWNT can be changed from *p*-type to *n*-type. According to
Bachtold et al. (1999), water molecules can also concentrate
onto the interface of the SiO$_2$ and nanotubes in the suspended
SWNT FET, and these water molecules could act as charge
traps. In the case of MWNTs, which usually have a metallic
nature of conductivity, the adsorption of water is also accompanied by a decrease in conductivity (Tsai et al. 2010; Cao
et al. 2011). Typical characteristics of MWNT-based devices
are shown in Figure 4.14. According to Tsai et al. (2010), this
is due to the adsorption of water molecules in the tube-to-tube
interface. Lekawa-Raus et al. (2015) suggested that the water
vapor is mainly adsorbed by the standard clustering mechanisms observed in other carbon materials, but the mechanisms
responsible for the improvement in electrical performance are
much more debatable.

There is an opinion that the metal/SWCNT junctions also
play an important role in the sensing mechanism (Leonard
and Tersoff 2000; Cui et al. 2003; Auvray et al. 2005; Zhang
et al. 2006; Bondavalli et al. 2009). According to Bondavalli
et al. (2009), the main sensing mechanism in CNT-based

sensors seems to be the modulation of the Schottky barrier
height at the contacts, due to the buildup of interface dipoles
that depend on the gas specie, but also on the chemical reactivity of the metal constituting the electrodes. This conclusion
is based on the results reported by Leonard and Tersoff (2000)
and Cui et al. (2003) who have shown that the interaction of
oxygen at the junction between the metal electrode and the
SWCNT changes the metal work function and so the Fermi-
level alignment. This means that the Fermi level at the contact
is not pinned by surface states as it happens for contacts of
most metals with conventional semiconductors (Si, GaAs,...),
and it is controlled by the metal work function.

Some papers state that capillary condensation can also
affect the properties of CNTs (Yeow and She 2006; Chen et al.
2009; Kyakuno et al. 2011). In particular, Chen et al. (2009)
believe that MWCNTs, where the tube diameter varies, are of
interest for improving the performance of humidity sensors at
low humidity due to capillary condensation. This allows more
nanotubes to become filled with water.

Typical configuration of such sensors is shown in
Figure 4.15. It has to be pointed out that for this kind of sensor, the research has essentially focused on SWCNTs because
MWCNTs are only metallic and therefore unsuitable to fabricate chemiresistors and transistors. Testing of such sensors
has shown that the sensitivity achieved was pretty good, and
also, removing the gas totally restored the initial resistance.

As a rule, to improve performance, carbon-based materi-
als are functionalized with different metals and metal oxides.
For example, Cao et al. (2011) showed that in the 11%–98%
RH range, the resistances of chemically treated and untreated
MWCNT-based humidity sensors increased by 120% and
28%, respectively.

### 4.4.3 DISADVANTAGES OF CNTS WHICH CAN LIMIT THEIR APPLICATION IN HUMIDITY SENSORS

So, a brief analysis of the results obtained indicates that CNTs
are really promising materials for gas- and humidity-sensor
applications (Li et al. 2008; Kalcher et al. 2009; Bondavalli

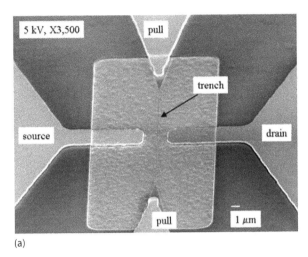

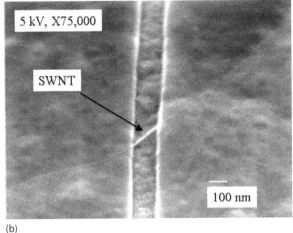

(a)                                                                          (b)

**FIGURE 4.15**   (a) Low; and (b) high-magnification SEM image of a CNT-based FET. (Reprinted with permission from Kaul, A.B. et al., *Nano Lett.*, 6, 2006, 942–947. Copyright 2006 American Chemical Society.)

et al. 2009). However, it is necessary to admit that the utilization of a single SWCNT presents some important issues (Bondavalli et al. 2009; Fam et al. 2011).

First, till now there has been no method that could fabricate only semiconducting SWCNTs. As a result, one cannot predict whether a SWCNT is metallic or semiconducting. In addition, it is known that to obtain the CNTs of high purity and uniformity is one of the big issues that still impact the applications of CNTs as gas-sensing material (Wang and Yeow 2009). The as-prepared CNTs are usually not pure and contain a lot of impurities, some of them, such as amorphous carbon and fullerenes, can hardly be completely removed from the raw materials, and the purity is difficult to be quantified. Thus, the measured physical and chemical properties of CNTs are peculiar to different research groups. Recently, some promising results on controlled synthesis of nanotubes in terms of morphology and diameter have been reported. However, chirality of the nanotube is difficult to control. Several strategies have been also reported to increase homogeneity. In particular, Arnold et al. (2006) proposed to differentiate CNTs using selective chemistry, which would involve the use of surface functionalization and or surfactants, which interacts with the surface of the CNT with specific chiralities and thereby sorting them. However, the overall cost of synthesis of pure CNTs strongly increases with the complexity of separation techniques adopted (Fam et al. 2011).

Second, it is quite laborious to identify the position of one single SWCNT on a sensor platform using standard methods. Proper manipulation techniques are required for applying a single tube or thin films of CNTs on the substrates that do not allow direct grow methods. Various proposals exist for their incorporation into devices in single-tube or thin-film architectures (Bachtold et al. 2001; Consales et al. 2008). However, though understanding that the realization of homogeneous thin films of CNTs with a controllable thickness and a tube size is an important basis for the future development of CNT-based devices for the sensor market, the development of

reasonable technologies for separation, selection tubes with similar diameter, and manipulation with nanotubes is still a task of great importance.

Third, considering that the CNTFET electrical characteristics are dependent on the individual SWCNT physical characteristics (the band gap in particular, which depends on the diameter for semiconductor specimens), it is very difficult to obtain reproducible devices. Depending on the preparation technique and process, the property and behavior of the sensors can vary significantly, which is very crucial for devices aimed for the sensor market. Therefore, the ability to synthesize identical and reproducible CNTs with consistent properties is very important for the application of CNTs in all areas (Wang and Yeow 2009).

We need to note that the previously mentioned disadvantages mainly relate to the sensors designed on the base of a single CNT. In the case of sensors such as SAW, QCM, capacitance, and optical sensors based on the using CNT networks (mats) (see Figure 4.16) or composites inclusive of CNTs, indicated disadvantages are not so important for sensor operation. Due to integral effect, there is no necessity to control parameters of individual CNTs in indicated devices.

There are also attempts to design FET-based or resistive sensors using SWCNT and MWCNT mats as channels (see Figure 4.17) (Snow et al. 2003; Kumar and Ramaprabhu 2006; Star et al. 2006; Chang et al. 2007; Cao et al. 2011; Mudimela et al. 2012). It has been established that SWCNT mats through a percolation effect can show semiconductor behavior even without separation on semiconducting and metallic CNTs. According to Bondavalli et al. (2009), only two conditions must be fulfilled: the distance between the two electrodes must be larger than the SWCNT length (otherwise metallic nanotubes could cause a short-circuit) and the real density of the SWCNT mat has to slightly exceed the percolation threshold for semiconducting specimens (remember that SWCNTs always come in mixtures of metallic (m) and semiconductor (s) specimens, with the approximate ratio s/m = 2; therefore, conditions can be found where the s-SWCNTs will

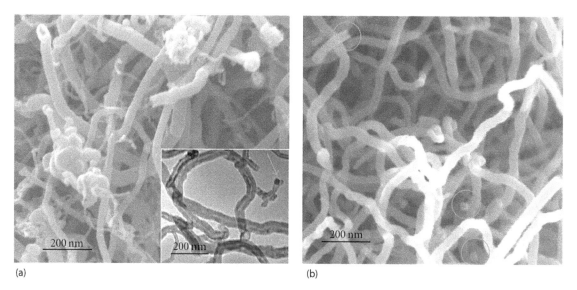

(a)                                                                          (b)

**FIGURE 4.16** SEM images of MWCNTs: (a) pristine MWCNTs; and (b) MWCNTs chemically treated in a mixture of 98% sulfuric acid and 79% nitric acid in ratio of 3:1; (insertion) TEM image of untreated carbon nanotubes. (Reprinted from Cao, C.L. et al., *J. Nanomater.*, 2011, 707303. Published by Hindawi as open access.)

(a)                                                                          (b)

**FIGURE 4.17** SEM images of (a) vertically aligned CNTs mat; and (b) CNTs trapped in castellated microelectrode gaps of sensors by positive dielectrophoresis (DEP). DEP is the electrokinetic motion of dielectrically polarized materials in non-uniform electric fields and has been used to manipulate CNTs for separation, orientation, and positioning of CNTs. (a: Reprinted from *Diamond Related Mater.*, 14, Huang, C.S. et al., Three-terminal CNTs gas sensor for $N_2$ detection, 1872–1875, Copyright 2005, with permission from Elsevier; b: Reprinted from *Sens. Actuators B*, 127, Suehiro, J. et al., Fabrication of interfaces between carbon nanotubes and catalytic palladium using dielectrophoresis and its application to hydrogen gas sensor, 505–511, Copyright 2007, with permission from Elsevier.)

percolate, whereas the m-SWCNTs will not, yielding an overall semiconductor behavior). In many works, including Novak et al. (2003) and Snow et al. (2003), this approach was realized with high efficiency.

Slow response and especially recovery process after interaction with the analyte is another disadvantage of SWCNT-based gas sensors. In dependence on measurement conditions, the recovery time was varied from 5 min to 12 hours (Kong et al. 2000; Liu et al. 2005; Benchirouf et al. 2013; Arunachalam et al. 2018). Examples of this sensor behavior are given in Figure 4.18. Unfortunately, the reasons for such a large difference in recovery times for different sensors is not clear. It is

important to note that the slow recovery process is inevitably accompanied by the appearance of a huge hysteresis (Kim et al. 2003) and a strong dependence of the parameters of SWCNTs from prehistory, which is very important for sensors intended for real use. According to Valentini et al. (2003, 2004), the fast recovery was observed only at 165°C. An important consequence of this study is that careful preparation of the SWNT should include high-temperature degassing, and that only dry, high-purity gases should be used in order to avoid artifacts when studying their effects on CNTs (Zahab et al. 2000). However, it was demonstrated that by integrating a microheater under the CNT's sensing layer or short exposure to UV

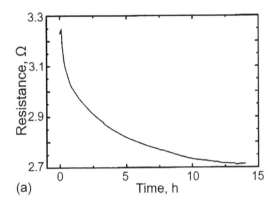
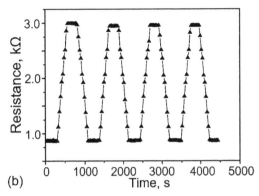

**FIGURE 4.18**  (a) The electrical resistivity variation of a compacted SWNT mat after the water volume injection (50 µL of distillated water) in a measurement chamber; and (b) resistivity variation of suspended CNTs during humidity cycles from 15% to 98% RH. The resistance of each device was allowed to stabilize for a few minutes before the beginning of each cycle. (a: Reprinted with permission from Zahab, A. et al., *Phys. Rev. B*, 62, 10000–10003, 2000. Copyright 2000 by the American Physical Society; b: Reprinted from Arunachalam, S. et al., *Sensors*, 18, 1655, 2018. Published by MDPI as open access.)

light, the response time of the sensor can be improved (Cho et al. 2005; Ueda et al. 2008). For example, Li et al. (2003b) reported that by using ultraviolet (UV) light, the recovery time was shortened to about 10 minutes. Continuous in-situ UV light illumination can also be used (Chen et al. 2012). It was established that the UV exposure decreases the desorption-energy barrier to ease the analyte desorption and thus contributes to the surface cleaning. According to Chen et al. (2012), the effect of in-situ UV light illumination is presumably reflected by: (1) affecting the background conductance, (2) cleaning the nanotube surface so that it is more accessible for gas adsorption, and (3) dynamically removing all adsorbed gas species from nanotubes. The first aspect shouldn't affect a sensor's performance dramatically. The second one has the potential to drastically enhance a sensor's sensitivity, while the third one will only reduce it. We need to recognize that though all these methods partly improved the recovery of the CNT-based sensors, but the recovery time was still not satisfactory. The experiment has shown that the recovery time could also be dramatically reduced, simply by applying a reverse bias for approximately 200 s (Hopkins and Lewis 2001). Researchers thought that the Coulomb interaction between the analyte and the negative charges induced in the channel by the gate voltage reduces the desorption barrier. The same method was employed by Chang et al. (2007).

Another way to improve the kinetics of the sensor response is the use of various treatments that affect the surface reactivity. For example, MWCNTs used by Liu et al. (2011) in humidity sensors were functionalized by $HNO_3$ at 140°C for 4 h to make them suspended with -OH and -COOH groups, which can make the CNTs hydrophilic. One can assume that it was this treatment that contributed to a decrease in response and recovery time of MWCNTs-$SiO_2$-based capacitance humidity sensors. Liu et al. (2011) reported that the response and recovery times in such sensors equaled 40 and 2 s, respectively.

Cao et al. (2011) also used CNT treatment in acid to improve humidity-sensor parameters. In the chemical-treatment process,

MWCNTs underwent an ultrasonic treatment for 150 minutes in a mixture of 98% sulfuric acid and 79% nitric acid in ratio of 3:1. The treated nanotubes were rinsed in deionized water and then let to dry. The thin-film humidity sensors were prepared by spraying MWCNTs dispersed in ethanol on the top of predeposited Au electrodes on quarts' plate. It was established that such treatment contributed to the increase in sensitivity and improved the stability of the sensor characteristics (see Figure 4.19).

Processing SWCNTs with ionic surfactants can also be used to control the parameters of humidity sensors. Evans et al. (2017) have shown that residual cationic and anionic surfactants induce a respective increase or decrease in the measured conductance across the randomly oriented SWCNT networks when exposed to water vapor. The approaches used to modify carbon-based materials will be discussed in more detail in Section 4.8 and Chapter 19.

The low solubility of CNTs is another factor-limiting CNT application (Tasis et al. 2003). Several attempts have been made in order to overcome this limitation (Star et al. 2002; Li et al. 2005; Backes et al. 2009). It was found that pristine CNTs are essentially insoluble, especially in polar solvents such as water. This has enabled the use of solution processing techniques, such as drop-casting, spin-casting, or spraying, which facilitates the fabrication of CNT-based devices. Moreover, the treatments used for solubility improvement often were accompanied by a strong change in the SWCNT's conductivity, which was not an admissible option for many applications. The development of novel methods that facilitate the processing of CNTs while having little impact on their electrical properties or providing the option to restore the conductivity in a subsequent step would therefore be desirable (Schnorr and Swager 2011).

According to (Pumera 2009), there is also a problem connected with features of CNT synthesis. As it is known, CNTs are typically grown from carbon-containing gas with the use of metallic-catalytic nanoparticles. It is well documented that such nanoparticles remain in the CNTs even after extensive purification procedures, leading to two very significant

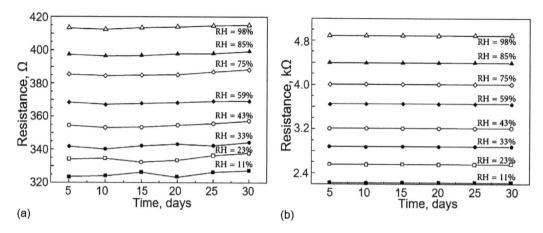

**FIGURE 4.19** Resistance variations with time for the sensors at various RH levels: (a) untreated; and (b) treated MWCNT humidity sensor. (Reprinted from Cao, C.L. et al., *J. Nanomater.*, 2011, 707303, 2011. Published by Hindawi as open access.)

problems (Pumera et al. 2007). It was shown that such residual metallic impurities are electrochemically active even when intercalated within the CNTs and that they can dominate the electrochemistry of CNTs (Liu et al. 2007). This is a significant problem for the construction of reliable sensors with reproducible parameters.

Interference from other vapors and gases, that is, low selectivity can be considered a disadvantage as well (Zhang and Zhang 2009). It was established that CNTs have high sensitivity to oxygen, $NO_2$, $NH_3$ and vapors of various organic solvents such as (Collins et al. 2000; Kong et al. 2000; Li et al. 2003a).

### 4.4.4 CNT-BASED COMPOSITES

One should note that CNTs are promising materials for preparing various composites (Barrau et al. 2003; Zeng 2003; Lee et al. 2013), which also can be used for humidity-sensor design. In this case, naturally, the most frequently used are

composites based on polymers, such as polyimide (PI) (Yoo et al. 2010; Tang et al. 2011), CNT/Nafion (Chen et al. 2005b; Su et al. 2006), hydroxyethyl cellulose (HEC) (Wang et al. 2014; Ma et al. 2015), poly(methyl methacrylate) (PMMA) (Su and Wang 2007), PVA (Fei et al. 2014), PEI (Li et al. 2012), vinyl-ethynyl-trimethyl-piperidole (VETP) (Saleem et al. 2010), poly-N-epoxypropylcarbazole (PEPC) (Shah et al. 2012), nanocrystalline cellulose (NCC) (Safari and van de Ven 2016), and the polyelectrolyte of QC-P4VP (Li et al. 2014). Studies have shown that CNT/polymer composites have the most evident advantages for humidity-sensor application. It was established that polymer-CNT composites combine the unique properties of nanotubes with the ease of processability of polymers. It should be noted that as a rule, CNT/polymer-based sensors are devices of the resistive type. Their typical characteristics are shown in Figure 4.20. It is seen that the performances of CNT/polymer composite-based humidity sensors may vary significantly. Unfortunately, to date, there is no comparative analysis of sensors made on the

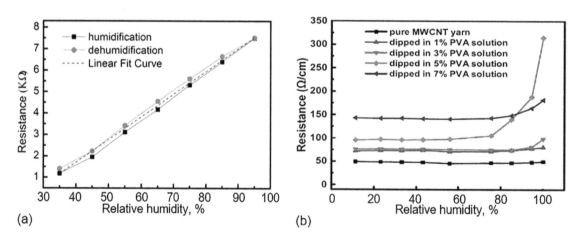

**FIGURE 4.20** (a) Hysteresis curve of the resistance change of the MWCNT/HEC composite film; and (b) typical resistance various as a function of relative humidity for CNTs-PVA samples with different composition. (a: Reprinted with permission from Wang, D. et al., *J. Mater. Res.*, 29, 2845–2853, 2014. Copyright 2014. Cambridge.com.; b: Reprinted from *Sens. Actuators B*, 230, Li, W. et al., A novel flexible humidity switch material based on multi-walled carbon nanotube/polyvinyl alcohol composite yarn, 528–535, 2016. Copyright 2016, with permission from Elsevier.)

basis of various polymers. Therefore, it is impossible to give preference to any particular polymer in the development of sensors. Too many parameters determine the characteristics of the sensors.

The features of the operation of CNT/polymer-composite sensors based on polymer swelling were discussed earlier in Section 4.2.1 and in Chapter 11 (Volume 2). Their feature is that for designing humidity-sensitive materials based on CNT/polymer composites, a very low fraction content of CNT is required. For example, according to Yoo et al. (2010), for achieving highly linear variation of the resistance with humidity, the concentration of CNTs in polyimide should be ~0.4%. In addition, the kinetics of the sensor response and hysteresis of such sensors are not controlled by adsorption-desorption processes on the surface of CNTs but by the specificity of the interaction of water vapor with the polymer. As a result, the response and recovery of the CNT-based humidity sensors can be fast (see Figure 4.21).

It is important to note that the dispersion of carbon nanofibers and nanotubes within a polymer matrix gives the possibility of overcoming the instability experienced with CB–polymer composites (Zhang et al. 2006; Ma et al. 2015). The instability of CB–polymer composites occurs because the nanosized CB particles tend to reaggregate when the composite absorbs vapor, which lowers the matrix viscosity and increases its volume. In contrast, dispersing CNFs and nanotubes in the polymer improves the vapor-sensing stability because their high aspect ratio resists movement within their polymer composites when vapor is absorbed and desorbed. Thus, the original electrical percolation pathways present in these composites are maintained after absorbed vapor has desorbed from the matrix. For CNT-based composites, such studies were carried out by Ma et al. (2015). They established that the HEC-CB-based humidity sensors had much higher sensitivity in comparison with HEC-CNT-based sensors. But

at the same time, the HEC-CNT-based sensors were characterized by better repeatability and stability under cyclic humidity changes, which is important for real applications.

As for other composite materials developed for use in humidity sensors, among them you can find composites such as MWCNT-SiO$_2$ (Liu et al. 2011), ZnO/MWCNTs/ZnO (Zhang et al. 2016a), CNT-MnO$_2$ (Jung et al. 2015), CNT-metal-organic frameworks (MOFs) (Chappanda et al. 2018), and SWCNT-Pt-P$_2$O$_5$ (Spiridon et al. 2017). In particular, Liu et al. (2011) have studied MWCNT-SiO$_2$ composites and established that the addition of CNTs in the SiO$_2$ matrix strongly increased the capacitance and the sensitivity to air humidity (see Figure 4.22a). The sharp increase in the capacitance value was observed between 0.5 and 1.5 wt%, where the capacitance changed from 97.4 to 9699 pF. According to Liu et al. (2011), the MWCNTs/SiO$_2$ films can be assumed as three-dimensional (3D) granular composites consisting of MWCNT molecules randomly dispersed in non-conducting SiO$_2$, as shown in Figure 4.22b. In the percolation model, the capacitance remains a stable value to a certain weight fraction of the CNTs. However, as the content of CNTs reaches a critical concentration or percolation threshold, some CNTs begin to get closer to each other (the distance between them decreases), which gives rise to the capacitance of the sensor. It was also established that the maximal effect of humidity influence is observed at the concentration of MWCNTs close to percolation threshold, which was ~1 wt%. The sensor with such composite has the highest sensitivity about 673 pF/% RH and the best linearity.

It is important to note that the sensing mechanism of devices based on CNT-inorganic material composites differs from the sensing mechanism observed in humidity sensors based on CNT-polymer composites. For example, according to Jung et al. (2015), MnO$_2$ coating on a CNT yarn serves as a catalyst for promoting the charge transfer between the

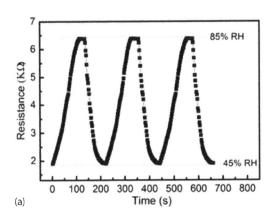

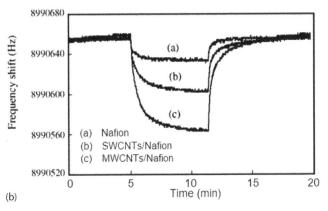

**FIGURE 4.21** (a) The response and recovery characteristic curve of the resistance change of the MWCNT/HEC composite film. Wang et al. (2014) believe that the actual response time and recovery time should be shorter because the measurement results were affected by the time delay of the humidity chamber; and (b) response of QCM humidity sensors developed using Nafion, SWCNTs/Nafion and MWCNTs/Nafion films to water vapors (469.3 ppmv). (a: Reprinted with permission from Wang, D. et al., *J. Mater. Res.*, 29, 2845–2853, 2014. Copyright 2014. Cambridge.com.; b: Reprinted from *Sens. Actuators B*, 115, Su, P.G. et al., A low humidity sensor made of quartz crystal microbalance coated with multi-walled carbon nanotubes/Nafion composite material films, 338–343, Copyright 2006 with permission from Elsevier.)

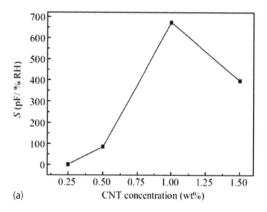 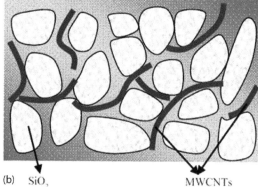

(a)  CNT concentration (wt%)          (b)  SiO₂              MWCNTs

**FIGURE 4.22** (a) Capacitance sensitivity of the sensors as a function of MWCNT concentration in SiO₂-MWCNTs composite; and (b) sketch map of the structure in MWCNTs/SiO₂ films. (Reprinted with permission from Liu, X. et al., *J. Semicond.*, 32, 034006, 2011. Copyright 2011. The Chinese Institute of Electronics.)

H₂O molecules and CNTs. In the case of the SWCNT-Pt-P₂O₅ composite (Spiridon et al. 2017), we have another situation because the humidity-sensitive properties are fully controlled by P₂O₅ (read Chapter 7, Volume 2).

## 4.5 GRAPHENE

Graphene is another carbon-based nanomaterial that can be promising for sensor applications (Kauffman and Star 2010; Ratinac et al. 2011; Basu and Bhattacharyya 2012; Gupta Chatterjee et al. 2015). Geim and Novoselev (2007) received the Nobel Prize in 2010 for their works related to graphene. Graphene is a two-dimensional (2D), single layer of sp²⁻ hybridized carbon that can be considered the "mother of all graphitic forms" of nanocarbon, including discussed earlier 1D CNTs (Geim and Novoselov 2007; Allen et al. 2010). Graphene has two atoms per unit cell. In graphene, carbon atoms are arranged in planar and hexagonal form shown in Figure 4.23. The crystalline or "flake" form of graphite consists of many graphene sheets stacked together. Its honeycomb structure has important consequences for the charge carriers (Marchenko et al. 2011). The π and π* bands of freestanding graphene form cones that touch each other in a single point signaling the presence of massless relativistic electrons (Morozov et al. 2008), which gives rise to outstanding transport properties. A high mobility of charge carriers and the ability to modify the electronic properties by doping (Rossi and Sarma 2008), by deformation (Huertas-Hernando et al. 2006), or by interaction with different substrates (Ran et al. 2009) places graphene among the most promising materials for future electronic devices.

### 4.5.1 SYNTHESIS AND DEPOSITION OF GRAPHENE LAYERS

At present, it was reported that methods such as chemical vapor deposition, reactive-ion etching, thermal decomposition of SiC, direct-current arc discharge with graphite rods in He atmospheres, chemical modification of graphite, and simple mechanical exfoliation, or "peeling off" layers from highly oriented pyrolytic graphite, can be applied for production of graphene (Allen et al. 2010; Choi et al. 2010; Singh

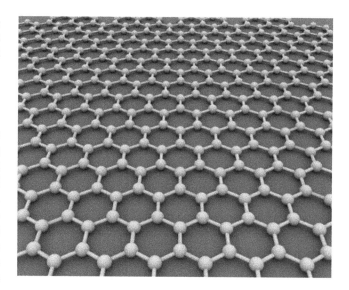

**FIGURE 4.23** Typical diagram illustrating the structure of graphene structure.

et al. 2011). Epitaxial graphene growth on the SiC substrates is also possible (Nomani et al. 2010). Mechanical exfoliation method is low cost, but the graphene produced is of poor quality with the limited area. It is particularly difficult and time-consuming to obtain a single-layer graphene in large scale with this method. The graphene obtained by epitaxial growth showed poor uniformity and contained a multitude of domains. Currently, however, the most popular method for graphene production relies on the chemical modification of graphite using the Hummers method, which involves the oxidation of graphite in the presence of strong acids and oxidants (Park and Ruoff 2009). In this case, oxidized graphite is cleaved via rapid thermal expansion or ultrasonic dispersion, and subsequently the graphene-oxide sheets were reduced to graphene. This method produces isolated, water-soluble graphite oxide sheets with many oxygen-containing defect sites (He et al. 1998). Graphite oxide can be transferred in graphene, also called reduced GO (RGO) or chemically converted graphene (CCG), by chemical reduction with aqueous hydrazine as a reducing agent (Stankovich et al. 2007).

**TABLE 4.5**
**Advantages and Disadvantages of Methods for Controlling the Size and the Shape of Graphene**

| Method | Defect Degree | Repeatability | Damage | Size Control | Layer Control | Shape Control | References |
|---|---|---|---|---|---|---|---|
| Nanoscale cutting | Low | Good | Low | 100 nm–1 mm | No | Yes | Mohanty et al. (2012) |
| Chemical control | Medium | Poor | Medium | 10–100 nm | No | No | Wang et al. (2012) |
| Chemical synthesis | Low | Good | N/A | 10 nm–100 μm | Yes | Yes | Wei and Liu (2010) |
| Differential separation | High | Good | Very low | 10–100 μm | No | No | Wang et al. (2011) |
| Density gradient separation | High | Good | Very low | 5 nm–5 μm | Yes | No | Green and Hersam (2009) |

*Source:* Zhang, G. and Sun, X., Size control methods and size-dependent properties of graphene, in Aliofkhazraei, M., Ali, N., Milne, W.I., Ozkan, C.S., Mitura, S., Gervasoni, J.L. (eds.), *Graphene Science Handbook, Size-Dependent Properties*, CRC Press, Boca Raton, FL, 27–40.

However, RGO does have limitations. A serious drawback of this method is that the oxidation process induces a variety of defects that would degrade the electronic properties of graphene. Therefore, at present increased interest is to prepare graphene using the CVD method (Li et al. 2009). It was established that CVD graphene tends to be more atomically smooth, whereas RGO usually has many oxygen-containing defect groups (Bagri et al. 2010). CVD-grown graphene can also show several orders of magnitude of lower resistivity, as compared to RGO (Li et al. 2009). In the review by Singh et al. (2011), one can find a description of all the previously mentioned methods. One can find additional information related to advantages and disadvantages of methods for controlling the size and the shape of graphene in Table 4.5.

Similar to CNTs, which can be functionalized with polymers, nanoparticles, or atomic dopants, different approaches toward the graphene functionalization have also been reported (Xu et al. 2008; Kauffman and Star 2010; Allen et al. 2010; Singh et al. 2011; Vedala et al. 2011). In particular, recently, N substitutionally doped graphene was first synthesized by a CVD method with the presence of $CH_4$ and $NH_3$ (Wei et al. 2009). As doping accompanies with the recombination of carbon atoms into graphene in the CVD process, dopant atoms can be substitutionally doped into the graphene lattice, which is hard to realize by other synthetic methods. The process of graphene doping will be discussed later in Section 4.8.3.

### 4.5.2 Graphene-Based Gas and Humidity Sensors

Taking into account unique properties of graphene, it was assumed that graphene-based devices should be viable candidates for the development of low-temperature gas (Arsat et al. 2009; Shafiei et al. 2010; Basu and Bhattacharyya 2012; Gupta Chatterjee et al. 2015; Varghese et al. 2015a, 2015b) and humidity sensors (Jung et al. 2008; Bi et al. 2013; Yao et al. 2014a; Wang et al. 2016a, 2016b). This assumption was based on the following facts.

First, graphene's electronic properties are strongly affected by the adsorption of molecules (Lin and Avouris 2008), a prerequisite for design of any type of gas sensors, including humidity sensors. It was established that the adsorption of gas molecules from the surrounding atmosphere is accompanied by doping of the graphene layers with electrons or holes depending on the nature of the adsorbed gas. As a result, by monitoring changes in resistivity, one can sense minute concentrations of certain gases present in the environment.

Second, the 2D structure of graphene constitutes an absolute maximum of the surface-area-to-volume ratio in a layered material, which is essential for high sensitivity. According to Pumera (2009), graphene has a theoretical surface area of 2630 m²/g, surpassing that of graphite (~10 m²/g), and is two times larger than that of CNTs (1315 m²/g).

Third, graphene has good long-term stability of parameters (Marchenko et al. 2011; Bi et al. 2013) (see Figure 4.24) and good compatibility with standard microelectronic technologies, such as the conventional lithographic process. (Berger et al. 2004; Shao et al. 2009). It should be noted that graphene is more suitable for device integration than CNTs because the planar nanostructure of the former makes it advantageous for use in standard microfabrication techniques. In addition, the recent improvements made to graphene-deposition methods have contributed to an increase in the applicability of graphene for device integration (Li et al. 2009; Reina et al. 2009).

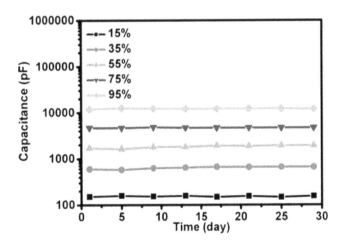

**FIGURE 4.24** Stability test of GO-based humidity sensors at fixed RH ($f$ = 1 kHz). (Reprinted by permission from Macmillan Publishers Ltd. *Sci. Rep.*, Bi, H. et al., 2013, copyright 2013.)

**TABLE 4.6**
**Comparison of Graphene-Based Humidity Sensors**

| Sensor Type | Material | RH Range (%) | Sensitivity | Response/Recovery Times, s | References |
|---|---|---|---|---|---|
| Capacitance | GO | 15–95 | Up to 46 pF/%RH | 10/41 | Bi et al. (2013) |
| | G/methyl red | 0–100 | 2869500% ($\Delta C/C$) | NA | Ali et al. (2016) |
| | RGO/SnO$_2$ | 10–95 | Up to 1600 pF/%RH | 6–102/6–9 | Zhang et al. (2016b) |
| | PVDF/G-Cu | 35–90 | 0.0371 pF/%RH | 20/NA | Hernández-Rivera et al. (2017) |
| | PVDF/G-Ag | | 0.0463 pF/% RH | 21/NA | |
| Resistive | RGO/SnO$_2$:Fe | 0–100 | 3.23 ($\Delta R/R$) | NA | Toloman et al. (2017) |
| | G | 4–84 | 65 ($\Delta R/R$) | 180/180 | Ghosh et al. (2009) |
| | G | 1–96 | 0.31 ($\Delta R/R/\%RH$) | 0.6/0.4 | Smith et al. (2015) |
| | GO | 30–80 | NA | 0.03/0.03 | Borini et al. (2013) |
| | G | 35–98 | 18.1 ($\Delta I/I$) | NA | Chen et al. (2014) |
| | RGO/SnO$_2$ | 11–97 | 0.15–0.45 ($\Delta R/R$) | NA | Zhang et al. (2016c) |
| Piezoresisitive | GO | 10–98 | 79.3 ($\Delta V/\%RH$) | 19/10 | Yao et al. (2012) |
| | Poly(dopamine)-treated GO/PVA | 30–95 | 0.4–2.3 (M$\Omega$/%RH) | NA | Hwang et al. (2014) |
| QCM | GO | 6–97 | 22.1 Hz/%RH | NA | Yao et al. (2011) |

The experiment confirmed the previously mentioned statements (see Table 4.6). In particular, Bi et al. (2013), while using the graphene films, developed capacitance-humidity sensors with improved sensitivity by a factor of 10 compared with conventional capacitance sensors, and their sensors had a reasonable response and recovery times. Studies have also shown that annealing was accompanied by the transformation of the properties of graphene, giving it hydrophobic properties, which significantly worsened the properties of humidity sensors.

For most of the gas-sensor applications, graphene synthesized by various methods was deposited on Si or Si/SiO$_2$ substrates and electrical contacts were prepared with Au/Ti or other metals, which provided good adhesion and ohmic contact with graphene (Schedin et al. 2007; Sundaram et al. 2008; Fowler et al. 2009). In addition, it was found that the role of electrode electrical contacts in the sensing mechanism

of graphene was minimum. The sensing mechanism was primarily attributed to the charge transfer at the graphene surface (Fowler et al. 2009).

One should note that due to stability requirements, GO is usually used in humidity sensors. In particular, Yao et al. (2014a) have shown that the stability of GO-based QCM humidity sensors was significantly higher than that of polymer-coated QCM sensors at any humidity condition. Especially at high-humidity points, the advantage of the stability of GO-based QCM sensor becomes more prominent. At that, RGO has superior conductivity than GO but inferior to the pristine graphene (Singh et al. 2011). Operating characteristics of the best graphene-based sensors are shown in Figures 4.25 and 4.26.

As for the mechanism of sensitivity to humidity, which is implemented in graphene-based sensors, in most of the works,

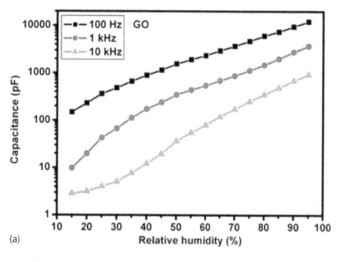
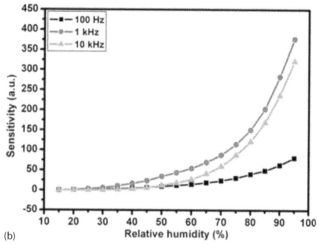

**FIGURE 4.25** Capacitance (a) of GO-based humidity sensor and sensitivity; and (b) versus RH at 100, 1, and 10 kHz. (Reprinted by permission from Macmillan Publishers Ltd. *Sci. Rep.*, Bi, H. et al., 2013, copyright 2013.)

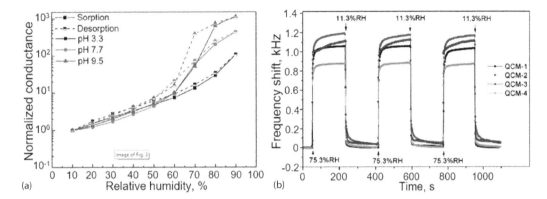

(a)

(b)

**FIGURE 4.26** (a) Normalized conductance of three different GO-based humidity sensors as a function of RH. Filled and half-filled symbols present the sorption and desorption process, respectively. GO was dispersed in deionized (DI) water to a concentration of 1 mg/mL and filtered through a membrane having 5 μm pores before use. The pH of GO solutions was adjusted to desired values using 0.1 M HCl or 0.1 M NaOH. The humidity-sensing layer was formed through drop-casting (10 μL) of the prepared GO solution on the sensor substrate with electrodes; and (b) repeatability of GO/PEI quartz crystal microbalance humidity sensors with three cycles. QCM devices consist of AT-cut quartz crystal (8 mm diameter) with a fundamental resonance frequency of 10 MHz and silver electrodes (5 mm diameter) on both sides. (a: Reprinted from *Sens. Actuators B*, 258, Park, E.U. et al., Correlation between the sensitivity and the hysteresis of humidity sensors based on graphene oxides, 255–262, Copyright 2018, with permission from Elsevier; b: Reprinted from *Sens. Actuators B*, 234, Yuan, Z. et al., Novel highly sensitive QCM humidity sensor with low hysteresis based on graphene oxide (GO)/poly(ethyleneimine) layered film, 145–154, Copyright 2016, with permission from Elsevier.)

it is believed that the work of graphene-based humidity sensors is based on the same principles that are implemented in other solid-state sensors (Bi et al. 2013). These mechanisms were discussed in Chapter 2.

The experiment showed that the detection limits of the graphene-based sensors ranged between the parts per billion (ppb) and parts per million (ppm) levels (Nomani et al. 2010). This and even higher levels of sensitivity are sought for industrial, environmental, and military monitoring. However, Schedin et al. (2007) believe that due to unique properties, graphene makes it possible to increase the sensitivity to its ultimate limit and detect individual dopants. They gave the following explanation of this statement. First, graphene is a strictly 2D material and, as such, has its whole volume, that is, all carbon atoms exposed to surface adsorbates, which maximizes their effect. Second, graphene is highly conductive, exhibiting metallic conductivity and, hence, low Johnson noise even in the limit of no-charge carriers, where a few extra electrons can cause a notable relative changes in the carrier concentration, *n*. As it was found, the mobility of electrons in graphene can be more than 100,000 cm²/ Vs at room temperature that is much higher than in other materials. For comparison, the mobility of electrons in silicon is ~1400 cm²/Vs. As a result, graphene has a resistivity of ~1.0 μΩ cm, which is about 35% less than the resistivity of copper. Third, graphene has few crystal defects (Novoselov et al. 2005; Geim and Novoselov 2007), which ensures a low level of excess (1/*f*) noise caused by their thermal switching. Fourth, graphene allows four-probe measurements on a single crystal device with electrical contacts that are ohmic and have low resistance (Schedin et al. 2007; Fowler et al. 2009). All of these features contribute to make a unique combination that maximizes the signal-to-noise ratio to a

level sufficient for detecting changes in a local concentration by less than one electron charge, *e*, at room temperature.

Graphene-based composite materials have been studied for humidity sensors as well (see Table 4.6). For example, Shafiei et al. (2010) and Zhang et al. (2016a, 2016b) used the RGO/SnO₂ composites (see Figure 4.27). Zhang et al. (2016b) showed that the sensitivity of the SnO₂/RGO composite was considerably higher than that of pure RGO. According to Zhang et al. (2016b), several possible reasons may be

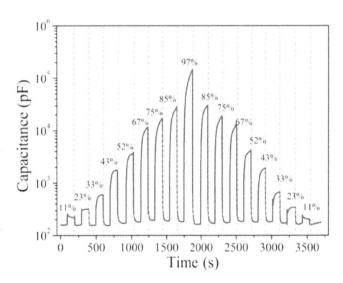

**FIGURE 4.27** Capacitance response of the SnO₂/RGO hybrid composite sensor under switching RH. (Reprinted from *Sens. Actuators B*, 225, Zhang, D. et al., Fabrication and characterization of an ultrasensitive humidity sensor based on metal oxide/graphene hybrid nanocomposite, 233–240, Copyright 2016, with permission from Elsevier.)

**TABLE 4.7**
**Flexible Graphene-Based Humidity Sensors**

| Sensor Type | Fabrication Method | Sensing Material | Working Range (% RH) | Sensitivity | Water Resistance | References |
|---|---|---|---|---|---|---|
| Impedance | TBLI | RGO | 10–90 | 6.3 Z/%RH | NO | Guo et al. (2012) |
| | LBL-anchored | RGO | 30–90 | 0.0423 log Z/%RH | Yes | Su and Chiou (2014) |
| | Sol-gel + self-assembly | GO/AuNPs | 20–90 | 0.0281 log Z/%RH | NA | Su et al. (2015) |
| Resistive | Hydrothermal | GQD | 1–100 | ~390 (R1%/ R100%) | NA | Hosseini et al. (2017) |
| Capacitance | Hydrothermal | $SnO_2$/RGO | 11–97 | 1604.89 pF/%RH | NA | Zhang et al. (2016b) |

LBL—layer-by-layer; GQD—graphene-quantum dots; TBLI—two-beam-laser interference

attributed for this enhancement in sensitivity. It is known that the ability of RGO to detect water molecules depends on its high surface-to-volume ratios and hydrophilic functional groups attached on its surface, such as hydroxyl and carboxyl groups. Zhang et al. (2016b) believe that the incorporation of $SnO_2$NPs into RGO sheets brings more active sites, such as vacancies and defects. Heterojunction may also be created at the interface of the two nanomaterials and contributes to the improvement in humidity sensing.

Experiments have shown that the flexible humidity sensors can also be designed on the basis of graphene (Dua et al. 2010; Guo et al. 2012). Parameters of such sensors are listed in Table 4.7.

### 4.5.3 DISADVANTAGES OF GRAPHENE, LIMITING ITS APPLICATION

Kauffman and Star (2010) concluded that there are several obstacles that should be overcome before graphene can compete with CNTs as the preferred carbon nanostructure for sensing platforms. According to Kauffman and Star (2010), the graphene-based sensor platforms in comparison with CNT or other nanowire-based sensors suffer from the following major disadvantages. First, the 2D nature of graphene inherently limits the sensor response. Second, graphene is a zero-band gap semiconductor, and it behaves as a semimetallic material (Geim and Novoselov 2007). It was established that it is very difficult to turn the graphene into an electrically conductive "off-state" because thermal promotion of charge carriers produces non-zero electrical conductance at any applied gate voltage in the FET structures. The absence of optical spectroscopy is other major limitation of graphene application as a sensor platform. Unlike SWNT, graphene does show UV-region absorbance (Liang et al. 2009). In addition, its luminescence is weak unless band gaps are created through chemical oxidation or size reduction (Gokus et al. 2009). Kauffman and Star (2010) believe that atomic doping of graphene (Boukhvalov and Katsnelson 2008) or decorating graphene with nanoparticles may improve the sensor response, and the development of spectroscopic techniques

for graphene will undoubtedly serve to help further the field of graphene-based gas sensors.

It is also necessary to take into account that it is nearly impossible to produce only single layers of graphene with current fabrication methods (Ratinac et al. 2011). Therefore, "graphene" can include anything from one to many layers, and the exact number of layers, $N$, critically affects properties, especially for low values of $N$. In particular, stacks with $N \geq 12$ or so tend to behave more like a thin-film graphite than graphene (Partoens and Peeters 2006). This means that the importance of knowing what sort of "graphene" you are working with is great.

In addition, one should take into account that graphene's single layers are not completely flat; instead, the flexible sheets have a tendency to fold, buckle, and corrugate (Ratinac et al. 2011). This flexibility is related to the out-of-plane phonons (flexural vibrations) that occur in soft membranes, which means that free-standing graphene tends to crumple (Castro Neto et al. 2009). Thus, the larger-scale distortions like folds and "pleats" (Novoselov et al. 2004) are seen to be an unavoidable by-product of graphite-cleaving techniques. When working with a soft membrane such as graphene, invariably some of the individual layers will fold and buckle during the process of mechanical peeling and subsequent solution deposition onto the substrate.

Some researchers have recently observed that inclusion of lithographic (photo or e-beam) steps in preparation of graphene can cause some negative effects on the sensing properties of graphene, due to the presence of residual polymers on the graphene surface. In the work by Dan et al. (2009), a cleaning process was demonstrated to remove the contamination on the sensor device structure, allowing an intrinsic chemical response of graphene-based sensors. The contamination layer was removed by a high-temperature cleaning process in a reducing ($H_2$/Ar) atmosphere (Dan et al. 2009).

In addition, we need to take into account that like CNT and other nanomaterials, the key challenge in synthesis and processing of bulk-quantity graphene sheets is aggregation (Singh et al. 2011). Unless well separated from each other, graphene tends to form irreversible agglomerates or even restack to form graphite through Van der Waals interactions.

The prevention of aggregation is essential for graphene sheets because most of their unique properties are only associated with individual sheets. This is the basis for a noticeable difference in the magnitude of the response for sensors fabricated in various laboratories using the same technology (Hernández-Rivera et al. 2017).

Bad selectivity of graphene-based sensors, which is typical for the most of solid-state gas sensors designed can also be considered as disadvantage of graphene-based sensors (Schedin et al. 2007; Pearce et al. 2011). It was established that similar to CNTs, pristine graphene interacts with numerous gases with large binding energies such as $NO_2$, $NH_3$, CO, $H_2$, $CO_2$, and $H_2O$ (see Figure 4.28). Detection of various vapors like nonanol, octanoic acid, trimethylamine, acetone, HCN, dimethyl methylphosphonate, dinitrotoluene, iodine, ethanol, and hydrazine hydrate has also been reported (Schedin et al. 2007; Robinson et al. 2008; Dan et al. 2009; Fowler et al. 2009; Allen et al. 2010).

Long response and recovery times, especially at room temperature (see Figure 4.28), are other disadvantages of graphene-based sensors (Varghese et al. 2015b; Chen et al. 2018). Schedin et al. (2007) found that the initial undoped state could be recovered only by annealing at 150°C or by short-time UV illumination, which should be considered as an alternative to thermal annealing. However, GO-based humidity sensors with fast response are also developed. In particular, Bi et al. (2013) reported that their capacitance-based humidity sensors had response and recovery times at 10–41 s, respectively. Bi et al. (2013) believed that the excellent response and recovery times are not only ascribed to the abundant hydrophilic functional groups on GO sheets but also to the large interlayer space in the GO films.

Based on the above, it can be concluded that despite the rush around the study of graphene, graphene-based devices will not appear on the humidity-sensor market in the near future. In addition, due to technological difficulties with separation and handling of graphene and nanotubes, the cost of such sensors significantly exceeds the cost of sensors based on polymers and ceramics.

## 4.6 NANODIAMOND PARTICLES

Nanodiamonds (NDs) are the members of the diverse structural family of nanocarbons discussed in this chapter. Therefore, it is anticipated that the attractive properties of NDs will be exploited in a similar manner to other carbon nanoparticles, in particular for the development of gas sensors of different types (Gi et al. 1997; Hayashi et al. 2001; Gurbuz et al. 2004; Chevallier et al. 2009; Davydova et al. 2012, 2014, 2018; Hribšek et al. 2010; Varga et al. 2015) and humidity (Davydova et al. 2012; Varga et al. 2015; Yao et al. 2014; Yao and Xue 2015; Kulha et al. 2016). In particular, based on diamond films were developed humidity sensors of various types, including QCM (Yao et al. 2014b; Varga et al. 2015; Kulha et al. 2016), impedance-based (Gi et al. 1997; Yao and Xue 2015) and resistive (Wang et al. 2007a; Kromka et al. 2010) humidity sensors. In the latter case, hydrogen-terminated nanocrystalline diamond or nanocomposites based on this material is generally used as the humidity-sensitive material (Sommer et al. 2008; Davydova et al. 2015).

It is known that diamond with a bandgap of 5.5 eV is a bonafide insulator in its undoped state, which does not respond to the changes in the $H_2O$-vapor concentration in the ambient air at all (see Figure 4.29). However, it was shown that it can exhibit a pronounced p-type surface conductivity of up to $10^{-5}\ \Omega^{-1}$ per square when the surface is terminated with hydrogen (Landstrass and Ravi 1989; Maki et al. 1992). In particular, it was established that both synthetic and natural diamonds become electrically conductive by treatment with hydrogen

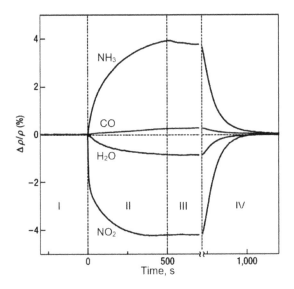

**FIGURE 4.28** Resistivity response of pristine graphene monocrystals to 1 ppm concentrations of different reducing and oxidizing gases. Regions: (*I*) response in vacuum before gas exposure, (*II*) exposure to 1 ppm of gases, (*III*) gas removed by vacuum, and (*IV*) gas desorption by annealing at 150°C. (Reprinted by permission from Macmillan Publishers Ltd. *Nat. Mater.*, 6, Schedin, F. et al., 2007, copyright 2007.)

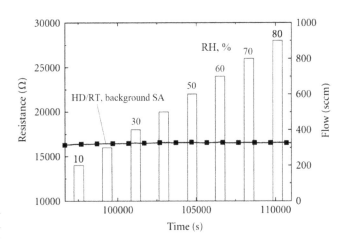

**FIGURE 4.29** Response of an HD sample toward water vapor. During these tests the HD sample was kept at room temperature. (Reprinted from Helwig, A. et al., *J. Sensors*, 2009, 620720, 2009. Published by Hindawi Publishing Corporation as open access.)

plasma and exploit the reproducible phenomenon that conductivity decreases with increasing RH. It is important that the H-terminated surface of intrinsic diamond (H-diamond) exhibits $p$-type semiconducting properties in the subsurface region without need of conventional doping (Hayashi et al. 1997). The best established model to explain this phenomenon and the operational principle of sensing properties of H-terminated surfaces (Wang et al. 2007b) is the transfer doping model of Maier et al. (2000). According to this model, a near-surface hole-accumulation layer with carrier densities in the range of $10^{12}$–$10^{13}$ cm$^{-2}$ forms via a transfer of diamond-valence electrons to an adsorbed surface species. It was found that such hydrogen-terminated nanocrystalline diamond is stable and responds to minimal variations in RH (Sommer et al. 2008). In addition, Kromka et al. (2010) have shown that similarly to other humidity-sensing materials, the increased surface-to-volume ratio due to the "porous" structure formation plays a crucial role in enhancing the sensitivity of the nanocrystalline diamond (NCD)-based gas and humidity sensors. The main advantages of the diamond include the following. On the one hand, it is chemically inert and thermally stable so it is considered a stable-sensor platform; while on the other hand, its carbon-terminated surface offers wide perspectives from organic chemistry and biochemistry for covalent attachment of specific receptors (Chevallier et al. 2009).

High pressure–high temperature synthesis can be used to obtain nanocrystalline diamond powders (Chevallier et al. 2009), while to form thin films, the CVD method, in particular, the microwave plasma-enhanced chemical vapor deposition process, should be preferred (Davydova et al. 2010, 2015; Possas-Abreu et al. 2017). Finally, all samples were exposed to hydrogen plasma in order to induce a $p$-type conductivity (Hayashi et al. 1997).

However, due to technological difficulties related to the synthesis of this material and sensor fabrication, and the absence of any improvement in the parameters of the gas and humidity sensors in comparison with other carbon materials (Helwig et al. 2009), the ND particles are used mainly in bioapplications (Schrand et al. 2009; Possas-Abreu et al. 2017). The benefits of using NDs in biomedical applications, including purification, sensing, imaging, and drug delivery, are based upon their desirable chemical, biological, and physical (optical, mechanical, electrical, thermal) properties (Table 4.8). In particular, it was established that diamond is biocompatible material. In addition, NDs are unique among the class of carbon nanoparticles because of their intrinsic hydrophilic surface, which is one of the many reasons that these nanocarbon particles are envisioned for biomolecular applications. The surface of ND particles contains a complex array of surface groups, including carboxylic acids, esters, ethers, lactones, and amines. Therefore, alterations in detonation of ND surface groups can produce a high density of chemical functionalities.

## 4.7 CARBON NITRIDE

### 4.7.1 CARBON NITRIDE AND ITS PROPERTIES

Carbon nitride is another representative of carbon-based materials, promising for humidity-sensor applications. Two modifications of this material are currently found: beta-carbon nitride ($\beta$-C$_3$N$_4$) and graphitic-carbon nitride (g-C$_3$N$_4$).

$\beta$-Carbon nitride is a solid diamond-like substance with the formula $\beta$-C$_3$N$_4$. The substance was first proposed in 1985 by Liu and Cohen (1989). They suggested that carbon and nitrogen atoms in a stable crystal lattice in 1:1.3 ratio can form a particularly short and strong bond (see Figure 4.30). According to calculations, this substance may be harder than diamond on the Mohs scale (Liu and Cohen 1989). The material has been considered difficult to produce and could not be

---

**TABLE 4.8**

**Physicochemical Properties of NDs Important for Biomedical Applications**

| Property | Characteristics |
|---|---|
| Structural | Small size of primary monocrystalline particles (~4–5 nm); availability of variable sizes and narrow size fractions; different forms (i.e., particulate, coating/film, substrate); large specific surface area (300–400 m$^2$/g); low porosity/permeability of films; high specific gravity (3.5 g/cm$^2$) |
| Chemical | Chemically resistant to degradation/corrosion, pH stability; high chemical purity; possible sp$^2$ carbon shells; numerous oxygen containing groups on surface; ease of surface functionalization (chemical, photochemical, mechanochemical, enzymatic, plasma- and laser-assisted methods); radiation/ozone resistance; large number of unpaired electrons on the surface |
| Biological | High biocompatibility, low toxicity; readily bind bio-active substances (i.e., proteins, DNA, etc.) with retained functionality; solid phase carrier |
| Optical | Photoluminescence: non-photobleaching, non-blinking, originates from N-vacancy defects; high refractive index, optical transparency; unique Raman spectral signal |
| Mechanical | High strength and hardness; fine abrasive |
| Electro-chemical | Electrochemical plating with metals; redox behavior of DND |
| Thermal | Can withstand very high/low temperatures |

*Source:* Schrand, A.M. et al., *Cr. Rev. Sol. St. Mater. Sci.*, 34, 18–74, 2009.

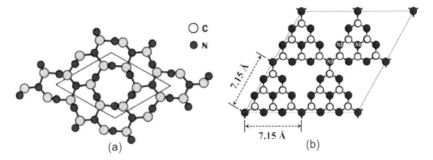

**FIGURE 4.30** Diagram illustrating the structure of one layer of 3D crystal lattice of (a) β-C$_3$N$_4$; and (b) g-C$_3$N$_4$. (Reprinted from *Appl. Catal. B*, 207, Zhu, B. et al., First principle investigation of halogen-doped monolayer g-C$_3$N$_4$ photocatalyst, 27–34, Copyright 2017, with permission from Elsevier.)

synthesized for many years. Recently, the production of beta-carbon nitride was achieved. For example, nanosized beta-carbon nitride crystals and nanorods of this material were prepared by means of an approach involving mechanochemical processing (Niu et al. 1993). However, one should note that till now, super-high hardness was not detected.

Graphitic or polymeric carbon nitride (g-C$_3$N$_4$) is a family of carbon nitride compounds with a general formula near to C$_3$N$_4$ (albeit typically with non-zero amounts of hydrogen) and the structure on the basis of heptazine cycles. Due to the special semiconductor properties, carbon nitrides show unexpected catalytic activity for a variety of reactions (Thomas et al. 2008; Mansor et al. 2016). Graphitic carbon nitride is a large band gap semiconductor (Mansor et al. 2016). According to McDermott et al. (2013), its bandgap is in the 2.5–2.8 eV range.

Graphitic carbon nitride can be made by polymerization of cyanamide, dicyandiamide, or melamine (Thomas et al. 2008). Graphitic carbon nitride can also be prepared by electrodeposition on the Si(100) substrate from a saturated acetone solution of cyanuric trichloride and melamine (ratio = 1:1.5) at room temperature (Li et al. 2003b). Well-crystallized graphitic carbon nitride nanocrystallites can be also prepared via benzene-thermal reaction between C$_3$N$_3$Cl$_3$ and NaNH$_2$ at 180–220°C for 8–12 h (Guo et al. 2003). Recently, a new method of synthesis of graphitic carbon nitrides by heating at 400°C–600°C of a mixture of melamine and uric acid in the presence of alumina has been reported. Alumina favored the deposition of the graphitic carbon nitrides layers on the exposed surface. This method can be assimilated to an in-situ CVD (Dante et al. 2011).

Many articles also cover amorphous carbon nitride (a-CN$_x$). As a rule, amorphous carbon nitride (a-CN$_x$) thin films, used for studies, were deposited using a radio-frequency plasma-enhanced chemical vapor deposition (rf-PECVD) technique (Zambov et al. 2000; Aziz et al. 2015). However, other methods can also be used for this purpose. For example, Lee and Lee (2007) used magnetron sputtering, and Zambov et al. (2000) used pulsed-laser deposition (PLD), combined with an RF discharge. One should note that exact amorphous-carbon nitride is commonly used in the design of humidity sensors. Low-density and refractive

index, high-dielectric constant and pronounced hygroscopicity as well as chemical inertness are specific properties of this material. Studies showed that the nitrogen atomic fraction (N/C) in a-CN$_x$ can be varied in a wide range depending on the method of deposition. In particular, Zambov et al. (2000) reported that N/C was at about 1 for PECVD films and lower for PLD coatings. Fourier transform infrared and XPS data also showed the presence of triple, double, and single bonds between carbon and nitrogen atoms in a-CN$_x$. It is important that despite a large number of efforts to synthesize β-C$_3$N$_4$ for over 10 years, most of the films obtained were also almost amorphous CN$_x$.

### 4.7.2 Humidity-Sensing Characteristics of Carbon Nitride

Typical characteristics of carbon-nitride-based humidity sensors are listed in Table 4.9. When developing the carbon-nitride-based humidity sensors, the focus was on devices of capacitive and resistive types. They had a typical configuration for all sensors of this type. As we noted earlier, predominantly amorphous carbon nitride (a-CNx) was used in these sensors as humidity-sensitive material.

Previously it was assumed that the use of carbon nitride in gas sensors can meet significant difficulties. On one hand, the perfect carbon nitride, β-C$_3$N$_4$ or a-C$_3$N$_4$, has been considered an extremely stable material that rarely reacts with any other gas or chemical at room temperature (Liu and Cohen 1989). On the other hand, it was established that this material is susceptible to hydrogen attacks, not only while the films were being deposited but also when they were exposed to the atmosphere (Boyd et al. 1995; Yoon et al. 2003). Hydrogen causes excessive stress that is high enough to swell the adhesion between the substrate and the film, due to the presence of moisture in the chamber wall as well as out of the chamber. This hydrogen attack can easily break or change the C≡N and C=N bonds to form C–H and N–H bonds (Lee et al. 2001). This is strongly undesirable behavior for the composition of crystalline-β-C$_3$N$_4$-aimed application in humidity sensors. However, it was later shown that defects caused by a hydrogen attack facilitate interaction with water and contribute to improved humidity-sensing characteristics. Therefore,

**TABLE 4.9**

**Humidity Sensors Based on Amorphous Carbon Nitride**

| Method of Deposition | Film Thickness | Sensor Type | RH Range | Sensor Signal | Reference |
|---|---|---|---|---|---|
| Plasma-chemical vapor deposition | 0.3–0.7 μm | Capacitive | 1%–70% RH | 2–5 pF/%RH | Zambov et al. (2000) |
| | 1–1.5 μm | Resistive | 9%–98% RH | $S=R_{9\%}/R_{98\%}$ ~3.4–4.3 | Awang et al. (2017); Aziz et al. (2015, 2017) |
| Pulsed-laser deposition combined with a modulated rf discharge | 0.3–0.7 μm | Capacitive | 1%–70% RH | 1.5–9 pF/%RH | Zambov et al. (2000) |
| Reactive RF magnetron sputtering | ~1.5 μm | Capacitive | 10%–90% RH | 4–20 pF/%RH | Lee and Lee (2004) |
| | 1.5–2 μm | Resistive | 5%–95% RH | 2.5–9 kΩ/%RH | Lee and Lee (2004, 2005a, 2005b, 2006) |
| | N/A | MOSFET | 20%–70% RH | 2.8 μA/%RH | Lee et al. (2008); Jung and Lee (2013) |
| Pyrolysis of nitrogen-rich precursor such as cyanamide | ~10 μm | Resistive | 11%–98% RH | $S=R_{11\%}/R_{98\%}$ ~$(0.8–2.3)\cdot10^4$ | Tomer et al. (2016) |
| In–SnO₂/meso-CN nanohybrid (pyrolysis) | ~10 μm | Resistive | 11%–98% RH | $S=R_{11\%}/R_{98\%}$ ~$(0.8–2.3)\cdot10^4$ | Malik et al. (2017) |

Lee and Lee (2004) believe that these undesired effects could be applied for fabricating humidity sensors. Figure 4.31 shows a schematic model for adsorption of water molecules. According to Lee and Lee (2005a, 2005b), when water molecules attack the carbon-nitride surface, hydroxyl groups bond with surface and interface carbon. The C≡N bonding breaks up, and the released nitrogen and carbon form C–HO and N–HO bonding. The C–HO bonding starts to form at the surface, due to the presence of excess carbon. In the case of paracyanogen (C=N), the carbon atom can accommodate two hydrogen atoms to become stabilized. In addition, C–H and N–H bonds formed in advance can return to C≡N and/or C=N bonds when in dry atmosphere. Therefore, the hydrogen-defected carbon-nitride films can provide dangling sites to capture water molecules physically or chemically.

Typical humidity-sensing characteristics of carbon-nitride-based sensors developed using thin-film technology are shown in Figure 4.32. These characteristics are similar to

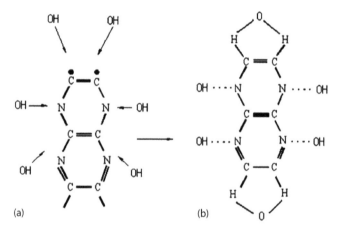

**FIGURE 4.31** Schematic model for adsorption of water molecules: (a) initial stage and (b) final stage of water-molecule adsorption. (Reprinted from Lee and Lee (2005). Published by Korean Phys. Soc. as open access.)

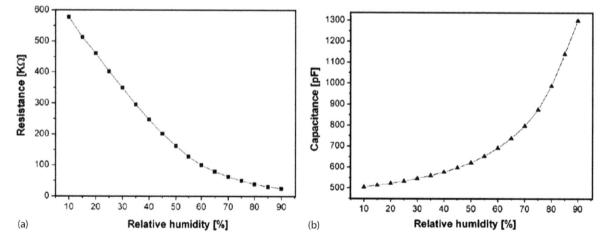

**FIGURE 4.32** (a) Resistance dependence, and (b) capacitance dependence on relative humidity for the a-*CN*ₓ-*based* sandwich-type sensors. CNₓ films were deposited in the atmosphere of pure nitrogen gas (99.999%) using a RF-magnetron-sputtering system with carbon target (99.99%). (Reprinted from Lee and Lee (2005). Published by Korean Phys. Soc. as open access.)

the characteristics of many previously considered solid-state humidity sensors. With an increase in humidity, a decrease in resistance and an increase in capacity occur. For carbon-nitride-based sensors, as well as for other sensors, the parameters are highly dependent on the deposition conditions and film porosity (Lee 2008; Awang et al. 2017). Response and recovery times of these sensors can be varied in the range of 60–90 s (Lee and Lee 2006). Typically, the hysteresis of characteristics did not exceed 10%RH (Lee and Lee 2005a, 2005b). After optimizing the technology, the hysteresis may decrease to 3%–4%RH (Lee and Lee 2006; Lee 2008).

One should note that the considered carbon-nitride-based sensors cannot compete in their parameters with the best samples of humidity sensors. However, Tomer et al. (2016) and Malik et al. (2017) have shown that humidity sensors with extremely high sensitivity can be designed based on carbon nitride. Humidity-sensing characteristics of the sensors developed by Tomer et al. (2016) are shown in Figure 4.33. For the manufacture of these sensors, Tomer et al. (2016) used 3D-replicated mesoporous g-C$_3$N$_4$ modified with well-dispersed catalytic Ag nanoparticles deposited on the

substrate by thick-film technology. g-CN was synthesized by pyrolysis of a nitrogen-rich precursor such as cyanamide using mesoporous silica KIT-6 as a templating support. Finally, the silica template was removed at room temperature using NH$_4$HF$_2$ solution. Tomer et al. (2016) reported that these sensors exhibited excellent sensitivity in the 11%–98% RH range while retaining high long-term stability, negligible hysteresis, and fast response. When the Ag content was 3 wt%, the impedance sensor showed more than 4.5 orders of magnitude change in the 11%–98% RH range. At that, the hysteresis was smaller than 1%RH. Compared to conventional resistive sensors based on metal oxides, a rapid response time (3 s) and recovery time (1.4 s) were observed in the 11%–98% RH range. According to Tomer et al. (2016), such impressive features originate from the planar morphology of g-CN as well as unique physical affinity and favorable electronic band positions of this material that facilitate the adsorption of water and charge transportation. As is known, a graphitic or polymeric carbon nitride (g-CN), generally presenting a 2D-layered structure promising for gas-sensor application. In addition, the 3D-ordered cubic mesostructure provides a high surface

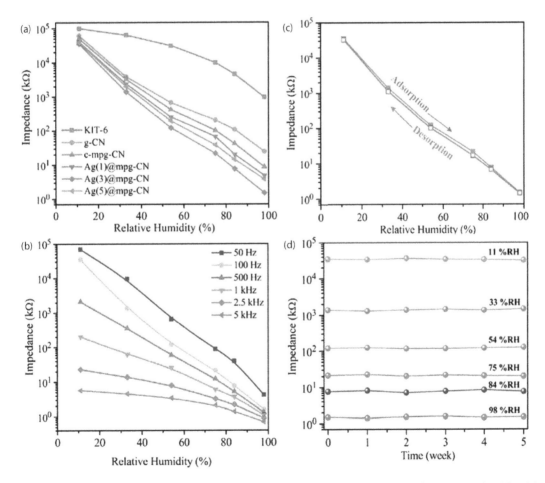

**FIGURE 4.33** (a) Humidity-sensing curves showing the decrease in impedance with the increase in %RH; (b) relationship of the impedance and relative humidity based on the Ag(3 wt%)@mpg-CN nanocomposite at various frequencies; (c) hysteresis curve showing adsorption–desorption responses measured in the 11%–98% RH range for Ag(3 wt%)@mpg-CN; and (d) the response of Ag(3 wt%)@mpg-CN monitored under different humidity conditions for 5 weeks. (Tomer, V.K. et al., *Nanoscale*, 8, 19794–19803, 2016. Reproduced by permission of The Royal Society of Chemistry.)

area thereby increasing the adsorption, transmission of charge carriers and desorption of water molecules across the sensor surfaces. Tomer et al. (2016) believe that mesoporous g-CN with Ag nanoparticles is demonstrated to provide an effective strategy in designing high-performance %RH sensors. It is important to note that to explain the observed humidity-sensing characteristics of carbon-nitride-based devices, the mechanism considered by us earlier for metal oxides is used. Approximately the same characteristics performed the sensors based on In–SnO$_2$-loaded cubic mesoporous graphitic carbon nitride (g-C$_3$N$_4$) (Malik et al. 2017). In–SnO$_2$/meso-CN nanohybrid was synthesized through a template inversion of mesoporous silica, KIT-6, developed by Tomer et al. (2016), using a nanocasting process. The optimized In–SnO$_2$/meso-CN nanohybrid exhibited an excellent response (5 orders change in impedance) in the 11%–98% RH range, high stability, negligible hysteresis (0.7%), and superfast response (3.5 s) and recovery (1.5 s).

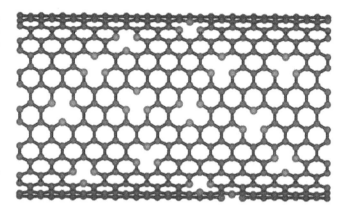

**FIGURE 4.34** Nitrogen-doped carbon nanotube scheme: a molecular model of carbon nanotubes containing pyridine-like and highly coordinated N-atoms. (With kind permission from Springer Science+Business Media: *Appl. Phys. A*, N-doping and coalescence of carbon nanotubes: Synthesis and electronic properties, 74, 2002, 355–361, Terrones, M. et al.)

## 4.8 SURFACE FUNCTIONALIZING OF CARBON NANOTUBES AND OTHER CARBON-BASED NANOMATERIALS

### 4.8.1 Carbon Nanotubes

As is known, in general, CNTs do not have sensing response to all gases and vapors, but only the ones with high-adsorption energy or that can interact with them. Pristine CNT-based gas sensors are currently limited to sense gases, such as NH$_3$, NO$_2$, SO$_2$, O$_2$, and NO. Gas molecules such as toxic gases (CO) and water, however, cannot be detected since they do not react (adsorb) with the surface of pure carbon SWCNTs surface (Terrones et al. 2004). For H$_2$ detection, bare CNT does not exhibit appreciable sensitivity as well. Experimental and theoretical studies established that the sensitivity of CNT sensors can be improved by doping the CNTs (Zhang and Zhang 2009). It was found that if the surface of the tube is doped with a donor or an acceptor, drastic changes in the electronic properties are observed. The N-doped nanotubes, for example, exhibit a higher reactivity toward reactants when compared to undoped tubes due to the introduction of nitrogen species and the structural irregularity of carbon hexagonal rings (see Figure 4.34). The N substitution reactions are also able to create radicals over nanotube surfaces, which can react with suitable reactants. As a result of the binding of the molecules to the doped locations—because of the presence of holes (B-doped tubes) or donors (N-doped tubes)—the surfaces became more reactive and sensitive to gas surrounding. Using first-principle calculations, Peng and Cho (2003) demonstrated that B- or N-doped SWCNT-based sensors can detect CO and water molecules, and more importantly, the response of these sensors can be controlled by adjusting the doping level of heteroatoms in a nanotube.

It is also important to point out that in spite of high sensitivity, CNT-based sensors are non-selective, which limits their use for sensing purposes in real samples. This means that mechanisms need to be developed to increase the selectivity of the detectors. As it was shown, the surface functionalizing is the most effective method, suitable for these purposes.

At present, there are two main approaches for the surface functionalizing of CNTs: covalent functionalization and non-covalent functionalization, depending on the types of linkages of the functional entities onto the nanotubes (Zhang et al. 2008). Several mechanisms of covalent and non-covalent modification of CNTs are illustrated in Figure 4.35. We need to note that altering the nanotube surface, besides influencing gas-sensing properties, strongly affects solubility properties, which can affect the ease of fabrication of CNT-based sensors. A review by Hirsh (2002), Balasubramanian and Burghard (2005), Shen et al. (2008), Zhang and Zhang (2009) discusses many of the functionalizations that have been commonly used. Currently, most covalently functionalized CNTs are based on esterification or amidation of carboxylic acid groups that are introduced on defect sites of the CNTs during acid treatment. Experiments showed that the CNT ends and sidewall "defect" have greater electrochemical activity (Banks and Compton 2005; Dumitrescu et al. 2009). The introducing defect sites along the sidewall of the CNTs can be carried out during the purification (oxidation) process as well (Zhang and Zhang 2009). The experiment showed that CNTs could possess structural defects: (1) five- or seven-membered rings within the carbon network; (2) sp$^3$-hybridized defects, with R=H and OH groups; (3) –COOH groups introduced by nanotube damage under oxidative conditions; and (4) open ends terminated with –COOH groups, or even other terminal groups such as –NO$_2$, –OH, –H, and =O (Hirsh 2002).

It is important that the oxidation in the gas or liquid phase could be used to increase the concentration of surface oxygen groups; while heating under inert atmosphere might be used to selectively remove some of these functions. It was shown that the gas-phase oxidation of the carbon mainly increased the concentration of hydroxyl and carbonyl surface groups, while oxidations in the liquid phase increased especially the concentration of carboxylic acids (Figueiredo et al. 1999).

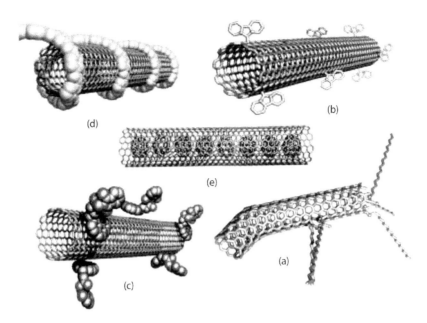

**FIGURE 4.35** Functionalization possibilities for SWNTs: (a) defect-group functionalization; (b) covalent sidewall functionalization; (c) non-covalent exohedral functionalization with surfactants; (d) non-covalent exohedral functionalization with polymers; and (e) endohedral functionalization with, for example, $C_{60}$. For methods B-E, the tubes are drawn in idealized fashion, but defects are found in real situations. (From Hirsh, A.: Functionalization of Single-Walled Carbon Nanotubes. *Angewandte Chemie International Edition*. 2002. 41. 1853–1859. Copyright Wiley-VCH Verlag GmbH & Co. KGaA. Reproduced with permission.)

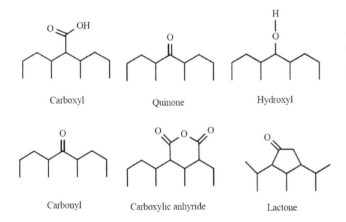

**FIGURE 4.36** Simplified schematic of some acidic surface groups bonded to aromatic rings on activated carbon.

Carboxyl, carbonyl, phenol, quinone, and lactone groups (Yang 2003) on carbon surfaces are shown in Figure 4.36.

Analysis of published results testifies that the treatment with strong acids, such as $HNO_3$, $H_2SO_4$, HCl, and oxidation agents, such as $KMnO_4/H_2SO_4$, $K_2Cr_2O_7/H_2SO_4$, and $OsO_4$, is one of the main steps in the CNT functionalizing. This treatment, usually used after CNTs synthesis, removes the end caps and may shorten the length of the CNTs. Acid treatment also adds oxide groups, primarily carboxylic acids, to the tube ends and defect sites. Further chemical reactions can be performed at these oxide groups to functionalize with the groups such as amides, thiols, or others. As an example, one can consider the polyaniline, which, after exposition to HCl, shows

a rapid drop in the resistance within a short period of time. According to Huang et al. (2003) and Virji et al. (2004), this doping of polyaniline is being achieved by protonation of the imine nitrogen by HCl. The charge created along the backbone by this protonation was counter-balanced by the resulting negatively charged chloride counterions. The change in conductivity was brought about by the formation of polarons (radical cations) that traveled along the polymer backbone. The mechanism for base dedoping of polyaniline is different than that for acid doping and may account for the slower response times and smaller resistance changes.

Some of modifiers, typically used for CNT functionalizing, are listed in Table 4.10. We would like to note that the approaches used for CNT functioning are similar to the approaches used for modification of polymer, activated carbon, fullerenes, and graphene surfaces.

Non-covalent functionalization is mainly based on supramolecular complexation using various adsorptive and wrapping forces, such as van der Waals and $\pi$-stacking interactions, without destruction of the physical properties of CNTs (Zhao and Stoddart 2009). It was established that functionalized CNT sensors often offer a higher sensitivity and a better selectivity compared with pristine CNT sensors (Qi et al. 2003; Valentini et al. 2003, 2004a, 2004b; Zhang et al. 2008; Zhang and Zhang 2009). This means that the doping and surface functionalization of CNTs may broaden the application range of CNT-based gas sensors.

As it is seen from the data presented in Table 4.10, the surface modification by metal nanoparticles, oxides, and polymers is also promising approach to optimizing gas-sensing characteristics of CNT-based devices. It is necessary to say

**TABLE 4.10**

**Modifiers Used for the Surface Functionalization of CNTs**

| Surface Modifiers | Examples |
|---|---|
| Functional groups | Carboxylic acid (–COOH) group; 3-aminopropyltriethoxysilane; amides; etc. |
| Polymers | PEI; Nafion; PABS; PPy; PDPA; PMMA; Poly (3-methylthiophene); polystyrene; PEG; POAS; poly(vinylacetate); polyisoprene; etc. |
| Metals | Pd; Pt; Ag; Au; Sn; Rh; Al |
| Metal complex | $Eu^{3+}$-containing dendrimer complex |
| Metal oxides | $SnO_2$; $WO_3$; $TiO_2$ |

*Source:* Jimenez-Cadena, G. et al., *Analyst*, 132, 1083–1099, 2007; Zhang, W.-D. and Zhang, W.-H., *J. Sensor*, 2009, 160698, 2009.

PABS, poly-(m-aminobenzene sulfonic acid); PEG, poly(ethylene glycol); PEI, polyethyleneimine; PDPA, polydiphenylamine; PMMA, poly(methyl methacrylate); POAS, Poly(o-anisidine); PPy, polypyrrole

that clusters of noble metals or metal oxides immobilized on carbon-based materials are quite common in modern chemistry, especially as hydrogenation catalysts. Methods such as wet or dry impregnation, deposition–precipitation, deposition–reduction, or ion-exchange protocols are routinely applied for these purposes (Toebes et al. 2001). Most of them rely on the treatment with aqueous solutions of suitable precursors, such as metal salts. It was found that the previously mentioned dopants attached to CNTs change the electronic properties of CNTs and therefore can be used for the surface functionalization of CNTs (Wang et al. 2007b; Zhang and Zhang 2009). For example, Qi et al. (2003) showed that noncovalently dropcoating of PEI onto SWCNTs changed the SWCNTs from *p*-type to *n*-type semiconductors.

It is necessary to note that the surface functionalizing also helps to resolve problems such as disentanglement of bundles, separation-purification, and dispersion-solubilization of CNTs, which are very important for preparation of CNT-based sensors and nanocomposites. As is known, synthetic chemistry primarily takes place in solvents. Thus, disentanglement and uniform dispersions of CNTs in several solvents have to be carried out in order to proceed with chemical reactions. Dispersion becomes difficult because CNTs are extremely resistant to wetting and are difficult to separate due to a strong van der

Waals interactions. Nevertheless, while using ultrasonic procedure in acids, it is possible to wet CNTs. Intercalated molecules play the role of disrupting and compensate the loss of van der Waals attractions between CNTs. In general, it was observed that ionic, covalent, non-covalent functionalization and the polymer-wrapping procedure could be effective for the achievement of the uniform CNT dispersion.

One can find more detailed analysis of the CNT-surface functionalizing in review papers prepared by Hirsh (2002), Liu (2005), Jimenez-Cadena et al. (2007), Kauffman and Star (2008), Zhang and Zhang (2009), Sun et al. (2011), and Schaetz et al. (2012).

### 4.8.2 CARBON BLACK AND FULLERENES

In closing, it is necessary to note that CB, fullerenes, and graphene can be functionalized using the same methods that were previously discussed (Tsubokawa 1992; Shen et al. 2008; Marques et al. 2011; Schaetz et al. 2012). Moreover, most of the indicated methods such as surface modifications by oxidation, grafting of polymers, treatment with surfactant, plasma treatment, and spattering were designed mainly for CB (Tsubokawa 1992). Different plasma treatment and the changes of related chemical functional groups on the surface of carbon are listed in Table 4.11. Due to using these treatments, the dispersibility of CB in solvents and compatibility in polymer matrices were markedly improved. In addition, the surface functionalizing of CB gave various functions to CB such as photosensitivity, bioactivity, cross-linking ability, and amphiphilic properties. For example, the vapor-grown carbon fibers were modified using $NH_3$, $O_2$, $CO_2$, $H_2O$, and HCOOH plasma gases to increase the wettability, and the results show that the oxidation strength was $O_2 > CO_2 > H_2O > HCOOH$ (Brüser et al. 2004).

The functionalization of fullerenes also gives positive effects. For example, the sensitivity of fullerene sensors toward polar vapors (e.g., ethanol and water) was enhanced by >50-fold through the deposition of a metal–fullerene-hybrid film containing both $C_{60}$ and aluminum due to a higher surface area and possibly metal-fullerene bonding (Grynko et al. 2009). UV exposure further enhanced the sensitivity of both pristine and $C_{60}$-Al hybrid films through the introduction of reactive sites on the $C_{60}$ surface (Grynko et al. 2009). The sensitivity of $C_{60}$-based QCM and SAW sensors toward polar and non-polar vapors such as volatile organic alcohols,

**TABLE 4.11**

**Related Chemical Groups Change at Different Plasma-Treatment Conditions**

| Plasma Gaseous | Increased Chemical Groups | Decreased Chemical Groups | References |
|---|---|---|---|
| $O_2$ | –C-OOH, C=O | –C-OH, C-O-C | Domingo-Garcia et al. (2000) |
| $N_2$ | –C-OH, C-O-C–, O=C-O, pyridine and quaternary nitrogen | –C=O (aromatic ring) | Brüser et al. (2004) |
| $NH_3$ | N-H | | Boudou et al. (2000) |
| $CO_2$ | –C-OOH, C=O | | Pai et al. (2006) |
| $H_2O$ | –C-OOH, C=O | | Pai et al. (2006) |

water vapors, aldehydes, and acids was enhanced by derivatizing the $C_{60}$ with supramolecular host compounds such as crown ethers and cryptands (Shih et al. 2001; Lin and Shih 2003). The proposed mechanism of this sensitivity enhancement involved a combination of enhanced chelation by the cryptand/crown ether as well as enhanced reactivity of the $C_{60}$ at the cryptand/crown ether binding site (Shih et al. 2001; Lin and Shih 2003). Another approach to generating supramolecular host compounds for vapor sensing with fullerenes involved liquid crystals (Dickert et al. 1997) where rigid linear (thermotropic liquid crystals) and globular (fullerenes) compounds formed a 1:1 stoichiometry sensing film, disturbing the close packing of both species and, thus, forming cavities and diffusion channels.

### 4.8.3 GRAPHENE

The same situation takes place with graphene. For example, the oxidation in an oxidizing acid ($HNO_3$ or $H_2SO_4$) is the most common approach, acceptable for graphene functionalizing. In such a treatment, it is possible to attach oxygen and hydroxyl (OH) or carboxyl (COOH) groups to graphene. When graphene is covered more or less uniformly with hydroxyl or carboxyl groups, the material is called graphene oxide, which is essentially a highly defective graphene sheet functionalized with oxygen groups (Bagri et al. 2010). It is important that GO, with a wide range of oxygen functional groups both on the basal planes and at the edges of GO sheets, becomes readily exfoliated in water. The previously mentioned reactive-oxygen functional groups of GO can then be chemically modified to produce homogeneous colloidal suspensions in various solvents and to influence the properties of graphene-based materials (Park et al. 2009). For these purposes, various methods of covalent and non-covalent functionalization can be applied (see Figure 4.37). It should be noted that these methods are similar to the approaches designed for CNT functionalizing.

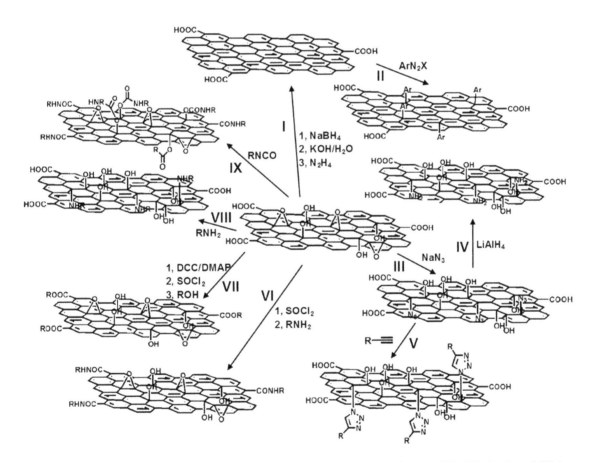

**FIGURE 4.37** Schematic showing various covalent functionalization chemistry of graphene or GO. *I*: Reduction of GO into graphene by various approaches. *II*: Covalent surface functionalization of reduced graphene via diazonium reaction. *III*: Functionalization of GO by the reaction between GO and sodium azide. *IV*: Reduction of azide functionalized GO (azide–GO) with $LiAlH_4$ resulting in the aminofunctionalized GO. *V*: Functionalization of azide–GO through click chemistry (R–ChCH/$CuSO_4$). *VI*: Modification of GO with long alkyl chains by the acylation reaction between the carboxyl acid groups of GO and alkylamine (after $SOCl_2$ activation of the COOH groups). *VII*: Esterification of GO by DCC chemistry or the acylation reaction between the carboxyl acid groups of GO and ROH alkylamine (after $SOCl_2$ activation of the COOH groups). *VIII*: Nucleophilic ring-opening reaction between the epoxy groups of GO and the amine groups of an amine-terminated organic molecular. *IX*: Treatment of GO with organic isocyanates leading to the derivatization of both the edge carboxyl and surface hydroxyl functional groups via formation of amides or carbamate esters (RNCO). (Loh, K.P. et al., *J. Mater. Chem.*, 20, 2277–2289, 2010. Reproduced by permission of Royal Society of Chemistry.)

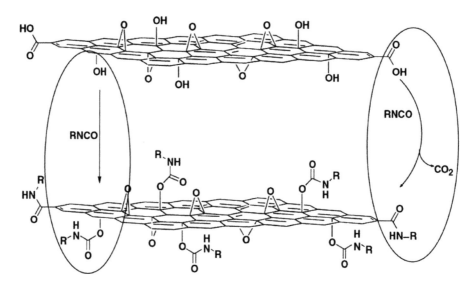

**FIGURE 4.38** Proposed reactions during the isocyanate treatment of GO where organic isocyanates react with the hydroxyl (*left oval*) and carboxyl groups (*right oval*) of graphene oxide sheets to form carbamate and amide functionalities, respectively. (Reprinted from *Carbon*, 44, Stankovich, S. et al., Synthesis and exfoliation of isocyanate-treated graphene oxide nanoplatelets, 3342–3347, Copyright 2006a, with permission from Elsevier.)

In particular, the treatment of isocyanates reduced the hydrophilicity of graphene oxide by forming amide and carbamate esters from the carboxyl and hydroxyl groups of graphene oxide, respectively. As a result, such isocyanate-derivatized GOs no longer exfoliate in water, but readily form stable dispersions in polar aprotic solvents (such as N, N-dimethylformamide), consisting of completely exfoliated, functionalized, individual graphene oxide sheets with thickness ~1 nm (Stankovich et al. 2006a). The proposed reaction is shown in Figure 4.38. This dispersion also facilitated the intimate mixing of the graphene oxide sheets with matrix polymers, providing a novel synthesis route to make graphene–polymer nanocomposites (Stankovich et al. 2006b). In particular, using the indicated graphene treatment, the polystyrene–graphene composites with percolation threshold near 0.1 vol.% were synthesized (Stankovich et al. 2006b). Such a low percolation threshold for a 3D isotropic case is evidently due to the extremely high aspect ratio of the graphene sheets and their excellent homogeneous dispersion in these composites.

In order to use carboxylic acid groups on graphene oxide to anchor other molecules, the carboxylic acid groups have been activated by thionyl chloride ($SOCl_2$), 1-ethyl-3-(3-dimethylaminopropyl)-carbodiimide, and N, N-dicyclohexylcarbodiimide (DCC or 2-(7-aza-1H-benzotriazole-1-yl)-1,1,3,3-tetramethyluronium hexa fluorophosphate) (Singh et al. 2011). The subsequent addition of nucleophilic species, such as amines or alcohols, produced covalently attached functional groups on graphene oxide via the formation of amides or esters. The attachment of hydrophobic long, aliphatic amine groups on hydrophilic graphene oxide improved the dispersability of modified graphene oxide in organic solvents (Niyogi et al. 2006). Subrahmanyam et al. (2008) have shown that soluble graphene layers in solvents such as $CCl_4$, $CH_2Cl_2$, and THF can

also be generated by the covalent attachment of alkyl chains to graphene layers by the reduction of graphite fluoride with alkyl lithium reagents.

In order to enhance the solubility of graphene-oxide nanosheets in water, the graphene-oxide nanosheets were functionalized with allylamine (Wang et al. 2009). Si and Samulski (2008) introduced a small number of p-phenyl-$SO_3H$ groups into the graphene oxide before it was fully reduced and the resulting graphene remained soluble in water and did not aggregate. The amine groups and hydroxyl groups on the basal plane of graphene oxide have also been used to attach polymers through either grafting-onto or grafting-from approaches. To grow a polymer from graphene oxide, an atom transfer radical polymerization initiator (i.e., a-bromoisobutyrylbromide) was attached to graphene surfaces (Fang et al. 2009). The following living polymerization produced graphene oxide with polymers that enhanced the compatibility of solvents and other polymer matrices.

Besides the carboxylic acid groups, the epoxy groups on the graphene oxide can be used to attach different functional groups through a ring-opening reaction. The preferred mechanism for this type of reaction involves a nucleophilic attack of amines on α-carbon. Various amine-ending chemicals such as octadecylamine (Wang et al. 2008), an ionic liquid 1-(3-aminopropyl)-3-methylimidazolium bromide (Yang et al. 2009) with an amine end group, and APTES have reacted with epoxy groups. This surface modification is of particular interest when attempting to disperse the GO in polar solvents (Yang et al. 2009).

Some of the most important routes of graphene-surface functionalizing, which are not based on the surface-bound oxygen and carbonyl moieties, are depicted in Figure 4.39. It is important that most $sp^2$-hybridized carbon scaffolds (including CNTs, fullerenes, and graphitic carbon shells) are amenable to these reactions.

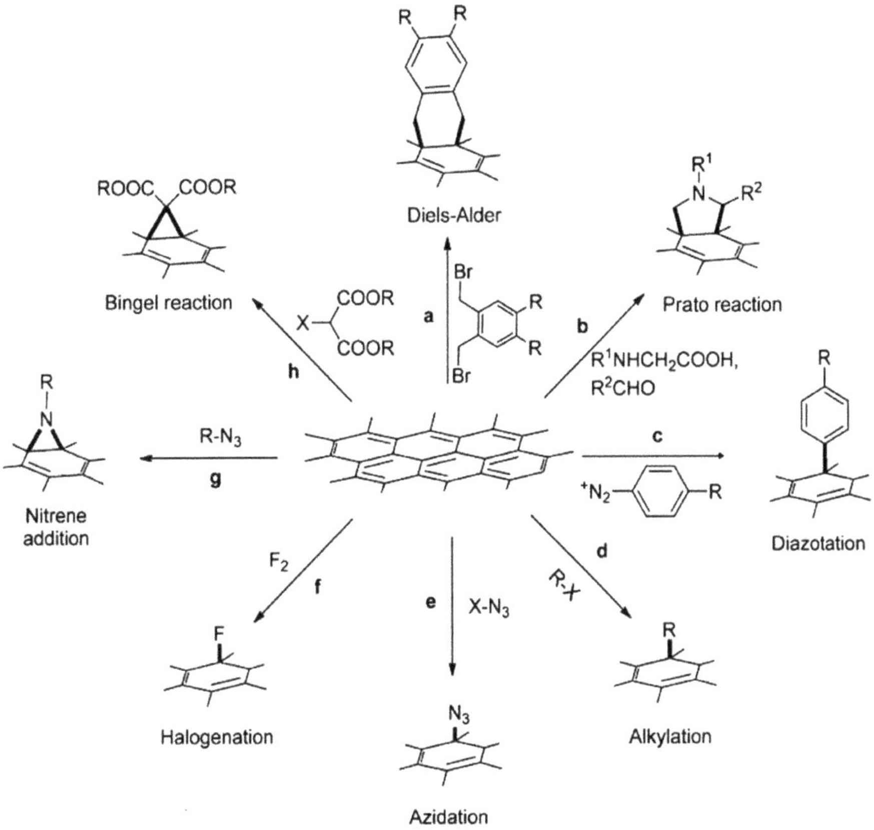

**FIGURE 4.39** Covalent functionalization of sp² -hybridized carbons: (a) Diels–Alder reaction; (b) Prato reaction; (c) diazonium chemistry; (d) alkylation of graphene oxide/activated graphenes; (e) azidation; (f) halogenations; (g) nitrene addition; and (h) Bingel reaction. (Reprinted with permission from Schaetz, A. et al., *ACS Catal.*, 2, 1267–1284, 2012. Copyright 2012 American Chemical Society.)

The non-covalent functionalization of graphene oxide utilizes the weak interactions (i.e., π–π interaction, Van der Waals interaction, and electrostatic interaction) between the graphene oxide and target molecules. Electrostatic interaction takes place mainly due to the fact that its surface is negatively charged due to the presence of oxygen functional groups. The sp² network on the graphene oxide provides π–π interactions with conjugated polymers and aromatic compounds that can stabilize reduced graphene oxide that resulted from chemical reduction and produce functional composite graphene–polymer materials, which can include poly(sodium 4-styrenesulfonate), sulfonated polyaniline, poly(3-hexylthiophene), conjugated polyelectrolyte, tetrasulfonate salt of copper phthalocyanine, porphyrin, and cellulose derivatives (Singh et al. 2011). Aromatic molecules have large aromatic planes and can anchor onto the reduced graphene-oxide surface without disturbing its electronic conjugation. This type of functionalization has some advantages in certain areas, such as chemical sensors and biomedical materials.

Plasma treatment and adsorption of atomic hydrogen on a graphene surface followed by its self-organization and hydrogen-island formation (Balog et al. 2010) can also be referred to in the context of graphene treatment by chemical methods.

It should be noted that most of the methods used to control the properties of carbon-based materials have not yet been tested for humidity sensors. Therefore, the study of the influence of these treatments on the electronic and humidity-sensitive properties of CNTs, graphene, and other carbon-based materials are urgently needed to move carbon-based devices to the sensor market.

## REFERENCES

Afify A.S., Ahmad S., Khushnood R.A., Jagdale P., Tulliani J.-M. (2017) Elaboration and characterization of novel humidity sensor based on micro-carbonized bamboo particles. *Sens. Actuators B* **239**, 1251–1256.

Ali Rafique M.M., Iqbal J. (2011) Production of carbon nanotubes by different routes—A review. *J. Encapsulation Adsorp. Sci.* **1**, 29–34.

Ali S., Hassan A., Hassan G., Bae J., Lee C.H. (2016) All-printed humidity sensor based on methyl-red/methyl-red composite with high sensitivity. *Carbon* **105**, 23–32.

Allen M.J., Tung V.C., Kaner R.B. (2010) Honeycomb carbon: A review of graphene. *Chem. Rev.* **110**, 132–145.

Arnold M.S., Green A.A., Hulvat J.F., Stupp S.I., Hersam M.C. (2006) Sorting carbon nanotubes by electronic structure using density differentiation. *Nat. Nanotechnol.* **1**, 60–65.

Arsat R., Breedo M., Shafiei M., Spizziri P.G., Gilje S., Kaner R.B., Kalantar-zadeh K., Wlodarski W. (2009) Graphene-like nanosheets for surface acoustic wave gas sensor applications. *Chem. Phys. Lett.* **467**(4–6), 344–347.

Arshak K., Moore E., Cavanagh L., Harris J., McConigly B., Cunniffe C., Lyons G., Clifford S. (2005) Determination of the electrical behaviour of surfactant treated polymer/carbon black composite gas sensors. *Composites A* **36**, 487–491.

Arunachalam S., Gupta A.A., Izquierdo R., Nabki F. (2018) Suspended carbon nanotubes for humidity sensing. *Sensors* **18**, 1655.

Auvray S., Borghetti J., Goffman M.F., Filoramo A., Derycke V., Bourgoin J.P. (2005) Carbon nanotubes transistor optimization by chemical control of nanotubes-metal interface. *Appl. Phys. Lett.* **84**(25), 5106–5108.

Awang R., Aziz N.F.H., Purhanudin N., Zalita Z. (2017) Characterization of a-CN$_x$ thin films prepared by RF-PECVD technique for humidity sensor. *Sains Malaysiana* **46**(3), 509–514.

Azam M.A., Zulkapli N.N., Nawi Z.M., Azren N.M. (2015) Systematic review of catalyst nanoparticles synthesized by solution process: Towards efficient carbon nanotube growth. *J. Sol-Gel Sci. Technol.* **73**, 484–500.

Aziz N.F.H., Ritikos R., Kamal S.A.A., Azman N.I., Awang R. (2015) Effect of RF power on the chemical bonding and humidity sensing properties of a-CN$_x$ thin films. *Adv. Mater. Res.* **1107**, 655–660.

Aziz S.A.A., Purhanudin N., Awang R. (2017) Chemical bonding and humidity sensing properties of amorphous carbon nitride (a-CN$_x$) by acetylene gas. *AIP Conference Proc.* **1838**, 020010.

Bachtold A., Hadley P., Nakanishi T., Dekker C. (2001) Logic circuits with carbon nanotube transistors. *Science* **294**(5545), 1317–1320.

Bachtold A., Strunk C., Salvetat J.P., Bonard J.M., Forro L., Nussbaumer T., Schonenberger C. (1999) Aharonov-Bohm oscillations in carbon nanotubes. *Nature* **397**, 673–675.

Backes C., Schmidt C.D., Hauke F., Bottcher C., Hirsch A. (2009) High population of individualized SWCNTs through the adsorption of water-soluble perylenes. *J. Am. Chem. Soc.* **131**, 2172–2184.

Baena J.R., Gallego M., Valcarcel M. (2002) Fullerenes in the analytical sciences. *Trends Anal. Chem.* **21**(3), 187–198.

Bagri A., Mattevi C., Acik M., Chabal Y.J., Chhowalla M., Shenoy V.B. (2010) Structural evolution during the reduction of chemically derived graphene oxide. *Nat. Chem.* **2**, 581–587.

Balasubramanian K., Burghard M. (2005) Chemically functionalized carbon nanotubes. *Small* **1**, 180–192.

Balog R., Jørgensen B., Nilsson L., Andersen M., Rienks E., Bianchi M., et al. (2010) Bandgap opening in graphene induced by patterned hydrogen adsorption. *Nat. Mater.* **9**, 315–319.

Banks C.E., Compton R.G. (2005) Exploring the electrocatalytic sites of carbon nanotubes for NADH detection: An edge plane pyrolytic graphite electrode study. *Analyst* **130**, 1232–1239.

Barkauskas J. (1997) Investigation of conductometric humidity sensors. *Talanta* **44**(6), 1107–1112.

Barrau S., Demont P., Peigney A. (2003) DC and AC conductivity of carbon nanotubes-polyepoxy composites. *Macromolecules* **36**, 5187.

Basu S., Bhattacharyya P. (2012) Recent developments on graphene and graphene oxide based solid state gas sensors. *Sens. Actuators B* **173**, 1–21.

Bauschlicher C.W., Ricca A. (2004) Binding of NH$_3$ to graphite and to (9,0) carbon nanotubes. *Phys. Rev. B* **70**, 115409.

Benchirouf A., Jalil A., Kanoun O. (2013) Humidity sensitivity of thin films based dispersed multi-walled carbon nanotubes. In: *Proceedings of 10th International Multi-Conference on Systems, Signals & Devices (SSD)*, Hammamet, Tunisia.

Berdinsky A.S., Shevtsov Y.V., Okotrub A.V., Trubin S.V., Chadderton L.T., Fink D., Lee J.H. (2000) Sensor properties of fullerene films and fullerene compounds with iodine. *Chem. Sustain. Dev.* **8**, 141–146.

Berger C., Song Z., Li T., Li X., Ogbazghi A.Y., Feng R., et al. (2004) Ultrathin epitaxial graphite: 2D electron gas properties and a route toward graphene-based nanoelectronics. *J. Phy. Chem. B* **108**, 19912–19916.

Bhargava C., Banga V.K., Singh Y. (2017) Reliability comparison of a fabricated humidity sensor using various artificial intelligence techniques. *Intern. J. Performab. Eng.* **13**(5), 577–586.

Bi H., Yin K., Xie X., Ji J., Wan S., Sun L., Terrones M., Dresselhaus M.S. (2013) Ultrahigh humidity sensitivity of graphene oxide. *Sci. Rep.* **3**, 2714.

Bondavalli P., Legagneux P., Pribat D. (2009) Carbon nanotubes based transistors as gas sensors: State of the art and critical review. *Sens. Actuators B* **140**, 304–318.

Borini S., White R., Wei D., Astley M., Hague S., Spigone E., et al. (2013) Ultrafast graphene oxide humidity sensors. *ACS Nano* **7**, 11166–11173.

Boudou J.P., Martinez-Alonzo A., Tascon J.M.D. (2000) Introduction of acidic groups at the surface of activated carbon by microwave induced oxygen plasma at low pressure. *Carbon* **38**, 1021–1029.

Boukhvalov D.W., Katsnelson M.I. (2008) Tuning the gap in bilayer graphene using chemical functionalization: Density functional calculations. *Phys. Rev. B* **78**, 085413.

Boyd K.J., Marton D., Todorov S.S., Al-Bayati A.H., Kulik J., Zuhr R.A., Rabalais J.W. (1995) Formation of C–N thin films by ion beam deposition. *J. Vac. Sci. Technol. A* **13**, 2110.

Brüser V., Heintze M., Brandl W., Marginean G., Bubert H. (2004) Surface modification of carbon nano fibres in low temperature plasmas. *Diamond Relat. Mater.* **13**, 1177–1181.

Bruzzi M., Miglio S., Scaringella M., Bongiorno G., Piseri P., Podesta A., Milani P. (2004) First study of humidity sensors based on nanostructured carbon films produced by supersonic cluster beam deposition. *Sens. Actuators B* **100**, 173.

Cao C.L., Hu C.G., Fang L., Wang S.X., Tian Y.S., Pan C.Y. (2011) Humidity sensor based on multi-walled carbon nanotube thin films. *J. Nanomater.* **2011**, 707303.

Castro Neto A.H., Guinea F., Peres N.M.R., Novoselov K.S., Geim A.K. (2009) The electronic properties of grapheme. *Rev. Mod. Phys.* **81**, 109–162.

Chang Y.W., Oh J.S., Yoo S.H., Choi H.H., Yoo K.-H. (2007) Electrically refreshable carbon nanotube-based gas sensors. *Nanotechnology* **18**, 435504.

Chao Y.C., Shih J.S. (1998) Adsorption study of organic molecules on fullerene with piezoelectric crystal detection system. *Anal. Chim. Acta* **374**, 39–46.

Chappanda K.N., Shekhah O., Yassine O., Patole S.P., Eddaoudi M., Salama K.N. (2018) The quest for highly sensitive QCM humidity sensors: The coating of CNT/MOF composite sensing films as case study. *Sens. Actuators B* **257**, 609–619.

Chen C., Wang X., Li M., Fan Y., Sun R. (2018) Humidity sensor based on reduced graphene oxide/lignosulfonate composite thin-film. *Sens. Actuators B* **255**, 1569–1576.

Chen G., Paronyan T.M., Pigos E.M., Harutyunyan A.R. (2012) Enhanced gas sensing in pristine carbon nanotubes under continuous ultraviolet light illumination. *Sci. Reports* **2**, 343.

Chen H.-J., Xue Q.-Z., Ma M., Zhou X.-Y. (2010) Capacitive humidity sensor based on amorphous carbon film/n-Si heterojunctions. *Sens. Actuators B* **150**, 487–489.

Chen H.W., Wu R.J., Chan K.H., Sun Y.L., Su P.G. (2005b) The application of CNT/Nafion composite material to low humidity sensing measurement. *Sens. Actuators B* **104**, 80–84.

Chen J., Tsubokawa N. (2000a) Novel gas sensor from polymer-grafted carbon black: Vapor response of electric R of conducting composites prepared from poly(ethylene-block-ethylene oxide)-grafted carbon black. *J. Appl. Poly. Sci.* **77**(11), 2437–2447.

Chen J., Tsubokawa N. (2000b) A novel gas sensor from polymer-grafted carbon black: Responsiveness of electric resistance of conducting composite from LDPE and PE-b-PEO-grafted carbon black in various vapors. *Polym. Adv. Technol.* **11**, 101–107.

Chen J., Tsubokawa N., Maekawa Y., Yoshida M. (2002) Vapor response properties of conducting composites prepared from crystalline oligomer-grafted carbon black. *Carbon* **40**, 1602–1605.

Chen M., Frueh J., Wang D., Lin X., Xie H., He Q. (2017) Polybenzoxazole nanofiber-reinforced moisture-responsive soft actuators. *Sci. Rep.* **7**, 769.

Chen M.-C., Hsu C.-L., Hsueh T.-J. (2014) Fabrication of humidity sensor based on bilayer graphene. *Electron Dev. Lett.* **35**, 590–592.

Chen R., Franklin N., Kong J., Cao J., Tombler T., Zhang Y., Dai H. (2001) Molecular photodesorption from single-walled carbon nanotubes. *Appl. Phys. Lett.* **79**, 2258–2260.

Chen S.G., Hu J.W., Zhang M.Q., Rong M.Z. (2005) Effects of temperature and vapor pressure on the gas sensing behavior of carbon black filled polyurethane composites. *Sens. Actuators B* **105**, 187–193.

Chen W.-P., Zhao Z.-G., Liu X.-W., Zhang Z.-X., Suo C.-G. (2009) A capacitive humidity sensor based on multi-wall carbon nanotubes (MWCNTs). *Sensors* **9**, 7431–7444.

Chen X., Chen X., Li N., Ding X., Zhao X. (2016) A QCM humidity sensors based on GO/Nafion composite films with enhanced sensitivity. *IEEE Sens.* **16**, 8874–8883.

Chen X., Wang R., Xu J., Yu D. (2004) TEM investigation on the growth mechanism of carbon nanotubes synthesized by hot-filament chemical vapor deposition. *Micron.* **35**(6), 455–460.

Chevallier E., Scorsone E., Bergonzo P. (2009) Modified diamond nanoparticles as sensitive coatings for chemical SAW sensors. *Procedia Chem.* **1**, 943–946.

Cho W.-S., Moon S.-I., Lee Y.-D., Lee Y.-H., Park J., Ju B.K. (2005) Multiwall carbon nanotube gas sensor fabricated using thermomechanical structure. *IEEE Electron. Dev. Lett.* **26**(7), 498–500.

Choi W., Lahiri I., Seelaboyina R., Kang Y.S. (2010) Synthesis of graphene and its applications: A review. *Crit. Rev. Sol. St. Mater. Sci.* **35**(1), 52–71.

Chu J., Peng X., Feng P., Sheng Y., Zhang J. (2013) Study of humidity sensors based on nanostructured carbon films produced by physical vapor deposition. *Sens. Actuators B* **178**, 508.

Collins P.G., Bradley K., Ishigami M., Zettl A. (2000) Extreme oxygen sensitivity of electronic properties of carbon nanotubes. *Science* **287**, 1801–1804.

Consales M., Cutolo A., Penza M., Aversa P., Giordano M., Cusano A. (2008) Fiber optic chemical nanosensors based on engineered single-walled carbon nanotubes. *J. Sensors* **2008**, 936074.

Cui X., Freitag M., Martel R., Brus L., Avouris P. (2003) Controlling energy-level alignment at carbon nanotubes/Au contact. *Nanoletters* **3**(6), 783–787.

Cusano A., Pisco M., Consales M., Cutolo A., Giordano M., Penza M., et al. (2006) Novel optochemical sensors based on hollow fibers and single walled carbon nanotubes. *IEEE Photonics Technol. Lett.* **18**(22), 2431–2433.

Dan Y., Lu Y., Kybert N.J., Luo Z., Johnson A.T.C. (2009) Intrinsic response of graphene vapor sensors. *Nano Lett.* **9**, 1472–1475.

Dante R.C., Martín-Ramos P., Correa-Guimaraes A., Martín-Gil J. (2011) Synthesis of graphitic carbon nitride by reaction of melamine and uric acid. *Mater. Chem. Phys.* **130**(3), 1094–1102.

Davydova M., Kromka A., Rezek B., Babchenko O., Stuchlik M., Hruska K. (2010) Fabrication of diamond nanorods for gas sensing applications. *Appl. Surf. Sci.* **256**, 5602–5605.

Davydova M., Kulha P., Babchenko O., Kromka A. (2015) Hydrogen-terminated diamond surface as gas sensing layer working at room temperature. In: *Proceedings of Nanoconference*, Brno, Czech Republic.

Davydova M., Kulha P., Laposa A., Hruska K., Demo P., Kromka A. (2014) Gas sensing properties of nanocrystalline diamond at room temperature. *Beilstein J. Nanotechnol.* **5**, 2339–2345.

Davydova M., Laposa A., Smarhak J., Kromka A., Neykova N., Nahlik J., et al. (2018) Gas-sensing behaviour of ZnO/diamond nanostructures. *Beilstein J. Nanotechnol.* **9**, 22–29.

Davydova M., Stuchlik M., Rezek B., Kromka A. (2012) Temperature enhanced gas sensing properties of diamond films. *Vacuum* **86**, 599–602.

Dickert F.L., Zenkel M.E., Bulst W.-E., Fischerauer G., Knauer U. (1997) Fullerene/liquid crystal mixtures as QMB- and SAW-coatings—Detection of diesel- and solvent-vapours. *Fresenius J. Anal. Chem.* **357**, 27–31.

Ding X., Chen X., Chen X., Zhao X., Li N. (2018) A QCM humidity sensor based on fullerene/graphene oxide nanocomposites with high quality factor. *Sens. Actuators B* **266**, 534–542.

Domingo-Garcia M., Lopez-Garzon F.J., Perez-Mendoza M. (2000) Effect of some oxidation treatments on the textural characteristics and surface chemical nature of an activated carbon. *J. Colloid Interface Sci.* **222**, 233–240.

Dong X.M., Fu R.W., Zhang M.Q., Zhang B., Li J.R., and Rong M.Z. (2003) Vapor induced variation in electrical performance of carbon black/poly(methyl methacrylate) composites prepared by polymerization filling. *Carbon* **41**, 371–374.

Dong X.M., Fu R.W., Zhang M.Q., Zhang B., Rong M.Z. (2004) Electrical R response of carbon black filled amorphous polymer composite sensors to organic vapors at low vapor concentrations. *Carbon* **42**(12–13), 2551–2559.

Dresselhaus M.S., Dresselhaus G., Eklund P.C. (1996) *Science of Fullerenes and Carbon Nanotubes*. Academic Press, New York.

Dresselhaus M.S., Lin Y.M., Rabin O., Jorio A., Souza Filho A.G., Pimenta M.A., et al. (2003) Nanowires and nanotubes. *Mater. Sci. Eng. C* **23**, 129–140.

Dua V., Surwade S.P., Ammu S., Agnihotra S.R., Jain S., Roberts K.E. (2010) All-organic vapor sensor using inkjet-printed reduced graphene oxide. *Angew. Chem. Int. Ed.* **49**, 2154–2157.

Dumitrescu I., Unwin P.R., Macpherson J.V. (2009) Electrochemistry at carbon nanotubes: Perspective and issues. *Chem. Commun.* **2009**(45), 6886–6901.

Evans G.P., Buckley D.J., Skipper N.T., Parkin I.P. (2017) Switchable changes in the conductance of single-walled carbon nanotube networks on exposure to water vapour. *Nanoscale* **9**(31), 11279–11287.

Fam D.W.H., Palaniappan A., Tok A.I.Y., Liedberg B., Moochhala S.M. (2011) A review on technological aspects influencing commercialization of carbon nanotube sensors. *Sens. Actuators B* **157**, 1–7.

Fang M., Wang K., Lu H., Yang Y., Nutt S. (2009) Covalent polymer functionalization of graphene nanosheets and mechanical properties of composites. *J. Mater. Chem.* **19**, 7098–7105.

Fei T., Jiang K., Jiang F., Mu R., Zhang T. (2014) Humidity switching properties of sensors based on multiwalled carbon nanotubes/polyvinyl alcohol composite films. *J. Appl. Polym. Sci.* **131**, 3972.

Figueiredo J.L., Pereira M.F.R., Freitas M.M.A., Orfao J.J.M. (1999) Modification of the surface chemistry of activated carbons. *Carbon* **37**, 1379–1389.

Fowler J.D., Allen M.J., Tung V.C., Yang Y., Kaner R.B., Weiller B.H. (2009) Practical chemical sensors from chemically derived graphene. *ACS Nano* **3**(2), 301–306.

Gautier L.A., Borgne V.L., Khakani M.A. (2016) Field emission properties of graphenated multi-wall carbon nanotubes grown by plasma enhanced chemical vapour deposition. *Carbon* **98**, 259–266.

Geim A.D., Novoselov K.S. (2007) The rise of graphene. *Nature Mater.* **6**, 183–191.

Ghadiry M., Gholami M., Lai C.K., Ahmad H., Chong W.Y. (2016) Ultra-sensitive humidity sensor based on optical properties of graphene oxide and nano-anatase TiO$_2$. *PLoS ONE* **11**, 0153949.

Ghosh A., Late D.J., Panchakarla L.S., Govindaraj A., Rao C.N.R. (2009) NO$_2$ and humidity sensing characteristics of few-layer graphenes. *J. Exp. Nanosci.* **4**, 313–322.

Gi R.S., Ishikawa T., Tanaka S., Kimura T., Akiba Y., Iida M. (1997) Possibility of realizing a gas sensor using surface conductive layer on diamond films. *Jpn. J. Appl. Phys.* **36**, Part 1, No. 4a, 2057–2060.

Gokus T., Nair R.R., Benetti A., Bohmler M., Lombardo A., Novoselov K.S., et al. (2009) Making graphene luminescent by oxygen plasma treatment. *ACS Nano* **3**, 3963–3968.

Green A.A., Hersam M.C. (2009) Solution phase production of graphene with controlled thickness via density differentiation. *Nano Lett.* **9**(12), 4031–4036.

Grynko D., Burlachenko J., Kukla O., Kruglenko I., Belyaev O. (2009) Fullerene and fullerene-aluminum nanostructured films as sensitive layers for gas sensors. *Semicond. Phys. Quantum Electron. Optoelectron.* **12**, 287–289.

Guo L., Jiang H.B., Shao R.Q., Zhang Y.L., Xie S.Y., Wanf J.N., et al. (2012) Two-beam-laser interference mediated reduction, patterning and nanostructuring of graphene oxide for the production of a flexible humidity sensing device. *Carbon* **50**, 1667–1673.

Guo Q.X., Xie Y., Wang X.J., Lv S.C., Hou T., Liu X.M. (2003) Characterization of well-crystallized graphitic carbon nitride nanocrystallites via a benzene-thermal route at low temperatures. *Chem. Phys. Lett.* **380**(1–2), 84–87.

Gupta Chatterjee S., Chatterjee S., Ray A.K., Chakraborty A.K. (2015) Graphene–metal oxide nanohybrids for toxic gas sensor: A review. *Sens. Actuators B* **221**, 1170–1181.

Gurbuz Y., Kang W.P., Davidson J.L., Kerns D.V. (2004) Diamond microelectronic gas sensor for detection of benzene and toluene. *Sens. Actuators B* **99**, 207–215.

Hayashi K., Sadanori Y., Hideyuki W., Takashi S., Hideyo O., Koji K. (1997) Investigation of the effect of hydrogen on electrical and optical properties in chemical vapor deposited on homoepitaxial diamond films. *J. Appl. Phys.* **81**, 744–753.

Hayashi K., Yokota Y., Tachibana T., Kobashi K., Fukunaga T., Takada T. (2001) Characteristics of diamond film gas sensors upon exposure to semiconductor doping gases. *J. Electrochem. Soc.* **148**(2), H17–H20.

He H., Klinowski J., Forster M., Lerf A. (1998) A new structural model for graphite oxide. *Chem. Phys. Lett.* **287**, 53–56.

Helwig A., Muller G., Sberveglieri G., Eickhoff M. (2009) On the low-temperature response of semiconductor gas sensors. *J. Sensors* **2009**, 620720.

Hernández-Rivera D., Rodríguez-Roldán G., Mora-Martínez R., Suaste-Gómez E. (2017) A capacitive humidity sensor based on an electrospun PVDF/Graphene membrane. *Sensors* **17**, 1009.

Hirsh A. (2002) Functionalization of single-walled carbon nanotubes. *Angew. Chem. Int. Ed.* **41**(11), 1853–1859.

Homma Y., Kobayashi Y., Ogino T., Ito R., Jung Y.J., Ajayan P.M. (2003) Role of transition metal catalysts in single-walled carbon nanotube growth in chemical vapor deposition. *J. Phys. Chem. B* **107**(44), 12161–12164.

Hopkins A.R., Lewis N.S. (2001) Detection and classification characteristics of arrays of carbon black/organic polymer composite chemiresistive vapor detectors for the nerve agent simulants dimethylmethylphosphonate and diisopropylmethylphosphonate. *Anal. Chem.* **73**,884–892.

Hossain M.E., Freund M.S., Jayas D.S., White N.D.G., Shafai C., Thomson D.J. (2012) Carbon black polymer sensor array for incipient grain spoilage monitoring. *Agric. Res.* **1**(1), 87–94.

Hosseini Z.S., Irajizad A., Ghiass M.A., Fardindoost S., Hatamie S. (2017) A new approach to flexible humidity sensors using graphene quantum dots. *J. Mater. Chem. C* **5**, 8966–8973.

Hribšek M.F., Ristic S.S., Radojkovic B.M. (2010) Diamond in surface acoustic wave sensors. *Acta Phys. Polonica A* **117**(5), 794–798.

Hu P.A., Zhang J., Li L., Wang Z., O'Neill W., Estrela P. (2010) Carbon nanostructure-based field-effect transistors for label-free chemical/biological sensors. *Sensors* **10**(5), 5133–5159.

Huang C.S., Huang B.R., Jang Y.H., Tsai M.S., Yeh C.Y. (2005) Three-terminal CNTs gas sensor for N$_2$ detection. *Diamond Related Mater.* **14**(11–12), 1872–1875.

Huang J.X., Virji S., Weiller B.H., Kaner R.B. (2003) Polyaniline nano fibers: Facile synthesis and chemical sensors. *J. Am. Chem. Soc.* **125**, 314–315.

Huertas-Hernando D., Guinea F., Brataas A. (2006) Spin-orbit coupling in curved graphene, fullerenes, nanotubes, and nanotube caps. *Phys. Rev. B* **74**, 155426.

Hwang S.-H., Kang D., Ruoff R.S., Shin H.S., Park Y.-B. (2014) Polyvinyl alcohol reinforced and toughened with poly(dopamine)-treated graphene oxide, and its use for humidity sensing. *ACS Nano* **8**, 6739–6747.

Iijima S. (1991) Helical microtubules of graphitic carbon. *Nature* **354**, 56–58.

Jiang K., Fei T., Jiang F., Wang G., Zhang T. (2014) A dew sensor based on modified carbon black and polyvinyl alcohol composites. *Sens. Actuators B* **192**, 658–663.

Jiang W.F., Xiao S.H., Feng C.Y., Li H.Y., Li X.J. (2007) Resistive humidity sensitivity of arrayed multi-wall carbon nanotube nests grown on arrayed nanoporous silicon pillars. *Sens. Actuators B* **125**, 651.

Jimenez-Cadena G., Riu J., Rius F.X. (2007) Gas sensors based on nanostructured materials. *Analyst* **132**, 1083–1099.

Jing H., Jiang Y., Du X., Tai H., Xie G. (2012) Humidity sensing properties of different single-walled carbon nanotube composite films fabricated by layer-by-layer self-assembly technique, *Appl. Phys. A* **109**, 111–118.

Journet C., Picher M., Jourdain V. (2012) Carbon nanotube synthesis: From large-scale production to atom-by-atom growth. *Nanotechnology* **23**, 142001.

Jung D., Kim J., Lee G.S. (2015) Enhanced humidity-sensing response of metal oxide coated carbon nanotube. *Sens. Actuators A* **223**, 11–17.

Jung I., Dikin D., Park S., Cai W., Mielke S.L., Ruoff R.S. (2008) Effect of water vapor on electrical properties of individual reduced graphene oxide sheets. *J. Phys. Chem. C* **112**, 20264.

Jung J., Lee S.P. (2013) Physical properties of nano-structured carbon nitride film for integrated humidity sensors. *J. Nanosci. Nanotechnol.* **13**(10), 7030–7032.

Kalcher K., Svancara I., Buzuk M., Vytras K., Walcarius A. (2009) Electrochemical sensors and biosensors based on heterogeneous carbon materials. *Monatsh. Chem.* **140**, 861–889.

Kauffman D.R., Star A. (2008) Carbon nanotube gas and vapor sensors. *Angew. Chem. Int. Ed.* **47**, 6550–6570.

Kauffman D.R., Star A. (2010) Graphene versus carbon nanotubes for chemical sensor and fuel cell applications. *Analyst* **135** (11), 2790–2797.

Kaul A.B., Wong E.W., Epp L., Hunt B.D. (2006) Electromechanical carbon nanotube switches for high-frequency applications. *Nano Lett.* **6**, 942–947.

Kim W., Javey A., Vermesh O., Wang Q., Li Y., Dai H. (2003) Hysteresis caused by water molecules in carbon nanotube field-effect transistors. *Nano Lett.* **3**, 193–198.

Kim Y.S., Ha S.-C., Yang Y., Kim Y.J., Cho S.M., Yang H., Kim Y.T. (2005) Portable electronic nose system based on the carbon black–polymer composite sensor array. *Sens. Actuators B* **108**, 285–291.

Kong J., Franklin N., Chou C., Pan S., Cho K.J., Dai H. (2000) Nanotube molecular wires as chemical sensors. *Science* **287**, 622–625.

Kromka A., Davydova M., Rezek B., Vanecek M., Stuchlik M., Exnar P., Kalbac M. (2010) Gas sensing properties of nanocrystalline diamond films. *Diamond Related Mater.* **19**, 196–200.

Kroto H.W., Heath J.R., O'Brien S.C., Curl R.F., Smalley R.E. (1985) $C_{60}$: Buckminsterfullerene. *Nature* **318**, 162–163.

Kuchibhatla S.V.N.T., Karakoti A.S., Bera D., Seal S. (2007) One dimensional nanostructured materials. *Prog. Mater. Sci.* **52**, 699–913.

Kulha P., Kroutil J., Laposa A., Prochazka V., Husak M. (2016) Quartz crystal microbalance gas sensor with ink-jet printed nano-diamond sensitive layer. *J. Electrical Eng.* **67**(1), 61–64.

Kumar M.K., Ramaprabhu S. (2006) Nanostructured Pt functionalized multiwalled carbon nanotube-based hydrogen sensor. *J. Phys. Chem. B* **110**, 11291–11298.

Kundrapu M., Li J., Shashurin A., Keidar M. (2012) A model of carbon nanotube synthesis in arc discharge plasmas. *J. Phys. D: Appl. Phys.* **45**, 315305.

Kyakuno H., Matsuda K., Yahiro H., Inami Y., Fukuoka T., Miyata Y., et al. (2011) Confined water inside single-walled carbon nanotubes: Global phase diagram and effect of finite length. *J. Chem. Phys.* **134**, 244501.

Lamb L.D., Huffman D.R. (1993) Fullerene production. *J. Phys. Chem. Solids* **54**(12), 1635–1643.

Landstrass M.I., Ravi K.V. (1989) Hydrogen passivation of electrically active defects in diamond. *Appl. Phys. Lett.* **55**(14), 1391–1393.

Le X., Wang X., Pang J., Liu Y., Fang B., Xu Z., et al. (2018) A high performance humidity sensor based on surface acoustic wave and graphene oxide on AlN/Si layered structure. *Sens. Actuators B* **255**, 2454–2461.

Lee J., Cho D., Jeong Y. (2013) A resistive-type sensor based on flexible multi-walled carbon nanotubes and composite films. *Solid-State Electron.* **87**, 80–84.

Lee J.G., Lee S.P. (2004) Nano-structured carbon nitride film for humidity sensor applications. *J. Korean Phys. Soc.* **45**(3), 619–622.

Lee J.G., Lee S.P. (2005a) Surface analysis and humidity-sensing properties of carbon nitride films for microsensors. *J. Korean Phys. Soc.* **47**, S429–S433.

Lee J.G., Lee S.P. (2005b) Humidity sensing properties of $CN_x$ film by RF magnetron sputtering system. *Sens. Actuators B* **108**, 450–454.

Lee J.G., Lee S.P. (2006) Impedance characteristics of carbon nitride films for humidity sensors. *Sens. Actuators B* **117**, 437–441.

Lee J.G., Lee S.P. (2007) Nano-structured carbon nitride films for microsensor applications. *Solid State Phenom.* **121–123**, 1199–1202.

Lee S.P. (2008) Synthesis and characterization of carbon nitride films for micro humidity sensors. *Sensors* **8**, 1508–1518.

Lee S.P., Kang J.B., Chowdhury S. (2001) Effect of hydrogen on carbon nitride film deposition. *J. Korean Phys. Soc.* **39**, S1–S6.

Lee S.P., Lee J.G., Chowdhury S. (2008) CMOS humidity sensor system using carbon nitride film as sensing materials. *Sensors* **8**, 2662–2672.

Lekawa-Raus A., Kurzepa L., Kozlowski G., Hopkins S.C., Wozniak M., Lukawski D., Glowacki B.A., Koziol K.K. (2015) Influence of atmospheric water vapour on electrical performance of carbon nanotube fibres. *Carbon* **87**, 18–28.

Leonard F., Tersoff J. (2000) Role of Fermi-level pinning in nanotubes Schottky diodes. *Phys. Rev. Lett.* **84**(20), 4693.

Li C., Cao C., Zhu H. (2003b) Preparation of graphitic carbon Nitride by electrodeposition. *Chinese Sci. Bull.* **48**(16), 1737–1740.

Li C., Thostenson E.T., Chou T.W. (2008) Sensors and actuators based on carbon nanotubes and their composites: A review. *Compos. Sci. Technol.* **68**, 1227–1249.

Li H., Cheng F., Duft A.M., Adronov A. (2005) Functionalization of single-walled carbon nanotubes with well-defined polystyrene by "click" coupling. *J. Am. Chem. Soc.* **127**, 14518–14524.

Li J., Lu Y., Ye Q., Cinke M., Han J., Meyyappan M. (2003a) Carbon nanotube sensors for gas and organic vapor detection. *Nano Lett.* **3**, 929–933.

Li K., Zhang C., Du Z., Li H., Zou W. (2012) Preparation of humidity-responsive antistatic carbon nanotube/PEI nanocomposites. *Synth. Met.* **162**, 2010.

Li Q., Chen F.J. (2012) Humidity sensor chips based on gold nanoparticles/PVA/Carbon black composite film. *Adv. Mater. Res.* **560–561**, 756–760.

Li W., Xu F., Sun L., Liu W., Qiu Y. (2016) A novel flexible humidity switch material based on multi-walled carbon nanotube/polyvinyl alcohol composite yarn. *Sens. Actuators B* **230**, 528–535.

Li X., Cai W., An J., Kim S., Nah J., Yang D., et al. (2009) Large-area synthesis of high-quality and uniform graphene films on copper foils. *Science* **324**, 1312–1314.

Li X., Chen X., Yu X., Chen X., Ding X., Zhao X. (2018) A high-sensitive humidity sensor based on water-soluble composite material of fullerene and graphene oxide. *IEEE Sens. J.* **18**(3), 962–966.

Li Y., Hong L., Chen Y., Wang H., Lu X., Yang M. (2007) Poly(4-vinylpyridine)/carbon black composite as a humidity sensor. *Sens. Actuators B* **123**, 554–559.

Li Y., Wu T.T., Yang M.J. (2014) Humidity sensors based on the composite of multi-walled carbon nanotubes and crosslinked polyelectrolyte with good sensitivity and capability of detecting low humidity. *Sens. Actuators B* **203**, 63–70.

Liang Y., Wu D., Feng X., Mullen K. (2009) Dispersion of graphene sheets in organic solvent supported by ionic interactions. *Adv. Mater.* **21**, 1679–1683.

Lin H.-B., Shih J.-S. (2003) Fullerene $C_{60}$-cryptand coated surface acoustic wave quartz crystal sensor for organic vapors. *Sens. Actuators B* **92**, 243–254.

Lin Y.M., Avouris P. (2008) Strong suppression of electrical noise in bilayer graphene nanodevices. *Nano Lett.* **8**, 2119–2125.

Liou W.-J., Lin N.-M. (2007) Nanohybrid $TiO_2$/carbon black sensor for $NO_2$ gas. *China Particuology* **5**, 225–229.

Liu A.Y., Cohen M.L. (1989) Prediction of new low compressibility solid. *Science* **245**(4920), 841—843.

Liu P. (2005) Modifications of carbon nanotubes with polymers. *Eur. Polym. J.* **41**, 2693–2703.

Liu X., Gurel V., Morris D., Murray D., Zhitkovich A., Kane A.B., Hurt R.H. (2007) Bioavailability of nickel in single-wall carbon nanotubes. *Adv. Mater.* **19**, 2790–2796.

Liu X., Luo Z., Han S., Tang T., Zhang D., Zhou C. (2005) Band engineering of carbon nanotube field-effect transistors via selected area chemical gating. *Appl. Phys. Lett.* **86**, 243501.

Liu X., Zhao Z., Li T., Wang X. (2011) Novel capacitance-type humidity sensor based on multi-wall carbon nanotube/$SiO_2$ composite films. *J. Semicond.* **32**(3), 034006.

Llobert E., Barbera-Brunet R., Etrillard C., Letard J.F., Debeda H. (2014) Humidity sensing properties of screen-printed carbon-black and Fe(II) spin crossover compound hybrid films. *Proc. Eng.* **87**, 132–135.

Loh K.P., Bao Q.L., Ang P.K., Yang J.X. (2010) The chemistry of graphene. *J. Mater. Chem.* **20**(12), 2277–2289.

Lonergan M.C., Severin E.J., Doleman B.J., Beaber S.A., Grubbs R.H., Lewis N.S. (1996) Array-based vapor sensing using chemically sensitive, carbon black-polymer resistors. *Chem. Mater.* **8**, 2298–2312.

Ma X., Ning H., Liu Y., Zhang J., Xu C. (2015) Highly sensitive humidity sensors made from composites of HEC filled by carbon nanofillers. *Mater. Technology. Adv. Perform. Mater.* **30**, 134–139.

Maier F., Riedel M., Mantel B., Ristein J., Ley L. (2000) Origin of surface conductivity in diamond. *Phys. Rev. Lett.* **85**(16), 3472–3475.

Maki T., Shikama S., Komori M., Sakaguchi Y., Sakuta K., Kobayashi T. (1992) Hydrogenating effect of single-crystal diamond surface. *Jpn. J. Appl. Phys.* **31**(10A), 1446–1449.

Malik R., Tomer V.K., Chaudhary V., Dahiya M.S., Sharma A., Nehra S.P., et al. (2017) An excellent humidity sensor based on In–$SnO_2$ loaded mesoporous graphitic carbon nitride. *J. Mater. Chem. A* **5**, 14134–14143.

Mamalis A.G., Vogtländer L.O.G., Markopoulos A. (2004) Nanotechnology and nanostructured materials: Trends in carbon nanotubes. *Precision Eng.* **28**, 16–30.

Mansor N., Miller T.S., Dedigama I., Jorge A.B., Jia J., Brázdová V., et al. (2016) Graphitic carbon nitride as a catalyst support in fuel cells and electrolyzers. *Electrochim. Acta* **222**, 44–57.

Marchenko D., Varykhalov A., Rybkin A., Shikin A.M., Rader O. (2011) Atmospheric stability and doping protection of noble-metal intercalated graphene on Ni(111). *Appl. Phys. Lett.* **98**, 122111.

Marques P.A.A.P., Gonçalves G., Cruz S., Almeida N., Singh M.K., Grácio J., Sousa A.C.M. (2011) Functionalized graphene nanocomposites. In: Hashim A. (ed.) *Advances in Nanocomposite Technology.* InTech, London pp. 247–272.

Mather P.J., Thomas K.M. (1997) Carbon black/high density polyethylene conducting composite materials. *J. Mater. Sci.* **32**, 1711–1715.

Matsuguchi M., Asahara K., Mizukami T. (2013) Highly sensitive toluene vapor sensors using carbon black/amino-functional copolymer composites. *J. Appl. Polym. Sci.* **127**(14), 2529–2535.

Matzger A.J., Lawrence C.E., Grubbs R.H., Lewis N.S. (2000) Combinatorial approaches to the synthesis of vapor detector arrays for use in an electronic nose. *J. Comb. Chem.* **2**, 301–304.

Mauter M.S., Elimelech M. (2008) Environmental applications of carbon-based nanomaterials. *Environ. Sci. Technol.* **42**(16), 5843–5859.

McDermott E.J., Wirnhier E., Schnick W., Virdi K.S., Scheu C., Kauffmann Y., et al. (2013) Band gap tuning in poly(triazine imide), a nonmetallic photocatalyst. *J. Phys. Chem. C* **117**, 8806–8812.

McLachlan D.S., Blaszkiewicz M., Newnham R.E. (1990) Electrical resistivity of composites. *J. Am. Ceram. Soc.* **73**(8), 2187–2203.

Meyyappan M. (2009) A review of plasma enhanced chemical vapour deposition of carbon nanotubes. *J. Phys. D: Appl. Phys.* **42**, 213001.

Mohanty N., Moore D., Xu Z., Sreeprasad T., Nagaraja A., Rodriguez A.A., Berry V. (2012) Nanotomy-based production of transferable and dispersible graphene nanostructures of controlled shape and size. *Nat. Commun.* **3**, 844.

Morozov S.V., Novoselov K.S., Katsnelson M.I., Schedin F., Elias D.C., Jaszczak J.A., Geim A.K. (2008) Giant intrinsic carrier mobilities in graphene and its bilayer. *Phys. Rev. Lett.* **100**, 016602.

Mudimela P.R., Grigoras K., Anoshkin I.A., Varpula A., Ermolov V., Anisimov A.S., et al. (2012) Single-walled carbon nanotube network field effect transistor as a humidity sensor. *J. Sensors* **2012**, 496546.

Munoz B.C., Steinthal G., Sunshine S. (1999) Conductive polymer-carbon black composites-based sensor arrays for use in an electronic nose. *Sens. Rev.* **19**(4), 300–305.

Niu C., Lu Y.Z., Lieber C.M. (1993) Experimental realization of the covalent solid carbon nitride. *Science* **261**(5119), 334–337.

Niyogi S., Bekyarova E., Itkis M.E., McWilliams J.L., Hamon M.A., Haddon R.C. (2006) Solution properties of graphite and graphene. *J. Am. Chem. Soc.* **128**(24), 7720–7721.

Nomani M.W.K., Shishir R., Qazi M., Diwan D., Shields V.B., Spencer M.G., et al. (2010) Highly sensitive and selective detection of $NO_2$ using epitaxial graphene on 6H-SiC. *Sens. Actuators B* **150**, 301–307.

Novak J.P., Snow E.S., Houser E.J., Park D., Stepnowski J.L., McGill R.A. (2003) Nerve agent detection using networks of single-walled carbon nanotubes. *Appl. Phys. Lett.* **83**(19), 4026–4029.

Novoselov K.S., Geim A.K., Morozov S.M., Jiang D., Zhang Y., Dubonos S.V., et al. (2004) Electric field effect in atomically thin carbon films. *Science* **306**(5696), 666–669.

Novoselov K.S., Jiang D., Schedin F., Booth T.J., Khotkevich V.V., Morozov S.V., Geim A.K. (2005) Two-dimensional atomic crystals. *PNAS* **102**, 10451.

Odom T.W., Huang J.L., Kim P., Lieber C.M. (1998) Atomic structure and electronic properties of single-walled carbon nanotubes. *Nature* **391**(6662), 62–64.

Okazaki M., Maruyama K., Tsuchida M., Tsubokawa N. (1999) A novel gas sensor from poly(ethylene glycol)-grafted carbon black. Responsibility of electrical resistance of poly(ethylene glycol)-grafted carbon black against humidity and solvent vapor. *Polymer J.* **31**(8) 672–676.

Pai Y.H., Ke J.H., Huang H.F., Lee C.M., Zen J.M., Shieu F.S. (2006) $CF_4$ plasma treatment for preparing gas diffusion layers in membrane electrode assemblies. *J. Power Sources* **161**, 275–281.

Park E.U., Choi B.I., Kim J.C., Woo S.-B., Kim Y.-G., Choi Y., Lee S.-W. (2018) Correlation between the sensitivity and the hysteresis of humidity sensors based on graphene oxides. *Sens. Actuators B* **258**, 255–262.

Park S., Dikin D.A., Nguyen S.B.T., Ruoff R.S. (2009) Graphene oxide sheets chemically cross-linked by polyallylamine. *J. Phys. Chem. C* **113**(36), 15801–15804.

Park S., Ruoff R.S. (2009) Chemical methods for the production of graphenes. *Nat. Nanotechnol.* **4**, 217–224.

Partoens B., Peeters F.M. (2006) From graphene to graphite: Electronic structure around the *K* point. *Phys. Rev. B* **74**, 075404.

Patel S.V., Jenkins M.W., Hughes R.C., Yelton W.G., Ricco A.J. (2000) Differentiation of chemical components in a binary solvent vapor mixture using carbon/polymer composite-based chemiresistors. *Anal. Chem.* **72**, 1532–1542.

Pati R., Zhang Y., Nayak S.K., Ajayan P.M. (2002) Effect of $H_2O$ adsorption on electron transport in a carbon nanotube. *Appl. Phys. Lett.* **81**(14), 2638–2640.

Pearce R., Iakimov T., Andersson M., Hultman, Lloyd Spetz A., Yakimova R. (2011) Epitaxially grown graphene-based gas sensors for ultrasensitive $NO_2$ detection. *Sens. Actuators B* **155**, 451–455.

Peng S., Cho K. (2000) Chemical control of nanotube electronics. *Nanotechnology* **11**, 57–60.

Peng S., Cho K. (2003) Ab initio study of doped carbon nanotube sensors. *Nano Lett.* **3**(4), 513–517.

Penza M., Anisimkin V.I. (1999) Surface acoustic wave humidity sensor using polyvinyl-alcohol film. *Sens. Actuators A* **76**(1), 162–166.

Penza M., Cassano G., Aversa P., Antolini F. (2004) Alcohol detection using carbon nanotubes acoustic and optical sensors. *Appl. Phys. Lett.* **85**(12), 2379–2381.

Possas-Abreu M., Ghassemi F., Rousseau L., Scorsone E., Descours E., Lissorgues G. (2017) Development of diamond and silicon MEMS sensor arrays with integrated readout for vapor detection. *Sensors* **17**, 1163.

Pumera M. (2009) Electrochemistry of graphene: New horizons for sensing and energy storage. *Chem. Rec.* **9**, 211–223.

Pumera M., Merkoci A., Alegret S. (2007) Carbon nanotube detectors for microchip CE: Comparative study of singlewall and multiwall carbon nanotube, and graphite powder films on glassy carbon, gold, and platinum electrode surfaces. *Electrophoresis* **28**(8), 1274–1280.

Qi P., Vermesh O., Grecu M., Javey A., Wang Q., Dai H. (2003) Toward large arrays of multiplex functionalized carbon nanotube sensors for highly sensitive and selective molecular detection. *Nano Lett.* **3**, 347–351.

Radeva E., Georgiev V., Spassov L., Koprinarov N., Kanev S. (1997) Humidity absorptive properties of thin fullerene layers studied by means of quartz microbalance. *Sens. Actuators B* **42**, 11–13.

Ramos M.V., Al-Jumaily A., Puli V.S. (2005) Conductive polymer-composite sensor for gas detection. In: *Proceedings of 1st International Conference on Sensing Technology*, Palmerston North, New Zealand, pp. 213–216.

Ran Q.S., Gao M.Z., Guan X.M., Wang Y., Yu Z.P. (2009) First-principles investigation on bonding formation and electronic structure of metal-graphene contacts. *Appl. Phys. Lett.* **94**, 103511.

Ratinac K.R., Yang W., Gooding J.J., Thordarson P., Braet F. (2011) Graphene and related materials in electrochemical sensing. *Electroanal.* **23**(4), 803–826.

Reina A., Jia X., Ho J., Nezich D., Son H., Bulovic V., et al. (2009) Large area, few-layer graphene films on arbitrary substrates by chemical vapor deposition. *Nano Lett.* **9**, 30–35.

Robinson J.T., Perkins F.K., Snow E.S., Wei Z., Sheehan P.E. (2008) Reduced graphene oxide molecular sensors. *Nano Lett.* **8**, 3137–3140.

Rossi E., Sarma S.D. (2008) Ground state of graphene in the presence of random charged impurities. *Phys. Rev. Lett.* **101**, 166803.

Ryan M.A., Shevade A.V., Zhou H., Homer M.L. (2004) Polymer–carbon black composite sensors in an electronic nose for air-quality monitoring. *MRS Bull.* **29**(10), 714–719.

Safari S., van de Ven T.G.M. (2016) Effect of water vapor adsorption on electrical properties of carbon nanotube/nanocrystalline cellulose composites. *ACS Appl. Mater. Interfaces* **8**(14), 9483–9489.

Saito R., Dresselhaus G., Dresselhaus M.S. (1998) *Physical Properties of Carbon Nanotubes.* Imperial College Press, London, UK.

Saito R., Fujita M., Dresselhaus G., Dresselhaus M.S. (1992) Electronic structure of graphene tubules based on $C_{60}$. *Phys. Rev. B* **46**, 1804–1811.

Saleem M., Karimov K.S., Karieva Z.M., Mateen A. (2010) Humidity sensing properties of CNT–OD–VETP nanocomposite films. *Phys. E* **43**, 28–32.

Sberveglieri G., Faglia G., Perego C., Nelli P., Marks R.N., Virgili T., et al. (1996) Hydrogen and humidity sensing properties of $C_{60}$ thin films. *Synth. Met.* **77**, 273–275.

Schaetz A., Zeltner M., Stark W.J. (2012) Carbon modifications and surfaces for catalytic organic transformations. *ACS Catal.* **2**, 1267–1284.

Schedin F., Geim A.K., Morozov S.V., Hill E.W., Blake P., Katsnelson M.I., Novoselov K.S. (2007) Detection of individual gas molecules adsorbed on grapheme. *Nat. Mater.* **6**, 652–655.

Schnorr J.M., Swager T.M. (2011) Emerging applications of carbon nanotubes. *Chem. Mater.* **23**, 646–657.

Schrand A.M., Ciftan Hens S.A., Shenderova O.A. (2009) Nanodiamond particles: Properties and perspectives for bio-applications. *Cr. Rev. Sol. St. Mater. Sci.* **34**, 18–74.

Shafiei M., Spizzirri P.G., Arsat R., Yu J., du Plessis J., Dubin S., et al. (2010) Platinum/graphene nanosheet/SiC contacts and their application for hydrogen gas sensing. *J. Phys. Chem. C* **114**, 13796–13801.

Shah M., Ahmad Z., Sulaiman K., Sayyad M.H. (2012) Carbon nanotubes' nanocomposite in humidity sensors. *Solid-State Electron.* **69**, 18–21.

Shao Q., Liu G., Teweldebrhan D., Balandin A.A., Roumyantes S., Shur M.S., Yan D. (2009) Flicker noise in bilayer graphene transistors. *IEEE Electron Device Lett.* **30**, 288–290.

Shen W., Li Z., Yiong L.Y. (2008) Surface chemical functional groups modification of porous carbon. *Recent Patents Chem. Eng.* **1**, 27–40.

Sheng L., Chen D., Chen Y. (2011) A surface acoustic wave humidity sensor with high sensitivity based on electrospun MWCNT/Nafion nanofiber films. *Nanotechnology* **22**, 265504.

Shih J.-S., Mao Y.-C., Sung M.-F., Gau G.-J., Chiou C.-S. (2001) Piezoelectric crystal membrane chemical sensors based on fullerene $C_{60}$. *Sens. Actuators B* **76**, 347–353.

Shim Y.-B., Park J.-H. (2000) Humidity sensor using chemically synthesized poly(1,5-diaminonaphthalene) doped with carbon. *J. Electrochem. Soc.* **147**(1), 381–385.

Si Y.C., Samulski E.T. (2008) Synthesis of water soluble graphene. *Nano Lett.* **8**, 1679–1682.

Singh V., Joung D., Zhai L., Das S., Khondaker S.I., Seal S. (2011) Graphene based materials: Past, present and future. *Prog. Mater. Sci.* **56**, 1178–1271.

Sinnott S.B., Andrews R. (2001) Carbon nanotubes: Synthesis, properties, and applications. *Crit. Rev. Solid State Mater. Sci.* **26**, 145–249.

Smith A.D., Elgammal K., Niklaus F., Delin A., Fischer A.C., Vaziri S., et al. (2015) Resistive graphene humidity sensors with rapid and direct electrical readout. *Nanoscale* **7**, 19099–19109.

Snow E., Perkins F., Houser E., Badescu S., Reinecke T. (2005) Chemical detection with a single-walled carbon nanotube capacitor. *Science* **307**, 1942–1945.

Snow E., Perkins F., Robinson J. (2006) Chemical vapor detection using single-walled carbon nanotubes. *Chem. Soc. Rev.* **35**, 790–798.

Snow E.S., Novak J.P., Campbell P.M., Park D. (2003) Random networks of carbon nanotubes as an electronic material. *Appl. Phys. Lett.* **82**(13), 2145–2147.

Sommer A.P., Zhu D., Fecht H.J. (2008) Genesis on diamonds. *Cryst. Growth Des.* **8**, 2628–2629.

Spiridon S.-I., Ionete E.I., Monea B.F., Ebrasu-Ion D., Enache S., Vaseashta A. (2017) Synthesis and characterization of integrated SWCNT-Pt-$P_2O_5$-based sensor platforms for absolute humidity measurement. In: *Proceedings of 2017 IEEE Sensors Applications Symposium (SAS)*, Glassboro, NJ.

Stankovich S., Dikin D.A., Dommett G.H.B., Kohlhaas K.M., Zimney E.J., Stach E.A., et al. (2006b) Graphene-based composite materials. *Nature* **442**, 282–286.

Stankovich S., Kikin D.A., Piner R.D., Kohlhaas K.A., Kleinhammes A., Jia Y., et al. (2007) Synthesis of graphene-based nanosheets via chemical reduction of exfoliated graphite oxide. *Carbon* **45**, 1558–1565.

Stankovich S., Piner R.D., Nguyen S.T., Ruoff R.S. (2006a) Synthesis and exfoliation of isocyanate-treated graphene oxide nanoplatelets. *Carbon* **44**, 3342–3347.

Star A., Joshi V., Skarupo S., Thomas D., Gabriel J.C.P. (2006) Gas sensor array based on metal-decorated carbon nanotubes. *J. Phys. Chem. B* **110**, 21014–21020.

Star A., Steuerman D.W., Heath J.R., Stoddart J.F. (2002) Starched carbon nanotubes. *Angew. Chem. Int. Ed.* **41**, 2508–2512.

Su P.-G., Chiou C.-F. (2014) Electrical and humidity-sensing properties of reduced graphene oxide thin film fabricated by layer-by-layer with covalent anchoring on flexible substrate. *Sens. Actuators B* **200**, 9–18.

Su P.-G., Shiu W.-L., Tsai M.-S. (2015) Flexible humidity sensor based on Au nanoparticles/graphene oxide/thiolated silica sol–gel film. *Sens. Actuators B* **216**, 467–475.

Su P.G., Sun Y.L., Lin C.C. (2006) A low humidity sensor made of quartz crystal microbalance coated with multi-walled carbon nanotubes/Nafion composite material films. *Sens. Actuators B* **115**, 338–343.

Su P.G., Wang C.S. (2007) In situ synthesized composite thin films of MWCNTs/PMMA doped with KOH as a resistive humidity sensor. *Sens. Actuators B* **124**, 303–308.

Subrahmanyam K.S., Vivekchand S.R.C., Govindaraj A., Rao C.N.R. (2008) A study of graphenes prepared by different methods: characterization, properties and solubilization. *J. Mater. Chem.* **18**, 1517–1523.

Suehiro J., Hidaka S., Yamane S., Imasaka K. (2007) Fabrication of interfaces between carbon nanotubes and catalytic palladium using dielectrophoresis and its application to hydrogen gas sensor. *Sens. Actuators B* **127**, 505–511.

Sun J.-T., Hong C.-Y., Pan C.-Y. (2011) Surface modification of carbon nanotubes with dendrimers or hyperbranched polymers. *Polym. Chem.* **2**, 998–1007.

Sundaram R.S., Navarro C.G., Balasubramaniam K., Burghard M., Kern K. (2008) Electrochemical modification of grapheme. *Adv. Mater.* **20**, 3050–3053.

Synowczyk A.W., Heinze J. (1993) Application of fullerenes as sensor materials. In: Kuzmany H., Mehring N., and Fink J. (eds.) *Electronic Properties of Fullerenes Springer Series in Solid State Sciences*, Vol. 117. Springer-Verlag, Berlin, Germany, pp. 73–77.

Tang Q.Y., Chan Y.C., Zhang K. (2011) Fast response resistive humidity sensitivity of polyimide/multiwall carbon nanotube composite films. *Sens. Actuators B* **152**, 99–106.

Tasis D., Tagmatarchis N., Georgakilas V., Prato M. (2003) Soluble carbon nanotubes. *Chem. Eur. J.* **9**, 4000–4008.

Terrones M., Ajayan P.M., Banhart F., Blase X., Carroll D.L., Charlier J.C., et al. (2002) N-doping and coalescence of carbon nanotubes: Synthesis and electronic properties. *Appl. Phys. A* **74**, 355–361.

Terrones M., Jorio A., Endo M., Rao A.M., Kim Y., Hayashi T., Terrones H., Charlier J.C., Dresselhaus G., Dresselhaus M.S. (2004) New direction in nanotube science. *Mater. Today* **7** (10), 30–45.

Thomas A., Fischer A., Goettmann F., Antonietti M., Müller J.-O., Schlögl R., Carlsson J.M. (2008) Graphitic carbon nitride materials: variation of structure and morphology and their use as metal-free catalysts. *J. Mater. Chem.* **18**(41), 4893–4908.

Toebes M.L., van Dillen J.A., de Jong K.P. (2001) Synthesis of supported palladium catalysts. *J. Mol. Catal. A* **173**, 75–98.

Toloman D., Popa A., Stan M., Socaci C., Biris A.R., Katona G., et al. (2017) Reduced graphene oxide decorated with Fe doped $SnO_2$ nanoparticles for humidity sensor. *Appl. Surf. Sci.* **402**, 410–417.

Tomer V.K., Thangaraj N., Gahlot S., Kailasam K. (2016) Cubic mesoporous Ag@CN: A high performance humidity sensor. *Nanoscale* **8**, 19794–19803.

Tsai J.T.H., Lu C.-C., Li J.G. (2010) Fabrication of humidity sensors by multi-walled carbon nanotubes. *J. Experiment. Nanosci.* **5** (4), 302–309.

Tsubokawa N. (1992) Functionalization of carbon black by surface grafting of polymers. *Prog. Polym. Sci.* **17**, 417–470.

Tsubokawa N., Yoshikawa S., Maruyama K., Ogasawara T., Saitoh K. (1997) Responsibility of electric resistance of polyethyleneimine-grafted carbon black against alcohol vapor and humidity. *Polym. Bull.* **39**, 217–224.

Ueda T., Bhuiyan M.M.H., Norimatsu H., Katsuki S., Ikegami T., Mitsugi F. (2008) Development of carbon nanotube-based gas sensors for $NO_x$ gas detection working at low temperature. *Physica E* **40**(7), 2272–2277.

Valentini L., Bavastrello V., Stura E., Armentano I., Nicolini C., Kenny J.M. (2004b) Sensors for inorganic vapor detection based on carbon nanotubes and poly(o-anisidine) nanocomposite material. *Chem. Phys. Lett.* **383**(5–6), 617–622.

Valentini L., Cantalini C., Armentano I., Kenny J.M., Lozzi L., Santucci S. (2004a) Highly sensitive and selective sensors based on carbon nanotubes thin films for molecular detection. *Diam. Relat. Mater.* **13**, 1301–1305.

Valentini L., Cantalini C., Lozzi L., Armentano I., Kenny J.M., Santucci S. (2003) Reversible oxidation effects on carbon nanotubes thin films for gas sensing applications. *Mater. Sci. Eng. C* **23**, 523–529.

Varga M., Laposa A., Kulha P., Kroutil J., Husak M., Kromka A. (2015) Quartz crystal microbalance gas sensor with nanocrystalline diamond sensitive layer. *Phys. Stat. Sol. B* **252**(11), 2591–2597.

Varghese O., Kichambre P., Gong D., Ong K., Dickey E., Grimes C. (2001) Gas sensing characteristics of multi-wall carbon nanotubes. *Sens. Actuators B* **81**, 32–41.

Varghese S.S., Lonkar S., Singh K.K., Swaminathan S., Abdala A. (2015a) Recent advances in graphene based gas sensors. *Sens. Actuators B* **218**, 160–183.

Varghese S.S., Varghese S.H., Swaminathan S., Singh K.K., Mittal V. (2015b) Two-Dimensional materials for sensing: Graphene and beyond. *Electronics* **4**, 651–687.

Vedala H., Sorescu D.C., Kotchey G.P., Star A. (2011) Chemical sensitivity of graphene edges decorated with metal nanoparticles. *Nano Lett.* **11**, 2342–2347.

Virji S., Huang J.X., Kaner R.B., Weiller B.H. (2004) Polyaniline nano fiber gas sensors: Examination of response mechanisms. *Nano Lett.* **4**, 491–496.

Wang D., Huang Y., Ma Y., Liu P., Liu C., Zhang Y. (2014) Research on highly sensitive humidity sensor based on Tr-MWCNT/ HEC composite films. *J. Mater. Res.* **29**(23), 2845–2853.

Wang D., Wang L., Dong X., Shi Z., Jin J. (2012) Chemically tailoring graphene oxides into fluorescent nanosheets for $Fe^{3+}$ ion detection. *Carbon* **50**(6), 2147–2154.

Wang G.X., Wang B., Park J., Yang J., Shen X.P., Yao J. (2009) Synthesis of enhanced hydrophilic and hydrophobic graphene oxide nanosheets by a solvothermal method. *Carbon* **47**, 68–72.

Wang Q., Qu S.L., Fu S.Y., Liu W.J., Li J.J., Gu C.Z. (2007a) Chemical gases sensing properties of diamond nanocone arrays formed by plasma etching. *J. Appl. Phys.* **102**, 103714.

Wang R., Zhang D., Sun W., Han Z., Liu C. (2007b) A novel aluminum-doped carbon nanotubes sensor for carbon monoxide. *J. Mol. Struct.* **806**(1–3), 93–97.

Wang S., Chia P.J., Chua L.L., Zhao L.H., Png R.Q., Sivaramakrishnan S., et al. (2008) Band-like transport in surface-functionalized highly solution-processable graphene nanosheets. *Adv. Mater.* **20**(18), 3440–3446.

Wang X., Bai H., Shi G. (2011) Size fractionation of graphene oxide sheets by pH-assisted selective sedimentation. *J. Am. Chem. Soc.* **133**(16), 6338–6342.

Wang Y., Shen C., Lou W., Shentu F. (2016a) Polarization-dependent humidity sensor based on an in-fiber Mach-Zehnder interferometer coated with graphene oxide. *Sens. Actuators B* **234**, 503–509.

Wang Y., Shen C., Lou W., Shentu F. (2016b) Fiber optic humidity sensor based on the graphene oxide/PVA composite film. *Opt. Commun.* **372**, 229–234.

Wang Y., Yeow J.T.W. (2009) A review of carbon nanotubes-based gas sensors. *J. Sensors* **2009**, 493904.

Wei D., Liu Y., Wang Y., Zhang H., Huang L., Yu G. (2009). Synthesis of N-doped graphene by chemical vapor deposition and its electrical properties. *Nano Lett.* **9**, 1752–1758.

Wei D., Liu, Y. (2010) Controllable synthesis of graphene and its applications. *Adv. Mater.* **22**(30), 3225–3241.

Wisitsoraat K.A.J., Tuantranont A., Lomas T. (2008) Humidity sensor utilizing multiwalled carbon nanotubes coated quartz crystal microbalance. In: *Proceedings of the 2nd IEEE International Nanoelectronics Conference*, INEC 2008, Shanghai, China, pp. 961–964.

Wu Z.-S., Zhou G., Yin L.-C., Ren W., Li F., Cheng H.-C. (2012) Graphene/metal oxide composite electrode materials for energy storage. *Nano Energy* **1**, 107–131.

Xie H., Yang Q., Sun X., Yang J., Huang Y. (2006) Gas sensor arrays based on polymer-carbon black to detect organic vapors at low concentration. *Sens. Actuators B* **113**, 887–891.

Xu C., Wang X., Zhu J. (2008) Graphene–metal particle nanocomposites. *J. Phys. Chem. C* **112**, 19841–19845.

Yang H., Shan C., Li F., Han D., Zhang Q., Niu L. (2009) Covalent functionalization of polydisperse chemically-converted graphene sheets with amine-terminated ionic liquid. *Chem. Commun.* **2009**(26), 3880–3882.

Yang R.T. (2003) *Adsorption*. Wiley, Hoboken, NJ.

Yao Y., Chen X., Guo H., Wu Z., Li X. (2012) Humidity sensing behaviours of graphene oxide-silicon bi-layer flexible structure. *Sens. Actuators B* **161**, 1053–1058.

Yao Y., Chen X., Ma W., Ling W. (2014b) Quartz crystal microbalance humidity sensors based on nanodiamond sensing films. *IEEE Trans. Nanotechnol.* **13**, 386–393.

Yao Y., Chen X.D., Guo H.H., Wu Z.Q. (2011) Graphene oxide thin film coated quartz crystal microbalance for humidity detection. *Appl. Surf. Sci.* **257**, 7778–7782.

Yao Y., Chen X.D., Li X.Y., Chen X.P., Li N. (2014) Investigation of the stability of QCM humidity sensor using graphene oxide as sensing films. *Sens. Actuators B* **191**, 779–783.

Yao Y., Xue Y.J. (2015) Impedance analysis of quartz crystal microbalance humidity sensors based on nanodiamond/ graphene oxide nanocomposite film. *Sens. Actuators B* **211**, 52–58.

Yeow J.T.W., She J.P.M. (2006) Carbon nanotube-enhanced capillary condensation for a capacitive humidity sensor. *Inst. Phys. Pub.* **17**, 5441–5448.

Yoo K.-P., Lim L.-T., Min N.-K., Lee M.J., Lee C.J., Park C.-W. (2010) Novel resistive-type humidity sensor based on multiwall carbon nanotube/polyimide composite films. *Sens. Actuators B* **145**, 120–125.

Yoon D.H., Suh S.J., Kim Y.T., Hong B., Jang G.E. (2003) Influence of hydrogen on a-SiC:H films deposited by RF-PECVD and annealing effect. *J. Korean Phys. Soc.* **42**, S943–S946.

Yuan Z., Tai H., Ye Z., Liu C., Xie G., Du X., Jiang Y. (2016) Novel highly sensitive QCM humidity sensor with low hysteresis based on graphene oxide (GO)/poly(ethyleneimine) layered film. *Sens. Actuators B* **234**, 145–154.

Zahab A., Spina L., Poncharal P. (2000) Water-vapor effect on the electrical conductivity of a single-walled carbon nanotube mat. *Phys. Rev. B* **62**(15), 10000–10003.

Zambov L.M., Popov C., Plass M.F., Bock A., Jelinek M., Lancok J., et al. (2000) Capacitance humidity sensor with carbon nitride detecting element. *Appl. Phys. A* **70**, 603–606.

Zeng H.C. (2003) Carbon nanotube-based nanocomposites. In: Nalwa H.S. (ed.) *Handbook of Organic Hybrid Materials and Nanocomposites*. American Scientific Publisher, Stevenson Ranch, CA, pp. 151–180.

Zhang B., Fu R., Zhang M., Dong X., Wang L., Pittman C.U. (2006) Gas sensitive vapour grown carbon nanofiber/polystyrene sensors. *Mater. Res. Bull.* **41**, 553–562.

Zhang D., Chang H., Li P., Liu R., Xue Q. (2016b) Fabrication and characterization of an ultrasensitive humidity sensor based on metal oxide/graphene hybrid nanocomposite. *Sens. Actuators B* **225**, 233–240.

Zhang D., Chang H., Liu R. (2016c) Humidity-sensing properties of one-step hydrothermally synthesized tin dioxide-decorated graphene nanocomposite on polyimide substrate. *J. Electron. Mater.* **45**, 4275–4281.

Zhang D., Yin N., Xia B., Sun Y., Liao Y., He Z., Hao S. (2016a) Humidity-sensing properties of hierarchical ZnO/MWCNTs/ ZnO nanocomposite film sensor based on electrostatic layer-by-layer self-assembly. *J. Mater. Sci. Mater. Electron.* **27**, 2481–2487.

Zhang G., Sun X. (2016) Size control methods and size-dependent properties of graphene. In: Aliofkhazraei M., Ali N., Milne W.I., Ozkan C.S., Mitura S., Gervasoni J.L. (eds.) *Graphene Science Handbook, Size-Dependent Properties*. CRC Press, Boca Raton, FL, pp. 27–40.

Zhang T., Mubeen S., Myung N.Y., Deshusses M.A. (2008) Recent progress in carbon nanotube-based gas sensors. *Nanotechnology* **19**, 332001.

Zhang W.-D., Zhang W.-H. (2009) Carbon nanotubes as active components for gas sensors. *J. Sensor* **2009**, 160698.

Zhang Y., Yu K., Xu R., Jiang D., Luo L., Zhu Z. (2005) Quartz crystal microbalance coated with carbon nanotube films used as humidity sensor. *Sens. Actuators A* **120**, 142–146.

Zhao J., Buldum A., Han J., Lu J. (2002) Gas molecule adsorption in carbon nanotubes and nanotube bundles. *Nanotechnology* **13**, 195–200.

Zhao J., Buldum A., Han J., Lu J.P. (2001) Gas molecules adsorption on carbon nanotubes. *Mater. Res. Soc. Proc.* **633**, 3.48.1–A13.48.6.

Zhao Y.L., Stoddart J.F. (2009) Noncovalent functionalization of single-walled carbon nanotubes. *Acc. Chem. Res.* **42**, 1161–1171.

Zhu B., Zhang J., Jiang C., Cheng B., Yu J. (2017) First principle investigation of halogen-doped monolayer g-$C_3N_4$ photocatalyst. *Appl. Catal. B* **207**, 27–34.

Ziegler D., Palmero P., Giorcelli M., Tagliaferro A., Tulliani J.-M. (2017) Biochars as innovative humidity sensing materials. *Chemosens.* **5**, 35.

# 5 Semiconductor-Based Humidity Sensors

## 5.1 CONVENTIONAL IV AND III-V SEMICONDUCTORS

Currently, conventional semiconductors such as Ge, Si, GaAs, and InP are widely used in microelectronics and optoelectronics (Mui et al. 1993; Yin and Tang 2007; Mokkapati and Jagadish 2009; Goley and Hudait 2014). However, it was found that these materials can also be the basis for the development of humidity sensors. Typically, these sensors are sensors of resistive and capacitance types (see Table 5.1). The experiment showed that for their manufacture it is possible to use materials of all types—epitaxial, nanocrystalline, porous, or 1D materials. But the same research found (Cheng et al. 2009; Salehi et al. 2006a, 2006b), that porous semiconductors exhibited much higher humidity sensitivity than the nonporous materials, such as epitaxial layers. Pore formation on semiconductor materials is usually achieved by electrochemical anodization using a suitable electrolyte. The use of nanocrystalline materials, especially 1D and 2D materials (Samà et al. 2017; Peng et al. 2018), instead of epitaxial films also has a positive effect (Chitara et al. 2010; Kerlau et al. 2006). As mentioned earlier, the reason for this optimizing effect is an increase in the actual surface area of the device structure and the formation of nanopores in which capillary condensation is possible.

For the synthesis of 1D materials such as nanowires (NWs) can be used a variety of methods. For example, in (Zhou et al. 2003) Silicon nanowires (SiNWs) were prepared by the oxide-assisted growth technique. Taghinejad et al. (2013) synthesized SiNWs using the vapor–liquid–solid (VLS) method. This technique and fabrication of the humidity sensor are depicted schematically in Figure 5.1. Li et al. (2009) and Tao et al. (2009) in preparing SiNWs used a chemical etching procedure. The etchant for SiNWs consisted of $AgNO_3$ (25 mM):HF (15%) = 1:1(volume ratio). After etching, all chips were taken out from the solution followed by rinsing with deionized (DI) water and diluted nitric acid (~30% in vol) in order to remove the residual silver particles on the surface of the SiNWs. Scanning electron microscope (SEM) images of SiNWs prepared using this method are shown in Figure 5.2. It is seen that the SiNWs are aligned perpendicularly to the bulk silicon substrate, and their average length is about 80 mm. It is important that the length of SiNWs can be adjusted by controlling the proper etching time. But you need to keep in mind that, like other humidity-sensitive materials, the control of microstructure of the used semiconductors is very important for sensing characteristics, and the pore-formation conditions should be carefully chosen to achieve a highly sensitive humidity sensor with a quick response (Salehi et al. 2006a). As for the diameter of SiNWs used for the manufacture of humidity sensors, it ranged from 20 to 25 nm (Zhou et al. 2003; Passi et al. 2011) up to 600 nm (Taghinejad et al. 2013).

As for the humidity-sensitive characteristics of humidity sensors based on conventional semiconductors, it is necessary to recognize that these devices, by a majority of parameters, as a rule, are significantly inferior to the previously considered humidity sensors based on metal oxides and polymers.

It is important to note that the sensitivity of capacitive NW-based sensors, in the manufacture of which were

---

**TABLE 5.1**
**RT Humidity Sensors Based On Conventional Semiconductors**

| Semiconductor | Form | Sensor Type | Electrodes | References |
|---|---|---|---|---|
| Ge | Nanowires | Resistive | Pt | Samà et al. (2017) |
| Si | Nanowires | Resistive | | Hsueh et al. (2011) |
| | Nanowires | Capacitance | Cu | Li et al. (2009) |
| | Nanowires | | Ag | Chen et al. (2011) |
| | Ni/SiNWs | | Cu | Tao et al. (2009) |
| | SiNWs:P | | Cu | Taghinejad et al. (2013) |
| GaAs | Porous | Schottky | Pd | Salehi et al. (2006a) |
| | | | Au | Salehi et al. (2006b) |
| InP | Epitaxial | Resistive | NA | Talazac et al. (2001) |
| | Porous/IONPs | Resistive | Ag/Cu | Cheng et al. (2009) |

IONPs—indium-oxide nanoparticles.

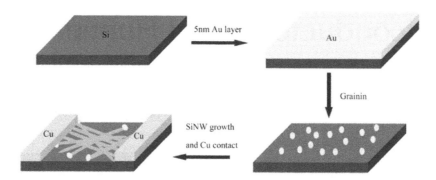

**FIGURE 5.1** Schematic drawing of SiNW-fabrication sequences and sensor preparation, starting with gold deposition and nano-grain formation. Once the growth is finalized, copper is used to serve as electrical contacts for the final electrical measurements. (Reprinted from *Sens. Actuators B*, 176, Taghinejad H. et al., Fabrication and modeling of high sensitivity humidity sensors based on doped silicon nanowires, 413–419, Copyright 2013, with permission from Elsevier.)

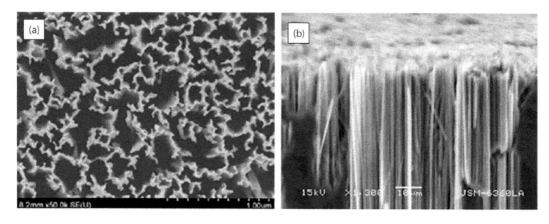

**FIGURE 5.2** SEM pictures of silicon nanowires (a) top-view image; and (b) cross-section image. (Reprinted with permission from *Phys. E Low Dimens. Syst. Nanostruct.*, 41, Li, H.L. et al., Investigation of capacitive humidity sensing behavior of silicon nanowires, 600–604, Copyright 2009, with permission from Elsevier.)

used NWs formed by chemical etching (Li et al. 2009; Chen et al. 2011), was much more than for sensors using NWs formed by the VLS technique (Taghinejad et al. 2013). The reason for this in the latter case is the low density of NWs in the sensitive layer. Taghinejad et al. (2013) believe that NW density can be increased by varying the growth pressure. But the possibilities of the VLS technique in the formation of high-density NWs are still very limited.

Typical humidity-sensitive characteristics of humidity sensors based on SiNWs, formed by chemical etching, are shown in Figure 5.3. The response and recovery time of such sensors typically ranged from 10 to 300 s, increasing with increasing humidity levels. The maximum hysteresis was also heavily dependent on sensor technology and ranged from 1.1% (Chen et al. 2011) up to 11% (Tao et al. 2009).

Low thermal and chemical stability should also be attributed to the shortcomings of these devices. Although it is necessary to recognize that SiNW-based capacitive sensors after additional heat treatment, contributing to the formation of an oxide layer on the surface of silicon, can be quite stable (Li et al. 2009; Chen et al. 2011). Test results performed by Chen et al. (2011) are shown in Figure 5.4.

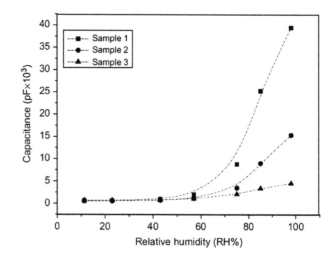

**FIGURE 5.3** The relationship between the capacitance variations in SiNW-based sensors and the corresponding relative humidity. For the samples 1–3, the etching time for the SiNWs was 60, 50, and 45 min, implying the different dimension of SiNWs resulted, respectively. (Reprinted with permission from *Phys. E Low Dimens. Syst. Nanostruct.*, 41 Li, H.L., et al., Investigation of capacitive humidity sensing behavior of silicon nanowires, 600–604, Copyright 2009, with permission from Elsevier.)

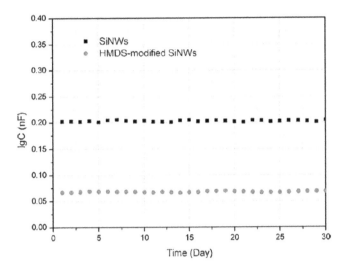

**FIGURE 5.4** The stability property of the SiNW-based humidity sensors in 30 days. The sensors were tested repeatedly under a fixed humidity level (75% RH). (Reprinted with permission from *Sens. Actuators B*, 156, Chen, X.J. et al., Humidity sensing behavior of silicon nanowires with hexamethyldisilazane modification, 631–636, Copyright 2011, with permission from Elsevier.)

## 5.2 GALLIUM AND BORON NITRIDES (WIDE-BAND SEMICONDUCTORS)

Gallium and boron nitrides are wide-band semiconductors of the III-V group (see Table 5.2). These are the most stable semiconductors of III-V group, and therefore, it is these semiconductors that have the maximum prospects for use in humidity sensors. Examples of the implementation of humidity sensors based on these materials are given in Table 5.3. It is seen that humidity sensors based on gallium and boron nitrides were the resistive type or had a Schottky diode structure.

### 5.2.1 GALLIUM NITRIDE

GaN in comparison with III-V compounds such as GaAs and InP exhibits the highest sensitivity to air humidity (Figure 5.5) and the stability of parameters (Figure 5.6). For example, GaN is not etched by any acid or base at temperatures below a few hundred degrees. Immunity to corrosion makes wide-band gap GaN very suitable for operation in chemically harsh environments. In addition, the presence of wide-band gap in GaN minimizes the generation of charge carriers due to undesirable

## TABLE 5.2
**Properties of Some Semiconductor Materials**

| Property | Si | GaAs | GaN | BN | SiC |
|---|---|---|---|---|---|
| Energy gap (eV) | 1.12 | 1.42 | 3.39 | 4.0–6.4 | 2.3–3.2 |
| Density (g/cm³) | 2.3 | 5.3 | 6.1 | 1.9–3.45 | 3.1 |
| Electron mobility (cm²/V·s) | 1450 | 8500 | 400–2000 | ~200 | 500–900 |
| Semi-insulating | No | Yes | Yes | Yes | Yes |
| Dielectric constant | 13.1 | 12.9 | 8.9–9.5 | 3.5–5.0 | 9.7 |
| Refractive index | 3.4 | 3.7 | 2.4–2.6 | 1.8–2.1 | 2.55 |
| Melting point (K) | 1687 | 1511 | >2700 | 3150–3400 | 3003 |
| Thermal conductivity (W/m·K) | 149 | 46–55 | 130–220 | 19–52 | 330–460 |
| Thermal expansion (μm/m·K) | 2.6 | 5.7 | 3–5.5 | 0.5–11 | 2.2–4.2 |
| Young's modulus (GPa) | 130–185 | 85.5 | 280–300 | 19.5–100 | 350–450 |
| Poisson ratio | 0.28 | 0.31 | 0.183 | 0.21–0.27 | 0.14–0.22 |

For pure materials at room temperature.

## TABLE 5.3
**RT Humidity Sensors Based on Gallium and Boron Nitrides**

| Semiconductor | Form | Sensor Type | Electrodes | References |
|---|---|---|---|---|
| GaN | Nanocrystalline | Resistive | Ti/Au | Chitara et al. (2010) |
| | Nanowires | Schottky | Ni | Peng et al. (2018) |
| AlGaN/GaN | Epitaxial | HEMT-based Schottky diode | Pt | Lo et al. (2010) |
| h-BN | Nanocrystalline | M-BN-Si, switching | Ag | Soltani et al. (2005) |
| BN:C, O | Nanocrystalline | Impedance | TiN | Aoki et al. (2007) |
| BN:Au | Nanotubes | Resistive | Ag/Ni | Yu et al. (2013) |
| h-BN | 3D structure | Resistive | Ag | Gautam et al. (2016) |

HEMT—high electron mobility transistor; M—metal.

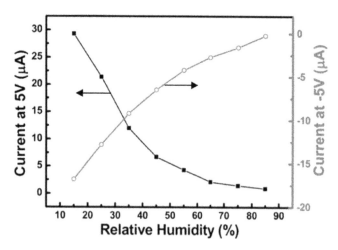

**FIGURE 5.5** The RH dependence of output current of GaN nanowire-array humidity sensor in dark at fixed bias voltage of 5 and −5 V, respectively. (Reprinted with permission from *Sens. Actuators B*, 256, Peng, M. et al., Ni-pattern guided GaN nanowire-array humidity sensor with high sensitivity enhanced by UV photoexcitation, 367–373, Copyright 2018, with permission from Elsevier.)

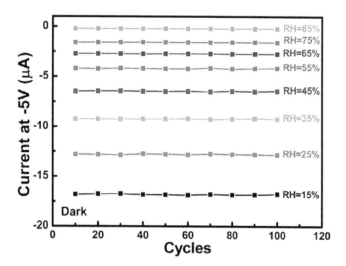

**FIGURE 5.6** Reproducibility of GaN-nanowire array humidity sensor at different RHs by cycling the RH conditions. (Reprinted from *Sens. Actuators B*, 256, Peng, M., et al., Ni-pattern guided GaN nanowire-array humidity sensor with high sensitivity enhanced by UV photoexcitation, 367–373, Copyright 2018, with permission from Elsevier.)

background optical or thermal excitation. This is very important for humidity-sensor operation at high temperature.

But it must be recognized that even this material cannot compete with metal oxides in the development of individual humidity sensors. The use of this material does not solve the problem of poor selectivity of solid-state humidity sensors as well. Like metal oxides, GaN-based sensors are sensitive to toxic gases and vapors of organic solvents (Lee et al. 2003; Kerlau et al. 2006; Prasad et al. 2017).

At the same time, it can be assumed that the need for GaN-based humidity sensors may appear in the development of integrated circuits based on GaN for high-power and

high-temperature electronics. The presence of efficient GaN UV light-emitted diodes (LEDs) also allows the development of hybrid monolithic structures, including a humidity sensor and LED. Studies have shown that the use of UV irradiation in the process of measuring humidity can have an optimizing effect on sensor parameters. In particular, Peng et al. (2018) have shown that UV photoexcitation strongly increases the response of GaN NW-array-humidity sensors (see Figure 5.7). With the increase of the UV optical power, the sensitivity to relative humidity (RH) sensitivity ($I_{RH15}/I_{RH85}$) of GaN sensors increased from 32 to 100 at 5 V and from 78 to 240 at −5 V, respectively. It is important that the response and recovery times also decreased under the influence of UV irradiation. Another advantage of nitride-based semiconductor devices according to Irokawa (2011) is utilization of AlGaN/GaN heterostructure. In an AlGaN/GaN heterostructure, the polarization-induced two-dimensional electron gas (2DEG) concentration at the AlGaN/GaN interface is extremely sensitive to the surface states. Any potential changes on the surface by adsorption of gas polar molecules would affect the surface potential and modulate the 2DEG density.

### 5.2.2 Boron Nitride

Boron nitride is an even a large-band-gap semiconductor (see Table 5.2), possessing very high chemical and thermal stability (Jiang et al. 2015). For example, Li et al. (2014b) reported that the monolayer boron nitride (BN) starts to oxidize at 700°C when heated in the air, which is much higher compared to the 250°C oxidation temperature of monolayer graphene. Liu et al. (2013) showed that ultrathin (2–5 nm) h-BN thin films are impervious to oxygen diffusion even at high temperatures and can be used as high-performance oxidant-resistance coatings for nickel up to 1100°C in

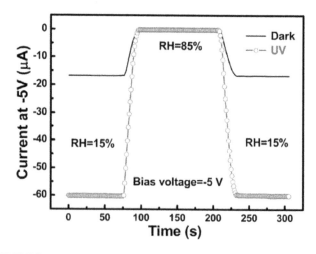

**FIGURE 5.7** The current response and recovery of GaN-nanowire array humidity sensor worked at −5 V under dark and UV power of 80 mW/cm² by switching the RH environment between 15% and 85%. (Reprinted from *Sens. Actuators B*, 256, Peng, M., et al., Ni-pattern guided GaN nanowire-array humidity sensor with high sensitivity enhanced by UV photoexcitation, 367–373, Copyright 2018, with permission from Elsevier.)

oxidizing atmospheres. Hou et al. (2014b) found lately that the thermal stability of BN nanofibers (400–1500 nm in diameter, 200–500 mm in length) is 250 K higher compared to powders with roughly the same size distribution (1250–1000 K), stating that the sample shape has an influence on the oxidation behavior. Wang et al. (2014) showed that the oxidation behavior of BN nanomaterials also depends on the specific surface area (SSA), reporting thermal stability of ~900°C for h-BN nanocrystals with average side length of ~210 nm, average thickness of ~70 nm, and SSA of 6 m²/g. At the same time, Kostoglou et al. (2015) have established that the h-BN platelets were stable up to ~1000°C, and their oxidation occurred in the temperature range between 1000°C and 1200°C, followed by the formation of boron oxide ($B_2O_3$). The above indicates that the BN-based humidity sensors must have a high stability of parameters.

As can be seen from Table 5.3 that humidity sensors based on BN were developed using materials with different crystallographic structure and morphology: nanocrystalline films (Soltani et al. 2005; Aoki et al. 2007), nanosheets (Sajjad and Feng 2014), nanotubes (Yu et al. 2013), and 3D structures (Gautam et al. 2016). For their synthesis, one can use different methods. Some of them were described in Jiang et al. (2015) and Kim et al. (2018). Unfortunately, the studies of BN-based humidity sensors were not comprehensive. Based on the published data, we can only say that BN-based humidity sensors with stable parameters (Yu et al. 2013; Gautam et al. 2016) can be made, and the resistance of these sensors can significantly and reversibly change under the influence of humidity (see Figures 5.8 and 5.9). At that, as for other solid-state humidity sensors, the high porosity and the surface area are responsible for enhancing the sensitivity of the sensor (Gautam et al. 2016).

However, from general considerations, the hexagonal 2D boron nitride (h-BN) should be the most promising for development of humidity sensors (Jiang et al. 2015). It possesses a honeycomb structure similar to graphene, but with alternating boron and nitrogen atoms. 2D h-BN-based humidity sensors have not been documented, but like other 2D materials, their properties enable BN sheets to be one of the potential sensing materials of the future. To date, BN sheets have been used to measure varying concentrations of oxygen, methane, hydrogen sulfide, and nitrogen-based gases (Ma 2015; Ayari et al. 2017; Yang et al. 2017).

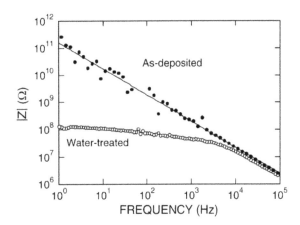

**FIGURE 5.8** Frequency dependence of impedance of BN:C film. BN:C films were synthesized by plasma-assisted chemical vapor deposition (PACVD). The water-treated samples were prepared in the atmosphere of around 100% in humidity. (Reprinted from *Diamond Rel. Mater.*, 16, Aoki, H., et al., Characterization of boron carbon nitride film for humidity sensor, 1300–1303, Copyright 2007, with permission from Elsevier.)

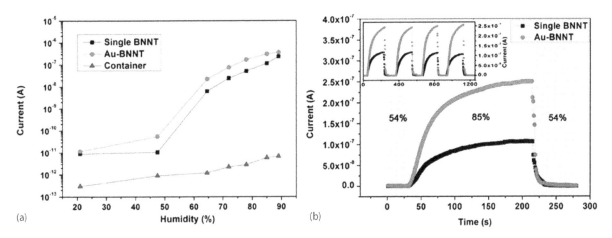

**FIGURE 5.9** (a) The I–RH curves of the container, a single BNNT, and Au-BNNT. Bias voltage of 10 V is applied during the test. c I–V curves of a single Au-BNNT in different RH levels from 47.5% to 89.0% RH measured at room temperature; (b) response and recovery characteristics of a single BNNT- and Au-BNNT-based humidity sensing units measured by switching the RH levels between RH = 54.0% and 85.0% (Inset is 4 cycles). Long bamboo-like BNNTs used in this work were synthesized by a ball-milling and subsequent high-temperature annealing processes. The resultant single BNNT was assembled between two Ni electrodes mounted onto an alumina ceramic plate. Silver paste was used for a good electrical contact between the BNNT and the electrodes. (Reprinted from *Electrochem. Commun.*, 30, Yu, Y., et al., Humidity sensing properties of single Au-decorated boron nitride nanotubes, 29–33, Copyright 2013, with permission from Elsevier.)

## 5.3 CHALCOGENIDES

Chalcogenides are other semiconductor compounds for use in humidity sensors. II–VI semiconductor compounds, such as ZnS, ZnSe, CdS, and CdSe are stable materials with high photosensitivity and good luminescence characteristics. Therefore, these materials are highly useful in making the thin-film solar cells, various devices of optoelectronics, photoresistors, LEDs, and phosphores. Nanoparticles of these materials are also used as quantum dots (Chun et al. 2011). Some parameters of II–VI compounds, important for applications are listed in Table 5.4.

As for gas- and humidity-sensor applications, the main advantage of these materials is their surface properties. Research has shown that II–VI semiconductor compounds do not have the surface Fermi-level pinning (see Figure 5.10). It has been generally recognized that semiconductors can be

**TABLE 5.4**
**Parameters of Chalcogenides Used in Gas Sensors**

| Semiconductor | $E_g$ (eV) | $T_{melting}$ (C) | $T_{sublimation}$ (C) | $T_{oxidation}$ (C) |
|---|---|---|---|---|
| ZnS | 3.54–3.91 | 1,827–1,850 | 1,178 | <500 |
| ZnSe | 2.7–2.721 | 1,525 | 800–900 | <300 |
| ZnTe | 2.25–2.27 | 1,238–1,290 | ~600 | |
| CdS | 2.42–2.46 | 1,475–1750 | 980 | <450–500 |
| CdSe | 1.74–1.751 | 1,268 | 600–700 | |
| CdTe | 1.49–1.51 | 1,092 | ~500 | |

*Source:* With kind permission from Springer: *Conventional Approaches*, Springer, New York, 2013, Korotcenkov, G.

classified into two broad groups based on the properties of the semiconductor–metal interface. "Ionic" materials display little or no Fermi-level stabilization at the interface; "covalent" materials display virtually complete stabilization. Figure 5.10 illustrates this statement. The slope $d\Phi_{Bn}/dX_M$, which is inversely correlated to the extent of Fermi-level stabilization, was found to be 1 (no Fermi-level stabilization) for "ionic" materials and 0.1 (nearly complete Fermi-level stabilization) for "covalent" materials. As we can see, for II–VI compounds, the slope $S$ varies from 0.3 for CdTe and CdSe to 1.0 for ZnS. This means that the surface Fermi-level position, that is, the band bending in these semiconductors can be changed over a wide range, contributing to the achievement of high-sensor response due to interaction with the gas surrounding, including wet air. For comparison, for semiconductors such as Si, GaAs, and InP, discussed in previous Section 5.1, the surface Fermi level is firmly pinned within their band gap due to the high surface state density. That is what causes the low sensitivity of the humidity sensors based on these materials. For information, for all metal oxides S is equal to 1.0.

It was established that II–VI semiconductor compounds can be used as sensing materials in various types of humidity sensors, including chemiresistors (Leung et al. 2008; Uzar et al. 2011; Okur et al. 2012; Demir et al. 2012; Jiang et al. 2012; Du et al. 2014; Liang and Liu 2014; Chen et al. 2018), QCM (Demir et al. 2011; Uzar et al. 2011; Okur et al. 2012), and heterojunction based (Hsueh et al. 2014). In these devices, they can be applied in the form of nanocrystalline layers (Demir et al. 2011, 2012), NW arrays (Uzar et al. 2011; Jiang et al. 2012; Okur et al. 2012), and nanocomposites (Chen et al. 2018). Operating characteristics of several II–VI-based humidity sensors with the best parameters are shown in Figures 5.11 through 5.13.

It is important that the sensitivity of II–VI-based gas and humidity sensors of the conductometric type for various semiconductors changes according to the value of $S$ from Figure 5.10. For example, in the couple CdS–CdSe, CdS-based sensors had much better sensitivity to vapors of water, ethanol, ammonia, acetone, and iodine (Nesheva et al. 2006). We also need to note that all regularities established for metal oxides can be applied for II–VI-based gas and humidity sensors as well (Lantto and Golovanov 1995). This means that for better sensitivity, II–VI semiconductor compounds should be porous with small both the grain size and diameter of NWs (Xu et al. 2011). The thinner films have also shown higher sensitivity and shorter response time (Afify and Battisha 2000). The increase of operating temperature decreases response and recovery times (Afify and Battisha 2000). Using NWs also helps to reduce response and recovery times. Du et al. (2014) reported that their humidity sensors based on individual CdS NWs had response and recovery times ~1.3 and 2.4 s, respectively. One should note that these sensors had much faster response than humidity sensors developed when using other materials (Fu et al. 2007; Kuang et al. 2007; Chang et al. 2010). Surface modification by noble metals improves the sensitivity of II–VI-semiconductor-compound-based humidity sensors as well. In particular, Liang and Liu (2014) established that

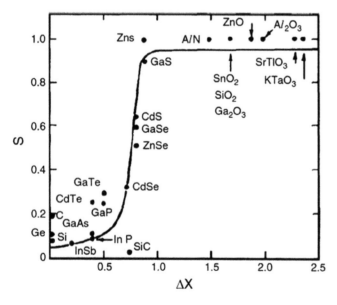

**FIGURE 5.10** Influence of electronegativity on the value of $S$ in $\Phi_{Bn} = S(\Delta X)$ dependence (Adapted with permission from Kurtin, S. et al., *Phys. Rev. Lett.* 22, 1433–1436, 1969, Copyright 1969, by the American Physical Society.)

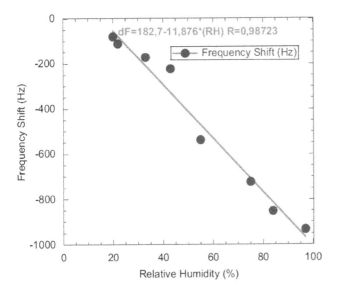

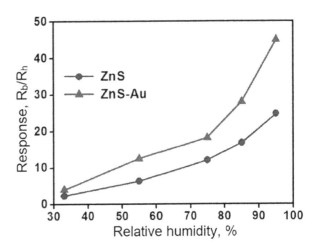

**FIGURE 5.11** The relationship between the frequency shifts of a ZnS-nanowire-coated QCM crystal and different relative humidities. ZnS nanowires with a diameter varied from about 60–120 nm were synthesized by evaporation of ZnS powder and transfer of ZnS vapor to lower-temperature regions using an inert gas as a carrier. The QCM sensor consisted of a AT-cut piezoelectric quartz crystal with resonance frequency of 8 MHz. ZnS nanowires were dispersed in ethanol using ultrasonic cleaner, and the solution was applied on the surface of a quartz crystal by the drop-casting technique. (Reprinted from *Physica E*, 44, Okur, S. et al., Synthesis and humidity sensing analysis of ZnS nanowires, 1103–1107, Copyright 2013, with permission from Elsevier.)

the ZnS-Au sensor exhibited considerably enhanced sensitivity compared with a pure ZnS nanocrystalline sphere sensor at various percent RH levels operated at room temperature.

However, in spite of the above results, one can assert that humidity sensors based on II–VI semiconductor compounds do not have perspectives in the humidity-sensor market. At first,

**FIGURE 5.13** Humidity-sensing responses of the ZnS and ZnS-Au sensors. ZnS spheres (50–60 nm) were synthesized using hydrothermal method. The Au ultrathin film was deposited onto the surfaces of the synthesized ZnS spheres using a DC sputtering system. ZnS nanoparticles formed nanospheres had a diameter of approximately 10–20 nm. (Reprinted from *Nanoscale Res. Lett.*, Synthesis and enhanced humidity detection response of nanoscale Au-particle-decorated ZnS spheres, 9, 647, 2014, Liang, Y.-C., and Liu, S.-L., Published by Springer as open access.)

they cannot replace metal-oxide gas sensors because they are sensitive to the same gases and vapors, have the same low selectivity, and the application of these materials does not give an improvement of operating characteristics in comparison with metal-oxide-based sensors. As a rule, the parameters of II–VI-based gas sensors are considerably worse. Second, the parameters of II–VI semiconductor compounds are highly dependent on lighting and temperature. In addition, despite high-melting temperatures, sublimation and especially irreversible oxidation occur at significantly lower temperatures (see Table 5.4) (Dimitrov and Boyanov 2000; Dimitrov et al.

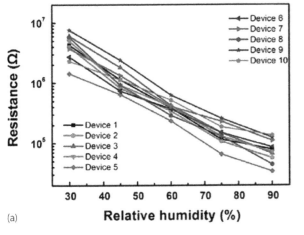

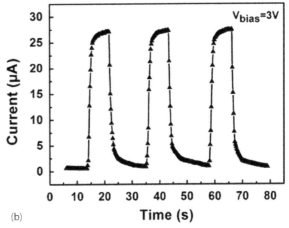

**FIGURE 5.12** (a) Relative-humidity influence on the resistances of sensors based on the individual CdS nanowire for ten devices; (b) current response of a CdS NW sensor to dynamic switches between dry air (5%) and moist air of 75% RH at $V_{bias}$ = 3 V. The CdS NWs were synthesized via a CVD method in a Cd-enriched atmosphere. The CdS NWs had the diameter in the range of 50–150 nm and the length of typically tens of micrometers. (Reprinted from *Mater. Lett.* 129, Du, L., et al., Synthesis of high-quality CdS nanowires and their application as humidity sensors, 46–49, Copyright 2014, with permission from Elsevier.)

2002). As a result, operating temperatures of II–VI-based gas and humidity sensors do not exceed 300°C–350°C for the most stable compounds ZnS and CdS. This means that II–VI semiconductor compounds cannot resolve the problems of selectivity and temporal instability of metal-oxide sensors. Too high resistance of the ZnS and CdS also creates difficulties in the development of conductometric humidity sensors (Fang et al. 2011).

## 5.4 DICHALCOGENIDES

### 5.4.1 FEATURES AND SYNTHESIS

Transition metal dichalcogenides (TMDCs), a class of 2D materials with the formula $MX_2$, where X is a chalcogen (S, Se, or Te), are layered materials with strong in-plane bonding and weak out-of-plane interactions, enabling exfoliation into 2D layers of a single-unit-cell thickness. Among the three chalcogen elements, sulfur is the most favorable atom since it is the most Earth-abundant among them and most of the $MS_2$ (transition metal disulfide) compounds are less toxic and more stable in air than $MSe_2$ and $MTe_2$. In these materials, the plane of metal atoms is sandwiched between two hexagonal planes of chalcogen atoms. In general, the thickness of each layer is about 0.6–0.7 nm (Andoshe et al. 2015). The layers are connected by weak van der Waals forces as the M-X bonds within layers are typically covalent. Although TMDCs have been studied for decades, recent advances in nanoscale materials characterization and device fabrication have opened up new opportunities for 2D layers of thin

TMDCs in nanoelectronics, optoelectronics (Wang et al. 2012a; Butler et al. 2013), and sensors applications (Varghese et al. 2015; Kim et al. 2017).

The experiment and simulations have shown that there are many interesting layer-dependent properties in 2D materials, which differ greatly from the properties of the bulk materials (Table 5.5). In particular, these family of materials have good optical transparency, excellent mechanical flexibility, great mechanical strength, and also peculiar electrical properties. For example, TMDCs such as $MoS_2$, $MoSe_2$, $WS_2$, and $WSe_2$ have sizable band gaps that change from indirect to direct in single layers. In particular, for $MoS_2$, the bulk indirect band gap of 1.3 eV increases to a direct band gap of 1.8 eV in a single-layer form (Kuc et al. 2011). The electronic structure of $MoS_2$ also enables valley polarization, which is not seen in bilayer $MoS_2$ (Mak et al. 2012). These unique properties allow these compounds to be used in the design of transistors, photodetectors, and electroluminescent devices. Due to the high surface area and the large number of reactive sites, 2D materials also provide a promising platform for the development of ultrahigh sensitive gas and humidity sensors (Varghese et al. 2015; Kim et al. 2017).

Reliable production of atomically thin, 2D TMDCs with uniform properties is essential for translating their new electronic and optical properties into applications. The review presented by Wang et al. (2012a) showed that for these purposes one can use various top-down and bottom-up methods (Figure 5.14). For example, atomically thin flakes of TMDCs can be peeled from their parent bulk crystals by micromechanical cleavage using adhesive tape (Mak et al. 2010),

## TABLE 5.5
## Summary of TMDC Materials and Properties

| Metal | S | | Se | | Te | |
|---|---|---|---|---|---|---|
| | Electronic Characteristics | References | Electronic Characteristics | References | Electronic Characteristics | References |
| Nb | Metal; superconducting; CDW | Beal et al. (1975) (E) | Metal; superconducting; CDW | Beal et al. (1975); Wilson et al. (1975) (E) | Metal | Ding et al. (2011) (T) |
| Ta | Metal; superconducting; CDW | Beal et al. (1975); Wilson et al. (1975) (E) | Metal; superconducting; CDW | Beal et al. (1975); Wilson et al. (1975) (E) | Metal | Ding et al. (2011) (T) |
| Mo | Semiconducting 1L: 1.8 eV | Mak et al. (2010) (E) | Semiconducting 1L: 1.5 eV | Liu et al. (2011) (T); | Semiconducting 1L: 1.1 eV | Liu et al. (2011) (T) |
| | Bulk: 1.2 eV | Kam and Parkinson 1982 (E) | Bulk: 1.1 eV | Kam and Parkinson (1982) (E) | Bulk: 1.0 eV | Gmelin (1994) (E) |
| W | Semiconducting 1L: 2.1 eV 1L: 1.9 eV | Kuc et al. 2011 (T) Liu et al. 2011 (T) | Semiconducting 1L: 1.7 eV | Ding et al. (2011) (T); | Semiconducting 1L: 1.1 eV | Ding et al. (2011) (T) |
| | Bulk: 1.4 eV | Kam and Parkinson (1982) (E) | Bulk: 1.2 eV | Kam and Parkinson (1982) (E) | | |

*Source:* Reprinted by permission from Macmillan Publishers Ltd. *Nat. Nanotechnol.*, Wang, Q.H. et al., 2012, copyright 2007.

The electronic characteristic of each material is listed as metallic, superconducting, semiconducting, or charge-density wave (CDW). For the semiconducting materials, the band-gap energies for monolayer (1L) and bulk forms are listed. The cited references are indicated as experimental (E) or theoretical (T) results.

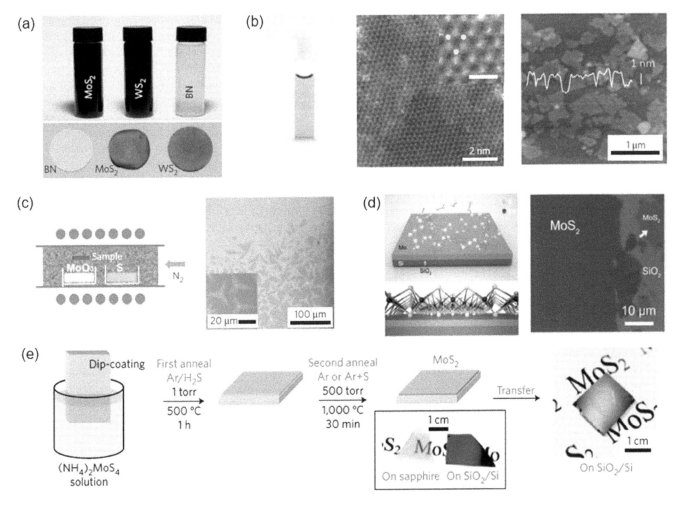

**FIGURE 5.14** Methods for synthesizing TMDC layers: (a) Stable suspensions of layered materials from liquid-phase exfoliation in solvents (top), and thin films derived from vacuum filtration of these suspensions (bottom); (b) suspension of lithium-intercalated and exfoliated $MoS_2$ in water (left). High-angle annular dark-field scanning transmission electron microscopy (HAADF-STEM) image of $MoS_2$ (middle) from the suspension in the left panel and atomic force microscopy (AFM) image of flakes of $MoS_2$ deposited on $SiO_2$ (right). The white line is a height profile taken at the position of the red line. The inset in the middle panel is a magnified view of the HAADF-STEM image, showing a honeycomb arrangement of $MoS_2$. Green and yellow dots represent Mo and S, respectively. Scale bar, 0.5 nm; (c) schematic of CVD of $MoS_2$ from solid S and $MoO_3$ precursors (left) and resulting $MoS_2$ films on $SiO_2$ (right). The red dots indicate the heating elements in the furnace. In this optical microscopy image, the lighter regions are $MoS_2$ and the darker regions are $SiO_2$; (d) CVD growth of $MoS_2$ from a solid layer of Mo on $SiO_2$ exposed to S vapor (top left), resulting in $MoS_2$ layers that are visible in optical microscopy (right). Bottom left: side-view schematic of an $MoS_2$ layer on the $Si/SiO_2$ substrate; and (e) CVD growth of $MoS_2$ from a dip-coated precursor on the substrate and growth in the presence of Ar gas and S vapor. (Reprinted by permission from Macmillan Publishers Ltd. *Nat. Nanotechnol.*, Wang, Q.H. et al., 2012, copyright 2007.)

applied to substrates and optically identified by the light interference (Benameur et al. 2011). Alternatively, TMDCs can be exfoliated by ultrasonication in appropriate liquids, including organic solvents, aqueous surfactant solutions, or solutions of polymers in solvents (Coleman et al. 2011).

As for bottom-up methods, the most promising is chemical vapor deposition (CVD) method. For example, CVD methods for growing atomically thin films of $MoS_2$ on insulating substrates use different solid precursors heated to high temperatures: sulfur powder and $MoO_3$ powder vaporized and co-deposited onto a nearby substrate (Lee et al 2012.); a thin layer of Mo metal deposited onto a wafer heated with solid sulfur (Zhan et al. 2012); and substrates dip-coated in a solution of

$(NH_4)_2MoS_4$ and heated in the presence of sulfur gas (Liu et al. 2012). Chemical preparation of TMDCs have also been demonstrated using hydrothermal synthesis (that is, growth of single crystals from an aqueous solution in an autoclave at high temperature and pressure (Peng et al. 2001).

### 5.4.2 DICHALCOGENIDE-BASED HUMIDITY SENSORS

When developing humidity sensors, various dichalcogenides such as $VS_2$ (Feng et al. 2012), $NbS_2$ (Divigalpitiya et al. 1990), $MoS_2$ (Zhang et al. 2014; Burman et al. 2016a, 2016b; Lei et al. 2016; Li et al. 2017a, 2017b; Lu et al. 2017; Zhao et al. 2017; Shin et al. 2018), $MoSe_2$ (Late et al. 2014), and $WS_2$

(Hou et al. 2014a; Jha and Guha 2016; Pawbake et al. 2016; Comini et al. 2017; Guan et al. 2017; Jha et al. 2017; Guo et al. 2017) were tested. However, the number of publications shows that the greatest use was found for $MoS_2$ and $WS_2$. According to Huo et al. (2014a), O'Brien et al. (2014), and Zhou et al. (2015), $WS_2$ possesses several advantages such as higher thermal stability, wider operation temperature range, and favorable band structure as compared to $MoS_2$.

As a rule, humidity sensors were made on the basis of exfoliated nanosheets of dichalcogenide materials (Zhang et al. 2014; Jha and Guha 2016; Comini et al. 2017), prepared using chemical ion intercalation and ultrasound-assistant liquid exfoliation methods (Zhou et al. 2011). Compared to the time-consuming and tough environment required for the ion-intercalation method, the ultrasound-assistant liquid method possesses advantages, such as low cost, ease of process, and potential for scale-up. The mechanical-cleavage method (manual mechanical exfoliation), often used in the initial stages of the study of graphene, can also be used for dichalcogenides, but this method is unsuited to mass production, and considerably decreases the reproducibility of the electronic device. Nanosheet solutions were then dropcast onto the sensor substrate, usually alumina or $SiO_2/Si$, with pre-patterned interdigitated electrodes (Donarelli et al. 2015). Nanoparticles synthesized by the CVD method are also used for humidity-sensor fabrication (Pawbake et al. 2016). At that, Pawbake et al. (2016) believe that the hotwire chemical vapor deposition (HW-CVD) and CVD are preferred methods for forming large-area, uniform TMDCs sheets and nanostructures for a wide range of device applications from the perspective of the repeatability and controllability. To date, humidity sensors of chemiresistive (Donarelli et al. 2015; Burman et al. 2016a), capacitive (Lei et al. 2016; Li et al. 2017a), impedance (Lu et al. 2017), FET (Late et al. 2012, Zhou et al. 2015; Zhao et al. 2017; Shin et al. 2018), and fiber-optic (Luo et al. 2016; Guan et al. 2017; Li et al. 2017b) types were developed on the basis of dichalcogenides. It should be noted that based on the parameters reported in published articles, that is, high sensitivity and quick response (see Table 5.6), dichalcogenides are very promising materials for humidity-sensor applications.

Besides dichalcogenides, humidity-sensitive materials have been also used the composites on their base, such as $MoS_2$/ graphene oxide (Burman et al. 2016a), $MoS_2@SnO_2$ (Lei et al. 2016), $MoS_2$/Ag (Li et al. 2017a, 2017b), $MoS_2$/ZnO (Lu et al. 2017), $WS_2$/GO (Comini et al. 2017; Jha et al. 2017). As a rule, the use of nanocomposites of optimal composition gave an improvement in the parameters of the sensors (see, for example, Figure 5.15 and Table 5.7). Regarding the mechanisms for ongoing improvement, in many cases they are debatable and more research is needed to determine them.

As for most solid-state humidity sensors, the sensing mechanism of 2D transition-metal-disulfide humidity sensors is based on grain boundary effects and charge transfer processes taking place due to water-vapor adsorption on the surface of dichalcogenides (read Chapter 2). It is believed that similarly to metal oxides, two conduction mechanisms take place simultaneously in dichalcogenides in humid atmosphere—one is the electron conduction through the $MoS_2$- or $WS_2$-sensing channel (water acts as an electron donor); and the second is the proton hopping through continuous water layers by the well-known Grotthuss mechanism (Chen and Lu 2005; Zhang et al. 2014; Burman et al. 2016a).

Yue et al. (2013) analyzed the most stable adsorption position, orientation, associated charge transfer, and the modification of electronic properties of the monolayer $MoS_2$ surface due to the adsorption of $H_2$, $O_2$, $H_2O$, $NH_3$, $NO$, $NO_2$, and $CO$ using first-principles calculation based on density functional theory (DFT). They found that all these molecules, acting as either electron donors or acceptors are only physisorbed on the $MoS_2$ surface, with small charge transfer and no significant alteration of the band structure upon molecule adsorption. They also observed significant modulation of the charge transfer by the application of a perpendicular electric field (Yue et al. 2013). Similarly, first-principles simulation has also been employed by Zhao et al. (2014) to investigate the adsorption of various gas molecules, such as $CO$, $CO_2$, $NH_3$, $NO$, $NO_2$, $CH_4$, $H_2O$, $N_2$, $O_2$, and $SO_2$ on monolayer $MoS_2$ by including van der Waals interactions between the gas molecules and $MoS_2$. They found that only $NO$, $NO_2$, and $SO_2$ could bind strongly to $MoS_2$ surface compared to other gas molecules, which was found to be in good agreement with experimental observations.

## TABLE 5.6

### Comparison in Response/Recovery Times Toward Various $MoS_2$- and $WS_2$-Based Humidity Sensors

| Type | Fabrication | Meas. Range | Response Time | Recovery Time | References |
|---|---|---|---|---|---|
| $MoS_2$/ZnO | Hydrothermal/chemical | 11%–95% RH | 1 s | 20 s | Lu et al. (2017) |
| $MoS_2$/$SnO_2$ | Hydrothermal | 0%–97% RH | 5 s | 13 s | Zhang et al. (2016) |
| $MoS_2$/rGO | Exfoliation | 25%–65% RH | 43 s | 37 s | Burman et al. (2016a) |
| $MoS_2$ | Exfoliation | 10%–60% RH | 9 s | 17 s | Zhang et al. (2014) |
| $MoS_2$ | Hydrothermal | 17%–89% RH | 140 s | 80 s | Tan et al. (2014) |
| $MoS_2$ | Exfoliation | 0%–40% RH | ~10 s | ~60 s | Zhao et al. (2017) |
| $WS_2$ | Exfoliation | 40%–80% RH | 13 s | 17 s | Jha and Guha (2016) |
| $WS_2$ | HW-CVD | 11%–97% RH | ~12 s | ~13 s | Pawbake et al. (2016) |
| $WS_2$/GO | Exfoliation | 40%–80% RH | ~25 s | ~29 s | Jha et al. (2017) |

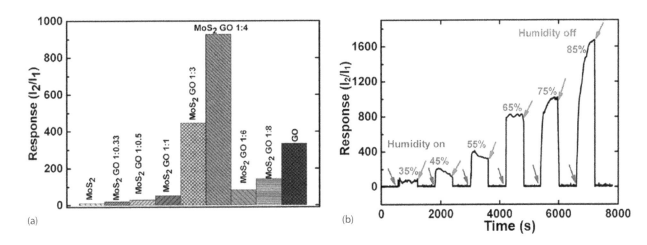

**FIGURE 5.15** (a) The comparative plot of response of different samples of MoS₂/graphene oxide nanocomposites at 65% RH; and (b) response of MoS₂/GO 1:4 sample for six different humidity levels. (Burman, D. et al., *RSC Adv.*, 6, 57424–57433, 2016, Reproduced by permission of RSC.)

**TABLE 5.7**

**Performances of MoS₂-Based Capacitance Humidity Sensors**

| Material | Meas. Range | Sensitivity | Response Time | References |
|---|---|---|---|---|
| $MoS_2$ | 17%–90% RH | 80.9 pF/%RH | 140 s | Tan et al. (2014) |
| $MoS_2$/Ag | 11%–97% RH | 21112 pF/%RH | ~1.5 s | Li et al. (2017) |

The same results were obtained for $WS_2$. The interactions of $NH_3$ and $H_2O$ molecules with monolayer $WS_2$ investigated by means of first-principles calculations indicated that both $NH_3$ and $H_2O$ molecules are physisorbed on monolayer $WS_2$ (Zhou et al. 2015). The results from Bader-charge analysis and plane-averaged differential charge density showed that $NH_3$ and $H_2O$ act as electron donor and acceptor, respectively, leading to *n*- and *p*-type doping, respectively. The charge transfer between the gas molecules and single-layer $WS_2$ was primarily determined from the mixing of the highest-occupied and lowest-unoccupied molecular orbital with the underlying $WS_2$ orbitals. They also provided a detailed explanation about the sensing mechanism of the $WS_2$-FET-based sensor toward $H_2O$. They found that in the case of $H_2O$ adsorption on monolayer WS2, the source-drain current of $WS_2$-FET-based gas sensors got suppressed by the electron trapping of the $H_2O$ molecule from the $WS_2$ channel (Zhou et al. 2015).

Since atomically 2D transition metal disulfides are transparent and flexible, Varghese et al. (2015) and Yang et al. (2016) believe that the application of 2D-transition-metal-disulfide-based sensors can find new areas that were not reached by the conventional chemoresistive humidity sensors. In addition, developing 3D nanostructures based on 2D transition-metal disulfides is promising to increase sensor response. It is important to note that compared to the previously reported metal-oxide sensors such as $SnO_2$, $MoS_2$ humidity sensors exhibit a faster response (Zhang et al. 2014; Lei et al. 2016; Li et al. 2017a, 2017b). As it is seen in Figure 5.16, response and recovery times can be smaller that 10–20 s. Li et al. (2017a, 2017b) reported that their sensors had response time smaller than 2 s. The excellent response speed of the $MoS_2$ sensor to humidity gas is attributed to its specific-surface property, in particular intrinsic hydrophobic property (Tahir et al. 2006), which accelerates the desorption process of water molecules on the surface, and weaker interaction between water molecule and $MoS_2$ surface (Lei et al. 2016).

As for the disadvantages of sensors based on 2D materials, they primarily include technological difficulties in manufacturing, low reproducibility of parameters, due to the peculiarities of the sensor-manufacturing technology, poor adhesion of the formed layers, and the effect of layer thickness on the electrophysical properties of dichalcogenides. As it is seen in Figures 5.16 and 5.17, the sensor response is highly dependent on exploitation conditions: the responses increased as the ultrasonic power and the time of exfoliation increased (Burman et al. 2016a; Jha and Guha 2016). Such sensors also perform low selectivity. Like the others solid-state humidity sensors, dichalcogenide-based sensors are sensitive not only to water vapor but also to reducing and oxidizing gases and organic solvent vapors (Donarelli et al. 2015; Varghese et al. 2015; Yang et al. 2016). It was established that under practical environmental conditions that includes oxygen, the properties of $MoS_2$ and $WS_2$ get strongly affected by the adsorption of oxygen, which leads to strong cross-sensitivity effects.

Improving the stability of sensor parameters is also a task to be addressed. As a rule, a noticeable change in the sensitivity of the sensors still occurs during operation (Figure 5.18).

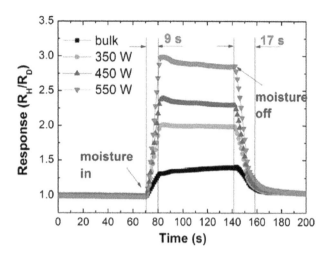

**FIGURE 5.16** Single-cycle response of the bulk and exfoliated MoS$_2$-based thin-film sensors to humidity gas with RH value about 60% in dependence on the ultrasonic powers used for exfoliation. (Reprinted from *Curr. Appl. Phys.*, 14, Zhang, S.-L. et al., Controlled exfoliation of molybdenum disulfide for developing thin film humidity sensor, 264–268, Copyright 2014, with permission from Elsevier.)

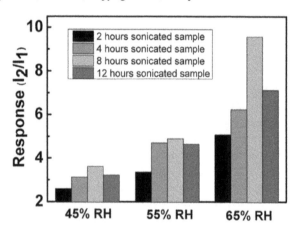

**FIGURE 5.17** Comparative response values of 2 hours, 4 hours, 8 hours, and 12 hours of sonicated MoS$_2$ samples to air humidity. (Burman, D. et al., *RSC Adv.*, 6, 57424–57433, 2016, Reproduced by permission of RSC.)

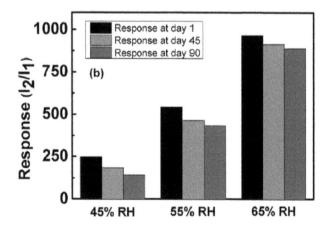

**FIGURE 5.18** Stability test of MoS$_2$/GO (1:4)-based humidity sensors. (Burman, D. et al., *RSC Adv.*, 6, 57424–57433, 2016, Reproduced by permission of RSC.)

## 5.5 SILICON CARBIDE (SiC)

Silicon carbide (SiC) is another wide-gap semiconductor that was considered promising for sensing applications (Wright and Horsfall 2007). Several selected parameters of SiC in comparison with Si and GaAs are presented in Table 5.1. One should note that SiC is a fairly well-studied material (Wright and Horsfall 2007). In fact, SiC exists in a large number of polytypes—different crystal structures built from the same Si–C subunit organized into a variety of stacking sequences. There are over a hundred of these polytypes known, but the preponderance of research and development has concentrated on only three: 3C, 6H, and 4H. One should note that all polytypes have identical planar arrangement. The difference lies in the stacking of the planes.

Among the previously noted polytypes, the 4H polytype is the most common for electronic devices due to its overall superior material properties, the suitability for which drives the investment in this area. The band gap of 4H SiC is 3.23 eV at room temperature (compared with 1.12 eV for silicon). The 3C–SiC polytype is more common for microelectromechanical (MEM)-based sensors due to the fact that it may be grown by chemical vapor deposition (CVD) or plasma enhanced CVD (PECVD) methods on standard Si wafers (thus reducing overall wafer cost compared with pure SiC technology). In addition, all forms of SiC have high radiation tolerance, a high thermal conductivity (better than copper), high hardness and Young's modulus (typically ~400 GPa compared with ~130 GPa for silicon), and a high critical electric field (in excess of 2 MV/cm) (Wright and Horsfall 2007). Due to the wide-band gap mentioned above, SiC is very thermally stable, and electronic devices can be operated at high temperatures (see Figure 5.19). This means that SiC-based devices can have a wider operating temperature range. Much better chemical, thermal, and mechanical stability in comparison with silicon

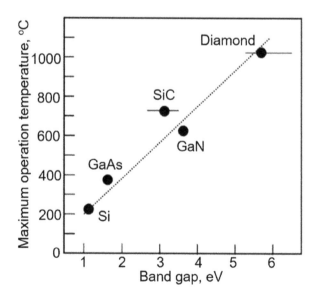

**FIGURE 5.19** Influence of the band gap on the maximum operating temperature of semiconductor devices. Data were extracted from Bogue (2002a, 2002b.)

make this semiconductor material more acceptable for the development of specific gas sensors aimed for operation in harsh environments. Immunity to corrosion makes SiC very suitable for operation in chemically harsh environments. Besides, the wide-band gap of SiC minimizes the generation of charge carriers due to undesirable background optical or thermal excitation. This is very important for gas-sensor operation at high temperature.

However, it should be noted that considered earlier in Section 5.2.1 GaN has the same advantages as SiC. Moreover, GaN has a unique advantage. A GaN gas or humidity sensor can be integrated with GaN-based optical devices or high-power, high-temperature electronic devices on the same chip. Another advantage of nitride-based semiconductor devices according to Irokawa (2011) is utilization of the AlGaN/GaN heterostructure. In an AlGaN/GaN heterostructure, the polarization-induced 2DEG concentration at the AlGaN/GaN interface is extremely sensitive to surface states. Any potential changes on the surface by adsorption of gas or vapor molecules would affect the surface potential and modulate the 2DEG density. Moreover, Irokawa (2011) believes that GaN has advantages in technology as well. First, GaN-based sensors, similarly to SiC, can be fabricated on the silicon substrate that contributes to a lower production costs. Second, the other promising wide-band gap semiconductor, SiC, has several technical drawbacks for a SiC device fabrication; the processing, particularly of SiC FETs (field-effect transistors), is inherently complicated, requiring high-temperature implantation and very high-temperature post-implantation annealing steps, leading to higher cost.

### 5.5.1 SiC-Based Humidity Sensors

The utilization of SiC-based devices as humidity sensors was started by Connolly et al. (2002, 2004a, 2004b) with the aim to develop a sensor able to operate in harsh chemical environments.

These studies have shown that membrane humidity sensors using porous SiC as the sensing element can be made.

Electrochemical etching was used to form the porous structure in 73% HF. Thin films of (p-type) SiC were deposited on standard p-type Si wafers, using PECVD, and doped with boron in situ. The thickness of the films was ~500 nm. Due to the very low etch rate of Al in 73% HF, it was able to use Al electrodes instead of Au, making the fabrication process of our sensors more cleanroom friendly. Testing of developed SiC-based capacitive humidity sensors has shown that the sensor response of these devices to the air-humidity change in the range of 10%–90% RH can reach 200%. Response times for the sensor were typically 2–3 min for a step change in RH of 10%–90%, and ~4 min for a step change in RH from 90% to 10%. It is seen that the parameters of developed SiC-based humidity sensors are significantly worse than the parameters of the sensors discussed earlier. In addition, the sensors had a large hysteresis, and moreover, this hysteresis got worse with the aging and burn-in treatments. However, at the same time, it was shown that SiC-based humidity sensors showed high resistance to the harsh environment effect—the outlet of a car exhaust and an ammonia atmosphere during several days.

More recent studies have shown that SiC-based humidity sensors can be significantly improved. In particular, Wang and Li (2010) developed carbonization technology of silicon nanoporous pillar array (Si-NPA). As a result, it was obtained the SiC/Si-NPA structure, which was used for elaboration a capacitance-humidity sensor. The capacitance–RH curves of SiC/Si-NPA measured with different signal frequencies of 20, 100, 1000, and 10,000 Hz are shown in Figure 5.20a. It is seen that at a testing frequency of 20 Hz and with the RH changed from 11% to 95%, the increase in the capacitance reaches over 1900%. The response and recovery times of the SiC/Si-NPA sensor were measured by switching over the environment humidity between RH = 11% and 95% under the frequency of 20 Hz, and the experimental curve is depicted in Figure 5.20b. It was shown that about 10 s were needed for the capacitance variation to reach 90% of the total variation during the vapor adsorption and desorption processes, respectively. Compared with the response and recovery times of 15 s

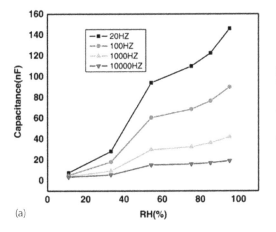
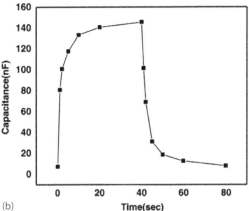

**FIGURE 5.20** (a) Capacitance–RH curve; and (b) room-temperature response and recovery times of the SiC/Si-NPA sensor. (Reprinted from *Mater. Lett.*, 64, Wang, H.Y. and Li X.J. Capacitive humidity-sensitivity of carbonized silicon nanoporous pillar array. 1268–1270, Copyright 2010, with permission from Elsevier.)

and 5 s of Si- NPA (Xu et al. 2005), the recovery rate of the SiC/Si-NPA sensor was a little slower maybe because of the decrease of the diameters of the microspores on the surface with the formation of SiC during the carbonization process. In contrast to the response and recovery times over 2 min reported for SiC-based sensors (Connolly et al. 2002), both the vapor adsorption and desorption processes of SiC/Si-NPA are rather rapid just lied on its pillar array and microspores structure, which might be an obvious technical predominance for fabricating practical SiC humidity sensors.

Further Wang et al. (2012b) also made an attempt to grow SiC NWs on the Si-NPA structures and thereby improve the parameters of the sensors. However, this approach did not give any improvement in sensor parameters compared to the approach described in (Wang and Li 2010). With the RH changing from 11% to 95%, the increase in capacitance was 960% at the measuring frequency of 100 Hz. The response and recovery times were measured to be ~105 s and 85 s, respectively, with a maximum humidity hysteresis of 4.5% at 75% RH. At the same time, testing for long-term stability has shown that compared with the capacitance–RH curve for freshly prepared samples, the curves measured after 6-month and 12-month storage remained almost unchanged. The capacitance zero drifts for the aged sensors were only ~0.7% and 0.9%, which were small enough to be neglected for practical detections. Considering the highly active surface of Si-NPA, which could easily be oxidized at room temperature even during its preparing process (Xu and Li 2008), the high stability of nw-SiC/Si-NPA sensors should come from the simultaneously occurred carbonization of the surface of Si-NPA during the high-temperature growing process of nw-SiC on the Si-NPA. It is the chemical inertness of SiC that facilitates nw-SiC/Si-NPA humidity sensor the high and long-term stability.

As for resistive humidity sensors based on SiC developed by Boukezzata et al. (2013, 2016) and Li et al. (2014a), then the results of their testing showed that the sensors have mediocre parameters in comparison with sensors developed on the basis of metal oxides and polymers. The change in the resistance of the SiC-based sensors developed by Boukezzata et al. (2013) is shown in Figure 5.21. The response and recovery times were found about 13 s and 20 min for RH varying from 5% to 86% and from 86% to 5%, respectively. The use of the SiC NWs for humidity-sensor fabrication (Li et al. 2014a) allowed significant improvement of the dynamics of the sensor response. Li et al. (2014a) reported that their sensors performed the response and recovery times ~5 and 26 s, correspondingly. However, the sensor response was small.

Regarding the mechanisms explaining the humidity-sensitive properties of SiC (Boukezzata et al. 2013, 2016; Li et al. 2014a), it should be noted that they are not different from the mechanisms used to interpret the results obtained on the basis of metal oxides and other solids (read Chapter 2).

As you can see, the use of SiC in the development of humidity sensors does not provide any improvement in comparison with the previously considered humidity sensors neither in sensitivity nor in selectivity. SiC-based sensors are not selective, as many other solid-state sensors (Bur et al.

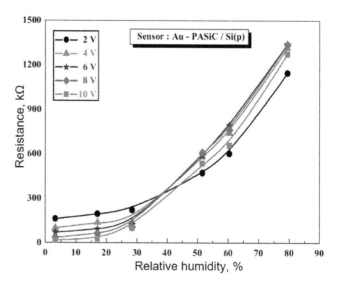

**FIGURE 5.21** Resistance versus RH of Au–PASiC/Si(p) sensor in dependence on bias voltage. The humidity-sensor structure was fabricated on an amorphous SiC thin films deposited on $p$-type silicon substrate with Au electrodes. Amorphous $\alpha$-$Si_{0.70}C_{0.30}$ films were prepared in DC magnetron co-sputtering system from 86 sprigs of a SiC (6HSiC) target. The thickness of the elaborated $\alpha$-$Si_{0.70}C_{0.30}$ films was 1.4 $\mu$m. Porous SiC was made by electrochemical etching in HF/ETG solution (1:1 by vol.) at an anodization current density (J) of 50 mA/cm$^2$ for 30 s. (Reprinted from *Sens. Actuators B*, 176, Boukezzata, A. et al., Investigation properties of Au–porous a-$Si_{0.70}C_{0.30}$ as humidity sensor, 1183–1190, Copyright 2013, Elsevier.)

2014). Therefore, it is difficult to expect such sensors to enter the market. Among the disadvantages of SiC-based humidity sensors, one can also point out a strong dependence of sensor parameters on the temperature. In addition, SiC still suffers from many technological difficulties that prevent near-term commercial application. For example, due to the high-etch resistance of SiC, wet-etching methods are not used generally in humidity-sensor fabrication. To improve reliability and control, dry-etching methods are preferred. For this purpose, the plasma-etch process using $SF_6/O_2$ can be used.

## REFERENCES

Afify H.H., Battisha I.K. (2000) Oxygen interaction with CdS based gas sensors by varying different preparation parameters. *Indian J. Pure Appl. Phys.* **38**(2), 119–126.

Andoshe D.M., Jeon J.-M., Kim S.Y., Jang H.W. (2015) Two-dimensional transition metal dichalcogenide nanomaterials for solar water splitting. *Electron. Mater. Lett.* **11**, 323–335.

Aoki H., Shima H., Kimura C., Sugino T. (2007) Characterization of boron carbon nitride film for humidity sensor. *Diamond Rel. Mater.* **16**, 1300–1303.

Ayari T., Bishop C., Jordan M.B., Sundaram S., Li X., Alam S., et al. (2017) Gas sensors boosted by two-dimensional h-BN enabled transfer on thin substrate foils: Towards wearable and portable applications. *Sci. Rep.* **7**, 15212.

Beal A.R., Hughes H.P., Liang W.Y. (1975) The reflectivity spectra of some group VA transition metal dichalcogenides. *J. Phys. C* **8**, 4236.

Benameur M.M., Radisavlievic B., Heron J.S., Sahoo S., Berger H., Kis A. (2011) Visibility of dichalcogenide nanolayers. *Nanotechnology* **22**, 125706.

Bogue R.W. (2002a) The role of materials in advanced sensor technology. *Sensor Rev.* **22**(4), 289–299.

Bogue R.W. (2002b) Advanced automotive sensors. *Sensor Rev.* **22**(2), 113–118.

Boukezzata A., Keffous A., Nezzal G., Gabouze N., Kechouane M., Zaafane K., et al. (2013) Investigation properties of Au–porous a-Si$_{0.70}$C$_{0.30}$ as humidity sensor. *Sens. Actuators B* **176**, 1183–1190.

Boukezzata A., Keffous A., Gabouze N., Guerbous L., Ouadjaout D., Menari H., Kechouane M. (2016) Surface modification of a-Si$_{0.60}$C$_{0.40}$:H films by Al-assisted photochemical etching: humidity sensing application. *Microsyst. Technol.* **22**, 935–941.

Bur C., Bastuck M., Schütze A., Juuti J., Lloyd Spetz A., Andersson M. (2014) Characterization of ash particles with a microheater and gas-sensitive SiC field-effect transistors. *J. Sens. Sens. Syst.* **3**, 305–313.

Burman D., Ghosh R., Santra S., Guha P.K. (2016a) Highly proton conducting MoS$_2$/graphene oxide nanocomposite based chemoresistive humidity sensor. *RSC Adv.* **6**, 57424–57433.

Burman D., Kumar Jha R., Santra S., Kumar Guha P. (2016b) Exfoliated MoS$_2$ based humidity sensing. *Adv. Mater. Proc.* **1**(2), 176–179.

Butler, S.Z., Hollen, S.M., Cao, L., Cui, Y., Gupta, J.A., Gutiérrez, H.R., et al. (2013) Progress, challenges, and opportunities in two-dimensional materials beyond graphene. *ACS Nano* **7**, 2898–2926.

Chang S.-P., Chang S.-J., Lu C.-Y., Li M.-J., Hsu C.-L., Chiou Y.-Z., et al. (2010) A ZnO nanowire-based humidity sensor. *Superlattices Microstruct.* **47**, 772–778.

Chen Q., Nie M., Guo Y. (2018) Controlled synthesis and humidity sensing properties of CdS/polyaniline composite based on CdAl layered double hydroxide. *Sens. Actuators B* **254**, 30–35.

Chen X.J., Zhang J., Wang Z.L., Yan Q., Hui S.C. (2011) Humidity sensing behavior of silicon nanowires with hexamethyldisilazane modification. *Sens. Actuators B* **156**, 631–636.

Chen Z., Lu C. (2005) Humidity sensors: A review of materials and mechanisms. *Sens. Lett.* **3**, 274–295.

Cheng D.D., Zheng M.J., Yao L.J., He S.H., Ma L., Shen W.Z., Kong X.Y. (2009) The fabrication and characteristics of indium-oxide covered porous InP. *Nanotechnology* **20**, 425302.

Chitara B., Late D.J., Krupanidhi S.B., Rao C.N.R. (2010) Room-temperature gas sensors based on gallium nitride nanoparticles. *Solid State Commun.* **150**, 2053–2056.

Chun J., Yang W., Kim J.S. (2011) Thermal stability of CdSe/ZnS quantum dot-based optical fiber temperature sensor. *Mol. Crystallogr. Liq. Crystallogr.* **538**(1), 333–340.

Coleman J. N., Lotya M., O'Neill A., Bergin S.D., King P.J., Khan U., et al. (2011) Two-dimensional nanosheets produced by liquid exfoliation of layered materials. *Science* **331**, 568–571.

Comini E., Sberveglieri G., Wisitsora-at A., Sofer Z., Mayorga Martinez C.C., Pumera M., Wlodarski W. (2017) GO/2D WS$_2$ based humidity sensor. *Proceedings* **1**, 469.

Connolly E.J., O'Halloran G.M., Pham H.T.M., Sarro P.M., French P.J. (2002) Comparison of porous silicon, porous polysilicon and porous silicon carbide as materials for humidity sensing applications. *Sens. Actuators A* **99**, 25–30.

Connolly E.J., Pham H.T.M., Groeneweg J., Sarro P.M., French P.J. (2004a) Relative humidity sensors using porous SiC membranes and Al electrodes. *Sens. Actuators B* **100**, 216–220.

Connolly E.J., Pham H.T.M., Groeneweg J., Sarro P.M., French P.J. (2004b) Silicon carbide membrane relative humidity sensor with aluminium electrodes. In: *Proceedings of IEEE Sensors Conference*, Maastricht, the Netherlands, pp. 193–196.

Demir R., Okur S., Seker M., Zor M. (2011) Humidity sensing properties of CdS nanoparticles synthesized by chemical bath deposition method. *Ind. Eng. Chem. Res.* 50, 5606–5610.

Demir R., Okur S., Seker M. (2012) Electrical characterization of CdS nanoparticles for humidity sensing applications. *Ind. Eng. Chem. Res.* **51**, 3309–3313.

Dimitrov R.I., Boyanov B.S. (2000) Oxidation of metal sulphides and determination of characteristic temperatures by DTA and TG. *J. Therm. Anal. Calorim.* **61**, 181–189.

Dimitrov R.I., Moldovanska N., Bonev I.K. (2002) Cadmium sulphide oxidation. *Thermochim. Acta* **385**, 41–49.

Ding Y., Wang Y., Ni J., Shi L., Shi S, Tang W. (2011) First principles study of structural, vibrational and electronic properties of graphene-like MX$_2$ (M = Mo, Nb, W, Ta; X = S, Se, Te) monolayers. *Physica B* **406**, 2254–2260.

Divigalpitiya W.M.R., Frindt R.F., Morrison S.R. (1990) Effect of humidity on spread NbS$_2$ films. *J. Phys. D: Appl. Phys.* **23**, 966–970.

Donarelli M., Prezioso S., Perrozzi F., Bisti F., Nardone M., Giancaterini L., et al. (2015) Response to NO$_2$ and other gases of resistive chemically exfoliated MoS$_2$-based gas sensors. *Sens. Actuators B* **207** Part A, 602–613.

Du L., Zhang Y., Lei Y., Zhao H. (2014) Synthesis of high-quality CdS nanowires and their application as humidity sensors. *Mater. Lett.* **129**, 46–49.

Fang X., Zhai T., Gautam U.K., Li L., Wua L., Bando Y., Golberg D. (2011) ZnS nanostructures: from synthesis to applications. *Prog. Mater. Sci.* **56**, 175–287.

Feng J., Peng L., Wu C., Sun X., Hu S., Lin C., et al. (2012) Giant moisture responsiveness of VS$_2$ ultrathin nanosheets for novel touchless positioning interface. *Adv. Mater.* **24**, 1969–1974.

Fu X., Wang C., Yu H., Wang Y., Wang T. (2007) Fast humidity sensors based on CeO$_2$ nanowires. *Nanotechnology* **18**, 145503–145506.

Gautam C., Tiwary C.S., Machado L.D., Jose S., Ozden S., Biradar S., et al. (2016) Synthesis and porous h-BN 3D architectures for effective humidity and gas sensors. *RSC Adv.* **6**, 87888.

Gmelin (1994) *Gmelin: Handbook of Inorganic and Organometallic Chemistry*, 8th edn, Vol. B7, Springer.

Goley P.S., Hudait M.K. (2014) Germanium based field-effect transistors: Challenges and opportunities. *Materials* **7**, 2301–2339.

Guan H., Xia K., Chen C., Luo., Tang J., Lu H., et al. (2017) Tungsten disulfide wrapped on micro fiber for enhanced humidity sensing. *Opt. Mater. Express* **7**(5), 1686–1696.

Guo H., Lan C., Zhou Z., Sun P., Wei D., Li C. (2017) Transparent, flexible, and stretchable WS$_2$ based humidity sensors for electronic skin. *Nanoscale* **9**, 6246.

Huo N., Yang S., Wei Z., Li S.-S., Xia J.-B., Li J. (2014a) Photoresponsive and gas sensing field-effect transistors based on multilayer WS$_2$ nanoflakes. *Sci. Rep.* **4**, 5209.

Hou X., Yu Z., Chou K.C. (2014b) Preparation and properties of hexagonal boron nitride fibers used as high temperature membrane filter. *MRS Bull.* **49**, 39–43.

Hsueh H.T., Hsueh T.J., Chang S.J., Hung F.Y., Weng W.Y., Hsu C.L., Dai B.T. (2011) Si nanowire-based humidity sensors prepared on glass substrate. *IEEE Sens. J.* **11**, 3036–3041.

Hsueh H.-T., Hsiao Y.-J., Lin Y.-D., Wu C.-L. (2014) Bifacial structures of ZnS humidity sensor and Cd-free CIGS photovoltaic cell as a self-powered device. *IEEE Electron. Dev. Lett.* **35**(12), 1272–1274.

Irokawa Y. (2011) Hydrogen sensors using nitride-based semiconductor diodes: The role of metal/semiconductor interfaces. *Sensors* **11**, 674–695.

Jiang P., Jie J., Yu Y., Wang Z., Xie C., Zhang X., et al. (2012) Aluminium-doped n-type ZnS nanowires as high-performance UV and humidity sensors. *J. Mater. Chem.* **22**, 6856.

Jiang X.-F., Weng Q., Wang X.-B., Li X., Zhang J., Golberg D., Bando Y. (2015) Recent progress on fabrications and applications of boron nitride nanomaterials: A review. *J. Mater. Sci. Technol.* **31**, 589–598.

Jha R.K. Burman D., Santra S., Guha P.R. (2017) WS$_2$/GO nanohybrids for enhanced relative humidity sensing at room temperature. *IEEE Sensors J.* **17**(22), 7340–7347.

Jha R.K, Guha P.K. (2016) Liquid exfoliated pristine WS$_2$ nanosheets for ultrasensitive and highly stable chemiresistive humidity sensors. *Nanotechnology* **27**, 475503.

Kam K.K., Parkinson B.A. (1982) Detailed photocurrent spectroscopy of the semiconducting group-VI transition-metal dichalcogenides. *J. Phys. Chem.* **86**, 463–467.

Kerlau M., Merdrignac-Conanec O., Reichel P., Barsan N., U. Weimar U. (2006) Preparation and characterization of gallium (oxy)nitride powders. Preliminary investigation as new gas sensor materials. *Sens. Actuators B* **115**, 4–11.

Kim J.H., Pham T.V., Hwang J.H., Kim C.S., Kim M.J. (2018) Boron nitride nanotubes: Synthesis and applications. *Nano Convergence* **5**, 17.

Kim T.H., Kim Y.H., Park S.Y., Kim S.Y., Jang H.W. (2017) Two-dimensional transition metal disulfides for chemoresistive gas sensing: Perspective and challenges. *Chemosensors* **5**, 15.

Korotcenkov G. (2013) *Handbook of Gas Sensor Materials*, Vol. 1: *Conventional Approaches.* Springer, New York.

Kostoglou N., Polychronopoulou K., Rebholz C. (2015) Thermal and chemical stability of hexagonal boron nitride (h-BN) nanoplatelets. *Vacuum* **112**, 42–45.

Kurtin S., McGill T.C., Mead C.A. (1969) Fundamental transition in the electronic nature of solids. *Phys. Rev. Lett.* **22**, 1433–1436.

Kuang Q., Lao C., Wang Z.L. Xie Z., Zheng L. (2007) High-sensitivity humidity sensor based on a single SnO$_2$ nanowire. *J. Am. Chem. Soc.* **129**, 6070–6071.

Kuc A., Zibouche N., Heine T. (2011) Influence of quantum confinement on the electronic structure of the transition metal sulfide TS$_2$. *Phys. Rev. B* **83**, 245213.

Lantto V., Golovanov V. (1995) A comparison of conductance behaviour between SnO$_2$ and CdS gas-sensitive films. *Sens. Actuators B* **24–25**, 614–618.

Late D.J., Doneux T., Bougouma M. (2014) Single-layer MoSe$_2$ based NH$_3$ gas sensor. *Appl. Phys. Lett.* **105**, 233103

Late D.J., Liu B., Matte H.S.S.R., Dravid V.P., Rao C.N.R. (2012) Hysteresis in single-layer MoS$_2$ field effect transistors. *ACS Nano* **6**, 5635–5641.

Lee D.-S., Lee J.-H., Lee Y.-H., Lee D.-D. (2003) GaN thin films as gas sensors. *Sens. Actuators B* **89**, 305–310.

Lee Y.-H., Z X.-Q., Zhang W., Chang M.-T., Lin C.-T., Chang K.-D., et al. (2012) Synthesis of large-area MoS$_2$ atomic layers with chemical vapor deposition. *Adv. Mater.* **24**, 2320–2325.

Lei X., Yu K., Li H., Tang Z., Guo B., Li J., et al. (2016) First-principle and experiment investigation of MoS$_2$@SnO$_2$ nanoheterogeneous structures with enhanced humidity sensing performance. *J. Appl. Phys.* **119**, 154303.

Leung Y.P., Choy W.C.H., Yuk T.I. (2008) Linearly resistive humidity sensor based on quasi one-dimensional ZnSe nanostructures. *Chem. Phys. Lett.* **457**, 198–201.

Li D., Lu H., Qiu W., Dong J., Guan H., Zhu W., et al. (2017b) Molybdenum disulfide nanosheets deposited on polished optical fiber for humidity sensing and human breath monitoring. *Opt. Express* **25**(23), 28407–28416.

Li G.-Y., Ma J., Peng, G. Chen W., Chu Z.-Y., Li Y.-H., Hu T.-J., Li X.-D. (2014a) Room-temperature humidity-sensing performance of SiC nanopaper. *ACS Appl. Mater. Interfaces* **6**, 22673–22679.

Li H.L., Zhang J., Tao B.R., Wan L.J., Gong W.L. (2009) Investigation of capacitive humidity sensing behavior of silicon nanowires. *Phys. E Low Dimens. Syst. Nanostruct.* **41**, 600–604.

Li L.H., Cervenka J., Watanabe K., Taniguchi T., Chen Y. (2014b) Strong oxidation resistance of atomically thin boron nitride nanosheets. *ACS Nano* **8**, 1457–1462.

Li N., Chen X.-D., Chen X.-P., Ding X., Zhao X. (2017a) Ultra-high sensitivity humidity sensor based on MoS$_2$/Ag composite films. *IEEE Electron. Dev. Lett.* **38**(6), 806–809.

Liang Y.-C., Liu S.-L. (2014) Synthesis and enhanced humidity detection response of nanoscale Au-particle-decorated ZnS spheres. *Nanoscale Res. Lett.* **9**, 647.

Liu K.-K., Zhang W., Lee Y.H., Lin Y.C., Chang M.T., Su C.Y., et al. (2012) Growth of large-area and highly crystalline MoS$_2$ thin layers on insulating substrates. *Nano Lett.* **12**, 1538–1544.

Liu L., Kumar S. B., Ouyang Y., Guo J. (2011) Performance limits of monolayer transition metal dichalcogenide transistors. *IEEE Trans. Electron Dev.* **58**, 3042–3047.

Liu Z., Gong Y., Zhou W., Ma L., Yu J., Carlos J., et al. (2013) Ultrathin high-temperature oxidation-resistant coating of hexagonal boron nitride. *Nat. Commun.* **4**, 2541–2548.

Lo C.F., Chang C.Y., Chu B.H., Pearton S.J., Dabiran A., Chow P.P., Ren F. (2010) Effect of humidity on hydrogen sensitivity of Pt-gated AlGaN/GaN high electron mobility transistor based sensors. *Appl. Phys. Lett.* **96**, 232106.

Lu Z., Gong Y., Li X., Zhang Y. (2017) MoS$_2$-modified ZnO quantum dots nanocomposite: Synthesis and ultrafast humidity response. *Appl. Surf. Sci.* **399**, 330–336.

Luo Y., Chen C., Xia K., Peng S., Guan H., Tang J., et al. (2016) Tungsten disulfide (WS$_2$) based all-fiber-optic humidity sensor. *Opt. Express* **24**(8), 8956–8966.

Ma S. (2015) Gas sensitivity of Cr doped BN sheets. *Appl. Mech. Mater.* **799–800**, 166–170.

Mak K.F., He K., Shan J., Heinz T.F. (2012) Control of valley polarization in monolayer MoS$_2$ by optical helicity. *Nature Nanotech.* **7**, 494–498.

Mak K.F., Lee C., Hone J., Shan J., Heinz T.F. (2010) Atomically thin MoS$_2$: A new direct-gap semiconductor. *Phys. Rev. Lett.* **105**, 13680.

Mokkapati S., Jagadish C. (2009) III-V compound SC for optoelectronic devices. *Mater. Today* **12**(4), 22–32.

Mui D.S.L., Wang Z., Morkoc H. (1993) A review of III–V semiconductor based metal-insulator-semiconductor structures and devices. *Thin Solid Films* **231**, 107–124.

Nesheva D., Aneva Z., Reynolds S., Main C., Fitzgerald A.G. (2006) Preparation of micro- and nanocrystalline CdSe and CdS thin films suitable for sensor applications. *J. Optoelectron. Adv. Mater.* **8**(6), 2120–2125.

O'Brien M., Lee K., Morrish R., Berner N.C., McEvoy N., Wolden C.A., Duesberg G.S. (2014) Plasma assisted synthesis of WS$_2$ for gas sensing applications. *Chem. Phys. Lett.* **615**, 6–10.

Okur S., Uzar N., Tekguzel N., Erol A., Arıkan M.C. (2012) Synthesis and humidity sensing analysis of ZnS nanowires. *Physica E* **44**, 1103–1107.

Passi V., Dubois E., Celle C., Clavaguera S., Simonato J.P., Raskin J.P. (2011) Functionalization of silicon nanowires for specific sensing. *ECS Trans.* **35**, 313–318.

Pawbake A.S., Waykar R.G., Late D.J., Jadka S.R. (2016) Highly transparent wafer-scale synthesis of crystalline WS$_2$ nanoparticle thin film for photodetector and humidity-sensing applications. *ACS Appl. Mater. Interfaces* **8**, 3359–3365.

Peng Y., Meng Z., Zhong C., Lu J., Yu W., Jia Y., Qian Y. (2001) Hydrothermal synthesis and characterization of single-molecular-layer MoS$_2$ and MoSe$_2$. *Chem. Lett.* **30**, 772–773.

Peng M., Zheng X., Ma Z., Chen H., Liu S., He Y., Li M. (2018) Ni-pattern guided GaN nanowire-array humidity sensor with high sensitivity enhanced by UV photoexcitation. *Sens. Actuators B* **256**, 367–373.

Prasad R.M., Lauterbach S., Kleebe H.-J., Merdrignac-Conanec O., Barsan N., Weimar U., Gurlo A. (2017) Response of gallium nitride chemiresistors to carbon monoxide is due to oxygen contamination. *ACS Sens.* **2**(6), 713–717.

Sajjad M., Feng P. (2014) Study the gas sensing properties of boron nitride nanosheets. *Mater. Res. Bull.* **49**, 35–38.

Salehi A., Kalantari D.J., Goshtasbi A. (2006b) Rapid response of Au/porous-GaAs humidity sensor at room temperature. In: *Proceedings of IEEE Conference on Optoelectronic and Microelectronic Materials and Devices, COMMAD 2006*, 6–8 Dec., Perth, WA, Australia, pp. 125–128.

Salehi A., Nikfarjam A., Kalantari D.J. (2006a) Highly sensitive humidity sensor using Pd/porous GaAs Schottky contact. *IEEE Sensors J.* **6**(6), 1415–1421.

Samà J., Seifner M.S., Domènech-Gil G., Santander J., Calaza C., Moreno M., Gràcia I., Barth S., Romano-Rodríguez A. (2017) Low temperature humidity sensor based on Ge nanowires selectively grown on suspended microhotplates. *Sens. Actuators B* **243**, 669–677.

Shin J., Hong Y., Wu M., Bae J.-H., Kwon H.-I., Park B.-G., Lee J.-H. (2018) An accurate and stable humidity sensing characteristic of Si FET-type humidity sensor with MoS₂ as a sensing layer by pulse measurement. *Sens. Actuators B* **258**, 574–579.

Soltani A., Thevenin P., Bakhtiar H., Bath A. (2005) Humidity effects on the electrical properties of hexagonal boron nitride thin films. *Thin Solid Films* **471**, 277–286.

Taghinejad H., Taghinejad M., Abdolahad M., Saeidi A., Mohajerzadeh S. (2013) Fabrication and modeling of high sensitivity humidity sensors based on doped silicon nanowires. *Sens. Actuators B* **176**, 413–419.

Tahir M., Zink N., Eberhardt M., Therese H., Kolb U., Theato P., Tremel W. (2006) Overcoming the insolubility of molybdenum disulfide nanoparticles through a high degree of sidewall functionalization using polymeric chelating ligands. *Angew. Chem. Int. Ed.* **45**, 4809–4815.

Talazac L., Brunet J., Battut V., Pauly A., Germain J.P., Pellier S., Soulier C. (2001) Air quality evaluation by monolithic InP-based resistive sensors. *Sens. Actuators B* **76**, 258–264.

Tan Y., Yu K., Yang T., Zhang Q., Cong W., Yin H., et al. (2014) The combinations of hollow MoS₂ micro nano-spheres: One-step synthesis, excellent photocatalytic and humidity sensing properties. *J. Mater. Chem. C* **2**(27), 5422–5430.

Tao B., Zhang J., Miao F., Li H., Wan L., Wang Y. (2009) Capacitive humidity sensors based on Ni/SiNWs nanocomposites. *Sens. Actuators B* **136**, 144–150.

Uzar N., Okur S., Arıkan M.C. (2011) Investigation of humidity sensing properties of ZnS nanowires synthesized by vapor liquid solid (VLS) technique. *Sens. Actuators A* **167**, 188–193.

Varghese S.S., Varghese S.H., Swaminathan S., Singh K.K., Mittal V. (2015) Two-dimensional materials for sensing: Graphene and beyond. *Electronics* **4**, 651–687.

Wang H.Y., Li X.J. (2010) Capacitive humidity-sensitivity of carbonized silicon nanoporous pillar array. *Mater. Lett.* **64**, 1268–1270.

Wang H.Y., Wang Y.Q., Hu Q.F., Li X.J. (2012b) Capacitive humidity sensing properties of SiC nanowires grown on silicon nanoporous pillar array. *Sens. Actuators B* **166–167**, 451–456.

Wang Q.H., Kalantar-Zadeh K., Kis A., Coleman J.N., Strano M.S. (2012a) Electronics and optoelectronics of two-dimensional transition metal dichalcogenides. *Nature Nanotechnol.* **7**, 699–712.

Wang L., Hang R., Xu Y., Guo C., Qian Y. (2014) From ultrathin nanosheets, triangular plates to nanocrystals with exposed (102) facets, a morphology and phase transformation of sp2 hybrid BN nanomaterials. *RSC Adv.* **4**, 14233–14240.

Wilson J.A., Di Salvo F.J., Mahajan S. (1975) Charge-density waves and superlattices in the metallic layered transition metal dichalcogenides. *Adv. Phys.* **24**, 117–201.

Wright N.G., Horsfall A.B. (2007) SiC sensors: A review. *J. Phys. D: Appl. Phys.* **40**, 6345–6354.

Xu H.J., Li X.J. (2008) Silicon nanoporous pillar array: A silicon hierarchical structure with high light absorption and triple-band photoluminescence. *Opt. Express* **16**, 2933–2941.

Xu L., Song H., Zhang T., Fan H., Fan L., Wang Y., Dong B., Bai X. (2011) A novel ethanol gas sensor-ZnS/cyclohexylamine hybrid nanowires. *J. Nanosci. Nanotechnol.* **11**(3), 2121–2125.

Xu Y.Y., Li X.J., He J.T., Hu X., Wang H.Y. (2005) Capacitive humidity sensing properties of hydrothermally-etched silicon nano-porous pillar array. *Sens. Actuators B* **105**, 219–222.

Yang S., Jiang C., Wei S.-H. (2017) Gas sensing in 2D materials. *Appl. Phys. Rev.* **4**, 021304.

Yang W., Gan L., Li H., Zhai T. (2016) Two-dimensional layered nanomaterials for gas-sensing applications. *Inorg. Chem. Front.* **3**, 433–451.

Yin Z., Tang X. (2007) A review of energy bandgap engineering in III–V semiconductor alloys for mid-infrared laser applications. *Solid-State Electron.* **51**(1), 6–15.

Yu Y., Chen H., Liu Y., Li L.H., Chen Y. (2013) Humidity sensing properties of single Au-decorated boron nitride nanotubes. *Electrochem. Commun.* **30**, 29–33.

Yue Q., Shao Z., Chang S., Li J. (2013) Adsorption of gas molecules on monolayer MoS₂ and effect of applied electric field. *Nanoscale Res. Lett.* **8**, 1–7.

Zhan Y., Liu Z., Najmaei S., Ajayan P.M., Lou J. (2012) Large-area vapor-phase growth and characterization of MoS₂ atomic layers on a SiO₂ substrate. *Small* **8**, 966–971.

Zhang D., Sun Y.E., Li P., Zhang Y. (2016) Facile fabrication of MoS₂-modified SnO₂ hybrid nanocomposite for ultrasensitive humidity sensing. *ACS Appl. Mater. Interfaces* **8**, 14142–14149.

Zhang S.-L., Choi H.-H., Yue H.-Y., Yang W.-C. (2014) Controlled exfoliation of molybdenum disulfide for developing thin film humidity sensor. *Curr. Appl. Phys.* **14**, 264–268.

Zhao J., Li N., Yu H., Wei Z., Liao M., Peng Chen P., et al. (2017) Highly sensitive MoS₂ humidity sensors array for noncontact sensation. *Adv. Mater.* **29**, 1702076.

Zhao S., Xue J., Kang W. (2014) Gas adsorption on MoS₂ monolayer from first-principles calculations. *Chem. Phys. Lett.* **595–596**, 35–42.

Zhou C.J., Yang W.H., Wu Y.P., Lin W., Zhu H.L. (2015) Theoretical study of the interaction of electron donor and acceptor molecules with monolayer WS₂. *J. Phys. D* **48**, 285303.

Zhou K.-G., Mao N.-N., Wang H.-X., Peng Y., Zhang H.-L. (2011) A mixed-solvent strategy for efficient exfoliation of inorganic graphene analogues. *Angew. Chem., Int. Ed. Engl.* **50**(46), 10839–10842.

Zhou X.T., Hu J.Q., Li C.P., Ma D.D.D., Lee C.S., Lee S.T. (2003) Silicon nanowires as chemical sensors. *Chem. Phys. Lett.* **369**, 220–224.

# 6 Porous Silicon

## 6.1 POROUS SILICON: GENERAL CONSIDERATION

The physical properties of porous silicon (PSi) have been discussed in detail in several good reviews (Feng and Tsu 1994; Cullis et al. 1997; Foll et al. 2006) and books (Canham 1997, 2014; Kochergin and Foll 2009; Sailor 2012; Santos 2014; Korotcenkov 2016a). According to the existing classification, PSi is subdivided into microporous ($R < 2$ nm), mesoporous (2 nm $< R <$ 50 nm), and macroporous ($R > 50$ nm) (Vial and Derrien 1995; Canham 1997; Cullis et al. 1997; Foll et al. 2002). The main parameter of any porous material is the index of porosity, which determines which part of the material volume is filled by pores. For PSi the index of porosity may lie in an unusually wide interval from 5% to 95%. Depending on the value of porosity and pore geometry, the surface area of PSi may range from 10 to 100 m$^2$/cm$^3$ for macroporous silicon, from 100 to 300 m$^2$/cm$^3$ for mesoporous silicon, and from 300 to 800 m$^2$/cm$^3$ for microporous silicon.

Structural studies of PSi have shown that this material is usually monocrystalline (Bomchil et al. 1989). It was established by numerous experiments that PSi, especially those with porosity lower than 35%–50%, and silicon in the substrate have the same crystallographic structure (Barla et al. 1984). The material with porosity more than 50% is completely depleted of charge carriers and characterized by a specific resistance of more than 10$^7$ $\Omega$·cm, whereas at the substrate this parameter is 1–10 $\Omega$·cm. Because PSi is a spongy structure with an extremely high specific-surface area, the electrical resistivity of the layer is high and sensitive to the ambient atmosphere. Therefore, the determination of exact values for the PSi resistivity is difficult (Ben-Chroin 1997). Among others, the resistivity depends on the quantum confinement, mobility and drift of the carriers, changes in the band structure and temperature, and the medium (gas or liquid) filling the pores (Parkhutik 1999). In addition, PSi has a low dielectric constant of around 2–3 in. vacuum that is especially important for capacitive sensors.

### 6.1.1 ADVANTAGES OF POROUS SILICON FOR HUMIDITY-SENSOR APPLICATION

PSi layers are very attractive from a sensor point of view because of a unique combination of crystalline structure: a large internal surface area of up to 200–500 m$^2$/cm$^2$, which is able to enhance the adsorbate effects, and high activity in surface chemical reactions. Several investigations show that electrical and optical characteristics of porous semiconductors may change considerably upon adsorption of molecules to their surfaces and/or by filling the pores (Feng and Tsu 1994;

Mares et al. 1995; Canham 1997, 2014; Cullis et al. 1997). This means that surface adsorption and capillary condensation effects in PSi layers can be used for development of effective sensor systems (Anderson et al. 1990; Parkhutik 1999).

Lauerhaas and Sailor (1993) and Di Francia et al. (1998) noted the following advantages of PSi over other porous materials, such as ceramics or nano- and polycrystalline films of metal oxides used for sensor design: (1) It is basically a crystalline material and thus, in principle, is perfectly compatible with common microelectronic processes devices; (2) It can be electrochemically fabricated in very simple and cheap equipment; (3) It can be produced in a large variety of morphologies, all exhibiting large values of surface-to-volume ratios. Other PSi advantages include the ability to create three-dimensional structures and design of multisensors, based on the use of various registration techniques for gas detection, for example, optical, electrical, and luminescent. Lower power consumption of PSi-based devices in comparison with metal-oxide gas sensors due to working at room temperature is another important advantage of those devices. Thus, because PSi sensors are based on silicon wafers, manufactured by using integrated circuit-production techniques and operated at room temperature using relatively low voltages, they can be used for producing compact and low-cost sensor systems on a chip, where both the sensing element and the read-out electronics can be effectively integrated on the same wafer. Some authors contend that the sensors based on PSi are so simple that they could ultimately be mass produced for pennies apiece. This means that those new sensors could be integrated into electronic equipment and used for building sensing arrays.

One should note that mentioned above properties of PSi are attractive not only for gas sensors (Huanca et al. 2008; De Stefano et al. 2016; Korotcenkov 2016b) but also for humidity sensors. Numerous research works have shown that many parameters of PSi such as the photoluminescence intensity, the capacity of the porous layer, the conductance, the reflection coefficient, infrared (IR) absorption, the resonance frequency of a Fabry-Pérot resonator made of the PSi, and so on, are sensitive to air humidity of surrounding atmosphere (Anderson et al. 1990; Ben-Chorin and Kux 1994a; Feng et al. 1994; Vial and Derrien 1995; Canham 1997; Blackwood and Akber 2006; Korotcenkov and Cho 2010a). This means that, based on its unique properties, PSi is a universal humidity-sensitive material suitable for developing all types of humidity sensors. Furthermore, both simple resistor-type structures (Di Francia et al. 2005) and more complex ones, such as transistors (Lazzerini et al. 2013) or heterostructures (Lundstrom et al. 1975), can be used for designing PSi-based solid-state humidity sensors (see Figure 6.1).

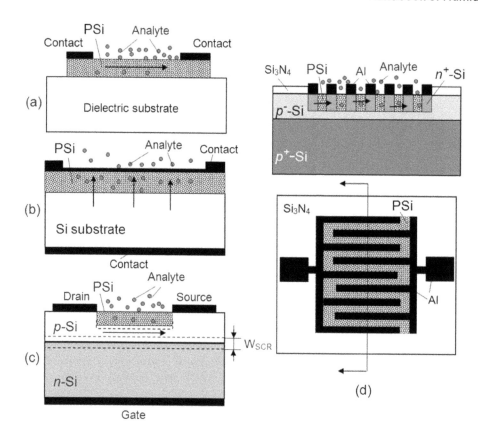

**FIGURE 6.1** Variants of PSi-based devices used for gas-sensing applications: (a, d) resistor-type structures, (b) Schottky barrier and *p-n* junction-type structure, and (c) transistor-type structure. (With kind permission from Taylor & Francis: *Porous Silicon: From Formation to Application*, Processing of porous silicon, in Korotcenkov, G. (ed.), Vol. 1: Formation and Properties, CRC Press, Boca Raton, FL, 2016b, 299–349, Korotcenkov, G. and Cho B.K.)

## 6.2 FEATURES OF SILICON POROSIFICATION

### 6.2.1 ELECTROCHEMICAL POROSIFICATION

PSi was first formed in the mid-1950s during research on the process of electrochemical polishing of the silicon surface in a water solution of hydrofluoric acid (HF). In such an experiment, a silicon wafer was the anode, whereas the cathode was a platinum electrode. During etching it was found that for certain conditions, such as a low density of anodic current (1–100 mA/cm²) and a high concentration of HF in electrolyte, instead of the process of electropolishing, the formation of porous layers was observed (Turner 1958). It was established that the necessary chemical reaction at the semiconductor–electrolyte interface is generally a mixture of direct dissolution, oxide formation, and oxide dissolution, with details sensitive to many parameters. The overall reaction may occur uniformly as electropolishing (often observed at high current densities), non-uniformly but everywhere on the surface, or highly localized including some self-organization. The last one corresponds to a pore-producing mode (Foll et al. 2006). So, porosification can be obtained by anodizing the silicon in a suitable electrolyte under suitable conditions.

Model notions of the porosification mechanism started being formed in the mid-1960s, but the common point of view has not been elaborated yet. Only a general knowledge about experimental parameters determining the PSi morphology and properties exists. Detailed discussions on suggested models of that process can be found in several articles (Smith and Collins 1992; Parkhutik 1999; Carstensen et al. 2000a; Chazalviel et al. 2000; Lehmann et al. 2000; Foll et al. 2002, 2006; Frey et al. 2007; Quiroga-González and Föll 2016). The variety of possible experimental conditions and the different morphologies that may be obtained (from microporous to macroporous) have been a formidable challenge for the understanding of the formation mechanisms. Though many details of the pore-formation mechanism are still unclear, a few general statements can be made. During pore formation, direct dissolution of the semiconductor almost always competes with oxidation plus subsequent dissolution of the oxide. The electrolyte therefore has to be able to dissolve the oxide.

Thus, from a general point of view, electrolytes acceptable for silicon porosification must contain chemical species allowing anodic Si dissolution with at least one of the two basic mechanisms such as direct dissolution and oxidation (followed by purely chemical dissolution of the oxide). It was established that various electrolytes can be used for Si porosification (Foll et al. 2002; Christophersen et al. 2003; Frey et al. 2007; Korotcenkov 2016c; Korotcenkov and Cho 2016a). However, until now, electrolytes on the basis of HF (15–40 wt.%) are the most utilized (Canham et al. 1996;

## TABLE 6.1
### Typical Parameters of Silicon Porosification Using Electrochemical Etching

| Substrate | Electrolyte | Parameters of Oxidation | Porosity, Thickness | References |
|---|---|---|---|---|
| $p$-Si (7–13 $\Omega$ cm) | HF: $C_2H_5OH$ = 1:1 | I = 200–250 mA·cm$^{-2}$ | 45%–60% $d\sim$15–40 $\mu$m | Baratto et al. (2001) |
| $p$-Si | HF: $C_2H_5OH$ = 1:1 | I = 20 mA·cm$^{-2}$ | | Racine et al. (1997) |
| $n$-Si | HF: $C_2H_5OH$ = 1:3 | I = 20 mA·cm$^{-2}$, t = 30 min, illumination (Nd:YAG laser) | 70%–80% | Koyama (2006) |
| $p$-Si (10 $\Omega$ cm) | HF: $C_2H_5OH$ = 1:2.5 | I = 5–20 mA·cm$^{-2}$ t = 20–120 min | 65%–67% | Holec et al. (2002) |
| $p$-Si (2–9 $\Omega$ cm) | HF: $H_2O$:$C_2H_5OH$ = 1:1:2 | I = 20–40 mA·cm$^{-2}$ | 57% | Hedrich et al. (2000) |
| $n$-Si (1 $\Omega$ cm) | HF: $C_2H_5OH$ = 7:3 | I$\sim$35 mA·cm$^{-2}$, t$\sim$25 min | $d\sim$ 80 $\mu$m | Di Francia et al. (1998) |
| $p$-Si | HF: $H_2O_2$:ethanol: $H_2O$ = (9–11):1:4:12 | Pt assisted electroless, t = 10 min | | Splinter et al. (2000) |
| $p$-Si (0.01–0.2 $\Omega$ cm) | HF: $C_2H_5OH$ = 1:1 | I$\sim$14 mA·cm$^{-2}$, t = 275 s I = 172 mA·cm$^{-2}$, t = 42 s | 33% 65% | Canham et al. (1996) |
| $p$-Si (0.02 $\Omega$ cm) | HF: $C_2H_5OH$ =1:2 | I = 20 mA·cm$^{-2}$ I = 75 mA·cm$^{-2}$ I = 150 mA·cm$^{-2}$ | 38% 62% 74% | Lysenko et al. (2002) |
| Si | 30% HF (including Triton X-100) | I = 30 mA·cm$^{-2}$, t$\sim$60 s | 65% | Connolly et al. (2002) |

*Source:* Reprinted with permission from Korotcenkov and Cho 2010b. Copyright 2010b: CRC Press.

Di Francia et al. 1998; Hedrich et al. 2000; Kordas et al. 2001; Connolly et al. 2002; Holec et al. 2002; Lysenko et al. 2002; Sharma et al. 2005). Table 6.1 gives some details of silicon electrochemical porosification (always for standard 49% HF and absolute ethanol; the substrate orientation is always {100}). The chemical reactions for HF-based electrolytes during silicon porosification can be written as (Lehmann 2002)

$$Si + 2HF + 2h^+ \Rightarrow SiF_2 + 2H^+ \tag{6.1}$$

$$Si + 2H_2O + 4h^+ \Rightarrow SiO_2 + 4H^+ \tag{6.2}$$

$$SiO_2 + 2HF_2^- + HF \Rightarrow SiF_6 + 2H_2O \tag{6.3}$$

As it follows from Table 6.1, a concentration of 15–40 wt.% of HF is a kind of standard for the fabrication of micro- and mesoporous silicon. The optimal concentration of the HF in an electrolyte depends on what kind of pores are to be etched and which parameter of the pores is to be optimized (Korotcenkov 2016d). Due to the hydrophobic character of the freshly etched (and H covered) Si surface, absolute ethanol ($C_2H_5OH$) is usually added because it might help to overcome surface tension problems (Bisi et al. 2000). In particular, ethyl alcohol (EtOH), added to this solution, reduces the surface tension at the silicon–solution interface. As a result, the HF dissolution in ethanol gives better surface wetting in comparison with water solution (Splinter et al. 2001). This is very important for the lateral homogeneity and the uniformity of the PSi layer in depth (Splinter et al. 2001). It was established that in order to be useful, the ethanol concentration should not be less than 15% (Bomchil et al. 1993). In addition, ethanol additions reduce the formation of hydrogen gas bubbles; this is essential for homogeneous layers, as it was pointed out before. In this case, the bubbles of gas produced during Si porosification are smaller and less persistent. Possibly ethanol acts as a surfactant and prevents bubbles from sticking to the silicon surface, but its action might be more complex than that because when a commercial surfactant was applied in the electrolyte instead of $C_2H_5OH$, the porosity could not exceed 70%.

A second prerequisite for the dissolution reaction and thereby pore formation in a semiconductor is holes. Summarizing various models, one can note the following. The Si surface in contact with HF water solutions becomes saturated by hydrogen and tends to be chemically inactive with respect to the electrolyte. If voltage is applied to the electrodes, the holes in the silicon wafer start migrating toward the silicon–electrolyte interface. At that, the Si atoms are freed from blocking by hydrogen, start interacting with ions and molecules of the electrolyte, and then transfer into solution. If electrolysis takes place at high-current density, a large number of holes would arrive at the electrode's surface. They move toward the interface in a uniform front and provide a reaction ability to almost every atom of Si. Microjets have larger surface than plane areas, so they dissolve faster. Thus, the surface of the silicon anode gradually flattens. This is a mode of electrochemical polishing. If electrolysis is made at a low current density, the number of holes is not sufficient for forming a uniform front, and therefore a local dissolution of silicon of the surface takes place. In general, at least one hole is needed to initiate the reaction chain at the interface that ultimately leads to the loss of one atom (or molecule in the case of compound semiconductors).

According to different models, initiation of the pore growth could begin at micro- and macrostructural defects, mechanically strained areas, or local perturbation of the surface's potential field. One of such model was developed by Allongue et al. (1995, 1997). Over time, the emerging pores continue to grow deep inside the electrode at the expense of the holes drifting to the tips of the pores, where the intensity of the electric field is higher. It is obvious that in silicon of $n$- and $p$-type the number of holes is different, and therefore the processes of porosification in both $n$-Si and $p$-Si have their own peculiarities. One of the major tricks for successful porosification experiments is to supply holes only locally. This trick works only in $n$-doped material, where the holes are the minority carriers. The holes then may be supplied by illumination or via electrical tunneling or avalanche breakdown of the material due to the high field strengths, as is the case for the pores in III–V compounds (Foll et al. 2006).

The holes in $p$-Si are the main charge carriers, and their concentration is $10^{14}$–$10^{18}$ cm$^{-3}$. In this case, as a rule, nanosized pores are formed. The pores are perpendicular to the surface channels with diameters of tens of nanometers with fine side derivations. In $n$-Si, where the main charge carriers are electrons, the hole concentration is extremely small ($10^2$–$10^6$ cm$^{-3}$). A necessary number of the holes can be obtained by photogeneration (by backlighting the Si electrode) or by avalanche generation (by anodizing in the area of high voltages). The structure of pores obtained in indicated conditions sufficiently differs from the structure of porous layers mentioned above and is characterized by the presence of pores of very large diameter. So, by changing the conditions of anodizing, one can form PSi with different pore morphology (geometry).

According to Chazalviel et al. (2000), existing theories can be divided into those focusing on the electrochemical aspects of the dissolution and those where the dominant role is played by the semiconducting character of silicon. According to one group of the authors, silicon can be made porous due to the specific features of silicon electrochemistry (Smith and Collins 1992; Parkhutik 1999; Zhang 2001; Lehmann 2002). Other authors have emphasized the role of the space charge in inducing various interface instabilities (Canham 1997; Chazalviel et al. 2000; Lehmann 2002). Both theories are able to make the qualitative prediction of the pore formation, but until now none has been worked out into a really quantitative

form that would correctly yield numerical figures, such as the porosity and characteristic structure sizes.

In recent years, attempts have been made to improve the understanding of the formation mechanism of porous semiconductors during the process of electrochemical etching (Foll et al. 2003a, 2003b; Christophersen et al. 2003; Outemzabet et al. 2004; Quiroga-González and Föll 2016). This required a more critical approach to the earlier presented models, designed for silicon.

For electrochemical porosification, a so-called electrochemical single or double cell is usually used (Bisi et al. 2000; Lammel and Renaud 2000; Korotcenkov 2016d). Examples of such cells are shown in Figure 6.2. The advantage compared to usual single cells is that no backside metallization of the wafer is needed. In the double cell, the contact is electrolytic, not metallic. An advantage of the double-cell arrangement is that it avoids a potential source of contamination of the PSi in any subsequent thermal and chemical processing (Halimaoui 1995). In addition, the uniformity of the processes is better than with metallic backside contacts. In the latter case, the forming of good backside contact with low resistance is an important requirement for obtaining a controlled etching process. However, a double cell has disadvantages as well. There is a risk that the overall reaction will be limited by the electrochemical reaction on the backside; such cases make double cells nontrivial setups.

For improvement conditions of the etching process, for control it was proposed to use the device called a potentiostat (see Figure 6.3a). A potentiostat allows controlling the voltage between the working electrode and the reference electrode directly. The working principle of a potentiostat is shown in Figure 6.3b.

## 6.2.2 OTHER METHODS OF SILICON POROSIFICATION

During the last decades it was established that besides conventional electrochemical etching, many other methods such as chemical stain etching, chemical vapor etching, spark processing, laser-induced etching, and reactive-ion etching

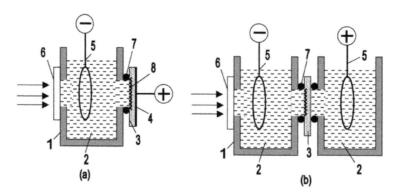

**FIGURE 6.2** Electrolytic cells for porous layer fabrication: (a) one-cell configuration, and (b) two-cell configuration. 1—Teflon bath, 2—electrolyte, 3—silicon wafer, 4—metal contact, 5—Pt electrode, 6—quartz window, 7—packing ring, 8—porous layer, arrows show the direction of illumination. (With kind permission from Taylor & Francis: *Porous Silicon: From Formation to Application*, Processing of porous silicon, in Korotcenkov, G. (ed.), Vol. 1: Formation and Properties, CRC Press, Boca Raton, FL, 2016b, 299–349, Korotcenkov, G. and Cho B.K.)

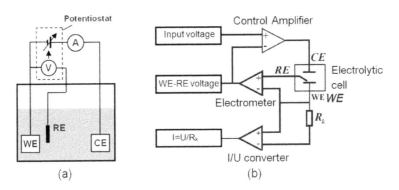

**FIGURE 6.3** (a) Experimental setup for performing electrochemical experiments operating in potentiostatic mode, and (b) a schematic representation of a three-electrode potentiostat. (Idea from http://www.porous-35.com/electrochemistry-semiconductors-10.html.)

can be used for Si porosification (Korotcenkov and Brinzari 2016). For example, the stain etching, in which the silicon sample is simply immersed in an HF-based solution, is the easiest way of PSi producing. It has been reported that the physical structure of those layers was similar to the one, fabricated by the anodization method. The resulting pores are in the range of 1 nm up to the micrometer. The stain etching is an electroless process and, therefore, it has several advantages in comparison with electrochemical etching. However, the layers, formed by this method, have low photoluminescence efficiency, deficient homogeneity, and poor reproducibility (Cullis et al. 1997).

HF spray and vapor-etching methods are also electroless processes. Therefore, vapor etching was investigated to address the difficulty of isolating metal contacts of devices from the electrolyte solution during anodic etching (Kalem and Yavuzcetin 2000). Spark erosion was also tested for preparing PSi (Hummel and Chang 1992). Noble-metal-assisted etching of Si is another method suitable to form PSi. This method is also a simple process that does not require the attachment of any electrodes and can be performed on objects of arbitrary shape and size (Cullis et al. 1997; Kolasinski 2005). Plasma etching can produce the pores as well. However, the field of applications of this method usually is limited by fabrication of macropores.

The laser-induced method of Si porosification has also small prospects to be introduced in mass production. It comes from disadvantages, such as the layer inhomogeneity over the area, the presence of optical radiation source, optical system, and the system of either scanning or sample's displacement. Irreproducibility of PSi-layer parameters is another disadvantage of this method (Koker and Kolasinski 2000). According to Koker and Kolasinski (2000), hydrocarbon contamination is one contributor to irreproducibility. A more detailed description of these methods can be found in reviews (Korotcenkov and Brinzari 2016) and books (Canham 2014; Korotcenkov 2016a) dedicated to PSi.

However, despite the wide variety of methods that can be used for silicon porosification, the electrochemical (anodic) etching is the most successful and used technique for producing pores in semiconductors. All previously mentioned methods allow forming PSi with its unique parameters, but none of them has reached a stage of maturity, similar to anodization. Besides, many methods have essential limitations on reproducibility, on attainment of required porosity and thickness of formed layer. Therefore, the attention of developers is attracted mainly to anodic oxidation, which is the most controllable process, and provides reproducible parameters. We must note that the standard method of electrochemical etching has appreciably more resources for fabrication of high-quality PSi layers in comparison with the above-mentioned methods, such as the stain etching, metal-assisted stain etching, laser-induced etching, and spark processing. Electrochemical etching is simple, inexpensive, and gives designers a sufficiently free hand for fabrication the PSi layer with required structure.

## 6.3 PSi CHARACTERIZATION

With regard to the characterization of PSi, it can be used for parameters such as pore size, pore density, specific surface area, and porosity (Korotcenkov 2016c). One should note that for PSi usually used in experiments, the porosity spans in unusually wide ranges from 5% to 95% (0%–30% is low porosity; 30%–70% is medium porosity, and 70%–95% is high porosity). Depending on the value of the porosity and pore geometry, the surface area of PSi also can change in wide range from 10 to 300–1000 $m^2/cm^3$. It should be mentioned that the surface area is not strictly proportional to the porosity as one might expect. As a rule, the dependence of surface area on the porosity has the maximum, and at a certain value of the porosity $\sim$50%–60% the surface area begins to decrease with increasing porosity (see Figure 6.4).

The structure of PSi also varies greatly depending on the etching conditions (see Figures 6.5 and 6.6). It is seen that the structure of PSi films can range from those holding micronsize pores to sponge-like layers with pores of several nanometers in diameter (Parkhutik 1999). Different structures and sizes of pores reflect differences in preparation conditions where everything is important: conductivity type, doping level, crystal orientation, construction of the electrolytic cell, current density, applied potential, temperature, illumination state of the semiconductor (front or backside), surface conditions (polished, rough, masked), sample preconditioning, and the type of electrolyte used (Lehmann 1993;

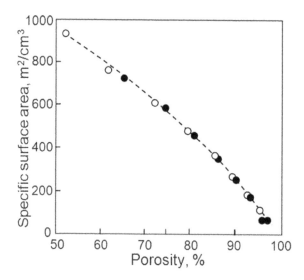

**FIGURE 6.4** Specific surface area as a function of PSi porosity. Layer thickness ~1 μm. (Data extracted from Halimaoui, A., Porous silicon formation by anodisation, in Canham, L.T. (ed.), *Properties of Porous Silicon*, IEE INSPEC, London, UK, 12–23, 1997.)

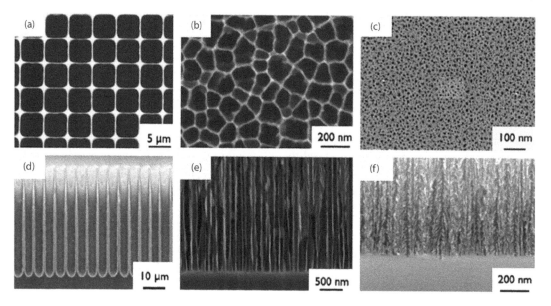

**FIGURE 6.5** Top views (a, b, c) and cross-sectional views (d, e, f) of PSi with different pore diameters: (a, d) ordered macropores, (b, e) medium-sized pores, and (c, f) mesopores. Actual average diameters of the pores, which are determined from the SEM images, are 20 nm (mesopore), 120 nm (medium-sized pore), and 5 μm (macropore), respectively. (Reprinted from *Electrochem. Commun.*, 10, Fukami, K., et al., (2008) Fine-tuning in size and surface morphology of rod-shaped polypyrrole using porous silicon as template. 56–60, Copyright 2008, with permission from Elsevier.)

**FIGURE 6.6** Types of pores formed in PSi: (a) closed pores at one end, (b) pores opened at both ends, and (c) pores with branching. (With kind from Taylor & Francis: *Porous Silicon: From Formation to Application*, Porous silicon characterization and application: General view, in Korotcenkov G. (ed.), Vol. 1: Formation and Properties. CRC Press, Boca Raton, FL, 2016b, 1–23, Korotcenkov, G; Allongue P. et al., *Thin Solid Films* **297**, 1–4, 1997.)

Parkhutik 1999; Bisi et al. 2000; Carstensen et al. 2000b; Kordas et al. 2001; Splinter et al. 2001; Christophersen et al. 2003; Foll et al. 2006; Kochergin and Foell 2006; Frey et al. 2007). For example, for either silicon with electron type of conductivity (n-Si), or strongly doped p-type silicon (p+-Si) the pores have perpendicular surface channels with diameter of tens of nanometers with fine side derivations. For the samples of weakly doped p-type silicon (p-Si) or for n-Si samples, being lightened, the structure is in the form of "sponge" or "coral." In such structures, the sizes of pores and nonetched areas can be very small, just a few nanometers. More detailed information about peculiarities of Si porosification and possible pore structure may be found in recent books and reviews (Feng and Tsu 1994; Vial and Derrien 1995; Canham 1997; Parkhutik 1999; Zhang 2001; Foll et al. 2002; Lehmann 2002).

Regarding detailed analysis of the influence of electrochemical etching conditions on porous semiconductor parameters such as the pore diameter, the pore depth, and the pore growth direction, one can find the answer to those questions in several published review articles (Bomchil et al. 1989; Lehmann and Gruning 1997; Pavesi 1997; Bisi et al. 2000; Christophersen et al. 2001; Korotcenkov and Cho 2010a, 2010b) and books (Canham 1997, 2014; Korotcenkov 2016 a,b,c,d). Some general regularities are summarized in Table 6.2 and Figures 6.7 through 6.9.

**TABLE 6.2**
**Effect of Anodization Parameters on PSi Formation**

|  | Parameters | | |
|---|---|---|---|
| **An Increase of** | **Porosity** | **Etching Rate** | **Critical Current** |
| HF concentration | Decreases | Decreases | Increases |
| Current density | Increases | Increases | — |
| Anodization time | Increases | Almost constant | — |
| Temperature | — | — | Increases |
| Wafer doping (p-type) | Decrease | Increases | Increases |
| Wafer doping (n-type) | Increases | Increases | |

*Source:* Reprinted with permission from *Surf. Sci. Rep.*, 38, Bisi, O. et al., Porous silicon: A quantum sponge structure for silicon based optoelectronics, 1–126, Copyright 2000, with permission from Elsevier.

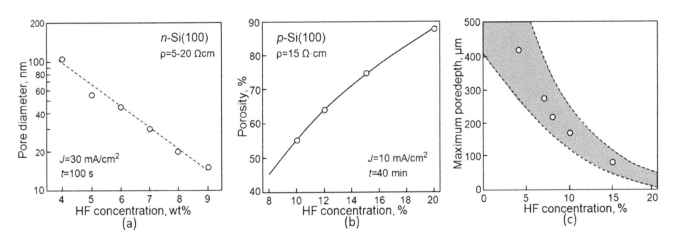

**FIGURE 6.7** (a) Diameter of mesopores on n-Si (100) versus HF concentration; (b) the porosity versus HF concentration (diluted with DMSO) for etching a p-type Si(100) wafer with a resistivity of 15 Ω cm. The etching time was 40 min with the current density of 10 mA/cm²; and (c) the maximum pore depth, shown by circles, which can be achieved in macroporous Si without degradation at the pore tips as a function of electrolyte concentration (with surfactant) for cylindrical pores. The shaded area shows the region of degradation for the pore geometries differing from cylindrical. ([a] Reprinted from Foll, H. et al., *J. Nanomater.*, 91635, 2006, Published by Hindawi Publishing Corporation as open access; [b] Reprinted from *Mater. Sci. Eng. B*, 118, Kan, P.Y.Y. and Finstad T.G., Oxidation of macroporous silicon for thick thermal insulation, 289–292, Copyright 2005, with permission from Elsevier; [c] Reprinted from *Thin Solid Films*, 297, Lehmann, V. and Gruning, U., The limits of macropore array fabrication, 13–17, Copyright 1997, with permission from Elsevier.)

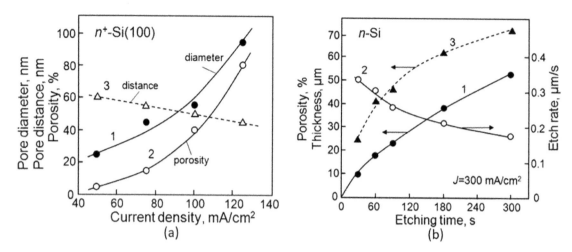

**FIGURE 6.8** (a) Relation between current density and pore-diameter (1), porosity (2) as well as pore distance (3) of the PSi-formation is achieved by control of the electrochemical parameters during anodization of a (100) $n+$-Si-wafer. The fabrication of the porous silicon template was performed in a 10 wt% aqueous HF-solution; (b) thickness (1) and porosity (3) of PSi layer, and etch rate (2) of the $n$-Si (100) [$n \sim (1.5-6.3) \times 10^{18}$ cm$^{-3}$] etched at 300 mA/cm$^2$ using electrolyte HF:ethanol = 1:1, as a function of etching time. ([a] From Rumpf, K. et al.: Transition metals specifically electrodeposited into porous silicon. *Phys. Stat. Sol. (c)*. 2009. 6. 1592–1595. Copyright Wiley-VCH Verlag GmbH & Co. KGaA. Reproduced with permission.; [b] Data extracted from Yaakob, S. et al., *J. Phys. Sci.*, 23, 17–31, 2012.)

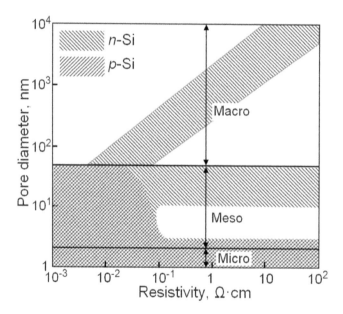

**FIGURE 6.9** The different pore sizes that form on $n$-type and $p$-type silicon substrates during anodization in hydrofluoric acid are shown as a function of the substrate resistivity. (Data extracted from Lehmann, V., *Mater. Lett.*, 28, 245–249, 1996.)

## 6.4 PSi-BASED HUMIDITY SENSORS

The experiment showed that due to the porous structure, PSi can be used for designing all types of humidity sensors.

### 6.4.1 CAPACITANCE HUMIDITY SENSORS

PSi-based humidity sensors operating on the capacitive principle have been demonstrated in Refs. (Kim et al. 2000a; Rittersma et al. 2000; Connolly et al. 2002, 2004a; Fujes et al. 2003; Björkqvist et al. 2004a, 2006; Xu et al. 2005; Islam and Saha 2007). Humidity sensors can be fabricated in both

one-sided and two-sided configurations. Electrodes with a comb structure cover only a small part of the porous layer surface. Typical parameters of humidity sensors based on a capacitance configuration are given in Table 6.3.

Experiments carried out in different labs have shown that PSi-based capacitance humidity sensors really can provide good sensitivity and reproducible signal throughout the entire relative humidity (RH) range. An example of the measuring abilities of a humidity sensor designed on the basis of PSi is presented in Figure 6.10. It is important that parameters of PSi-based sensors depend on the construction and technology used and can vary in the wide range. For example, Kim

**TABLE 6.3**
**Typical Parameters of Capacitance-Type Humidity Sensors Based on Porous Silicon**

| Material | Analyte | Concentration | Sensitivity[a] | Linearity | References |
|---|---|---|---|---|---|
| SiC | Humidity | 20%–90% RH | 1.32 | No | Connolly et al. (2004b) |
| PSiC | Humidity | 10%–90% RH | | No | Connolly et al. (2004a) |
| PSiC | Humidity | 10%–75% RH | ~2 | No | Connolly et al. (2002) |
| PSi | Humidity | 11%–94% RH | ~15 | No | Xu et al. (2005) |
| PSi | Humidity | 10%–90% RH | | Yes | O'Halloran et al. (1999) |
| PSi | Humidity | 10%–90% RH | ~25 | No | Connolly et al. (2002) |
| PSi | Humidity | 10%–95% RH | ~50 | No | Fujes et al. (2003) |
| PSi | Humidity | 5%–85% RH | ~49 | No | Björkqvist et al. (2004a) |

[a]  Sensitivity calculated as a ratio, $S = C(RH = max)/C(RH = min)$.

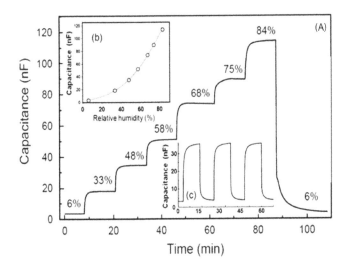

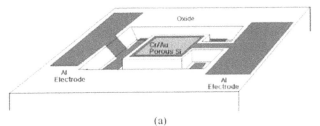

(a)

(b)

**FIGURE 6.10**   (a) Dynamic response of the capacitance of a TC-PSi humidity sensor. Electrical parameters were measured using 85 Hz frequency; (b) dependence of the capacitance on relative humidity measured using 85 Hz frequency; and (c) repeatability of a TC-PSi humidity sensor. Relative humidity is changed between 6% and 58%. Capacitance was measured using 120 Hz frequency. (Reprinted with permission from Björkqvist et al. (2004a). Copyright 2004: Elsevier.)

**FIGURE 6.11**   (a) Schematic diagram of the top view, and (b) a photograph of the sensor developed by Kim et al. (2000a). (Reprinted with permission from Kim et al. (2000a). Copyright 2000: IOP.)

et al. (2000a, 2000b) reported that their PSi-based capacitive sensors have shown a large increase in capacitance of over 300% for the range of 25%–95% RH at room temperature. For comparison, in the polymer-based humidity sensors, the capacitance increases by only a few tens of percents. Sensor, developed by Kim et al. (2000a, 2000b) was fabricated monolithically to be compatible with the typical IC-process technology except for the formation of PSi layer. Figure 6.11a shows a schematic diagram for the top view of this sensor, and Figure 6.11b a photograph of the completed sensor with the sensitive area of $2.3 \times 2.3$ mm² in the middle.

It was shown that in a general case the increase in the PSi-based structure capacitance being measured with the increase in the air humidity is connected with water-molecule adsorption by the porous layer, which leads to the increase in the permittivity of the layer on one hand, and to the increase in

the surface states density at the Si–SiO₂ interface on the other hand (Fastykovsky and Mogilnitsky 1999; Kim et al. 2000b). Prevalence of one effect over the other in the capacitance being measured, depends on the air humidity and electrical bias mode at the metal-oxide semiconductor (MOS) structure and determines the linearity of the capacitance versus air humidity characteristics.

In addition to adsorption–chemisorption processes, capillary-condensation phenomenon is also involved in the mechanism of humidity sensing (Adamson and Gast 1997). The typical pore size in Si layers used for sensors is in the range of 2–4 to 12–15 nm. Therefore, water transport in PSi is described by Knudsen diffusion. Therefore, the performance of a PSi-based humidity sensor is determined by its nano- and microscopic dimensions, including pore size, thickness of the porous layer, the size distribution of the surface structural unit, the regularity of the surface morphology, and the distance between the electrodes (Kim et al. 2000b; Di Francia et al. 2002). This fact indicates that the regularity and controllability of porous semiconductor structures are of great importance in sensor application. For example, due to water condensation in nanosize pores, it is difficult to prepare rapid humidity sensors on the basis of nanoporous material (Björkqvist et al. 2004a).

Yarkin (2003) and Björkqvist et al. (2004b) established that the characteristics of PSi-based humidity sensors are also strongly affected by the hydrophilic/hydrophobic properties of the walls. It has been assumed that only the hydrophilic regions (water adsorbing sites) on a sample adsorb polar water molecules through hydrogen bonding, whereas the hydrophobic regions, which contain very weak dispersion forces, do not adsorb water (Raghavan et al. 2001). It should be noted that the functional groups of hydrophilic surfaces, such as $NH_2$, OH, and COOH, are highly hydrogen bonded and tend to associate weakly with other chemical species, except strong electron donors such as water. For $SiO_2$ surface, two-thirds of the SiOH groups are hydrogen bonded, and water preferentially reacts at these sites (Iler 1979). Once water is adsorbed, the site then has free active hydrogen for adsorption through hydrogen bonding. This water-induced increase of the hydrogen bonding sites explains the interaction increase with increasing RH for the silicon substrate.

It is well known that the hydrogen-covered silicon surface is hydrophobic, whereas the surface covered with imperfect oxide is hydrophilic (Okorn-Schmidt 1999). Based on the Kelvin equation, the Kelvin radius becomes smaller when the surface becomes more hydrophobic. Modification of the contact angle (hygroscopicity) of the PSi surface is possible when using various treatment temperatures (Buckley et al. 2003). For example, it was established that the oxidation strongly promotes capillary condensation of water vapors in the pores. Modification of surface properties may be used for influencing the operating characteristics of humidity sensors since the

condensation of water on the hydrophobic surface occurs at a higher values of relative air humidity. However, as Björkqvist et al. (2006) established, in that case, the configuration of sensors must be optimized because the hydrophilic electrodes also affect water condensation in porous medium. This effect can lead to abnormal sensor behavior.

### 6.4.2 RESISTIVE HUMIDITY SENSORS

Resistive-type humidity sensors based on PSi were also designed and tested by Baratto et al. (2002), Di Francia et al. (2005), Wang and Yeow (2008), and Jalkanen et al. (2012). Measurement of the electrical conductivity of direct current (DC) and alternating current (AC) of PSi showed that the properties of PSi change dramatically as a result of the environmental impact. At that, Schechter et al. (1995) established that this effect is most pronounced for a hydrophilic surface. At the same time, PSi with a hydrophobic surface can have no response to water vapor. Water vapor affected the conductivity of hydrophobic PSi only at a high humidity level and prolonged exposure. It was also found that the humidity response can be easily changed by a simple pre-treatment or aging of the samples.

There are several approaches to explain the behavior of the current in PSi with the adsorbed gas (Schechter et al. 1995; Green and Kathirgamanathan 2002). The most popular is associated with changes of the dielectric constant due to the condensation of the gas in the pores. Moreover, the formation of a condensed film can introduce a parallel ionic conductivity. Nevertheless, this hypothesis has been ruled out by Ben-Chorin and Kux (1994). Indeed, they observed that the variation of conductivity versus the methanol pressure started below the condensation conditions. According to Archer et al. (2005), due to interaction of gas molecules with the PSi surface, the porous layer becomes a charged layer that can modulate the field in the Si channel (see Figure 6.12). This modulation can be carried out by two mechanisms: (1) the change in the space charge region by the charge redistribution, and (2) the change in the width of the conductive channel. However, the adsorption mechanism and its influence on the carrier concentration in PSi are still not clear (Schechter

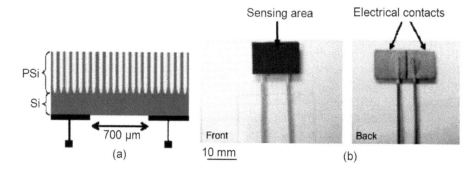

**FIGURE 6.12** (a) Schematic cross-sectional view of a porous silicon sensor. The electrical contacts are placed on the back part of the layer (c-Si) by aluminum evaporation or colloidal silver paint 700 μm apart; and (b) pictures of the front and backsides of the device. (Reprinted from *Sens. Actuators B.*, 106, Archer M. et al., Electrical porous silicon chemical sensor for detection of organic solvents, 347–357, Copyright 2005, with permission from Elsevier.)

et al. 1995; Barillaro et al. 2000). From our point of view, the absence of a realistic model of PSi-based gas sensors is connected with the lack of understanding of the conductivity mechanism in porous materials. At present, too many assumptions are being used for this phenomenon explanation (Yamana et al. 1990; Ben-Chorin and Kux 1994, Ben-Chorin et al. 1994; Fejfar et al. 1995; Lubianiker and Balberg 1998; Lee et al. 2000; Shi et al. 2000; Remaki et al. 2001; Garrone et al. 2003; Sukach et al. 2003), which is probably a consequence of the presence of many factors influencing the current percolation in porous structures.

Experiments carried out by Jalkanen et al. (2012) have shown that optimized PSi-based resistive humidity sensors can provide good sensitivity and reproducible signals throughout the entire RH range (see Figure 6.13a). In addition, negligibly small hysteresis (Figure 6.13b), accompanied by relatively fast response and recovery times was also demonstrated. Jalkanen et al. (2012) believe that the low hysteresis, observed in Figure 6.13b, is a result of applied surface functionalization, which minimized the influence of capillary condensation and increased the role of surface effects.

It is important to note that monitoring the conductance variation of porous semiconductors is the simplest and cheapest way to realize a humidity sensor with PSi. However, some difficulties need to be overcome. For example, a reliable contact on PSi using a microwelder cannot be obtained due to the fragility of the material.

The optimization of the sensor performance also needs a better control of the thickness and porosity of the PSi membrane and its wetting ability. Salgado et al. (2006) concluded that a very thin PSi layer is not enough to detect some appreciable resistance changes and for thick PSi layers the diffusion time of the molecules plays an important role to have fast responses.

The same result, concerning the strong sensitivity dependence of the porous microstructure as well as of the porous

layer thickness, was obtained by Pancheri et al. (2003, 2004). They observed a large microstructural transformation in PSi even for an HF concentration change between 13% and 15%. Pancheri et al. (2003, 2004) assumed that the observed different sensitivity most likely originated in the degree of interconnection between the conducting microchannels. In the structures with lower degree of interconnection, high-resistance paths, which are highly sensitive to gas, dominate the overall resistance, as in a series arrangement. On the other hand, in structures with higher interconnections, low resistance, and insensitive paths frequently allow local bypasses across the sensitive paths (parallel arrangement). Thus, in the latter case, the sensitivity is dramatically inhibited.

### 6.4.3 OTHER TYPES OF PSi-BASED HUMIDITY SENSORS

The effect of humidity on PSi-based heterostructures were reported in Strikha et al. (2001), Yarkin (2003), Litovchenko et al. (2004), and Kayahan (2015). For example, a three-orders-of-magnitude increase of conductivity in an Au/PSi/c-Si diode was observed by Yamana et al. (1990) when the RH of the ambient atmosphere increased from 10% to 100%. Pd/PSi/c-Si structures also showed sensitivity to water vapor. For explanation of the observed effects, a well-known model designed for Pd-semiconductor gas sensors is commonly used (Lundstrom et al. 1975, 1989; Johansson et al. 1998). According to this model, the sensor response is determined by the change of the potential barrier height caused by a change of gas environment. Hydrogen or water molecules rapidly dissociate into hydrogen atoms and hydroxyl species by the catalytic property of the Pd metal, and then the hydrogen atoms penetrate through the Pd metal with a high diffusion coefficient. These hydrogen atoms are trapped at the interface between Pd metal and semiconductor and, due to the forming a dipole layer, give rise to an electrical polarization and the increase in the height of Schottky barrier (Johansson et al. 1998; Litovchenko et al. 2004).

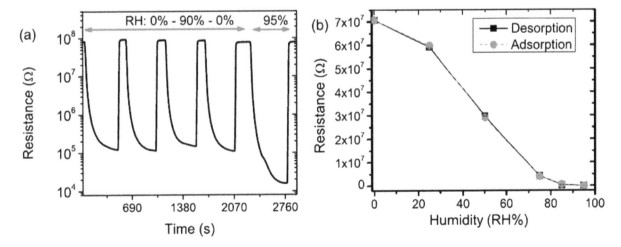

**FIGURE 6.13** (a) The sensor response to RH modulation between 0% and 90%. Final humidity pulse to 95 RH% demonstrates that the sensitivity is well below 5 RH%, even for higher humidity values; (b) the sensor response to a stepwise relative humidity sweep ranging from 95 to 0 RH%, and the corresponding hysteresis curve determined from the measurement. (Reprinted with permission from Jalkanen, T. et al., *Appl. Phys. Lett.* 101, 263110, 2012. Copyright 2012 by the American Institute of Physics.)

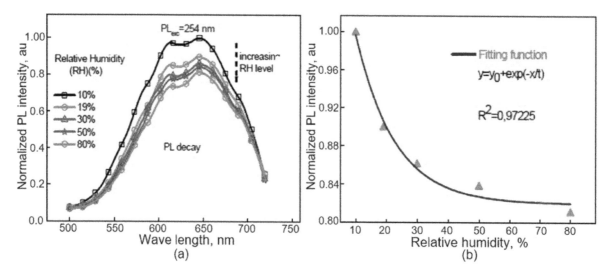

**FIGURE 6.14**   (a) PL spectra, and (b) the PL intensity changes of porous silicon with increasing RH levels. (Reprinted with permission from Kayahan, E., *Acta Phys. Polonica A*, 127, 1397–1399, 2015. Copyright 2010 by the American Institute of Physics.)

However, the above-mentioned model is too simplified. Due to the porous structure of both metal film and semiconductor, inside the pores we have a semiconductor surface uncovered by metal, which also can participate in the gas-sensing phenomena. This means that direct gas-PSi interactions must be taken into account for a correct explanation of the observed effects. We believe that such sensitivity of PSi-based sensors is conditioned by the porosity of the metal layer and the participation of the uncovered PSi surface in the gas-sensing effects. Moreover, we believe that the modulation of the PSi conductivity along the pores, due to gas interaction with PSi surface inside pores, plays a dominant role in achieving extremely high sensitivity of PSi-based humidity sensors.

The water vapor influence on the photoluminescence properties of PSi can be also used for humidity measurement (see Figure 6.14). This effect was demonstrated in Maruyama and Ohtani (1994), Baratto et al. (2002), and Kayahan (2015). However, it should be noted that this method is not very convenient to use and does not provide the necessary reproducibility.

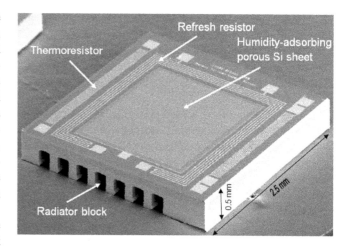

**FIGURE 6.15**   SEM micrograph of the PSi humidity sensor of capacitive type. (Reprinted with permission from *Sens. Mater.*, 12, Rittersma, Z.M. and Benecke, W., A humidity sensor featuring a porous silicon capacitor with an integrated refresh resistor, 35–55, Copyright 2000, with permission from Elsevier.)

### 6.4.4   MICROMACHINED PSi-BASED HUMIDITY SENSORS

Another advantage of PSi as a sensor material is the possibility to integrate it with already existing silicon technology. CMOS compatibility of PSi enables the development of microminiaturized humidity sensors with improved sensitivity and reduced size. It is this characteristic of PSi that made it possible to develop on its basis a capacitive sensor that has a shorter recovery time after interaction with water vapor. The humidity sensors realized using micromachining technology (see Figure 6.15) usually consist of three parts (Rittersma et al. 2000, Rittersma and Benecke 2000): (1) a humidity-sensitive capacitor with a PSi dielectric, (2) two integrated thermoresistors, and (3) a refresh resistor. The capacitor is formed between a meshed top electrode and a low-Ohmic backside of the sensor. Refreshing can take considerable electrical power,

and hence for rapid refreshing it would be advantageous to fabricate the capacitor such that it is thermally insulated from the bulk silicon. For this purpose, the part of the back surface may be etched with KOH or reactive ion etching using micromachining technology. This technology will be discussed in details in Chapter 15. The choice of metal resistors was made in view of a possible economical process sequence and ease of packaging: placed on the surface of the sensor, the resistors and bondpads can be fabricated in a single metallization step together with the top electrode of the capacitor (Rittersma and Benecke 2000). Because the latter contains small meshes, the PSi dielectric can be formed as the final step in the fabrication sequence.

Further, the described approach was used in the development of sensors with improved functionality (Das et al. 2003).

For this, the PSi-based capacitance humidity sensor has been integrated with a phase detection electronic circuit. The PSi sensor has been fabricated with a membrane-type contact structure for better capacitive sensing and a built-in planar micro hot plate at the top surface for restoration and processing convenience. It has been shown that simple electronics improves the behavior of the sensor. In particular, a simple phase detector circuit allowed to measure and monitor the capacitive changes with accuracy and resolution (better than 0.1%) and good linearity over the entire range of humidity (15%–98%). The response time was less than a few seconds.

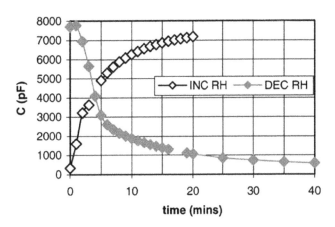

## 6.5 LIMITATIONS OF PSi-BASED HUMIDITY SENSORS

However, research has shown that in the most cases PSi-based humidity sensors efficiently operate only in the limited range of RH (Connolly et al. 2002, 2004a; Fujes et al. 2003; Di Francia et al. 2005, Xu et al. 2005). For example, analysis of fabricated devices showed that porous semiconductor-based humidity sensors in capacitance configuration efficiently operate only in the range of RH higher than 10–20% (O'Halloran et al. 1999; Connolly et al. 2002, 2004a, 2004b; Salonen et al. 2002; Füjes et al. 2003; Xu et al. 2005; Björkqvist et al. 2004a, 2004b, 2006). As indicated earlier, the capacity effects are attributed to the variation of the dielectric constant due to the filling of the pores by the gaseous species.

The low selectivity of PSi-based humidity sensors is also a disadvantage of such kinds of sensors. Studies have shown that such sensors, in addition to response to air humidity, are highly sensitive to vapors of organic solvents and oxidizing gases such as $NO_2$ and $O_3$. As a result, this kind of device is not able to identify the components of a complex mixture containing water vapors and gas. However, this disadvantage is peculiar to all adsorption type sensors, including metal-oxide-based gas sensors (Korotcenkov 2005, 2007).

**FIGURE 6.16** Response-time data for porous polysilicon (after burn-in); open symbols correspond to a change in RH from 10% to 90%, and the filled symbols a change in RH from 90% to 10%. (Reprinted with permission from *Sens. Actuators A*, 99, Connolly, E.J. et al., Comparison of porous silicon, porous polysilicon and porous silicon carbide as materials for humidity sensing applications, 25–30, Copyright 2002, with permission from Elsevier.)

Slow response, due to the limited mass transportation and capillary condensation, is another disadvantage of PSi-based humidity sensors (Connolly et al. 2004b). From the results presented in Figure 6.16, one can see that the response and recovery processes in PSi-based humidity sensors really are very slow (Connolly et al. 2002). According to Connolly et al. (2002), even the fastest SiC-based humidity sensors showed the response time about 120 s when the RH changed from 10% to 90%. As shown by Fujes et al. (2003) an excessive thickness of the PSi layer would make the absorption and desorption processes much more difficult and lead to a long response time. Therefore, the optimization of a porous-layer thickness is needed for achieving optimal parameters of humidity sensors.

Poor linearity of operating characteristics (Wang et al. 2010) can be added to the disadvantages of porous-semiconductor-based RH sensors (see Figure 6.17b). Large hysteresis

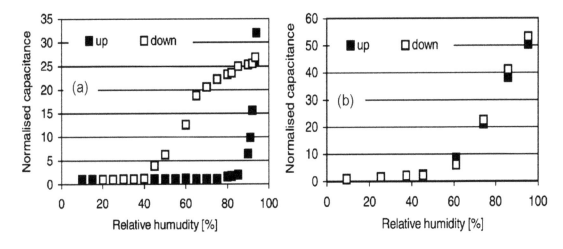

**FIGURE 6.17** Hysteresis effect in as-prepared PSi-based humidity sensors (a) without any heating, and (b) with refreshing before each measurement (20 s, 120 mW). Up and down indicate the direction of humidity change. (Reprinted with permission from *Sens. Actuators B*, 95, Füjes, P., et al., Porous silicon-based humidity sensor with interdigital electrodes and internal heaters, 140–144, Copyright 2003, with permission from Elsevier.)

is also a common problem in humidity sensing based on adsorption (see Figure 6.17a). It is believed that larger hysteresis is due to the intrinsic properties of the material causing capillary condensation of the water vapor to occur in the nanometer scale pores.

The aging of PSi and long-term drift of sensor parameters is also a significant disadvantage of PSi-based humidity sensors (Maruyama and Ohtani 1994; Islam and Saha 2007). For example, Kim et al. (2001) found that the exposure of freshly formed PSi films to humid atmosphere (30 Torr) at room temperature results in a gradual increase of the photoluminescence (PL) yield during the first two days of storage. Simultaneously, the maximum spectral yield shifts toward red. The initial luminescence intensity may be restored by evacuating the PSi sample at a temperature of 150°C. Treatment of PSi in humid atmosphere in excess of two days results in irreversible decrease of PL intensity. It was established that the aging time may require more than 5–10 days (Setzu et al. 1998). It was established that when a PSi is stored in oxidizing environments like water and $H_2O_2$, the formation of Si-O-Si bonds replacing Si-$H_x$ bonds takes place (Hossain et al. 2000). The same effect was observed during the PSi surface carbonization as well (Mahmoudi et al. 2007). It was observed that the rate of oxidation of the PSi layer depends on the concentration of OH− and holes in the valence band of the PSi layer.

The fabrication of stable contacts with low resistance is a great problem for PSi as well (Lewis et al. 2005). It was established that the physical methods of metal deposition do not minimize a contact resistance due to a small contact area, whereas during chemical methods of metal deposition a strong degradation of gas-sensing characteristics could occur. Some technical approaches for resolving these problems were proposed (Gole et al. 2000; Tucci et al. 2004; Lewis et al. 2005). In particular, in research carried out by Lewis et al. (2005), the optimization of technology of contact forming allowed a considerable improvement of the operating characteristics of PSi-based sensors, including their stability. Using a designed metallization process, Gole and his collaborators (2000) dramatically reduced the resistance of the electrodes, allowing their PSi-based sensors to operate between 1 and 10 mV. This is very good result, because sensors based on PSi that have been built before usually required an operating voltage of as much as 3–5 V due to the high resistance in the electrodes connected to the PSi.

The other important disadvantage of the devices based on PSi is strong nonreproducibility of the properties of PSi. The varieties of the morphology of PSi are too high, and the dependence of its parameters on the layer structure is too strong (Foll et al. 2002). One can understand how relevant the problem of the irreproducibility of parameters of porous films is from the fact that the samples prepared by different research groups are hardly comparable even if the preparation conditions are apparently the same (Parkhutik 1999).

As follows from the above, humidity sensors on the basis of PSi have disadvantages that can restrain their commercialization (Connolly et al. 2004b). However, research carried out during recent years has shown that operating parameters of

PSi-based humidity sensors can be considerably improved. For example, Xu et al. (2005) showed that for PSi-based humidity sensors, prepared by hydrothermal etching, when the RH changed from 11% to 95%, the capacitance showed an increment over 1500% at a frequency of 100 Hz.

## 6.6 APPROACHES TO OPTIMIZATION OF PSi-BASED HUMIDITY SENSORS

### 6.6.1 SENSITIVITY

As follows from the analysis of the operation of PSi-based sensors, the sensitivity of a PSi sensor is determined by the amount of adsorbed and condensed analyte in the pores, and by the refractive index and dielectric constant of the analyte (Anderson et al. 2003; Salem et al. 2006). For a liquid analyte, there is full infiltration of the porous matrix and the PSi sensor acts as a refractometer (see Figure 6.18) or capacitor with increased capacitance.

However, for a gas-phase analyte, the spectral shift and electroconductivity of PSi are mainly determined by the effects of surface adsorption, which govern the amount of adsorbed analyte, and not the refractive index or dielectric constant of the material. For example, Ruminski et al. (2010) established that despite the fact that isopropanol and heptane have similar refractive indexes, the response of optical sensors toward 500 ppm of isopropanol was ~9 times as large as the response toward 500 ppm of heptane. Only in the case of capillary condensation of vapors into a PSi matrix, the refractive index and dielectric constant of the material become the factors determining the sensor response.

However, for a gas-phase analyte, the signal of PSi sensors is determined mainly by surface adsorption effects, which govern the amount of analyte adsorbed, rather than the refractive index of the material. For example, Ruminski et al. (2010) established that despite the fact that isopropanol and heptane

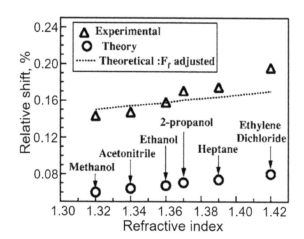

**FIGURE 6.18** Graph of theoretical and observed relative shifts in the resonant wavelength of anodically oxidized porous silicon rugate filter (PSRF) vs. the refractive index of the liquid immersed into the pores. (Reprinted with permission from Salem, M.S. et al., *J. Appl. Phys.*, 100, 083520, 2006, Copyright 2010 by the American Institute of Physics.)

have similar refractive indexes, the response of optical sensor towards 500 ppm of isopropanol was ~9 times as large as the response toward 500 ppm of heptane. Only in the case of capillary condensation of vapors into the PSi matrix, the refractive index of the material becomes a factor that determines the response of optical sensors. As it was indicated earlier in Chapter 2, microcapillary condensation is a nanoscale phenomenon that provides an additional means of increasing sensitivity by spontaneously concentrating volatile molecules (Casanova et al. 2007).

The experiment showed that the extent of both adsorption and capillary condensation of vapors into PSi optical sensors are influenced by the surface affinity of analytes to the porous matrix. For example, freshly etched PSi possesses a hydride-terminated surface that is generally unstable and does not provide good binding sites for organic and water vapors (Ruminski et al. 2010). Thus, freshly etched PSi is generally modified by diverse chemical treatments, surface functionalization, to provide more stable surfaces and to impart sensitivity or specificity for the intended analyte (Buriak 2002; Gao et al. 2000, 2002a, 2002b; Ruminski et al. 2010). In particular, it was found that the surface chemistry could be changed by thermal oxidation, silanization, carbonization, or functionalization by the covalent binding of functional groups (O'Halloran et al. 1997; Buriak et al. 1998; Stewart and Buriak 1998; Salonen et al. 2000; Björkqvist et al. 2004a, 2004b; Mahmoudi et al. 2007). For example, carbonization changes the originally hydrophobic PSi surface to hydrophilic, thus improving its humidity sensing properties. Moreover, the sensitivity of TC-PSi is presumably better due to its larger specific surface area. A recent report on TC-PSi humidity sensor showed good sensitivity (see Figure 6.10a) and repeatability of the sensor (Figure 6.10c) (Björkqvist et al. 2004a).

Surface modification with metals and metal oxides can also be used to improve sensor performance. For example, it was established that the performance of PSi organic vapor sensors based on PL quenching can also be strongly improved after surface modification by $HfO_2$ (Gan et al. 2014) and $LaF_3$ (Mou et al. 2011) layers. At optimal thickness of the covering, enhanced PL and a slightly decreased PL lifetime were achieved. However, with the increasing thickness of the $LaF_3$ layer, PL intensity of PSi was decreasing along with a small blue shift. According to Gan et al. (2014), the compact coating of an $HfO_2$ layer effectively passivates the surface states and induces the Purcell effect, which suppresses the nonradiative recombination rates and increases the radiative recombination rates. Mou et al. (2011) believe that the $LaF_3$ layer prevents oxidation of PSi.

Decreasing the pore size of the PSi sensor is also a key factor of improving its sensitivity toward water vapor. Although analyte vapors will adsorb onto the surface of meso and macroporous silicon, nanoscale pores possess an additional capability to concentrate a water vapor, known as microcapillary condensation (read Chapter 2). Taking into account that the water in liquid form has the maximum effect on dielectric constant and refractive index, it becomes obvious that the sensitivity of the capacitive and refractometric optical sensors is controlled by the pore size and the porosity of the PSi. The smaller the pore radius, the lower the partial pressure at which water-vapor condensation can occur at a given temperature (Adamson 1990; Israelachvili 1992). The greater the porosity and open pores volume are, the greater is the amount of condensed water or organic solvent, and the greater is the range of capacity and refractive index change when interacting with various vapors. Exactly capillary condensation is responsible for the visibly observable, large (~100 nm) spectral shifts of the rugate stop band upon water-vapor exposure. Strong increase in capacitance in humid atmosphere is also the result of capillarity condensation. It is believed that the volume fraction of pores in sensing material designed for capacitive sensors should be as high as 45%. As it was shown in Section 6.3, the pore size can be controlled by changing both the current density during silicon porosification, and the dopant level of the crystalline silicon wafer. Chemical stain etching (Kolasinski 2005), which degrades PSi without the application of an electrochemical bias, has also been used to create microporous layers with small pore diameters.

The pore size distribution is also an important parameter. According to the prediction of Kelvin equation, water vapor starts to condense at room temperature in mesopores of size 2 nm around 15% RH and continues to around 100 nm under saturated atmosphere. Because of that, the pore size in the range of 2–10 nm is optimal for effective humidity sensing in the low RH range, whereas the material with larger (20–100 nm) pores is more preferable for the design of humidity sensors for high RH values (Seiyama et al. 1983; Connolly et al. 2002). This means that sensors with a large pore size have a low sensitivity in the region of low concentration of analyte and they, as a rule, begin to feel vapors usually at a humidity level exceeding 60%–80% RH. Therefore, optimal samples of PSi, developed for sensors designed to perform measurements in a wide range of concentrations of detected gases and vapors, should have a specific distribution of pore sizes. For example, for humidity sensors, experiment has shown that the best results were obtained for samples with a wide distribution of pore sizes from 2 nm up to radii, in which capillary condensation cannot take place at any humidity at their operating temperatures (Shimizu et al. 1985).

As for the specific surface area and pore volume, it is important to note that the specific surface area is the principal microstructure for sensing water vapors under low concentration, while pore volume dominates at high concentration of analyte. For example, at low humidity, water adsorption on the sample surface is likely to be the dominant factor for electrical conductivity. Consequently, a higher surface area will provide more sites for water adsorption and will give more charge carriers for electrical conductivity. Under high RH, for example, above 80%, as a rule, mesopores are already filled with water, and therefore it is precisely the pore volume determines the total quantities of water condensed in the metal-oxide matrix, that is, the magnitude of the change in dielectric constant.

It should be noted that the specific surface area is directly related to the porosity of the material. The higher the porosity, the greater the specific surface area. But this regularity is performed only up to a certain amount of porosity; exceeding this threshold value is accompanied by a decrease in the specific surface area. As a rule, the reduction of a specific surface area begins when the porosity exceeds 50%–60% (Halimaoui 1997).

The experiment showed that temperature control also gives possibility to influence the sensitivity of PSi-based sensors. In particular, Zangooie et al. (1999) demonstrated that vapor capture is promoted at lower temperatures, resulting in an increase of the ellipsometric response. Furthermore, decreasing the temperature resulted in the shifts in the ellipsometric parameters occurring at lower partial pressures.

## 6.6.2 SELECTIVITY

In principle, PSi-based sensors are not selective sensors. This means that it is not possible to quantify a complex gas–water vapor mixture in one measurement, as it happened in gas chromatography. The experiment showed that the improvement of selectivity can be achieved only via using a number of special approaches. They are as follows:

*Lock and key approach.* In this approach to specificity, PSi can be conjugated to a specific capture probe for a target species (Rickert et al. 1996; Albert et al. 2000). Specificity comes with a trade-off, since a separate sensor must be constructed for each new target analyte and capture probes that tightly bind their targets can have poor reversibility (Rickert et al. 1996). This approach is mainly used in PSi-based biosensors to specifically bind a target analyte such as a protein.

*Surface reactivity control via surface functionalizing.* The surface of PSi has been extensively tailored due to the large library of reactions available to modify the silicon–hydrogen and silicon–silicon bonds (Song and Sailor 1999; Salem et al. 2006; Ruminski et al. 2010). Methods include hydrosilylation (Buriak 1999), radical coupling (Lewis 1998), Grignard reactions (Chazalviel and Ozanam 1997), electrochemical reduction of alkyl halides (Gurtner et al. 1999), and thermal carbonization with acetylene (Salonen et al. 2000, 2004). As a result of these treatments, the PSi surface can directly react with some chemical species (Rocchia et al. 2003; Ruminski et al. 2010). Selectivity PSi interaction with analyte can be also improved by loading with catalysts that react with specific compounds. For example, Litovchenko et al. (2017) have found that nanostructured catalysts based on nanoclusters of transition metals (W, Pd, Cu) and their oxides deposited on PSi are characterized by an enhanced activity with respect to the adsorption and catalytic decomposition of $H_2S$, $H_2$, and $H_2O$ molecules.

*Physical control of surface reactions.* As was shown, the chemical composition of the surface of PSi has a profound effect on the sorption of analyte vapor. But sorption can also be modulated physically by heating or cooling the porous layer. Thermal cycling techniques for distinguishing organic vapors have been used with metal oxides and other types of sensors (Heilig et al. 1997; Nakata et al. 1998; Lee and Reedy 1999) but have not been widely studied with PSi. There are only two papers related to this topic (King et al. 2011; Kelly et al. 2011). The necessity of the spectral curves measurement limits the ability of measurements in dynamic mode.

De Stefano et al. (2003, 2004) and Letant and Sailor (2001), who studied PSi-based optical sensors, used another approach for selective detection of vapors. When using a time-resolved measurement of the red shift of PSi microcavity (MC) resonance peak, De Stefano et al. (2003, 2004) demonstrated that the MC red shift is sensitive to hydrocarbon vapors and specific to their molecular weight. Letant and Sailor (2001) applied time-resolved reflectivity to discriminate two vapors (acetone and methanol) and to identify their components in a binary gas mixture. Of course, this approach can be used for identification of water vapor in gas–water–vapor mixture.

As the experience of using metal-oxide gas sensors has shown, an improvement in the selectivity of the sensor response can also be achieved through changes in operating temperatures (Korotcenkov, and Cho 2013). Gases and vapors can be selectively recognized by changes in operating temperature, since different gases and vapors due to the particular combination of adsorption/desorption parameters and reaction rates can have different temperature profiles of sensor response. This means that by changing operation temperature it is possible to create conditions when the response to one gas will exceed considerably the response to other one. However, one should admit that this approach cannot completely resolve the problem of low selectivity of PSi-based gas sensors, since temperature profiles of sensor signal are too broad and there are too many gases, which have similar $S = f(T_{oper})$ dependences. In addition, as applied to PSi-based gas sensors, this mode of operation is undesirable, since operation at elevated temperatures enhances the degradation processes and the temporal drift of the characteristics of these devices.

*Size exclusion.* The ability to create layers of PSi with different pore geometries led to molecular selectivity by size exclusion. However, this approach was used mostly in biosensors, where the detected molecules are large. For example, Orosco et al. (2009) designed a double layer of mesoporous silicon for this purpose in enzyme sensors. Collins et al. (2002) showed that in addition to stacks of PSi layers with different pore sizes, lateral pore diameter gradients of PSi can also be used to separate molecules by size. For gas sensors, this approach has limited capabilities, since the pore diameter should be too small for effective molecular selectivity by size exclusion.

*Using of sensors arrays.* This approach is the basis for the development of various electronic noses (Letant and Sailor 2000). Usually in this approach, arrays are used that vary the material or coating of each sensor element, but use a single transduction methodology. In the case of PSi-based sensors, the change in the pore size can replace the use of different materials. In particular, this approach was realized by Islam et al. (2006). They demonstrate the discrimination of organic vapors using an array of PSi sensors with different porosity and pore morphology. It was shown that every element in this sensor arrays have specific response to analyte. Therefore, processing of the sensor arrays responses using specialized computer software can create multidimensional response profiles which can be viewed as a digital "smellprint" or "fingerprint" that represents the chemical complexity of the gas mixture (Oton et al. 2003; Hutter and Ruschin 2012). Hutter et al. (2011) also used sensor array for gas detection. However, to get a specific response, they used PSi monolayers infiltrated with pH-sensitive dye (bromocresol purple) at different concentrations. The transduction mechanism was the reflectance change due to a spectral shift of the dye-absorbance band reducing the reflectance intensity. The correlation of the array output (vector of four signals) with an expected gradient-like response (owing to the gradient of the dye concentration) enabled the detection of ammonia, minimizing false alarms from the humidity presence and unstable illumination.

The recognition can be further enhanced either by increasing the number of sensors in the array or by using the data of the response and the recovery time, which are unique for each vapor. Park et al. (2010) showed a different approach to the development of sensors, suitable for use in the electronic nose. They produced PSi samples with gradually varying porosity. The technology was based on inserting a Si wafer gradually (or by stages) into the HF solution during the anodization process to obtain the pore-size and gradient of the layer thickness or various multilayers, on a single substrate. Park et al. (2010) believe that PSi structures with a lateral pore gradient distribution can be used in various areas, including matrix in optical electronic nose systems.

*Multiparameters sensing.* This approach uses simultaneously several differing transduction methods. In particular, multiparameters control can combine responses from sensors that use different operating principles, like reflectance sensing with PSi, photoluminescence quenching, Fourier transform IR spectroscopy, Raman spectroscopy, fluorescence detection, ion mobility spectroscopy, mass spectroscopy, surface Plasmon resonance, or other methods (Letant et al. 2000; Mulloni and Pavesi 2000; Baratto et al. 2002; Park et al. 2010). Therefore, having access to several gas properties, multiparametric PSi sensors are expected to have better gas selectivity than single-parameter sensors. For instance, simultaneously monitoring thin-film reflectance and PL enabled Letant et al. (2000) to discriminate among a large number of vapors, with the results being comparable to the performance of a commercial electronic nose. In addition to selectivity and sensitivity, this provides a way to rule out false positives, which increases the detection capability of the sensor. This approach was successfully used for $NO_2$ detection in humid air. Oton et al. (2003) showed that simultaneous measurement of electrical conductivity and the index of PSi refraction gives the possibility for independent control of $NO_2$ and air humidity. As established by Oton et al. (2003), the effective index of refraction of the porous layer depends on the water content inside the porous layer but not on the $NO_2$ concentration (see Figure 6.19).

This approach has received further development in the research of Baratto et al. (2002), who suggested comparing the sensor response as a result of the gas influence on the photoluminescence intensity, electrical conduction, and effective refractive index for resolving the problem of the selectivity of gas sensors based on PSi. It was shown that in a PSi microcavity, the photoluminescence intensity, the electrical conduction, and the resonance peak of the cavity are affected by the presence of $NO_2$, ethanol, and water in different ways (see Table 6.4).

One can see that $NO_2$ increases DC current and PL intensity, but no peak shift is produced. Humidity and ethanol cause a peak shift and decrease in DC current. The peak shift is higher with ethanol than with water due to the higher value of refractive index, whereas the conductance decrease is higher with humidity than with ethanol. In contrast with the case of nitrogen dioxide, the peak intensity variation is small for both humidity and ethanol. Thus, the use of PSi in multiparametric sensors can help to discriminate the presence of $NO_2$, which is a dangerous pollutant, from interfering gases like humidity and ethanol. The decrease in DC current may be associated with a decrease in $NO_2$ concentration or with an increase in RH. If a concomitant peak shift and intensity variation is observed, it happens due to the presence of humidity. If no peak shift is observed, the variation must be ascribed to $NO_2$ concentration variations. The same discrimination can be obtained with ethanol. An appropriate pattern recognition step could improve preliminary results of gas species identification and quantification. Unfortunately, any appreciable success in this direction was not achieved.

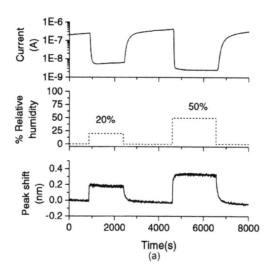
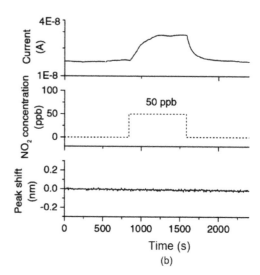

**FIGURE 6.19** Simultaneous measurement of optical and electrical response to humidity and NO$_2$. The electrical current is measured on the 30-μm deep monolayer employing the three-terminal configuration, and the reflectance peak shift is measured on the microcavity: (a) response to two different humidity values in the absence of NO$_2$, and (b) response to 50 ppb of NO$_2$ in the presence of 20% humid air. (From Oton, C.J. et al.: Multiparametric porous silicon gas sensors with improved quality and sensitivity. *Phys. Stat. Sol. (a)*. 2003. 197. 523–527. Copyright Wiley-VCH Verlag GmbH & Co. KGaA. Reproduced with permission.)

---

**TABLE 6.4**

**Sensor Signals That Can Be Used for Gas or Vapor Recognition. PSi Layers Were Formed Using Highly *p*-Type Doped Si Substrates**

|          | DC Current | PL Intensity | Peak Shift |
|----------|-----------|--------------|------------|
| NO$_2$   | Increase  | Decrease     | None       |
| Humidity | Decrease  | Varies       | Red shift  |
| Ethanol  | Decrease  | Varies       | Red shift  |

*Source:* Reprinted from Baratto, C. et al., *Sensors*, 2, 121–126, 2012.

---

***Impedance spectroscopy.*** Studies have shown, that selectivity of the sensor response can be improved by using specific measurement methods. Ibraimov et al. (2016) on the example of polar and non-polar vapors showed that the selectivity of a single PSi sensor can be achieved by the impedance measurements within the frequency range $10^3$–$10^5$ Hz. It was shown that the capacitance of the structure and its frequency dispersion are different under the influence of different gases and vapors, providing a unique agent signature. This difference can be attributed to the difference in the dipole moments, molar masses, and viscous lag of the analyte molecules in condensed liquids (Archer et al. 2005; Ozdemir and Gole 2007; Gole et al. 2007). As a result, it is possible to distinguish different gases and vapors based on the amplitude of the capacitance change and on the shape of the dispersion curve. At that, the highest change in the capacitance was found for the vapors of polar solvents: acetonitrile, ethanol, and methanol. The non-polar vapors such as toluene and chloroform displayed a smaller but still easily measurable

change of the capacitance in comparison with air. The use of conductivity measurement was not effective because the conductance was less sensitive to the vapors, especially at low frequencies. The characteristic time of response to different vapors gives an additional parameter to sense gases and vapors selectively. For example, it was established that the characteristic rise times for vapors of polar liquids were much smaller than the decay time. At that, the adsorption times were similar for all these agents, while the decay times were different, providing an additional parameter to distinguish the gases and vapors. It was also found that the most pronounced effect was observed on the frequency 1 kHz.

### 6.6.3  THE RATE OF RESPONSE

Based on the general concepts, it is clear that the kinetics of the sensor response of devices based on PSi is controlled by adsorption-desorption, diffusion, and capillary condensation of the analyte in the nanoporous matrix. This means that the main structural parameters affecting the kinetics of the sensor response are thickness, pore size, and porosity of sensitive material. For example, an experiment showed that sensors with a pore size above 100 nm in diameter usually exhibit a fast response, but in the case when the pore size is smaller 10 nm in diameter, the response and especially recovery may be long. As demonstrated by Fujikura et al. (2000), an excessive thickness of the porous layer also makes the absorption and desorption processes much more difficult and leads to a long response time. Letant and Sailor (2001), who studied single-layer PSi interferometer, also confirmed that too much thickness of the layer leads to an increase in the condensation time in the pores and, therefore, to a slower sensor response (see Figure 6.20a). They also found that an increase in porosity, achieved through an increase in pore size, also contributes to a decrease in response time (see Figure 6.20b). Therefore, it is quite clear that the most effective method of reducing the time of sensor response and recovery is to increase the pore diameter and reduce the thickness of the porous layer itself. Using this approach, Xu et al. (2005) developed humidity sensor with response and recovery times equaled about 15 and 5 s, respectively. Xu et al. (2005) believed that the fast response to humidity of PSi-based sensors might be due to the regular morphology and suitable thickness of the sensing layer, obtained during optimized hydrothermal etching process. However, this approach is not always acceptable, since when it is used, an improvement in the kinetics of the sensor response will be accompanied by a sharp decrease in the sensitivity of the sensor. As it was mentioned above, the presence of micropores causes a slow response time, while the absence of micropores is accompanied by a decrease in sensitivity (read Section 6.6.1).

We must not forget that the temperature mode of operation also affects the slowdown of the kinetics of the PSi-based sensor response. As is known, adsorption, desorption, and diffusion are activation processes and an increase in temperature is the easiest way to speed up these processes. For example, it was found that both sensitivity and recovery time extremes can be improved through the heating the PSi structures up to 200°C (Björkqvist et al. 2006). However, this approach with respect to PSi-based sensors is used very rarely, since the ability to operate at room temperature is one of the important advantages of these sensors. Therefore, with respect to PSi-based sensors, a slightly different approach is used, which was proposed by Björkqvist et al. (2006). They found that with periodical refreshing by heating up to 200°C, the sensitivity of PSi-based humidity sensors was nearly the same as without heating, but the recovery was faster and the hysteresis was negligible. This means that the application of chemical sensor refreshing before every measuring step is highly recommended. At the same time, studies have shown that the use of internal heating filaments made it possible to reach the desired temperature values faster around the active part of the chip than for heaters placed outside of the PSi structure (Fujes et al. 2003). King (2010) showed that a thermal pulse can also be successfully applied to PSi-sensors to refresh the sensor response after exposure to low-volatility vapor analytes. It was established that thermal pulse with heating already up to 160°C rapidly refreshes the oxidized silica sensors to their initial baseline.

### 6.6.4 HYSTERESIS

As noted earlier, large hysteresis is also a common problem for PSi sensors based on adsorption. Typical hysteresis effect observed in capacitive sensors is shown in Figure 6.17a. It was found that hysteresis is strong in hydrophilic samples. Therefore, it becomes clear that when designing sensors with minimal hysteresis, it is necessary to give preference to PSi with hydrophobic properties.

Björkqvist et al. (2004a) established that the hysteresis is also related to the size and geometry of the pores and how the presence of moisture changes this geometry. Widening of the pore size in PSi structures might solve the hysteresis problem. For example, a thermal carbonized (TC) PSi-based humidity

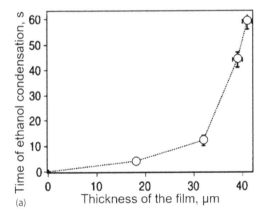

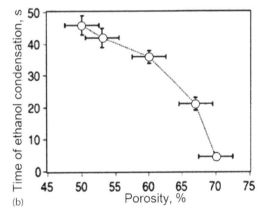

(a) (b)

**FIGURE 6.20** Time of condensation of ethanol in the pores (a) versus thickness, and (b) porosity of the PSi film. The samples were etched from the same silicon substrate in a solution of saturated (49%) aqueous HF/ethanol 3:1 (v/v) with a current density of 50 mA/cm$^2$ and then oxidized by a 20-min ozone treatment. The increase in porosity was achieved by increasing the pore size in the formed material by dissolution in aqueous HF solution (25 vol.%). (From Letant, S.E. and Sailor, M.J.: Molecular identification by time-resolved interferometry in a porous silicon film. *Adv. Mater.* 2001. 13. 335–338. Copyright Wiley-VCH Verlag GmbH & Co. KGaA. Reproduced with permission.)

sensor with larger pores designed by Björkqvist et al. (2006) showed only slight hysteresis above 75% RH. Unfortunately, the increase in the pore size reduces the sensitivity too. Because of that, the pore size in the range 2–10 nm is optimal for effective humidity sensing in the low RH range, whereas the material with larger (20–100 nm) pores is more preferable for the design of humidity sensors for high RH values (Seiyama et al. 1983; Connolly et al. 2002).

Björkqvist et al. (2004a) established that the hysteresis is also related to the size and geometry of the pores. They believe that the widening of the pore size in PSi structures might solve the hysteresis problem. For example, a TC PSi-based humidity sensor with larger pores designed by Björkqvist et al. (2006) showed only slight hysteresis above 75% RH (see Figure 6.21a). Unfortunately, as it was indicated above in Section 6.6.1, the increase in the pore size also reduces the response of PSi sensors.

Jalkanen et al. (2012) showed that thermal hydrocarbonization (THC) of PSi also gives a strong decrease in hysteresis. In particular, negligibly small hysteresis (Figure 6.21b), accompanied by relatively fast response and recovery times was demonstrated for PSi-based humidity sensors. To achieve such results, the surface of the PSi layer was stabilized by THC. To impart hydrophilicity to the surface, the THC PSi films were functionalized with undecylenic acid by immersing the films in a solution of undecylenic acid for 12 h at 110°C. Jalkanen et al. (2012) believe that the low hysteresis of THC PSi-based humidity sensors, observed in Figure 6.21b, is the result of applied surface functionalization, which minimizes the effect of capillary condensation and increases the role of surface effects.

It is important to note that in each of these cases, reduced hysteresis is mainly based on a decrease in the amount of analyte condensed in the pores Björkqvist et al. (2006). Unfortunately, in a capacitive-type and refractometric optical sensors, the sensor signal of which largely depends on the amount of analyte condensed in the pores, this also means lower sensitivity. However, the experiment showed that the

sensitivity of the PSi-vapor sensor even after the proposed modifications is still sufficient for accurate measurements.

Björkqvist et al. (2006) showed that the continuous heating of the PSi-based sensors also reduced the hysteresis drastically. However, Füjes et al. (2003) confirmed that the continuous heating reduces the hysteresis drastically, but they have also found that continuous heating also decreases sensitivity. Therefore, they proposed to use periodic refreshment. In this case, the sensitivity is almost the same as without heating, but the hysteresis was insignificant. The results of their studies related to capacitive humidity sensors are presented in Figure 6.10.

### 6.6.5 LINEARITY

The requirement of good linearity of sensing characteristics is usually imposed on capacitive and refractometric optical sensors designed to work in a wide range of concentrations of detected gases and vapors. As can be seen from the data shown in Figures 6.10b, 6.17, and 6.21 capacitive sensors typically have characteristics that are substantially non-linear.

Since the most significant change in the properties of PSi occurs during the condensation of water and organic solvent vapor in the pores, it becomes quite obvious that the linearity of the dependence of the capacitance of structures on air humidity or vapors of organic solvents is possible only with a certain distribution of pore diameters in PSi. This distribution of pore diameters should provide capillarity condensation at various concentrations of detected gas or vapor. As mentioned earlier, this situation can be realized in PSi, in which the pore diameters vary according to a certain law in a wide range from 2 nm to the size of a micrometer. However, it should be recognized that it is very difficult to technologically implement this situation, and the necessary size distribution of pores can be achieved only under a successful combination of circumstances. O'Halloran et al. (1999) believe that the linearity of sensing characteristics of capacitive humidity sensors can be improved by decreasing the current density used to form the

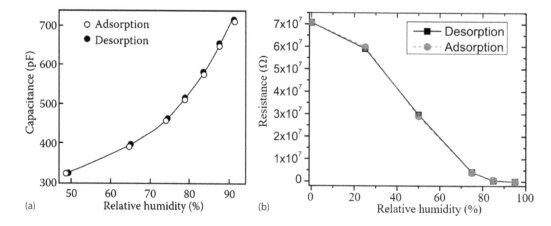

**FIGURE 6.21** (a) Hysteresis loop for PSi-based humidity sensors with larger pores. Humidity conditions are changed every 4 min. Data extracted from Björkqvist et al. (2006); (b) conductometric response of PSi-based sensor to a step-wise relative humidity sweep ranging from 95 to 0 RH%. (Reprinted with permission from Jalkanen, T. et al., *Appl. Phys. Lett.* 101, 263110, 2012. Copyright 2012 by the American Institute of Physics.)

porous layer. It was shown that when using indicated specific parameters of silicon porosification, it is really possible to fabricate a sensor with linear $C = f(RH)$ dependence over the humidity range from 10% to 90% RH. However, PSi layers formed at such low-current density values showed a long response time and poor sensitivity in comparison to the layers formed using higher values of current density. Therefore, an analysis of possible approaches to improve the linearity of the output sensor signal shows that the simplest solution is to use electronic circuits that allow linearizing the output characteristics of the sensors. In particular, this approach was used by Das et al. (2003).

Wang and Yeo (2008) used a different approach for linearization. To eliminate the capillary condensation, which gives the maximum contribution to the non-linearity, they used ordered $n$-type macroporous silicon with a pore diameter from 1.5 μm to 6 μm. As a result, they received conductometric humidity sensors with good linearity in the range of 0%–92% RH. However, the sensitivity of such sensors was very low. When the humidity in the specified range was changed, the resistance of the sensors changed only by 8%. The same result was obtained when using macroporous silicon in the development of capacitive humidity sensors (Wang and Yeo 2008). These results were presented in Figure 6.22. It was shown that in the range of 0%–80% RH, sensor response has linear dependence. However, this change is small.

## 6.6.6 STABILITY

Instability of PSi-based optical sensors' parameters has three main reasons. First, there is an aging of PSi. Due to its high chemical activity, PSi is easily oxidized. For example, it is well known that at room temperature and ambient

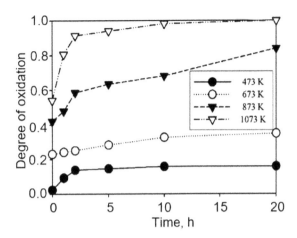

**FIGURE 6.23** Oxidation extent of 70% porosity and 30-μm thick porous membranes versus time and temperature as measured for the amount of oxide formed during the heating and cooling periods. (Reprinted with permission from Pap, A.E. et al., *J. Phys. Chem. B*, 108, 12744–12747, 2004. Copyright 1997 American Chemical Society.)

environment about 7 nm of Si, a native oxide can be formed with time. Oxidation, as reported by Björkqvist et al. (2003), is a very slow process at room temperature and thus leads to continuous changes in the structure of PSi and its physical parameters. Therefore, the properties of highly PSi and PSi-based devices are not stable in time, especially at initial stages of exploitation after manufacturing (see Figure 6.23), and can be significantly influenced not only by fabrication and drying conditions, but even by the manner in which it is stored, prior to examination or use (Table 6.5). As stated previously, PSi slowly reacts with the ambient air and consequently its chemical composition and its properties evolve continuously with the storage time. The effect of "aging" on both the composition and the structure of PSi are now well documented (Zhu et al. 1992; Canham 1997; Lee and Tu 2007). Changes in material properties, such as electrical resistivity and strain, and optical properties, such as refractive index and photoluminescence, accompany the aging process.

### 6.6.6.1 Stabilization through Oxidation

The experiment showed that the oxidation of PSi is the most effective method for resolving these problems (Salem et al. 2006; Korotcenkov and Cho 2016b). Commonly used modes of thermal oxidation of PSi are presented in Table 6.6. In particular, the use of pre-aged treatment by oxidation was one of elaborations suggested for aging effect prevention. It was found that oxidized PSi may be more stable and may contain fewer interface states affecting the Fermi-level pinning. For example, it was established that PSi could be converted to porous $SiO_2$ within 60 min at 950°C under a water vapor atmosphere. Connolly et al. (2002) proposed using the burn-in process, which involves heating the device to ~55°C, and repeated cycling of the RH between 5 and 95% for sensor parameter stabilization. It was found that this treatment improved humidity sensor stability as well as reduced hysteresis effects dramatically. Pancheri et al. (2004) showed that

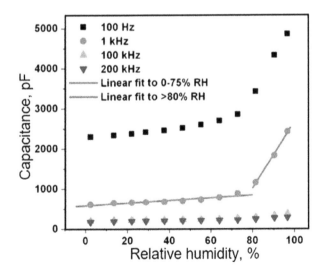

**FIGURE 6.22** Capacitance response of macroporous PSi-based sensor to air humidity at different measurement frequencies. The porous structured used in this sensor had a pore size of 4 μm and pore depth of 97 μm. (Reprinted with permission from *Sens. Actuators B*, 149, Wang, Y. et al., A capacitive humidity sensor based on ordered macroporous silicon with thin film surface coating, 136–142, Copyright 2010, with permission from Elsevier.)

## TABLE 6.5
### Effects of Varying Storage Conditions on Properties of PSi

| Storage Conditions | Major Effect |
|---|---|
| Ambient air (15 min–15 months) | Contaminated native oxide growth |
| Air, vacuum | Changes in layer strain |
| HF, ethanol, freon, ether | Lowest carbon levels for HF storage |
| $N_2$, $H_2$, forming gas, $O_2$ (min–hr) | Widely varying PL stability |
| Dry $N_2$, then UHV | Avoids photostimulated oxidation |
| Vacuum ($10^{-6}$ torr) | Carbon and oxygen pickup |
| Vacuum ($10^{-3}$ torr) | Heavy hydrocarbon contamination |
| Transport under propanol (<1 day) | Minimize oxidation by reducing air exposure |
| Ethanol storage and removal in UHV ($<2 \times 10^{-9}$ torr) | Minimize oxidation by completely avoiding airexposure |
| Cooled ethylene glycol | Green PL retained |
| Plastic and glass containment vessels | Blue PL due to plastic boxes outgassing |

*Source:* Data extracted from Canham L. (ed.), *Properties of Porous Silicon*, INSPEC, London, UK, 1997.

## TABLE 6.6
### Technological Parameters, Usually Used for Thermal Oxidation of Porous Si

| 1 step | 2 Step | 3 Step |
|---|---|---|
| Dry air, 800°C, 20 h, | | |
| Wet $O_2$; 1050°C; 3 h | | |
| Dry $O_2$, 1000°C, 60 s | | |
| Wet $O_2$; 300°C; 1 h | Wet $O_2$; 900°C, 1 h | |
| Dry $O_2$; 400°C; 30 min | Dry $O_2$; 1000°C; 1 h | Wet $O_2$, 1000°C; 30 min |
| Dry $O_2$; 300°C | Wet $O_2$; 850°C | Wet $O_2$; 1100°C |

*Source:* With kind permission from CRC Press: *Crit. Rev. Sol. St. Mater. Sci.*, Silicon porosification: State of the art 35, 2010b, 153–260, Korotcenkov, G. and Cho, B.K.

in aged PSi samples, the changes in resistivity of the porous layer, associated with exposure to $NO_2$, have improved the reversibility compared to fresh samples. The stabilization of the current is also faster in the aged than in the fresh samples. It is seen that the baseline shift (1) is absent in the aged sample and significant in the fresh sample, and the signal stabilization (2) is good in the aged, thin sample and poor in the fresh, thick sample. The same results were observed by Massera et al. (2004). They also proposed prolonged exposure to high concentrations of $NO_2$ for stabilization of sensor parameters. It was established that after such treatments, including a 10-h exposure to 2 ppm of $NO_2$ at 20% RH in synthetic air, followed by 2 weeks of stabilization in the ambient air, the devices were stable in the sub-ppm range and their electrical characteristics were greatly improved. The sensors had faster dynamics and reduced hysteresis. Both results suggest that the sample aging is a potentially good strategy for use of PSi conductometric sensors because reversibility and signal stabilization are still major unsolved limitations for such applications of freshly etched PSi-based sensors.

For the same purposes, Ben-Chorin and Kux (1994) proposed storing porous samples in a highly oxidizing medium like $H_2O_2$ for 48 h. Ruminski et al. (2010) established that the sensors with ozone oxidized PSi layer also had stable characteristics. It was found that initial oxidation in an $H_2O_2$ solution could stabilize the PSi layer to a greater extent. Holec et al. (2002) showed that methyl 10-undeacetonate derivatization of the PSi surface increases the PL-time stability of PSi-based sensors as well. Boiling in $HNO_3$ can also be used for PSi properties stabilization (Kochergin and Foell 2006).

It is important to note that silicon oxide has a refractive index lower than crystalline silicon, so that the average refractive index of oxidized PSi films strongly changes (Salem et al. 2006). This means that reflectance spectrum will be changed during PSi oxidation (Figure 6.24). In addition, the pore size may also change during the heat-treatment process. Pore enlargement caused by the higher surface treatment temperature is clearly seen from Figure 6.25, which depicts the pore size distributions of free-standing PSi films etched with the current density of 10 mA/cm². The lower treatment

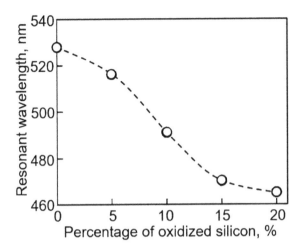

**FIGURE 6.24** Simulated resonant wavelength of oxidized porous silicon rugate filter (PSRF) at different volume fractions of oxidized silicon. (Reprinted with permission from Salem, M.S. et al., *J. Appl. Phys.*, 100, 083520, 2006, Copyright 2010 by the American Institute of Physics.)

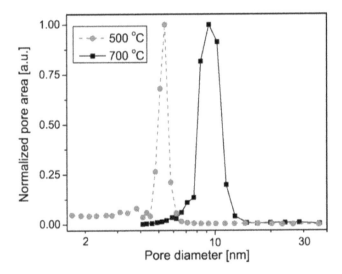

**FIGURE 6.25** Pore-size distribution for free-standing PSi films etched with the current density of 10 mA/cm². The higher treatment temperature used for sensor stabilization, leads to considerable pore enlargement. (Reproduced with permission from *Sens. Actuators B*, 147, Jalkanen, T. et al., Electro-optical porous silicon gas sensor with enhanced selectivity, 100–104, Copyright 2010, with permission from Elsevier.)

temperature of 500°C does not increase significantly the average pore diameter value (Salonen et al. 2009). However, when the temperature is raised up to 700°C, annealing starts to affect the pore morphology and the difference in the obtained pore sizes is obvious (Jalkanen et al. 2010). The design of the sensors should take this into account, since the increased pore diameter has a noticeable effect on the adsorption behavior of ambient gases infiltrating into the pores; in this case, the capillary condensation can only be observed for higher gas concentrations.

### 6.6.6.2 Stabilization through Thermal Carbonization and Nitridation

The long-term stability studies of differently stabilized PSi samples have shown that TC (Salonen et al. 2002; Björkqvist et al. 2004a) and THC of PSi (Jalkanen et al. 2012) are even a more efficient stabilizing method than thermal oxidation (Björkqvist et al. 2003, 2006). Characteristics of TC-PSi humidity sensor have been shown in Figure 6.10. A thermally carbonized PSi surface has been found stable in humid air and even in harsh environments. Thermal carbonization is a stabilizing method that exploits the dissociation of acetylene at high temperatures (Salonen et al. 2002). One of the biggest advantages of carbonization or hydrosilylation is that the material remains mesoporous and the surface area is diminished only slightly, whereas most oxidation methods will delete practically all mesopores from the sample (Björkqvist et al. 2003, 2004a; Salonen et al. 2004).

Instead of using oxidation or carbon-based stabilization methods, nitridation can be applied when passivating the PSi surface. With nitridation a highly stable surface and durable material is achieved, but the loss of surface area in the pores is sharply increased, especially during high-temperature treatments (Sailor 2014). No doubts, such effect is disadvantage of this method for sensor applications. Usually the routes for producing nitride passivation involve either $NH_3$ or $N_2$ atmospheres and relatively high temperatures (Morazzani et al. 1996; Lai et al. 2011).

### 6.6.6.3 Stabilization through Surface Functionalizing

Surface functionalizing by metal oxides can also promote the increase in stability of PSi-based sensor performance. For example, the coating of $HfO_2$ (Gan et al. 2014) and $Ta_2O_5$ (Wang et al. 2010) modifies the surface properties and thus improves the sensing stability of the PSi-based devices. In such sensors, metal oxide acts as an adsorption enhancement layer, while PSi provides a large surface area. At the same time, Mou et al. (2011) established that $LaF_3$ coatings do not have such pronounced stabilizing properties. Moreover, it was observed that all the coated PSi samples showed degradation in the PL intensity with time, but annealing at 400°C could recover and stabilize the degraded PL.

The deposition of polymer layers on the porous structure had similar effects. Vrkoslav et al. (2005) showed that impregnation of PSi with cobalt phthalocyanine (Co(II)Pc) is an effective way to improve the stability of the PL quenching response. According to the results obtained by Vrkoslav et al. (2005), the protection of the PSi surface with Co(II) Pc results in a substantial increase of resistance against slow ambient temperature oxidation. Regarding the accelerating of PL quenching in modified PSi, Vrkoslav et al. (2005) believed that the reduction of the pore size and the increase of the PSi surface polarity due to a Co(II)Pc impregnation were the main reasons for these effects. Hedrich et al. (2000) established that methyl 10-undeacetonate derivatization of the PSi surface also increased the PL time stability of PSi-based sensors. However, as a rule, the better stability of sensor parameters is paid for by a decrease in the sensor response. For example, a PSi-surface stabilization by Co(II)Pc was accompanied by a decrease of sensitivity by a factor of 1.5–2.2.

### 6.6.6.4 Temperature Stabilization

Considering the activation nature of many processes involved in gas-sensitive effects, the change in temperature of the environment in the measurement process also contributes to instability of the PSi-based sensors parameters. As is known, ambient temperature during the year can change in a very wide range. For example, in North Europe ambient temperature during the year can range from −30°C to +30°C. Therefore, in order to eliminate this effect, it is desirable to stabilize the temperature even for sensors operating at room temperature. This of course complicates the measuring system, but significantly improves the accuracy and reproducibility of measurements. The most advanced version of such a sensor is shown in Figure 6.15. This sensor includes an integrated heater and temperature sensor.

### 6.6.6.5 Drift Compensation

Summing up this review, it is necessary to recognize that the problem of instability of PSi-based sensor parameters has not yet been resolved. None of the previously mentioned techniques succeeded in complete prevention of the aging effect. Even after aging, sensors had temporal drift (Islam and Saha 2007). Lewis et al. (2005) reported that their devices continued to work months after their initial fabrication without requiring cleaning or recalibration. However, to test the sensor, Lewis et al. (2005) used dry nitrogen as a carrier gas. That is, the studies were conducted in the absence of humidity and oxygen, which have the strongest effect on the parameters of PSi. But, even in these conditions, temporal drift of sensor characteristics was present.

Islam and Saha (2007) believed that the problem of temporal drift could be partially resolved through compensation, using soft computing techniques. It can be assumed that drift compensation using additional processing algorithms is the easiest way to improve the stability of sensor parameters. This approach to mitigate the negative effect of gas sensor drift is not new and has been discussed in many articles related to metal-oxide-based sensors (Di Carlo and Falasconi 2012; Vergara et al. 2012; Korotcenkov and Cho 2014). However, to implement this approach, we need a more fundamental understanding of the nature of the drift effect in PSi sensors, which requires additional research. According to Pancheri et al. (2004), further efforts should be directed toward the development of an oxidation treatment, which could reproduce the effect of aging on PSi conductivity. Unfortunately, it is almost impossible to take into account all possible factors for building a mathematical model necessary to compensate the aging effect of PSi-based sensors using soft computing techniques. Too many hard-to-predict factors affect the aging process of sensors. Besides, this approach requires the ability to perfectly reproduce fabricated sensors. However, the experiment

shows that even two identical PSi-based sensors (same label) with the same history have different performance and different time stability.

## 6.7 CONCLUSIONS AND FUTURE TRENDS

Numerous studies conducted in recent decades have shown that PSi has great potential as an inexpensive platform for developing highly sensitive gas and vapor sensors. However, it must be recognized that the realization of these opportunities is encountering considerable difficulties, connected with insufficiently high stability of the parameters of PSi, low selectivity, and reproducibility of the sensor signal. As follows from our consideration, in recent years, significant progress has been made in improving the parameters of PSi-based sensors. In particular, modifications made to the manufacture of PSi-based sensors led to the creation of much more sensitive and reversible gas sensors than previously reported. As a result, there were hopes for the development of truly stable and reversible sensors capable of detecting various gases and vapors. But in order to realize these hopes, fully further studies are necessary. For example, a detailed classification of the structural peculiarities of PSi depending on the conditions of porosification and the performance of the PSi-based devices will be an important step toward both a deeper understanding of the nature of gas-sensing effects on PSi and the successful commercialization of gas and vapor sensors fabricated on its base. Korotcenkov and Cho (2010a) believe that establishing a correlation between the electrophysical properties of the initial material (bulk Si) and the parameters of PSi-based sensors can also provide additional information needed to find ways to optimize sensor parameters. The search for new approaches to stabilizing the parameters of PSi-based sensors and improving the selectivity of their response is also necessary, since the existing methods do not provide the required stability of the characteristics and the selectivity of their response.

## REFERENCES

Adamson A.W. (1990) *Physical Chemistry of Surfaces.* John Wiley & Sons, New York.

Adamson A.W., Gast A.P. (1997) *Physical Chemistry of Surface.* Wiley, New York.

Albert K.J., Lewis N.S., Schauer C.L., Sotzing G.A., Stitzel S.E., Vaid T.P., Walt D.R. (2000) Cross-reactive chemical sensor arrays. *Chem. Rev.* **100**(7), 2595–2626.

Allongue P. (1997). Porous silicon formation mechanisms. In: Canham L. (ed.) *Properties of Porous Silicon.* INSPEC, London, pp. 3–11.

Allongue P., Henry de Villeneuve C., Bernard M.C., Peo J.E., Boutry-Forveille A., Levy-Clement C. (1997) Relationship between porous silicon formation and hydrogen incorporation. *Thin Solid Films* **297**, 1–4.

Allongue P., Kieling V., Gerischer H. (1995), Etching mechanism and atomic structure of H-Si(111) surface prepared in NH$_4$F. *Electrochim. Acta.* **40**(10), 1353–1360.

Anderson R.C., Muller R.S., Tobias C.W. (1990) Investigation of porous silicon for vapour sensing. *Sens. Actuators A* 21–23, 835–839.

Anderson M.A., Tinsley-Bown A., Allcock P., Perkins E.A., Snow P., Hollings M., et al. (2003). Sensitivity of the optical properties of porous silicon layers to the refractive index of liquid in the pores. *Phys. Stat. Sol. (a)* **197**(2), 528–533.

Archer M., Christophersen M., Fauchet P.M. (2005) Electrical porous silicon chemical sensor for detection of organic solvents. *Sens. Actuators B* **106**, 347–357.

Barla K., Herino R., Bomchil G., Pfister J.C., Freund A. (1984) Determination of lattice parameter and elastic properties of porous silicon by x-ray diffraction. *J. Crystal Growth* **68**, 727–732.

Baratto C., Faglia G., Comini E., Sberveglieri G., Taroni A., La Ferrara V., et al. (2001). A novel porous silicon sensor for detection of sub-ppm NO₂ concentrations. *Sens. Actuators B* **77**, 62–66.

Baratto C., Faglia G., Sberveglieri G., Gaburro Z., Pancheri L., Oton C., et al. (2002) Multiparametric porous silicon sensors. *Sensors* **2**, 121–126.

Barillaro G., Diligenti A., Nannini A., Pieri F. (2000) Organic vapours detection using an integrated porous silicon device. In: *Proceeding of the 14th European Conference on Solid-State Transducers, EUROSENSORS XIV*, August 27–30, 2000, Copenhagen, Denmark. Abstract M2P39, pp. 197–199.

Ben-Chroin M. (1997) Resistivity of porous silicon. In: Canham L. (ed.) *Properties of Porous Silicon*. INSPEC, London, pp. 38 and 165.

Ben-Chorin M., Kux A. (1994) Adsorbate effects on photoluminescence and electrical conductivity of porous silicon. *Appl. Phys. Lett.* **67**, 481–483.

Ben-Chorin M., Moller F., Koch F. (1994) Nonlinear electrical transport in porous silicon. *Phys. Rev. B* **49**, 2981–2984.

Bisi O., Ossicini S., Pavesi L. (2000) Porous silicon: A quantum sponge structure for silicon based optoelectronics. *Surf. Sci. Rep.* **38**, 1–126.

Björkqvist M., Paski J., Salonen J., Lehto V.-P. (2006) Studies on hysteresis reduction in thermally carbonized porous silicon humidity sensor. *IEEE Sensors J.* **6**(3), 542–547.

Björkqvist M., Salonen J., Laine E. (2004b) Humidity behavior of thermally carbonized porous silicon. *Appl. Surf. Sci.* **222**, 269–274.

Björkqvist M., Salonen J., Laine E. (2004b) Humidity behavior of thermally carbonized porous silicon. *Appl. Surf. Sci.* **222**, 269–274.

Björkqvist M., Salonen J., Paski J., Laine E. (2003) Comparison of stabilizing treatments on porous silicon for sensor applications. *Phys. Stat. Sol. (a)* **197**, 374–377.

Björkqvist M., Salonen J., Paski J., Laine E. (2004a) Characterization of thermally carbonized porous silicon humidity sensor. *Sens. Actuators A* **112**, 244–247.

Blackwood D.J. and Akber M.F.B.M. (2006) In-situ electrochemical functionalization of porous silicon. *J. Electrochem. Soc.* **153**, G976–G980.

Bomchil G., Halimaoui A., Herino R. (1989) Porous silicon: The material and its applications in silicon-on-insulator technologies. *Appl. Surf. Sci.* **41/42**, 604–613.

Bomchil G., Halimaoui A., Sagnes I., Badoz P.A., Berbezier I., Perret P., et al. (1993) Porous silicon: Material properties, visible photo and electroluminescence. *Appl. Surf. Sci.* **65/66**, 394–407.

Buckley D.N., O'Dwyer C., Harvey E., Melly T., Serantoni M., Sutton D., Newcomb S.B. (2003) Anodic behavior of InP: film growth, porous layer formation and current oscillations. In: *Proceeding of 203 Meeting of Electrochemical Society*, April 27–May 2, 2003, Paris, France. Abstract 776. The Electrochemical Society, Inc. NJ, U.S.

Buriak J.M. and Allen M.J. (1998) Lewis acid mediated functionalization of porous silicon with substituted alkenes and alkynes. *J. Am. Chem. Soc.* **120**, 1339–1340.

Buriak J.M. (1999) Silicon-carbon bonds on porous silicon surfaces. *Adv. Mater.* **11**, 265–267.

Buriak J.M. (2002) Organometallic chemistry on silicon and germanium surfaces. *Chem. Rev.* **102**(5), 1272–1308.

Canham L. (ed.) (1997) *Properties of Porous Silicon*. INSPEC, London, UK.

Canham L. (ed.) (2014) *Handbook of Porous Silicon*. Springer, Switzerland.

Canham, L.T., Cox, T.I., Loni, A., Simons, A.J. (1996) Progress towards silicon optoelectronics using porous silicon technology. *Appl. Surf. Sci.* **102**, 436–441.

Carstensen J., Christophersen M., Föll H. (2000a) Pore formation mechanisms for the Si-HF system. *Mater. Sci. Eng. B* **69–70**, 23–28.

Carstensen J., Christophersen M., Hasse G., Föll H. (2000b) Parameter dependence of pore formation in silicon with a model of local current bursts. *Phys. Stat. Sol. (a)* **182**, 63–69.

Casanova F., Chiang C.E., Li C.P., Schuller I.K. (2007) Direct observation of cooperative effects in capillary condensation: The hysteretic origin. *Appl. Phys. Lett.* **91**, 243103.

Chazalviel J.-N. (1987) Surface methoxylation as the key factor for the good performance of n-Si/methanol photoelectrochemical cells. *J. Electroanal. Chem.* **233**, 37–48.

Chazalviel J.N., Ozanam F. (1997) Surface modification of porous silicon. In: Canham L. (ed.) *Properties of Porous Silicon*. INSPEC, London, UK, p. 59.

Chazalviel J.N., Wehrspohn R.B., Ozanam F. (2000) Electrochemical preparation of porous semiconductors: From phenomenology to understanding. *Mater. Sci. Eng. B* **69–70**, 1–10.

Christophersen M., Carstensen J., Voigt K., Foll H. (2003) Organic and aqueous electrolytes used for etching macro- and mesoporous silicon. *Phys. Stat. Sol. (a)* **197**(1), 34–38.

Christophersen M., Merz P., Quenzer J., Carstensen J., Föll H. (2001) Deep electrochemical trench etching with organic hydrofluoric electrolyte. *Sens. Actuators A* **88**, 241–245.

Connolly E.J., O'Halloran G.M., Pham H.T.M., Sarro P.M., French P.J. (2002) Comparison of porous silicon, porous polysilicon and porous silicon carbide as materials for humidity sensing applications. *Sens. Actuators A* **99**, 25–30.

Connolly E.J., Pham H.T.M., Groeneweg J., Sarro P.M., French P.J. (2004a) Relative humidity sensors using porous SiC membranes and Al electrodes. *Sens. Actuators B* **100**, 216–220.

Connolly E.J., Timmer B., Pham H.T.M., Groeneweg J., Sarro P.M., Olthuis W., et al. (2004b) Porous SiC as an ammonia sensor. In: *Proceedings of IEEE Sensors*, Vol. 1, Article number M2P-P.21, pp. 178–181.

Collins B.E., Dancil K.P., Abbi G., Sailor M.J. (2002) Determining protein size using an electrochemically machined pore gradient in silicon. *Adv. Funct. Mater.* **12**, 187–191.

Cullis A.G., Canham L.T., Calcott P.D.G. (1997) The structural and luminescence properties of porous silicon. *J. Appl. Phys.* **82**, 909–965.

Das J., Dey S., Hossain S.M., Rittersma Z.M.C., Saha H. (2003) Hygrometer comprising a porous silicon humidity sensor with phase-detection electronics. *IEEE Sensor J.* **3**(4), 414–420.

De Stefano L., Moretti L., Rendina I., Rossi A.M. (2003) Porous silicon microcavities for optical hydrocarbons detection. *Sens. Actuators A* **104**, 179–182.

De Stefano L., Moretti L., Rendina I., Rossi A.M. (2004) Time-resolved sensing of chemical species in porous silicon optical microcavity. *Sens. Actuators B* **100**, 168–172.

De Stefano L., Rea I., Caliò A., Politi J., Terracciano M., Korotcenkov G. (2016) Porous silicon-based optical chemical sensors. In: Korotcenkov G. (ed.) *Porous Silicon: From Formation to Application.* Vol. 2: *Biomedical and Sensor Applications,* CRC Press, Boca Raton, FL, pp. 69–94.

Di Carlo S., Falasconi M. (2012) Drift correction methods for gas chemical sensors. In: W. Wang (ed.) *Artificial Olfaction Systems: Techniques and Challenges.* InTech, New York, pp. 305–326.

Di Francia G., Castaldo A., Massera E., Nasti I., Quercia L., Rea I. (2005) A very sensitive porous silicon based humidity sensor. *Sens. Actuators B* **111–112,** 135–139.

Di Francia, G., De Filippo, F., La Ferrara, V. et al. (1998) Porous silicon layers for the detection at RT of low concentrations of vapours from organic compounds. In: *Proceeding of European Conference Eurosensors XII,* September 13–16, 1998, Southampton, UK. IOP, Bristol, Vol. 1, pp. 544–547.

Di Francia G., Noce M.D., Ferrara V.L., Lancellotti L., Morvillo P., Quercia L. (2002) Nanostructured porous silicon for gas sensor application. *Mater. Sci. Technol.* **18,** 767–771.

Fastykovsky P.P., Mogilnitsky A.A. (1999) Effect of air humidity on the metal-oxide-semiconductor tunnel structures' capacitance. *Sens. Actuators B* **57,** 51–55.

Fejfar A., Pelant I., Sípek E., Kočka J., Juška G. (1995) Transport study of self-supporting porous silicon. *Appl. Phys. Lett.* **66,** 1098–1100.

Feng Z.C., Tsu R. (eds.) (1994) *Porous Silicon.* Word Science, Singapore.

Feng Z.C., Wee A.T.S., Tan K.L. (1994). Surface and optical analyses of porous silicon membranes. *J. Phys. D* **27,** 1968–1975.

Foll H., Carstensen J., Frey S. (2006) Porous and nanoporous semiconductors and emerging applications. *J. Nanomater.* **2006,** 91635 (1–10).

Foll H., Christophersen M., Carstensen J., Hasse G. (2002) Formation and application of porous silicon. *Mater. Sci. Eng. R* **39,** 93–142.

Foll H., Langa S., Carstensen J., Christophersen M., Tiginyanu I.M. (2003b). Pores in III-V semiconductors. *Adv. Mater.,* **15**(3), 183–198.

Foll H., Langa S., Carstensen J., Lolkes S., Christophersen M., Tiginyanu I.M. (2003a) Engineering porous III-Vs, III-Vs REVIEW. *Adv. Semicond. Mag.* **16**(7), 42–43.

Frey S., Keipert S., Chazalviel J.-N., Ozanam F., Carstensen J., Foll H. (2007) Electrochemical formation of porous silica: Toward an understanding of the mechanisms. *Phys. Stat. Sol. (a)* **204**(5), 1250–1254.

Fujikura H., Liu A., Hamamatsu A., Sato T., Hasegawa H. (2000). Electrochemical formation of uniform and straight nano-pore arrays on (001) InP surfaces and their photoluminescence characterizations. *Jpn. J. Appl. Phys.* **39,** 4616–4620.

Fukami K., Harraz F.A., Yamauchi T., Sakka T., Ogata Y.H. (2008) Fine-tuning in size and surface morphology of rod-shaped polypyrrole using porous silicon as template. *Electrochem. Commun.* **10,** 56–60.

Füjes P., Kovacs A., Ducso Cs., Adam M., Muller B., Mescheder U. (2003) Porous silicon-based humidity sensor with interdigital electrodes and internal heaters. *Sens. Actuators B* **95,** 140–144.

Gan L., He H., Sun L., Ye Z. (2014) Improved photoluminescence and sensing stability of porous silicon nanowires by surface passivation. *Phys. Chem. Chem. Phys.* **16,** 890–894.

Gao J., Gao T., Sailor M.J. (2000) A porous silicon vapor sensor based on laser interferometry. *Appl. Phys. Lett.* **77,** 901–903.

Gao J., Gao T., Li Y.Y., Sailor M.J. (2002a) Vapor sensors based on optical interferometry from oxidized microporous silicon films. *Langmuir* **18,** 2229–2233.

Gao T., Gao J., Sailor M.J. (2002b) Tuning the response and stability of thin film mesoporous silicon vapor sensors by surface modification. *Langmuir* **18,** 9953–9957.

Garrone E., Borini S., Rivolo P., Boarino L., Geobaldo F., Amato G. (2003) Porous silicon in NO$_2$: A chemisorption mechanism for enhanced electrical conductivity. *Phys. Stat. Sol. (a)* **197,** 103–106.

Gole J.L., Lewis S., Seungwoo L. (2007) Nanostructures and porous silicon: Activity at interfaces in sensors and photocatalytic reactors. *Phys. Stat. Sol. A.* **204,** 1417–1422.

Gole J.L., Seals L.T., Lillehei P.T. (2000) Patterned metallization of porous silicon from electroless solution for direct electrical contact. *J. Electrochem. Soc.* **147,** 3785–3789.

Green S., Kathirgamanathan P. (2002) Effect of oxygen on the surface conductance of porous silicon: Towards room temperature sensor applications. *Mater. Lett.* **52,** 106–113.

Gurtner C., Wun A.W., Sailor M.J. (1999) Surface modification of porous silicon by electrochemical reduction of organo halides. *Angew. Chem. Intern. Ed.* **38**(13–14), 1966–1968.

Halimaoui A. (1995) Porous silicon: Material processing, properties and applications. In: Vial J.C. and Derrien J. (eds.) *Porous Silicon Science and Technology.* Springer-Verlag, New York.

Halimaoui A. (1997) Porous silicon formation by anodisation. In: Canham L.T. (ed.) *Properties of Porous Silicon.* IEE INSPEC, London, UK, pp. 12–23.

Hedrich F., Billat S., Lang W. (2000). Structuring of membrane sensors using sacrificial porous silicon. *Sens. Actuators B* **84,** 315–323.

Heilig A., Barsan N., Weimar U., Schweizer-Berberich M., Gardner J.W., Gopel W. (1997) Gas identification by modulating temperatures of SnO$_2$-based thick film sensors. *Sens. Actuators B* **43,** 45–51.

Holec H., Chvojka T., Jelinek I., Jindřicha J., Němeca I., Pelant I., et al. (2002) Determination of sensoric parameters of porous silicon in sensing of organic vapours. *Mater. Sci. Eng. C* **19,** 251–254.

Hossain S., Chakraborty S., Dutta S.K., Das J., Saha H. (2000) Stability in photoluminescence of porous silicon. *J. Lumin.* **91,** 195–202.

Huanca D.R., Ramirez-Fernandez F.J., Salcedo W.J. (2008). Porous silicon optical cavity structure applied to high sensitivity organic solvent sensor. *Microelectron. J.* **39,** 499–506.

Hummel R.E., Chang S.-S. (1992) Novel technique for preparing porous silicon. *Appl. Phys. Lett.* **61,** 1965–1967.

Hutter T., Horesh M., Ruschin S. (2011) Method for increasing reliability in gas detection based on indicator gradient in a sensor array. *Sens. Actuators B* **152,** 29–36.

Hutter T., Ruschin S. (2012) Some methods for improving the reliability of optical porous silicon sensors. In: Wang W. (ed.) *Advances in Chemical Sensors.* IntechOpen, pp. 47–62.

Ibraimov M.K., Sagidolda Y., Rumyantsev S.L., Zhanabaev Z.Zh., Shur M.S. (2016) Selective gas sensor using porous silicon. *Sensor Lett.* **14**(6), 588–591.

Iler R.K. (1979) *The Chemistry of Silica.* John Wiley, New York, p. 622.

Islam T., Das J., Saha H. (2006) Porous silicon based organic vapor sensor array for E-nose applications. In: *Proceedings of 5th IEEE Sensors Conference, Sensors 2006,* 22–25 October, Daegu, Korea, pp. 1085–1088.

Islam T., Saha H. (2007) Study of long-term drift of a porous silicon humidity sensor and its compensation using ANN technique. *Sens. Actuators A* **133,** 472–479.

Israelachvili J.N. (1992) *Intermolecular and Surface Forces.* 2nd edn. Academic Press, London, UK.

Jalkanen T., Makila E., Maattanen A., Tuura J., Kaasalainen M., Lehto V.-P., et al. (2012) Porous silicon micro- and nanoparticles for printed humidity sensors. *Appl. Phys. Lett.* **101,** 263110.

Jalkanen T., Tuura K., Mäkilä E., Salonen J. (2010) Electro-optical porous silicon gas sensor with enhanced selectivity. *Sens. Actuators B* **147**, 100–104.

Johansson M., Lundstrom I., Ekedahl L.-G. (1998) Bridging the pressure gap for palladium metal-insulator-semiconductor hydrogen sensors in oxygen containing environments. *J. Appl. Phys.* **84**(1), 44–51.

Kalem S., Yavuzcetin O. (2000) Possibility of fabricating light-emitting porous silicon from gas phase etchants. *Opt. Express* **6**(1), 7–11.

Kan P.Y.Y., Finstad T.G. (2005) Oxidation of macroporous silicon for thick thermal insulation. *Mater. Sci. Eng. B* **118**, 289–292.

Kayahan E. (2015) Porous silicon based humidity sensor. *Acta Phys. Polonica A* **127**(4), 1397–1399.

Kelly T.L., Sega A.G., Sailor M.J. (2011) Identification and quantification of organic vapors by time-resolved diffusion in stacked mesoporous photonic crystals. *Nano Lett.* **11**, 3169–3173.

Kim S.-J., Jeon B.-H., Choi K.-S., Min N.-K. (2000a) Capacitive porous silicon sensors for measurement of low alcohol gas concentration at room temperature. *J. Solid State Electrochem.* **4**, 363–366.

Kim S.J., Lee S.H., Lee C.J. (2001) Organic vapour sensing by current response of porous silicon layer. *J. Phys. D: Appl. Phys.* **34**, 3505–3509.

Kim S.J., Park J.Y., Lee S.H., Yi S.H. (2000b) Humidity sensors using porous silicon layer with mesa structure. *J. Phys. D: Appl. Phys.* **33**, 1781–1784.

King B.H. (2010) Design and manipulation of 1-D rugate photonic crystals of porous silicon for chemical sensing applications, PhD thesis, University of California, San Diego.

King B.H., Wong T., Sailor M.J. (2011) Detection of pure chemical vapors in a thermally cycled porous silica photonic crystal. *Langmuir* **27**(13), 8576–8585.

Kochergin V.R., Foell H. (2006) Novel optical elements made from porous Si. *Mater. Sci. Eng. R* **52**, 93–140.

Kochergin V., Foll H. (2009) *Porous Semiconductors: Optical Properties and Applications*. Springer, Berlin, Germany.

Koker L., Kolasinski K.W. (2000). Applications of a novel method for determining the rate of production of photochemical porous silicon. *Mater. Sci. Eng. B* **69–70**, 132–135.

Kolasinski K.W. (2005) Silicon nanostructures from electroless electrochemical etching. *Curr. Opin. Solid State Mater. Sci.* **9**(1–2), 73–83.

Kordas K., Remes J., Beke S., Hu T., Leppavuorim S. (2001) Manufacturing of porous silicon: Porosity and thickness dependence on electrolyte composition. *Appl. Surf. Sci.* **178**, 190–193.

Korotcenkov G. (2005) Gas response control through structural and chemical modification of metal oxides: State of the art and approaches. *Sens. Actuators B* **107**, 209–232.

Korotcenkov G. (2007) Metal oxides for solid state gas sensors. What determines our choice? *Mater. Sci. Eng. B* **139**, 1–23.

Korotcenkov G. (2016d) Silicon porosification: Approaches to PSi parameters control. In: Korotcenkov G. (ed.) *Porous Silicon: From Formation to Application*. Vol. 1: *Formation and Properties*. CRC Press, Boca Raton, FL, pp. 69–121.

Korotcenkov G. (2016c) Porous silicon characterization and application: General view. In: Korotcenkov G. (ed.) *Porous Silicon: From Formation to Application*. Vol. 1: *Formation and Properties*. CRC Press, Boca Raton, FL, pp. 1–23.

Korotcenkov G. (2016b) Solid state gas and vapor sensors based on porous silicon. In: Korotcenkov G. (ed.) *Porous Silicon: From Formation to Application*. Vol. 2: *Biomedical and Sensor Applications*. CRC Press, Boca Raton, FL, pp. 3–43.

Korotcenkov G. (ed.) (2016a) *Porous Silicon: From Formation to Application*. Vols. 1–3. CRC Press, Boca Raton, FL.

Korotcenkov G., Brinzari V. (2016) Alternative methods of silicon porosification and properties of these PSi layers. In: Korotcenkov G. (ed.) *Porous Silicon: From Formation to Application*. Vol. 1: *Formation and Properties*. CRC Press, Boca Raton, FL, pp. 245–284.

Korotcenkov G., Cho B.K. (2010b) Silicon porosification: State of the art. *Crit. Rev. Sol. St. Mater. Sci.* **35**(3) (2010) 153–260.

Korotcenkov G., Cho B.K. (2010a) Porous semiconductors: Advanced material for gas sensor applications. *Crit. Rev. Sol. St. Mater. Sci.* **35**(1), 1–37.

Korotcenkov G., Cho B.K. (2013) Engineering approaches to improvement operating characteristics of conductometric gas sensors. Part 1: Improvement of sensor sensitivity and selectivity. *Sens. Actuators B* **188**, 709–728.

Korotcenkov G., Cho B.K. (2014) Engineering approaches to improvement operating characteristics of conductometric gas sensors. Part 2: Decrease of dissipated (consumable) power and improvement stability and reliability. *Sens. Actuators B* **198**, 316–341.

Korotcenkov G., Cho B.K. (2016b) Processing of porous silicon. In: Korotcenkov G. (ed.) *Porous Silicon: From Formation to Application*. Vol. 1: *Formation and Properties*. CRC Press, Boca Raton, FL, pp. 299–349.

Korotcenkov G., Cho B.K. (2016a): Technology of Si porous layer fabrication using anodic etching: General scientific and technical issues. In: Korotcenkov G. (ed.) *Porous Silicon: From Formation to Application*. Vol. 1: *Formation and Properties*, CRC Press, Boca Raton, FL, pp. 43–68.

Koyama H. (2006) Strong photoluminescence anisotropy in porous silicon layers prepared by polarized-light-assisted anodization. *Solid State Commun.* **138**, 567–570.

Lai M., Parish G., Liu Y., Dell J.M., Keating A.J. (2011) Development of an alkaline-compatible porous-silicon photolithographic process. *J. Microelectromech. Syst.* **20**, 418–423.

Lammel G., Renaud P. (2000) Free-standing, mobile 3D porous silicon microstructures. *Sens. Actuators A* **85**, 356–360.

Lauerhaas J.M., Sailor M.J. (1993) Chemical modification of the photoluminescence quenching of porous silicon. *Science* **261**, 1567–1568.

Lazzerini G.M., Strambini L.M., Barillaro G. (2013) Self-tuning porous silicon chemitransistor gas sensors. In: *Proceedings of IEEE Sensors Conference, Sensors 2013*, 3–6 November, Baltimore, MD.

Lee A.P., Reedy B.J. (1999) Temperature modulation in semiconductor gas sensing. *Sens. Actuators B* **60**, 35–42.

Lee M.-K., Tu H.-F. (2007) Stabilizing light emission of porous silicon by *in-situ* treatment. *Jpn. J. Appl. Phys.* **46**, 2901–2903.

Lee W.H., Lee C., Kwon Y.H., Hong C.Y., Cho H.Y. (2000) Deep level defects in porous silicon. *Sol. St. Commun.* **113**, 519–522.

Lehmann V. (1993). The physics of macropore formation in low doped *n*-type silicon. *J. Electrochem. Soc.* **140**(10), 2836–2843.

Lehmann V. (1996) Developments in porous silicon research. *Mater. Lett.* **28**, 245–249.

Lehmann V. (2002) *Electrochemistry of Silicon*. Wiley-VCH, Weinheim, Germany.

Lehmann V., Gruning U. (1997) The limits of macropore array fabrication. *Thin Solid Films* **297**, 13–17.

Lehmann V., Stengl R., Luigart A. (2000) On the morphology and the electrochemical formation mechanism of mesoporous silicon. *Mater. Sci. Eng. B* **69–70**, 11–22.

Letant S.E., Content S., Tan T.T., Zenhausern F., Sailor M.J. (2000) Integration of porous silicon chips in an electronic artificial nose. *Sens. Actuators B* **69**, 193–198.

Letant S.E., Sailor M.J. (2000) Detection of HF gas with a porous silicon interferometer. *Adv. Mater.* **12**(5), 355–359.

Letant S.E., Sailor M.J. (2001) Molecular identification by time-resolved interferometry in a porous silicon film. *Adv. Mater.* **13**, 335–338.

Lewis N.S. (1998) Progress in understanding electron-transfer reactions at semiconductor/liquid interfaces. *J. Phys. Chem. B* **102**(25), 4843–4855.

Lewis S.E., DeBoer J.R., Gole J.L., Hesketh P.J. (2005) Sensitive, selective, and analytical improvements to a porous silicon gas sensor. *Sens. Actuators B* **110**, 54–65.

Litovchenko V.G., Gorbanyuk T.I., Solntsev V.S., Evtukh A.A. (2004) Mechanism of hydrogen, oxygen and humidity sensing by Cu/Pd-porous silicon–silicon structures. *Appl. Surf. Sci.* **234**, 262–267.

Litovchenko V.G., Gorbanyuk T.I., Solntsev V.S. (2017) Mechanism of adsorption-catalytic activity at the nanostructured surface of silicon doped with clusters of transition metals and their oxides. *Ukr. J. Phys.* **62**(7), 605–614.

Lubianiker Y., Balberg I. (1998) A comparative study of the Meyer–Neldel rule in porous silicon and hydrogenated amorphous silicon. *J. Non-Cryst. Solids* **227–230**, 180–184.

Lundstrom L., Armgarth M., Petersson L.G. (1989) Physics with catalytic metal gate chemical sensors. *CRC Crit. Rev. Sol. State Mater. Sci.* **15**, 201–278.

Lundstrom L., Shivaraman M.S., Svensson C.M., Lundkvist L. (1975) A hydrogen–sensitive MOS field–effect transistor. *Appl. Phys. Lett.* **26**, 55–57.

Lysenko V., Perichon S., Remaki B., Barbier D. (2002). Thermal isolation in microsystems with porous silicon. *Sens. Actuators A* **99**, 13–24.

Mahmoudi B., Gabouze N., Haddadi M., Mahmoudi Br., Cheraga H., Beldjilali K., Dahmane D. (2007) The effect of annealing on the sensing properties of porous silicon gas sensor: Use of screen-printed contacts. *Sens. Actuators B* **123**, 680–684.

Mares J.J., Kristofik J., Hulicius E. (1995) Influence of humidity on transport in porous silicon. *Thin Solid Films* **255**, 272–275.

Maruyama T., Ohtani S. (1994) Photoluminescence of porous silicon exposed to ambient air. *Appl. Phys. Lett.* **65**(11), 1346–1348.

Massera E., Nasti I., Quercia L., Rea I., Di Francia G. (2004) Improvement of stability and recovery time in porous-silicon-based NO$_2$ sensor. *Sens. Actuators B* **102**, 195–197.

Morazzani V., Cantin J., Ortega C., Pajot B., Rahbi R., Rosenbauer M., von Bardeleben H., Vazsonyi E. (1996) Thermal nitridation of p-type porous silicon in ammonia. *Thin Solid Films* **276**, 32–35.

Mou S.S., Islam Md. J., Md. Ismail A.B. (2011) Photoluminescence properties of LaF$_3$-coated porous silicon. *Mater. Sci. Appl.* **2**, 649–653.

Mulloni V., Pavesi L. (2000). Electrochemically oxidised porous silicon microcavities. *Mater. Sci. Eng. B* **69–70**, 59–65.

Nakata S., Nakasuji M., Ojima N., Kitora M. (1998) Characteristic nonlinear responses for gas species on the surface of different semiconductor gas sensors. *Appl. Surf. Sci.* **135**, 285–292.

O'Halloran G.M., Kuhl M., Trimp P.J., French P.J. (1997) The effect of additives on the adsorption properties of porous silicon. *Sens. Actuators A* **61**, 415–420.

O'Halloran G.M., van der Vlist W., Sarro P.M., French P.J. (1999) Influence of the formation parameters on the humidity sensing characteristics of a capacitive humidity sensor based on porous silicon. In: *CD Proceedings of the 13th European Conference on Solid-State Transducers, EUROSENSORS XIII*, The Hague, September 12–15, Abstract 4P13, 117–120.

Okorn-Schmidt H.F. (1999) Characterization of silicon surface preparation processes for advanced gate dielectrics. *IBM J. Res. Dev.* **43**, 351–365.

Orosco M.M., Pacholski C., Sailor M.J. (2009) Real-time monitoring of enzyme activity in a mesoporous silicon double layer. *Nature Nanotech.* **4**, 255–258.

Oton C.J., Pancheri L., Gaburro Z., Pavesi L., Baratto C., Faglia G., Sberveglieri G. (2003) Multiparametric porous silicon gas sensors with improved quality and sensitivity. *Phys. Stat. Sol. (a)* **197**, 523–527.

Outemzabet R., Cherkaoui M., Ozanam F., Gabouze N., Kesri N., Chazalviel J.N. (2004) Anisotropy in the anodic dissolution of silicon elucidated by in situ infrared spectroscopy. *J. Electroanal. Chem.* **563**, 3–8.

Ozdemir S., Gole J.L. (2007) The potential of porous silicon gas sensors. *Curr. Opin. Solid State Mater. Sci.* **11**, 92–100.

Parkhutik V. (1999) Porous silicon-mechanisms of growth and applications. *Sol. St. El.* **43**, 1121–1141.

Pancheri L., Oton C.J., Gaburro Z., Soncini G., Pavesi L. (2003) Very sensitive porous silicon NO$_2$ sensor. *Sens. Actuators B* **89**, 237–239.

Pancheri L., Oton C.J., Gaburro Z., Soncini G., Pavesi L. (2004) Improved reversibility in aged porous silicon NO$_2$ sensors. *Sens. Actuators B* **97**, 45–48.

Pap A.E., Kordas K., George T.F., Leppavuori S. (2004) Thermal oxidation of porous silicon: Study on reaction kinetics. *J. Phys. Chem. B* **108**, 12744–12747.

Park S.., Lee, K.W., Kim Y.Y. (2010) A technique for the fabrication of various multiparametric porous silicon samples on a single substrate. *Thin Solid Films* **518**, 2860–2863.

Pavesi L. (1997) Porous silicon dielectric multilayers and microcavities. *La Rivista del Nuovo Cimento* **20**, 1–76.

Quiroga-González E., Föll H. (2016) Fundamentals of silicon porosification via electrochemical etching. In: Korotcenkov G. (ed.) *Porous Silicon: From Formation to Application.* Vol. 1: *Formation and Properties.* CRC Press, Boca Raton, FL, pp. 30–45.

Racine G.A., Genolet G., Clerc P.A., Despont M., Vettiger P., De Rooij N.F. (1997). Porous silicon sacrificial layer technique for the fabrication of free standing membrane resonators and cantilever arrays. In: *CD Proceeding of the 11th European Conference on Solid State Transducers Eurosensors XI*, September 21–24, 1997, Warsaw, Poland, Vol. 1, pp. 285–288.

Raghavan D., Gu X., Nguyen T., Vanlandingham M. (2001) Characterization of chemical heterogeneity in polymer systems using hydrolysis and tapping-mode atomic force microscopy. *J. Polymer Sci. B: Polymer Phys.* **39**, 1460–1470.

Remaki B., Perichon S., Lysenko V., Barbier D. (2001) Electrical transport in porous silicon from improved complex impedance analysis. In: Fauchet P.M., Buriak J.M., Canham L.T., Koshida N., White B.E. Jr. (eds.) *Microcrystalline and Nanocrystalline Semiconductors, MRS Proc.* Vol. 638, MRS, PA, USA, pp. F321–E326.

Rickert J., Weiss T., Gopel W. (1996) Self-assembled monolayers for chemical sensors: molecular recognition by immobilized supramolecular structure. *Sens. Actuators B* **31**, 45–50.

Rittersma Z.M., Benecke W. (2000) A humidity sensor featuring a porous silicon capacitor with an integrated refresh resistor. *Sens. Mater.* **12**(1), 35–55.

Rittersma Z.M., Splinter A., Bodecker A., Benecke W. (2000) A novel surface-micromachined capacitive porous silicon humidity sensor. *Sens. Actuators B* **68**, 210–217.

Rocchia M., Garrone E., Geobaldo F., Boarino L., Sailor M.J. (2003) Sensing CO$_2$ in a chemically modified porous silicon film. *Phys. Stat. Sol. (a)* **197**, 365–369.

Ruminski A.M., King B.H., Salonen J., Snyder J.L., Sailore M.J. (2010 Porous silicon-based optical microsensors for volatile organic analytes: Effect of surface chemistry on stability and specificity. *Adv. Mater.* **20**(17), 2874–2883.

Rumpf K., Granitzer P., Poelt P., Krenn H. (2009) Transition metals specifically electrodeposited into porous silicon. *Phys. Stat. Sol. (c)* **6**(7), 1592–1595.

Sailor M.J (ed.) (2012) *Porous Silicon in Practice: Preparation, Characterization and Applications.* Wiley-VCH Verlag GmbH.

Sailor M.J. (2014) Chemical reactivity and surface chemistry of porous silicon. In: L. Canham, (ed.) *Handbook Porous Silicon*, Springer, Zug Heidelberg, Switzerland, pp. 355–380.

Salem M.S., Sailor M.J., Harraz F.A., Sakka T., Ogata Y.H. (2006) Electrochemical stabilization of porous silicon multilayers for sensing various chemical compounds. *J. Appl. Phys.* **100**(8), 083520.

Salgado G.G., Becerril T.D., Santiesteban H.J., Andrés E.R. (2006) Porous silicon organic vapor sensor. *Opt. Mater.* **29**, 51–55.

Salonen J., Björkqvist M., Laine E., Niinisto L. (2004) Stabilization of porous silicon surface by thermal decomposition of acetylene. *Appl. Surf. Sci.* **225**, 389–394.

Salonen J., Laine E., Niinisto L. (2002) Thermal carbonization of porous silicon surface by acetylene. *J. Appl. Phys.* **91**(1), 456–461.

Salonen J., Lehto V.P., Björkqvist M., Laine E., Niinistö L. (2000) Studies of thermally-carbonized porous silicon surfaces. *Phys. Stat. Sol. (a)* **182**, 123–126.

Salonen J., Mäkilä E., Riikonen J., Heikkilä T., Lehto V.-P. (2009) Controlled enlargement of pores by annealing of porous silicon. *Phys. Status Solidi (a)* **206**, 1313–1317.

Santos H.A. (ed.) (2014) *Porous Silicon for Biomedical Applications.* Woodhead Publishing, Cambridge, UK.

Schechter I., Ben-Chorin M., Kux A. (1995) Gas sensing properties of porous silicon. *Anal Chem.* **67**, 3727–3732.

Seiyama T., Yamazoe N., Arai H. (1983) Ceramic humidity sensors. *Sens. Actuators* **4**, 85–96.

Setzu S., Letant S., Solsona P., Romenstain R., Vial, J.C. (1998) Improvement of the luminescence in p-type as-prepared or dye impregnated porous silicon microcavities. *J. Lumin.* **80**, 129–132.

Sharma S.N., Sharma R.K., Lakshmikumar S.T. (2005) Role of an electrolyte and substrate on the stability of porous silicon. *Physica E* **28**, 264–272.

Shi F.G., Mikrajuddin, Okuyama K. (2000) Electrical conduction in porous silicon: Temperature dependence. *Microelectron. J.* **31**, 187–191.

Shimizu Y., Aral H., Seiyama T. (1985) Theoretical studies on the impedance-humidity characteristics of ceramic humidity sensors. *Sens. Actuators* **7**, 11–22.

Smith R.L., Collins S.D. (1992) Porous silicon formation mechanisms. *J. Appl. Phys.* **71**(8), R1–R22.

Song J.H., Sailor M.J. (1999) Chemical modification of crystalline porous silicon surfaces. *Comm. Inorg. Chem.* **21**, 69–84.

Splinter A., Sturmann J., Benecke W. (2000). New porous silicon formation technology using internal current generation with galvanic elements. In: *CD Proceeding of the 13th European Conference on Solid-State Transducers, EUROSENSORS XIIV*, August 27–30, 2000, Copenhagen, Denmark, Abstract T2P03, pp. 423–426.

Splinter A., Bartels O., Benecke W. (2001) Thick porous silicon formation using implanted mask technology. *Sens. Actuators B* **76**, 354–360.

Stewart M.P., Buriak J.M. (1998) Photopatterned hydrosilylation on porous silicon. *Angew Chem. Int. Ed. Eng.* **37**, 3257–3259.

Strikha V., Skryshevsky V., Polishchuk V., Souteyrand E., Martin J.R. (2001) A study of moisture effects on Ti/porous silicon/silicon Schottky barrier. *J. Porous Mater.* **7**, 111–114.

Sukach G.A., Oleksenko P.F., Smertenko P.S., Evstigneev A.M., Bogoslovskaya A.B. (2003) Study of charge flow mechanisms in metal-porous silicon structures by photoluminescent and electrophysical techniques. In: Svechnikov S.V., Valakh M.Y. (eds.) *Optics and Photonics: Optical Diagnostics of Materials and Devices for Opto-, Micro-, and Quantum Electronics*, *Proc. SPIE*, Vol. 5024, SPIE, pp. 72–79.

Tucci M., La Ferrara V., Della Noce M., Massera E., Quercia L. (2004) Bias enhanced sensitivity in amorphous/porous silicon heterojunction gas sensors. *Non-Cryst. Solids* **338–340**, 776–779.

Turner D.R. (1958) Electropolishing silicon in hydrofluoric. *J. Electrochem. Soc.* **105**, 402–408.

Vergara A., Vembu S., Ayhan T., Ryan M.A., Homer M.L., Huertaa R. (2012) Chemical gas sensor drift compensation using classifier ensembles. *Sens. Actuators B* **166–167**, 320–329.

Vial J.C. and Derrien J. (eds.) (1995) *Porous Silicon: Science and Technology*, Les Editions de Physique, Les Ulis. Springer, Berlin, Germany.

Vrkoslav V., Jelınek I., Matocha M., Kral V., Dian J. (2005) Photoluminescence from porous silicon impregnated with cobalt phthalocyanine. *Mater. Sci. Eng. C* **25**, 645–649.

Wang Y., Park S., Yeow J.T.W., Langner A., Müller F. (2010) A capacitive humidity sensor based on ordered macroporous silicon with thin film surface coating. *Sens. Actuators B* **149**, 136–142.

Wang Y., Yeow J.T.W. (2008) A $HfO_2$ thin film enhanced porous silicon humidity sensor. In: *Proceedings of IEEE Sensors Conference*, 26–29 October, Lecce, Italy, pp. 819–822.

Xu Y.Y., Li X.J., He J.T., Hai X.H., Wang Y. (2005) Capacitive humidity sensing properties of hydrothermally-etched silicon nano-porous pillar array. *Sens. Actuators B* **105**, 219–222.

Yaakob S., Abu Bakar M., Ismail J., Abu Bakar N.H.H., Ibrahim K. (2012) The formation and morphology of highly doped *n*-type porous silicon: Effect of short etching time at high current density and evidence of simultaneous chemical and electrochemical dissolutions. *J. Phys. Sci.* **23**(2), 17–31.

Yamana M., Kashiwazaki N., Kinoshita A., Nakano T., Yamamoto M., Walton, C.W. (1990) Porous silicon oxide layer formation by the electrochemical treatment of a porous silicon layer. *J. Electrochem. Soc.* **137**, 2925.

Yarkin D.G. (2003) Impedance of humidity sensitive metal/porous silicon/*n*-Si structures. *Sens. Actuators A* **107**, 1–6.

Zangooie S., Jansson R., Arwin H. (1999) Investigation of optical anisotropy of refractive-index-profiled porous silicon employing generalized ellipsometry. *Mater. Res.* **14**, 4167–4175.

Zhang X.G. (2001) *Electrochemistry of Silicon and Its Oxide.* Kluwer Academic/Plenum, New York.

Zhu W.X., Gao Y.X., Zhang L.Z., Mao J.C., Zhang B.R., Duan J.Q., Qin G.G.(1992) Time evolution of the localized vibrational mode infrared absorption of porous silicon in air. *Superlattices Microstruct.* **12**(3), 409–412.

# 7 Mesoporous Silica and Its Prospects for Humidity Sensor Application

## 7.1 CLASSIFICATION OF POROUS MATERIALS

As shown in previous chapters, humidity-sensitive materials should be porous with controlled pore sizes. Unfortunately, the synthesis of materials with a given pore size is a difficult technological problem. Therefore, in the development of humidity sensors, as well as in the development of various gas sensors, great interest is shown in materials with inherent porosity, primarily in materials with mesoporous and microporous structure. According to the International Union of Pure and Applied Chemistry (IUPAC), the prefix meso- refers to a region 2–50 nm, macro- is a region >50 nm, and micro- is a region <2 nm. Several examples of materials with different porosity are presented in Table 7.1. The small mesopores and micropores limit the kinds of ions and molecules that can be admitted to the interior of the materials. In addition, control over the pore size offers the possibility of molecular sieving or molecular selectivity. Mesoporosity and microporosity can also endow a material with a high surface area exceeding 1,000 $m^2/g$ and pore volume greater than 1 $cm^3/g$. This greatly expands the potential of the materials for application to adsorption and as a support for immobilized catalytic or sensing moieties (Moos et al. 2006, 2009; Xu et al. 2006; Carrington and Xue 2007; Slowing et al. 2007; Basabe-Desmonts et al. 2007; Ariga et al. 2007; Walcarius 2008; Melde et al. 2008; Sahner et al. 2008; Zheng et al. 2012; Kresge and Roth 2013; Wagner et al. 2013; Amonette and Matyas 2017).

## 7.2 MESOPOROUS SILICA

There are two types of mesoporous silica: random mesoporous structures and ordered mesoporous structures (Galarneau et al. 2001). Mesoporous silica, especially those exhibiting ordered pore systems and uniform pore diameters, have shown great potential for sensing applications in recent years (Melde et al. 2008).

Sol-gel chemistry is frequently employed in designing random mesoporous structures of silicates (Brinker and Scherer 1990; Corma 1997). Liquid silicon alkoxide (Si(OR)$_4$) precursors are hydrolyzed and condensed to form siloxane bridges, a process that is often described as inorganic polymerization and is represented below:

**TABLE 7.1**
**Different Types of Porous Materials**

| Pore Size | Type of Material | Examples | Pore Size Range (nm) |
|---|---|---|---|
| >50 nm | Macroporous | Porous glasses | >50 |
| 2–50 nm | Mesoporous | Pillared-layered clays | 1, 10 |
| | | Silica | |
| | |    M41S | 1.6–10 |
| | |    SBA-15 | 8–10 |
| | |    SBA-16 | 5 |
| | |    Diatom biosilica | 2–50 |
| | | Mesoporous alumina | 2 |
| <2 nm | Microporous | Zeolites | <1.42 |
| | | Activated carbon | 0.6 |
| | | ZSM-5 | 0.45–0.6 |
| | | Zeolite A | 0.3–0.45 |
| | | Beta and Mordernite-Zeolites | 0.6–0.8 |
| | | Faujasite | 0.74 |
| | | Cloverite | 0.6–1.32 |

*Source:* Data extracted from Naik, B., and Ghosh, N.N., *Recent Pat. Nanotechnol.*, 3, 213–224, 2009.

$$\text{Hydrolysis: } Si(OR)_4 + nH_2O \rightarrow HO_n - Si(OR)_{4-n} + nROH$$

$$\text{Condensation: } (RO)_3 Si - OH + HO - Si(OR)_3 \rightarrow$$

$$(RO)_3 Si - O - Si(OR)_3 + H_2O \text{ and / or}$$

$$(RO)_3 Si - OR + HO - Si(OR)_3 \rightarrow$$

$$(RO)_3 Si - O - Si(OR)_3 + ROH$$

$$(7.1)$$

The most commonly used precursors are tetraethoxysilane (TEOS) and tetramethoxysilane (TMOS). A colloidal sol of condensed silicate species can eventually interconnect as an immobile three-dimensional network encompassing the space of its reaction container. Drying a gel under ambient conditions or with heat will typically cause shrinkage as solvent leaves the micropores of the silicate network. This type of material is called a xerogel. Alternatively, supercritical drying can be applied to remove solvent yielding a product that is more similar to the size and shape of the original gel. Such aerogels may have low solid volume fractions near 1% and, therefore, very high pore volumes. The use of basic pH and an excess of water can result in particulate precipitation. Gels can also be deposited allowing for the generation of thin films or membranes. The isoelectric point of silica is in the pH range 1–3. This value determines the surface charge of a condensing silicate or material in solution due to protonation and deprotonation of silanol groups (Si–OH).

Ordered mesoporous materials are made with a combination of using self-assembled surfactants as template and simultaneous sol-gel condensation around template (micelles) (Corma 1997; Galarneau et al. 2001). Surfactants are organic molecules that are comprised of two parts with different polarity. One part is a hydrocarbon chain (often referred to as polymer tail), which is nonpolar and hence hydrophobic and lipophilic, whereas the other is polar and hydrophilic (often called hydrophilic head). Because of such a molecular structure, surfactants tend to enrich at the surface of a solution or interface between aqueous and hydrocarbon solvents, so that the hydrophilic head can turn toward the aqueous solution, resulting in a reduction of the surface or interface energy. Such concentration segregation is spontaneous and thermodynamically favorable. Surfactant molecules can be generally classified into four families, and they are known as anionic, cationic, nonionic, and amphoteric surfactants, which are briefly discussed below:

1. Typical anionic surfactants are sulfonated compounds with a general formula R-SO$_3$Na, and sulfated compounds of R-OSO$_3$Na, with R being an alkyl chain consisting of 11–21 carbon atoms.
2. Cationic surfactants are commonly comprised of an alkyl hydrophobic tail and a methyl-ammonium ionic compound head, such as cetyl trimethyl ammonium bromide (CTAB), $C_{16}H_{33}N(CH_3)_3Br$

and cetyl trimethyl ammonium chloride (CTAC), $C_{16}H_{33}N(CH_3)_3Cl$.
3. Nonionic surfactants do not dissociate into ions when dissolved in a solvent as both anionic and cationic surfactant. Their hydrophilic head is a polar group such as ether, R–O–R, alcohol, R–OH, carbonyl, R–CO–R, and amine, R–NH–R.
4. Amphoteric surfactants have properties similar to either nonionic surfactants or ionic surfactants. Examples are betaines and phospholipids.

When surfactants dissolve into a solvent, forming a solution, the surface energy of the solution will decrease rapidly and linearly with an increasing concentration. This decrease is due to the preferential enrichment and the ordered arrangement of surface of surfactant molecules on the solution surface, that is, hydrophilic heads inside the aqueous solution and/or away from nonpolar solution or air. However, such a decrease stops when a critical concentration is reached, and the surface energy remains constant with further increase in the surfactant concentration. This critical concentration is termed as the *critical micelle concentration*, or CMC. Below the CMC, the surface energy decreases due to an increased coverage of surfactant molecules on the surface as the concentration increases. At the CMC, the surface has been fully covered with the surfactant molecules. Above the CMC, further addition of surfactant molecules leads to phase segregation and formation of colloidal aggregates, or micelles (Mittal and Fendler 1982). The initial micelles are spherical and individually dispersed in the solution (Figure 7.1a) and would transfer to a cylindrical rod shape (Figure 7.1b), with further increased surfactant concentration. Continued increase of surfactant concentration results in an ordered parallel hexagonal packing of cylindrical micelles (Figure 7.1c). At a still higher concentration, lamellar micelles would form (Figure 7.1d).

The process of ordered mesoporous structures synthesis is conceptually straightforward and can be briefly described below (Corma 1997). Surfactants with a certain molecule length are dissolved into a polar solvent with a concentration exceeding its CMC, mostly at a concentration, at which hexagonal or cubic packing of cylindrical micelles is formed. At the same time, the precursors for the formation of silica are also dissolved into the same solvent, together with other necessary chemicals such as catalysts. Inside the solution, several processes proceed simultaneously. Surfactants segregate and form micelles, whereas precursors undergo hydrolysis and condensation around the micelles simultaneously. Surfactants are often removed by calcination, or burning, to produce molecular sieves with narrow pore-size distributions and highly ordered mesostructures. When extraction of templates is used instead of calcinations, organic functional groups can be incorporated into the materials during synthesis.

The calcination is usually performed at temperatures 500°C–550°C (Wang et al. 2005; Tu et al. 2012). The processes occurring during calcination process can be understood

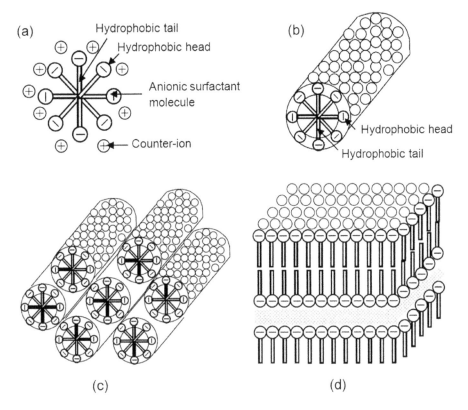

FIGURE 7.1 Schematics of various micelles formed at various surfactant concentrations above the CMC.

on the basis of the infrared (IR) spectroscopy data shown in Figure 7.2, where infrared spectra are compared for both as-prepared silica aerogel samples and 500°C calcined ones. According to Wang et al. (2005), the characteristic difference between two spectra is an indication for thermal removal of surface bonded species. In analyzed spectra, two peaks at 2981 and 1376 cm⁻¹ are assigned to C–H asymmetric stretching and symmetric deformation of $CH_2$ species. Absorption bands at 2907 and 1469 cm⁻¹ are characteristics of C–H asymmetric stretching and deformation of $CH_2$ species. Absorbance bands in the lower wavenumber region 1300–400 cm⁻¹ are used to study the lattice vibrations of silica. Three sharp peaks at 1080, 795, and 460 cm⁻¹ are primary characteristics of the asymmetric stretching, symmetric, and bending modes of ≡Si–O–Si lattice vibrations in a silica skeletal network (Yoda and Ohshima 1999). The absorption band at 960 cm⁻¹ is attributed to a stretching mode of free surface silanols (≡Si–OH). On the base of IR spectra, it was concluded that the presence of ethoxy groups ($OCH_2CH_3$) on the surface of as-prepared silica is due to the incomplete hydrolysis of TEOS in the sol-gel process (Hong et al. 1998). Subsequent annealing of the silica sample up to 500°C in the air causes the hydrocarbon residue to be completely burned out (curve [b] in Figure 7.2). Wang et al. (2005) believe that at the same the surface species appear, which provides a hydrophilic capability for adsorbing water through a hydrogen-bonding force. It is seen that the hydroxyl peak became intense after the silica sample was thermally treated at 500°C, suggesting that the amount of surface OH groups was increased after removing

ethyl hydrocarbon species. This consideration is also consistently associated with the increase in spectrum intensity for the broad band, centered at 3459 cm⁻¹ and shouldered at 3243 cm⁻¹ and the vibration band at 1636 cm⁻¹ (curves [b] versus [a] in Figure 7.2), as they are well known as bulk water

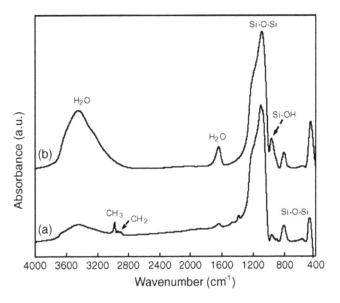

FIGURE 7.2 Infrared spectra of silica aerogels: (a) crude sample as-prepared, and (b) calcined sample at 500°C. (Reprinted from *Sens. Actuators B*, 107, Wang, C.-T. et al., Humidity sensors based on silica nanoparticle aerogel thin films, 402–410, Copyright 2005, with permission from Elsevier.)

on the surface (Li and Willey 1997). Generally, making quantitative determination of analytes by IR spectra without using an internal standard is difficult. However, based on an equal intensity of the Si–O–Si band at 1080 cm$^{-1}$ in the two spectra measured under the exactly same circumstance, it is still realizable to conclude that the silica aerogel surface becomes more hydrophilic to water and polar molecules after thermal treatment (Wang and Willey 2001).

The first ordered mesoporous structure was realized in 1992 with the reports of the M41S materials (Kresge et al. 1992; Beck et al. 1992), followed by the introduction of FSM-16 (Inagaki et al. 1993). The syntheses of M41S materials employs cationic alkylammonium surfactants in amounts above their critical micelle concentrations. These surfactants cooperatively assemble with inorganic precursors to produce a silicate matrix. The most well-known representatives of this class include the silica solids MCM-41 (with a hexagonal arrangement of the mesopores, space group $p6$ mm), MCM-48 (with a cubic arrangement of the mesopores, space group $Ia3d$), and MCM-50 (with a laminar structure, space group $p2$) (see Figure 7.3). Since the first reports of M41S, many different surfactants, precursors, and combinations of the two have been studied. A schematic model for the formation of the mesoporous materials from kanemite is shown in Figure 7.4. The silicate layers of kanemite can wind around the exchanged alkyltrimethylammonium ions. This causes the condensation of silanol groups on the adjacent silicate layers of kanemite, since the silicate layers have the flexibility to wind due to its single-layered structure.

Additional information regarding synthesis and properties of mesoporous silica, including as-synthesized and functionalized materials, one can find in review papers of Huo et al. (1996), Corma (1997), Soler-Illia et al. (2002),

Hoffmann et al. (2006), Naik and Ghosh (2009), Carroll and Anderson (2011), Wu et al. 2013.

It is necessary to note that the described approach to synthesis of mesoporous-silica-based on the sol-gel process can be used for synthesizing other metal oxides, including conductive metal oxides, designed for gas-sensor applications (read Chapter 2).

## 7.3 ADVANTAGES OF SILICA FOR SENSOR APPLICATIONS

Silica is an attractive material for many sensing applications because of its high surface area, stability over a fairly wide range of pH (excluding alkaline), relative inertness in many environments, transparency in the UV-visible spectrum, and its morphological control (Melde et al. 2008; Amonette and Matyas 2017). Moreover, the silane chemistry gives possibility to covalently attach molecular probes to the pore walls. Therefore, spectrophotometrically active molecular probes can be entrapped in sol-gel glass and applied for heterogeneous detection of analytes in the solution or gas. Bulk materials may be applied as synthesized (e.g., batch adsorption of an analyte from solution) or as part of a surface coating. Gels can be used to form monolithic materials or thin films on a wide variety of substrates by spin and dip-coating techniques. They can be deposited as or embedded in a specialized coating on an electrode and active area of various gas-sensing devices. The pore size can be controlled and the surface properties can be altered (e.g., grafting hydrophobic groups) to encourage the entrance of a specific analyte over that of similar species.

Depending on the method of drying, silica can be in the form of xerogel or aerogel. A xerogel is a solid, formed from a gel by drying with unhindered shrinkage. Xerogels usually

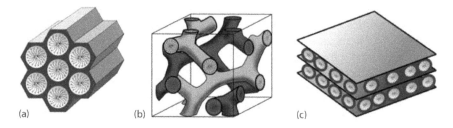

**FIGURE 7.3** Structures of the mesoporous silica material: (a) SBA-15, (b) KIT-6, and (c) MCM-50 (lamellar). (Kresge, C. T., and Roth, W. J., *Chem. Soc. Rev.*, 42, 3663–3670, 2013. Reproduced by permission of The Royal Society of Chemistry.)

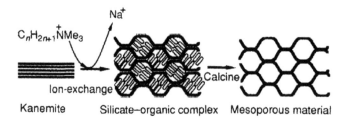

**FIGURE 7.4** Schematic model for the formation of the mesoporous material from kanemite. (Inagaki, S. et al., *J. Chem. Soc. Chem. Commun.*, 8, 680–682, 1993. Reproduced by permission of The Royal Society of Chemistry.)

retain high porosity (15%–50%) and enormous surface area (150–900 m²/g), along with very small pore size (1–10 nm). When solvent removal occurs under supercritical conditions (Aegerter et al. 2011), the network does not shrink and a highly porous, extremely low-density material with low thermal conductivity known as an aerogel is produced.

Silica aerogels have special interest for sensing applications, because silica aerogels have many properties that make them ideal as sensor platforms for analysis of gaseous species (Walcarius and Collinson 2009; Bonnaud et al. 2016; Amonette and Matyas 2017). They have porosities as high as 99%, and thus are mostly gas permeable. They have an index of refraction nearly identical to air, which ensures ready transmission of light through their volume. They have large internal surface areas for the sorption of analytes and for the attachment of functional groups to attract and detect those analytes. Their surfaces can be made hydrophilic or hydrophobic, depending on the nature of the analyte and the environment in which they are used. Because they are synthesized by a sol-gel process, it is easy to incorporate a wide variety of chromophores, fluorophores, and even biological receptors (including enzymes and bacteria) into their structure, making the potential range of their application limited only by the imagination of scientists and engineers, designing them. They can be prepared in powder, thin layer, and monolithic forms, allowing great flexibility in their design. Finally, they are amenable to hybridization with organic polymers to give them mechanical strength, varying degrees of electrical conductivity, and chemical durability, as needed. As a result, aerogel-based sensors have better sensitivity. In particular, Wang and Wu (2006) tested aerogel- and xerogel-based humidity sensors and found that the xerogel-coated sensors showed only slight impedance response to changes in relative humidity (RH) (see Figure 7.5). At the same time, the

aerogel-coated sensors showed an impedance drop of about 700 kΩ at 1 kHz frequency and 25°C in going from 20% to 90% RH. The authors concluded that the aerogel thickness and the pore structure were the two primary factors, affecting sensor response to humidity.

## 7.4 SILICA-BASED HUMIDITY SENSORS

### 7.4.1 Approaches to Humidity Sensor Design

At present, there are several approaches to mesoporous silica using in humidity sensors. In particular, humidity sensors can be designed on the base of silicates (Innocenzi et al. 2001, 2005; Domansky et al. 2001; Falcaro et al. 2004; Bertolo et al. 2005). Water molecules interact with hydroxyl sites, providing a base for physisorption of water layers as RH increases. For a dry surface at relatively low humidity, the conductance occurs through a proton "hopping" between the adsorption sites. At higher humidity, water concentrates to form multilayers or condenses to fill a pore. Proton mobility, therefore, becomes more facile, and the conductivity increases with protons moving from molecule to molecule (Grotthus chain reaction model). A mesopore structure increases the surface area and the number of hydroxyl groups available for water adsorption. Factors that affect the sensor response, include the size and accessibility of mesopores, the film thickness, the number of hydroxyl sites, and organic matter within the pores. Organic matter refers to residual surfactant from a templating process or polymer introduced either during or after synthesis.

Typically, humidity sensors are being developed based on ordered mesoporous silica of SBA-15 type. However, ordered mesoporous silica of other types such as SBA-16, KIT-6, MCM-41, and HMS can also be used. These studies pointed out that the factors affecting the humidity-sensing response of porous silica include accessibility of inner mesopores, processing condition, calcination temperature, film thickness, working voltage, working frequency, and electrode (Bearzotti et al. 2003, 2004; Tongpool and Jindasuwan 2005; Tu et al. 2012). Confirmation of this statement can be found in Figure 7.6. It can be seen that the sensor response depends on the type of mesoporous silica used and the conditions for subsequent processing of the synthesized material. Tu et al. (2012) compared humidity sensors fabricated using SBA-15, SBA-16, and crystal-like SBA-16 and showed that crystal-like SBA16-based sensors had maximum sensitivity. It was also found that after microwave digestion (MWD) treatment, used for copolymer template removal, the humidity-sensitive property of mesoporous silica has been greatly improved; the impedance of crystal-like SBA-16 after MWD treatment changed by more than three orders of magnitude over the whole humidity range (11%–95% RH) and, the impedance changes were several orders of magnitude greater than that of the calcined sample. According to Tu et al. (2012), a MWD is an effective technique to remove organic surfactant, which shows many advantages over other methods such as lower structural shrinkage, larger pores, and additional pores in the silica wall (Tian et al. 2002). Tu et al. (2012) believe that this treatment allows

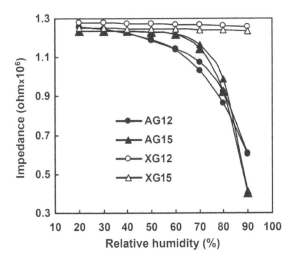

**FIGURE 7.5** Impedance responses and hysteresis characteristics of aerogel and xerogel silica film sensors as a function relative humidity, measured at 25°C and 1 kHz. (Reprinted from *Thin Solid Films* 496, Wang, C.-T. and Wu, C.-L., Electrical sensing properties of silica aerogel thin films to humidity. 658–664 Copyright 2006, with permission from Elsevier.)

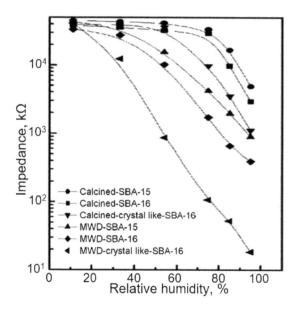

**FIGURE 7.6** Humidity-sensitive properties of mesoporous silica. (Reprinted from *Sens. Actuators B*, 136, Tu, J. et al., Humidity sensitive property of Li-doped 3D periodic mesoporous silica SBA-16, 392–398, Copyright 2009, with permission from Elsevier.)

for large numbers of hydroxyl groups reserving on the surface, which may improve its humidity-sensitive property. Tu et al. (2012) also established that the MWD-treated samples possess both larger main mesopores and additional connective pores within their pore walls, which makes this material much more suitable for mass transfer than the calcined counterpart. As a result, these sensors were fast enough. For humidity sensors based on the MWD treated crystal-like SBA-16, the response time was 40 s when RH increased from 11% to 95%, and the recovery time was about 100 s.

The experiment showed that silica-based hybrid materials are also very effective materials for fabrication of resistive or impedance humidity sensors. Examples of such sensors are listed in Table 7.2.

It is seen that hybrid silica-based materials can be prepared using ionic compounds, metal oxides, and polymers. At that, regardless of composition, composite-based sensors demonstrate good parameters both in terms of sensitivity and stability, and in rate of response. Typical characteristics of such sensors are shown in Figure 7.7. Sensing mechanisms, explaining the operation of silica-ionic-compound-based sensors (Falcaro et al. 2004; Tomer and Duhan 2015) are identical to the mechanisms considered in the analysis of

**TABLE 7.2**

**Comparison of Sensor Response Measured in the Range of 11%–98% RH for Different Mesoporous Silica-Based Resistive Humidity Sensors Reported in Literature**

| Sensing Material | Post Heat Treatment | Sensitivity Orders of Magnitude | Resp./Rec. Times (s) | References |
|---|---|---|---|---|
| Li/SBA-15 | Yes (550°C) | Three | 21/51 | Geng et al. (2007) |
| Li/SBA-16 | No | Four–five | 40/100 | Tu et al. (2009) |
| Li/MCM-41 | No | Two–three | | Wang et al. (2008) |
| LiCl/KIT-6 | Dried at 60°C | Three | 15/28 | Zhao et al. (2015) |
| KCl/MCM-41 | Yes (100°C) | Three–four | | Liu et al. (2010) |
| KCl/SBA-15 | Yes (550°C) | Five-six | 10/25 | Zhang et al. (2012) |
| K₂CO₃/SBA-15 | Yes (550°C) | Five | 15/50 | Yuan et al. (2009) |
| K₂CO₃/SiO₂ | Dried at 100°C | Four | 10/38 | Guo et al. (2016) |
| NaCl/SBA-15 | Yes (550°C) | Four–five | 20/65 | Geng et al. (2015) |
| NaCl/KIT-6 | No | Five | 47/150 | He et al. (2016) |
| NaCl/HMS | No | Five | 18/184 | Ge et al. (2017) |
| MgO/SBA-15 | Yes (550°C) | Three–four | 10/20 | Wang et al. (2010a) |
| ZnO/SBA-15 | Yes (550°C) | Three–four | | Yuan et al. (2010) |
| ZnO/SBA-15 | Yes (550°C) | Five | 17/18 | Tomer et al. (2015) |
| Fe/SBA-15 | Yes (550°C) | Four | 20/50 | Qi et al. (2008) |
| Fe₂O₃/silica | Yes (550°C) | Four | 20/40 | Yuan et al. (2011) |
| TiO₂/SBA-15 | Yes (600°C) | Five | 14/19 | Tomer and Duhan (2015a) |
| WO₃/SBA-15 | Yes (600°C) | Four–five | 18/25 | Tomer and Duhan (2015b) |
| SnO₂/SBA-15 | Yes (550°C) | Five-six | 15/21 | Tomer and Duhan (2015c) |
| SnO₂:Ag/SBA-15 | Yes (550°C) | Five | 5/8 | Tomer et al. (2016) |
| Ag/SBA-15 | | Five | 100/125 | Tomer et al. (2014) |
| NiO-PPY/SBA-15 | | Three–four | 45/90 | Wang et al. (2010b) |
| PSS/SBA-15 | Dried at 60°C | Three | 5/106 | Zhao et al. (2017a) |
| Propyl-S-DMC/SBA-15 | Dried at 60°C | Three | 11/61 | Zhao et al. (2017b) |

Abbreviations: DMC—methacrylatoethyl trimethyl ammonium chloride; HMS—hexagonal mesoporous silica; MCM—mesoporous *silica* materials; Propyl-S-DMC—propyl-SH modified by DMC; PSS—vinyl modified by sodium p-styrenesulfonate; SBA—Santa Barbara Amorphous.

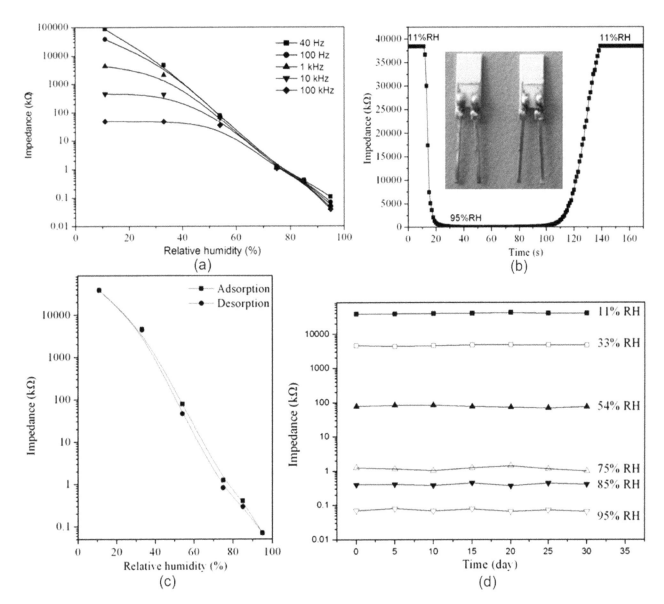

**FIGURE 7.7** (a) Relationship of impedance and RH at different frequencies based on K-SBA-15(0.5); (b) response and recovery properties of K-SBA-15(0.5); (c) hysteresis of K-SBA-15(0.5) sensor measured at100 Hz; and (d) stability of the K-SBA-15(0.5) sensor measured at100 Hz. Tested sensors were manufactured using the following process: The mixture of KCl and SBA-15 was ground for 1 h. The grounded mixtures were calcined at 550°C for 6 h, the products were designated as K-SBA-15(X), where X is the mass fraction of KCl in 1 g of SBA-15. Then K-SBA-15(X) powder was ground and mixed with deionized water in a weight ratio of 100:5 to form a dilute paste. The paste was screen-printed on a ceramic substrate (6 mm × 3 mm, 0.5 mm thick) with five pairs of Ag-Pd interdigital electrodes. To form a film with a thickness of about 10 μm, and then the film was dried in vacuum at 90°C for 5 h. In order to improve the sensor antipollution, we made a solution of 0.1 g of ethyl cellulose in ethylester acetate (4 mL), which was screen-printed on the surface of the sensitive film as the protective layer. Finally, the humidity sensor was obtained after aging at RH of 95% and a voltage of 1 V and 100 Hz for 24 h. (Reprinted from *J. Phys. Chem. Solids*, 73, Zhang W. et al., Humidity sensitive properties of K-doped mesoporous silica SBA-15, 517–522, Copyright 2012, with permission from Elsevier.)

metal-oxide sensors (read Chapter 2). There you will find also a description of the nature of the observed optimizing effect. In particular, according to the ion transport model of Casalbore-Miceli et al. (2000), KCl, NaCl, or LiCl may dissolve in the adsorbed water, and the dissociated ions ($K^+$, $Na^+$, $Li^+$ and $Cl^-$) from it can play the role of conduction carriers.

As for silica-metal-oxide nanocomposites, then Tomer and Duhan (2015) have found that for direct synthesis method, the uniform dispersion of $TiO_2$ nanoparticles in SBA-15 and unavailability of chocked pores provides for optimal composition better opportunities for free movement of charge carriers across its surface.

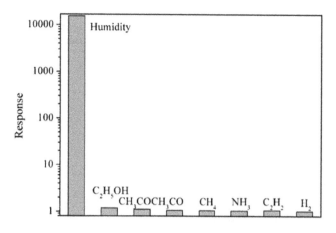

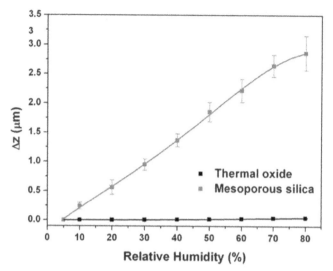

**FIGURE 7.8** The response of Fe-SBA-15 sensor to humidity (from 11% to 95% RH) and different gases (1000 ppm) at room temperature measured at 100 Hz. Mesoporous silica SBA-15 was synthesized using triblock copolymer P123. The sensing materials of Fe-doped SBA-15 were prepared by mixing of $Fe(NO_3)_3$ and SBA-15, and grounding in an agate mortar for 1 h. (Reprinted with permission from Qi et al. (2008). Copyright 2008: Elsevier.)

**FIGURE 7.9** Static evaluation of the mesoporous silica and bare cantilever responses under relative humidity variations. (Reprinted from Stassi S. et al., *J. Mater. Chem.*, *C*3, 12507, 2015, Reproduced by permission of The Royal Society of Chemistry.)

It is important that through the formation of nanocomposites one can also achieve a significant improvement in the selectivity of humidity sensors. For example, Qi et al. (2008) have found that Fe-SBA-15 sensors are almost insensitive to $C_2H_5OH$, $CH_3COCH_3$, CO, $CH_4$, $NH_3$, $C_2H_2$, and $H_2$ (see Figure 7.8).

Optical and fiber optical sensors are another direction for mesoporous silica application (Balkose et al. 1998; Melde et al. 2008; Viegas et al. 2009; Mallik et al. 2017). The principles of such a sensor operation were discussed in detail in Volume 1 of our series.

Experiments have shown that other types of humidity sensors can also use mesoporous silica as a sensing layer. For example, silica-based humidity-sensing films were incorporated in surface acoustic wave (SAW) (Tang et al. 2015; Tsukahara et al. 2017), QCM (Vashist and Vashist 2011; Zhu et al. 2014), capacitive (Wagner et al. 2011), and cantilever (Stassi et al. 2015)-based humidity sensors. At that, Stassi et al. (2015) have shown that a dip-coating deposition of mesoporous silica thin film with an accessible pore size of approximately 8 nm contributed to an increase in sensitivity to RH up to two orders of magnitude with respect to bare cantilevers (see Figure 7.9). The microcantilever (MC) arrays used in this work were prepared using a standard combination of surface and bulk micromachining processes, starting from a silicon-on-insulator (SOI) wafer. The mesoporous silica was deposited by dip coating the silicon-micromachined dice in a silica precursor solution, prepared by an acid-catalyzed process (Lee et al. 2012).

The same optimizing effect was observed for SAW- and QCM-based sensors. Moreover, as it is seen in Figure 7.10, the use of silica gives a significantly greater optimizing effect when compared to metal-oxide films (Tang et al. 2015). Apparently, this is due to the greater porosity and surface area of the silica layer.

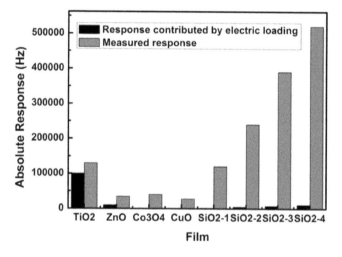

**FIGURE 7.10** Measured absolute frequency shifts and calculated absolute frequency shifts contributed by the electric loading effect during humidity measurement. The humidity increased from 30% to 93%. The $SiO_2$ and semiconducting oxide films were fabricated using sol-gel technology. Synthesized sols were coated onto the entire quartz substrates using a spin-coating method. (Reprinted from *Sens. Actuators B* 215, Tang Y. et al., Highly sensitive surface acoustic wave (SAW) humidity sensors based on sol–gel $SiO_2$ films: Investigations on the sensing property and mechanism, 283–291, Copyright (2015), with permission from Elsevier.)

It is important to note that silica-based humidity sensors, regardless of their type, demonstrate good reversibility, reproducibility, and stability of the parameters (see Figures 7.7 and 7.11), which is crucial to the practical application.

A mesoporous silica can also be employed as a hard template for the synthesis of a mesoporous material of a composition valuable for humidity sensing (Melde et al. 2008). Ordered, well-defined mesostructures are particularly suited

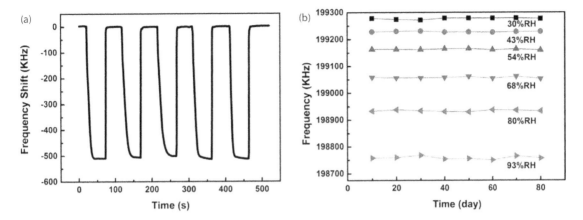

**FIGURE 7.11** (a) Frequency response of the SAW sensors based on the SiO$_2$ film when the RH increases from 30% to 93% for five cycles; and (b) frequency of the sensor in circumstances with various RH values for 80 days. (Reprinted from *Sens. Actuators B* 215, Tang Y. et al., Highly sensitive surface acoustic wave (SAW) humidity sensors based on sol–gel SiO$_2$ films: Investigations on the sensing property and mechanism, 283–291, Copyright (2015), with permission from Elsevier.)

to these applications. In particular, mesoporous silicates have been used to template carbons, metals, and metal oxides. The silicate framework is usually removed following templating by dissolving with hydrofluoric acid or a strong base.

### 7.4.2 Features of Humidity Sensor Fabrication

The experiment showed that calcination temperature has a large effect on the response of the templated thin film to humidity with variations resulting from differences in the number of silanol groups on the surface and the amount of surfactant left in the mesopores (Melde et al. 2008). Several studies have compared thin films prepared using templating techniques and calcination between 150°C and 550°C (Innocenzi et al. 2001, 2005; Falcaro et al. 2004; Bertolo et al. 2005). It has been found that lower calcination temperatures resulted in residual surfactant within the pores of the materials. When calcinations below 300°C were used, enhanced sensitivity to low RH values resulted. Calcinations at higher temperatures provided higher saturation levels. It has been speculated that the residual surfactant in the low-calcination-temperature materials is the strongest contributing factor to the differences observed.

The pore size also plays a role in the effectiveness of silicate materials used for sensing changes in RH. A comparison of nontemplated to templated materials demonstrated an increase in the current of several orders of magnitude (Bearzotti et al. 2003, 2004). Silica aerogels (surface area 866 m$^2$/g; average pore size of 20.5 nm; and pore volume 2.83 cm$^3$/g) were compared to a xerogel of lower porosity (surface area 709 m$^2$/g, average pore diameter 6.7 nm, and pore volume 1.29 cm$^3$/g) (Wang et al. 2005). The aerogels were much more sensitive to RH, as measured by impedance and capacitance, than the xerogel.

The pore size in silica aerogels and their texture depends on many parameters of the technological process used to synthesize them (Hæreid et al. 1995; He et al. 2009; Pang et al. 2013). Low-density silica aerogels require supercritical drying

to prevent the collapse of the pores during the removal of solution from the gels. But if the structure of silica aerogel is not strong enough, even drying in the supercritical fluids, the silica aerogel will shrink strongly or produce severe cracks. On the other hand, low strength limits the large-scale using of silica aerogel in many fields. Many efforts, thus, were paid to increase the strength of silica aerogel in order to improve its practical performances. It was established that the modulus of rupture of silica aerogel can be increased by heat-treating the wet gel in hot water (Hæreid et al. 1995) or aging in water/ethanol mixture (Einarsrud et al. 2001). According to Hæreid et al. (1995), prolonging the aging time can increase the skeletal strength of silica gel while increasing the aging temperature can shorten the aging period. Hæreid et al. (1995) believe that aging in a water solution can promote the dissolution and reprecipitation process of small silica onto the contact point of particles, which will increase the neck area and thus enhance the skeletal strength of silica gel. Reichenauer (2004) attributed this heat-treatment effect to an Ostwald ripening process, which can enhance the backbone strength and thus prevent the collapse or cracking of silica aerogel during drying process. Aging in a mother solution or adding an additional monomer before or after the gel point are also reported to be able to improve the strength and stiffness of the wet gel by rebuilding the network via further hydrolysis, condensation, and specific reaction of these monomers. So, silica aerogel cured in methanol exhibits high elastic modulus and hardness (Hæreid et al. 1996; Lucas et al. 2004). Rigacci et al. (2004) have shown that the increase in strength and stiffness can also be achieved after aging the silica gel in a polyethoxydisiloxane/ethanol solution. Thus, controlling the composition of the aging solution allows one to tune the texture of silica gel, and the gels with required pore size and more homogeneous pore size distribution can be obtained. In particular, the primary particle size and pore size can usually be increased and homogenized during the aging process. For example, He et al. (2009) offered for this purpose aging in a 100°C autoclave with a TEOS/ethanol-mixed solution and aging in ethanol at room temperature for the same time. According to He et al. (2009), high temperature and pressure aging for wet silica gel can promote the following processes: the dissolution and reprecipitation of silica, the further hydrolysis and condensation of unhydrolyzed $\equiv Si-O-C_2H_5$ groups, and the esterification of silanols by TEOS from the aging solution. Hence, after supercritical $CO_2$ drying, it leads to the silica aerogel having strong skeletal strength, large primary particle size, and average pore size, and thus high specific-pore volume and good monolithic performance. He et al. (2009) have found that aging at room temperature in ethanol results after supercritical $CO_2$ drying in silica aerogel with low average pore size, weak skeletal strength, and thus low specific pore volume and poor monolithic appearance due to the high shrinkage during the drying process. Silica aerogel aged in 100°C autoclave shows hydrophobic characteristics while that aged at room temperature exhibits more hydrophilic feature (see Figure 7.12). TEOS from the aging solution makes the backbone of silica wet gel much stronger, but easily results in the remaining of small-sized pores.

The accessibility of the mesopores is another important parameter for achievement-required detection efficiency (Melde et al. 2008). For example, a three-dimensional mesopore system should be more useful for sensing in a thin film than a two-dimensional mesostructure with channels running parallel to the plane of the film. A mesostructure that is highly ordered is less likely to obstruct an analyte with "dead ends" that are more likely in disordered structures. Recent developments in the synthesis of hierarchically structured materials, combining improved accessibility with function, should benefit sensing applications. Some examples of macrostructured-ordered mesoporous silicates include close-packed colloidal crystals of mesoporous silicas (Yang et al. 2000), inverse opals of mesoporous silica (Holland et al.

1999), and microphase-separation-induced macroporous-mesoporous silica and organosilica monoliths (Nakanishi and Kanamori 2005; Nakanishi et al. 2008). Macropores can facilitate diffusion-enhancing access to the mesopores. This feature should aid in material sensitivity and response and recovery time. Based on reports of periodic mesoporous organosilicas with crystal-like order within the pore walls (Inagaki et al. 2002; Kapoor et al. 2003), it may be possible to design macro-meso-microstructured materials with high specific adsorption capacities bearing functional moieties.

In addition, selection of the electrode materials employed in these types of sensing applications was found to be also important. Chromium and gold electrodes provided the most stable current responses while titanium and aluminum showed decreasing responses over extended operation periods (Bearzotti 2008).

## REFERENCES

Aegerter M.A., Leventis N., Koebel M.M. (2011) *Aerogels Handbook*. Springer, New York.

Amonette J.E., Matyas J. (2017) Functionalized silica aerogels for gas-phase purification, sensing, and catalysis: A review. *Micropor. Mesopor. Mat.* **250**, 100–119.

Ariga K., Vinu A., Hill J.P., Mori T. (2007) Coordination chemistry and supramolecular chemistry in mesoporous nanospace. *Coord. Chem. Rev.* **251**, 2562–2591.

Balkose D., Ulutan S., Ozkan F.C., Celebi S., Ulku S. (1998) Dynamics of water vapor adsorption on humidity-indicating silica gel. *Appl. Surf. Sci.* **134**, 39–46.

Basabe-Desmonts L., Reinhoudt D.N., Crego-Calama M. (2007) Design of fluorescent materials for chemical sensing. *Chem. Soc. Rev.* **36**, 993–1017.

Bearzotti A. (2008) Influence of metal electrodes on the response of humidity sensors coated with mesoporous silica. *J. Phys. D: Appl. Phys.* **41**, 1–5.

Bearzotti A., Bertolo J.M., Innocenzi P., Falcaro P., Traversa E. (2003) Relative humidity and alcohol sensors based on mesoporous silica thin films as synthesized from block copolymers. *Sens. Actuators B* **95**, 107–110.

Bearzotti A., Bertolo J.M., Innocenzi P., Falcaro P., Traversa E. (2004) Humidity sensors based on mesoporous silica thin films synthesized by block copolymers. *J. Eur. Cer. Soc.* 24, 1969–1972.

Beck J.S., Vartuli J.C., Roth W.J., Leonowicz M.E., Kresge C.T., Schmitt K.D., et al. (1992) A new family of mesoporous molecular sieves prepared with liquid crystal templates. *J. Am. Chem. Soc.* **114**, 10834–10843.

Bertolo J.M., Bearzotti A., Generosi A., Palummo L., Albertini V.R. (2005) X-rays and electrical characterizations of ordered mesostructured silica thin films used as sensing membranes. *Sens. Actuators B* **111–112**, 145–149.

Bonnaud P.A., Miura R., Suzuki A., Miyamoto N., Hatakeyama N., Miyamoto A. (2016) Confined water in nanoporous silica: Application to humidity sensors. In: *Proceedings of the 16th International Conference on Nanotechnology*, Sendai, Japan, August 22–25, pp. 270–272.

Brinker C.J., Scherer G.W. (1990) *Sol–Gel Science: The Physics and Chemistry of Sol–Gel Processing*. Academic, San Diego, CA.

Carrington N.A., Xue Z.L. (2007) Inorganic sensing using organofunctional sol–gel materials. *Acc. Chem. Res.* **40**, 343–350.

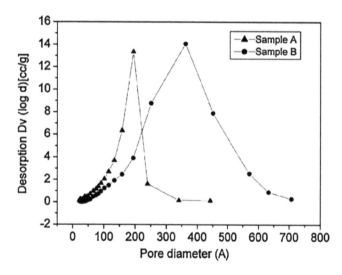

**FIGURE 7.12** Pore-size distribution of silica aerogel aged in ethanol at room temperature (sample A) and in autoclave at 100°C (sample B). (Reprinted from *J. Mater. Proces. Technol.* 209, He F. et al., Modified aging process for silica aerogel. 1621–1626, Copyright 2009, with permission from Elsevier.)

Carroll M.K., Anderson A.M. (2011) Aerogels as platforms for chemical sensors. In: Aegerter M.A., Leventis N., Koebel M.M. (eds.) *Aerogels Handbook*. Springer, New York, pp. 637–650.

Casalbore-Miceli G., Yang M.J., Camaioni N., Mari C.-M., Li Y., Sun H., Ling M. (2000) Investigations on the ion transport mechanism in conducting polymer films. *Solid State Ionics* **131**, 311–321.

Corma A. (1997) From microporous to mesoporous molecular sieve materials and their use in catalysis. *Chem. Rev.* **97**, 2373–2419.

Domansky K., Liu J., Wang L.-Q., Engelhard M.H., Baskaran S. (2001) Chemical sensors based on dielectric response of functionalized mesoporous silica films. *J. Mater. Res.* **16**, 2810–2816.

Einarsrud M.-A., Nilsen E., Rigacci A., Pajonk G.M., Buathier S., Valette D., et al. (2001) Strengthening of silica gels and aerogels by washing and aging processes. *J. Non-Cryst. Solids* **285**, 1–7.

Falcaro P., Bertolo J.M., Innocenzi P., Amenitsch H., Bearzotti A. (2004) Ordered mesostructured silica films: Effect of pore surface on its sensing properties. *J. Sol-Gel Sci. Tech.* **32**, 107–110.

Galarneau A., Di Renzo F., Fajula F., Vedrine J. (eds.) (2001) *Zeolites and Mesoporous Materials at the Dawn of the 21st Century*. Elsevier, Amsterdam, The Netherlands.

Ge S., He X., Jia L., Duan L., Zhang S., Zhang Q., Geng W. (2017) Facile fabrication of NaCl-added mesoporous silica HMS composite and its humidity responding performance. *J. Sol-Gel Sci. Technol.* **82**, 635–642.

Geng W., Wang R., Li X., Zou Y., Zhang T., Tu J., et al. (2007) Humidity sensitive property of Li-doped mesoporous silica SBA-15. *Sens. Actuators B* **127**, 323–329.

Geng W., Zhou L., Duan L., Gu J., Zhang Q. (2015) Humidity sensing property of NaCl-added mesoporous silica synthesized by a facile way with low energy cost. *Int. J. Appl. Ceram. Technol.* **12**(1), 169–175.

Guo L., Chu X. Gao X., Wang C., Chi Y., Yang X. (2016) AC humidity sensing properties of mesoporous $K_2CO_3$-$SiO_2$ composite materials. *J. Chem.* 2016, 3508307.

Hæreid S., Anderson J., Einarsrud M.A., Hua D.W., Smith D.M. (1995) Thermal and temporal aging of TMOS-based aerogel precursors in water. *J. Non-Cryst. Solids* **185**, 221–226.

Hæreid S., Nilsen E., Einarsrud M.A. (1996) Properties of silica gels aged in TEOS. *J. Non-Cryst. Solids* **204**, 228–234.

He F., Zhao H., Qu X., Zhang C., Qiu W. (2009) Modified aging process for silica aerogel. *J. Mater. Proces. Technol.* **209**, 1621–1626.

He X., Geng W., Zhang B., Jia L., Duan L., Zhang Q. (2016) Ultrahigh humidity sensitivity of NaCl-added 3D mesoporous silica KIT-6 and its sensing mechanism. *RSC Adv.* **6**, 38391.

Hoffmann F., Cornelius M., Morell J., Froba M. (2006) Silica-based mesoporous organic–inorganic hybrid materials. *Angew Chem. Int. Ed.* **45**, 3216–3251.

Holland B.T., Blanford C.F., Do T., Stein A. (1999) Synthesis of highly ordered, three-dimensional, macroporous structures of amorphous or crystalline inorganic oxides, phosphates, and hybrid composites. *Chem. Mater.* **11**, 795–805.

Hong J.K., Kim H.R., Park H.H. (1998) The effect of sol-viscosity on the sol–gel derived low density $SiO_2$ xerogel film for inter-metal dielectric application. *Thin Solid Films* **332**, 449–454.

Huo Q., Margolese D.I., Stucky G. (1996) Surfactant control of phases in the synthesis of mesoporous silica-based materials. *Chem. Mater.* **8**, 1147–1160.

Inagaki S., Fukushima Y., Kuroda K. (1993) Synthesis of highly ordered mesoporous materials from a layered polysilicate. *J. Chem. Soc. Chem. Commun.* **8**, 680–682.

Inagaki S., Guan S., Ohsuna T., Terasaki O. (2002) An ordered mesoporous organosilica hybrid material with a crystal-like wall structure. *Nature* **416**, 304–307.

Innocenzi P., Falcaro P., Bertolo J.M., Bearzotti A., Amenitsch H. (2005) Electrical responses of silica mesostructured films to changes in environmental humidity and processing conditions. *J. Non-Cryst. Solids* **351**, 1980–1986.

Innocenzi P., Martucci A., Guglielmi M., Bearzotti A., Traversa E. (2001) Electrical and structural characterisation of mesoporous silica thin films as humidity sensors. *Sens. Actuators B* **76**, 299–303.

Kapoor M.P., Yang Q., Inagaki S. (2003) Self-assembly of biphenylene-bridged hybrid mesoporous solid with molecular-scale periodicity in the pore walls. *J. Am. Chem. Soc.* **124**, 15176–15177.

Kresge C.T., Leonowicz M.E., Roth W.J., Vartuli J.C., Beck J.S. (1992) Ordered mesoporous molecular sieves synthesized by a liquid-crystal template mechanism. *Nature* **359**, 710–712.

Kresge C.T., Roth W.J. (2013) The discovery of mesoporous molecular sieves from the twenty-year perspective. *Chem. Soc. Rev.* **42**, 3663–3670.

Lee H.J., Park K.K., Kupnik M., Melosh N.A., Khuri-Yakub B.T. (2012) Mesoporous thin film on highly-sensitive resonant chemical sensor for relative humidity and $CO_2$ detection. *Anal. Chem.* **84**, 3063–3066.

Li W., Willey R.J. (1997) Stability of hydroxyl and methoxy surface groups on silica aerogels. *J. Non-Cryst. Solids* **212**, 243–249.

Liu L., Kou L.Y., Zhong Z.C., Wang L.Y., Liu L.F., Li W. (2010) Preparation and humidity sensing properties of KCl/MCM-41 composite. *Chin. Phys. Lett.* **27**, 050701.

Lucas E.M., Doescher M.S., Ebenstein D.M., Wahl K.J., Rolison D.R. (2004) Silica aerogels with enhanced durability, 30-nm mean pore-size, and improved immersibility in liquids. *J. Non-Cryst. Solids* **350**, 244–252.

Mallik A.K., Farrell G., Wu Q., Semenova N.Y. (2017) Study of the influence of the agarose hydrogel layer thickness on sensitivity of the coated silica microsphere resonator to humidity. *Appl. Opt.* **56**(14), 4065–4069.

Melde B.J., Johnson B.J., Charles P.T. (2008) Mesoporous silicate materials in sensing. *Sensors* **8**, 5202–5228.

Mittal K.L., Fendler E.J. (eds.) (1982) *Solution Behavior of Surfactants*. Plenum, New York.

Moos R., Sahner K., Fleischer M., Guth U., Barsan N., Weimar U. (2009) Solid state gas sensor research in Germany—A status report. *Sensors* **9**, 4323–4365.

Moos R., Sahner K., Hagen G., Dubbe A. (2006) Zeolites for sensors for reducing gases. *Rare Metal Mater. Eng. Suppl. B* **35**, 447–451.

Naik B., Ghosh N.N. (2009) A review on chemical methodologies for preparation of mesoporous silica and alumina based materials. *Recent Pat. Nanotechnol.* **3**, 213–224.

Nakanishi K., Amatani T., Yano S., Kodaira T. (2008) Multiscale templating of siloxane gels via polymerization-induced phases separation. *Chem. Mater.* **20**, 1108–1115.

Nakanishi K., Kanamori K. (2005) Organic-inorganic hybrid poly(silsesquioxane) monoliths with controlled macro- and mesopores. *J. Mater. Chem.* **15**, 3776–3786.

Pang S.C., Kho S.Y., Chin S.F. (2013) Tailoring microstructure of silica xerogels via a facile synthesis approach. *J. Mater. Environ. Sci.* **4**(5), 744–751.

Qi Q., Zhang T., Zheng X., Wan, L. (2008) Preparation and humidity sensing properties of Fe-doped mesoporous silica SBA-15. *Sens. Actuators B* **135**, 255–261.

Reichenauer G. (2004) Thermal aging of silica gels in water. *J. Non-Cryst. Solids* **350**, 189–195.

Rigacci A., Einarsrud M.A., Nilsen E., Pirard R., Dolle F.E., Chevalier B. (2004) Improvement of the silica aerogel strengthening process for scaling-up monolithic tile production. *J. Non-Cryst. Solids* **350**, 196–201.

Sahner K., Hagen G., Schönauer D., Reiß S., Moos R. (2008) Zeolites—Versatile materials for gas sensors. *Solid State Ionics* **179**(40), 2416–2423.

Slowing I.I., Trewyn B.G., Giri S., Lin V.S.-Y. (2007) Mesoporous silica nanoparticles for drug delivery and biosensing applications. *Adv. Funct. Mater.* **17**, 1225–1236.

Soler-Illia G.J., Sanchez C., Lebeau B., Patarin J. (2002) Chemical strategies to design textured materials: From microporous and mesoporous oxides, to nanonetworks and hierarchical structures. *Chem. Rev.* **102**, 4093–4138.

Stassi S., Cauda V., Fiorilli S., Ricciardi C. (2015) Surface area enhancement by mesoporous silica deposition on microcantilever sensors for small molecule detection. *J. Mater. Chem. C* **3**, 12507.

Tang Y., Li Z., Ma J., Wang L., Yang J., Du B., Yu Q., Zu X. (2015) Highly sensitive surface acoustic wave (SAW) humidity sensors based on sol–gel $SiO_2$ films: Investigations on the sensing property and mechanism. *Sens. Actuators B* **215**, 283–291.

Tian B.Z., Liu X.Y., Yu C.Z., Gao F., Luo Q., Xie S.H., et al. (2002) Microwave assisted template removal of siliceous porous materials. *Chem. Commun.* **11**, 1186–1187.

Tomer K., Adhyapak P.V., Duhan S., Mulla I.S. (2014) Humidity sensing properties of Ag-loaded mesoporous silica SBA-15 nanocomposites prepared via hydrothermal process, *Micropor. Mesopor. Mat.* **197**, 140–147.

Tomer V.K., Duhan S. (2015a) Nano titania loaded mesoporous silica: Preparation and application as high performance humidity sensor. *Sens. Actuators B* **220**, 192–200.

Tomer V.K., Duhan S. (2015b) Highly sensitive and stable relative humidity sensors based on $WO_3$ modified mesoporous silica. *Appl. Phys. Lett.* **106**, 063105.

Tomer V.K., Duhan S. (2015c) In-situ synthesis of $SnO_2$/SBA-15 hybrid nanocomposite as highly efficient humidity sensor. *Sens. Actuators B* **212**, 517–525.

Tomer V.K., Duhan S. (2016) A facile nanocasting synthesis of mesoporous Ag-doped $SnO_2$ nanostructures with enhanced humidity sensing performance. *Sens. Actuators B* **223**, 750–760.

Tomer V.K., Duhan S., Malik R., Nehra S.P., Devi S. (2015) A novel highly sensitive humidity sensor based on ZnO/SBA-15 hybrid nanocomposite. *J. Am. Ceram. Soc.* **98**, 3719–3725.

Tongpool R., Jindasuwan S. (2005) Sol–gel processed iron oxide–silica nanocomposite films as room-temperature humidity sensors. *Sens. Actuators B* **106**, 523–528.

Tsukahara Y., Tsuji T., Oizumi T., Akao S., Takeda N., Yamanaka K. (2017) A thermodynamic consideration on the mechanism of ultrasensitive moisture sensing by amorphous silica. *Int. J. Thermophys.* **38**, 108.

Tu J., Li N., Geng W., Wang R., Lai X., Cao Y., et al. (2012) Study on a type of mesoporous silica humidity sensing material. *Sens. Actuators B* **166–167**, 658–664.

Tu J., Wang R., Geng W., Lai X., Zhang T., Li N., et al. (2009) Humidity sensitive property of Li-doped 3D periodic mesoporous silica SBA-16. *Sens. Actuators B* **136**, 392–398.

Vashist S.K., Vashist P. (2011) Recent advances in quartz crystal microbalance-based sensors. *J. Sensors* **2011**, 571405.

Viegas D., Goicoechea J., Santos J.L., Araújo F.M., Ferreira L.A., Arregui F.J., Matias I.R. (2009) Sensitivity improvement of a humidity sensor based on silica nanospheres on a long-period fiber grating. *Sensors* **9**, 519–527.

Wagner T., Haffer S., Weinberger C., Klaus D., Tiemann M. (2013) Mesoporous materials as gas sensors. *Chem. Soc. Rev.* **42**, 4036–4053.

Wagner T., Krotzky S., Weiß A., Sauerwald T., Kohl C.-D., Roggenbuck J., Tiemann M. (2011) A high temperature capacitive humidity sensor based on mesoporous silica. *Sensors* **11**, 3135–3144.

Walcarius A. (2008) Electroanalytical applications of microporous zeolites and mesoporous (organo)silicas: Recent trends. *Electroanal.* **20**(7), 711–738.

Walcarius A., Collinson M.M. (2009) Analytical chemistry with silica sol-gels: Traditional routes to new materials for chemical analysis. *Annu. Rev. Anal. Chem.* **2**, 121–143.

Wang C.T., Willey R.J. (2001) Mechanistic aspects of methanol partial oxidation over supported iron oxide aerogels. *J. Catal.* **202**, 211–219.

Wang C.-T., Wu C.-L. (2006) Electrical sensing properties of silica aerogel thin films to humidity. *Thin Solid Films* **496**, 658–664.

Wang C.-T., Wu C.-L., Chen I.-C., Huang Y.-H. (2005) Humidity sensors based on silica nanoparticle aerogel thin films. *Sens. Actuators B* **107**, 402–410.

Wang L., Li D., Wang R., He Y., Qi Q., Wang Y., Zhang T. (2008) Study on humidity sensing property based on Li-doped mesoporous silica MCM-41. *Sens. Actuators B* **133**, 622–627.

Wang R., Liu X., He Y., Yuan Q., Li X., Lu G., Zhang T. (2010a) The humidity-sensitive property of MgO-SBA-15 composites in one-pot synthesis. *Sens. Actuators B* **145**, 386–393.

Wang R., Zhang T., He Y., Li X., Geng W., Tu J., Yuan Q. (2010b) Direct-current and alternating-current analysis of the humidity-sensing properties of nickel oxide doped polypyrrole encapsulated in mesoporous silica SBA-15. *J. Appl. Polym. Sci.* **115**, 3474.

Wu S.-H., Mou C.-Y., Lin H.-P. (2013) Synthesis of mesoporous silica nanoparticles. *Chem. Soc. Rev.* **42**, 3862–3875.

Xu X., Wang J., Long Y. (2006) Zeolite-based materials for gas sensors. *Sensors* **6**, 1751–1764.

Yang S.M., Coombs N., Ozin G.A. (2000) Micromolding in inverted polymer opals (MIPO): Synthesis of hexagonal mesoporous silica opals. *Adv. Mater.* **12**, 1940–1944.

Yoda S., Ohshima S. (1999) Supercritical drying media modification for silica aerogel preparation. *J. Non-Cryst. Solids* **248**, 224–234.

Yuan Q., Geng W., Li N., Tu J., Wang R., Zhang T., Li X. (2009) Study on humidity sensitive property of $K_2CO_3$-SBA-15 composites. *Appl. Surf. Sci.* **256**, 280–283.

Yuan Q., Li N., Geng W., Chi Y., Tu J., Li X., Shao C. (2011) Humidity sensing properties of mesoporous iron oxide/silica composite prepared via hydrothermal process. *Sens. Actuators B* **160**, 334–340

Yuan Q., Li N., Tu J., Li X., Wang R., Zhang T., Shao C. (2010) Preparation and humidity sensitive property of mesoporous ZnO–$SiO_2$ composite. *Sens. Actuators B* **149**, 413–419.

Zhang W., Wang R., Zhang Q., Li J. (2012) Humidity sensitive properties of K-doped mesoporous silica SBA-15. *J. Phys. Chem. Solids* **73**, 517–522.

Zhao H., Liu S., Wang R., Zhang T. (2015) Humidity-sensing properties of LiCl-loaded 3D cubic mesoporous silica KIT-6 composites. *Mater. Lett.* **147**, 54–57.

Zhao H., Zhang T., Dai J., Jiang K., Fei T. (2017b) Preparation of hydrophilic organic groups modified mesoporous silica materials and their humidity sensitive properties. *Sens. Actuators B* **240**, 681–688.

Zhao H., Zhang T., Qi R., Dai J., Liu S., Fei T., Lu G. (2017a) Organic-inorganic hybrid materials based on mesoporous silica derivatives for humidity sensing. *Sens. Actuators B* **248**, 803–811.

Zheng Y., Li X., Dutta P.K. (2012) Exploitation of unique properties of zeolites in the development of gas sensors. *Sensors* **12**, 5170–5194.

Zhu Y., Chen J., Li H., Zhu Y., Xu J. (2014) Synthesis of mesoporous $SnO_2$–$SiO_2$ composites and their application as quartz crystal microbalance humidity sensor. *Sens. Actuators B* **193**, 320–325.

# 8 Aluminosilicate (Zeolites)-Based Humidity Sensors

## 8.1 ZEOLITES AND FEATURES OF THEIR SYNTHESIS

Zeolites or aluminosilicates are another group of compounds that have generated some interest (Rolison 1990; Walcarius 1999; Kulprathipanja 2010; Wales et al. 2015). Zeolites are three-dimensional (3D), microporous, crystalline, hydrated aluminosilicates of the alkaline and alkaline-earth metals. They are very ordered, well-defined, uniform, and the complex structure is an open framework based on sharing oxygen atoms by connecting $SiO_4$ and $AlO_4$ tetrahedra infinitely to outspread the 3D network. A high degree of open porosity gives rise to an exceptionally high surface area. Chemically, they are represented by the empirical formula:

$$M_{x/m}\left[(AlO_2)_x(SiO_2)_y\right]zH_2O, \qquad (8.1)$$

where M is the cation with valence $m$, $z$ is the number of water molecules in each unit cell, and $x$ and $y$ are integers (Jacobs 1977). Each $AlO_4$ tetrahedron in the framework bears a net negative charge, which is balanced by an extra-framework cation. It is generally observed that no twofold $AlO_4$ can be interconnected by sharing their junction in the zeolite framework. Therefore, the Si/Al ratio of a zeolite framework is always greater than 1.

Great interest to zeolites is not surprising considering that they comprise a microporous open framework structure which is accessible to certain guest molecules. The initial building blocks of the zeolite crystal lattice are $AlO_4$ or $SiO_4$ tetrahedra that are interconnected via oxygen bridges. Thus, a well-defined, 3D framework is created in which micro- and mesopores are linked by channels (see Figure 8.1). The intracrystalline channels or voids can be 1D, 2D, or 3D (Flanigen et al. 2010). Intracrystalline channels or interconnected voids are occupied by the cations and water molecules. The cations are mobile and ordinarily undergo ion exchange. The water in many zeolites may be removed reversibly, generally by the application of heat, which leaves intact a crystalline host structure permeated by the micropores and voids, which may amount to 50% of the crystals by volume.

Zeolites are usually synthesized under hydrothermal conditions in the low temperature range (70°C–300°C), from solutions of sodium aluminate, sodium silicate, or sodium hydroxide (Yu 2007). The precise zeolite formed is determined by the reactants used and the particular synthesis conditions used, such as temperature, time, pH (usually pH > 10), and templating ion. The templating ion is usually a cation around, which the aluminosilicate lattice is formed, so that the tunnel size is determined by the templating cation (Kulprathipanja 2010). The zeolite synthesis is carried out with inorganic as well as organic precursors. The inorganic precursors yielded more hydroxylated surfaces whereas the organic precursors easily incorporated the metals into the network. Temperatures higher 200°C–300°C often gives denser materials. The addition of fluoride to the reactive gel led to more perfect and larger crystals of known molecular sieve structures as well as new structures and compositions (Villaescusa and Camblor 2003). The fluoride ion also is reported to serve as a template or structure-directing agent (SDA) in some cases. Fluoride addition extends the synthesis regime into the acidic pH region.

Zeolites are classified according to following criteria; chemical composition (Si/Al ratio), framework structures, pore-opening diameter, and pore dimensions of zeolite. The details of classification as per the previously mentioned criteria are summarized in Tables 8.1 through 8.3.

At present, more than 200 distinct framework structures of zeolites are known, in addition to 40 natural ones. Today, in addition to the natural types of zeolites, a huge number of synthetic ones is available (Auerbach et al. 2003; Xu et al. 2007). Several of them are listed in Table 8.4 and shown in Figure 8.1. Some of the more important zeolite types, most of which have been used in commercial applications, include the zeolite minerals mordenite, chabazite, erionite, and clinoptilolite, the synthetic zeolite types A, X, Y, L, "Zeolon" mordenite, ZSM-5, beta, and MCM-22, and the zeolites F and W. They exhibit pore sizes from 0.3 to 1.0 nm and pore volumes from about 0.10 to 0.35 cm³/g. Typical zeolite pore sizes include: (1) small pore zeolites with eight-ring pores, free diameters of 0.30–0.45 nm (e.g., zeolite A); (2) medium pore zeolites with 10-ring pores, 0.45–0.60 nm in free diameter (ZSM-5); (3) large-pore zeolites with 12-ring pores of 0.6–0.8 nm [e.g., zeolites X, Y (see Figure 8.2)]; and (4) extra-large-pore zeolites with 14-ring pores (e.g., UTD-1) (Flanigen et al. 2010). It is necessary to note that the zeolite framework should be viewed as somewhat flexible, with the size and shape of the framework and pore responding to the changes in temperature and guest species. For example, ZSM-5 with sorbed neopentane has a near-circular pore of 0.62 nm, but with substituted aromatics as the guest species, the pore assumes an elliptical shape, 0.45–0.70 nm in diameter.

It was established that the ratio of silicon to aluminum atoms in the lattice, $x/y$, is a very important parameter for zeolite characterization. The more aluminum-based units are

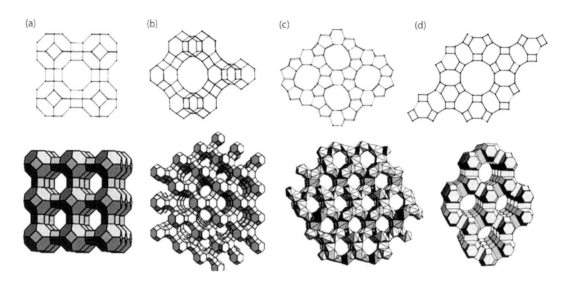

**FIGURE 8.1**   Representative zeolite frameworks, (with pore openings): (a) zeolite A (3D, 4.2 Å); (b) zeolite Y (3D, 7.4 Å); (c) Zeolite L (1D, 7.1 Å); and (d) ZSM-5 (silicalite) (2D, 5.3 × 5.6 Å, 5.1 × 5.5 Å): D—dimensions of channel system. (Reprinted from Zheng, Y. et al., *Sensors*, 12, 5170–5194, 2012. Published by MDPI as open access.)

### TABLE 8.1
### Classification of Zeolite According to Si/Al Ratio

| Class | Si/Al Ratio | Examples |
|---|---|---|
| Minimum SiO$_2$ | 1.0–1.5 | Faujasite; Zeolite A, X |
| In-between SiO$_2$ | 1.5–5 | Erionite; Clinoptilolite; Chabazite (CHA); Zeolite L, Y |
| Maximum SiO$_2$ | 5—several hundred | MFI; zeolite-β (BEA); NCL-1 |
| All Silica Molecular Sieves | Several thousand infinity | Silicate-I (ZSM-5); Silicate-II (ZSM-11); Si-ZSM-48 |

*Source:* Data extracted from Auerbach, S.M. et al., *Handbook of Zeolite Science and Technology.* Marcel Dekker, New York, 2003; Scott, M.A. et al., *Handbook of Zeolite Science and Technology*, CRC Press, New York, 2003; Xu, X. et al., *Sensors*, 6, 1751–1764, 2006; Cejka, J. et al., *Introduction to Zeolite Science and Practice*, 3rd edn., Studies in Surface Science and Catalysis, 168. Elsevier, Amsterdam, the Netherlands, 2007.)

### TABLE 8.2
### Classification of Zeolites According to Pore Size

| Types | Pore Size | Pore Diameter | Examples |
|---|---|---|---|
| Small-pore zeolites | 8MR | 0.3–0.5 nm | CHA; LTA |
| Medium-pore zeolites | 10 MR | 0.5–0.6 nm | ZSM-5 |
| Large-pore zeolites | 12 MR | 0.6–0.9 nm | FAU |
| Extra-large-pore zeolites | 14 MR–18 MR | >0.9 nm | UDT-1(14–ring); VPI-5(18-ring); Cloverite(20-ring) |

*Source:* Data extracted from Auerbach, S.M. et al., *Handbook of Zeolite Science and Technology*, Marcel Dekker, New York, 2003; Scott, M.A. et al., *Handbook of Zeolite Science and Technology*, CRC Press, New York, 2003; Xu, X. et al., *Sensors*, 6, 1751–1764, 2006; Cejka, J. et al., *Introduction to Zeolite Science and Practice*, 3rd edn., Studies in Surface Science and Catalysis, 168. Elsevier, Amsterdam, the Netherlands, 2007.)

## TABLE 8.3
## Zeolite Classification as Per Pore Dimensionality

| 1D | 2D | 3D |
|---|---|---|
| BEA (Beta) | MFI (ZSM-5) | FAU (faujasite) |
| MOR (Mordenite) | FER (Ferrierite) | MEL (ZSM-11) |
| LTL (Zeolite L) | | LTA (Linde Type A), |
| | | ANA (Analcime) |
| | | CHA (Chabazite) |

*Source:* Data extracted from Auerbach, S.M. et al., *Handbook of Zeolite Science and Technology*, Marcel Dekker, New York, 2003; Scott, M.A. et al., *Handbook of Zeolite Science and Technology*, CRC Press, New York, 2003; Xu, X. et al., *Sensors*, 6, 1751–1764, 2006; Cejka, J. et al., *Introduction to Zeolite Science and Practice*, 3rd edn., Studies in Surface Science and Catalysis, 168. Elsevier, Amsterdam, the Netherlands, 2007.)

## TABLE 8.4
## Typical Oxide Formula of Some Synthetic Zeolites

| Framework | Cationic Form | Formula of Typical Unit Cell | Window | Effective Channel Diameter (nm) |
|---|---|---|---|---|
| A | Na | $Na_{12}[(AlO_2)_{12}(SiO_2)_{12}]$ | 8-ring | 0.38 |
| | Ca | $Ca_2Na_2[(AlO_2)_{12}(SiO_2)_{12}]$ | | 0.44 |
| | K | $K_{12}[(AlO_2)_{12}(SiO_2)_{12}]$ | | 0.29 |
| X | Na | $Na_{86}[(AlO_2)_{86}(SiO_2)_{106}]$ | 12-ring | 0.84 |
| | Ca | $Ca_{40}Na_6[(AlO_2)_{86}(SiO_2)_{106}]$ | | 0.80 |
| | Sr, Ba | $Sr_{21}Ba_{22}[(AlO_2)_{86}(SiO_2)_{106}]$ | | 0.80 |
| Y | Na | $Na_{56}[(AlO_2)_{56}(SiO_2)_{136}]$ | 12-ring | 0.80 |
| | K | $K_{56}[(AlO_2)_{56}(SiO_2)_{136}]$ | | |
| | $NH_4$ | $(NH_4)_{39}(Na)_{10}[(AlO_2)_{49}(SiO_2)_{143}]$ | | |
| Mordenite (MOR) | Ag | $Ag_8[(AlO_2)_8(SiO_2)_{40}]$ | 12-ring | 0.70 |
| | H | $H_8[(AlO_2)_8(SiO_2)_{40}]$ | | |
| ZSM-5 | Na | $Na_3[(AlO_2)_3(SiO_2)_{93}]$ | 10-ring | 0.60 |
| Silicate | - | $(SiO_2)_{96}$ | 10-ring | 0.60 |

*Source:* Data extracted from Auerbach, S.M. et al., *Handbook of Zeolite Science and Technology*, Marcel Dekker, New York, 2003.)

present in the zeolite lattice, that is, the higher $y$, the more cations are needed for charge compensation. As a consequence, zeolites with high aluminum content are highly polar materials with excellent ion-exchange capacity and potential for high ion conductivity. This means that by changing the framework Si/Al ratio of the zeolite, the ion-exchange capacity and conductivity, the interaction between the zeolite and the adsorbed molecules, and the modification of hydrophilic or hydrophobic properties, can all be changed (Xu et al. 2006, 2007). The zeolites with low silica contents are hydrophilic and are usually used as drying agents for absorbing steam, whereas hydrophobic high-siliceous zeolites are used for absorbing organic molecules from humid air or water. Thus, varying the framework Si/Al ratio of zeolites greatly changes the adsorption selectivity toward molecules with different polarity. In addition, the aluminum ions form catalytically active sites, since they may either act as Bronstedt or Lewis acids. Such sites are of high interest in a number of catalyzed organic reactions. So, the $x/y$ ratio indicates the amount of acidic centers per unit cell as well as the content of mobile cations.

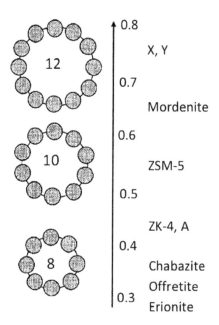

**FIGURE 8.2** Dimensions of the pores of different zeolites (in nm). (With kind permission from Springer Science+Business Media: *Catal. Lett.*, Adaptation of the porosity of zeolites for shape selective reactions, 22, 1993, 107–121, Ribeiro, F.R.)

## 8.2 ADVANTAGES FOR GAS-SENSOR APPLICATIONS

As it was shown before, zeolites have several physical and structural features that can be exploited for gas sensing (Zheng et al. 2012; Wales et al. 2015). In particular, their surface and structural properties can be easily modified, which makes them ideal candidates for the selective adsorption of various volatile hydrocarbons and small organic molecules (Pejcic et al. 2007). For example, various post-synthesis steps were developed to chemically modify the zeolite by incorporating catalytically active metal clusters, such as Pt, Fe, or Cu, or by anchoring organic dyes within the pore structure. The surface properties (viz., hydrophobicity, hydrophilicity, and binding to reactant molecules) of zeolites bearing negative surface charge can be varied also by organic functionalization of their internal and external surfaces, which can improve their affinity to absorb water and other cations. Such modifications result in an alteration in the surface properties so much that the hydrophilic zeolites are converted into hydrophobic zeolites, which can absorb molecular diameters (e.g., organic cations) larger then water (Scott et al. 2003; Jha and Singh 2011). The internal microporosity of zeolites that gives rise to the high surface area also provides sites for adsorption of molecules. The negative aluminosilicate framework of zeolites necessitates the presence of neutralizing ion-exchangeable cations within the framework. These cations can influence adsorption, diffusion, and catalytic properties of zeolites, and thereby influence sensing behavior. The extra framework cations are bound electrostatically at preferential sites and can perform energy activated motion between these sites. The presence of a guest molecule within the zeolite can

interfere with this motion and has been used as the basis of sensing. Interaction of the molecule with the cation is manifested in the change in impedance/capacitance as measured by the frequency dependent impedance spectra. Upon adsorption, mass changes as well as optical/electrical properties are altered, which can be used for the sensor transduction. Selectivity toward analytes has also been observed by adsorption of species within a zeolite. Microporous spaces within the zeolite can be served as a host for guest species. So, when used in sensor elements, zeolites can take various roles, which fall into two major categories (see Figure 8.3).

In a great number of cases, zeolites are used as auxiliary elements. They may act either as a framework to stabilize the sensor material, as a filter layers (either catalytic or size-restrictive) to enhance selectivity of a sensitive film, or as a preconcentrator of specific analytes from diluted solutions. For example, due to excellent chemical and thermal stability, zeolites can be used as a substrate to prepare compounds and devices with desirable fundamental physical and chemical properties (Xu et al. 2006). For example, inorganic or organic compounds, metal and metal-organic compounds and their clusters can be assembled into the pores and the cages in zeolites. Some nano-sized metal or metal-oxide particles have been successfully inserted into the caves and the pores or highly dispersed on the external surface of zeolites.

The second group encompasses devices in which the zeolite itself is the main functional material leading to a sensor effect. Such detection principles rely directly on adsorptive, catalytic, or conductive properties of one specific zeolite that are subject to well-defined changes, depending on the composition of the surrounding gaseous (Alberti and Fetting 1994; Xu et al. 2006; Sahner et al. 2008). For example, the encapsulating ruthenium (II) complexes inside zeolite supercages gives possibility to design oxygen sensors, the fixing of zeolites on the surface of conductometric, quartz crystal microbalance, or surface acoustic wave (SAW)-based sensors also improves sensitivity of these devices, etc. (Zhou et al. 2003; Moos et al. 2009; Zheng et al. 2012).

It was established that better selectivity can also be achieved through a correct choice of both the geometrical properties, i.e., pore sizes and types of porous network, and acidic properties of zeolites, which can be controlled, for example, by the Si/Al ratio and the nature and quantity of the compensation cations (Ribeiro 1993; Mann et al. 2005; Moos et al. 2009; Satsuma et al. 2011). As it is known, only molecules of certain size are able to be absorbed by a given zeolite material, or pass through its pores, while molecules of bigger size cannot (see Table 8.4). For example, Satsuma et al. (2011) studied the effect of zeolite acidity and pore diameter of zeolites on the conductivity response to base molecules, such as water, acetonitrile, ammonia, benzonitrile, pyridine, aniline, and trimethylamine. They established that for smaller molecules, having molecular diameter of less than 0.4 nm (water, acetonitrile, and ammonia), a good correlation was observed between the response magnitude and the proton affinity of base molecules. It was indicated that acid–base interaction is the major factor when molecular diameter is smaller enough than zeolite pore diameter. As for

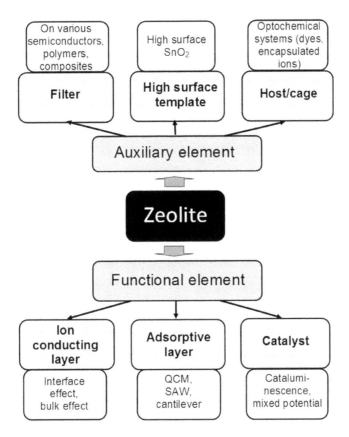

**FIGURE 8.3** Classification of zeolite-based gas sensors. (Reprinted from *Solid State Ionics, 179*, Sahner, K. et al., Zeolites—versatile materials for gas sensors, 2416–2423, Copyright 2008, with permission from Elsevier.)

larger molecules (benzonitrile, pyridine, aniline, and triethylamine) in which the molecular diameter is nearly the pore diameter of zeolites, the response magnitudes were lower than those of smaller molecules. Quantification of adsorbed species by in-situ Fourier Transform Infrared Spectroscopy (FTIR) revealed that the response magnitude for larger molecules significantly depends on the mobility of migrating species controlled by relative size of molecules to zeolite pore diameter. On practice, the size of the pores can be chosen from large-pore zeolites, like zeolite Y and ZSM-20, which are very useful in the transformation of large molecules, to intermediate pore zeolites, like ZSM-5, which already presents shape selectivity toward molecules such as ramified paraffins, down to small-pore zeolites that have large applications in the separation of very small molecules, like oxygen and nitrogen. When using such an approach, ammonia gas sensors have been developed as well (Rodriguez-Gonzalez et al. 2005; Moos et al. 2009). Various composites with specific properties also can be designed on the base of zeolites. For example, Chuapradit et al. (2005) reported about polyaniline/zeolite composites responsive to CO.

One should note that the use of one or another zeolite depends not only on the control of the pore structure and intrazeolite chemistry but also on the ability to control the morphology and pre-shaping, for example, the growth of thin zeolite films with required parameters. For example, for numerous synthetic strategies, the main goal is to obtain adhesive layers of zeolites on various substrates, such as noble and non-noble metals, glass, ceramics, silicon, and even 3D substrates (FEZA 2011). The experiment showed that the porous zeolite films can be deposited on the gas-sensor platform in a continuous and crack-free manner. At present, the following three approaches are usually applied for deposition of zeolite films (Wales et al. 2015):

1. Spin (dip) coating (thickness in the range of 200–1,000 nm)
2. Hydrothermal growth of seeded zeolite layers (thickness of 500–5,000 nm)
3. Screen printing based on self-bonded zeolite crystals with/without additives (thickness in the range of 2,000–10,000 nm).

As for the other properties of zeolites and their gas-sensor applications, then it should be noted that so far, a number of review papers were published on zeolite-based gas sensors (Moos et al. 2006, 2009; Xu et al. 2006; Walcarius 2008; Sahner et al. 2008; Zheng et al. 2012; Wales et al. 2015). They focus on various aspects of this materials class and mostly classify the devices according to the analytes to be detected or the type of sensor transduction. Description of other applications of zeolite-based materials one can find in Corma (1997), Davis (2002), Xu et al. (2007), and Pina et al. (2011).

## 8.3 ZEOLITE-BASED HUMIDITY SENSORS

Table 8.5 summarizes the types of humidity sensors developed based on zeolite. One can see that zeolites can be used in all types of humidity sensors (Wales et al. 2015). In particular, Li et al. (2000, 2004) developed the resistive LiCl/NaY-based humidity sensors, which showed good humidity-sensitive performance with a change of four orders of magnitude in electrical conductivity over the whole range of relative humidity. In addition, they also showed satisfactory reversibility responses to environmental moisture changes. Li et al. (2004) employed a porous-zeolite support to improve the stability and durability of LiCl at high humidity. Indeed, when the humidity exceeds 40% of relative humidity (RH), the conductivity response of LiCl is no longer linear, and thus it becomes difficult to use the material as a humidity sensor. The LiCl was dispersed into a NaY zeolite to obtain a linear response to up to 75% RH. The authors claimed that the well-defined cavities of NaY-type zeolites are more suitable as host materials for preparation of humidity-sensitive composite materials than those with channel structure, such as ZSM-5 ormordenite zeolites. Zou et al. (2001a, 2001b, 2004) similarly reported that the loading of LiCl into stilbite resulted in an electrically conductive material, exhibiting very high humidity sensitivities with a linear response in conductivity with a humidity change from 0% to 63% and fast response times for adsorption and desorption.

Changes in zeolite conductivity due to extra-framework cations contained in the framework have also been employed for the concept of humidity sensors with measurements based on the changes in either impedance or capacitance. In particular, Neumeier et al. (2008) and Satsuma et al. (2011) studied the impedance of the H-form of ZSM-5 as a function of water concentration (0–300 ppm) in $H_2$. ZSM-5 zeolite was chosen for its small pore size (0.55 nm), which improved water selectivity, and also for its acidic properties, due to the acid–base chemistry, involved in impedance measurements. Neumeier et al. (2008) have shown that the zeolite-modified sensors can be extremely robust; H-ZSM-5 based humidity sensors were capable to work under harsh conditions [reducing atmosphere ($H_2$) and temperature up to 600°C], which is not possible with humidity-sensors based on metal oxides or polymers, and provided a reversible and linear response (Figure 8.4).

Zeolite-based capacitance humidity sensors were developed by Alcantara et al. (2017) and Urbiztondo et al. (2011). Alcantara et al. (2007) fabricated interdigital capacitive humidity sensors based on ZSM-5 zeolite using $Al_2O_3$ ceramic substrates with electrode gaps of 20 µm. The results showed that the sensor was capable of good performance (detection limit of ~7.32% RH) and was suitable for use under a broader range of environmental conditions (from 39% to 96% RH). Even greater sensitivity was demonstrated by capacitive sensors developed by Urbiztondo et al. (2011). They studied humidity sensors with zeolite films consisting

### TABLE 8.5
### Zeolite-Based Humidity Sensors

| Sensor Type | Zeolite Type | References |
|---|---|---|
| Resistive | LiCl-ZSM-39, LiCl-ZSM-5 | Zou et al. (2004) |
| | LiCl/H-STI | Zou et al. (2001a, 2001b, 2004) |
| | LiCl/NaY | Li et al. (2000, 2004) |
| Capacitive | NaY | Alberti et al. (1994, 1997); Plog et al. (1995) |
| | Nb-TMS-1 | Kinsel and Balkus (1998) |
| | Zeolite A | Urbiztondo et al. (2011) |
| | ZSM-5 | Alcantara et al. (2017) |
| Impedance | NaY | Kurzweil et al. (1995) |
| | LTA | Alcantara et al. (2017) |
| | H-ZSM-5 | Neumeier et al. (2008) |
| | LTA | Lopes et al. (2017) |
| QCM | NaY | Yan and Bein (1992) |
| | NaA | |
| | LTA | Mintova et al. (2001) |
| SAW | NAY | Bein and Brown (1989) |
| | NaA | |
| Microwave | Fe-Zeolite | Bogner et al. (2017) |
| Microcantilever | Zeolite | Scandella et al. (1998) |
| Optical and fiberoptic | NaY/Ru-dipy | Meier et al. (1995) |
| | Methylene blue-H-MOR | Zanjanchi and Sohrabnezhad (2005); Sohrabnezhad et al. (2007) |
| | Nile Red in Y zeolite | Meinershagen and Bein (1999); Pelleiro et al. (2007) |
| | Ag- LTA(Na) | Coutino-Gonzalez et al. (2016) |

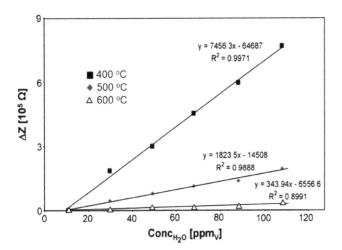

FIGURE 8.4 Water concentration dependence of impedance difference of H-ZSM-5 type zeolite in hydrogen at 400°C, 500°C, and 600°C. A H-ZSM-5-type zeolite (M80) sensor device was fabricated by thick-film technology on a comb-like platinum electrode on an $Al_2O_3$ substrate. (Reprinted from *Sens. Actuators B*, 134, Neumeier, S. et al., Zeolite based trace humidity sensor for high temperature applications in hydrogen atmosphere, 171–174, Copyright 2008, with permission from Elsevier.)

of different zeolites with Si/Al ratios ranging from 1.5 (zeolite A) to infinite (silicalite), and found that the performance of the sensor has been related to the electrical properties of the zeolites (relative permittivity, $\varepsilon_r$), which in turn is a function of their Si/Al ratio (see Figure 8.5). It was established that the most hydrophilic zeolite used (Si/Al ratio of 1.5) gave the lowest detection limit (0.45 ppm) and the shortest response time, in contrast with the behavior shown by the

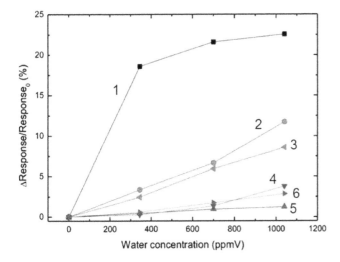

FIGURE 8.5 The increase of sensor capacitance with humidity concentration for different zeolite coatings on IDCs at room temperature. Interdigital capacitors (IDCs) had electrode gaps of 50 (1, 2, 6) or 10 μm (3, 4, 5): 1—zeolite A (LTA); 2, 4—zeolite Y (FAU) crystals; 3, 5—zeolite β (BEA); 6—silicalite (MFI). (Reprinted from *Sens. Actuators B*, 157, Urbiztondo, M. et al., Zeolite-coated interdigital capacitors for humidity sensing, 450–459, Copyright 2011, with permission from Elsevier.)

sensor coated with silicalite (no Al), where the detection limit was 1800 ppmV. It was also found that for zeolite coatings with a Si/Al ratio of 30 or higher, only small changes in capacitance are observed as the water concentration increases from 0 to 1000 ppm.

Mintova et al. (2001) reported that high sensitivity, good reversibility, and long life of the film at low steam concentrations also performed zeolite based QCM humidity sensors. They tested LTA (zeolite A) and BEA (zeolite β) type zeolites and found that due to the large pore size channels (0.67 nm) (Mintova and Bein 2001), beta polymorph A-based sensors exhibited more interference from alkanes compared to zeolite A-based sensors. Larger resonant frequency shifts were measured with the zeolite A coated QCM sensor, e.g., 2000 Hz after 150 ppm water exposure for zeolite A, compared to 400 Hz for beta polymorph A. However, the response and recovery times were faster for the beta polymorph A zeolite sensor, which was rationalised on the basis of its larger pore channel diameter and higher hydrophobicity due to a higher Si/Al ratio. Sasaki et al. (2002) also designed the QCM sensor based on zeolite A and have shown that such sensors can be used as humidity sensors. The example of zeolite-coated QCMs is shown in Figure 8.6.

Zeolite-based optical sensors with acceptable parameters were developed by Zanjanchi and Sohrabnezhad (2005) and Sohrabnezhad et al. (2007). Methylene blue (MB) encapsulated into the protonated MOR-type (mordenite) zeolite (H-MOR) via ion-exchange reaction. It was believed that the porous structure of the zeolites allows the encapsulation of organic chromophores, providing a protective environment that increases their stability. The mechanism of the sensor was based on the protonation or deprotonation of the dye molecules, which are associated with desorption or adsorption of water molecules by the zeolite, respectively. It was established that the sensor presented a linear response range from 9% to 92% RH with good stability and reversibility. The sensor

FIGURE 8.6 A gas-sensing unit with two zeolite-coated QCMs. (Reprinted from *Micropor. Mesopor. Mat.*, 144, Pina, M.P. et al., Zeolite films and membranes. Emerging applications, 19–27, Copyright 2011, with permission from Elsevier.)

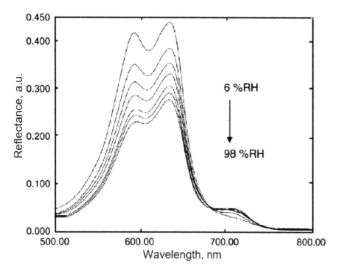

**FIGURE 8.7** Diffuse reflectance spectra of the zeolite, MB–HMOR, exposed to different humidity in the range of 6%–98% RH. To obtain the protonated form of the zeolite (HMOR), the as-synthesized sample was stirred in the solution of HCl at room temperature for 24 h. Obtained samples were immersed in an aqueous solution of new methylene blue. (Reprinted from *Mater. Lett.*, 61, Sohrabnezhad, S. et al., New methylene blue incorporated in mordenite zeolite as humidity sensor material, 2311–2314, Copyright 2007, with permission from Elsevier.)

could operate at two bands, either of 650 or of 745 nm, but a higher sensitivity of the measurement performance was exhibited at 650 nm (see Figure 8.7). The sensor provided relatively fast responses, about 2 min in the adsorption step and about 4 min in the desorption step. Another example of an optical humidity detector utilized solvatochromic Nile Red dye encapsulated in a NaY zeolite (Pellejero et al. 2007). This device had a detection limit of 200 ppm humidity, a response time of around 4 min and discrimination against hexane was also demonstrated. One further example of an optical detector involved impregnation of silver on ZSM-5-type zeolite, where the authors showed that Ag+ species in the Ag-ZSM-5 zeolite exhibited stable optical properties under high-temperature treatment (Sazama et al. 2008). The behavior of the 37,700 cm⁻¹ band was followed with in-situ UV-vis spectroscopy. Above 150°C, this band was found to be sensitive to the presence of water vapor in the gas stream (2.8% $H_2O$ in He), disappearing in the presence of water vapor while full intensity could be recovered under water-free conditions.

High sensitivity was also demonstrated by the cantilever-based humidity sensors, elaborated on the base of zeolites. In particular, Scandella et al. (1998) demonstrated water detection in the nanogram range by attaching a few ZSM-5 crystals to a silicon cantilever (Berger et al. 1997). Pina et al. (2011) believe that the sensitivity of zeolite-based sensors can be even higher. The cantilevers just discussed consist of a zeolite layer on the top of a silicon cantilever. In this case, the Si cantilever is the resonant element and the zeolite layer on the top of it acts as a target for the objective molecules. It is obvious that in this case most of the cantilever mass (the silicon support) is not sensitive toward the

gas-phase molecules. Pina et al. (2011) hopes that through the development of the technology of zeolite film patterning (Pellejero et al. 2008) it is possible to develop self-supported zeolite structures and open the way to zeolite-only cantilevers (Agusti et al. 2010).

The experiment showed that metal-oxide-based sensors can be further modified with zeolite films to improve their response to gas analytes compared with uncoated metal oxides. In particular, a zeolite layer can be used as an adsorbent filter to enhance the selectivity of the metal oxide by retaining analytes and/or interfering species. The example of such approach was reported by Vilaseca et al. (2007). They developed conductometric sensor with $SnO_2$ as a sensing layer and zeolite A and ZSM-5 zeolites as filters. Results showed that zeolite A improved the selectivity toward $H_2O$ by the elimination of the response to propane, hydrogen, and methane while continuing to respond significantly to ethanol and CO. On the other hand, the ZSM-5-type silicalite-coated sensor gave enhanced response to both $H_2O$ and $H_2$.

## 8.4 LIMITATIONS OF ZEOLITE APPLICATION IN HUMIDITY SENSORS

However, one should note that interest in zeolites as humidity-sensitive materials is very limited. At first, due to a small-pore size the diffusion limitation will be present in kinetics of zeolite-based humidity sensors. This means that the response and especially the recovery will be slow. For example, for optochemical humidity sensors based on Nile Red-Y composite due to the hydrophilic nature of the zeolite employed as a host, the recovery times were around 35 min (Pellejero et al. 2007). The same effect was observed by Urbiztondo et al. (2011), who tested the capacitive humidity sensors. They established that the strong adsorption of water in the most hydrophilic zeolites resulted in longer recovery times at room temperature, which means that temperature cycling (with a high-temperature desorption step) would be needed in most sensing scenarios. Large hysteresis is also a feature of zeolite-based sensors. This means that such sensors cannot be used for in-situ measurements. It is believed that the using very thin zeolitic films (nanometer dimensions) can only improve the response and recovery times. However, experiments carried out by Mintova et al. (2001) have shown that even thicknesses less than 100 nm do not provide fast response and recovery (see Figure 8.8).

The experiment also showed that zeolite-based sensors are very sensitive to prehistory. Therefore, when using them immediately before measurements it is necessary to do degassing, that is, through thermal treatment exercise zeolites activation. This greatly complicates and delays the measurement process. Otherwise, you can get data that does not correspond to reality (see Figure 8.9).

In addition, we also need to take into account that the main problems with the use of zeolites with small sizes (nano-sizes) are their low synthesis yields and inconsistent reproducibility.

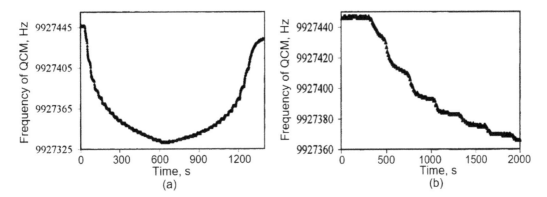

**FIGURE 8.8** Frequency response of QCM-based sensor upon sorption and desorption of water vapor as a function of time at 23°C on LTA zeolite with thickness of 65 nm: (a) sorption/desorption at different water concentrations (30–180 ppm); and (b) sorption of low water concentration (up to 45 ppm). (Reprinted from *Chem. Mater.*, 13, Mintova, S. et al., Humidity sensing with ultrathin LTA-type molecular sieve films grown on piezoelectric devices, 901–905, Copyright 2001, with permission from Elsevier.)

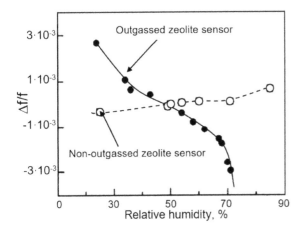

**FIGURE 8.9** Cantilever-based humidity sensor using ZSM-5 single crystals. (Reprinted from *Micropor. Mesopor. Mat.*, 21, Scandella, L. et al., Combination of single crystal zeolites and microfabrication: Two applications towards zeolite nanodevices, 403–409, Copyright 1998, with permission from Elsevier.)

Chiral zeolites have better reproducibility, but the difficulties in resolving enantiopure zeolites still persist. The lack of selectivity can also be attributed to the disadvantages of these sensors.

## REFERENCES

Agusti J., Pellejero I., Abadal G., Murillo G., Urbiztondo M.A., Sesé J. et al. (2010) Optical vibrometer for mechanical properties characterization of silicalite-only cantilever based sensors. *Microelectr. Eng.* **87**, 1207–1209.

Alberti K., Fetting F. (1994) Zeolites as sensitive materials for dielectric gas sensors. *Sens. Actuators B* **21**, 39–50.

Alberti K., Haas J., Plog C., Fetting F. (1997) Zeolite coated interdigital capacitors as a new type of gas sensor. *Catal. Today* **8**, 509–513.

Alcantara G.P., Ribeiro L.E.B., Alves A.F., Andrade C.M.G., Fruett F. (2017) Humidity sensor based on zeolite for application under environmental conditions. *Micropor. Mesopor. Mat.* **247**, 38–45.

Auerbach S.M., Carrado K.A., Dutta P.K. (2003) *Handbook of Zeolite Science and Technology.* Marcel Dekker, New York.

Bein T., Brown K. (1989) Molecular sieve sensors for selective detection at the nanogram level. *J. Am. Chem. Soc.* **111**, 7640–7641.

Berger R., Gerber C.H., Lang H.P., Gimzewski J.K. (1997) Micromechanics: A toolbox for femtoscale science: "Towards a laboratory on a tip." *Microelectr. Eng.* **35**, 373–379.

Bogner A., Steiner C., Walter S., Kita J., Hagen G., Moos R. (2017) Planar microstrip ring resonators for microwave-based gas sensing: Design aspects and initial transducers for humidity and ammonia sensing. *Sensors* **17**, 2422.

Cejka J., Van Bekkum H., Corma A., and Schuth F. (eds.) (2007) *Introduction to Zeolite Science and Practice*, 3rd edn., Studies in Surface Science and Catalysis, Vol. 168. Elsevier, Amsterdam, the Netherlands.

Chuapradit C., Wannatong L.R., Chotpattananont D., Hiamtup P., Sirivat A., Schwank J. (2005) Polyaniline/zeolite LTA composites and electrical conductivity response towards CO. *Polymer* **46**, 947–953.

Corma A. (1997) From microporous to mesoporous molecular sieve materials and their use in catalysis. *Chem. Rev.* **97**, 2373–2419.

Coutino-Gonzalez E., Baekelant W., Dieu B., Roeffaers M.B.J., Hofkens J. (2016) Nanostructured Ag-zeolite composites as luminescence-based humidity sensors. *J. Visualized Exp.* **117**, e54674.

Davis M.E. (2002) Ordered porous materials for emerging applications. *Nature* **417**, 813–821.

FEZA (2011) *Proceedings of 5th FEZA School on Zeolites*, Valencia, Spain, p.245.

Flanigen E.M., Broach R.W., Wilson S.T. (2010) Introduction. In: Kulprathipanja S. (ed.) *Zeolites in Industrial Separation and Catalysis.* Wiley-VCH Verlag, Weinheim, Germany.

Jacobs P.A. (1977) *Carboniogenic Activity of Zeolites.* Elsevier Scientific, New York.

Jha B., Singh D.N. (2011) A review on synthesis, characterization and industrial application of fly ash zeolites. *J. Mater. Edu.* **33**(1–2), 65–132.

Kinsel M.E., Balkus K.J., Jr.(1998) Mesoporous molecular sieve thin films. *Studies Surf. Sci. Catal.* **117**, 111–118.

Kulprathipanja S. (ed.) (2010) *Zeolites in Industrial Separation and Catalysis.* Wiley-VCH Verlag, Weinheim, Germany.

Kurzweil P., Maunz W., Plog C. (1995) Impedance of zeolite-based gas sensors. *Sens. Actuators B* **24–25**, 653–656.

Li N., Li X.T., Zhang T., Qiu S.L., Zhu G.S., Zheng W.T., Yu W.X. (2004) Host-guest composite materials of LiCl/NaY with wide range of humidity sensitivity. *Mater. Lett.* **58**, 1535–1539.

Li X.T., Shao C.L., Ding H., Zhang T., Li N., Qiu S.L., Xiao F.S. (2000) Preparation and humidity sensitive property of LiCl/ NaY composite materials. *Chem. J. Chin. Univ.-Chin.* **21**, 1167–1170.

Lopes M., Sakai A., Bosio L., Alcantara I., Andrade C. (2017) Development and investigation of three different topologies of devices for sensing of moisture traces. In: *Proceedings of International Conference on Electronic, Control, Automation and Mechanical Engineering (ECAME 2017)*, 19–20 November, Sanya, China, pp. 426–430.

Mann D.P., Paraskeva T., Pratt K.F.E., Parkin I.P., Williams D.E. (2005) Metal oxide semiconductor gas sensors utilizing a Cr-zeolite catalytic layer for improved selectivity. *Meas. Sci. Technol.* **16**, 1193–1200.

Meier B., Werner T., Klimant I., Wolfbeis O.S. (1995) Novel oxygen sensor material based on a ruthenium bipyridyl complex encapsulated in zeolite Y: Dramatic differences in the efficiency of luminescence quenching by oxygen on going from surface-adsorbed to zeolite-encapsulated fluorophores. *Sens. Actuators B* **29**, 240–245.

Meinershagen J., Bein T. (1999) Optical sensing in nanopores. Encapsulation of the solvatochromic Dye Nile Red in zeolites. *J. Am. Chem. Soc.* **121**, 448–449.

Mintova S., Bein T. (2001) Nanosized zeolite films for vapor-sensing applications. *Micropor. Mesopor. Mat.* **50**, 159–166.

Mintova S., Mo S., Bein T. (2001) Humidity sensing with ultrathin LTA-type molecular sieve films grown on piezoelectric devices. *Chem. Mater.* **13**, 901–905.

Moos R., Sahner K., Fleischer M., Guth U., Barsan N., Weimar U. (2009) Solid state gas sensor research in Germany—A status report. *Sensors* **9**, 4323–4365.

Moos R., Sahner K., Hagen G., Dubbe A. (2006) Zeolites for sensors for reducing gases. *Rare Metal Mater. Eng. Suppl. B* **35**, 447–451.

Neumeier S., Echterhof T., Bolling R., Pfeifer H., Simon U. (2008) Zeolite based trace humidity sensor for high temperature applications in hydrogen atmosphere. *Sens. Actuators B* **134**, 171–174.

Pejcic B., Eadington P., Ross A. (2007) Environmental monitoring of hydrocarbons: A chemical sensor perspective. *Environ. Sci. Technol.* **41**(18), 6333–6342.

Pellejero I., Urbiztondo M., Izquierdo D., Irusta S., Salinas I., Pina M.P. (2007) An optochemical humidity sensor based on immobilized Nile Red in Y Zeolite. *Ind. Eng. Chem. Res.* **46**(8), 2335–2341.

Pellejero I., Urbiztondo M.A., Villarroya M., Sesé J., Pina M.P., Santamaría J. (2008) Development of etching processes for the micropatterning of silicalite films. *Micropor. Mesopor. Mat.* **114**, 110–120.

Pina M.P., Mallada R., Arruebo M., Urbiztondo M., Navascués N., de la Iglesia O., Santamaria J. (2011) Zeolite films and membranes. Emerging applications. *Micropor. Mesopor. Mat.* **144**, 19–27.

Plog C., Maunz W., Kurzweil P., Obermeier E., Scheibe C. (1995) Combustion gas sensitivity of zeolite layers on thin-film capacitors. *Sens. Actuators B* **24–25**, 403–406.

Ribeiro F.R. (1993) Adaptation of the porosity of zeolites for shape selective reactions. *Catal. Lett.* **22**, 107–121.

Rodriguez-Gonzalez L., Franke M.E., Simon U. (2005) Electrical detection of different amines with proton-conductive H-ZSM-5. *Stud. Surf. Sci. Catal.* **158**, 2049–2056.

Rolison D.R. (1990) Zeolite-modified electrodes and electrode-modified zeolites. *Chem. Rev.* **90**(5), 867–878.

Sahner K., Hagen G., Schönauer D., Reiß S., Moos R. (2008) Zeolites—Versatile materials for gas sensors. *Solid State Ionics* **179**(40), 2416–2423.

Sasaki I., Tsuchiya H., Nishioka M., Sadakata M., Okubo T. (2002) Gas sensing with zeolite-coated quartz crystal microbalances—Principal component analysis approach. *Sens. Actuators B* **86**, 26–33.

Satsuma A., Yang D., Shimizu K.I. (2011) Effect of acidity and pore diameter of zeolites on detection of base molecules by zeolite thick film sensor. *Micropor. Mesopor. Mat.* **141**, 20–25.

Sazama P., Jirglova H., Dedecek J. (2008) Ag-ZSM-5 zeolite as high-temperature water-vapor sensor material. *Mater. Lett.* **62**, 4239–4241.

Scandella L., Binder G., Mezzacasa T., Gobrecht J., Berger R., Lang H.P., et al. (1998) Combination of single crystal zeolites and microfabrication: Two applications towards zeolite nanodevices. *Micropor. Mesopor. Mat.* **21**, 403–409.

Scott M.A., Kathleen A.C., Dutta P.K. (2003) *Handbook of Zeolite Science and Technology.* CRC Press, New York.

Sohrabnezhad S., Pourahmad A., Sadjadi M.A. (2007) New methylene blue incorporated in mordenite zeolite as humidity sensor material. *Mater. Lett.* **61**, 2311–2314.

Urbiztondo M., Pellejero I., Rodríguez Á., Pina M.P., Santamaría R.S. (2011) Zeolite-coated interdigital capacitors for humidity sensing. *Sens. Actuators B* **157**, 450–459.

Vilaseca M., Coronas J., Cirera A., Cornet A., Morante J.R., Santamaria J. (2007) Gas detection with SnO₂ sensors modified by zeolite films. *Sens. Actuators B* **124**, 99–110.

Villaescusa L., Camblor M. (2003) The fluoride route to new zeolites. *Recent Res. Dev. Chem.* **1**, 93–141.

Walcarius A. (1999) Zeolite-modified electrodes in electrochemical chemistry. *Anal. Chim. Acta* **384**, 1–16.

Walcarius A. (2008) Electroanalytical applications of microporous zeolites and mesoporous (organo)silicas: Recent trends. *Electroanal.* **20**(7), 711–738.

Wales D.J., Grand J., Ting, V.P. Burke R.D., Edler K.J., Bowen C.R., S., Burrows A.D. (2015) Gas sensing using porous materials for automotive applications. *Chem. Soc. Rev.* **44**, 4290–4321.

Xu R., Pang W., Yu J., Huo Q., Chen J. (2007) *Chemistry of Zeolites and Related Porous Materials: Synthesis and Structure.* John Wiley & Sons, Singapore.

Xu X., Wang J., Long Y. (2006) Zeolite-based materials for gas sensors. *Sensors* **6**, 1751–1764.

Yan Y., Bein T. (1992) Molecular recognition on acoustic wave devices: Sorption in chemically anchored zeolite monolayers. *J. Phys. Chem.* **96**, 9387–9393.

Yu J. (2007) Synthesis of zeolites. In: Cejka J., Van Bekkum H., Corma A., and Schuth F. (eds.) *Introduction to Zeolite Science and Practice*, 3rd edn., Studies in Surface Science and Catalysis, Vol. 168. Elsevier, Amsterdam, the Netherlands, pp. 39–103.

Zanjanchi M.A., Sohrabnezhad S. (2005) Evaluation of methylene blue incorporated in zeolite for construction of an optical humidity sensor. *Sens. Actuators B* **105**, 502–507.

Zheng Y., Li X., Dutta P.K. (2012) Exploitation of unique properties of zeolites in the development of gas sensors. *Sensors* **12**, 5170–5194.

Zhou J., Li P., Zhang S., Long Y., Zhou F., Huang Y. et al. (2003) Zeolite-modified microcantilever gas sensor for indoor air quality control. *Sens. Actuators B* **94**, 337–342.

Zou J., He H.Y., Dong J.P., Long Y.C. (2001a) A novel LiCl/H-STI zeolite guest/host assembly material with superior humidity sensitivity: Fabrication and characterization. *Chem. Lett.* **8**, 810–811.

Zou J., He H.Y., Dong J.P., Long Y.C. (2004) A guest/host material of LiCl/H-STI (stilbite) zeolite assembly: Preparation, characterization and humidity-sensitive properties. *J. Mater. Chem.* **14**, 2405–2411.

Zou J., Luo Z.L., Jiang Z.Y., Long Y.C. (2001b) Studies on $C_xN$ natural zeolite IV. Research on the humidity-sensitive property of the novel LiCl/H STI guest/host material. *Acta Chim. Sin.* **59**, 862–866.

# 9 Metal Phosphate-Based Humidity Sensitive Materials

## 9.1 METAL PHOSPHATES

Metal phosphates are another group of microporous and mesoporous crystalline materials, which is of interest for use in humidity sensors (Cheetham et al. 1999; Rajic et al. 2000; Yu and Xu 2003; Xu et al. 2007). In general, the compounds such as aluminophosphates ($AlPO_4$) and silicoaluminophosphates (SAPO) belong to the group of zeolites (Murcia 2013). However, we decided to consider these materials separately. The first representatives of this group of materials were aluminophosphates, synthesized in 1982 (Wilson et al. 1982). This discovery opened up the field for a large number of open-framework inorganic materials. At present, the aluminophosphate-based family of microporous compounds has over 200 members. As in aluminosilicate structures, aluminophosphates avoid Al-O-Al bonds (Loewenstein 1954), and additionally P-O-P bonds, which leads to even-numbered rings within the structures.

The general formula of aluminophosphates can be expressed as $[(AlO_2)_x(PO_2)_x \cdot yH_2O]$, indicating that, unlike most zeolites, the aluminophosphate molecular sieves are ordered with an Al:P ratio that is always unity. As a result, the AlPO structures are based on a strict alternation of $[AlO_4]^-$ and $[PO_4]^+$ tetrahedra leading to a non-charged framework in contrast to most zeolitic frameworks, and similar to all-silica zeolites (see Figure 9.1). However, the charge can be induced into AlPO frameworks, either by an Al/P—ratio of >1 or by partial replacement of aluminum and/or phosphorus by silicon or other elements. The substitution of Al by divalent metal ions generates Brønsted-acid sites (acidic bridging–OH groups) as well as Lewis-acid sites (anionic vacancies deriving from missing lattice oxygens) in the aluminophosphate lattice. The incorporation of a transition metal cation, which can easily change its oxidation number, also creates a redox active site. The coupling of acidic with redox properties opens up the routes toward the shape selective bifunctional catalysis and to the design of novel catalysts. In addition, the incorporation of heteroatoms into aluminophosphates has played an important role in enhancing the diversity of structures and compositions of microporous compounds and molecular sieves (Weckhuysen et al. 1999; Rajic 2005; Li et al. 2012). Metal hetero elements normally substitute Al atoms in the framework (SM1 substitution) whereas Si atoms predominantly substitute P atoms (SM2). Additionally, Si can adopt both Al and P positions (SM3), which leads to silicon islands. If other metals are incorporated, the compounds are called

SAPO-n (S = Si), MeAPO-n (Me = Fe, Mg, Mn, Zn, Co, etc.), MeASO-n, ElAPO-n (El = Ba, Ga, Ge, Li, As, etc.), REPO$_4$ (RE = La, Sm, Gd), and ElAPSO-n (*n* refers to a distinct structure type).

The experiment showed that the discovery of AlPOs led to a greater flexibility in the possible frameworks compared to the traditional zeolite structure. In particular, in addition to incorporation, many new framework elements (V, Cr, Ni, Li, Be, B, Mg, Ti, Mn, Fe, Co, Zn, Ga, As, Cu) (Weckhuysen et al. 1999; Rajic 2005; Li et al. 2012), the opportunity to synthesize materials with pore sizes larger than 12-membered rings appeared (Davis et al. 1988; Davis 2002; Wang et al. 2018). For instance, VPI-5 ($(H_2O)_{42}[Al_{18}P_{18}O_{72}]$) is an aluminophosphate with 18-membered-ring apertures (1.27 × 1.27 nm) (Davis et al. 1988) (Figure 9.2), and JDF-20 ($[Et_3NH]_2[Al_5P_6O_{24}H] \cdot 2H_2O$) is a microporous aluminophosphate with the largest aperture size (20-membered ring, 1.45 × 0.62 nm). In addition, the primary building unit in aluminophosphates is not only limited to a tetrahedral coordination (Pastore et al. 2005; Yu and Xu 2006). As a result, some AlPO-based molecular sieves have the structure analog to known zeolites, but many compounds have no zeolite counterpart. At present, large-, medium-, and small-pore $AlPO_{4-n}$ molecular sieves were prepared on the basis of aluminophosphate-based microporous compounds (Xu et al. 2007).

## 9.2 SYNTHESIS AND MECHANISM OF MOLECULAR SIEVE GROWTH

Usually, aluminophosphates were synthesized through the crystallization of Al, P, and other element sources together, under hydrothermal or solvothermal conditions (Cundy and Cox 2005; Rajic 2005; Xu et al. 2007; Van Heyden 2008; Murcia 2013). However, a technique combining hydrothermal and microwave heating can also be employed for the preparation of aluminophosphates (Girnus et al. 1995; Lohse et al. 1996). The application of this technique reduces the crystallization time from a few days to a few minutes. As in illustration, the crystallizations of MnAPO-*n* and MnAPSO-*n* (*n* = 5, 44) are completed in 30 minutes (Lohse et al. 1996). At the same time, the formation of the first large $AlPO_4$-5 crystals is observed after only 60 s (Girnus et al. 1995). It is assumed that the microwaves destroy the hydrogen bridges between water molecules causing the fast-gel dissolution and the formation of Al-O-P building blocks.

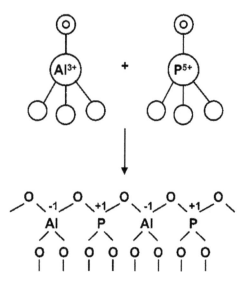

**FIGURE 9.1** Assembly of a microporous aluminophosphate material. (From Weckhuysen, B.M. et al.: Transition metal ions in microporous crystalline aluminophosphates: Isomorphous substitution. *Eur. J. Inorg. Chem.* 4. 5652577. 1999. Copyright Wiley-VCH Verlag GmbH & Co. KGaA. Reproduced with permission.)

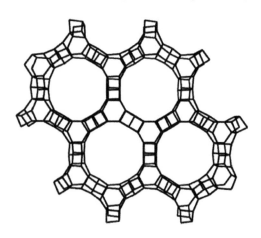

**FIGURE 9.2** Perspective t-atom representation of aluminophosphate molecular sieves, VPI-5, showing the 18-ring pore openings along [001]. (Reprinted from DZS (2007) Database of Zeolite Structures, http://www.iza-structure.org/databases/.)

As the solution chemistries of silicon and phosphorus are different, the synthesis conditions of zeolites and AlPOs are also different. Traditionally, zeolites are synthesized by mixing appropriate sources for aluminum and silicon with a cation source, and most often an organic template in an alkaline media, whereas AlPOs can be obtained within a broad pH ranging. Table 9.1 summarizes some of the aluminophosphates and organic templates used during their synthesis.

In the case of SAPO, the reactants, containing aluminum, phosphorus, and eventually silicon are mixed with a suitable template (Figure 9.3). Afterward, the aqueous reaction mixture is heated in a sealed autoclave at temperatures usually between 100°C and 200°C. For some time after raising the synthesis temperature, the reactants remain amorphous or in their pre-synthesis phase. After this "induction period," crystalline products can be detected. Gradually, all precursor materials get converted into an approximately equal mass of zeolite/aluminophosphate crystals. The products are recovered by filtration or centrifugation, washing, and drying.

The framework elements are typically introduced in the solution in an oxidic forms or salts (see Table 9.2), and sometimes the metal organic molecular precursors are used, which are hydrolyzed in solution to an oxidic species. Phosphorus can be introduced in the form of phosphorus oxide, but more commonly phosphoric acid is used. The precursors contain Al/Si/P-O bonds, which are converted to Si-O-Al bonds during the hydrothermal reaction by hydrolysis and condensation reactions. The bond types of educts and products are very similar in the bond energy; therefore, the overall free-energy change during the synthesis is small, and the process is usually kinetically controlled. Thus, the desired product is frequently metastable, and care needs to be taken that an elongated synthesis time does not lead to redissolution of the product.

As for the mechanisms of formation of the microporous structure, that is, the mechanisms by which the transformation from the left to the right side in Figure 9.3 proceeds, then the main ones are covered in Cundy and Cox (2005) and Van Heyden (2008). It should be noted that most of the investigations were concerned with zeolite synthesis, and only relatively few with aluminophosphates (Oliver et al. 1998; Taulelle 2001).

**TABLE 9.1**

**Examples of Templates Used for Synthesis of Aluminum-Phosphate-Based Microporous Materials**

| Reaction Directing Agent (Template) | Aluminum Phosphate Synthesized Using this Template |
|---|---|
| Et$_3$N (Triethylamine, N(CH$_2$CH$_3$)$_3$) | AlPO$_4$-5 (pH = 1); SAlPO$_4$-5; AlPO$_4$-8, SAlPO$_4$-8, MnAlPO$_4$-8; AlPO$_4$-25; Fe$^{III}$AlPO$_4$-31; Co$^{III}$(Mn$^{III}$)AlPO$_4$-36 (pH = 13); Co$^{II}$(Co$^{III}$)AlPO$_4$-36 (pH = 13); AlPO$_4$-41 |
| Pr$_3$N (Tripropylamine, (CH$_3$CH$_2$CH$_2$)$_3$N) | AlPO$_4$-5 |
| [Et$_4$N]OH (Tetraethylammonium, C$_8$H$_{20}$N$^+$) | FeAlPO$_4$-5 |
| (i-Pr)$_2$NH (Di-isopropylamine) | AlPO$_4$-11, AlPO$_4$-11; SAlPO$_4$-11; CoAlPO$_4$-11; MnAlPO$_4$-11 |
| Di-n-propylamine, C$_6$H$_{15}$N | AlPO4-11; VPI-5; MAPO-39; CoAPO-43; CoAPO-50; SAPO-n (n = 5,11,31) |
| (iPr2)EtN (Diisopropylethylamine, [(CH$_3$)$_2$CH]$_2$NC$_2$H$_5$) | AlPO$_4$-18; MAlPO$_4$-18, M = Mg, Mn, Co, Ni, Zn |
| Hexamethylenetriamine, (NH$_2$(CH$_2$)$_6$NH(CH$_2$)$_6$NH$_2$) | (NH$_4$)$_{0.88}$(H$_3$O)$_{0.12}$AlPO$_4$(OH)$_{0.33}$F$_{0.67}$; AlPO$_4$-CJ2 |
| Ethylene or propylene oxide (C$_2$H$_4$O or CH$_3$CHCH$_2$O) | AlPO$_4$-Al$_2$O$_3$ (APAl) |

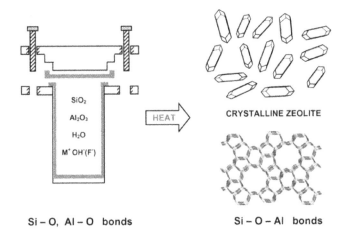

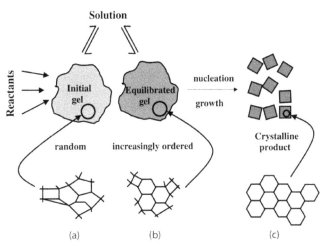

**FIGURE 9.3** Hydrothermal zeolite synthesis. The starting materials (Si-O and Al-O bonds) are converted by an aqueous mineralizing medium (OH and/or F) into the crystalline product (Si-O-Al bonds), whose microporosity is defined by the crystal structure. (Reprinted from *Micropor. Mesopor. Mat.*, 82, Cundy, C.S., and Cox, P.A., The hydrothermal synthesis of zeolites: Precursors, intermediates and reaction mechanism, 1–78, Copyright 2005, with permission from Elsevier.)

**FIGURE 9.4** The evolution of order from the (a) primary amorphous phase, through the (b) secondary amorphous phase, to the (c) crystalline product. (Reprinted from *Micropor. Mesopor. Mat.*, 82, Cundy, C.S., and Cox, P.A., The hydrothermal synthesis of zeolites: Precursors, intermediates and reaction mechanism, 1–78, Copyright 2005, with permission from Elsevier.)

## TABLE 9.2
## Precursor Used for Metal Phosphate Synthesis

| Metal Phosphate | Precursor Used for Metal Phosphate Synthesis |
| --- | --- |
| Al | $Al_2O_3$; $Al(OH)_3$; $(Al(NO_3)_3 \cdot 9H_2O)$ |
| Zr | $Zr(SO_4)_2$; $ZrOCl_2 \cdot 8H_2O$; $Zr(OC_3H_7)_4$; $Zr(OC_4H_9)_4$ |
| Ti | $Ti(OC_3H_7)_4$; $TiCl_4$; $Ti(OMe)_4$; $Ti_4(OCH_3)_{16}$ |
| Fe | $FeCl_3$; $Fe_2O_3$; $FeCl_2 \cdot 3H_2O$; $Fe(NO_3)_3 \cdot 9H_2O$ |
| Zn | $Zn(NO_3)_2 \cdot 6H_2O$ |
| Co | $Co(NO_3)_2 \cdot 6H_2O$ |
| Ni | $Ni(NO_3)_2 \cdot 6H_2O$ |
| Cu | $Cu(NO_3)_2 \cdot 3H_2O$ |
| Bi | $Bi(NO_3)_3 \cdot 5H_2O$ |
| B | $H_3BO_3$ |

This is important, because in general, the crystallization mechanism varies for either zeolites or aluminophosphates, just as each specific synthesis mixture will result in different reactions under the given conditions. Nonetheless, there are some general mechanistic considerations, which are accepted for most of the different molecular sieve systems. The overall process is generally divided into three different periods, namely, the induction, nucleation, and growth period. At the induction stage, a formation of a gel phase and the generation of so-called secondary building units (SBU) in the liquid phase within and around the gel particles takes place (Figure 9.4). Typically, these units comprise a small number of alternating connected $TO_4$ tetrahedra. The moment when the first particles with a long-range order appear in the synthesis mixture is called nucleation. Nucleation proceeds by the assembly of a certain number of SBUs to a periodically ordered domain of a certain size. Therefore, a certain degree

of supersaturation of the SBUs must be present in the mixture. After nucleation, the individual nuclei grow by further addition of nutrients from the solution. Predominantly, these nutrients are delivered in the form of monomeric units to the surface of the crystals, where they find the positions with maximum bonding possibilities, for example, at topological steps on the zeolite surface.

Certainly, the presented mechanism represents an idealized case, while in each system the situation can vary to a certain degree. For example, in most synthesis mixtures, the different steps will not be separated strictly in time over the whole synthesis volume. Formed nuclei will tend to grow straight after nucleation, and nucleation can still proceed in other parts of the system. This is more likely to happen in large and unstirred batches. Especially during and after the growth phase, other processes can proceed depending on composition and temperature. As zeolite formation is a dynamic process, redissolution processes can occur and lead to recrystallization of other zeolite or dense phases, or large crystals grow at the expense of smaller crystals (Ostwald ripening). Additionally, individual crystals can grow together and form agglomerates, or they can aggregate depending on their surface charge.

The template species remain occluded in the as-synthesized products, and it is necessary to remove them in order to make the aluminophosphate framework porous for further use. This can be achieved thermally by calcination in air or oxygen, usually at 400°C–600°C. It was found that the temperature at which the template removal occurs depends on the incorporated metal (Rajic 2005). Calcination may sometimes be accompanied by a transformation of one crystalline phase into another. A typical example is the transformation of $AlPO_4$-21 to $AlPO_4$-25 (Jelinek et al. 1991).

It should be noted that the formation of other metal phosphates is subject to the same regularities identified for aluminophosphates (Rajic 2005; Lin and Ding 2013).

## 9.3 METAL PHOSPHATE-BASED HUMIDITY SENSORS

Metal phosphates tested as the humidity sensitive materials are listed in Table 9.3. It is seen that a large amount of metal phosphates has been tested. At the same time, based on them, the researchers tried to develop humidity sensors of various types from resistive and capacitive to optical and fiber optic. Humidity-sensitive layers in these devices were formed using various methods of ceramics, thick-film and thin-film technologies, including the spraying layer-by-layer (LbL) technique (Hernaez et al. 2013), pulsed-laser ablation (Munoz et al. 1998), tape casting (Sheng et al. 2012), and spin coating (Szendrei et al. 2017).

The results of the studies showed that sensors based on metal phosphates may have a sensitivity inherent to the best samples of humidity sensors developed on the basis of other materials. So, capacitive humidity sensors with the cubic bismuth phosphate structure, developed by Sheng et al. (2012), revealed linear capacitance variations of four orders, namely, from 1.1 pF to 12,908 pF over the entire RH range (30%–95%RH) (see Figure 9.5). Like the sensing mechanisms of previous types, the pore properties of capacitive-type sensors are responsible for their humidity-sensing performance, which are controlled by the physisorption and chemisorption of water-vapor molecules as well as capillary condensation (Girnus et al. 1995; Malla and Komarneni 1995; Balkus et al. 1997a, 1997b; Newalkar et al. 1998; Tsutsumi et al. 1999). In particular, Newalkar et al. (1998) interpreted the water condensation phenomenon as induced by the structural characteristics of the aluminophosphate framework porosity.

Mulla et al. (1998) have shown that boron phosphate is a potential candidate for the resistive humidity sensors.

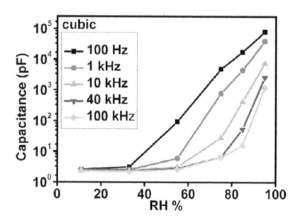

**FIGURE 9.5** Capacitance vs. relative humidity at frequencies of 100 Hz, 1 kHz, 10 kHz, 40 kHz, and 100 kHz for cubic bismuth phosphate. Cubic sillenite-type bismuth phosphates were synthesized via hydrothermal methods. (Reprinted from. *Sens. Actuators B*, 166–167, Sheng, M. et al., Humidity sensing properties of bismuth phosphates, 642–649, Copyright 2012, with permission from Elsevier.)

A resistance of boron phosphate films changed from $10^{11}$ to $10^6$ $\Omega$ with change in relative humidity from 35% to 90% (Figure 9.6). At that, this dependence was characterized by the linear behavior. It was established that in boron phosphate, humidity sensitivity largely depends on the optimization of the preparation parameters and more significantly the variation of boron/phosphorus ratio. About the same sensitivity had amorphous-zirconium-phosphate-based sensors developed by Sadaoka et al. (1988). Sadaoka et al. (1988) have shown that the humidity-impedance characteristics of zirconium-phosphate-based humidity sensors improved when the acidic protons of

## TABLE 9.3
### Metal Phosphates Used for Humidity Sensor Development

| Sensor Type | Phosphate | References |
|---|---|---|
| Conductometric, resistive | Nasicon, $Na_{1+x}Zr_2Si_xP_{3-x}O_{12}$ | Sadaoka and Sakai (1985) |
| | Boron phosphate, $BPO_4$ | Mulla et al. (1998) |
| | Rare earth phosphates, $REPO_4$ (RE = La, Sm, Gd) | Hanna et al. (2007) |
| Capacitance | Aluminum phosphate, $AlPO_4$-5 | Balkus et al. (1997a) |
| | Iron aluminophosphate, FeAPO-5 | Munoz et al. (1998) |
| | Vanadium aluminum phosphate (VAPO-5); Manganese aluminum phosphate (MnAPO-5); Magnesium aluminum phosphate (MAPO-36) | Balkus et al. (1997b) |
| | *Bismuth phosphates*, $BiPO_4$ | Sheng et al. (2012) |
| Impedance | Antimony phosphate, $HSb(PO_4)_2$ | Miura et al. (1988) |
| | Zirconium phosphate, $Zr(HPO_4)_2 \cdot H_2O$; | Sadaoka et al. (1988); Feng et al. (1999) |
| | $Zr(LiPO_4)_2 \cdot 2H_2O$; $Zr(NaPO_4)_2 \cdot 3H_2O$; $Zr(KPO_4)_2 \cdot 3H_2O$ | Sadaoka et al. (1988) |
| | Lithium titanium phosphate, $LiTi_2(PO_4)_3$ | Nocun (2000) |
| | Hydroxyapatite, $[Ca_{10}(PO_4)_6(OH)_2$; HAp] | Hontsu et al. (2010) |
| | Aluminum phosphate, $AlPO_4$-5 | Chen et al. (2015) |
| | Titanium hydrogen phosphate, $Ti^{IV}Ti^{IV}(HPO_4)_4$ | Mileo et al. (2017) |
| Optical | Antimony phosphate, $HSbP_2O_8$ | Ganter et al. (2016) |
| Fluorescent | 2D antimony phosphate ($H_3Sb_3P_2O_{14}$) | Szendrei et al. (2017) |
| Fiber optic | Poly(sodium phosphate) | Hernaez et al. (2013) |

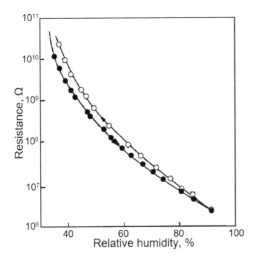

**FIGURE 9.6** Change in BPO₄ resistance vs. relative humidity. The solid-state reaction between metaphosphate and boric acid was used to prepare BPO₄. BPO₄ were calcined at 350°C. Reprinted with permission from *Sens. Actuators A*, 69, Mulla, I.S. et al. Humidity sensing properties of boron phosphate, 72–76, Copyright 1998, with permission from Elsevier.)

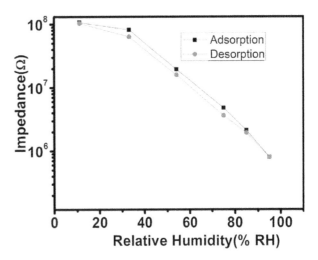

**FIGURE 9.7** Humidity hysteresis characteristic of AlPO₄-5 zeolite. AlPO₄-5 was prepared through a hydrothermal method. (Reprinted from *Sens. Actuators B*, 212, Chen, Y. et al., Humidity sensor based on AlPO₄-5 zeolite with high responsivity and its sensing mechanism, 242–247, Copyright 2015, with permission from Elsevier.)

the sample were replaced with alkali cations. In particular, it was established that for calcined zircon, the impedance in a humid atmosphere was lowered when the sample was calcined with monobasic alkali phosphate as compared with the sample calcined with phosphoric acid, while the surface area hardly depended on the additives. In the case of amorphous-zirconium-phosphate-based samples, replacing the acidic protons with alkali cations such as Na and K made the impedance lower and less dependent on the calcination temperature, while it decreased the surface area and the response time. Similar effects of alkali cations on impedance were also observed with crystalline zirconium phosphate.

An even greater sensitivity was demonstrated by the impedance-humidity sensors based on a three-dimensional titanium phosphate, $Ti^{IV}Ti^{IV}(HPO_4)_4$. It was found that this solid shows a spectacular increase of the conductivity by nine orders of magnitude at 90°C, when the relative humidity goes from 0% to 95% (Mileo et al. 2017). The molecular simulations revealed that the POH groups in the structure of a three-dimensional titanium hydrogen phosphate act as a proton source, and the proton is transported from one pore to another through the hydrogen bond network formed by water molecules and the bridging O of the solid. According to Mileo et al. (2017), this material is one of the best porous solids reported to date in terms of efficiency for proton transfer with a resulting activation energy (0.13 eV) as low as that reported for Nafion commercially used as electrolyte membrane in fuel cells. Its conductivity attains $1.2 \times 10^{-3}$ S/cm at room temperature and 95% relative humidity. Chen et al. (2015) have shown that AlPO₄-5-based resistive humidity sensors had good linearity over an RH range from 11% to 95% with the hysteresis smaller than 3% (see Figure 9.7).

It is also important to note that aluminophosphate-based humidity sensors are fairly stable devices. For example, it was established that in some cases calcined AlPO₄-34 is thermally

stable up to 1000°C (Rajic et al. 1997; Tuel et al. 2000). It was also shown that aluminophosphate-based humidity sensors are able to work both at room and at elevated temperatures (Miura et al. 1988). Thus, Miura et al. (1988) reported that a proton-conductive antimony phosphate ($HSb(PO_4)_2$ can operate at a medium temperature. The sensor could detect a water-vapor pressure ($P_{H_2O}$) in the range from 0.05 to 4.1 kPa at 150°C–250°C. Its humidity sensitivity (impedance change) was relatively high, especially in the low $P_{H_2O}$ range. Feng et al. (1999) have found that the conductivity of polycrystalline $\gamma$-Zr(HPO₄)₂·2H₂O is also highly sensitive to humidity in the temperature range of 120°C–200°C. The results of long-term stability tests are shown in Figure 9.8. It is seen that

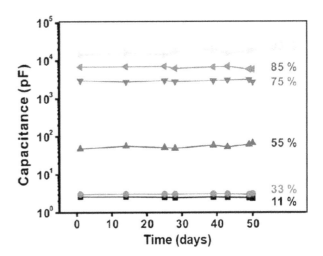

**FIGURE 9.8** Long-term stability of cubic bismuth phosphate-based sensors at various RH values over 50 days. (Reprinted from *Sens. Actuators B*, 166–167, Sheng, M. et al., Humidity sensing properties of bismuth phosphates, 642–649, Copyright 2012, with permission from Elsevier.)

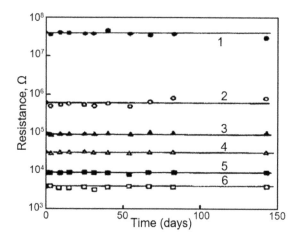

FIGURE 9.9 The stability test of zirconium phosphate, $Zr(KPO_4)_2$, when it is kept in an ambient atmosphere: 1—0%RH; 2—20%RH; 3—40%RH; 4—60%RH; 5—80%RH; 6—90%RH. (Reprinted from *Sens. Actuators*, 13, Sadaoka, Y. et al., Humidity sensor using zirconium phosphates and silicates. Improvements of humidity sensitivity, 147–157, Copyright 1988, with permission from Elsevier.)

metal phosphate-based sensors, especially zirconium-phosphate-based sensors (Figure 9.9), retain their characteristics unchanged during the whole test, even at high humidity.

## 9.4 LIMITATIONS

As for the limitations that may arise when using metal phosphates in humidity sensors, they are mainly related to technological difficulties appearing from the reproducible synthesis of these materials. For example, unlike metal oxides, apart from the temperature and time, the type of metal phosphate structure formed depends on the control of a number of variables, including the reactant-gel composition, the individual characteristics of reactants, the type of template, and pH. Most of these variables are not independent and influence one another during the crystallization process. As an example, two different crystal structures (Rajic 2005) can simultaneously crystallize from one in the same reaction mixture (their chemical formulas differing only in the water content). It was also found that several organic amines are known to produce different framework structures by slightly changing the synthesis variables (Rajic 2005). Transition-metal complexes may also act as structure-directing agents.

Slow response and recovery are also characteristic for metal phosphate-based humidity sensors (see Figure 9.10). This behavior is associated with a small pore diameter. Although it should be recognized that under certain conditions, the aluminophosphate humidity sensors can be quite fast. According to Chen et al. (2015), the $AlPO_4$-5-based resistive humidity sensors at the frequency of 100 Hz had response and recovery times about 2 s. Chen et al. (2015) believe that the good behavior can be attributed to the particular structure of the $AlPO_4$-5 zeolite. The $AlPO_4$-5 zeolite consists of four- and six-membered rings of alternating phosphate and aluminum ions bridged by oxygen. These secondary rings are arranged to form primary twelve-membered rings and produce one-dimensional channels of

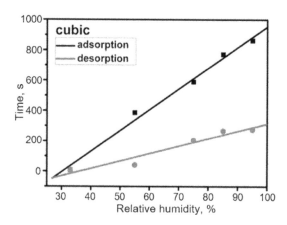

FIGURE 9.10 Response and recovery times vs. RH for cubic bismuth phosphate-based capacitance sensors. (Reprinted from *Sens. Actuators B*, 166–167, Sheng, M. et al., Humidity sensing properties of bismuth phosphates, 642–649, Copyright 2012, with permission from Elsevier.)

0.73 nm in diameter (Balkus et al. 1997a). Usually water sorption takes place in the six-membered-ring channels first via energetic considerations of structural defects, and then water molecules fill the main 12-membered ring channels via a "capillary condensation" mechanism. The initial water sorption could be driven by the affinity of the framework, and then chains of water molecule inside the large 12-membered cavities are formed by hydrogen bonds (Floquet et al. 2004). Due to the one-dimensional microstructures, water molecules are easily adsorbed on the large surface of the material with current carriers transferring fast. Besides, the large space inside cavities makes water molecules pass through easily, which contributes to the observed response and recovery properties. According to Brunauer-Emmett-Teller (BET) measurements, the surface area of the $AlPO_4$-5 zeolite synthesized by Chen et al. (2015) was 4.2 $m^2$/g. Most of the pores fell into the size range of 1.7–38 nm, and the average pore diameter of the $AlPO_4$-5 zeolite was about 16.2 nm. Apparently, it is the presence of large pores and a large scatter in their diameter that ensures acceptable linearity and good speed.

In addition, as well as other solid-state humidity sensors, metal phosphate-based humidity sensors are not selective. For example, $AlPO_4$-5-, MnAPO-5-, and MgAPO-5-based sensors exhibited a significant change in capacitance upon exposure also to small molecules such as $CO_2$, CO, and $N_2$ (Balkus et al. 1997a, 1997b). In particular, significant responses to $CO_2$ and CO were registered at levels as low as 50 and 100 ppm, respectively.

## REFERENCES

Balkus K.J., Jr., Ball L.J., Gimon-Kinsel M.E., Anthony J.M., Gnade B.E. (1997b) A capacitance-type chemical sensor that employs VAPO-5, MnAPO-5 and MAPO-36 molecular sieves as the dielectric phase. *Sen. Actuators B* **42**, 67–79.

Balkus K.J., Jr., Ball L.J., Gnade B.E., Anthony J.M. (1997a) A capacitance type chemical sensor based on $AlPO_4$-5 molecular sieves. *Chem. Mater.* **9**, 380–386.

Cheetham A.K., Ferey G., Loiseau T. (1999) Open-framework inorganic materials. *Angew. Chem. Int. Ed.* **38**, 3268–3292.

Chen Y., Zhang Y., Li D., Gao F., Feng C., Wen S., Ruan S. (2015) Humidity sensor based on AlPO$_4$-5 zeolite with high responsivity and its sensing mechanism. *Sens. Actuators B* **212**, 242–247.

Cundy C.S., Cox P.A. (2005) The hydrothermal synthesis of zeolites: Precursors, intermediates and reaction mechanism. *Micropor. Mesopor. Mat.* **82**, 1–78.

Davis M.E. (2002) Ordered porous materials for emerging applications. *Nature* **417**, 813–821.

Davis M.E., Saldarriaga C., Montes C., Garces J.M., Crowder C. (1988) A molecular sieve with eighteen—Membered rings. *Nature* **331**, 698–699.

DZS (2007) Database of Zeolite Structures, http://www.iza-structure.org/databases/.

Feng S.-H., Zhao H., Li L.-S., Greenblatt M. (1999) Synthesis, proton conductivity and humidity sensing property of γ-type zirconium hydrogen phosphate, Zr(HPO$_4$)$_2$·2H$_2$O. *Chem. Res. Chin. Univ.* **15**(1), 5–9.

Floquet N., Coulomb J.P., Dufau N., Andre G. (2004) Structure and dynamics of confined water in AlPO4-5 zeolite. *J. Phys. Chem. B* **108**, 13107–13115.

Ganter P., Szendrei K., Lotsch B.V. (2016) Humidity sensing: Towards the nanosheet-based photonic nose: Vapor recognition and trace water sensing with antimony phosphate thin film devices. *Adv. Mater.* **28**(34), 7294–7294.

Girnus I., Jancke K., Vetter R., Richter-Mendau J., Caro J. (1995) Large AlPO$_4$-5 crystals by microwave heating. *Zeolites* **15**, 33–39.

Hanna A.A., Ashraf A.M.M., Ali F., El-Sayed M.A. (2007) Rare earth phosphates as humidity sensors. *Phosphorus Res. Bull.* **21**, 78–83.

Hernaez M., Lopez-Torres D., Elosua C., Matias I.R., Arregui F.J. (2013) Sensitivity enhancement of a humidity sensor based on poly(sodium phosphate) and poly(allylamine hydrochloride). In: *Proceedings of IEEE Sensors Conference*, November 3–6, Baltimore, MD.

Hontsu S., Nakamori M., Nishikawa H., Kusunoki M. (2010) Characteristics of a humidity sensor using a Na-doped hydroxyapatite thin film. *Mem. Faculty B.O.S.T. Kinki Univ.* **26**, 87–91.

Jelinek R., Chmelka B.F., Wu Y., Grandinetti P.J., Pines A., Barrie P.J., Klinowski J. (1991) Study of the aluminophosphate AlPO$_4$-21 and AlPO$_4$-25 by aluminum-27 double-rotation NMR. *J. Am. Chem. Soc.* **113**, 4097–4101.

Li J., Yu J., Xu R. (2012) Progress in heteroatom-containing aluminophosphate molecular sieves. *Proc. R. Soc. A* **468**, 1955–1967.

Lin R., Ding Y. (2013) A review on the synthesis and applications of mesostructured transition metal phosphates. *Materials* **6**, 217–243.

Loewenstein W. (1954) The distribution of aluminum in the tetrahedra of silicates and aluminates. *Am. Mineral.* **39**, 92–96.

Lohse U., Bruckner A., Schreirer E., Bertam R., Janchen J., Fricke R. (1996) Microwave synthesis of MnAPO, MnAPSO-44 and MnAPO-5-stability of Mn$^{2+}$ ions on framework positions. *Microporous Mater.* **7**, 139–149.

Malla P.B., Komarneni S. (1995) Zeolite, effect of pore size on the chemical removal of organic template molecules from synthetic molecular sieves. *Zeolites* **15**(4), 324–332.

Mileo P.G.M., Kundu T., Semino R., Benoit V., Steunou N., Llewellyn P.L., et al. (2017) Highly efficient proton conduction in a three-dimensional titanium hydrogen phosphate. *Chem. Mater.* **29**(17), 7263–7271.

Miura N., Mizuno H., Yamazoe N. (1988) Humidity sensor using antimony phosphate operative at a medium temperature of 150°C–250°C. *Jpn. J. Appl. Phys.* **27**(5A), L931.

Mulla I.S., Chaudhary V.A., Vijayamohanan K. (1998) Humidity sensing properties of boron phosphate. *Sens. Actuators A* **69**, 72–76.

Munoz T., Jr., Balkus K.J., Jr. (1998) Preparation of FeAPO-5 molecular sieve thin films and application as a capacitive type humidity sensor. *Chem. Mater.* **10**, 4114–4122.

Murcia A.B. (2013) Ordered porous nanomaterials: The merit of small. *Nanotechnology* 2013, 257047.

Newalkar B.L., Jasra R.V., Kamath V., Bhat S.G.T. (1998) Sorption of water in aluminophosphate molecular sieve AlPO$_4$-5. *Micropor. Mesopor. Mat.* **20**, 129–137.

Nocun M. (2000) Humidity sensor based on porous glass-ceramic. *Opt. Applicata* **30**(4), 613–618.

Oliver S., Kuperman A., Ozin G.A. (1998) A new model for aluminophosphate formation: Transformation of a linear chain aluminophosphate to chain, layer, and framework structures. *Angew. Chem., Int. Ed.* **37**, 46–62.

Pastore H.O., Coluccia S., Marchese L. (2005) Porous aluminophosphates: From molecular sieves to designed acid catalysts. *Annu. Rev. Mater. Res.* **35**, 351–395.

Rajic N. (2005) Open-framework aluminophosphates: synthesis, characterization and transition metal modifications. *J. Serb. Chem. Soc.* **70**(3), 371–391.

Rajic N., Gabrovsek R., Kaucic V. (2000) Thermal investigation of two FePO materials prepared in the presence of 1,2-diaminoethane. *Thermochim. Acta* **359**, 119–122.

Rajic N., Gabrovsek R., Ristic A., Kaucic V. (1997) Thermal investigations of some AlPO and MeAPO materials prepared in the presence of HF. *Thermochim. Acta* **306**, 31–36.

Sadaoka Y., Sakai Y. (1985) Humidity sensor using sintered zircon with alkali-phosphate. *J. Mater. Sci.* **20**(8), 3027–3033.

Sadaoka Y., Sakai Y., Mitsu S. (1988) Humidity sensor using zirconium phosphates and silicates. Improvements of humidity sensitivity. *Sens. Actuators* **13**, 147–157.

Sheng M., Gu L., Kontic R., Zhou Y., Zheng K. Chen G., et al. (2012). Humidity sensing properties of bismuth phosphates. *Sens. Actuators B* **166–167**, 642–649.

Szendrei K., Jiménez-Solano A., Lozano G., Lotsch B.V., Míguez H. (2017) Fluorescent humidity sensors based on photonic resonators. *Adv. Optical Mater.* **5**, 1700663.

Taulelle F. (2001) Crystallogenesis of microporous metallophosphates. *Curr. Opin. Solid State Mater. Sci.* **5**, 397–405.

Tsutsumi K., Mizoe K., Chubachi K. (1999) Colloid adsorption characteristics and surface free energy of AlPO$_4$-5. *Polym. Sci.* **277**, 83–90.

Tuel A., Caldarelli S., Meden A., McCusker L.B., Baerlocher Ch., Ristic A., et al. (2000) NMR characterization and rietveld refinement of the structure of rehydrated AlPO$_4$-34. *J. Phys. Chem.* **104**, 5697–5705.

Van Heyden H. (2008) (Silico-)aluminophosphates: Synthesis and application for heat storage and transformation. PhD thesis. Ludwig-Maximilians-Universität München, Germany.

Wang Y., Zou X., Sun L., Rong H., Zhu G. (2018) A zeolite-like aluminophosphate membrane with molecular-sieving property for water desalination. *Chem. Sci.* **9**, 2533–2539.

Weckhuysen B.M., Ramachandra Rao R., Martens J.A., Schoonheydt R.A. (1999) Transition metal ions in microporous crystalline aluminophosphates: Isomorphous substitution. *Eur. J. Inorg. Chem.* 4, 565–577.

Wilson S.T., Lok B.M., Messina C.A., Cannan T.R., Flanigen E.M. (1982) Aluminophosphate molecular sieves: A new class of microporous crystalline inorganic solids. *J. Am. Chem. Soc.* **104**, 1146–1147.

Xu R., Pang W., Yu J., Huo Q., Chen J. (2007) *Chemistry of Zeolites and Related Porous Materials: Synthesis and Structure*. John Wiley & Sons, Singapore.

Yu J., Xu R. (2006) Insight into the construction of open-framework aluminophosphates. *Chem. Soc. Rev.* **35**, 593–604.

Yu J.H., Xu R.R. (2003) Rich structure chemistry in the aluminophosphate family. *Acc. Chem. Res.* **36**, 481–490.

# 10 Black Phosphorus and Phosphorene-Based Humidity Sensors

## 10.1 BLACK PHOSPHORUS

Phosphorus is well known for a commonly used material in explosives and matches. Its structural instability and toxicity have been keeping physicists and chemists from applying outstanding electronic optical properties of the black phosphorus (BP) in various fields (Fletcher and Galambos 1963; Island et al. 2015). However, recently, almost a century after its first discovery, BP was rediscovered as a promising two-dimensional (2D) material, phosphorene. Due to its distinct properties compared with other 2D materials, 2D BP is extensively studied with purpose to estimate their prospects for various applications (Park and Sohn 2007; Li et al. 2014; Sa et al. 2014; Xia et al. 2014; Abbas et al. 2015; Shen et al. 2015; Zhou et al. 2016; Lee et al. 2016a; Gusmão et al. 2017; Jain et al. 2017; Lin et al. 2017; Yi et al. 2017; Choi et al. 2018; Irshad et al. 2018; Wang et al. 2018; Wu et al. 2018; Yang et al. 2018; Korotcenkov 2019). Experiment and simulations have shown that 2D BP devoid of many of the shortcomings inherent in other 2D materials. In the case of graphene, this is an absence of the band gap, which reduces graphene applicability in semiconducting and optoelectronic devices. The 2D layer form of silicon, which is called as silicene, has the main issue on its environmental stability. When exposed to air, however, silicene is readily oxidized without identified reasons. Transition metal dichalcogenides (TMDCs) have direct band-gap and strong-light absorption that can be utilized in the optoelectronic devices. However, the Mo- and W-based TMDCs that are mainly used have relatively high band-gap energies. Thus, they are only proper for the absorption of limited portion of visible lights for optoelectronic applications. Furthermore, TMDCs have relatively low mobilities when compared to graphene. These drawbacks hinder the improvement of device performance using TMDCs for electronic applications and broad the applications, which require different band-gap range.

## 10.2 CRYSTAL AND ELECTRONIC STRUCTURES OF BP

BP forms a layered structure with single element, phosphorus, just like graphene (Du et al. 2010; Wei and Peng 2014). The top view of the hexagonal structure of BP with bond angles 102.09° and 96.34° is shown in Figure 10.1b. However, the BP structure has several differences when compared with other layered materials consisting of Group IV elements. In its layered structure, phosphorus atoms have five valence electrons that have configuration of $3s^23p^3$ valence shell. To build the bond with other atoms, each phosphorus atom goes through hybridization and forms $sp^3$ hybridized orbitals. These orbitals make covalent bonding with adjacent four phosphorus atoms, and this constructs a puckered structure (see Figure 10.1a). Because of its puckered structure, a single layer of BP consists of two atomic layers and includes two kinds of interatomic bonds. A shorter bond with a length of 0.2224 nm connects phosphorus in the same layer, and a longer bond with a length of 0.2244 nm connects phosphorus at the top and bottom in the single layer (Du et al. 2010; Appalakondaiah et al. 2012). With the schematic diagram of the BP in Figure 10.1a, it is possible to confirm its layered structure with space of 0.53 nm. Each layer interacts with other layers with weak van der Waals forces. Thus, BP has two types of bonds, namely, covalent intralayer bonding and weak interlayer van der Waals bonding (Aldave et al. 2016). In addition, there are two distinctive directions at the edge of BP, zigzag and armchair. These two

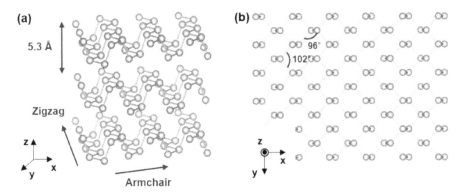

**FIGURE 10.1** (a, b) Schematic diagram of the crystal structure of black phosphorus. The system is relaxed using density functional theory calculation. (Reprinted with permission from Zhang, C. et al., *J. Phys. Chem. C* **113**, 18823–18826, 2009. Copyright 2009 American Chemical Society.)

**TABLE 10.1**

**A Compendium of Electronic and Mechanical Properties of Phosphorene, Graphene, and MoS$_2$**

| Material type | Phosphorene | Graphene | MoS$_2$ |
|---|---|---|---|
| Conduction type | Semiconductor, ambipolar | Semimetal | Semiconductor, $n$-type |
| Band gap (eV) | 0.3–2.0 | 0 | 1.2–1.8 |
| Carrier mobility (cm$^2$/V·s) | 600–1,000 | 200,000 | 200 |
| On/off Ratio | $10^3$–$10^5$ | 5.5–44 | $10^6$–$10^8$ |
| Thermal conductance (W/m·K) | 10–36 | 2,000–5,000 | 34.5–52 |
| Thermoelectric figure of merit, ZT | 1–2.5 | 0 | 0.4 |
| Strain to failure (%) | 24–32 | 19.4–38 | 19.5–36 |
| Young's modulus (GPa) | 35–166 | 1,000 | 270 ± 100 |

*Source:* Reproduced from *Mater. Sci. Eng. B*, 221, Khandelwal, A. et al., Phosphorene—The two-dimensional black phosphorous: Properties, synthesis and applications, 17–34, Copyright 2017, with permission from Elsevier.

different directions cause strong in-plane anisotropy in the BP properties such as effective masses, phonon dispersions, thermal transport, magnetic properties, etc.

Electrophysical properties of BP in comparison with other 2D materials are listed in Table 10.1. Bulk BP is a direct band-gap $p$-type semiconductor with good electrical conductivity ($\approx 10^2$ S/m), reasonable density (2.69 g/cm$^3$), and an intrinsic energy gap of $\approx 0.34$ eV (Sun et al. 2014). This semiconductor also exhibits great electrical properties with electron and hole mobilities of 220 and 350 cm$^2$/V s, respectively (Appalakondaiah et al. 2012). In addition, BP has three crystalline phases, namely, orthorhombic, rhombohedral, and simple cubic phases (Liu et al. 2015a). Under high pressure, semiconducting orthorhombic BP, which is stable under ambient conditions, can transform to a semimetallic rhombohedral structure at 5.5 GPa at room temperature (Iwasaki et al. 1986). With increasing pressures, the distances among the individual layers decrease faster than the intralayer atomic separations because the interlayer van der Waals coupling is relatively weaker (Morita 1986). Under a much higher pressure (10 GPa), the semimetallic rhombohedral structure converts further into metallic cubic phase because of an internal distortion (Ahuja 2003).

Regarding monolayer of BP (also termed "phosphorene"), then the description of the electronic structure and electron-transport properties of this material one can find in Takao and Morita (1981), Asahina and Morita (1984), Du et al. (2010), Castellanos-Gomez et al. (2014), Han et al. (2014), Rudenko and Katsnelson (2014), Wu et al. (2015), and Xu et al. (2015). Figure 10.2 shows the band gap structures of the monolayer, bilayer, and multilayer BP, constructed by density functional theory calculation with HSE06 hybrid functional, and determined by photoluminescence (PL) measurements. It was established that the band gap of BP is layer dependent, and monolayer BP has a direct band gap of ~2.0 eV at the Γ point of the first Brillouin zone (Li et al. 2014). Liu et al. (2014) initially measured the experimental band-gap value of phosphorene. The PL studies performed in the visible region, proved that the band gap in phosphorene (1.45 eV) was really significantly larger than in bulk BP (0.3 eV). Other data are shown in Figure 10.2.

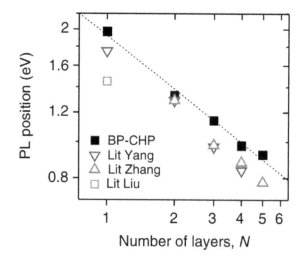

**FIGURE 10.2** Photoluminescence of black phosphorous nanosheets. Position of PL lines plotted versus layer number: 1—BP nanosheets were exfoliated by sonication in solvent of N-cyclohexyl-2-pyrrolidone (CHP); 2 to 4—data for mechanically cleaved BP nanosheets taken from Yang et al. (2014), Zhang et al. (2014), and Liu et al. (2014). (Reprinted by permission from Macmillan Publishers Limited Liquid exfoliation of solvent-stabilised black phosphorus: applications beyond electronics. *Nature Commun.*, Hanlon D. et al. 2015, copyright 2015.)

Simulations also showed that the shapes of the band are similar to each other and that they have the same translational structural symmetry and bond interactions (Carvalho et al. 2016; Rahman et al. 2016). Compared with other 2D materials, the electronic structure of BP has some distinct properties (see Figure 10.3). BP has a direct band gap at the Γ point of the Brillouin zone regardless of its number of layers. In the case of TMDCs, they have direct band gaps at the $K$ point only for the single layers, and commonly they have indirect band gaps. Thus, depending on the thickness, BP may have band-gap values ranging from 0.3 eV in the bulk to 2.0 eV in a single layer (Asahina and Morita 1984). It indicates that the tunable band gap of BP supplies the gap between graphene (0 eV) and transition metal chalcogenides (1.0–2.0 eV) (Wang et al. 2012;

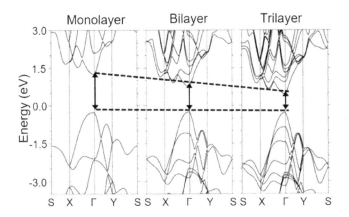

**FIGURE 10.3** Electronic structures of monolayer, bilayer, and trilayer black phosphorus calculated using first principles. Direct band gap of each layer is presented with black arrow at gamma position. The energy is scaled with respect to the Fermi energy. (Reproduced from Cai et al. (2014). Published by IOP Publishing LTD as open access.)

Lv et al. 2015). Notably, the energy range of BP covers the near-infrared (IR) and mid-IR parts of the electromagnetic wave spectrum, suggesting a promising future in the fields of IR and thermal conductivity sensors.

Qiao et al. (2014) predicted the effective mass and mobility of charge carriers in phosphorene with first-principle calculations. It was established that effective masses of both electron and hole decrease, when the number of BP layers increase. Compared with each other, the hole has the lower effective mass and, thus, the conductivity of the hole surpasses that of the electron, making the hole the major carrier in BP. However, it was established that *n*-type BP can be achieved by doping. In particular, Xu et al. (2015) have shown that substituting Te atoms to P atoms let doped BP had *n*-type conductivity. Other studies have found that the addition of Cu adatoms (Koenig et al. 2016) or chemical doping with benzyl viologen (Yu et al. 2016) can also be used as an effective electron doping strategy to obtain *n*-type conductivity. The ability to realize both *n*-type and *p*-type conductivity is a unique property of phosphorene (Das et al. 2014b). Most of the 2D semiconductors, known till now, are *n*-type semiconductors. It is also important to note that the field-effect mobility of phosphorene (almost 1,000 cm²/V·s along armchair direction) was lower than graphene (Schwierz 2010) but much higher than typical silicon-based devices (about 500 cm²/V·s).

More details on the properties of BP and phosphorene can be found in the reviews prepared by Lee et al. (2016a), Akhtar et al. (2017), Jain et al. (2017), Khandelwal et al. (2017), Yi et al. (2017), and Wu et al. (2018).

## 10.3 SYNTHESIS OF BLACK PHOSPHORUS AND PHOSPHORENE

### 10.3.1 BULK BLACK PHOSPHORUS

BP can be synthesized by a variety of methods. Its first synthesis was done by Bridgman (1914), by converting white phosphorus to BP at 373 K under 1.2 GPa within 5–30 min. In the 1980s, several Japanese groups made some important contributions to optimize this method, which significantly improved the quality and size of BP single crystals (Maruyama et al. 1981; Shirotani 1982; Endo et al. 1982). In particular, in 1981, Maruyama et al. (1981) found the method to synthesize BP in comparatively moderate pressure, by using the solution of white phosphorus and liquid bismuth at 573 K under 0.5 MPa. Different from the synthesis methods, using white phosphorus as mentioned, recently the researchers began to select red phosphorus (RP) as a raw material for BP fabrication. In 2007, Lange et al. (2007) discovered that BP can be prepared from RP via the addition of gold, tin, and tin (IV) iodide in small amounts under low pressure condition at 873 K.

Krebs and Schultze-Gebhardt (1955) first reported a "low-pressure" synthesis of BP using the catalytic action of Hg on WP at high temperatures, and this "low-pressure" method was tremendously promoted by Nilges et al. (2008). This low-pressure route method included a chemical vapor transport reaction method using Au, Sn, RP, and SnI₄ as reactant sources. Then, in 2014, Köpf et al. (2014) showed that BP can be grown by a short-way transport reaction with RP, tin, and tin (IV) iodide additives. The mixture of Sn, SnI₄, and RP was located in the furnace at the temperature of 650°C. Then the mixture was cooled down to the 550°C during 7.5 h for the synthesis of BP. In contrast to the previous method by Lange et al. (2007), they succeeded to reduce the amount of side phases such as Au₃SnP₇, AuSn, or Au₂P₃. One should note that "low-pressure" method allowed to produce high quality BP (Liu et al. 2015a).

Phase transformation from RP to BP has also occurred under the high-energy mechanical milling (HEMM) condition, which also provides high pressures and temperatures. Park et al. (2007) reported the synthesis of BP by HEMM using steel balls and milling vessel for 54 h.

### 10.3.2 PHOSPHORENE

Synthesis methods used for phosphorene preparing are listed in Table 10.2. Typical SEM images of bulk BP and phosphorene are shown in Figure 10.4. One should note that the first few layers of phosphorene were prepared by mechanical exfoliation, similar to graphene and other 2D materials (Li et al. 2014; Liu et al. 2014). The experiment showed that mechanical exfoliation can produce BP in the range from 50 layers to even two layers. As mentioned above, BP is a kind of layered material, and the interlayer interaction is dominated by weak van der Waals forces. Such a structural feature satisfies the applicable range of the top-down procedure. In several studies, phosphorene was fabricated by mechanical exfoliation for electronic devices such as field effect transistors (FETs) (Xia et al. 2014). However, one single-layer phosphorus was hard to obtain. Lu et al. (2014) have shown that the combination of mechanical cleavage and plasma ablation is promising for controlling the layer number of BP. They reported that using Ar⁺ plasma it is possible to thin few-layer BP to monolayer phosphorene. It was established that the BP nanosheets (NS) afforded by this technique are not only controllable

**TABLE 10.2**

**Methods Used for Phosphorene Preparation**

| Process | Precursor | Treatment | Thickness (nm) | References |
|---|---|---|---|---|
| Mechanical exfoliation | BP | Scotch tape on $SiO_2$/Si | 0.7–6 | Dhanabalan et al. (2017) |
| ME-PDMS | BP | Scotch tape on $SiO_2$/Si, curved PDMS | 1.6–2.8 | Andres et al. (2014) |
| Hydrothermal | RP, $NH_4F$ | Teflon-lined autoclave, 200°C | 3 | Zhao et al. (2017) |
| Electrochemical exfoliation | BP, Pt, $Na_2SO_4$ | Voltage of +7 V was applied across electrode for 90 min | 1.4–10 | Erande et al. (2016) |
| Plasma assisted exfoliation | BP | $Ar^+$ plasma at 30 W, the pressure of 30 Pa, 20 s | 2–10 | Lu et al. (2014) |
| LPE | BP, organic solvent/ water/ionic liquids | Bath sonication for 24–48 h/tip sonication for 2–4 h, centrifugation at 2000–1000 rpm for 30 min | 0.7–6 | Dhanabalan et al. (2017) |
| CVD | BP thin film over Si (substrate) $SnI_4$, Sn | Tube furnace, 950°C | 3.4 | Joshua et al. (2016) |
| Pulsed layer deposition | BP | KrF ($\lambda$: 248 nm, $\nu$: 5 Hz), 150°C, vacuum chamber | N/A | Yang et al. (2015) |

*Abbreviations:* BP—black phosphorous, CVD—chemical vacuum deposition, LPE—liquid phase exfoliation, ME—mechanical exfoliation, PDMS—polydimethylsiloxane, RP—red phosphorous.

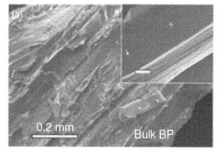 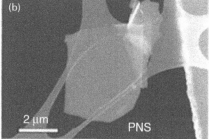

**FIGURE 10.4** (a) Scanning electron microscopy (SEM) image of (a) bulk BP. The inset is a magnified image showing the layered structure; scale bar, 10 μm; and (b) SEM image of exfoliated phosphorene nanosheets. (Reprinted by permission from Macmillan Publishers Ltd. *Nature Commun.* Cui S., et al., 2015, copyright 2015.)

and homogeneous, but also highly crystalline. The plasma-assisted thickness control of BP flakes was also demonstrated by Jia et al. (2015) and Lee et al. (2016b).

Although mechanical exfoliation is quite useful for fundamental research (Huo et al. 2015), the large-scale production of phosphorene is still difficult. This approach is labor intensive and time-consuming. Besides, this method does not have possibility to systematic control of the shape, size and the thickness of BP layer. A further disadvantage is the residual organic contamination resulted from the adhesive tapes. In order to avoid this limitation, a liquid phase exfoliation was reported as a reliable method to produce low-size flakes of few- and single-layer BP in greater quantities (Brent et al. 2014; Guo et al. 2015; Huo et al. 2015; Yasaei et al. 2015a). In fact, liquid exfoliation is just a general concept, because there are different mechanisms for different approaches. As a result, there are many methods that implement this approach. However, the main methods of liquid exfoliation include ion intercalation, ion exchange, and sonication-assisted exfoliation (Nicolosi et al. 2013). Of these methods,

the most common in the formation of 2D BP is the method of sonication-assisted exfoliation. In this case, the stable dispersions of BP are achieved via ultra-sonication, wherein collapsing cavitation bubbles yielded intense tensile and shear stress fields that exfoliated and fragmented bulk crystals. Liquid exfoliation seems very simple. NSs can be made just by putting the bulk materials into some solvents and using proper ultrasound treatments. However, there are so many solvents available it is very important to choose the appropriate solvent. In particular, Brent et al. (2014) and Guo et al. (2015) produced phosphorene by 24 h sonication of bulk BP in a basic N-methyl-2-pyrrolidone (NMP) solution. The atomic force microscopy (AFM) results suggested that the large NSs consisted of three to five layers. The monolayer BP NSs were also obtained. Almost the same time, Yasaei et al. (2015a) attempted several solvents, covering a wide range of surface tensions and polar interaction parameters. They found that the aprotic and polar solvents like dimethylformamide (DMF) and dimethyl sulfoxide (DMSO) were appropriate for the exfoliation and dispersion.

After 6-h sonication at 130 W for 0.2 mg BP ground bulk in 10 mL solvents (DMF and DMSO) and 30-min centrifugation at 2,000 r/min, they obtained uniform and stable dispersions with concentrations up to 0.01 mg/mL. Hanlon et al. (2015) exfoliated BP in the solvent N-cyclohexyl-2-pyrrolidone (CHP). The sonication time lasted 5 h, and then, the samples were centrifuged at 1,000 r/min for 180 min. By this way, the concentration of BP NSs achieved as high as 1 mg/mL. AFM results showed that 70% of observed NS thickness was less than 10 layers. For the size of the samples, it was interesting that the dispersions contained very small flakes (average length was 130 nm) and very large flakes (average length was 2.3 μm).

The above discussions show that the optimal solvents are organic solvents, which have a high boiling point and are difficult to remove. In addition, they are toxic. For this reason, exfoliation in the water phase has attracted a considerable interest. Recently, Wang et al. (2015b) successfully prepared ultrathin BP NSs by exfoliating the bulk counterpart in water. With the consideration of the extreme sensitivity of mechanically exfoliated BP to moisture, it seems to be a surprise that BP NSs can be exfoliated in water. However, it is worth to be noted that the water-exfoliated BP NSs kept stable in water for weeks without obvious degradation in the dark and preserved their NS morphology well (Wang et al. 2015b).

Apart from mechanical exfoliation and liquid-phase exfoliation (LPE), another interesting top-down approach is electrochemical exfoliation. Recently, Erande et al. (2015, 2016) successfully prepared an atomically thin layer of BP NSs by electrochemical exfoliation. The BP crystal was fixed on a highly conductive metal electrode, and the platinum wire was used as a counter electrode. Once the voltage was applied between the two electrodes, ·OH and ·O radicals would be produced around the BP crystal. The radicals were inserted into the space between consecutive layers of BP crystal and weakened the van der Waals interaction. As a result, the BP interlayer separated from BP crystal and was dispersed into

the electrolyte solution. Mayorga-Martinez et al. (2016) used a similar approach to prepare BP nanoparticles (NPs) by electrochemical exfoliation. The as-prepared BP macroparticles solution was transferred to the two-platinum electrodes electrochemical system. The induced potential difference on the BP macroparticles at opposite sides of the particle resulted in the fragmentation of the BP macroparticles into NPs.

As compared with top-down methods, the bottom-up methods are infrequently used. Chemical vacuum deposition (CVD) is one of the typical bottom-up methods used to fabricate 2D nanomaterials. This method is performed on a given substrate with precursors at high temperature. Another widely available bottom-up method for preparation of 2D materials is wet-chemical methods (Yoo et al. 2014; Zhang et al. 2016). The chemical synthesis of phosphorene has many advantages such as good crystallinity and high purity. Although chemical bottom-up methods have not made a breakthrough, the rapid development of chemical vapor deposition method creates a promising future for phosphorene chemical growth (Bagheri et al. 2016). The requirement for high quality, strong controllability and efficiency will further increase with the development of the field of nanoelectronics. More detailed description of mentioned above methods one can find in (Huo et al. 2015; Akhtar et al. 2017; Jain et al. 2017; Khandelwal et al. 2017; Lin et al. 2017; Yi et al. 2017; Wu et al. 2018).

## 10.4 BLACK PHOSPHORUS-BASED HUMIDITY SENSORS

Numerous studies in recent years have shown that BP shows a great potential for humidity sensor applications (Korotcenkov 2019). It was established that the best BP-based sensors have an ultrasensitive (Table 10.3) and selective response (see Figure 10.5) toward humid air with the trace-level detection capability. Most of these sensors were fabricated using 2D BP prepared by the method of LPE. The typical view of the BP-based humidity sensors is shown in Figure 10.6.

**TABLE 10.3**
**The List of the Performances of Different Humidity Sensors Based on BP**

| Materials | Transduction Methods | Sensitivity | Response Time (s) | Recovery Time (s) | Test Range (%RH) | References |
|---|---|---|---|---|---|---|
| BP flakes | Resistive | ~4 orders from 10%RH to 85%RH | <1 | ~1–2 | 10–85 | Yasaei et al. (2015b) |
| BP flakes | Resistive | 521% at 97%RH | 101 | 26 | 11–97 | Erande et al. (2016) |
| BP flakes | Resistive | 99.17% at 97.3%RH | 255 | 10 | 11–97 | Late (2016) |
| BP quantum dots | Resistive | ~4 orders from 10%RH to 90%RH | — | — | 10–90 | Zhu et al. (2016) |
| BP/graphene | Resistive | 43.4% at 70%RH. | 9 | 30 | 15–70 | Phan et al. (2017) |
| BP flakes | SIW resonator | 197.67 kHz/%RH | — | — | 11–97 | Chen et al. (2017) |
| | | 5.82 MHz/%RH (at RH>84%) | — | — | 11–97 | Chen and Xu (2018) |
| BP flakes | QCM | 82.7 Hz/pg at 90%RH | 14 | 10 | 10–90 | Walia et al. (2017); Yao et al. (2017) |

*Source:* Reproduced from *Nano Today*, 20, Yang, A. et al., 2018 Recent advances in phosphorene as a sensing material, 13–32, 2018. Copyright 2018, with permission from Elsevier.)

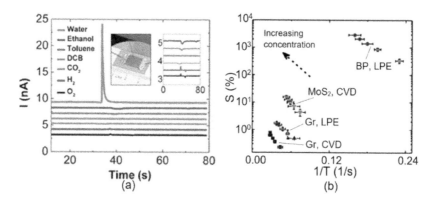

**FIGURE 10.5** (a) Response of the BP sensor to different analytes. The inset (right) magnifies the same curves for clarity. The inset (left) shows the schematic of a typical BP film sensor fabricated; (b) comparison of sensitivities of a typical BP film, graphene, and $MoS_2$ sensors. (a) (Reprinted with permission from Yasaei, P., et al., *Adv. Mater.*, 27, 1887–1892, 2015a. Copyright [2015a] American Chemical Society; Yasaei, P., et al., *ACS Nano*, 9, 9898–9905, 2015b. Copyright [2015a] American Chemical Society.). (b) (Reproduced from *Mater. Sci. Eng. B*, 221, Khandelwal, A. et al., Phosphorene—The two-dimensional black phosphorous: Properties, synthesis and applications, 17–34, Copyright 2017, with permission from Elsevier.). (From Yasaei P., et al., *Adv. Mater*, 27, 1887–1892, 2015a; Yasaei P., et al., *ACS Nano*, 9, 9898–9905, 2015b.)

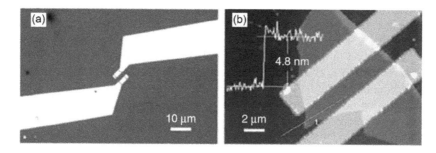

**FIGURE 10.6** (a) Optical microscopy; and (b) AFM images of the phosphorene nanosheet (PNS)-based FET sensor showing that the PNS electrically bridges the gold electrodes. The profile in (b) indicates the PNS has a thickness of 4.8 nm. (Reprinted by permission from Macmillan Publishers Ltd. *Nature Commun.* Cui S., et al., 2015, copyright 2015.)

**FIGURE 10.7** Schematic of fabrication of the suspended structure via dry transfer. Suspended BP flakes were fabricated on two high posts formed by the etched $SiO_2$ layer. (Reproduced from *Sens. Actuators B*, 250, Lee G. et al., Suspended black phosphorus nanosheet gas sensors, 569–573, Copyright 2017, with permission from Elsevier.)

As a rule, these sensors had the FET configuration. For their fabrication, various approaches can be used. For example, Figure 10.7 shows a schematic of the procedure for fabricating a suspended BP device structure. BP flakes mechanically exfoliated from the bulk BP crystals (Smart Elements) were transferred onto source-drain electrodes via a dry-transfer technique (Gel-Pak gel-film), where the electrodes were pre-patterned on a $SiO_2$/Si substrate. A drop-casting method can also be used for sensor fabrication (Yao et al. 2017). For these purposes, the BP NSs dispersion is used.

Sensors that most clearly demonstrate the advantages of the BP-based humidity sensors were developed by Yasaei et al. (2015). BP films used in this study were prepared using LPE through sonication of the ground BP powder in DMF or DMSO. After the synthesis of BP films, electrical contacts were established using gold paste or gallium-indium eutectic. The sensing performance of the fabricated films was studied in a dynamic condition using the pulse-injection method. Current modulation with respect to time plot is shown in Figure 10.17a. Nearly, a five-fold increase in the drain current was observed upon injection of water vapor, whereas the response to other analytes was much smaller. Also, direct flow of hydrogen, oxygen, and carbon dioxide gasses induces a negligible response in the sensor. This selective response of BP toward water vapor makes it viable for practical humidity detection as it overcomes cross-sensitivity and false positive issues. The film also shows full recovery of the response within 1–5 s of injection, depending upon the volume of water vapor injected. Yasaei et al. (2015) also compared the water-vapor-sensing characteristics of BP films to those of graphene and molybdenum disulfide ($MoS_2$). Figure 10.17b shows the sensitivity S [where $S = ((I-I_0)/I_0)\%$] of different sensors with respect to reciprocal recovery time ($1/T$) upon injection of water vapor. Under identical conditions, BP sensors exhibited higher sensitivity (by 2 orders of magnitude) and faster recovery (twofold faster) compared to other tested nanomaterial-based sensors. Also, in static sensing experiments, the drain current of the sensor increases by ~4 orders of magnitude as the relative humidity (RH) varies from 10% to 85%. This level of sensitivity compares with the highest levels ever reported for humidity sensing (read previous chapters).

Erande et al. (2016) also fabricated a BP humidity sensor using stacked BP flakes. The BP films were prepared by electrochemical exfoliation method. However, the sensor response of these sensors was small in comparison with data reported by Yasaei et al. (2015). The resistance of the BP sensor decreased by ~85% with increasing humidity from 11% to 97%. The response time and the recovery time were also longer ~101 s and ~26 s, respectively. It should be note that the humidity sensors previously mentioned had quite thick BP films (a few μm), which is not optimal in terms of achieving maximum sensitivity. This experimentally showed by Late (2016), when studying the humidity-sensing performances of BP NSs with different thickness. Late (2016) prepared three kinds of BP films by LPE with different centrifugal speeds, namely, 3000, 5000, and 10,000 rpm. Compared to the other two samples, the BP sample obtained by 10,000 rpm was thinner and exhibited better humidity-sensing performance with faster response and recovery speed.

It is important that for the $NO_2$ sensor, a significant increase in sensitivity with a decrease in BP thickness is also observed (see Figure 10.8). At that, the sensitivity dramatically increased when the phosphorene nanosheet (PNS) was thinner than ~10 nm. Interestingly, that the growth in the sensitivity was observed with decreasing phosphorene NS thickness until 4.8 nm, after which the sensitivity dramatically decreased. Such behavior Cui et al. (2015) explained by the effect of thickness on the band-gap value. Physically, this dependence arises from the fact that a larger band-gap semiconductor has poorer ability to attract the gas molecules due to its lower carrier concentration, whereas the conductivity change is less probable for a smaller band-gap semiconductor due to its high carrier concentration. Therefore, there would be an optimum band-gap range for the highest sensitivity by balancing these two effects.

Besides BP flakes, black phosphorus quantum dots (BP QDs) were also used to fabricate humidity sensor. Zhu et al. (2016) reported an ultrafast approach for preparing BP QDs by the LPE method using a household kitchen blender. The humidity sensor was fabricated with the prepared the BP QDs spin-coated on interdigital electrodes. The resistance of the sensor decreased by ~4 orders of magnitude with RH varied from 10% to 90%. More importantly, the response characteristics of the sensor were only slightly changed after 66 h operation in the high humidity level (90%), while a negligible change was found in the moderate humidity level (35%).

As for the mechanisms explaining the sensitivity of BP to air humidity, then as a rule, they do not differ from the mechanisms developed for metal oxides and other solid-state materials (Erande et al. 2016).

Chen et al. (2017) developed a passive microwave substrate-integrated waveguide (SIW) resonator humidity sensor operated at microwave frequencies using BP as sensitive layers. The reflection coefficient of the SIW sensor was measured by a vector network analyzer. The humidity sensitivity of the sensor with BP layer was 197.67 kHz/%RH, about 40 times larger than that of the sensor without BP sensitive layers.

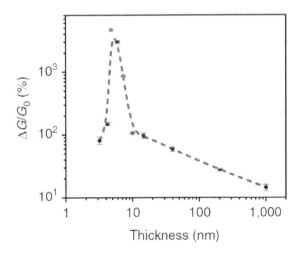

**FIGURE 10.8** The influence of phosphorene nanosheets thickness on the conductivity response to $NO_2$ (500 ppb) in dry air. (Reprinted by permission from Macmillan Publishers Ltd. *Nature Commun.* Cui S., et al., 2015, copyright 2015.)

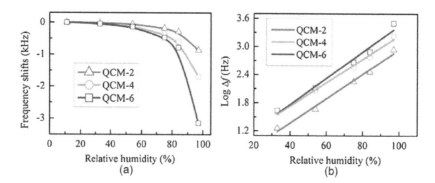

**FIGURE 10.9** (a) Frequency response of the black phosphorous-based QCM humidity sensors as a function of humidity; and (b) logarithmic fitting curve of Log (Df) versus humidity for all the sensors. (Reprinted from *Sens. Actuators B*, 244, Yao, Y., et al., Novel QCM humidity sensors using stacked black phosphorus nanosheets as sensing film, 259–264, Copyright 2017, with permission from Elsevier.)

Yao et al. (2017) have shown that the QCM humidity sensors based on BP can also be developed. In this reported work, the BP NSs were synthesized using a traditional liquid exfoliation method. From Figure 10.9 it has been inferred that the resonance frequency decreases as the humidity level is increased. As the humidity is varied from 11.3% to 97.3%, the change in the resonance frequency for the QCM-2, QCM-4, and QCM-6 sensor types (with a relative increase in the amount of BP deposition in QCM) is 863 Hz, 1698 Hz, and 3145 Hz, respectively. This reveals that the resonance frequency response of the BP-based QCM sensors is strongly related to the amount of BP NSs deposited on the QCM. Increased sensitivity of QCM-6 was ascribed by the increased number of water adsorption sites provided by increased deposition of BP on the QCM transducer. The response time of three types of BP sensors (QCM-2, QCM-4, and QCM-6) was 14 s, 23 s, and 29 s, respectively. The recovery time of all three sensors was recorded as 10 s. It is important that the moisture hysteresis for this sensor was observed to be under 4%. In addition, Yao et al. (2017) reported that developed sensors were stable during 4 weeks at a high humidity level 97.3%, and the drift, similar to a conductivity drift observed by Yasaei et al. (2015b), has not been observed. The BP conductometric sensor exhibited a good stability without obvious drifts in its sensing characteristics after 3 months' exposure to atmosphere only with relatively low humidity (25°C and 25 ± 12% RH). This means that the QCM-based BP sensors are more suitable for humidity detection in wide range of RH in comparison with conductometric-based BP sensors.

## 10.5 LIMITATIONS

### 10.5.1 THE TUNABILITY OF BLACK PHOSPHORUS PROPERTIES

No doubt that BP has unique humidity-sensitive properties, which makes this material promising for use in humidity sensors. But at the same time, BP has a number of disadvantages, which significantly limit its use in real devices. These include the technological difficulties in forming humidity-sensitive layers with a given thickness, a strong dependence of the properties on both the synthesis and subsequent processing conditions, and

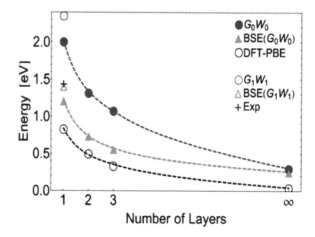

**FIGURE 10.10** Band gap with increasing number of layers are calculated by various methods and fitting curves are presented with dashed lines. The experimental value is brought from (Dai et al. 2014). (Reprinted with permission from Tran, V. et al., *Phys. Rev. B* **89**, 235319, 2014. Copyright 2014 American Chemical Society.)

the layer thickness (Cai et al. 2014; Das et al. 2014a; Kim et al. 2015; Carvalho et al. 2016; Lee et al. 2016a; Korotcenkov 2019). This thickness dependence is clearly identified with calculation data in Figure 10.10. To figure out the band gaps of monolayer, bilayer, trilayer, and bulk BP, several calculation methods were employed. From ab initio calculations with the GW approximation, the band gap of BP varies from 2 eV for monolayer to approximately 0.3 eV for bulk (Tran et al. 2014). The dependence of the thickness of the band gap is due to the charge carriers' quantum confinement effect, in the out-of-plane direction. TMDCs also have band gaps varying with thickness, but this dependence is not so strong as in BP (Conley et al. 2013).

Rodin et al. (2014) and Wang et al. (2015a) suggested that the strain can also induce a modification of the band gap. By the Density Functional Theory and the Tight-Binding Theory for monolayer BP, they have shown that as the strain increases, the original direct band gap of BP changes into an indirect band gap and, at last, it shows the zero band gap of a metal. Liu et al. (2014) calculated the strain dependence of the band gap of BP as well. According to these calculations, in case of negative strain, both armchair and zigzag direction strains lead to the band-gap decrease. On the other hand, when the positive strain is applied, zigzag-direction strain consistently increases the band gap while the band gap abruptly decreases with more than 5% of the armchair-direction strain. Moreover, Liu et al. (2014) reported that a compressive strain of 5% triggered the transition from direct to indirect band gap. The stacking pattern of BP is another factor that is possible to deal with the band gap. Dai and Zeng (2014) have found that resulting from different stackings, calculated band gaps of bilayer BP have the range of 0.8 eV–1.05 eV.

In recent years, interest has also been shown in phosphorene oxide (Wang et al. 2015a). Phosphorene oxide, by analogy to graphene oxide (GO), is expected to have novel chemical and electronic properties and may provide an alternative route to the synthesis and application of phosphorene. In particular, it was established that stoichiometric and non-stoichiometric phosphorene oxide configurations to be stable at ambient conditions, thus suggesting that the oxygen adsorption may not degrade the phosphorene. However, Wang et al. (2015a) have found that the nature of the band gap of the phosphorene oxides depends on the degree of functionalization of phosphorene; an indirect gap is predicted for the non-stoichiometric configurations, whereas a direct gap is predicted for the stoichiometric phosphorene oxide.

Taking into account the strong interrelation of the electronic parameters of the phosphorene with its adsorption and catalytic properties, it becomes clear that all of the above makes it difficult to manufacture phosphorene-based sensors with reproducible parameters. But the main disadvantage of BP, limiting its use in humidity sensors, is still the instability of their parameters (Hanlon et al. 2015; Ma et al. 2015; Khandelwal et al. 2017; Irshad et al. 2018; Wu et al. 2018). Bulk BP is stable at atmospheric conditions for a few months, but exfoliated 2D BP shows a relatively high reactivity and air instability (Ling et al. 2015). After a few days, the BP flakes degraded, and large water droplets on the surface were observed. As a result, the sensors ceased to function (see Figure 10.11a). In addition to the sharp deterioration in the

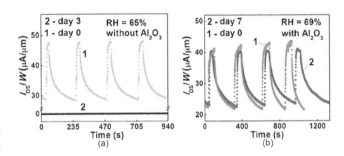

**FIGURE 10.12** Long-term stability of the BP humidity sensors: (a) dynamic sensing response for the unencapsulated BP sensor right after fabrication (orange) and after being stored in ambient for 3 days (black), (black line RH ~21%, orange line RH ~65%); and (b) dynamic-sensing response for the encapsulated BP sensor right after fabrication (red) and after being stored in ambient for 7 days (blue), (blue line RH ~21%, red line RH ~69%). (Reprinted with permission from Miao J. et al., *ACS Appl. Mater. Inter.* 9, 2017. Copyright 2017 American Chemical Society.)

sensor response, the degradation of BP can be quantified by Raman spectroscopy (Figure 10.12a). The Raman peak intensity decreased gradually after continuous exposure in the air (Favron et al. 2015; Miao et al. 2017). Similarly, the degradation of BP can be revealed by AFM.

## 10.5.2 Instability of Black Phosphorus

In most previous studies, researchers simply attributed the degradation of 2D BP in ambient conditions to moisture and strong hydrophilicity. However, the degradation mechanism of 2D BP is controversial and not entirely clear (Jain et al. 2017; Wu et al. 2018; Korotcenkov 2019). The experiment showed that there are many factors leading to instability of the parameters of BP. But the most important are degradations by oxygen, light, water, and temperature (Huang et al. 2016; Lee et al. 2016a; Wang et al. 2016). In Figure 10.13, it is possible to identify the progress of the degradation of BP in the ambient

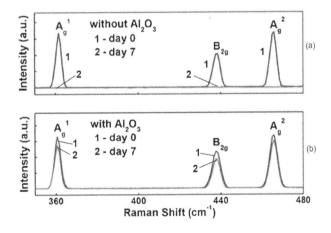

**FIGURE 10.11** (a) Raman spectra of a BP flake without Al₂O₃ capsulation after exfoliation (blue line) and after 7 days under ambient condition (red line); and (b) Raman spectra of a 6-nm-thick Al₂O₃ encapsulated BP flake immediately after exfoliation (blue line) and after 7 days under ambient condition (red line). %). (Reprinted with permission from Miao J. et al., *ACS Appl. Mater. Inter.* 9, 2017. Copyright 2017 American Chemical Society.)

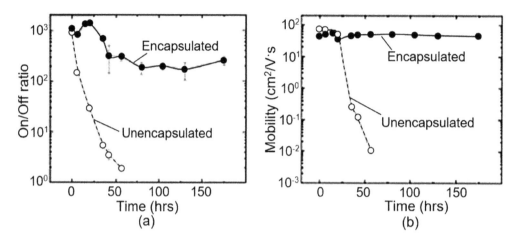

**FIGURE 10.13** On/off ratio (a) and hole mobility (b) of the black phosphorus field-effect transistor for encapsulated one and unencapsulated one. (Reprinted with permission from Wood, J. D. et al., *Nano Lett.* 14, 6964–6970, 2014. Copyright 2014 American Chemical Society.)

condition. Given the specifics of the work of humidity sensors, the presence of degradations caused by oxygen and water molecules are especially critical for their performance.

The interaction with oxygen has been established by both calculation and experiment. During interaction, the $O_2$ molecule easily dissociates and chemisorbed at the surface of BP (Ziletti et al. 2015; Wang et al. 2016). The dissociation energy of oxygen is exothermic (–4.07 eV) and has a barrier of 0.54 eV (Wang et al. 2016). As a result of the reaction, interstitial oxygen can also occur, and both interstitial and chemisorbed oxygen atoms deteriorate the properties of BP devices by initiating degradation. In particular, due to interaction with oxygen, huge structural deformation occurs and deep donor/acceptor levels are introduced in the gap of BP. The levels caused by oxygen bridge, accelerate the recombination at the surface of BP (Ziletti et al. 2015). In addition, because of the difference in electronegativity between oxygen and phosphorus, oxygen-adsorbed phosphorus has increased hydrophilicity, and this leads to the degradation by water molecule. As it was shown before, BP has strong affinity for water molecule. For example, Island et al. (2015) have shown that when BP left in the ambient condition, its volume during several days increased at 200% due to the condensation of moisture from the air. According to Wang et al. (2016), $H_2O$ will not strongly interact with pristine phosphorene; however, an exothermic reaction could occur if phosphorene is first oxidized. The pathway of oxidation first, followed by exothermic reaction with water is the most likely route for the chemical degradation of phosphorene-based devices in the air. It was also found that the strong dipole–dipole interaction between water molecule and phosphorus causes a significant distortion on the BP structure. In addition, Island et al. (2015) have established that long-term exposure to ambient conditions resulted in a layer-by-layer etching process of BP flakes; flakes can be etched down to single-layer (phosphorene) thicknesses. For example, continuous measurements of BP FETs in the air have shown the degradation and breakdown of the channel material after several days due to the layer-by-layer etching process.

Studies established that light and temperature also contribute to the degradation of phosphorene. Favron et al. (2015) have found that with the oxygen- and water-existent condition, the photoinduced degradation of the phosphorene was aroused. Moreover, the oxidation rate was proportional to the oxygen concentration and light intensity. As for thermal stability of the 2D BP, the studies made by Liu et al. (2015b) have shown that the decomposition of BP was observed at ~400°C in vacuum, in contrast to the 550°C bulk BP sublimation temperature. The decomposition initiates with an eye-shaped crack and at last amorphous RP remains.

In recent years, many studies have been carried out aimed at stabilizing the properties of BP (Li et al. 2015; Wu et al. 2018; Korotcenkov 2019). In particular, to prevent detrimental effect of the degradation, Wood et al. (2014) and Miao et al. (2017) made the passivation on the BP layer by the $AlO_x$ overlayer. Encapsulation layers such as h-BN (Cai et al. 2015), polymer (Koenig et al. 2014), $SiO_2$ (Wan et al. 2015), $MoS_2$ (Son et al. 2017), GO (Xing et al. 2017), and graphene (Cai et al. 2015) were also applied as barriers to protect 2D BP from its structure and chemical degradation. As we can see in Figures 10.12 and 10.13, passivated (encapsulated) BP maintains its properties well compared with uncapsulated one. According to Miao et al. (2017), the BP device encapsulated a 6-nm-thick $Al_2O_3$ layer did not display any noticeable change after being stored in ambient condition for over 5 days. Thus, it seems obvious that passivation on the BP layer blocks the fast degradation by oxygen and water molecule. However, if this approach gives good results for field-effect transistors, then surface passivation has limited application for humidity sensors. For gas-sensing applications, BP has to be in direct contact with the environment. Otherwise, there is a sharp drop in sensitivity (see Figure 10.14). Abbas et al. (2015) suggest using thicker BP flakes that are not subject to such rapid degradation. But even in this case, the time of stable operation is very limited and does not meet the requirements for devices intended for the sensor market. As noted earlier, even bulk BP is stable at atmospheric conditions for only a few months.

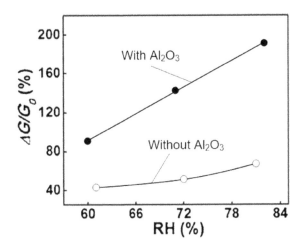

FIGURE 10.14 Conductivity response of the BP humidity sensors without and with $Al_2O_3$ layer as a function of RH levels. (Reprinted with permission from Miao, J. et al., *ACS Appl. Mater. Inter.* 9, 2017. Copyright 2017 American Chemical Society.)

A recent approach to overcome the difficulty related to the corrosion of BP under ambient conditions has been reported by Tan et al. (2018). This research work suggested

an alternative solution to improve the stability of BP by intercalating alkali metal hydride in multi-layered BP. Thus, the X-ray photoelectron spectroscopy (XPS), low-energy electron diffraction (LEED), and chemical and transport studies of synthesized Quasi-monolayer LiH-BP revealed that LiH intercalation had reduced the reactivity of BP toward oxygen by donating electrons to electron trap sites in BP, consequently neutralizing the effect of hole oxidizers in BP. Moreover, the LiH-BP holds good crystallinity and showed carrier mobility up to ~800 cm$^2$/V·s, even after ambient exposure. AFM and Raman studies also have shown the stability of intercalated BP after exposure to the surrounding atmosphere for 1 month. However, these studies do not provide an answer to the question, how will such structures behave in humidity sensors? With this regard, the studies carried out by Phan et al. (2017) give a more specific answer to this question. They have shown that the stability of the BP-based humidity sensors can be significantly improved through the formation of BP/graphene hybrid structures (see Figure 10.15). As it is seen, the sensor based on BP-graphene had good repeatability and non-degradation after 1 hour. However, these sensors are also subject to aging with longer operation (see Figure 10.16).

Studies have shown that to preserve the properties of BP flakes, they must be stored in solutions. However, even in

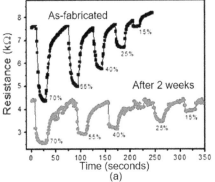

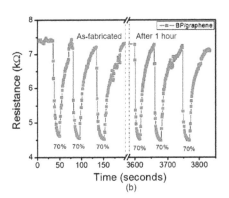

FIGURE 10.15 Transient response and estimated stability of the humidity sensor after 1 hour based on (a) BP only; and (b) BP/graphene heterojunction. (Reprinted by permission from Macmillan Publishers Ltd. *Sci. Rep.* Phan D.-T., et al., 2017, copyright 2017.)

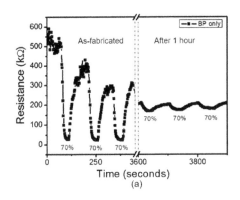

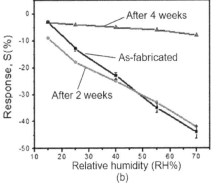

FIGURE 10.16 (a) Transient response of the BP/graphene humidity sensor as-fabricated and after 2 weeks; and (b) aging influence on sensor response of BP/graphene humidity sensor. (Reprinted by permission from Macmillan Publishers Ltd. *Sci. Rep.* Phan D.-T., et al., 2017, copyright 2017.)

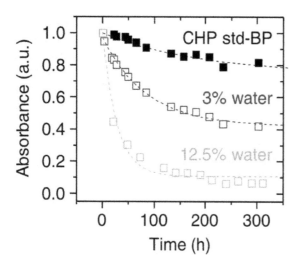

**FIGURE 10.17** The stability of absorbance (465 nm) by black phosphorous nanosheets exfoliated in different conditions: 1—Fitted time-dependent data for BP nanosheets exfoliated in CHP, and 2, 3—data for BP exfoliated in CHP solution with 3 vol% and 12.5 vol% of water added. (Reprinted by permission from Macmillan Publishers Limited Liquid exfoliation of solvent-stabilised black phosphorus: applications beyond electronics. *Nature Commun.*, Hanlon D. et al. 2015, copyright 2015.)

this case, BP flakes change properties, albeit at a slower rate. According to Hanlon et al. (2015), the most stable characteristics were performed by the BP NSs exfoliated in dry, deoxygenated N-cyclohexyl-2-pyrrolidone. The BP in the aqueous environment degrades most rapidly (see Figure 10.17), although still with considerably longer time constants than micromechanically cleaved BP.

## REFERENCES

Abbas A.N., Liu B., Chen L., Ma Y., Cong S., Aroonyadet N., et al. (2015) Black phosphorus gas sensors. *ACS Nano* **9**, 5618–5624.

Ahuja R. (2003) Calculated high pressure crystal structure transformations for phosphorus. *Phys. Status Solidi B* **235**, 282–287.

Akhtar M., Anderson G., Zhao R., Alruqi A., Mroczkowska J.E., Sumanasekera G., Jasinski J.B. (2017) Recent advances in synthesis, properties, and applications of phosphorene. *Npj 2D Mater. Appl.* **1**(1), 5.

Aldave S.H., Yogeesh M.N., Zhu W.N., Kim J., Sonde S.S., Nayak A.P., Akinwande D. (2016) Characterization and sonochemical synthesis of black phosphorus from red phosphorus. *2D Mater.* **3**(1), 014007.

Andres C.-G., Vicarelli L., Prada E., Island J.O., Narasimha-Acharya K.L., Blanter S.I., et al. (2014) Isolation and characterization of few-layer black phosphorus. *2D Mater.* **1**(2), 025001.

Appalakondaiah S., Vaitheeswaran G., Lebegue S., Christensen N.E., Svane A. (2012) Effect of van der Waals interactions on the structural and elastic properties of black phosphorus. *Phys. Rev. B* **86**, 035105.

Asahina H., Morita A. (1984) Band structure and optical properties of black phosphorus. *J. Phys. C: Solid State Phys.* **17**(11), 1839–1852.

Bagheri S., Mansouri N., Aghaie E. (2016) Phosphorene: A new competitor for graphene. *Int. J. Hydrogen Energy* **41**(7), 4085–4095.

Brent J.R., Savjani N., Lewis E.A., Haigh S.J., Lewis D.J., O'Brien P. (2014) Production of few-layer phosphorene by liquid exfoliation of black phosphorus. *Chem. Commun.* **50**, 13338–13341.

Bridgman P.W. (1914) Two new modifications of phosphorus. *J. Am. Chem. Soc.* **36**, 1344–1363.

Cai Y., Zhang G., Zhang Y.-W. (2014) Layer-dependent band alignment and work function of few-layer phosphorene. *Sci. Rep.* **4**, 6677.

Cai Y.Q., Zhang G., Zhang Y.W. (2015) Electronic properties of phosphorene/graphene and phosphorene/hexagonal boron nitride heterostructures. *J. Phys. Chem. C* **119**, 13929–13936.

Carvalho A., Wang M., Zhu X., Rodin A.S., Su H., Neto A.H.C. (2016) Phosphorene: From theory to applications. *Nat. Rev. Mater.* **1**, 16061.

Castellanos-Gomez A., Vicarelli L., Prada E., Island J.O., Narasimha-Acharya K., Blanter S.I., et al. (2014) Isolation and characterization of few-layer black phosphorus. *2D Mater.* **1**, 025001.

Chen C.-M., Xu J. (2018) A miniaturized evanescent mode HMSIW humidity sensor. *Int. J. Microwave Wireless Technol.* **10**(1), 87–91.

Chen C.M., Xu J., Yao Y. (2017) SIW resonator humidity sensor based on layered black phosphorus. *Electron. Lett.* **53**, 249–251.

Choi J.R., Yong K.W., Choi J.Y., Nilghaz A., Lin Y., Xu J., Lu X. (2018) Black phosphorus and its biomedical applications. *Theranostics* **8**(4), 1005–1026.

Conley H.J., Wang B., Ziegler J.I., Haglund R.F., Jr., Pantelides S.T., Bolotin K.I. (2013) Bandgap engineering of strained monolayer and bilayer MoS$_2$. *Nano Lett.* **13**, 3626–3630.

Cui S., Pu H., Wells S.A., Wen Z., Mao S., Chang J., et al. (2015) Ultrahigh sensitivity and layer-dependent sensing performance of phosphorene-based gas sensors. *Nature Commun.* **6**, 8632.

Dai J., Zeng X.C. (2014) Bilayer phosphorene: Effect of stacking order on bandgap and its potential applications in thin-film solar cells. *J. Phys. Chem. Lett.* **5**, 1289–1293.

Das S., Demarteau M., Roelofs A. (2014b) Ambipolar phosphorene field effect transistor. *ACS Nano* **8**, 11730–11738.

Das S., Zhang W., Demarteau M., Hoffmann A., Dubey M., Roelofs A. (2014a) Tunable transport gap in phosphorene. *Nano Lett.* **14**, 5733–5739.

Dhanabalan S.C., Ponraj J.S., Guo Z., Li S., Bao Q., Zhang H. (2017) Emerging trends in phosphorene fabrication towards next generation devices. *Adv. Sci.* **4**, 1600305.

Du Y., Ouyang C., Shi S., Lei M. (2010) Ab initio studies on atomic and electronic structures of black phosphorus. *J. Appl. Phys.* **107**, 093718.

Endo S., Akahama Y., Terada S., Narita S. (1982) Growth of large single crystals of black phosphorus under high pressure. *Jpn. J. Appl. Phys.* **21**(8), L482–L484.

Erande M.B., Pawar M.S., Late D.J. (2016) Humidity sensing and photodetection behavior of electrochemically exfoliated atomically thin-layered black phosphorus nanosheets. *ACS Appl. Mater. Interfaces* **8**, 11548–11556.

Erande M.B., Suryawanshi S.R., More M.A., Late D.J. (2015) Electrochemically exfoliated black phosphorus nanosheets—Prospective field emitters. *Eur. J. Inorg. Chem.* **19**, 3102–3107.

Favron A., Gaufrès E., Fossard F., Phaneuf-L'Heureux A.-L., Tang N.Y., Lévesque P.L., et al. (2015) Photooxidation and quantum confinement effects in exfoliated black phosphorus. *Nat. Mater.* **14**, 826–832.

Fletcher G.F., Galambos J.T. (1963) Phosphorus poisoning in humans. *Arch. Int. Med.* **112**, 846–852.

Guo Z., Zhang H., Lu S., Wang Z., Tang S., Shao J., et al. (2015) From black phosphorus to phosphorene: Basic solvent exfoliation, evolution of Raman scattering, and applications to ultrafast photonics. *Adv. Funct. Mater.* **25**, 6996–7002.

Gusmão R., Sofer Z., Pumera M. (2017) Black phosphorus rediscovered: From bulk material to monolayers. *Angew Chem. Int. Ed. Engl.* **56**(28), 8052–8072.

Han C., Yao M., Bai X., Miao L., Zhu F., Guan D., et al. (2014) Electronic structure of black phosphorus studied by angle-resolved photoemission spectroscopy. *Phys. Rev. B* **90**, 085101.

Hanlon D., Backes C., Doherty E., Cucinotta C.S., Berner N.C., Boland C., et al. (2015) Liquid exfoliation of solvent-stabilised black phosphorus: Applications beyond electronics. *Nature Commun.* **6**, 8563.

Huang Y., Qiao J., He K., Bliznakov S., Sutter E., Chen X., et al. (2016) Degradation of black phosphorus: The role of oxygen and water. *Chem. Mater.* **28**(22), 8330–8339.

Huo C., Yan Z., Song X., Zeng H. (2015) 2D materials via liquid exfoliation: A review on fabrication and applications. *Sci. Bull.* **60**(23), 1994–2008.

Irshad R., Tahir K., Li B., Sher Z., Ali J., Nazir S. (2018) A revival of 2D materials, phosphorene: Its application as sensors. *J. Industrial Eng. Chem.* **64**(25), 60–69.

Island J.O., Steele G.A., van der Zant H.S., Castellanos-Gomez A. (2015) Environmental instability of few-layer black phosphorus. *2D Mater.* **2**, 011002.

Iwasaki H., Kikegawa T., Fujimura T., Endo S., Akahama Y., Akai T., et al. (1986) Synchrotron radiation diffraction study of phase transitions in phosphorus at high pressures and temperatures. *Physica B+C* **139**, 301–304.

Jia J., Jang S.K., Lai S., Xu J., Choi Y.J., Park J.H., Lee S. (2015) Plasma-treated thickness-controlled two-dimensional black phosphorus and its electronic transport properties. *ACS Nano* **9**, 8729–8736.

Jain R., Narayan R., Sasikala S.P., Lee K.E., Jung H.J., Kim S.O. (2017) Phosphorene for energy and catalytic application—Filling the gap between graphene and 2D metal chalcogenides. *2D Mater.* **4**, 042006.

Joshua B.S., Daniel H., Hai-Feng J. (2016) Growth of 2D black phosphorus film from chemical vapor deposition. *Nanotechnology* **27**, 215602.

Khandelwal A., Mani K., Karigerasi M.H., Lahiri I. (2017) Phosphorene—The two-dimensional black phosphorous: Properties, synthesis and applications. *Mater. Sci. Eng. B* **221**, 17–34.

Kim J., Baik S.S., Ryu S.H., Sohn Y., Park S., Park B.-G., et al. (2015) Observation of tunable band gap and anisotropic dirac semimetal state in black phosphorus. *Science* **349**, 723–726.

Koenig S.P., Doganov R.A., Schmidt H., Neto A.H.C., Ozyilmaz B. (2014) Electric field effect in ultrathin black phosphorus. *Appl. Phys. Lett.* **104**, 103106.

Koenig S.P., Doganov R.A., Seixas L., Carvalho A., Tan J.Y., et al. (2016) Electron doping of ultrathin black phosphorus with Cu adatoms. *Nano Lett.* **16**, 2145–2151.

Köpf M., Eckstein N., Pfister D., Grotz C., Krüger I., Greiwe M., et al. (2014) Access and in situ growth of phosphorene-precursor black phosphorus. *J. Cryst. Growth* **405**, 6–10.

Korotcenkov G. (2019) Black phosphorus—New nanostructured material for humidity sensors: Achievements and limitations. *Sensors* (MDPI) **19**, 1010.

Krebs H., Schultze-Gebhardt F. (1955) Über die struktur und eigenschaften der halbmetalle. VII. Neubestimmung der struktur des glasigen selens nach verbesserten röntgenographischen methoden. *Acta Crystallogr.* **8**(7), 412–419.

Lange S., Schmidt P., Nilges T. (2007) Au$_3$SnP$_7$@black phosphorus: An easy access to black phosphorus. *Inorg. Chem.* **46**, 4028–4035.

Late D.J. (2016) Liquid exfoliation of black phosphorus nanosheets and its application as humidity sensor. *Micropor. Mesopor. Mat.* **225**, 494–503.

Lee G., Kim S., Jung S., Jang S., Kim J. (2017) Suspended black phosphorus nanosheet gas sensors. *Sens. Actuators B* **250**, 569–573.

Lee T.H., Kim S.Y., Jang H.W. (2016a) Black phosphorus: Critical review and potential for water splitting photocatalyst. *Nanomaterials* **6**, 194.

Lee G., Lee J.Y., Lee G.H., Kim J. (2016b) Tuning the thickness of black phosphorus *via* ion bombardment-free plasma etching for device performance improvement. *J. Mater. Chem. C* **4**, 6234–6239.

Li L., Yu Y., Ye G.J., Ge Q.Q., Ou X.D., Wu H., et al. (2014) Black phosphorus field-effect transistors. *Nat. Nanotechnol.* **9**, 372–377.

Li P., Zhang D., Liu J., Chang H., Sun Y., Yin, N. (2015) Air-stable black phosphorus devices for ion sensing. *ACS Appl. Mater. Interfaces* **7**(44), 24396.

Lin S., Chui Y., Li Y., Lau S.P. (2017) Liquid-phase exfoliation of black phosphorus and its applications. *FlatChem.* **2**, 15–37.

Ling X., Wang H., Huang S.X., Xia F.N., Dresselhaus M.S. (2015) The renaissance of black phosphorus. *Proc. Natl. Acad. Sci. USA* **112**, 4523–4530.

Liu H., Du Y.C., Deng Y.X., Ye P.D. (2015a) Semiconducting black phosphorus: Synthesis, transport properties and electronic applications. *Chem. Soc. Rev.* **44**(9), 2732–2743.

Liu H., Neal A.T., Zhu Z., Luo Z., Xu X.F., Tománek D., Ye P.D. (2014) Phosphorene: An unexplored 2D semiconductor with a high hole mobility. *ACS Nano* **8**(4), 4033–4041.

Liu X., Wood J.D., Chen K.-S., Cho E., Hersam M.C. (2015b) In situ thermal decomposition of exfoliated two-dimensional black phosphorus. *J. Phys. Chem. Lett.* **6**, 773–778.

Lu W.L., Nan H.Y., Hong J.H., Chen Y.M., Zhu C., Liang Z., et al. (2014) Plasma-assisted fabrication of monolayer phosphorene and its Raman characterization. *Nano Res.* **7**(6), 853–859.

Lv R.T., Robinson J.A., Schaak R.E., Sun D., Sun Y.F., Mallouk T.E., Terrones M. (2015) Transition metal dichalcogenides and beyond: Synthesis, properties, and applications of single and few-layer nanosheets. *Acc. Chem. Res.* **48**(1), 56–64.

Ma X., Lu W., Chen B., Zhong D., Huang L., Dong L., et al. (2015) Performance change of few layer black phosphorus transistors in ambient. *AIP Advances* **5**, 107112.

Maruyama Y., Suzuki S., Kobayashi K., Tanuma S. (1981) Synthesis and some properties of black phosphorus single crystals. *Physica B+C* **105**(1–3), 99–102.

Mayorga Martinez C.C., Latiff N.M., Eng A.Y.S., Sofer Z., Pumera M. (2016) Black phosphorus nanoparticle labels for immunoassays via hydrogen evolution reaction mediation. *Anal. Chem.* **88**, 10074–10079.

Miao J., Cai L., Zhang S., Nah J., Yeom J., Wang C. (2017) Air-stable humidity sensor using few-layer black phosphorus. *ACS Appl. Mater. Interfaces* **9**(11), 10019.

Morita A. (1986) Semiconducting black phosphorus. *Appl. Phys. A: Mater. Sci. Process.* **39**, 227–242.

Nicolosi V., Chhowalla M.., Kanatzidis MG., Strano M.S., Coleman J.N. (2013) Liquid exfoliation of layered materials. *Science* **340**, 1226419.

Nilges T., Kersting M., Pfeifer T. (2008) A fast low-pressure transport route to large black phosphorus single crystals. *J. Solid State Chem.* **181**(8), 1707–1711.

Park C.M., Sohn H.J. (2007) Black phosphorus and its composite for lithium rechargeable batteries. *Adv. Mater.* **19**, 2465–2468.

Phan D.-T., Park I., Park A.-R., Park C.-M., Jeon K.-J. (2017) Black P/graphene hybrid: A fast response humidity sensor with good reversibility and stability. *Sci. Rep.* **7**, 10561.

Qiao J., Kong X., Hu Z.-X., Yang F., Ji W. (2014) High-mobility transport anisotropy and linear dichroism in few-layer black phosphorus. *Nat. Commun.* **5**, 4475.

Rahman M.Z., Kwong C.W., Davey K., Qiao S.Z. (2016) 2D phosphorene as a water splitting photocatalyst: Fundamentals to applications. *Energy Environ. Sci.* **9**, 709–728.

Rodin A.S., Carvalho A., Castro Neto A.H. (2014) Strain-induced gap modification in black phosphorus. *Phys. Rev. Lett.* **112**(17), 176801.

Rudenko A.N., Katsnelson M.I. (2014) Quasiparticle band structure and tight-binding model for single-and bilayer black phosphorus. *Phys. Rev. B* **89**, 201408.

Sa B., Li Y.-L., Qi J., Ahuja R., Sun Z. (2014) Strain engineering for phosphorene: The potential application as a photocatalyst. *J. Phys. Chem. C* **118**, 26560–26568.

Schwierz F. (2010) Graphene transistors. *Nat. Nanotechnol.* **5**, 487–496.

Shen Z., Sun S., Wang W., Liu J., Liu Z., Jimmy C.Y. (2015) A black-red phosphorus heterostructure for efficient visible-light-driven photocatalysis. *J. Mater. Chem. A* **3**, 3285–3288.

Shirotani I. (1982) Growth of large single crystals of black phosphorus at high pressures and temperatures, and its electrical properties. *Mol. Cryst. Liq. Cryst.* **86**(1), 203–211.

Son Y., Kozawa D., Liu A.T., Koman V.B., Wang Q.H., Strano M.S. (2017) A study of bilayer phosphorene stability under $MoS_2$-passivation. *2D Mater.* **4**, 025091.

Sun J., Zheng G.Y., Lee H.W., Liu N., Wang H.T., Yao H.B., et al. (2014) Formation of stable phosphorus–carbon bond for enhanced performance in black phosphorus nanoparticle–graphite composite battery anodes. *Nano Lett.* **14**, 4573–4580.

Takao Y., Morita A. (1981) Electronic structure of black phosphorus: Tight binding approach. *Physica B+C* **105**, 93–98.

Tan S.J.R., Abdelwahab I., Chu L., Poh S.M., Liu Y., Lu J., et al. (2018) Quasi-monolayer black phosphorus with high mobility and air stability. *Adv. Mater.* **30**(6), 1704619.

Tran V., Soklaski R., Liang Y., Yang L. (2014) Layer-controlled band gap and anisotropic excitons in few-layer black phosphorus. *Phys. Rev. B* **89**, 235319.

Walia S., Sabri Y., Ahmed T., Field M.R., Ramanathan R., Arash A., et al. (2017) Defining the role of humidity in the ambient degradation of few-layer black phosphorus. *2D Mater.* **4**, 015025.

Wan B.S., Yang B.C., Wang Y., Zhang J.Y., Zeng Z.M., Liu Z.Y., et al. (2015) Enhanced stability of black phosphorus field-effect transistors with $SiO_2$ passivation. *Nanotechnology* **26**(6), 453702.

Wang G., Pandey R., Karna S.P. (2015a) Phosphorene oxide: Stability and electronic properties of a novel two-dimensional material. *Nanoscale* **7**, 524–531.

Wang G., Slough W.J., Pandey R., Karna S.P. (2016) Degradation of phosphorene in air: Understanding at atomic level. *2D Mater.* **3**, 025011.

Wang H., Yang X., Shao W., Chen S., Xie J., Zhang X., et al. (2015b) Ultrathin black phosphorus nanosheets for efficient singlet oxygen generation. *J. Am. Chem. Soc.* **137**, 11376–11382.

Wang Q.H., Kalantar-Zadeh K., Kis A., Coleman J.N., Strano M.S. (2012) Electronics and optoelectronics of two-dimensional transition metal dichalcogenides. *Nat. Nanotechnol.* **7**(11), 699–712.

Wang W., Xie G., Luo J. (2018) Black phosphorus as a new lubricant. *Friction* **6**(1), 116–142.

Wei Q., Peng X. (2014) Superior mechanical flexibility of phosphorene and few-layer black phosphorus. *Appl. Phys. Lett.* **104**, 251915.

Wood J.D., Wells S.A., Jariwala D., Chen K.-S., Cho E., Sangwan V.K., et al. (2014) Effective passivation of exfoliated black phosphorus transistors against ambient degradation. *Nano Lett.* **14**, 6964–6970.

Wu R.J., Topsakal M., Low T., Robbins M.C., Haratipour N., Jeong J.S., et al. (2015) Atomic and electronic structure of exfoliated black phosphorus. *J. Vac. Sci. Technol. A* **33**, 060604.

Wu S., K Hui K.S., Hui K.N. (2018) 2D black phosphorus: From preparation to applications for electrochemical energy storage. *Adv. Sci.* **5**, 1700491.

Xia F., Wang H., Jia Y. (2014) Rediscovering black phosphorus as an anisotropic layered material for optoelectronics and electronics. *Nat. Commun.* **5**, 4458.

Xing C., Jing G., Liang X., Qiu M., Li Z., Cao R., et al. (2017) Graphene oxide/black phosphorus nanoflake aerogels with robust thermo-stability and significantly enhanced photothermal properties in air. *Nanoscale* **9**, 8096.

Xu Y., Dai J., Zeng X.C. (2015) Electron-transport properties of few-layer black phosphorus. *J. Phys. Chem. Lett.* **6**, 1996–2002.

Yang A., Wang D., Wang X., Zhang D., Koratkar N., Rong M. (2018) Recent advances in phosphorene as a sensing material. *Nano Today* **20**, 13–32.

Yang J., Xu R., Pei J., Myint Y.W., Wang F., Wang Z., et al. (2014) Unambiguous identification of monolayer phosphorene by phase-shifting interferometry. Preprint at http://arxiv.org/abs/1412.6701.

Yang Z., Hao J., Yuan S., Lin S., Yau H.M., Dai J., Lau S.P. (2015) Field-effect transistors based on amorphous black phosphorus ultrathin films by pulsed laser deposition. *Adv. Mater.* **27**, 3748–3754.

Yao Y., Zhang H., Sun J., Ma W., Li L., Li W., Du J. (2017) Novel QCM humidity sensors using stacked black phosphorus nanosheets as sensing film. *Sens. Actuators B* **244**, 259–264.

Yasaei P., Behranginia A., Foroozan T., Asadi M., Kim K., Khalili-Araghi F., Salehi-Khojin A. (2015b) Stable and selective humidity sensing using stacked black phosphorus flakes. *ACS Nano* **9**, 9898–9905.

Yasaei P., Kumar B., Foroozan T., Wang C., Asadi M., Tuschel D., et al. (2015a) A. high-quality black phosphorus atomic layers by liquid-phase exfoliation. *Adv. Mater.* **27**, 1887–1892.

Yi Y., Yu X.-F., Zhou W., Wang J., Chu P.K. (2017) Two-dimensional black phosphorus: Synthesis, modification, properties, and applications. *Mater. Sci. Eng. R* **120**, 1–33.

Yoo D., Kim M., Jeong S., Han J., Cheon J. (2014) Chemical synthetic strategy for single-layer transition-metal chalcogenides. *J. Am. Chem. Soc.* **136**, 14670–14673.

Yu X., Zhang S., Zeng H., Wang Q.J. (2016) Lateral black phosphorene P-N junctions formed via chemical doping for high performance near-infrared photodetector. *Nano Energy* **25**, 34–41.

Zhang C., Lian J., Yi W., Jiang Y., Liu L., Hu H., et al. (2009) Surface structures of black phosphorus investigated with scanning tunneling microscopy. *J. Phys. Chem. C* **113**, 18823–18826.

Zhang S., Yang J., Xu R., Wang F., Li W., Ghufran M., et al. (2014) Extraordinary photoluminescence and strong temperature/angle-dependent Raman responses in few-layer phosphorene. *ACS Nano* **8**, 9590–9596.

Zhang Y., Rui X., Tang Y., Liu Y., Wei J., Chen S., et al. (2016) Wet-chemical processing of phosphorus composite nanosheets for high-rate and high-capacity Lithium-ion batteries. *Adv. Energy Mater.* **6**(10), 1502409.

Zhao G., Wang T., Shao Y., Wu Y., Huang B., Hao X. (2017) A novel mild phase-transition to prepare black phosphorus nanosheets with excellent energy applications. *Small* **13**, 1602243.

Zhou L. Zhang J., Zhuo Z., Kou L., Ma W., Shao B., et al. (2016) Novel excitonic solar cells in phosphorrene-TiO$_2$ heterostructures with extraordinary charge separation efficiency. *J. Phys. Chem. Lett.* **7**, 1880–1887.

Zhu C., Feng X., Zhang L., Li M., Chen J., Xu S., et al. (2016) Ultrafast preparation of black phosphorus quantum dots for efficient humidity sensing. *Chem. Eur. J.* **22**, 7357–7362.

Ziletti A., Carvalho A., Campbell D.K., Coker D.F., Neto A.C. (2015) Oxygen defects in phosphorene. *Phys. Rev. Lett.* **114**, 046801.

# 11 Metal-Organic Framework-Based Humidity Sensors

## 11.1 METAL-ORGANIC FRAMEWORKS

### 11.1.1 GENERAL CONSIDERATION

Among various porous materials, the metal organic frameworks (MOFs), also called porous coordination polymers (PCPs) (Rowsell and Yaghi 2004; Fang et al. 2010), are a class of ultraporous hybrid organic-inorganic crystalline supramolecular materials with exceptionally high accessible surface area due to the ordered framework, produced by the inorganic nodes coordinated by organic bridging ligands (Fang et al. 2010; MacGillivray 2010; Meek et al. 2011). It is important to note that the synthesis of MOFs goes through the formation the building blocks of metal ions and organic linkers, the so-called secondary building unit (SBUs) (Yaghi et al. 2003). The formation of the SBUs imposes the precise disposition of the links. In this way, the pore aperture of MOFs can be controlled on the angstrom level through the gradual increase in the number of atoms in the organic links used in the MOF design.

Depending on the metal ion and its oxidation state, coordination numbers could commonly be 2–6 for transition metals, or 6–12 for lanthanides. Different coordination numbers result in various geometries, which can be linear, T- or Y-shaped, tetrahedral, square-planar, square-pyramidal, trigonal-bipyramidal, octahedral, trigonalprismatic, pentagonal-bipyramidal, or polyhedral coordination geometry, and the corresponding distorted forms (Kitagawa et al. 2004). Besides crystallinity, one great advantage of MOFs is that, given a starting framework geometry, it is possible to build frameworks that have the same topology, but that differ by the presence of functional groups and by the size of the organic building blocks. This concept, called isoreticularity (Eddaoudi et al. 2002; Cavka et al. 2008; Garibay and Cohen 2010), allows one to tune the pore size of the material and adds the possibility of introducing functional groups within the framework. Moreover, if two or more isoreticular organic linkers are employed, frameworks bearing different functionalities that are randomly and homogeneously distributed within the framework are produced by exploiting the concept of multi-variable or mixed MOFs (MTV-MOFs or MIXMOFs) (Burrows et al. 2008; Kleist et al. 2009; Deng et al. 2010).

The potential to construct porous structures of coordination polymers by the coordination bonds was initially proposed in 1989 by Hoskins and Robson (1989). However, it took almost 10 years to realize the first few porous MOFs with permanent porosity, established by gas adsorption studies (Kondo et al. 1997), as exemplified by MOF-5 in 1999 with significantly high surface area of greater than 3000 $m^2$/g (Li et al. 1999;

Chui et al. 1999). The availability of various building blocks of metal ions and organic linkers, the so-called SBUs (Yaghi et al. 2003), makes possible to prepare an infinite number of new MOFs with diverse structures, topologies, and porosity. For example, there are papers where the surface area around 10,000 $m^2$/g have been reported (Furukawa et al. 2010). Several examples of MOFs with their characterization are presented in Table 11.1. The typical structures of MOFs are shown in Figure 11.1.

It is known that mesoporous silica (Chapter 7), porous carbon (Chapter 4), and other related materials also can have very large apertures (up to 100 nm), and their pore size can be varied in the scale of a few nanometers. However, the surface areas of MOFs are much higher as compared with those found in mentioned above materials, thus providing more readily available surfaces for interaction with guest molecules. In addition, the pore size in MOFs can be changed in bigger range. At that, in contrast to nanoporous materials such as zeolites and carbon nanotubes, the MOFs have ability to tune the structure and functionality of MOFs directly during synthesis. This tunability is significantly different from that of traditional zeolites whose pores are confined by rigid tetrahedral oxide skeletons that are difficult to alter.

### 11.1.2 MOF SYNTHESIS AND STRUCTURAL ENGINEERING

MOFs can be made by attaching inorganic metal ions to organic linkers via strong chemical bonds, as discovered by Yaghi and Li in the late 1990s (Zhou et al. 2012). It was established that various synthetic methods, including microwave, electrochemical, mechanochemical, ultrasonic, and high-throughput syntheses can be applied to synthesize MOFs and microstructure engineering (Meek et al. 2011; Liu et al. 2017). However, it was found that MOFs as well as zeolites are produced almost exclusively by hydrothermal or solvothermal techniques, where the crystals are slowly grown from a hot solution of metal precursor, such as metal nitrates and bridging ligands (Li et al. 1999; Yaghi and Li 1995; Yaghi et al. 2003; Pichon et al. 2006). We need to note that MOFs in comparison with zeolites have very little changes in synthetic technique. Ligands (see Table 11.2), the organic units used for MOFs synthesis, are typically mono-, di-, tri-, or tetravalent ones. This means that the pores can be tuned by the organic linkers of different length and/or space. Thus, similar to the synthesis of organic copolymers, the building blocks of a MOF should be chosen carefully Rowsell and Yaghi (2004). Whereas the nature and concentration of the monomers in an organic polymer determine

**TABLE 11.1**
**Characteristic Data of Several MOFs**

| MOFs | Formula | BET Surface Area (m²/g) | Pore/Channel Diameter (Å) | Window Diameter (Å) | Open Metal Sites | Thermostability (°C) | Moisture |
|---|---|---|---|---|---|---|---|
| MOF-74 | $M_2$(2,5-DOT) (M is $Zn^{2+}$, $Mg^{2+}$) | | 10, 14 | | | 300 | |
| IRMOF-74-I to -XI | $M_2$(2,5-DOT) (M is $Zn^{2+}$, $Mg^{2+}$) | 1350–2510 | 10, 14–85, 98 | | | 300 | |
| MIL-101(Cr) | $Cr_3O(H_2O)_2F(BDC)_3$ | 2736–2907 | 29, 34 | 12, 16 | Yes | 300–330 | Yes |
| MIL-100(Cr) | $Cr_3O(H_2O)_2F(BTC)_2$ | 1595 | 25, 29 | 5.6, 8.6 | Yes | 350 | Yes |
| MOF-5, IRMOF-1 | $Zn_4O(BDC)_3$ | 630–2900 | 11, 15 | 7.5, 11.2 | No | 400–480 | No |
| HKUST-1, MOF-199 | $Cu_3(BTC)_2$ | 1000–1458 | 12 | 8, 9 | Yes | 280 | No |
| IRMOF-3 | $Zn_4O(NH_2\text{-}BDC)_3$ | 1957 | | 9.6 | No | 320 | No |
| ZIF-8 | Zn(2-methylimidazole)$_2$ | 1504 | 11.4 | 3.4 | No | 380–550 | Yes |
| MIL-53(Al) | $Al^{III}(OH)(BDC)$ | 940–1038 | 8.5 | 8.5 | No | 330 | Yes |
| MIL-47(V) | $V^{IV}O(BDC)$ | 800 | 8.5 | 8.5 | No | 350 | Yes |
| ZIF-7 | Zn(benzimidazolate)$_2$ | | 4.3 | 2.9 | No | 480 | Yes |
| Copper(II) isonicotinate | $Cu(4\text{-}C_5H_4N\text{-}COO)_2(H_2O)_4$ | 146 | | | No | | Yes |

*Source:* Data extracted from Gu, Z.-Y. et al., *Acc. Chem. Res.* 45, 734–745, 2012. and Deng, H. et al., *Science*, 336, 1018–1023, 2012.
*Abbreviation:* BDC—terephthalic acid; BTC—1,3,5-benzenetricarboxylate; DOT = dioxidoterephthalate.

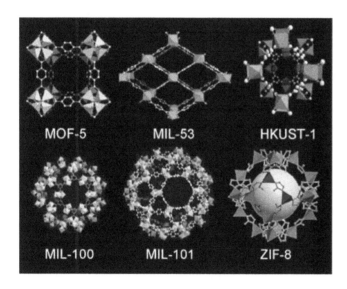

**FIGURE 11.1** Structures of typical MOFs used in analytical chemistry. (Reprinted with permission from Gu, Z.-Y. et al., *Acc. Chem. Res.*, 45, 734–745, 2012. Copyright 2012 American Chemical Society.)

its processability, physical and optical characteristics, it is the network connectivity of the building units that largely determines the properties of a MOF, for example, the definition of large channels available for the passage of molecules. The choice of metal has a significant effect on the structure and properties of the MOF as well. For example, the metal's coordination preference influences the size and the shape of pores by dictating how many ligands can bind to the metal and in which orientation. Thus, interchangeable linkers and coordinating metal ions offer great flexibility in the framework

design (see Figure 11.2). This allows judicious manipulation of the pore or channel sizes, surface area, and type of metal sites in the MOFs (Li et al. 1999; Deng et al. 2012; Eddaoudi et al. 2002; Rowsell and Yaghi 2004; Rosseinsky 2004; Sudik et al. 2005; Kitagawa et al. 2006; Allendorf and Stavila 2015). Rowsell and Yaghi (2004) believe that several factors must be borne in mind when approaching the synthesis of a new metal-organic framework, aside from the geometric principles that are considered during its design. By far the most important is the maintenance of the integrity of the building blocks. Quite often a great deal of efforts has been expended on the synthesis of a novel organic link and conditions must be found that are mild enough to maintain the functionality and conformation of this moiety, yet reactive enough to establish the metal–organic bonds. The inclusion of chiral centers or reactive sites within an open framework are also active goals for generating functional materials (Rowsell and Yaghi 2004). It was found that the pore surfaces can be functionalized by the immobilization of functional sites, such as $-NH_2$ and $-OH$, into their isostructural MOFs (Eddaoudi et al. 2002; Chen et al. 2010a, 2010b; Vaidhyanathan et al. 2010).

Zeolite synthesis often makes the use of a variety of templates, or structure-directing compounds, and a few examples of templating, particularly by organic anions, are seen in the MOF literature as well. A particular templating approach that is useful for MOFs intended for gas storage is the use of metal-binding solvents such as N,N-diethylformamide and water. In these cases, metal sites are exposed when the solvent is fully evacuated, allowing hydrogen to bind at these sites. A solvent-free synthesis of a range of crystalline MOFs also can be used (Pichon et al. 2006).

One important issue during MOFs preparation is the activation of MOFs after synthesis. The solvents used during

**TABLE 11.2**
**Common Ligands in MOFs**

| Common Name | IUPAC Name | Chemical Formula | Structural Formula |
|---|---|---|---|
| | | **Bidentate Carboxylics** | |
| Oxalic acid | Ethanedioic acid | HOOC-COOH | |
| Malonic acid | Propanedioic acid | HOOC-(CH$_2$)-COOH | |
| Succinic acid | Butanedioic acid | HOOC-(CH$_2$)$_2$-COOH | |
| Glutaric acid | Pentanedioic acid | HOOC-(CH$_2$)$_3$-COOH | |
| Phthalic acid | Benzene-1,2-dicarboxylic acid $o$-phthalic acid | C$_6$H$_4$(COOH)$_2$ | |
| Isophthalic acid | Benzene-1,3-dicarboxylic acid $m$-phthalic acid | C$_6$H$_4$(COOH)$_2$ | |
| Terephthalic acid | Benzene-1,4-dicarboxylic acid $p$-phthalic acid | C$_6$H$_4$(COOH)$_2$ | |
| | | **Tridentate Carboxylates** | |
| Citric Acid | 2-Hydroxy-1,2,3-propanetricarboxylic acid | (HOOC)CH$_2$C(OH)(COOH)CH$_2$(COOH) | |
| Trimesic acid | Benzene-1,3,5-tricarboxylic acid | C$_9$H$_6$O$_6$ | |
| | | **Azoles** | |
| 1,2,3-Triazole | 1$H$-1,2,3-triazole | C$_2$H$_3$N$_3$ | |

*(Continued)*

**TABLE 11.2 (*Continued*) Common Ligands in MOFs**

| Common Name | IUPAC Name | Chemical Formula | Structural Formula |
|---|---|---|---|
| Pyrrodiazole | 1*H*-1,2,4-triazole | $C_2H_3N_3$ | |
| **Other** | | | |
| Squaric acid | 3,4-Dihydroxy-3-cyclobutene-1,2-dione | $C_4H_2O_4$ | |

*Source:* Modified from Almeida Paz, F.A., et al., *Chem. Soc. Rev.*, 41, 1088–1110, 2012; Dhaka, S., et al., *Coord. Chem. Rev.* 380, 330–352, 2019; Pullen, S., and Clever, G. H., *Acc. Chem. Res.* 51, 3052–3064, 2018.

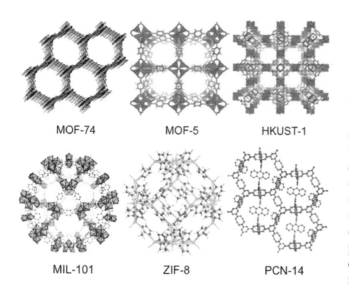

**FIGURE 11.2** Crystal structures of common MOFs. (Allendorf, M.D., and Stavila, V., *CrystEngComm.*, 17, 229–246, 2015. Reproduced by permission of The Royal Society of Chemistry.)

synthesis often remain in the pores of the materials, and usually activation by heating is required to remove these solvent molecules. Studies have shown that activation at elevated temperatures can cause a sample decomposition, whereas activation at lower temperatures greatly minimizes the danger of reducing metal ions (Liu et al. 2007).

Post-synthetic functionalization of MOFs opens up another dimension of structural possibilities that might not be achieved by conventional synthesis (Ranocchiari et al. 2012). This post-synthetic modification of MOFs gives a large additional variety of materials with different chemical and physical properties (Wang et al. 2009; Cohen 2011; Tanabe and Cohen 2009; Sun et al. 2013). A great deal of recent works explores covalent modification of the bridging ligands. Of particular interest to MOFs for gas sensors are modifications

which expose metal sites. This has been demonstrated with post-synthetic coordination of additional metal ions to the sites on the bridging ligands, and addition and removal of metal atoms to the metal site.

Thus, due to the variety of existing metal ions and organic linkers, limitless combinations of MOFs can be developed to suit the targeted application (Rowsell and Yaghi 2004; Shekhah et al. 2011; Deng et al. 2012; Liu et al. 2017). This has led to the development of well over 20 000 MOFs in the past two decades (Furukawa et al. 2013). Thus, depending on the need of a specific application, MOF materials can be designed with specific structures, textural characteristics, and dimensions to meet the demands of the applications. In particular, depending on the organic and inorganic units, structurally, MOFs can be prepared in one-dimension (1D), two-dimension (2D), or three-dimension (3D). The experiment showed that owing to fascinating structures and unusual properties, such as permanent nanoscale porosity (up to 90% free volume), high surface area, tunable pore size, adjustable internal surface properties, good thermostability, uniform structured cavities, and unique sieving properties, MOFs have great potential for diverse applications in clean energy as storage media for gases such as hydrogen and methane, and as high-capacity adsorbents to meet various separation needs. Applications of MOFs in membranes, thin film devices, catalysis, analytical chemistry, chemical sensors, and biomedical imaging are also increasingly gaining importance (Eddaoudi et al. 2002; Sudik et al. 2005; Li and Yang 2007; Kuppler et al. 2009; Thomas 2009; Gu et al. 2010, 2012; Horcajada et al. 2010; Allendorf et al. 2011; Shekhah et al. 2011; He et al. 2012; Shah et al. 2012; Liu et al. 2017; Salunkhe et al. 2017). Gas- and humidity-sensor development is also an important application area for MOFs. Several recent reviews, summarizing the gas-adsorption isotherms and gas-sensing applications of MOFs, one can find in Thomas (2009), Chen et al. (2010a), Fang et al. (2010), Allendorf et al. (2011), Keskin and Kizilel (2011), Meek et al. (2011), Shekhah et al. (2011), Khoshaman and Bahreyni (2012), Kreno et al. (2012),

Venkatasubramanian et al. (2012), and Lei et al. (2014). It is important to note that MOFs, due to great similarity to zeolites, can have the same area of applications. However, compared to crystalline and microporous fully inorganic zeolites, MOFs have much broader synthetic flexibility facilitated by the coordination environment provided by the metal ion and the geometry of the organic "linker" groups (Meek et al. 2011; Liu et al. 2017). But at the same time we need to recognize that MOFs have worse stability in comparison with zeolites. This limits the application of these materials in high-temperature devices.

## 11.2 HUMIDITY-SENSOR APPLICATIONS

### 11.2.1 ADVANTAGES FOR GAS-SENSOR APPLICATIONS

To date, four kinds of MOF-based materials have been used as sensors: crystalline MOF materials, MOF-based functional composites, nanoscale MOFs (NMOFs), and MOF membranes. Several general growth methods for these materials have been developed in recent years, and several excellent review articles on this topic are available (Shekhah et al. 2007, 2011; Masoomi and Morsali 2012; Stavila et al. 2014; Heinke et al. 2015; Li et al. 2015; Wales et al. 2015; Zhang et al. 2015; Liu et al. 2017). Studies of these materials have shown that MOFs have the following special features, important for sensor applications (Yi et al. 2016):

1. High surface areas would concentrate analytes to high levels, which enhances detective sensitivity. For example, not unlike other nanoporous materials, MOFs have the ability to adsorb large quantities

of water. It is reported that HKUST-1 can adsorb as much as 40 wt.% water (Chui et al. 1999; Wang et al. 2002);

2. Specific functional sites (open metal sites, Lewis acidic/basic sites, and tunable pore sizes) that can realize specific recognition with unprecedented selectivity through host–guest interactions or size exclusion; and

3. The presence of an organic component within the structure creates many opportunities to synthetically modify the pore environment with respect to both size and chemical properties. Flexible porosity or frameworks enable the reversible uptake and release of substrates that increases regeneration and recycling;

4. Unique set of parameters that makes MOFs attractive for the sensors, using different platforms (Mintova et al. 2001; Allendorf et al. 2008).

### 11.2.2 MOF-BASED HUMIDITY SENSORS

Generally speaking, the unique properties of MOFs make these materials attractive for measuring humidity using various transduction principles. Table 11.3 summarizes the results of research related to application of MOFs in humidity sensors. However, taking into account the unusually high surface area (>3000–6000 $m^2$/g) (Ferey et al. 2005; Furukawa et al. 2010) and high porosity, one can assume that selective sorptive layers in mass sensitive humidity sensors (Biemmi et al. 2008; Ameloot et al. 2009; Zybaylo et al. 2010; Khoshaman and Bahreyni 2012) are the best areas for MOFs applications. Mass-sensitive humidity sensors, such as surface acoustic

## TABLE 11.3
## Examples of MOF Applications in Humidity Sensors

| Sensor Type/$T_{oper}$ | MOF | Characterization | References |
|---|---|---|---|
| Optical, refractometry-based | Cu(II)-BTC$_2$(H$_2$O)$_3$ (HKUST-1) | Sensitive to various vapors and gases | Lu et al. (2011) |
| Optical, fluorescence-based | CuMOF | | Yu et al. (2014) |
| | LnMOF | | Yu et al. (2012) |
| | MgMOF | | Douvali et al. (2015a) |
| Microcantilever | Cu-BTC | | Prestipino et al. (2006) |
| | Cu(II)-BTC$_2$(H$_2$O)$_3$ (HKUST-1) | Sensitive only to water | Allendorf et al. (2008) |
| QCM | Cu$_3$(BTC)$_2$(H$_2$O)$_3$·xH$_2$O (HKUST-1) | Good sensitivity. Hysteresis is absent at $T_{oper}$=343 K | Biemmi et al. (2008) |
| | [Cu$_3$(BTC)$_2$ (HKUST-1) | | Ameloot et al. (2009) |
| SAW | Cu-BTC | 0.28–14,800 ppm range | Robinson et al. (2012) |
| Capacitance | Cu$_3$(BTC)$_2$ | | Liu et al. (2011) |
| Impedance | Fe-BTC; Cu-BTC; Al-BDC | Sensitivity only for hydrophilic gases | Achmann et al. (2009) |
| | MOF NPs (NH$_2$-MIL-125(Ti)) | | Zhang et al. (2013) |
| Work function | Cu-BTC | Sensitive to various vapors and gases | Davydovskaya et al. (2013) |

BTC—1,3,5-benzenetricarboxylate; BDC—1,4-benzenedicarboxylate.

wave (SAW) sensors, quartz crystal microbalances (QCMs), and microcantilevers (MCLs) are the devices that can transduce mass or molecular adsorption changes into vibration frequency or mechanical energy signals (read Volume 2). In each case, a signal detection requires the analyte to be adsorbed onto the surface of the sensor. Highly porous materials such as MOFs should be inherently sensitive for the gas or vapor detection using mass sensitive devices because they effectively concentrate analyte molecules at higher levels than are present in the external atmosphere. In particular, water uptake as high as 41 wt.% has been reported (Biemmi et al. 2008). In addition, the MOF thin films can be easily deposited on QCMs, SAW-based devices, and MCLs, since their hydroxylated surface provides reactive sites for binding linkers and metal ions.

### 11.2.2.1 Mass-Sensitive Sensors

The typical response of MOF-based QCM humidity sensors is shown in Figure 11.3. Ameloot et al. (2009) directly coated HKUST-1 on the top of QCM substrates by electrochemical deposition that allowed tuning the size of the densely packed crystallites. The water-sorption capacity of the films was found to be 25–30 wt.%.

Robinson et al. (2012) designed a thin film of Cu-BTC to functionalize a SAW device for water-vapor detection. They have shown that a thin film of Cu-BTC, grown on a SAW sensor, can be used to detect water vapor at sub-ppmv concentrations, demonstrating that functionalizing devices with MOF materials can enable highly sensitive humidity detection. It was found that the MOF-coated SAWs exhibited the response to water-vapor concentrations, spanning 4 orders of magnitude from −85°C to +10°C frost point (0.28–14,800 ppm at local elevation). The response was also fast (seconds) and reproducible. These results demonstrated 3 orders of magnitude better response to humidity using

HKUST-1 as compared to the same coating on QCMs-based devices discussed before (Biemmi et al. 2008; Ameloot et al. 2009). In addition, no cross-sensitivity to $N_2$, $O_2$, $CO_2$, Ar, or methane was observed. Robinson et al. (2012) accomplished this by covalently binding Cu-BTC to the surface hydroxyl groups on the quartz surface. The dependence of the SAW-sensor response on the film thickness was also determined. Cu-BTC layers varying from 75 to 350 nm were grown using 10, 20, 30, 40, 50, 60, and 100 LBL cycles. The response to water vapor as a function of the number of coating cycles, shown in Figure 11.4, indicates that there is an optimal thickness, above which the sensor response saturates. For coatings of 40–100 cycles, corresponding to 150, 175, 200, and 350-nm film thicknesses, respectively, the response to low humidity (−50°C to −40°C FP) did not appreciably increase after 50 cycles. This is evidently due to poor coupling of the portions of the film far from the SAW surface to the acoustic waves. SAWs with 10–30 coating cycles are not shown here, as their responses were significantly lower.

Robinson et al. (2012) have also simulated water adsorption isotherm for Cu-BTC (see Figure 11.5). It is seen that the initial uptake of $H_2O$ below −3°C frost point or 5,700 ppmv is linear with respect to vapor pressure. According to Robinson et al. (2012), this region corresponds to adsorption at open Cu-sites. Robinson et al. (2012) believe that above the initial linear uptake regime (at about 7°C frost point or 10,000 ppmv), some cages in the MOF structures begin to fill with $H_2O$ molecules, and adsorption of $H_2O$ is no longer linear with respect to pressure. A multi-step isotherm for $H_2O$ uptake in Cu-BTC has been shown previously by experiment (Biemmi et al. 2008; Kusgens et al. 2009) with the initial stage assigned to adsorption at open Cu-sites, followed by pore-filling. The saturation process initially manifests as the complete filling of individual pores rather than an even

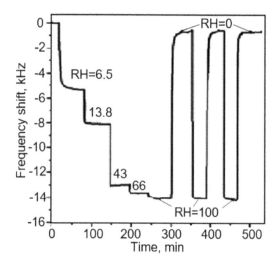

**FIGURE 11.3** The signal of the QCM sensor induced by adsorption of water from nitrogen streams at different relative humidity values, illustrating sensor reversibility and reproducibility. [$Cu_3(BTC)_2$] layer was electrochemically grown on the surface of QCM. (Reprinted with permission from Ameloot, R. et al., *Chem. Mater.*, 21, 2580–2582, 2009. Copyright 2009 American Chemical Society.)

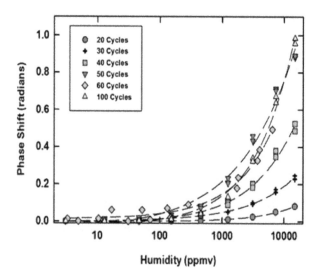

**FIGURE 11.4** The response of Cu-BTC SAWs with a different number of coating cycles to various humidity levels. (Reprinted with permission from Robinson, A.L. et al., *Anal. Chem.*, 84, 7043–7051, 2012. Copyright 2012 American Chemical Society.)

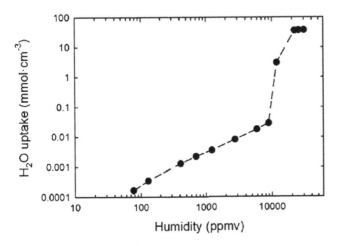

**FIGURE 11.5** Water-adsorption isotherm calculated from GCMC simulation. (Reprinted with permission from Robinson, A.L. et al., *Anal. Chem.*, 84, 7043–7051, 2012. Copyright 2012 American Chemical Society.)

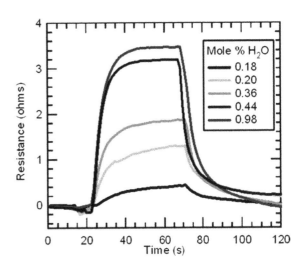

**FIGURE 11.6** Temporal response of the piezoresistive sensor to water vapor diluted in $N_2$. The device coated with HKUST-1 responds rapidly to water vapor but has no response to $N_2$ or $O_2$. (Reprinted with permission from Allendorf, M.D. et al., *J. Am. Chem. Soc.*, 130, 14404–14410, 2008. Copyright 2008 American Chemical Society.)

distribution of $H_2O$ molecules throughout the MOF structure. This indicates that hydrogen bonding between $H_2O$ molecules is energetically more favorable than $H_2O$–MOF interactions at "intermediate" water-vapor pressures (−3°C to 16°C frost point, equal to 5,700–22,000 ppmv, between the linear initial uptake region and complete saturation of the structure). Thus, clustering of water molecules is favored over an even distribution. Similar behavior has been reported for $H_2O$ adsorption in PCN-14 (a MOF with Cu paddle-wheels and multi-pore structure, similar to Cu-BTC) at 10 mbar (Zeitler et al. 2011). Similar conclusions were made by Canivet et al. (2014) on the basis of the results of their studies of CPO-27 (also known as MOF-74), UiO-66, and mesoporous MIL-101.They concluded that there are three main types of water adsorption mechanism: (1) adsorption on metallic clusters, which modifies the first coordination sphere of the metal ion (chemisorption); (2) layer/cluster adsorption; and (3) capillary condensation.

An MCL-based device is another possibility to develop MOF-based humidity sensor. MCLs detect the presence of analyte(s) by one of two transduction mechanisms: modification of the cantilever oscillation frequency as a result of the mass uptake (dynamic mode) and strain-induced bending (static deflection mode) (read Chapter 13, Volume 2). In the dynamic mode, changes in the oscillation frequency of the sensor are typically detected optically, while in static mode, adsorption produces strain at the coating-MCL interface, causing deflection of the cantilever beam that can be detected either optically or by using a built-in piezoresistive sensor. The experiment showed that the structural flexibility of MOFs is an advantage for chemical detection using static MCL, because even small changes in unit cell dimensions can result in a large tensile or compressive stresses at the interface between the cantilever and a MOF thin film (Fletcher et al. 2005; Kitagawa and Matsuda 2007). Temporal response to water vapor diluted in $N_2$ of the piezoresistive cantilever-based sensor with Cu-BTC sensing layer is shown in Figure 11.6. However, Prestipino et al. (2006) established that Cu-BTC

does not exhibit very large adsorption-induced structural changes; removal of the coordinated waters changes the unit cell dimension of this cubic MOF by only 0.12 Å. This means that other MOFs with better mechanical properties are required for high-sensitive cantilever-based humidity sensors. More recently, Venkatasubramanian et al. (2012) explored optimization of MOF coating properties and MCL design to optimize this sensing platform. HKUST-1-based piezoelectric MCL humidity sensors were developed by Allendorf et al. (2008). The reason for HKUST choice was that small distortions of the HKUST-1 framework occur upon adsorption of analytes in the pores. Allendorf et al. (2008) established that the energy of molecular adsorption within a porous MOF structure could be transformed to mechanical energy, which could be utilized to create a responsive, reversible, and selective sensor. The authors tested the response of the device to water vapor in nitrogen carrier gas with HKUST-1 in either the hydrated as-synthesised state or the "activated" dehydrated state. The dehydrated state was achieved by passing current through the piezoresistive elements so that temperature reached 50°C whilst under a flow of dry nitrogen gas for 2 h. It was found that the sensing response to water vapor did not significantly differ whether the HKUST-1 was hydrated or dehydrated. Measurements were performed by exposing the sensor to a flow of water vapor at various concentrations (0%–1%) in nitrogen-carrier gas at room temperature and pressure. The sensor response was non-linear within the range of water vapor concentrations investigated. The authors reported a measurable change in the resistance for adsorption after 0.5 s upon exposure to moist nitrogen gas, though this was the shortest measurement interval used. The time constant for desorption was ~10 s. Allendorf et al. (2011) believe that MOF application in MCL-based devices is limited only by the development of MOFs with high chemical selectivity and the ability to grow these onto the desired substrate.

## 11.2.2.2 Other Types of Humidity Sensors

However, it was established that the application of MOFs for detection of gas and water molecules sensing using other transduction principles such as capacitance (Liu et al. 2011), impedometric (Achmann et al. 2009; Zhang et al. 2013), optical (Lu and Hupp 2010; Kreno et al. 2010; Chen and Ma 2012; Hinterholzinger et al. 2012), and mechanical (Goeders et al. 1997; Allendorf et al. 2008; Venkatasubramanian et al. 2010) is also possible. In particular, Liu et al. (2011) utilized a method of homogeneous nucleation to prepare a large-scale continuous thin film of HKUST-1 on a polished copper slice. The film sensor exhibited high sensitivity and repeatability based on the humidity capacitance signal in the reverse order of humidification and desiccation processes for relative humidity (RH) values of 11.3%, 32.8%, 57.6%, 75.3%, and 84.3%, with a rapid response and recovery time (Figure 11.7).

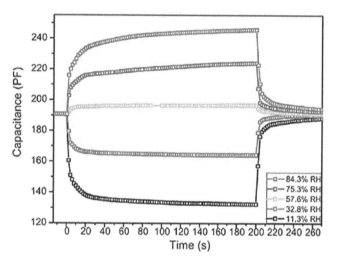

**FIGURE 11.7** Temporal response to different RH values of the $Cu_3(BTC)_2$ film sensor. (Liu, J. et al., *J. Mater. Chem.*, 21, 3775–3778, 2011, Reproduced by permission of The Royal Society of Chemistry.)

Refractometry-based optical humidity sensors were developed and tested by Lu et al. (2011). Figure 11.8a shows the shifting of the stop band up to ~16 nm induced by the adsorption of different analytes. Because the refractive index change increases with increasing amounts of adsorbed analyte, the shift also depends predictably on analyte concentration. By converting these stop-band shifts to normalized volume fractions of adsorbates, Lu and co-workers were able to construct adsorption curves on the basis of their optical measurements (Figure 11.8b). Given the ability of HKUST-1 to sorb a wide range of analytes, it is not surprising that the sensor also responded to other vapors and gases, including argon, carbon dioxide, ethane, ethylene, and ambient air. Shifts in the stop band as small as 0.015 nm were resolvable, yielding limits of detection as low as 2.6, 0.5, and 0.3 ppm for water, carbon disulfide, and ethanol, respectively.

The MOF-based sensors, operating through changes to fluorescence signals have also been successfully applied to humidity measurements. Yu et al. (2014) reported a naked-eye colorimetric sensor for humidity with CuMOF. The color of these crystals changed to red–brown from bright yellow above 33% RH, and the luminescent intensity was gradually quenched by encapsulated water molecules after different times owing to nonradiative energy transitions (Figure 11.9). Another LnMOF humidity sensor was also synthesized by them with the heteroatom-rich mixed-ligand system of pyridyl-4,5-imidazole dicarboxylic acid and oxalic acid (Yu et al. 2012a). The switching on and off of fluorescence for $Ln^{3+}$ emitters can be controlled by adsorbed and desorbed water based on O-H vibrational quenching. Later, Douvali et al. (2015a, 2015b) reported that MgMOF as an efficient sensor, capable of rapidly detecting traces (0.05%–5% v/v) of water in different organic solvents based on the sensing mechanism turn-on fluorescent enhancement. Eu and Tb MOFs were also studied (Wales et al. 2015). It was found that the terbium MOF showed the greatest change in luminescence intensity when the dehydrated form was rehydrated upon exposure to

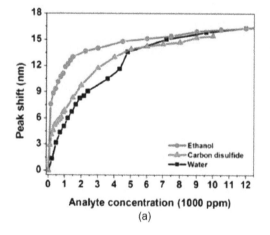

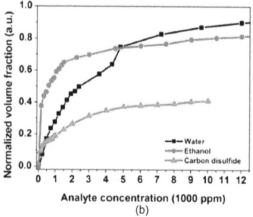

**FIGURE 11.8** (a) Dependence of stop-band peak shift of the MOF-SCC thin film on the vapor concentrations of different solvents; and (b) normalized vapor adsorption isotherms of different solvents by MOF-SCC. (From Lu, Z.-Z. et al.: Solvatochromic behavior of a nanotubular metal–organic framework for sensing small molecules. *J. Am. Chem. Soc.* 2011. 133. 4172–4174. Copyright Wiley-VCH Verlag GmbH & Co. KGaA. Reproduced with permission.)

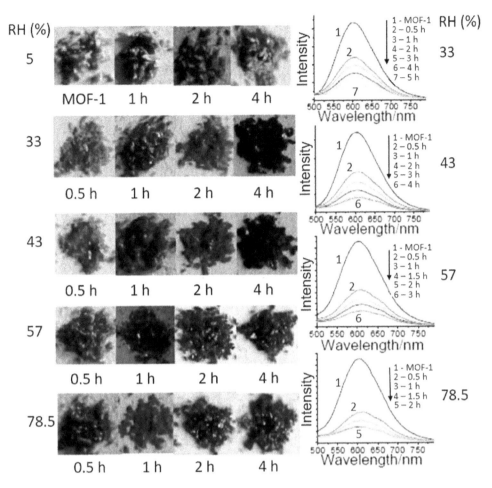

FIGURE 11.9 Left: Photographs are showing the color changes of the bulk crystal samples of Cu^I-MOF in different atmospheres with different RH. Right: The corresponding solid-state emission spectra of Cu^I-MOF in different atmospheres with different RH (33%–78.5%). (From Yu, Y. et al., *Chem. Commun.*, 50, 1444–1446, 2014. Reproduced by permission of The Royal Society of Chemistry.)

known quantities of water vapor in the air, that is, different RH levels. However, the response time of the terbium MOF probe was slow even in high RH environments. For recovery of the luminescence emission to an intensity equivalent to the luminescence emission intensity of the hydrated form, the dehydrated form had to be exposed to 85% RH for 9 h. Other results related to optical MOF-based humidity sensors one can find in (Lee et al. 2011; Wang et al. 2011; Ibarra et al. 2012; Ellern et al. 2013).

Davydovskaya et al. (2012) utilized HKUST-1 as the active sensing material for water and other analytes. The device was a Kelvin probe work function-based sensor that consisted of an alumina substrate coated on the top side with a ~2 μm TiN back electrode and a screen-printed platinum-resistive heater element on the bottom side. On the top of the TiN back electrode was a drop cast film of HKUST-1 as the sensing layer (Figure 11.10). Using this Kelvin probe setup, the measured signal represented the difference in the work function of the oscillating gold paddle and the sensing layer. As the gold paddle is not sensitive to the gases used, the signal represents the changes of the electronic structure of the sample under investigation. Humidity-sensing measurements were

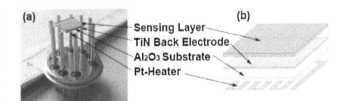

FIGURE 11.10 (a) Photograph of the Kelvin probe-type sensor fabricated by Fleischer et al.; and (b) schematic of the Kelvin probe highlighting the different layers that comprise the device. (Reprinted from *Sens. Actuators B*, 187, Davydovskaya, P. et al., Work function based gas sensing with Cu-BTC metal-organic framework for selective aldehyde detection, 142–146, Copyright 2012, with permission from Elsevier.)

performed in synthetic air (20% $O_2$ and 80% $N_2$) at a flow rate of 1 L min^{-1} at 25°C, and the humidity was studied in the range 0%–50%RH. It was found that the sensing response was reversible and stable with a decrease in work function ($\Delta\Phi$) of 5.5 mV per percent increase in RH. In addition, it was found that there was only a small influence of variation in humidity on the work function response of the TiN sample; the

HKUST-1 layer imparted sensitivity to water vapor. However, the authors also demonstrated the sensitivity of the device to aldehydes and thus the device suffers from cross sensitivity. No response time was given for sensing of water and the thermal stability of the sensor device was also not indicated.

Although electrical and electrochemical methods have been widely used in humidity sensors based on solid electrolytes, chemiresistive metal oxides, and metal-oxide semiconductor field-effect transistors (read Volume 2), they have been minimally explored for MOFs. This is most likely because the majority of MOFs are insulating (Kreno et al. 2012). To our knowledge, only two groups have reported measurement of MOF electrical properties as a means of humidity sensing. For example, Achmann et al. (2009) found that Al-BDC responded to humidity changes, but significant drift of the baseline led to irreproducible measurements. At the same time, M-BTC (M-Al, Fe, Cu) MOF did not show any signal to $O_2$ (10%vol.), $CO_2$ (10%vol.), $C_3H_8$ (1,000 ppm), NO (1,000 ppm), and $H_2$ (1,000 ppm). Noticeable effects were observed only for hydrophilic gases like ethanol (0%–18%vol.), methanol (0%–35%vol.), and water (0%–2.5%vol.), which were applied. Achmann et al. (2009) have found that the sensor responded similarly to ethanol and methanol with sensitivity increasing in the following order: methanol <ethanol <water. According to Achmann et al. (2009) their impedance sensor can detect water vapor at concentrations as low as 0.25 vol%. Responses to water vapors are shown in Figure 11.11. It is seen that devices using the Fe-BTC MOF responded reproducibly to water vapor with the absolute value of the complex impedance, |Z|, decreasing linearly with increasing water vapor concentration, $c(H_2O)$, at the lowest measured temperature, 120°C (see Figure 11.11). At higher temperatures, however, the sensitivity to changes in water concentration

decreased, and the dependence changed to an exponential decay. Achmann et al. (2009) believe that the Fe-BTC sensor can fill a need for humidity sensing, where current sensors fail, at water concentrations below 10% and at low temperature. However, it is also seen that even in this case impedometric gas sensors had low sensitivity. Zhang et al. (2013) have also reported that amine-functionalized titanium MOF NPs, $[Ti_8O_8(OH)_4(abdc)_6]$ [$NH_2$-MIL-125(Ti)], can be used as sensing elements in the impedance humidity sensor. A film of $NH_2$-MIL-125(Ti) was deposited onto five pairs of Ag–Pd interdigitated electrodes on a ceramic substrate. Upon exposure to RH ranging from 11% to 95% RH, the largest magnitude change in impedance was measured at 100 Hz. Within this humidity range, the impedance decreased from 4.5 MΩ at 11% RH to ~124 Ω at 95% RH in a non-linear manner. The $t_{90}$ response time of the sensor, upon increasing concentration of water vapor, was 45 s, and the $t_{90}$ recovery time was ~50 s.

### 11.2.3 Features of MOF-Based Sensor Fabrication

Many signal transduction schemes require a physical interface between the MOF and a device. This generally involves fabricating the MOF as a thin or thick film on a surface. The increasing interest in utilizing MOFs as gas or humidity sensors has led to a surge of interest in preparing MOF thin films and their pattering (Zacher et al. 2009, 2011). The MOF film growth and patterning has been recently reviewed (Zacher et al. 2009; Farrusseng 2011; Shekhah et al. 2007, 2011; Falcaro et al. 2014).

As it was indicated before, the MOF crystals are produced by a process of self-assembly, which allows (under the proper conditions) for the spontaneous formation of ordered lattices.

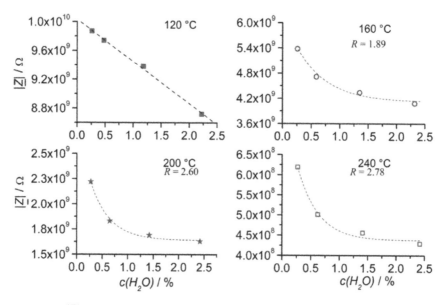

**FIGURE 11.11** Response curves (|Z| vs $H_2O$ concentration) of a planar Fe-BTC interdigital electrode sensor at different temperatures: 120°C, 160°C, 200°C, and 240°C. The sensor characteristics changed from a linear dependence at 120°C to a behavior that can be approximated by exponential decay. Measurement frequency: 1 Hz. (Reprinted from Achmann, S. et al., *Sensors*, 9, 1574–1589, 2009. Published by MDPI AG as open access.)

This bottom-up approach enables the growth of hybrid crystals with complex supramolecular architectures. However, achieving control over the spatial localisation of the self-assembly sites is a challenging task (Aizenberg et al. 1999), which remains a major scientific goal for the development of MOF-based technology (Falcaro et al. 2012). To address this issue, a number of different approaches have been proposed to control the position of these ultraporous crystals. These approaches, used for localization of MOFs during sensor fabrication, are listed in Table 11.4. Their detailed descriptions, including advantages and shortcomings, can be found in (Falcaro et al. 2014).

According to Kreno et al. (2012), most commonly, MOF films have been synthesized using the bottom-up approach directly on the surface of interest from the appropriate molecular and ionic precursors. Typically, the surface is a metal, metal oxide, glass, or silicon. Film formation can sometimes be accomplished by simply placing a platform in a reactor with the MOF precursors. These direct growth approaches often require functionalization of the surface with a self-assembled monolayer or seeding of the growth with small MOF crystals to nucleate film formation. In some cases, MOF films can be grown by one molecular or ionic layer at a time by sequential immersions of the substrate in solutions of the metal and organic precursors (Zacher et al. 2007). Functional groups on the surface (e.g., terminal components of self-assembled monolayers) may nucleate the MOF growth in a specific crystallographic direction, leading to preferentially oriented films (Biemmi et al. 2007; Zacher et al. 2007; Schoedel et al. 2010). Examples of indicated approaches one can find in Biemmi et al. (2008), Ameloot et al. (2009), and Robinson et al. (2012). Biemmi et al. (2008) demonstrated selective growth of $Cu_3(BTC)_2$ on functionalized QCM gold electrodes. The preparation of the MOF thin film was achieved by direct growth on a 11-mercaptoundecanol self-assembled monolayer (SAM). Ameloot et al. (2009), for fabrication of MOF-coated QCM, have used electrochemically synthesized Cu-BTC films, grown directly on the device. At the same time, Robinson et al. (2012) have used the layer-by-layer (LBL) deposition

## TABLE 11.4
## Permanent Localization of MOFs: Patterns

| Classification | Patterning Approach | MOF | Pattern Thickness | Pattern Resolution/ Gap Size |
|---|---|---|---|---|
| **Bottom-up** | | | | |
| *Surface functionalization* | LPE | ZIF-8 | ~700 nm | ~2 μm |
| | AFM | HKUST-1 | ~60 nm | ~15 μm |
| | Gel-layer | NH2-MIL-88B(Fe) | 40–550 nm | N/A |
| | LPE | $Cu_2(ndc)_2(dabco)$ | N/A | μm range |
| *Electrochemical deposition* | Anodic deposition | HKUST-1 | 1–20 μm | ~100 μm |
| | Precision milling and anodic deposition | HKUST-1 | 5–15 μm | μm range |
| | Galvanic displacement | HKUST-1 | N/A | 20 μm |
| *Nucleating agents* | Heterogeneous seeding with lithography | MOF-5 | N/A | ~5 μm |
| *Contact printing* | EISA combined with μCP | HKUST-1; MOF-5 | N/A | μm range |
| | MIMIC | Zn-[4,40-di(4-pyridyl)cyanostilbene] | 2 μm | μm range |
| | Pen-type lithography | HKUST-1; $Cd_3[Co(CN)_6]_2$; $Zn_3[Co(CN)_6]_2$; | <1 μm | μm range |
| | μmCP (click printing) | $Mn_3[Co(CN)_6]_2$; $Ag_3[Co(CN)_6]$ MOF-5 | ~40 nm | μm range |
| *Microfluidics* | Microfluidics | HKUST-1 | N/A | μm range |
| | Microfluidics with LbL | HKUST-1 | ~550 nm | μm rang |
| | Microfluidics | HKUST-1 spheres | D ~1–2 μm | ~400 μm |
| *Conversion from ceramics* | Pseudomorphic replication mechanism | [Al(OH)(ndc)]n | 0.2–1 μm | 200 nm |
| | Combined mechanisms | HKUST-1 | N/A | 10 μm |
| | N/A | ZIF-8 | ~1–5 μm | 15 μm |
| *Ink-jet printing and spray coating* | Ink-jet printing | HKUST-1 | ~6 μm | μm range |
| | LbL, spray coating | HKUST-1 | ~1 μm | μm range |
| **Top-down** | | | | |
| *Photolithography* | Deep X-ray lithography | ZIF-9 | 5 μm | 25 μm |
| | UV lithography | ZIF-8 | 200 nm | μm range |
| | UV lithography & Imprinting | $NH_2$-MIL-53(Al); ZIF-67($Co(Im)_2$), and ZIF-8 | N/A | 5 μm |

*Source:* Falcaro, P. et al., *Chem. Soc. Rev.*, 43, 5513–5560, 2014. Reprinted with permission from Royal Society of Chemistry.

**FIGURE 11.12** Surface functionalisation by microcontact printing (mCP): (a–c) a lithographed stamp is inked with solution containing the functional units; (d–f) the solution is then transferred to the substrate by placing the stamp in contact with the substrate; and (g and h) the solvent is then allowed to evaporate, producing the self-assembled monolayer (SAM). (From Falcaro, P. et al., *Chem. Soc. Rev.*, 43, 5513–5560, 2014. Reproduced by permission of The Royal Society of Chemistry.)

method for deposition of the MOF film. An example of the implementation of bottom-up approach using microcontact printing (μCP) method is shown in Figure 11.12.

The second approach to making a film based on top-down patterning is to first synthesize small MOF particles and subsequently deposit them on a surface using methods of thick-film technology. This has been demonstrated, for example, for Cr-MIL-101, where a suspension of monodisperse nanoparticles was obtained by microwave heating and then layered onto Si wafers via repetitive dip coating (Demessence et al. 2009). Similar techniques have been used for ZIF-8 film formation

(Lu and Hupp 2010), where a particle growth may nucleate on the surface while also incorporating particles formed initially homogeneously in solution; however, the exact mechanism is unknown. According to Kreno et al. (2012), these methods are remarkably reproducible with regard to both average film thickness and uniformity of film thickness. An unusual, but useful, third approach involves MOF film formation within the spatial constraints of a gel layer (Schoedel et al. 2010).

In contrast to bottom-up patterning methods, top-down approaches can take advantage of the wide range of protocols, available for synthesising thin films and powders, as the patterning step is applied after the MOF material has been produced (Falcaro et al. 2014). These techniques can be particularly powerful in situations where a MOF, that can only be produced using harsh synthesis conditions, needs to be incorporated within a delicate device. Top-down approaches for permanently localizing MOF are just beginning to emerge and are currently based on photolithographic methods (Dimitrakakis et al. 2012; Lu et al. 2012; Doherty et al. 2013). Although only a few methods have been reported to date, the pattern features are summarized in Table 11.4. Examples of their realization are shown in Figure 11.13.

Although the suitability of top-down approaches is currently being investigated for MOF-based device production, they do present some advantages over the bottom-up protocols. In particular, this technology is compatible with an industrial lithographic technique commonly used for microfabrication, namely photolithography. The good versatility (compatible with several different MOFs) and potentially high production rates of photolithography and imprinting make these approaches very attractive for device fabrication. Although this particular research stream is promising, the type of research required to progress the field involves the combination of multi-disciplinary expertise, not readily available in most research teams. This requirement may slow the rate of progress (Falcaro et al. 2014).

## 11.2.4 How to Optimize the Sensor Performance?

Important elements to consider when optimizing the performance and utility of humidity sensors are sensitivity, selectivity, response time, materials stability, and reproducibility (Kreno et al. 2012). As it was shown before, the development of highly porous nanostructured materials with required chemical and electrophysical properties is crucial for achieving high performance of humidity sensors. Operation of humidity sensors is based on the adsorption of water vapor, and therefore parameters of humidity sensors are strongly governed by several parameters of sensing materials, including porosity, pore size, crystallinity, surface reactivity, as well as the surface area. The sensitivity depends also on the sorption capacity of the MOF. From this point of view, highly porous MOFs are excellent humidity sensitive materials as they have an extremely great surface area and the ability to adsorb large quantities of water; for example, it is reported that HKUST-1 can adsorb as much as 40 wt.% water (Chui et al. 1999; Wang et al. 2002).

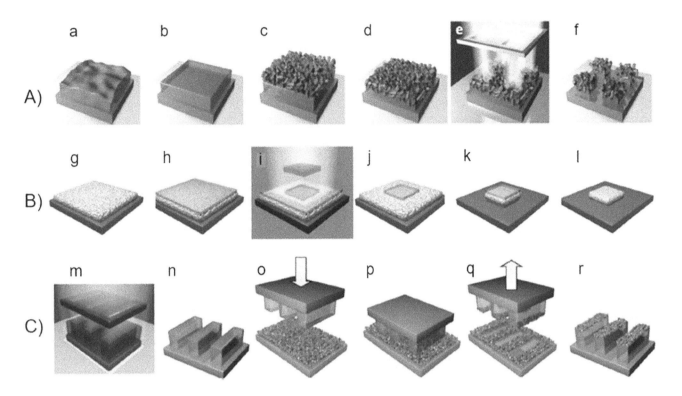

**FIGURE 11.13** Schematic illustration of the photolithography and imprinting techniques for the formation of MOF patterns developed by Dimitrakakis et al. (2012) (A), Lu et al. (2012) (B), and Doherty et al. (2013) (C): A—(a) prehydrolyzed PhTES-based solution is drop-cast on silicon wafers, and dried (b); (c) presynthesized MOF powder is spread on the film; (d) the film is heated to soften out, thereby anchoring the MOF particles to the substrate; (e–f) a photo mask positioned on top of the film is exposed to synchrotron X-rays. The unexposed region is etched off, leaving a well-defined pattern with a superficial layer of MOFs. B—(g) a substrate is immersed in MOF precursor solution, forming a MOF film; (h) a photoresist was then spin-coated on top of the MOF film, and (i and j) covered with a shadow mask and exposed to UV-light; (k) the subsequent immersion in a base solution reveals the pattern exposing the MOF film in controlled locations (areas exposed to the UV-light); (l) a further etching process removes the exposed MOF film. C—(m and n) a UV-lithographed film of photoresist is prepared and pressed (o) onto a pre-prepared MOF film; (p and q) the two films are separated, (r) resulting in a patterned MOF surface. (Falcaro, P. et al., *Chem. Soc. Rev.*, 43, 5513–5560, 2014. Reproduced by permission of The Royal Society of Chemistry.)

In rigid MOFs, uptake is controlled primarily by the adsorbate-pore surface interaction and steric interactions (Li et al. 2009), and depends on the strength of analyte binding to the MOF (stronger binding translates into lower detection limits). The experiment showed that in MOFs the interaction with water vapors is accomplished through several types of interactions that include van der Waals interactions of the framework surface with gases, coordination of the gas molecules to the central metal ion, and hydrogen bonding of the framework surface with gases (Kuppler et al. 2009; MacGillivray 2010; Chen et al. 2010a, 2010b; Zacher et al. 2011). Sensitivity depends also on the dynamics of analyte transport within the MOF. Exceptionally sluggish transport can lead to long response times that are difficult to distinguish from baseline drift. It is worth noting that, all else being equal, small pores will adsorb gas or vapor analytes more strongly than will large ones, and thereby enhance sensitivity.

An important role in sensor performance is also played by the adhesion of the humidity-sensitive layer. It is known that in each of mass-sensitive sensors, the MOF film must be tightly attached to the device surface to obtain good sensitivity. Robinson et al. (2012) have found that attaching the MOF film to the device through strong covalent bonds to the silicon dioxide surface, rather than via a SAM attached to the gold electrodes, provides also a much more thermally stable interface. This allows the device to be heated to temperatures necessary to quickly remove adsorbed water and regenerate the sensor. Film morphology may also influence sensitivity. For example, currently available methods for growing MOF films produced polycrystalline films, which if they are very rough will scatter acoustic waves to a greater extent than amorphous films that lack grain boundaries, reducing the sensitivity of SAW and QCM devices (Kreno et al. 2012). Regarding the effect of the roughness of the MOF films on humidity-sensor parameters, then further studies in this direction are required. It is clear that for these purposes a versatile coating methods, in which both the film thickness and morphology can be controlled, are needed.

The potential selectivity of MOF materials for specific analytes, or classes of analytes, is substantial, but, as yet, is not highly developed. Among possible mechanisms of molecular (analyte) selectivity, the most intuitive is size exclusion (molecular sieving) wherein atoms or molecules that are smaller than the MOF's apertures can be adsorbed, but larger molecules cannot (Dinca and Long 2005). As is known, the pore apertures dictate the size of the molecules that may enter the pores, which provide the surface and space to carry out these functions (Deng et al. 2012). Another source of selectivity is chemically specific interactions of the adsorbate with the MOF internal surface, for example, via hydrogen bonding, Mulliken-type electron donor/acceptor interactions, or formation of coordinate-covalent bonds.

As for the accurate architectural design of MOF materials on a nanometer scale, it could be realized through aperture size adjustment, cage modification, and functional group post-decoration (Eddaoudi et al. 2002; Serre et al. 2007; Lin et al. 2013). In particular, rational adjustment of the aperture size can be achieved through: (1) metal ion substitution, (2) rational design and functionalization of ligands, and (3) preferred orientation control. Cage modification, which can be categorized into in-situ and post-cage modification, respectively, represents a new concept in the field of MOF membranes. In-situ cage modification refers to embedding guest modifiers in cages during the synthesis of MOF materials. One important consideration for designing pore and aperture size is the tendency for MOFs with lengthier struts (which should produce larger apertures) to catenate, thereby yielding smaller pores (Farha et al. 2009; Deshpande et al. 2010). Pore dimensions can also be modulated by removing non-structural ligands (e.g., coordinated solvent molecules) from framework nodes or replacing node-coordinated solvent molecules with larger or smaller ligands (Farha et al. 2008).

Often, the desired functionality can be incorporated at the MOF-synthesis stage (Lin et al. 2006). Some functional groups cannot be readily incorporated during materials synthesis due to their tendency to coordinate to the MOF corners and form undesired structures. To overcome this problem, several methods entailing post-synthetic modification of MOFs have been developed. Some involve alteration or addition of functional groups on struts (Wang and Cohen 2007; Gadzikwa et al. 2009), others are node-based and entail binding pore-modifying molecules at coordinatively unsaturated metal sites at the MOF (Farha et al. 2008; Hwang et al. 2008). Last, certain potential analytes may be preferentially adsorbed if they favorably interact with open metal sites in the MOF. Thus, MOFs with immobilizing functional sites such as Lewis basic or acidic and open metal sites within porous MOFs to introduce their specific interaction with gas molecules can give great abilities for specific and unique gas molecular recognition (Allendorf et al. 2008). Based on the nature of interactions, functional groups post-decoration could be categorized into physical and chemical post-decoration, respectively. Physical post-decoration is based on physical interactions between MOF layers and functional groups. Therefore, surface properties of MOF layers will not exert significant influence on the choice of modifying agents. Post-modification was also proved effective for patching defects in MOF membranes. Chemical post-decoration relies on chemical-bonding interactions between MOFs and modifying agents so that modifying agents should be deliberately designed and selected based on the functionality of

MOF materials. Compared with physical post-decoration, chemical post-decoration was enabled to significantly change adsorptive selectivity, aperture size, and grain boundary structures of MOF membranes simultaneously. It is important that in contrast to accurate manipulation of both the aperture size and cage of MOF membranes, which emphasized on optimization of their diffusive selectivity, functional group post-decoration of MOF membranes enabled enhanced adsorptive selectivity and sealed grain boundary defects (Liu et al. 2017). One should also note that when designing MOFs for selective sensing applications, the existing literature on gas separation by MOFs (Li et al. 2009) can offer insight into what structures and functionalities may be useful.

The final requirements of rapid response time and sensor regeneration are dependent on sorption kinetics and thermodynamics. Because most guests are physisorbed, MOF sensors should be recyclable simply by subjecting the material to dynamic vacuum, if necessary, at slightly elevated temperature (Kreno et al. 2012). The response rate is governed by the rate of guest diffusion within the pores and by MOF particle size or film thickness. It is worth noting that diffusion-based mass transport resistance effects could have significant negative consequences for sensor response time. (Recall that diffusion times increase as the square of diffusion distance.) Bearing this in mind, both thin film and bulk crystalline MOF sensors should be designed with small enough dimensions to ensure rapid analyte uptake and equilibration. For example, Kreno et al. (2012) expect that dense, pinhole-free films with thicknesses on the order of 100 nm will provide sufficient analyte adsorption to be detected by QCM, SAW, and MCL devices with adequate response times. Several studies of molecule diffusion within MOFs have been reported, based on either theoretical modeling (Skoulidas and Sholl 2005; Haldoupis et al. 2010) or experimental measurements (Stallmach et al. 2006; Song et al. 2010; Zybaylo et al. 2010). In particular, Song et al. (2010) have shown that diffusion times can also be shortened by increasing MOF aperture sizes.

## 11.2.5 LIMITATIONS

In spite of extraordinary properties of MOFs, we have to recognize that the exploration of porous MOFs for sensing functions is still at a very early stage. Therefore, till now we do not have MOFs-based gas and humidity sensors acceptable for sensor market. New approaches to film deposition and sensor fabrication are required, and new effective methods should be designed to reduce the amount of drift and improve the reproducibility in order to make them suitable for long-term applications in environmental conditions (Khoshaman and Bahreyni 2012).

The experiment showed that the main limitation of MOF's application in humidity sensors is limited long-term stability (see Figure 11.14). It was established that many MOFs tend to decompose once exposed to humid air (Huang et al. 2003; Greathouse and Allendorf 2006; Kaye et al. 2007; Li and Yang 2007; Kusgens et al. 2009; Kizzie et al. 2011; De Coste et al. 2013a; Tan et al. 2015). For example, IRMOF-1 degrades in

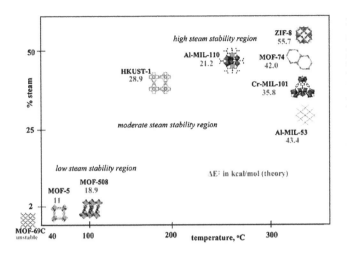

**FIGURE 11.14** Steam stability map of MOFs. The position of the structure for a given MOF represents its maximum structural stability as probed by XRD measurements, while the energy of activation for ligand displacement by a water molecule as determined by molecular modeling is represented by the magenta numbers (in kcal/mol). (Reprinted with permission from Low, J.J. et al., *J. Am. Chem. Soc.*, 131, 15834–15842, 2009. Copyright 2009 American Chemical Society.)

the presence of small amounts of water at room temperature. Kusgens et al. (2009) suggested that the structure of Cu-BTC (HKUST-1), widely used in the development of humidity sensors (Table 11.3), should be stable only in low RH environments. MOF-177, which has a very high surface area (>5000 m²/g), quickly loses its original crystalline structure upon exposure to air (within 3 days) resulting in a drastic decrease in its surface area (30 m²/g) (Li and Yang 2007). Moreover, the class of MOF described by the general formula M(bdc)(ted)$_{0.5}$ [M = Zn, Ni; bdc = 1,4-benzenedicarboxylate; ted = triethylenediamine], which has excellent absorbent properties for a variety of gases, only remains stable in humidity levels below 30% and fully decomposes in 60% humidity at room temperature (Liang et al. 2010). For some MOFs, the gas uptake capacity is greatly reduced after hydration, although their crystalline structure is maintained (Kizzie et al. 2011; Schoenecker et al. 2012). The most striking example is MOF-74, which consists of a 3D honeycomb lattice with 1D pores and contains a high density of coordinatively unsaturated metal sites, which are available for binding small molecules (Yu et al. 2012b). Thermal regeneration under a vacuum is sufficient to remove the adsorbed water molecules and restore the crystalline structure, even after hydration under high humidity (70%). However, only 16% of its original adsorption capacity is recovered after the hydration–regeneration process for Mg-MOF-74 (Sumida et al. 2012).

The knowledge concerning the MOF degradation mechanism by water vapor is rather poor. However, it was found that due to a relatively weak coordination bond between the metal SBU and the ligands, MOFs are susceptible to reaction with moisture; water could either displace the bound ligand, leading to the collapse of the MOF structure, or block the binding sites and prevent the adsorption of other target molecules (Sumida et al. 2012; Burtch et al. 2014; Canivet et al. 2014; Tan et al. 2015). Theoretical modeling, including molecular dynamics and quantum mechanical calculations, have offered a promising way to unveil the detailed mechanism of water interaction with building units of MOFs and possible reaction pathways (Greathouse and Allendorf 2006; Han et al. 2010; De Toni et al. 2012). In particular, Greathouse and Allendorf (2006) used empirical force fields and molecular dynamics to infer that the reaction is initiated by a direct attack of MOF-5 by water molecules in the possible way: oxygen of water replaces one of ZnO$_4$ tetrahedra, leading to the release of the ligand. This ligand-displacement mechanism was also proposed by De Toni et al. (2012) modeling-based first-principles calculations. It was found that at higher loading, a water cluster can be formed at the Zn$_4$O site and stabilize the water-bound state. This structure rapidly transforms into a linker-displaced state, where water has fully displaced one arm of a linker. Earlier, Han et al. (2010) used a reactive force field (ReaxFF) approach to perform a detailed study of the hydrolysis of IRMOF-1 (MOF-5), IRMOF-10, and MOF-74, and found that above a certain amount of loading, for instance, >4.5 wt.% H$_2$O in IR-MOF-1, the water molecules are randomly distributed in the MOF. Then, after interacting with the Zn–O moiety in an SBU, the H$_2$O molecule dissociates into H$^+$ and OH$^-$. The OH attaches to the Zn, while the H attaches to the oxygen atoms of the 1,3,5-benzenetricarboxylate (1,3,5-BTC). After this, the IR-MOF-1 loses its original structure and collapses. Similar results were found for MOF-74, where the MOF loses its original structure after the dissociation of the water molecules leads to the H$^+$ and OH$^-$ to bond to the O atoms of the linker and the metal centers, respectively. The two mechanisms (hydrolysis and ligand displacement) are summarized by the following equations (Low et al. 2009; Canivet et al. 2014):

$$ML + H_2O \rightarrow M(OH_2) + L \tag{11.1}$$

$$ML + H_2O \rightarrow M(OH) + LH \tag{11.2}$$

Both mechanisms were established from computational chemistry and confirmed experimentally (Low et al. 2009). The ligand displacement reaction involves the insertion of a water molecule into the M–O metal–ligand bond of the framework. This leads to the formation of a hydrated cation and to the release of a free ligand Equation (11.1). In contrast, during the hydrolysis reaction, the metal–ligand bond is broken and water dissociates to form a hydroxylated cation and a free protonated ligand Equation (11.2).

There are several approaches to resolving this stability problem. The main approach to tackling the stability of MOFs has focused on the modification of the linkers to have a hydrophobic character, such as in the case of FMOF-1 (Fluorous Metal-Organic Frameworks) (Yang et al. 2011) and ZIF (zeolitic imidazolate frameworks) materials (Kusgens et al. 2009; Lu and Hupp 2010). Experimental studies have shown that at present the most stable materials were MIL-110/-101, CPO-27 (also known as MOF-74), ZIF-8, and Zr-MOFs such as UiO-66

(Low et al. 2009; Schoenecker et al. 2012; Canivet et al. 2014). Unfortunately, the FMOF-1 and ZIF materials cannot be used in all devices. In addition, such materials have low sensitivity to air humidity. Another approach is based on the enhancing the metal–linker bond by changing the metal center, for example, using zirconium and nickel instead of copper or zinc (Cavka et al. 2008). It was found that the stability and decomposition reaction pathway of MOFs upon exposure to water vapor critically depends on their structure and the specific metal cation in the building units. For example, Ni(bdc)(ted)$_{0.5}$ remains stable under conditions where all other M(bdc)(ted)$_{0.5}$ materials are chemically attacked. However, although Zr-MOFs such as UiO-66 were found to be extremely stable in the presence of water, isostructural UiO-67, MOF-805,and MOF-806 made of biphenyldicarboxylate or bipyridinedicarboxylate ligands (instead of benzenedicarboxylate in UiO-66) are unstable in water vapor (De Coste et al. 2013b; Furukawa et al. 2014). The instability of the extended biaryl-based Zr-MOF derivatives was attributed to the torsional strain undergone by the crystal leading to its structural collapse. Similar adsorption isotherms found for Al-MIL-100 seem to indicate instability or at least partial collapse of the structure (Furukawa et al. 2014). Using powder XRD (PXRD), Dietzel and co-workers (2006) pointed out that exposure to oxygen from ambient air during water desorption/adsorption initiates the degradation of Ni/Mg-CPO-27.

An alternative approach explored is installing protective groups to prevent the access of water to the metal center, albeit on the cost of surface area (Jasuja et al. 2012; Makal et al. 2013). These approaches are described in more detail by Tan et al. (2015).

Other limitations of MOFs-based gas sensors are the followings (Allendorf et al. 2011; Kreno et al. 2012):

1. Expensive technology, that is, high cost;
2. Extremely slow growth rates. Typically requiring up to a day to produce films on the order of 100-nm thick;
3. Low thermal stability: most MOFs are already decomposed at temperatures around 350°C–400°C. Only some of them are stable up to temperatures above 500°C (Cavka et al. 2008);
4. The absence of large-scale manufacturing. Present methods of growing bulk samples of MOFs produce no more than ~1 g of material; and
5. Low selectivity. Though the potential selectivity of MOF materials for specific analytes, or classes of analytes, is substantial, but, as yet, is not highly developed.

## 11.3 PROSPECTS OF MOF FOR STRUCTURAL ENGINEERING OF HUMIDITY-SENSITIVE MATERIALS

As it is shown before, MOF is promising area of structural engineering of humidity sensitive materials. It was established that MOFs can be used as sacrificial templates to derive various porous nanomaterials by applying different

thermal and/or chemical treatments (Salunkhe et al. 2017). For instance, the heat treatment of MOFs in an inert atmosphere can produce highly porous carbons with high specific surface areas (~3000 m$^2$/g), given that chemical etching is conducted to remove the surface metal ions. Compared with carbonaceous materials, fabricated using conventional precursors, MOF-derived carbons often exhibit controllable porous architectures, pore volumes, and surface areas. Typically, the carbonization of MOFs to produce porous carbons can be performed directly without ("direct carbonization") or with secondary precursors ("indirect carbonization") (Liu et al. 2008; Kaneti et al. 2017). The former is more advantageous, as it is more facile and involves only a single calcination step. To date, various porous carbon materials with controlled morphologies from 0D to 3D have been successfully derived from the carbonization of MOFs, highlighting the versatility of MOFs as precursors (Xia et al. 2013; Pachfule et al. 2016). On the other hand, the direct-heat treatment of MOFs in the air can lead to their decomposition into their corresponding metal oxides (Salunkhe et al. 2015) (see Figure 11.15). As MOFs are made up of metal ion centers coordinated to organic linkers, their controlled heating in the air can give rise to porous metal oxides due to the oxidation of the metal ions and the release of gaseous CO$_2$ and NO$_2$ originating from the decomposition of organic linkers during the heat treatment. Typically, the derived metal-oxide nanostructures possess the structures with the same shape and porosity of the parent MOFs (with optimized heating conditions in nitrogen (N$_2$) and/or air atmosphere), as well as higher surface areas compared to metal oxides produced by other methods, often with readily developed porous architectures (Salunkhe et al. 2017). Moreover, the variations in annealing temperature and time can provide useful control over the composition, surface area, and pore-size distribution of MOF-derived metal oxides (Xia et al. 2015; Kaneti et al. 2017). Apart from pure metal-oxide nanostructures, porous metal-oxide nanocomposites can also be fabricated through the direct-heat treatments of bimetallic MOFs or core–shell MOF composites at high temperatures in the air or through a two-step annealing, first in N$_2$, then in air. The latter method is particularly beneficial for preserving

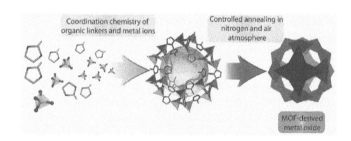

**FIGURE 11.15** Schematic illustration, showing the fabrication of porous metal-oxide nanostructures with large surface area and high porosity from metal-organic frameworks (MOFs) composed of metal ions and organic linkers as precursors through a two-step annealing in nitrogen (N$_2$) and air, respectively. (Reprinted with permission from Salunkhe, R.R. et al., *ACS Nano*, 11, 5293–5308, 2017. Copyright 2017 American Chemical Society.)

the original structure of the parent MOFs and, therefore, for achieving higher surface area and porosity. Thus, the highly tunable properties of MOF-derived metal oxides along with the possibilities to functionalizing them with various surface active dopants make them attractive candidates for next-generation humidity sensors.

# REFERENCES

Achmann S., Hagen G., Kita J., Malkowsky I.M., Kiener C., Moos R. (2009) Metal–organic frameworks for sensing applications in the gas phase. *Sensors* **9**, 1574–1589.

Aizenberg J., Black A.J., Whitesides G.M. (1999) Control of crystal nucleation by patterned self–assembled monolayers. *Nature* **398**, 495–498.

Allendorf M.D., Houk R.J.T., Andruszkiewicz L., Talin A.A., Pikarsky J., Choudhury A., et al. (2008) Stress–induced chemical detection using flexible metal–organic frameworks. *J. Am. Chem. Soc.* **130**, 14404–14410.

Allendorf M.D., Schwartzberg A., Stavila V., Talin A.A. (2011) A roadmap to implementing metal–organic frameworks in electronic devices: Challenges and critical directions. *Chem. Eur. J.* **17**, 11372–11388.

Allendorf M.D., Stavila V. (2015) Crystal engineering, structure–function relationships, and the future of metal–organic frameworks. *CrystEngComm.* **17**, 229–246.

Almeida Paz F.A., Klinowski J., Vilela S.M.F., Tome J.P.C., Cavaleiro J.A.S., Rocha J. (2012) Ligand design for functional metal–organic frameworks. *Chem. Soc. Rev.* **41**, 1088–1110.

Ameloot R., Stappers L., Fransaer J., Alaerts L., Sels B.F., De Vos D.E. (2009) Patterned growth of metal–organic framework coatings by electrochemical synthesis. *Chem. Mater.* 21 2580–2582.

Biemmi E., Darga A., Stock N., Bein T. (2008) Direct growth of $Cu_3(BTC)_2(H_2O)_3 \cdot xH_2O$ thin films on modified QCM–gold electrodes – Water sorption isotherms. *Micropor. Mesopor. Mat.* **114**, 380–386.

Biemmi E., Scherb C., Bein T. (2007) Oriented growth of the metal organic framework $Cu_3(BTC)_2(H_2O)_3 \cdot xH_2O$ tunable with functionalized self–assembled monolayers. *J. Am. Chem. Soc.* **129**, 8054–8055.

Burrows A.D., Frost C.G., Mahon M. F., Richardson C. (2008) Post–synthetic modification of tagged metal–organic frameworks. *Angew. Chem.* 47, 8610–8614.

Burtch N.C., Jasuja H., Walton K.S. (2014) Water stability and adsorption in metal–organic frameworks. *Chem. Rev.* **114**, 10575–10612.

Canivet J., Fateeva A., Guo Y., Coasne B., Farrusseng D. (2014) Water adsorption in MOFs: Fundamentals and applications. *Chem. Soc. Rev.* **43**, 5594–5617.

Cavka J.H., Jakobsen S., Olsbye U., Guillou N., Lamberti C., Bordiga S., Lillerud K.P. (2008) A new zirconium inorganic building brick forming metal organic frameworks with exceptional stability. *J. Am. Chem. Soc.* **130**, 13850–13851.

Chen B., Xiang S., Qian G. (2010a) Metal–organic frameworks with functional pores for recognition of small molecules. *Acc. Chem. Res.* **43**, 1115–1124.

Chen Y., Ma S. (2012) Microporous lanthanide metal–organic frameworks. *Rev. Inorg. Chem.* **32**(2–4), 81–100.

Chen Z., Xiang S., Arman H.D., Li P., Zhao D., Chen B. (2010b) A microporous metal–organic framework with immobilized–OH functional groups within the pore surfaces for selective gas sorption. *Eur. J. Inorg. Chem.* **24**, 3745–3749.

Chui S.S.-Y., Lo S.M.-F., Charmant J.P.H., Orpen A.G., Williams I.D. (1999) A chemically functionalizable nanoporous material $[Cu_3(TMA)_2(H_2O)_3]_n$. *Science* **283**, 1148–1150.

Cohen S.M. (2011) Postsynthetic methods for the functionalization of metal–organic frameworks. *Chem. Rev.* **112**, 970–1000.

Davydovskaya P., Pohle R., Tawil A., Fleischer M. (2012) Work function based gas sensing with Cu-BTC metal-organic framework for selective aldehyde detection. *Sens. Actuators B* **187**, 142–146.

De Coste J.B., Peterson G.W., Jasuja H., Glover T.G., Huang Y.-G., Walton K.S. (2013b) Stability and degradation mechanisms of metal–organic frameworks containing the $Zr_6O_4(OH)_4$ secondary building unit. *J. Mater. Chem. A* **1**, 5642–5650.

De Coste J.B., Peterson G.W., Schindler B.J., Killops K.L., Browe M.A., Mahle J.J. (2013a) The effect of water adsorption on the structure of the carboxylate containing metal–organic frameworks Cu–BTC, Mg–MOF–74, and UiO–66. *J. Mater. Chem.* **1**, 11922–11932.

De Toni M., Jonchiere R., Pullumbi P., Coudert F.-X., Fuchs A.H. (2012) How can a hydrophobic MOF be water-unstable? Insight into the hydration mechanism of IRMOFs. *Chem Phys Chem.* **13**, 3497–3503.

Demessence A., Horcajada P., Serre C., Boissiere C., Grosso D., Sanchez C., Ferey G. (2009) Elaboration and properties of hierarchically structured optical thin films of MIL -101(Cr). *Chem. Commun.* **46**, 7149–7151.

Deng H., Doonan C.J., Furukawa H., Ferreira R.B., Towne J., Knobler C.B., Wang B., Yaghi O.M. (2010) Multiple functional groups of varying ratios in metal–organic frameworks. *Science* **327**, 846–850.

Deng H., Grunder S., Cordova K.E., Valente C., Furukawa H., Hmadeh M., et al. (2012) Large–pore apertures in a series of metal–organic frameworks. *Science* **336**(6084), 1018–1023.

Deshpande R.K., Minnaar J.L., Telfer S.G. (2010) Thermolabile groups in metal–organic frameworks: Suppression of network interpenetration, post-synthetic cavity expansion, and protection of reactive functional groups. *Angew. Chem. Int. Ed.* **49**, 4598–4602.

Dhaka S., Kumar R., Deep A., Kurade M.B., Ji S.-W., Jeon B.-H. (2019) Metal–organic frameworks (MOFs) for the removal of emerging contaminants from aquatic environments. *Coord. Chem. Rev.* **380**, 330–352.

Dietzel P.D.C., Panella B., Hirscher M., Blom R., Fjellvag H. (2006) Hydrogen adsorption in a nickel based coordination polymer with open metal sites in the cylindrical cavities of the desolvated framework. *Chem. Commun.* **9**, 959–961.

Dimitrakakis C., Marmiroli B., Amenitsch H., Malfatti L., Innocenzi P., Grenci G., et al. (2012) Top–down patterning of zeolitic imidazolate framework composite thin films by deep X–ray lithography. *Chem. Commun.* **48**, 7483–7485.

Dinca M., Long J.R. (2005) Strong $H_2$ binding and selective gas adsorption within the microporous coordination Ssid $Mg_3(O_2C-C_{10}H_6-CO_2)_3$. *J. Am. Chem. Soc.* **127**, 9376–9377.

Doherty C.M., Grenci G., Ricco R., Mardel J.I., Reboul J., Furukawa S., et al. (2013) Combining UV lithography and an imprinting technique for patterning metal-organic frameworks. *Adv. Mater.* **25**, 4701–4705.

Douvali A., Tsipis A.C., Eliseeva S.V., Petoud S., Papaefstathiou G.S., Malliakas C.D., et al. (2015a) Turn–on luminescence sensing and real–time detection of traces of water in organic solvents by a flexible metal–organic framework. *Angew. Chem. Int. Ed.* **54**, 1651–1656.

Douvali A., Tsipis A.C., Eliseeva S.V., Petoud S., Papaefstathiou G.S., Malliakas C.D., et al. (2015b) Turn-on luminescence sensing and real-time detection of traces of water in organic solvents by a flexible metal–organic framework. *Angew. Chem.* **127**, 1671–1676.

Eddaoudi M., Kim J., Rosi N., Vodak D., Wachter J., Keeffe M.O., Yaghi O.M. (2002) Systematic design of pore size and functionality in isoreticular MOFs and their application in methane storage. *Science* **295**, 469–472.

Ellern I., Venkatasubramanian A., Lee J.H., Hesketh P.J., Stavilla V., Allendorf M.D., Robinson A.L. (2013) HKUST-1 coated piezoresistive microcantilever array for volatile organic compound sensing. *Micro Nano Lett.* **8**(11), 766–769.

Falcaro P., Buso D., Hill A.J., Doherty C.M. (2012) Patterning techniques for metal organic frameworks. *Adv. Mater.* **24**, 3153–3168.

Falcaro P., Ricco R., Doherty C.M., Liang K., Hill A.J., Styles M.J. (2014) MOF positioning technology and device fabrication. *Chem. Soc. Rev.* **43**, 5513–5560.

Fang Q.-R., Makal T.A., Young M.D., Zhou H.-C. (2010) Recent advances in the study of mesoporous metal–organic frameworks. *Commun. Inorg. Chem.* **31**, 165–195.

Farha O.K., Malliakas C.D., Kanatzidis M.G., Hupp J.T. (2009) Control over catenation in metal–organic frameworks via rational design of the organic building block. *J. Am. Chem. Soc.* **132**, 950–952.

Farha O.K., Mulfort K.L., Hupp J.T. (2008) An example of node–based post–assembly elaboration of a hydrogen–sorbing, metal–organic framework material. *Inorg. Chem.* **47**, 10223–10225.

Farrusseng D. (ed.) (2011) *Metal–Organic Frameworks: Applications from Catalysis to Gas Storage.* Wiley–VCH, New York.

Ferey C., Mellot–Draznieks C., Serre C., Millange F., Dutour J., Surble S., Margiolaki I.A (2005) Chromium terephthalate-based solid with unusually large pore volumes and surface area. *Science* **309**, 2040–2042.

Fletcher A.J., Thomas K.M., Rosseinsky M.J. (2005) Flexibility in metal–organic framework materials: Impact on sorption properties. *J. Solid State Chem.* **178**, 2491–2510.

Furukawa H., Cordova K.E., O'Keeffe M., Yaghi O.M. (2013) The chemistry and applications of metal–organic frameworks. *Science* **341**, 1230444.

Furukawa H., Gandara F., Zhang Y.-B., Jiang J., Queen W.L., Hudson M.R., Yaghi O.M. (2014) Water adsorption in porous metal–organic frameworks and related materials. *J. Am. Chem. Soc.* **136**, 4369–4381.

Furukawa H., Ko N., Go Y.B., Aratani N., Choi S.B., Choi E., et al. (2010) Ultrahigh porosity in metal–organic frameworks. *Science* **329**, 424–428.

Gadzikwa T., Farha O.K., Mulfort K.L., Hupp J.T., Nguyen S.T (2009) A Zn-based, pillared paddlewheel MOF containing free carboxylic acids via covalent post-synthesis elaboration. *Chem. Commun.* **25**, 3720–3722.

Garibay S.J., Cohen S.M. (2010) Isoreticular synthesis and modification of frameworks with the UiO–66 topology. *Chem. Commun.* **46**, 7700–7702.

Goeders K.M., Colton J.S., Bottomley L.A. (1997) Microcantilevers: Sensing chemical interactions via mechanical motion. *Chem. Rev.* **108**, 522–542.

Greathouse J.A., Allendorf M.D. (2006) The interaction of water with MOF–5 simulated by molecular dynamics. *J. Am. Chem. Soc.* **128**(40), 10678–10679.

Gu Z.-Y., Wang G., Yan X.-P. (2010) MOF–5 metal–organic framework as sorbent for in–field sampling and preconcentration in combination with thermal desorption GC/MS for determination of atmospheric formaldehyde. *Anal. Chem.* **82**, 1365–1370.

Gu Z.-Y., Yang C.-X., Chang N., Yan X.-P. (2012) Metal–organic frameworks for analytical chemistry: From sample collection to chromatographic separation. *Acc. Chem. Res.* **45**(5), 734–745.

Haldoupis E., Nair S., Sholl D.S. (2010) Efficient calculation of diffusion limitations in metal organic framework materials: A tool for identifying materials for kinetic separations. *J. Am. Chem. Soc.* **132**, 7528–7539.

Han S.S., Choi S.-H., van Duin A.C.T. (2010) Molecular dynamics simulations of stability of metal–organic frameworks against $H_2O$ using the $Re_{ax}FF$ reactive force field. *Chem. Commun.* **46**, 5713–5715.

He Y., Zhang Z., Xiang S., Fronczek F.R., Krishna R., Chen B. (2012) A microporous metal–organic framework for highly selective separation of acetylene, ethylene, and ethane from methane at room temperature. *Chem. Eur. J.* **18**, 613–619.

Heinke L., Tu M., Wannapaiboon S., Fischer R.A., Woll C. (2015) Surface–mounted metal–organic frameworks for applications in sensing and separation. *Micropor. Mesopor. Mat.* **216**, 200–215.

Hinterholzinger F.M., Ranft A., Feckl H., Bein T., Lotsch B.T. (2012) One–dimensional metal–organic framework photonic crystals used as platforms for vapor sensing. *J. Mater. Chem.* **22**, 10356–10362.

Horcajada P., Chalati T., Serre C., Gillet B., Sebrie C., Baati T., et al. (2010) Porous metal–organic–framework nanoscale carriers as a potential platform for drug delivery and imaging. *Nat. Mater.* **9**, 172–178.

Hoskins B.F., Robson R. (1989) Infinite polymeric frameworks consisting of three dimensionally linked rod–like segments. *J. Am. Chem. Soc.* **111**, 5962–5964.

Huang L.M., Wang H.T., Chen J.X., Wang Z.B., Sun J.Y., Zhao D.Y., Yan Y.S. (2003) Synthesis, morphology control, and properties of porous metal–organic coordination polymers. *Micropor. Mesopor. Mat.* **58**, 105–114.

Hwang Y.K., Hong D.-Y., Chang J.-S., Jhung S.H., Seo Y.-K., Kim J., et al. (2008) Amine grafting on coordinatively unsaturated metal centers of MOFs: Consequences for catalysis and metal encapsulation. *Angew. Chem. Int. Ed.* **47**, 4144–4148.

Ibarra I.A., Hesterberg W., Holliday B.J., Lynch V.M., Humphrey S.M. (2012) Gas sorption and luminescence properties of a terbium(III)-phosphine oxide coordination material with two-dimensional pore topology. *Dalton Trans.* **41**, 8003–8009.

Jasuja H., Huang Y.-G., Walton K.S. (2012) Adjusting the stability of metal–organic frameworks under humid conditions by ligand functionalization. *Langmuir* **28**, 16874–16880.

Kaneti Y.V., Tang J., Salunkhe R.R., Jiang X., Yu A., Wu K.C.W., Yamauchi Y. (2017) Nanoarchitectured design of porous materials and nanocomposites from metal–organic frameworks. *Adv. Mater.* **29**, 1604898.

Kaye S.S., Dailly A., Yaghi O.M., Long J.R. (2007) Impact of preparation and handling on the hydrogen storage properties of $Zn_4O(1,4-benzenedicarboxylate)_3$ (MOF-5). *J. Am. Chem. Soc.* **129**, 14176–14177.

Keskin S., Kizilel S. (2011) Biomedical applications of metal organic frameworks. *Ind. Eng. Chem. Res.* **50**, 1799–1812.

Khoshaman A.H., Bahreyni B. (2012) Application of metal organic framework crystals for sensing of volatile organic gases. *Sens. Actuators B* **162**, 114–119.

Kitagawa S., Kitaura R., Noro S. (2004) Functional porous coordination polymers. *Angew. Chem. Int. Ed.* **43**, 2334–2375.

Kitagawa S., Matsuda R. (2007) Chemistry of coordination space of porous coordination polymers. *Coord. Chem. Rev.* **251**, 2490–2509.

Kitagawa S., Noro S.-I., Nakamura T. (2006) Pore surface engineering of microporous coordination polymers. *Chem. Commun.* **7**, 701–707.

Kizzie A.C., Wong–Foy A.G., Matzger A.J. (2011) Effect of humidity on the performance of microporous coordination polymers as adsorbents for $CO_2$ capture. *Langmuir* **27**, 6368–6373.

Kleist W., Jutz F., Maciejewski M., Baiker A. (2009) Mixed-linker metal-organic frameworks as catalysts for the synthesis of propylene carbonate from propylene oxide and $CO_2$. *Eur. J. Inorg. Chem.* (24), 3552–3561.

Kondo M., Yoshitomi T., Seki K., Matsuzaka H., Kitagawa S. (1997) Three–dimensional framework with channeling cavities for small molecules: $\{[M_2(4,4'–bpy)_3(NO_3)_4]\cdot xH_2O\}_n$ (M = Co, Ni, Zn). *Angew. Chem. Int. Ed. Engl.* **36**, 1725–1727.)

Kreno L.E., Leong K., Farha O.K., Allendorf M., Van Duyne R.P., Hupp J.T. (2012) Metal–organic framework materials as chemical sensors. *Chem. Rev.* **112**, 1105–1125.

Kuppler R.J., Timmons D.J., Fang Q.-R., Li J.-R., Makal T.A., Young M.D., et al. (2009) Potential applications of metal–organic frameworks. *Coord. Chem. Rev.* **253**, 3042–3066.

Kusgens P., Rose M., Senkovska I., Frode H., Henschel A., Siegle S., Kaskel S. (2009) Characterization of metal–organic frameworks by water adsorption. *Micropor. Mesopor. Mater.* **120**, 325–330.

Lee T., Liu Z.X., Lee H.L. (2011) A biomimetic nose by microcrystals and oriented films of luminescent porous metal–organic frameworks. *Cryst. Growth Des.* **11**, 4146–4154.

Lei J., Qian R., Ling P., Cui L., Ju H. (2014) Design and sensing applications of metal–organic framework composites. *TrAC Trends Anal. Chem.* **58**, 71–78.

Li H., Eddaoudi M., O'Keeffe M., Yaghi O.M. (1999) Design and synthesis of an exceptionally stable and highly porous metal–organic framework. *Nature* **402**, 276–279.

Li J.R., Kuppler R.J., Zhou H.C. (2009) Selective gas adsorption and separation in metalorganic frameworks. *Chem. Soc. Rev.* **38**, 1477–1504.

Li W., Zhang Y., Li Q., Zhang G. (2015) Metal–organic framework composite membranes: Synthesis and separation applications. *Chem. Eng. Sci.* **135**, 232–257.

Li Y., Yang R.T. (2007) Gas adsorption and storage in metal–organic framework MOF–177. *Langmuir* **23**, 12937–12944.

Liang Z., Marshall M., Chaffee A.A. (2010) $CO_2$ adsorption, selectivity and water tolerance of pillared–layer metal organic frameworks. *Micropor. Mesopor. Mat.* **132**, 305–310.

Lin R.-B., Li F., Liu S.-Y., Qi X.-L., Zhang J.-P. Chen X.-M. (2013) A noble-metal-free porous coordination framework with exceptional sensing efficiency for oxygen. *Angew. Chem. Int. Ed.* **52**, 13429–13433.

Lin X., Blake A.J., Wilson C., Sun X.Z., Champness N.R., George M.W., et al. (2006) A porous framework polymer based on a Zinc(II) 4,4'–Bipyridine–2,6,2',6'–tetracarboxylate: Synthesis, structure, and "zeolite–like" behaviors. *J. Am. Chem. Soc.* **128**, 10745–10753.

Liu B., Shioyama H., Akita T., Xu Q. (2008) Metal–organic framework as a template for porous carbon synthesis. *J. Am. Chem. Soc.* **130**, 5390–5391.

Liu J., Culp J.T., Natesakhawat S., Bockrath B.C., Zande B., Sankar S.G., et al. (2007) Experimental and theoretical studies of gas adsorption in $Cu_3(BTC)_2$: An effective activation procedure. *J. Phys. Chem. C* **111**, 9305–9313.

Liu J., Sun F., Zhang F., Wang Z., Zhang R., Wang C., Qiu S. (2011) *In situ* growth of continuous thin metal–organic framework film for capacitive humidity sensing. *J. Mater. Chem.* **21**, 3775–3778.

Liu Y., Ban Y., Yang W. (2017) Microstructural engineering and architectural design of metal–organic framework membranes. *Adv. Mater.* **29**, 1606949.

Low J.J., Benin A.I., Jakubczak P., Abrahamian J.F., Faheem S.A., Willis R.R. (2009) Virtual high throughput screening confirmed experimentally: Porous coordination polymer hydration. *J. Am. Chem. Soc.* **131**, 15834–15842.

Lu G., Farha O.K., Zhang W., Huo F., Hupp J.T. (2012) Engineering ZIF-8 thin films for hybrid MOF-based devices. *Adv. Mater.* **24**, 3970–3974.

Lu G., Hupp J.T. (2010) Metal–organic frameworks as sensors: A ZIF–8 based Fabry–Pérot device as a selective sensor for chemical vapors and gases. *J. Am. Chem. Soc.* **132**, 7832–7833.

Lu Z.-Z., Zhang R., Li Y.-Z., Guo Z.-J., Zheng H.-G. (2011) Solvatochromic behavior of a nanotubular metal–organic framework for sensing small molecules. *J. Am. Chem. Soc.* **133**, 4172–4174.

MacGillivray L.R. (ed.) (2010) *Metal–Organic Frameworks: Design and Application.* Wiley, Hoboken, NJ.

Makal T.A., Wang X., Zhou H.-C. (2013) Tuning the moisture and thermal stability of metal–organic frameworks through incorporation of pendant hydrophobic groups. *Cryst. Growth Des.* **13**, 4760–4768.

Masoomi M.Y., Morsali A. (2012) Applications of metal–organic coordination polymers as precursors for preparation of nano–materials. *Coord. Chem. Rev.* **256**, 2921–2943.

Meek S.T., Greathouse J.A., Allendorf M.D. (2011) Metal–organic frameworks: A rapidly growing class of versatile nanoporous materials. *Adv. Mater.* **23**, 249–267.

Mintova S., Mo S.Y., Bein T. (2001) Humidity sensing with ultra-thin LTA–type molecular sieve films grown on piezoelectric devices. *Chem. Mater.* **13**, 901–905.

Pachfule P., Shinde D., Majumder M., Xu Q. (2016) Fabrication of carbon nanorods and graphene nanoribbons from a metal–organic framework. *Nat. Chem.* **8**, 718–724.

Pichon A., Lazuen–Garay A., James S.L. (2006) Solvent–free synthesis of a microporous metal organic framework. *CrystEngComm.* **8**, 211–214.

Prestipino C., Regli L., Vitillo J.G., Bonino F., Damin A., Lamberti C., et al. (2006) Local structure of framework Cu(II) in HKUST–1metal-organic framework: Spectroscopic characterization upon activation and interaction with adsorbates. *Chem. Mater.* **18**, 1337–1346.

Pullen S., Clever G.H. (2018) Mixed-ligand metal–organic frameworks and heteroleptic coordination cages as multifunctional scaffolds—A comparison. *Acc. Chem. Res.* **51**(12), 3052–3064.

Ranocchiari M., Lothschütz C., Grolimund D., van Bokhoven J.A. (2012) Single–atom active sites on metal–organic frameworks. *Proc. R. Soc. A* **468**(2143), 1985–1999.

Robinson A.L., Stavila V., Zeitler T.R., White M.I., Thornberg S.M., Greathouse J.A., Allendorf M.D. (2012) Ultrasensitive humidity detection using metalorganic framework–coated microsensors. *Anal. Chem.* **84**, 7043–7051.

Rosseinsky M.J. (2004) Recent developments in metal–organic framework chemistry: Design, discovery, permanent porosity and flexibility. *Micropor. Mesopor. Mat.* **73**, 15–30.

Rowsell J.L.C., Yaghi O. (2004) Metal–organic frameworks: A new class of porous materials. *Micropor. Mesopor. Mat.* **73**, 3–14.

Salunkhe R.R., Kaneti Y.V., Yamauchi Y. (2017) Metal–organic framework–derived nanoporous metal oxides toward supercapacitor applications: Progress and prospects. *ACS Nano* **11**(6), 5293–5308.

Salunkhe R.R., Tang J., Kamachi Y., Nakato T., Kim J.H., Yamauchi Y. (2015) Asymmetric supercapacitors using 3D nanoporous carbon and cobalt oxide electrodes synthesized from a single metal–organic framework. *ACS Nano* **9**, 6288–6296.

Schoedel A., Scherb C., Bein T. (2010) Oriented nanoscale films of metal–organic frameworks by room–temperature gel–layer synthesis. *Angew. Chem. Int. Ed.* **49**, 7225–7228.

Schoenecker P.M., Carson C.G., Jasuja H., Flemming C.J.J., Walton K.S. (2012) Effect of water adsorption on retention of structure and surface area of metal–organic frameworks. *Ind. Eng. Chem. Res.* **51**, 6513–6519.

Serre C., Mellot-Draznieks C., Surblé S., Audebrand N., Filinchuk Y., Férey G. (2007) Role of solvent-host interactions that lead to very large swelling of hybrid frameworks. *Science*, **315**, 1828–1831.

Shah M., McCarthy M.C., Sachdeva S., Lee A.K., Jeong H.-K. (2012) Current status of metal–organic framework membranes for gas separations: Promises and challenges. *Ind. Eng. Chem. Res.* **51**, 2179–2199.

Shekhah O., Liu J., Fischer R.A., Woll C. (2011) MOF thin films: Existing and future applications. *Chem. Soc. Rev.* **40**, 1081–1106.

Shekhah O., Wang H., Kowarik S., Schreiber F., Paulus M., Tolan M., et al. (2007) Step–by–step route for the synthesis of metal–organic frameworks. *J. Am. Chem. Soc.* **129**, 15118–15119.

Skoulidas A.I., Sholl D.S. (2005) Self–diffusion and transport diffusion of light gases in metal–organic framework materials assessed using molecular dynamics simulations. *J. Phys. Chem. B* **109**, 15760–15768.

Song F., Wang C., Falkowski J.M., Ma L., Lin W. (2010) Isoreticular chiral metal–organic frameworks for asymmetric alkene epoxidation: Tuning catalytic activity by controlling framework catenation and varying open channel sizes. *J. Am. Chem. Soc.* **132**, 15390–15398.

Stallmach F., Groger S., Kunzel V., Karger J., Yaghi O.M., Hesse M., Muller U. (2006) NMR studies on the diffusion of hydrocarbons on the metal-organic framework material MOF-5. *Angew. Chem. Int. Ed.* **45**, 2123–2126.

Stavila V., Talin A.A., Allendorf M.D. (2014) MOF–based electronic and opto–electronic devices. *Chem. Soc. Rev.* **43**, 5994–6010.

Sudik A.C., Millward A.R., Ockwig N.W., Côté A.P., Kim J., Yaghi O.M. (2005) Design, synthesis, structure, and gas ($N_2$, Ar, $CO_2$, $CH_4$, and $H_2$) sorption properties of porous metal–organic tetrahedral and heterocuboidal polyhedra. *J. Am. Chem. Soc.* **127**, 7110–7118.

Sumida K., Rogow D.L., Mason J.A., McDonald T.M., Bloch E.D., Herm Z.R., et al. (2012) Carbon dioxide capture in metal–organic frameworks. *Chem. Rev.* **112**, 724–781.

Sun F., Yin Z., Wang Q.Q., Sun D., Zeng M.H., Kurmoo M. (2013) Tandem postsynthetic modification of a metal–organic framework by thermal elimination and subsequent bromination: Effects on absorption properties and photoluminescence. *Angew. Chem. Int. Ed.* **52**, 4538–4543.

Tan K., Nijem N., Gao Y., Zuluaga S., Li J., Thonhauser T., Chabal Y.J. (2015) Water interactions in metal organic frameworks. *CrystEngComm.* **17**, 247–260.

Tanabe K.K., Cohen S.M. (2009) Engineering a metal–organic framework catalyst by using postsynthetic modification. *Angew. Chem. Int. Ed.* **48**, 7424–7427.

Thomas K.M. (2009) Adsorption and desorption of hydrogen on metal-organic framework materials for storage applications: Comparison with other nanoporous materials. *Dalton Trans.* (9), 1487–1505.

Vaidhyanathan R., Iremonger S.S., Shimizu G.K.H., Boyd P.G., Alavi S., Woo T.K. (2010) Direct observation and quantification of $CO_2$ binding within an amine–functionalized nanoporous solid. *Science* **330**, 650–653.

Venkatasubramanian A., Lee J.-H., Houk R.J., Allendorf M.D., Nair S. (2010) Characterization of HKUST–1 crystals and their application to MEMS microcantilever array sensors. *ECS Trans.* **33**(8), 229–238.

Venkatasubramanian A., Lee J.-H., Stavila V., Robinson A., Allendorf M.D., Hesketh P.J. (2012) MOF @ MEMS: Design optimization for high sensitivity chemical detection. *Sens. Actuators B* **168**, 256–262.

Wales D.J., Grand J., Ting, V.P. Burke R.D., Edler K.J., Bowen C.R., Mintova S., Burrows A.D. (2015) Gas sensing using porous materials for automotive applications. *Chem. Soc. Rev.* **44**, 4290–4321.

Wang C.-C., Yang C.-C., Chung W.-C., Lee G.-H., Ho M.-L., Yu Y.-C., et al. (2011) A new coordination polymer exhibiting unique 2D hydrogen-bonded ($H_2O$)$_{16}$ ring formation and water-dependent luminescence properties. *Chem. Eur. J.* **17**, 9232–9241.

Wang L., Reis A., Seifert A., Philippi T., Ernst S., Jia M., Thiel W.R. (2009) A simple procedure for the covalent grafting of triphenylphosphine ligands on silica: Application in the palladium catalyzed Suzuki reaction. *Dalton Trans.* **17**, 3315–3320.

Wang Q.M., Shen D.M., Bulow M., Lau M.L., Deng S.G., Fitch F.R., et al. (2002) Metallo–organic molecular sieve for gas separation and purification. *Micropor. Mesopor. Mat.* **55**, 217–230.

Wang Z., Cohen S.M. (2007) Post–synthetic covalent modification of a neutral metal–organic framework. *J. Am. Chem. Soc.* **129**, 12368–12369.

Xia W., Mahmood A., Zou R., Xu Q. (2015) Metal–organic frameworks and their derived nanostructures for electrochemical energy storage and conversion. *Energy Environ. Sci.* **8**, 1837–1866.

Xia W., Qiu B., Xia D., Zou R. (2013) Facile preparation of hierarchically porous carbons from metal–organic gels and their application in energy storage. *Sci. Rep.* **3**, 1935.

Yaghi O.M., Li H.L. (1995) Hydrothermal synthesis of a metal organic framework containing large rectangular channels. *J. Am. Chem. Soc.* **117**, 10401–10402.

Yaghi O.M., O'Keeffe M.O., Ockwig N.W., Chae H.K., Eddaoudi M., Kim J. (2003) Reticular synthesis and the design of new materials. *Nature* **423**, 705–714.

Yang C., Kaipa U., Mather Q.Z., Wang X., Nesterov V., Venero A.F., Omary M.A. (2011) Fluorous metal–organic frameworks with superior adsorption and hydrophobic properties toward oil spill cleanup and hydrocarbon storage. *J. Am. Chem. Soc.* **133**, 18094–18097.

Yi F.-Y., Chen D., Wu M.-K., Han L., Jiang H.-J. (2016) Chemical sensors based on metal–organic frameworks. *Chem Plus Chem.* **81**, 675–690.

Yu K., Kiesling K., Schmidt J.R. (2012b) Trace flue gas contaminants poison coordinatively unsaturated metal–organic frameworks: Implications for $CO_2$ adsorption and separation. *J. Phys. Chem. C* **116**, 20480–20488.

Yu Y., Ma J.-P., Dong Y.-B. (2012a) Luminescent humidity sensors based on porous Ln$^{3+}$–MOFs. *CrystEngComm.* **14**, 7157–7160.

Yu Y., Zhang X.-M., Ma J.-P., Liu Q.-K., Wang P., Dong Y.-B. (2014) Cu(I)–MOF: Naked-eye colorimetric sensor for humidity and formaldehyde in single–crystal-to-single–crystal fashion. *Chem. Commun.* **50**, 1444–1446.

Zacher D., Baunemann A., Hermes S., Fischer R.A. (2007) Deposition of microcrystalline [Cu$_3$(btc)$_2$] and [Zn$_2$(bdc)$_2$(dabco)] at alumina and silica surfaces modified with patterned self–assembled organic monolayers: Evidence of surface selective and oriented growth. *J. Mater. Chem.* **17**, 2785–2792.

Zacher D., Schmid R., Woll C., Fischer R.A. (2011) Surface chemistry of metal–organic frameworks at the liquid–solid interface. *Angew. Chem. Int. Ed.* **50**, 176–199.

Zacher D., Shekhah O., Woll C., Fischer R.A. (2009) Thin films of metal–organic frameworks. *Chem. Soc. Rev.* **38**, 1418–1429.

Zeitler T.R., Allendorf M.D., Greathouse J.A. (2011) Grand canonical Monte Carlo simulation of low–pressure methane adsorption in nanoporous framework materials for sensing applications. *J. Phys. Chem. C* **116**, 3492–3502.

Zhang X., Wang W., Hu Z., Wang G., Uvdal K. (2015) Coordination polymers for energy transfer: Preparations, properties, sensing applications, and perspectives. *Coord. Chem. Rev.* **284**, 206–235.

Zhang Y., Chen Y., Zhang Y., Cong H., Fu B., Wen S., Ruan S. (2013) A novel humidity sensor based on NH$_2$–MIL–125(Ti) metal organic framework with high responsiveness. *J. Nanopart. Res.* **15**, 2014.

Zhou H.-C., Long J.R., Yaghi O.M. (2012) Introduction to metal–organic frameworks. *Chem. Rev.* **112**, 673–674.

Zhou H.C., Long J.R., Yaghi O.M. (2012) Introduction to metal–organic frameworks. *Chem. Rev.* **112**, 673–674.

Zybaylo O., Shekhah O., Wang H., Tafipolsky M., Schmid R., Johannsmann D., Woll C. (2010) A novel method to measure diffusion coefficients in porous metal–organic frameworks. *Phys. Chem. Chem. Phys.* **12**, 8092–8097.

# 12 Supramolecular Materials

Supramolecular materials are architectures consisting of molecules that are able to self-assemble into larger constructs (Vögtle 1996; Lehn 2007; Kurth 2008). Polymers as well as small molecules are able to form self-assembled structures. In nature, self-assembly is essential in many processes (see, e.g., the cell membrane or the formation of cell adhesions), and frequently the assembly and disassembly of supramolecular materials is controlled enzymatically. One should note that the design and creation of regularly shaped nanoscale objects, which can serve as building blocks of supramolecular materials, is an extremely important goal in materials science. At that, it is important that the form and/or function of the final supramolecular material differs from that of the original building blocks. Controlling the spatial organization of such objects could deliver materials with defined chemical and topographical nanoscale features, potentially leading to novel physical properties or integrating several properties in a single material, making them useful for various applications, such as drug delivery, gas separation, and chemical sensing (Stupp et al. 1998; Palacios et al. 2007; Kurth 2008; Hahn and Gianneschi 2011; Sharma and Cragg 2011; Kumar et al. 2017).

The driving force behind the self-assembly of the building blocks can be based either on the phase segregation due to hydrophilic/hydrophobic interactions (e.g., block copolymers) (Amir et al. 2009) or on the formation of well-ordered secondary structures such as β-sheets through hydrogen-bonds (Mutter and Vuileumier 1989; Mutter et al. 1991). A combination of both mechanisms can be accomplished using conjugates of polymers with oligo- or polypeptides (Zubin et al. 2002; Castelletto et al. 2010; De Graaf et al. 2012).

The most famous representatives of the supramolecular materials are metal-organic frameworks (MOFs), zeolites, cavitands, and MCs (Vögtle 1996; Biedermann and De Cola 2018). MOFs and zeolites were covered in previous chapters. Therefore, in this chapter we will give a brief description of other representatives of supramolecular materials.

## 12.1 CAVITANDS

### 12.1.1 CHARACTERIZATION

Cavitands, synthetic organic compounds with a container shape, are extremely interesting and versatile molecular receptors (Cram 1983). Possible structures of cavitands are shown in Figure 12.1. The specific interactions between the cavitand and guest molecules are mainly based on their bucket-like conformation. The cavity of the cavitand allows it to engage in host-guest chemistry with guest molecules of a complementary shape and size. As it is seen in Figure 12.1, cavitands have

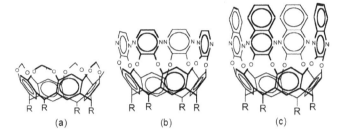

**FIGURE 12.1** Three cavitands: (a) methylene-bridged (Me-Cav), (b) pyrazine-bridged (Py-Cav), and (c) quinoxalines-bridged (Qx-Cav). (Reprinted from *Sens. Actuators B*, 97, Feresenbet, E. et al., Optical sensing of the selective interaction of aromatic vapors with cavitands, 211–220, Copyright 2004, with permission from Elsevier.)

different shapes and sizes of their cavities, which are easy to modify, and therefore they have different complexing abilities with target molecules.

Phosphonate cavitands represent one class of molecular receptors that have been studied in detail to understand the factors that lead to selective binding using alcohols as model analyte vapors.

It was established that these factors include: (1) simultaneous hydrogen bonding with a P=O group and CH-π interactions with the π-basic cavity; (2) a rigid cavity that provides a permanent free volume for the analyte around the inward-facing P=O groups, which is essential for effective hydrogen bonding; and (3) a network of energetically equivalent hydrogen-bonding sites available to the analyte (Melegari et al. 2008). The nonspecific dispersion interactions can be much stronger than the specific interactions and can depend on the chain length of sensed alcohols and their concentration (Pinalli et al. 2004). Thus, the main specific interactions responsible for recognition are H-bonding, CH-π, and dipole–dipole interactions (Hartmann et al. 1994; Dickert and Schuster 1995). This strategy for selectivity-enhancement is fundamentally different from that used in other chemically selective coatings (e.g., polymers) that rely on the solubility of the targets with the coating layer.

Examples of cavitands include cyclodextrins, calixarenes, pillarrenes, and cucurburils. However, in gas sensors, usually cyclodextrins and calixarenes are used. These compounds are a relatively new family of ion receptors that are receiving increasing attention due to their ease of synthesis and multiple sites for structural modification. For example, the calyx[*n*]arenes are a class of cyclooligomers or cyclic supramolecules based on a hydroxyalkylation product of a phenol and an aldehyde. The *n* in calix[*n*]arenes represents the number of aryl units in the macrocyclic ring, which are linked to each other through methylene bridges. For example, a calix[4]arene has four units in the ring and a calix[6]arene has six.

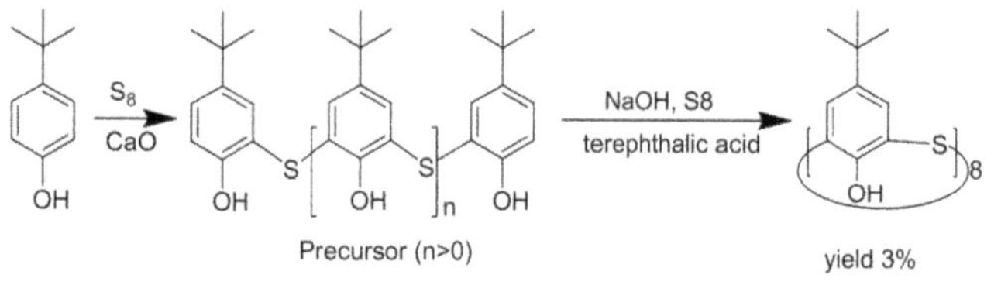

**FIGURE 12.2**  Synthesis of *p-tert*-butyl-octathiacalix[8]arene.

A great number of calyx[*n*]arenes varying in the shape and diameter of the nanocavity (cylinder, truncated cone) as well as in the type of peripheral functional groups have been developed (Parker 1996). One of examples of calixarenes is shown in Figure 12.2. It was found that calixarenes form cavities of various diameters and are able to capture metal ions and organic molecules into these cavities ("host-guest" complexation) (Gutsche 1989). Forster et al. (1991) have shown that the nature of the ionophoric activity displayed by derivatized calixarenes is strongly dependent on the cavity size, which can be conveniently altered by varying the reaction conditions and the number of phenyl units (*n* = 4–20) in the macrocycle. Taking into account the synthetic flexibility of the calyx[*n*] arenes allowing direct control the number of phenolic units present within a single supramolecule and the nature and length of the spacer units, we obtain possibility to manage ionophoric activity of calixarenes over this cavity size. More detailed information about synthesis, properties, and application of calix[*n*]arenes in chemical sensing one can find in Gutsche (1989), Chawla et al. (2011), Sharma and Cragg (2011), and Gardiner (2016).

Cyclodextrins (CDs) possess the same features. CDs are naturally occurring cyclic oligosaccharides, which have a rigid torus shape with an inner hydrophobic cavity and an outer hydrophilic one (Szetjli 1998). Similar to calix[*n*]arenes, they possess a remarkable ability to form inclusion complexes with host molecules. The inner diameter of the cavity, hydrophobic properties and the weak van der Waals forces are the factors that decide the bonding between CD molecules and guest molecules. Moreover, sulfated β-cyclodextrin, in particular, shows a high solubility in water and an anionic behavior in aqueous solution, so it can be electrochemically incorporated in a polymer matrix during an oxidative process. The structure of β-cyclodextrin is shown in Figure 12.3.

### 12.1.2 CAVITANDS AS A MATERIAL FOR GAS AND ELECTROCHEMICAL SENSORS

The previously mentioned features of calixarenes and cyclodextrins reveal wide possibilities of using cavitand films as sensitive layers for various kinds of sensors (Sharma and Cragg 2011; Kumar et al. 2017). It is believed that sensing films based on cavitands promise responses with improved selectivity due to the presence of organic hosts with enforced cavities whose rigid organization vastly enhances their ability to selectively bind guests (Cram 1983; Schierbaum et al. 1994). The experiment showed that various derivatives with differing selectivities for various gas molecules can be prepared via functionally

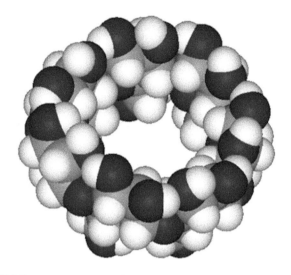

**FIGURE 12.3**  Model of β-cyclodextrin. (Modified from https://commons.wikimedia.org/wiki/File:Beta-cyclodextrin3D.png and https://helix.northwestern.edu/blog/2015/06/take-whiff-chemistry.)

modifying either the upper and/or lower rims (Diamond and Nolan 2001; Rudkevich 2007). For example, calixarene derivatives can be incorporated into plasticized poly(vinylchloride) membranes to produce functioning ion-selective electrodes (ISEs) and fluorescence-based matrix (Forster et al. 1991; Ludwig and Dzung 2002) in electrochemical and optical gas sensors. It was found that calixarenes are also useful building platforms in the design of multichromophoric systems in which photoinduced phenomena (electron charge and proton transfers, excimer formation and resonance energy transfer) are controlled by ions. The outstanding selectivities offered by calixarene-based ligands is of major interest for sensing ions (Valeur and Leray 2007; Sharma and Cragg 2011). This possibility was realized in $Cl_2$, $NH_3$, HCl, and $NO_2$ calixarene-based gas sensors (Lavrik et al. 1996; Rudkevich 2007; Ohira et al. 2009). In particular, Grady et al. (1997) designed calixarene-based optical sensors for gaseous ammonia detection in the fish samples. The optical detector was based on a calix[4] arene to which a nitrophenylazophenol chromophore was attached. Similarly, it has been observed that the calixarene derivatives respond very prominently to chloroform vapors (Wang et al. 2002). The specific and selective formation of a colored complex of alkylated calixarenes has been utilized to develop a fiber-optic-based colorimetric $NO_2$ sensor (Ohira et al. 2009). A long period grating (LPG) modified with a mesoporous film infused with a calixarene (CA[8] or CA[4]) as a functional compound was also employed for the detection of individual volatile organic compounds (VOCs) and their mixtures (Hromadka et al. 2017). When the response in individual VOCs was investigated, an initial response of the sensor was observed in less than 30 s, and the sensors exhibited difference responses to individual VOCs and mixtures. Maffei et al. (2011) established that fluorescent phosphonate cavitands are also good basis for designing selective optical sensing of alcohols vapors. Moreover, they demonstrated that it is possible to achieve high selectivity in chemical vapor sensing by harnessing the binding specificity of a cavitand receptor.

Shenoy (2005) has shown that, by using a real-time, label-free, optical technique called surface plasmon resonance (SPR), refractive index changes induced by analyte-cavitand interactions provide selective signals for sensitive chemical-vapor detection as well. For sensor fabrication, cavitand solutions (0.38 mM) in chloroform were spin-coated onto surface plasmon resonance substrates (50-nm thick gold-coated cover glass). Feresenbet et al. (2004) established that the methylene-bridged cavitands (Me-Cav) (see Figure 9.1) with shallow cavities do not complex aromatic vapors, and the pyrazine-bridged cavitands (Py-Cav) show intermediate selectivity, whereas the quinoxalines-bridged cavitands (Qx-Cav) with the deepest cavities show the best selectivity for aromatic vapors. A comparison with polymer coatings polyepichlorhydrin (PECH) and polyisobutylene (PIB) shows that cavitands have higher selectivity despite the fact that the polymer coatings are more than twice the thickness of the spin-coated cavitand films Shenoy (2005).

However, the application of calixarenes and other cavitands such as the phosphorus-bridged ones or β-cyclodextrin in biosensors (Oh et al. 2005) and low-temperature gas sensors such as quartz crystal microbalance (QCM)- and surface acoustic wave (SAW)-based sensors, is also possible (Grady et al. 1997; Dickert et al. 1997; Rosler et al. 1998; Li and Ma 2000; Mermer et al. 2012; Kumar et al. 2017). For example, Hartmann et al. (1996) and Kalchenko et al. (2002) reported about cavitand-based QCM sensors array aimed for detection of volatile organic vapors including pentane, heptane, benzene, chloroform, methanol, acetonitrile, tetrachloroethylene, diethylamine, ethanol, and nitrobenzene (see Figure 12.4). Dickert et al. (1997) have shown that in combination with aliphatic spacers, this material fulfills the required demands of high sensitivity and short response times, and the detection of solvents in the gas phase to 2.5 ppm can be realized. Moreover, according to Dickert et al. (1997), the molecular

structure of calixarenes could be modified to tune density and porosity of the coating to the special requirements of SAW and QCM devices. For example, this could be realized on the one hand by using various aldehydes to create different basic calix[4]resorcinarene cavities with variable spacers, such as alkyl chains or alkyl thiolates for self-assembling on gold surfaces and, on the other hand, via bridging of two resorcine molecules of the cavitand by forming cyclic ethers (Davis and Stirling 1995). Sensitivity and selectivity of the host molecules could be varied within a large range in this way. Pinalli et al. (2004) have shown that phosphate and phosphonate cavitands are also sensitive to ethanol, methanol, and benzene vapors. The same result was observed for β-cyclodextrin-based QCM sensors (Wang et al. 2001; Palaniappan et al. 2006). In this study, alkenyl-β-CD was used as the sensing material because guest molecules such as benzene bind tightly to β-CD when compared to α-CD or γ-CD. In addition, it was found that the presence of alkenyl groups in β-CD ensures a covalent attachment of β-CD to the silica matrix used (Palaniappan et al. 2006). Clathrate materials that crystallize in phases with channels or cavities containing solvent molecules can also be used as sensing materials (Ehlen et al. 1993; Finklea et al. 1998; Yakimova et al. 2008; Cha et al. 2009). It was shown that these materials were ~100 times more sensitive to VOCs than polymer-coated thickness shear mode (TSM) devices at low concentrations (Finklea et al. 1998).

### 12.1.3 Cavitand-Based Humidity Sensors

Reports related to the development of humidity sensors based on calixarene molecules also exist in the literature (Okur et al. 2010; Su et al. 2013). Most of these are sensors based on the QCM technique, using calix[4]arene as a humidity-sensitive material (see Table 12.1).

Typical characteristics of calix[4]arene-based QCM humidity sensors are shown in Figure 12.5. It is seen that the adsorption and desorption between 22% and 75% relative humidity (RH) are nearly linear, but at higher RH values the frequency is increasing quickly. In addition, it was established that calix[4]arene films showed extremely fast adsorption response against to humidity changes. This feature is most probably sourced from both carboxylate and sulfonate groups. Water molecules can easily make strong complex with calix[4]arene from both upper and lower rims, which are functionalized eight carboxylate and sulfonate groups. As a result, the desorption process is slower than the rate of adsorption, and the large hysteresis appears on the humidity-sensitive characteristics. To describe the adsorption and desorption kinetics of gas-vapor molecules onto organic or inorganic films, the Langmuir adsorption isotherm model (Gregg and Sing 1967) is frequently used (Okur et al. 2010).

Su et al. (2013) have shown that the functionalization of calix[4]arenes can improve the surface properties of calix[4]arenes and the sensitivity of calix[4]arenes to humidity. It was established that the 5-(4′-nitrophenyl)azo-25,26,27-tribenzoyloxy-28-hydroxycalix[4]arene (NTBHC) thin film

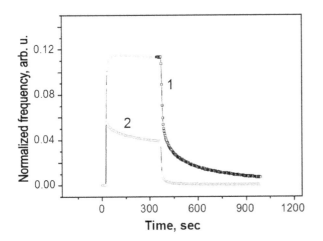

**FIGURE 12.4** Experimental response of calixarene-based QCM sensors to injection of acetone vapor: sensitive layers are: 1—calix[4] arene, 2—calix[4]arene containing single O=P(OPri)Ph functional group. (Reprinted from Kalchenko, V.I. et al., *Mater. Sci. (Poland)*, 20, 73–88, 2002. Published by Wrocław University of Science and Technology as open access.)

**TABLE 12.1**

**Examples of Calixarene-Based Humidity Sensors**

| Sensor Type | Sensing Material | Working Range | Sensitivity | Hysteresis | References |
|---|---|---|---|---|---|
| QCM | Calix[4]arene derivatives with both carboxylate and sulfonate groups | 22–84%RH | ~3.3 Hz/%RH | 18%RH | Okur et al. (2010) |
| | Calix[4]arene derivatives with both benzoyloxy and nitrophenylazo groups | 20–90%RH | ~7 Hz/%RH | 2%RH | Su et al. (2013) |
| R, C | 5,11,17,23-tetra-ter-butyl-25,27-dehydrazinamidcarbonilmetoxy-26,28--dehydroxy-calix[4]aren -CNTs | 17–85%RH | $\Delta R \sim 190\ \Omega$ | NA | Özbek et al. (2013) |
| | Calix[n]arene sulfonic acids | 0–90%RH | $\sigma \cdot 10^{-5} - 2 \cdot 10^{-1}$ S/cm | NA | Shmygleva et al. (2017) |

R—resistive, C—conductometric.

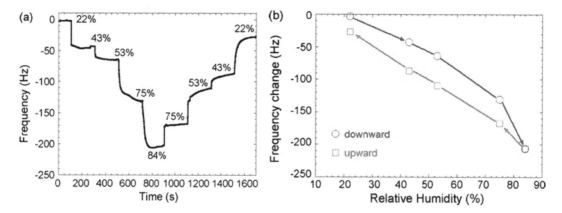

**FIGURE 12.5** The response of calix[4]arene-based QCM sensors to the change of relative humidity in the range of 22% and 84% RH: (a) – F = f(RH) and (b) – $\Delta$F = f(RH). Calix[4]arene films with 300 nm thickness were obtained by a drop-cast technique. (Reprinted from *Sens. Actuators B*, 145, Okur, S. et al., Humidity adsorption kinetics of water soluble calix[4]arene derivatives measured using QCM technique, 93–97, Copyright 2010 with permission from Elsevier.)

that was coated on the QCM electrode had high sensitivity (especially at low RH levels), acceptable linearity of the relationship between the frequency change [$\Delta$f (Hz)] and RH in the range of 20%–90%, and negligible hysteresis (within 2% RH), and a long-term stability measured at 25°C. $NO_2$, CO, $NH_3$, $H_2$, and $C_2H_4$ gases had also no significant cross-sensitivity effects on the humidity sensor at 60% RH.

As can be seen from Table 12.1, resistive humidity sensors can also be developed based on calixarenes. For these purposes, one can use calix[n]arene sulfonic acids (Shmygleva et al. 2017), as well as calix[n]arene-CNTs composites (Özbek et al. 2013). Shmygleva et al. (2017) have shown that calix(n) arene sulfonic acids hydrates had high proton conductivity up to $10^{-1}$ S/cm. The effect of humidity on conductivity in the calix[n]arene sulfonic acid films is shown in Figure 12.6. As in other resistive humidity sensors, water molecules are involved in a proton transport in calixarenes due to the formation and destruction of bound proton forms. It was found that all known sulfonates are hydrates, which can contain up to 16 water molecules per molecule of calixarene (Shmygleva et al. 2017).

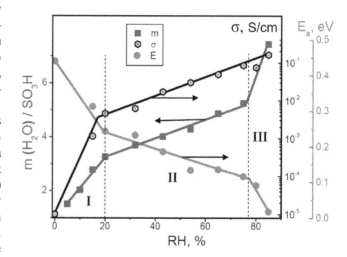

**FIGURE 12.6** Water content (squares), proton conductivity (hexagons), and activation energy (circles) dependencies on RH for calix[6]arene sulfonic acids at 25°C. (Reprinted from Shmygleva, L. et al., *Int. J. Electrochem. Sci.*, 12, 4056–4076, 2017. Published by ESG publisher as open access.)

### 12.1.4 LIMITATIONS

However, despite the seemingly undeniable advantages of cavitands for the development of humidity sensors, their real use is very much limited. The experiment showed that the use of compounds with a container shape does not give any noticeable improvement in the characteristics of humidity sensors in comparison with conventional polymers. Moreover, cavitand-based devices have advantages and disadvantages similar to advantages and disadvantages of polymer-based devices. In particular, cavitand-based sensors are low-temperature devices. For example, the melting temperatures of calix[$n$]arenes are varied usually from 200°C to 450°C. Experiments carried out with cavitand-based QCM sensors have shown that the increase of operating temperature is accompanied by a strong decrease in the sensor signal (see Figure 12.7). Ferrari et al. (2004) believe that observed temperature dependence of response is due to the partition coefficient behavior. Another characteristic feature of calixarenes and tetrathiacalixarenes, which limits technological possibilities of sensor fabrication, is their insolubility in water as well as in aqueous bases and their very low solubility in organic solvents. The solubility of calixarenes can be substantially modified only via derivatization (Asfari and Vicens 1988).

It is worth noting that the response in cavitand-based gas sensors is usually slow. Response and recovery times for various sensors vary in the range from 10 sec to 10 min (Ferrari et al. 2004). Liu et al. (2005) established that the recovery during the $N_2$ purging cycle can be even irreversible at room temperature.

In addition, we have to recognize that despite all attempts, a fully specific supramolecular sensor, in which nonspecific interactions and competitive binding by undesired analytes

have been eliminated, has not been yet obtained. Contrary to expectations, the mere presence of a cavity in the molecules, which forms the sensitive layer, does not guarantee sensing selectivity. This fact has been clearly demonstrated by Grate et al. (1996) some years ago by comparing the selectivity patterns of polymers with those of cavitands, cyclodextrins, and cyclophanes toward a set of analytes. It was found that selectivity response patterns of cavitands to small molecules can be similar to common amorphous and crystalline polymers (Grate et al. 1996), indicating that cavitand sensors can respond not only to molecules with an ideal fit in the cavity but also to sorbed molecules occupying both intracavity and intercavity sites (Dickert and Schuster 1995; Grate 2000; Pirondini and Dalcanale 2007; Schneider 2009; Kumar et al. 2017). This means that cavitand-based sensors are not selective. In particular, it was found that calixarene-based sensors are sensitive to ammonia gas, volatile amines, aromatic nitrogenous molecules, acidic gases (HCl, $NO_x$, $CO_2$), organic vapors, gaseous aromatic compounds, toxic gases, $H_2$, etc. (Kumar et al. 2017). Hromadka et al. (2017) also reported that calixarene-based sensors cannot discriminate between individual VOCs. Grate et al. (1996) have shown that the binding and selectivity in the examples cited are governed primarily by general dispersion interactions and not by specific oriented interactions, which could lead to a molecular recognition. Nevertheless, the presence of a preorganized cavity in cavitands does promise an advantage in sensitivity compared to amorphous polymers, especially if applied to the sensor in multilayers (Grate 2000). Moreover, Pirondini and Dalcanale (2007) believe that a truly specific receptor for a given molecule can be designed and prepared. According to Pirondini and Dalcanale (2007), two different strategies can be envisioned to avoid nonspecific interactions. From the receptor side, the challenge is to design a host incorporating a suitable transduction group (i.e., a chromophore), which can be activated exclusively by the molecular recognition event. Alternatively, the collective behavior of self-organizing materials can be tapped to amplify the molecular recognition phenomena at the macroscopic level.

As in the case of other humidity-sensitive materials, the sensitivity and kinetics of the response of cavitand-based sensors strongly depends on the morphology of the formed layer (Betti 2007). Technological difficulties are an additional limitation in the application of calixarenes. One should note that calixarenes with required properties are difficult to produce in required quantities, because random polymerization occurs inside of complex mixtures of linear and cyclic oligomers with different numbers of repeating units. Finely tuned starting materials and tightly controlled reaction conditions are necessary for reproducible synthesis. It was established that in such conditions prominent calix[n]arenes (called major calixarenes, when $n = 4, 6$, or 8) can be prepared in excellent yields with high purity. However, the current level of technology provides the obtaining of these materials at multigram scale only, which is not enough for mass use. Calix[$n$]arenes containing five and seven aryl units (called minor calixarenes), have been obtained in low yields. Calix[$n$]arenes containing

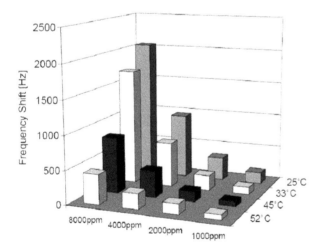

**FIGURE 12.7** Steady-state frequency shift (absolute value) versus toluene concentrations at different substrate temperatures for a (Me-Cav)-based QCM sensor. (Reprinted with *Sens. Actuators B*, 103, Ferrari, M. et al., Cavitand-coated PZT resonant piezo-layer sensors: Properties, structure, and comparison with QCM sensors at different temperatures under exposure to organic vapors, 240–246, Copyright 2004 with permission from Elsevier.)

more than eight aryl units (referred to as higher calixarenes), have been discovered only recently and have not yet become readily available.

## 12.2 METALLO-COMPLEXES

### 12.2.1 CHARACTERIZATION

Since transition metals are able to establish reversible interactions with other atoms, they can be exploited to form metallo-complexes (MCs), which may act as receptors for different types of analytes. It was shown that macrocyclic compounds, such as crown ethers, cyclodextrins, calixarenes, cyclophanes, cavitands, cryptands, spherands, carcerands, cyclopeptides, and other structurally related species can be incorporated in these metallosupramolecules (Atwood et al. 1996). Metal phthalocyanines (MPc), aromatic macrocyclic compounds, which have semiconductor properties, also can be referred to this class of materials (Schollhorn et al. 1998; Fietzek et al. 1999; Ceyhan et al. 2006). Phthalocyanine (Pc) ligands can coordinate with various metal ions, and the central metals can interact with small molecules through a coordination bond (see Figure 12.8). The phthalocyanines are stable up to 450°C; at this temperature, the materials decompose but do not melt. Because of their high-decomposition temperature, they can be vacuum evaporated to produce thin films (Karimov et al. 2008; Aziz et al. 2010). Porphyrins molecules can be also assembled into nanostructures using several methods (Kosal et al. 2002; Medforth et al. 2009). Several reviews are available, which analyze the performance of porphyrins and cyanines in gas sensing (Di Natale et al. 1998, 2007; Ozturk et al. 2009; Nardis et al. 2011; Trogler 2012). In metalloporphyrins, metallophthalocyanines, and related macrocycles, gas sensing is accomplished either by π-stacking of the gas into organized layers of the flat macrocycles or by gas coordination to the metal center without the cavity inclusion. In particular, metalloporphyrins provide several mechanisms of gas response, including hydrogen bonding, polarization, polarity interactions, metal-center-coordination interactions, and molecular arrangements (Di Natale et al. 2007; Nardis et al. 2011).

Harbeck et al. (2011) and Sen et al. (2011) have shown that metal complexes of *vic*-dioximes can be characterized as candidate materials for VOC sensing with sorption-based chemical gas sensors as well. The *vic*-dioximes are known to form stable complexes with a variety of metals such as $Ni^{2+}$, $Pd^{2+}$, $Cu^{2+}$, $Co^{2+}$, $Zn^{2+}$, or $Cd^{2+}$. Furthermore, they can be also modified easily in the substituent structure.

Metallo-dendrimers are representatives of MCs as well (Albrecht van Koten 1999; Hwang et al. 2007; Satija et al. 2011). A possible configuration of metallo-dendrimers is shown in Figure 12.9. The metal in these compounds can be positioned at the infrastructure's core, in connectors between branching centers or act as terminal groups. Metal centers can also be integrated as structural auxiliary points within the dendritic framework by their incorporation after dendrimer construction. This means that there are numerous combinations in MCs. It is important that all these supramolecules possess a specific host–guest behavior with different luminescent or electronic properties, which can be exploited for sensing purposes. At that, the necessary components can be incorporated into MCs according to the needs of the analyte to generate analytically useful and observable signals.

### 12.2.2 ADVANTAGES FOR GAS-SENSOR APPLICATIONS

No doubt that MCs are preferable for cation recognition (De Silva et al. 1997; Bergonzi et al. 1998; Bren 2001). However, the application of MCs for detection of gas molecule using different transduction principles such as optical (Del Bianco et al. 1993; De Silva et al. 1997; Albrecht van Koten 1999; Elosua et al. 2006), and mechanical (Nieuwenhuizen and Harteveld 1994; Benkstein et al. 2000; Kimura et al. 2010) is also promising. Studies have shown that the MC conductometric gas sensors based mainly on MPcs can also be developed. Detailed analysis of MPc-based conductometric gas sensors one can find in the papers published by Wright (1989), Schollhorn et al. (1998), and Germain et al. (1998). We need to say that MPcs, which are organic semiconductors (Wright 1989), can also be used to design gas-sensitive thin-film transistors. For explanation of MPc gas sensitivity, the approach based on gas molecule interaction with the π-electron network of the Pcs is usually used (Roisin et al. 1992). For example, in the case of VOC

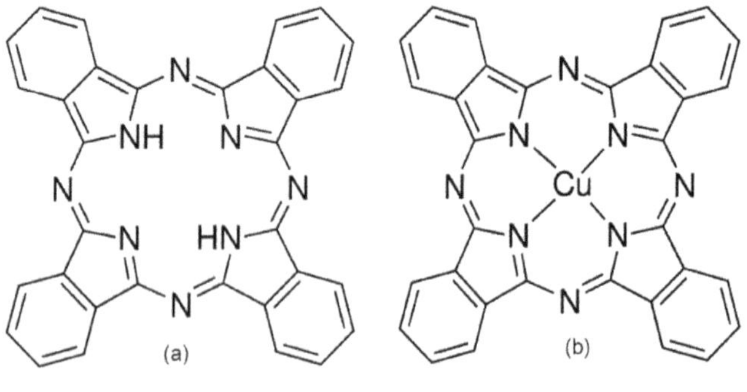

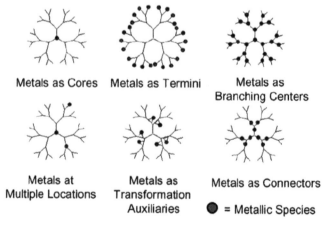

Metals as Cores    Metals as Termini    Metals as Branching Centers

Metals at Multiple Locations    Metals as Transformation Auxiliaries    Metals as Connectors

⬤ = Metallic Species

**FIGURE 12.9** The different roles metals can play in metallo-dendrimers. (Hwang, S.-H. et al., *New J. Chem.*, 31, 1192–1217. Reproduced by permission of The Royal Society of Chemistry.)

**FIGURE 12.8** Structural formula of (a) 29H,31H-Phthalocyanine ($H_2Pc$), and (b) Copper(II) 2,3-naphthalocyanine (CuPc).

sensors, the VOCs would play an electron acceptor role. When the VOC molecules interact with the π-electron network of the phthalocyanines, it causes the transfer of an electron from the phthalocyanine ring to the VOC molecule (Schollhorn et al. 1998). Thus, the induced positive holes on the film surface give rise to an increase in the *p*-type conductivity of the film. It was found, however, that MCs that contain transition metal centers are especially convenient for fluorescence sensors, as they undergo fast and kinetically uncomplicated one-electron redox changes (Bergonzi et al. 1998). Moreover, transition metals tend to participate in luminescence quenching of both the electron transfer, eT, and energy transfer, ET, varieties (see Figure 12.10). Finally, the potential of the metal-centered redox couple can be modulated by varying the nature of the hosting coordinative environment. One possible example of MCs applied in luminescence- and QCM-based VOCs sensors is shown in Figure 12.11.

Testing of MC-based gas sensors has shown that first, the MCs are generally more sensitive and responsive on electro- and photochemical stimuli compared to metal-free organic macrocyclic molecules (Kumar et al. 2008). According to Albrecht van Koten (1999), a proper choice of the (transition) metal center and the corresponding ligand array is crucial for gas-sensor design since the metal center generally exhibits a high selectivity for particular substances. This enables the preparation of detector materials of a high selectivity. Second, various functionalities can easily be introduced into MC structure by employing functional ligands. For example, the experiment showed that a wide range of chromophores, fluorophores, and redox-active functionalities have been successfully incorporated into supramolecular frameworks (Holliday and Mirkin 2001). Furthermore, ligand tuning in organometallic complexes

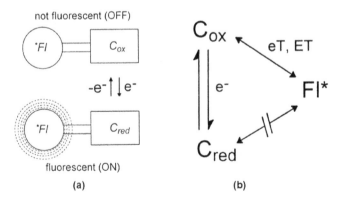

**FIGURE 12.10** (a) The multicomponent approach to the design of a redox switch of fluorescence. Switch efficiency requires that the control unit C in its oxidized form, $C_{ox}$, quenches the proximate photo-excited fluorophore Fl* and the reduced form $C_{red}$ does not (OFF/ON switch). The other favourable ON/OFF situation can be obtained when $C_{red}$ quenches Fl* and $C_{ox}$ does not. (b) The basis of an OFF/ON redox switch of fluorescence. Either an energy transfer (ET) or an electron transfer (eT) mechanism can be responsible for the quenching of the light emitting fragment Fl*, in a multicomponent redox unit-spacer-fluorophore system C~Fl. (Reprinted with *Coord. Chem. Rev.*, 170, Bergonzi, R. et al., Molecular switches of fluorescence operating through metal centred redox couples, 31–46, Copyright 1998 with permission from Elsevier.)

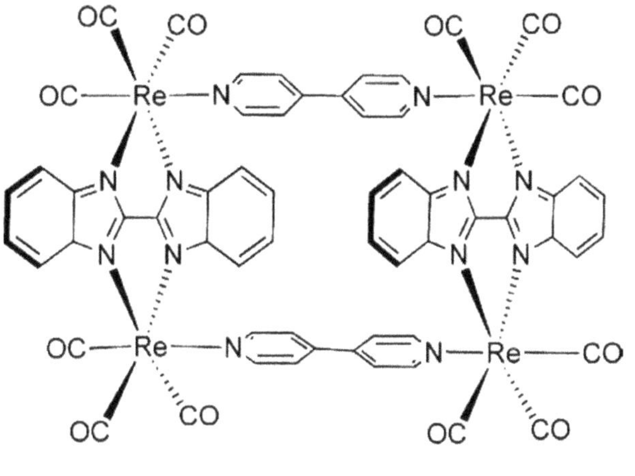

**FIGURE 12.11** Metallo-complex sensitive to volatile organic compounds (VOCs). (Reprinted with *Coord. Chem. Rev.*, 252, Kumar, A. et al., Directed assembly metallocyclic supramolecular systems for molecular recognition and chemical sensing, 922–939, Copyright 2008 with permission from Elsevier.)

can be used as a method to modify and optimize selectivity and sensitivity of the detector units through electronic and steric effects. Ligand fragments that have minor or no influence on the sensor activity may also serve as potential anchoring points to immobilize the sensor sites on an appropriate support, for example, on dendrimers or polymers (Albrecht van Koten 1999). In addition, they provide sites for the introduction of signal transduction and amplification devices. Through a suitable fixation of the sensing unit, recovery of the sensors by common separation techniques is facilitated. Thus, one can assume that due to specific interactions in MCs, an MC can provide new opportunities to develop novel devices with improved sensitivity and selectivity (Jimenez-Cadena et al. 2007).

In addition, MCs may play the role of catalysts (Hwang et al. 2007). In particular, an MPc and its derivatives are shown to catalyze reactions like the reduction of oxygen and $CO_2$ (Ceyhan et al. 2006). The catalytic behavior of MPc complexes is associated with the redox activity of the central metal, which undergoes oxidation and reduction. The values of the oxidation or reduction potentials of the central metal in MPc thus strongly influence the catalytic behavior of the complexes, which is also essential for the design of more efficient gas sensors. One can find a more detailed description of MCs in review papers by De Silva et al. (1997), Leininger et al. (2000), Donilfo and Hupp (2001), Holliday and Mirkin (2001), Kumar et al. (2008).

Detailed discussions of gas adsorption on MPCs one can find in the review by Wright (1989). Analyzing the response of QCM sensor with an MPc-sensing layer, Kimura et al. (2010) have shown that the incorporation of sterically protected MPcs within the polymer brushes on QCMs allows them to work as molecular receptors to recognize the chemical properties of VOC vapors based on their size and polarity. The selectivity and sensitivity of the sensing layer on QCMs can be tuned by modifying both molecular recognition receptors, which depend on the structure of peripheral substituents and the central metals and polymer brushes. In particular, they found that sulfonated CoPc shows a stronger affinity for acetone, ethanol, and pyridine vapors relative to other MPcs (NiPc, CuPc, ZnPc). These coordinative VOC vapors can form a weak coordination interaction with the central metal, and the formation of coordination bonds enhances the selectivity of sensors.

However, we need to recognize that indicated sensors in comparison with conventional metal-oxide gas sensors have low sensitivity to many specific gases and slow response and recovery processes even at $T_{oper} = 180°C$. Sadaoka et al. (1990a, 1990b) found that post-deposition annealing at 330°C could shorten the response time. But this improvement is not cardinal. The failure to return to baseline after the VOC vapors or other gas molecules are removed originates from the strong bonding of the gas molecules on the surface. This appreciably limits an application of these materials in the devices for the sensor market.

### 12.2.3 Humidity-Sensor Applications of Metallo-Complexes

Testing of the gas-sensitive characteristics of the MCs showed that the humidity of the surrounding atmosphere has a significant impact on its parameters, especially at low temperatures (Belghachi and Collins 1990). This means that MCs can be used as humidity-sensitive materials. Examples of such use are given in Table 12.2.

It can be seen that when developing humidity sensors, optical-, QCM- and resistive-based sensors were tested. However, the main research focused on the development of the conductometric

sensors based on organic semiconductors, such as MPcs. It was established that phthalocyanines are stable up to temperatures around 400°C and have high sensitivity to humidity and gases (Snow and Barger 1989; Belghachi and Collins 1990; Řeboun and Hamáček 2008; Kubersky et al. 2010). The advantage of phthalocyanines is their modifiability (McKeown 1998). In particular, intrinsic conductivity of the material can be changed by changing of the central atom. Over 70 metals can be used as the central metal atom (Al, Cu, Ni or other metals) in humidity-sensitive materials. It was also established that the other features of phthalocyanines like dissolvability and sensitivity can be changed through substituents. The benzene ring of phthalocyanines can be modified by means of several different substituents. It was established that the sulfonated groups cause better sensitivity to humidity and the material is easily dissolvable in water after substitution (Kubersky et al. 2010). Sulfonated phthalocyanine are hydrophilic materials. Water molecules are absorbed within layers of phthalocyanines simultaneously with the dissociation of sulfonated groups. These phenomena cause an increase in conductivity. At the same time, sulfamidic phthalocyanines are rather hydrophobic materials. The sulfamidic groups cause better sensitivity to other gases, and the material is not dissolvable in water after substitution of the benzene ring (Skothem and Reynolds 2007; Jiang 2009).

As a rule, humidity sensors are based on an organic sensitive layer that is applied on the substrate with interdigital electrodes. Easy depositing of thin organic layers is the main advantage when comparing to conventional materials.

### TABLE 12.2
### Metallo-Complexes Used as Humidity Sensitive Materials

| Sensor Type | Metallo-Complex | References |
|---|---|---|
| Resistive, conductive, impedance | Metal phtalocyanines | Snow and Barger (1989); Řeboun and Hamáček (2008) |
| | Nickel phthalocyanine, NiPc | Belghachi and Collins (1990); Ahmad et al. (2017) |
| | Cobalt tetrasulfonated phthalocyanine, CoTsPc | Centurion et al. (2012) |
| | Aluminum phthalocyanine chloride (AlPcCl) | Chani et al. (2012) |
| | The octafluoro copper phthalocyanine, $Cu(F_8Pc)$ | Wannebroucq et al. (2018) |
| | Amine terminated polyamidoamine dendrimer—AuNPs | Su and Shiu (2012) |
| | Coronene tetracarboxylate-dodecyl methyl viologen | Mogera et al. (2014) |
| | Ni(II) ions and bis(2,9-dimethyl-1,10-phenanthroline) | Pandey et al. (2014) |
| | DMBHPET | Azmer et al. (2015) |
| | SIM | Yan et al. (2017) |
| Capacitance | Copper phthalocyanine, CuPc | Karimov et al. (2008) |
| | Nickel phthalocyanine, NiPc | Shah et al. (2008); Ahmad et al. (2017) |
| | Vanadyl phthalocyanine (VOPc) | Aziz et al. (2010, 2012) |
| | PEPC+NiPc+ZnO | Karimov et al. (2010) |
| | PEPC+NiPc+$Cu_2O$ | Ahmad et al. (2013) |
| | DMBHPET | Azmer et al. (2015) |
| | TMBHPET | Al-sehemi (2015) |
| Optical | H-bonded cholesteric liquid crystal (CLC) polymer film | Herzer et al. (2012) |
| QCM | Amine terminated polyamidoamine dendrimer—AuNPs | Su and Tzou (2012) |

DMBHPET—(E)-2-(4-(2-(3,4-dimethoxybenzeylidene)hydrazinyl)phenyl)ethane-1,1,2-tricarbonitrile; PEPC—poly-$N$-epoxypropylcarbazole; SIM is prepared by ionic self-assembly in aqueous solutions of electroactive 2,2'-azino-bis(3-ethylbenzothiazoline-6-sulfonic acid) (ABTS) and 1,10-bis(3-methyl-imidazolium-1-yl) decane ($C_{10}(mim)_2$); TMBHPET—(E)-2-{4-[2-(3,4,5-trimethoxybenzylidene) hydrazinyl]phenyl}ethylene-1,1,2-tricarbonitrile.

The organic material can be deposited by using of various methods. The most usual methods are spin coating and dip coating. In this case, the thicknesses of these layers were in the range of 50–100 nm. As a result of testing, Belghachi and Collins (1990) have found that NiPc demonstrated much greater conductivity response to humidity than the other metal Pcs studied. Kubersky et al. (2010) also believed that the sulfonated NiPc is the best material for humidity sensors even though it is less sensitive to humidity (see Figure 12.12). Its impedance-humidity characteristic was exponential, and the maximum value of impedance was lower than 1 MΩ. The values of the impedance did not change within the large range of amplitude. This means that measurement and evaluation are simpler when comparing to the other materials. In addition, it was established that NiPc thin layers had almost the same response measured in the temperature range of 20°C–50°C. At the same time, Chani et al. (2012) believe that aluminum phthalocyanine chloride (AlPcCl) is a material optimal for conductometric humidity sensors. They established that for the change in RH from 20% to 92%, the change in resistance was ~2 × 10³ and 1.3 × 10³ times, respectively, for the sensors having 50- and 100-nm thick films. Moreover, they found that the 1 h annealing of these samples at 100°C resulted in the increase in average sensitivity up to 30%–40%. It was also observed that annealing results in minimization of hysteresis and the reduction of the recovery time ($\tau_{rec}$) up to 63% and 70% in the sensors with 50- and 100-nm thick organic film, respectively, while the response ($\tau_{res}$) time was 10 s for both the sensors.

As for the other sensors, there are significantly fewer results, which does not allow for any comparative analysis. We can only conclude that phthalocyanines demonstrate good performance as humidity sensitive material. Karimov et al. (2008) reported that the capacitance of copper phthalocyanine (CuPc)-based humidity sensors increased from 15 to 17 pF at 35% RH to 2,450–4,300 at RH = 92%, that is, the capacitances of the hygrometers increased by 160–250 times (on average

≈200 time) with increasing humidity. Approximately the same sensitivity was performed by NiPc (Ahmad et al. 2017) and vanadium (IV) oxide phthalocyanine (VOPc)-based (Aziz et al. 2010, 2012) humidity sensors. It is observed that the capacitance of the samples increased to 250, 225, and 110 times for vanadyl phthalocyanine films, having thicknesses of 50, 100, and 150 nm, respectively, with the RH being varied over 0%–95% range. At that, as seen in Figure 12.13, the maximum capacity change was observed at RH > 50%. Sensors developed on the basis of organic-inorganic composite were also highly sensitive. (PEPC+NiPc+Cu₂O). Ahmad et al. (2013) reported that the maximum sensitivity ~31.6 pF/%RH at 100 Hz in capacitive mode of operation has been attained. When the humidity level was increased from 40% to 97%, the response time for these sensors was 13 seconds and when the RH was decreased from 97% to 40%, the recovery time was 15 seconds. For comparison, the TMBHPET ((E)-2-{4-[2-(3,4,5-trimethoxybenzylidene)hydrazinyl]phenyl}ethyl-ene-1,1,2-tricarbonitrile) organic-dye-based humidity sensor developed by Al-sehemi (2015) had a maximum sensitivity ~46 fF/%RH at 1 KHz measured in the range of 20%–90%RH. According to Aziz et al. (2010, 2012), observed capacitance response in metal phthalocyanine-based sensors is attributed to polarization due to the adsorption of water molecules and the charge carriers transfer.

However, it was found that the phthalocyanine films after a long-time humidity exposure (RH~90%) are close to quasi-liquid phase, they are spread on the electrode system, and can be washed out from the substrate. To achieve the appropriate stability of the organic film, Řeboun and Hamáček (2008) proposed to mix water-soluble phthalocyanines with an acrylate resin. They found that with the concentration of acrylate resin in phthalocyanine 1:3 or higher, the long-term stability of electrical parameters is sufficient for humidity changes detection in the range from 20% to 90%.

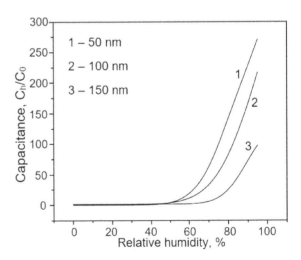

FIGURE 12.13 Capacitance versus RH relationships for the Ag/VOPc/Ag capacitive sensors for the VOPc thicknesses of (1) 50, (2) 100, and (3) 150 nm. (Reprinted with permission from Aziz, F. et al., *J. Semicond.*, 31, 114002, 2010. Copyright 2010 Chinese Institute of Electronics & the Institute of Semiconductors of the Chinese Academy of Sciences.)

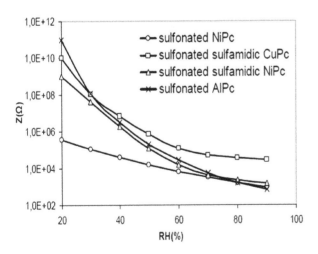

FIGURE 12.12 Impedance-humidity characteristics. (From Kubersky, P. et al., Phthalocyanine layers for humidity detection, in: *Proceedings of 33rd International Spring Seminar on Electronics Technology, ISSE 2010*, May 12–16, Warsaw, Poland, pp. 14–17.)

### 12.2.4 Approaches to Improving Gas-Sensor Parameters and Limitations

According to Schollhorn et al. (1998), the selectivity and sensitivity of metal complexes to gases could be improved by molecular engineering, which can be made in three ways:

1. *The change of the macrocycle structure: porphyrines and phthalocyanines.* No structure could be found yet that fulfills at once the requirements of high sensitivity to a gas, selectivity to a gas toward interferents, and good reversibility. However, owing to their properties (good chemical stability for instance), phthalocyanines appear as good materials for gas-sensing purposes.
2. *The change of central metal ion M.* The exchange of the metal in MPcs results in variations of sensitivity to gases that can be assigned both to changes in oxidation potentials and to morphological modifications. Unfortunately, the selectivity toward gases cannot be improved significantly this way.
3. *The change of peripheral substituents of the macrocycle.* Addition of electron-withdrawing (resp. -donating) substituents on a metallo phthalocyanine macrocycle decreases (resp. increases) the electron density of the conjugated cycle and increases (resp. decreases) the oxidation potential of the macrocycle. The sensitivity to reducing gases (resp. oxidizing gases) should be increased by substitution of the cycle by electron-withdrawing (resp. -donating) groups.

### 12.2.5 Limitations

As we previously mentioned, in principle, MCs are promising materials for designing selective gas sensors, since each receptor can be designed to interact with a specific analyte. However, despite the indisputable advantages of metal complexes for sensor applications, their real use, especially in the development of humidity sensors, is very limited. Therefore, till now, such humidity sensors were not developed. This is primarily due to the fact that the synthesis of these MCs can be expensive. It is known that as the scale and complexity of target molecules increases, the stepwise synthesis of large discrete supermolecules from molecular building blocks becomes increasingly difficult, often low-yielding, and specific to only a few approaches. The synthesis of large symmetric structures is long and requires careful consideration of the entropic and enthalpic costs involved. This process includes the time required for the linear, stepwise synthesis of complex macromolecules and molecular assemblies composed of hundreds or even thousands of subunits, as well as a decrease of the overall yield in multistep reactions. The purification of the larger molecules also still poses some problems (Leininger et al. 2000). This means that the reproducibility of sensor parameters can be low. Advantages and disadvantages of various approaches, such as the directional-bonding approach, symmetry-interaction approach, and weak-link synthetic strategy, usually used for

synthesis of MCs, were reviewed in detail by Holliday and Mirkin (2001). Low sensitivity, despite the declared ability to achieve high selectivity, is also a limiting factor. As was established, MCs are sensitive to $NH_3$ (Wannebroucq et al. 2018), $SO_2$ (Albrecht and van Koten 1999), $NO_2$ (Belghachi and Collins 1990), toxic chemicals (Ishihara et al. 2016), and various vapors (Dzugan et al. 2014). In addition, the problem related to sensor stability also is present for discussed sensors. For example, Elosua et al. (2006) reported about 3 months' stability only for optical-gas sensors designed on the base of the (Au-Ag) $[Au_2Ag_2(C_6F_5)_4(C_6H_5N)_2]$ complex.

## REFERENCES

Ahmad Z., Touati F., Zafar Q., Mutbar Shah M. (2017) Integrated capacitive and resistive humidity transduction via surface type nickel phthalocyanine based sensor. *Int. J. Electrochem. Sci.* **12**, 3012–3019.

Ahmad Z., Zafar Q., Sulaiman K., Akram R., Karimov K.S. (2013) A humidity sensing organic–inorganic composite for environmental monitoring. *Sensors* **13**, 3615–3624.

Albrecht M., van Koten G. (1999) Gas sensor materials based on metallodendrimers. *Adv. Mater.* **11**(2), 171–174.

Al-Sehemi A. (2015) Sensing performance optimization by tuning electrical properties of organic (D-π-A) dye-based humidity sensor. In: *Proceedings of 16th Tetrahedron Symposium*, Asia Edition, Shanghai, China.

Amir R.J., Zhong S., Pochan D.J., Hawker C.J. (2009) Enzymatically triggered self-assembly of block copolymers. *J. Am. Chem. Soc.* **131**(39), 13949–13951.

Asfari Z., Vicens J. (1988) Preparation of series of calix[6]arenes and calix[8]arenes derived from *p-n*-alkylphenols. *Tetrahedron Lett.* **29**(22), 2659–2660.

Atwood J.L., Davis J.E.D., McNicol D.D., Vogtle F., Lehn J.-M. (eds.) (1996) *Comprehensive Supramolecular Chemistry*, Vols. 1–11, Pergamon, Oxford, UK.

Aziz F., Hassan Sayyad M., Sulaiman K., Majlis B., Karimov K.S., Ahmad Z., Sugandi G. (2012) Influence of humidity conditions on the capacitive and resistive response of an Al/VOPc/Pt co-planar humidity sensor. *Meas. Sci. Technol.* **23**, 014001.

Aziz F., Sayyad M., Karimov K.S., Saleem M., Ahmad Z., Khan S.M. (2010) Characterization of vanadyl phthalocyanine based surface-type capacitive humidity sensors. *J. Semicond.* **31**, 114002.

Azmer M.I., Ahmad Z., Sulaiman K., Al-sehemi A. (2015) Humidity dependent electrical properties of an organic material DMBHPET. *Measurement* **61**, 180–184.

Belghachi A., Collins R.A. (1990) The effects of humidity on phthalocyanine $NO_2$ and $NH_3$ sensors. *J. Phys. D: Appl. Phys.* **23**, 223.

Benkstein K.D., Hupp J.T., Stern C.L. (2000) Luminescent mesoporous molecular materials based on neutral tetrametallic rectangles. *Angew. Chem. Int. Ed.* **39**, 2891–2893.

Bergonzi R., Fabbrizzi L., Licchelli M., Mangano C. (1998) Molecular switches of fluorescence operating through metal centred redox couples. *Coord. Chem. Rev.* **170**, 31–46.

Betti P. (2007) Cavitand receptors for environmental monitoring. PhD Thesis. Università degli Studi di Parma, Italy.

Biedermann F., De Cola L. (2018) Porous supramolecular materials: The importance of emptiness. *Supramolecular Chem.* **30**(3), 166–168.

Bren V.A. (2001) Fluorescent and photochromic chemosensors. *Rus. Chem. Rev.* **70**(12), 1017–1036.

Castelletto V., Hamley I.W., Perez J., Abezgauz L., Danino, D. (2010) Fibrillar superstructure from extended nanotapes formed by a collagen-stimulating peptide. *Chem. Commun.* **46**, 9185–9187.

Centurion L.M.P.C., Moreira W.C., Zucolotto V. (2012) Tailoring molecular architectures with cobalt tetrasulfonated phthalocyanine: Immobilization in layer-by-layer films and sensing applications. *J. Nanosci. Nanotechnol.* **12**, 2399–2405.

Ceyhan T., Altındal A., Erbil M.K., Bekaroglu O. (2006) Synthesis, characterization, conduction and gas sensing properties of novel multinuclear metallo phthalocyanines (Zn, Co) with alkylthio substituents. *Polyhedron* **25**, 737–746.

Cha J.-H., Lee W., Lee H. (2009) Hydrogen gas sensor based on proton-conducting clathrate hydrate. *Angew. Chem. Int. Ed.* **48**, 8687–8690.

Chani M.T.S., Karimov K.S., Ahmad Khalid A., Raza K., Umer Farooq M., Zafar Q. (2012) Humidity sensors based on aluminium phthalocyanine chloride thin films. *Physica E* **45**, 77–81.

Chawla H.M., Pant N., Kumar S., Black D.St. C., Kumar N. (2011) Calixarene-based materials for chemical sensors. In: G. Korotcenkov (ed.) *Chemical Sensors: Fundamentals of Sensing Materials.* Vol. 3. *Polymers and other Materials.* Momentum Press, New York, pp. 117–200.

Cram D.J. (1983) Cavitands: Organic hosts with enforced cavities. *Science* **219**(4589), 1177–1183.

Davis F., Stirling C.J.M. (1995) Spontaneous multilayering of calyx[4]resorcinarenes. *J. Am. Chem. Soc.* **117**, 10385–10386.

De Graaf A.J., Mastrobattista E., Vermonden T., van Nostrum C.F., Rijkers D.T.S., Liskamp R.M.J., Hennink W.E. (2012) Thermosensitive peptide-hybrid ABC block copolymers obtained by ATRP: Synthesis, self-assembly, and enzymatic degradation. *Macromolecules* **45**(2), 842–851.

De Silva A.P., Gunaratne H.Q.N., Gunnlaugsson T., Huxley A.J.M., McCoy C.P., Rademacher J.T., Rice T.E. (1997) Signaling recognition events with fluorescent sensors and switches. *Chem. Rev.* **97**, 1515–1566.

Del Bianco A., Baldini F., Bacci M., Klimant I., Wolfbeis O.S. (1993) A new kind of oxygen sensitive transducer based on an immobilised metallo-organic compound. *Sens. Actuators B* **11**, 347–350.

Di Natale C., Macagnano A., Repole G., Saggio G., D'Amico A., Paolesse R., Boschi T. (1998) The exploitation of metalloporphyrins as chemically interactive material in chemical sensors. *Mater. Sci. Eng. C* **5**, 209–215.

Di Natale C., Paolesse R., D'Amico A. (2007) Metalloporphyrins based artificial olfactory receptors. *Sens. Actuators B* **121**, 238–246.

Diamond D., Nolan K. (2001) Calixarenes: Designer ligands for chemical sensors. *Anal. Chem.* **73**(1), 22–29A.

Dickert F.L., Balumler U.P.A., Stathopulos H. (1997) Mass-sensitive solvent vapor detection with calix[4]resorcinarenes: Tuning sensitivity and predicting sensor effects. *Anal. Chem.* **69**, 1000–1005.

Dickert F.L., Schuster O. (1995) Supramolecular detection of solvent vapours with calixarenes: Mass-sensitive sensors, molecular mechanics and BET studies. *Mikrochim. Acta* **119**(1–2), 55–62.

Donilfo P.H., Hupp J.T. (2001) Supramolecular coordination chemistry and functional microporous molecular materials. *Chem. Mater.* **13**, 3113–3125.

Dzugan T., Kroupa M., Reboun J. (2014) Sensitivity of organic humidity sensor element on organic vapours. *Proc. Eng.* **69**, 962–967.

Ehlen A., Wimmer C., Weber E., Bargon J. (1993) Organic clathrate-forming compounds as highly selective sensor coatings for the gravimetric detection of solvent vapors. *Angew. Chem. Int. Ed.* **32**, 110–112.

Elosua C., Bariain C., Matias I., Arregui F., Luquin A., Laguna M. (2006) Volatile alcoholic compounds fibre optic nano sensor. *Sens. Actuators B* **115**, 444–449.

Feresenbet E., Dalcanale E., Dulcey C., Shenoy D.K. (2004) Optical sensing of the selective interaction of aromatic vapors with cavitands. *Sens. Actuators B* **97**, 211–220.

Ferrari M., Ferrari V., Marioli D., Taroni A., Suman M., Dalcanale E. (2004) Cavitand-coated PZT resonant piezo-layer sensors: Properties, structure, and comparison with QCM sensors at different temperatures under exposure to organic vapors. *Sens. Actuators B* **103**, 240–246.

Fietzek C., Bodenhofer K., Haisch P., Hees M., Hanack M., Steinbrecher S., Zhou F., Plies E., Gopel W. (1999) Soluble phthalocyanines as coatings for quartz-microbalances: Specific and unspecific sorption of volatile organic compounds. *Sens. Actuators B* **57**, 88–98.

Finklea H.O., Phillippi M.A., Lompert E., Grate J.W. (1998) Highly sorbent films derived from $Ni(SCN)_2(4\text{-picoline})_4$ for the detection of chlorinated and aromatic hydrocarbons with quartz crystal microbalance sensors. *Anal. Chem.* **70**, 1268–1276.

Forster R., Cadogan A., Diaz M.T., Diamond D., Harris S.J., McKervey M.A. (1991) Calixarenes as active agents for chemical sensors. *Sens. Actuators B* **4** (3), 325–331.

Gardiner W.H. (2016) Investigations into the synthesis of upper-rim functionalised calix[4]arenes and their applications. PhD Thesis. School of Chemistry, University of East Anglia, Norwich, UK.

Germain J.P., Pauly A., Maleysson C., Blanc J.P., Schollhorn B. (1998) Influence of peripheral electron-withdrawing substituents on the conductivity of zinc phthalocyanine in the presence of gases. Part 2: Oxidizing gases. *Thin Solid Films* **333**, 235–239.

Grady T., Butler T., MacCraith B.D., Diamond D., McKervey M.A. (1997) Optical sensor for gaseous ammonia with tunable sensitivity. *Analyst* **122**(8), 803–806.

Grate J.W. (2000) Acoustic wave microsensor arrays for vapor sensing. *Chem. Rev.* **100**, 2627–2647.

Grate J.W., Patrash S.J., Abraham M.H., Du C.M. (1996) Selective vapor sorption by polymers and cavitands on acoustic wave sensors: Is this molecular recognition? *Anal. Chem.* **68**, 913–917.

Gregg S.J., Sing, K.S.W. (1967) *Adsorption Surface Area and Porosity.* Academic Press, London, UK.

Gutsche C.D. (1989) *Calixarenes.* Royal Society of Chemistry, Cambridge, UK.

Hahn M.E., Gianneschi N.C. (2011) Enzyme-directed assembly and manipulation of organic nanomaterials. *Chem. Commun.* **47** (43), 11814–11821.

Harbeck N., Sen Z., Gurol I., Gumus G., Musluoglu E., Ahsen V., Ozturk Z.Z. (2011) Vic-dioximes: A new class of sensitive materials for chemical gas sensors. *Sens. Actuators B* **56**, 673–679.

Hartmann J., Auge J., Hauptmann P. (1994) Using the quartz crystal microbalance principles for gas detection with reversible and irreversible sensors. *Sens. Actuators B* **18–19**, 429–433.

Hartmann J., Hauptmann P., Levi S., Dalcanale E. (1996) Chemical sensing with cavitands: Influence of cavity shape and dimensions on the detection of solvent vapors. *Sens. Actuators B* **35–36**, 154–157.

Herzer N., Guneysu H., Davies D.J., Yildirim D., Vaccaro A.R., Broer D.J., et al. (2012) Printable optical sensors based on H-bonded supramolecular cholesteric liquid crystal networks. *J. Am. Chem. Soc.* **134** (18), 7608–7611.

Holliday B.J., Mirkin C.A. (2001) Strategies for the construction of supramolecular compounds through coordination chemistry. *Angew. Chem. Int. Ed.* **40**, 2022–2043.

Hromadka J., Korposh S., Partridge M., James S.W., Davis F., Crump D., Tatam R.P. (2017) Volatile organic compounds sensing using optical fibre long period grating with mesoporous nanoscale coating. *Sensors* **17**, 205.

Hwang S.-H., Shreiner C.D., Moorefield C.N., Newkome G.R. (2007) Recent progress and applications for metallodendrimers. *New J. Chem.* **31**, 1192–1217.

Ishihara S., Azzarelli J.M., Krikorian M., Swager T.M. (2016) Ultratrace detection of toxic chemicals: Triggered disassembly of supramolecular nanotube trappers. *J. Am. Chem. Soc.* **138**(26), 8221–8227.

Jiang J. (ed.) (2009) *Functional Phthalocyanine Molecular Materials.* Springer, Berlin, Germany.

Jimenez-Cadena G., Riu J., Rius F.X. (2007) Gas sensors based on nanostructured materials. *Analyst* **132**, 1083–1099.

Kalchenko V.I., Koshets I.A., Matsas E.P., Kopulov O.N., Solovyov A., Kazantseva Z.I., Shitshov Yu.M. (2002) Calixarene-based QCM sensors array and its response to volatile organic vapours. *Mater. Sci. (Poland)* **20** (3), 73–88.

Karimov K.S., Cheong K.Y., Saleem M., Murtaza I., Farooq M., Noor A.F.M. (2010) Ag/PEPC/NiPc/ZnO/Ag thin film capacitive and resistive humidity sensors. *J. Semicon.* **31**, 054002.

Karimov K.S., Qazi I., Khan T.A., Draper P.H., Khalid F.A., Mahroof-Tahir M. (2008) Humidity and illumination organic semiconductor copper phthalocyanine sensor for environmental monitoring. *Environ. Monit. Assess.* **141**, 323–328.

Kimura M., Sugawara M., Sato S., Fukawa T., Mihara T. (2010) Volatile organic compound sensing by quartz crystal microbalances coated with nanostructured macromolecular metal complexes. *Chem. Asian J.* **5**, 869–876.

Kosal M.E., Chou J.-H., Wilson S.R., Suslick K.S. (2002) A functional zeolite analogue assembled from metalloporphyrins. *Nat. Mater.* **1**, 118–121.

Kubersky P., Hamacek A., Dzugan T., Kroupa M., Cengery J., Reboun J. (2010) Phthalocyanine layers for humidity detection. In: *Proceedings of 33rd Int. Spring Seminar on Electronics Technology, ISSE 2010*, May 12–16, Warsaw, Poland, pp. 14–17.

Kumar A., Sun S.-S., Lees A.J. (2008) Directed assembly metallocyclic supramolecular systems for molecular recognition and chemical sensing. *Coord. Chem. Rev.* **252**, 922–939.

Kumar S., Chawla S., Chinlun Zou M.C. (2017) Calixarenes based materials for gas sensing applications: a review. *J. Incl. Phenom. Macrocycl. Chem.* **88**, 129–158.

Kurth D.G. (2008) Metallo-supramolecular modules as a paradigm for materials science. *Sci. Technol. Adv. Mater.* **9**, 014103.

Lavrik N.V., DeRossi D., Kazantseva Z.I., Nabok A.V., Nesterenko B.A., Piletsky S.A., et al. (1996) Composite polyaniline/calixarene Langmuir-Blodgett films for gas sensing. *Nanotechnology* **7**(4), 315–319.

Lehn J.-M. (2007) From supramolecular chemistry towards constitutional dynamic chemistry and adaptive chemistry. *Chem. Soc. Rev.* **36**, 151–160.

Leininger S., Olenyuk B., Stang P.J. (2000) Self-assembly of discrete cyclic nanostructures mediated by transition metals. *Chem. Rev.* **100**, 853–908.

Li D.Q., Ma M. (2000) Surface acoustic wave microsensors based on cyclodextrin coatings. *Sens. Actuators B* **69**, 75–84.

Liu C.J., Lin J.T., Wang S.H., Jiang J.C., Lin L.G. (2005) Chromogenic calixarene sensors for amine detection. *Sens. Actuators B* **108**, 521–527.

Ludwig R., Dzung N.T.K. (2002) Calixarene-based molecules for cation recognition. *Sensors* **2**, 397–416.

Maffei F., Betti P., Genovese D., Montalti M., Prodi L., De Zorzi R., Geremia S., Dalcanale E. (2011) Highly selective chemical vapor sensing by molecular recognition: Specific detection of C1–C4 alcohols with a fluorescent phosphonate cavitand. *Angew. Chem. Int. Ed.* **50**, 4654–4657.

McKeown N.B. (1998) *Phthalocyanine Materials: Synthesis, Structure, and Function.* Cambridge University Press, Cambridge, UK.

Medforth C.J., Wang Z., Martin K.E., Song Y., Jacobsen J.L., Shelnutt J.A. (2009) Self-assembled porphyrin nanostructures. *Chem. Commun.* **2009**, 7261–7277.

Melegari M., Suman M., Pirondini L., Moiani D., Massera C., Ugozzoli F., et al. (2008) Supramolecular sensing with phosphonate cavitands. *Chem.—Eur. J.* **14**(19), 5772–5779.

Mermer Ö., Okur S., Sümer F., Özbek C., Sayın S., Yılmaz M. (2012) Gas sensing properties of carbon nanotubes modified with calixarene molecules measured by QCM techniques. *Acta Phys. Polonica A* **121**, 240–242.

Mogera U., Sagade A.A., George S.J., Kulkarni G.U. (2014) Ultrafast response humidity sensor using supramolecular nanofibre and its application in monitoring breath humidity and flow. *Sci. Rep.* **4**, 4103.

Mutter, M., Gassmann, R., Buttkus, U. Altmann, K.H. (1991) Switch peptides: pH-induced α-Helix to β-sheet transitions of bisamphiphilic oligopeptides. *Angew. Chem. Int. Ed. Engl.* **30**, 1514–1516.

Mutter M., Vuileumier S. (1989) A chemical approach to protein design-template-assembled synthetic proteins (TASP). *Angew. Chem.* **28**, 535–676.

Nardis S., Pomarico G., Tortora L., Capuano R., D'Amico A., Di Natale C., Paolesse R. (2011) Sensing mechanisms of supramolecular porphyrin aggregates: A teamwork task for the detection of gaseous analytes. *J. Mater. Chem.* **21**, 18638–18644.

Nieuwenhuizen M.S., Harteveld J.L.N. (1994) An automated SAW gas sensor testing system. *Sens. Actuators A* **44**, 219–229.

Oh S.W., Moon J.D., Lim H.J., Park S.Y., Kim T., Park J., et al. (2005) Calixarene derivative as a tool for highly sensitive detection and oriented immobilization of proteins in a microarray format through noncovalent molecular interaction. *FASEB J.* **19** (10), 1335–1337.

Ohira S.I., Wanigasekar E., Rudkevich D.M., Dasgupta P.K. (2009) Sensing parts per million levels of gaseous $NO_2$ by a optical fiber transducer based on calix[4]arenes. *Talanta* **77**(5), 1814–1820.

Okur S., Kus M., Ozel F., Yilmaz M. (2010) Humidity adsorption kinetics of water soluble calix[4]arene derivatives measured using QCM technique. *Sens. Actuators B* **145**, 93–97.

Özbek C., Culcular E., Okur S., Yilmaz M., Kurt M. (2013) Electrical characterization of interdigitated humidity sensors based on CNT modified calixarene molecules. *Acta Phys. Polonica A* **123**(2), 461–463.

Ozturk Z.Z., Kilinc N., Atilla D., Gurek A.G., Ahsen V. (2009) Recent studies of chemical sensors based on phthalocyanines. *J. Porphyrins Phthalocyanines* **13**, 1179–1187.

Palacios M.A., Nishiyabu R., Marquez M., Anzenbacher P. Jr. (2007) Supramolecular chemistry approach to the design of a high-resolution sensor array for multianion detection in water. *J. Am. Chem. Soc.* **129**, 7538–7544.

Palaniappan A., Li X., Tay F.E.H., Li J., Su X. (2006) Cyclodextrin functionalized mesoporous silica films on quartz crystal microbalance for enhanced gas sensing. *Sens. Actuators B* **119**, 220–226.

Pandey R.K., M Hossain Md. D., Moriyama S., Higuchi M. (2014) Real-time humidity-sensing properties of ionically conductive Ni(II)-based metallo-supramolecular polymers. *J. Mater. Chem. A* **2**, 7754–7758.

Parker D. (ed.) (1996) *Macrocycle Synthesis. A practical Approach.* Oxford University Press, Oxford, UK.

Pinalli R., Suman M., Dalcanale E. (2004) Cavitands at work: From molecular recognition to supramolecular sensors. *Eur. J. Org. Chem.* **2004**, 451–462.

Pirondini L., Dalcanale E. (2007) Molecular recognition at the gas–solid interface: A powerful tool for chemical sensing. *Chem. Soc. Rev.* **36**, 695–706.

Řeboun J., Hamacek A. (2008) Organic materials for humidity sensors. In: *Digest of Articles.* Technical University, Budapešť, pp. 481–485.

Roisin P., Wright J.D., Nolte R.J.M., Sielcken O.E., Thorpe S.C. (1992) Gas-sensing properties of semiconducting films of crown-ether-substituted phthalocyanines. *J. Mater. Chem.* **2**, 131-137.

Rosler S., Lucklum R., Borngraber R., Hartmann J., and Hauptmann P. (1998) Sensor system for the detection of organic pollutants in water by thickness shear mode resonators. *Sens. Actuators B* **48**, 415–424

Rudkevich D.M. (2007) Progress in supramolecular chemistry of gases. *Eur. J. Org. Chem.* **20**, 3255–3270.

Sadaoka Y., Jones T.A., Gopel W. (1990a) Fast $NO_2$ detection at room temperature with optimized lead phthalocyanine thin film structure. *Sens. Actuators B* **1**, 148–153.

Sadaoka Y., Jones T.A., Revell G.S., Gopel W. (1990b) Effects of morphology on $NO_2$ detection in air at room temperature with phthalocyanine thin films. *J. Mater. Sci.* **25**, 5257–5268.

Satija J., Sai V.V.R., Mukherji S. (2011) Dendrimers in biosensors: Concept and applications. *J. Mater. Chem.* **21**, 14367–14386.

Schierbaum K.D., Weiss T., Thoden Van Velzen E.U., Engbersen J.F.J., Reinhoudt DN, Göpel W. (1994) Molecular recognition by self-assembled monolayers of cavitand receptors. *Science* **265**, 1413–1415.

Schneider H.-J. (2009) Binding mechanisms in supramolecular complexes. *Angew. Chem. Int. Ed.* **48**, 3924–3977.

Schollhorn B., Germain J.P., Pauly A., Maleysson C., Blanc J.P. (1998) Influence of peripheral electron-withdrawing substituents on the conductivity of zinc phthalocyanine in the presence of gases. Part 1: Reducing gases. *Thin Solid Films* **326**, 245–250.

Sen Z., Gumus G., Gurola I., Musluoglu E., Ozturk Z.Z., Harbecka M. (2011) Metal complexes of *vic*-dioximes for chemical gas sensing. *Sens. Actuators B* **160**, 1203–1209.

Shah M., Sayyad M.H., Karimov Kh.S. (2008) Fabrication and study of nickel phthalocyanine based surface-type capacitive sensors. *Proc. World Acad. Sci. Eng. Technol.* **2008**, *43*, 392.

Sharma K., Cragg P.J. (2011) Calixarene based chemical sensors. *Chem. Sens.* **1** (9), 1–18.

Shenoy D.K. (2005) Cavitands: Container molecules for surface Plasmon resonance (SPR)-based chemical vapor detection. *Mater. Sci. Technol.* 2005 NRL Review, 171–173.

Shmygleva L., Slesarenko N., Chernyak A., Sanginov E., Karelin A., Pisareva A., et al. (2017) Effect of calixarene sulfonic acids hydration on their proton transport properties. *Int. J. Electrochem. Sci.* **12**, 4056–4076.

Skothem T.A., Reynolds J.R. (eds.) (2007) *Handbook of Conducting Polymers.* CRC Press, Boca Raton, FL.

Snow A.W., Barger W.R. (1989) Phthalocyanine films in chemical sensors. In: Kadish K.M., Smith K.M., Guilard R. (eds.) *Phthalocyanines—Properties and Applications.* VCHm, New York, pp. 343–392.

Stupp S.I., Keser M., Tew G.N. (1998) Functionalized supramolecular materials. *Polymer* **39**(19), 4505–4508.

Su P.-G., Lin L.-G., Tzou W.-H. (2013) Humidity sensing properties of calix[4]arene and functionalized calix[4]arene measured using a quartz-crystal microbalance. *Sens. Actuators B* **181**, 795–801.

Su P.-G., Shiu C.-C. (2012) Electrical and sensing properties of a flexible humidity sensor made of polyamidoamine dendrimer-Au nanoparticles. *Sens. Actuators B* **165**, 151–156.

Su P.-G., Tzou W.-H. (2012) Low-humidity sensing properties of PAMAM dendrimer and PAMAM–Au nanoparticles measured by a quartz-crystal microbalance. *Sens. Actuators A* **179**, 44–49.

Szetjli J. (1998) Introduction and general overview of cyclodextrin chemistry. *Chem. Rev.* **98**, 1743–1753.

Trogler W.C. (2012) Chemical sensing with semiconducting metal phthalocyanines. *Struct. Bond* **142**, 91–118.

Valeur B., Leray I. (2007) Ion-responsive supramolecular fluorescent systems based on multichromophoric calixarenes: A review. *Inorg. Chim. Acta* **360**, 765–774.

Vögtle F. (ed.) (1996) *Comprehensive Supramolecular Chemistry.* Pergamon, Oxford, UK.

Wang C., Chen F., He X.-W. (2002) Kinetic detection of benzene/chloroform and toluene/chloroform vapors using a single quartz piezoelectric crystal coated with calix[6]arene. *Anal. Chim. Acta* **464**, 57–64.

Wang C., Chen F., He X.W., Kang S.Z., You C.C., Liu Y. (2001) Cyclodextrin derivative-coated quartz crystal microbalances for alcohol sensing and application as methanol sensors. *Analyst* **126**, 1716–1720.

Wannebroucq A., Ouedraogo S., Meunier-Prest R., Suisse J.-M., Bayo M., Bouvet M. (2018) On the interest of ambipolar materials for gas sensing. *Sens. Actuators B* **258**, 657–664.

Wright J.D. (1989) Gas adsorption on phthalocyanines and its effects on electrical properties. *Prog. Surf. Sci.* **31**, 1–60.

Yakimova L.S., Ziganshin M.A., Sidorov V.A., Kovalev V.V., Shokova E.A., Tafeenko V.A., Gorbatchuk V.V. (2008) Molecular recognition of organic vapors by adamantylcalix[4]arene in QCM sensor using partial binding reversibility. *J. Phys. Chem. B* **112**, 15569–15575.

Yan H., Zhang L., Yu P., Mao L. (2017) Sensitive and fast humidity sensor based on a redox conducting supramolecular ionic material for respiration monitoring. *Anal. Chem.* **89**(1), 996–1001.

Zubin E.M., Romanova E.A., Oretskaya T.S. (2002) Modern methods for the synthesis of peptide-oligonucleotide conjugates. *Russ. Chem. Rev.* **71**(3) 239–264.

# 13 Biomaterials as Sensing Elements of Humidity Sensors

Currently, bioengineering is considered one of the most promising areas of research. Therefore, it is quite natural that research has emerged, aimed at the use of biomaterials in the development of humidity sensors. A list of biomaterials used in humidity sensors is given in Table 13.1. As you can see, the most commonly used are peptides.

## 13.1 PROTEINS AS HUMIDITY-SENSITIVE MATERIAL

Consequently, proteins are considered polarizable materials, which can be characterized by dielectric spectroscopy (Bonincontro et al. 1997). Frequency is another critical factor, influencing the behavior of dielectric materials such as proteins, particularly their polarization. With increasing frequency, polarizations will no longer be able to follow the quick electric field reversals, resulting in the dropping out of constituents, while others still prevail (Figueiró et al. 2006). This results in a drop of the dielectric permittivity and dielectric loss values of the protein. Bonincontro and Risuleo (2003) suggest that protein solutions, exposed to an electric field in the radio frequency range (3 kHz to 300 GHz), generally show dielectric relaxation due to orientation polarization (dipole-dipole interactions). There are also several reports on protein sensitivity in contact with water vapor, oxygen, $NH_3$, carbon dioxide (Gontard et al. 1996; Mujica-Paz et al. 1997; McAlpine et al. 2008) and ethanol gas (Yoshizawa et al. 2014). Taking into account their electrical properties, dielectric properties, and their sensitivity to environmental factors such as water vapor and carbon dioxide, it is hypothesized that proteins can be used as bio-sensors for the measurement of a physical, chemical, biological, or any other parameter, giving a response in terms of resistance, capacitance, voltage, or current (Mukhopadhyay 2013).

The relation between dielectric properties of proteins and environmental changes was recently taken into account to develop a humidity sensor. Bibi et al. (2016b) investigated a thin wheat-gluten-protein film, deposited on a designed and manufactured interdigital capacitor as a humidity sensor in food packaging on the frequency range from 30 to 1000 MHz. When relative humidity (RH) increased from 20% to 95% RH at 25°C, the authors showed dielectric permittivity and loss change from 5.0 to 9.8 and from 0.4 to 1.5, respectively. The dielectric permittivity and loss were very sensitive to RH and increased exponentially due to the comparable exponential increase of wheat-gluten water content versus RH. The authors attested that the dependency of both dielectric permittivity and dielectric loss of wheat gluten on RH offered the possibility to use the protein for monitoring Rh in packed food products. Bibi et al. (2016b) compared the capacitance sensitivity and the hysteresis of different humidity sensors found in the literature, with their device designed with wheat gluten. They have found that although some materials (metal oxides, including porous aluminum oxide) have higher sensitivity performances, wheat gluten has a very good sensitivity value on a very wide range of RH and has some advantage over the other materials, being a natural polymer and a by-product of the agrifood industry. This ensures a much lower cost than the others. In addition to this, wheat gluten has a very low hysteresis, indicating that the measurements are reversible on several RH sorption and desorption cycles.

## 13.2 COLLAGEN-BASED HUMIDITY SENSORS

Shapardanis et al. (2014) used a different peptide to develop humidity sensors, namely collagen. Coming from the bones, tendons, and skin of bovine and pigs, this by-product of the meat and leather industry is plentiful, inexpensive, and simple to process. Almost 30% of the protein found in animals is collagen (Burgeson and Nimni 1992). This is due to collagen's importance in developing the body's architecture, such

## TABLE 13.1
## Biomaterials Used in Humidity Sensors

| Material | Sensor Type | References |
|---|---|---|
| DNA-metal ion biomatrix | Optical | Yamada and Sugiyama (2008) |
| Dopamine coated Au NPs | Resistive | Wang et al. (2012); Lee et al. (2014) |
| The Hygrophila Spinosa T. Anders (HST) seeds-$SnO_2$ | Resistive | Kennedy et al. (2015) |
| Onion membrane | Impedance | Sajid et al. (2016) |
| Proteins (gluten, collagen, gelatin, etc.) | Capacitive | Shapardanis et al. (2014); Bibi et al. (2016a, 2016b) |
| | Resistive | McAlpine et al. (2008) |
| Polysaccharide | Impedance | Liakos et al. (2017) |

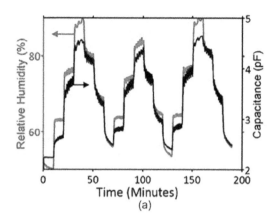
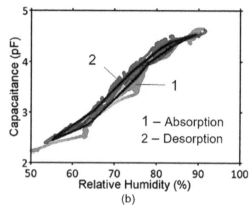

**FIGURE 13.1** Humidity-sensing characteristics of collagen-based humidity sensors. The devices were tested using a commercial RH chamber at a constant 36°C. The devices were tested at 10% RH intervals for 10 minutes from 50% RH to 90% RH, then back to 50% RH: (a) the overall data that compares the RH as measured by the commercial data logger to the capacitance of the sensor; and (b) the device's dependence on RH with distinctions between absorption and desorption. (Reprinted from Shapardanis, S. et al., *AIP Adv.*, 4, 127132, 2014. Published by AIP as open access.)

as bones and skin, while also providing mechanical support, such as connective tissues. The physical properties of collagen are derived from its hierarchical structure, starting at the molecular level (Shoulders and Raines 2009). The capacitive RH sensors were built using a coplanar interdigitated electrode geometry with the gelatin dielectric patterned between and below the comb-like electrodes. Operating characteristics of these sensors are shown in Figure 13.1. Testing these sensors revealed that collagen-based sensors are found to have a sensitivity to moisture in the range of 0.065 pF/%RH. Absorption and desorption times were found to be 20 and 8 seconds, respectively. Shapardanis et al. (2014) believe that gelatin, a network of denatured collagen molecules (Duhamel et al. 2002), which is inexpensive, widely available, and easy to process, can be an effective dielectric-sensing polymer for capacitive-type RH sensors, as a good alternative to more conventional polymers.

## 13.3 OTHER BIOMATERIALS

Shapardanis et al. (2014) and Bibi et al. (2016a) believe that the development of a natural polymers could be an innovation in the bio-sensor technology. In spite of its complexity, it would have a significant financial and environmental potential for bio-sensor development, assuming that detection of fluctuations in dielectric properties, influenced by changes in environmental conditions, is possible. One can agree with Bibi et al. (2016a), that the use of biomaterials in the development of humidity sensors is a new innovative approach. However, it should be noted that the prospects of its development are very limited due to the necessity of the presence on the sensor market of devices with very high manufacturability and stability of parameters.

As for other biomaterials, studies have shown that these materials also have great potential for the development of various devices, including humidity sensors. According to Yamada and Sugiuama (2008), the DNA-metal ion

$(Co^{2+}$ or $Ni^{2+})$ biomatrix may have a potential for the novel optical RH sensor with being flexible, low cost, nonhazardous, and environmentally benign. These matrices indicated the different color with the change of RH (see Figures 13.1 and 13.4). Especially, the color of $Co^{2+}$-containing biomatrix changed from blue to red with the increase of RH. These biomatrices were constructed by the electrostatic interaction to the phosphate group and the coordination interaction to the nucleic-acid base (Figures 13.2 and 13.3).

The results, obtained by Lee et al. (2014), deserve attention. This study presented a simple process for producing the resistance-based humidity sensors, utilizing dopamine (DA)-coated gold nanoparticles (AuNPs) as the sensing material. Highly hydrophilic DA biomolecules are physically bonded to

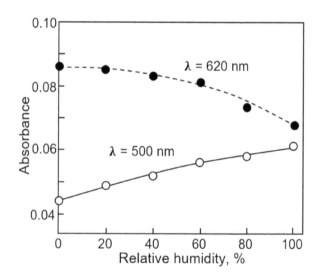

**FIGURE 13.2** Absorbance at 620 nm (●) and 500 nm (○) of DNA-$Co^{2+}$ biomatrix under various RH conditions. The error bar represents the error of ±5%. These measurements of absorption spectra were performed at room temperature. (Reprinted with permission from Macmillan Publishers Ltd. *Polymer J.*, Yamada, M. and Sugiyama, T, Copyright 2008: Springer Nature.)

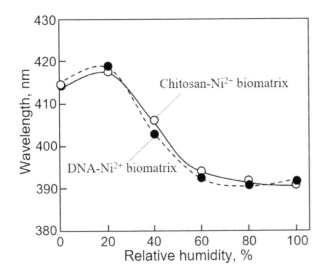

**FIGURE 13.3** The shift in maximum absorption peak of biopolymer-$Ni^{2+}$ biomatrix under various RH conditions. The RH range is from 0% to 100%. ●, DNA-$Ni^{2+}$ biomatrix; ○, chitosan-$Ni^{2+}$ biomatrix. These measurements of absorption spectra were performed at room temperature. (Reprinted with permission from Macmillan Publishers Ltd. *Polymer J.*, Yamada, M. and Sugiyama, T, Copyright 2008: Springer Nature.)

4–6 nm AuNPs to enhance the humidity sensing performance. Lee et al. (2014) have shown that the DA-coated AuNPs have good humidity-sensing performance in the range of 20%–90% RH (see Figure 13.4). It is seen that AuNPs with DA concentration below 100 μM show small variations of resistance between 30% and 90% RH, while DA-AuNPs concentration

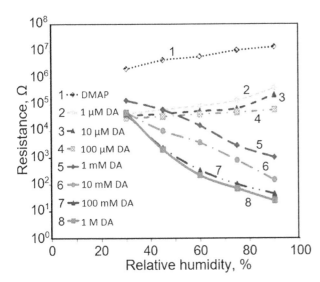

**FIGURE 13.4** Measured resistance of the humidity sensors coated with the AuNPs treated by different concentration of DA solutions. (Reprinted with *Sens. Actuators B*, 191, Lee, H.-C. et al., High-performance humidity sensors utilizing dopamine biomolecule-coated gold nanoparticles, 204–210, Copyright 2014 with permission from Elsevier.)

above 1 mM may become a candidate for humidity sensing. The developed humidity sensor also showed fast response time for water adsorption (5 s) and desorption (10 s). Moreover, a three-day long-term measurement at low-, medium-, and high-humidity ranges also showed that the developed sensor has good stability.

Other studies carried out by Yamada and Sugiyama (2008), Kennedy et al. (2015), and Sajid et al. (2016) are also interesting, but unlikely to be of interest for practical use. It is hard to imagine that sensors with onion membrane will appear on the sensor market.

## REFERENCES

Bibi F., Guillaume C., Vena A., Gontard N., Sorli B. (2016b) Wheat gluten, a bio-polymer layer to monitor relative humidity in food packaging: Electric and dielectric characterization. *Sens. Actuators A Phys.* **247**, 355–367.

Bibi F., Villain M., Guillaume C., Sorli B., Gontard N. (2016a) A review: Origins of the dielectric properties of proteins and potential development as bio-sensors. *Sensors* **16**, 1232.

Bonincontro A., Mari C., Mengoni M., Risuleo G. (1997) A study of the dielectric properties of *E. coli* ribosomal RNA and proteins in solution. *Biophys. Chem.* **67**, 43–50.

Bonincontro A., Risuleo G. (2003) Dielectric spectroscopy as a probe for the investigation of conformational properties of proteins. *Spectrochim. Acta A Mol. Biomol. Spectrosc.* **59**, 2677–2684.

Burgeson R., Nimni M. (1992) Collagen types: Molecular structure and tissue distribution. *Clin. Orthop. Relat. R* **282**, 250–272.

Duhamel C., Hellio D., Djabourov M. (2002) All gelatin networks: 1. Biodiversity and physical chemistry. *Langmuir* **18**, 7208–7217.

Figueiró S., Macêdo A.A., Melo M.R., Freitas A.L., Moreira R., de Oliveira R., et al. (2006) On the dielectric behaviour of collagen-algal sulfated polysaccharide blends: Effect of glutaraldehyde crosslinking. *Biophys. Chem.* **120**, 154–159.

Gontard N., Thibault R., Cuq B., Guilbert S., Breton J.F. (1996) Influence of relative humidity and film composition on oxygen and carbon dioxide permeabilities of edible films. *J. Agric. Food Chem.* **44**, 1064–1069.

Kennedy L.J., Umapathy M.J., Aruldoss U. (2015) Synthesis of simple and novel biocomposite doped nanocrystalline tin oxide and its humidity sensing properties. *Measurement* **67**, 1–9.

Lee H.-C., Wang C.-Y., Lin C.-H. (2014) High-performance humidity sensors utilizing dopamine biomolecule-coated gold nanoparticles. *Sens. Actuators B* **191**, 204–210.

Liakos I.L., Mondini A., Filippeschi C., Mattoli V., Tramacere F., Mazzolai B. (2017) Towards ultra-responsive biodegradable polysaccharide humidity sensors. *Mater. Today Chem.* **6**, 1–12.

McAlpine M.C., Agnew H.D., Rohde R.D., Blanco M., Ahmad H., Stuparu A.D., et al. (2008) Peptide-nanowire hybrid materials for selective sensing of small molecules. *J. Am. Chem. Soc.* **130**, 9583–9589.

Mujica-Paz H., Gontard N. (1997) Oxygen and carbon dioxide permeability of wheat gluten film: Effect of relative humidity and temperature. *J. Agric. Food Chem.* **80**, 4101–4105.

Mukhopadhyay S.C. (2013) *Intelligent Sensing, Instrumentation and Measurements*. Springer: Berlin, Germany.

Sajid M., Aziz S., Kim G.B., Kim S.W., Jo J., Choi K.H. (2016) Bio-compatible organic humidity sensor transferred to arbitrary surfaces fabricated using single cell-thick onion membrane as both the substrate and sensing layer. *Sci. Rep.* **6**, 30065.

Shapardanis S., Hudpeth M., Kaya T. (2014) Gelatin as a new humidity sensing material: Characterization and limitations. *AIP Adv.* **4**, 127132.

Shoulders M., Raines R. (2009) Collagen structure and stability. *Annu. Rev. Biochem.* **78**, 929–958.

Wang C.-Y., Lee H.-C., Lin C.-H. (2012) High performance humidity sensors based on dopamine biomolecules coated gold-nanoparticles. *IEEE Sensor Conference, Sensors* 2012, October 28–31, Taipei, Taiwan.

Yamada M., Sugiyama T. (2008) Utilization of DNA-metal ion biomatrix as a relative humidity sensor. *Polymer J.* **40**(4), 327–331.

Yoshizawa S., Arakawa T., Shiraki K. (2014) Dependence of ethanol effects on protein charges. *Int. J. Biol. Macromol.* **68**, 169–172.

# Section II

## Sensor Technologies and Related Materials

# 14 Substrates and Electrodes in Humidity Sensors

## 14.1 CONVENTIONAL SUBSTRATES FOR HUMIDITY SENSORS

As for the substrates used in humidity sensors, they are subject to standard requirements: the substrate to be used in humidity sensors preferably should have good dielectric properties, excellent stability, and should be chemically inert. There must be no undesirable reactions between the substrate and the humidity-sensitive layer during the production and operation of the sensor. In addition, good adhesion of the humidity-sensitive layer, deposited on these substrates, must be ensured. The thermal coefficient of expansion, which is an important parameter for metal-oxide gas sensors, in the case of humidity sensors operating at room temperature, is not so significant. However, in the case of using high-temperature processes, a large difference in thermal coefficient of expansion can affect the quality of the formed coatings.

Experience has shown that taking into account manufacturing concerns, the substrates should also provide the following (Prudenziati 1994): (1) mechanical strength: it is important for easy handling and basic for several sensors; (2) surface finish: it is important for good line definition and printed film uniformity; (3) good planeness: there should be a minimum distortion of the substrate. An excess can result in screening problems in thick-films and photo-processing problems in thin-films; (4) the absence of visual defects: surface defects such as small pits or burrs can result in circuit defects such as open circuits and pin holes; (5) good compatibility of materials: the surface should be chemically and physically compatible with the chemicals and material used in fabrication. Generally speaking, the chemical inertness of ceramic substrates is particularly useful for sensors designed to work in very aggressive media; (6) low cost of producing large quantities; and (7) small tolerance: should be possible to achieve the suitable tolerances in overall dimensions.

Substrates for planar humidity sensors are most often developed from alumina ceramics, but other materials might also be used. Comparison of materials suitable for application as substrates in humidity sensors is present in Table 14.1. Ceramics based on $Al_2O_3$ are generally preferred, since this compound has excellent temperature stability as well as excellent dielectric and temperature conductance properties (William and Boardman 1975; Qu 1996; Kita et al. 2015). Therefore, alumina ceramics have been extensively used as substrate material for hybrid microelectronics and in microelectronics packaging. Moreover, $Al_2O_3$ is characterized by good chemical stability when exposed to many gas atmospheres, good

### TABLE 14.1
### Parameters of Materials That Can Be Used as Humidity Sensor Platforms

| Parameter | $Al_2O_3$ | $SiO_2$ (quartz) | Si |
|---|---|---|---|
| Melting point, °C | 2072 | 1600 | 1414 |
| Thermal conductivity, W/m°K (20°C) | 35 | 1.3 | 149 |
| Refractive index | 1.77 | 1.52 | 3.42 |
| Density, g/cm³ | 3.95 | 2.65 | 2.3 |
| Volume resistivity, ohm·cm (20°C) | $>10^{14}$ | $10^{12}$–$10^{16}$ | — |
| Dielectric constant (1 MHz) | 9.8 | 3.9 | 11.9 |
| Coefficient of thermal expansion, $10^{-6}$/°C (100°C) | 8.4 | 0.55 | 2.6 |
| Maximum use temperature, °C | 1750 | 1100 | 300 |

adhesion to many gas-sensitive materials, and a relatively modest price. The elastic moduli and mechanical strength of alumina ceramic are high, which means that it is one of the strongest refractory oxides. Quartz and glass substrates can also be used (Hsueh et al. 2011; Zafar and Sulaiman 2016). Glass slides are easily available in good quality. Glass exhibits attractive dielectric and optical properties. Since glass is transparent for visible light, it is particularly suited for devices with optical detection principles. If the sensor production process is low temperature, they can be used—even if it is difficult to produce reliable contacts. Single-crystal quartz with its hexagonal lattice structure is piezoelectric and is, therefore, used as substrate material for acoustic-wave devices (see Chapter 17). Last but not least, glasses are chemically inert and suitable for high-temperature applications. A number of micromachining techniques, such as isotropic wet etching or anisotropic dry etching, have been developed to structure glass (Hierlemann 2004). However, glass slides were mostly used in very early prototyping stages, normally not in series production.

Planar humidity sensors can be also developed using oxidized silicon (Si) substrates. The thin-film technologies usually require the substrates with very flat surfaces. Si is the standard substrate material for integrated circuit (IC) fabrication and, thus, the most common substrate material in microfabrication (read Chapter 15). Besides its favorable electrical properties, the single-crystal silicon also has excellent physical properties (mechanical strength, thermal conductivity), which enable the design of micromechanical structures. In particular, the Si substrates are used for fabrication of various sensing platforms, such as membranes and cantilevers (read Chapter 16). The use of Si substrate material enables also the co-integration of transducers and circuitry, which is used to advantage, for example, in realizing complementary metal-oxide semiconductor (CMOS)-based microsystems (Hierlemann 2004). When using Si substrates, the sensors, integrated with heaters and electrical circuits, can be fabricated.

One should note that polymers have also been more and more applied over the last years as an inexpensive substrate material for humidity sensors. Due to the cost advantage, disposable devices and devices for specific applications can be developed using flexible polymer substrates. However, we will not consider these substrates in this chapter. The materials used for the manufacture of flexible substrates and the properties of devices developed on the basis of such substrates are discussed in sufficient detail in Chapter 22 included in Volume 2.

## 14.2 ELECTRODES

It is clear that in most cases, the electrodes of the humidity sensors are similar to the electrodes of an electronic device, which deliver current flow or electric power supply without loss from the external power sources to the device. Therefore, strong mechanical adhesion and small contact resistance in addition to high stability are the most important requirements of materials intended for electrode fabrication (Korotcenkov 2007; Lee 2017). The study showed that degradation of contacts takes place mainly due to diffusion processes, occurring at the electrode–sensing material interface or the electrode interaction with substrate and surrounding atmosphere (Meixner and Lampe 1996). This means that electrode materials should be chemically and mechanically stable on the substrates, the

connection to the lead-out terminals should be easy, the sensing film should not be damaged during electrode formation, and the technology of their deposition should provide the contact geometry required for the operation of the sensor. From the point of view of chemical and thermodynamic stability, the best option for the manufacture of electrodes is the use of noble metals and their alloys. Table 14.2 summarizes the properties of noble metals used in semiconductor gas sensors.

Among noble metals, in characterizing a maximum stability in the air over a wide range of temperatures, the most intensive aging and degradation effects become apparent for Ag contacts. Silver can easily move if the electric field applied to the electrode at operating temperature is higher than 300°C. Gold also diffuses quickly at high temperatures. Platinum (Pt)-based contacts are the most stable (Park and Kim 1999; Esch et al. 2000). Pt is not oxidized at high temperatures; it has low diffusivity and is resistive to corrosive gases (Fleischer et al. 1994; Dziedzic et al. 1997). In addition, Pt has a good ohmic contact with most metal oxides (MOXs). Qu (1996) has also shown that Pt electrodes prepared using thick-film technology had more porous structure in comparison with Ag- and Ag/Pd-based electrodes, that is a very important factor for humidity sensors. As a result, humidity sensors using Pt electrodes had higher sensitivity and faster response. However, even for the most stable metals there are some limitations. For example, a potential reason for possible parameter changes in contacts is bad adhesion of Pt. Therefore, an intermediate thin film of an "adhesion" metal such as chromium (Cr), titanium (Ti), or tungsten (W), 10–100 nm thick, is deposited between the MOX and electrode layers. Good adhesion materials for platinum (Pt) films over substrates used have been found to be tantalum (Ta) (Hoefer et al. 1994) and TiN (Michel et al. 1995). However, it was established that using a

## TABLE 14.2
### Properties of Printing Metals and Alloys for Electrode of Semiconductor Gas Sensors

| Materials | Electrical Properties | Advantages | Disadvantages |
|---|---|---|---|
| Silver | High conductivity. Compatible with resistor and dielectric system. Resistivity: $1.59 \cdot 10^{-8}$ $\Omega \cdot$m | Least expensive; Good bond strength. | Tendency to migrate over the surface of insulants and resistors under high humidity. |
| Gold | High conductivity and reliability. Resistivity: $2.44 \cdot 10^{-8}$ $\Omega \cdot$m | Alloy with tin may be made without the use of flux | High cost. Unsuitability for solder joining. |
| Platinum | Use where extreme resistance to molten solder and to bond strength degradation by solder is required. Resistivity: $11.0 \cdot 10^{-8}$ $\Omega \cdot$m | Available wire, flat plate, and tube. Large range of size. Usable at high temperature. | Most expensive. |
| Palladium-Silver | Compatible with resistor and dielectric system. Sheet resistance: 0.01–0.04 $\Omega$/sq | Suitable for ultrasonic wire bonding. | The possibility of silver migration under high humidity |
| Platinum-Silver | Alternative to Pd-Ag. Sheet resistance: 0.01–0.04 $\Omega$/sq | Suitable for ultrasonic wire bonding. | Not recommended for hybrid applications involving ultrasonic wire bonding. |
| Platinum-Gold | Compatible with most thick film materials. Sheet resistance: 0.08–0.1 $\Omega$/sq | Excellent solderability. Suitable for both wire and die bonding. | High cost. Rather high electrical resistivity. |
| Palladium-Gold | Similar properties to Pt-Au. Sheet resistance: 0.04–0.10 $\Omega$/sq | Less expensive than Pt-Au. | Inferior solder leach resistance and solder ageing than Pt-Au |

*Source:* Data extracted from Holmes, P. and Loasby, R., *Handbook of Thick Film Technology*, Electrochemical Publications, Glasgow, UK, 1976; Lee, S.P., *Sensors*, 17, 683, 2017.

sublayer does not resolve all the problems. No doubt that the use of Ti/Pt layers as the electrical contacts, which are more stable than Ti/Au or Ti/Pd/Au, might help to prevent the rather fast degradation of contacts (Sozza et al. 2004). In the case of microhotplate fabrication using micromachining technology, we have a more complicated situation because besides the thermodynamic stability of the materials used, we need to take into account the compatibility of these materials with Si technology. A more detailed analysis of electrode materials acceptable for microhotplate fabrication was carried out by Furjes et al. (2002). Materials acceptable for realization of microhotplates are presented in Table 14.3.

As seen in Table 14.3, there are several alternatives for fabrication electrodes in micromachining devices; however, according to Furjes et al. (2002), the materials with ideal properties cannot be found when having to consider their high-thermal and low-electrical conductivity, unfavorable thermal coefficient of resistance (TCR) dependence vs. temperature, high residual stress, chemical reactivity with Si at high temperature, and non-compatibility with Si processing. Most of the materials are soluble in HF as used in porous Si processing (in the case of Pt, the adhesive Ti layer); therefore, the wires have to be encapsulated by another $SiN_{1.05}$ layer. Unfortunately, the high-deposition temperature of the second $SiN_{1.05}$ layer makes the application of aluminum (Al) impossible. NiFe is far from being a CMOS-compatible material at all. According to Furjes et al. (2002), the materials presented in Table 14.3 have the following advantages and shortcomings:

- *Ti/Pt.* The excellent chemical compatibility and low resistivity of Pt offer the most attractive solution for wiring. The generated heat is negligible in the Pt wires, while the heat transfer can be kept at a low level by the formation of a minimized cross section in the suspension beams. Nevertheless, the high-operational temperature of the contacts between the Si heater and Pt results in continuous deterioration by the formation and lateral creeping of a silicide phase.

A conductive diffusion barrier between Si and Pt, such as TiN, may eliminate this problem.
- *Polycrystalline silicon.* Although the contacts between polycrystalline silicon and the filament are inherently perfect, the high resistance in combination with the high-positive TCR of the poly-Si limits the operation temperature of the structure. Increasing the input power (and temperature), the resistance of the polysilicon wires on the suspension beams may dominate, resulting in malfunction of the device.
- *$TiSi_2$.* $TiSi_2$ and other refractory silicides are among the most promising materials because of their low resistivity, low contact resistance, stability at high temperature, and chemical compatibility. Due to the superior electromigration resistance of $TiSi_2$, thin layers can be formed; therefore, the high stress and the thermal loss via wiring can be reduced to an acceptable level.

Shim et al. (2011) have shown that conductive metal oxides such as indium tin oxide (ITO) can also be used as electrode materials in humidity sensors. They established that the $WO_3$ or $SnO_2$ thin-film sensors with ITO interdigitated electrodes (IDEs) on glass substrates displayed higher responses than sensors with Pt IDEs, attributed to the low-resistance ohmic contacts between the electrode (ITO) and the sensing material ($WO_3$ or $SnO_2$).

Of course, in the case of humidity sensors operated at room temperature, we do not have such strong requirements from electrode materials, and therefore other materials, which are not as stable as Pt, can be used. Parameters of some metals used in the manufacture of humidity sensors are listed in Table 14.4. In particular, gold (AU), Al, copper (Cu), silver (Ag), Cr, nickel (Ni), Ti, as well as carbon (C) and Si can also be used in humidity sensors as electrode materials (Wang et al. 2011; Zafar and Sulaiman 2016). For example, gold (front and back) electrodes are used in most conventional quartz crystal microbalances (QCMs). The two metallizations, most commonly used to fabricate transducers on surface acoustic wave (SAW) devices, are gold-on-chromium and aluminum (Ballantine et al. 1997). Au is often chosen for gas detection applications because of its inertness and resistance to corrosion. Unfortunately, the inertness of Au also prevents its adhesion to quartz and other oxides utilized for acoustic wave (AW) device substrates. Therefore, an underlayer of Cr (2–10 nm thick) is utilized to promote the adhesion of Au to the substrate: the electropositive (reactive) nature of Cr allows it to form strong bonds with oxide surfaces, while alloying between the Cr and Au chemically binds the two metal layers tightly together. Care must be taken not to expose a freshly deposited Cr layer to oxygen (air) before the Au is deposited, as a chromium-oxide layer will form instantaneously, preventing adhesion of the Au to the Cr. At elevated temperatures (ca. 300°C and above), Cr and Au interdiffuse; the unfortunate result of this is that the conductivity of the Au layer decreases significantly, eventually rendering the metallization too resistive for use. This problem can be partially circumvented by substitution of Ti for Cr as an adhesion layer, although Ti is more difficult to deposit.

## TABLE 14.3
## Guide for Selection of Materials for Wiring

| Parameter | Al | Ti/Pt | Polycrystalline Si | $TiSi_2$ | NiFe |
|---|---|---|---|---|---|
| Heat transfer | a | a | a | a | a |
| Resistivity and TCR | ++ | ++ | – | + | ++ |
| Contact (if any) | ++ | –(TiN) | ++ | + | – |
| Stress | + | +/a | + | a/– | a |
| e⁻ migration | a | a | + | + | a |
| Chemical compatibility | – | + | ++ | a | – |

*Source:* Reprinted with permission from Furjes, P. et al., Materials and processing for realization of micro-hotplates operated at elevated temperature, *J. Micromech. Microeng.*, 12, 425–429, 2002. Copyright 2002, Institute of Physics.

++, excellent; +, good; a, acceptable; –, not acceptable.

**TABLE 14.4**

**Properties of Metallic Materials Used for Humidity Sensors Fabrication**

| Property | Al | Au | Cr | Fe | Cu | Ni | Pt |
|---|---|---|---|---|---|---|---|
| Density, $\rho_m$ (kg/m³) | 2700 | 19320 | 7190 | 7860 | 8960 | 8910 | 21450 |
| Melting point, $T_{mp}$ (°C) | 659 | 1095 | 1765 | 1535 | 1083 | 1453 | 1760 |
| Boiling point, $T_{bp}$ (°C) | 2467 | 2807 | 2672 | 2750 | 2567 | 2732 | 3827 |
| Electrical conductivity, $\sigma$ ($10^3$ S/cm) | 377 | 488 | 79 | 112 | 607 | 146 | 94 |
| Temperature coefficient of resistance, $\alpha_r$ ($10^{-4}$/K) | 39 | 34 | 30 | 50 | 39 | 60 | 30 |
| Work function, $\phi$ (eV) | 4.3 | 5.1 | 4.5 | 4.2 | 4.5 | 5.3 | 5.6 |
| Thermal conductivity, $\kappa$ (W/mK) | 236 | 319 | 97 | 84 | 403 | 94 | 72 |
| Specific heat capacity, $C_p$ (J/K kg) | 904 | 129 | 448 | 449 | 385 | 456 | 133 |
| Linear expansivity, $\alpha_l$ ($10^{-6}$ K⁻¹) | 23.1 | 14.2 | 4.9 | 11.8 | 16.5 | 13.4 | 8.8 |
| Young's modulus, $E_m$ (GPa) | 70 | 78 | 279 | 152 | 130 | 219 | 168 |
| Yield strength, Y (MPa) | 50 | 200 | — | 160 | 150 | 148 | <14 |
| Poisson's ratio, $\nu$ | 0.35 | 0.44 | 0.21 | 0.27 | 0.34 | 0.31 | 0.38 |

*Source:* Data extracted from James, A.M. and Lord, M.P. (eds.), *MacMillans Chemical and Physical Data*, MacMillans Press, London, UK, 1992; Lide, D.R., *Handbook of Chemistry and Physics*, CRC Press, Boca Raton, 2004. These values are intended only as a guide and we recommend, wherever possible, validation against other sources. Measurements taken at 20°C where appropriate unless stated otherwise.

Aluminum has the advantage that it adheres well to common oxide substrates, it is easy to deposit, it is only 17% less conductive (for an equivalent thickness) than Au, and it is far less dense. The lower density is significant because reflections of AWs from Au interdigital transducer (IDT) fingers in delay-line applications can cause appreciable pass band ripple in the IDT frequency response. Al's main disadvantage is the relative ease with which it corrodes; this problem is sometimes addressed, particularly for (non-sensor) commercial applications of SAW devices, by passivating the Al using a relatively impermeable layer of a material such as $Si_3N_4$ or AIN.

Although it is not yet common for AW devices, other areas of microelectronics, including integrated humidity sensors, have demonstrated the utility of more exotic metallization, such as Pt–Ti, for demanding, high-temperature applications; this combination would also be very corrosion resistant, though the relatively high density and poor conductivity of Pt are less than optimal for SAW devices.

Despite a fairly large set of metals that can be used in the manufacture of humidity sensors, even in this case the selection of electrode materials requires attention. For example, it was established that the electrical sensitivity of quartz resonators working at high frequencies (e.g., mesa-shaped QCMs) depends not only on the geometry but also on the electrode material. Recently it was found that in mesa-shaped quartz resonators operated in the frequency range 1.4–3.4 GHz, the electromagnetic radiation losses in Al electrodes were significantly greater (by an order of magnitude) than their viscoelastic losses at frequencies higher than 2.3 GHz (Yong et al. 2009). In these experiments it was found that at high frequencies, the vibrating Al electrodes worked as an emitter of electromagnetic waves.

Deposition of metals can be accomplished using various methods, such as thermal evaporation, electron-beam-induced evaporation, and sputtering. However, films prepared using methods of thick-film technology, which are performed by screen printing, are also employed in many cases (Qu 1996). The most commonly used materials for printed electrodes are noble metals, such as Ag, Au, Pt, and palladium. The alloys of these metals are also widely used (Holmes and Loasby 1976)

As for the film thickness, there are no general rules, as each type of humidity sensor has its own requirements for the electrodes used. For example, in planar-capacitive sensors with parallel-plate structure and in field-effect transistors, the upper electrode must be porous and, at the same time, well conductive. Therefore, in such sensors it is preferred to use Au with a thickness of 10–20 nm, which at such thicknesses already has low resistance, but remains permeable to water vapor (Lee and Park 1996). At the same time, in capacitive sensors with interdigital electrodes, the electrodes have to be as thick as possible (Markevicius et al. 2012). SAW-based sensors have their own requirements. The metal film used to make the IDT must be thick enough to offer low electrical resistance and thin enough so that it does not present an excessive mechanical load to the AW. Therefore, typical IDTs have thickness at ~100–200 nm.

## 14.3 MATERIALS FOR HEATER FABRICATION

The large part of the humidity sensors requires temperature stabilization or periodic cleaning through high-temperature annealing. So, the heater is an important part of these devices, which influences their operating characteristics. For instance, the resistive humidity sensors have a clearly expressed sensitivity dependence on operating temperature. Therefore, in the absence of temperature control, even small changes in the resistance of a heater will be accompanied by a change in the sensors' temperature mode. This means that electrode materials used for heaters should be very stable, because they need to provide maximal stability while working at high temperatures. In addition, the heating element material is preferred to have a linear relationship (Eq. 14.1) between its resistance and temperature within the operation range:

$$R = R_0(1 + \alpha(T - T_0)), \tag{14.1}$$

where $T_0$ is the baseline temperature, $R_0$ is the resistance at the baseline temperature, $\alpha$ is the temperature coefficient of resistance, $T$ is the operating temperature, and $R$ is the resistance at the operating temperature. The linear behavior assisted the measurement and control of temperature and simplified the signal processing steps.

To achieve acceptable heating uniformity, heaters for planar sensors are conventionally fabricated in the shape of meander, double spiral, and drive-wheel made of various metals or alloys. However, the meander shape is, thus far, the most extensively studied one due to its simple geometry. The most commonly utilized material is Pt deposited as thin or thick layers (e.g., Dziedzic et al. 1997; Mailly et al. 2001; Mo et al. 2001; Aslam et al. 2004; Dai 2007; Liu et al. 2018). Pt/Ti and Pt/tantalum sandwiches are often chosen for the reasons discussed in the previous section. As we wrote before Pt has excellent chemical properties. It is corrosion resistant and can be operated at elevated temperatures for a long period of time without changing its physical properties. However, for operations above 500°C, Ti is not recommended, because it will diffuse into the Pt layer and form precipitates on the grain boundaries (Ababneh et al. 2017), whereas tantalum shows better behavior due to its function as a diffusion barrier (Tiggelaar et al. 2009; Marasso et al. 2016). Ceramic adhesion films have also been investigated. Ababneh et al. (2017) reported that the titanium-dioxide adhesion layer would not diffuse into Pt for temperatures up to 800°C. Halder et al. (2007) obtained stable performance of electrodes on a platinized substrate at 1000°C, with aluminum oxide as the adhesion layer, due to an increase in grain size. Although $TiO_2$ and $Al_2O_3$ show much better adhesion performance, their applications are restricted by the feasibility of integration into the CMOS process due to their high-temperature deposition conditions.

In some works related to the heater manufacturing, it was suggested to use polysilicon (Vincenzi et al. 2001), NiCr or Si-Ni-Cr alloys (Korotchenkov et al. 1999), SiC (Chen and Mehregany 2007), TiN (Creemer et al. 2008), and $RuO_2$ (Dziedzic et al. 1994; Jelenkovic et al. 2003; Bai et al. 2006). The main advantage of $RuO_2$ is its weak dependence on the temperature. For comparison, the resistance of the Pt film approximately doubles, when it reaches temperatures of 350°C–400°C. Although testing has shown that at $T > 350°C–400°C$ the Pt heater has more stable characteristics. Nevertheless, Pt heaters can be used to monitor temperature low-price sensors. In addition, while using polysilicon, some problems with formation

of stable contacts may occur; while using $RuO_2$, other problems may arise. For example, in an ambient hydrogen atmosphere at temperatures as low as 150°C–250°C, $RuO_2$ tends to reduce to ruthenium (Jelenkovic et al. 2003). In addition to single-layer adhesion films, a Cr/CrN/Pt/CrN/Cr multilayer approach was demonstrated by Chang and Hsihe (2016), who show improved adhesion and structural stability up to 480°C. However, the multilayer approach has not been widely applied, because it introduces extra depositing and etching steps to the fabrication processes.

Doped polysilicon is another widely adopted heater material with linear resistance-temperature relations (Vincenzi et al. 2001), It is a fully CMOS-compatible material, and the adhesion problem no long exists. However, polysilicon is only suitable for operating temperatures below 500°C, beyond which the recrystallization of polysilicon will cause drift in resistance (Ehmann et al. 2001). Special packaging techniques, such as inert gas sealing, are required to alleviate this problem (Samaeifar et al. 2015). However, this technology is not preferred for commercialization due to its higher cost. For microhotplates operating above the stability point of Pt and polysilicon, molybdenum (Mele et al. 2012; Rao et al. 2017) and tungsten (Ali et al. 2008; Wang and Yu 2015; Shao et al. 2016) heaters were studied for applications in harsh conditions. Other than these two metals, Creemer et al. (2008) investigated microhotplates based on CMOS-compatible titanium nitride (TiN) heater and operating temperature can reach up to 700°C, but the high stress of TiN decreases the yield of the device. The materials above are relatively high cost and mainly used in applications above 300°C. More affordable materials, such as Ni, serve as better alternatives for operating temperature below 300°C (Bhattacharyya et al. 2008). Table 14.5 summarizes the maximum temperature and CMOS compatibility of the various heater materials. Doped polysilicon and tungsten (W) are preferred for monolithic CMOS-MEMS devices due to their compatibility with the processes (Gardner et al. 2002), and their temperature range can cover most operating temperatures of the metal oxides.

To avoid catalytic interaction of gases with a heater made of noble metal, the films are frequently coated with a thin,

**TABLE 14.5**

**Comparison of Maximum Temperature and Complementary Metal-Oxide Semiconductor (CMOS) Compatibility Among Various Heater Materials**

| Heater Material | Maximum Temperature (°C) | CMOS Compatibility |
| --- | --- | --- |
| Pt | 600 | No |
| Doped polysilicon | 500 | Yes |
| Ni | 300 | No |
| Mo | 800 | No |
| W | 700 | Yes |
| TiN | 700 | Yes |

*Source:* Reprinted from Liu, H. et al., *Micromachines*, 9, 557, 2018. Published by MDPI as open access.

chemically inert layer of $SiO_2$. Such passivation very often serves as a support for further functional layers in top-down microelectronic technologies. It is necessary to note that the use of passivation of electrode materials allows reducing requirements to their thermodynamic stability. In particular, the indicated approach is used in microhotplate fabrication. As a result, the most microhotplate designers consider polycrystalline Si doped with boron (B) or phosphorus (P) impurities to be a very appropriate material for making heaters and temperature sensors, because with capsulation covering it is stable up to 1000°C (Panchapakesan et al. 2001; Hwang et al. 2011), and it does not contaminate the IC processing equipment. Moreover, when the poly-Si layer is doped with B at a concentration close to $1 \times 10^{20}$ atoms/cm$^3$, its electrical resistance becomes almost independent on the temperature. Therefore, if the poly-Si is employed as a heater, it generates a uniform heat flow, which is not affected by temperature variations. When the poly-Si layer is doped with B at a concentration close to $2 \times 10^{19}$ atoms/cm$^3$, its electrical resistance is a linear function of temperature, which makes it possible to design a reliable thermometer. B is frequently selected as a dopant, because the amount can be accurately controlled in order to predict the electrical resistivity of the poly-Si layer. Following implantation, the structures are annealed to let the B diffuse and to recrystallize the Si layer. No doubt that $TiSi_2$, discussed in the previous section, can be used for making heaters as well.

## REFERENCES

Ababneh A., Al-Omari A.N., Dagamseh A.M.K., Tantawi M., Pauly C., Mucklich F., et al. (2017) Electrical and morphological characterization of platinum thin-films with various adhesion layers for high temperature applications. *Microsyst. Technol.* **23**, 703–709.

Ali S.Z., Udrea F., Milne W.I., Gardner J.W. (2008) Tungsten-based SOI microhotplates for smart gas sensors. *J. Microelectromech. Syst.* **17**, 1408–1417.

Aslam M., Gregory C., Hatfield J.V. (2004) Polyimide membrane for micro-heated gas sensor array. *Sens. Actuators B* **103**, 153–157.

Bai Z., Wang A., Xie C. (2006) Laser grooving of $Al_2O_3$ plate by a pulsed Nd:YAG laser: Characteristics and application to the manufacture of gas sensors array heater. *Mater. Sci. Eng. A* **435–436**, 418–424.

Ballantine D.S. Jr, White R.M., Martin S.J., Ricco A.J., Frye G.C., Zellers E.T., Wohltjen H. (1997) *Acoustic Wave Sensors: Theory, Design, and Physico-Chemical Applications.* Academic, San Diego, CA.

Bhattacharyya P., Basu P.K., Mondal B., Saha H. (2008) A low power MEMS gas sensor based on nanocrystalline ZnO thin films for sensing methane. *Microelectron. Reliab.* **48**, 1772–1779.

Chang W.-Y., Hsihe Y.-S. (2016) Multilayer microheater based on glass substrate using MEMS technology. *Microelectron. Eng.* **149**, 25–30.

Chen L., Mehregany M. (2007) Exploring silicon carbide for thermal infrared radiators. In: *Proceedings of the 6th IEEE Sensors Conference*, Atlanta, GA, October 28–31, 2007, pp. 620–623.

Creemer J.F., Briand D., Zandbergen H.W., Vlist W., Boer C.R., Rooij N.F., Sarro P.M. (2008) Microhotplates with TiN heaters. *Sens. Actuators A* **148**, 416–421.

Dai C.L. (2007) A capacitive humidity sensor integrated with micro heater and ring oscillator circuit fabricated by CMOS-MEMS technique. *Sens. Actuators B* **122**, 375–380.

Dziedzic A., Golonka L.J., Kozlowski J., Licznerski B.W., Nitsch K. (1997) Thick-film resistive temperature sensors. *Meas. Sci. Technol.* **8**, 78–85.

Dziedzic A., Golonka L.J., Licznerski B.W., Hielscher G. (1994) Heaters for gas sensors from thick conductive or resistive films. *Sens. Actuators B* **18–19**, 535–539.

Ehmann M., Ruther P., von Arx M., Paul O. (2001) Operation and short-term drift of polysilicon-heated CMOS microstructures at temperatures up to 1200 K. *J. Micromech. Microeng.* **11**, 397–401.

Esch H., Huyberechts G., Mertens R., Maes G., Manca J., DeCeuninck W., De Schepper L. (2000) The stability of Pt heater and temperature sensing elements for silicon integrated tin oxide gas sensors. *Sens. Actuators B* **65**, 190–192.

Fleischer M., Hollbauer L., Meixner H. (1994) Effect of the sensor structure on the stability of $Ga_2O_3$ sensors for reducing gases. *Sens. Actuators B* **18–19**, 119–124.

Furjes P., Vizvary Z., Adam M., Barsony I., Morrissey A., Ducso C. (2002) Materials and processing for realization of microhotplates operated at elevated temperature. *J. Micromech. Microeng.* **12**, 425–429.

Gardner J.W., Varadan V.K., Awadelkarim O.O. (2002) *Microsensors, MEMS, and Smart Devices.* John Wiley & Sons, Chichester, UK.

Halder S., Schneller T., Waser R. (2007) Enhanced stability of platinized silicon substrates using an unconventional adhesion layer deposited by CSD for high temperature dielectric thin film deposition. *Appl. Phys. A* **87**, 705–708.

Hierlemann A. (2004) *Integrated Chemical Microsensor Systems in CMOS Technology.* Springer, Berlin, Germany.

Hoefer U., Kuhner G., Schweizer W., Sulz G., Steiner K. (1994) CO and $CO_2$ thin film $SnO_2$ gas sensors on Si substrates. *Sens. Actuators B* **22**, 115–119.

Holmes P., Loasby R. (1976) *Handbook of Thick Film Technology.* Electrochemical Publications, Glasgow, UK.

Hsueh H.T., Hsueh T.J., Chang S.J., Hung F.Y., Tsai T.Y., Weng W.Y., Hsu C.L., Dai B.T. (2011) CuO nanowire-based humidity sensors prepared on glass substrate. *Sens. Actuators B* **156**, 906–911.

Hwang W.-J., Shin K.-S., Roh J.-H., Lee D.-S., Choa S.-H. (2011) Development of micro-heaters with optimized temperature compensation design for gas sensors. *Sensors* **11**, 2580–2591.

James A.M., Lord M.P. (eds.) (1992) *MacMillans Chemical and Physical Data.* MacMillans Press, London, UK.

Jelenkovic E.V., Tong K.Y., Cheung W.Y., Wong S.P. (2003) Degradation of $RuO_2$ thin films in hydrogen atmosphere at temperatures between 150°C and 250°C. *J. Microelectron. Reliab.* **43**, 49–55.

Kita J., Engelbrecht A., Schubert F., Groß A., Rettig F., Moos R. (2015) Some practical points to consider with respect to thermal conductivity and electrical resistivity of ceramic substrates for high-temperature gas sensors. *Sens. Actuators B* **213**, 541–546.

Korotcenkov G. (2007) Practical aspects in design of one-electrode semiconductor gas sensors: Status report. *Sens. Actuators B* **121**, 664–678.

Korotchenkov G.S., Dmitriev S.V., Brynzari V.I. (1999) Processes development for low cost and low power consuming $SnO_2$ thin film gas sensors (TFGS). *Sens. Actuators B* **54**, 202–209.

Lee S.P. (2017) Electrodes for semiconductor gas sensors. *Sensors* **17**, 683.

Lee S.P., Park K.J. (1996) Humidity sensitive field effect transistors. *Sens. Actuators B* **35–36**, 80–84.

Lide D.R. (2004) *Handbook of Chemistry and Physics.* CRC Press, Boca Raton, FL.

Liu H., Zhang L., Li K.H.H., Tan O.K. (2018) Microhotplates for metal oxide semiconductor gas sensor applications—Towards the CMOS-MEMS monolithic approach. *Micromachines* **9**, 557.

Mailly F., Giani A., Bonnot R., Temple-Boyer P., Pascal-Delannoy F., Foucaran A., Boyer A. (2001) Anemometer with hot platinum thin film. *Sens. Actuators A* **94**, 32–38.

Marasso S.L., Tommasi A., Perrone D., Cocuzza M., Mosca R., Villani M., et al. (2016) A new method to integrate ZnO nano-tetrapods on MEMS micro-hotplates for large scale gas sensor production. *Nanotechnology* **27**(38), 385503.

Markevicius V., Navikas D., Valinevicius A., Andriukaitis D., Cepenas M. (2012) The soil moisture content determination using interdigital sensor. *Electronika Electrotech.* **18**(10), 25–28.

Meixner H., Lampe U. (1996) Metal oxide sensors. *Sens. Actuators B* **33**, 198–202.

Mele L., Santagata F., Iervolino E., Mihailovic M., Rossi T., Tran A.T., et al. (2012) A molybdenum MEMS microhotplate for high-temperature operation. *Sens. Actuators A* **188**, 173–180.

Michel H.-J., Michel H.-J., Leiste H., Halbritter J. (1995) Structural and electrical characterization of PVD-deposited $SnO_2$ films for gas-sensor application. *Sens. Actuators B* **24–25**, 568–572.

Mo Y.W., Okawa Y., Tajima M., Nakai T., Yoshiike N., Katukawa K. (2001) Micro-machined gas sensor array based on metal film micro-heater. *Sens. Actuators B* **79**, 175–181.

Panchapakesan B., DeVoe D.L., Widmaier M.R., Cavicchi R., Semancik S. (2001) Nanoparticle engineering and control of tin oxide microstructures for chemical microsensor applications. *Nanotechnology* **12**, 336–349.

Park J.H., Kim K.H. (1999) Improvement of long-term stability in $SnO_2$-based gas sensor for monitoring offensive odor. *Sens. Actuators B* **56**, 50–58.

Prudenziati M. (ed.). (1994) *Thick Film Sensors* (Middelhoek S. series ed.) *Handbook of Sensors and Actuators*, Vol. 1. Elsevier, Amsterdam, the Netherlands.

Qu W. (1996) Effect of electrode materials on the sensitive properties of the thick-film ceramic humidity sensor. *Solid State Ionics* **83**, 257–262.

Rao L.L.R., Singha M.K., Subramaniam K.M., Jampana N., Asokan S. (2017) Molybdenum microheaters for MEMS-based gas sensor applications: Fabrication, electro-thermo-mechanical and response characterization. *IEEE Sens. J.* **17**, 22–29.

Samaeifar F., Hajghassem H., Afifi A., Abdollahi H. (2015) Implementation of high-performance MEMS platinum micro-hotplate. *Sens. Rev.* **35**, 116–124.

Shao F., Fan J.D., Hernandez-Ramirez F., Fabrega C., Andreu T., Cabot A., et al. (2016) $NH_3$ sensing with self-assembled ZnO-nanowire µHP sensors in isothermal and temperature-pulsed mode. *Sens. Actuators B* **226**, 110–117.

Shim Y.-S., Moon H.G., Kim D.H., Jang H.W., Kang C.-Y., Yoon Y.S., Yoon S.-J. (2011) Transparent conducting oxide electrodes for novel metal oxide gas sensors. *Sens. Actuators B* **160**, 357–363.

Sozza A., Dua C., Kerlain A., Brylinski C., Zanoni E. (2004) Long-term reliability of Ti–Pt–Au metallization system for Schottky contact and first-level metallization on SiC MESFET, *Microelectron. Reliab.* **44**, 1109–1113.

Tiggelaar R.M., Sanders R.G.R., Groenland A.W., Gardeniers J.G.E. (2009) Stability of thin platinum films implemented in high-temperature microdevices. *Sens. Actuators A* **152**, 39–47.

Vincenzi D., Butturi M.A., Guidi V., Carotta M.C., Martinelli G., Guarnieri V., et al. (2001) Development of a low-power thick-film gas sensor deposited by screen-printing technique onto a micromachined hotplate. *Sens. Actuators B* **77**, 95–99.

Wang J., Yu J. (2015) Multifunctional platform with CMOS-compatible tungsten microhotplate for pirani, temperature, and gas sensor. *Micromachines* **6**, 1597–1605.

Wang L., He Y., Hu J., Qi Q., Zhang T. (2011) DC humidity sensing properties of $BaTiO_3$ nanofiber sensors with different electrode materials. *Sens. Actuators B* **153**, 460–464.

Wang J., Yu J. (2015) Multifunctional platform with CMOS-compatible tungsten microhotplate for pirani, temperature, and gas sensor. *Micromachines* **6**, 1597–1605.

William W., Boardman J. (1975) Semiconductor gas sensor and method therefore. U.S. Patent 3901067.

Yong Y.K., Patel M., Vig J., Ballato A. (2009) Effects of electromagnetic radiation on the Q of quartz resonators. *IEEE Trans. Ultrason. Ferroelectr. Freq. Control.* **56**, 353–360.

Zafar Q., Sulaiman K. (2016) Utility of PCDTBT polymer for the superior sensing parameters of electrical response based relative humidity sensor. *React. Funct. Polymers* **105**, 45–51.

# 15 Fundamentals of Microfabrication Technologies

Microtechnology and microfabrication processes are used to produce devices with dimensions ranging from micrometers to millimeters. Microfabrication processes can be effectively applied to yield a single device or thousands of devices such as actuators or different sensors (Eaton and Smith 1997; Bausells 2015; Loizeau et al. 2015). The so-called "batch processing," that is, the fabrication of many devices in parallel, does not only lead to a tremendous cost reduction but also enables the production of array structures or large device series with minute-fabrication tolerances (Hierlemann 2004). In high-volume production, the advantage of batch processing is paramount, and the high development and setup costs amortize. In Volume 2 of this series it was shown that microfabrication techniques can be also used to produce humidity sensors (Fenner and Zdankiewicz 2001). Such approach allows one to significantly improve sensor characteristics in comparison to conventionally fabricated devices, and to develop devices with new functionality that cannot be realized in conventional fabrication technology. Key advantages of microfabricated humidity sensors include the small size of the device and sampling volume, and high reproducibility of transducer/sensor characteristics due to the precise geometric control in the fabrication steps.

## 15.1 CMOS PROCESS AND ITS USE IN THE MANUFACTURE OF HUMIDITY SENSORS

### 15.1.1 CMOS Process: General Consideration

Integrated circuit (IC) fabrication processes, including complementary metal-oxide semiconductor (CMOS) processes, are the most important microfabrication processes. CMOS technology is used in microprocessors, microcontrollers, static RAM, and other digital logic circuits. CMOS technology is also used for several analog circuits, such as sensors, data converters, and highly integrated transceivers for many types of communication. The success of CMOS-technology, which is one of the enabling technologies of the information age, clearly demonstrates the efficiency of microfabrication technologies. Typical microelectronic CMOS-chips generally consist of a substrate, the transistor components, the metal layers, and a passivation layer on top (Hierlemann 2004). The substrate is a silicon wafer, the thickness of which depends on the wafer size: 525 $\mu$m for a four-inch wafer and 850 $\mu$m for an eight-inch wafer. *Silicon* is the standard substrate material for IC fabrication and, hence, the most common substrate material in microfabrication in general (Brand 2005). It is supplied as single-crystal wafers with diameters between 100 and 300 mm. In addition to its favorable electrical properties, single-crystal silicon also has excellent mechanical properties, which enable the design of micromechanical structures: (1) crystalline silicon is a hard and brittle material that deforms elastically until it reaches its yield strength, at which point it breaks; (2) tensile yield strength = 7 GPa (~1500 lb suspended from 1 mm²); (3) Young's Modulus near that of stainless steel - {100} = 130 GPa; {110} = 169 GPa; {111} = 188 GPa; (4) mechanical properties uniform, no intrinsic stress; (5) mechanical integrity up to 500°C; (6) good thermal conductor; (7) low thermal expansion coefficient; and (8) high piezoresistivity.

CMOS processes for digital electronics typically use low-doped (doping concentration in the $10^{16}$ cm$^{-3}$ range) silicon wafers, whereas processes for mixed-signal or analog electronics are often based on high-doped (doping concentration in the $10^{19}$ cm$^{-3}$ range) wafers with a low-doped epitaxial layer to minimize latch-up. The choice of the substrate material might already require a compromise between the requirements for the micromechanical part and the on-chip electronics. For example, the fabrication of membrane structures is typically based on anisotropic silicon etching in a potassium hydroxide (KOH) solution (read Section 15.2.1). High $p$-type doping ($N_A \geq 10^{19}$ cm$^{-3}$) substantially reduces the silicon etch rates in KOH solutions, thus preventing the use of highly $p$-doped CMOS substrates in combination with KOH etching.

The fabrication of ICs using CMOS or BiCMOS technology is based on four basic microfabrication techniques: deposition, patterning, doping, and etching (Brand 2005). In standard CMOS IC, up to eight metal layers consisting of aluminum (down to a feature size of 0.18 $\mu$m) or copper (0.13 and 0.09 $\mu$m CMOS) are used to wire the electronic components and to establish connections to the outside world (bondpads). Intermetal oxide (Si-oxide) layers are used as electrical insulator between the different metal layers. Finally, silicon nitride, silicon oxinitride, or silicon-oxide layers passivate the device and protect the electronics. Photolithography and chemical wet and plasma etching are used to pattern and form the silicon platform and measurement structures. Oxidation and epitaxy may also be involved in the CMOS process. Figure 15.1 illustrates how these techniques are combined to build up an IC layer by layer: a thin film, such as an insulating silicon dioxide film, is deposited on the substrate, a silicon wafer. A light-sensitive photoresist layer is then deposited on top and patterned using photolithography. Finally, the pattern is transferred from the photoresist layer to the silicon dioxide layer by an etching process. After removing the

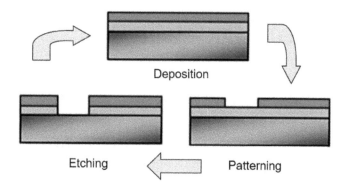

**FIGURE 15.1** Main fabrication technologies involved in CMOS process.

remaining photoresist, the next layer is deposited and structured, and so on. Doping of a semiconductor material by ion implantation, the key step for the fabrication of diodes and transistors, can be performed directly after photolithography, that is using a photoresist layer as a mask or after patterning an implantation mask (e.g., a silicon dioxide layer).

It is important to note that indicated standard processing steps originating from semiconductor technology can be used for fabrication of solid-state humidity sensors discussed in Volume 2 (Dennis et al. 2015). Humidity sensors of this type are fabricated using standard CMOS process with some form of post-CMOS micromachining or other additional post-CMOS processes to complete the sensor design. A hybrid CMOS process is also can be used to deposit, pattern, and activate the humidity-sensing element. Thick-film technology uses screen-printing techniques to apply and pattern conducting pastes or inks onto ceramic substrates to form interconnects and passive components such as resistors (Prudenziati and Morten 1986). Typical film thickness is usually greater than 5 μm. Table 15.1 summarizes the key features of each technology (Prudenziati and Morten 1986; Prudenziati 1994; Fenner and Zdankiewicz 2001). One should note that material purity and process cleanliness during CMOS-based sensor fabrication are critical. Therefore, semiconductor-class tooling is required for CMOS process realization because contaminants will adversely affect the mechanical and electrical properties of the sensing structures, resulting in parts that fail to meet product specifications.

Taking into account technological processes used for humidity-sensor fabrication, one can highlight the following types of CMOS-based humidity sensors (Dennis et al. 2015):

- CMOS-based humidity sensors
- CMOS-MEMS-based humidity sensors

As dielectric layers and masking coatings in the CMOS processes $SiO_2$, $Si_3N_4$, or $Al_2O_3$, can be used. Their parameters are given in Table 15.2.

### 15.1.2 CMOS-BASED HUMIDITY SENSORS

Miniaturization trends have necessitated the fabrication of resistive and capacitive humidity sensors, as well as field effect or acoustic wave (AW)-based humidity sensors, using CMOS process technology and some additional post-CMOS steps, such as drop-coating or deposition of sensitive materials on the CMOS die. The working principle of CMOS-based humidity sensors is similar to conventional humidity sensors in that they convert information about the humidity level in the air into electrical signals. The changes in the sensitive layer are detected by the respective transducer and translated into current or voltage output. The design and fabrication of different types of CMOS-based humidity sensors have been reported by many researchers (Dai 2002; Gu et al. 2004; Nizhnik et al. 2011; Deng et al. 2014).

### 15.1.3 CMOS-MEMS-BASED HUMIDITY SENSORS

The term MEMS is an acronym of microelectromechanical systems that have both mechanical and electronic components. However, the label MEMS usually is being used to describe both a category of micromechatronic devices and the processes used when manufacturing them. In Japan, MEMS are more commonly known as micromachines, and in European countries, MEMS are more commonly referred to as microsystems technology (MST). Some MEMS don't even have mechanical parts, yet they are classified as MEMS because they are miniaturize structures used in conventional machinery, such as springs, cantilevers, channels, cavities, holes, and membranes. MEMS does offer a challenge in the area of how to effectively package devices that require more than an electrical contact to

**TABLE 15.1**
**Comparison of Microsensor Manufacturing Technologies**

| Manufacturing Technology | Film Thickness (μm) | Critical Dimension (μm) | Aspect Ratio | Topology | Device Size (μm) |
|---|---|---|---|---|---|
| CMOS | <1 | 0.17 | 2:1 | <1 | 1 |
| MEMS | 2–6 | 1.0 | 6:1 | 4–10 | 100 |
| Thick film | >5 | >500 | >50:1 | >200 | >1000 |

*Source:* Data extracted from Fenner, R. and Zdankiewicz, E., *IEEE Sensors J.*, 1, 309–317; Brand, O., Fabrication technology, in: *Advanced Micro and Nanosystems, Vol. 2. CMOS: MEMS*, Baltes, H., Brand, O., Fedder, G.K., Hierold, C., Korvink, J., and Tabata, O., WILEY-VCH Verlag GmbH & Co. KGaA, Weinheim, Germany, pp. 1–64, 2005.

**TABLE 15.2**
**Properties of Dielectric Materials Used for Humidity Sensors Fabrication**

| Property | $Al_2O_3$ | $SiO_2$ | $Si_3N_4$ |
|---|---|---|---|
| Density, $\rho_m$ (kg/m³) | 3965 | 2200 | 3100 |
| Melting point, $T_{mp}$ (°C) | 2045 | 1713 | 1900 |
| Boiling point, $T_{bp}$ (°C) | 2980 | 2230 | — |
| Thermal conductivity, $\kappa$ (W/mK) | 28–35 | 1.4 | 20 |
| Specific heat capacity, $C_p$ (J/K kg) | 730 | 700 | 600–800 |
| Temperature expansivity, $\alpha_l$ ($10^{-6}K^{-1}$) | ~8.0 | 0.4 | 3.3 |
| Dielectric constant, $\varepsilon_r$ | 9–10 | 3.8 | ~7.5 |
| Refractive index | ~1.77 | 1.46 | 2.05 |
| Energy band gap, $E_v$ | 18–23 | 9 | ~5 |
| Young's modulus, $E_m$ (GPa) | — | 57–85 | 304 |
| Electrical breakdown strength (MV/cm) | 10–35 | 10 | 2.3 |

*Source:* James, A.M. and Lord, M.P. (eds.), *MacMillans Chemical and Physical Data*, MacMillans Press, London, UK, 1992; Lide, D.R., *Handbook of Chemistry and Physics*, CRC Press, Boca Raton, FL, 2004. These values are intended only as a guide and we recommend, wherever possible, validation against other sources.

the out of package. A MEMS usually is designed to achieve a certain engineering functions by electromechanical or electrochemical means. The core element in MEMS generally consists of two principal components: a sensing or actuating element and a signal transudation unit. Therefore, MEMS devices may also be referred to as transducers. MEMS are constructed with both IC-based fabrication techniques and other mechanical fabrication techniques (Gabriel 1995; Madou 1997; Chen et al. 2001; Tao and Bin 2002) such as micromachining ones (Mehregany and Zorman 2001; Tang 2001).

As it was indicated, MEMS is a technology that integrates mechanical elements, sensors, actuators, and electrical and electronic components on a common silicon substrate with feature sizes ranging from millimeters to micrometers. The most significant advantage of MEMS is their ability to communicate easily with electrical elements in semiconductor chips. Furthermore, there are many other advantages for MEMS, such as small size, low-power consumption, low cost, easy integration into systems, and possibility of use for array fabrication. The integration of micromachining processes with CMOS technology can be accomplished in different ways. The additional process steps (or process modules) can either precede the standard CMOS process sequence (pre-CMOS), or they can be performed in between the regular CMOS steps (intra-CMOS) or after the completion of the CMOS process (post-CMOS). In the case of post-CMOS micromachining, the microstructures are built from either the CMOS layers themselves or from additional layers deposited on top of the CMOS wafer. When CMOS layers are used in MEMS devices as structural layers and post-CMOS micromachining is used to release these structures, the resulting devices are known as CMOS-MEMS devices (Lazarus et al. 2010; Yang et al. 2011). The design and fabrication of different types CMOS-MEMS humidity sensors have been reported by many researchers (Dai 2007; Kim et al. 2010; Lazarus et al. 2010; Yang et al. 2011; Dennis et al. 2015).

Some of the primary mechanical elements utilized in the development of CMOS-MEMS humidity sensors included microsized cantilevers and membranes or plates.

## 15.2 FEATURES OF MICROMACHINING PROCESSES

As it was shown in Volume 2, micromachined structures such as membranes and cantilevers are widely used in humidity sensors. Membranes are required for thermal isolation in thermal-conductivity humidity sensors, as well as for fabrication-resistive piezoelectric and surface acoustic wave (SAW) humidity sensors and whereas cantilevers can be used as resonant structures for mass-sensitive humidity sensors. More details on micromachining techniques can be found in dedicated books on microsystem technology (Madou 1997; Hsu 2002). The micromachining techniques are categorized into bulk micromachining, surface micromachining, and Lithographie, Galvanoformung, Abformung (LIGA) micromachining processes.

### 15.2.1 BULK MICROMACHINING

Bulk micromachining is the most used of the two principal silicon micromachining technologies (Lang et al. 1994a, 1994b; Lang 1996; Gardner et al. 2002). It emerged in the early 1960s and has been used since then in the fabrication of many different microstructures. Bulk micromachining is utilized in the manufacture of the majority of commercial devices—almost all pressure sensors and silicon valves and 90 percent of silicon acceleration sensors. The term *bulk micromachining* expresses the fact that this type of micromachining is used to realize micromechanical structures within the bulk of a single-crystal silicon (SCS) wafer by selectively removing the wafer material. The microstructures fabricated using bulk micromachining may cover the thickness range from submicrons to the thickness of the full wafer (200–500 μm) and the lateral size ranges from microns to the full diameter of a wafer (75–200 mm). Etching is the key technological step for bulk micromachining. Bulk micromachining techniques can be classified into *isotropic* and *anisotropic* etching techniques (structure geometry), or into wet- and dry-etching techniques (liquid or plasma and reactive ion etching) (Hierlemann 2004). In the case of isotropic etching, the same etch rate applies to all directions (Figure 15.2a), whereas in the case of anisotropic etching, the substrate is preferentially etched away along certain crystal planes while it is preserved in other directions (Figure 15.2b). Various etch-stop techniques are an important element of these technologies.

### 15.2.1.1 Wet Etching

Wet chemical etching is widely used in semiconductor processing (Gardner et al. 2002). It is used for lapping and polishing to give an optically flat and damage-free surface and to remove contamination that results from wafer handling and storing. Most importantly, it is used in the fabrication of discrete devices and ICs of relatively large dimensions to delineate patterns and to open windows in insulating materials.

### 15.2.1.1.1 Isotropic and Orientation-Dependent Wet Etching

It is worth noting that in standard semiconductor processing, most of the wet-etching processes are isotropic, that is, unaffected by crystallographic orientation. The most common *isotropic wet* silicon etchant is HNA, a mixture of hydrofluoric acid (HF), nitric acid ($HNO_3$), and acetic acid ($CH_3COOH$): Nitric acid oxidizes the silicon surface, and hydrofluoric acid etches the grown silicon-dioxide layer (Hierlemann 2004). The acetic acid controls the dissociation of $HNO_3$, which provides the oxidation of the silicon. The etch rates and the resulting surface quality strongly depend on the chemical composition (Table 15.3). A mixture of HF and $HNO_3$ can be diluted with water or acetic acid. However, $CH_3COOH$ is preferred, because it prevents $HNO_3$ dissociation.

The etching reaction of silicon with this etchant can be described by Eq. (15.1). The etching process actually occurs in several steps. First step, nitric acid oxidizes the silicon (Eq. 15.2). In the second step, the newly formed silicon dioxide is etched by the HF (Eq. 15.3).

$$HNO_3\left(aq\right) + Si\left(s\right) + 6HF\left(aq\right) \rightarrow H_2SiF_6\left(aq\right)$$
$$+ HNO_2\left(aq\right) + H_2O\left(l\right) + H_2\left(g\right) \tag{15.1}$$

$$HNO_3\left(aq\right) + H_2O\left(l\right) + Si\left(s\right) \rightarrow SiO_2\left(s\right)$$
$$\text{i.} \quad + HNO_2\left(aq\right) + H_2\left(g\right) \tag{15.2}$$

$$SiO_2\left(s\right) + 6HF\left(aq\right) \rightarrow H_2SiF_6\left(aq\right)$$
$$\text{ii.} \quad + 2\,H_2O\left(l\right) \tag{15.3}$$

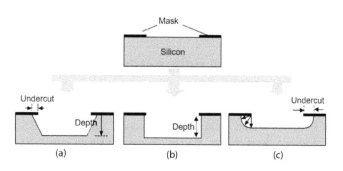

**FIGURE 15.2** Bulk micromachining techniques: (a) wet anisotropic etching, (b) anisotropic DRIE etching, and (c) isotropic etching.

**TABLE 15.3**

**Isotropic Etching of Si, $SiO_2$, and $Si_3N_4$ ($T = 22°C–25°C$)**

| Etchant | Typical Composition (mL) | Application | Etch rate (μm/min) | Masking Films Etch Rate (nm/min) |
|---|---|---|---|---|
| $HF/HNO_3$/(water or $CH_3COOH$) | 10/30/80 | Si | 0.7–3.0 | $SiO_2$ (30) |
| | 25/50/25 | Si | 40 | $Si_3N_4$ |
| | 9/75/30 | Si | 7.0 | $SiO_2$ (70) |
| $HF/HNO_3/HC_2H_3O_2$ | 8/75/17 | Si | 5 | Photoresist $Si_3N_4$ |
| $HF/HNO_3/H_2O$ | 1/50/20 | Poly-Si | 0.31 | Photoresist $Si_3N_4$ (0.2) |
| HF | 48% | $SiO_2$ | 0.020–2.0 | |
| *BOE* (HF/$NH_4F/H_2O$) | 28/113 g/170 | $SiO_2$ | 0.1–0.5 | Photoresist $Si_3N_4$ |
| HF | 48% | $Si_3N_4$ | $7–10 \times 10^{-3}$ | |
| $HF/NH_4F/H_2O$ | 45/200 g/300 | $Si_3N_4$ | $\sim4–7 \times 10^{-3}$ | |
| $H_3PO_4$ (180°C) | 90% | $Si_3N_4/SiO_2$ | $\sim10 \times 10^{-3}$ ($Si_3N_4$) $\sim1 \times 10^{-3}$ ($SiO_2$) | |

BOE—buffered-oxide etch.

**TABLE 15.4**

**Typical Solutions for Metal Wet Etching**

| Metals | Etchants |
|---|---|
| Al | $H_3PO_4$:$HNO_3$:$H_2O$ |
| W, TiW | $H_2O_2$:$H_2O$ |
| Cu | $HNO_3$:$H_2O$ (1:1) |
| Ni | $HNO_3$:$CH_3COOH$:$H_2SO_4$ |
| Au | $KI$:$I_2$:$H_2O$ |
| Pt, Au | $HNO_3$:$HCl$ (1:3) "aqua regia" |

As for the etching of metals used in the manufacture of humidity sensors, the typical etchants used for these purposes are listed in Table 15.4.

However, some wet etchants are orientation dependent, that is, they have the property of dissolving a given crystal plane of a semiconductor much faster than other planes (see Table 15.5). *Anisotropic wet etching* of silicon is the most common micromachining technique and is used to release, for example, membranes and cantilevers for sensors of various types. It was established that in diamond and zinc-blende lattices, the (111) plane is more closely packed than the (100) plane and, hence, for any given etchant, the etch-rate is expected to be slower. It is known that the etch grooves are limited by crystal planes, along which etching proceeds at slowest speed, that is, the (111) planes. In case of (100) silicon wafers, the (111) planes are intersecting the wafer surface at an angle of 54.7°, so that the typical pyramid-shape etch grooves are formed.

A commonly used orientation-dependent etch for silicon consists of a mixture of KOH in water and isopropyl alcohol. Various concentration of KOH can be used; 20–40 wt % is common. For this solution, the ratio of the etch rates for the (100) and (110) planes to the (111) plane are very high and can achieve up to 400:1 and 600:1, respectively. KOH solution is very stable, yields reproducible etching results, is relatively inexpensive, and is, therefore, the most common anisotropic-wet-etching chemical in industrial manufacturing. The disadvantages of KOH include the relatively high $SiO_2$– and Al-etch rates, which require a protection of fabricated structures during etching. The etch rate on thermally grown silicon oxide is ~1.4 nm/min. This means that an etching during 10 hours consumes 0.84 μm of oxide. At the same time, etch rate on silicon nitride prepared using low pressure chemical vapor deposition (LPCVD) method is negligible. Therefore, silicon nitride films are often used as a mask during etching by KOH-based solution. Etching with KOH is typically performed from the back side of the wafer, with the front side protected by a mechanical cover and/or a protective film.

Alternative silicon wet etchants are ammonium hydroxide compounds, such as tetramethyl ammonium hydroxide (TMAH), and ethylene diamine/pyrochatechol (EDP) solutions. TMAN is characterized by the etch rate ~10 μm/min (90°C). At that, $SiO_2$ is virtually unreactive. Some EDP formulations, such as EDP type S, exhibit relatively low Al-etch rates. In addition, etch rate for silicon nitride and silicon oxide is almost negligible; even native oxide becomes important in processing. Such parameters of EDP solutions render them suitable for releasing microstructures on the front side of CMOS-wafer; highly directional selective etching allows using cheap oxide as mask. However, this solution is chemically unstable. Etch rate and color changes with time after exposure to oxygen. More detailed discussions of wet etching of silicon can be found in Madou (1997), Gardner et al. (2002) and Brand (2005).

Unlike isotopic wet etching, rate-limited reactions are preferred for anisotropic wet etching. In this case, it is easier to control the process and it is more repeatable. In the case of diffusion limited reaction, the reaction products may remain on the surface. Other differences between anisotropic wet etching and isotopic wet etching can be found in Table 15.6.

As it was indicated before, mask materials for anisotropic silicon etchants are silicon dioxide and silicon nitride.

**TABLE 15.5**

**Etching Characteristics of Different Anisotropic Wet Etchants for Single-Crystalline Silicon**

| Etchant (typical composition) | $T$, °C | Etch Rate (μm/min) | Anisotropic Etch Rate Ratio [(100)/(111)] | Dopant Dependence | Masking Films (etch rate of mask) |
|---|---|---|---|---|---|
| EDP (Ethylenediamine/ pyrocatechol/water: 750 mL/120 g/100 mL) | 115 | 9.75–1.25 | 35:1 | >$7 \times 10^{19}$ cm$^{-3}$ boron reduces etch rate by about 50 | $SiO_2$ (0.2 nm/min); $Si_3N_4$ (0.1 nm/min); Au, Cr, Ag, Cu, Ta |
| KOH (Potassium hydroxide/ (water or isopropyl); 50 g/100 mL) | 50–85 | 1.0–1.4 | 400:1 | >$10^{20}$ cm$^{-3}$ boron reduces etch rate by about 20 | $Si_3N_4$; $SiO_2$ (1.4 nm/min) |
| $H_2N_4$ (Tetrazadiene/(water or isopropyl): 100 mL/100 mL) | 100 | 2.0 | — | No dependence | $SiO_2$; Al |
| NaOH (Sodium hydroxide/ water: 10 g/100 mL) | 65 | 0.25–1.0 | — | >$3 \times 10^{20}$ cm$^{-3}$ boron reduces etch rate by about 10 | $Si_3N_4$; $SiO_2$ (0.7 nm/min) |
| TMAN (Tetramethyl ammonium/(water or isopropyl): 220 g/780 mL) | 90 | 1.0 | 100:1 | >$10^{20}$ cm$^{-3}$ boron reduces etch rate | $SiO_2$; $Si_3N_4$ |

**TABLE 15.6**

**Features of Isotopic and Anisotropic Wet Si Etching**

| Isotopic Wet Etching | Anisotropic Wet Etching |
|---|---|
| • All crystallographic directions are etched at the same rate. This means that etch rate is the same in all directions | • Etch rate is different for different crystal plane directions |
| • Etchants are usually acids | • Etchants are usually alkaline |
| • Etch temperature: 20...50°C | • Elevated etch temperatures (70°C–120°C) |
| • Isotropy is due to the fast chemical reactions. Reaction is diffusion-limited | • Reaction is rate-limited |
| • Etch rates from X μm/min to XX μm/min | • Different theories propose for anisotropy |
| • Strong undercutting | • Etch depths depend on geometry |
| | • Undercutting also depends on geometry |
| | • Low etch rate (ca. 1 μm/min) |
| | • Small mask undercutting |

*Source:* Gardner, J.W. et al., *Microsensors, MEMS, and Smart Devices*, John Wiley & Sons, Chichester, UK, 2002; Anoop, P.A.B. et al., A review of various wet etching techniques used in micro fabrication for real estate consumption, in: *Proceedings of International Conference on Innovations in Intelligent Instrumentation, Optimization and Signal Processing "ICIIIOSP-2013"*, pp. 26–31, 2013.

It is important to note, that "convex" corners of the etch mask are underetched in case of (100) silicon substrates, leading to, for example, completely underetched cantilever structures. The etch rates in preferentially etched crystal directions such as the (100) and the (110) direction, and the ratio of the etching rates in different crystal directions strongly depend on the exact chemical composition of the etching solution and the process temperature. Figure 15.3 shows orientation-dependent etching of (100)-oriented silicon through patterned silicon dioxide (SiO₂), which acts as a mask. Precise V-grooves, in which the edges are (111) planes at an angle of approximately 55° from the (100) surface, can be realized by the etching. If the etching time is short, or the window in the mask is sufficiently large, U-shaped grooves could also be realized. The width of the bottom surface, *w*, is given by the $w_0$.

### 15.2.1.1.2 Etch-Stop Techniques

Reliable etch-stop techniques are very important for achieving reproducible etching results. The simplest method is to control the etching time based on the known etching rate (Figure 15.4). This is a very simple but inaccurate method, as the etching rate varies with the chemical condition of the etchant and geometrical factors limiting the agitation of the etch. Typical accuracy is ±20 μm. Periodic control of the depth of the etched cavity in appropriate time intervals until desired depth is reached increases accuracy. Uneven etching depth from cavity to cavity due to chemical and geometrical factors is still a problem. Typical accuracy for this method is ±10 μm.

As already mentioned, wet-anisotropic-silicon etchants "stop" etching, that is, the etch rate is reduced by at least 1–4 orders of magnitude, as soon as a (111) silicon plane or a silicon dioxide/nitride layer is reached. Selectivity ratio of etchants (Eq. 15.4) to SiO₂ and Si₄N₄ is presented in Table 15.7. The higher the selectivity ratio, the better the mask material is. This means that the formation of such layers in the required place can be used as etch-stop technique.

$$\text{Selectivity ratio} = \frac{\text{Etching rate of silicon}}{\text{Etching rate of the material}}. \quad (15.4)$$

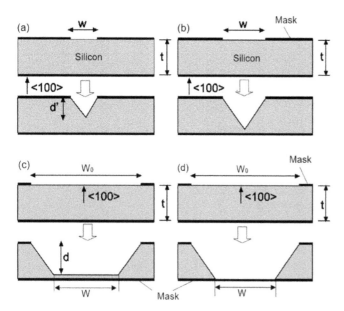

**FIGURE 15.3** (a, b, c, d) Possible variants of anisotropic etching of (100) crystal silicon.

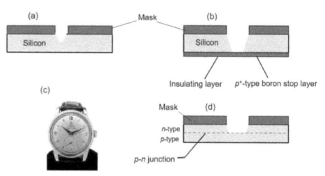

**FIGURE 15.4** Techniques to actively stop the etching process: (a) self-limiting etch or V-groove method, (b) insulator etch stop, (c) etch-rate control or timed etch stop, and (d) etch-stop via doping or the boron etch stop.

## TABLE 15.7
### Selectivity Ratios for Si Etching Using $SiO_2$ and $Si_3N_4$ Stop Layers

| Stop Layer | Etchants | Selectivity Ratios |
|---|---|---|
| $SiO_2$ | KOH | $10^3$ |
|  | TMAN | $10^3$–$10^4$ |
|  | EDP | $10^3$–$10^4$ |
| $Si_3N_4$ | KOH | $10^4$ |
|  | TMAN | $10^3$–$10^4$ |
|  | EDP | $10^4$ |

EDP—ethylene-diamine and pyrocatecol; TMAH—tetramethyl ammonium hydroxide.

In addition, the etch rate is greatly reduced in highly boron-doped regions (doping concentration $\geq 10^{20}$ cm$^{-3}$) (Greenwood 1969; Bohg 1971; Gardner et al. 2002). Doping influence on the rate of silicon etching is shown in Table 15.8. A heavily boron-doped layer, usually used for silicon membranes fabrication, can be epitaxially grown or formed by the diffusion or implantation of boron into a lightly doped substrate. The main benefits of the high-boron etch-stop are the independence of crystal orientation, the smooth-surface finish, and the possibilities it offers for fabricating released structures with arbitrary-lateral geometry in a single etch step. Typical accuracy of this method is $\pm 3$ $\mu$m. On the other hand, the high levels of boron required are known to introduce considerable mechanical stress into the material; this may even cause buckling or even fracture in a diaphragm or other double-clamped structures. Moreover, the introduction of electrical components for sensing purposes into these microstructures, such as the implantation of piezoresistors, is inhibited by the excessive background doping. The latter consideration constitutes an important limitation to the applicability of the high-boron-dose etch-stop. Usually this problem is solving by depositing an epitaxial layer atop the stop layer, with appropriate doping as substrate material for integrated devices (Figure 15.4).

The etching can also be stopped at a $p$-n-junction using a so-called electrochemical etch-stop technique (ECE) (Figure 15.5). Etching stops if the applied potential exceeds a threshold value, called *passivation potential*. This method

has been extensively used to release silicon membranes and $n$-well structures. ECE relies on the passivation of silicon surfaces through application of a sufficiently high-anodic potential with respect to the potential of the etching solution. In this case, one can use very light doping compared to boron etch-stop. This technology is OK with CMOS standards for integrated circuit fabrication. Consequently, this *an electrochemical etch-stop* is currently the most widely used etch-stop technique.

### 15.2.1.2 Dry Etching
Reactive ion (plasma) etching (see Figure 15.6) can be also used for micromachining. In a high-intensity AC electric field, gas molecules will be deassociated and form very active

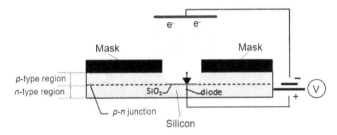

**FIGURE 15.5** "Reverse bias" voltage applied to p-n junction keeps current from flowing.

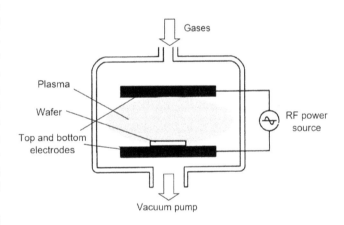

**FIGURE 15.6** Chamber for dry etching: typical parameters: RF frequency, 13.56 MHz; RF power, from tens of watts to kilowatts; pressure, from mTorr to >100 Torr.

## TABLE 15.8
### Dopant-Dependent Etch Rates of Selected Silicon Wet Etchants

| Etchant (diluent) | Temperature (°C) | (100) etch Rate ($\mu$m/min) for Boron Doping $<<10^{19}$ cm$^{-3}$ | Etch Rate ($\mu$m/min) for Boron Doping ~$10^{20}$ cm$^{-3}$ |
|---|---|---|---|
| EDP ($H_2O$) | 115 | 0.75 | 0.015 |
| KOH ($H_2O$) | 85 | 1.4 | 0.07 |
| NaOH ($H_2O$) | 65 | 0.25–1.0 | 0.025–0.1 |

*Source:* Gardner, J.W. et al.: *Microsensors, MEMS, and smart devices*, 2002. Copyright Wiley-VCH Verlag GmbH & Co. KGaA. Reproduced with permission.

radicals, such as Cl⁻ or F⁻, which can react with silicon or other materials to remove the materials.

*Isotropic dry etching* of silicon is done with xenon difluoride, $XeF_2$. As it is known, spontaneous chemical etching without ion bombardment is a process of isotropic etching of silicon (Ibbotson et al. 1984). The etch rate during Si etching using $XeF_2$ can vary from 1 to 3 μm/min up to 40 μm/min. This vapor-phase etching method exhibits excellent etch selectivity with respect to aluminum, Cr, silicon dioxide, silicon nitride, TiN, and photoresist, all of which can be used as etch masks. The $XeF_2$ etch's selectivity to silicon nitride is better than 100:1. The selectivity to silicon dioxide is reported to be better than 10,000:1. The $XeF_2$ silicon etch rates depend on the loading (size of the overall silicon surface exposed to the etchant) with typical values of approximately 1 μm/min.

*Anisotropic dry etching* of silicon is usually carried out as reactive-ion etching (RIE) with plasma-assisted etching systems. RIE is the most common dry etching method, which combines the effects of chemically active gaseous radicals and physical ion bombardment (Rangelow 1996, 2001; Elwenspoek and Jansen 1998; Franssila and Sainiemi 2008). Due to the combinations of reactive neutrals and an ion bombardment, the etch rate may be 10 times greater than that obtained by considering these contributions separately. By controlling the process parameters, such as process gases and process pressure, the etching can be rendered either isotropic or anisotropic. The dry-etching anisotropy originates from experimental parameters such as the direction of the ion bombardment, and is, therefore, independent of the crystal orientation of the substrate material.

Deep reactive ion etching (DRIE) is considered an extension of RIE, but DRIE enables the fabrication of deeper and narrower structures with a higher etch rate than conventional RIE. DRIE reactors are also equipped with two power sources, an inductive coupled plasma (ICP) source for high-density plasma generation, and a capacitive coupled plasma (CCP) source for controlling the ion energies. However, unlike the traditional RIE system, DRIE employs sidewall passivation to enhance process anisotropy. The passivation

layer improves the directionality of the etching. The layer is removed from horizontal surfaces by sputtering, but sidewalls remain protected. DRIE techniques are typically utilized to create sidewalls that are as vertical as possible. DRIE can be used to produce tilted sidewalls as well. By controlling the amount of passivation during the process, both positively and negatively tapered sidewalls are attainable. If passivation is not used, DRIE is capable of producing completely isotropic etch profiles, which is especially beneficial for the release of freestanding structures (Sainiemi 2009).

### 15.2.1.2.1 Features of Reactive Ion Etching

Detailed analysis of RIE can be found in published review articles and books (Winters 1978; Oehrlein 1990; Jansen et al. 1996, 2009; Elwenspoek and Jansen 1998; Kiihamaki and Franssila 1999; Cardinaud et al. 2000; Rangelow 2001; Franssila and Sainiemi 2008; Sainiemi 2009). Analysis of the processes taking place during RIE can be found there as well. According to Li et al. (1994), these processes include: (1) mass transportation of reactive species from plasma to the substrate surface; (2) adsorption of the reactive species to the substrate surface; (3) chemical reaction of the species with the substrate material; (4) desorption of the reaction product from the substrate surface; and (5) removal of the reaction product from the system. Some typical characteristics of the RIE process are in Table 15.9.

For RIE, $SF_6$ (Arens-Fischer et al. 2000), $Cl_2$ (Fischer and Chou 1993), or gas mixtures such as $SF_6/O_2$, $SF_6/C_4F_8$, $SF_6/CHF_3$ (Tserepi et al. 2003) and $Cl_2/BCl_3/H_2$ (French et al. 1997) are usually being used. The use of $SF_6/O_2$ plasma at various flow rates makes it possible to vary the etching directionality between isotropic and anisotropic ones, while $Cl_2$ offers lower etching rates but provides much better profile control (Yoo 2010). During $SF_6/O_2$ plasma etching, F* radicals provide the chemical etching of silicon materials by formation of volatile $SiF_4$. $O_2$ plasma produces O* radicals for sidewall passivation with $Si_xO_yF_z$, which helps to control the etching profiles. A low amount of oxygen in $CF_4$ or $SF_6$ plasma also increases the etch rate of silicon because the recombination

---

**TABLE 15.9**

**Typical Technological Parameters of Si Reactive Ion Etching**

| Plasma | Protection Layer | The Parameter of the RIE-Process |
|---|---|---|
| $SF_6/O_2$ | Photoresist; Al; $SiO_2$; $Al_2O_3$ | Plasmalab System 100 reactor (Oxford Instruments), ICP, CCP-13.56 MHz, |
| $SF_6$ | Ti (100 nm) | Pressure: 20 μ bar; power: 70 W; voltage: 240 V and gas flow: 20 ml/min. |
| $Cl_2$ | Cr (50 nm) | $Cl_2$ and $SiCl_4$ flow rates, 76.6 and 13.3 seem respectively, a power density of 0.32 W/cm², and a pressure of 40 mTorr. |
| $SF_6$; $SF_6/C_4F_8$; $SF_6/CHF_3$ | — | High Density Plasma (HDP) reactor (Micromachining Etching Tool of Alcatel) and in a Reactive Ion Etching (RIE) reactor (NE330 of Nextral) |
| $Cl_2/BCl_3/H_2$ | — | Plasma Therm PK-1250 parallel-plate RIE Power 200–400 W; Pressure 20–40 mtorr |
| $Cl_2(Br_2)/BCl_3$; $SF_6/O_2/CHF_3$ | Cr; Ni; Al; or $SiO_2$ | RF power density: 0.05 W/cm²; ion current density: 0.35 mA/cm² |

*Source:* With kind permission from Taylor & Francis: *Crit. Rev. Sol. St. Mater. Sci.*, Silicon porosification: State of the art, 35, 2010a, 153–260, Korotcenkov, G. and Cho B.K. Copyright 2010. CRC Press.

of fluorine radicals with $CF^{3+}$ or $SF^{5+}$ ions is reduced, which increases the amount of free fluorine (Mogab et al. 1978). Too high oxygen concentration in plasma results in a thick passivation layer, which leads to a reduced etch rate and formation of silicon nanograss (or black silicon, silicon nanoturf, and columnar microstructures) (Jansen et al. 1995).

Formation of a non-volatile passivation layer on the substrate surface during the RIE process has been reported, especially when exploiting $CHF_3$ (or $CF_4/H_2$) plasma or plasma composed of molecules that contain fluorine (e.g., $SF_6$ or $CF_4$) and oxygen (Jansen et al. 1996). In the case of $CHF_3$ plasma, the passivation layer is composed of carbon and fluorine. The role of hydrogen is to catalyze the formation of polymeric precursors such as CF. Hydrogen also reduces the density of free fluorine radicals by forming HF. The quality of the passivation layer strongly depends on the process temperature as well. At cryogenic temperatures, the passivation layer is more stable and less oxygen is required for its formation (Jansen et al. 2009). Examples of RIE process chemistry are shown in Figure 15.7.

### 15.2.1.2.2 Deep Reactive Ion Etching

There are two main DRIE processes. The most common one is the Bosch process (Larmer and Schilp 1996), which is also known as "switched process" or "time multiplexed process." The other one is known as a "cryogenic process" due to its low process temperature (Tachi et al. 1988; De Boer et al. 2002). The Bosch process utilizes a separate passivation step followed by an etching step, while in the cryogenic process the passivation occurs simultaneously with the etching.

The Bosch process is the most widely used DRIE technique. It provides maximal etching rate, which attains 50 μm/min. The processing of masked silicon wafer starts with a short etching step that utilizes $SF_6$ plasma. After this etching step, a thin fluorocarbon film is deposited on the wafer. The fluorocarbon film passivates the surface and prevents etching. Octofluoro cyclobutane ($C_4F_8$) is commonly used in the passivation step. It generates ($CF_2$) $n$ radicals and results in a Teflon-like soft polymer film. At the beginning of the next short etching step, the fluorocarbon film is removed from horizontal surfaces. The $SF_6$ etching step is not anisotropic, but the polymer still etches preferentially from the horizontal surfaces due to directional ion bombardment, while the vertical sidewalls remain protected. The repetition of etching

and passivation cycles results in almost vertical sidewalls (Kiihamaki 2005). In using this technology, many different MEMS devices were fabricated.

The drawback of the process is the scalloping of the sidewalls due to the alternating etching and passivation steps (Andersson et al. 2000). The sidewall roughness can be reduced by shortening the duration of the etching and passivation steps (Sainiemi 2009) or by postprocessing: thermal oxidation followed by oxide etching (Matthews and Judy 2006) or annealing in a hydrogen atmosphere at high temperature (Lee and Wu 2006) reduces the size of the scallops. In the Bosch process, it is also important to have an adequate ratio between ions and radicals. A relative ion concentration that is too high degrades the sidewall profiles. The reactor issues are discussed in a more detailed manner by Walker (2001).

The cryogenic DRIE process does not have separate etching and passivation steps, as they both occur simultaneously. The etching is performed in $SF_6/O_2$ plasma. At cryogenic temperatures ($T < -100°C$), a passivating $SiO_xF_y$ layer forms on the top of the silicon surface (Dussart et al. 2004; Mellhaoui et al. 2005), which again is sputtered away from horizontal surfaces by directional ion bombardment. The substrate temperature plays a key role in cryogenic processes. Therefore, the possibility of controlling the substrate temperature accurately at very low temperatures is crucial. When the temperature is fixed, the thickness of the passivation layer is mainly determined by the $O_2$ flow rate. Too low oxygen flow results in the failing of the passivation layer and isotropic etching profiles, whereas too high oxygen content in the plasma leads to over-passivation, a reduction to the silicon etch rate, and the creation of black silicon (De Boer et al. 2002; Suni et al. 2008; Sainiemi 2009).

Changing the $SF_6/O_2$ ratio is the most convenient way to optimize the passivation layer thickness and, ultimately, the sidewall angles. The etch rate of silicon is mainly dependent on $SF_6$ flow rate and the power of the ICP source. Higher $SF_6$ flow and ICP power increase the quantity of free fluorine radicals that result in the higher etch rate of silicon. The etch rate of the masking material is mainly dependent on the ion energies that are determined by CCP source. The ions must have sufficient energy to remove the passivating layer from horizontal surfaces, but when a certain threshold is reached, an increase in CCP power only increases the etch rate of masking material and undercutting (Sainiemi 2009).

The main advantage of the cryogenic process over the Bosch process is smooth sidewalls. The sidewall quality of structures etched using cryogenic DRIE is superior to the Bosch process. Cryogenic processes typically have higher selectivity than Bosch processes because ions that are at a low energy are already enough to sputter the thin passivation layer (De Boer et al. 2002). However, the etch rate during cryogenic processes is almost 10 times smaller than during Bosch process.

We need to note that the RIE process is complicated enough. In order to obtain high aspect ratio, 2D- and 3D-structures (see e.g., Figure 15.8) in silicon with reactive ion etching, several process conditions should be controllable and properly

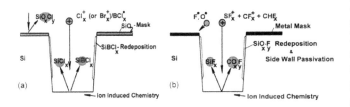

**FIGURE 15.7** The chemistry during the etching of Si in (a) $Cl_2$ (or $Br_2$)/$BCl_3$ plasma, and (b) $SF_6/O_2/CHF_3$ plasma. (Reprinted from *Microel. Eng.*, 27, Rangelow, I.W., Hudek, P., MEMS fabrication by lithography and reactive ion etching (LIRIE), 471–474, Copyright (1995), with permission from Elsevier.)

**FIGURE 15.8** SEM pictures of (a) a silicon chip containing two arrays of membrane-type sensors, and (b) a close view of a cantilever-based sensor fabricated on the same wafer. Their microfabrication process has been presented in (Loizeau et al. 2013). The size of a chip is 13 mm × 10 mm, on which two arrays of sensors, in total 16 sensors, are designed. (Reprinted from *Sens. Actuators A*, 228, Loizeau, F. et al., Comparing membrane- and cantilever-based surface stress sensors for reproducibility, 9–15, Copyright (2015), with permission from Elsevier.)

chosen. They are (Rangelow 1996, 2001): (1) control the ratio of ion flux to radical flux and their density and energy; (2) choice of plasma chemistry; (3) mechanisms for forming sidewall passivants (sidewall passivation engineering) and good knowledge about the surface kinetics; and (4) control of the substrate temperature in a wide range (–100°C to +150°C).

### 15.2.1.2.3 Masking during Reactive Ion Etching

The mask pattern is an important element of RIE technology. The etching processes applied to etch required material should not affect the mask. Pattern transfer requires that a substrate material should be preferentially etched with respect to the masking layer. This parameter is called the etch selectivity. It means that the mask material should be stable under the etching conditions.

Thin photoresist (1 μm) is a good mask material for the RIE of silicon, if only shallow features (<100 μm) are required, and the etching is performed at room temperature (Jansen et al. 1996). Thicker photoresist masks (>1.5 μm) make it possible to create deeper structures, but the use of thick photoresist is typically not desirable due to line width limitations and possible cracking problems (Walker 2001). Photoresists

are also unable to tolerate harsh wet-etching conditions such as heated KOH solutions. The photoresist masks also have a quite limited temperature range because high and low temperatures are known to harm the resist (Walter 1997; Walker 2001; Sainiemi 2009).

If photoresist cannot be used during the silicon-etching step, a hard mask is needed. Usually, hard masks are utilized when deep structures are required. Typically, the selectivity of hard mask materials is at least 1 order of magnitude higher than the selectivity of photoresists. The temperature range permitted by hard masks is also much greater than in the case of photoresists. Hard-mask materials do not suffer from cracking because their coefficients of thermal expansion (CTEs) are better matched to the CTE of silicon (Sainiemi 2009). Hard-mask materials also have better mechanical properties than photoresists. The obvious drawback of all hard masks is the increased amount of complexity in the process because extra deposition and etching steps are required. The isotopic etching of hard-mask material also results in poor dimensional control (Rakhshandehroo and Pang 1996). Deposition of a hard mask may also require the inclusion of high-temperature steps in the process.

The most common hard-mask material is silicon dioxide. The popularity of silicon dioxide is based on its well-explored material properties and designed growth, deposition, and etching techniques. The etch rate of $SiO_2$ during DRIE of silicon is very low, which makes the creation of deep structures possible. According to French et al. (1997), the ratio of etch rates $Si/SiO_2$ could change from 370 for etching in $SF_6/O_2$ up to 20 for etching in $Cl_2/BCl_3/H_2$. The thermal growth of $SiO_2$ requires temperatures around 1000°C, which limits its use. Plasma-enhanced chemical vapor deposition (PECVD) can be done at considerably lower temperatures (ca. 300°C). In plasma etching, the etch rate of PECVD oxide is comparable with thermal oxide (Buhler et al. 1997).

After the lithography, the $SiO_2$ layer is etched and the photoresist is removed. The patterned $SiO_2$ layer now acts as an etch mask. The thickness chosen for this dielectric mask is a compromise and usually is around 100–200 nm. The thicker the mask, the longer it withstands plasma and the deeper one can etch the semiconductor. The thinner the mask, the thinner resist one can use to pattern it and thus achieve higher resolution (Krauss and De La Rue 1999). The temperature range allowed by the $SiO_2$ mask is much wider in comparison with standard photoresists. The stability of $SiO_2$ during the wet etching of silicon is also reasonable. Silicon dioxide has not been reported to inflict surface roughening, changes in etch rate, or pronounced undercutting like some of the metal masks.

Other common hard-mask materials used during DRIE include metals (Tian et al. 2000). Many metal masks such as aluminum and nickel offer easy deposition at room temperature by sputtering or evaporation. Aluminum is commonly utilized as an etch mask because of its wide availability, reasonable price, and the fact that it can easily be etched anisotropically in chlorine-based plasmas (Fedynyshyn et al. 1987a; Mansano et al. 1996; Boufnichel et al. 2005). It was established that metal masks are even more selective than $SiO_2$, but some metals have been reported to affect etch rate (Fedynyshyn et al. 1987a, 1987b), undercutting (Mansano et al. 1996; Boufnichel et al. 2005), and the surface quality of the etched features (Fedynyshyn et al. 1987a; Fleischman et al. 1998). Fedynyshyn et al. (1987a, 1987b) noted that the etch rate of silicon increased in fluorine-containing plasmas when aluminum was used as an etch mask. According to Sainiemi (2009), aluminum catalyzes the generation of free fluorine radicals and thus increases the etch rate. It is also known that using aluminum masks can be accompanied by formation of micro- or nanograss on the etched silicon surfaces due to sputtering and redeposition of the aluminum on the etch field (Fedynyshyn et al. 1987a; Fleischman et al. 1998).

Metal oxides, such as aluminum and titanium oxides (Dekker et al. 2006; Sainiemi 2009), and silicon nitride can be used. Amorphous alumina ($Al_2O_3$) combines the good properties of aluminum and silicon dioxide. This material has extreme selectivity (66,000:1), a fully conformal deposition profile, a deposition temperature of 85°C, and it does not inflict micromasking (Dekker et al. 2006; Chekurov et al. 2007; Grigoras et al. 2007; Sainiemi 2009). The pattern transfer to the $Al_2O_3$ layer can be done accurately, even when isotropic wet etchants such as phosphoric acid or HF are used, because an alumina layer that is just a few nanometers thick is enough for through-wafer etching. According to Chekurov et al. (2007), $Al_2O_3$ mask only a few angstroms thick ($d \sim 0.3$–$0.9$ nm) is required for etching a thickness, $h$, which is defined by the processed structure (typically $1 < h < 20$ µm). Dekker et al. (2006) has shown that for $TiO_2$, the etch rate using high-frequency operation is roughly 10 times higher than that of $Al_2O_3$. In addition, $TiO_2$ etch rates were not as reliable as the $Al_2O_3$ etch rates (more scatter). In addition, $TiO_2$ is less resistant and appears to suffer more from chemical attack.

Silicon nitride is an excellent masking material in KOH etching, but it is consumed quite rapidly during DRIE of silicon. Nevertheless, silicon nitride is also sometimes utilized as an etch mask for DRIE due to its low built-in stresses. Therefore, nitride is well suited for membrane applications (Leivo and Pekola 1998; Zhang et al. 2000).

One should note that the dry-etching process in comparison with wet etching has both advantages and disadvantages indicated in Table 15.10, and therefore the choice of etching method must proceed from the tasks that need to be solved.

### 15.2.1.3 Examples of the Manufacture of Membranes and Cantilevers

It should be noted that currently a large number of technological methods were developed, allowing the production of membranes and cantilevers. Some variants of these approaches to the manufacture of membranes and cantilevers, based on the principles of bulk micromachining, are shown in Figures 15.9 and 15.10.

In more detail, the processes of manufacturing membranes and cantilevers, as applied to humidity sensors, will be discussed in Chapter 16.

### 15.2.2 Surface Micromachining

Surface micromachining comprises a number of techniques to produce microstructures from the thin films, previously deposited onto a substrate, and is based on a sacrificial-layer method (see, e.g., Zeitschel et al. 1999; Gardner et al. 2002). In contrast to bulk micromachining, surface micromachining leaves the substrate intact. The sacrificial layer is deposited over the substrate surface and covered by a second layer (mechanical layer), which is intended to form a membrane.

### TABLE 15.10
### Wet and Dry Etching Comparison

| Parameters | Dry Etching | Wet Etching |
|---|---|---|
| Directionality | Good for most materials | Only with single crystal materials (aspect ratio up to 100) |
| Production-automation | Good | Poor |
| Environment impact | Low | High |
| Masking film adherence | Not as critical | Vey critical |
| Selectivity | Poor | Very good |
| Materials to be etched | Only certain materials | All |
| Process scale up | Difficult | Easy |
| Cleanliness | Conditionally clean | Good to very good |
| Critical dimensional control | Very good (<0.1 µm) | Poor |
| Equipment cost | Expensive | Less expensive |
| Typical etch rate | Slow (0.1 µm/min) to fast (6 µm/min) | Fast (1 µm/min and up) |
| Operational parameters | Many | Few |
| Control of etch rate | Good in case of slow etch | Difficult |

*Source:* Gardner, J.W. et al., *Microsensors, MEMS, and Smart Devices*, John Wiley & Sons, Chichester, UK, 2002; Anoop, P.A.B. et al., A review of various wet etching techniques used in micro fabrication for real estate consumption, in: *Proceedings of International Conference on Innovations in Intelligent Instrumentation, Optimization and Signal Processing "ICIIIOSP-2013,"* pp. 26–31, 2013.

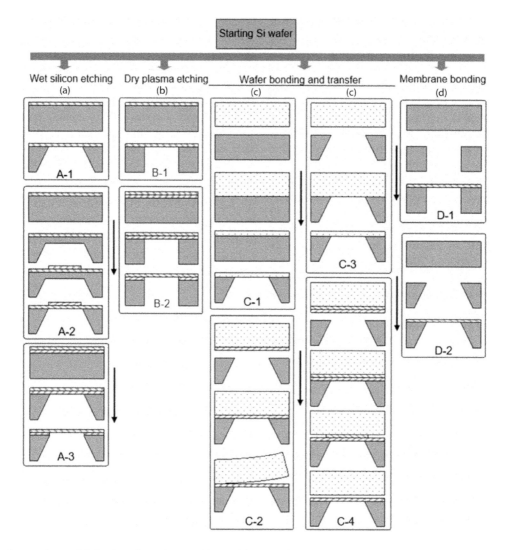

**FIGURE 15.9** Approaches to fabrication of membranes: (a) wet silicon etching, (b) dry plasma etching, (c) wafer bonding and transfer, and (d) membrane bonding.

Then the sacrificial layer is removed by etching, leaving a cavity beneath the membrane layer. The thickness of the sacrificial layer determines the distance of the structural parts from the substrate surface. Clamped beams, cantilevers, microbridges, or microchannels can be fabricated this way. This basic process is illustrated in Figure 15.11. The process can be further expanded to produce multiple mechanical layers to yield more complex mechanical structures.

As it is seen, surface micromachining, unlike bulk micromachining, creates 3D microstructures by adding material to the substrate. This means that there is a little waste of substrate materials. Moreover, added materials may not be the same as the substrate material. This provides flexibility in the manufacture of devices.

The drawback of this method is the limited distance between the membrane and substrate, which is typically several micrometers according to the thickness of the sacrificial layer. Although this distance is usually sufficient for other functional units, such as micromotors, gas sensors often require the thickness of the sacrificial layer to be of the

order of several dozens of micrometers to reduce heat transfer to the substrate. In addition, drying process is more critical during surface micromachining. The gap between each layer is only several micrometers, and therefore this process can be accompanied by stiction (see Figure 15.12). The stiction phenomenon is the collapsing of the layers supported by the sacrificial layers once they are removed by etching. It is the most serious technical problem in surface micromachining. Stiction may occur in the example of the cantilever beam fabricated by surface micromachining. Once stiction takes place, there is little chance to separate the parts again. The stiction is induced by several mechanisms: (1) van der Waals forces, (2) the structure bent by the surface tension of the solution during drying, and (3) the structure bonded to the substrate due to the chemical reaction between the contact area. For resolving this problem, there are the following solutions: (1) reduce the contact area, (2) reduce the drying time (use isopropyl alcohol (IPA) instead of water), (3) dry or supercritical release, (4) add supporting structures, and (5) use harmonic excitation.

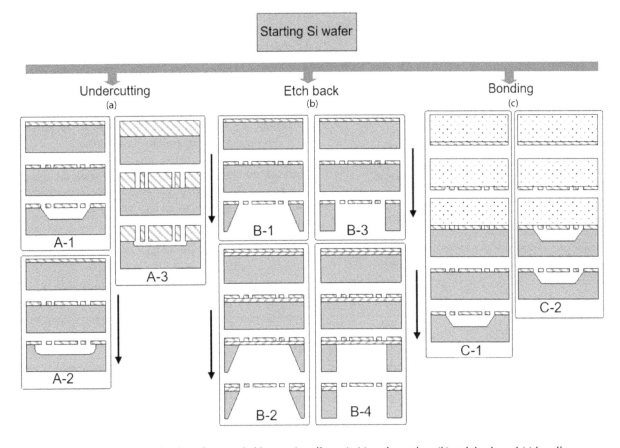

**FIGURE 15.10**    Approaches to fabrication of suspended beams (cantilevers): (a) undercutting, (b) etch back, and (c) bonding.

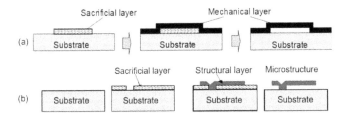

**FIGURE 15.11**    Diagram illustrating surface micromachining: (a) CVD $SiO_2$, poly-Si, etching; and (b) $SiO_2$, poly-Si, etching.

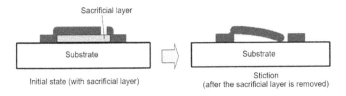

**FIGURE 15.12**    Diagram illustrating the stiction effect.

An important consideration in the fabrication of an ideal mechanical microstructure is that it is without any residual mechanical stress. This means that surface micromachining requires specific technological parameters of film deposition and a compatible set of structural materials, sacrificial materials, and chemical etchants (Gardner et al. 2002). The structural materials must possess the physical and chemical properties that are suitable for the desired application. In addition, the structural materials must have appropriate mechanical properties, such as high yield and fracture strengths, minimal creep and fatigue, and good wear-resistance. The sacrificial materials must also have good mechanical properties to avoid device failure during the fabrication process. These properties include good adhesion and low residual stress to eliminate device failure by delamination and/or cracking. The etchants must have excellent etch selectivity and must be able to etch off the sacrificial materials without affecting the structural ones. In addition, the etchants must also have appropriate viscosity and surface-tension characteristics. One should note that there are a number of choices for both sacrificial and mechanical layers, and some examples are listed in Table 15.11. Similar to KOH etching, the TMAH-based etchant is commonly used for fast removal of sacrificial layer and silicon micromachining. Other disadvantages of surface micromachining include the need to use additional deposition processes and multiple masks, that are expensive and time consuming.

### 15.2.2.1  Features of Surface Micromachining Processes

#### 15.2.2.1.1  Polycrystalline Silicon and Silicon Dioxide

When polycrystalline silicon is used as the structural layer, sacrificial-layer technology normally employs silicon dioxide ($SiO_2$) as the sacrificial material. The poly-Si usually is

**TABLE 15.11**

**Examples of Sacrificial and Mechanical Layers with a Suitable Etchant**

| Sacrificial Layer | Mechanical Layer | Sacrificial Etchant |
|---|---|---|
| Silicon dioxide | Polysilicon, silicon nitride, | HF-based |
| Silicon dioxide | silicon carbide | Pad etch, 73% |
| Polysilicon | Aluminium | HF |
| Polysilicon | Silicon nitride, silicon carbide | KOH-based |
| Resist, polymers | Silicon dioxide | TMAH-based |
| | Aluminium, silicon carbide | Acetone, oxygen plasma |

*Source:* With kind permission from Springer Science+Business Media: *Advances Materials and Technologies for Micro/Nano Devices, Sensors and Actuators*, Smart sensors: Advantages and pitfalls, 2010, 249–259, French, P.J. Copyright 2010: Springer.

deposited by LPCVD, while the $SiO_2$ layer is thermally grown (or LPCVD). The oxide is readily dissolved in HF solution without the poly-Si being affected. Silicon nitride is often used together with this material system for electrical insulation. The advantages of this material system include the following (Gardner et al. 2002):

1. Both poly-Si and $SiO_2$ are used in Integrated circuit (IC) processing; therefore, their deposition technologies are readily available.
2. Poly-Si has excellent mechanical properties and can be doped for various electrical applications. Doping not only modifies the electrical properties but can also modify the mechanical properties of poly-Si. For example, the maximum mechanically sound length of a freestanding beam is significantly larger for phosphorus-doped as compared with undoped poly-Si8. However, in most cases, the maximum length attainable is limited by the tendency of the beam to stick to the substrate.
3. The oxide can be thermally grown and deposited by CVD over a wide range of temperatures (from about 200°C to 1200°C), which is very useful for various processing requirements. However, the quality of the oxide will vary with deposition temperature.
4. The material system is compatible with 1C processing. Both poly-Si and $SiO_2$ are standard materials for IC devices. This commonality makes them highly desirable in sacrificial-layer-technology applications that demand integrated electronics.

### 15.2.2.1.2   Polyimide and Aluminum

In this second material system, the polymer "polyimide" is used for the structural material, whereas aluminum is used for the sacrificial material. Acid-based aluminum etchants

are used to dissolve the aluminum sacrificial layer. The three main advantages of this material system are as follows (Gardner et al. 2002):

1. Polyimide has a small elastic modulus, which is about 50 times smaller than that of polycrystalline silicon.
2. Polyimide can take large strains before fracture.
3. Both polyimide and aluminum can be prepared at relatively low temperatures (<400°C).

However, the main disadvantage of this material system lies with polyimide in that it has unfavorable viscoelastic characteristics (i.e., it tends to creep), and so devices may exhibit considerable parametric drift.

### 15.2.2.1.3   Silicon Nitride/Polycrystalline Silicon and Tungsten/Silicon Dioxide

In the third material system of silicon nitride/poly-Si, silicon nitride is used as the structural material and poly-Si as the sacrificial material. For this material system, silicon anisotropic etchants such as KOH and EDP are used to dissolve the poly-Si.

In the fourth material system of tungsten/oxide, tungsten deposited by CVD is used as the structural material with the oxide as the sacrificial material. Here again, HF solution is used to remove the sacrificial oxide. In addition to the listed materials, phosphosilicate (PSG) and boronphosphosilicate (BPSG) glasses can also be used as sacrificial layers. For their etching, one can use the solution 1:1 $HF:H_2O$ + 1:1 $HCl:H_2O$. The etching rate is specified in Table 15.12.

## 15.2.3   PROCESSES USING BOTH BULK AND SURFACE MICROMACHINING

It is clearly possible to fabricate a variety of microsensor and MEMS devices, including humidity sensors, using either solely bulk-micromachining techniques or surface-micromachining techniques. However, all such devices suffer from limitations that are inherent in one or the other of these two techniques (see Table 15.13). Therefore, recently there have been many developments using the advantages of the fabrication possibilities offered by bulk and surface-micromachining techniques. This is undoubtedly a promising direction, since combining the two techniques in fabricating MEMS devices opens up new opportunities for the fabrication of a new class

**TABLE 15.12**

**Etching Rates for Sacrificial Layers**

| Sacrificial Layers | Lateral Etch Rate (μm/min) |
|---|---|
| CVD $SiO_2$ (densified at 1050°C for 30 min) | 0.6170 |
| Ion-implanted $SiO_2$ (at $8 \cdot 10^{15}$ cm$^{-2}$, 50 KeV) | 0,8330 |
| Phosphosilicate glass (PSG) | 1.1330 |
| Borophosphosilicate glass (BPSG) | 4.1670 |

**TABLE 15.13**

**Comparison Between Bulk and Surface Micromachining Technologies**

| Bulk Micromachining | | Surface Micromachining | |
|---|---|---|---|
| Rugged structures that can withstand vibration and shock | Large die areas that give it high cost | Small die areas that makes it cheaper | Less-rugged structures with respect to vibration and shock |
| Large mass/area (suitable for capacitive sensors) | Not fully integrated with IC process | Fits well within IC process | Small mass/area, which would typically reduce sensitivity |
| Well-characterized material (i.e., Si) | Limited structural geometry possible | Wider range of structural geometry | Some of the materials are not very well understood |

*Source:* Gardner, J.W. et al.: *Microsensors, MEMS, and smart devices*, 2002. Copyright Wiley-VCH Verlag GmbH & Co. KGaA. Reproduced with permission.

of MEMS devices that are not possible to fabricate using either of the technique alone.

For example, using a surface-micromachining one can form an active element, while a silicon cap to protect against external influences can be formed using bulk micromachining. The general concept of a sensing die with moving parts that needs protection is shown in Figure 15.13.

### 15.2.4 OTHER MICROMACHINING TECHNIQUES

#### 15.2.4.1 PSi-Based Micromachining

Lang and co-workers (Lang et al. 1994a, 1994b; Lang 1996) have shown that silicon porosification technology can be also used in micromachining processes. Detailed discussion of the process of silicon porosification is presented in (Korotcenkov and Cho 2010a, 2015; Korotcenkov 2015; Korotcenkov and Brinzari 2015) and in the Chapter 6 of the present series. The typical processes of membrane and cantilever fabrication using porous silicon (PSi) sacrificial layers are presented in

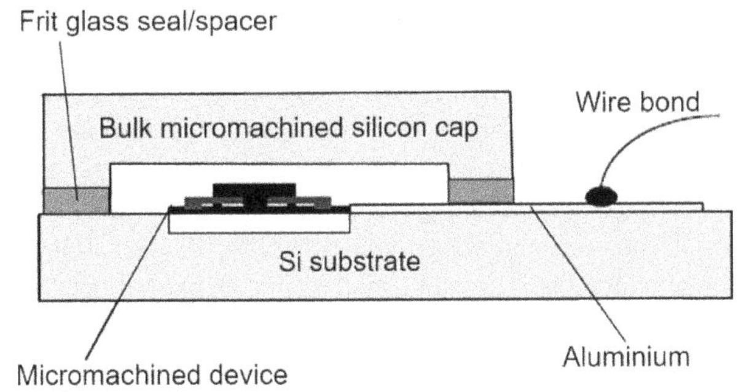

**FIGURE 15.13** Combined surface and bulk micromachining for protection of the sensor from contamination and external influences. Sensing membrane or cantilever formed by surface micromachining are sealed inside a protective cavity formed by the silicon substrate, a bulk micromachined top (cover) wafer. The top-wafer design provides a physical protection and access to the bond pads. When bonded, the cavity that protects the sensor is achieved. One can find a description of the wafer bonding in the Section 15.3. (Idea from Frank, R., *Understanding Smart Sensors*, Artech House, Boston, MA, 2013.)

Figure 15.14 (Korotcenkov and Cho 2010a, 2010b). According to the research (Steiner and Lang 1995; Dusco et al. 1997; Splinter et al. 2001), for application of PSi as a sacrificial layer material, it is necessary to form pores with sizes in the range between 4 and 50 nm (mesoporous silicon). These pore sizes guarantee a physically stable layer. If the dimensions are too small, the porous structure becomes unstable. The removal of PSi for sacrificial layer application can be done with highly diluted alkaline solutions, especially KOH or NaOH at room temperature in a concentration of 0.5–5 wt.%. High-density plasma etching can be used for this purpose as well (Canham et al. 1996).

According to Lang and co-workers (Lang et al. 1994a, 1994b; Lang 1996) and Hedrich et al. (2000), PSi technology has the following specific features:

- Very thick layers can be produced that extend over the whole wafer thickness. This property is employed to develop sensor membranes that need a sacrificial layer thicker than 50 μm, which cannot be obtained using such conventional layers as $SiO_2$ or doped silicate glasses. Due to the large thickness, PSi offers options for creating a wide air gap between the membrane and the substrate.
- Chemical reactions occurring at internal surfaces of the PSi layer are very fast. This permits quick-and-easy removal of the sacrificial layer, even at room temperature, with low-concentration alkaline solutions rather than the aggressive HF employed in other cases.
- Back-side lithography is not necessary, which makes packaging the final sensor and the use of adhesives easier, gives higher yields, and provides better mechanical stability of the wafer.
- There are no geometric restrictions caused by the crystallographic orientation of the substrate. Unlike anisotropic etching, the geometry of the PSi layers is not limited to certain planes, and so they can be placed locally on a wafer with controlled undercutting. This is a considerable advantage in the development of special sensors, for example,

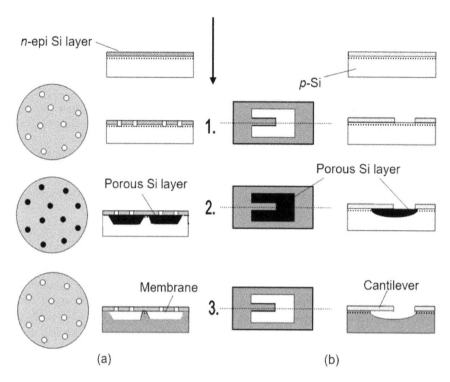

**FIGURE 15.14**  Process sequence for the fabrication of (a) membrane, and (b) cantilever using PSi as a sacrificial layer: 1: photolithography; 2: Si porosification; 3: PSi etching. (With kind permission from Taylor & Francis: *Crit. Rev. Sol. St. Mater. Sci.*, Silicon porosification: State of the art, 35, 2010a, 153–260, Korotcenkov, G. and Cho B.K. Copyright 2010. CRC Press.)

for optical applications where circular designs are required.

- Any PSi layer geometry can be created via appropriate surface-masking layers.
- The PSi layers can be further covered with other thin films, by such as epitaxy processes, which are widely employed in CMOS technologies.

Thus, surface micromachining using PSi layers allows the production of chips of small sizes at low cost. However, the PSi layers are subjected to stresses and stick to the substrates because of the narrow gap, which may limit their application (Lang 1996). In addition, the PSi structures cannot be made by lateral undercutting, and the fabrication of free-standing planes requires holes to be made for the entrance of an etchant. The removal of PSi to make a sacrificial layer is conventionally carried out with a highly dilute alkaline solution, such as KOH or NaOH, at a concentration of 0.5–5 wt% at room temperature or by employing a high-density plasma (Tserepi et al. 2003). Etching the PSi sacrificial layer, made with low-concentration alkali at room temperature makes it possible to retain other chip components. This is especially important when the sacrificial layer is removed as the last step in chip production.

### 15.2.4.2   Lasers in Micromachining

In addition to chemical etching, lasers are used to perform critical trimming and thin-film cutting in semiconductor and sensor processing (Sun and Swenson 2002). The flexibility of laser programming systems allows their usage in marking, thin-film removal, milling, and hole drilling (Gower 2000). The hole diameters and close spacing are achieved without causing fracturing or material degradation. Lasers also provide noncontact residue-free machining in semiconductor products, including sensors (Frank 2013). The precise value of the thin-film resistors in interface circuits is accomplished by interactive laser trimming (Sun and Swenson 2002). Also, lasers can vaporize the material (ablation) using high-power density. Lasers have also been investigated as a means of extending the bulk micromachining process (Alavi et al. 1991). Figure 15.15 shows that either <110> or <100> wafers can be processed using a combination of photolithography, laser melting, and anisotropic etching. A deeper and wider etch occurs in the area that has been damaged. The grooved shape or microchannel obtained by this process has been used to precisely position fibers and spheric lenses in hybrid microoptical devices without requiring additional bonding or capturing techniques.

### 15.2.4.3   LIGA Micromachining

A technique that allows overcoming the two-dimensionality of surface micromachining is the LIGA process. The term LIGA is an acronym for German term in "*Lithography* (Lithographie), *electroforming* (Galvanoformung), and *molding* (Abformung)." The technology was developed in Germany as a method for separation of uranium isotopes using miniaturized nozzles (Ehrfeld et al. 1988). The LIGA process is radically different from silicon-based

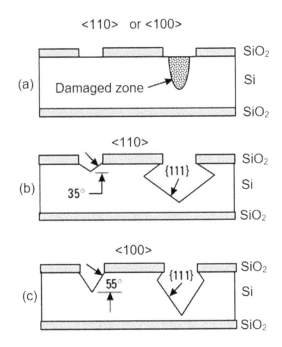

**FIGURE 15.15** Influence of the damaged zones (a) formed by laser beam on the results of anisotropic etching of (110) (b) and (100) (c) silicon wafers in alkaline solution. (Idea from Alavi, M. et al., Laser machining for fabrication of new microstructures, in: *Digest of Technical Papers Presented in International Conference on Solid-State Sensors and Actuators*, TRANSDUCERS'91, June, 24–27, San Francisco, CA, pp. 512–515.)

micromanufacturing: (1) LIGA can produce microstructures that have high aspect ratio; (2) there is no restriction on using silicon or silicon compounds as substrate. Nickel is a common material for LIGA products; and (3) it is easier to be produced in large volumes.

When using LIGA technology, it is possible to produce a microstructure with a height ranging from a few to hundreds of microns, like bulk and surface micromachining relies on lithographic patterning. But instead of ultraviolet light streaming through a photolithographic mask, this process utilizes high-energy X-ray that penetrates several hundred microns into a thick layer of polymer. Exposed areas are stripped away with a developing chemical, leaving a template that can be filled with nickel or another material by electrodeposition (Bacher et al., 1994). What remains may be either a structural element or the master for a molding process. As with surface micromachining, LIGA structures can be processed to etch away an underlying sacrificial layer, leaving suspended or movable structures on a substrate (Hruby 2001). The entire process can be carried out on the surface of a silicon chip, giving LIGA a degree of compatibility with microelectronics (Stadler and Ajmera 2002). The biggest limitation of this technology is the availability of high-energy synchrotrons for the X-ray generation. There are, for instance, no more than ten synchrotrons in the USA. Synchrotron radiation (X-ray) source is a very expensive facility. Advantages and disadvantages of the LIGA process in comparison with previously

**TABLE 15.14**
**Advantages and Disadvantages of Micromachining Processes**

| Technology | Features |
| --- | --- |
| Bulk micromachining | • Less expensive in the process, but material loss is high. |
| | • Suitable for microstructures with simple geometry. |
| | • Limited to low-aspect ratio in geometry. |
| Surface micromachining | • Requires the building of layers of materials over the substrate. |
| | • Complex masking design and productions. |
| | • Etching of sacrificial layers is necessary—not always easy and wasteful. |
| | • The process is tedious and more expensive. |
| | • There are serious engineering problems such as interfacial stresses and stiction. |
| | • Major advantages: |
| | • Not constrained by the thickness of silicon wafers. |
| | • Wide choices of thin film materials to be used. |
| | • Suitable for complex geometry such as microvalves and actuators. |
| The LIGA process | • Most expensive in initial capital costs. |
| | • Requires special synchrotron radiation facility for deep X-ray lithography. |
| | • Microinjection molding technology and facility for mass productions. |
| | • Major advantages are: |
| | • Virtually unlimited aspect ratio of the microstructure geometry. |
| | • Flexible in microstructure configurations and geometry. |
| | • The only technique that allows the production of metallic microstructures. |

*Source:* Madou, M., *Fundamentals of Microfabrication: The Science of Miniaturization*, CRC Press, Boca Raton, FL, 2003; Saile, V. et al., (eds.), *LIGA and its Applications*, Wiley-VCH, Weinheim, Germany, 2009.

**TABLE 15.15**
**Comparison of Bulk, Surface, and LIGA Micromachining Processes**

| Capability | Bulk | Surface | LIGA |
|---|---|---|---|
| Planar geometry | Rectangular | Unrestricted | Unrestricted |
| Min. planar feature size | 2 × depth | 1 μm | 3 μm |
| Sidewall features | 54.74° slope | Dry etch limited | 0.2 μm runout over 400 μm |
| Surface and edge definitions | Excellent | Mostly adequate | Very good |
| Material properties | Very well controlled | Mostly adequate | Well controlled |
| Integration with electronics | Demonstrated | Demonstrated | Difficult |
| Capital investment and cost | Low | Moderate | High |
| Published knowledge | Very high | High | Low |

*Source:* Frank, R., *Understanding Smart Sensors*, Artech House, Boston, MA, 2013.

reviewed processes of surface and bulk micromachining processes are listed in Tables 15.14 and 15.15.

### 15.2.4.3.1 Major Steps in LIGA Process

Major steps in LIGA process are shown in Figures 15.16 and 15.17. As for the materials suitable for use in this process, they are listed in Table 15.16. Poly(methyl methacrylate) (PMMA) is the most popular photoresist for the LIGA process, but other polymers are also available (see Table 15.17).

## 15.3 WAFER BONDING

In addition to micromachining, different types of wafer bonding are needed to produce more complex sensing structures

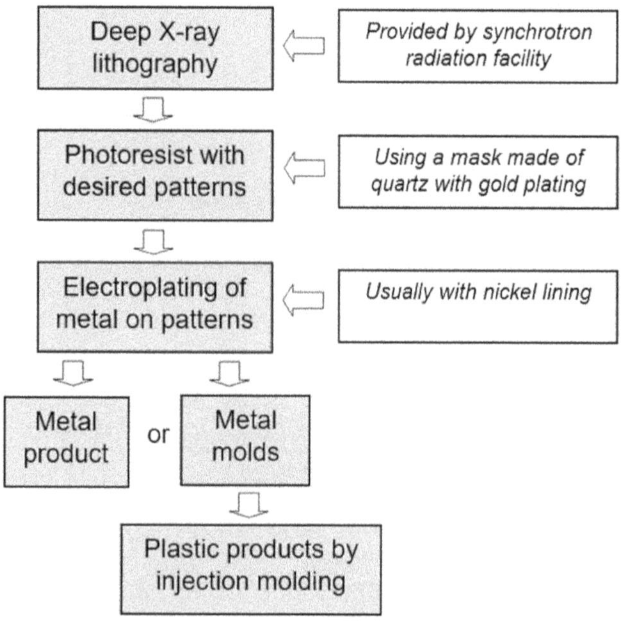

**FIGURE 15.16** Major steps in the LIGA process.

(Peterson and Barth 1989; Frank 2013). The attachment of silicon to a second silicon wafer or silicon to glass is an important aspect of semiconductor sensors. For example, such a need arises when using silicon-on-insulator (SOI) wafers with buried structures (cavities, etc.), heterogeneous integration, fabrication of multilayer devices and 2D structures, and with packaging sensors, especially of the MEMS type (Maszara et al. 1988; Gösele and Tong 1998; Tong and Gösele 1999; Suni et al. 2006a, 2006b). One of the possible options is shown in Figure 15.18. In fact, a clear dependence on wafer bonding as the enabling technology for high-volume MEMS has been identified (Mirza and Ayon 1998). Several different approaches to wafer bonding are possible.

### 15.3.1 DIRECT BONDING (HIGH TEMPERATURE)

Direct-wafer bonding or fusion bonding generally means any joining of two materials without an intermediate layer or external force, including an electrical field. In principle, most materials bond together if their surfaces are flat, smooth, and clean. The principle of this method is simple: two flat, clean, and smooth wafer surfaces are brought into contact and form a weak bonding based on physical forces (Gösele and Tong 1998; Tong and Gösele 1999). The physical forces can be van der Waals forces, capillary forces, or electrostatic forces. The wafer pair is then annealed at high temperature (in the case of hydrophilic Si at >1000°C), and the physical forces are converted to chemical bonds. In the case of silicon, high-temperature bonding falls into two categories: hydrophilic bonding, in which the bonded surfaces are silicon dioxide, and hydrophobic bonding, in which the surfaces are silicon.

Hydrophilic high-temperature bonding is used commercially, for example, in SOI wafer manufacturing. The hydrophilicity of the surface can be enhanced with various methods, of which the most popular is warm SC-1 (1:1:5 $NH_3:H_2O_2:H_2O$ solution). In the case of hydrophilic bonding before bonding, both silicon wafers are usually treated in a solution, such as boiling nitric acid or sulfuric peroxide (Peterson and Barth 1989). This step covers the surface of both wafers with a few

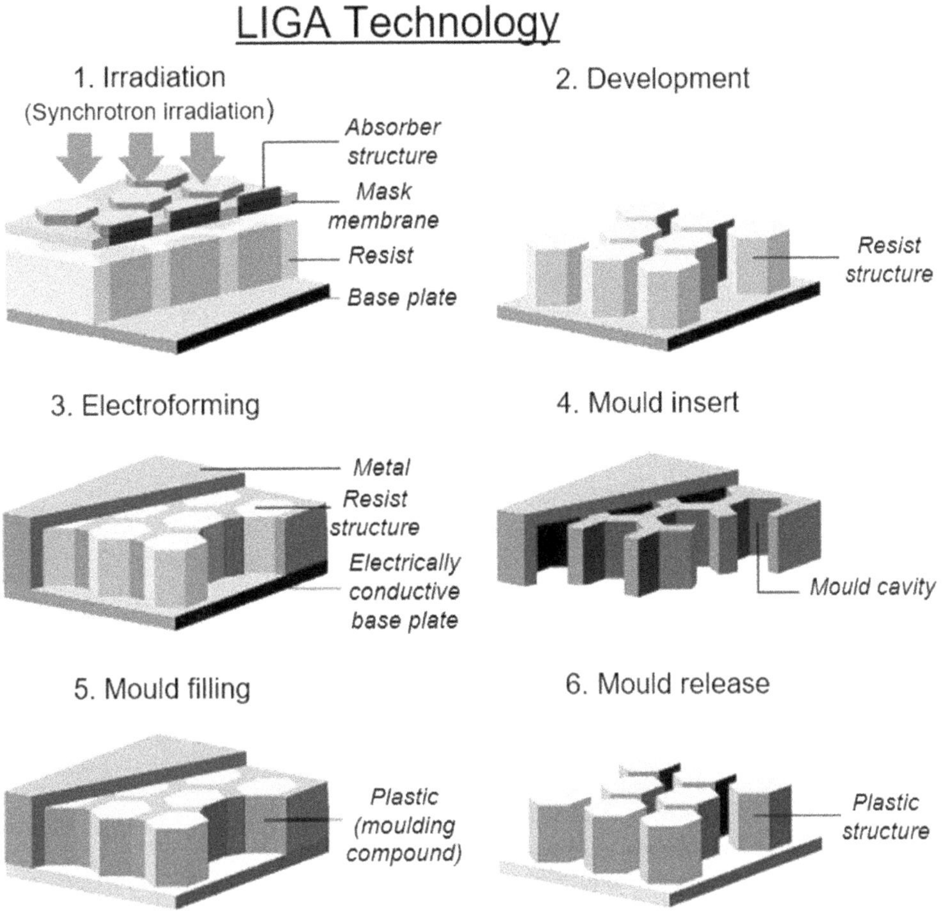

## LIGA Technology

### 1. Irradiation
(Synchrotron irradiation)

Absorber structure

Mask membrane

Resist

Base plate

### 2. Development

Resist structure

### 3. Electroforming

Metal

Resist structure

Electrically conductive base plate

### 4. Mould insert

Mould cavity

### 5. Mould filling

Plastic (moulding compound)

### 6. Mould release

Plastic structure

**FIGURE 15.17** Schematic illustration of the LIGA process. (Adapted from IMM LIGA Technology page. Available online at http://www. imm-mainz.de/.)

### TABLE 15.16
### Requirements to Materials Used in LIGA Process

| Materials | Requirements |
|---|---|
| Materials for substrates | • Substrates in LIGA process must be electrically conductive to facilitate subsequent electroplating over photoresist mold.<br>• Metals such as: steel, copper plates, titanium, and nickel, or<br>• Silicon with thin titanium or silver/chrome top layer; glass with thin metal layers. |
| Photoresist materials | • Must be sensitive to X-ray radiation.<br>• Must have high resolution and resistance to dry and wet etching.<br>• Must have thermal stability up to 140°C.<br>• The unexposed part must be absolutely insoluble during development.<br>• Good adhesion to substrate during electroplating. |

*Source:* Madou, M., *Fundamentals of Microfabrication: The Science of Miniaturization*, CRC Press, Boca Raton, FL, 2003; Saile, V. et al., (eds.), *LIGA and its Applications*, Wiley-VCH, Weinheim, Germany, 2009.

monolayers of reactive hydroxyl molecules. Initial contact of the wafers holds them together through a strong surface tension. Subsequent processing at temperatures from 900°C to 1100°C drives off the hydroxyl molecules. The remaining oxygen reacts with the silicon to form silicon dioxide and fuses the two surfaces. Silicon-fusion bonding can be used to reduce the size of a micromachined structure.

Hydrophobic bonding also provides a strong bond. However, the process requires high purity and flatness of the surface (<1 nm roughness) and high vacuum. Otherwise, high-temperature treatments are required. During annealing at >700°C, surface diffusion of silicon takes place and closes the microgaps between the surfaces (Tong and Gösele 1999). The problem with the bonding process is the presence of hydrogen at the bonded interface, which may cause voids. Tong et al. (1994) have also shown that hydrophilic bonding provides a significantly stronger bond strength at low temperatures, while hydrophobic bonding is more effective at higher temperatures.

The main disadvantage of direct bonding is the high processing temperature. Therefore, this method cannot be used if the wafers are preprocessed, contain humidity-sensitive materials or components, or have different thermal expansion

**TABLE 15.17**
**Polymers Used as Photoresists in LIGA Process**

| Parameter | PMMA | POM | PAS | PMI | PLG |
|---|---|---|---|---|---|
| Sensitivity | Bad | Good | Excellent | Reasonable | Reasonable |
| Resolution | Excellent | Reasonable | Very bad | Good | Excellent |
| Sidewall smoothness | Excellent | Very bad | Very bad | Good | Excellent |
| Stress corrosion | Bad | Excellent | Good | Very bad | Excellent |
| Adhesion on substrate | Good | Good | Good | Bad | Good |

*Source:* Madou, M., *Fundamentals of Microfabrication: The Science of Miniaturization*, CRC Press, Boca Raton, FL, 2003; Saile, V. et al., (eds.), *LIGA and its Applications*, Wiley-VCH, Weinheim, Germany, 2009.
PMMA—Poly(methyl methacrylate); POM—Polyoxymethylene; PAS—p-Aminosalicylic acid; PMI—Polymethacrylimide; PLG—Polypropylene glycols.

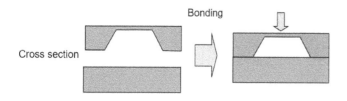

**FIGURE 15.18** Illustration of the wafer-bonding process.

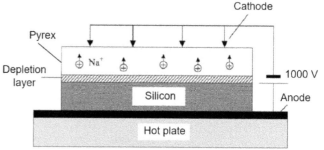

**FIGURE 15.19** Diagram illustrating anodic bonding.

coefficients. Using plasma activation can significantly reduce the processing temperature (Moriceau et al. 2005). The most often reported plasma gases are argon, nitrogen, and oxygen. For example, Bengtsson and Amirfeiz (2000) and Farrens et al. (1995) reported that after such plasma treatment a strong bonding was achieved even at low temperatures (<400°C). Different chemical activation methods have also been published (Plössl Kräuter 1999; Tong and Gösele 1999), but their effectiveness is not as good as plasma activation.

### 15.3.2 ANODIC BONDING

Electrostatic, or anodic, bonding is a process used to attach a silicon top wafer to a glass substrate and also to attach silicon to silicon, glass to glass, and glass to metal. Anodic bonding is widely used for bonding glass substrate to other conductive materials due to its good bond quality. It can serve as a hermetic and mechanical connection between glass and metal substrates or as a connection between glass and semiconductor substrates (Henmi et al. 1994; Rogers and Kowal 1995; Wei et al. 2003). During anodic bonding, the mobile sodium ions from glass move toward the electrode, leaving a negatively charged region into the glass wafer, and electrostatic forces pull the silicon and glass tightly together (see Figure 15.19). Then, the electrochemical reaction takes place and covalent bonds are formed between the glass and silicon. Anodic bonding is less sensitive to the surface roughness than direct bonding, but it requires that one of the wafers is alkali glass. It also requires high voltage and moderate temperatures. One should also note that coefficients of thermal expansion of silicon and glass must be matching; otherwise, cracking upon cooling will be observed, and only certain glasses such as Pyrex and Borofloat are suitable for anodic bonding.

The effect of temperature on the unbonded area for different structures is plotted in Figure 15.20. Studies performed by

Wei (2006) showed that for Si-to-glass anodic bonding, the unbonded area decreases markedly from 1.55% to 0.13% when the bonding temperature is increased from 200°C to 300°C. However, the bubble-free interface cannot be achieved. For a Si-to-glass bonding, the unbonded area is largely reduced. The unbonded area decreases from 0.4% to 0.13% when the bonding temperature is increased from 200°C to 225°C. When the bonding temperature is increased to higher than 250°C and when the voltage is increased beyond 600 V, the total wafer area becomes bonded together. For glass-to-glass bonding, a bubble-free interface can be achieved when the bonding temperature is higher than 275°C. The bond strength also increases with an increase in the bonding temperature (see Figure 15.20b). Therefore, in anodic bonding, the substrates are typically heated to a temperature between 350°C and 450°C when 500 V or more is applied across the structure. In the case of silicon-to-silicon bonding, low-temperature anodic bonding (T~400°C) is only possible when using the intermediate layer bonding approach. Wei et al. (2006) established that with careful control of the cleaning and bonding processes, for direct silicon-to-silicon bonding at temperature of 400°C only bond efficiency of about 90% and bond strength of about 10 MPa can be achieved.

### 15.3.3 BONDING USING INTERMEDIATE LAYERS

Currently, various low-temperature wafer-bonding processes, using intermediate layers such as sodium-rich glass (glass frit), adhesive paste, and metal have been reported. The intermediate layers with thicknesses in the range of several

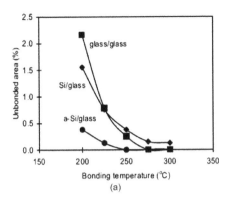
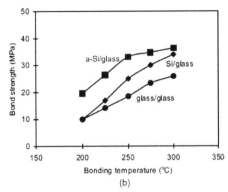

**FIGURE 15.20** (a) Unbonded area versus bonding temperature at voltage of 600 V, bonding force of 200 N, bonding time of 10 minutes, and vacuum of 1 Pa; and (b) bond strength versus bonding temperature. (Reprinted with permission from Wei, J., Wafer bonding techniques for microsystem packaging, *J. Phys.: Conf. Series*, 34, 943–948, 2006. Copyright 2006, Institute of Physics.)

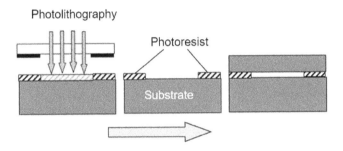

**FIGURE 15.21** Diagram illustrating the process of wafer bonding using photoresist as an adhesive.

microns usually have been deposited on the surface of bottom (constraint) wafer, which is then thermocompression-bonded to the top wafer (see Figure 15.21).

Typical adhesives are different polymers such as photoresists, polyimides, and benzocyclobutane (BCB) (Niklaus et al. 2001, 2006; Pan et al. 2005). Adhesive bonding requires only low-temperature annealing. The bonding process is also cheap, and the requirements to surface smoothness are low. Moreover, this process applicable to almost any material. The disadvantages of these materials are long-term instability and a limited temperature range for their use. Besides, this technique may generate problems such as outgassing, low-positioning accuracy, poor long-term reliability, and uncertain bond quality. However, the adhesive bonding is commonly used in applications where hermeticity is not needed.

Metallic bonding uses an intermediate metal layer between the wafers. It includes thermocompression, solder, and eutectic bonding. In thermocompression bonding, two metallic surfaces (usually gold–gold) are joined using high pressure and intermediate or low temperatures. In solder bonding, the solder balls are fabricated first by electroplating on one of the wafers, then the solder balls are brought into contact with contact pads on another wafer. The bonding is finished with low-temperature solder reflow (Tilmans et al. 2000). In eutectic bonding, typical metal systems are AuSi, AlGe, and AuSn (Enoksson et al. 2005; Kim and Lee 2006). Metallic bonding provides a hermetic seal

at relatively low temperatures (200°C–400°C). Metallic bonding, however, has limitations regarding wafer topography, and it is difficult to use it as a wafer-level bonding process. Metal intermediate layers increase also the difficulty of leading out the electrical circuit from a sealed cavity.

In glass-frit bonding, a glass-paste layer is used as the bonding medium. It usually consists of finely ground (grain size less than 15 μm) lead or lead-silicate glass and organic binder (Knechtel 2005). The temperature needed is above the softening temperature of the glass used, usually still below 450°C. The typical process of glass-frit bonding is shown in Figure 15.22. The advantages of the method are insensitivity to surface roughness, usability with most materials, and cheap price. The disadvantages are incompatibility with IC technology (due to Pb) and the requirement for a wide bonding rim (>200 μm, usually 500 μm). In addition, sodium ions can degrade the performance of electronic devices.

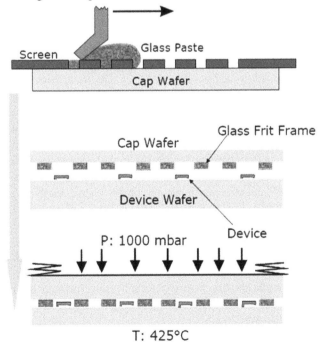

**FIGURE 15.22** Typical process of glass-frit bonding.

**TABLE 15.18**

**Comparison of Typical Wafer Bonding Process Conditions**

| Process | Temperature (°C) | Pressure (Bar) | Voltage (V) | Surface Roughness (nm) | Precise Gaps | Hermetic Seal | Vacuum Level during Bonding (Torr) |
|---|---|---|---|---|---|---|---|
| Silicon-silicon (glass frit) | 400–500 | 1 | N/A | N/A | No | Yes | 10 |
| Anodic | 300–500 | N/A | 100–1,000 | 20 | Yes | Yes | $10^{-5}$ |
| Silicon fusion | 1,000 | N/A | N/A | 0.5 | Yes | Yes | $10^{-3}$ |

*Source:* Mirza, A.R., Ayon A.A., Silicon wafer bonding: Key to MEMS high-volume manufacturing, *Sensors* (Peterborough), 15, 24–33, 1998; Frank, R., *Understanding Smart Sensors*, Artech House, Boston, MA, 2013.

**TABLE 15.19**

**Bonding Techniques Used in Microelectronics and Their Advantages and Drawbacks**

| Technique | | Advantages | Drawbacks |
|---|---|---|---|
| Bonding without interlayer | Anodic | Hermetic; strong bond | High-voltage; bond time; sodium glass; flat surface required |
| | Direct | Hermetic; strong bond | High-temperature; very flat surface required |
| | Direct (surface-activated) | Hermetic; decrease in temperature; varies | Varies |
| Metallic interlayer | Eutectic | Hermetic; strong bond | Specific metals required; flat surface required |
| | Thermocompression | Hermetic; not-flat surface is OK | High forces; specific metals required; |
| | Solder | Hermetic; self-aligning | Specific metals required; solder flow possible |
| Insulating interlayer | Glass frit | Hermetic, common in MEMS; not-flat surface is OK | Large area, medium to high temperature |
| | Adhesive | Versatile; not-flat surface is OK | Non-hermetic |

*Source:* Suni, T., Direct wafer bonding for MEMS and microelectronics, Ph.D. Thesis, Helsinki University of Technology, Espoo, Finland, 2006c.

To avoid the above problems, sol-gel amorphous silica layers can be used (Barbe et al. 2000). This approach has several advantages. First, sol-gel processing is essentially a low-temperature process as the conversion of Si-OH groups into Si-O-Si units can occur at low temperatures (Eaton et al. 1994). Second, it is possible to deposit a thin, uniform and homogeneous layer of amorphous silica on the substrates. This effectively smoothes the initial wafer surface and enables rough surfaces to be bonded. Third, the sol-gel films can be coated both inside or outside of complex shapes over large areas. Finally, sol-gel processing is simple and low cost (Sokoll et al. 1991). Sol-gel intermediate layer bonding is attractive in the applications such as transferring a CMOS circuit to a foreign substrate, combining MEMS and microelectronic circuits on silicon, combining photonic devices such as lasers and detectors with driver or amplifier circuits, or monolithic integration, which describes the integration of disparate devices onto a single chip (London et al. 1999). The devices inside the wafers usually cannot endure high temperatures and high voltages, which is used in direct bonding and anodic bonding, respectively. Sol-gel intermediate layer bonding can provide a stable and insulated interlayer for these applications (Wei 2006).

A comparison of the main methods used for wafer bonding is given in Tables 15.18 and 15.19.

## 15.4 SUMMARY

Our consideration shows that there are many different microfabrication processes and technological approaches to produce devices with specified parameters. Each of these technologies has its advantages and disadvantages. Therefore, each developer, based on the available capabilities and requirements for the device being developed, must choose the process that is most suitable for the implementation of his plans.

## REFERENCES

Alavi M., Buttgenbach S., Schumacher A., Wagner H.J. (1991) Laser machining for fabrication of new microstructures. In: *Digest of Technical Papers Presented in International Conference on Solid-State Sensors and Actuators*, TRANSDUCERS'91, June 24–27, San Francisco, CA, pp. 512–515.

Andersson H., van der Wijngaart W., Enoksson P., Stemme G. (2000) Micromachined flow-through filter-chamber for chemical reactions on beads. *Sens. Actuators B* **67**, 203–208.

Anoop P.A.B., Grace J., Manu C.M. (2013) A review of various wet etching techniques used in micro fabrication for real estate consumption. In: *Proceedings of International Conference on Innovations in Intelligent Instrumentation, Optimization and Signal Processing "ICIIIOSP-2013"*, Foundation of Computer Science, New York, pp. 26–31.

Arens-Fischer R., Kruger M., Thonissen M., Ganse V., Hunkel D., Marso M., Lüth H. (2000) Formation of porous silicon filter structures with different properties on small areas. *J. Porous Mater.* **7**, 223–225.

Bacher W., Menz W., Mohr J. (1994) The LIGA technique and its potential for microsystems. In: *Proceedings of the 20th International Conference on Industrial Electronics, Control, and Instrumentation IECON '94*, Piscataway, NJ, Vol. 3, pp. 1466–1471.

Barbe C.J., Cassidy D.J., Triani G., Latella B.A., Mitchell D.R.G., Finnie K.S., et al. (2000) Low temperature bonding of ceramics by sol-gel processing. *J. Sol-gel Sci. Tech.* **19**, 321–324.

Bausells J. (2015) Piezoresistive cantilevers for nanomechanical sensing. *Microelectronic Eng.* **145**, 9–20.

Bengtsson S., Amirfeiz P. (2000) Room temperature wafer bonding of silicon, oxidized silicon and crystalline quartz. *J. Electron/Mater.* **29** (7), 909–915.

Bohg A. (1971) Ethylene diamine-pyrocatechol-water mixture shows etching anomaly in boron-doped silicon. *J. Electrochem. Soc.* **118**, 401–402.

Boufnichel M., Lefaucheux P., Aachboun S., Dussart R., Ranson P. (2005) Origin, control and elimination of undercut in silicon deep plasma etching in the cryogenic process. *Microelectron. Eng.* **77**, 327–336.

Brand O. (2005) Fabrication technology. In: Baltes H., Brand O., Fedder G.K., Hierold C., Korvink J., Tabata O. (eds.) *Advanced Micro and Nanosystems.* Vol. 2. *CMOS – MEMS*, Wiley-VCH Verlag GmbH & Co. KGaA, Weinheim, Germany, pp. 1–64.

Buhler J., Steiner F.-P., Baltes H. (1997) Silicon dioxide sacrificial layer etching in surface micromachining. *J. Micromech. Microeng.* **7**, R1–R13.

Canham L.T., Cox T.I., Loni A., Simons A.J. (1996) Progress towards silicon optoelectronics using porous silicon technology. *Appl. Surf. Sci.* **102**, 436.

Cardinaud C., Peignon M.C., Tessier P.Y. (2000) Plasma etching: Principles, mechanisms, application to micro- and nano-technologies. *Appl. Surf. Sci.* **164**, 72–83.

Chekurov N., Koskenvuori M., Airaksinen V.-M., Tittonen I. (2007) Atomic layer deposition enhanced rapid dry fabrication of micromechanical devices with cryogenic deep reactive ion etching. *J. Micromech. Microeng.* **17**, 1731–1736.

Chen J., Zou J., Liu C. (2001) A review of MEMS fabrication technologies. *J. Chin. Soc. Mech. Eng.* **22**, 459–475.

Dai C.L. (2002) A capacitive humidity sensor integrated with micro heater and ring oscillator circuit fabricated by CMOS–MEMS technique. *Sens. Actuators B* **122**, 375–380.

De Boer J.M., Gardeniers J.G.E., Jansen H.V., Smulders E., Gilde M.-J., Roelofs G., et al. (2002) Guidelines for etching silicon MEMS structures using fluorine high-density plasmas at cryogenic temperatures. *J. Microelectromech. Syst.* **11**, 385–401.

Dekker J., Kolari K., Puurunen R.L. (2006) Inductively coupled plasma etching of amorphous $Al_2O_3$ and $TiO_2$ mask layers grown by atomic layer deposition. *J. Vac. Sci. Technol. B* **24**, 2350–2355.

Deng F., He Y., Zhang C., Feng W. (2014) A CMOS humidity sensor for passive RFID sensing applications. *Sensors* **14**, 8728–8739.

Dennis J.-O., Ahmed A.-Y., Khir M.-H. (2015) Fabrication and characterization of a CMOS-MEMS humidity sensor. *Sensors* **15**, 16674–16687.

Dusco C., Vazzonyi E., Adam M., Szabo I., Barsony I., Gardeniers J.G.E., Van den Berg A. (1997) Porous silicon bulk micromachining for thermally isolated membrane formation. *Sens. Actuators A* **60**, 235–239.

Dussart R., Boufnichel M., Marcos G., Lefaucheux P., Basillais A., Benoit R., et al. (2004) Passivation mechanisms in cryogenic $SF_6/O_2$ etching process. *J. Micromech. Microeng.* **14**, 190–196.

Eaton W.P., Risbud S.H., Smith R.L. (1994) Silicon wafer-to-wafer bonding at T<200°C with polymethylmethacrylate. *Appl. Phys. Lett.* **65**, 439.

Eaton W.P., Smith J.H. (1997) Micromachined pressure sensors: Review and recent developments. *Smart Mater. Struct.* **6**, 530–539.

Ehrfeld W., Gotz F., Munchmeyer D., Schelb W., Schmidt D. (1988) LIGA process: Sensor construction techniques via X-ray lithography. In: *IEEE Technical Digest of Solid-State Sensor and Actuator Workshop*, Hilton Head Island, SC, pp. 1–4.

Elwenspoek M., Jansen H. (1998) *Silicon Micromachining.* Cambridge University Press, Cambridge, UK.

Enoksson P., Rusu C., Sanz-Velasco A., Bring M., Nafiri A., Bengtsson S. (2005) Wafer bonding for MEMS. In: Hobart K.D., Bengtsson S., Baumgart H., Suga T., Hunt C.E. (eds.) *Semiconductor Wafer Bonding: Science, Technology and Applications VIII.* The Electrochemical Society, Princeton, NJ, pp. 157–172.

Farrens S.N., Dekker J.R., Smith J.K., Roberds B.E. (1995) Chemical free room temperature wafer to wafer direct bonding. *J. Electrochem. Soc.* **142**(11), 3949–3955.

Fedynyshyn T.H., Grynkewich G.W., Hook T.B., Liu M.-D., Ma T.-P. (1987a) The effect of aluminium versus photoresist masking on the etching rates of silicon and silicon dioxide in $CF_4/O_2$ plasmas. *J. Electrochem. Soc.* **134**, 206–209.

Fedynyshyn T.H., Grynkewich G.W., Ma T.-P. (1987b) Mask dependent etch rates II. The effect of aluminium versus photoresist masking on the etching rates of silicon and silicon dioxide in fluorine containing plasmas. *J. Electrochem. Soc.* **134**, 2580–2585.

Fenner R., Zdankiewicz E. (2001) Micromachined water vapor sensors: A review of sensing technologies. *IEEE Sensors J.* **1**(4), 309–317.

Fischer P.B., Chou S.Y. (1993) Sub-50 nm high aspect-ratio silicon pillars, ridges, and trenches fabricated using ultrahigh resolution electron beam lithography and reactive ion etching. *Appl. Phys. Lett.* **62**(12), 1414.

Fleischman A.J., Zorman C.A., Mehregany M. (1998) Etching of 3C-SiC using $CHF_3/O_2$ and $CHF_3/O_2/He$ plasmas at 1.75 Torr. *J. Vac. Sci. Technol. B* **16**, 536–543.

Frank R. (2013) *Understanding Smart Sensors.* Artech House, Boston, MA.

Franssila S., Sainiemi L. (2008) Reactive ion etching. In: Li D. (ed.) *Encyclopedia of Micro and Nanofluidics.* Springer, New York, pp. 1772–1781.

French P.J. (2010) Smart sensors: Advantages and pitfalls. In: Gusev E., Garfunkel E., Dibeikin A. (eds.) *Advances Materials and Technologies for Micro/Nano Devices, Sensors and Actuators*, Springer, Dordrecht, the Netherlands, pp. 249–259.

French P.J., Gennissen P.T.J., Sarro P.M. (1997) New silicon micromachining techniques for microsystems. *Sens. Actuators A* **62**, 652–662.

Gabriel K.J. (1995) Engineering microscopic machines. *Sci. Am.* **273**, 118–121.

Gardner J.W., Varadan V.K., Awadelkarim O.O. (2002) *Microsensors, MEMS, and Smart Devices.* John Wiley & Sons, Chichester, UK.

Gösele U., Tong Q.-Y. (1998) Semiconductor wafer bonding. *Annu. Rev. Mater. Sci.* **28**, 215–241.

Gower M.C. (2000) Industrial applications of laser micromachining. *Opt. Exp.* **7**(2), 56–67.

Greenwood J.C. (1969) Ethylene diamine-cathechol-water mixture shows preferential etching of p-n junction. *J. Electrochem. Soc.* **116**, 1325–1326.

Grigoras K., Sainiemi L., Tiilikainen J., Saynatjoki A., Airaksinen V.-M., Franssila S. (2007) Application of ultra-thin aluminum oxide etch mask made by atomic layer deposition technique. *J. Phys. Conf. Series.* **61**, 369–373.

Gu L., Huang Q.A., Qin M. (2004) A novel capacitive-type humidity sensor using CMOS fabrication technology. *Sens. Actuators B* **99**, 491–498.

Hedrich F., Billat S., Lang W. (2000) Structuring of membrane sensors using sacrificial porous silicon. *Sens. Actuators A* **84**, 315–323.

Henmi H., Shoji S., Shoji Y., Yoshimi K., Esashi M. (1994) Vacuum packaging for microsensors by glass-silicon anodic bonding. *Sens. Actuators A* **43**, 243–248.

Hierlemann A. (2004) *Integrated Chemical Microsensor Systems in CMOS Technology*. Springer, Germany.

Hruby J. (2001) LIGA technologies and applications. *MRS Bull.* **26** (4), 337–340.

Hsu T.R. (2002) *MEMS & Microsystems Design and Manufacturing*. McGraw-Hill, New York.

Ibbotson D.E., Flamm D.L., Mucha J.A., Donnelly V.M. (1984) Comparison of $XeF_2$ and F-atom reactions with Si and $SiO_2$. *Am. Inst. Phys.* **44**(12), 1129–1131.

James A.M., Lord M.P. (eds.) (1992) *MacMillans Chemical and Physical Data*. MacMillans Press, London, UK.

Jansen H., de Boer M., Legtenberg R., Elwenspoek M. (1995) The black silicon method: A universal method for determining the parameter setting of a fluorine-based reactive ion etcher in deep silicon trench etching with profile control. *J. Micromech. Microeng.* **5**, 115–120.

Jansen H., Gardeniers H., de Boer M., Elwenspoek M., Fluitman J. (1996) A survey on the active ion etching of silicon in microtechnology. *J. Micromech. Microeng.* **6**, 14–28.

Jansen H.V., de Boer M.J., Unnikrishnan S., Louwerse M.C., Elwenspoek M.C. (2009) Black silicon method: A review on high speed and selective plasma etching of silicon with profile control: An in-depth comparison between Bosch and cryostat DRIE processes as roadmap to next generation equipment. *J.Micromech. Microeng.* **19**, 033001 (1–41).

Kiihamaki J. (2005) *Fabrication of SOI Micromechanical Devices*. Otamedia Oy, Espoo, Finland.

Kiihamaki J., Franssila S. (1999) Pattern shape effects and artefacts in deep silicon etching. *J. Vac. Sci. Technol. A* **17**, 2280–2285.

Kim J., Lee CC. (2006) Fluxless wafer bonding with Sn-rich Sn-Au dual-layer structure. *Mater. Sci. Eng. A* **417**, 143–148.

Kim J.H., Hong S.M., Moon B.M., Kim K. (2010) High-performance capacitive humidity sensor with novel electrode and polyimide layer based on MEMS technology. *Microsys. Technol.* **16**(12), 2017–2021.

Knechtel R. (2005) Glass frit bonding: A universal technology for wafer level encapsulation and packaging. *Microsyst. Technol.* **12**, 63–68.

Korotcenkov G. (2015) Silicon porosification: Approaches to PSi parameters control. In: Korotcenkov G. (ed.) *Porous Silicon: From Formation to Application*. Vol. 1: *Formation and Properties*, CRC Press, Boca Raton, FL, pp. 69–121.

Korotcenkov G., Brinzari V. (2015) Alternative methods of silicon porosification and properties of these PSi layers. In: Korotcenkov G. (ed.) *Porous Silicon: From Formation to Application*. Vol. 1: *Formation and Properties*, CRC Press, Boca Raton, FL, pp. 245–284.

Korotcenkov G., Cho B.K. (2010a) Silicon porosification: State of the art. *Crit. Rev. Sol. St. Mater. Sci.* **35**(3), 153–260.

Korotcenkov G., Cho B.K. (2010b) Porous semiconductors: Advanced material for gas sensor applications. *Crit. Rev. Sol. St. Mater. Sci.* **35**(1), 1–37.

Korotcenkov G., Cho B.K. (2015) Technology of Si porous layer fabrication using anodic etching: General scientific and technical issues. In: Korotcenkov G. (ed.) *Porous Silicon: From Formation to Application*. Vol. 1: *Formation and Properties*, CRC Press, Boca Raton, FL, pp. 43–68.

Krauss T.F., De La Rue R.M. (1999) Photonic crystals in the optical regime—Past, present and future. *Prog. Quant. Electron.* **23**, 51–96.

Lang W. (1996) Silicon microstructuring technology. *Mater. Sci. Eng. R* **17**, 1–55.

Lang W., Steiner P., Richter A., Marusczyk K., Weimann G., Sandmaier H. (1994b) Applications of porous silicon as a sacrificial layer. *Sens. Actuators A* **43**, 239–242.

Lang W., Steiner P., Schaber U., Richter A. (1994a) A thin film bolometer using porous silicon technology. *Sens. Actuators A* **43**, 185–187.

Larmer F., Schilp A. (1996) Method for anisotropically etching silicon. US-Patent 5,501,893.

Lazarus N., Bedair S.S., Lo C.C., Fedder G.K. (2010) CMOS-MEMS capacitive humidity sensor. *J. Microelectromech. Syst.* **19**, 183–191.

Lee M.C.M., Wu M.C. (2006) Thermal annealing in hydrogen for 3-D profile transformation on siliconon-insulator and sidewall roughness reduction. *J. Microelectromech. Syst.* **15**, 338–343.

Leivo M.M., Pekola J.P. (1998) Thermal characteristics of silicon nitride membranes at sub-Kelvin temperatures. *Appl. Phys. Lett.* **72**, 1305.

Li Y.X., Wolffenbuttel M.R., French P.J., Laros M., Sarro P.M., Wolffenbuttel R.F. (1994) Reactive ion applications. *Sens. Actuators A* **41–42**, 317–323.

Lide D.R. (2004) *Handbook of Chemistry and Physics*. CRC Press, Boca Raton, FL.

Loizeau F., Lang H.P., Akiyama T., Gautsch S., Vettiger P., Yoshikawa G., de Rooij N.F. (2013) Piezoresistive membrane-type surface stress sensor arranged in arrays for cancer diagnosis through breath analysis. In: *Proceedings of IEEE International Conference of Micro Electro Mechanical System* 1, pp. 621–624.

Loizeau F., Lang H.P., Akiyama T., Gautsch S., Vettiger P., Yoshikawa G., de Rooij N.F. (2015) Comparing membrane- and cantilever-based surface stress sensors for reproducibility. *Sens. Actuators A* **228**, 9–15.

London J.M., Loomis A.H., Ahadian J.F., Fonstad C.G. (1999) Preparation of silicon-on-gallium arsenide wafers for monolithic optoelectronic integration. *IEEE Photonics Tech. Lett.* **11**, 958–960.

Madou M. (1997) *Fundamentals of Microfabrication*. CRC Press, Boca Raton, FL.

Madou M. (2003) *Fundamentals of Microfabrication: The Science of Miniaturization*. CRC Press, Boca Raton, FL.

Mansano R.D., Verdonck P., Maciel H.S. (1996) Deep trench etching in silicon with fluorine containing plasmas. *Appl. Surf. Sci.* **100/101**, 583–586.

Maszara W.P., Goetz G., Caviglia A., McKitterick J.B. (1988) Bonding of silicon wafers for silicon-on-insulator. *J. Appl. Phys.* **64**(10), 4943–4950.

Matthews B., Judy J.W. (2006) Design and fabrication of a micromachined planar patch-clamp substrate with integrated microfluidics for single-cell measurements. *J. Microelectromech. Syst.* **15**, 214–222.

Mehregany M., Zorman, C.A. (2001) Micromachining techniques for advanced SiC MEMS. In: Argarwal A., Skowronski M., Cooper J.A., Janzen E. (eds.) *MRS Online Proceedings* Vol. 640. doi: 10.1557/PROC-640-H4.3.

Mellhaoui X., Dussart R., Tillocher T., Lefaucheux P., Ranson P. (2005) $SiO_xFy$ passivation layer in silicon cryoetching. *J. Appl. Phys.* **98**, 104901-1–10.

Mirza A.R., Ayon A.A. (1998) Silicon wafer bonding: Key to MEMS high-volume manufacturing. *Sensors (Peterborough)* **15**(12), 24–33.

Mogab C.J., Adams A.C., Flamm D.L. (1978) Plasma etching of Si and $SiO_2$—The effect of oxygen additions to $CF_4$ plasmas. *J. Appl. Phys.* **49**, 3796.

Moriceau H., Rieutord F., Morales C., Sartori S., Charvet A.M. (2005) Surface plasma activation before direct wafer bonding: A short review and recent results. In: Hobart K.D., Bengtsson S., Baumgart H., Suga T., Hunt C.E. (eds.) *Semiconductor Wafer Bonding: Science, Technology and Applications VIII.* The Electrochemical Society, Pennington, NJ, pp. 34–49.

Niklaus F., Andersson H., Enoksson P., Stemme G. (2001) Low temperature full wafer adhesive bonding of structured wafers. *Sens. Actuators A* **91**, 235–241.

Niklaus F., Kumar R.J., McMahon J.J., Yu J., Lu J.-Q., TCale T.S., Gutmann R.J. (2006) Adhesive wafer bonding using partially cured benzocyclobutene for three-dimensional integration. *J. Electrochem. Soc.* **153**(4), G291–G295.

Nizhnik O., Higuchi K., Maenaka K. (2011) A standard CMOS humidity sensor without post-processing. *Sensors* **11**, 6197–6202.

Oehrlein G.S. (1990) Reactive etching. In: Rossnagel S.M. (ed.) *Handbook of Plasma Processing Technology: Reactive Ion Etching.* Noyes, Park Ridge, NJ, pp. 196–232.

Pan C.T., Cheng P.J., Chen M.F., Yen C.K. (2005) Intermediate wafer level bonding and interface behavior. *Microelectron. Reliab.* **45**, 657–663.

Peterson K., Barth P. (1989) Silicon fusion bonding: Revolutionary tool for silicon sensors and microstructures. *WESCON Technical Papers* **33**, 220–224.

Plössl A., Kräuter G. (1999) Wafer direct bonding: Tailoring adhesion between brittle materials. *Mater. Sci. Eng. R* **25**, 1–88.

Prudenziati M. (ed.) (1994) *Thick Film Sensors* (Middelhoek S. (Ser. ed.) *Handbook of Sensors and Actuators).* Elsevier, New York.

Prudenziati M., Morten B. (1986) Thick-film sensors: An overview. *Sens. Actuators* **10**, 65–92.

Rakhshandehroo M.R., Pang S.W. (1996) Fabrication of Si field emitters by dry etching and mask erosion. *J. Vac. Sci. Technol. B* **14**, 612–616.

Rangelow I.W. (1996) *Deep Etching of Silicon.* Oficyna Wydawnicza Politekchniki, Wroclav, Poland.

Rangelow I.W. (2001) Dry etching-based silicon micro-machining for MEMS. *Vacuum* **62**, 279–291.

Rangelow I.W., Hudek P. (1995) MEMS fabrication by lithography and reactive ion etching (LIRIE). *Microel. Eng.* **27**, 471–474.

Rogers T., Kowal J. (1995) Selection of glass, anodic bonding conditions and material compatibility for silicon-glass capacitive sensors. *Sens. Actuators A* **46–47**, 113–120.

Saile V., Wallrabe U., Tabata O., Korvink J.G. (eds.) (2009) *LIGA and Its Applications.* Wiley-VCH, Weinheim, Germany.

Sainiemi L. (2009) Cryogenic deep reactive ion etching of silicon micro and nanostructures. PhD Thesis, Helsinki University of Technology, Espoo, Finland.

Sokoll R., Tiller H.J., Hoyer T. (1991) Thermal desorption and infrared studies of sol-gel derived $SiO_2$ coatings on Si wafer. *J. Electrochem. Soc.* **138**, 2150–2153.

Splinter A., Bartels O., Benecke W. (2001) Thick porous silicon formation using implanted mask technology. *Sens. Actuators B* **76**, 354–360.

Stadler S., Ajmera P.K. (2002) Integration of LIGA structures with CMOS circuitry. *Sensors Mater.* **14**, 151–166.

Steiner P., Lang W. (1995) Micromachining applications of porous silicon. *Thin Solid Films* **255**, 52–58.

Sun Y., Swenson E.J. (2002) Laser micromachining in the microelectronics industry. *Proc. SPIE* **4915**, 17–22.

Suni N., Haapala M., Makinen A., Sainiemi L., Franssila S., Färm E., et al. (2008) Rapid and simple method for selective surface patterning with electric discharge. *Angew. Chem. Int. Ed.* **47**, 7442–7445.

Suni T. (2006c) Direct wafer bonding for MEMS and microelectronics. PhD Thesis, Helsinki University of Technology, Espoo, Finland.

Suni T., Henttinen K., Dekker J., Luoto H., Kulawski M., Mäkinen J., Mutikainen R. (2006b) Silicon-on-insulator wafers with buried cavities. *J. Electrochem. Soc.* **153**(4), G299–G303.

Suni T., Henttinen K., Lipsanen A., Dekker J., Luoto H., Kulawski M. (2006a) Wafer scale packaging of MEMS by using plasma-activated wafer bonding. *J. Electrochem. Soc.* **153**(1), G78–G82.

Tachi S., Kazunori K., Okudaira S. (1988) Low-temperature reactive ion etching and microwave plasma etching of silicon. *Appl. Phys. Lett.* **52**, 616.

Tang W.C. (2001) Surface micromachining: A brief introduction. *MRS Bull.* **26**, 289–290.

Tao Z.H., Bin Z. (2002) Microelectromechanical system technology and its application. *Electron. Components Mater.* **21**, 28–30.

Tian W.-C., Weigold J.W., Pang S.W. (2000) Comparison of $Cl_2$ and F-based dry etching for high aspect ratio Si microstructures etched with an inductively coupled plasma source. *J. Vac. Sci. Technol. B* **18**, 1890–1896.

Tilmans H.A.C., Van der Peer M.D.J., Beyne E. (2000) The indent reflow sealing (IRS) technique: A method for the fabrication of sealed cavities for MEMS devices. *J. Microelectromech. Syst.* **9**(2), 206–217.

Tong Q.-Y., Cha G., Gafiteanu R., Gosele U. (1994) Low temperature wafer direct bonding, *J. Microelectromech. Syst.* **3**, 29–35.

Tong Q.-Y., Gösele U. (1999) *Semiconductor Wafer Bonding: Surface and Technology.* John Wiley & Sons, New York.

Tserepi A., Tsamis C., Gogolides E., Nassiopoulou A.G. (2003) Dry etching of porous silicon in high density plasmas. *Phys. Stat. Sol. (a)* **197**(1), 163–167.

Walker M.J. (2001) Comparison of Bosch and cryogenic processes for patterning high aspect ratio features in silicon. *Proc. SPIE*, **4407**, 89–99.

Walter L. (1997) Photoresist damage in reactive ion etching. *J. Electrochem. Soc.* **144**, 2150–2154.

Wei J. (2006) Wafer bonding techniques for microsystem packaging. *J. Phys. Conf. Series* **34**, 943–948.

Wei J., Xie H., Nai M.L., Wong C.K., Lee L.C. (2003) Low temperature wafer anodic bonding. *J. Micromech. Microeng.* **13**, 217–222.

Winters H.F. (1978) The role of chemisorption in plasma etching. *J. Appl. Phys.* **49**, 5165.

Yang T.Y., Huang J.J., Liu C.Y., Wang H.Y. (2011) A CMOS-MEMS humidity sensor. In: *Proceedings of the International Conference on Circuits, System and Simulation (ICCSS),* Bangkk, Thailand, May 28–29, pp. 212–217.

Yoo J. (2010) Reactive ion etching (RIE) technique for application in crystalline silicon solar cells. *Solar Energy* **84**(4), 730–734.

Zeitschel A., Friedberger A., Welser W., Muller G. (1999) Breaking the isotropy of porous silicon formation by current focusing. *Sens. Actuators A* **74**, 113–117.

Zhang T.-Y., Su Y.-J., Qian C.-F., Zhao M.-H., Chen L.-Q. (2000) Microbridge testing of silicon nitride thin films deposited on silicon wafers. *Acta Mater.* **48**, 2843–2857.

# 16 Micromachining Platforms for Humidity Sensors and Examples of Their Fabrication

As shown in the Volume 2 (Chapter 13) and in Vashist and Korotcenkov (2011), membranes and cantilevers are important components of many humidity sensors, largely determining their sensitivity and power consumption. As a rule, membranes and cantilevers are made using the micromachining techniques described in the previous chapter. Based on the requirements for the sensor, membranes and cantilevers can be formed by different materials with different physical and physicochemical properties. Therefore, the process of their manufacture is not unified. Each type of membrane and cantilever has its own specific manufacturing and its own specific set of processes. Let's look at some of them.

## 16.1 MICROMACHINING MEMBRANES

Suspended membranes are the basis for the development of mass-sensitive humidity sensors and micro-hotplates, designed to heat the sensors in order to restore the original characteristics and reduce the recovery time after interaction with water vapor (read Chapters 12 and 13, Volume 2). As a rule, the interdigital metal electrodes are patterned over the membrane and covered by the humidity-sensing layer using technologies described in Chapter 15. The membrane, therefore, needs to be very robust to allow deposition of the humidity-sensing material and heating to required temperatures. This means that the materials chosen for the membrane of the micro-hotplate should in theory combine low-thermal conductivity—in other words, small thickness—with high mechanical strength—in other words, large thickness (Horrillo et al. 1999). The first requirement provides a reduction in the power necessary to achieve the required temperature, which is critical for portable devices. The second requirement requires reliable production, particularly when the sensing layer is deposited and annealed.

Scanning electron microscope (SEM) views of several hotplates used for gas-sensor design are presented in Figure 16.1. The best parameters of these hotplates are presented in Table 16.1. It is seen that indicated platforms achieve temperatures in the range of 100°C–800°C with extremely small power of 3–30 mW (Han et al. 2001; Furjes et al. 2004).

To fabricate the membranes, including hotplates, one can use either silicon bulk or surface micromachining (Mlcak et al. 1994; Lang 1996; Simon et al. 2001). Examples of membrane fabrication using bulk micromachining are shown in Figure 16.2.

The microscopic photo of the fabricated film bulk acoustic resonator (FBAR) chip using processes listed in Figure 16.3b is shown in Figure 16.3a. It is seen that humidity sensor has embedded microheater and temperature sensor, which significantly improves sensor performance.

Another option of a membrane-based platform with built-in complementary metal-oxide semiconductor (CMOS) microheater was developed by Santra et al. (2015) for resistive humidity sensors. A cross-section view and an optical image of this sensor structure are shown in Figure 16.4. A key component of this resistive humidity sensor is the CMOS microelectromechanical systems (MEMS)-based micro-hotplate ($\mu$HP) structure. The silicon die had a size $1 \times 1$ mm. It was designed using a 1.0 $\mu$m silicon-on-insulator (SOI) CMOS process technology and fabricated in a commercial foundry, followed by deep reactive-ion etching (DRIE) to release the thin membrane. The processing employs SOI wafers with 1 $\mu$m buried oxide, 0.25 $\mu$m SOI layer, and three metallization layers. The $\mu$HP structure typically consists of an embedded, 0.3 $\mu$m thick, resistive tungsten microheater (metal layer 1), 0.3-$\mu$m thick heat-spreader plate (metal layer 2), and a top gold layer for interdigitated sensing electrodes (IDEs). The diameters of the circular heater and membrane structures are 250 and 600 $\mu$m, respectively. The heater was fabricated during the CMOS process while the top gold electrodes and corresponding tracks are deposited as a post-CMOS process in the same commercial foundry. Use of tungsten in the heater allows the device to operate at a very high temperature (up to 750°C), if required, for example, when oxide-based sensing materials are used. Gold was used as the electrode material because of its chemical inertness (and hence, unchanged conductivity over prolonged use under various temperature and humidity conditions) compared to commonly used aluminum in this SOI process. The silicon underneath the $\mu$HP was etched away, using the dioxide layer as the etch stopper, at a wafer level by DRIE technique. This forms a 4.5 $\mu$m silicon dioxide: $SiO_2$ (4 $\mu$m)/silicon nitride: $Si_3N_4$ (0.5 $\mu$m) membrane structure onto which the microheater and electrodes were suspended. The membrane structure reduced DC-power consumption of the sensing device to <5 mW when used for humidity sensing. The heating temperature was uniformly confined over the microheater region, due to the buried heat spreader. The temperature decreased rapidly away from the heater region and was at close to room temperature at the membrane rim, allowing reliable temperature, independent on-chip circuit performance.

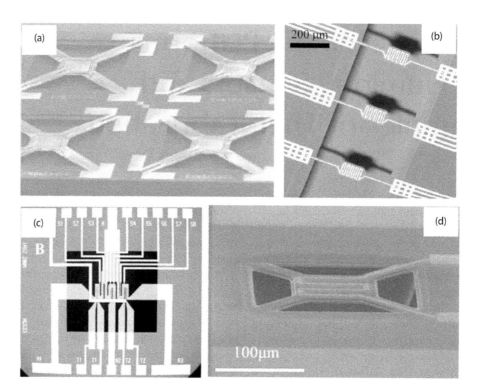

**FIGURE 16.1** (a) SEM view of thermally decoupled membrane array. (Reprinted from *Sens. Actuators B*, 76, Splinter, A. et al., Thick porous silicon formation using implanted mask technology, 354–360, Copyright 2001, with permission from Elsevier.) (b) SEM view of a free-standing sensor platform fabricated using a sacrificial layer of porous silicon. (Reprinted from *Superlattice Microstruct.*, 35, Furjes, P. et al., Thermal characterisation of micro-hotplates used in sensor structures, 455–464, Copyright 2004, with permission from Elsevier.) (c) Optical microscope photograph of a three-section heater sensor array. (Reprinted from *Sens. Actuators B*, 134, Francioso, L. et al., Linear temperature microhotplate gas sensor array for automotive cabin air quality monitoring, 660–665, Copyright 2008, with permission from Elsevier.) (d) Suspended porous silicon microhotplate with a Pt heater. The thickness of the membrane is 4 μm. The depth of the cavity under the membrane is more than 60 μm. (Reprinted from *Sens. Actuators B*, 95, Tsamis, C. et al., Thermal properties of suspended porous silicon micro-hotplates for sensor applications, 78–82, Copyright 2003, with permission from Elsevier.)

**TABLE 16.1**
**The Thermal Characterization of Designed Micro-Hotplates**

| References | Parameter | | | |
|---|---|---|---|---|
| | Dimension of the Sensor Element (μm²) | Maximum Temperature (°C) | Dissipation Power, (°C/mW) | $t_T$ (ms)[a] |
| Furjes et al. (2004) | 100 × 100 | 550 | 27 | 4.2 |
| Tsamis et al. (2003) | 100 × 100 | 600 | 20 | — |
| Triantafyllopoulou et al. (2006) | 100 × 100 | 900 | 180–340 | — |

[a] Time necessary to obtain maximum operating temperature.

It is worth noting that the micromachining technology allows the manufacture of substantially more complex membrane configurations. For example, Dennis et al. (2015), using standard 0.35 μm CMOS process technology and post-CMOS micromachining technology based on dry etching, fabricated a platform for a humidity sensor with an embedded microheater, operated in dynamic mode. Figure 16.5a shows a schematic of the top view of the device and consists of a plate, supported by four beams with each beam having two sections of different widths of 7 and 19 μm but the same length of 240 μm each.

The stator comb-fingers are anchored to the substrate while the rotor comb-fingers are attached to the 400 × 400 μm plate and also to the thick 19-μm beam (Beam 2). After completion of the CMOS process, before the dry-etching step, the eight-inch wafer is back-grinded to a thickness of 350 μm from the initial thickness of 750 μm. All metal layers are aluminum (Al), whereas silicon dioxide ($SiO_2$) is used as the dielectric layer, and tungsten is used for the vias. The CMOS-MEMS sensor was actuated by applying an AC-current through the embedded microheater. These vibrations produce stress at the

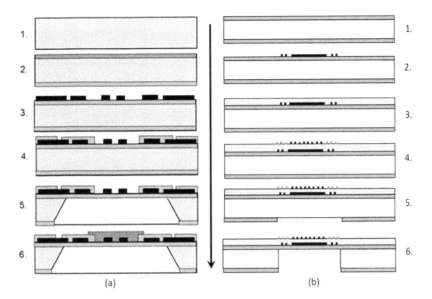

**FIGURE 16.2** (a) Fabrication of a micromachined resistive humidity sensor using bulk etching: 1, Si substrate; 2, deep oxidation of Si substrate and deposition of additional dielectric layer (Si₃N₄); 3, metallization to form heaters and metal contacts; 4, passivation; 5, back-side etching; 6, deposition of humidity-sensing layer. (With kind permission from Springer Science+Business Media: *Solid State Gas Sensing*, Electrical-based gas sensing, 2009a, 47–107, Comini, E. et al. Copyright 2009 Springer; and (b) Fabrication of FBAR sensor chip: 1, first 1.5 μm silicon nitride film was deposited on the silicon substrate by low-pressure chemical vapor deposition (LPCVD); 2, Pt film for bottom electrode and resistor heater was deposited on top surface of the silicon nitride film by physical vapor deposition (PVD) and patterned; 3, then 1.2-μm ZnO film was sputtered on the top surface of the chip and patterned; 4, Pt film for top electrode and resistor temperature sensor was deposited on surface of the ZnO films and patterned; 5, on the back of the chip, silicon-nitride film in suspending area was etched by reactive ion etching (RIE); 6, with the patterned silicon nitride film as a mask, silicon was etched from the back by deep reactive ion etching (DRIE), until it reached the top silicon nitride film. (Reprinted from Zhang, M. et al., *Micromachines*, 8, 25, 2017. Published by MDPI as open access.)

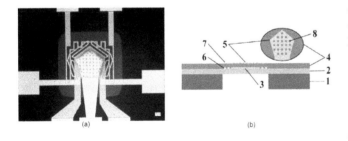

**FIGURE 16.3** (a) The microscopic photo of the fabricated FBAR chip; (b) cross-section of the FBAR chip with detail of the micro-trough-holes in top electrode: 1—silicon substrate; 2—silicon nitride support film; 3—bottom electrode; 4—ZnO piezoelectric film; 5—top electrode; 6—resistor heater; 7—resistor temperature sensor; 8—micro through hole. (Reprinted from Zhang, M. et al., *Micromachines*, 8, 25, 2017. Published by MDPI as open access.)

anchor points of the four supporting beams, and this stress changes the resistance of the piezoresistors (PZRs), which can be measured using a Wheatstone bridge circuit.

In Chapter 13 (Volume 2), it was shown that microcantilevers are very promising sensor platforms, which give possibility to improve sensitivity of capacitance and various mass-sensitive devices. The recently developed nanocantilevers present a major technology breakthrough that have given rise to nanoelectromechanical systems used for humidity

detection (Lavrik et al. 2004; Lang, Gerber 2008; Goeders et al. 2008; Lang 2009; Vashist and Korotcenkov 2011). Microcantilevers generally used for sensing applications are some hundreds of micrometers long, some tens of micrometers wide, and some hundreds of nanometers thin. The spring constant of the cantilever is dependent on the shape and dimensions of the microcantilever. Increasing the cantilever length decreases the resonance frequency. The length-to-thickness ratio is directly proportional to the sensitivity of the cantilever.

As it was mentioned before, microcantilevers can have different compositions and various shapes (see Figure 16.6). The shape of the cantilever often depends on the detection technique (Goeders et al. 2008). For example, square pads on the ends of cantilever beams are used with capacitive detection systems to increase sensitivity, because the measured capacitance is proportional to the surface area of the parallel plates. Piezoresistive cantilevers are often U-shaped, with components of the Wheatstone bridge circuit manufactured at their base points. For optical detection schemes, rectangular, paddle, and T-shaped cantilevers are quite common.

Cantilevers are normally made of silicon or related materials such as silicon nitride that have a high Young's modulus. However, it is known that cantilever sensitivity depends critically on the spring constant of material used, that is, Young's module. It has long been established that the lower

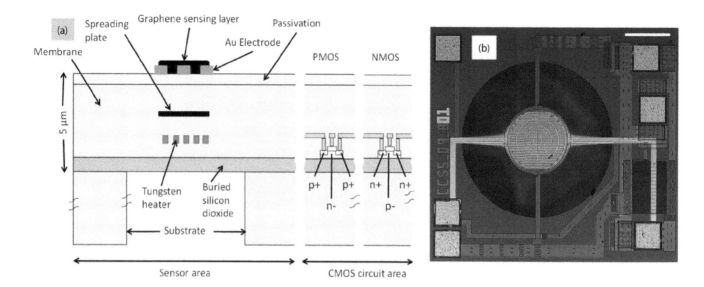

**FIGURE 16.4** (a) Cross-sectional view (not to scale); and (b) optical micrograph of the CMOS device (the scale bar at top right is 200 μm). (Reprinted by permission from Macmillan Publishers Ltd., *Sci. Rep.*, Santra, S. et al., 2015, Copyright 2015.)

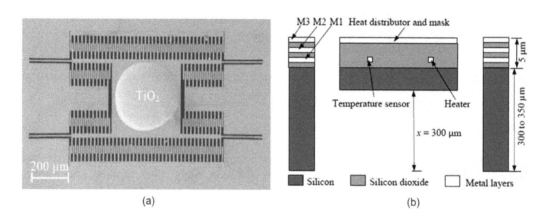

**FIGURE 16.5** (a) FESEM image of CMOS-MEMS device with TiO₂ paste deposited on its plate; (b) cross-sectional view shows CMOS layers and the substrate. (Reprinted from Dennis, J.-O. et al., *Sensors*, 15, 16674–16687, 2015. Published by MDPI as open access.)

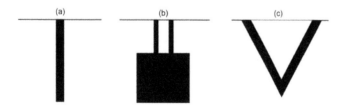

**FIGURE 16.6** Different types of microcantilevers (top view): (a) rectangular; (b) double-legged; and (c) triangular. (Reprinted from Vashist, S.K., *J. Nanotechnol.*, 3, 1–15, 2007. Published by AZom. Com Pty Ltd. as open access.)

**TABLE 16.2**
**Material Properties of Si and Several Polymers. RT Measurements**

| Material | Young's Modulus (GPa) | Poisson's Ratio |
|---|---|---|
| Si | 169 | 0.3 |
| Polyimide | 3.3 | 0.35 |
| SU-8 | 3.6 | 0.33 |

*Source:* Fragakis, J. et al., *J. Phys. Confer. Ser.*, 10, 305–308, 2005; Chung, S. et al., *Electron. Mater. Lett.*, 2, 175–181, 2006.

the spring constant of the cantilever, the higher the sensitivity. This means that a cantilever made of a softer material will be more sensitive for static-deflection measurements in comparison with Si-based cantilevers. Young's modulus for several polymers acceptable for cantilever design is listed in Table 16.2.

## 16.2 SILICON-BASED MICROCANTILEVERS

Silicon-based microcantilevers are made using well-established fabrication techniques, which generally include a thin-layer deposition, photolithographic patterning, etching,

Fabrication diagram | Mask | Optical microscope images

(a) Silicon oxide lithography and Boron diffusion

(b) Opening contacts

(c) Metal deposition (Al or Au)

(d) Cantilever outline - etching

(e) Cantilever release - etching

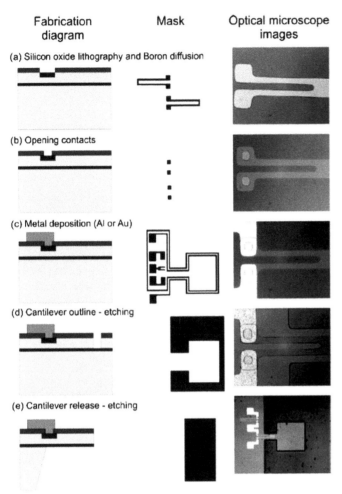

**FIGURE 16.7** (Left) Scheme of the microcantilever fabrication process using bulk micromachining: (a) oxidation, aperture and doping; (b) contacts opening; (c) aluminum or gold deposition; (d) front-face etching; and (e) back-side etching. (Middle) Configuration of masks used during microcantilever fabrication. (Right) Pictures showing the top view of the structures after (a–e) processes. (Reprinted with permission from *Sens. Actuators B*, 137, Urbiztondo, M.A. et al., Zeolite-modified cantilevers for the sensing of nitrotoluene vapors, 608–616, Copyright 2009, with permission from Elsevier.)

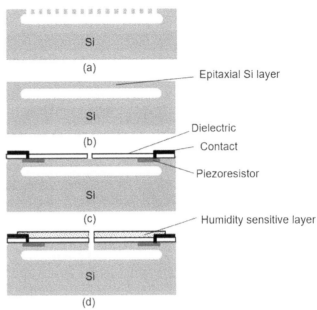

**FIGURE 16.8** Fabrication process of the microcantilever-piezoresistive humidity sensor: (a) Pre-process of the substrate, which includes Si porosification with subsequent removal of PSi; (b) epitaxial growth of Si layer; (c) structure after CMOS process; and (d) structure after post-process. (Reprinted from Huang, J.-Q. et al., *Micromachines*, 6, 1569–1576, 2015. Published by MDPI as open access.)

**TABLE 16.3**
**Details of the Materials and Fabrication Process**

| Layer | Material | Thickness (μm) |
|---|---|---|
| Epitaxial layer | Single crystal silicon | 8 |
| Dielectric | TEOS (tetraethoxysilane)/SiN$_x$ | 0.3/0.1 |
| Metal | AlCu | 1.9 |
| Humidity sensitive material | Polyimide (PI) | 3 |

*Source:* Huang, J.-Q. et al., *Micromachines*, 6, 1569–1576, 2015. Published by MDPI as open access.

and micromachining (surface and bulk), discussed before in Chapter 15. In particular, Si-cantilever sensor arrays can be microfabricated using a dry-etching SOI fabrication technique, developed in the Micro-/Nanomechanics Department at IBM's Zurich Research Laboratory. One example of a microcantilever fabrication process is shown in Figure 16.7, where a scheme of the masks used and a top view of the cantilever during the different stages are also provided. Briefly, a five-step photolithography procedure was used to first achieve the development of the strain gauges and electrical contacts, and then to define the geometry of the microcantilever.

An example of manufacturing of a microcantilever-based piezoresistive humidity sensor, using surface micromachining process, is shown in Figure 16.8. Fabrication started with an *n*-type (100) Si wafer. Before the CMOS process, a two-step etching process was performed as shown in Figure 16.8a. The size of the cantilevers was 400 μm wide by 100 μm long. Details of the materials and fabrication process are presented in Table 16.3. Compared with a bulk-fabricated microcantilever piezoresistive humidity sensor, the structure of sensors fabricated using surface micromachining was more compact.

The processing technologies previously discussed enable batch fabrication of cantilevers with high yield, good reproducibility, and low cost. Thus, it is possible to make thousands of microcantilevers in an array format on a wafer (see Figure 16.9). A more detailed description of microcantilever fabrication using commercial CMOS technology can be

**FIGURE 16.9** SEM image of an array of eight silicon microcantilevers fabricated at the Micro-and Nano-mechanics Group, IBM Zurich Research Laboratories. Each cantilever is 1 μm thick, 500 μm long, and 100 μm wide, with a pitch of 250 μm, spring constant 0.02 N/m. (Reprinted from *Sens. Actuators B, 77*, Battiston, F.M. et al., A chemical sensor based on a microfabricated cantilever array with simultaneous resonance-frequency and bending readout, 122–131, Copyright 2001, with permission from Elsevier.)

found in the literature (Hierlemann et al. 2000; Lavrik et al. 2004; Martorelli 2008; Boisen and Thundat 2009).

When the piezoresistive technique is applied to monitor-cantilever deflection, the cantilever is fabricated with an integrated resistor, having piezoresistive properties, and its resistance changes as a function of cantilever bending (Boisen and Thundat 2009). The extent of cantilever bending can be sensitively measured by a simple electrical measurement. The commonly fabricated piezoresistors are based on boron-doped silicon layers. The doped silicon layers are then capped with insulating materials. To enhance sensitivity, the piezoresistive layer of the cantilever should be placed in the region of maximum stress, preferably on the surface. One such cantilever is shown in Figure 16.10. For a first approximation, the resulting relative change in the resistance is directly proportional to the stress, and therefore, to the deflection.

The silicon resistor is defined in microcrystalline or single-crystal silicon and encapsulated usually in silicon nitride. The thickness of the deposited silicon nitride on either side of the cantilever is adjusted so that the neutral axis of the cantilever lies inside the silicon-nitride layer, rather than in the silicon layer. This asymmetry in material composition ensures that the resistor is placed close to one of the cantilever surfaces for optimal sensitivity. In addition, the silicon nitride serves as efficient electrical insulation for the resistor and ensures that the device can be operated in liquids. The signal-to-noise ratio of the piezoresistive cantilever depends highly on the doping and the crystallinity of the silicon layer. The best performance is clearly found for single-crystalline resistors, which can detect deflections below 1.0-nm resolution (www.cantion.com).

The experiment showed that microcantilevers and membranes can also be fabricated using polymers. It was found that microcantilevers and membranes, fabricated from polymers, inherently possess readily tailorable mechanical and

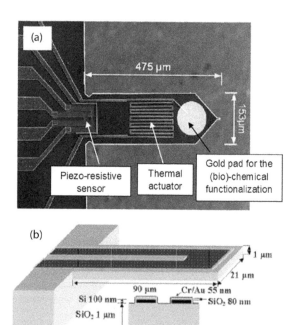

**FIGURE 16.10** Thermally driven piezoresistive cantilevers: (a) top view and (b) schematic diagram. (a) (Reprinted from *Microelectron. Eng.*, 86, Woszczyna, M. et al., Thermally driven piezoresistive cantilevers for shear-force microscopy, 1212–1215, Copyright 2009, with permission from Elsevier.) (b) (Reprinted with permission from Li, P. et al., A single-sided micromachined piezoresistive SiO$_2$ cantilever sensor for ultra-sensitive detection of gaseous chemicals, *J. Micromech. Microeng.*, 16, 2539–2546, 2006. Copyright 2006, Institute of Physics.)

chemical properties, which can be used to improve sensor performance (Goeders et al. 2008; Chen et al. 2017; Steffens et al. 2018). However, in this chapter we will not consider the specifics of the manufacture of such microcantilevers and membranes. These issues will be addressed in Chapter 19.

## 16.3 FUNCTIONALIZATION OF CANTILEVER AND MEMBRANE SURFACES

It is essential that the surfaces of the cantilever and membranes should be coated in a proper way to provide suitable receptor surfaces for the molecules to be detected (Lang and Gerber 2008). Such coatings should be specific, homogeneous, stable, and reproducible. The choice of reactive group depends on the cantilever surface composition; the choice of terminal group depends on the specific chemical interaction desired to attract the analyte to the cantilever surface (Goeders et al. 2008).

For static-mode measurements, one side of the cantilever should be passivated to block unwanted adsorption. Often, for this purpose, the cantilever's lower side is passivated using silane chemistry for coupling an inert surface, such as polyethylene glycol silane. Silanization is performed first on the silicon microcantilever. Then, a humidity-sensing material is deposited on the top side of the cantilever, leaving the lower side unchanged (Lang and Gerber 2008). In this case, the concentration of the water vapor in the air is proportional to the extent of deflection. When all surfaces of the cantilever are coated with a humidity-sensitive material, then the concentration of water vapor in the air is proportional to the change in resonance frequency.

No doubt that the coating method used for surface functionalizing of nanocantilevers should be fast, reproducible, reliable, and allow one or both cantilever surfaces to be coated separately (Lang 2009; Vashist and Korotcenkov 2011). Thin metallic or ceramic films can be applied to the desired surfaces of the cantilever using standard film deposition techniques, such as thermal evaporation or sputtering. To prevent delamination, an intermediate adhesion-promoting layer is often employed. For example, during fabrication of cantilevers with a gold covering deposited by evaporation, a thin underlayer of titanium or chromium is used to promote adhesion of the gold film onto the silicon cantilever. The disadvantage of all these methods is that they are suitable only for coating large areas, not individual cantilevers in an array, unless shadow masks are used. Such masks need to be accurately aligned to the cantilever structures, which is a time-consuming process.

Other methods to coat cantilevers use manual placement of particles onto the cantilever, which requires skillful handling of tiny samples. For example, cantilevers can be coated by directly pipetting solutions of the probe molecules onto the cantilevers or by employing airbrush spraying and shadow masks to coat the cantilevers separately. Microdropping may also be used for deposition of different sensing layers (Urbiztondo et al. 2009). All these methods, however, have only limited reproducibility and are very time-consuming if a large number of cantilever arrays should be coated.

It was found, however, that for coating with polymer layers, ink-jet spotting or inject printing is more convenient, as it is possible to coat only the upper or lower surface (read Chapter 18). The method is also appropriate for coating many cantilever sensor arrays in a rapid and reliable way (Bietsch et al. 2004a, 2004b). Inkjet spotting allows a cantilever to be coated within seconds and yields very homogeneous, reproducibly deposited layers of well-controlled thickness (Lange et al. 2002; Savran et al. 2003). This technique has also been applied to functionalize a polymer-coated microcantilever array (Bietsch et al. 2004b).

Spin-coating technology can also be used for deposition of polymer films on the surface of silicon cantilever (Steffens et al. 2014, 2018). Using this method, Steffens et al. (2014, 2018) fabricated silicon-cantilever-based humidity sensors with polyaniline, poly(o-ethoxyaniline), and polypyrrole as a sensitive layer. This process included the following steps: After spinning at 500 rpm for 8 s, a polymer solution was deposited on the microcantilever surface. The spinning rate was increased to 1,000 rpm for 10 s and 3,000 rpm for 1 min. Afterward, the coated microcantilever sensors were dried in a vacuum desiccator for 12 h at room temperature. Next, the sensitive polymer layer was doped with 1 M HCl (hydrochloric acid). The results of the Si functionalization by the polymer layer are shown in Figure 16.11. These results

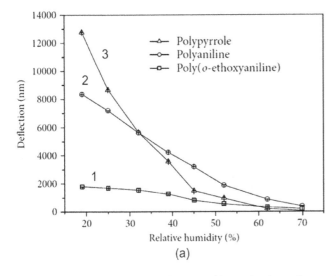

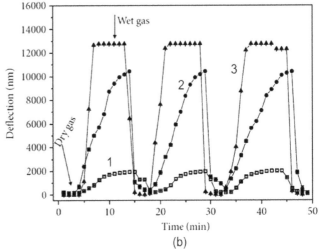

(a)

(b)

**FIGURE 16.11** Si-cantilever deflection with conducting polymers (polyaniline, poly(o-ethoxyaniline) and polypyrrole) as a function of (a) relative humidity, and (b) time. (Reprinted from Steffens, C. et al., *Scanning*, 4782685, 2018. Published by Hindawi as open access.)

demonstrated that the cantilever sensor functionalized with polypyrrole, presented the higher sensitivity in the range of low relative humidity (RH), higher reversibility and detection limit to humidity, and faster response. The response-time measurements showed 3 seconds. The hysteresis of the sensor was found as 1.23% RH.

# REFERENCES

Battiston, F.M., Ramseyer J.-P., Lang H.P., Baller M.K., Gerber Ch., Gimzewski J.K., Meye E., Güntherodt H.-J. (2001) A chemical sensor based on a microfabricated cantilever array with simultaneous resonance-frequency and bending readout. *Sens. Actuators B* **77**(1–2), 122–131.

Bietsch A., Hegner M., Lang H.P., Gerber C. (2004a) Inkjet deposition of alkanethiolate monolayers and DNA oligonucleotides on gold: Evaluation of spot uniformity by wet etching. *Langmuir* **20**, 5119–5122.

Bietsch A., Zhang J., Hegner M., Lang H.P., Gerber C. (2004b) Rapid functionalization of cantilever array sensors by inkjet printing. *Nanotechnology* **15**, 873–880.

Boisen A., Thundat T. (2009) Design and fabrication of cantilever array biosensors. *Mater. Today* **12**, 32–38.

Chen M., Frueh J., Wang D., Lin X., Xie H., He Q. (2017) Polybenzoxazole nanofiber-reinforced moisture-responsive soft actuators. *Sci. Rep.* **7**, 769.

Chung S., Makhar S., Ackler H., Park S. (2006) Material characterization of carbon-nanotube-reinforced polymer composite. *Electron. Mater. Lett.* **2**(3), 175–181.

Comini E., Faglia G., Sberveglieri G. (2009a) Electrical-based gas sensing. In: Comini E., Faglia G., Sberveglieri G. (eds.) *Solid State Gas Sensing*, Springer–Verlag, New York, pp. 47–107.

Dennis J.-O., Ahmed A.-Y., Khir M.-H. (2015) Fabrication and characterization of a CMOS-MEMS humidity sensor. *Sensors* **15**, 16674–16687.

Fragakis J., Chatzandroulis S., Papadimitriou D., Tsamis C. (2005) Simulation of capacitive type bimorph humidity sensors. *J. Phys. Confer. Ser.* **10**, 305–308.

Francioso L., Forleo A., Taurino A.M., Siciliano P., Lorenzelli L., Guarnieri V., et al. (2008) Linear temperature microhotplate gas sensor array for automotive cabin air quality monitoring. *Sens. Actuators B* **134**, 660–665.

Furjes P., Ducso C., Adam M., Zettner J., Barsony I. (2004) Thermal characterisation of micro-hotplates used in sensor structures. *Superlattice Microstruct.* **35**, 455–464.

Goeders K.M., Colton J.S., Bottomley L.A. (2008) Microcantilevers: Sensing chemical interactions via mechanical motion. *Chem. Rev.* **108**, 522–542.

Han P.G., Wong H., Poon M.C. (2001) Sensitivity and stability of porous polycrystalline silicon gas sensor. *Colloids Surf. A* **179**, 171–175.

Hierlemann A., Lange D., Hagleitner C., Kerness N., Koll A., Brand O., Baltes H. (2000) Application-specific sensor systems based on CMOS chemical microsensors. *Sens. Actuators B* **70**, 2–11.

Horrillo M., Sayago I., Ares L., Rodrigo J., Gutirrez J., Gotz A., et al. (1999) Detection of low $NO_2$ concentrations with low power micromachined tin oxide gas sensors. *Sens. Actuators B* **58**, 325–332.

Huang J.-Q., Li F., Zhao M., Wang K. (2015) Surface micromachined CMOS MEMS humidity sensor. *Micromachines* **6**, 1569–1576.

Lang H.P. (2009) Cantilever-based gas sensing. In: Comini E., Faglia G., Sberveglieri G. (eds.) *Solid State Gas Sensing*, Springer Science+Business Media, New York, pp. 305–328.

Lang H.P., Gerber Ch. (2008) Microcantilever sensors. *Top. Curr. Chem.* **285**, 1–27.

Lang W. (1996) Silicon microstructuring technology. *Mater. Sci. Eng. R* **17**, 1–55.

Lange D., Hagleitner C., Hierlemann A., Brand O., Baltes H. (2002) Complementary metal oxide semiconductor cantilever arrays on a single chip: Mass-sensitive detection of volatile organic compounds. *Anal. Chem.* **74**, 3084–3095.

Lavrik N.V., Sepaniak M.J., Datskosa P.G. (2004) Cantilever transducers as a platform for chemical and biological sensors. *Rev. Sci. Instrum.* **75**(7), 2229–2253.

Li P., Li X. (2006) A single-sided micromachined piezoresistive $SiO_2$ cantilever sensor for ultra-sensitive detection of gaseous chemicals. *J. Micromech. Microeng.* **16**, 2539–2546.

Martorelli J.V. (2008) Monolithic CMOS-MEMS resonant beams for ultrasensitive mass detection. PhD thesis, Universitat Autonoma de Barcelona, Spain.

Mlcak R., Tuller H.L., Greiff P., Sohn J., Niles L. (1994) Photoassisted electrochemical micromachining of silicon in HF electrolytes. *Sens. Actuators A* **40**, 49–55.

Santra S., Hu G., Howe R.C.T., De Luca A., Ali S.Z., Udrea F., et al. (2015) CMOS integration of inkjet-printed graphene for humidity sensing. *Sci. Rep.* **5**, 17374.

Savran C.A., Burg T.P., Fritz J., Manalis S.R. (2003) Microfabricated mechanical inherently differential readout. *Appl. Phys. Lett.* **83**, 1659–1661.

Simon I., Bârsan N., Bauer M., Weimar U. (2001) Micromachined metal oxide gas sensors: opportunities to improve sensor performance. *Sens. Actuators B* **73**, 1–26.

Splinter A., Bartels O., Benecke W. (2001) Thick porous silicon formation using implanted mask technology. *Sens. Actuators B* **76**, 354–360.

Steffens C., Brezolin A.N., Steffens J. (2018) Conducting polymer-based cantilever sensors for detection humidity. *Scanning* **2018**, 4782685.

Steffens C., Leite F.L., Manzoli A., Sandoval R.D., Fatibello O., Herrmann P.S.P. (2014) Microcantilever sensors coated with doped polyaniline for the detection of water vapour. *Scanning* **36**(3), 311–316.

Triantafyllopoulou R., Chatzandroulis S., Tsamis C., Tserepi A. (2006) Alternative micro-hotplate design for low power sensor arrays. *Microelectron. Eng.* **83**, 1189–1191.

Tsamis C., Nassiopoulou A.G., Tserepi A. (2003) Thermal properties of suspended porous silicon micro-hotplates for sensor applications. *Sens. Actuators B* **95**, 78–82.

Urbiztondo M.A., Pellejero I., Villarroya M., Sese J., Pina M.P., Dufour I., Santamaria J. (2009) Zeolite-modified cantilevers for the sensing of nitrotoluene vapors. *Sens. Actuators B* **137**, 608–616.

Vashist S.K. (2007) A review of microcantilevers for sensing applications. *J. Nanotechnol.* **3**, 1–15.

Vashist S.K., Korotcenkov G. (2011) Microcantilever-based chemical sensors. In: G. Korotcenkov (ed.) *Chemical Sensors: Comprehensive Sensor Technologies Volume 4: Solid-State Devices*, Momentum Press, New York, pp. 321–376.

Woszczyna M., Gotszalk T., Zawierucha P., Zielony M., Ivanow T., Ivanowa K., et al. (2009) Thermally driven piezoresistive cantilevers for shear-force microscopy. *Microelectron. Eng.* **86**, 1212–1215.

Zhang M., Du L., Fang Z., Zhao Z. (2017) A sensitivity-enhanced film bulk acoustic resonator gas sensor with an oscillator circuit and its detection application. *Micromachines* **8**, 25.

# 17 Platforms and Materials for QCM and SAW-Based Humidity Sensors

Piezoelectric-based humidity sensors, such as surface acoustic wave (SAW), quartz crystal microbalance (QCM) or bulk acoustic wave (BAW), and cantilever-based devices, create a certain class of gas sensors widely used in various applications (Korotcenkov 2011). As humidity sensors, the resonators are coated with layers, which selectively absorb or adsorb water vapor and thereby induce a mass change that is then detected via a shift in the resonant frequency of the device (read Chapter 13, Volume 2). In all these acoustic wave (AW)-based devices, piezoelectric materials play one of the main roles. These materials are needed to generate the AW that propagates along the surface in SAW devices or throughout the bulk of the structure in BAW devices. The main feature of piezoelectric materials is the ability to couple mechanical strain to electrical polarization (the direct piezoelectric effect). They may also reversibly and rapidly change their shape under the influence of an alternating electric field (the reverse piezoelectric effect). Due to this phenomenon, piezoelectric materials are widely used as transducers to convert electrical to mechanical energy and vice versa. Studies have shown that such an effect only occurs in crystals that lack a center of inversion symmetry (Ballantine et al. 1996).

## 17.1 PIEZOELECTRIC MATERIALS

Usually, $\alpha$-quartz ($SiO_2$) and nonstoichiometric lithium niobate ($LiNbO_3$) are used in QCM- and SAW-based devices as a piezoelectric material (Jakubik et al. 2002). These materials are chosen on the basis of the device's design specifications for operating temperature, fractional bandwidth, and insertion loss.

Quartz is a piezoelectric material that is widespread in nature (Rosen et al. 1992). Quartz, a crystalline form of $SiO_2$, presents an excellent crystal quality, and this material can be very weakly sensitive to parasitic temperature effects (compared to other piezoelectric materials) according to its cut. In addition, thermodynamically, quartz is very stable. These properties explain its widespread use as a platform for AW sensors (Wang et al. 2006; Fanget et al. 2011). Though it is far from easy to grow, quartz's widespread use has led to its availability in comparatively large sizes and large quantities from several suppliers. Artificially grown quartz crystals show high stability over long times at room temperature and good resistance to adverse chemical conditions. In addition, various crystallographic orientations that minimize the temperature coefficient (near-room temperature) of AW devices have been characterized over the years and are now widely available; these include the ST cut for AW devices and the AT and BT cuts for "thickness-shear-mode" (TSM) resonators or QCM

sensors. AT-cut crystals are used, in particular, as transducers in QCM devices, since they demonstrate high-frequency stability in microbalance applications (Lucklum and Eichelbaum 2007). For this cut, the electromechanical coupling coefficient is very small. This feature of the AT-cut devices is a great advantage, since the resonator response frequency can be monitored for only the fundamental mode and overtones (Rosen et al. 1992), the resonance curve is sharp and narrow, while the probability for spurious resonances separated from the main resonance is relatively small (Yoon et al. 2009). BT-cut devices also satisfy this criterion, and their resonance frequencies are therefore stable at room temperature (Lucklum and Eichelbaum 2007). Micromachining allows fabrication of thinly cut crystals with a thickness of less than 10 $\mu$m and a diameter smaller than 100 $\mu$m (Tadigadapa and Mateti 2009). Eigen frequencies of AT-cut crystals are typically in the range 5–20 MHz, but higher frequencies (up to 300 MHz) are also available.

Lithium niobate ($LiNbO_3$) has a relatively large electromechanical coupling coefficient ($k^2$) that allows two-port AW devices to utilize interdigital transducers comprised of nearly an order of magnitude fewer pairs of fingers than quartz. The coupling coefficient represents the ratio of the mechanical (electrical) energy converted to the input electrical (mechanical) energy for the piezoelectric material (Schwartz et al. 2004). The orientation of the coupling coefficient is important and must be distinguished clearly. The planar coupling coefficient, $k_p$, describes the radial coupling in a thin disc, when the electrical field is applied through the thickness, and the thickness coupling coefficient, $k_t$, is identical to $k_{33}$ when the element is clamped laterally (Setter 2002). Unfortunately, $LiNbO_3$ suffers from a very large temperature coefficient—approximately 80 ppm/°C for a SAW device with propagation in the Z direction on Y-cut $LiNbO_3$. The consequences of this are that $LiNbO_3$ is thermally quite fragile and that exceptional temperature stability is necessary when using $LiNbO_3$ to detect anything other than temperature changes (Ballantine et al. 1997).

Among the most common and important piezo materials, which can be used in QCM and SAW-based sensors, are also zinc oxide (ZnO) (Özgür et al. 2005; Lang 2005), lithium tantalate ($LiTaO_3$) (Fechete et al. 2006), lead zirconate titanate (PZT) (Ferrari et al. 2000; Lee et al. 2004, 2006), and aluminum nitrate (AlN) films (Tadigadapa and Mateti 2009).

Zinc oxide (ZnO) is a wide-band gap chemically stable semiconductor of interesting physical properties that can be changed in a controlled way during chemical preparation. ZnO is both piezoelectric (Özgür et al. 2005) and pyroelectric (Lang 2005). The high wettability of ZnO is employed in superhydrophobic coatings, which is important for humidity sensors (Zhou et al. 2007). An advantage of the ZnO-based resonators is that it is possible to fabricate them in a solidly mounted resonator (SMR) configuration and to integrate them with integrated circuit (IC) systems. ZnO films can be deposited using various methods, from metal-organic chemical vapor deposition (MOCVD) to laser-pulse deposition and spray pyrolysis. The ZnO microcrystals can be oriented in the process of the structure growth with the crystallite $c$ axis (001 direction) either perpendicular to or aligned in the substrate plane. Due to this property, textured ZnO films (Yanagitani et al. 2007) can be used in SAW, Love mode, and BAW types of devices as active elements or as wave-guiding layers (Uchino and Ito 2009). It is believed that ZnO materials offer lower cost, easier fabrication, and highly sensitive sensor components for SAW and "electronic nose" devices in comparison with quartz crystals (Lozano et al. 2006).

Lead zirconate titanate ($Pb(Zr, Ti)O_3$ (PZT), $Pb(La, Zr, Ti)O_3$, or $Pb(Zr, Ti, Zn)O_y$) is a piezoelectric ceramic material, which demonstrates an exceptional combination of pyroelectric, piezoelectric, and inverse piezoelectric effects, high-temperature stability, and large electromechanical coupling coefficients (Ferrari et al. 2000; Lee et al. 2004, 2006). The sensitivity of PZT is double that of quartz. These properties enable the development through advanced engineering of PZT-based microelectromechanical systems (MEMS), including micromachined cantilevers (Ferrari 2004; Lee et al. 2004). Piezoelectric polymers, polyvinylidene fluoride or polyvinylidene difluoride (PVDF), generally known as piezo-films, allow for extremely low-cost sensors but generally do not provide acceptable stability for many applications. The comparison of piezoelectric properties of several major piezoelectric ceramics discussed above is given in Table 17.1. In this table, PZT 4: hard PZT (PZT doped with acceptor ions, such as $K^+$ or $Na^+$ at the A site, or $Fe^{3+}$, $Al^{3+}$ or $Mn^{3+}$ at the B site); PZT 5H: soft PZT (PZT doped with donor ions, such as $La^{3+}$ at the A site, or $Nb^{5+}$ or $Sb^{5+}$ at the B site); LF4T:

$(K_{0.44}Na_{0.52}Li_{0.04})(Nb_{0.86}Ta_{0.10}Sb_{0.04})O_3$; PVDF: piezoelectric polymer synthesized using copolymerization of vinilydene difluoride with trifluoroethylene (TrFE).

According to Gerfers et al. (2010), the AlN films, as far as the piezoelectric materials which are concerned (see Table 17.2), have been rather less investigated than PZT and ZnO films because of their smaller piezoelectric constant. However, their temperature/humidity stability (Lakin et al. 2000), higher signal-to-noise ratio (Trolier-Mckinstry and Muralt 2004) and the compatibility with complementary metal-oxide semiconductor (CMOS) processing are making them more popular in the field of nanotechnology (Gerfers et al. 2010). In particular, AlN, as well as of ZnO films due to their compatibility with IC components and lower costs, are the most promising materials for application in the film bulk acoustic resonator (FBAR) technology. Since extremely thin piezoelectric films are readily fabricated, both ZnO and AlN have been used to make bulk resonators that operate at much higher frequencies (e.g., 100–3,000 MHz) (Roy and Basu 2002) than TSM resonators made from quartz disks, which typically resonate at 5–20 MHz (Ballantine et al. 1997). Moreover, AlN is a large-band-gap material (6 eV) with a large resistivity. On the contrary, ZnO is a semiconductor

**TABLE 17.2**

**Typical Piezoelectric Properties of AlN, ZnO, and PZT Thin Films**

| Property | AlN | ZnO | PZT Thin Film |
|---|---|---|---|
| Density, g/cm$^3$ | 3.26 | 5.68 | 7.5–7.6 |
| $k_t^2$, % | 6.5 | 9 | 7–15 |
| $Q_m$ @ 2 GHz | 2490 | 1770 | — |
| $k_t^2 Q_m$ @ 2 GHz | 160 | 160 | — |
| $e_{31,f}$, C/m$^2$ | −1.05 | −1.0 | −8 to 12 |
| $d_{33,f}$, pm/V | 3.9 | 5.9 | 60–130 |

*Source:* Tadigadapa S. and Mateti K., *Meas. Sci. Technol.*, 20, 092001–30, 2009.

**TABLE 17.1**

**Comparison of Piezoelectric and Related Properties of Several Important Piezoelectric Materials**

| Parameter | Quartz | BaTiO$_3$ | PbTiO$_3$:Sm | PZT 4 | PZT 5H | LF4T | PVDF |
|---|---|---|---|---|---|---|---|
| $d_{33}$ (pC/N) | 2.3 | 190 | 65 | 289 | 593 | 410 | 33 |
| $g_{33}$ ($10^{-3}$ Vm/N) | 57.8 | 12.6 | 42 | 26.1 | 19.7 | 20.2 | 380 |
| $k_t$ | 0.09 | 0.38 | 0.50 | 0.51 | 0.50 | 0.45 | 0.30 |
| $k_p$ | 5 | 0.33 | 0.03 | 0.58 | 0.65 | 0.6 | 6 |
| $\varepsilon^x_{33}/\varepsilon_0$ | | 1700 | 175 | 1300 | 3400 | 2300 | 3–10 |
| $Q_m$ | | >10$^5$ | 900 | 500 | 65 | — | — |
| $T_c$ (°C) | — | 120 | 355 | 328 | 193 | 253 | |

*Source:* Kholkin, A.L. et al., Smart ferroelectric ceramics for transducer applications, Chapter 9.1, in Schwartz M. (ed.) *Smart Materials*, CRC Press, Boca Raton, FL, 2009.

$d_{33}$: piezoelectric charge sensor constant; $g_{33}$: piezoelectric voltage constant; $k$: electromechanical coupling factor; $\varepsilon_{33}/\varepsilon_0$: dielectric constant; $Q_m$: mechanical quality factor; $T_c$: Curie temperature.

and a piezoelectric material with a band gap of 3 eV that holds the intrinsic risk of increased conductivity because of off-stoichiometry. This low DC resistivity due to high conductivity turns into a high-dielectric loss at low frequencies, which is especially harmful for sensor and actuators operating at frequencies lower than 10 kHz (Trolier-Mckinstry and Muralt 2004). Due to these reasons, AlN has the upper hand over ZnO, which means that AlN is better suited for deflection devices (especially for sensors), while ZnO is better suited for longitudinal BAW generation, as it yields larger coupling coefficients for longitudinal BAW generation.

Between PZT and AlN, PZT displays a dielectric constant which is 100 times higher than in AlN. Moreover, its (PZT) piezoelectric coefficients are almost 10 times higher than that of AlN. But, for a given strain, PZT produces an almost 10 times lower electric field than AlN, which makes AlN more suitable for sensor applications than PZT. So, for sensor applications, where an output parameter is voltage, AlN is more suitable than PZT, whereas for energy harvesting applications, PZT can be suitable because PZT produces more energy than AlN due to its high electromechanical coupling.

Table 17.3 shows some important material parameters that can be applied in SAW devices. The fundamental parameters controlling operating characteristics of SAW-based sensors are the SAW velocity, the temperature coefficients of delay (TCD), the electromechanical coupling factor, and the propagation loss. The free surface velocity, $V_0$, of the material is a function of the cut angle and the propagative direction. The TCD is an indication of the frequency shift expected from a transducer due to a temperature change and is also a function of the cut angle and the propagation direction. In SAW applications, the coupling factor $kp^2$ relates to the maximum bandwidth obtainable, and the amount of a signal loss between input and output that determines the fractional bandwidth vs. minimum insertion loss for a given material

and a filter. The propagation loss, one of the major factors determining the insertion loss of the device, is caused by the wave scattering by crystalline defects and surface irregularities. Materials that have high electromechanical coupling factors combined with small TCD times are likely to be required.

## 17.2 HIGH-TEMPERATURE AW DEVICES

High-temperature sensors are desired for the automotive, aerospace, and energy industries. Specifically, high-temperature sensors are critical for the development of high-performance engines, as well as for engine control and health assessment, in which the sensors must be able to operate reliably under harsh environments, where the sensors often need to be close to the engine component of interest for adequate sensitivity, for fuel combustion efficiency studies, and engine prognostic and engine health management. Other applications that benefit from high-temperature sensors include power plants, for the structure health monitoring of the furnace components or reactor systems. However, it was established that the materials discussed in the previous section cannot be used at elevated temperatures (see Table 17.4). For example, the phase transition at 573°C and the strong decreasing resonator quality factor (Q-factor) of quartz permit their use up to around 450°C. The phase transitions in general lead to instability and irreversibility of the properties with temperature. The chemical instability of nonstoichiometric lithium niobate, that is, its tendency to decomposition, limits its application temperature to about 450°C (Fachberger et al. 2004). The decrease of electrical resistivity of $LiNbO_3$ at high temperature (Smith and Welsh 1971), which contributes to the charge drift, interfering with piezoelectrically induced charges, also limits the application at high temperatures of the materials discussed above. Many factors, including the previously mentioned ones, together with increased attenuation of AWs and dielectric

### TABLE 17.3
### Parameters of Main SAW Materials

| Material | Cut–Propagation Direction | $k^2$ (%) | TCD (ppm/C) | $V_0$ (m/s) | $\varepsilon_r$ |
|---|---|---|---|---|---|
| **Single Crystal** | | | | | |
| Quartz | ST—X | 0.16 | 0 | 3158 | 4.5 |
| $LiNbO_3$ | 128°Y—X | 5.5 | −74 | 3960 | 35 |
| $LiTaO_3$ | X112°—Y | 0.75 | −18 | 3290 | 42 |
| $Li_2B_4O_7$ | (110) —<001> | 0.8 | 0 | 3467 | 9.5 |
| **Ceramic** | | | | | |
| PZT-In($Li_{3/5}W_{2/5}$)$O_3$ | | 1.0 | 10 | 2270 | 690 |
| (Pb, Nd)(Ti, Mn, In)$O_3$ | | 2.6 | <1 | 2554 | 225 |
| **Thin Film** | | | | | |
| ZnO/glass | | 0.64 | −15 | 3150 | 8.5 |
| ZnO/Sapphire | | 1.0 | −30 | 5000 | 8.5 |

*Source:* Uchino, K. and Ito, Y., Smart ceramics: Transducers, sensors, and actuators, Chapter 9.2, in: Schwartz M. (ed.), *Smart Materials*, CRC Press, Boca Raton, FL, 2009.

## TABLE 17.4

### Operation Temperature Limits of Piezoelectric Materials

| Material | Temperature Limit (°C) | Reasons |
|---|---|---|
| $Li_2B_4O_7$ | 230 | Excessive ionic conductivity |
| $LiNbO_3$ | 300 | Decomposition |
| $LiTaO_3$ | 300 | Decomposition |
| $\alpha$-$SiO_2$ ($\alpha$-quartz) | 573 | Phase transformation |
| $AlPO_4$ | 588 | Phase transformation |
| $GaPO_4$ | 970 | Phase transformation |
| AlN | ~1000 | Oxidation resistance |
| $La_3Ga_5SiO_{14}$ | 1470 | Melting point |

*Source:* Reprinted with permission from Fritze, H., High-temperature bulk acoustic wave sensors, *Meas. Sci. Technol.*, 22, 012002, 2011. Copyright 2011, Institute of Physics.

losses, must be considered when choosing a material appropriate for a particular high-temperature application.

Commonly used ferroelectric and piezoelectric materials, which can be applied at high temperature, are listed in Table 17.5. The materials are categorized into different groups, including ferroelectric materials with perovskite, tungsten bronze, Aurivillius and perovskite layer structures, and nonferroelectric single crystals (Zhang et al. 2011). The sensitivities of indicated piezoelectric crystals vs. their usage temperature range are shown in Figure 13.4. Perovskite ferroelectric ceramics, such as PZT, have high piezoelectric

sensitivity. Although the electrical resistivity remains sufficiently high up to Curie temperature $T_C$, the usage is generally limited to 1/2 $T_C$, above which aging occurs rapidly (Setter 2002). Ferroelectric ceramics with Aurivillius and/or perovskite-layer structures possess Curie temperatures >600°C. However, their application is limited due to low piezoelectric properties and reduced electrical resistivity at elevated temperatures (Setter 2002; Ye 2008). The mineral tourmaline has the same limitations.

It was found that langasite ($La_3Ga_5SiO_{14}$, LGS) and gallium orthophosphate ($GaPO_4$), which are also piezoelectric materials, have much better characteristics at high temperatures (Smythe et al. 2000; Fritze and Tuller 2001). Experiments have shown that the phase transformation of $GaPO_4$ limits its operation to temperatures below 970°C, whereas LGS has been shown to exhibit BAW vibrations up to at least 1,400°C (Schulz et al. 2009). Both materials exhibited very good Q-factors at elevated temperatures. The most critical issues for the indicated materials are stoichiometry changes due to, for example, low-oxygen partial pressures and high losses. According to Schulz et al. (2009), the mechanical loss in langasite is significantly impacted by the electrical conductivity due to the piezoelectric coupling. The effect of the piezoelectric coupling on the loss is negligible for gallium phosphate, since it shows an extremely low electrical conductivity. Besides, gallium phosphate has minimal dependence of resonance frequency on the temperature (see Figure 17.1). However, the poor availability of $GaPO_4$ single crystals and an expensive, time-intensive growth process limit its large-scale applications. LGS has better manufacturability and is therefore a more promising candidate

## TABLE 17.5

### Properties of Various High-Temperature Piezoelectric Crystals and Commercially Available Ferroelectric Materials

| Material | Structure | $T_C/T_{melt}$ (°C) | $\varepsilon_r$ | $d_{33}$ (pC/N) |
|---|---|---|---|---|
| **Ferroelectric Ceramics/Single Crystals** | | | | |
| $BaTiO_3$ | Perovskite | 115 | 1700 | 190 |
| PZT-5A | Perovskite | 330 | 2000 | 400 |
| $PbTiO_3$ (modified) | Perovskite | 470 | 200 | 60 |
| $PbNb_2O_6$ (modified) | Tungsten bronze | 500 | 300 | 85 |
| $Bi_4Ti_3O_{12}$ | Aurivillius | 675 | 180 | 20 |
| $CaBi_4Ti_4O_{12}$ | Aurivillius | 800 | 150 | 14 |
| $La_2Ti_2O_7$ | Perovskite layer | 1460 | 60 | 2.6 |
| $LiTaO_3$ (crystal) | Trigonal, 3 m | 720 | 43 | 6 |
| $LiNbO_3$ (crystal) | Trigonal, 3 m | 1150 | 28 | 6 |
| **Piezoelectric Single Crystals** | | | | |
| $\alpha$-Quartz | Trigonal, 32 | 570 | 4.5 | ~3 |
| Tourmaline | Trigonal, 3 m | 1100–1200 | 6 | ~2 |
| $GaPO_4$ | Trigonal, 32 | 1650 | 6–7 | 4–5 |
| Langasite ($La_3Ga_5SiO_{14}$) | Trigonal, 32 | ~1470 | 16–20 | 4–7 |
| Oxyborate (ReCOB) | Monoclinic m-point group | ~1500 | 11–12 | 5–7 |

*Source:* Reprinted from *J. Crystal Growth*, 318, Zhang, S. et al., High-temperature ReCOB piezocrystals: Recent developments, 884–889, Copyright 2011, with permission from Elsevier.

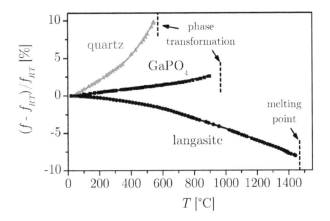

FIGURE 17.1 Temperature-dependent resonance frequency of langasite, gallium phosphate, and quartz BAW resonators normalized using the resonance frequency at room temperature $f_{RT}$. (Reprinted with permission from Fritze, H., High-temperature bulk acoustic wave sensors, *Meas. Sci. Technol.*, 22, 012002, 2011. Copyright 2011, Institute of Physics.)

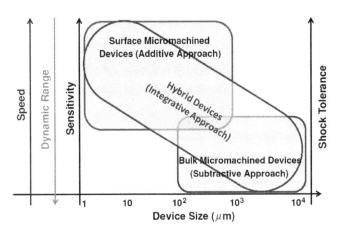

FIGURE 17.2 Three approaches toward fabricating piezoelectric MEMS devices. (Reprinted with permission from Tadigadapa, S. and Mateti, K., Piezoelectric MEMS sensors: State-of-the-art and perspectives. *Meas. Sci. Technol.*, 20, 092001-30, 2009. Copyright 2009, Institute of Physics.)

for the design of high-temperature gas sensors (Tuller 2003; Richter et al. 2006; Fritze et al. 2006). In addition, it has been found that small donator doping suppresses the electrical conductivity and the losses in LGS in the entire temperature range of measurement. The main problems during LGS-based sensors operation are related to instability of oxygen vacancies and water influence. Fritze et al. (2006) established that, in the absence of hydrogen, the formation of oxygen vacancies limits the stable operation of langasite at fairly low oxygen partial pressures, for example, at about 10–24 and 10–36 bar at 800°C and 600°C, respectively. The oxygen-vacancy concentration remains almost constant, and a stable resonance frequency is expected in this range. However, water vapor results in a frequency shift at higher oxygen partial pressure, starting below about 10–15 and 10–20 bar at 800°C and 600°C, respectively.

Zhang et al. (2011) believe that oxyborates $ReCa_4O(BO_3)_3$ (Re: rare earth element; abbreviated as ReCOB) are also promising piezoelectric crystals for high-temperature-sensing applications. Large-sized and high-quality crystals can be readily grown by the Czochralski growth method. In contrast to quartz and $GaPO_4$ crystals, no phase transformations are observed prior to their melting points, being in the order of ~1,500°C. Moreover, the resistivity of ReCOB was found to be ~$2 \times 10^8$ Ohm cm at 800°C, 2 orders higher than langasite.

## 17.3 MICROMACHINING QCM-BASED SENSORS

There are essentially three approaches to realizing piezoelectric MEMS devices: (1) deposition of piezoelectric thin films on silicon substrates with appropriate insulating and conducting layers, followed by the surface or silicon bulk micromachining to realize the micromachined transducers ("additive approach"); (2) direct-bulk micromachining of a single crystal or polycrystalline piezoelectrics and piezoceramics, which are thereafter appropriately electroded to realize micromachined transducers ("subtractive approach"); and (3) integrate-micromachined structures in silicon via bonding techniques

onto bulk piezoelectric substrates ("integrative approach") (Tadigadapa and Mateti 2009).

Depending upon the exact physical size of the desired mechanical sensors, each of these approaches offers specific advantages as illustrated schematically in Figure 17.2. As indicated before, there are many piezoelectric materials that can be used for gas- and humidity-sensor design. However, the experiment showed that for MEMS devices we have limitations in the application of several of these materials. It was established that the deposition of the piezoelectric material in the form of thin films from 1/10th of a micron to several tens of microns in the thickness with piezoelectric properties approaching those of the corresponding bulk materials is critical to the realization of a piezoelectric micromachined transducer. Unfortunately, the materials such as quartz, langasite, lithium niobate, and lithium tantalate are only available as bulk single crystals, and currently there are no effective processes for their deposition in a single-crystal thin-film form (Tadigadapa and Mateti 2009). This means that these materials can be integrated in micromachined transducers only through direct-bulk micromachining or hybrid integration methods.

At the same time, ZnO, AlN, and PZT do not have technological limitations during film deposition. Thin films of these materials can be prepared using various methods, including sputtering, chemical-solution methods, screen printing, and atomic-layer deposition technique. It was found that the properties of thin-film piezoelectric materials depend upon (1) stoichiometry, (2) film morphology, (3) film density, (4) impurities, and (5) defects. In order to obtain a large response to mechanical deformations, piezoelectric films have to be grown with a textured structure with a high degree of alignment of the piezoelectric (poling) axis. In addition to the conditions used for the growth process, the orientation and quality of the substrate material is also found to have a strong influence on the nucleation and subsequent growth of the piezoelectric films.

As it follows from the results presented in Table 17.2, aluminum nitride is one of the promising thin-film piezoelectric materials due to its high-ultrasonic velocity and fairly large piezoelectric coupling factors. Furthermore, AlN is CMOS compatible, a key consideration for on-chip integration of electronics with MEMS devices. The high-ultrasonic velocity is useful in BAW for higher frequencies because of the reduced film thickness, corresponding to the fundamental mode $\lambda/2$ thickness, which scales with $1/f$ and is only a few hundred nanometers above 5 GHz. The longitudinal material coupling coefficient for $c$-axis oriented AlN was estimated by Lakin (1991) as 6.5%–7%, respectively, 7.8% for ZnO. The slightly higher coupling coefficient of ZnO is offset by its higher temperature coefficient of frequency (TCF) and low acoustic velocity. Microstructural and stoichiometric considerations limit the use of PZT to high deposition temperatures, while CMOS-contamination requirement limits the use of ZnO in CMOS-compatible processes. PZT has a very high dielectric constant compared to AlN and ZnO.

As for the features of the manufacturing technology of microminiaturized AW sensors, which include thin-film resonators (TFRs), the thin-film bulk acoustic resonator (TFBAR) and SMR structures, then a description of the technological approaches used for these purposes (wet and dry etching, bulk and surface micromachining, etc.) one can find in the review by Tadigadapa and Mateti (2009). It is important to note that these are the same operations that were discussed earlier in Chapters 15 and 16. The features of TFBAR- and SMR-based humidity sensors were considered in Volume 2 (Chapter 12) of our series. Two typical structures of TFBAR-based sensors are shown in Figure 17.3. The first (Figure 17.3a) shows the electrodes and the thin piezoelectric film, deposited on the top of an optional insulating layer (typically silicon nitride), supported by a substrate. A portion of the substrate is removed, typically by a wet-chemical-etching

process, thereby defining the resonator. The top electrode may be patterned in a ground–signal–ground configuration, with the bottom electrode serving as the electrical ground plane (Lakin 2005). The second TFBAR structure (Figure 17.3b) shows the vibrating membrane suspended over an air gap. However, for humidity-sensing applications, the first configuration is the most widely utilized. In most gas-sensing applications, a sensitive layer is deposited on one side of the structure. For the supporting substrate with the etched cavity, the sensitive layer is typically deposited in the region of this etched cavity as shown in Figure 17.3a. However, for the air gap-based structure, the sensitive layer is deposited on the top side. The fundamental difference between FBAR and QCM or TSM is in the operating frequency. FBARs have a fundamental resonant frequency around 1 GHz due to the thinner piezoelectric layer. This also means higher mass sensitivity.

More detailed description of different TFBARs structures, fabricated using several technology approaches and piezoelectric materials, one can find in Gabl et al. (2003), Penza et al. (2008) Wingqvist et al. (2008), Tukkiniemi et al. (2009), Gerfers et al. (2010), Wang et al. (2011), Johnston et al. (2012), García-Gancedo et al. (2013), and Zhang et al. (2017). For example, the technology, developed by García-Gancedo et al. (2013) for fabrication devices shown in Figure 17.4 included the following steps and processes.

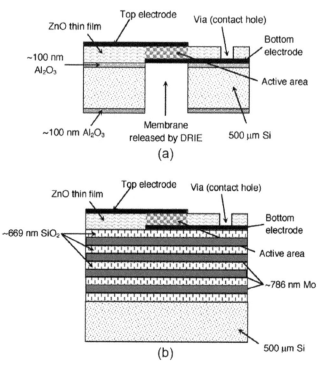

(a)

(b)

FIGURE 17.4  Schematic view of the devices fabricated by García-Gancedo et al. 2013: (a) FBAR; and (b) SMR. The FBAR and SMR devices fabricated for this work consist of a reactively sputtered thin film of ZnO (~1.2 μm) sandwiched between Cr/Au electrodes. (Reprinted from *Sens. Actuators B* 183, García-Gancedo, L. et al., Direct comparison of the gravimetric responsivities of ZnO-based FBARs and SMRs, 136–143, Copyright 2013, with permission from Elsevier.)

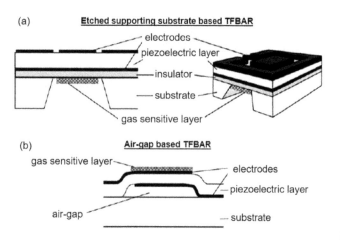

FIGURE 17.3  (a) Cross-section and diagram of a TFBAR with etched supporting substrate; and (b) diagram of TFBAR having the vibrating part suspended over an air gap. These structures were realized using bulk (a) and surface micromachining (b) technologies. (With kind permission from Springer Science+Business Media: *Solid State Gas Sensing*, Acoustic wave gas and vapor sensors, 2009, 261–304, Ippolito, S.J. et al. Copyright 2009, Springer.)

The TFBARs fabrication process commenced by patterning and sputtering a thin layer (~100 nm) of $Al_2O_3$ on the bottom of a double-side polished (100)-oriented Si substrate. This layer acts as a hard mask for the forthcoming deep reactive ion etching (DRIE) process and defines the dimensions of the membrane. A second $Al_2O_3$ layer (100 nm) was reactively sputtered on the top of the wafer. This layer doubles as a support for the ZnO membrane and as an etch barrier for the forthcoming DRIE process. The SMR fabrication process commenced by growing seven alternating low- and high-acoustic impedance layers of porous $SiO_2$ (669 nm) and Mo (786 nm) on a single-side polished (100)-oriented substrate to form an acoustic mirror, centered at ~2 GHz with ~1 GHz bandwidth. The layers forming the Bragg reflector were deposited on a standard magnetron sputtering system, and although it is not possible to measure the thickness of each layer individually other than with a precalibrated thickness monitor within the sputtering chamber, it is estimated that each layer may have an error in thickness and homogeneity of up to several nm. For both the TFBARs and SMRs, the bottom electrodes (4/50 nm Cr/Au) were patterned through a standard photolithography process and thermally evaporated on the top $Al_2O_3$ layer (for the TFBARs) or the top $SiO_2$ layer (for the SMRs). The ZnO piezoelectric films were then reactively sputtered at high rates (~50 nm/min) and at room temperature from a 4-in diameter metallic Zn target in a high-target utilization sputtering (HiTUS) system. This technique yields high-quality films with very low stress and defect density and a homogeneity better than 99% over a 6-in diameter area.

Another approach to fabrication of TFBAR resonator structure, developed by Gerfers et al. (2010), is shown in Figure 17.5. In this structure, a piezoelectric film is sandwiched between two electrode films. The substrates were 6-in diameter (100)-silicon wafers. The process started with a silicon trench etch, followed by trench fill using silicon dioxide as a sacrificial layer. The surface was then planarized by a chemical-mechanical polishing (CMP) step. It is important to have a very smooth surface for the following piezoelectric-film-stack deposition. A bottom Mo layer (0.32 μm) was sputtered and patterned to define the electrode area. An AlN layer (1.1 μm) was deposited by reactive sputtering. A second Mo layer was deposited and patterned to define top electrode area. The AlN layer was dry etched to open contact windows on the bottom Mo electrodes and form release holes. Finally, the sacrificial layer was etched with buffered HF (BOE) to release the membrane. Figure 17.5b shows optical and cross-section SEM micrographs of a resonator after completion of the fabrication.

It is important to note that structurally AW-based gas sensors and humidity sensors are identical. Only the sensing layers used to increase the sensitivity to a particular gas or vapor differ. Therefore, the technologies developed for gas sensors can be used in the manufacture of humidity sensors. One of the TFBARs designed for humidity measurement is shown in Figure 17.6 (Zhang et al. 2015, 2017). The ZnO film with nanostructure acted as both the resonant layer and sensitive layer. Micro through-holes, with size of $10 \times 10$ μm$^2$, were made in the top electrode to provide paths for gas reaching the ZnO layer directly and improve sensitivity of the sensor. In addition, thermal modules, a temperature sensor

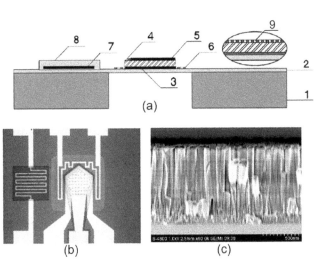

**FIGURE 17.6** (a) Cross section of the FBAR device with detail of the through-hole array: 1- Si substrate; 2—support film (1.5 μm silicon nitride); 3—bottom electrode; 4—ZnO resonance film (1.2 μm); 5—top Pt electrode; 6—Pt resistance heater; 7—resistance temperature sensor; 8—insulated film; 9—micro through-hole. All the structures were on a $SiN_x$-insulated support film and a monocrystalline silicon substrate. Silicon in center of the back of substrate was etched by deep reactive ion etching (RIE), until it reached the top silicon nitride film, forming a suspending area; (b) microscopic photo of the FBAR chip; and (c) cross section of ZnO film by SEM. (Reprinted from *Procedia Eng.*, 120, Zhang, M. et al., Micro through-hole array in top electrode of film bulk acoustic resonator for sensitivity improving as humidity sensor, 663–666, Copyright 2015, with permission from Elsevier.)

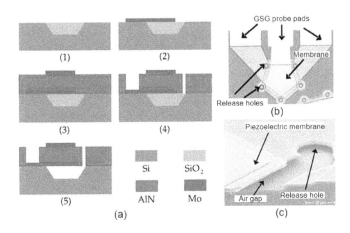

**FIGURE 17.5** (a) Fabrication process flow of AlN piezoelectric resonators, and (b, c) images of resonator after fabrication: (b) Topview, and (c) SEM cross-section. (Reprinted from Gerfers, F. et al., Sputtered AlN thin films for piezoelectric MEMS devices: FBAR resonators and accelerometers, in: Swart J.W. (ed.), *Solid State Circuits Technologies*, INTECH, Rijeka, Croatia, pp. 333–352. Published by INTECH as open access.)

and a heater, were added to monitor and control working temperature of the resonator. As a result, high temperature was available to enhance thermal desorption. The testing of this humidity sensor in the range of 30%–90%RH has shown that the micro through-hole array design made sensitivity rise to 18.6 ppm/%RH, namely 21.2 kHz/%RH, which was 18 times higher than for the samples without micro holes. TFBAR-based humidity sensors, designed by Giangu et al. (2014), showed even higher sensitivity The sensitivity of the TFBAR humidity sensor reached ~43 kHz/% RH with a resonance frequency of 6 GHz. TFBAR devices were manufactured on a 0.5-μm-thin GaN membrane using 50 nm Mo metallization. As a humidity-sensitive material, it was used as a composite-containing polyvinyl alcohol and titanium-dioxide nanoparticles with wide-diameter distribution, ranging from 80 to 1000 nm. GaN, applied in designed sensor, as well as AlN, has good piezoelectric properties, but only recently it became possible to use this material for SAW and FBAR device manufacturing. The quality of the wide-band gap GaN-thin layers, grown or deposited on SiC, on sapphire and also on Si substrate, greatly improved in recent years. GaN compatibility with advanced micromachining and nanoprocessing technologies allows one to use it to increase the resonance frequency of various acoustic devices (Müller et al. 2006, 2009, 2010), including humidity sensors (Cismaru et al. 2012). The sound velocity calculated, neglecting the influence of metallic electrodes, was in the range of 6900–7350 m/s (Bougrov et al. 2001; Müller et al. 2009).

But, it is necessary to recognize that despite growing success in solving TFBAR technological problems, unresolved issues remain. The problem of optimizing the resonator characteristics quality factor $Q$ and electromechanical coupling factor $k_t^2$ may be solved by improving the film quality of the piezoelectric material, but they are also influenced by the design of the resonator and the layer properties of the materials surrounding the piezoelectric film. As is known, good piezoelectric properties can be achieved by having good stoichiometry and morphology of the materials. To achieve these, the piezoelectric materials need proper seed layers as well as proper control of the nucleation, growth, and crystallization processes. Solutions are also needed for maintaining a uniform film thickness for frequency control. This may be achieved either by improving the deposition tools or by developing techniques for individual sensor trimming. Applied to AlN and ZnO films, this problem is being solved by using rf or dc sputtering. These technologies were developed specifically for TFBAR fabrication in 1980 by a number of groups. These films are now readily produced with optimized piezoelectric properties and very high thickness uniformity on 4–8″ substrates with standard industrial sputtering machines. As for GaN, the piezoelectric GaN film can be grown on silicon by MOCVD or MBE techniques (Müller et al. 2009). These processes ensure the high quality of the deposited layer and offer the possibility of monolithic integration with High Electron Mobility Transistors (HEMT) manufactured on the same chip.

## REFERENCES

Ballantine D.S. Jr, White R.M., Martin S.J., Ricco A.J., Frye G.C., Zellers E.T., Wohltjen H. (1997) *Acoustic Wave Sensors: Theory, Design, and Physico-Chemical Applications.* Academic, San Diego, CA.

Ballantine D.S. Jr, White R.M., Martin S.J., Ricco A.J., Zellers E.T., Frye G.C., Wohltjen H. (1996) Acoustic wave sensors: Theory, design, and physico-chemical applications. In: Levy M., Stern R. (eds.) *Applications of Modern Acoustics.* Academic Press, San Diego, CA.

Bougrov V., Levinshtein M.E., Rumyantsev S.L., Zubrilov A. (2001) Gallium Nitride (GaN). In: M.E. Levinshtein, S.L. Rumyantsev, M.S. Shur (eds.) *Properties of Advanced Semiconductor Materials GaN, AlN, InN, BN, SiC, SiGe.* Wiley, New York, pp. 1–30.

Cismaru A., Konstantinidis G., Szacilowski K., Müller A., Stefanescu A., Stavrinidis A., et al. (2012) Gas and humidity sensors based on high frequency acoustic devices manufactured on GaN/Si. In: *RF MEMS Technologies: Recent Developments, Series in Micro and Nanoengineering*, Vol. 21. Editura Academiei Romane, Bucharest, Romania, Vol. 21, pp. 263–270.

Fachberger R., Bruckner G., Knoll G., Hauser R., Biniasch J., Reindl L. (2004) Applicability of LiNbO$_3$, langasite and GaPO$_4$ in high temperature SAW sensors operating at radio frequencies. *IEEE Trans. Ultrason. Ferroelectr. Freq. Control* **51**, 1427–1431.

Fanget S., Hentz S., Puget P., Arcamone J., Matheron M., Colinet E., Andreucci P., Duraffourg L., Myers E., Roukes M.L. (2011) Gas sensors based on gravimetric detection—A review. *Sens. Actuators B* **160**, 804–821.

Fechete A.C., Wlodarski W., Kalantar-Zadeh K., Holland A.S., Antoszewski J., Kaciulis S., Pandolfi L. (2006) SAW-based gas sensors with rf sputtered InO$_x$ and PECVD SiN$_x$ films: Response to H$_2$ and O$_3$ gases. *Sens. Actuators B* **118**, 362–367.

Ferrari V. (2004) Acoustic-wave piezoelectric and pyroelectric sensors based on PZT thick films. In: Yurish S.Y., Gomes M.T.S.R. (eds.) *Smart Sensors and MEMS*, Vol. 181, NATO science series II: Mathematics, physics and chemistry. Springer, New York, pp. 125–154.

Ferrari V., Marioli D., Taroni A., Ranucci E. (2000) Multisensor array of mass microbalances for chemical detection based on resonant piezo-layers of screen-printed PZT. *Sens. Actuators B* **68**, 81–87.

Fritze H. (2011) High-temperature bulk acoustic wave sensors. *Meas. Sci. Technol.* **22**, 012002.

Fritze H., Schulz M., She H., Tuller H.L. (2006) Sensor application-related defect chemistry and electromechanical properties of langasite. *Solid State Ionics* **177**, 2313–2316.

Fritze H., Tuller H.L. (2001) Langasite for high temperature bulk acoustic wave applications. *J Appl. Phys. Lett.* **78**, 976–977.

Gabl R., Green E., Schreiter M., Feucht H.D., Zeininger H., Primig R., et al. (2003) Novel integrated FBAR sensors: A universal technology platform for bio- and gas-detection. In: *Proceedings of IEEE Sensors Conference*, October 22–24, Toronto, Canada, pp. 1184–1188.

García-Gancedo L., Pedrós J., Iborra E., Clement M., Zhao X.B., Olivares J., et al. (2013) Direct comparison of the gravimetric responsivities of ZnO-based FBARs and SMRs. *Sens. Actuators B* **183**, 136–143.

Gerfers F., Kohlstadt P.M., Ginsburg E., He M.Y., Samara-Rubio D., Manoli Y., Wang L. (2010) Sputtered AlN thin films for piezoelectric MEMS devices: FBAR resonators and accelerometers. In: Swart J.W. (ed.) *Solid State Circuits Technologies.* INTECH, Rijeka, Croatia, pp. 333–352.

Giangu I., Buiculescu V., Konstantinidis G., Szaciáowski K., Stefanescu A., Bechtold F., et al. (2014) Acoustic wave sensing devices and their LTCC packaging. In: *Proceedings of the International IEEE Semiconductor Conference*, Sinaia, Romania, October 13–15, pp. 147–150.

Ippolito S.J., Trinchi A., Powell D.A., Wlodarski W. (2009) Acoustic wave gas and vapor sensors. In: Comini E., Faglia G., Sberveglieri G. (eds.) *Solid State Gas Sensing*. Springer, New York, pp. 261–304.

Jakubik W.P., Urbaczyk M.W., Kochowski S., Bodzenta J. (2002) Bilayer structure for hydrogen detection in a surface acoustic wave sensor system. *Sens. Actuators B* **82**, 265–271.

Johnston M.L., Edrees H., Kymissis I., Shepard K.L. (2012) Integrated VOC vapor sensing on FBAR-CMOS array. In: *Proceedings of Symposium on MEMS*, Paris, France, January 29–February 2, pp. 846–849.

Kholkin A.L., Kiselev D.A., Kholkine L.A., Safari A. (2009) Smart ferroelectric ceramics for transducer applications, Chapter 9.1. In: Schwartz M. (ed.) *Smart Materials*. CRC Press, Boca Raton, FL.

Korotcenkov G. (ed.) (2011) *Chemical Sensors: Comprehensive Sensor Technologies*, Volume 4: *Solid State Devices*. Momentum, New York.

Lakin K.M. (1991) Fundamental properties of thin film resonators. In: *Proceedings of the 45th Annual Symposium on Frequency Control*, Vol. 1. May 29–31, Los Angeles, CA, pp. 201–206.

Lakin K.M. (2005) Thin film resonator technology. *IEEE Trans Ultrason. Ferroelectr.* **52**, 707–716.

Lakin K.M., McCarron K.T., McDonald J.F. (2000) Temperature compensated bulk acoustic thin film resonators. In: *Proceedings of IEEE Ultrasonics Symposium*, October 22–25, San Juan, Puerto Rico, pp. 855–858.

Lang S.B. (2005) Pyroelectricity: From ancient curiosity to modern imaging tool. *Phys. Today* **58**, 31–36.

Lee J.H., Yoon K.H., Hwang K.S., Park J., Ahn S., Kim T.S. (2004) Label free novel electrical detection using micromachined PZT monolithic thin film cantilever for the detection of C-reactive protein. *Biosens. Bioelectron.* **20**, 269–275.

Lee Y., Lim G., Moon W. (2006) A self-excited micro cantilever biosensor actuated by PZT using the mass micro balancing technique. *Sens. Actuators A* **130–131**, 105–110.

Lozano J., Fernandez M.J., Fontecha J.L., Aleixandre M., Santos J.P., Sayago I., et al. (2006) Wine classification with a zinc oxide SAW sensor array. *Sens. Actuators B* **120**, 166–171.

Lucklum R., Eichelbaum F. (2007) Interface circuits for QCM sensors. In: Steinem C. and Janshoff A. (eds.) *Piezoelectric Sensors*. Springer, New York, pp. 3–47.

Müller A., Neculoiu D., Konstantinidis G., Deligeorgis G., Dinescu A., Stavrinidis A., et al. (2010) SAW devices manufactured on GaN/Si for frequencies beyond 5 GHz. *IEEE Electron Dev. Lett.* **31**(12), 1398–1400.

Müller A., Neculoiu D., Konstantinidis G., Stavrinidis A., Vasilache D., Cismaru A., et al. (2009) 6.3 GHz film bulk acoustic resonator structures based on a Gallium Nitride/Silicon thin membrane. *IEEE Electron Dev. Lett.* **30**(8), 799–801.

Müller A., Neculoiu D., Vasilache D., Dascalu D., Konstantinidis G., Kosopoulos A., Adikimenakis A., et al. (2006) GaN micromachined FBAR structures for microwave applications. *Superlattice. Microst.* **40**, 426–431.

Özgür Ü., Alivov Ya. I., Liu C., Teke A., Reshchikov M.A., Doğan S., Avrutin V., Cho S.-J., Morkoç H. (2005) A comprehensive review of ZnO materials and devices. *J. Appl. Phys.* **98**, 041301.

Penza M., Aversa P., Cassano G., Suriano D., Wlodarski W. (2008) Thin-film bulk-acoustic-resonator gas sensor functionalized with a nanocomposite Langmuir–Blodgett layer of carbon nanotubes. *IEEE Trans. Electron Dev.* **55**(5), 1237–1243.

Richter D., Fritze H., Schneider T., Hauptmann P., Bauersfeld N., Kramer K.D., Wiesner K., Fleischer M., Karle G., Schubert A. (2006) Integrated high temperature gas sensor system based on bulk acoustic wave resonators. *Sens. Actuators B* **118**, 466–471.

Rosen C.Z., Hiremath B.V., Newnham R. (eds.) (1992) *Piezoelectricity*. Springer-Verlag, New York.

Roy S., Basu S. (2002) Improved zinc oxide film for gas sensor applications. *Bull. Mater. Sci.* **25**, 513–515.

Schulz M., Sauerwald J., Richter D., Fritze H. (2009) Electromechanical properties and defect chemistry of high-temperature piezoelectric materials. *Ionics* **15**, 157–161.

Schwartz R.W., Ballato J., Haertling G.H. (2004) Piezoelectric and electro-optic ceramics. In: Buchanan R.C. (ed.) *Ceramic Materials for Electronics*. Dekker, New York, pp. 207–315.

Setter N. (ed.) (2002) *Piezoelectric Materials in Devices*. EPFL Swiss Federal Institute of Technology, Lausanne, Switzerland.

Smith R.T., Welsh F.S. (1971) Temperature dependence of the elastic, piezoelectric, and dielectric constants of lithium tantalate and lithium niobate. *J. Appl. Phys.* **42**, 2219–2230

Smythe R., Helmbold R.C., Hague G.E., Snow K.A. (2000) Langasite, langanite, and langatate bulk-wave Y-cut resonators. *IEEE Trans. Ultrason. Ferroelectr. Freq. Control.* **47**, 355–360.

Tadigadapa S., Mateti K. (2009) Piezoelectric MEMS sensors: State-of-the-art and perspectives. *Meas. Sci. Technol.* **20**, 092001–092030.

Trolier-Mckinstry S., Muralt P. (2004) Thin film piezoelectrics for MEMS. *J. Electroceramics* **12**, 7–17.

Tukkiniemi K., Rantala A., Nirschl M., Pitzer D., Huber T., Schreiter M. (2009) Fully integrated FBAR sensor matrix for mass detection. *Procedia Chem.* **1**, 1051–1054.

Tuller H.L. (2003) Defect engineering: Design tools for solid state electrochemical devices. *Electrochim. Acta* **48**, 2879–2887.

Uchino K., Ito Y. (2009) Smart ceramics: Transducers, sensors, and actuators, Chapter 9.2. In: Schwartz M. (ed.) *Smart Materials*. CRC Press, Boca Raton, FL.

Wang J.-J., Chen D., Zhang L.-Y., Xu Y. (2011) Thin-film-bulk-acoustic-resonator gas sensor for the detection of organophosphate vapor detection. In: *Proceedings of Symposium on Piezoelectricity, Acoustic Waves and Device Applications (SPAWDA)*, December 9–11, Shenzhen, China. doi: 10.1109/SPAWDA.2011.6167205.

Wang W., He S., Li S., Pan Y. (2006) Enhanced sensitivity of SAW gas sensor based on high frequency stability oscillator. *Smart Mater. Struct.* **15**, 1525–1530.

Wingqvist G., Yantchev V., Katardjiev I. (2008) Mass sensitivity of multilayer thin film resonant BAW sensors. *Sens. Actuators A* **148**, 88–95.

Yanagitani T., Kiuchi M., Matsukawa M., Watanabe Y. (2007) Characteristics of pure-shear mode BAW resonators consisting of (1120) textured ZnO films. *IEEE Trans. Ultrason. Ferroelectr. Freq. Control* **54**, 1680–1686.

Ye Z.G. (ed.) (2008) *Handbook of Advanced Dielectric Piezoelectric and Ferroelectric Mater: Synthesis Characterization and Applications*. Woodhead, Cambridge, UK.

Yoon S.M., Cho N.J., Kanazawa K. (2009) Analyzing spur-distorted impedance spectra for the QCM. *J. Sensors* **2009**, 259746.

Zhang M., Du L., Fang Z., Zhao Z. (2015) Micro through-hole array in top electrode of film bulk acoustic resonator for sensitivity improving as humidity sensor. *Procedia Eng.* **120**, 663–666.

Zhang M., Du L., Fang Z., Zhao Z. (2017) A sensitivity-enhanced film bulk acoustic resonator gas sensor with an oscillator circuit and its detection application. *Micromachines* **8**, 25.

Zhang S., Yu F., Xia R., Fei Y., Frantz E., Zhao X., Yuan D., Chai B.H.T., Snyder D., Shrout T.R. (2011) High-temperature ReCOB piezocrystals: Recent developments. *J. Crystal Growth* **318**, 884–889.

Zhou X., Zhang J., Jiang T., Wang X., Zhu Z. (2007) Humidity detection by nanostructured ZnO: A wireless quartz crystal microbalance investigation. *Sens. Actuators A* **135**, 209–214.

# 18 Technologies Suitable for Fabrication of Humidity Sensing Layers

## General Consideration

## 18.1 INTRODUCTION: ACCEPTABLE TECHNOLOGIES AND THEIR ADVANTAGES AND LIMITATIONS

Making high-quality humidity sensitive materials is one of the most important tasks of the technology designed for humidity-sensor fabrication. As shown in previous chapters, humidity-sensitive materials used in humidity sensors need to fulfill a range of requirements, related to the crystallographic structure, chemical composition, morphology, electrophysical properties, catalytic activity, porosity, and so on. As it was shown in previous chapters, the materials for humidity sensors can come in a variety of forms, including films, ceramics, or powders. Their structure may be amorphous, glassy, nano-crystalline, polycrystalline, single crystalline, or epitaxial. They may be either dense or porous. These materials may be elementary substances, complex compounds, or composites. Polymers, metals, dielectrics, and semiconductors can also be used as materials for humidity sensors. They may be either organic or inorganic in nature.

This vast amount of variation indicates that it is impossible to produce such a wide range of materials using just one method. The possible differences in the physical-chemical properties of the materials are too great; so too are the resulting differences in the conditions required for the synthesis and deposition of these materials. Therefore, for preparing humidity-sensing materials with required properties we have to use various methods such as vacuum deposition, sputtering, sol-gel process, and spray pyrolysis. The short description of these methods is presented in Tables 18.1 and 18.2. These techniques differ in deposition rates, substrate temperature during deposition, precursor materials, the necessary equipment, expenditure, and in the quality of the resulting films. For example, Figure 18.1 gives the comparison of two widely used methods for forming humidity-sensitive materials. More detailed analysis of the methods, acceptable for application in humidity sensor fabrication, one can find in a vast array of quality reviews devoted to this subject (Brinker and Scherer 1990; Randhaw 1991; Hitchman and Jensen 1993; Hecht et al. 1994; Bunshah 1994; Brinker et al. 1996; Glocker and Shah 1995; Nenov and Yordanov 1996; Huczko 2000; Arthur 2002;

Willmott 2004; Tay et al. 2006; Vahlas et al. 2006; Van Tassel and Randall 2006; Vayssiere 2007; Viswanathan et al. 2006; Christen and Eres 2008; Jaworek and Sobszyk 2008; Milchev 2008; Tiemann 2008; Korotcenkov and Cho 2010). Most of the recommended articles deal with technologies related to gas-sensitive materials preparation. However, gas sensors and humidity sensors in most cases have the same principles of operation and use the same materials. Therefore, the recommendations of these reviews are well founded.

For the formation and synthesis of humidity-sensitive materials, specific methods such as template-based synthesis, electro-spinning, and Langmuir-Blodgett (LB) process can also be used (read Table 18.3).

It is clear that the choice of a method acceptable for sensor-material synthesis or deposition in the design and manufacture of a sensor is a difficult task. We need to analyze many different factors, including the type of technology, which will be used for sensor fabrication: ceramic, thick-film, or thin-film technologies. No doubt that every technology has advantages and disadvantages. Table 18.4 gives a comparison of several of these methods.

## 18.2 CERAMIC TECHNOLOGY

Ceramic elements have been investigated since the 1950s and are still employed as gas chemiresistors (Vandrish 1996). The simplest ceramic sensor design is a sintered block arrangement. The block is prepared by sintering metal-oxide powders, with the electrode wires embedded into the block. The techniques used to produce ceramic humidity sensors are similar to those used to fabricate conductometric gas sensors, thermistors, resistors, and electrodes from ceramics. The regimes of annealing and high-temperature processing are optimized for each material to provide either dense ceramic for solid electrolyte electrochemical-humidity sensors or porous ceramics for other types of humidity sensors. As indicated above, in the porous structure, it is necessary to increase both the surface-to-volume ratio and the water-vapor penetrability of humidity-sensing matrix. The final shape of the ceramic element may be a sphere, tablet, or cylinder, which contains a heater and electrical contact wires.

**TABLE 18.1**

**Vapor-Phase Methods Used During Humidity-Sensor Fabrication**

| Method | Description |
|---|---|
| **Physical vapor deposition (PVD)** | PVD covers a number of deposition technologies in which material is released from a source and transferred to the substrate. The two most important technologies are evaporation and sputtering. |
| **Vacuum evaporation** | *Vacuum evaporation* (including sublimation) is a PVD process where material from a thermal vaporization source reaches the substrate without collision with gas molecules in the space between the source and substrate. The vacuum is required to allow the molecules to evaporate freely in the chamber, and then subsequently condense on all surfaces. |
| | There are two popular evaporation technologies, which are e-beam evaporation and resistive evaporation, each referring to the heating method. In e-beam evaporation, an electron beam is aimed at the source material causing local heating and evaporation. In resistive evaporation, usually a tungsten boat, containing the source material, is heated electrically with a high current to make the material evaporate. |
| **Molecular beam epitaxy** | *Epitaxial techniques* are techniques of arranging atoms in single-crystal fashion on crystalline substrates so that the lattice of the newly grown film duplicates that of the substrate. If the film is of the same material as the substrate, the process is called homoepitaxy, epitaxy, or simply epi. In molecular beam epitaxy (MBE), the heated single-crystal sample is placed in an ultrahigh vacuum ($10^{-10}$ Torr) in the path of streams of atoms from heated cells that contain the materials of interest. These atomic streams impinge on the surface, creating layers whose structure is controlled by the crystal structure of the surface, the thermodynamics of the constituents, and the sample temperature. The deposition rate of MBE is very low. |
| **Sputtering** | *Sputter deposition* is a method of depositing thin films by sputtering material from a "target." Sputtering is a term used to describe the mechanism in which atoms are ejected from the surface of a material when that surface is stuck by sufficiency energetic ions generated by low-pressure gas plasma, usually an argon plasma. The sputtering takes place at much lower temperature than evaporation. Sputtered atoms ejected into the gas phase in vapor form are not in their thermodynamic equilibrium state and tend to condense on all surfaces in the vacuum chamber including the substrate. |
| | There are several modifications of sputtering. However, the basic principle of sputtering is the same for all sputtering technologies. The differences typically relate to the manner in which the ion bombardment of the target is realized. |
| **Reactive sputtering** | *Reactive sputtering* is a modification of conventional sputtering process. During this process, reactive gas is introduced into the sputtering chamber in addition to the argon plasma. As a result, the compound, which is formed by the elements of that gas combining with the sputter material, can be deposited. |
| **Laser ablation** | *Laser ablation* is the process of removing material from a solid surface by irradiating it with a laser beam. The ablation process takes place in a vacuum chamber, either in a vacuum or in the presence of some background gas. In the case of oxide films, oxygen is the most common background gas. |
| | At low-laser flux, the material is heated by the absorbed laser energy and evaporates or sublimates. At high-laser flux, the material is typically converted to a plasma. As a result, a supersonic jet of particles (plume) with composition similar to composition of material is ejected normal to the target surface. The plume, similar to the rocket exhaust, expands away from the target with a strong forward-directed velocity distribution of the different particles. The ablated species condense on the substrate placed opposite to the target. |
| **Chemical vapor deposition (CVD)** | *Chemical vapor deposition* refers to the formation of a non-volatile solid material from the reaction of chemical reactants, called precursors, being in the vapor phase in the right constituents. A reaction chamber is used for this process, into which the reactant gases are introduced to decompose and react with the substrate to form thin film or powders. |
| | There are several main classification schemes for chemical vapor deposition processes. These include classification by the pressure (atmospheric, low-pressure, or ultrahigh high vacuum), characteristics of the vapor (aerosol or direct liquid injection), or plasma processing type (microwave plasma-assisted deposition, plasma-enhanced deposition, remote plasma-enhanced deposition). |

*Source:* Korotcenkov, G., *Handbook of Gas Sensor Materials*, Springer, New York, 2013.

**TABLE 18.2**
**Liquid-Phase Methods Used for Humidity-Sensing Material Synthesis and Deposition**

| Method | Description |
|---|---|
| Sol-gel method | The *sol-gel process* is a wet-chemical technique used for the fabrication of both glassy and ceramic materials. In this process, the sol (or solution) evolves gradually toward the formation of a gel-like network containing both a liquid phase and a solid phase. In other words, the sol-gel process is the formation of an oxide network through polycondensation reactions of a molecular precursor in a liquid. A "sol" is a stable dispersion of colloidal particles or polymers in a solvent. These particles may be amorphous or crystalline. A "gel" consists of a three-dimensional continuous network, which encloses a liquid phase. In a colloidal gel, the network is built from agglomeration of colloidal particles. Typical precursors are metal alkoxides and metal chlorides, which undergo hydrolysis and polycondensation reactions to form a colloid. The basic structure or morphology of the solid phase can range anywhere from discrete colloidal particles to continuous chain-like polymer networks. |
| Chemical deposition | Unit species of material to be deposited is applied in liquid/solution form. Substrates act as a physical support and no reaction. Deposition carried out at lower temperatures ($<100°C$) typically atmospheric pressures |
| Chemical solution deposition (CSD) | The *chemical solution deposition* (CSD) method used for preparation of oxide films comprises the deposition of a liquid sol on a substrate and the conversion of gel films to resulting ceramic films via heat treatment. |
| Liquid phase deposition (LPD) | *Liquid phase deposition* (LPD) is a method for the "non-electrochemical" production of polycrystalline ceramic films at low temperatures. LPD, along with other aqueous solution methods [chemical bath deposition (CBD), successive ion layer adsorption, and reaction (SILAR) and electroless deposition (ED) with catalyst] has developed as a potential substitute for vapor-phase and chemical-precursor systems. The method involves immersion of a substrate in an aqueous solution containing a precursor species (commonly a fluoro-anion), which hydrolyzes slowly to produce a supersaturated solution of the desired oxide, which then precipitates preferentially on the substrate surface, producing a conformal coating. |
| Electrochemical deposition (ECD) *(electroplating)* | The *electrochemical deposition* process, which is typically restricted to electrically conductive materials and is carried out in a liquid solution of ions (electrolyte), is well suited to make films of metals such as copper, gold, nickel, etc. The films can be made in any thickness from $<0.1$ to $>100$ μm. Other materials including metal oxides can be deposited as well. There are basically two technologies for plating: *electroplating* and *electroless* plating. During *electroplating*, when an electrical potential is applied between a conducting area on the substrate and a counterelectrode (usually platinum) in the liquid, a chemical redox process takes place, resulting in the formation of a layer of material on the substrate and usually some gas generation at the counter electrode. This method, although more complicated, allows for more operator control. In the *electroless* plating process, a more complex chemical solution is used, in which deposition happens spontaneously on any surface, which forms a sufficiently high electrochemical potential with the solution. The deposition is from a solution containing a metal salt and a reducing agent as well as various other additives, such as stabilizers and surfactants. This process is desirable since it does not require conductive substrates, any external electrical potential, and contact to the substrate during processing. Unfortunately, it is also more difficult to control with regards to film thickness and uniformity |
| Electrophoretic deposition (EPD) | *Electrophoretic deposition* is a particulate forming process. It is a high-level efficient process for production of films or coatings on electrically conducting objects from colloidal suspensions. It begins with a dispersed powder material in a solvent and uses an electric field to move the powder particles into a desired arrangement on an electrode surface. Deposition on the electrode occurs via particle coagulation. The technique allows depositing thin and thick films, and the shaping of bulk objects with metallic, polymeric, or ceramic particles. There are four defining characteristics of EPD: (1) it begins with particles that are well dispersed and able to move independently in solvent suspension; (2) the particles have a surface charge due to electrochemical equilibrium with the solvent; where the particles would normally be electrically neutral, a compound might be bonded to them to give them an electrical charge in suspension; (3) there is electrophoretic motion of the particles in the bulk of the suspension; and (4) a rigid (finite shear strength) deposition of the particles is formed on the deposition electrode |

*Source:* With kind permission from Springer Science+Business Media: *Handbook of Gas Sensor Materials*, 2013, Korotcenkov, G. Copyright 2013, Springer.

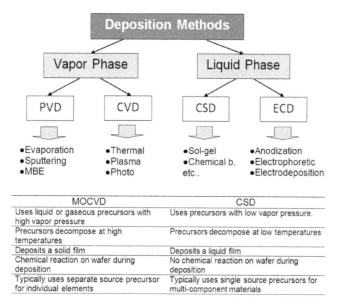

| MOCVD | CSD |
|---|---|
| Uses liquid or gaseous precursors with high vapor pressure | Uses precursors with low vapor pressure. |
| Precursors decompose at high temperatures | Precursors decompose at low temperatures |
| Deposits a solid film | Deposits a liquid film |
| Chemical reaction on wafer during deposition | No chemical reaction on wafer during deposition |
| Typically uses separate source precursor for individual elements | Typically uses single source precursors for multi-component materials |

**FIGURE 18.1** Comparison of main deposition methods. (With kind permission from Springer Science+Business Media: *Handbook of Gas Sensor Materials*, 2013, Korotcenkov, G.. Copyright 2013 Springer.)

Metal-oxide powders for ceramic humidity sensors can be synthesized using various methods such as sol-gel, flame pyrolysis, and hydrothermal synthesis. Many of these technologies offer some specific advantages for fabricating nanoparticles, such as simplicity, flexibility, low cost, ease of use on large substrates, and the ability to modify the composition by the addition of dopants and modifiers. These methods are well described in reviews and books (Comini et al. 2009a; Korotcenkov and Cho 2010). These processes will be briefly discussed in Chapter 20. Modification of the humidity-sensing properties of these ceramics is possible by changing the parameters of synthesis and sintering (Song and Park 1994). By varying synthesis conditions, the characteristic size of the crystallites in the ceramic samples might be reduced down to 4 nm (Yu et al. 1997), which promotes high gas sensitivity (read Chapter 2 in Volume 1). Ball milling is an efficient operation that can also be used to adjust the parameters of ceramic sensors (Choi et al. 2005). Powders obtained using chemical methods due to strong agglomeration usually have low gas permittivity; ball milling destroys dense agglomerates and makes the gas-sensing "3D" matrix more homogeneous (Yamazoe and Miura 1992).

## TABLE 18.3
## Specific Methods Used for Humidity-Sensitive Materials Preparing

| Method | Description |
|---|---|
| *Template-based synthesis* | *Template-based synthesis* involves the fabrication of the desired material within the pores or channels of a nanoporous template. A template may be defined as a central structure within which a network forms in such a way that removal of the template creates a filled cavity with morphological and/or stereochemical features related to those of the template. Track-etch membranes, porous alumina, and other nanoporous structures have been characterized as templates. Electrochemical and electroless depositions, chemical polymerization, sol-gel deposition, and chemical vapor deposition have been presented as major template synthetic strategies. *Template-based synthesis* can be used to prepare nanostructures of conductive polymers, metals, metal oxides, semiconductors, carbons, and other solid matter. |
| | If the templates that are used have cylindrical pores of uniform diameter, monodisperse nanocylinders of the desired material are obtained within the voids of the template material. Depending on the operating parameters, these nanocylinders may be solid (a nanorod) or hollow (a nanotubule). The nanostructures can remain inside the pores of the templates, or they can be freed and collected as an ensemble of free nanoparticles. Alternatively, they can protrude from the surface like the bristles of a brush. |
| *Electrospinning* | *Electrospinning* is a simple method for generating nanofibers made of polymers, ceramics, and composites. In the electrospinning process a high voltage is used to create an electrically charged jet of solution, mainly polymer-based, which dries or solidifies to leave a very fine (typically on the micro or nanoscale) fiber. One electrode is placed into the spinning solution and the other attached to a collector. The electric field is subjected to the end of a capillary tube that contains the liquid (polymer) fluid held by its surface tension. When a sufficiently high voltage is applied to a liquid droplet, the body of the liquid becomes charged, and electrostatic repulsion counteracts the surface tension and the droplet is stretched; at a critical point, a stream of liquid erupts from the surface. The process of electrospinning does not require the use of coagulation chemistry or high temperatures to produce solid threads from solution. However, for preparing metal-oxide fibers, the following annealing is required. Electrospinning from molten precursors is also practiced; this method ensures that no solvent can be carried over into the final product. |
| | Successful electrospinning requires the use of an appropriate solvent and polymer system to prepare solutions exhibiting the desired viscoelastic behavior. The traditional setup for electrospinning works well for most conventional polymers, but it cannot be easily applied to polymers with limited solubilities (e.g., conjugated polymers) or low molecular weights. |
| *Langmuir–Blodgett film* | In the *Langmuir–Blodgett* (LB) process, a monolayer of film forming molecules (stearic acid is often used as a model molecule) on an aqueous surface is compressed into a compact floating film and transferred to a solid substrate by passing the substrate through the water surface. |

*Source:* Korotcenkov, G., *Handbook of Gas Sensor Materials*, Springer, New York, 2013.

## TABLE 18.4
## Advantages and Disadvantages of Synthesis and Deposition Methods Usually Used During Humidity-Sensor Fabrication

### Vacuum Evaporation

**Advantages**

- High-purity films can be deposited from high-purity source material.
- Source of material to be vaporized may be a solid in any form and purity.
- The line-of-sight trajectory and limited-area sources allow the use of masks to define areas of deposition on the substrate and shutters between the source and substrate to prevent deposition when not desired.
- Deposition-rate monitoring and control are relatively easy.
- It is the least expensive of the PVD processes.

**Disadvantages**

- Large-volume vacuum chambers are generally required to keep an appreciable distance between the hot source and the substrate.
- Inconstancy of evaporation rates during the deposition process.
- A considerable difference in composition between the evaporated and deposited materials Therefore, many compounds and alloy compositions, due to stoichiometry problem, can only be deposited with difficulty. The same problems take place for materials that have high melting temperature and low saturated vapor pressure.
- Line-of-sight trajectories and limited-area sources result in poor film-thickness uniformity and in poor surface coverage on complex surfaces.
- Possible contamination from the evaporator.
- The need to periodically load the evaporator.

### Sputtering

**Advantages**

- Low-defect-density films of almost all materials used in gas sensors, including high-melting-point materials can be grown on unheated substrates.
- The opportunity to synthesize compounds and alloys that cannot be obtained by thermal evaporation of materials in a high vacuum.
- The high adhesion of a film.
- A high coefficient of use of the sputtered material.
- Becoming homogeneous through thickness coverings.
- The opportunity to create apparatus and production lines for continuous operation.

**Disadvantages**

- Difficulties in plasma stabilization, particularly at low pressure and in large areas.
- The rate's dependence on electrical power.
- A large particle energy resulting in surface damage due to surface bombardment. This problem is particularly important for the deposition of polymer materials and films, which cannot be thermally processed for recovery to the initial state.

### Laser-Ablation

**Advantages**

- PLD permits precise control of deposition of multilayer structures *in situ* in one technological step.
- Any metal, ceramic, alloy, or intermetallic compound, as well as fully reacted metals, can be deposited on virtually any substrate, including plastic, paper, metals, and ceramics.
- Films can be deposited at either low (including room temperature) or high temperature.
- The method is very convenient for the quick preparation and study of new sensing materials.
- Process is compatible with oxygen and other reactive gases.
- PLD technology creates the possibility for controlled deposition of ultrathin coverings, which is very attractive for surface modification of the materials for gas sensors

**Disadvantages**

- Low productivity of the method.
- Relatively high investment costs.
- The composition and thickness depend on too many deposition conditions, such as wavelength, energy and shape of the laser pulse, focusing geometry, process atmosphere, and substrate temperature.
- Difficulties in the deposition of thick layers.
- Difficulties in attaining the necessary stoichiometry of materials containing volatile components.
- Difficulties in scaling up to large wafers.
- Due to repeated interaction of the laser beam with the target, structural changes occur on the surface with craters forming. Therefore, the composition and properties of the deposited material will depend on the duration of the deposition process.

### Chemical Vapor Deposition

**Advantages**

- CVD has the ability to coat complex shapes internally and externally because it is a non-line-of-sight process with strong throwing power.
- CVD can produce uniform films with strong reproducibility and adhesion at reasonably high deposition rates.
- CVD can produce multilayered coatings and coatings for a variety of metals, alloys, and compounds.
- CVD provides the ability to control crystal structure, surface morphology, and the orientation of the products by controlling the process parameters.
- Coatings are dense, and their purity can be controlled.

**Disadvantages**

- Up-front capital costs can be high, with complex handling, safety, and automatic systems.
- High-temperature requirements may limit substrate choices.
- Some substrates can be attacked by the coating gases.
- Poor adhesion or lack of metallurgical bonding is possible.
- Masking portions that are not to be coated can be difficult.
- The difficulty with stoichiometry control of multicomponent materials is possible.

*(Continued)*

**TABLE 18.4** (*Continued*)

**Advantages and Disadvantages of Synthesis and Deposition Methods Usually Used During Humidity-Sensor Fabrication**

**Deposition from Aerosol Phase**

**Advantages**

- The required equipment for aerosol deposition is simple and safe, and the process is straightforward, rapid, reliable, and inexpensive.
- The deposition of films does not require a vacuum at any stage. Deposition can be carried out at atmospheric pressure.
- It is possible to use simple, less expensive, and less toxic precursors, which do not possess the high pressure of vapor under saturation. This creates the possibility to apply some salts and metal-organic compounds that cannot be used with standard methods such as CVD and MOCVD.
- The deposition rate, the thickness, and the composition of the films can be easily controlled. It offers an extremely simple way to dope films with virtually any element in any proportion, merely by adding it in some form to the spray solution.

**Disadvantages**

- Fairly low reproducibility and the homogeneity of the film thickness distribution over the substrate's area. Especially if the substrate used is not flat.
- Other technological problems are associated with the plotting of sensitive material on small areas.
- Low effectiveness in the use of the precursor during the film-deposition process.

**Sol-Gel Technique**

**Advantages**

- Sol-gel is low-cost wet-chemical technology, which offers the possibility to prepare solids with predetermined structure, including thick, porous ceramics needed for gas sensors, by varying the experimental conditions.
- Multicomponent compounds and doped materials may be prepared with a controlled stoichiometry by mixing sols of different compounds and using multiple different dopants.
- There is possibility of independent control over porosity, crystal structure, and grain size.
- This method can easily shape materials into complex geometries in a gel state. So, materials in different shapes as films, fibers, powders, and bulk could be obtained.
- Very small quantities of raw material can be involved and hence the cost of metal-organic precursors is not a consideration.
- The possibility to synthesis of ceramic material at a temperature close to room temperature opens the opportunity of incorporating volatile components or soft dopants, such as fluorescent dye molecules and organic chromophores, in synthesized matrix.

**Disadvantages**

- Weak bonding and as result low adhesion and low-wear resistance. Therefore, the sol-gel technique is very substrate-dependent, and the thermal expansion mismatch prevents its wide application.
- Large shrinkage during drying processing, which is accompanied by the cracking effect, increased with film thickness increasing.
- The presence of residual hydroxyl and residual carbon.
- High cost of raw materials.
- Long processing time.

*Source:* With kind permission from Springer Science+Business Media: *Handbook of Gas Sensor Materials*, 2013, Korotcenkov, G. Copyright 2013 Springer.

Despite complex geometry and structure, the ceramic sensors have rather high-humidity sensitivity and long-term stability of functional properties. The high sensitivity seems to be a result of the small size of crystallites that constitute the powders and high porosity of the humidity-sensing matrix. The long-term stability seems to be a result of thoroughly developed aging procedures. An additional advantage of the ceramics is their rather low production cost, even in small quantities.

The major disadvantages are the difficulty of reliably producing sensors with the same parameters on a large scale, large thermal sluggishness, insufficient mechanical durability, and little compatibility with most microelectronic devices fabricated using planar technologies. In addition, conventional ceramic elements are fairly large in size. As a result, nowadays, the ceramic technology is practically not used in the manufacture of humidity sensors.

## 18.3 PLANAR SENSORS

At present, planar constructions of devices are generally considered to be better than ceramics for the development of humidity sensors. These structures can be fabricated by a number of microelectronic protocols as "single-sided" or "double-sided" designs. The choice depends mostly on how to deposit the main functional elements of the sensor, the heater, and the humidity-sensitive layer in the case of chemiresistors. These elements may be deposited on the same (front) side of the substrate ("single-sided" design) or on different sides of the substrate ("double-sided" design) (Oyabu 1982; Schierbaum et al. 1990). The double-sided construction requires employing two-sided photolithography or other deposition protocol and makes it difficult to wire the sensor at the housing. However, this type of design allows smaller chip size and decreased power loss. Another

**TABLE 18.5**

**Conventional Techniques for Fabricating Planar Gas Sensors**

| Thick-Film Technologies | Thin-Film Technologies |
|---|---|
| Sol-gel | Chemical vapor deposition (thermal, plasma, laser-induced) |
| Flame pyrolysis | |
| Precipitation | Sputtering (reactive, cathode) |
| Screen printing | Evaporation (reactive, thermal, arc, laser) |
| Dip coating | Spray pyrolysis |
| Drop coating | Molecular-beam epitaxy |
| Spin coating | Electroplating |
| | Ion plating |
| | Photolithography |
| | Etching |

advantage of this construction is the option to apply different materials when forming the heater and contact electrodes.

Planar humidity sensors can be fabricated via thin- or thick-film technologies. These technologies involve various protocols (see Table 18.5), which result in different structural and functional properties of the sensors. However, the general architecture of thin-film and thick-film sensors is similar. Planar sensors are very robust; the temperature homogeneity over the sensing layer is good.

Thick- and thin-film planar humidity sensors are generally favored compared to sintered ceramic-based sensors, especially in the terms of technological adaptability, reproducibility of parameters, and the rate of response. However, as for the ceramic elements, the humidity sensitivity is also highly dependent on the film porosity, film thickness, operating temperature, the presence of additives, and crystallite size (Korotcenkov 2005, 2008). One should note that thick-film and thin-film technologies are frequently combined

to produce planar sensors, especially in laboratory investigations. For example, in resistive humidity sensors, the humidity-sensing metal-oxide layers may be prepared via thin-film technologies (CVD, sputtering, etc.), while the heaters are formed with thick-film technology.

### 18.3.1 Thick-Film Technology

Thick-film technology is one of the main technologies used for humidity-sensor fabrication (Prudenziati 1994; Ihokura and Watson 1994; Galan-Vidal et al. 1995; White and Turner 1997; Grundler 2007; Bakrania and Wooldridge 2009; Patil 2011). The fabrication processes, utilizing thick-film technologies, have been developed over 40 years and are rather specific to the particular devices (Janata 1989; Madou and Morrison 1989; Moseley et al. 1991). Ability to form a humidity-sensing layer with required composition and porosity gives possibility to design humidity sensors with high-operating characteristics. In many ways, the microstructure of the thick film compares with that of ceramics and is a function of grain sintering conditions. As a rule, the films prepared by thick-film technology are porous. Reproducibility of sensors fabricated by thick-film technology is better than for ceramic elements.

There are several approaches used in thick-film technology (Agnew 1973; Holmes and Loasby 1976; Prudenziati 1994; Heule and Gauckler 2001; Lee et al. 2007b). They are dip coating, drop coating, spin coating, spray coating, flame pyrolysis, conformal coverage with thermoplastic transfer molding, and screen printing (Heule and Gauckler 2001; Lee et al. 2007b). A brief description of these methods is presented in Table 18.6. One can find the description of the sol-gel method in Chapter 20 of our series. It is important to note that during the formation of the films of many

**TABLE 18.6**

**Methods of Thick-Film Technology Usually Used for Preparing Humidity-Sensitive Layers**

| Method | Description |
|---|---|
| Casting | Casting is a simple technology that can be used for a variety of materials (mostly polymers). In this process, the material to be deposited is dissolved in liquid form in a solvent. The material can be applied to the substrate by spraying or spinning. Once the solvent is evaporated, a thin film of the material remains on the substrate. The thicknesses that can be cast on a substrate range all the way from a single monolayer of molecules (adhesion promotion) to tens of micrometers. The control on film thickness depends on exact conditions but can be sustained within ±10% in a wide range. Delamination and cracking can occur if liquid film is too thick. This method gives a more uniform and a more reproducible membrane than dip coating. |
| Spin coating | In the spin-coating process, the substrate spins around an axis, which should be perpendicular to the coating area (see Figure 18.2a). The quality of the coating depends on the rotation velocity, rheological parameters of the coating liquid, and surrounding atmosphere. The coating thickness varies between several hundreds of nanometers and up to 10 micrometers. Desired thickness obtained by precursor dilution, spin speed, and number of layers. Equipment similar to that of spin-coat tracks used for photoresist deposition. |
| Spray coating | Precursor is atomized to form fine aerosol, which then is deposited on a slowly rotating wafer (see Figure 18.2b). Deposition enhanced by electrostatic charging of aerosol. Desired thickness is controlled by adjusting deposition time and number of layers. The coating step is suitable for establishing an in-line process. |
| Slip casting | Slip casting is a technique in which a suspension (slip) is poured into a porous mold (generally made of plaster). The mold's pores absorb the liquid, and particles are compacted on the mold walls by capillary forces, i.e., solidify, producing parts of uniform thickness. Once dried to the leather-hard stage, the molds are opened, and the cast object removed to dry completely before firing. |

*(Continued)*

**TABLE 18.6 (*Continued*)**

**Methods of Thick-Film Technology Usually Used for Preparing Humidity-Sensitive Layers**

| Method | Description |
|---|---|
| Tape casting | The tape casting is a technique for continuous production of ceramic or other tapes according to the "doctor-blade principle." In this process, a suspension of ceramic, metal, or polymer particles in an organic solvent or water, mixed together with strengthening plasticizers and/or binders, can be used. The slip is cast onto a precisely machined stone plate, on which the carrier film is moved smoothly and without perturbations. By the doctor blade, the slurry is spread homogeneously on the surface of the tape. Drying and firing are final stages of the actual tape forming. |
| Dip coating | Dip-coating techniques can be described as a process where the substrate to be coated is immersed in a liquid and then withdrawn with a well-defined withdrawal speed under controlled temperature and atmospheric conditions (see Figure 18.2c). The coating thickness is mainly defined by the withdrawal speed, by the solid content, and the viscosity of the liquid. |
| | The applied coating may remain wet for several minutes until the solvent evaporates. This process can be accelerated by heated drying. In addition, the coating be exposed to various thermal, UV, or IR treatments for stabilization. |
| Screen printing | Screen printing is a printing technique that uses a woven mesh to support an ink-blocking stencil. The paste (ink) used is a mixture of the material of interest, an organic binder, and a solvent. The attached stencil forms open areas of mesh that transfer ink or other printable materials, which can be pressed through the mesh as a sharp-edged image onto a substrate. A roller or squeegee is moved across the screen stencil, forcing or pumping ink past the threads of the woven mesh in the open areas. After printing, the wet films are allowed to settle for 15–30 min to flatten the surface and are dried. This removes the solvents from the paste. Subsequent firing burns off the organic binder, metallic particles are reduced or oxidized, and glass particles are sintered. It can be used to print on a wide variety of substrates, including paper, paperboard, plastics, glass, metals, fabrics, and many other materials. |
| Inkjet printing | Inkjet technologies, which are based on the 2D printer technique of using a jet to deposit tiny drops of ink onto substrate, are perfectly suited to controllably dispense small and precise amounts of "liquid" to precise locations. The available inkjet technologies include: (1) continuous inkjet; (2) drop-on-demand inkjet; (3) thermal inkjet; and (4) piezo inkjet. |
| | The "liquid" materials can encompass low- to high-viscosity fluids, colloidal suspensions, frits, metallic suspensions, and almost any other material that can be dispersed in a liquid-carrier material. The carrier material can be aqueous- or non-aqueous-based solvent material. When printed, liquid drops of these materials instantly cool and solidify to form a layer of the part. Usually inkjet printing is accompanied by thermal treatment. |

*Source:* Korotcenkov, G., *Handbook of Gas Sensor Materials*, Springer, New York, 2013.

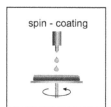 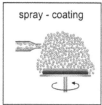 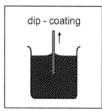

**FIGURE 18.2** Illustration of the commonly used coating techniques. (Reprinted from *Comptes Rendus Chimie*, 7, Schwartz, R.W. et al., Chemical solution deposition of electronic oxide films, 433–461, Copyright 2004, with permission from Elsevier.)

humidity-sensitive materials (e.g., metal oxide, carbon, zeolites, MOFs, etc.) by thick-film technology methods, as a rule, one can use powders, fibers, or various 1D and 2D presynthesized structures.

During the formation of films by the indicated methods, the substance used passes through several stages, including gel formation, drying, and crystallization. These stages are shown in Figure 18.3. One can find more detailed description of the processes, taking place in the films during their formation using methods of thick-film technology, in Schwartz et al. (2004).

### 18.3.1.1 Screen Printing

One should note that screen printing is one of the most important thick-film deposition methods (White and Turner 1997). This method is similar to that used for traditional silkscreen printing (see Figure 18.4). The main difference lies in the screen materials and the degree of sophistication of the printing machine. Special thick-film pastes (or inks) can be formulated to paint or print an active layer onto a substrate. To formulate the paste, finely milled metal oxides or other sensing materials are combined with small amounts of special additives, such as binders, resins, and organic solvents, used as an organic vehicle to form a printable paste. The paste is spread on the substrate by means of a screen made from mesh mounted on a metallic frame. The mesh is coated with an ultraviolet (UV) sensitive emulsion onto which the required pattern can be formed photographically. The finished stencil has open mesh areas through which the desired pattern can be printed and is held in position at a distance of around 0.5 mm from substrate surface of the in the screen-printing machine. The ink is placed on the opposite side of the screen and a squeegee traverses the screen under pressure, thereby bringing it into contact with the substrate and also forcing the ink through the open areas of the mesh

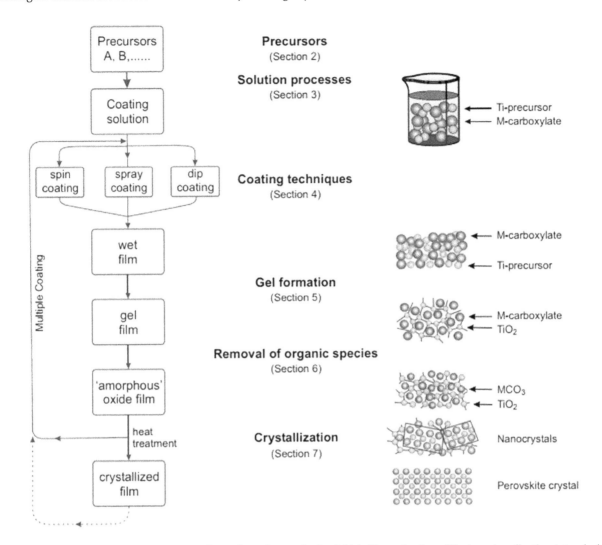

**FIGURE 18.3** Flow chart of a typical process formation using methods of thick-film technology. The bars describe the status during the procedure of film forming, while the arrows indicate the treatment and the internal processes. On the right-hand side, the structural status of the molecules and the ions, respectively, for the evolution of a metal titanate perovskite is schematically depicted. (Reprinted from *Comptes Rendus Chimie*, 7, Schwartz, R.W. et al., Chemical solution deposition of electronic oxide films, 433–461, Copyright 2004, with permission from Elsevier.)

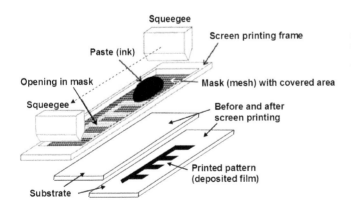

**FIGURE 18.4** Manufacture of thick-film structures by screen printing. (Idea from Grundler, P., *Chemical Sensors: An Introduction for Scientists and Engineers*, Springer-Verlag, Berlin, Germany, 2007.)

(see Figure 18.5). The required circuit pattern is thus left on the substrate. The process is sometimes referred to as being "off-contact" printing, since the screen only contacts the printed surface at the point where the squeegee passes over it.

One should note that an "on-contact" (or "in-contact") process, so-called stencil printing, also is possible (Figure 18.6). The stencil is a metal mask that rests directly in contact with the surface of the board. In stencil printing, the board is moved into contact with the stencil before the squeegee starts to move. When the squeegee has *completed* its stroke, the board and stencil are then separated vertically, which releases the paste from the stencil, producing well-defined edges to the print. It is usual, but not essential, for the stencil to remain fixed and the board to be raised for printing and lowered at a controlled speed afterward. The process depends on the

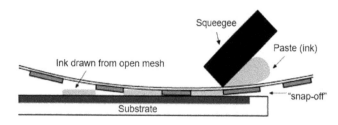

**FIGURE 18.5** Schematic illustration of the screen-printing process. (Based on http://gwent.org/gem_screen_printing.html; https://screenprintingtechniques.wordpress.com/2015/03/29/.)

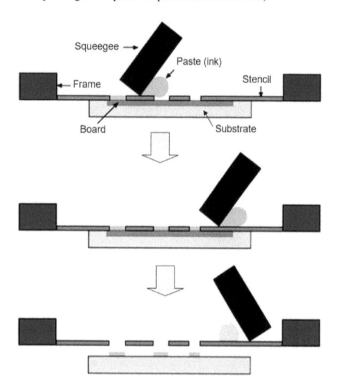

**FIGURE 18.6** Schematic illustration of stencil printing ("on contact" printing). (Based on http://gwent.org/gem_screen_printing.html; https://screenprintingtechniques.wordpress.com/2015/03/29/)

interaction of several factors: (1) the stencil and squeegee, (2) the board material and pad design, (3) the physical properties of the solder paste, and (4) the printing operation. If one of these factors is incorrect, printing quality will be poor, and therefore the printer itself is only one of the decisive factors in the whole process. The process window (the region where process values can vary but still produce good results) can be enlarged by careful choice of materials and design.

To remove organic solvents, the layers are dried with an infrared belt drier or at conventional oven at temperatures of around 150°C. After drying, the adhesion of the layers to the substrate is enhanced. In general, further annealing of the thick films is performed at high temperatures, in the range of 300°C–1200°C. During this step, frequently called firing, the glass component melts, the fine powders sinters, and the overall layer of a solid composite material attaches to the substrate. The thickness of the fabricated layer depends on the paste viscosity and the size of the apertures in the mesh

(Madou and Morrison 1989; Comini et al. 2009a). The thermal treatment has to be gradual to minimize the temperature-induced stresses, which appear under sudden heating and cooling (Moseley and Tofield 1987; Sberveglieri 1992a). These steps can be repeated using materials appropriate for fabricating specific areas of the device. Standard printable thick-film materials for resistive heaters, and conductor lines may also be fabricated by thick-film technology on the substrate before or after the sensor layer (normally pastes of noble metals like Pt and Au).

The thickness of the layers fabricated by thick-film technologies is commonly in the range of 2–300 μm. The lowest resolution limit of screen printing is typically about 50 μm. Of course, this accuracy does not always suit the developers of sensors, and therefore various laboratories, are continuing intensive research toward finding new methods to reduce the characteristic size of sensors fabricated by thick-film technologies. One method, which was developed recently, is to cast a suspension into appropriate photoresist structures (Schoenholzer et al. 2000). Another method, which is more cost-efficient, uses a soft lithography of liquid materials (Xia and Whitesides 1998). The use of liquid inorganic precursor polymers has already been demonstrated for the fabrication of microstructured ceramics (Yang et al. 2001) and other hierarchically ordered oxides (Yang et al. 1998). In the soft lithography process, the micropatterns are transferred by casting a silicone rubber, poly-dimethylsiloxane (PDMS), against the master structure. The PDMS is then peeled off, cut, and used as a mold that forms microcapillaries on a substrate, which can be filled with a liquid. This technique is referred to as micromolding in capillaries (MIMIC). The most striking result of using PDMS as the mold material is that this elastomeric material readily establishes a reversible conformal contact on the molecular level to a variety of substrates, thus sealing the capillaries optimally. Additionally, the master structures may be reused many times to cast PDMS molds. It is important that the MIMIC technique does not require clean-room conditions. Experiments carried out by Heule et al. (2001) have shown that one can use a colloidal dispersion of ceramic powders with solid contents of 0.1–40 vol% to form the film. These methods enabled the fabrication of microstructured ceramic lines with a spatial resolution of 10 μm, which can be integrated into a miniaturized metal-oxide (MOX)-based gas sensor (Heule and Gauckler 2001).

### 18.3.1.1.1 Features of Materials Used for Screen-Printing Technology

The performance of humidity sensors, fabricated by thick-film technology, depends critically on the applied materials, primarily on the pastes employed to make the humidity-sensing layer and electrodes. The contact electrodes, heater, and sensing layer are formed by sequential deposition of the corresponding pastes on the substrate with subsequent annealing. Therefore, the substances used to fabricate for example resistive-humidity sensors can be either fine metal powders to prepare electrodes, heaters, and temperature sensors or fine-grained MOX powders and polymers to prepare humidity-sensitive layers. In choosing these materials, important considerations are good adhesion, similar coefficients of expansion to prevent stress-related damage during sudden heating and cooling, the ability to retain their characteristics throughout the fabrication process, easy availability, and low cost, among others. Furthermore, the paste and composite materials used in the sensor device have to provide adequate and reliable electrical contacts for solid–solid interfaces.

The screen defines the pattern of the printed film and also meters the amount of paste that is deposited. It is a very important part of the screen-printing equipment and is essentially a stencil through which the paste is forced during the printing process. The most common type of the screen consists of a frame, normally cast aluminum, onto which a finely woven mesh is stretched.

The mesh itself is usually based on a plain-weave pattern. Some important properties of the screen mesh are: the size and density of the strands (usually quoted in terms of lines per inch), the tension, the orientation, and the material. In addition, the mesh material must make sure that the printed deposit is uniform. The mesh material must be precisely woven and have uniform mesh apertures. The fabric should also be flexible enough to enable good contact all over the substrate. The fabric needs to be resilient so that the mesh returns to its original position after the printing stroke. The squeegee itself is in contact with the fabric for most of the printing stroke, so the finish of the fabric must be slippery and smooth so that the resistance to the squeegee is minimum. The mesh material must also be chemically stable and very resistant to attack from the various solvents and other chemicals used in the thick-film process. Finally, the fabric must have an economic working lifetime and suitable mechanical support for the emulsion.

At present, the three most common materials used for thick-film screen meshes are polyester, nylon, and stainless steel. They all have their own advantages and disadvantages. Polyester is flexible and can be used for printing onto uneven surfaces. It is also much more resilient than nylon and stainless steel. The elastic properties of polyester are better than those of stainless steel but cannot match those of nylon. The registration and definition properties are generally good. Polyester screens have a long lifetime and give low squeegee wear. Nylon is the most elastic of the three screen fabrics, which can be an advantage in certain circumstances. Unfortunately, this also means that the open areas of the screen tend to deform during a print stroke, thereby resulting in elongated images. Another disadvantage of nylon is that its low resilience means that high-viscosity inks should not be used. The mesh tends to stick to the substrate and not peel oil, which results in poor print quality. The main advantages of stainless-steel screens are that they produce high standards of line definition and registration and also good control of ink deposition. The mesh filaments can be drawn finer than those of nylon or polyester and can therefore provide a higher percentage of open area for a given mesh count. Stainless steel is ideal for printing onto flat surfaces, but because of its poor flexibility and resilience, it is much more difficult to use on uneven surfaces. It is

generally recommended that stainless steel screens be used for printing on small areas where high definition and registration are needed.

The squeegee is essentially a flexible blade, whose function is to transfer the paste through the screen and onto the substrate. During printing, the squeegee forces ink through the open areas of the mesh and, by virtue of the surface tension between the film and substrate, the required pattern is transferred to the substrate as the screen and substrate separate. The squeegee's shape, the material, and the pressure are all factors that dictate the life of the screen and the squeegee. Clearly, the squeegee must be resistant to the solvents and inks used in thick-film processing. Polyurethane and neoprene are common materials used for squeegee fabrication.

As a rule, the basic constituents of thick-film ink are: (1) active material, (2) a low, softening temperature glass frit, and (3) organic carrier. Of course the inks can contain other components such as catalytically active noble metals (Pd, Pt, Au, Ag) or metal oxides ($Al_2O_3$, $SiO_2$, etc.), which can be added for improvement porosity, catalytic activity, and stability. The glass frit acts as a binder, which holds the active particles together and bonds the film to the substrate. During the firing cycle, the glass melts, typically migrates to the substrate–metal interface, and forms a mechanical key at the film-to-substrate interface. As a result, the film is firmly bonded to the substrate. The glass frit also provides a suitable matrix for the active material of the film. The organic carrier gives the paste the desired viscosity for screen printing. The characteristics required of the ink viscosity depend on the printing stage. When the ink is being forced through the screen, viscosity must be low. After printing, however, viscosity must he high, because the film must retain its printed geometry and not run. For ideal (Newtonian) fluids, the viscosity is independent of the shear rate and only varies with temperature. However, for a thick-film paste, the viscosity must change with the pressure applied.

The active material in inks for the humidity-sensing-layer forming is essentially finely divided powders of metal oxides with a typical particle size in the range from of a few nanometers to a few microns. However, nowadays powders with nanometric grain size are very commonly used for humidity-sensor fabrication, because better sensing properties of the active layer could be obtained.

Conductor pastes can contain a precious metal or metal alloy like silver, platinum, gold, palladium, and their alloys. After processing, the metallic particles fuse to form continuous electrical paths through the carrier glass. Sheet resistivities of the order of 10 m$\Omega$/cm$^2$ are typical. In sensor applications, conducting inks have an important function in the formation of electrode patterns, which range from simple rectangular structures to interdigitated pairs. Platinum is also used for resistance thermometry and for combined heaters and thermometers in gas sensing, where control at different temperatures is an important technology. The dielectric pastes usually have ceramic powders as the active material. Dielectric inks are mainly used as insulants, either at crossovers or when a conducting substrate has been employed.

Choosing the glass content for thick-film components is mainly dictated by the following requirements:

- The linear coefficient of thermal expansion (CTE) should be as close as possible to the CTE of the substrate. A slightly lower CTE is tolerated, since the glass is put slightly in compression, that is, in a condition safer than the opposite, which creates tensile strains.
- The softening point ($T_S$) should fall between 400°C and 600°C. In these circumstances, the viscosity of the glass will be low enough at the peak temperatures of the common firing cycles (850°C to 1000°C) for a continuous glassy matrix to form in a few minutes, and/or for there to be a liquid phase, which assists the sintering of metal particles and the required reactions with the other components of the paste.
- No ions should be in motion under electric bias if mass transport and the degradation of the properties of the film are to be avoided. These conditions are prerequisites for the stability of the thick-film layers at relatively high temperatures.

It was established that pure silica does not meet these requirements (mainly because the softening point is too high and the CTE is too low). Adding $Bi_2O_3$ decreases the $T_S$ value and increases the CTE but makes the glass too "short" (namely, the viscosity $\eta$ changes too rapidly when the firing temperature is changed); moreover, $Bi_2O_3$ may induce devitrification. Alkaline or alkaline-earth cations, widely used in the glass industry as a means for controlling the viscosity and workability of silicate glasses, cannot be used in large quantities because of their consistent ion mobilities. The experience showed that lead is the most suitable additive for achievement required properties of the glass. PbO may be added to silica in large fractions without devitrification. Its cation exhibits low mobility in the glass and makes it very stable up to relatively high temperatures. Therefore, lead borosilicate glasses ($Pb/B_2O_3/SiO_2$ with $Bi_2O_3$ as fluxing agent) are often used as a glass frit in inks designed for gas-sensor fabrication.

The crucial point here is that the specific composition of the glass (e.g., the relative amounts of $SiO_2$, $Bi_2O_3$, PbO, $B_2O_3$) greatly affects the dissolution kinetics of Al and Be (from alumina and beryllia substrates), the reactions with the active ingredients of the ink, and the reproducibility of the glass batches because it affects the degree of volatilization of Pb from the melted glass during the preparation of both the frit and the components. Usually the concentration of these additives does not exceed 5%. In addition, it is necessary to take into account that in excessively high-firing temperature profiles, glass may "float" on the conductor surface and result in poor solderability, poor adhesion, and wire-bond ability. The chemical and physical properties of the substrate also play an important role in determining how the glass binder phase wets the substrate surface and is attached to it. Changes in the chemical composition of the surface, grain size, and substrate smoothness can lead to a wide variation in the characteristics of thick-film conductors.

For obtaining required thick-film adhesion, the oxides (fritless conductors) can also be used. Small amounts (0.1%–1%) of chemically active oxides, such as CuO, CdO, or NiO, added to the paste, react high temperatures with alumina substrates to form oxides like $CuAlO_2$, which provide adhesion. A combination of glass and oxides also is used, but the amount of glass in this case is much lower than in the fritted systems. Adhesion is believed to be related to a combination of the effects, mentioned above.

"Organic vehicle" is a generic term that describes a blend of volatile solvents and polymers or resins together with a surfactant, which are needed to provide a homogeneous suspension of the particles of the functional materials and a rheology that is suitable for the printing of the film configuration. The vehicle is a temporary, sacrificial ingredient, which should be completely removed in the following steps of the process, during which the microstructure of the deposits is formed. The composition of the organic vehicle determines (or helps to determine) the shelf life of the paste, its drying rate on the screen, the change in printability with ambient temperature, the resolution of fine lines, some electrical properties of the fired films, and their cosmetic appearance. The cooperative effects and the relationships between the properties (density, wettability, surface energies, viscosity etc.) of the organic vehicle and inorganic constituents of the paste contribute to the static and dynamic properties of the paste. Moreover, the solvents should not be so volatile that after a short time they leave a hard unprintable paste on the screen mesh, but at the same time, they should evaporate early in the drying phase of the process. Finally, the fluid properties (rheology) of the vehicle should prevent penetration through the screen when the paste is at rest but should enable printing to be fast under the squeeze pressure and the film to settle quickly in the desired configuration on the substrate without bleeding or smearing from the defined geometry.

Organic vehicle formulations are usually blends of cellulose-type resins or cellulose acetate (the polymeric viscosifier) and solvents like terpineol, butylcarbitol, ethylcellulose, cyclohexanone, or ethylene glycol. Several examples one can see in Table 18.7. The relative fractions of these ingredients vary according to the types and amounts of the inorganic materials of the paste. The ratio of the inorganic to organic part usually is kept in the past at (70–75)/(25–30). These simple vehicles perform well with many solid systems in a variety of printing conditions and environments. However, the requirements for fine resolution of lines, speed of fast printing, smooth surfaces and the like stimulated the development of complex vehicles.

We need to note that the composition of the ink used for humidity-sensor fabrication and the size of inorganic particles added in the past have determining influence on operating characteristics of humidity sensors designed, because exactly the composition controls the porosity and surface chemistry of the gas-sensing matrix (Bakrania and Wooldridge 2009). For example, it is necessary to take into account that the excess of the glass in the ink decreases the porosity, that is, the gas penetrability of the gas-sensing layer (Willett et al. 1998). Therefore, in many cases, the ink does not contain

**TABLE 18.7**

**Examples of Organic Vehicles (Binder + Solvent) Used in Thick-Film Technology for Preparing Metal-Oxide Films**

| Organic Carrier | References |
|---|---|
| Terpineol HVS100: Ethylcellulose: Texaphor 963: Rilanit spez = 89.3: 1.2: 6.0: 3.5 wt% | Ivanov (2004) |
| Terpineol HVS100: Ethylcellulose: Texaphor 963: Rilanit spez: Disponil = 85.5: 1.1: 5.7: 3.4: 4.3 wt% | Ivanov (2004) |
| 8% ethyl cellulose and 92% butyl carbitol acetate | Jayadev Dayan et al. (1998) |
| $\beta$-terpeneol: butyl carbitol acetate: ethyl cellulose | Nitta and Haradome (1979) |
| Ethyl cellulose | Zhang et al. (2012) |
| Ethyl cellulose: butyl cellulose: butyl carbitol acetate: terpineol | Deore et al. (2011) |
| Diethyl glycol monobutyl: $\alpha$-terpinol | Yadava et al. (2010) |
| Deionized water: $\alpha$-terpinol | Choi et al. (2005) |
| $\alpha$-terpineol: ethyl cellulose = 89: 11 wt% | Lee et al. (2007a) |
| Isopropyl alcohol: hydroxylpropyl cellulose | |

the glass. The presence of other components such as CuO, added for improvement of the film adhesion, can also be accompanied by undesirable effects, because CuO promotes the metal-oxide film densification during high-temperature annealing. Moreover, the additives can be incorporated in the lattice of metal oxides with proper changes in their bulk and surface properties. This means that the change of the ink composition will be always accompanied by the change of gas sensors. For example, Ivanov (2004) has studied the influence of film-adhesion promoters such as $Bi_2O_3$ and $Cu_2O$ on the response of $SnO_2$ and $WO_3$-based sensors and found that tungsten-oxide sensors containing only $Bi_2O_3$ were more sensitive to nitrogen dioxide than to ammonia when operated at low temperatures. At the same time, $WO_3$ sensors with $Bi_2O_3$ and $Cu_2O$ were found to be very sensitive to ammonia. Their selectivity to ammonia can be improved by increasing the amount of $Cu_2O$ in the ink. The same regularity was observed for $SnO_2$ sensors as well. However, it was established that the inclusion of $Cu_2O$ into the past increases the response to the water vapor. Moreover, tin-oxide sensors, containing only $Bi_2O_3$, were found to be sensitive and very selective to ethanol vapors. Thus, it is seen that the composition of the past influences both the magnitude of the response and the response time.

### 18.3.1.2 Other Printing Technologies

Gravure printing is also a very promising method for depositing colloidal nanoparticle-based layers with high resolution (Figure 18.7). A detailed description of this method is given elsewhere (Kraus et al. 2007), where it was shown that this approach allows the possibility of achieving resolution in the submicrometer range.

However, in many cases, especially during micromachining technology, when the prefabricated micro-hotplates are used, lithographic processing and screen printing of the sensing layers are not possible (Spannhake et al. 2009). The sensing layers, therefore, need to be deposited over shadow masks, which should allow placing the layers exclusively onto predefined spots between the electrodes.

This requires only moderate precision, so the masks can be made of mechanically stiff ceramics, which are conventionally produced in a microelectromechanical systems (MEMS) foundry in the form of micromachined silicon wafers.

Another method, promising for local depositing a humidity-sensitive layer, is based on the use of drop coating (Figure 18.8). This method allows placing small volumes of

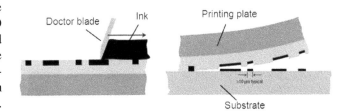

**FIGURE 18.7** Conventional gravure printing technology. The doctor blade fills the recessed features of a printing plate with ink. Then particles dispersed in the ink are transferred from the plate onto the substrate. (Reprinted by permission from Macmillan Publishers Ltd., *Nature Nanotechnol.*, Kraus, T. et al., 2007, Copyright 2007: Nature.)

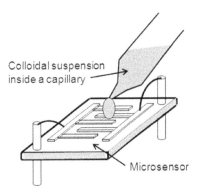

**FIGURE 18.8** Deposition of MOX sol onto a microheater using the drop-dispenser technique. (Idea from Spannhake, J. et al., Micro-fabrication of gas sensors, in: Comini E., Faglia G., and Sberveglieri G. (eds.), *Solid State Gas Sensing*, Springer-Verlag, Berlin, Germany, 1–46, 2009.)

humidity-sensitive material as a sol on the top of the hotplate following by thermal annealing to form a gel (Vincenzi et al. 2001; Guidi et al. 2002; Francioso et al. 2006; Epifani et al. 2007). Because of the good thermal insulation, the center regions of the hotplates can be heated to temperatures of 300°C–700°C, which is necessary to transform the colloidal suspension into a solid film.

Direct writing of powders via ink jets is another promising technology for local deposition of MOXs on micro-hotplates or cantilevers. This method offers the possibility of combining the advantages of thick films and micromachining. A general review of the state of the art in ink-jet printing of various materials can be found elsewhere (Calvert 2001; Zhao et al. 2002; Bietsch et al. 2004). The scheme of the method is drawn in Figure 18.9. Inkjet printing is a non-contact technique, which does not require any masks for design and repeated production of microscale patterns. Usually ink-jet printing involves the drop formation of polymer or metal-oxide-laden inks through the use of a piezoelectric-dispensing printhead. The printhead consists of a glass capillary surrounded by a piezoelectric material. Through an applied voltage, the piezoelectric material provides compression on the glass capillary and droplets of picoliter volume can be produced depending on the orifice of the glass capillary. There are several properties that must be controlled during ink-jet printing including, but not limited to, solids loading, density, viscosity, surface tension, and evaporation rate of the ink as well as printing settings of applied voltage, vacuum level, and

orifice diameter. These properties must be regulated to produce consistent drops of a specific volume that correspond to the amount of material deposited on the miniaturized sensor substrate. The physical size of the micro-hotplate also dictates the positioning of the drops produced by the ink-jet printer must be precise and reproducible. In contrast to parallel processes such as photolithography, the ink-jet printing makes it possible to fabricate each sample individually, although this contradicts the requirements of mass-scale production. Still, modern computer-driven setups could help to build such individual devices with the required automation and accuracy. Moreover, this method enables one to fabricate high-quality patterns on a big variety of flexible paper- or polymer-based substrates and suits for the inexpensive production of sensor modules for microelectromechanical systems. Such ink-jet technology has been utilized to print aqueous 2D MOX suspensions with a size of ~37 μm (Windle and Derby 1999), which is sufficient to design sensors using micromachining (Liu 1995).

It is worth saying that at present inkjet printing is used mostly for deposition of polymer-based materials. However, sensors with acceptable parameters fabricated using inkjet printing of carbon- and metal-oxide-based sensing layers, are also presented (Santra et al. 2015; Rieu et al. 2016). Examples of various types of inkjet printing humidity sensors are listed in Table 18.8.

Conducting materials can also be deposited using inject printing. These materials can be metallic or polymeric based. There is a range of metallic inks commercially available for inject printing purposes. Sun Chemical Inc. makes stable silver nanoparticle solutions for inkjet printing purposes. Sigma-Aldrich is another well-known company for making gold functional inks for the purpose of inkjet printing. A numbers of studies are being conducted with the focus on making other metallic solutions to be suitable for different fabrication methods. Other types of conductive inks are based on intrinsically conducting polymers (ICPs) such as PEDOT:PSS. Some of the conductive inks that are available commercially for the inject-printing process are listed Table 18.9.

### 18.3.1.3 Advantages and Disadvantages of Thick-Film Technology

Barsan and Weimar (2001) considered the thick-film technology an excellent technique for fabricating many types of gas sensors, since it allows one to make the highly porous sensitive layers of nanostructured films. In general, thick-film technology has several significant advantages for the development of sensors:

1. Flexibility to develop various sensor constructions
2. Extensive choice of materials
3. Easy integration with electronic circuits
4. Low cost even in low-volume production
5. The possibility to use automatic fabrication processes
6. Compatibility with other (micro)electronic devices

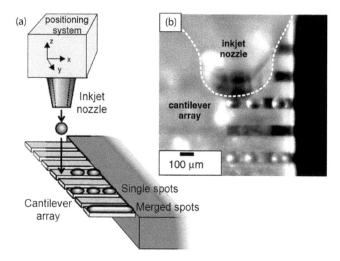

**FIGURE 18.9** Inkjet printing of individual droplets onto a cantilever array (a) as a scheme, and (b) as seen by a video camera. A positioning system allows accurate placement of single droplets onto selected cantilevers. When deposited with a small pitch, the droplets merge into a continuous layer covering the entire cantilever length. For demonstration, three droplets of water are deposited onto selected cantilevers. Owing to the oblique view of the camera, only the central cantilever is in focus. (Reprinted with permission from Bietsch, A. et al., Rapid functionalization of cantilever array sensors by inkjet printing, *Nanotechnology*, 15, 873–880, 2004. Copyright 2004, Institute of Physics.)

## TABLE 18.8
### Comparison of Different Types of Inkjet Printing Humidity Sensors

| Type of Sensor | Active Material | Substrate | Features | Sensitivity | References |
|---|---|---|---|---|---|
| Resistive | SPS:PEDOT | PET | NP/RGO electrodes | 0-98%RH, $\tau_{res}$~39s | Yuan et al. (2016) |
| Resistive (RFID) | Copper-acetate | Kodak photo paper | Ag IDE configuration | 0%–45%RH | Quddious et al. (2016) |
| Resistive | UV-curable polymers | Alumina | Au IDE configuration | 30%–90%RH, $\tau_{res}$<400 s | Cho et al. (2008) |
| Resistive | Graphene | CMOS Si-micromachining | Au IDE configuration | 10%–80%RH, 0.25%/%RH, $\tau_{res}$~6–16 s | Santra et al. (2015) |
| Capacitive | Paper (Lumi silk) | Paper (Lumi silk) | Ag IDE configuration | 40%–100%RH, 2 pF/RH% | Gaspar et al. (2017) |
| Capacitive | Polyimide | Polyimide substrate | Ag IDE configuration | 4.5 fF/%RH | Rivadeneyra et al. (2014) |
| Capacitive | Cellulose acetate butyrate | PET | Ag IDE configuration | 10%–80%RH, 3.2 fF/%RH, $\tau_{res}$~320 s | Molina-Lopez et al. (2013) |
| Capacitive | Polymeric core—shell particles | PET | Ag IDE configuration | 5%–90%RH, 0.4–0.6 pF/%RH | Starke et al. (2012) |
| Piezoresistive microcantilever | PEDOT:PSS s | Si3N4 /polysilicon/ SiO2 | Polysilicon resistor | 20%–70%RH, ~$3 \cdot 10^{-3}$ $\Omega$/%RH | Sappat et al. (2011) |
| Cantilever | PSS, PAH | Silicon | Deflection; frequency shift | 5%–80%RH, ~1 Hz/%RH | Toda et al. (2014) |

GO—graphene oxidized; IDE—interdigital electrodes; NP—nanoparticles; REID—Radio frequency identification technology; PAH—poly(allylamine hydrochloride; PEDOT—poly(3,4-ethylenedioxythiophene); PET—polyethylene terephthalate; PSS—polystyrene sulfonated acid; SPS—Sulfonate polystyrene.

## TABLE 18.9
### Commercially Available Conductive Inks for Inject Printing

| Name | Material | Manufacturer |
|---|---|---|
| NanoGold | Au nanoparticle | Sigma Aldrich |
| NanoSilver | Ag nanoparticle | Sun Chemical |
| Clevios™ PH 1000 | PEDOT:PSS | Heraeus |
| Polyaniline (Emeraldine salt) | PANI nanoparticles | Sigma Aldrich |

PEDOT:PSS—poly(3,4-ethylenedioxythiophene) polystyrene sulfonate.

In addition, the composition of thick films can be functionalized via incorporation of numerous catalytic promoter impurities, and they can be used to make mesoporous structures with high specific area (Moseley and Tofield 1987; Madou and Morrison 1989; Ihokura and Watson 1994; Morrison 1994; Bârsan et al. 1999; Marek et al. 2003; Graf et al. 2004).

Compared to ceramic elements, thick films are more mechanically reliable and can be mounted in standard housings for integrated micromechanical schemes. Compared to thin films, thick films are less sensitive to the quality of the substrates. The parameters of thick-film-based sensors are more reliable than those of ceramic elements and, in general, the gas sensitivity is higher. The operating regime of these sensors can be maintained to keep constant either the working current or the working voltage. If this regime is chosen at the maximum temperature dependence of gas sensitivity, these sensors are easily employed as gas alarms.

Although thick-film technology is an attractive option for sensor fabrication, it has some drawbacks:

1. The different materials employed are not very compatible and depend strongly on the manufacturing process.
2. These processes require complicated curing cycles under high temperatures carried out over long time frames.
3. The thick-film sensors are still not very reproducible in series.
4. The resolution of this technology is limited to a possible minimum line width of 100 µm (Sinner-Hettenbach 2000).
5. The surface of screen-printed layers is rather rough. In addition, thick-film heaters are not always very reproducible, with variations in their temperature coefficient of resistance and nominal resistance.

The cracking effect observed in the formed films during the heat-treatment process is another strong disadvantage of the thick-film technology. In addition, the organics trapped by the thick coating often cause a failure during the thermal process. It is important to note that this disadvantage is characteristic of all films manufactured using thick-film technology. The consequences of the cracking effect in $In_2O_3$ ceramics are shown in the SEM images presented in Figure 18.10. Atkinson and Guppy (1991) observed that the crack spacing increased with film thickness and attributed this behavior to a mechanism in which partial delamination accompanies crack propagation. Therefore, since the crack-free property is the essential requirement for protective coatings, the maximum coating

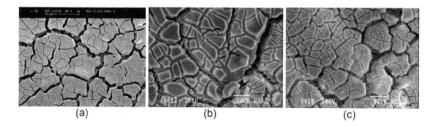

**FIGURE 18.10** Cracking effects in $In_2O_3$ ceramics prepared by the sol-gel method: (a) $In_2O_3$:Bi, (b) $In_2O_3$:Mn, and (c) $In_2O_3$:Cu. (Reprinted from *Sens. Actuators* B, 103, Korotcenkov, G. et al., Gas sensing characteristics of one-electrode gas sensors on the base of doped $In_2O_3$ ceramics, 13–22, Copyright 2004, with permission from Elsevier.)

thickness in this case is limited to 0.5 μm (Olding et al. 2001). However, for sensor applications, the presence of the cracking effect is not a factor, which limits the sensing properties. Therefore, in gas and humidity sensors, film thickness may exceed 100 μm. The presence of cracks improves gas permeability and helps to reduce response time.

It is worth noting that some of the mentioned disadvantages of thick films and ceramic elements—such as reliability of sensor characteristics under production in series, long times to stabilize the functional properties, and rather high power consumption of sensors fabricated on conventional platform—are substantially improved by the application of thin-film microelectronic technologies.

### 18.3.2 THIN-FILM TECHNOLOGY

Compared to the technologies employed to fabricate thick films and ceramics, the development of thin-film sensor elements is based on well-managed deposition processes (Wu et al. 1993; Liu 1995). Humidity-sensitive films are produced via all the available thin-film technologies, among which one can note thermal evaporation and sputtering (Stryhal et al. 2002; Saadeddin et al. 2007), laser ablation (Phillips et al. 1996), spray pyrolysis (Tiburcio-Silver and Sanchez-Juarez 2004), chemical vapor deposition (Heilig et al. 1999), and atomic-layer deposition (ALD) (Takada 2001). A brief description of these methods can be found in Table 18.1. Each method has advantages and disadvantages. For example, spray pyrolysis and chemical vapor deposition seem to be the cost-effective techniques that are attractive for the production of inexpensive sensors (Sberveglieri et al. 1993; Labeau et al. 1993; Rumyantseva et al. 1996; Brousse and Schleich 1996; Olvera 1996). Spray pyrolysis is quite flexible in terms of materials and structures that can be used to design gas sensors and to deposit composite materials (Korotcenkov et al. 2001b). ALD allows one to deposit highly homogeneous thin films with excellent coverage and thickness control. Films grown with ALD are usually dense, pinhole-free, and extremely conformal to the underlying substrate. Furthermore, control of the film thickness at the monolayer level can be easily achieved by simply counting the number of ALD cycles (Göpel and Reinhardt 1996). Reactive sputtering yields high reproducibility (Lalauze et al. 1991) and long-term stability of the functional properties of the film (Sayago et al. 1995b). Magnetron sputtering allows one to vary the crystalline structure of the film from amorphous

to single-crystal (epitaxial) just by changing the substrate temperature and the rate of deposition (LeGore et al. 1997; Kissin et al. 1999a), whereas the film stoichiometry is driven by the amount of oxygen in the vacuum chamber (Williams and Coles 1993, 1995; DiGiulio et al. 1996; Miccoci et al. 1996; Kissin et al. 1999b). The minimum crystallite size in the gas-sensitive polycrystalline MOX thin films is now down to 3 nm (Barbi et al. 1995). According to Demarne and Grisel (1993), the DC magnetron sputtering allows one to use lower temperatures than radio-frequency sputtering, which is a definite advantage for Si substrates. A more detailed comparison of selected deposition methods is presented in Tables 18.4 and 18.10. Table 18.10 compares two main deposition methods of thin-film technology such as physical vapor deposition (PVD) and chemical vapor deposition (CVD), frequently used in the design of gas sensors for film deposition. The same comparison one can make for CVD and chemical solution deposition (CSD) methods (see Table 18.11).

#### 18.3.2.1 Features of the Morphology of the Films Deposited Using Methods of Thin-Film Technology

Sensors prepared using thin-film technologies have humidity-sensitive layers of typical thickness up to 1 μm. The thin-film technologies allow one to reliably control the physical properties, such as the thickness, morphology, microstructure, and stoichiometry of the humidity-sensitive layers, which is extremely important, since these parameters govern the overall sensor performance (Korotcenkov 2008). Therefore, these deposition techniques are of great interest to both research laboratories and manufacturers. The conventional planar thin-film technologies employ masks made by photolithography with a resolution of 1–10 μm, which may be brought down to 10–100 nm by using ion-beam or electron-beam lithography. These techniques can be used to fabricate gas sensors of a few millimeters in size. Another option is to form multisensor arrays on a single substrate, which extends the long-term performance of the sensor as well as allowing selective analyzation of gases using the "electronic nose" concept (Gardner et al. 1995; Althainz et al. 1996).

The morphology of MOX films, deposited using thin-film technologies, is much more diverse than that of the films made using thick-film technologies. For example, as has been established, that "thin-" and "thick-" film gas sensors have completely different dependencies of the grain size

## TABLE 18.10

### Comparison Between Chemical Vapor Deposition and Physical-Vapor-Deposition-Coating Techniques

| Chemical Vapor Deposition (CVD) | Physical Vapor Deposition (PVD) |
|---|---|
| **Sophisticated reactor and/or vacuum system** | **Sophisticated reactor and vacuum system** |
| Simpler deposition rigs with no vacuum system has been adopted in variants of CVD, AACVD, ESAVD, FACVD, and CCVD | |
| ***Expensive techniques*** for LPCVD, PECVD, PACVD, MOCVD, EVD, ALE, UHVCVD; | ***Expensive techniques*** |
| ***Relatively low-cost techniques*** for AACVD and FACVD | |
| ***Non-line-of-sight process*** | ***Line-of-sight process*** |
| Therefore, it can coat complex-shaped components and deposit coating with good conformal coverage. | Therefore, it has difficulty coating complex-shaped components and producing conformal coverage. |
| ***Tend to use volatile/toxic chemical precursors*** | ***Tend to use expensive sintered solid targets/sources*** |
| Less volatile/more environmentally friendly precursors have been adopted in variants of CVD such as AACVD, ESAVD, and CCVD | Potential difficulties in large-area deposition and varying the composition or stoichiometry of the deposits |
| ***Multisource precursors*** tend to produce nonstoichiometric films | ***Both single and multiple targets*** do not guarantee the deposition of stoichiometric films because different elements will evaporate or sputter at different rates, except with the laser-ablation method |
| ***Single-source precursors*** (AACVD, PICVD) have overcome such problems. | |
| ***High deposition temperatures*** in conventional CVD. | ***Low to medium deposition temperatures*** |
| ***Low to medium deposition temperatures*** can be achieved using variants of CVD such as PECVD, PACVD, MOCVD, AACVD, ESAVD | |

*Source:* Reprinted from *Prog. Mater. Sci.*, 48, Choy, K.L., Chemical vapour deposition of coatings, 57–170, Copyright (2003), with permission from Elsevier. CCVD, combustion chemical vapor deposition; MOCVD, metal-organic-assisted CVD; PECVD, plasma-enhanced CVD; FACVD, flame-assisted CVD; AACVD, aerosol-assisted CVD; ESAVD, electrostatic-atomization CVD; LPCVD, low-pressure CVD; APCVD, atmospheric-pressure CVD; PACVD, photo-assisted CVD; TACVD, thermal-activated CVD; EVD, electrochemical vapor deposition; RTCVD, rapid thermal CVD; UHVCVD, ultrahigh-vacuum CVD; ALE, atomic-layer epitaxy; PICVD, pulsed-injection CVD.

## TABLE 18.11

### Comparative Characterization of CVD and CSD Methods

| | MOCVD | CSD |
|---|---|---|
| *Merits* | • Control of microstructure and hence properties.<br>• Compatible with MEMS technology.<br>• Mature systems have good thickness control and repeatability.<br>• Excellent conformal coating and thickness scaling.<br>• In-situ deposition. | • Simple, inexpensive means of film deposition.<br>• Low-temperature process.<br>• Rapid means for studying film-substrate interaction, grain-growth behavior, effect of dopants, etc.<br>• Precise stoichiometry control.<br>• Deposition of multicomponent materials without any problems. |
| *Limitations* | • More sophisticated and expensive than CSD technique.<br>• Deposition of multicomponent materials requires strong control.<br>• Not very flexible to variation of deposition parameters.<br>• Long cycle of development. | • Conformal coating and thickness scaling.<br>• Crack formation and delamination of material during drying process.<br>• Additional thermal treatment is necessary<br>• Deposition of continuous films with nm thickness. |

*Source:* With kind permission from Springer Science+Business Media: *Handbook of Gas Sensor Materials*, Korotcenkov, G. Copyright 2013 Springer.

on the film thickness. In both ceramic and thick-film sensors, the size of the metal-oxide grains does not depend on the thickness of the sensing layer. In the materials prepared using thick film and ceramic technologies, the method of powder preparation (precursor material, aging time, pH, etc.) and the sintering temperature are the main parameters that control the grain size (Risti et al. 2002; Vuong et al. 2004; Amjoud et al. 2005). In contrast, for thin-film sensors, which

were fabricated using metal-oxide deposition at temperatures higher 200°C–300°C, the grain size of the metal oxides is usually determined directly by the thickness of the deposited film (Korotcenkov et al. 2005a, 2005b). Increasing the film thickness leads to a larger grain size (see Figure 18.11). The morphology of the films deposited using thin-film technologies depends on the number of factors. For example, it was found that the grain size of $SnO_2$ and $In_2O_3$ films deposited by spray

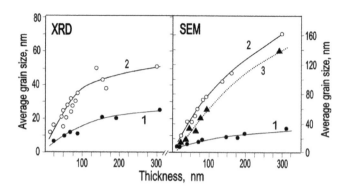

**FIGURE 18.11** Grain-size dependence on the thickness of films prepared by pyrolysis at different temperatures, calculated using (a) XRD data, and (b) SEM images: (1) $T_{pyr}$ = 350°C–375°C; (2) $T_{pyr}$ = 450°C–475°C; (3) $T_{pyr}$ = 510°C–535°C. (Reprinted from *Thin Solid Films*, 471, Korotcenkov, G. et al., Faceting characterization of $SnO_2$ nanocrystals deposited by spray pyrolysis from $SnCl_4$-$5H_2O$ water solution, 310–319, Copyright 2005, with permission from Elsevier.)

pyrolysis depends on such deposition parameters as the pyrolysis temperature, the film thickness, the distance between the atomizer and the substrate, and the properties of the precursor employed (Palatnik et al. 1972; Brinzari et al. 2002; Korotcenkov et al. 2001a, 2005a, 2005b; Chakrapani 2016). Such distinctions make it impossible to apply regularities established for sensors, fabricated by thick-film technology, to those, fabricated by thin-film technology, making the present research necessary.

The same effect is observed in other MOXs (Bender et al. 2002; Kiriakidis et al. 2007), made by various methods as magnetron sputtering (Bender et al. 2002; Suchea et al. 2006; Kiriakidis et al. 2007), pulsed-laser deposition (Dolbec et al. 2003), and plasma-enhanced chemical vapor

deposition (PECVD) (Huang et al. 2006). With these techniques, the grain size also depends on the total gas pressure and the oxygen partial pressure in the deposition chamber. For example, it was found that changing the oxygen partial pressure from 10 to 200 mTorr during the $SnO_2$-pulsed laser deposition leads to the increase in the $SnO_2$ grain size from 3 to 10 nm (Dolbec et al. 2003).

The two-dimensionality of the grains that compose the oxide films is another feature of thin-film morphology, which should be considered when analyzing the influence of the grain size on the gas-sensing effect. In many cases, the MOX films deposited by standard thin-film methods have a columnar structure (see Figure 18.12) (Korotcenkov et al. 2004, 2005a, 2005b). This means that the in-plane size of the grains may differ significantly from the grain size measured in the growth direction. Ordinary analytical methods, employed for analyzing the film structure and the grain size, such as X-ray diffraction (XRD), scanning electron microscopy (SEM), and atomic force microscopy (AFM), provide mainly in-plane grain sizes. It is worth noting that the two-dimensionality effects are stronger in thicker films. The grained structure (see Figure 18.12a) that is typical for MOX, using ceramic or thick-film technology, is usually observed only in thin films deposited at low temperatures (Korotcenkov et al. 2005b). With spray pyrolysis, for example, the grained structure of $SnO_2$ films was observed only when the deposition temperatures did not exceed 400°C.

A detailed study also showed that, despite the significantly smaller thickness, the metal-oxide films obtained using thin-film technology do not have the porosity found in sensors, fabricated by "thick" or "ceramic" technologies. Typical SEM images of intergrain boundary in the $SnO_2$ films deposited at temperatures higher 350°C are shown in Figure 18.13. It is seen that the necks between grains are absent. As it was shown

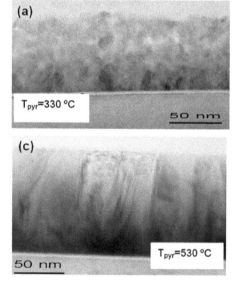

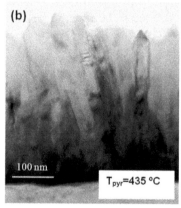

**FIGURE 18.12** TEM cross-section micrographs of $SnO_2$ films deposited by spray pyrolysis: (a) grained structure, $T_{pyr}$ ~ 330°C–350°C, $d$ ~ 70–80 nm; (b, c) columnar structure, (b) $T_{pyr}$ ~ 475°C, $d$ ~ 300 nm, (c) $T_{pyr}$ ~ 510°C, $d$ ~ 75–100 nm. (Reprinted from *Thin Solid Films*, 471, Korotcenkov, G. et al., Faceting characterization of $SnO_2$ nanocrystals deposited by spray pyrolysis from $SnCl_4$-$5H_2O$ water solution, 310–319, Copyright 2005, with permission from Elsevier.)

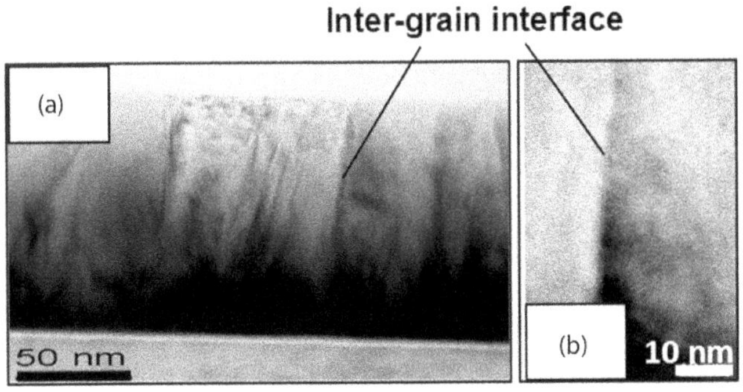

**Inter-grain interface**

FIGURE 18.13 (a,b) SEM images of the SnO$_2$ films deposited at $T_{pyr}$ = 520°C ($d$~120 nm). (Reprinted from *Sens. Actuators B*, 142, Korotcenkov, G. et al., Thin film SnO$_2$-based gas sensors: Film thickness influence, 321–330, Copyright 2009, with permission from Elsevier.)

before, the SnO$_2$ films deposited at temperatures, exceeding 350°C, have columnar structure in which the grains can grow through the entire film thickness (see Figure 18.12). This means that films deposited by thin-film technology have a larger contact area between crystallites. Moreover, the comparison of the images of the films, having different thickness, shows that the area of indicated contacts increases considerably with the film growth. Korotcenkov and co-workers (Korotcenkov et al. 2007a, 2007b; Korotcenkov and Cho 2009) have shown that kinetics of sensor response in thin-film devices is being controlled by adsorption–desorption processes and the surface diffusion in intercrystalline space. In the frame of this model, the time required for the diffusion of oxygen or oxygen vacancies into intercrystallite space should increase along with the film-thickness growth. The specific character of such film structure is shown in Figure 18.14. It is important to note that the processes mentioned above begin to dominate in the kinetics of the gas-sensing effect only when the film thickness exceeds 60–80 nm. Experiments have shown that in films with a thickness of less than 60–80 nm, there is no any diffusion limitation in the kinetics of the sensor response (Korotcenkov and Cho 2009). In this case, the sensitivity is mainly determined by the effectiveness of surface reactions.

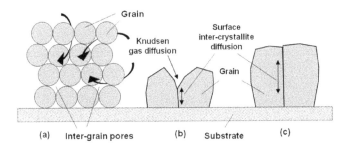

FIGURE 18.14 Models, illustrating diffusion processes dominating in the films formed using (a) thick-film, and (b and c) thin-film technologies. (Reprinted from *Sens. Actuators B*, 142, Korotcenkov, G. et al., Thin film SnO$_2$-based gas sensors: Film thickness influence, 321–330, Copyright 2009, with permission from Elsevier.)

## 18.4 SUMMARY

Thus, a brief overview of technologies suitable for the manufacture of humidity sensors shows that we do not have the ideal technology, and we cannot name the best method for the deposition of the humidity-sensing layer. The choice of the method optimal for each specific technical application should take into account the properties of the deposited material, the sensor design, and the possible consequences for the sensor parameters when applying the selected method. For example, the findings of Will et al. (2000) regarding the methods that can be applied to produce metal-oxide films can be used for this purpose (see Table 18.12).

Based on the analysis conducted in this section, it can be concluded that only simple stoichiometric compounds can be deposited using standard CVD and PVD methods, as each component has to be evaporated at different temperatures due to their different vapor pressures. Moreover, the CVD process also applies very toxic precursors. The constituents have to be deposited from independently controlled sources, adding complexity to the system. At the same time, PVD is a line-of-sight process, meaning that it has difficulty in coating complex-shaped components.

Other methods also have limitations. Magnetron sputtering of multicomponent materials requires expensive sintered solid targets and produces surface damage. Ceramic-powder methods and spray-pyrolysis methods, on the other hand, have the potential to be good candidates for complicated stoichiometric compositions or for mixtures of materials. However, sol-gel-coated films tend to crack and have a thickness limitation for each layer, meaning that the process needs to be repeated to obtain the required thickness. Spray pyrolysis has limitations in repeatability. The investment cost for CVD and PVD apparatus is high compared to the droplet and powder techniques, whereas the setup for spray pyrolysis is inexpensive and simple. Standard CVD methods, however, have the advantage of being able to coat large areas uniformly, and they feature easily controlled deposition rates and film thicknesses. On the other hand, all liquid-precursor methods, such as sol-gel and slurry coating, are time-, labor-, and energy-intensive because coating and drying/sintering have to be repeated in order to avoid crack formation.

According to consideration of general requirements to sensor technology, one can expect that the ideal method in any application should meet the following criteria:

- Compatibility with the process of manufacture of the chemical sensor
- No impairment of, or effect on, the properties of the bulk materials used in the device
- Ability to deposit the required type of material with the required thickness and structure
- Improvement in the quality of the designed sensor
- Ability to coat the engineering components uniformly with respect to both size and shape
- Cost-effectiveness in terms of the cost of the substrate, depositing material, and coating technique
- Ecologically clean and safe for attending personnel

**TABLE 18.12**

**Comparison of Methods Acceptable for Producing Metal-Oxide Films**

| Technique | Film Characteristics | | | Process Features |
| | Structure | Deposition Rate or Thickness | Cost | Characteristics and Limitations |
|---|---|---|---|---|
| **Vapor Phase** | | | | |
| Thermal spray technologies | | 100–500 μm/h | | High deposition rates, various compositions, possible thick and porous coatings, high temperatures necessary |
| EVD | C | 3–50 μm/h | Exp. | High-reaction temperatures necessary, corrosive gases equipment and processing costs |
| CVD | C | 1–10 μm/h | Exp. | Various precursor materials possible, high-reaction equipment temperatures necessary, corrosive gases |
| PVD (RF and magnetron) sputtering) | A to P | 0.25–2.5 μm/h | Exp. | Tailor-made films, dense and crack-free, low films equipment deposition temperatures, multipurpose technique, relatively low deposition rate |
| Laser ablation | A to P | | Exp. | Intermediate deposition temperatures, difficult equipment upscaling, time-sharing of laser, relatively low deposition rate |
| Spray pyrolysis | A to P | 5–60 μm/h | Ec. | Robust technology, upsealing possible, easy control of polycrystalline parameters, corrosive salts, post-thermal treatment usually necessary |
| **Liquid Phase** | | | | |
| Sol-gel, liquid precursor route processes | P | 0.5–1 μm for each coating | Ec. | Various precursors possible, very thin films, low temperature sintering, coating and drying/ heating have to be repeated 5–10 times, crack formation during drying, many process parameters |
| **Solid Phase** | | | | |
| Tape casting | P | 25–200 μm | | Robust technology, upscaling possible, crack formation |
| Slip casting and slurry coating | P | 25–200 μm | Ec. | Robust technology, crack formation, slow |
| Tape calendering | P | 5–200 μm | | Upscaling possible, co-calendering possible |
| EPD | P | 1–200 μm | | Short formation time, little restriction to shape of substrate, suitable for mass production, high deposition rates, inhomogeneous thickness |
| Transfer printing | P | 5–100 μm | Ec. | Robust technology, rough-substrate surfaces possible, adhesion on smooth substrates difficult |
| Screen printing | P | 10–100 μm | Ec. | Robust technology, upscaling possible, crack formation |

*Source:* Reprinted from *Solid State Ionics*, 131, Will, J., Fabrication of thin electrolytes for second-generation solid oxide fuel cells, 79–96, Copyright 2000, with permission from Elsevier.
A—amorphous; C—columnar; Ec—economical; Exp.—expensive; EPD—electrophoretic deposition; EVD—electrochemical vapor deposition; P—polycrystalline

It is obvious that at present there is no ideal method that would meet all the requirements. Therefore, in practice it is always necessary to search for the best compromise and choose a method that meets as many requirements as possible. At the same time, these requirements may change significantly during the development stage of the device or in the process of industrial development. Properties that may be preferred during development, such as multi-functionality, the ability to vary the parameters of the technological process and the deposited material, and the speed of the reorganization of the technological process, are very different from the properties required during industrial processing, such as compatibility with basic technological processes, reproducibility, productiveness, and cost. Such differences in requirements are inevitable, and they must be taken into account during research aimed at developing devices designed for use in the gas-sensor market.

## REFERENCES

Agnew J. (1973) *Thick Film Technology: Fundamentals and Applications in Microelectronics.* Hayden Book, Rochelle Park, NJ.
Althainz P., Goschnick J., Ehrmann S., Ache H.J. (1996) Multisensor microsystem for contaminants in air. *Sens. Actuators B* **33**, 72–76.

Amjoud M., Rhouta B., Alimoussa A., Hajji L., Mezzane D., Ahamdane H. (2005) Effect of pH adjustment in sol-gel synthesis route on grain size of tin dioxide intended for gas sensors application. *Phys. Chem. News* **22**, 120–124.

Arthur J.A. (2002) Molecular beam epitaxy. *Surf. Sci.* **500**, 189–217.

Atkinson A., Guppy R.M. (1991) Mechanical stability of sol-gel films. *J. Mater. Sci.* **26**(14), 3869–3873.

Bakrania S.D., Wooldridge M.S. (2009) The effects of two thick film deposition methods on tin dioxide gas sensor performance. *Sensors* **9**, 6853–6868.

Barbi G.B., Santos J.P., Serrini P., Gibson P.N., Horrillo M.C., Manes L. (1995) Ultrafine grain-size tin-oxide films for carbon monoxide monitoring in urban environments. *Sens. Actuators B* **25**, 559–563.

Bârsan N., Schweizer-Berberich M., Gopel W. (1999) Fundamental and practical aspects in the design of nanoscaled $SnO_2$ gas sensors. A status report. *Fresen. J. Anal. Chem.* **365**, 287–304.

Bârsan N., Weimar U. (2001) Conduction model of metal oxide gas sensors. *J. Electroceram.* **7**, 143–167.

Bender M., Gagaoudakis E., Douloufakis E., Natsakou E., Katsarakis N., Cimalla V., Kiriakidis G., Fortunato E., Nunes P., Marques A., Martins R. (2002) Production and characterization of zinc oxide thin films for room temperature ozone sensing. *Thin Solid Films* **418**, 45–50.

Bietsch A., Zhang J., Hegner M., Lang H.P., Gerber C. (2004) Rapid functionalization of cantilever array sensors by inkjet printing. *Nanotechnology* **15**, 873–880.

Brinker C.J., Hurd A.J., Schunk P.R., Ashely C.S, Cairncross R.A., Samuel J., Chen K.S., Scotto C., Schwartz R.A. (1996) Sol-gel derived ceramic films—Fundamentals and applications. In: Stern K. (ed.) *Metallurgical and Ceramic Protective Coatings.* Chapman & Hall, London, UK, pp. 112–151.

Brinker C.J., Scherer G.W. (1990) *Sol-Gel Science: The Physics and Chemistry of Sol-Gel Processing.* Academic Press, San Diego, CA.

Brinzari V., Korotcenkov G., Schwank J., Lantto V., Saukko S., Golovanov V. (2002) Morphological rank of nano-scale tin dioxide films deposited by spray pyrolysis from $SnC_{14}·5H_2O$ water solution. *Thin Solid Films* **408**, 51–58.

Brousse T., Schleich D.M. (1996) Sprayed and thermally evaporated $SnO_2$ thin films for ethanol sensors. *Sens. Actuators B* **31**, 77–79.

Bunshah R.F. (ed.) (1994) *Handbook of Deposition Technologies for Film and Coatings: Science, Technology and Applications.* Noyes Publications, Park Ridge, NJ.

Calvert P. (2001) Inkjet printing for materials and devices. *Chem. Mater.* **13**, 3299–3305.

Chakrapani V. (2016) Modulation of stoichiometry, morphology and composition of transition metal oxide nanostructures through hot wire chemical vapor deposition. *J. Mater. Res.* **31**(1), 17–27.

Cho N., Lim T., Jeon Y., Gong M. (2008) Inkjet printing of polymeric resistance humidity sensor using UV-curable electrolyte inks. *Macromol. Res.* **16**, 149–154.

Choi U.-S., Shimanoe K., Yamazoe N. (2005) Influences of ball-milling time on gas-sensing properties of $Co_3O_4$–$SnO_2$ composites. *Sens. Actuators B* **107**, 516–522.

Choy K.L. (2003) Chemical vapour deposition of coatings. *Prog. Mater. Sci.* **48**(2), 57–170.

Christen H.M., Eres G. (2008) Recent advances in pulsed-laser deposition of complex oxides. *J. Phys.: Condens. Matter* **20**, 264005.

Comini E., Faglia G., Sberveglieri G. (2009a) Electrical-based gas sensing. In: Comini E., Faglia G., Sberveglieri G. (eds.), *Solid State Gas Sensing.* Springer–Verlag, New York, pp. 47–107.

Demarne V., Grisel A. (1993) A new $SnO_2$ low temperature deposition technique for integrated gas sensors. *Sens. Actuators B* **15–16**, 63–67.

Deore M.K., Gaikwad V.B., Jain G.H. (2011) LPG gas sensing properties of CuO loaded ZnO thick film resistors. In: *Proceedings of Fifth International Conference on Sensing Technology*, Palmerston North, New Zealand, November 28–December 1, IEEE, pp. 233–238.

DiGiulio M., Serra A., Tepore A., Rella R., Siciliano P., Mirenghi L. (1996) Influence of the deposition parameters on the physical properties of tin oxide thin films. *Mater. Sci. Forum* **203**, 143–148.

Dolbec R., El Khakani M.A., Serventi A.M., Saint-Jacques R.G. (2003) Influence of the nanostructural characteristics on the gas sensing properties of pulsed laser deposited tin oxide thin films. *Sens. Actuators B* **93**, 566–571.

Epifani M., Francioso L., Siciliano P., Helwig A., Mueller G., Dıraz R., et al. (2007) $SnO_2$ thin films from metalorganic precursors: Synthesis, characterization, microelectronic processing and gas-sensing properties. *Sens. Actuators B* **124**, 217–226.

Francioso L., Russo M., Taurino A.M., Siciliano P. (2006) Micrometric patterning process of sol–gel $SnO_2$, $In_2O_3$ and $WO_3$ thin film for gas sensing applications: Towards silicon technology integration. *Sens. Actuators B* **119**, 159–166.

Galan-Vidal C.A., Munoz J., Dominguez C., Alegret S. (1995) Chemical sensors, biosensors and thick-film technology. *Trends Anal. Chem.* **14**(5), 225–231.

Gardner J.W., Pike A., de Rooij N.F., Koudelka-Hep M., Clerc P.A., Hierlemann A., Gopel W. (1995) Integrated array sensor for detecting organic solvents. *Sens. Actuators B* **26–27**, 135–139.

Gaspar C., Olkkonen J., Passoja S., Smolander M. (2017) Paper as active layer in inkjet-printed capacitive humidity sensors. *Sensors* **17**, 1464.

Glocker D.A., Shah I. (eds.) (1995) *Handbook of Thin Film Process Technology.* Institute of Physics Publishing, Bristol, UK.

Göpel W., Reinhardt G. (1996) Metal oxide sensors: New devices through tailoring interfaces on the atomic scale. In: Baltes H., Göpel W., Hesse J. (eds.), *Sensors Update. Sensor Technology—Applications Markets*, Vol. 1. VCH, Weinheim, Germany, pp. 49–120.

Graf M., Barrettino D., Zimmermann M., Hierlemann A., Baltes H., Hahn S., et al. (2004). CMOS monolithic metal-oxide sensor system comprising a micro hotplate and associated circuitry. *IEEE Sens. J.* **4**(1), 9–16.

Grundler P. (2007) *Chemical Sensors: An Introduction for Scientists and Engineers.* Springer-Verlag, Berlin, Germany.

Guidi V., Butturi M.A., Carotta M.C., Cavicchi B., Ferroni M., Malagu C., et al. (2002) Gas sensing through thick film technology. *Sens. Actuators B* **84**, 72–77.

Hecht G., Richter F., Hahn J. (eds.) (1994) *Thin Films.* DGM Informationgessellschaft, Oberursel, Germany.

Heilig A., Bârsan N., Weimar U., Gopel W. (1999) Selectivity enhancement of $SnO_2$ gas sensors: Simultaneous monitoring of resistances and temperatures. *Sens. Actuators B* **58**, 302–309.

Heule M., Gauckler L.J. (2001) Gas sensors fabricated from ceramic suspensions by micromolding in capillaries. *Adv. Mater.* **13**, 1790–1793.

Heule M., Meier L., Gauckler L.J. (2001) Micropatterning of ceramics on substrates towards gas sensing applications. *Mater. Res. Soc. Symp. Proc.* **657**, EE9.4.

Hitchman M.L., Jensen K.F. (1993) *CVD: Principles and Applications.* Academic Press, San Diego, CA.

Holmes P.J., Loasby R.G. (eds.) (1976) *Handbook of Thick Film Technology.* Electrochemical Publications, Ayr, UK.

Huang H., Tan O.K., Lee Y.C., Tse M.S. (2006) Preparation and characterization of nanocrystalline $SnO_2$ thin films by PECVD. *J. Crystal Growth* **288**, 70–74.

Huczko A. (2000) Template-based synthesis of nanomaterials. *Appl. Phys. A* **70**, 365–376.

Ihokura K., Watson J. (1994) *The Stannic Oxide Gas Sensor: Principle and Application.* CRC Press, Boca Raton, FL.

Ivanov P.T. (2004) *Design, Fabrication and Characterization of Thick-Film Gas Sensors.* PhD Thesis, University Rovira i Virgili, Tarragona, Spain.

Janata J. (1989) *Principle of Chemical Sensors.* Plenum Press, NewYork.

Jaworek A., Sobczyk A.T. (2008) Electrospraying route to nanotechnology: An overview. *J. Electrostatics* **66**, 197–219.

Jayadev Dayan N., Sainkar S.R., Karekar R.N., Aiyer R.C. (1998) Formulation and characterization of ZnO:Sb thick-film gas sensors. *Thin Solid Films* **325**, 254–258.

Kiriakidis G., Suchea M., Christoulakis S., Horvath P., Kitsopoulos T., Stoemenos J. (2007) Structural characterization of ZnO thin films deposited by dc magnetron sputtering. *Thin Solid Films* **515**, 8577–8581.

Kissin V.V., Voroshilov S.A., Sysoev V.V. (1999a) Oxygen flow effect on gas sensitivity properties of tin oxide film prepared by r.f. sputtering. *Sens. Actuators B* **55**, 55–59.

Kissin V.V., Voroshilov S.A., Sysoev V.V. (1999b) A comparative study of $SnO_2$ and $SnO_2$:Cu thin films for gas sensor applications. *Thin Solid Films* **348**, 307–314.

Korotcenkov G. (2005) Gas response control through structural and chemical modifications of metal oxide films: State of the art and approaches. *Sens. Actuators B* **107**, 209–232.

Korotcenkov G. (2008) The role of morphology and crystallographic structure of metal oxides in response of conductometric-type gas sensors. *Mater. Sci. Eng. R* **61**, 1–39.

Korotcenkov G. (2013) *Handbook of Gas Sensor Materials.* Springer, New York.

Korotcenkov G., Boris I., Brinzari V., Luchkovsky Yu., Karkotsky G., Golovanov V., et al. (2004b) Gas sensing characteristics of one-electrode gas sensors on the base of doped $In_2O_3$ ceramics. *Sens. Actuators* B **103**(1–2), 13–22.

Korotcenkov G., Brinzari V., DiBattista M., Schwank J., Vasiliev A. (2001a) Peculiarities of $SnO_2$ thin film deposition by spray pyrolysis for gas sensor application. *Sens. Actuators B* **77**, 244–252.

Korotcenkov G., Brinzari V., Ivanov M., Cerneavschi A., Rodriguez J., Cirera A., et al. (2005a) Structural stability of $In_2O_3$ films deposited by spray pyrolysis during thermal annealing. *Thin Solid Films* **479**, 38–51.

Korotcenkov G., Brinzari V., Schwank J., Cerneavschi A. (2001b) Possibilities of aerosol technology for deposition of $SnO_2$-based films with improved gas sensing characteristics. *J. Mater. Sci. Eng. C* **19**, 73–77.

Korotcenkov G., Brinzari V., Stetter J.R., Blinov I., Blaja V. (2007b) The nature of processes controlling the kinetics of indium oxide-based thin film gas sensor response. *Sens. Actuators B* **128**, 51–63.

Korotcenkov G., Cerneavschi A., Brinzari V., Vasiliev A., Cornet A., Morante J.R., et al. (2004a) $In_2O_3$ films deposited by spray pyrolysis as a material for ozone gas sensors. *Sens. Actuators B* **99**, 304–310.

Korotcenkov G., Cho B.K. (2009) Thin film $SnO_2$-based gas sensors: Film thickness influence. *Sens. Actuators B* **142**, 321–330.

Korotcenkov G., Cho B.K. (2010) Methods of sensing materials synthesis and deposition. In: Korotcenkov G. (ed.) *Chemical Sensors: Fundamentals of Sensing Materials.* Vol. 1. *General Approaches.* Momentum Press, New York, pp. 214–303.

Korotcenkov G., Cornet A., Rossinyol E., Arbiol J., Brinzari V., Blinov Y. (2005b) Faceting characterization of $SnO_2$ nanocrystals deposited by spray pyrolysis from $SnCl_4$-5$H_2O$ water solution. *Thin Solid Films* **471**, 310–319.

Korotcenkov G., Ivanov M., Blinov I., Stetter J.R. (2007a) Kinetics of $In_2O_3$-based thin film gas sensor response: The role of "redox" and adsorption/desorption processes in gas sensing effects. *Thin Solid Films* **515**(7–8), 3987–3996.

Kraus T., Malaquin L., Schmid H., Riess W., Spencer N.D., Wolf H. (2007) Nanoparticle printing with single-particle resolution. *Nature Nanotechnol.* **2**, 570–576.

Labeau M., Gautheron B., Delabouglise G., Pena J., Ragel V., Varela A., et al. (1993) Synthesis, structure and gas sensitivity properties of pure and doped $SnO_2$. *Sens. Actuators B* **15–16**, 379–389.

Lalauze R., Breuli P., Pijolat C. (1991) Thin films for gas sensors. *Sens. Actuators B* **3**, 175–182.

Lee D.-H., Chang Y.-J., Stickle W., Chang C.-H. (2007a) Functional porous tin oxide thin films fabricated by inkjet printing process. *Electrochem. Solid-State Lett.* **10**(11), K51–K54.

Lee S., Lee G.-G., Kim J., Kang S.-J.L. (2007b) A novel process for fabrication of $SnO_2$-based thick film gas sensors. *Sens. Actuators B* **123**, 331–335.

LeGore L.J., Greenwood O.D., Paulus J.W., Frankel D.J., Lad R.J. (1997) Controlled growth of $WO_3$ films. *J. Vacuum Sci. Technol. A* **15**, 1223–1227.

Liu C. (1995) Development of chemical sensors using microfabrication and micromachining techniques. *Mater. Chem. Phys.* **42**, 87–90.

Madou, M. J., Morrison S.R. (1989) *Chemical Sensing with Solid State Devices.* Academic Press, Boston, MA.

Marek J., Trah H.-P., Suzuki Y., Yokomori I. (eds.) (2003) *Sensors for Automotive Technology.* VCH, Weinheim, Germany.

Miccoci G., Serra A., Siciliano P., Tepore A., Ali-Adib Z. (1996) CO sensing characteristics of reactively sputtered $SnO_2$ thin films prepared under different oxygen partial pressure values. *Vacuum* **47**, 1175–1177.

Milchev A. (2008) Electrocrystallization: Nucleation and growth of nano-clusters on solid surfaces. *Russ. J. Electrochem.* **44**, 619–645.

Molina-Lopez F., Vasquez Quintero A., Mattana G., Briand D., de Rooij N.F. (2013) Large-area compatible fabrication and encapsulation of inkjet-printed humidity sensors on flexible foils with integrated thermal compensation. *J. Micromech. Microeng.* **23**, 025012.

Morrison S.R. (1994) Chemical sensors. In: Sze S.M. (ed.), *Semiconductor Sensors.* John Wiley & Sons, New York, pp. 404–408.

Moseley P.T., Norris J.O.W., Williams D.E. (1991) *Techniques and Mechanisms in Gas Sensing.* Adam Hilger, Bristol, UK.

Moseley P.T., Tofield B.C. (eds.) (1987) *Solid State Gas Sensors.* Adam Hilger, Bristol, UK.

Nenov T.G., Yordanov S.P. (1996) *Ceramic Sensors: Technology and Applications.* Technomic, Basel, Switzerland.

Nitta M., Haradome M. (1979) Thick-film CO gas sensors. *IEEE Trans. El. Dev. ED* **26**(3), 247–249.

Olding T., Sayer M., Barrow D. (2001) Ceramic sol–gel composite coatings for electrical insulation. *Thin Solid Films* **398–399**, 581–586.

Olvera M.L., Maldonaldo A., Asomoza R. (1996) Characterization of a thin film tin oxide gas sensor deposited by chemical spraying. *AIP Conf. Proc.* **378**, 376–381.

Oyabu T. (1982) Sensing characteristic of $SnO_2$ thin film gas sensors. *J. Appl. Phys.* **53**, 2785–2787.

Palatnik L.S., Fuks M.I., Kosevich V.M. (1972) *Mechanism of Formation and Substructure of Condensed Films*. Science Press, Moscow (in Russian).

Patil A. (2011) *ZnO Thick Films Gas Sensor: Electrical, Structural and Gas Sensing Characteristics with Different Dopants*. LAP Lambert Academic Publishing, Saarbrücken, Germany.

Phillips H.M., Li Y., Bi X., Zhang B. (1996) Reactive pulsed laser deposition and laser induced crystallization of $SnO_2$ transparent conducting thin films. *Appl. Phys. A* **63**, 347–351.

Prudenziati M. (ed.) (1994) *Thick Film Sensors* (Middelhoek S. (series ed.) *Handbook of Sensors and Actuators*, Vol. 1). Elsevier, Amsterdam, the Netherlands.

Quddious A., Yang S., Khan M.M., Tahir F.A., Shamim A., Salama K.N., Cheema H.M. (2016) Disposable, paper-based, inkjet-printed humidity and $H_2S$ gas sensor for passive sensing applications. *Sensors* **16**, 2073.

Randhaw H. (1991) Review of plasma-assisted deposition processes. *Thin Solid Films* **196**, 329–349.

Rieu M., Camara M., Tournier G., Viricelle J.-P., Pijolat C., De Rooij N.F., et al. (2016) Fully inkjet printed $SnO_2$ gas sensor on plastic substrate. *Sens. Actuators B* **236**, 1091–1097.

Risti M., Ivanda M., Popovi S., Musi S. (2002) Dependence of nanocrystalline $SnO_2$ particle size on synthesis route. *J. Non-Crystal. Solids* **303**, 270–280.

Rivadeneyra A., Fernández-Salmerón J., Agudo M., López-Villanueva J.A., Capitan-Vallvey L.F., Palma A.J. (2014) Design and characterization of a low thermal drift capacitive humidity sensor by inkjet-printing. *Sens. Actuators B* **195**, 123–131.

Rumyantseva M.N., Labeau M., Senateur J.P., Delabouglise G., Boulova M.N., Gaskov A.M. (1996) Influence of copper on sensor properties of tin dioxide films in $H_2S$. *Mater. Sci. Eng. B*, **41**, 228–234.

Saadeddin I., Pecquenard B., Manaud J.P., Decourt R., Abrugère C., Buffeteau T., Campet G. (2007) Synthesis and characterization of single- and co-doped $SnO_2$ thin films for optoelectronic applications. *Appl. Surf. Sci.* **253**, 5240–5249.

Santra S., Hu G., Howe R.C.T., De Luca A., Ali S.Z., Udrea F., et al. (2015) CMOS integration of inkjet-printed graphene for humidity sensing. *Sci. Rep.* **5**, 17374.

Sappat A., Wisitsoraat A., Sriprachuabwong C., Jaruwongrungsee K., Lomas T., Tuantranont A. (2011) Humidity sensor based on piezoresistive microcantilever with inkjet printed PEDOT/PSS sensing layers. In: *Proceedings of the 8th International Conference on Electrical Engineering, Electronics, Computer, Telecommunications and Information Technology (ECTI)*, May 17–19, Khon Kaen, Thailand, pp. 34–37.

Sayago I., Gutierrer F.J., Ares L., Robla J.I., Horrillo M.C., Getino J., et al. (1995a) The effect of additives in tin oxide on the sensitivity and selectivity to $NO_x$ and CO. *Sens. Actuators B* **26**, 19–23.

Sayago I., Gutierrer J., Ares L., Robla J.I., Horrillo M.C., Getino J., Rino J., Agapito J.A. (1995b) Long-term reliability of sensors for detection of nitrogen oxides. *Sens. Actuators B* **26**, 56–58.

Sberveglieri G. (ed.) (1992a) *Gas Sensors*. Kluwer, Dordrecht, the Netherlands.

Sberveglieri G., Nelli P., Benussi G.P., Depero L.E., Zocchi M., Rossetto G., Zanella P. (1993) Enhanced response to methane for $SnO_2$ thin films prepared with the CVD technique. *Sens. Actuators B* **15–16**, 334–337.

Schierbaum K.D., Vaihinger S., Gopel W. (1990) Prototype structure for systematic investigations of thin-film gas sensors. *Sens. Actuators B* **1**, 171–175.

Schoenholzer U., Hummel R., Gauckler L.J. (2000) Microfabrication of ceramics by filling of photoresist molds. *Adv. Mater.* **12**, 1261–1263.

Schwartz R.W., Schneller T., Waser R. (2004) Chemical solution deposition of electronic oxide films. *Comptes Rendus Chimie* **7**, 433–461.

Sinner-Hettenbach M. (2000) $SnO_2$ (110) and nano-$SnO_2$: Characterization by surface analytical techniques. PhD thesis, University of Tübingen, Tübingen, Germany.

Song K.-H., Park S.J. (1994) Factors determining the carbon monoxide sensing properties of tin oxide thick films calcined at different temperatures *J. Am. Ceram. Soc.* **77**, 2935–2939.

Spannhake J., Helwig A., Schulz O., Muller G. (2009) Microfabrication of gas sensors. In: Comini E., Faglia G., and Sberveglieri G. (eds.) *Solid State Gas Sensing*. Springer-Verlag, Berlin, Germany, pp. 1–46.

Starke E., Türke A., Schneider M., Fischer W.-J. (2012) Setup and properties of a fully inkjet printed humidity sensor on PET substrate. In: *Proceedings of IEEE Sensor Conference*, October 28–31, Taipei, Taiwan. doi:10.1109/ICSENS.2012.6411259.

Stryhal Z., Pavlik J., Novak S., Mackova A., Perina V., Veltruska K. (2002) Investigations of $SnO_2$ thin films prepared by plasma oxidation. *Vacuum* **67**, 665–671.

Suchea M., Katsarakis N., Christoulakis S., Nikolopoulou S., Kiriakidis G. (2006) Low temperature indium oxide gas sensors. *Sens. Actuators B* **118**, 135–141.

Takada T. (2001) A temperature drop on exposure to reducing gases for various metal oxide thin films. *Sens. Actuators B* **77**, 307–311.

Tay B.K., Zhao Z.W., Chua D.H.C. (2006) Review of metal oxide films deposited by filtered cathodic vacuum arc technique. *Mater. Sci. Eng. R* **52**, 1–48.

Tiburcio-Silver A., Sanchez-Juarez A. (2004) Regeneration processes study on spray-pyrolyzed $SnO_2$ thin films exposed to CO-loaded air. *Sens. Actuators B* **102**, 174–177.

Tiemann M. (2008) Repeated templating. *Chem. Mater.* **20**, 961–971.

Toda M., Chen Y., Nett S.K., Itakura A.N., Gutmann J., Rü Berger R. (2014) Thin polyelectrolyte multilayers made by inkjet printing and their characterization by nanomechanical cantilever sensors. *J. Phys. Chem. C* **118**, 8071–8078.

Vahlas C., Caussat B., Serp P., Angelopoulos G.N. (2006) Principles and applications of CVD powder technology. *Mater. Sci. Eng. R* **53**, 1–72.

Van Tassel J.J., Randall C.A. (2006) Mechanism of electrophoretic deposition. In: Boccaccini A.R., Van der Biest O., Clasen R. (eds.) *Electrophoretic Deposition: Fundamentals and Applications*. Trans. Tech. Publications, Zurich, Switzerland.

Vandrish G. (1996) Ceramic applications in gas and humidity sensors. *Key Eng. Mater.* **122–124**, 185–224.

Vayssieres L. (2007) An aqueous solution approach to advanced metal oxide arrays on substrates. *Appl. Phys. A* **89**, 1–8.

Vincenzi D., Butturi M.A., Stefancich M., Malagu C., Guidi V., Carotta M.C., et al. (2001) Low-power thick-film gas sensor obtained by a combination of screen printing and micromachining techniques. *Thin Solid Films* **391**, 288–292.

Viswanathan V., Laha T., Balani K., Agarwal A., Seal S. (2006) Challenges and advances in nanocomposite processing techniques. *Mater. Sci. Eng. R* **54**, 121–285.

Vuong D.D., Sakai G., Shimanoe K., Yamazoe N. (2004) Preparation of grain size-controlled tin oxide sols by hydrothermal treatment for thin film sensor application. *Sens. Actuators B* **103**, 386–391.

White N.M., Turner J.D. (1997) Thick-film sensors: Past, present and future. *Meas. Sci. Technol.* **8**, 1–20.

Will J., Mitterdorfer A., Kleinlogel C., Perednis D., Gauckler L.J. (2000) Fabrication of thin electrolytes for second-generation solid oxide fuel cells. *Solid State Ionics* **131**, 79–96.

Willett M.J., Burganos V.N., Tsakiroglou C.D., Payatakes A.C. (1998) Gas sensing and structural properties of various pre-treated nanopowders tin (IV) oxide samples. *Sens. Actuators B* **53**, 76–90.

Williams G., Coles G.S.V. (1993) $NO_x$ response of tin-dioxide-based gas sensors. *Sens. Actuators B* **15–16**, 349–353.

Williams G., Coles G.S.V. (1995) The influence of deposition parameters on the performance of tin dioxide $NO_2$ sensors prepared by radio-frequency magnetron sputtering. *Sens. Actuators B* **25**, 469–473.

Willmott P.R. (2004) Deposition of complex multielemental thin films. *Prog. Surf. Sci.* **76**, 163–217.

Windle J., Derby B. (1999) Ink jet printing of PZT aqueous ceramic suspensions. *J. Mater. Sci. Lett.* **18**, 87–90.

Wu Q., Lee K.-M., Lin C.-C. (1993) Development of chemical sensors using microfabrication and micromachining techniques. *Sens. Actuators B* **13–14**, 1–6.

Xia Y., Whitesides G. M. (1998) Soft lithography. *Angew. Chem. Int. Ed.* **37**, 550–575.

Yamazoe N., Miura N. (1992) Some basic aspects of semiconductor gas sensors. In: Yamauchi S. (ed.) *Chemical Sensor Technology*, Vol. 4. Elsevier, Amsterdam, the Netherlands.

Yang H., Deschatelets P., Brittain S.T., Whitesides G.M. (2001) Fabrication of high performance ceramic microstructures from a polymeric precursor using soft lithography. *Adv. Mater.* **13**, 54–58.

Yang P., Deng T., Zhao D., Feng P., Pine D., Chmelka B.F., Whitesides G.M., Stucky G.D. (1998) Hierarchically ordered oxides. *Science* **282**, 2244–2246.

Yu K.N., Xiong X., Liu Y., Xiong C. (1997) Microstructural change of nano-$SnO_2$ grain assemblages with the annealing temperature. *Phys. Rev. B* **55**, 2666–2671.

Yuan Y., Peng B., Chi H., Li C., Liu R., Liu X. (2016) Layer-by-layer inkjet printing SPS:PEDOT NP/RGO composite film for flexible humidity sensors. *RSC Adv.* 6, 113298–113306

Zhang M.L., Song J.P., Yuan Z.H., Zheng C. (2012) Response improvement for $In_2O_3$–$TiO_2$ thick film gas sensors. *Curr. Appl. Phys.* **12**(3) 678–683.

Zhao X., Evans J.R.G., Edirisinghe M.J. (2002) Direct ink-jet printing of vertical walls. *J. Am. Ceram. Soc.* **85**, 2113–2115.

# 19 Polymer Technologies

## 19.1 INTRODUCTION

There are two main options for incorporating polymers into humidity sensors (Gardner and Bartlett 1995; Kumar and Sharma 1998; Harsanyi 1995, 2000):

1. Preprocessed polymer films are synthesized and shaped by extrusion, stretched into sheet forms, and covered by metal film separately from the sensor structures. They can then be attached to inorganic sensor surfaces (Loo et al. 2002). Illustration of this process is presented in Figure 19.1.
2. Polymerization occurs directly on the sensor surfaces. The synthesis and shaping process occur on the sensor surface.

No doubt that the second method is more progressive and is therefore the more commonly used technique for the fabrication of thin polymer layers used in the majority of humidity sensors.

## 19.2 METHODS OF POLYMER SYNTHESIS

There is no single method for synthesizing polymers aimed for humidity sensors. Most polymers, except ionomeric polymers, can be synthesized using standard methods of polymerization, both conventional and specific routes, including the polycondensation process and metal-catalyzed polymerization techniques, so-called chain-growth polymerization and step-growth polymerization (Flory 1953; Adhikari and Kar 2010). In chain-growth polymerization, an initiator reacts with a monomer molecule to create a reactive site, and the reactive site then reacts with successive monomer molecules to yield the polymer. A homopolymer that forms via chain-growth polymerization usually forms from one monomer; copolymers result from chain-growth polymerization of two or more monomers with the same type of reactive functional site (olefinic unsaturation). This type of polymerization is popular for the monomers that have double bonds, which can act as the reactive functional site during polymerization (Odian 2004). The formation of polyethylene from ethylene in the presence of an initiator is an example of chain-growth polymerization. Step-growth polymerization begins when one monomer with two reactive functional groups reacts with another monomer, containing functional groups of another type such that a small by-product molecule leaves the chain. Polymerization usually proceeds by reactions between two different reactive functional groups, for example, hydroxyl and carboxyl groups, isocyanate and hydroxyl groups, amine and acid groups, etc. So, according to the pair of functional groups in the monomers, a number

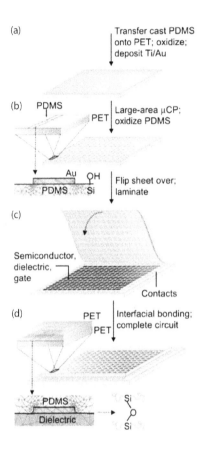

**FIGURE 19.1** Schematic illustration of steps for building embedded plastic circuits by lamination: (a) the fabrication begins by transfer casting a thin film of the elastomer polydimethylsiloxane (PDMS; 25–50 µm thick) onto indium tin oxide (ITOMylar; ~175 µm thick), oxidizing the exposed PDMS surface, and depositing uniform layers of Ti (~1 nm, adhesion promoter) and Au (15–20 nm); (b) contact printing (CP) on the gold-coated PDMS film defines the transistor sourcedrain electrodes and related interconnections. The Insets provide magnified views of a pair of sourcedrain electrodes and a side profile of an electrode. Plasma oxidation produces hydroxyl groups on the exposed surface of the PDMS; (c) aligning and laminating this sheet to a bottom plastic substrate (PET; ~175 µm thick) that supports the semiconductor, dielectric, and gate levels produces a complete circuit with top contact transistors. In this case, the contacts at the edges of the bottom sheet enable electrical connection to the embedded circuit; and (d) the Insets provide schematic views of a typical transistor and a side profile of a laminated electrode. Finite-element modeling qualitatively illustrates conformal contact, or wetting, of the PDMS layer against the dielectric. This wetting is a critical aspect of the lamination approach; it leads to extremely good electrical contacts between the two sheets. Chemical reaction of –Si–OH groups on the PDMS surface with those on the dielectric yield –Si–O–Si– bonds that provide a robust mechanical seal. (Reprinted with permission from Loo, Y.-L. et al., 2002. Soft, conformable electrical contacts for organic semiconductors: High-resolution plastic circuits by lamination. *Proc. Nat. Acad. Sci. U.S.A.*, 99, 10252–10256. Copyright 2006 National Academy of Sciences, U.S.A.)

of different chemical reactions may be used to synthesize polymeric materials by step polymerization, for example, esterification, amidation, the formation of urethanes, and aromatic substitution (Odian 2004).

One should note that indicated types of polymerization can be realized using different approaches. Polymers may be synthesized by any of the following techniques:

- Chemical or radical polymerization
- Electrochemical polymerization
- Photochemical polymerization
- Metathesis polymerization
- Concentrated emulsion polymerization
- Inclusion polymerization
- Solid-state polymerization
- Plasma polymerization
- Pyrolysis
- Soluble-precursor polymer preparation

These methods are described in detail in various comprehensive reviews (Malkin and Siling 1991; Skotheim 1986; Kumar and Sharma 1998; Gurunathan et al. 1999; Malinauskas 2001; Reisinger and Hillmyer 2002; Odian 2004; Wu et al. 2012).

Chemical polymerization (oxidative coupling) is followed by the oxidation of monomers to a cation radical and their coupling to form dications. The repetition of this process generates a polymer, and all classes of conjugated polymers can be synthesized using this technique. Chemical polymerization is conducted with relatively strong chemical oxidants. The reaction is controlled by the concentration and oxidizing power of the oxidant. Chemical polymerization occurs in the bulk of the solution, and the resulting polymers precipitate as insoluble solids. The polymer formed by chemical synthesis is generally a black powder. Among all of the above-listed techniques, chemical polymerization is the most used for preparing large amounts of conductive polymers, such as polypyrrole, polythiophene, polyfuran, polyisothionaptha-lene, polyindole, polyaniline, polycarbazole, poly p-phenylene vinylene (PPV), and polypyrene (Gurunathan et al. 1999). Synthesis of polymers by oxidative coupling polymerization is a very easy and simple method (Malinauskas 2001). The procedure involves simple mixing of monomer and oxidant in aqueous or organic protonic acid solution. Commonly used oxidants are ammonium peroxydisulfate (APS), ferric chloride, hydrogen peroxide, permanganate or bichromate anions, cerium sulfate, etc. A comprehensive picture of the fundamentals of chemical polymerization may be obtained from some review reports (Waltman and Bargon 1986; Okamoto and Nakano 1994; Toshima and Hara 1995; Feast et al. 1996; Smith 1998; Corrigan et al. 2016; Ren et al. 2016; Pan et al. 2018).

Another chemical method used in the preparation of polymers is chemical vapor deposition. This method uses reagents in the gaseous phase, with the polymer being formed on a substrate present in the reagent vapor. An alternative is to have only the monomer in gaseous form, with the oxidant being present in liquid form on the surface of the substrate to be coated.

Electrochemical polymerization is usually carried out on the working electrode in a single compartment cell with a three-electrode configuration, including a working electrode (generally Pt, but may vary according to the final requirements), a reference electrode (saturated calomel electrode), and a secondary electrode (Pt, Ni, or C). Generally, organic solvents are used in electrochemical polymerization, but aqueous solutions have been employed as well. In electrochemical polymerization, the electrochemically active groups are either built into the polymer structure or are added as a pendant group. Electrochemical polymerization can be carried out potentiometrically using a suitable power supply (potentio-galvanostat). Generally, potentiostatic conditions are recommended to obtain thin films, while galvanostatic conditions are recommended to obtain thick films. The advantage of this method is that precise flow control and the rate of film deposition can be maintained by varying the potential/current conditions of the working electrode in the system. The voltage potential is the most important parameter in controlling the polymerization process. For electrochemical oxidation, a certain electropolymerization potential (EP) must be applied to the solution for the monomer to be oxidized. Table 19.1 gives the peak oxidation potentials for some of the aromatic compounds that can produce conducting polymers using the electrochemical technique (Miller 1982). Table 19.1 shows that the electrochemically polymerizable monomers reported to date have relatively lower anodic-oxidation potential peaks, that is, oxidation potentials of less than 2.1 V. No polymerization occurs below the EP. Above the EP, the rate of polymerization increases with the potential, which may be affected by a number of parameters, including monomer concentration, electrolyte concentration, and

**TABLE 19.1**

**Electrochemical Data for Some Heterocyclic and Aromatic Monomers Used for Electrochemical Polymerization**

| Monomer | Oxidation Potential (V) |
| --- | --- |
| Pyrrole | 0.6–0.8 |
| Bipyrrole | 0.55 |
| Terpyrrole | 0.26 |
| Thiophene | 2.07 |
| Biothiophene | 1.31 |
| Terthiophene | 1.05 |
| Azulene | 0.91 |
| Pyrene | 1.30 |
| Carbazole | 1.82 |
| Fluorene | 1.62 |
| Fluorabthene | 1.83 |
| Aniline | 0.71 |

the nature of the electrode. For example, the EP for pyrrole is generally between 0.6 and 0.8 V. During pyrrole polymerization, lower current densities lead to the formation of a more crystalline polymer with fewer cross-linkages, while higher current densities produce rougher films. Using this technique, a variety of conductive polymers has been generated, such as polypyrrole, polythiophene, polyaniline, polyphenylene oxide pyrrole, and polyaniline/polymeric acid composite (Sato et al. 1986; Martin et al. 1993). The degree of polymer doping during electrochemical polymerization depends on the dopant concentration, the amount of charge passed, and the voltage applied. For example, the oxidation state of the polymer can be varied electrochemically by cycling the potential between oxidized, conducting, and the neutral, insulating state, or by using suitable redox compounds. Varying the composition of the polymerization medium also leads to a change in the conductivity of the polymer (Miasik et al. 1986; Pei and Inganas 1993). The electrochemical polymerization is a preferred technique for polymer-film electrodes, thin-layer sensors, in microtechnology, etc., because the control of potential is a prerequisite for polymer-film deposition on the anode during synthesis.

Photochemical polymerization takes place in the presence of UV irradiation (Chatani et al. 2014). This technique utilizes photons to initiate a polymerization reaction in the presence of photosensitizers. Pyrrole has recently been photopolymerized using a ruthenium(II) complex as the photosensitizer. Under photoirradiation, Ru(II) is oxidized to Ru(III), and the polymerization is initiated by a one-electron-transfer oxidation process. Polypyrrole (Ppy) films can be obtained through the photosensitized polymerization of pyrrole using a copper complex as the photosensitizer. Photopolymerization of benzo(C)thiophene has been carried out in acetonitrile using $CCl_4$ and tetrabutylammonium bromide. Photosensitive polymers have a unique advantage in processing, as polymerization and shaping can occur simultaneously with UV illumination. Photosensitive polymers can be applied and patterned with the same technology used by photoresists.

Plasma polymerization is a technique that is used in the preparation of ultrathin uniform layers (5–10 nm) that adhere strongly to an appropriate substrate. An electric glow discharge is used to create low-temperature "cold" plasma (Saravanan et al. 2004). A schematic diagram of the apparatus used for plasma polymerization is provided in Figure 19.2. The device used for plasma polymerization is commonly constructed with an anode, a mesh cathode, and a monomer supply ring. The stage is cooled by water because the probability that the monomer radicals will stick to the wafer decreases at high temperature, and the plasma polymerization rate decreases along with the temperature. The monomer liquid is cooled at 0°C to maintain the vapor pressure. Argon gas is supplied to the chamber through the holes in the anode and is ionized between the anode and the mesh cathode. The advantage of this technique is that it eliminates a number of steps needed in the conventional coating process (Favia and De

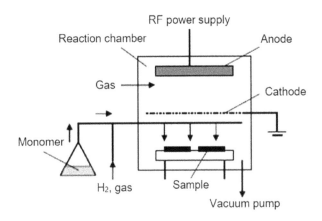

**FIGURE 19.2** Typical configuration of device for polymer film deposition by reactive sputtering. (With kind permission from Springer Science+Business Media: *Handbook of Gas Sensor Materials*, Springer, New York, 2013, Korotcenkov, G.)

Agostino 1998). Plasma polymerization or electropolymerization of monomers are preferred when the polymers are difficult to process because of insolubility or infusibility.

Metathesis polymerization is unique, differing from all other polymerizations in that all the double bonds in the monomer remain in the polymer. It is a natural outgrowth of Ziegler–Natta polymerization in that the catalysts used are similar, and often identical. This usually involves a transition-metal compound plus an organometallic alkylating agent. Metathesis polymerization is further divided into three classes: ring-opening metathesis of cyclo-olefins (ROMP); metathesis of alkynes, acyclic or cyclic; and metathesis of diolefins (Sutthasupa et al. 2010; Martinez et al. 2014). By far, the greatest amount of work has been done on ROMP.

Pyrolysis is probably one of the oldest approaches utilized to synthesize conductive polymers by heating the polymer to form extended aromatic structures, thus eliminating heteroatoms (Garcia-Nunez et al. 2017). The product of polymer hydrolysis can be a film, a powder, or a fiber, depending on the form and nature of the standing polymer and the pyrolysis condition.

One should take into account that polymer properties like morphology, electroactivity and conductivity are very much dependent on the polymerization conditions used, including the monomer concentration, solvent, electrolyte, temperature, and the potential/current applied.

## 19.3 PREPARATION OF POROUS POLYMERS

As for the porosity of polymeric materials, they are also controlled by the conditions of synthesis and the modes of subsequent treatments (Wu et al. 2012). The past several years have witnessed an expansion of various methodologies directed at preparing porous polymers, including direct templating, block copolymer self-assembly, and direct synthesis methodologies. Each of these methodologies has its own strengths and limitations, as summarized in Table 19.2. The direct templating

**TABLE 19.2**

**Summary of the Pore Generation, Major Advantages, and Limitations for the Main Methodologies of the Preparation of Porous Polymers**

| Methodology | Pore Generation | | Major Advantages | Major Limitations |
|---|---|---|---|---|
| Direct templating | Removal of templates | | The synthesis is relatively easy to carry out, since the template structure is fixed; precise control over meso-/macropore structure can be realized; and the introduction of smart functionality into the framework can be achieved | Sacrificial templates are needed, leading to the difficulty in scaling up production and the limited utility of materials; products lack micropores |
| Block copolymer self-assembly | Removal of sacrificial component; morphology reconstruction; vesiculation | | Porous polymers show tailored pore size, well-defined pore architectures and long-range order; their framework and pore surface can demonstrate sophisticated structures and smart or stimuli-responsive properties | The synthesis is often hard to scale up when using expensive and lab-synthesized amphiphilic block copolymers; there is a lack of micropores |
| Direct synthesis | Microporous polymers | Linkage of polymerizable monomer building blocks or sometimes hyper-cross-linkable polymers | Permanent porosity with high surface area can be achieved; the ore structure can be tailored by designing monomers with targeted structures | Strict requirements on the monomer structures and synthetic routes are needed; raw materials are expensive in many cases; fine design and control over meso-/macroporous structure is unavailable |
| | Meso-/macroporous polymers | Reaction-induced phase separation | Relatively precise control over meso-/macroporous network structure can be realized | There is a loss of micropores |
| | Hierarchical porous polymers | Combination of hypercross-linking with phase separation | Simultaneous construction of hierarchical micro-, meso- and macropores is achieved | Precise design and control over hierarchical organization needs to be improved |
| High internal phase-emulsion polymerization | Removal of internal phase | | Porosity is very high, typically 74 ~ 95 vol%; macroporosity with a unique void and window structure is 3D highly interconnected | Surface area is usually low, and mechanic strength is often weak |
| Interfacial polymerization | Inheritance of confined space from micelle | | It is especially suitable for synthesizing nanocapsules | Definite control over pore geometry is relatively difficult |
| Breath figures | Removal of condensed water droplet templates | | The preparation is simple; honeycomb patterned porous polymer films with ordered structure can be obtained | Preparation of ordered porous films with pore sizes below 100 nm is a challenge |

*Source:* Reprinted with permission from Wu, D. et al., *Chem. Rev.*, 112, 3959–4015, 2012. Copyright 2012 American Chemical Society.

methodology is a simple and versatile approach for preparation of porous polymers (Yang and Zhao 2005; Thomas et al. 2008). It involves a casting or molding process, which essentially follows the same design concept using a predesigned mold (i.e., template) to prepare plastic bottles but is scaled down to the nanometer range. The block copolymer self-assembly methodology is very useful for making mesoporous or macroporous polymers, especially materials with long-range order due to the microphase separation of incompatible blocks yielding mesoscale structures (Hillmyer 2005; Olson et al. 2008). The direct synthesis methodology can directly generate pores during solution polymerization, followed by removal of the solvent from the pores (Al-Muhtaseb and Ritter 2003; Mastalerz 2008; Jiang and Cooper 2010). Microporous polymers with extremely high surface area and permanent porosity that persists even in the dry state can be prepared by direct-synthesis procedures, for example, polymerization of rigid and contorted monomer building blocks that inhibit space-efficient packing (Cooper 2009; McKeown and Budd 2010). Mesoporous

and/or macroporous (meso-/macroporous) polymers can be fabricated by reaction-induced phase-separation procedures (Al-Muhtaseb and Ritter 2003; Kanamori et al. 2006). In addition to these three types of frequently used methodologies, high-internal-phase emulsion polymerization (Kimmins and Cameron 2011), interfacial polymerization (Sun and Deng 2005; Li et al. 2010), breath figures (Bunz 2006), and some relatively infrequently utilized but important methods have also been developed. A detailed description of these methods is given in a review prepared by Wu et al. (2012). However, one should note that the development of new methods and approaches should be continued, despite the significant progress in preparation of porous polymers.

## 19.4 FABRICATION OF POLYMER FILMS

The deposition of polymer as a sensing layer on the surface of sensor platform is a very tricky process, since the sensor sensitivity depends on the thickness, chemical composition, crystallinity, conductivity of the coated polymer, etc. At present, thin and thick polymer films can be deposited using a variety of techniques that vary in complexity and applicability. The present state of the art of polymer deposition is based on extensive research and development, and numerous research articles are available in various sensor-related journals and books (Chrisey and Hubler 1994; Gardner and Bartlett 1995; Matsumoto et al. 1998; Pique et al. 2003; Adhikari and Kar 2010).

### 19.4.1 Solution-Based Methods

The choice of deposition technique depends on the physicochemical properties of the material, the film-quality requirements, and the substrate being coated. The simplest methods involve the application of a liquid-solution polymer in a volatile solvent (Harsanyi 1995). These techniques were considered in Chapter 18. Methods of polymer film deposition, usually used for humidity sensors fabrication are presented in Table 19.3.

As it is seen, methods of thick-film technology such as spin coating, dip coating, and drop coating discussed in Chapter 18 are the most common methods of forming a polymer film in the manufacture of humidity sensors. In *spin coating* or *spin casting*, a dilute polymer solution droplet is placed on the surface of a rotating electrode. Because of the rotation, excess solution is spun off the surface and a very thin polymer film is formed. Following the same procedure, multiple layers can be formed until the desired thickness is obtained. This procedure provides pinhole-free thin films. *Drop coating or solvent evaporation uses another approach.* Applying a drop of polymer solution of the required consistency onto the sensor surface, followed by solvent evaporation, creates a film. The quality and thickness of a polymer film formed by this manual technique depends on the personal skill of the research worker, but the method is advantageous, because the thickness of the humidity sensitive polymer film can be known from the original concentration of polymer solution and droplet volume. *Dip coating* consists of immersing the electrode material in a solution of the polymer in a suitable solvent for a sufficient period to allow spontaneous film formation onto the substrate by adsorption. The film thickness may be controlled by adjusting the polymer solution viscosity and the speed of withdrawing the electrode from the solution, followed by solvent evaporation to form the polymer film on the sensor surface. *Screen printing* is a widely used technology in processing polymer composite materials available in the paste form. This method is capable of printing all the polymer components but has limited feature size resolution (75 μm or larger).

Polymer films can also be deposited using *ink-jet printing*, discussed before in Chapter 18. Resolution on the order of 25 μm can be achieved without surface modifications, while hydrophobic dewetting patterns can be used to obtain resolutions approaching 200 nm (Wang et al. 2004a). The efficiency of solution deposition methods such as the dipping and spin coating depend on the solubility of the conductive polymers, previously synthesized by the chemical polymerization technique. In this case, the surface to be coated is enriched either with a monomer or an oxidizing agent, and is then treated with a solution of either oxidizer or monomer, respectively. A major advantage of this process is that the polymerization occurs almost exclusively at the surface; no bulk polymerization takes place in the solution. For some polymer materials, the surface can be enriched with a monomer by its sorption from the solution. Enrichment of the surface by an oxidizer

---

### TABLE 19.3
### Methods of Polymer Films Forming Used for Humidity-Sensor Fabrication

| Method of Thin Film Forming | Polymer | Thickness (μm) | References |
|---|---|---|---|
| Spin coating | BCP of PS-b-P4VP; BEHP-co-MEH:PPV; PDA; PEDOT:PSS; PI; PVA; VTTBNc | 0.14–15 | Boltshauser et al. (1991); Qui et al. (2001); Sajid et al. (2017); Ahmad et al. (2018); Aziz et al. (2018); Li et al. (2018); Wang et al. (2018); Zafar et al. (2018) |
| Drop coating | Pan; PEDOT:PSS; PETMP-DVB; PMPS | 0.3–65 | McGovern et al. (2005b); Kus and Okur (2009); Xiao et al. (2018) |
| Dip coating (Immersing) | AEPAB/ST/HEMA/BPA; PDVB; PEO; SPBI; PODS; 4-VP-co-BuMA | N/A | Sadaoka et al. (1986); Li et al. (2005); Gong et al. (2010); Fei et al. (2014); Park and Gong (2017); Wang et al. (2018) |
| Electro-spraying | PVA | ~1 | Ahmad et al. (2018) |
| Ink-jet printing | PI | N/A | Lazarus et al. (2010) |
| Layer-by-layer | PSS | 0.05–0.5 | Zhang et al. (2017) |

*Abbreviations*: AEPAB—[2-(Acryloyloxy)ethyl] trimethyl ammoniumchloride; BCP—block copolymers; BEHP-co-MEH:PPV—Poly{[2-[2′,5′-bis(2″-ethylhexyloxy)phenyl]-1,4- phenylenevinylene]-co-[2-methoxy-5-(2′-ethylhexyloxy)-1,4- phenylenevinylene]}; BPA—4-Acryloxybenzophenone; DVD—divinylbenzene; HEMA—2-hydroxyethylmethacrylate; PAn—polyaniline; PDA—polydopamine; PDVB—poly(divinylbenzene); PEDOT:PSS—poly(3,4-ethylenedioxythiophene)/poly(4-styrene sulfonate); PEO—Poly(ethylene oxide); PETMP—Pentaerythritol tetrakis(3-mercaptopropionate); PI—polyimide; PMPS—polysquaraine (poly-(1-methylpyrrol-2-ylsquaraine); PODS—polymerized *n*-octadecylsiloxane; PS-b-P4VP—poly(styrene)-b-poly(4-vinylpyridine); PSS—poly(sodium styrene sulfonate); PVA—Poly-vinyl Alcohol; SPBI—sulfonated polybenzimidazoles; ST—Styrene; VTTBNc—Vanadyl 2,11,20,29-Tetra Tert-Butyl 2,3 naphthalocyanine; 4-VP-co-BuMA—poly(4-vinylpyridine-co-butyl methacrylate).

can be achieved either by using an ion-exchange mechanism or by deposition of an insoluble layer of oxidizer.

A part of polymers formed by chemical polymerization can deposit spontaneously on the surface of various materials immersed in the polymerization solution. The distribution of the resulting polymers between the precipitated and deposited forms depends on many variables and varies within a broad range. To coat materials with a polymer layer, it is desirable to shift this distribution toward the surface-deposited form, whereas bulk polymerization should be diminished as much as possible. This can usually be achieved by choosing appropriate reaction conditions, such as the concentration of the solution components, the concentration ratio of oxidant to monomer, the reaction temperature, and an appropriate treatment of the surface of the material to be coated by conducting polymers. Although a bulk polymerization cannot be completely suppressed, a reasonably high yield of surface-deposited polymers can be achieved by adjusting the reaction conditions (Malinauskas 2001).

Electrochemical polymerization discussed above can also be used for preparing polymer films. Using this method, polymers can be easily deposited on various substrates by simple techniques. A monomer solution is oxidized or reduced to an activated form that leads to a polymer film formed directly on the sensor surface. This procedure results in few pinholes, since polymerization would be accentuated at exposed (pinhole) sites at the electrode surface. Unless the polymer film itself is redox-active, electrode passivation occurs and further film growth is prevented. In the case of electrode preparation by direct electrochemical deposition of the conducting polymer, knowledge of the kinetics of the electrodeposition process is also of utmost importance in order to obtain proper sensing function of the electrode. Generally, a mixed material is deposited on the surface, containing electrochemically active and conducting as well as inactive and insulating parts, if the polymerization conditions are not carefully optimized (Otero and Rodriguez 1994). Along with the above-mentioned influencing parameters on polymer growth during electrodeposition, the effect of the electrode material and its surface properties need attention.

Of course, humidity sensitive layer can be formed using the Langmuir–Blodgett (LB) technique. This technique is based on the transfer of an insoluble polymer monolayer on a substrate with a hydrophilic surface, as it is raised from the liquid covered by this polymer monolayer (Gaines 1966; Osada and DeRossi 2000). It is also possible to dip a substrate with a hydrophobic surface in water covered by a polymer monolayer and then slowly draw it back out. The possibility of depositing ordered films with known and controlled thickness (in the range ±2.5 nm) is the main advantage multimolecular layers and architectures with high perfection, different layer symmetries, and molecular orientations. This method, however, has very low technological effectiveness, making its application in real chemical-sensor-fabrication processes unlikely. The number of polymers that can be used for preparing polymer films with the LB method is also very limited.

The main disadvantage of solution-based process is that it is limited by materials that can be covered or enriched with a layer of either monomer or oxidizer in a separate stage, preceding the surface polymerization. In addition, in many cases the post-processing treatments are required to improve molecular ordering and grain sizes of the thin polymer films. It is also often difficult to purify the polymers and achieve good molecular ordering over large-area substrates. Another major concern for solution processing methods is the effect of the solvent on underlying organic features, is requiring chemically compatible materials. For this reason, dry processing methods are being developed (Blanchet et al. 2003).

## 19.4.2 Dry Methods

The most well-known dry deposition methods used for polymeric films preparation are vacuum-deposition technologies. These polymer-deposition techniques involve *in-situ* polymerization on a substrate surface affected by various factors. The following are the examples of vacuum-deposition processes (Skotheim 1986; Harsanyi 2000):

- Vacuum pyrolysis consisting of sublimation, a pyrolysis, and a deposition-polymerization process;
- Vacuum polymerization stimulated by electron bombardment;
- Vacuum polymerization initiated by UV irradiation;
- Vacuum evaporation using either a resistance-heated solid polymer source or, more effectively, an electron beam;
- Radio-frequency sputtering of a polymer target in a plasma composed of polymer fragments with argon added to the plasma; and
- Plasma or glow-discharge polymerization of monomer gases or vapors.

Vacuum deposition is usually used for small molecules and oligomers. The layers can be deposited on any substrates that cannot be damaged by the vacuum processes. It is somewhat costly because of expensive equipment and low deposition throughput, but this method can produce thin polymer films that have high density, good electrophysical properties, thermal stability, and insolubility in organic solvents, acids, and alkalis. Examples of organic semiconductor films that have been deposited in this manner are oligothiophene and oligofluorene derivatives, metallophthalocyanines, and acenes (pentacene and tetracene). Polymer properties can be improved by controlling the deposition rate and temperature, which affect the morphology of the films. Modification of the interface between the substrate and the organic layer and post-deposition treatments, such as annealing, also improve molecular ordering.

As it follows from our discussions, most of the currently available techniques, however, are not advanced enough to be capable of depositing in a continuous process of thin, uniform, and solvent-free polymer thin films without a simultaneous influence on their chemical integrity and physicochemical properties. Furthermore, most of these techniques are not appropriate for the fabrication of multilayers, since they rely on the application of a solvent solution containing the material of interest, which may dissolve any previously deposited layers. Therefore, intensive research has been carried out over the last few years, attempting to perfect the methods for the deposition of polymer films. The most successful approach so far has been a modification of the pulsed-laser deposition (PLD) method. Previous work with UV PLD showed the ability of this technique to deposit thin films of various types of polymer materials. However, it also has been established that the PLD of polymers in a standard variant is limited to a small class of materials.

The recently developed matrix-assisted pulsed-laser evaporation (MAPLE) method considerably extends the possibilities of PLD for the deposition of polymeric materials. MAPLE differs from PLD in the way the target is prepared and in the laser-energy regime under which the laser–material interactions at the target take place (Pique et al. 2003). The MAPLE deposition process has been used successfully to deposit various types of polymer and organic materials, including chemoselective polymers. MAPLE offers significant advantages for the fabrication of chemical sensors, since it allows deposition of solvent-free chemoselective polymers on a variety of substrate surfaces. For example, using MAPLE, highly uniform films of siloxane fluoroalcohol (SXFA) have been deposited on the surface of surface acoustic wave (SAW) resonators. The performance of these MAPLE-coated sensors was comparable to that of spray-coated SAW devices.

It is necessary to note that in the chemical vapour deposition (CVD) method with laser activation, the method of cooling the substrate during polymer deposition is also effective. This cooling method creates conditions that prevent degradation of the polymer and can significantly increase the deposition rate. For example, by irradiating a cooled substrate with an excimer laser in an organic gas environment, polymethylmethacrylate (PMMA) films have been selectively deposited with a high deposition rate (Takashima et al. 1994).

Another dry printing method for depositing conducting polymers is based on thermal imaging technique (Blanchet et al. 2003). The printing process is illustrated in Figure 19.3. This method, developed for the ablative transfer of a patterned layer onto a flexible receiver layer, has resolution down to 5 μm. It is important that this technology can be used to process successive layers without the use of a solvent that could degrade the underlying organic layers. It needs to be highlighted that the appearance of effective dry methods for polymer deposition is an important achievement, since standard wet processes do not contribute to the integration of devices in modern semiconductor processing (Harsanyi 2000). However, it must be recognized that dry deposition methods are almost never used in the manufacture of humidity sensors.

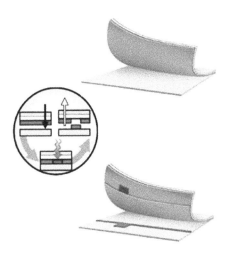

**FIGURE 19.3** Illustration of the printing process. The two flexible films, a multilayer donor and a receiver are held together by vacuum. The laser beam is focused onto a thin absorbing layer that converts light into heat, an optional ejection layer placed directly underneath, and a DNNSA–PANI/SWNT-conducting layer coated on top. The heat generated at the metal interface decomposes the surrounding organics creating a gas bubble that when expanding propels the conducting layer onto the receiver. After imaging is completed, the donor and receiver films are separated. (Reprinted with permission from Blanchet, G.B. et al., *Appl. Phys. Lett.*, 82, 463–465, 2003. Copyright 2003 by the American Institute of Physics.)

## 19.5 BULK AND STRUCTURE MODIFICATION OF POLYMERS

As it was shown in Volume 2, performances of humidity sensors depend on a large number of parameters that describe not only chemical and electrophysical properties, but also structural and mechanical properties. It is important to note that for each type of device the combination of these properties is specific. This means that methods are needed that allow these properties to be controlled over a wide range. The experiment showed that polymer technology has such capabilities.

As a rule, polymeric materials are not a single polymer, but they contain various chemicals. Some of these additives are present in minute amounts, while others are the main amounts of the total composition. It is these additives that make it possible to control the properties of the formed polymeric materials, including physical and/or chemical behavior (Carraher 2008). For example, appropriate substitutions in the monomer can improve the stability in the air of the electrochemically produced polymers. Although prepared under identical electrochemical conditions, the substitution of a methyl group, for example, poly(3-methyl thiophene), has shown better air stability than that of polythiophene (Tourillon and Garnies 1982; Bryce et al. 1987). In addition to its increased stability in the air, poly(3-methyl thiophene) has also shown improved conductivity over that of the parent polythiophene (Waltman et al. 1983). Unfortunately, such an increase in electrical conductivity has not been observed in a pyrrole system after alkyl-group substitution.

The methyl-substituted polypyrroles had lower conductivities than the parent polypyrrole (Diaz et al. 1982).

However, the additives discussed above are specific additives used in the synthesis of individual polymers. At the same time, there are additives that are used in the synthesis of almost all polymeric materials. These are solvents, cross-linkers, initiators, and plasticizers. Let's consider their role in the formation of polymeric materials in more detail.

### 19.5.1 SOLVENTS AND THEIR ROLE IN THE FORMATION OF PORES

As is known, gas penetrability and specific surface area are the most important parameters of polymers aimed for application in humidity sensors. This means that polymers synthesized should be porous. Experiments have shown that these parameters of polymers are controlled mainly by the solvent (Pichon and Chapuis-Hugon 2008). The solvent serves to bring all the components in the polymerization—the template, the functional monomer(s), the cross-linker, and the initiator—into one phase, and therefore just the physical and chemical characteristics of the solvent determine both the accuracy of the assembly between the template and the monomer, and the creating the pores in polymers (Cormack and Elorza 2004). For this reason, the solvent is commonly referred to as the "porogen." When polymerization occurs, solvent molecules occupy space in the polymer network and create the pores required to allow the diffusion of the template out of the network, and its subsequent diffusion back into the polymer, during recognition. This means that the porosity of the polymers can be determined by the overall concentration of monomers and cross-linkers in the solution. A favorable solvent for molecular imprinting polymers (MIPs) will create well-developed pores within the network and increase the total pore volume. However, a large amount of solvent can ultimately lead to the formation of microspheres and nanospheres instead of a large and stable cross-linked network (Bergmann and Peppas 2008).

Aside from its dual role as a solvent and a pore-forming agent, the solvent in a covalent polymerization must also be judiciously chosen so that it simultaneously maximizes the likelihood of complex formation between the template and the functional monomers. More specifically, the use of a highly thermodynamic solvent tends to result in polymers with well-developed pore structures and high specific surface areas, while the use of a thermodynamically poor solvent leads to polymers with poorly developed pore structures and low specific surface areas. Historically, chloroform has been used as a highly thermodynamic solvent; other solvents that have been investigated include dimethyl sulfoxide, toluene, carbon tetrachloride, n-hexane, and benzene (Bergmann and Pappes 2008). Several examples of solvents typically used for preparing polymer films are listed in Table 19.4. It was established that chloroform is a really good solvent for the polymer, while other solvents, such as carbon tetrachloride and n-hexane, are pore solvents (Odian 2004). The inherent viscosity of the polymer increases as the reaction mixture contains a larger proportion of chloroform.

**TABLE 19.4**
**Solvents Typically Used for Preparing Polymer Films**

| Solvent | Polymer |
| --- | --- |
| Toluene/ethanol | Ethyl cellulose (EC); poly(vinyl chloride-co-isobutyl vinyl ether) (PVC-iBVE); poly(4-tert.-butyl styrene) (PTBS); polystyrene (PS); poly(vinyl methyl ketone) (PVMK) |
| Tetrahydrofurane | Poly(tetrafluor ethylene-covinylidenfluorid-co-propylene) (PFE-VFP); poly(4-vinyl phenol) (PVPh); poly(vinyl chloride) (PVC) |
| Chloroform | Poly(styrene-co-acrylonitrile) (PSAN); cellulose acetate (CAc); polysulfone (PSu); poly(bisphenol A carbonate) (PC); poly(methyl methacrylate) (PMMA) |
| Dimethylformamide | Poly(acrylonitrile) (PAN) |

A solvent must be chosen that will not interfere with the template–monomer complex. If the solvent does indeed create an interaction, it could inhibit the formation of the imprinted sites. As a result, many imprinting systems shun polar solvents and instead utilize nonpolar solvents in order to maximize attraction of the template by the functional monomers (Bergmann and Peppas 2008). In this context, water is an especially poor solvent choice for MIPs because of its highly polar nature. Water is both a hydrogen-bond donor and a hydrogen acceptor. Thus, many hydrogen bonds that are formed during covalent imprinting can be destroyed by the sheer amount of water molecules present. In addition, many cross-linking agents that are soluble in water have little structural integrity, thus limiting these materials in extraction applications, such as high-pressure liquid chromatography. However, if hydrophobic forces are being used to drive the complexation, then water could well be the solvent of choice. In this case, polar solvents are preferred because their use helps to stabilize hydrogen bonds. Supercritical $CO_2$ has been proposed as an alternative porogen for MIP production (Alexander et al. 2006).

Finally, it is necessary to know that the better the reaction medium as a solvent for the polymer, the longer the polymer stays in solution and the larger the polymer molecular weight. With a solvent medium that is a poor solvent for the polymer, the molecular weight is limited by precipitation. In addition to the effect of a solvent on the course of a polymerization, the solvent is a poor or good solvent for the polymer, and solvents affect polymerization rates and molecular weights due to preferential salvation or other specific interactions with either the reactants or transition state of the reaction or both.

### 19.5.2 CROSS-LINKERS

In polymers, the cross-linker fulfills three major functions (Cormack and Elorza 2004). First, the cross-linker is important in controlling the morphology of the polymer matrix, whether it is a gel type, macroporous, or a microgel powder. In particular, from a polymerization point of view, high cross-link ratios are generally preferred to access permanently porous (macroporous) materials. Second, it serves to stabilize the binding sites in imprinted polymers. In other words, cross-links can act to lock in "memory" preventing free-chain movement. As it is known, most successful MIP networks involve hydrogel components, such as polyethylene glycol (PEG) or hydroxyethylmethacrylate (HEMA), that absorb large amounts of water, causing them to swell exponentially in volume. In an MIP system, this can lead to further swelling in the template cavities (Pichon and Chapuis-Hugon 2008). Finally, it imparts adequate mechanical stability to the polymer matrix. Cross-linking can also be effected through application of heat, mechanically, through exposure to ionizing radiation and nonionizing (such as microwave) radiation, through exposure to active chemical agents, or through any combination of these. Chemical cross-linking generally renders the material insoluble.

A number of known cross-linkers are commercially available and a few of which are capable of simultaneously complexing with a template and thus acting as functional monomers (see Figure 19.4). Ethylene glycol dimethacrylate (EDMA) is the most commonly used cross-linker for methacrylate-based systems, primarily because it provides materials with mechanical and thermal stability, good wettability in most rebinding media, and rapid mass transfer with strong recognition properties (Alexander et al. 2006; Sellergren et al. 2009). With the exception of trimethacrylate monomers, such as trimethylolpropane trimethacrylate (Kempe and Mosbach 1995), no other cross-linking monomers provide similar recognition properties for such a large variety of target templates. Divinylbenzene (DVB) has proven to be a superior matrix monomer for some templates, but it is most commonly used in combination with other polymerization formats, such as emulsion, precipitation, or suspension. Polar protic cross-linking monomers, such as methylenediacrylamide (Hart and Shea 2001) and pentaerythritoltrimethacrylate (Manesiotis et al. 2005), have been useful for imprinting and applications in more polar solvents.

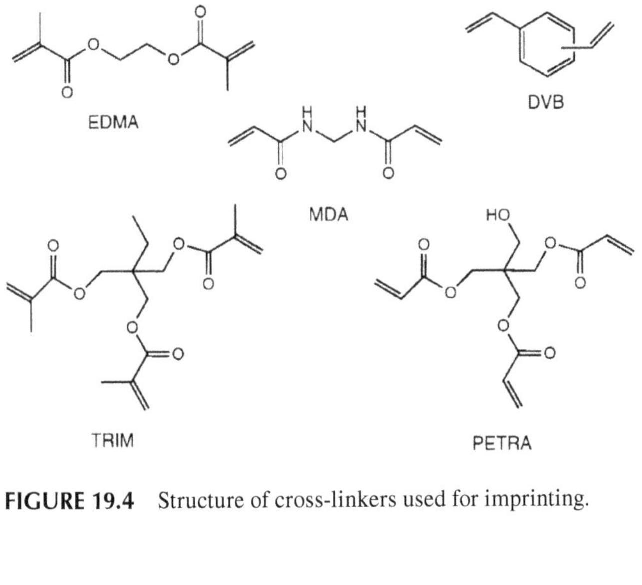

**FIGURE 19.4** Structure of cross-linkers used for imprinting.

However, we need to note that the cross-linking agent and its concentration must be very carefully chosen. For example, according to Pichon and Chapuis-Hugon (2008), the choice of cross-linker is important to the imprinting binding sites obtained, since the binding capacity of the polymer increases in relation to the degree of cross-linking. The nature of the interactions developed by the cross-linker must be taken into consideration as well. In addition, the mole ratio of functional monomers to the cross-linker must also be taken into consideration so that the functional cavities are sufficiently spaced to allow the individual binding pockets to swell (Pichon and Chapuis-Hugon 2008). If there is too little cross-linking, the template cavities will be too close to each other, thus creating a larger, less recognitive pore. However, the amount of cross-linking must not be so high as to limit diffusion of the template into the network. Polymers with cross-link ratios in excess of 80% are often the norm. For the same reason that the reactivity ratios of functional monomers need to be matched in a cocktail polymerization to ensure smooth incorporation of the comonomers, and the reactivity ratio of the cross-linker should ideally be matched to that of the functional monomer(s).

We also need to take into account that in some humidity-sensor applications, where the swelling effect plays the main role, the using of cross-linkers can be limited. Cross-linking increases the strength of the cross-linked material but decreases its flexibility and increases its brittleness. Most chemical cross-linking is not easily reversible.

The application of cross-linker can give other effects as well. For example, Matsuguchi et al. (2003) analyzed the influence of cross-linker on the characteristics of quartz-crystal-microbalance-based $SO_2$ sensors and found that the sensors with cross-linked structure had lower sorption ability but faster sorption/desorption rates. As is known, the last one is very important for sensors designed for *in-situ* measurements. However, it must be borne in mind that as a rule the optimizing effect is observed only at certain concentration of cross-linker.

Chen et al. (2006) have shown that cross-linking agent (isophorone diisocyanate) added to the carbon black (CB)/waterborne polyurethane (WPU) composite latexes can both weaken the unwanted negative vapor-coefficient effect and improve reproducibility of CB/WPU composites acting as gas-sensing materials in the environment of organic solvent vapors. Moreover, the maximum magnitude of the response and response rate of the composites are significantly increased after cross-linking treatment as well. Chen et al. (2006) believe that the mechanism responsible for the sensing behavior of the composites remains unchanged even though the matrix polymer has been cross-linked, while the interaction between CB particles and the matrix polymer is enhanced due to the appearance of the cross-linking structure. As a result, the movement of the fillers in the swollen matrix is localized, which ensures the reversible breakdown and establishment of the conductive networks throughout the composites.

### 19.5.3 INITIATORS

Functional initiators find application in conventional radical polymerization for the synthesis of various polymers. The initiator functionality can be considered by reason of the presence of functional end groups, such as hydroxyl and carboxyl, or azo and perester bonds, which undergo dissociation to the alkyl, alkoxy, or acyloxy radicals under the influence of temperature or irradiation and initiate the polymerization (Moad and Solomon 1995). A *free radical* is simply a molecule with an unpaired electron. The tendency for this free radical to gain an additional electron in order to form a pair makes it highly reactive so that it breaks the bond on another molecule by stealing an electron. As a result, the formation of two additional molecules with an unpaired election (which are another free radicals) takes place. Free radicals are often created by the division of a molecule (known as an *initiator*) into two fragments along a single bond. The following diagram in Figure 19.5 shows the formation of a radical from its initiator, in this case benzoyl peroxide.

During the activation of the initiator consisting of a heterocyclic, arylsubstituted, or with aryl-ring-fused sulfonium salt, a carbon-sulfur bond is broken via a ring-opening reaction leading to the formation of a sulfide and a carbocation (carbenium ion) within the same molecule. The functionality of initiators is utilized in various ways to achieve a specific goal. In particular, thermal decomposition of azo initiators such as azo-bis-isobutyronitrile is the most commonly used source of free radicals in the formation of both DVB- and (meth) acrylate-based MIPs. The photochemical decomposition of this compound allows MIPs to be prepared at a low temperature and with a resulting increase in separation efficiency of the polymers (Alexander et al. 2006). Azo-dialkyl peroxides, azo-diacyl peroxides, azo-perestersazo-hydroperoxides, etc. are other examples of initiators used for radical polymerization (Pabin-Szafko et al. 2005). The first three groups of azoperoxy compounds can play a role of bifunctional initiators in generation of block copolymers. They can also play a role of traditional initiator in radical polymerization of just one type of monomers. Azo-diacyl peroxides and azo-peresters were tested as initiators in styrene and acrylamide polymerization processes and in preparation of block copolymers from vinyl and acrylic monomers (Czech et al. 2008).

It is worth noting that since the activation does not lead to fragmentation of the initiator-molecule into smaller molecules, no molecular sulfur-containing decomposition products form that otherwise would evaporate or migrate from the polymer-causing bad smell.

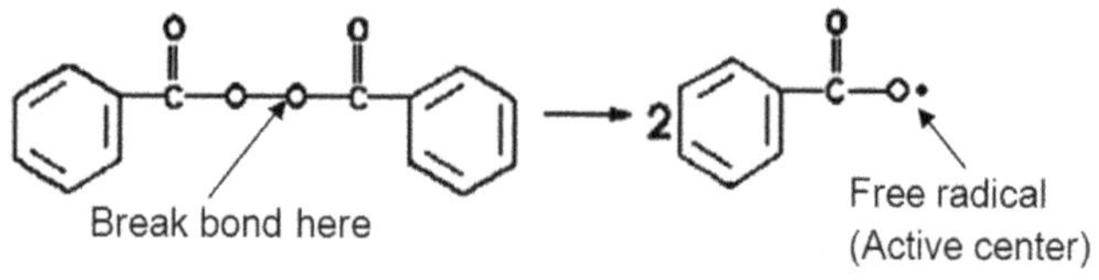

FIGURE 19.5 Diagram illustrating the forming of free radicals.

### 19.5.4 PLASTICIZERS

A plasticizer is a material incorporated into a plastic to increase its workability and flexibility. The addition of a plasticizer may lower the melt viscosity, elastic modulus, and glass transition temperature ($T_g$). Plasticization can occur through the addition of an external chemical agent or may be incorporated within the polymer itself (Carraher 2008). Internal plasticization can be produced through copolymerization, giving a more flexible polymer backbone, or by grafting another polymer onto a given polymer backbone. Thus, poly(vinylchloride-co-vinyl acetate) is internally plasticized because of the increased flexibility brought about by the change in the structure of the polymer chain. The presence of bulky groups in the polymer chain increases segmental motion and the placement of such groups through grafting acts as an internal plasticizer. Internal plasticization achieves its end goal at least in part by discouraging association between polymer chains. External plasticization is achieved by incorporating a plasticizing agent into a polymer through mixing and/or heating.

Plasticizers used should be relatively nonvolatile, nonmobile, inert, inexpensive, nontoxic, and compatible with the system to be plasticized. They can be divided based on their solvating power and compatibility (Carraher 2008). Primary plasticizers are used as either the sole plasticizer or the major plasticizer with the effect of being compatible with some solvating nature. Secondary plasticizers are materials that are usually blended with a primary plasticizer to improve certain performance, such as flame or mildew resistance, or to reduce the cost. The separation between primary and secondary plasticizers is sometimes arbitrary.

The three main chemical groups of plasticizers are phthalate esters, trimellitate esters, and adipate esters. Most plasticizers are classified as general-purpose, performance, or specialty plasticizers (Carraher 2008). General-purpose plasticizers are those that offer good performance inexpensively. Most plasticizers belong to this group. Performance plasticizers offer added performance over general-purpose plasticizers generally with added cost. Performance plasticizers include fast-solvating materials such as butyl benzyl phthalate and dihexyl phthalate; low-temperature plasticizers such as di-n-undecylphthalate and di-2-ethylhexyladipate; and so-called permanent plasticizers such as tri-2-ethyl hexyl trimellitate, triisononyl trimellitate, and diisodecyl phthalate. The plasticizers tributyl phosphate (TBP), tris(2-ethylhexyl)phosphate (TOP), 2-(octyloxy)benzonitrile (OBN), and 2-nitrophenyl octyl ether (NPOE) are also used to plasticize the polymers aimed for gas-sensor fabrication (Apostolidis 2004). Specialty plasticizers include materials that provide important properties such as reduced migration, improved stress–strain behavior, flame resistance, and increased stabilization. In all cases, performance is varied through the introduction of different alcohols into the final plasticizer product. There is a balance between compatibility and migration. Generally, the larger the ester group, the lesser the migration, up to a point where compatibility becomes a problem and a limiting factor.

Regarding plasticizers influence on performance of polymer-based gas sensors, we can say that this effect depends on many factors, including technological route used, the type of polymer, and target gas. For example, Apostolidis (2004) found that although the plasticizers did not coercively increase sensitivity of polystyrene-based optical sensors, they enhanced performance by reducing response time. As we previously mentioned, the permeability $P$ of a gas into a polymer is linked to the diffusion coefficient $D$ and the solubility $S$ of a gas in a particular polymer. Thus, increasing the plasticizer content in a polymer matrix usually leads to an increase of the diffusion coefficient of the gas into a particular polymer. As a result, the response time of the sensor material is affected, and the response is more rapid than that of a nonplasticized matrix. Positive effect of plasticizers influence was also observed for polyvinyl chloride (PVC)-based ammonia sensors. The sensitivity of PVC was increased dramatically upon plastification. The addition of plasticizers, for example, TBP or TOP, enabled an enhancement of oxygen sensitivity of the poly(4-tert-butylstyrene) (PTBS)-based materials as well. The same effect of sensitivity increase was observed in optical carbon dioxide sensors. However, in other combinations, we can have the opposite effect. The resulting matrix, for example, may offer lower permeability than the nonplasticized due to over compensation of this effect by a reduced solubility of the target gas in the softened matrix. Thus, the observed decrease in sensitivity can be attributed to a decrease of solubility of the target gas with increasing plasticizer content. In particular, it was found that OBN and NPOE at concentrations of 10 % (w/w) in the PTBS- and polystyrene (PS)-matrix caused a decrease in sensitivity of the oxygen sensors. In addition, Apostolidis (2004) found that the introduction of plasticizers into the polymer matrix increased the temperature dependence of the $CO_2$ sensor parameter.

## 19.6  APPROACHES TO POLYMER FUNCTIONALIZING

As a rule, most polymeric materials have a hydrophobic, chemically inert surface, untreated nonpolar polymer surfaces often have adverse problems in adhesion, coating, painting, coloring, lamination, packaging, colloid stabilization, etc. To solve these problems, an enormous number of basic and applied researches in various fields using different innovative techniques, including chemical and physical processes, have been devoted to the functionalizing of polymeric materials (Garbassi et al. 1994; Chan 1994; Hoffman 1996; Uyama et al. 1998; Kang and Zhang 2000; Kato et al. 2003).

### 19.6.1  Polymer Doping

Studies have shown that many different approaches can be used for polymer functioning (Skotheim 1986; Nalwa 1997). The doping is the most common process of conducting polymer modification. The concept of doping is the unique, central, underlying, and unifying theme that distinguishes conducting

polymers from all other types of polymers (Chiang et al. 1977; MacDiarmid 2002). During the doping process, an organic polymer, either an insulator or semiconductor having a small conductivity, typically in the range from $10^{-10}$ to $10^{-5}$ S/cm, can be converted to a polymer, which is in the "metallic" conducting regime (~$1$–$10^4$ S/cm). The controlled addition of known, usually small ($\leq 10\%$), nonstoichiometric quantities of chemical species results in dramatic changes in the electronic, electrical, magnetic, optical, and structural properties of the polymer. In the doped state, the backbone of a conducting polymer consists of a delocalized $\pi$-system. In the undoped state, the polymer may have a conjugated backbone such as in trans-$(CH)_x$, which is retained in a modified form after doping, or it may have a nonconjugated backbone, as in polyaniline (leucoemeraldine base form), which becomes truly conjugated only after $p$-doping, or a nonconjugated structure as in the emeraldine base form of polyaniline, which becomes conjugated only after protonic acid doping. Doping is reversible to produce the original polymer with little or no degradation of the polymer backbone. Both doping and undoping processes, involving dopant counterions, which stabilize the doped state, may be carried out chemically or electrochemically (Kanatzidis 1990). Transitory doping by methods which introduce no dopant ions are also known (Ziemelis et al. 1991).

In general, the doping includes the following processes.

#### 19.6.1.1  Redox Doping (Ion Doping)

All conductive polymers can undergo either $p$- and/or $n$- redox doping by chemical and/or electrochemical processes. Dopants used for selected polymers are listed in Table 19.5. The $p$-doping, that is, partial oxidation of the $\pi$-backbone of an organic polymer, was first discovered by treating trans-$(CH)_x$ with an oxidizing agent such as iodine. The $n$-doping, that

**TABLE 19.5**
**Conductivities of Conductive Polymers with Selected Dopants**

| Polymer | Doping Materials | Conductivity (s/cm) |
|---|---|---|
| Polyacetylene | $I_2$, $Br_2$, Li, Na, $AsF_5$ | $10^4$ |
| Polypyrrole | $BF_4^-$, $ClO_4^-$, tosylate | $500$–$7.5 \times 10^3$ |
| Polythiophene | $BF_4^-$, $ClO_4^-$, tosylate, $FeCl_4^-$ | $10^3$ |
| Poly(3-alkylthiophene) | $BF_4^-$, $ClO_4^-$, $FeCl_4^-$ | $10^3$–$10^4$ |
| Polyphenylenesulfide | $AsF_5$ | $500$ |
| Polyphenylene-vinylene | $AsF_5$ | $10^4$ |
| Polythienylene-vinylene | $AsF_5$ | $2.7 \times 10^3$ |
| Polyphenylene | $AsF_5$, Li, K | $10^3$ |
| Polyisothi-anaphthene | $BF_4^-$, $ClO_4^-$ | $50$ |
| Polyazulene | $BF_4^-$, $ClO_4^-$ | $1$ |
| Polyfuran | $BF_4^-$, $ClO_4^-$ | $100$ |
| Polyaniline | HCl | $200$ |

*Source:* Reprinted from *Eur. Polym. J.*, 34, Kumar, D. and Sharma, R.C., Advances in conductive polymers, 1053–1060, Copyright 1998, with permission from Elsevier.

is, partial reduction of the backbone π-system of an organic polymer, was also discovered using *trans*-(CH)ₓ by treating it with a reducing agent (Chiang et al. 1977). These doping processes change the number of electrons associated with the polymer backbone; hence, the conductivity of the polymer is also changed. For example, undoped polyacetylene is silvery, insoluble, and intractable, with a conductivity similar to that of semiconductors. However, when it was weakly oxidized by compounds such as iodine it turned a golden color, and its conductivity increased to about $10^4$ S/m. The same situation takes place with other conjugated polymers.

Doping of polymers for the manufacture of electronic devices, including conductometric sensors, has been an important area of research since the 1980s. Studies have shown that typical oxidizing dopants, which can be used for polymer modification, besides iodine, include chlorine, bromine, arsenic pentachloride, iron(III) chloride, and $NOPF_6$. Many experiments have been done using halogen dopants (Cl, Br, I) since the doping of the polymer with these elements can be easily achieved by a simple chemical process. A typical reductive dopant is liquid sodium amalgam or preferably sodium naphthalide. The main criteria for using the additives mentioned above is their ability to oxidize or reduce a polymer without reducing its stability or regardless of whether they are able to initiate side reactions that inhibit the ability of polymers to conduct electricity. An example of the latter is the doping of a conjugated polymer with bromine and chlorine. It has been shown that if the high electron affinity of Cl and Br allows to obtain a stable charge transfer complex (CT-complex) between these halogens and polymers, in return there is often a progressive attack of the polymer backbone by the halogen (Safoula et al. 1999, 2001). For example, bromine is too powerful oxidant and adds across the double bonds to form sp3 carbons. As a result, the chlorine and bromine doping induce partial degradation of the polymer. The same problem may also occur with $NOPF_6$ if left too long. Therefore, it appears that iodine doping, even if it is less stable, is more promising because it does not induce any polymer degradation. After iodine doping only, CT-complex formation takes place (Napo et al. 1999; Safoula et al. 2001).

### 19.6.1.2 Photo and Charge-Injection Doping

When intrinsically conducting polymers (ICPs) are exposed to radiation, whose energy is greater than the band-gap energy of the polymer, electrons hop across the gap, and the polymer undergoes photo-doping. However, the charge carriers disappear rapidly due to recombination of electrons and holes when irradiation is discontinued. Charge-injection doping is most conveniently carried out using a metal/insulator/semiconductor configuration, involving a metal and a conducting polymer, separated by a thin layer of a high-dielectric-strength insulator.

### 19.6.1.3 Nonredox Doping

In this process, the number of electrons associated with the polymer backbone does not change during the doping process but instead the energy levels of these electrons are rearranged during doping. An example of this occurs when an ICP such

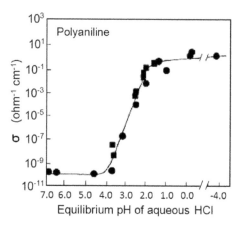

**FIGURE 19.6** Conductivity of emeraldine base as a function of pH of HCl dopant solution as it undergoes protonic acid doping. (Reprinted from *Synth. Met.*, 13, Chiang, J.C. and Macdiarmid, A.G., Polyaniline: Protonic acid doping of the emeraldine from to the metallic regime, 193–205, Copyright 1986, with permission from Elsevier.)

as the emeraldine base form of polyaniline becomes highly conductive by immersing the polymer in an acid. The emeraldine base interacts with aqueous protonic acids to produce the protonated emeraldine, which is highly conductive (Chiang and Macdiarmid 1986; Macdiarmid et al. 1987). This process can increase the conductivity of the polymer by several orders of magnitude. The example of such influence is shown in Figure 19.6.

### 19.6.2 POLYMER GRAFTING

The grafting is other approach used for polymers functioning (Uchida and Ikada 1996; Ranby 1999; Kang and Zhang 2000; Zhao and Brittain 2000). Among the surface-modification techniques developed to date, surface grafting has emerged as a simple, useful, and versatile approach to improve surface properties of polymers for many applications. According to Gopal et al. (2007), the grafting has the following advantages: (1) the ability to modify the polymer surface to have distinct properties through the choice of different monomers; (2) the controllable introduction of graft chains with a high density and exact localization to the surface, without affecting the bulk properties; and (3) long-term chemical stability, which is assured by covalent attachment of graft chains (Gopal et al. 2007). The latter factor contrasts with physically coated polymer chains that can in principle be removed rather easily. The experiment showed that surface grafting provides versatile techniques for introducing functional groups such as amine, imine, hydroxyl, carboxylic acid, sulfonate, and epoxide onto a broad range of conventional polymeric substrates, most of which have a nonpolar, less reactive surface (see Table 19.6). As is known, hydroxyl, amine, carboxyl, and sulfone groups are hydrophilic functional groups (Van der Bruggen 2009). Usually functional groups are localized on the side chains. A typical polymer consists of a backbone,

## TABLE 19.6

### Methods and Functional Groups Used for Functionalizing of Polymer Used in Gas Sensors

| Backbone | Side Chains |
| --- | --- |
| Polypyrrole (PPy) | Alkyl; alkoxy; hydroxyalkyl; carboxyalkyl; alkyl sulfonic acid; amine; ester group |
| Polyaniline (PANI) | Alkoxy; sulfonic acid; phenyl; boronate |
| Polythiophene (PTh) | Alkyl; alkoxy; ester group; alkthio; carboxyl alkyl |
| Poly(3,4ethylenedioxy thiophene); (PEDOT) | Alkoxy; ether group |
| Poly(acryl acid) (PA) | Amine |
| Poly(diethylyl benzene) (PEB) | Alkoxy |

*Source:* Bai, H. and Shi, G. *Sensors*, 7, 267–307, 2007.

typically made up of a main chain of long strands of monomer units—from ten to millions—and side chains. A side chain is simply a relatively short branch of the polymer molecule, usually several atoms or groups of atoms, that are connected to the polymer backbone. There may be a few or many of them. Sometimes even the branches (side chains) have branches (side chains). The presence of these side chains can affect the physical properties of a polymer. For example, high-density polyethylene, with its near-absence of side chains, is harder, more abrasion resistant, and will withstand higher temperatures, compared to low-density polyethylene that has numerous molecular branches or side chains. The functional groups introduced with help of side chains can be utilized to further reaction with small or large molecules through covalent or noncovalent linkage. Functionalization is achieved by either direct grafting of functional monomer or postderivatization of graft chains.

The introduction of grafts to the backbones of conducting polymers has two effects (Bai and Shi 2007). First, most of the side chains are able to increase the solubility of conducting polymers. This makes them be processed into the sensing film by LB technology, spin coating, ink printing or other solution-assistant method. Second, some functional chains can adjust the physical and chemical properties of conducting polymers, such as space between molecules (Li et al. 2006) or dipole moments (Torsi et al. 2003), or bring additional interactions with analytes, which may enhance the response, shorten the response time, or produce new sensitivity to other gases (Ruangchuay et al. 2004).

The techniques to initiate grafting are: (1) chemical, (2) photochemical and/or via high-energy radiation, (3) the use of a plasma, and (4) enzymatic (Nady et al. 2001; Kato et al. 2003; Van der Bruggen 2009). The choice for a specific grafting technique depends on the chemical structure of the polymer and the desired characteristics after surface modification. For example, modification of polymer surfaces can be rapidly and cleanly achieved by plasma treatment due to the possibility of the formation of various active species on the surface of polyethylene, polypropylene, polytetrafluoroethylene, etc. By

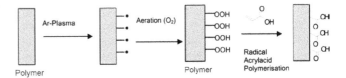

**FIGURE 19.7** Schematical representation of polyacrylic acid (PAAc) graft modification by Ar plasma treatment. (Reprinted from *Desalination*, 189, Tyszler, D. et al., Reduced fouling tendencies of ultrafiltration membranes in wastewater treatment by plasma modification, 119–129, Copyright 2006, with permission from Elsevier.)

variation of plasma-treatment parameters, the surfaces with different properties can be obtained. Possible gases include $CF_4$, Ar, $O_2$, $H_2$, He, Ne, $N_2$, and $CO_2$, in addition to $H_2O$. The surface is bombarded with ionized-plasma components to generate radical sites. Bonds that can be attacked by radicals are C–C, C–H, and C–S bonds, with the exclusion of the aromatic C–H and C–C bonds. This is similar to photodegradation. The generated radicals can subsequently react with gas molecules (depending on the plasma), schematically shown for Ar in Figure 19.7. Remaining radical sites bind with oxygen or nitrogen after contact with the air.

### 19.6.3 SURFACE MODIFICATION VIA PLASMA TREATMENT

The experiment showed that the surface modification is also possible using plasma-treatment techniques. It was found that the plasma treatment activates the polymer surface to facilitate the deposition of other polymers, having active functionality, or the polymer can be used directly as the detecting layer for a chemical-sensing device. For example, Van der Bruggen (2009), analyzing chemical modification of polymer membranes, have noted that $CO_2$-plasma treatment leads to the incorporation of oxygen in the membrane surface in the form of carbonyl, acid, and ester groups, yielding an increase in hydrophilicity. Modification of polymer surfaces with $CO_2$ plasmas in general leads to a surface oxidation and the formation of hydrophilic surfaces. A fast reaction was observed; the treated membranes had a better fouling resistance. Continued plasma treatment, however, resulted in membrane degradation. $H_2O$ plasma treatment also leads to the incorporation of oxygen, containing functional groups on the surface. $O_2$ plasmas have a similar effect, with reported functional groups mainly being hydroxyl, carbonyl, and carboxyl groups (Supriya and Claus 2004). Nitrogen-containing plasma systems, on the other hand, yield amine, imine, amide, and nitrile functional groups on the membrane surface. Through post reactions after contact with air, an oxygen compounds can also be present.

However, plasma treatment is generally slow and expensive in fiber applications. Plasma treatment also results in deposition of a macromolecular structure, graft polymerization, etching, roughening, and cross-linking. In addition to surface modifications, plasma treatment under certain conditions can contribute to an increase in the crystallinity of the film and even to a new crystallization with a concentration gradient depending on the film thickness. The effects of plasma treatment on the surface of polymeric materials were discussed in detail comparing the energies of plasma with those of the chemical bonds in polymers (Nakamatsu and Delgado-Aparicio 1997). Chemical modification gives possibility to introduce various functional groups on the polymer surface using various reactions, such as sulfonation, chloromethylation, aminomethylation, and lithiation (Guiver and Apsimon 1988; Michael et al. 1989; Breitbach et al. 1991). The main challenge for modification by chemical treatment of commercial polymer membranes is that the modification agent may partly block the pores of the membranes. Even if the modified membranes are less prone to fouling, the total flux after modification is generally smaller than before modification (Anke Nabe et al. 1997).

It is worth noting that the possibility of keeping bulk properties of polymers without changes during their modification is a good feature of the grafting. The performance of polymeric materials in many applications, especially in gas sensors, relies largely upon the combination of bulk (e.g., mechanical) properties in combination with the properties of their surfaces. However, polymers very often do not possess the surface properties needed for these applications. Vice versa, those polymers that have good surface properties frequently do not possess the mechanical properties that are critical for their successful application. Due to this dilemma, (surface) modification of polymers without changing the bulk properties has been a topical aim in research for many years, mostly, because surface modification provides a potentially easier route than, for example, polymer blending to obtain new polymer properties, important for humidity-sensor design (Kato et al. 2003; Gopal et al. 2007).

## 19.7 ADHESION OF POLYMER FILMS AND HOW TO IMPROVE THIS PARAMETER

Adhesive bonds are dependent on the chemistry of the interface, and thus the control of surface chemistry is critical to bond quality (Kinloch 1987; Hart-Smith 1983; Awaja et al. 2009). High-performance polymers very often do not possess the desired surface properties for strong adhesive bonding. They are hydrophobic in nature, and in general exhibit insufficient adhesive bond strength due to relatively low surface-free energy (SFE) (Hart-Smith et al. 1998). It has been established that for good adhesion, polymer and substrate need to have special surface properties such as polarity and high surface-free energy (SFE) (Bhowmik et al. 2006). Roughness on the surface is also important in achieving higher adhesion strength, as it would obtain the strongest and most durable bonds. Therefore, the most important step in adhesive bonding of polymers to inorganic substrate is substrate surface modification, which not only prevents or removes contaminants that can adversely affect bonding but also helps in creating chemically active sites on the surface to maximize bond strength. For example, the covering substrate surfaces with a thin layer of a silane adhesion promoter improves the adhesion between silicon dioxide and polyimides. Silane groups usually have two different reactive groups. One group is reactive to the substrate and the other to the polymer (Ebnesajjad 2014). The experiment has shown that other techniques that form covalent coupling also work well. In particular, glycidyl methacrylate (GMA) chains, which contain epoxy functionality, can be grafted onto the surface of the silicon dioxide before deposition of the polyamic acid. On curing, the GMA chains react with the polyimide. Solvent-based primers such as toluene and xylene have also been used as pre-treatments to enhance adhesion of polymers, despite the fact that the by-products are a chemical hazard (Cheng et al. 2006).

Several conventional methods such as thermal treatment and mechanical treatment can also be used to improve polymer-substrate bond (Bhatnagar 2017). Such an approach is especially effective for the polymer–polymer interface, where to form strong polymer–polymer adhesion it is necessary for the network to be continuous across the interface. This continuity can be formed by chain interdiffusion, if the materials are sufficiently miscible, or by chemical reaction to form coupling chains at the interface. As it is known, thermal and mechanical treatments promote these reactions. However, modification of the substrate surface via plasma treatment is the most effective method to improve the adhesive bonds (Shenton et al. 2001; Awaja et al. 2009; Wolf and Sparavigna 2010).

Plasma contains active species, such as electrons, ions, radicals, photons, and so on, that initiate chemical and physical modifications on the substrate surface (Weikart and Yasuda 2000; Larrieu et al. 2005). Ions and electrons present in plasma break polymer chains and chemical bonds due to their high kinetic energy. At the same time, free radicals in plasma modify the chemical properties of substrate surface by introducing functional groups on the surface (Yasuda 1985). Thus, when the material whose surface is to be modified is placed in the plasma chamber, these energetic particles collide with the surface of the material and cause molecular disruptions. This leads to a drastic modification of the structure and properties of the surface (Kaplan et al. 1993) that depends on the composition of the surface and the gas used.

There are mainly two types of plasmas: thermal plasma and cold plasma. Thermal plasma is used to destroy solid, liquid, and gaseous toxic halogenated and hazardous substances or to generate anticorrosion, thermal barrier, anti-wear coatings, and so on. Cold plasmas are used for surface modifications of materials, ranging from simple topographical changes to the creation of surface chemistries and coatings that are radically different from the bulk material (Bonizzoni and Vassallo 2002). Cold plasmas are generated by glow discharges at reduced pressures of 0.01–10 torr, and pressure of around 1 torr is sufficient for surface modification of polymers and solid substrates (Liston et al. 1993). The low-pressure plasma using glow discharge and atmospheric-pressure plasma using corona discharge are convenient methods for surface modification of substrates to enhance their adhesion characteristics.

It was established that only short plasma treatment times are required to increase the bond strength between two substrates. This form of surface treatment allows for modification or tailoring of surface properties without changing the overall bulk properties of the polymer and is generally environmentally friendly. Plasma treatment of surfaces often induces the formation of oxygen-containing functional groups, such as hydroxyl groups, resulting in increased surface wetting and improved adhesion (Ooij et al. 1995; Süzer et al. 1999; Mutel et al. 2000). One takes into account that the resulting surface changes depends on the composition of the surface and the gas used. Gases or mixtures of gases, used for plasma treatment of polymers, can include N, Ar, $O_2$, He, nitrous oxide, water vapor, carbon dioxide, methane, ammonia, and others. Each gas produces a unique plasma composition and results in different surface properties.

There are also a variety of surface treatment techniques for improving adhesion other than chemical and plasma treatments described above. For example, corona discharge, as for plasma treatment, introduces oxygen-containing polar groups to the surface and improves the surface energy and adhesion strength (Sun et al. 1999). Irradiation treatments can also be used to increase the adhesion characteristics of the surfaces (Lawrence and Li 2001).

It is important to note that the same treatments can be used to improve adhesion of metal films on the surface of polymers (Pappas et al. 1991; Li et al. 2004; Langowskia 2011). For example, Pappas et al. (1991) have found that *in-situ* low energy $Ar^+$ and/or $O_2^+$ ion bombardment of the polymer surface prior to metal deposition provides major improvement in the adhesion of the metal layer. Li et al. (2004) established that the $O_2/CF_4$ plasma was more effective than the Ar or $O_2$ plasma on modifying the polysiloxane-based polymer surface, in terms of increasing the wettability and roughness of the polymer surface, which is of great benefit to improving the adhesion between Au films and the polymer. The insertion of two adhesion layers, $Al_2O_3/Al$, between the polymer and the Au film was also shown to be useful for improving the adhesion between them (see Table 19.7).

One can find more detail information related to polymer adhesion and approaches to adhesion improvement in Chang et al. (1990), Brown (2000), and Awaja et al. (2009).

## 19.8 THE ROLE OF POLYMER FUNCTIONALIZATION IN HUMIDITY-SENSING EFFECT

It is important to note that the previously mentioned studies related to a polymer structure modification and functionalization mainly have not been focused on the design of humidity-sensing material. As a result, we did not find a comprehensive analysis and generalization of the effects of doping or structure and surface-modification influence on the operating characteristics of polymer-based humidity sensors. We have only individual papers related to this topic of research, which indicate that polymers doped with different ions may give distinct responses to a specific analyte (Bai and Shi 2007). In addition, these studies testify that the great importance for humidity-sensing characteristics have the nature of dopants (Brie et al. 1996; Van and Potje-Kamloth 2001), the molecular sizes of dopants (De Souza et al. 2001), the doping levels (Kawai et al. 1998; Nicho et al. 2001; Ruangchuay et al. 2004), and the structural features of polymer films formed (Li et al. 2014a). In particular, it was found that the addition of bulky $-CF_3$ groups in polyimide (PI) resulted in a good gas permeability and selectivity; the presence of a bulky $-CF_3$ group in the polyimide structure has been shown to impart steric congestion, which led to producing soluble PIs with more free volume (Tena et al. 2015). This also decreased the overall dielectric constant of the polymer. Jain et al. (2003) have found that camphosulfonic acid-doped polyaniline (PANI) shows the best response comparing with those doped with diphenyl phosphate and maleic acid (Mac) when detecting water vapor. Hong et al. (2004) reported that strong acid dopants resulted in better reversibility of PANI-based chemiresistor, while they performed a worse response. A strong difference in sensing performances between $Cl^-$, $SO_4^{-2}$ and $NO_3^-$ doped PPy composites was also observed by Guernion et al. (2002).

## TABLE 19.7
**Pull-Off Strength of the Au Film Deposited on the Surface of Polysiloxane Based Polymer**

| Sample No. | Sample Structure | Processing Gases for the Plasma-Treatment (min) | Time of Plasma Treatment | Pull-Off Strength for Au Film (N/mm²) |
|---|---|---|---|---|
| A | Polymer/Au | – | 0 | 1.2 |
| B | Polymer/Au | Ar | 10 | 3.2 |
| C | Polymer/Au | $O_2$ | 5 | 5.5 |
| D | Polymer/Au | $O_2/CF_4 = 2:1$ | 1 | 8.5 |
| E | Polymer/$Al_2O_3$/Au | $O_2/CF_4 = 2:1$ | 1 | 2.3 |
| F | Polymer/$Al_2O_3$/Al/Au | $O_2/CF_4 = 2:1$ | 1 | 14.7 |
| G[a] | Polymer/$Al_2O_3$/Al/Au | $O_2/CF_4 = 2:1$ | 1 | >35 |

*Source:* Reprinted from *Appl. Surf. Sci.*, 233, Li, W.T. et al., Significant improvement of adhesion between gold thin films and a polymer, 227–233, Copyright 2004, with permission from Elsevier.

[a] Annealed at 150°C for 2 h.

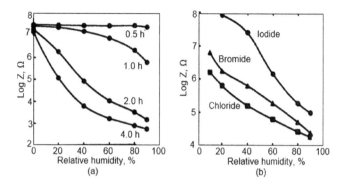

**FIGURE 19.8** (a) Impedance of the sulfonated porous polyethylene films as a function of humidity. The sulfonation reaction time are indicated in the figure; (b) impedance of polytetrafluoroethylene-graft-quaternized polyvinyipyridine as a function of humidity: (●) Iodide, (▲) bromide, (■) chloride. (Reprinted from *Sens. Actuators B*, 35–36, Sakai, Y. et al., Humidity sensors based on polymer thin films, 85–90, Copyright 1996, with permission from Elsevier.)

In particular, Figure 19.8 illustrates the influence of doping (grafting) on sensing characteristics of polymer-based conductometric humidity sensors designed by Sakai et al. (1996).

It was also established that surface modification and doping strongly affect the temporal stability of the functionalized polymer surface. For example, De Leeuw et al. (1997) reported that many polymer-based devices were not designed due to the lack of environmentally stable *n*-type-doped conducting polymers. Research has shown that some heavily *p*-type doped polymers are stable under ambient conditions, for example, poly- pyrrole and polyaniline. Conjugated materials that are stable as an undoped or very slightly doped *p*-type doped material are also known, for example, polythiophene. Until now no *n*-type-doped conjugated polymers, having similar stability, are known. The difficulty to arrive at such a polymer is related to the well-known instability of organic anions, especially of carbanions. Typically, these ions are easily oxidized in contact with air or water. This means that design and synthesis of an *n*-type conducting polymer with high stability is not trivial. As we know, temporal stability is one of the most important problems in the field of gas-sensor design. In addition, we need to take into account that in general the doping of polymers is reversible process, and therefore for charge stabilization on the polymer backbone a counter "dopant" ions should be introduced in polymer during all chemical and electrochemical *p*- and *n*-doping processes (MacDiarmid 2001).

Adhesion of polymer films to substrates also depends on the bulk and surface modification. It was established that adhesive forces can be increased by using special adhesion prompters or by plasma treatment of the substrate surface. For example, currently commercially available polyimides already contain an integrated adhesion promoter: Pimel G-X Grade (Asahi 1994), Pyralin PI 2700 (DuPont 1994), and probimide 7000, 7500 (OCG 1994). By chemical modification of the substrate surface, the PANI could be grafted onto Si substrate with good adherence as well (Chen et al. 2001). The pretreatment of electrode surface sometimes is also able

to optimize the contact resistance, as well as the sensitivity. Cui and Martin (2003) have shown that on a rough Au electrode surface, the electrodeposited PPy adhered much better than on a smooth Au electrode. Of course, the pretreatment of polymer sensing film, including the heating, may also affect the performance of the humidity sensors.

Analyzing the absence of any generalizations in the field of the influence of doping and surface modification on the performance of humidity sensors based on polymers, one should recognize that this situation is quite explainable, since the results that need to be analyzed are specific for every experiment. It was established that consequences of surface modification are strongly dependent on the polymer used and the modification method selected for the surface functioning. All this naturally creates difficulties for generalizing the results obtained in this area. Therefore, to compare the effectiveness of the influence of various methods of surface functionalizing on the properties of polymers, we use the results of the analysis performed by Nady et al. (2001) for polymer-based membranes, aimed for electrochemical gas sensors. Usually, the aims of surface modification of a membrane are the following: (1) minimization of undesired interactions (adsorption or adhesion, or in more general terms membrane fouling) that reduce the performance as described previously, and (2) improvement of the selectivity or even the formation of entirely novel separation functions (Ulbricht 2006). An overall comparison between the different methods used for surface modification of polymer membrane is presented in Table 19.8. Here we must also keep in mind that it is not always right to interpret and compare results, because one method of modification can influence many parameters simultaneously. Thus, Table 19.8 gives a general impression only.

According to Nady et al. (2001), all the surface modification methods mentioned earlier allow modification without affecting the bulk properties too much when appropriate conditions are selected; mostly, the flux is similar to the base membrane or slightly lower. Complete and seemingly permanent hydrophilic modification of poly(arylsulfone) membranes is achieved by blending and photoinduced grafting. A few studies showed that chemically redox initiation-grafting could be successfully applied to polyethersulfone ultrafiltration membranes (Belfer et al. 2000; Reddy et al. 2005). However, chemical treatment usually employs harsh treatment; often it may lead to undesirable surface changes and contamination, and may not be the best choice in environmental terms. Plasma treatment is probably one of the most versatile poly (arylsulfone) membrane-surface-treatment techniques. For example, simple inert gas, nitrogen, or oxygen plasmas have been used to increase the surface hydrophilicity of membranes, and a water-plasma treatment have successfully rendered asymmetric polysulfone membranes permanently hydrophilic. However, its high costs and technical complexity remain drawbacks for large-scale use. Combination of two or three modification techniques is complex in terms of cost effectiveness and environmental drawbacks but could lead to multi-functional membranes that are of great interest for "membranes of the future." Such membranes may need more functions than "only" providing a selective barrier with high

**TABLE 19.8**

**Advantages and Disadvantages of Surface Modification Methods Applied to Polymer Membranes**

| Modification Method | Flux after Modification | Simplicity/Versatility | Reproducibility | Environmental Aspects | Cost Effectiveness |
|---|---|---|---|---|---|
| Coating | L | H | H | H | H |
| Blending | H | E | H | H | H |
| Composite | H | H | L | H | H |
| Chemical | L | H | L | L | H |
| Grafting initiated by: | | | | | |
| *Chemical* | H | H | H | L | H |
| *Photochemical* | H | H | H | H | H |
| *Radiation* | H | L | L | L | L |
| *Plasma* | H | L | L | H | L |
| Combined methods | L | L | H | L | L |

*Source:* Reprinted from *Desalination*, 275, Nady, N. et al., Modification methods for poly(arylsulfone) membranes: A mini-review focusing on surface modification, 1–9, Copyright 2001, with permission from Elsevier.

*Abbreviations:*   L—low, H—high, E—excellent.

performance (flux and stability). To be complete, it should be noted that all mentioned methods influence membrane smoothness/roughness (Rana and Matsuura 2010).

## 19.9 POLYMER-BASED NANOCOMPOSITES AND THEIR APPLICATION IN HUMIDITY SENSORS

As it was shown in Volume 2 (Chapter 11), the polymer-based composite materials have attracted interest of a number of researchers developing humidity sensors. Many different nanocomposites can be prepared using polymers (Rajesh et al. 2010). However, in humidity sensors, mainly polymer–polymer, polymer–metal, polymer–metal oxide and polymer–carbon nanocomposites are used. The interest to nanocomposites is due to the synergistic and hybrid properties of composite materials. For example, conductive components, dispersed in electrically insulating polymer matrix, may be responsible for its conductivity while the polymer may promote specific interaction with the target gas or vapor. Ease of processability of organic polymers combined with the better mechanical and specific electrophysical properties

of nanoparticles has led to the fabrication of many devices with improved parameters. In particular, it was found that the transition to polymer-based nanocomposites provides additional opportunities for the design of highly effective humidity sensors as well. Ahmadizadegan (2016) established that the polyimide/ZnO nanocomposite films showed better thermal stability associated with high glass-transition temperatures, high char yields (higher than 80% at 800°C in nitrogen), and increased gas permeation in comparison with the pristine polyimide membrane. As is known, gas permeation largely determines the sensitivity and speed of humidity sensors. The experiment also showed that various additives such as metal oxides, carbon, noble metals, and polymers can also be used to improve the adsorption of water vapors by the polymer-based matrix, increase the number of interacting sites with the analyte, increase the intra- and interchain mobility of the charge in conducting polymer chains, or even change the affinity of the composite for the target gas. Furthermore, they can be used as a transducer matrix to convert the ionic charge into an electronic one, as a mediator, and sometimes even as a catalyst. Examples of polymer-based nanocomposites applied for humidity sensors fabrication are listed in Table 19.9.

**TABLE 19.9**

**Polymer-Based Composites Used for Humidity Sensor Fabrication and Methods of Their Preparing**

| Type of Composite | Composites | References |
|---|---|---|
| Polymer-polymer | CE-PPy; PANI-PVA; PANI-PVP; PI-PU | McGovern et al. (2005a); Lavine et al. (2006); Li et al. (2012); Mahadeva et al. (2011); Bae et al. (2017) |
| Polymer-metal oxide | PAni-ZnO; PPy-Ta$_2$O$_5$; PPy-ZnO; PEDOT:PSS-AZO; PEDOT:PSS-SiO$_2$; PSDA-ZnO | Shukla et al. (2012); Lee et al. (2013); Chaluvaraju et al. (2016); Najjar and Nematdoust (2016); Zor and Cankurtaran (2016) |
| Polymer-carbon | DAN-CB; PEGAs-CNT; PVDF-G; PDDA-GO; PSSNa-GO; PI-CNTs | Shim and Park (2000); Lee et al. (2008); Li et al. (2014b); Çiğil et al. (2017); Hernández-Rivera et al. (2017) |
| Polymer-metal | PVA-Ag | Power et al. (2010) |

*Abbreviations:*   AZO—aluminum zinc oxide; CB—carbon black; CE—cellulose; CNT—carbon nanotube; DAN—Poly(1,5-diaminonaphthalene); G—graphene; GO—graphene oxide; PANI—polyaniline; PDDA—poly(diallydimethyldiammonium chloride); PEGAs—Polyethylene glycol acrylate; PPy—polypyrrole; PSDA—poly(diphenylamine sulfonic acid); PSSNa—polystyrenesulfonate; PVDF—polyvinylidene fluoride; PU—polyurethane-urea.

By now, a large number of methods for synthesis and fabrication of polymer-based composites have been demonstrated (Bein and Stucky 1996; Malinauskas et al. 2005; Karnicka et al. 2005; Hussain et al. 2006; Koo 2006; Xiao and Li 2008; Hanemann and Szabó 2010; Rajesh et al. 2010; Sattler et al. 2010; Ruiz-Hitzky et al. 2011; Oliveira and Machado 2013; Khan et al. 2017). Figure 19.9 makes a summary of the basic methods used to prepare nanocomposites.

The simplest and effective approach to composite preparation is the introduction of nanoparticles or nanotubes into polymer matrices or simply mixing them with a polymer matrix (see Figure 19.10). In this case, the inorganic nanoparticles are first synthesized and then introduced in the polymer solution or melt. This is the so-called *ex-situ* approach. The *ex-situ* approach is the most general one, since there are no limitations on the kinds of nanoparticles and polymers that can be used (Rozenberg and Tenne 2008; Hanemann and Szabó 2010). However, one should take into

account that such approach requires blending or mixing the components with the polymer in the solution or in melt form, but many of polymers are not fusible and they are generally insoluble in common solvents. In addition, homogeneous dispersion of inorganic nanoparticles in the polymer matrix are difficult to obtain. Therefore, the most important step in fabrication of polymer-based nanocomposites, using this approach, is the dispersion of filler in the matrix. Various mechano-chemical approaches, including sonication by ultrasound, can be used for this purpose. However, the scope of such approaches for dispersing the nanoparticles is limited by reaggregation of the individual nanoparticles and establishment of an equilibrium state under certain conditions, which determines the size distribution of the agglomerate of the dispersed nanoparticles. Therefore, many *ex-situ* processes generally suffer extremely from the high agglomeration tendency of nanoparticles, as it is rather difficult to destroy the nanoparticle agglomerates even using high external shear forces. Other limitations are related to temperature conditions and the limited stability of some types of inorganic nanoparticles to mechanical impacts (Rozenberg and Tenne 2008; Hanemann and Szabó 2010). The preparation of nanocomposites by diffusion of dispersed colloids in polymer films is also possible. This method is suited only for rather insoluble polymers with good swelling behavior or for very thin films.

Surface functionalization is one of the most effective approaches to improving the dispersion of inorganic particles. In particular, it was found that several powders of surface-modified colloids have been found to disperse well in liquids. Such dispersions can be mixed with dissolved polymers, and subsequently nanocomposites can readily be obtained by casting followed by evaporation of the solvent or by spin coating. This approach is widely used in the preparation of composites based on carbon-based materials. For example, noncovalent modifications of carbon nanotubes (CNTs) can be used to wrap polymers around the nanotubes. At the same time, it was found that a surface hydrophobization using physisorption or chemisorption causes a steric stabilization, enabling a repulsive interaction of the particles (Krishnamoorti 2007), which prevents reagglomeration of dispersed particles. The treatment of the hydroxyl-terminated metal oxides with organosilanes yields via chemisorption a hydrophobic surface (Krishnamoorti 2007; Smits et al. 2008). If the organosilanes carry a reactive, polymerizable functionality, the surface modified particle can be attached to the resulting polymer backbone or network. A short review dealing with these grafting techniques was published by Rong et al. (2006). Dispersants or surfactants are amphiphilic molecules with a polar and a nonpolar molecular moiety. They are attached physically (physisorption) via van der Waals forces or hydrogen bridges to the particles forming a hydrophobic surface as well. A comprehensive overview of the different surfactants and the application possibilities was given by Karsa (2003).

Particles coated with a polymer shell are also considerably more stable against aggregation because of a large decrease of their surface energy in comparison with bare

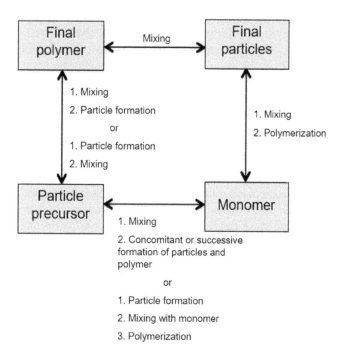

**FIGURE 19.9** Summary of nanocomposites preparation methods. (Idea from Kickelbick, G., *Hybrid Materials: Synthesis, Characterization, and Applications*, Wiley-VCH, Weinheim, Germany, 2007.)

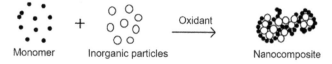

**FIGURE 19.10** Formation of nanocomposites. (Reprinted from Rajesh et al., Nanocomposites: From fabrication to chemical sensor application, in: Korotcenkov, G. (ed.), *Chemical Sensors: Fundamentals of Sensing Materials*, Vol. 2: Nanostructured Materials, Momentum Press, New York, pp. 335–368, 2010. With permission.)

particles. There are many techniques for the deposition of nanometer-sized conducting polymer layers onto different substrates, including nanosized ones (Malinauskas 2001; Chen et al. 2006). Such a polymer shell can be obtained by first synthesizing the inorganic nanoparticles and then dispersing them in a polymer solution. For example, metal nanoparticles embedded into a conducting polymer matrix can be easily obtained by the chemical reduction of metal ions from their salt solution at the conducting polymer/solution interface. Such a process of the polymer shell formation on preformed inorganic cores can also be realized by polymerization of the desired monomer with inorganic nanoparticles dispersed in it. Finally, the polymer-coated inorganic nanoparticles are precipitated into a nonsolvating phase. Then the nanocomposite material is formed. This affords a simple technique for the preparation of nanocomposites by direct mixing of particles and polymers (Caseri 2003). The presence of such a shell on the surface of inorganic nanoparticle increases the compatibility of the particles in the polymer matrix and makes it easier to disperse them (Malinauskas et al. 2005).

In some cases, the process of the polymer-based nanocomposite formation and nanoparticle preparation can be combined into one process or performed as a series of consecutive processes in one reactor (the *in-situ* approach). In the *in-situ* methods, nanocomposites are generated inside a polymer matrix by precursors, which are transformed into the desired nanoparticles by appropriate reactions. *In-situ* approaches are currently getting much attention because of their obvious technological advantages over *ex-situ* methods (Rozenberg and Tenne 2008). One-step synthesis leads to improved compatibility of the filler and the polymer matrix and enhanced dispersion of the filler.

Up to now, *in-situ* formation of particles in a liquid medium has probably been the most widely used method for the preparation of polymer nanocomposites, containing isotropic inorganic particles (Wang et al. 2004b). Commonly, soluble inorganic or organometallic compounds are converted by chemical reactions to colloids in water or organic solvents. The polymer may be already present during colloid synthesis or may be added afterward. The particle dispersion can be destabilized or stabilized by the polymer, depending on the system. In the former case, the nanocomposite forms spontaneously by co-precipitation after colloid formation; in the latter case, nanocomposites can be obtained by addition of a solvent that acts as a coprecipitation agent, by casting followed by solvent evaporation, or by spin coating (Caseri 2003). Sol-gel method is one of the varieties of *in-situ* methods (Koo 2006). The term sol-gel is associated to two relations steps, sol and gel. Sol is a colloidal suspension of solid nanoparticles in monomer solution, and gel is the 3D interconnecting network formed between phases. In this method, solid nanoparticles are dispersed in the monomer solution, forming a colloidal suspension of solid nanoparticles (sol). They form an interconnecting network between phases (gel) by polymerization reactions followed by the hydrolysis procedure. The polymer nanoparticle 3D network extends

throughout the liquid. The polymer serves as a nucleating agent and promotes the growth of layered crystals. As the crystals grow, the polymer is seeped between layers, and thus the nanocomposite is formed.

Inorganic particles can also be prepared *in situ* in solid polymer matrices, for example, by thermal decomposition of incorporated precursors, reaction of incorporated compounds with gaseous species, or when the polymer films, containing an incorporated precursor, are immersed in liquids, containing the reactive species required for the formation of the desired colloid. If solid reaction by-products arise from the particle synthesis, they can be embedded in the nanocomposites, hence the formation of volatile reaction side products, which are able to leave polymer matrices, should be preferred if possible (Caseri 2003).

It is important to note that the use of nanocomposites can significantly improve the parameters of humidity sensors, but this improvement is possible only with the correct choice of the composition of the composite and the operating conditions of the sensors. For successful application of composites in humidity sensors, two important factors need to be addressed: functionality and processability. These properties are related to the aspect ratio of the two components in the matrix, the interfacial adhesion between the additives, electrical conductivity, and structure (Hatchett and Josowicz 2008). For example, PANI/polyvinyl alcohol (PVA) composites appear to be very promising for humidity-sensor design (McGovern et al. 2005a). The cross-linked PVA additive may serve a protective role against thermal and environmental degradation of the conducting polymer blend (Bai and Shi 2007). However, it was found that the composites with more than 20% by weight of PANI in PVA were immiscible, and only when less than 10% of PANI was present, the composite was miscible. In addition, it was found that in general, processed PANI/PVA composite films (150 μm thick) did not show any change with respect to relative humidity between 10% and 60%. The change in resistance of PANI/PVA composite was observed only when the PANI/PVA composite was exposed to relative humidity between 60% and 90% (McGovern et al. 2005a). Moreover, a relatively small increase in the minority component PANI has a dramatic influence on the overall properties of the composite. It should be noted that a particularly strong dependence of properties on the composition is observed for composites based on carbon-based materials. For example, as it is seen in Figure 19.11, for composite PI-CNTs, when the concentration of CNTs in the composite varies in the range of 0–0.4 wt%, the nature of the influence of humidity on the resistance can change dramatically.

The same situation is typical for other nanocomposites. For example, Figure 19.12 shows results obtained by Bae et al. (2017) for the composite polyimide (PI)-polyurethaneurea (PU) block copolymers. This study established that the sensor sensitivity to humidity increased sharply with increasing PU content from 0 to 25 wt %, and then increased a little. The hysteresis of the sensor decreased sharply with increasing PU content up to

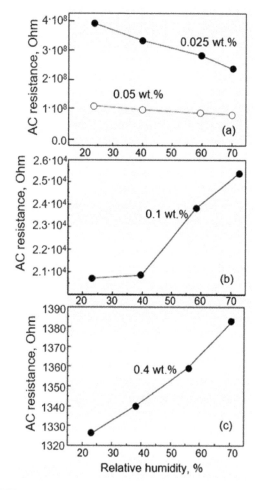

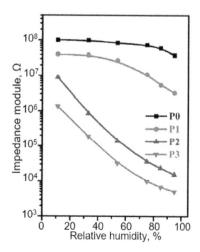

FIGURE 19.13 Impedance modules of MPOSS:DVB:DMC-based composites with various composition at different RH (1 V AC, 1 kHz): P0—MPOSS:DVB:DMC=1: 4: 0; P1—1: 3: 2; P2—1: 2.75: 2.5; and P3—1: 3.25: 1.5. (Reprinted from *Sens. Actuators B*, 253, Dai, J. et al., LiCl loaded cross-linked polymer composites by click reaction for humidity sensing, 361–367, Copyright 2017, with permission from Elsevier.)

FIGURE 19.11 Humidity dependence of the AC resistance of the devices with different MWCNT contents: (a) low MWCNT contents; (b, c) higher MWCNT contents. (Data extracted from Lee, M.J. et al., Humidity sensing characteristics of plasma functionalized multiwall carbon nanotube-polyimide composite films, in: *Proceedings of IEEE SENSORS Conference*, October 26–29, Lecce, Italy, pp. 430–433, 2008.)

50 wt %, and then decreased a little. However, the same studies have shown that the thermal stability and mechanical properties of the copolymer decreased markedly with increasing PU content. The effect of the composition of the composite mercaptopropyl polyhedral oligomeric silsesquioxane (MPOSS)/1,4-divinylbenzene (DVB)/methacrylatoethyl trimethyl ammonium chloride (DMC) on the humidity sensitive characteristics is shown in Figure 19.13. It is seen that depending on the composition, the impedance response can be change considerably. It is a pity that in most of the works the composition of the composite, which is really optimal in terms of its humidity-sensing characteristics, is not determined.

It is important to note that there are other approaches to the development of humidity sensors based on polymer-based composites. In particular, Chen et al. (2017) proposed humidity sensors designed using CNTs/PVA-based bilayer actuators (Figure 19.14). They established that the PBONF-reinforced CNT/PVA bilayer actuators can unsymmetrically adsorb and desorb water, resulting in a reversible deformation. More importantly, the actuators show a pronounced increase of conductivity due to the deformation induced by the moisture change, which allows the integration of a moisture-sensitive actuator and a humidity sensor. Upon changing the environmental humidity, the actuators can respond by the deformation for shielding and report the humidity change in a visual manner (Figure 19.14e, f), which has been demonstrated by a tweezer and a curtain. According to Chen et al. (2017), such nanofiber-reinforced bilayer actuators with the sensing capability should hold considerable promise for the applications, such as soft robots, sensors, intelligent switches, integrated devices, and material storage.

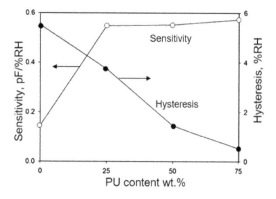

FIGURE 19.12 The effect of PU content on the sensitivity and hysteresis of capacitive humidity sensors prepared using PI-PU copolymers at 25°C. (From Bae, Y.-M. et al.: Polyimide-polyurethane/urea block copolymers for highly sensitive humidity sensor with low hysteresis. *J. Appl. Polym. Sci.* 2017. 2017. 44973. Copyright Wiley-VCH Verlag GmbH & Co. KGaA. Reproduced with permission.)

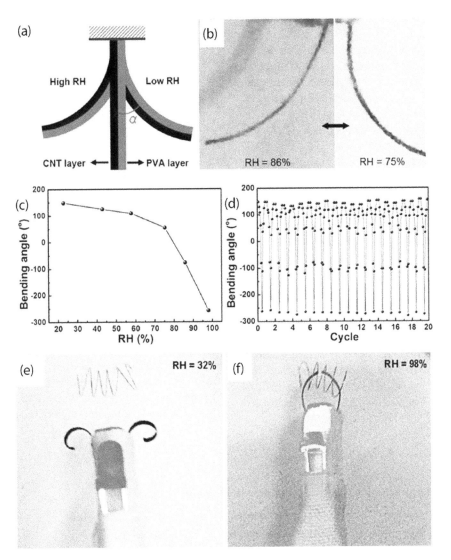

**FIGURE 19.14** Moisture-responsive properties of the PBONF-reinforced CNT/PVA bilayer actuators: (a) scheme indicating the definition of the bending angle $\alpha$; (b) photographs of the PBONF-reinforced CNT/PVA bilayer actuator at RH = 75% and RH = 86%; (c) bending angle of the actuator as a function of RH; (d) reversible deformation of the actuator when the RH is repeatedly changed for 20 cycles; and (e, f) photographs illustrating that an electrical tweezer grips a copper wire, closes the circuit, and lightens up the LED bulb at the same time, when the RH is switched from (c) 32% to (d) 98%. The tweezer consists of two PBONF-reinforced CNT/PVA strips. (Reprinted by permission from Macmillan Publishers Ltd. *Sci. Reports*, Chen, M. et al., 2017, Copyright 2017.)

## 19.10 POLYMER-BASED MICROMACHINING

In Chapter 15, we considered the micromachining process based on silicon technology. However, the experiment has shown that the micromachining technology can also be based on the polymer approach. It was established that due to specific mechanical and chemical properties of polymers, the polymer-based micromachining technology has specific features not available to silicon-based micromachining (Goeders et al. 2008). Micromachined polymers may be employed as structural or functional elements as well as soft, flexible substrates that contain other devices. This versatility is afforded by the development of a wide range of processing techniques, unique for polymer materials. Simple polymer structural elements can be photopatterned or casted, eliminating the need for complicated etching steps and lithographic masking required in silicon processing. All this allows to implement a large range of MEMS devices with improved parameters (Kim and Meng 2016). For example, to alter the stiffness of silicon-based cantilevers, their geometry must be changed or a rigid coating must be applied to the surface. In contrast, the stiffness of polymeric cantilevers requires only a change in material. In this way, microcantilevers with the same geometry but different properties can be produced. Thus, the development of polymer micromachining technologies that complement traditional silicon approaches has enabled the broadening of microelectromechanical system (MEMS) applications. The capabilities of these technologies have opened the door to alternative and potentially more cost-effective manufacturing options to produce highly flexible structures and substrates with tailorable bulk and surface properties, important for various applications, including humidity sensors.

At present, many polymers have been explored in the research literature for MEMS applications (Curran et al. 1998; Jiguet et al. 2004; Kim and Meng 2016; Chen et al. 2017). They include SU-8, polyimide, Parylene, polydimethylsiloxane (PDMS), liquid-crystal polymers (LCPs), cyclic-olefin polymers (COPs), polymethyl methacrylate (PMMA or plexiglass), polycarbonate (PC), polystyrene (PS), and CNT/PVA bilayer actuators. These polymers have their own specific combination of material properties, processing conditions, and performance demands that drives selection of a particular polymer for a device. However, among the polymers mentioned above, SU-8, polyimide, and Parylene are currently the most used for manufacturing membranes, cantilevers, and other structural elements on hybrid silicon-polymer devices such as gas and humidity sensors. Compared to the other polymers, these three are compatible with more standard microfabrication techniques, that is, photolithography and wet/dry etches, which have motivated a large effort within the community to develop novel strategies for processing and device construction. Comparative characteristic of these polymers is shown in Table 19.10. Additional information about these polymers and their application areas within MEMS can be found in reviews (Grayson et al. 2004; Ziaie et al. 2004; Liu 2007; Becker and Gartner 2008; Kim et al. 2008; Kim and Meng 2016).

## 19.10.1 SU-8-Based Micromachining Technology

As is seen in Table 19.8, SU-8, polyimide, and Parylene really have great potential for making various MEMS elements. However, the polymer most commonly used for manufacturing microcantilevers and membranes, acceptable for the use of humidity sensors, is SU-8 (Jiguet et al. 2004; Johansson et al. 2005; Lukes and Dickensheets 2013). SU-8, a photopolymerizable epoxy-acrylate polymer (glycidyl-ether-bisphenol-A novolac), is a negative photosensitive polymer used as a structuring material in MEMS and for fabrication purposes within the micro-total-analysis-system (lTAS) area (Carlier et al. 2004). SU-8 is highly transparent in the UV region, allowing fabrication of relatively thick (hundreds of micrometers) structures with nearly vertical side walls. The SU-8 photoresist is most commonly exposed with conventional UV (350–400 nm) radiation, although i-line (365 nm) is the recommended wavelength. SU-8 may also be exposed with e-beam or X-ray radiation. As it can be seen from its formula, the SU-8 polymer has quite low molecular weight (~7000) and thus, when non-cross-linked can easily be dissolved by a number of solvents (e.g., propylene-glycol-methyl ether, gamma-butyrol-acetone, and methyl iso-butyl ketone). Typically, the lithography of SU-8 involves a set of processing steps similar to standard thick photoresists (Kim and Meng 2016): (1) deposition on a substrate (usually via spinning); (2) a softbake to evaporate

## TABLE 19.10
## Overview of Polymers Promising for Fabrication of MEMS Elements

| Polymer | Properties Summary | Fabrication Overview | Examples of Applications | Fabrication Challenges |
|---|---|---|---|---|
| SU-8 | • High thermal and chemical stability;<br>• Tunable properties;<br>• Low optical loss | • Thick photo-patternable films;<br>• Multiple exposure steps and single development;<br>• Backside exposure;<br>• SU-8 to polymer bonding;<br>• Etching is difficult and rates are slow; mechanical processes necessary | • Structural molds for soft lithography;<br>• Microfluidics;<br>• Microneedles;<br>• Optical waveguides;<br>• Neural probes | • Properties of film dependent on processing parameters;<br>• Importance of exposure on dimensional accuracy and resolution;<br>• Large stresses between SU-8 and substrate |
| Polyimide | • High thermal and chemical stability;<br>• Low moisture absorption;<br>• Tailorable film via chemical modification;<br>• Photosensitized formulations;<br>• Biocompatible | • Pattern photosensitive polyimides;<br>• Dry etching of using oxygen or fluorine chemistries;<br>• Hot embossing capability;<br>• Surface modification using plasma | • Flexible microelectrode arrays;<br>• Sensors;<br>• Microchannels;<br>• Lab-on-a-tube technology;<br>• Self-assembly via imidization shrinkage | • Poor adhesion onto materials;<br>• Significant shrinkage during imidization process |
| Parylene C | • Chemical inertness;<br>• Uniform and conformal CVD;<br>• Gas phase, pinhole-free deposition;<br>• Low intrinsic stress;<br>• High transmittance in visible spectrum;<br>• Biocompatible | • Deposition onto structured surfaces and liquids;<br>• Dry etching of using oxygen or fluorine chemistries;<br>• Hot embossing and thermal forming;<br>• Surface modification using plasma | • Microchannels;<br>• Sensors;<br>• Neural probes;<br>• Cuff electrodes;<br>• Retinal and cochlear implants | • Sensitive to temperature (low $T_g$);<br>• Adhesion issues between Parylene-Parylene and Parylene-metal in soaking conditions |

*Source:* Reprinted with permission from Kim, B.J. and Meng, E., Review of polymer MEMS micromachining, *J. Micromech. Microeng.*, 26, 013001, 2016. Copyright 2016, Institute of Physics.

the solvent; (3) exposure to cross-link the polymer: the exposure of this polymer to UV light generates a strong photoacid, which protonates the epoxy groups of the monomer and starts a cross-linking reaction to create a highly cross-linked polymer (Martinez-Duarte 2014); (4) post-exposure bake to finalize the cross-linking; and (5) development to reveal the cross-linked structure. After exposition and developing, its highly cross-linked structure gives it high stability to chemicals and radiation damage. Cured cross-linked SU-8 shows very low levels of outgassing. The last one is good for gas and humidity-sensor design. Parameters of SU-8 are listed in Table 19.11.

One should note that the fabrication of polymer-based MEMS elements, including cantilevers and membranes is inexpensive and fast. Polymeric microcantilevers and membranes can be fabricated in a variety of ways; the method used is determined by the type of polymer to be used. For example, the process for fabricating SU-8-based cantilevers is quite similar to that used for silicon-based cantilevers. The fabrication process is shown in Figure 19.15. A thin film of SU-8 is deposited onto a wafer by spin coating. Photolithography then is used to define the regions that will comprise both the cantilever and the chip to which it is attached. The unwanted material is removed and the polymer cantilevers are released from the substrate by immersion in an appropriate solvent. SU-8 cantilevers have been made into arrays for optical lever (Calleja et al. 2005; Ransley et al. 2006) and piezoresistive (Johansson et al. 2006) detection schemes.

An example of a polymer SU-8-based cantilever with integrated gold resistor is shown in Figure 19.16. The gold resistors serve as low-noise piezoresistors; the signal-to-noise ratio is comparable to the value for silicon-nitride cantilevers with single-crystal resistors (Johansson et al. 2005). The relative resistance change of gold is approximately 40 times smaller than for single-crystal silicon, and new materials are therefore being investigated as possible candidates for even better piezoresistors. The challenge is to find a material that is soft while having a large change in the resistance upon deflection. Moreover, the electrical noise in the system should be low (Boisen and Thundat 2009).

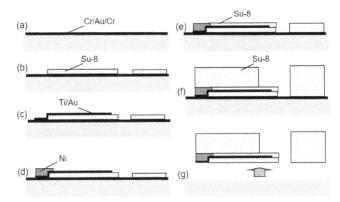

**FIGURE 19.15** Polymer cantilever fabrication process: (a) evaporation of 5/50/50 nm of Cr/Au/Cr as release layer on a Si wafer; (b) patterning of 1.5-μm-thick SU-8 for the cantilever; (c) evaporation of 5/60 nm Ti/Au for the resistors and wires; (d) electroplating approximately 5 μm of Ni for the contact pads; (e) patterning 3.5-μm-thick SU-8 for encapsulation of the resistors; (f) patterning 200-μm-thick SU-8 for the microfluidic channel; and (g) release of the chip from the wafer by etching the Cr layer. (Reprinted from *Sens. Actuators A*, 123–124, Johansson, A. et al., SU-8 cantilever sensor system with integrated readout, 111–115, Copyright 2005, with permission from Elsevier.)

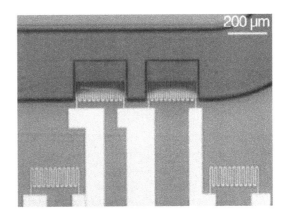

**FIGURE 19.16** Image of polymer (SU-8) cantilevers with integrated gold resistors. These devices are softer than similar silicon devices and are much faster to fabricate. However, polymers with the same stability as silicon still need to be identified. (Reprinted from *Sens. Actuators A*, 123–124, Johansson, A. et al., SU-8 cantilever sensor system with integrated readout, 111–115, Copyright 2005, with permission from Elsevier.)

### TABLE 19.11
### Properties of SU-8

| Properties | SU-8 2000 | SU-8 3000 |
|---|---|---|
| Aspect ratio | 10:1 | 5:1 |
| 5% weight loss temperature (°C) | 279 | |
| Coefficient of thermal expansion (ppm/°C) | 52 | 52 |
| Young's modulus (GPa) | 2.0 | 2.0 |
| Tensile strength (MPa) | 60 | 73 |
| Elongation (%) | 6.5 | 4.8 |
| Dielectric constant (1 GHz, 50% RH) | 4.1 | 3.2 |
| Dielectric strength (V/μm) | 112 | 115 |
| Volume resistivity (Ω cm) | $0.3–2 \times 10^{17}$ | $2–5 \times 10^{16}$ |
| Shelf life (years) | 1 | 1 |

*Source:* Courbat, J. et al., *J. Micromech. Microeng.*, 20, 055026, 2010.

With SU-8, it is also possible to utilize multiple exposure steps on multiple layers that are subsequently released in a single development (Mata et al. 2006; Zhu et al. 2012). For this method, the first SU-8 layer is spun on and exposed. Then, instead of developing, a second layer of SU-8 is spun on and exposed. SU-8 layers are repeatedly applied and exposed until the final layer of the device is processed. In the final step, all unexposed regions would be developed away in a single development step, greatly simplifying the fabrication process (Figure 19.17a). Unconventional UV exposure methods can be used to form structures that are difficult to obtain using standard MEMS materials. For example, to create protective

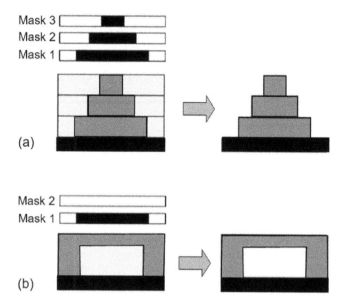

**FIGURE 19.17** (a) Schematic of multiple exposure steps of SU-8, where a layer of SU-8 is spun on and then exposed with Mask 1. Then a second layer of SU-8 is spun on and exposed with Mask 2, and a last layer of SU-8 is spun on and exposed with Mask 3. Following development of the pyramid structure is released; and (b) schematic of forming a cap on SU-8 posts using two masks: Mask 1 is used to expose a thick SU-8 layer to create the posts and Mask 2 is used to underexposure the top surface of the layer to form the cap. (Reprinted with permission from Kim, B.J. and Meng, E., Review of polymer MEMS micromachining, *J. Micromech. Microeng.*, 26, 013001, 2016. Copyright 2016, Institute of Physics.)

sealing caps at the wafer-level, partial exposure (i.e., underexposure) of SU-8 was used to cross-link the top of a cap, while the region underneath was not exposed (Zine-El-Abidine and Okoniewski 2009). Illustration of this process is shown in Figure 19.17b. Unexposed SU-8 was developed away through access ports patterned on the lid. A final lithographic step then sealed the holes and package.

The cross-linked nature of SU-8 imparts chemical stability, which in turn makes removal of cured SU-8 difficult and requires strong chemistries. Typical wet-etch recipes for SU-8 include hot 1-methyl-2-pyrrolidone (NMP), piranha etch ($H_2SO_4$, $H_2SO_4/H_2O_2$), and fuming $HNO_3$ (Lorenz et al. 1997). Ozone solution can also remove SU-8 (Yanagida et al. 2011), but removal rates are slow; the rate can be improved by using ozone steam (Yoshida et al. 2014). However, the ozone solution, as well as other stronger solutions, may not be compatible with other materials present in a device; nickel corrosion was reported after exposure to ozone (Yoshida et al. 2014). Dry etching of SU-8 is also possible, but with slow etch rates (Hong et al. 2004).

SU-8 films can be separated from the substrate to create free-film devices. The typical approach is to use release layers consisting of metal films such as aluminum (Al) (Daniel 2001), titanium (Ti) (Dellmann et al. 1998), copper (Cu) (Seidemann et al. 2002), chromium (Cr) (Truong and Nguyen 2004), or chromium/gold/chromium (Cr/Au/Cr) (Johansson

et al. 2005) that are later removed by wet etching. A sacrificial layer of gold by itself cannot be used, as SU-8 has been found to have poor adhesion to gold (Song et al. 2003). SU-8 can be released from silicon substrates directly (Truong and Nguyen 2004). However, when using potassium hydroxide (KOH) to etch silicon to free the final SU-8 device, the heat required to achieve a reasonable etch rate can introduce added internal stresses within the SU-8, which can lead to microcracks or warpage (Lau et al. 2013). Other approaches to fabrication of SU-8-based MEMS elements are discussed in Kim and Meng (2016).

### 19.10.2 Polyimide-Based Micromachining

As it was shown by Frazier et al. (1994), Fragakis et al. (2005), and Wilson and Atkinson (2007), polyimides also can be used for design of MEMS elements such as cantilevers and membranes. Parameters of polyimide were considered earlier in Chapter 22 (Volume 2). Electrical and mechanical properties are reviewed in Pedersen et al. (1998a), and characterization of the mechanical properties in thin films was performed in Kim and Allen (1999). Now, polyimides are in both photopatternable and nonphotopatternable versions. Key characteristics of polyimides include high glass transition temperature, high thermal stability (up to 400°C), low dielectric constant, high mechanical strength, low modulus, low moisture absorption, chemical stability, and solvent resistance (Frazier 1995). This versatile class of polymers can be linear (aliphatic) or cyclic (aromatic) in structure, and the cured material can exhibit both thermoset or thermoplastic behavior (Gad-el-Hak 2006). Synthesis generally starts with a polyamic acid precursor that is imidized at elevated temperatures (typically 300°C–500°C) in a nitrogen environment to form the final polyimide structure (Barth et al. 1985). The imidization process involves solvent removal and subsequent ring closure in aromatic versions (Pedersen et al. 1998a). The polyamic acid precursor is soluble in polar inorganic solvents, including n-methylpyrrolidone (NMP), dimethyl formamide (DMF), and dimethylsulfoxide (DMSO) (Pedersen et al. 1998a).

Polyimides for specific MEMS applications were briefly reviewed in Frazier et al. (1994), Frazier (1995), and Kim and Meng (2016). The chemical and thermal stability of polyimides make them an attractive sacrificial layer material (Bagolini et al. 2002). Because of its chemical structure, polyimide can accept various degrees of chemical modification, allowing it to be tailored for various applications (Wilson and Atkinson 2007; Othman et al. 2008), including piezoresistive sensing applications (Frazier and Allen 1993a). Silicon/polyimide-based cantilevers can also be used in capacitive type bimorph humidity sensors (Fragakis et al. 2005). The top-surface view of such cantilever is shown in Figure 19.18. During operation, the polyimide layer swells as it is exposed to ambient humidity; the induced compressive stress decreases in value and the cantilever tends to return to its horizontal position, effectively increasing the device capacitance.

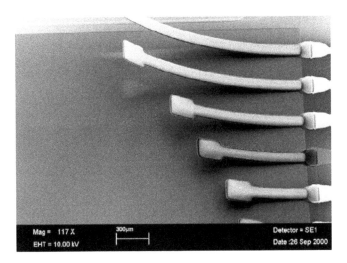

between the two electrodes. The air gap in the device is created by means of a sacrificial-layer technique in which an aluminum layer is deposited and subsequently etched away. The opening to the sacrificial layer on the backside is made by combined etching of the silicon with KOH and reactive ion etching (RIE). Since the whole sensor fabrication process contains only low temperature (<300°C) IC-compatible steps, the process can be carried out as a post-process on a substrate where integrated circuits have already been completed.

Similar to SU-8 films, cured polyimides are difficult to remove by wet etching, but removal using hot bases and very strong acids has been reported (Pedersen et al. 1998a). A combination of sulfuric acid and hydrogen peroxide was successful in removing 1-$\mu$m thick sacrificial polyimide islands with good selectivity against silicon nitride and oxide (Memmi et al. 2002). Ozone solutions have also been explored (Yanagida et al. 2013). The removal of uncured polyimides (i.e., not imidized) by wet etching has been accomplished in KOH solutions (5%–30%) (Frazier and Allen 1993b; Frazier et al. 1994). More practical polyimide removal is achieved by conventional dry-etching techniques. Commonly used plasmas are formed from $O_2$ gas alone (Schmidt et al. 1988) or in combination with CF4, CHF3, and SF6 gases (Frazier et al. 1994). Al is most commonly used as a hard mask (Frazier et al. 1994; Bagolini et al. 2002); Cr/Au (Kim and Allen 1992), PECVD silicon nitride (Suh et al. 1997), oxide (Bagolini et al. 2002), and silicon carbide (SiC) (Bagolini et al. 2002) have also been explored. Plasma etching has also been used for removal the polyimide sacrificial layers (Ma et al. 2009).

Release from Si substrates can be achieved by undercutting in an HF:HNO$_3$ (1:1) etch (Shamma-Donoghue et al. 1982),

**FIGURE 19.18** Top-surface view of the polyimide-covered cantilevers. The devices are bent upward due to the polyimide precursor shrinking after curing. (Reprinted with permission from Fragakis, J. et al., Simulation of capacitive type bimorph humidity sensors, *J. Phys. Confer. Ser.*, 10, 305–308, 2005. Copyright 2005, Institute of Physics.)

Membrane-based versions of capacitive sensor and the process of their fabrication are shown in Figure 19.19. This structure was developed by Pedersen et al. (1997) as a pressure sensor. However, the same structure can be used and as a humidity sensor. The sensor basically consists of polyimide diaphragm under which a metal electrode is placed. On the silicon substrate, a counter electrode is deposited, and a second polyimide layer is used to provide electrical insulation

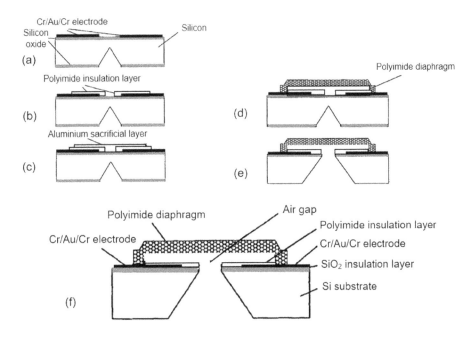

**FIGURE 19.19** (a–e) Sensor fabrication process; and (f) cross-sectional view of the membrane-based capacitive sensor. (Reprinted from *Sens. Actuators A*, 63, Pedersen, M. et al., An IC-compatible polyimide pressure sensor with capacitive readout, 163–168, Copyright 1997, with permission from Elsevier.)

or from $SiO_x$ sacrificial layers by undercutting with HF (Rousche et al. 2001). For -OH terminated $SiO_2$ surfaces (e.g., oxidized Si or Pyrex wafers), polyimide films can be released by immersion in hot deionized (DI) water followed by buffered HF (BHF). It is hypothesized that the relatively rapid rate of film release is attributed to BHF induced breakage of hydrogen bonds between the polyimide and -OH terminated $SiO_2$ surfaces and not etching alone (Engel et al. 2003).

Many metal sacrificial layers have also been used, including Al wet release with a mixture of phosphoric-acetic-nitric acids and water (Schmidt et al. 1988; Pedersen et al. 1997), anodic dissolution in sodium chloride (Metz et al. 2004), electrochemical erosion (Jiang et al. 2013), thick electroplated Cu (ferric chloride release of 15–50 µm thick films) (Kim and Allen 1992), Cr ($HCl:H_2O=1$: 1 etch) (Frazier and Allen 1993b), and Ti (removal in dilute HF) (Li et al. 2001; McNamara et al. 2005). Polymers that decompose at high temperatures such as polycarbonates (Metz et al. 2004) and polynorbonenes (Bhusari et al. 2001) can also be used as sacrificial layers. These materials are removed at elevated temperatures of 300°C and 370°C–425°C, respectively, in microchannels or sealed cavities.

The poor adhesion of polyimide onto certain materials, as well as vice versa, has been noted as a processing challenge. Excellent adhesion to Cr without adhesion promotion has been reported, but for other substrates such as Si, Si-derivatives, Al, and Cu, it is possible to apply silane-based adhesion promotion agents (Pedersen et al. 1998b). Adhesion of thin film metals to polyimide has also been shown to be problematic (e.g., Al) (Bagolini et al. 2002). One technique to improve metal adhesion is through the use of adhesive layers such as $SiO_x$ for Al (Bagolini et al. 2002) or Ti for Au (Bae et al. 2012) and Pt (Marton et al. 2015). Others have noted that $O_2$ plasma RIE roughening of polyimide surfaces can promote metal-to-polyimide (Nakamura et al. 1996) and polyimide-to-polyimide adhesion (Rousche et al. 2001).

As for Parylene C, we will not consider this polymer, since Parylene devices, both structural and free film, are largely associated with bioMEMS due to the outstanding biocompatibility and optimal material properties of the polymer for biomedical applications. This information can be found in (Kim and Meng 2016).

Injection molding, an economical mass production technique, has also been used to fabricate microcantilevers out of thermoplastic polymers (McFarland et al. 2004, 2005a, 2005b). In this process, a molten polymer such as polypropylene is forced under pressure into a steel cavity (mold); the shape of the cavity defines the dimensions of both the base and the cantilever(s). Injection-molded microcantilevers have been shown to be of equal caliber to commercial silicon microcantilevers. McFarland and co-workers (2004, 2005) specified in detail the fabrication of injection-molded microcantilevers. One should note that despite their advantages over silicon-based cantilever arrays, polymeric cantilever arrays are not commercially available.

## REFERENCES

Adhikari B., Kar P. (2010) Polymers in chemical sensors. In: Korotcenkov G. (ed.) *Chemical Sensors: Fundamentals of Sensing Materials*. Vol. 3: *Polymers and Other Materials*. Momentum Press, New York, pp. 1–76.

Ahmad Z., Abbas M., Gunawan I., Shakoor R.A., Ubaid F., Touati F. (2018) Electro-sprayed PVA coating with texture-enriched surface morphology for augmented humidity sensing. *Prog. Organic Coatings* **117**, 7–9.

Ahmadizadegan H. (2016) Synthesis and gas transport properties of novel functional polyimide/ZnO nanocomposite thin film membranes. *RSC Adv.* **6**, 106778–106789.

Alexander C., Andersson H.S., Andersson L.I., Ansell R.J., Kirsch N., Nicholls I.A., et al. (2006) Molecular imprinting science and technology: A survey of the literature for the years up to and including 2003. *J. Mol. Recognit.* **19**, 106–180.

Al-Muhtaseb S.A., Ritter J.A. (2003) Preparation and properties of resorcinol–formaldehyde organic and carbon gels. *Adv. Mater.* **15**, 101–114.

Anke Nabe A., Staude E., Belfort G. (1997) Surface modification of polysulfone ultrafiltration membranes and fouling by BSA solutions. *J. Membr. Sci.* **133**, 57–72.

Apostolidis A. (2004) Combinatorial approach for development of optical gas sensors: Concept and application of high-throughput experimentation. PhD Thesis, University of Regensburg, Regensburg, Germany.

Asahi (1994) Tokyo, Japan, Technical data sheet for Asahi Chemical "Pimel TL-500, G-7000 and IX Grade, Asahi Chemical Co., Ltd, Tokyo, Japan.

Awaja F., Gilbert M., Kelly G., Fox B., Pigram P.J. (2009) Adhesion of polymers. *Prog. Polymer Sci.* **34**, 948–968.

Aziz F., Bakar A.A., Ahmad Z., Bawazeer T.M., Alsenany N., Alsoufi M.S., Supangat A. (2018) Template-assisted growth of nanoporous VTTBNc films: Morphology and moisture sensitivity studies. *Mater. Lett.* **211**, 195–198.

Bae S.H., Che J.-H., Seo J.-M., Jeong J., Kim E.T., Lee S.W., et al. (2012) *In vitro* biocompatibility of various polymer-based microelectrode arrays for retinal prosthesis. *Invest. Ophthalmol. Vis. Sci.* **53**, 2653–2657.

Bae Y.-M., Youn Lee Y.-H., Kim H.-S., Don Lee D.-J., Kim S.Y., Kim H.-D. (2017) Polyimide-polyurethane/urea block copolymers for highly sensitive humidity sensor with low hysteresis. *J. Appl. Polym. Sci.* **2017**, 44973.

Bagolini A., Pakula L., Scholtes T.L.M., Pham H.T.M., French P.J., Sarro P.M. (2002) Polyimide sacrificial layer and novel materials for post-processing surface micromachining. *J. Micromech. Microeng.* **12**, 385–389.

Bai H., Shi G. (2007) Revised gas sensor based on conducting polymer. *Sensors* **7**, 267–307.

Barth P.W., Bernards S.L., Angell J.B. (1985) Flexible circuit and sensor arrays fabricated by monolithic silicon technology. *IEEE Trans. Electron Devices* **32**, 1202–1205.

Becker H., Gartner C. (2008) Polymer microfabrication technologies for microfluidic systems. *Anal. Bioanal. Chem.* **390**, 89–111.

Bein T., Stucky G.D. (eds.) (1996) Special issue on nanostructured materials and references therein. *Chem. Mater.* **8**, 1569–2194.

Belfer S., Fainchtain R., Purinson Y., Kedem O. (2000) Surface characterization by FTIR-ATR spectroscopy of polyethersulfone membranes-unmodified, modified and protein fouled. *J. Membr. Sci.* **172**, 113–124.

Bergmann N.M., Peppas N.A. (2008) Molecularly imprinted polymers with specific recognition for macromolecules and proteins. *Prog. Polym. Sci.* **33**, 271–288.

Bhatnagar N. (2017) Plasma surface treatment to enhance adhesive bonding. In: *Handbook of Adhesive Technology*. CRC Press, Boca Raton, FL, pp. 67–94.

Bhowmik S., Bonin H.W., Bui V.T., Weir R.D. (2006) Modification of high-performance polymer composite through high-energy radiation and low-pressure plasma for aerospace and space applications. *J. Appl. Polym. Sci.* **102**, 1959–1967.

Bhusari D., Reed H.A., Wedlake M., Padovani A.M., Allen S.A.B., Kohl P. (2001) Fabrication of air-channel structures for microfluidic, microelectromechanical, and microelectronic applications. *J. Microelectromech. Syst.* **10**, 400–408.

Blanchet G.B., Loo Y.-L., Rogers J.A., Gao F., Fincher C.R. (2003) Large area, high resolution, dry printing of conducting polymers for organic electronics. *Appl. Phys. Lett.* **82**(3), 463–465.

Boisen A., Thundat T. (2009) Design and fabrication of cantilever array biosensors. *Mater Today* **12**, 32–38.

Boltshauser T, Chandran L., Balks H., Bose F., Steiner D. (1991) Humidity sensing properties and electrical permittivity of new photosensitive polyimides. *Sens. Actuators B* **5**, 161–164.

Bonizzoni G., Vassallo E. (2002) Plasma physics and technology: Industrial applications. *Vacuum* **64**, 327–336.

Breitbach L., Hinke E., Staude E. (1991) Heterogeneous functionalizing of polysulfone membranes. *Angew. Makromol. Chem.* **184**, 183–196.

Brie M., Turcu R., Neamtu C., Pruneanu S. (1996) The effect of initial conductivity and doping anions on gas sensitivity of conducting polypyrrole films to NH$_3$. *Sens. Actuators B* **37**, 119–122.

Brown H.R. (2000) Polymer adhesion. *Mater. Forum* **24**, 49–58,

Bryce M.R., Chissel A., Kathirgamanathan P., Parker D., Smith N.R.M. (1987) Soluble, conducting polymers from 3-substituted thiophenes and pyrroles. *J. Chem. Soc. Chem. Commun.* **6**, 466–467.

Bunz U.H.F. (2006) Breath figures as a dynamic templating method for polymers and nanomaterials. *Adv. Mater.* **18**, 973–989.

Calleja M., Nordstrom M., Alvarez M., Tamayo J., Lechuga L.M., Boisen A. (2005) Highly sensitive polymer-based cantilever-sensors for DNA detection. *Ultramicroscopy* **105**, 215–222.

Carlier J., Arscott S., Thomy V., Fourrier J.C., Caron F., Camart J.C., Druon C., Tabourier P. (2004) Integrated microfluidics based on multi-layered SU-8 for mass spectrometry analysis. *J. Micromech. Microeng.* **1**, 619–624.

Carraher C.E., Jr. (2008) *Polymer Chemistry*. Taylor & Francis Group, Boca Raton, FL.

Caseri W. (2003) Nanocomposites. In: Yang P. (ed.) *The Chemistry of Nanostructured Materials*. World Scientific, Singapore.

Chaluvaraju B.V., Ganiger S.K., Murugendrappa M.V. (2016) Thermo-electric power and humidity sensing studies of the polypyrrole/tantalum pentoxide composites. *J. Mater. Sci.: Mater Electron* **27**, 1044–1055.

Chan C.-M. (1994) *Polymer Surface Modification and Characterization*. Hanser, Munich, Germany.

Chang C.-A., Kim Y.-K., Schrott A.G. (1990) Adhesion studies of metals on fluorocarbon polymer films. *J. Vac. Sci. Technol. A* **8**(4), 3304–3309.

Chatani S., Kloxin C.J., Bowman C.N. (2014) The power of light in polymer science: Photochemical processes to manipulate polymer formation, structure, and properties. *Polym. Chem.* **5**, 2187–2201.

Chen M., Frueh J., Wang D., Lin X., Xie H., He Q. (2017) Polybenzoxazole nanofiber-reinforced moisture-responsive soft actuators. *Sci. Reports* **7**, 769.

Chen W., Li C.M., Chen P., Sun C.Q. (2006) Electrosynthesis and characterization of polypyrrole/Au nanocomposite. *Electrochim. Acta* **52**, 1082–1086.

Chen Y.J., Kang E.T., Neoh K.G., Tan K.L. (2001) Oxidative graft polymerization of aniline on modified Si(100) surface. *Macromolecules* **34**, 3133–3141.

Cheng C., Liye Z., Zhan R.-J. (2006) Surface modification of polymer fibre by the new atmospheric pressure cold plasma jet. *Surf. Coatings Technol.* **200**(24), 6659–6665.

Chiang C.K., Fincher Jr. C.R., Park Y.W., Heeger A.J., Shirakawa H., Louis E.J., MacDiarmid A.G. (1977) Electrical conductivity in doped polyacetylene. *Phys. Rev. Lett.* **39**, 1098–1101.

Chiang J.C., Macdiarmid A.G. (1986) Polyaniline: Protonic acid doping of the emeraldine from to the metallic regime. *Synth. Met.* **13**, 193–205.

Chrisey D., Hubler G. (eds.) (1994) *Pulsed Laser Deposition of Thin Films*. Wiley, New York.

Çiğil A.B., Cankurtaran H., Kahraman M.V. (2017) Photocrosslinked thiolene-based hybrid polymeric sensor for humidity detection. *React. Funct. Polym.* **114**, 75–85.

Cooper A.I. (2009) Conjugated microporous polymers. *Adv. Mater.* **21**, 1291–1295.

Cormack P.A.G., Elorza A.Z. (2004) Molecularly imprinted polymers: Synthesis and characterisation. *J. Chromatogr. B* **804**, 173–182.

Corrigan N., Shanmugam S., Xu J., Boyer C. (2016) Photocatalysis in organic and polymer synthesis. *Chem. Soc. Rev.* **45**, 6165–6212.

Courbat J., Briand D., de Rooij N.F. (2010) Foil level packaging of a chemical gas sensor. *J. Micromech. Microeng.* **20**, 055026.

Cui X.Y., Martin D.C. (2003) Fuzzy gold electrodes for lowering impedance and improving adhesion with electrodeposited conducting polymer films. *Sens. Actuators A* **103**, 384–394.

Curran S., Ajayan P.M., Blau W., Carroll D.L., Coleman J.N., Dalton A., et al. (1998) A composite from poly(*m*-phenylenevinylene-*co*-2,5-dioctoxy-*p*-phenylenevinylene) and carbon nanotubes: A novel material for molecular optoelectronics. *Adv. Mater.* **10**, 1091–1093.

Czech Z., Butwin A., Herko E., Hefczyc B., Zawadiak J. (2008) Novel azo-peresters radical initiators used for the synthesis of acrylic pressure-sensitive adhesives. *eXPRESS Polymer Lett.* **2**(4), 277–283.

Dai J., Zhang T., Qi R., Zhao H., Fei T., Lu G. (2017) LiCl loaded cross-linked polymer composites by click reaction for humidity sensing. *Sens. Actuators B* **253**, 361–367.

Daniel J.H. (2001) Micro-electro-mechanical system fabrication technology applied to large area x-ray image sensor arrays. *J. Vac. Sci. Technol. A* **19**, 1219–1223.

De Leeuw D.M., Simenon M.M.J., Brown A.R., Einerhand R.E.E. (1997) Stability of *n*-type doped conducting polymers and consequences for polymeric microelectronic devices. *Synth. Met.* **87**, 53–59.

De Souza J.E.G., dos Santos F.L., Barros-Neto B., dos Santos C.G., de Melo C.P. (2001) Polypyrrole thin films gas sensors. *Synth. Met.* **119**, 383–384

Dellmann L., Roth S., Beuret C., Racine G.-A., Lorenz H., Despont M., et al. (1998) Fabrication process of high aspect ratio elastic and SU-8 structures for piezoelectric motor applications. *Sens. Actuators A* **70**, 42–47.

Diaz A.F., Castillo J., Kanazawa K.K., Logan J.A., Salmon M., Fajardo O. (1982) Conducting poly-N-alkylpyrrole polymer films. *J. Electroanal. Chem.* **133**(2), 233–239.

DuPont (1994) Technical data sheet for Dupont "Pyralin PI 2700" series. Wilmington, DE.

Ebnesajjad S. (2014) Surface treatment and bonding of ceramics. In: *Surface Treatment of Materials for Adhesive Bonding* (2nd edn.). Elsevier, Waltham, pp. 283–299.

Engel J., Chen J., Liu C. (2003) Development of polyimide flexible tactile sensor skin. *J. Micromech. Microeng.* **13**, 359–366.

Favia P., De Agostino R. (1998) Plasma treatments and plasma deposition of polymers for biomedical applications. *Surf. Coat. Technol.* **98**, 1102–1106.

Feast W.J., Tsibouklis J., Pouwer K.L., Groenendaal L., Meijer E.W. (1996) Synthesis, processing and material properties of conjugated polymers. *Polymer* **37**(22), 5017–5047.

Fei T., Jiang K., Liu S., Zhang T. (2014) Humidity sensors based on Li-loaded nanoporous polymers. *Sens. Actuators B* **190**, 523–528.

Flory P.J. (1953) *Principles of Polymer Chemistry.* Cornell University Press, Ithaca, NY, chap. 2.

Fragakis J., Chatzandroulis S., Papadimitriou D., Tsamis C. (2005) Simulation of capacitive type bimorph humidity sensors. *J. Phys. Confer. Ser.* **10**, 305–308.

Frazier A.B. (1995) Recent applications of polyimide to micromachining technology. *IEEE Trans. Ind. Electron.* **42**, 442–448.

Frazier A.B., Ahn C.H., Allen M.G. (1994) Development of micromachined devices using polyimide-based processes. *Sens. Actuators A* **45**, 47–55.

Frazier A.B., Allen M.G. (1993a) Piezoresistive graphite polyimide thin-films for micromachining applications. *J. Appl. Phys.* **73**, 4428–4433.

Frazier A.B., Allen M.G. (1993b) Metallic microstructures fabricated using photosensitive polyimide electroplating molds. *J. Microelectromech. Syst.* **2**, 87–94.

Gad-el-Hak M. (2006) *MEMS: Design and Fabrication*, 2nd edn. CRC Press, Boca Raton, FL.

Gaines G.L. Jr. (1966) *Insoluble Monolayers at Liquid-Gas Interfaces.* Wiley-Interscience, New York.

Garbassi F, Morra M, Occhiello E. (1994) *Polymer Surfaces.* Wiley, Chichester, UK.

Garcia-Nunez J.A., Pelaez-Samaniego M.R., Garcia-Perez M.E., Fonts I., Abrego J., Westerhof R.J.M., Garcia-Perez M. (2017) Historical developments of pyrolysis reactors: A review. *Energy Fuels* **31** (6), 5751–5775.

Gardner J.W., Bartlett P.N. (1995) Application of conducting polymer technology in microsystems. *Sens. Actuators A* **51**, 57–66.

Goeders K.M., Colton J.S., Bottomley L.A. (2008) Microcantilevers: Sensing chemical interactions via mechanical motion. *Chem. Rev.* **108**, 522–542.

Gong M.-S., Kim J.U., Kim J.-G. (2010) Preparation of water-durable humidity sensor by attachment of polyelectrolyte membrane to electrode substrate by photochemical crosslinking reaction. *Sens. Actuators B* **147**, 539–547.

Gopal R., Zuwei M., Kaur S., Ramakrishna S. (2007) Surface modification and application of functionalized polymer nanofibers. In: Mansoori G.A., George Th. F., Assoufid L., Zhang G. (eds.), *Molecular Building Blocks for Nanotechnology: From Diamondiods to Nanoscale Materials and Applications.* E-Publishing, Springerlink.com, Berlin, Germany, pp. 72–91.

Grayson A.C.R, Shawgp R.S., Johnson A.M., Flynn N.T., Li Y., Cima M.J., Langer R. (2004) A BioMEMS review: MEMS technology for physiologically integrated devices. *Proc. IEEE* **92**, 6–21.

Guernion N., Costello B.P.J.D., Ratcliffe N.M. (2002) The synthesis of 3-octadecyl- and 3-docosylpyrrole, their polymerisation and incorporation into novel composite gas sensitive resistors. *Synth. Met.* **128**, 139–147.

Guiver M.D., Apsimon J.W. (1988) The modification of polysulfone by metalation. *Polym. Sci. C: Polym. Lett.* **26**, 123–127

Gurunathan K., Murugan A.V., Marimuthu R., Mulik U.P., Amalnerkar D.P. (1999) Electrochemically synthesized conducting polymeric materials for applications towards technology in electronics, optoelectronics and energy storage devices. *Mater. Chem. Phys.* **61**(3), 173–191.

Hanemann T., Szabó D.V. (2010) Polymer-nanoparticle composites: From synthesis to modern applications. *Materials* 3, 3468–3517.

Harsanyi G. (1995) Polymeric sensing films: New horizons in sensorics? *Sens. Actuators A* **46–47**, 85–88.

Harsanyi G. (2000) *Sensors in Biomedical Applications: Fundamentals, Technology and Applications.* Technomic, Basel, Switzerland.

Hart B.R., Shea K.J. (2001) Synthetic peptide receptors: Molecularly imprinted polymers for the recognition of peptides using peptide-metal interactions. *J. Am. Chem. Soc.* **123**, 2072–2073.

Hart-Smith J.J. (1983) Adhesive layer thickness and porosity criteria for bonded joints, Final report AFWAL-TR-82-4172I, Materials Laboratory (AFWAL/MLBC), Long Beach, California.

Hart-Smith L.J., Brown D., Wong S. (1998) Surface preparations for ensuring that the glue will stick in bonded composite structures. In: Peters S.T. (ed.) *Handbook of Composites.* Springer, Boston, MA, pp. 667–685.

Hatchett D.W., Josowicz M. (2008) Composites of intrinsically conducting polymers as sensing nanomaterials. *Chem. Rev.* **108**, 746–769.

Hernández-Rivera D., Rodríguez-Roldán G., Mora-Martínez R., Suaste-Gómez E. (2017) A capacitive humidity sensor based on an electrospun PVDF/Graphene membrane. *Sensors* **17**, 1009.

Hillmyer M.A. (2005) Nanoporous materials from block copolymer precursors. *Adv. Polym. Sci.* **190**, 137–181.

Hoffman A.S. (1996) Surface modification of polymers: Physical, chemical, mechanical and biological methods. *Macromol. Symp.* **101**, 443–454.

Hong G., Holmes A.S., Heaton M.E. (2004) SU8 resist plasma etching and its optimization. *Microsyst. Technol.* **10**, 357–359.

Hussain F., Hojjati M., Okamoto M., Gorga R.E. (2006) Review article: Polymer-matrix nanocomposites, processing, manufacturing, and application: An overview. *J. Composite Mater.* **40**(17), 1511–1574.

Jain S., Chakane S., Samui A.B., Krishnamurthy V.N., Bhoraskar S.V. (2003) Humidity sensing with weak acid-doped polyaniline and its composites. *Sens. Actuators B* **96**, 124–129.

Jiang J.-X., Cooper A.I. (2010) Microporous organic polymers: Design, synthesis, and function. *Top. Curr. Chem.* **293**, 1–33.

Jiang X., Sui X., Lu Y., Yan Y., Zhou C., Li L., et al. (2013) *In vitro* and *in vivo* evaluation of a photosensitive polyimide thin-film microelectrode array suitable for epiretinal stimulation. *J. NeuroEng. Rehabil.* **10**, 48.

Jiguet S., Bertsch A., Hofmann H., Renaud P. (2004) SU8-silver photosensitive nanocomposite. *Adv. Eng. Mater.* **6**, 719–724.

Johansson A., Blagoi G., Boisen A. (2006) Polymeric cantilever-based biosensors with integrated readout. *Appl. Phys. Lett.* **89**, 173505.

Johansson A., Calleja M., Rasmussen P.A., Boisen A. (2005) SU-8 cantilever sensor system with integrated readout. *Sens. Actuators A*, **123–124**, 111–115.

Kanamori K., Nakanishi K., Hanada T. (2006) Rigid macroporous poly(divinylbenzene) monoliths with a well-defined bicontinuous morphology prepared by living radical polymerization. *Adv. Mater.* **18**, 2407–2411.

Kanatzidis M.G. (1990) Polymeric electrical conductors. *Chem. Eng. News.* **68**(49), 36–54.

Kang E.T., Zhang Y. (2000) Surface modification of fluoropolymers via molecular design. *Adv. Mater.* **12**, 1481–1494.

Kaplan S.L., Lopata E.S., Smith J. (1993) Plasma processes and adhesive bonding of polytetrafluoroethylene. *Surf. Interf. Anal.* **20**, 331–336.

Karnicka K., Chojak M., Miecznikowski K., Skunik M., Baranowska B., Kolary A., et al. (2005) Polyoxometallates as inorganic templates for electrocatalytic network films of ultrathin conducting polymers and platinum nanoparticles. *Bioelectrochemistry* **66**, 79–87.

Karsa D.R. (2003) *Surfactants in Polymers, Coatings, Inks and Adhesives.* Blackwell Publishing, Oxford, UK.

Kato K., Uchida E., Kang E.-T., Uyama Y., Ikada Y. (2003) Polymer surface with graft chains. *Prog. Polym. Sci.* **28**, 209–259.

Kawai T., Kojima S., Tanaka F., Yoshino K. (1998) Electrical property of poly(3-octyloxythiophene) and its gas sensor application. *Jpn. J. Appl. Phys. Part 1* **37**, 6237–6241.

Kempe M., Mosbach K. (1995) Receptor binding mimetics: A novel molecularly imprinted polymer receptor. *Tetrahedr. Lett.* **36**, 3563–3566.

Khan W.S., Hamadneh N.F., Khan W.A. (2017) Polymer nanocomposites—Synthesis techniques, classification and properties. In: Di Sia P. (ed.) *Science and Applications of Tailored Nanostructures.* One Central Press, Cheshire, UK, pp. 50–67.

Kickelbick G. (2007) *Hybrid Materials: Synthesis, Characterization, and Applications.* Wiley-VCH, Weinheim.

Kim B.J., Meng E. (2016) Review of polymer MEMS micromachining. *J. Micromech. Microeng.* 26, 013001.

Kim P., Kwon K.W., Park M.C., Lee S.H., Kim S.M., Suh K.Y. (2008) Soft lithography for microfluidics: A review. *Biochip J.* **2**, 1–11.

Kim Y.W., Allen M.G. (1992) Single-layer and multilayer surface-micromachined platforms using electroplated sacrificial layers. *Sens. Actuators* A **35**, 61–68.

Kim Y.W., Allen M.G. (1999) *In situ* measurement of mechanical properties of polyimide films using micromachined resonant string structures. *IEEE Trans. Compon. Packaging Technol.* **22**, 282–290.

Kimmins S.D., Cameron N.R. (2011) Functional porous polymers by emulsion templating: Recent advances. *Adv. Funct. Mater.* **21**, 211–225.

Kinloch A.J. (1987) *Adhesion and Adhesives: Science & Technology.* Chapman and Hall, New York.

Koo J.H. (2006) *Polymer Nanocomposites: Processing, Characterization and Applications.* McGraw-Hill, New York.

Korotcenkov G. (2013) *Handbook of Gas Sensor Materials.* Springer, New York.

Krishnamoorti R. (2007) Strategies for dispersing nanoparticles in polymers. *Mater. Res. Bull.* **32**, 341–347.

Kumar D., Sharma R.C. (1998) Advances in conductive polymers. *Eur. Polym. J.* **34** (8), 1053–1060.

Kus M., Okur S. (2009) Electrical characterization of PEDOT:PSS beyond humidity saturation. *Sens. Actuators B* **143**, 177–181.

Langowskia H.-C. (2011) Surface modification of polymer films for improved adhesion of deposited metal layers. *J. Adhesion Sci. Technol.* **25** (1–3), 223–243.

Larrieu J., Held B., Martinez H., Tison Y. (2005) Ageing of a tactic and isotactic polystyrene thin films treated by oxygen DC pulsed plasma. *Surf. Coat. Technol.* **200**, 2310–2316.

Lau K.H., Giridhar A., Harikrishnan S., Satyanarayana N., Sinha S.K. (2013) Releasing high aspect ratio SU-8 microstructures using AZ photoresist as a sacrificial layer on metallized Si substrates. *Microsyst. Technol.* **19**, 1863–1871.

Lavine B.K., Kaval N., Westover D.J., Oxenford L. (2006) New approaches to chemical sensing-sensors based on polymer swelling. *Anal. Lett.* **39**, 1773–1783.

Lawrence J., Li L. (2001) Modification of the wettability characteristics of polymethyl methacrylate (PMMA) by means of $CO_2$, Nd:YAG, excimer and high power diode laser radiation. *Mater. Sci. Eng. A* **303**(1–2), 142–149.

Lazarus N., Bedair S.S., Lo C.-C., Fedder G.K. (2010) CMOS-MEMS capacitive humidity sensor. *J. Microelectromech. Systems* **19**(1), 183–191.

Lee C.-H., Chuang W.-Y., Lin S.-H., Wu W.-J., Lin C.-T. (2013) A printable humidity sensing material based on conductive polymer and nanoparticles composites. *Jpn. J. Appl. Phys.* **52**, 05DA08.

Lee M.J., Lee C.-J., Singh V.R., Yoo K.-P., Min N.-K. (2008) Humidity sensing characteristics of plasma functionalized multiwall carbon nanotube-polyimide composite films. In: *Proceedings of IEEE SENSORS Conference*, October 26–29, Lecce, Italy, pp. 430–433.

Li B., Sauve G., Iovu M.C., Jeffries-El M., Zhang R., Cooper J., et al. (2006) Volatile organic compound detection using nanostructured copolymers. *Nano Lett.* **6**, 1598–1602.

Li H., Zhang S., Gong C., Liang Y., Qi Z., Li Y. (2014a) Novel high $T_g$, organosoluble poly(etherimide)s containing 4,5-diazafluorene unit: Synthesis and characterization. *Eur. Polym. J.* 2014, **54**, 128–137.

Li M.H., Wu J.J., Gianchandani Y.G. (2001) Surface micromachined polyimide scanning thermocouple probes *J. Microelectromech. Syst.* **10**, 3–9.

Li W., Yoon J.A., Matyjaszewski K. (2010) Dual-reactive surfactant used for synthesis of functional nanocapsules in miniemulsion. *J. Am. Chem. Soc.* **132**, 7823–7825.

Li W.T., Charters R.B., Luther-Davies B., Mar L. (2004) Significant improvement of adhesion between gold thin films and a polymer. *Appl. Surf. Sci.* **233**, 227–233.

Li Y., Deng C., Yang M. (2012) A novel surface acoustic wave-impedance humidity sensor based on the composite of polyaniline and poly(vinyl alcohol) with a capability of detecting low humidity. *Sens. Actuators B* **165**, 7–12.

Li Y., Deng C., Yang M. (2014b) Facilely prepared composites of polyelectrolytes and graphene as the sensing materials for the detection of very low humidity. *Sens. Actuators B* **194**, 51–58.

Li Y., Yang M.J., She Y. (2005) Humidity sensitive properties of crosslinked and quaternized poly(4-vinylpyridine-co-butyl methacrylate). *Sens. Actuators B* **107**, 252–257.

Li Y., Zhao H., Jiao M., Yang M. (2018) Sulphonated polystyrene-b-poly(4-vinylpyridine) with nanostructures induced by phase separation as promising humidity sensitive material. *Sens. Actuators B* **257**, 1118–1127.

Liston E.M., Martinu L., Wertheimer M.R. (1993) Plasma surface modification of polymers for improved adhesion: A critical review. *J. Adhesion Sci. Technol.* **7**, 1077–1089.

Liu C. (2007) Recent developments in polymer MEMS. *Adv. Mater.* **19**, 3783–3790.

Loo Y.-L., Someya T., Baldwin K.W., Bao Z., Ho P., Dodabalapur A., et al. (2002) Soft, conformable electrical contacts for organic semiconductors: High-resolution plastic circuits by lamination. *Proc. Nat. Acad. Sci. U.S.A.* **99**, 10252–10256.

Lorenz H., Despont M., Fahrni N., LaBianca N., Renaud P., Vettiger P. (1997) SU-8: A low-cost negative resist for MEMS. *J. Micromech. Microeng.* **7**, 121–124.

Lukes S.J., Dickensheets D.L. (2013) SU-8 2002 surface micromachined deformable membrane mirrors. *J. Microelectromech. Syst.* **22**(1), 94–106.

Ma S., Li Y., Sun X., Yu X., Jin Y. (2009) Study of polyimide as sacrificial layer with $O_2$ plasma releasing for its application in MEMS capacitive FPA fabrication. In: *Proceedings of Int. Conf. on Electronic Packaging Technology & High Density Packaging*, August 10–13, Beijing, China, pp. 526–529.

MacDiarmid A.G. (2001) "Synthetic metals": A novel role for organic polymers. *Curr. Appl. Phys.* **1**, 269–279.

MacDiarmid A.G. (2002) Synthetic metals: A novel role for organic polymers. *Synthetic Metals* **125**, 11–22.

Macdiarmid A.G., Chiang J.C., Richter A.F., Epstein A.J. (1987) Polyaniline: A new concept in conducting polymers. *Synth. Met.* **18**, 285–290.

Mahadeva S.K., Yun S., Kim J. (2011) Flexible humidity and temperature sensor based on cellulose–polypyrrole nanocomposite. *Sens. Actuators A* **165**, 194–199.

Malinauskas A. (2001) Chemical deposition of conducting polymers. *Polymer* **42**, 3957–3972.

Malinauskas A., Malinauskiene J., Ramanavicius A. (2005) Conducting polymer-based nanostructurized materials: Electrochemical aspects. *Nanotechnology* **16**, R51–R62.

Malkin A.Y., Siling M.I. (1991) Scientific principles of present-day and future technologies of synthesis and processing polycondensation polymers. *Rev. Polymer Sci.* **33**, 2135–2160.

Manesiotis P., Hall A.J., Courtois J., Irgum K., Sellergren B. (2005) An artificial riboflavin receptor prepared by a template analogue imprinting strategy. *Angew. Chem. Int. Ed.* **44**, 3902–3906.

Martin C.R., Parthasarathy R., Menon V. (1993) Template synthesis of electronically conductive polymers—A new route for achieving higher electronic conductivities. *Synth. Met.* **55** (2–3), 1165–1170.

Martinez H., Ren N., Matta M.E., Hillmyer M.A. (2014) Ring-opening metathesis polymerization of 8-membered cyclic olefins. *Polym. Chem.* **5**, 3507–3532.

Martinez-Duarte R. (2014) SU-8 Photolithography as a toolbox for carbon MEMS. *Micromachines* **5**, 766–782.

Marton G., Kiss M., Orbán G., Pongrácz A., Ulbert I. (2015) A polymer-based spiky microelectrode array for electrocorticography. *Microsyst. Technol.* **21**, 619–624.

Mastalerz M. (2008) The next generation of shape-persistent zeolite analogues: Covalent organic frameworks. *Angew. Chem., Int. Ed.* **47**, 445–447.

Mata A., Fleischman A.J., Roy S. (2006) Fabrication of multi-layer SU-8 microstructures. *J. Micromech. Microeng.* **16**, 276–284.

Matsuguchi M., Sakurada K., Sakai Y. (2003) Effect of crosslinked structure on $SO_2$ gas sorption properties in amino-functional copolymers. *J. Appl. Polymer Sci.* **88**, 2982–2987.

Matsumoto Y., Yoshida K., Ishida M. (1998) A novel deposition technique for fluorocarbon films and its applications for bulk- and surface-micromachined devices. *Sens. Actuators A* **66**, 308–314.

McFarland A.W., Colton J.S. (2005a) Role of material microstructure in plate stiffness with relevance to microcantilever sensors. *J. Micromech. Microeng.* **15**, 1060–1067.

McFarland A.W., Colton J.S. (2005b) Chemical sensing with micromolded plastic microcantilevers. *J. Microelectromech. Syst.* **14**, 1375–1385.

McFarland A.W. Poggi M.A. Bottomley L.A., Colton J.S. (2004) Injection moulding of high aspect ratio micron-scale thickness polymeric microcantilevers. *Nanotechnology* **15**, 1628–1632.

McGovern S.T., Spinks G.M., Wallace G.G. (2005a) Micro-humidity sensors based on a processable polyaniline blend. *Sens. Actuators B* **107**, 657–665.

McGovern S.T., Spinks G.M., Wallace G.G. (2005b) Highly processable method for the construction of miniature conducting polymer moisture sensors. *Proc. SPIE* **5649**, 607–615.

McKeown N.B., Budd P.M. (2010) Exploitation of intrinsic microporosity in polymer-based materials. *Macromolecules* **43**, 5163–5176.

McNamara S., Basu A.S., Lee J.-H., Gianchandani Y.D. (2005) Ultracompliant thermal probe array for scanning non-planar surfaces without force feedback. *J. Micromech. Microeng.* **15**, 237–243.

Memmi D., Foglietti V., Cianci E., Caliano G., Pappalardo M. (2002) Fabrication of capacitive micromechanical ultrasonic transducers by low temperature process. *Sens. Actuators A* **99**, 85–91.

Metz S., Bertsch A., Renaud Ph. (2004) Polyimide and SU-8 microfluidic devices manufactured by heat-depolymerizable sacrificial material technique. *Lab Chip* **4**, 114–120.

Miasik J., Hooper A., Tofield B. (1986) Conducting polymer gas sensors. *J. Chem. Soc. Faraday Trans. 1* **82** (4), 1117–1127.

Michael D., Guiver M.D., Kutowy O., Apsimon J.W. (1989) Functional group polysulphones by bromination–metalation. *Polymer* **30**, 1137–1142.

Miller J.S. (ed.) (1982) *Catalysis and Electrocatalysis.* Am. Chem. Soc. Symp. Ser. Vol. 192. American Chemical Society, Washington, DC.

Moad G., Solomon D.H. (1995) *The Chemistry of Free Radical Polymerization.* Elsevier Science, Oxford, UK.

Mutel B., Grimblot J., Dessaux O., Goudmand P. (2000) XPS investigations of nitrogen-plasma-treated polypropylene in a reactor coupled to the spectrometer. *Surf. Interface Analysis* **30** (1), 401–406.

Nady N., Franssen M.C.R., Zuilhof H., Mohy Eldin M.S., Boom R., Schroën K. (2001) Modification methods for poly(arylsulfone) membranes: A mini-review focusing on surface modification. *Desalination* **275**, 1–9.

Najjar R., Nematdoust S. (2016) A resistive-type humidity sensor based on polypyrrole and ZnO nanoparticles: Hybrid polymers I nanocomposites. *RSC Adv.* **6**, 112129.

Nakamatsu K.J., Delgado-Aparicio V.L.F. (1997) Modificacion de superficies de polymeros con plasma. *Revista Plasticos Modernos* **74**, 262–268.

Nakamura Y., Suzuki Y., Watanab Y. (1996) Effect of oxygen plasma etching on adhesion between polyimide films and metal. *Thin Solid Films* **291**, 367–369.

Nalwa H.S. (1997) *Handbook of Organic Conductive Materials and Polymers.* Wiley, New York.

Napo K., Safoula G., Bernede J.C., D'Almeida K., Touirhi S., Alimi K., Barreau A. (1999) Influence of the iodine doping process on the properties of organic and inorganic polymer thin films. *Polym. Degrad. Stabil.* **66**, 257–262.

Nicho M.E., Trejo M., Garcia-Valenzuela A., Saniger J.M., Palacios J., Hu H. (2001) Polyaniline composite coatings interrogated by a nulling optical-transmittance bridge for sensing low concentrations of ammonia gas. *Sens. Actuators B* **76**, 18–24.

OCG (1994) *Technical data sheet for Ciba Geigy "Probimide 7000, 7500 and 400*, OCG Microelectronics Materials AG, Basel, Switzerland.

Odian G. (2004) *Principles of Polymerization*, 4th ed. John Wiley, Hoboken, NJ.

Okamoto Y., Nakano T. (1994) Asymmetric polymerization. *Chem. Rev.* **94** (2), 349–372.

Oliveira M., Machado A.V. (2013) Preparation of polymer-based nanocomposites by different routes. In: Wang X. (ed.) *Nanocomposites: Synthesis, Characterization and Applications.* NOVA Science Publishers, New York.

Olson D.A., Chen L., Hillmyer M.A. (2008) Templating nanoporous polymers with ordered block copolymers. *Chem. Mater.* **20**, 869–890.

Ooij W.J., Surman D., Yasuda H.K. (1995) Plasma-polymerized coatings of trimethylsilane deposited on cold-rolled steel substrates Part 2. Effect of deposition conditions on corrosion performance. *Prog. Org. Coatings* **25**, 319–337.

Osada Y., DeRossi D.E. (eds.) (2000) *Polymer Sensors and Actuators.* Springer-Verlag, Berlin, Germany.

Otero T.F., Rodriguez J. (1994) Parallel kinetic studies of the electro-generation of conducting polymers: Mixed materials, composition and properties control. *Electrochim. Acta* **39**(2), 245–253.

Othman M.B.H., Ahmad Z., Akil H.M. (2008) Fabrication of nanoporous polyimide of low dielectric constant. In: Proceedings of *33rd IEEE/CPMT International Electronic Manufacturing Technology Symposium*, November 4–6, Penang, Malaysia, pp. 1–4.

Pabin-Szafko B., Wisniewska E., Szafko J. (2005): Functional azo-initiators—Synthesis and molecular characteristics. *Polimery* **50**, 271–278.

Pan X., Fantin M., Yuan F., Matyjaszewski K. (2018) Externally controlled atom transfer radical polymerization. *Chem. Soc. Rev.* **47**, 5457–5490.

Pappas D.I., Cuomo J.J., Sachdev K.G. (1991) Studies of adhesion of metal films to polyimide. *J. Vac. Sci. Technol. A* **9**(5), 2704–2708.

Park K.-J., Gong M.-S. (2017) A water durable resistive humidity sensor based on rigid sulfonated polybenzimidazole and their properties. *Sens. Actuators B* **246**, 53–60.

Pedersen M., Meijerink M.G.H., Olthuis W., Bergveld P. (1997) An IC-compatible polyimide pressure sensor with capacitive readout. *Sens. Actuators A* **63**, 163–168.

Pedersen M., Olthuis W., Bergveld P. (1998a) High-performance condenser microphone with fully integrated CMOS amplifier and DC-DC voltage converter. *J. Microelectromech. Syst.* **7**, 387–394.

Pedersen M., Olthuis W., Bergveld P. (1998b) Development and fabrication of capacitive sensors in polyimide. *Sensors Mater.* **10**, 1–20.

Pei Q., Inganas O. (1993) Conjugated polymers as smart materials, gas sensors and actuators using bending beams. *Synth. Met.* **57**(1), 3730–3735.

Pichon V., Chapuis-Hugon F. (2008) Role of molecularly imprinted polymers for selective determination of environmental pollutants—A review. *Anal. Chim. Acta* **622**, 48–61.

Pique A., Auyeung R.C.Y., Stepnowsk J.L., Weir D.W., Arnold C.B., McGill R.A., Chrisey D.B. (2003) Laser processing of polymer thin films for chemical sensor applications. *Surf. Coat. Technol.* **163–164**, 293–299.

Power A., Betts T., Cassidy J. (2010) Silver nanoparticle polymer composite based humidity sensor. *Analyst* **135**, 1645–1652.

Qui Y.Y., Azeredo-Leme C., Alcacer L.R., Franca J.E. (2001) A CMOS humidity sensor with on-chip calibration. *Sens. Actuators A* **92**, 80–87.

Rajesh, Ahuja T., Kumar D. (2010) Nanocomposites: From fabrication to chemical sensor application. In: Korotcenkov G. (ed.) *Chemical Sensors: Fundamentals of Sensing Materials.* Vol. 2: *Nanostructured Materials.* Momentum Press. New York, pp. 335–368.

Rana D., Matsuura T. (2010) Surface modifications for antifouling membranes. *Chem. Rev.* **110**, 2448–2471.

Ranby B. (1999) Surface modification and lamination of polymers by photografting. *Int. J. Adhes. Adhes.* **19**, 337–343.

Ransley J.H.T., Watari M., Sukumaran D., McKendry R.A., Seshia A.A. (2006) SU8 bio-chemical sensor microarrays. *Microelectron. Eng.* **83**, 1621–1625.

Reddy A.V.R., Trivedi J.J., Devmurari C.V., Mohan D.J., Singh P., Rao A.P., et al. (2005) Fouling resistant membranes in desalination and water recovery. *Desalination* **183**, 301–306.

Reisinger J.J., Hillmyer M.A. (2002) Synthesis of fluorinated polymers by chemical modification. *Prog. Polym. Sci.* **27**, 971–1005.

Ren J.M., McKenzie T.G., Fu Q., Wong E.H.H., Xu J., An Z., et al. (2016) Star polymers. *Chem. Rev.* **116**(12), 6743–6836.

Rong M.Z., Zhang M.Q., Ruan W.H. (2006) Surface modification of nanoscale fillers for improving properties of polymer nanocomposites: A review. *Mater. Sci. Technol.* **22**, 787–796.

Rousche P J., Pellinen D.S., Pivin D.P., Jr., Williams J.C., Vetter R.J., Kipke D.R. (2001) Flexible polyimide-based intracortical electrode arrays with bioactive capability. *IEEE Trans. Biomed. Eng.* **48**, 361–371,

Rozenberg B.A., Tenne R. (2008) Polymer-assisted fabrication of nanoparticles and nanocomposites. *Prog. Polym. Sci.* **33**, 40–112.

Ruangchuay L., Sirivat A., Schwank J. (2004) Electrical conductivity response of polypyrrole to acetone vapor: Effect of dopant anions and interaction mechanisms. *Synth. Met.* **140**, 15–21.

Ruiz-Hitzky E., Aranda P., Darder M., Ogawa M. (2011) Hybrid and biohybrid silicate based materials: Molecular vs block-assembling bottom-up processes. *Chem. Soc. Rev.* **40**, 801–828.

Sadaoka Y., Sakai Y., Akiayama H. (1986) A humidity sensor using alkali salt poly(ethylene oxide) hybrid films. *J. Mater. Sci.* **21**, 235–240.

Safoula G., Napo K., Bernede J.C., Touihri S., Alimi K. (2001) Electrical conductivity of halogen doped poly-(N-vinylcarbazole) thin films. *Eur. Polym. J.* **37**, 843–849.

Safoula G., Touihri S., Bernede J.C., Jamali M., Rabiller C., Molinie Ph., Napo K. (1999) Properties of the complex salt obtained by doping the poly(N-vinylcarbazole) with bromine. *Polymer* **40**, 531–539.

Sajid M., Kim H.B., Yang Y.J., Jo J., Choi K.H. (2017) Highly sensitive BEHP-co-MEH:PPV + Poly(acrylic acid) partial sodium salt based relative humidity sensor. *Sens. Actuators B* **246**, 809–818.

Sakai Y., Sadaoka Y., Matsuguchi M. (1996) Humidity sensors based on polymer thin films. *Sens. Actuators B* **35–36**, 85–90.

Saravanan S., Mathai J., Venkatachalam S., Anantharaman M.R. (2004) Low K thin films based on RF plasma-polymerized aniline. *New J. Phys.* **6**, 64–68.

Sato, M., Tanaka S., Kacriyama K. (1986) Electrochemical preparation of conducting poly(3-methylthiophene): Comparison with polythiophene and poly(3-ethylthiophene). *Synth. Met.* **14**(4), 279–288.

Sattler K.D., Pricl S., Posocco P., Scocchi G., Fermeglia M. (2010) Polymer–clay nanocomposites. In: Sattler K.D. (ed.) *Handbook of Nanophysics, Functional Nanomaterials.* CRC Press, Boca Raton, FL, Chapter 3.

Schmidt M.A., Howe R.T., Senturia S.D., Aritonidis H. (1988) Design and calibration of a microfabricated floating-element shear-stress sensor. *IEEE Trans. Electron Devices* **35**, 750–757.

Seidemann V., Rabe J., Feldmann M., Büttgenbach S. (2002) SU8-micromechanical structures with *in situ* fabricated movable parts. *Microsyst. Technol.* **8**, 348–350.

Sellergren B., Schillinger E., Lanza F. (2009) Experimental combinatorial methods in molecular imprinting. In: Potyrailo R.A., Mirsky V.M. (eds.) *Combinatorial Methods for Chemical and Biological Sensors*. Springer Science+Business Media, New York, pp. 173–200.

Shamma-Donoghue S.A., May G.A., Cotter N.E., White R.L. Simmons F.B. (1982) Thin-film multielectrode arrays for a cochlear prosthesis. *IEEE Trans. Electron Devices* **29**, 136–144.

Shenton M.J., Lovell-Hoare M.C., Stevens G.C. (2001) Adhesion enhancement of polymer surfaces by atmospheric plasma treatment. *J. Phys. D: Appl. Phys.* **34**, 2754–2760.

Shim Y.-B., Park J.-H. (2000) Humidity sensor using chemically synthesized Poly(1,5-diaminonaphthalene) doped with carbon. *J. Electrochem. Soc.* **147**(1), 381–385.

Shukla S.K., Khati V., Minakshi, Bharadavaja A., Shekhar A., Tiwari A. (2012) Fabrication of electro-chemical humidity sensor based on zinc oxide/polyaniline nanocomposites. *Adv. Mat. Lett.* **3**(5), 421–425.

Skotheim T.A. (1986) *Handbook of Conducting Polymers, Vols. 1 and 2*. Marcel Dekker, New York.

Smith J.D.S. (1998) Intrinsically electrically conducting polymers. Synthesis, characterization, and their applications. *Prog. Polym. Sci.* **23**, 57–79.

Smits V., Chevalier P., Deheunynck D., Miller S. (2008) A new filler dispersion technology. *Reinf. Plast.* **12**, 37–73.

Song H.-C., Oh M.-C., Ahn S.-W., Steier W.H. (2003) Flexible low-voltage electro-optic polymer modulators. *Appl. Phys. Lett.* **82**, 4432–4434.

Suh J.W., Glander S.F., Darling R.B., Storment C.W., Kovacs G.T.A. (1997) Organic thermal and electrostatic ciliary microactuator array for object manipulation. *Sens. Actuators* A **58**, 51–60.

Sun C., Zhang D., Wadsworth L.C. (1999) Corona treatment of polyolefin films—A review. *Adv. Polymer Technol.* **18**(2), 171–180.

Sun Q., Deng Y. (2005) In situ synthesis of temperature-sensitive hollow microspheres via interfacial polymerization. *J. Am. Chem. Soc.* **127**, 8274–8275.

Supriya L., Claus R.O. (2004) Fabrication of electrodes for polymer actuators and sensors via self-assembly. In: *Proceedings of IEEE Sensor Conference*, October 24–27, Vienna, Austria, Vol. 2, pp. 619–622.

Sutthasupa S., Shiotsuki M., Sanda F. (2010) Recent advances in ring-opening metathesis polymerization, and application to synthesis of functional materials. *Polymer J.* **42**, 905–915.

Süzer S., Argun A., Vatansever O., Aral O. (1999) XPS and water contact angle measurements on aged and corona-treated PP. *J. Appl. Polymer Sci.* **74**(7), 1846–1850.

Takashima K., Minami K., Esashib M., Nishizawa J. (1994) Laser projection CVD using the low temperature condensation method. *Appl. Surf. Sci.* **79/81**, 366–374.

Tena A., Fernandez A.M., Viuda M., Palacio L., Pradanos P., Lozano A.E., et al. (2015) Advances in the design of co-poly(etherimide) membranes for $CO_2$ separations. Influence of aromatic rigidity on crystallinity, phase segregation and gas transport. *Eur. Polym. J.* **62**, 130–138.

Thomas A. Goettmann F. Antonietti M. (2008) Hard templates for soft materials: Creating nanostructured organic materials. *Chem. Mater.* **20**, 738–755.

Torsi L., Tanese M.C., Cioffi N., Gallazzi M.C., Sabbatini L., Zambonin P.G., et al. (2003) Side-chain role in chemically sensing conducting polymer field effect transistors. *J. Phys. Chem. B* **107**, 7589–7594.

Toshima N., Hara S. (1995) Direct synthesis of conducting polymers from simple monomers. *Prog. Polym. Sci.* **20**, 155–183.

Tourillon G., Garnies F. (1982) New electrochemically generated organic conducting polymers. *J. Electroanal. Chem.* **135**(1), 173–178.

Truong T.Q., Nguyen N.T. (2004) A polymeric piezoelectric micropump based on lamination technology. *J. Micromech. Microeng.* **14**, 632–638.

Tyszler D., Zytner R.G., Batsch A., Brugger A., Geissler S., Zhou H., et al. (2006) Reduced fouling tendencies of ultrafiltration membranes in wastewater treatment by plasma modification. *Desalination* **189**, 119–129.

Uchida E., Ikada Y. (1996) Surface modification of polymers by UV-induced graft polymerization. *Curr. Trends Polym. Sci.* **1**, 135–146.

Ulbricht U. (2006) Advanced functional polymer membranes. *Polymer* **47**, 2217–2262.

Uyama Y., Kato K., Ikada Y. (1998) Surface modification of polymers by grafting. *Adv. Polym. Sci.* **137**, 1–39.

Van C.N., Potje-Kamloth K. (2001) Electrical and $NO_x$ gas sensing properties of metallophthalocyanine-doped polypyrrole/silicon heterojunctions. *Thin Solid Films* **392**, 113–121.

Van der Bruggen B. (2009) Chemical modification of polyethersulfone nanofiltration membranes: A review. *J. Appl. Polymer Sci.* **114**, 630–642.

Waltman R.J., Bargon J. (1986) Electrically conducting polymers: A review of the electropolymerization reaction, of the effects of chemical structure on polymer film properties, and of applications towards technology. *Can. J. Chem.* **64**(1), 76–95.

Waltman R.J., Bargon J., Diaz A.F. (1983) Electrochemical studies of some conducting polythiophene films. *J. Phys. Chem.* **87**(8), 1459–1463.

Wang J., Chan S., Carlson R.R., Luo Y., Ge G., Ries R.S., Heath J.R., Tseng H-R. (2004b) Electrochemically fabricated polyaniline nanoframework electrode junctions that function as resistive sensors. *Nano Lett.* **4**, 1693–1697.

Wang J.Z., Zheng Z.H., Li H.W., Huck W.T.S., Sirringhaus H. (2004a) Dewetting of conducting polymer inkjet droplets on patterned surfaces. *Nat. Mater.* **3**, 171–176.

Wang L., Yu Y., Xiang Q., Xu J., Cheng Z., Xu J. (2018) PODS-covered PDA film based formaldehyde sensor for avoiding humidity false response. *Sens. Actuators B* **255**, 2704–2712.

Weikart C.M., Yasuda H.K. (2000) Modification, degradation, and stability of polymeric surfaces treated with reactive plasmas. *J. Polym. Sci., Part B: Polym. Phys.* **38**, 3028–3042.

Wilson W.C., Atkinson G.M. (2007) Review of polyimides used in the manufacturing of micro systems. *Technical Memorandum* NASA/TM-2007-2148702007.

Wolf R., Sparavigna A.C. (2010) Role of plasma surface treatments on wetting and adhesion. *Engineering* **2**, 397–402.

Wu D., Xu F., Sun B., Fu R., He H., Matyjaszewski K. (2012) Design and preparation of porous polymers. *Chem. Rev.* **112**, 3959–4015.

Xiao X., Zhang Q.-J., He J.-H., Xu Q.-F., Li H., Li N.-J., et al. (2018) Polysquaraines: Novel humidity sensor materials with ultra-high sensitivity and good reversibility. *Sens. Actuators B* **255**, 1147–1152.

Xiao Y., Li C.M. (2008) Nanocomposites: From fabrications to electrochemical bioapplications. *Electroanal.* **20**(6), 648–662.

Yanagida H., Esashi M., Tanaka S. (2013) Simple removal technology of chemically stable polymer in MEMS using ozone solution. *J. Microelectromech. Syst.* **22**(1), 87–93.

Yanagida H., Yoshida S., Esashi M., Tanaka S. (2011) Simple removal technology using ozone solution for chemically-stable polymer used for MEMS. In: *Proceedings of IEEE 24th Int. Conf. on Micro Electro Mechanical Systems*, January 23–27, Cancun, Mexico, pp. 324–327.

Yang H., Zhao D. (2005) Synthesis of replica mesostructures by the nanocasting strategy. *J. Mater. Chem.* **15**, 1217–1231.

Yasuda H.K. (1985) Modification of polymer surfaces by plasma treatment and plasma polymerization. In: Lee L.-H. (ed.) *Polymer Wear and Its Control. ACS Symposium Series*, Vol. 287. American Chemical Society, Washington, DC, pp. 89–102.

Yoshida S., Esashi M., Tanaka S. (2014) Development of UV-assisted ozone steam etching and investigation of its usability for SU-8 removal. *J. Micromech. Microeng.* **24**, 035007.

Zafar Q., Abdulla S.M., Azmer M.I., Najeeb M.A., Qadir K.W., Sulaiman K. (2018) Influence of relative humidity on the electrical response of PEDOT:PSS based organic field-effect transistor. *Sens. Actuators B* **255**, 2652–2656.

Zhang D., Jiang C., Sun Y., Zhou Q. (2017) Layer-by-layer self-assembly of tricobalt tetroxide-polymer nanocomposite toward high-performance humidity-sensing. *J. Alloys Compounds* **711**, 652–658.

Zhao B., Brittain W.J. (2000) Polymer brushes: Surface-immobilized macromolecules. *Prog. Polym. Sci.* **25**, 677–710.

Zhu J., Cao Y., Li Y., Chen X., Chen D. (2012) Integrating process and novel sacrificial layer fabricating technique based on diluted SU-8 resist. *Microelectron. Eng.* **93**, 56–60.

Ziaie B., Baldi A., Lei M., Gu Y., Siegel R.A. (2004) Hard and soft micromachining for BioMEMS: Review of techniques and examples of applications in microfluidics and drug delivery. *Adv. Drug Deliv. Rev.* **56**, 145–172.

Ziemelis K.E., Hussain A.T., Bradley D.D.C., Friend R.H., Rilhe J., Wegner G. (1991) Optical spectroscopy of field-induced charge in poly(3-hexyl thienylene) metal-insulator-semiconductor structures: Evidence for polarons. *Phys. Rev. Lett.* **66**, 2231–2234.

Zine-El-Abidine I., Okoniewski M. (2009) A low temperature SU-8 based wafer-level hermetic packaging for MEMS devices. *IEEE Trans. Adv. Packaging* **32**, 448–452.

Zor F.D., Cankurtaran H. (2016) Impedimetric humidity sensor based on nanohybrid composite of conducting poly(diphenylamine sulfonic acid). *J. Sensors* **2016**, 5479092.

# 20 Synthesis of Humidity-Sensitive Metal Oxides

## Powder Technologies

### 20.1 METHODS ACCEPTABLE FOR SYNTHESIS OF HUMIDITY-SENSITIVE METAL OXIDES

Currently, metal oxides can be synthesized using two main approaches:

1. Solution-based synthesis, which includes electrocrystallization, electrophoretic deposition, chemical-bath deposition, selective ionic-layer absorption and reaction, salts solution, chemical precipitation/co-precipitation, microemulsion, sol-gel technology, hydrothermal/solvothermal synthesis, hydrothermal oxidation, and microwave-assisted synthesis (Somiya and Roy 2000; Niesen and De Guire 2001; Mao et al. 2007).
2. Vapor-phase synthesis (dry processes), which includes methods such as a physical vapor deposition (PVD), chemical vapor deposition (CVD), metal organic CVD (MOCVD), spray pyrolysis, and flame pyrolysis (Choy 2003).

In principle, during the development of humidity sensors, it is possible to use metal oxides synthesized by all the above methods (see Table 20.1). The main thing is that metal oxides meet the requirements for materials intended for this application (read Volume 2 of our series). However, in this chapter we will briefly review only the methods most commonly used in the manufacture of humidity-sensitive metal oxides, or the methods best suited for mass production. In this case, the focus will be on methods for the synthesis of metal oxides intended for use in thick-film technology. This technology, based on screen-printing, spin-coating, deep-coating, and drop-coating techniques, is currently dominant in the manufacture of humidity sensors. Comparative characteristics of these methods are given in Table 20.2. When considering this table, one should keep in mind that it was compiled on the basis of data obtained more than 30 years ago. The basic laws of the formation of metal-oxide films by methods of thin-film technology such as thermal and electron-beam evaporation, sputtering, laser ablation, or pulsed-laser deposition (PLD), were considered earlier in Chapter 18. A detailed description of these methods can also be found in Korotcenkov and Cho (2010).

---

## TABLE 20.1

### Examples of Different Techniques Used for Synthesis of Metal Oxides Designed for Application in Humidity Sensors

| Method of Synthesis/ Deposition | Metal Oxides | Sensor Type | References |
|---|---|---|---|
| Sol-gel | $In_2O_3$; $SnO_2$; $In_2O_3$:Sn; $Fe_2O_3$; ZnO; $BaTiO_3$; $Al_2O_3$; | R, I, O, C | Yuk and Troczynski (2003); Mahboob et al. (2016); Babu and Vadivel (2017); Üzar et al. (2017); Yadav et al. (2017) |
| Solid-state reaction | $MgFe_2O_4$; $MgO-Al_2O_3$; $MgFe_{1.8}Mn_{0.2}O_4$; $Mg_{0.5}Cu_{0.5}Fe_{1.8}Ga_{0.2}O_4$; $Mg_{0.5}Zn_{0.5}Fe_2O_4$; | R | Rezlescu et al. (2005); Klym et al. (2016) |
| Metal–organic decomposition (MOD) | $Bi_{3.25}La_{0.75}Ti_3O_{12}$ | I | Zhang et al. (2017a) |
| Co-precipitation | $Ni_{1-x}Co_xMoO_4$ | R | Jeseentharani et al. (2018) |
| Hydrothermal/solvothermal | $SnO_2$-$Mn_3O_4$; $VO_2$; $CeO_2$; $SnO_2$ | I, R, C | Xie et al. (2015); Feng et al. (2017); Evans et al. (2018); Khan et al. (2018) |
| One-step wet chemical method | $TiO_2/SrTiO_3$; $WO_3$; $BaMO_3$ | I, R, C | Viviani et al. (2001); Ramkumar and Rajarajan (2017); Zhang et al. (2017b) |
| Solution immersion | ZnO:Sn | R | Ismail et al. (2018) |
| Microwave assisted | $Zn_2SnO_4$ | R | Parthibavarman et al. (2012) |
| Reactive sputtering | $WO_3$:Fe; ZnO | R | Narimani et al. (2016); Piloto et al. (2018) |
| Electrochemical anodization | $Al_2O_3$ | C | Juhász and Mizsei (2009); Kim et al. (2009) |
| CVD | $SnO_2$; $WO_3$ | R | Shafura et al. (2013); Jadkar et al. (2017) |

*Abbreviations*: R—resistive, I—impedance; O—optical, fiber optical; C—capacitive.

---

**TABLE 20.2**
**Metal-Oxide Powder Synthesis Route Comparison**

| | Synthesis Method | | | | | |
|---|---|---|---|---|---|---|
| Parameter | Solid-State Reaction | Co-precipitation | Sol-gel | Spray Pyrolysis | Emulsion Synthesis | Hydrothermal Synthesis |
| Compositional control | Poor | Good | Excellent | Excellent | Excellent | Excellent |
| Morphology control | Poor | Moderate | Moderate | Excellent | Excellent | Good |
| Powder reactivity | Poor | Good | Good | Good | Good | Good |
| Particle size (nm) | >1000 | >10 | >10 | >10 | >100 | >100 |
| Purity (%) | <99.5 | <99.5 | >99.9 | >99.9 | >99.9 | >99.5 |
| Agglomeration | Moderate | High | Moderate | Low | Low | Low |
| Calcination step | Yes | Yes | Yes | No | Yes | No |
| Milling step | Yes | Yes | Yes | No | Yes | No |
| Costs | Low-moderate | Moderate | Moderate-high | High | Moderate | Moderate |

*Source:* Data extracted from Dawson, W.J., *Am. Ceram. Soc. Bull.*, 67, 1673–1678, 1988; Cousin, P. and Ross, R.A., *Mater. Sci. Eng. A*, 130, 119–125, 1990.

## 20.2 WET CHEMICAL METHODS OF METAL-OXIDE SYNTHESIS

### 20.2.1 CO-PRECIPITATION METHODS

Co-precipitation from solution is one of the oldest techniques for the preparation of metal oxides. This process involves dissolving a salt precursor, usually a chloride, oxychloride, or nitrate such as $AlCl_3$ to make $Al_2O_3$ (Figure 20.1). The corresponding metal hydroxides usually form and precipitate in water by adding a basic solution (the precipitating agent) such as NaOH, $NH_3$, KOH, $NH_4OH$, or $Na_2CO_3$ to the solution (Stankic et al. 2016). As soon as the critical concentration of species in the solution is reached, a short burst of nucleation occurs, followed by a growth phase. Precipitation includes several processes that occur simultaneously including initial nucleation (formation of small crystallites), growth (aggregation), coarsening, and agglomeration. Several parameters, such as pH, mixing rates, temperature, and concentration, have to be controlled to produce satisfactory results (Reddy 2011). Then, the particles are allowed to age for hours to days, which ensures the growth of particles. Finally, the precipitated product with low solubility in the solvent is separated from the liquid by filtration. The resulting chloride salts, that is, NaCl or $NH_4Cl$, are washed away and the hydroxide is calcined after filtration and washing to obtain the final oxide powder. The precursor powder undergoes thermal treatment at elevated temperature for several hours to determine the final crystalline structure and morphology of the nanomaterial (read Chapter 18). Thus, co-precipitation is comprised of two main steps: (1) a chemical synthesis in the liquid phase that determines the chemical composition, and (2) a thermal treatment that determines the crystal structure and morphology (Mirzaei and Neri 2016). This method has been employed in synthesizing metal oxides like ZnO, $MnO_2$, $BiVO_4$, MgO, $SnO_2$, $Al_2O_3$, and $Y_2O_3$ (Stankic et al. 2016). This method is also useful in preparing composites of different oxides by co-precipitation of the corresponding hydroxides in the same solution (Reddy 2011).

The advantage of this method is its low cost, simple water-based reaction, flexibility, mild-reaction conditions, and size control. The composition control and purity of the resulting product are good. One disadvantage of this method is the difficulty to control the particle size and size distribution. Different rates of precipitation of each individual compound may lead to microscopic inhomogeneity. Moreover, very often, fast (uncontrolled) precipitation takes place, resulting in large particles. In addition, aggregates are generally formed, as with other solution techniques. Therefore, the use of surfactants is a common practice to prevent agglomeration, which also affects the particle size of the composites obtained by this technique (Jadhav et al. 2009).

### 20.2.2 SONOCHEMICAL METHOD

In sonochemical methods, the solution of the starting material (e.g., metallic salts) is subjected to a stream of intensified ultrasonic vibrations (between 20 kHz and 10 MHz), which breaks the chemical bonds of the compounds (Suslick 1990).

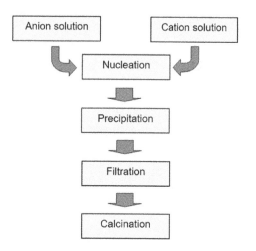

**FIGURE 20.1** Diagram illustrating co-precipitation method.

The physical phenomenon responsible for the sonochemical process is acoustic cavitation. The ultrasound waves pass through the solution, causing alternate compression and relaxation. This leads to acoustic cavitation, that is, formation, growth, and implosive collapse of bubbles in the liquid. In addition, the change in pressure creates microscopic bubbles that implode violently, leading to emergence of shock waves within the gas phase of the collapsing bubbles. These bubbles are in the nanometer size range. Cumulatively, the effect of millions of bubbles collapsing produces an excessive amount of energy that is released in the solution. Transient temperatures of ~5000 K, pressure of ~1800 atm, and cooling rates above 1010 K/s have been recorded at the localized cavitational implosion hotspots (Suslick 1990). The excessively high rate of cooling process is found to affect the formation and crystallization of the obtained products (Gendanken 2003). The products could be either amorphous or crystalline depending on the temperature in the ring region of the bubble.

This method has been used to synthesize a wide range of metal-oxide nanomaterials (Bang and Suslick 2010). Examples of reported metal oxides synthesized by this method include $TiO_2$, $ZnO$, $CeO_2$, $MoO_3$, $V_2O_5$, $In_2O_3$, $ZnFe_2O_4$, $ZnAl_2O_4$, $ZnGa_2O_4$, and $Fe_3O_4$ (Stankic et al. 2016). The advantages associated with sonochemical methods include uniform-size distribution, a higher surface area, faster reaction time, and improved phase purity of the metal-oxide nanoparticles as observed by various research groups.

## 20.2.3 MICROEMULSION TECHNIQUE

Microemulsions or micelles (including reverse micelles) represent an approach based on the formation of micro/nanoreaction vessels for the preparation of nanoparticles, and they have received considerable interest in recent years (Capek 2004; Eastoe et al. 2006; D'Souza and Richards 2007). This technique uses an inorganic phase in water-in-oil microemulsions, which are isotropic liquid media with nanosized water droplets that are dispersed in a continuous oil phase. In general, microemulsions consist of, at least, a ternary mixture of water, a surfactant, or a mixture of surface-active agents and oil. The classic examples for emulsifiers are sodium dodecyl sulfate and aerosol bis(2-ethylhexyl)sulfosuccinate. Surfactants such as pentadecaoxyethylene nonylphenylether, decaoxyethylene nonylphenyl ether, poly(oxyethylene)5nonyl phenolether, and many others, that are commercially available, can also be used in these processes. The surfactant (emulsifier) molecule stabilizes the water droplets, which have polar heads and nonpolar organic tails. The organic (hydrophobic) portion faces toward the oil phase, and the polar (hydrophilic) group toward water. In diluted water (or oil) solutions, the emulsifier dissolves and exists as a monomer, but when its concentration exceeds a certain limit, called the critical micelle concentration, the molecules of emulsifier associate spontaneously to form aggregates called micelles. These micro-water droplets then form nanoreactors for the formation of nanoparticles.

A typical method for the preparation of metal-oxide nanoparticles within micelles consists of forming two

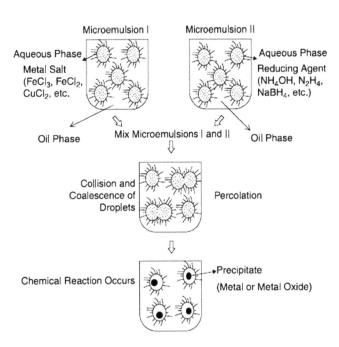

**FIGURE 20.2** Proposed mechanism for the formation of metal particles by the microemulsions approach. (Reprinted from *Adv. Colloid Interface Sci.*, 110, Capek, I., Preparation of metal nanoparticles in water-in-oil (w/o) microemulsions, 49–74, Copyright 2004, with permission from Elsevier; Scheme 1.)

microemulsions, one with the metal salt of interest and the other with the reducing or oxide containing agent and mixing them together. A schematic diagram is shown in Figure 20.2. When two different reactants mix, the interchange of the reactants takes place due to the collision of water microdroplets. Once a nucleus forms with the minimum number of atoms, the growth process starts. Nanoparticle synthesis inside the micelles can be achieved by different methods, including hydrolysis of reactive precursors, such as alkoxides, and precipitation reactions of metal salts (Hartl et al. 1997). The reaction (reduction, nucleation, and growth) takes place inside the droplet, which controls the final size of the particles. The interchange of the reactant is very fast so that for the most commonly used microemulsions, it occurs just during the mixing process. The reactant concentration has a major influence on the reduction rate. The interchange of nuclei between two microdroplets does not take place due to the special restrictions from the emulsifier. Once the particle inside the droplets attains its full size, the surfactant molecules attach to the metal surface, thus stabilizing and preventing further growth. Solvent removal and subsequent calcination leads to the final product. This method was used to synthesize iron-oxide nanoparticles, $NiO$, $CeO_2$, $TiO_2$, $ZnO$, $SnO_2$, $CuO$, and many other oxides and composites (Song and Kim 2000; Stankic et al. 2016).

The ability to control the formation of different kinds of core–shell structures with subnanometric resolution is seen as a major benefit of this technique (Stankic et al. 2005). A literature survey depicts that the ultra-fine nanoparticles in the size range between 2 and 50 nm can be easily prepared by this method

(Song and Kim 2000). In addition, the method also provides the ability to manipulate the size and shape of nanoparticles by adjusting parameters such as concentration and type of surfactant, the type of continuous phase, the concentration of precursors, and molar ratio of water to surfactant (Solans et al. 2005).

Disadvantages include low production yields and the need to use large amount of solvents and surfactants: this method involves the necessity of several washing processes and further stabilization treatment due to aggregation of the produced nanoparticles (Wu et al. 2008). Modifications have been incorporated to overcome these disadvantages. For instance, reverse microemulsion technique has been developed (Eastoe et al. 2006). The synthesis of metal oxides from reverse micelles is similar in most aspects to the synthesis of metal oxides in aqueous phase by the precipitation process. For example, precipitation of hydroxides can be done by addition of bases like $NH_4OH$ or $NaOH$ to a reverse micelle solution, containing aqueous metal ions at the micelle cores. Another approach was used to produce monodisperse spherical ZnO nanoparticles. ZnO nanoparticles were not directly produced in the microemulsion, but by the thermal decomposition of zinc glycerolate microemulsion product during subsequent calcination process (Yildirim and Durucan 2010). The modified technique prevented agglomeration whereas the calcination temperature and concentration of surfactant could be varied in order to tune the particle size and morphology of the ZnO nanoparticles, respectively.

## 20.2.4 THE SOL-GEL PROCESSING

Sol-gel techniques have long been known for the preparations of metal oxides and have been described in several books and reviews (Ivanovskaya 2000; Sakka 2003; Niederberger 2007; Corriu and Anh 2009; Guglielmi et al. 2014). A *sol* is a dispersion of solid particles (~0.01–1 µm) in a liquid in which only Brownian motion suspends the particles. In a *gel*, liquid and solid are dispersed within each other, presenting a solid network containing liquid components.

It has been established that, through the sol-gel process, homogeneous inorganic oxide materials with desirable properties of hardness, optical transparency, chemical durability, tailored porosity, and thermal resistance can be produced at room temperatures. (Hench and Ulrich 1984; Brinker et al. 1984, 1991; Brinker and Scherer 1990; Warren et al. 1990; Pope et al. 1995; Komarneni et al. 1998). The surfactant-free process involves the simple wet-chemical reaction based on hydrolysis of metal reactive precursors, usually alkoxides in an alcoholic solution, and condensation leading to formation of sol, which through the process of aging results in formation of an integrated network of metal hydroxide as gel. The gel is a polymer of a three-dimensional skeleton surrounding interconnected pores. Different reactions can create the cross-linkages that result in the gelation of the solution (Zeigler and Fearon 1990; Brinker and Scherer 1990; Corriu and Anh 2009). According to Brinker and Scherer (1990), sol-gel polymerization occurs in three stages. First, the monomers are polymerized to form particles, followed by the growth of particles, and finally the particles are

linked into chains. Networks then extend throughout the liquid medium, thickening into a gel. The characteristics and properties of a sol-gel inorganic network are related to a number of factors that affect the rate of hydrolysis and condensation reactions: the pH level, the temperature and time of the reaction, the reagent concentrations, the nature and concentration of the catalyst, the $H_2O/M$ molar ratio (R), the aging temperature, and the drying time. Among these factors, the pH level, the nature and concentration of the catalyst, the $H_2O/M$ molar ratio (R), and the temperature have been identified as being the most important (Brinker and Scherer 1990). Although, hydrolysis can occur without the addition of an external catalyst, it is more rapid and complete when a catalyst is employed. Mineral acids (HCl) and ammonia are used most often; however, other catalysts may be used as well, including acetic acid, KOH, amines, KF, and HF.

Removal of the solvents and appropriate drying of the gel is an important step that results in an ultra-fine powder of the metal hydroxide. When the sol is cast into a mold, a wet gel forms. With further drying, the gel is converted into dense particles. If the liquid in a wet gel is removed under a supercritical condition, a highly porous and extremely low-density material, called an aerogel, is obtained. Heat treatment of the hydroxide is a final step that leads to the corresponding ultra-fine powder of the metal oxide. Nanoparticles in the gel are often amorphous, and the final step of thermal treatment imparts the desired crystalline structure to the particles, although it also leads to some agglomeration (Corriu and Anh 2009). Depending on the heat-treatment procedure, the final product may end up in the form of a nanometer scale powder, bulk material, or oxygen-deficient metal oxides. The size, shape, and structure of final product are greatly influenced by the reaction parameters.

Thus, the sol-gel process has seven steps (Klein 1994): (1) preparing of required solution; (2) formation of a stable metal precursor solution (sol); (3) formation of a gel by a poly-condensation reaction (gel); (4) aging the gel for hours to days resulting in the expulsion of solvent, Ostwald ripening, and formation of a solid mass; (5) drying the gel of any liquids; (6) dehydration and surface stabilization; and (7) heat treatment of the gels at high temperatures to generate crystalline nanoparticles. Details of the sol-gel process are discussed more extensively in several excellent review articles (Livage et al. 1988; Brinker and Scherer 1990; Minh and Takahashi 1995; Brinker et al. 1996; Bagwell and Messing 1996; Livage 1997; Narendar and Messing 1997; Troczynski and Yang 2001; Olding et al. 2001). A diagram illustration of the sol-gel process is shown in Figure 20.3.

It is worth noting that, in addition to inorganic precursors [salts such as $In(NO_3)_3$], metal-organic precursors may also be used in the sol-gel process. Taurino et al. (2003) believe that the choice between these two classes should be based on a number of factors. First, metal organic precursors are more expensive, and the sol preparation requires sometimes the use of organic solvents (Niederberger 2007), such as 2-methoxyethanol, that require careful handling and disposal due to their combined toxicity and high vapor pressure (Corriu and Anh 2009). At the same time, handling of inorganic precursors does not generally require any special

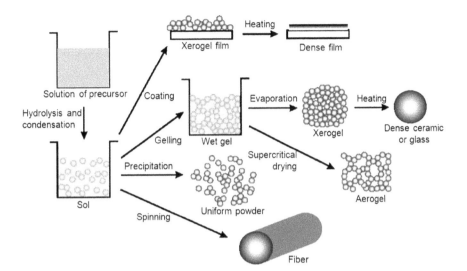

**FIGURE 20.3** Typical stages of sol-gel route. (Idea from Klein, L.C. (ed.), *Sol-Gel Optics: Processing and Applications*, Kluwer Academic Publishers, Boston, MA, 1994; http://www.chemat.com.)

equipment such as a glove-box. Their chemistry is also more extensively known and more easily manipulated than that of metal-organic precursors, and the spin coating of their solutions does not require any special conditions such as low moisture. Additionally, the solutions have long-term stability against the increase of viscosity. On the other hand, films prepared from solutions based on metal-organic precursors are much more uniform than those using inorganic precursors, and they can be easily deposited onto various substrates, including silicon. Metalorganic precursors are also free from possible contaminants such as chlorine or sulfur.

Using sol-gel processing, a lot of different metal oxides like $TiO_2$, $ZnO$, $MgO$, $CuO$, $ZrO_2$, $Nb_2O_5$, $Al_2O_3$, $TiO_2$, $SnO_2$, $In_2O_3$, $Fe_2O_3$, various nanocomposites, and doped metal oxides were synthesized (see Table 20.3) (Ivanovskaya 2000; Ba et al. 2006; Stankic et al. 2016).

The modification of sol-gel processing consisted of employing ultrasonic conditions to overcome the effects of high temperature conditions (up to 600°C), which could lead to oxidation of the products. Under the effect of ultrasound vibrations, high temperatures and pressures could instantaneously be generated and then dissipated in the local environment of the particles avoiding oxidation (Corr et al. 2004). In addition, it has been reported that in the frame of sol-gel technology supercritical fluids (hydrothermal approach) can be used to synthesize nanoparticles like $TiO_2$, $ZrO_2$, $Al_2O_3$, $TiO_2$-$SiO_2$, $SiO_2$-$Al_2O_3$, and $ZrO_2$/$TiO_2$ hybrid oxide nanotubes (Sui and Charpentier 2012).

The sol-gel method presents various advantages. First of all, the mixing of starting reagents being at the atomic/molecular level, results in faster reaction times at lower temperatures, which helps to reduce the interdiffusion from one phase into the other and the formation of parasitic phases. Since the process begins with a relatively homogeneous mixture, the resulting product is a uniform ultra-fine porous powder. In addition, it provides the particle size to be tuned by simply varying the gelation time (Cannas et al. 2004).

The sol-gel method is also very attractive for the synthesis of multicomponent metal oxides, containing more than one metal, since the slow reaction kinetics allows good structural engineering of the final product. Another advantage is that the reactions are conducted at low temperatures or at room temperature.

Sol-gel processing also has the advantage that it can also be expanded to accommodate an industrial-scale production. In addition, the sol-gel procedure also allows coating of substrates with complex shapes on the nanometer to micrometer scale, which some commonly used coating procedures cannot accomplish. In addition, sol-gel is suitable for preparing thick, porous ceramics necessary for various sensor applications because the reaction proceeds with the coexisting solvent phase in which evaporation leaves numerous cavities.

Despite these advantages, the sol-gel technique has never reached its full industrial potential due to several limitations, including weak bonding, low wear resistance, difficulty in controlling porosity, and cracking effect. The consequences of the cracking effect in $In_2O_3$ ceramics are shown in the scanning electron microscopy images presented in Figure 18.10. Atkinson and Guppy (1991) observed that the crack spacing increased with film thickness and attributed this behavior to a mechanism in which partial delamination accompanies crack propagation. Therefore, because of the crack-free property of coatings, the maximum coating thickness in this case is limited to 0.5 μm (Olding et al. 2001). However, one should note that for sensor applications, the presence of the cracking effect is not a factor that limits the sensing properties. Therefore, the thickness of the humidity-sensitive films can exceed 100 μm in gas and humidity sensors. The presence of cracks improves gas permeability and helps to reduce the response time.

A more important disadvantage of the films prepared by the sol-gel method for use in sensors is the strong dependence of the quality of these layers on the properties of the substrate. A large difference in the coefficients of thermal expansion of the substrate and the formed layer leads to the destruction

**TABLE 20.3**

**Some Metal Oxides Prepared by the Sol-Gel Method and Their Parameters**

| Metal Oxide | Precursor, Temperature of Hydrolysis | Temperature and Time of Calcination | Grain Size |
|---|---|---|---|
| $SnO_2$ | $[Sn(NMe_2)_2]_2$, 135°C | 600°C, 6 h | 10–20 nm |
| $SnO_2$ | $SnCl_4$, 150°C | 700°C, 1 h | ~10 nm |
| $In_2O_3$ | $InCl_3$, 110°C | 600°C, 1 h | 20–27 nm |
| $In_2O_3$ | $In(NO_3)_3$, 60°C | 500°C–800°C, 1 h | 18–50 nm |
| $In_2O_3$ | $In(OAc)_3$, 3 days | 400°C–700°C, 30 min | 23–37 nm |
| $WO_3$ | $W(OC_2H_5)_6$, RT, 1 h | 400°C–700°C | |
| $WO_3$ | $(NH_4)_{10}W_{12}O_{41}$ $5H_2O$ | 300°C–1000°C, 3–5 h | |
| $TiO_2$ | $Ti(OC_2H_5)_4$, RT, 1 h | 350°C–500°C, 4 h | 10–15 nm |
| $TiO_2$ | $TiCl_4$ | 750°C, 2 h | Whiskers |
| $Ta_2O_5$ | $TaCl_5$, 24 h | 600°C–700°C, 4 h | ~50 nm |
| $Fe_2O_3$ | $Fe(OC_2H_4OCH_3)_3$, 90°C | 400°C, 2 h | ~5–30 nm |
| $ZrO_2$ | $Zr(acac)$, 220°C | 600°C, 4 h | ~20–60 nm |
| $TiO_2$-$WO_3$ | $Ti(OC_3H_7)_4 + H_2WO_4$ | 700°C–900°C | |
| $TiO_2$-$SnO_2$ | $Ti(OC_3H_7)_4 + SnCl_2$ | 600°C–700°C | 10–50 nm |
| $BaSnO_3$ | $Ba(OH)_2 + K_2SnO_3$ $3H_2O$ | 1000°C | 200 nm |

*Source:* Reprinted from Korotcenkov, G. and Cho, B.K., Synthesis and deposition of sensor materials, in: *Chemical Sensors: Fundamentals of Sensing Materials*, vol. 1: *General Approaches*, Korotcenkov, G. (ed.), Momentum Press, New York, pp. 215–304, 2010. With permission.

of the coating. In addition, the organics trapped by the thick coating often caused failure during the thermal process. It is important to note that this disadvantage is inherent in all films manufactured by thick-film technology.

## 20.2.5 Hydrothermal/Solvothermal Techniques

Hydrothermal/solvothermal synthesis is a potentially superior method for low-cost production of metal oxides, including advanced mixed oxides and nanocomposites for various applications, because complex oxide powders are formed directly (Somiya and Roy 2000; D'Souza and Richards 2007; Wang et al. 2003, 2007a, 2007b, 2007c; Hayashi and Hakuta 2010; Modeshia and Walton 2010; Feng and Li 2011; Köseoğlu et al. 2011; Yang et al. 2013; Meng et al. 2016). This solution chemical process can be easily differentiated from other processes, such as sol-gel and co-precipitation, by the temperatures and especially the pressures involved. The hydrothermal/solvothermal method involves the heterogeneous chemical reaction in a solvent (aqueous or nonaqueous), occurring above room temperature and at pressure more than 1.0 atm in a closed system (Byrappa and Yoshimura 2012). When the reaction is performed using water as the solvent, the method is called hydrothermal synthesis. The metal complexes are decomposed thermally either by boiling the contents in an inert atmosphere or by using an autoclave. The chemical reaction is carried out for several (typically 6–48) hours leading to the nucleation and growth of nanoparticles. The temperatures during hydrothermal synthesis lie between the boiling point of water and its critical temperature (374°C), whereas pressures can go as high as 15 MPa. Currently, the production of various metal-oxide particles such as $TiO_2$, $K_2Ti_6O_{13}$,

$K_4Nb_6O_{17}$, $KNbO_3$, $KTiNbO_3$, $KTaO_3$, $Zn_2SiO_4$:Mn, $ZrO_2$, $AlOOH$, $Al_2O_3$, $Ba(Sr)Ti(Zr)O_3$, $Ca_{0.8}Sr_{0.2}Ti_{1-x}FeO_{3-x}$, $Cr_2O_3$, YSZ, $(Fe, In)_2O_3$(ITO), $LiFePO_4$, $(Ce, Zr)O_2$, $In_2O_3$, $YVO_4$, $(Co, Cu, Ni)(Fe, Co)_2O_4$, $Fe_2O_3$, YAG, $ErOOH$, $CuAlO_2$, ZnO, $LiMn_2O_4$, $La_xNi_yO_3$, $SnO_2$, and $(Ca, Mg)(PO_4)_3$ has been demonstrated by hydrothermal batch (Hayashi and Hakuta 2010). Solvothermal synthesis uses other solvents. For example, nanoparticles of MgO, $TiO_2$, $MnFe_2O_4$, $CoFe_2O_4$, and $Fe_3O_4$ have been synthesized using polyol as the solvent (Kim et al. 2007; Subramania et al. 2007; Wang 2008). One should note that the calcination step required by earlier techniques can be eliminated in many cases. However, many of the other steps employed in the previous wet methods can be used in hydrothermal/solvothermal processes. So, hydrothermal/solvothermal synthesis with the exception of heat treatment contains the same steps as the sol-gel process. Hydrothermal/solvothermal synthesis generally produces nanoparticles with crystalline structure that are relatively uncontaminated and thus do not require purification or post-treatment annealing but may have a wider size distribution if special treatment for size control is not applied. Several examples of metal-oxide nanoparticles synthesized using hydrothermal method are presented in Table 20.4. Other advantages are the use of inexpensive raw materials, such as oxides, hydroxides, chlorides, nitrates, and organometallic complex. Elimination of impurities, associated with milling and achievement of very fine and highly reactive anhydrous crystalline powders, is also an advantage of this method.

A new trend in hydrothermal/solvothermal synthesis is to use this technique in combination with microwave (Verma et al. 2004; Meng et al. 2016; Mirzaei and Neri 2016), sol-gel (Li et al. 2005) that can not only vary the physiochemical

**TABLE 20.4**

**Metal-Oxide Particles Produced by the Supercritical Hydrothermal Synthesis Technique**

| Metal Salt Used as Starting Material | Product | Particle Size, nm |
|---|---|---|
| $Al(NO_3)_3$ | $\gamma$-AlOOH, $\gamma$-$Al_2O_3$ | 4–1000 |
| $Fe(NO_3)_3$ | $\alpha$-$Fe_2O_3$ | 16–50 |
| $Fe_2(SO_4)_3$ | $\alpha$-$Fe_2O_3$ | ~50 |
| $FeCl_2$ | $\alpha$-$Fe_2O_3$ | ~50 |
| $Fe(NH_4)_2H(C_6H_5O_7)_2$ | $Fe_3O_4$ | ~50 |
| $Co(NO_3)_2$ | $Co_3O_4$ | ~100 |
| $Ni(NO_3)_2$ | NiO | ~200 |
| $Zn(CH_3CO_2)_2$, $H_2O_2$ | ZnO | 39–320 |
| $ZrOCl_2$ | $ZrO_2$ (cubic) | ~10 |
| $Ti(SO_4)_2$ | $TiO_2$ | ~20 |
| $TiCl_4$ | $TiO_2$ | ~20 |
| $Ti(OC_3H_7)_4$ | $TiO_2$ | 20–115 |
| $Ce(NO_3)_3$ | $CeO_2$ | 5–300 |
| $Ce(NO_3)_3$, NaOH | $CeO_2$ | 3–8 |
| $ZrO(NO_3)_2$, $ZrO(Ac)_2$ | $ZrO_2$ | ~7 |
| $SnCl_2$, $InCl_3$ | $In_2O_3$, $SnO_2$, ITO | <10 |
| $Fe(NO_3)_3$, $Ba(NO_3)_2$ | $BaO{\cdot}6Fe_2O_3$ | 50–1000 |
| $Li(NO_3)$, $Co(NO_3)_2$ | $LiCoO_2$ | 40–400 |
| $Nb_2O_5$, KOH | $KNbO_3$ | 7–42 |
| $Ta_2O_5$, KOH | $KTaO_3$ | 1000–10,000 |
| $Ti(OC_3H_7)_4$, KOH | $K_2Ti_6O_{13}$ | 12–20 |

*Source:* Data extracted from Adschiri, T. et al., *Ind. Eng. Chem. Res.*, 39, 4901–4907, 2000; Hayashi, H. and Hakuta, Y., *Materials*, 3, 3794–3817, 2010.

and structural properties of the materials, but in addition to that may lead to the formation of the single-phased materials with enhanced stability (Li et al. 2005). In addition, simply changing the temperature, time, and pressure of the reaction, you can control the particle size, composition, stoichiometry, and particle shape (Adschiri et al. 2000; Wang et al. 2007a, 2007b; Hayashi and Hakuta 2010; Modeshia and Walton 2010). Moreover, because the properties of the solvent mixture as well as those of the reactants differ intrinsically under high temperature and pressure conditions from their corresponding properties under ambient conditions, the overall result is that this type of processing allows you to more manipulate a large set of experimental variables in the synthesis of high-quality nanomaterials (Mao et al. 2007; Modeshia and Walton 2010). It was observed that basicity and hydrolysis ratio of the reacting medium together with the steric or electrostatic stabilization of the reactive molecules also affect the nucleation and growth steps, which in turn control the particle size, shape, composition, and crystal structure of the particles. For instance, varying the hydrolysis ratio allows to synthesize either metal or (oxy)hydroxide or oxide nanoparticles (Feldmann and Jungk 2001). To modify the size and properties, the use of surfactants, capping agents, or mineralizers is common practice (Wang et al. 2007a, 2007b; Akhtar et al. 2008; Köseoğlu et al. 2011). For example, by using a zwitterionic surfactant, smaller ZnO particle sizes were obtained as compared with

those obtained from surfactant-free hydrothermal reaction (Maneedaeng 2015). A suitable capping agent or stabilizer, such as a long-chain amine, thiol, and trioctylphospine oxide, can be added to the reaction contents at a suitable point to hinder the growth of the particles and, hence, provide their stabilization against agglomeration. The stabilizers also help in dissolution of the particles in different solvents. The addition of mineralizers also helps to reduce the size of crystallites. Typical mineralizers are hydroxides (NaOH, KOH, LiOH), carbonates ($Na_2CO_3$), and halides (NaF, KF, NaCl, KCl, LiCl). Different mineralizers result in crystals of different sizes and shapes (Somiya and Roy 2000). As a result, this method allows for the preparation of ultra-small nanoparticles (<5 nm). In particular, using this method, Ren et al. (2012) synthesized $2.5 \times 4.3$ nm $TiO_2$ nanoparticles, and Soultanidis et al. (2012) synthesized $1.6 \pm 0.3$ nm $WO_x$ nanoparticles. In the latter case, it was shown that the use of reducing/oxidizing agents may strongly affect both, the size (use of an oxidizing agent led to particles with diameters smaller than 1 nm) and the shape (use of a reducing agent led to rod-shaped nanoparticles). Tian et al. (2011) have shown that adjusting the ratio of reducing agent and solvent can tune the particle size of magnetite nanoparticles from ~6 to 1 nm, while iron-oxide nanostructures could be produced in different morphologies—such as, nanocubes (Wang et al. 2007c) and hollow spheres (Titirici et al. 2006)—by this synthesis route.

The major obstacle in developing the hydrothermal/solvothermal synthesis routes into rational synthesis, where new materials are prepared in a predictive manner, thus lies in the fact that more detailed understanding of the complex solution chemistry, which cannot necessarily be generalized for all systems, is required (Modeshia and Walton 2010).

### 20.2.6 MICROWAVE-ASSISTED METHOD

Microwave irradiation is electromagnetic irradiation in the frequency range 0.3–300 GHz, (wavelengths of 1 mm–1 m). Therefore, the microwave region of the electromagnetic spectrum lies between infrared and radio frequencies. Microwave-assisted method has been of increasing interest, as it is relatively low energy and time consuming (Panda et al. 2006; Tompsett et al. 2006; Lagashettya et al. 2007; Baghbanzadeh et al. 2011; Mirzaei and Neri 2016). As it stated above, in spite of benefits of wet chemical routes for synthesis of nano metal oxides, as a rule, they need long time for synthesis. Microwave-assisted processing attracted a great deal of attention due to its advantages to supply a higher synthesis rate, resulting in superior traditional heating. Indeed, since microwaves can penetrate the material and supply energy, heat can be generated throughout the volume of the material, resulting in volumetric heating. Microwave-assisted methods involve quick and uniform heating of the reaction medium with no temperature gradients through two mechanisms: dipolar polarization and ionic conduction. The ability to elevate the temperature of a reaction well above the boiling point of the solvent increases the speed of reactions by a factor of 10–1000. As a result, the reaction times are reduced from a few hours to several minutes or even seconds without compromising the particle purity or size (Roberts and Strauss 2005).

Faster reaction rates favor rapid nucleation and formation of small, highly monodisperse particles. Yields are also generally higher, and the technique may provide a means of synthesizing compounds that are not available conventionally (Schanche 2003). Additionally, the method can lead to the synthesis of materials with smaller particle size, narrow particle size distribution, high purity, and enhanced physicochemical properties (Motshekga et al. 2012). Furthermore, microwave methods are unique in providing scaled-up processes without suffering thermal gradient effects, thus leading to a potentially industrially important advancement in the large-scale synthesis of nanomaterials (Panda et al. 2006). Therefore, it's not then a surprise that microwave processes have been used to synthesize different highly crystalline nanoparticles of pure and mixed metal oxides $Fe_3O_4$, ZnO, $TiO_2$, MnO, $Fe_3O_4$, $SnO_2$, NiO, $PtO_2$, CoO, $Mn_2O_3$, MgO, $CeO_2$ CuO, $Zn_2SnO_4$, $LaFe_{0.7}Zn_{0.3}O_3$, $LiFePO_4$, $NiFe_2O_4$, and its composites (Bilecka et al. 2008; Polshettiwar et al. 2009; Goharshadi et al. 2011; Prado-Gonjal et al. 2015; Mirzaei and Neri 2016; Stankic et al. 2016).

Presently, microwave heating seems the most promising way to achieve short synthesis times. Microwave energy is precisely controllable and can be turned on and off instantly, eliminating the need for warm-up and cool-down. In addition, automation allows control over the reaction conditions and hence facilitates manipulation of particle size, morphology, and crystallinity (Bilecka and Niederberger 2010). The choice of starting metal-oxides precursor (as acetates, chlorides, isopropyls) and solvents (as ethylene glycol, benzene) can govern reaction success, particle size, and crystal structure (Bilecka et al. 2008).

Limitations of the microwave-assisted processes are associated with the instrumental apparatus itself. The possibility of varying the reaction conditions by finely tuning/controlling the irradiation power and the temperature are the main drawbacks that may, in some cases, hinder reproducibility, especially when not laboratory-designed microwave ovens are used (Mirzaei and Neri 2016). At last, it is noteworthy that papers so far reported in the literature regarding metal oxides synthesized by microwave results lacking of comprehensive investigations feels. Many researchers just synthesized the materials by microwave irradiation and did not focus their attention on the understanding of the effect of microwave irradiation parameters, such as irradiation time and microwave power, on their final properties.

### 20.2.7 CALCINATION OF METAL HYDROXIDES SYNTHESIZED BY WET METHODS

Research has shown that stabilized metal hydroxides are usually synthesized during the sol-gel process or precipitation method. This means that the deposited materials are very far from thermodynamic equilibrium. All of these products are very unstable at temperatures over 150°C and do not have a crystalline structure. Furthermore, depending on the formation procedure, the material also presents traces of other chemical species, such as chlorides or traces of organic precursors. Therefore, thermal treatment (calcination) is required to transfer these hydroxides in their stoichiometric oxide form. The modes of thermal treatment commonly used in the calcination of some oxide phases are presented in Table 20.3 (Tahar et al. 1997; Chung et al. 1998; Sangaletti et al. 1999; Li et al. 1999; Nayral et al. 2000; Ivanovskaya 2000; Llobet et al. 2000; Yue and Gao 2000; Leiti et al. 2001; Zakrzewska 2001; Kaya et al. 2002).

For the majority of oxides, thermal annealing changes the properties of the oxides in a consistent manner. These changes generally are as follows:

- The concentration of chlorine and traces of other precursors decreases as the annealing temperature increases. This concentration falls below the detection limit of X-ray photoelectron spectroscopy, which is ~0.1%, at temperatures above 600°C–700°C.
- As a rule, the crystalline phase determined by X-ray diffraction appears at temperatures above 400°C. When the annealing temperature is increased, the crystallinity of the samples also increases. This process is accompanied by grain-size growth, in which strong increases have been shown experimentally at temperatures above 450°C (Zhang and Liu 2000;

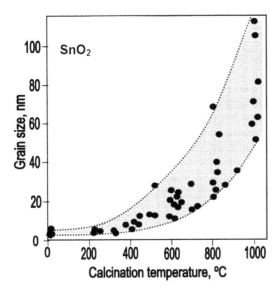

**FIGURE 20.4** Influence of calcination temperature on the size of $SnO_2$ powders prepared by the sol-gel method. (Reprinted from Korotcenkov, G. and Cho, B.K., Synthesis and deposition of sensor materials, in: *Chemical Sensors: Fundamentals of Sensing Materials*, vol. 1: *General Approaches*, Korotcenkov, G. (ed.), Momentum Press, New York, pp. 215–304, 2010. With permission.)

Cirera et al. 2000; Leite et al. 2001; Cabot et al. 2001). In treatments at about 1000°C, the grains size may be as large as 100–150 nm (see Figure 20.4). Taking into account the straight correlation between the surface activity of a material and the grain size (Leite et al. 2001), this metal-oxide behavior requires a careful choice in the method by which materials are stabilized using sol-gel technology.

- At low calcination temperatures, nanoparticles are characterized by a quasi-spherical size, strong disorder, and water content, which progressively disappear with increasing calcination temperature. When the transition temperature is above 450°C and changes occur very quickly, practically no water is found, meaning that the nanoparticles are of good crystalline quality and are faceted (Dieguez et al. 1999b, 2000).
- As a rule, the majority of the particles are in a highly agglomerated state (Kaya et al. 2002; Nayral et al. 2000; Dieguez et al. 2000);
- Nanoparticles prepared using low calcination temperature are characterized by a strong distortion in the crystalline structure. Mechanical strains observed in metal-oxide powders are taken off during the annealing process at T > 400°C–600°C (Dieguez et al. 1999b; Cirera et al. 2000).
- The introduction of different additives for sol stabilization can change the particulars of oxide phase formation (Ivanovskaya 2000; Cabot et al. 2001) because the additives are incorporated into the sol micelle, and further, into the crystalline-phase structures. Additives introduced to change the electrophysical and catalytic properties of oxides

(Sangaletti et al. 1999; Cabot et al. 2001) show the same influence. The fundamentals of the processes taking place in nanoceramics during sintering are given by Lu (2008).

## 20.3 SYNTHESIS IN THE VAPOR PHASE

Nanoscaled powders are the main materials used to prepare ceramics and thick films for a variety of applications, including humidity-sensor design. As shown previously, the various wet chemical methods can be used for synthesis of such materials (Hench and Ulrich 1984; Lavernia and Wu 1996; Niesen and De Guire 2001; Vahlas et al. 2006). The challenge in synthesizing nanostructure powders using wet-chemical routes is to control and engineer the physical properties (size, shape, composition, etc.) of the starting powders, since these affect the properties of the final products. From an industrial point of view, however, cost-effective and less complicated synthesis techniques are required for large-scale applications. Therefore, the various gas-phase synthesis methods have been developed. In all of these techniques, the primary products are nanosized powders, which are subsequently transformed by various consolidation techniques to materials in either bulk form or coatings, with or without porosity. Some of the most widely used techniques are summarized schematically in Figure 20.5. The powder synthesis processes shown are (1) gas processing condensation (GPC), (2) chemical vapor condensation (CVC), (3) microwave plasma processing (MPP), and (4) low-pressure combustion flame synthesis (CFS). These processes can all be interchanged and used in similar chamber designs.

### 20.3.1 Gas Processing Condensation

In GPC, a metallic or inorganic material, such as a suboxide, is vaporized using thermal evaporation in an atmosphere of another inert gas (at 1–50 mbar) (Hahn 1997). Clusters form in the vicinity of the source by homogeneous nucleation in the gas phase and grow through coalescence and incorporation of atoms from the gas phase. The cluster or particle size depends critically on the residence time of the particles in the growth regime and can be influenced by the gas pressure, the type of inert gas, and the evaporation rate/vapor pressure of the evaporating material. The average particle size of the nanoparticles can be increased by increasing the gas pressure, vapor pressure, and mass of the inert gas.

Log-normal size distributions have been found experimentally and explained theoretically by the growth mechanisms of the particles. Even in more complex processes, such as low-pressure combustion flame synthesis, in which a number of chemical reactions are involved, the size distributions are log-normal. This method, however, can only be used in a system designed for gas flow (GFC gas-flow condensation). In such a system, a dynamic vacuum is generated by continuous pumping, and the inlet of gas takes place via a mass flow controller. One major advantage over convectional gas flow is the improved control of the particle sizes. The particle-size distribution in gas flow systems shift toward smaller average values, with an appreciable reduction in the standard deviation of the distribution.

Nanocrystalline oxide powders are formed through controlled postoxidation of the primary nanoparticles of a pure metal (e.g., Ti to $TiO_2$) or a suboxide (e.g., ZrO to $ZrO_2$). In some cases, direct evaporation of oxides, namely, $SnO_2$, $Al_2O_3$, $CeO_2$, has resulted in average particle sizes as small as 3 nm with essentially no agglomeration. The velocity of the postoxidation process, influenced by the heat of the reaction, determines the degree of agglomeration of the primary powder particles and is essential for the evolution of the microstructure during compaction and sintering. Since the reaction is performed in a gas flow, there is no agglomeration during the reaction of the isolated particles, which is a clear advantage of this method.

### 20.3.2 Chemical Vapor Condensation

The CVC process, shown schematically in Figure 20.5, was originally designed to adjust the parameter field during synthesis in order to suppress film formation and enhance homogeneous nucleation of the particles in a gas flow. In a certain range of residence time, both particle and film formation can be obtained. The residence time of the precursor molecules can be adjusted by changing the gas-flow rate, the pressure difference between the precursor delivery system and the main chamber, and the temperature of the hot-wall reactor. These changes result in prolific production of nanosized particles of metals and ceramics instead of thin films, as in CVD

processing (Tschope et al. 1997). The mechanisms involved in the process of particle formation for two sources, thermal evaporation and electrospraying, are discussed by Nakaso et al. (2003) and presented schematically in Figure 20.6.

In the simplest form of the CVC process, a metal-organic precursor is introduced into the hot zone of the reactor using a mass flow controller. In spite of the increased number of factors controlling the CVC process in comparison to CFC, a wider range of ceramics, including nitrides and carbides, can be synthesized.

In addition to the formation of single-phase nanoparticles, the CVC reactor allows the synthesis of mixtures of (1) nanoparticles of two phases or doped nanoparticles by supplying two precursors at the front end of the reactor, and (2) coated nanoparticles, for example, $n$-$ZrO_2$ powders coated with $n$-$Al_2O_3$ or vice versa, by supplying a second precursor in a second stage of the reactor. In the latter case, nanoparticles formed by homogeneous nucleation are coated by heterogeneous nucleation in a second stage of the reactor. Because the CVC processing is continuous, production capabilities are much greater than with GPC processing. The microstructure of nanoparticles and the properties of materials obtained by CVC have been identical to the GPC-prepared powders for all oxides studied.

### 20.3.3 Microwave Plasma Processing

MPP is similar to the previously discussed CVC method but employs plasma instead of high temperature for the decomposition of the metal-organic precursors. In comparison to

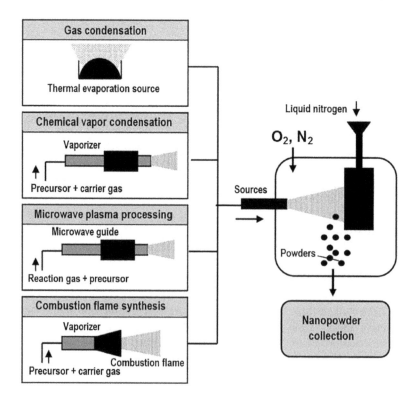

**FIGURE 20.5** Techniques for gas-phase synthesis of nanocrystalline powders. (Reprinted from *NanoStruct. Mater.*, 9, Hahn, H., Gas phase synthesis of nanocrystalline materials, 3–12, Copyright 1997, with permission from Elsevier.)

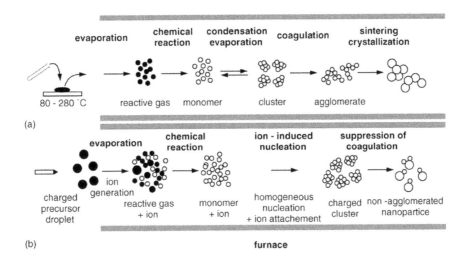

**FIGURE 20.6** Mechanisms in the process of particle formation: (a) homogeneous thermal CVD and (b) electrospray CVD. (Reprinted from *Mater. Sci. Eng. R*, **53**, Vahlas, C. et al., Principles and applications of CVD powder technology, 1–72, Copyright 2006, with permission from Elsevier.)

thermal activation, the major advantage of plasma-assisted pyrolysis is the low-temperature reaction, which reduces the tendency for agglomeration of the primary particles.

### 20.3.4 COMBUSTION FLAME SYNTHESIS

The CFS process involves the combustion of liquid or gaseous precursors injected/delivered into diffused or premixed flames, where the liquid precursor either decomposes or vaporizes and then undergoes a chemical reaction and/or combustion in the flame. The flame source and the combustion process provide the required thermal environment for vaporization, decomposition, and chemical reaction. The fuel for the CFS process can be hydrogen or a hydrocarbon ($C_2H_2$), mixed with $O_2$. The use of a hydrocarbon often leads to the formation of soot, whereas the combustion of hydrogen is a faster process and produces no condensed species. The flame temperature is usually high, typically 1727°C–2727°C, which often causes a homogeneous gas-phase reaction to occur, leading to the deposition of powders. Therefore, CFS is widely used in industrial applications to produce large quantities of powders from metal-chloride precursors in hydrocarbon flames (Hampikian and Carter 1999; Carter 1999). The main process parameters that can be optimized to control the crystal structure and particle size are the flame temperature and its distribution, the choice of precursor and its residence time in the flame, and the ratio of precursor to fuel (Bunshah 1994; Vincenzini 1995; Carter 1999). Additives may also be introduced into the flame to alter the size, phase, and shape of the powders.

A typical CFS system is shown in Figure 20.7. It consists of a water-cooled vacuum chamber continuously pumped with a roughing pump. A constant dynamic pressure is maintained with a closed-loop pressure controller. Inside the chamber, a burner and a substrate are rigidly fixed to a mount to allow the distance between the burner and substrate to be varied. The fuel flows upstream along with vapors of the precursor

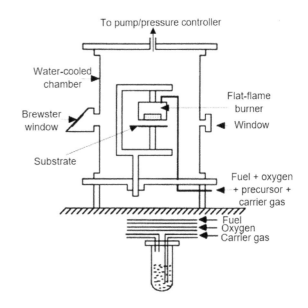

**FIGURE 20.7** Schematic of burner–substrate assembly for combustion-flame chemical vapor condensation. (Reprinted from *NanoStruct. Mater.*, **11**, Skandan, G. et al., Synthesis of oxide nanoparticles in low pressure flames, 149–158, Copyright 1999, with permission from Elsevier.)

carried by a carrier gas, such as helium or another inert gas. As the configuration in Figure 20.7 indicates, the flow is highly one-dimensional, and the temperature and chemical species concentrations vary only in the axial direction (i.e., perpendicular to the burner face). The precursor pyrolyzes in the preheat zone of the flame, and particles condense as the flame gases cool upon approaching the substrate. The high degree of uniformity in these flames ensures that the particle-size distribution in the deposit is narrow, since all particles experience a similar time/temperature history. The chamber may be sited on an optical table, which houses a tunable-dye-laser system, to probe the flame for the gas temperature and radical species concentrations.

The CFS method presents the following advantages:

- The high flame temperatures allow using volatile and less volatile chemical precursors to form a chemical vapor.
- The formation of the reaction product is a single-step process with no postprocessing treatments, such as calcination.
- The mixing of reactants on a molecular scale takes place quickly, thus significantly reducing the processing time and enabling better control of the stoichiometry of the multicomponent powders.
- The vaporization, decomposition, and chemical reactions occur rapidly, leading to a high deposition rate.
- CFS technology is a relatively low-cost compared to conventional methods and can be performed in an open atmosphere without the need for a sophisticated reactor or vacuum system.

The main drawback to the CFS method is the large temperature fluctuation of the flame source during deposition, due to the large temperature gradient in the flame. Efforts have been made to minimize the instability of the flame temperature by developing specially designed burners, such as the counterflow flame burner and the reduced-pressure flat-flame burner, to produce a very flat and uniform flame in the horizontal plane (Skandan et al. 1999; Choy 2003). Such burners allow for better control over the microstructure, the particle size, and its distribution, leading to the fabrication of nanocrystalline powders that are difficult to produce using the conventional CFS method (Glumac et al. 1999).

At the same time, however, the high flame temperature is another problem of conventional CFS. The high gas pressure during the reaction produces highly agglomerated powders, which is disadvantageous for subsequent processing. Therefore, the low-pressure combustion flame synthesis method (LPCFS) has been developed over the past several years. The basic concept of LPCFS is to extend the pressure range to the pressures used in gas-phase synthesis and thus to reduce or avoid agglomeration.

Additional information regarding the application of this method for the formation of complex-shaped metal-oxide nanomicrostructures can be found in the review by Mishra et al. (2014). I was shown that facile and cost-effective flame transport synthesis allows controlled growth of different nanostructures and their interconnected networks in a scalable process.

### 20.3.5 NANOPOWDERS COLLECTION

A number of collection devices was used to separate the nanoparticles from the gas (Choy 2003; Vahlas et al. 2006). Traditionally, a rotating cylindrical device cooled with liquid nitrogen has been employed for particle collection. The nanoparticles are subsequently removed from the surface of the cylinder with a scraper, usually in the form of a metallic plate. However, the easiest way to collect nanopowders is to use a mechanical filter with a small pore size. The disadvantage of this method is that parts of the filter material are often incorporated in the nanocrystalline powder, which leads to the appearance of impurities and defects in the bulk of the nanocrystalline material. Nanoparticles ranging from 2–50 nm in size can be extracted from the gas flow using thermophoretic forces due to the applied constant temperature gradient, and then loosely deposited on the surface of the device for collecting as a low-density powder without agglomeration. In addition, various devices developed for aerosol science and pollution control can be used to efficiently collect nanoparticles. Among these techniques, corona discharge and electrostatic devices (Karthikeyan et al. 1997) are the most appropriate for highly effective large-scale separation and continuous processing. In all of these techniques, the nanopowders can be taken out of the system and further processed by mixing, pressing, and sintering to obtain bulk nanocrystalline materials with variable microstructures.

## 20.4 MECHANICAL MILLING OF POWDERS: MECHANO-CHEMICAL METHOD

For some applications, especially solid-state gas humidity sensors, it is necessary to use nanoscaled grains with sizes smaller than 5–10 nm (Nayral et al. 2000; Leite et al. 2001). In principle, this is possible, since during the synthesis, crystallites with a size of 2–20 nm are formed. In many cases, however, the powders synthesized using dry and wet methods are agglomerated and are large in size up to several micrometers. Mechanical milling is an effective method of resolving this problem (Hadjipanayis and Siegel 1994; Koch 1997).

Mechanical attrition, or ball milling, which induces heavy cyclic deformation in powders, is a technique that produces nanostructures not by cluster assembly, but by the structural decomposition of coarser-grained structures as the result of severe plastic deformation. Using ball milling has become a popular method for producing nanocrystalline materials because of its simplicity, the relatively inexpensive equipment (on the laboratory scale) needed, and the applicability to essentially all classes of materials. The most commonly cited advantage is the possibility for easily scaling up material to tonnage quantities for various applications. On the other hand, the most commonly noted problems are contamination from milling media and/or atmosphere and the need in many applications to consolidate the powder product without coarsening the nanocrystalline microstructure. In fact, the contamination problem is often presented as a reason to dismiss this method, at least for some materials. If steel balls and containers are used, iron contamination becomes a problem. This is the most serious problem for the highly energetic mills and depends on the mechanical behavior of the powder being milled as well as its chemical affinity for the milling media. Lower-energy mills result in much less, often negligible, iron contamination. Other milling media, such as tungsten carbide or ceramics, may be used, but contamination is also possible from such media (Koch 1997). Surfactants (process control agents) may also be used to minimize contamination.

At longer milling times, the grain size decreases steadily. In principle, using ball milling, one can reduce the size of grains to a crystallite size. But in reality, the minimum grain size obtainable by milling has been attributed to a balance between the defect/dislocation structure introduced by the plastic deformation of milling and its recovery through thermal processes. Hadjipanayis and Siegel (1994) summarizes the phenomenology of the development of a nanocrystalline microstructure by mechanical milling into the following three stages:

Stage 1: Deformation is localized in shear bands, containing a high dislocation density.
Stage 2: Dislocation annihilation/recombination/rearrangement forms a cell/subgrain structure with nanoscale dimensions. Further milling extends this structure throughout the sample.
Stage 3: The orientation of the grains becomes random.

It was found that the energy of the mill is not critical for the final microstructure, but the kinetics of the process, of course, depends on the energy. The amount of time needed to attain the same microstructure can be several orders of magnitude longer in low-energy mills than in high-energy mills. It was also assumed that the total strain introduced by milling, rather than the milling energy or ball–powder–ball collision frequency, is responsible for determining the nanocrystalline grain size (Koch 1997). Thus, as a result of mechanical milling, additional mechanical tensions appear in the material. The appearance of these mechanical tensions and the change of microstructure can be an important additional factor that influences gas-sensing characteristics. Mechanical tensions should therefore be considered during the development of procedures for sensor fabrication. These tensions are only relieved through thermal treatment at $T \approx 600°C–800°C$ (Cirera et al. 2000). Thus, an additional thermal treatment after mechanical milling may be required.

It should be noted that ball milling is also the basis of the mechano-chemical method (Figure 20.8) used to prepare mainly metal-oxide nanocomposites. Although binary oxides can also be synthesized (Avvakumov and Karakchiev 2004; Tsuzuki and McCormick 2004; Šepelák et al. 2012). The mechano-chemical method is the combination of mechanical and chemical phenomena on a nanoscale solid material. Here nanomaterials are synthesized by mechanical activation and in this method, ball milling is a widely used technique, wherein the powder mixture is placed in a ball mill and is subjected to high-energy collision from the balls, and thus mechanical force is used to achieve chemical processing and transformation. Mechano-chemical processing is characterized by the repeated welding, deformation, and fracture of the mixture of reactants. Chemical reactions occur at the interfaces of the nanometer-sized grains that are continuously regenerated during milling (Koch 1991). As a consequence, chemical reactions, which usually require high temperatures due to separation of the reaction phases by product phases, can occur at low temperatures in a ball mill without the need for external heating (Yang and McCormick 1998; McCormick et al. 2001).

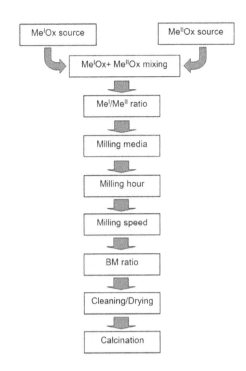

**FIGURE 20.8** Diagram illustrating the mechano-chemical method.

Contamination, long processing time, no control on particle morphology, agglomerates, and residual strain in the crystallized phase are other disadvantages of high-energy ball milling process. However, the method is famous for its results, various applications, and potential scientific values. The method of nanocomposite preparing consists mainly of mixing two oxides taken in required MeOx(I)/MeOx(II) ratio. Metal oxide can be synthesized by various methods. The highlight of this method is choosing a mechanical milling with selected milling media, such as zirconia and stainless steel, with particular speed and duration as well as critically maintaining the ball-mass ratio. After ball grinding, the samples are heat treated.

## REFERENCES

Adschiri T., Hakuta Y., Arai K. (2000) Hydrothermal synthesis of metal oxide fine particles at supercritical conditions. *Ind. Eng. Chem. Res.* **39**, 4901–4907.

Akhtar M.S., Khan M.A., Jeon M.S., Yang O.B. (2008) Controlled synthesis of various ZnO nanostructured materials by capping agents-assisted hydrothermal method for dye-sensitized solar cells. *Electrochim. Acta* **53**(27), 7869–7874.

Atkinson A., Guppy R.M. (1991) Mechanical stability of sol-gel films. *J. Mater. Sci.* **26**, 3869–3875.

Avvakumov E.G., Karakchiev L.G. (2004) Mechanochemical synthesis as a method for the preparation of nanodisperse particles of oxide materials. *Chem. Sustainable Development.* **12**, 287–291.

Ba J., Fattakhova Rohlfing D., Feldhoff A., Brezesinski T., Djerdj I., Wark M., Niederberger M. (2006) Nonaqueous synthesis of uniform indium tin oxide nanocrystals and their electrical conductivity in dependence of the tin oxide concentration. *Chem. Mater.* **18**, 2848–2854.

Babu B.M., Vadivel S. (2017) High performance humidity sensing properties of indium tin oxide (ITO) thin films by sol–gel spin coating method. *J. Mater. Sci.: Mater. Electron.* **28**, 2442–2447.

Baghbanzadeh M., Carbone L., Cozzoli P.D., Kappe C.O. (2011) Microwave assisted synthesis of colloidal inorganic nanocrystals. *Angew Chem.* **50**, 11312–11359.

Bagwell R.B., Messing G.L. (1996) Critical factors in the production of sol–gel derived porous alumina. *Key Eng. Mater.* **115**, 45–64.

Bang J.H., Suslick K.S. (2010) Applications of ultrasound to the synthesis of nanostructured materials. *Adv. Mater.* **22**, 1039–1059.

Bilecka I., Djerdj I., Niederberger M. (2008) One-minute synthesis of crystalline binary and ternary metal oxide nanoparticles. *Chem. Commun.* **7**, 886–888.

Bilecka I., Niederberger M. (2010) Microwave chemistry for inorganic nanomaterials synthesis. *Nanoscale.* **2**, 1358–1374.

Brinker C.J, Clark D.E., Ullrich D.R. (eds.) (1984) *Better Ceramics Through Chemistry.* Elsevier-North Holland, New York.

Brinker C.J., Hurd A.J., Schunk P.R., Ashely C.S, Cairncross R.A., Samuel J., Chen K.S., Scotto C., Schwartz R.A. (1996) Sol-gel derived ceramic films—Fundamentals and applications. In: Stern K. (ed.) *Metallurgical and Ceramic Protective Coatings.* Chapman & Hall, London, UK, pp. 112–151.

Brinker C.J., Scherer G.W. (1990) *Sol-Gel Science: The Physics and Chemistry of Sol-Gel Processing.* Academic Press, San Diego, CA.

Brinker C.R., Hurd A.J., Frye G.C., Schunk P.R., Ashley C.S. (1991) Sol-gel thin film formation. *J. Ceram. Soc. Japan* **99**, 862–877.

Bunshah R.F. (ed.) (1994) *Handbook of Deposition Technologies for Films and Coatings*, 2nd ed. Noyes, Park Ridge, NJ.

Byrappa K., Yoshimura M. (2012) *Handbook of Hydrothermal Technology.* Elsevier, New York.

Cabot A., Dieguez A., Romano-Rodriguez A., Morante J., Barsan N. (2001) Influence of the catalytic introduction procedure on the nano-SnO$_2$ gas sensor performance. Where and how stay the catalytic atoms? *Sens. Actuators B* **79**, 98–106.

Cannas C., Musinu A., Navarra G., Piccaluga G. (2004) Structural investigation of Fe$_2$O$_3$–SiO$_2$ nanocomposites through radial distribution functions analysis. *Phys. Chem. Chem. Phys.* **6**, 3530–3534.

Capek I. (2004) Preparation of metal nanoparticles in water-in-oil (w/o) microemulsions. *Adv. Colloid Interface Sci.* **110**(1–2), 49–74.

Carter W.B. (1999) Flame assisted deposition of oxide coatings. In: Dahotre N.B., Sudarshan T.S. (eds.) *Intermetallic and Ceramic Coatings.* Marcel Dekker, New York, pp. 233–266.

Choy K.L. (2003) Chemical vapour deposition of coatings. *Prog. Mater. Sci.* **48**(2), 57–170.

Chung W.K., Sakai G., Shimanoe K., Miura N., Lee D.D., Yamazoe N. (1998) Preparation of indium oxide thin film by spin-coating method and its gas-sensing properties. *Sens. Actuators B* **46**, 139–145.

Cirera A., Cornet A., Morante J.R., Olaizola S.M., Castano E., Garcia J. (2000) Comparative structural study between sputtered and liquid pyrolysis nanocrystalline SnO$_2$. *Mater. Sci. Eng. B* **69–70**, 406–410.

Corr S.A., Gun'ko Y.K., Douvalis A.P., Venkatesan M., Gunning R.D. (2004) Magnetite nanocrystals from a single source metallorganic precursor: Metallorganic chemistry vs. biogeneric bacteria. *J. Mater. Chem.* **14**, 944–946.

Corriu R., Anh N.T. (2009) *Molecular Chemistry of Sol-gel Derived Nanomaterials.* John Wiley & Sons, Chichester, UK.

Cousin P., Ross R.A. (1990) Preparation of mixed oxides: A review. *Mater. Sci. Eng. A* **130**, 119–125.

D'Souza L., Richards R. (2007) Synthesis of metal-oxide nanoparticles: Liquid–solid transformations. In: Rodríguez J.A., Fernández-Garcia M. (eds.) *Synthesis, Properties, and Applications of Oxide Nanomaterials.* John Wiley & Sons, Hoboken, pp. 81–117.

Dawson W.J. (1988) Hydrothermal synthesis of advanced ceramic powder. *Am. Ceram. Soc. Bull.* **67**(10), 1673–1678.

Dieguez A., Romano-Rodriguez A., Alay J.L., Morante J.R., Barsan N., Kappler J., et al., (2000) Parameter optimization in SnO$_2$ gas sensors for NO$_2$ detection with low cross-sensitivity to CO: Sol-gel preparation, film preparation. Powder calcination, doping and grinding. *Sens. Actuators B* **65**, 166–168.

Dieguez A., Romano-Rodriguez A., Morante J.R., Kappler J., Barsan N., Gopel W. (1999b) Nanoparticle engineering for gas sensor optimisation: Improved sol-gel fabricated nanocrystalline SnO$_2$ thick film gas sensor for NO$_2$ detection by calcination, catalytic metal introduction and grinding treatments. *Sens. Actuators B* **60**, 125–137.

Eastoe J., Hollamby M.J., Hudson L. (2006) Recent advances in nanoparticle synthesis with reversed micelles. *Adv. Colloid Interface Sci.* **128–130**, 5–15.

Evans G.P., Powell M.J., Johnson I.D., Howard D.P., Bauer D., Darr J.A., Parkin I.P. (2018) Room temperature vanadium dioxide–carbon nanotube gas sensors made via continuous hydrothermal flow synthesis. *Sens. Actuators B* **255**, 1119–1129.

Feldmann C., Jungk H.O. (2001) Polyol-mediated preparation of nanoscale oxide particles we thank Jacqueline Merikhi and Gerd Much for carrying out the scanning electron microscopy (SEM) and the atomic force microscopy (AFM) investigations, respectively. *Angew Chem.* **40**, 359–362.

Feng H., Li C., Li T., Diao F., Xin T., Liu B., Wang Y. (2017) Three-dimensional hierarchical SnO$_2$ dodecahedral nanocrystals with enhanced humidity sensing properties. *Sens. Actuators B* **243**, 704–714.

Feng S., Li G. (2011) Hydrothermal and solvothermal syntheses. In: Xu R., Pang W., Huo Q. (eds.) *Modern Inorganic Synthetic Chemistry.* Elsevier, Amsterdam, the Netherlands, pp. 63–93.

Gendanken A. (2003) Sonochemistry and its applications in nanochemistry. *Curr. Sci.* **85**, 1720–1722.

Glumac N.G., Skandan G., Chen Y.J., Kear B.H. (1999) Particle size control during flat flame synthesis of nanophase oxide powders. *NanoStruct. Mater.* **12**, 253–258.

Goharshadi E.K., Samiee S., Nancarrow P. (2011) Fabrication of cerium oxide nanoparticles: Characterization and optical properties. *J. Colloid Interface Sci.* **356**, 473–480.

Guglielmi M., Kickelbic G., Martucci A. (eds.) (2014) *Sol-Gel Nanocomposites.* Springer-Verlag, New York.

Hadjipanayis G.C., Siegel R.W. (eds.) (1994) *Nanophase Materials.* Kluwer, Dordrecht, the Netherlands.

Hahn H. (1997) Gas phase synthesis of nanocrystalline materials. *NanoStruct. Mater.* **9**, 3–12.

Hampikian J.M., Carter W.B. (1999) The combustion chemical vapor deposition of high temperature materials. *Mater. Sci. Eng. A* **267**, 7–18.

Hartl W., Beck C., Roth M., Meyer F., Hempelmann R. (1997) Nanocrystalline metals and oxides II: Reverse microemulsions. *Ber. Bunsen-Ges. Phys. Chem.* **101**(11), 1714.

Hayashi H., Hakuta Y. (2010) Hydrothermal synthesis of metal oxide nanoparticles in supercritical water. *Materials* **3**, 3794–3817.

Hench L.L., Ulrich D.R. (eds.) (1984) *Ultrastructure Processing of Ceramics, Glasses, and Composites.* Wiley, New York.

Ismail A.S., Mamat M.H., Yusoff M.M., Malek M.F., Zoolfakar A.S., Rani R.A. et al. (2018) Enhanced humidity sensing performance using Sn-doped ZnO nanorod Array/SnO$_2$ nanowire heteronetwork fabricated via two-step solution immersion. *Mater. Lett.* **210**, 258–262.

Ivanovskaya M. (2000) Ceramic and film metal oxide sensors obtained by sol-gel method: Structural features and gas-sensitive properties. *Electron. Technol.* **33**(1/2), 108–112.

Jadhav A.P., Kim C.W., Cha H.G., Pawar A.U., Jadhav N.A., Pal U. et al. (2009) Effect of different surfactants on the size control and optical properties of $Y_2O_3$:$Eu^{3+}$ nanoparticles prepared by co-precipitation method. *J. Phys. Chem. C* **113**(31), 13600–13604.

Jadkar V., Pawbake A., Waykar R., Jadhavar A., Date A., Late D. et al. (2017) Synthesis of $\gamma$-$WO_3$ thin films by hot wire-CVD and investigation of its humidity sensing properties. *Phys. Status Solidi (a)* **214**(5), 1600717.

Jeseentharani V., Dayalan A., Nagaraja K.S. (2018) Nanocrystalline composites of transition metal molybdate ($Ni_{1-x}Co_xMoO_4$; $x = 0, 0.3, 0.5, 0.7, 1$) synthesized by a co-precipitation method as humidity sensors and their photoluminescence properties. *J. Phys. Chem. Solids* **115**, 75–83.

Juhász L., Mizsei J. (2009) Humidity sensor structures with thin film porous alumina for on-chip integration. *Thin Solid Films* **517**, 6198–6201.

Karthikeyan J., Berndt C.C., Tikkannen J., Wang J.Y., King A.H., Herman H. (1997) Preparation of nanophase materials by thermal spray processing of liquid precursors. *NanoStruct. Mater.* **9**, 137–140.

Kaya C., He J.Y., Gu X., Butler E.G. (2002) Nanostructured ceramic powders by hydrothermal synthesis and their applications. *Microporous Mesoporous Mater.* **54**, 37–49.

Khan S.B., Karimov Kh. S., Din A., Akhtar K. (2018) A multimodal impedimetric sensor for humidity and mechanical pressure using a nanosized $SnO_2$-$Mn_3O_4$ mixed oxide. *Microchim. Acta* **185**, 24.

Kim D.H., Kang J.W., Kim T.R., Kim E.J., Im J.S., Kim J. (2007) A polyol-mediated synthesis of titania-based nanoparticles and their electrochemical properties. *J. Nanosci. Nanotech.* **7**, 3954–3958.

Kim Y., Jung B., Lee H., Kim H., Lee K., Hyuncul Park H. (2009) Capacitive humidity sensor design based on anodic aluminum oxide. *Sens. Actuators B* **141**, 441–446.

Klein L.C. (ed.) (1994) *Sol-Gel Optics: Processing and Applications.* Kluwer Academic Publishers, Boston, MA.

Klym H., Ingram A., Shpotyuk O., Hadzaman I., Hotra O., KostivYu. (2016) Nanostructural free-volume effects in humidity-sensitive $MgO$-$Al_2O_3$ ceramics for sensor applications. *JMEPEG* **25**, 866–873.

Koch C.C. (1991) Mechanical milling and alloying. In: Cahn R.W., Haasen P., Kramer E.J. (eds.) *Materials Science and Technology*, Vol. 15. VCH Verlagsgesellschaft GmbH, Weinheim, Germany, pp. 193–245.

Koch C.C. (1997) Synthesis of nanostructured materials by mechanical milling: Problems and opportunities. *Nanostuct. Mater.* **9**, 13–22.

Komarneni S., Sakka S., Phule P.P., Laine R.M. (eds.) (1998) *Sol-Gel Synthesis and Processing.* Ceramic Transactions, Vol. 95. Wiley, New York.

Korotcenkov G., Cho B.K. (2010) Synthesis and deposition of sensor materials. In: *Chemical Sensors: Fundamentals of Sensing Materials*, vol. 1: *General Approaches*, G. Korotcenkov (ed.) Momentum Press, New York, pp. 215–304.

Köseoğlu Y., Bay M., Tan M., Baykal A., Sözeri H., Topkaya R. et al. (2011) Magnetic and dielectric properties of $Mn_{0.2}Ni_{0.8}Fe_2O_4$ nanoparticles synthesized by PEG-assisted hydrothermal method. *J. Nanoparticle Res.* **13**(5), 2235–2244.

Lagashetty A., Havanoor V., Basavaraja S., Balaji S.D., Venkataraman A. (2007) Microwave-assisted route for synthesis of nanosized metal oxides. *Sci. Technol. Adv. Mater.* **8**, 484–493.

Lavernia E., Wu Y. (1996) *Spray Atomization and Deposition.* Wiley, Chichester, UK.

Leite E.R., Cerri J.A., Longo E., Valera J.A., Paskocima C.A. (2001) Sintering of ultrafine undoped $SnO_2$ powders. *J. Eur. Ceram. Soc.* **21**, 669–675.

Li G.L. Wang G.H., Hong J.M. (1999) Morphologies of rutile form $TiO_2$ twins crystals. *J. Mater. Sci. Lett.* **18**, 1243–1246.

Li Z., Hou B., Xu Y., Wu D., Sun Y., Hu W., et al. (2005) Comparative study of sol–gel-hydrothermal and sol–gel synthesis of tita-nia–silica composite nanoparticles. *J. Solid State Chem.* **178**(5), 1395–1405.

Livage J. (1997) Sol-gel processes. *Solid State Mater. Sci.* **2**, 132–136.

Livage J., Henry M., Sanchez C. (1988) Sol-gel chemistry of transition metal oxides. *Prog. Solid State Chem.* **18**, 259–342.

Llobet E., Molas G., Molinas P., Calderer J., Vilanova X., Brezmes J. et al. (2000) Fabrication of highly selective tungsten oxide ammonia sensors. *J. Electrochem. Soc.* **147**(3), 776–779.

Lu K. (2008) Sintering of nanoceramics. *Int. Mater. Rev.* **53**, 21–38.

Mahboob Md. R., Zargar Z.H., Islam T. (2016) A sensitive and highly linear capacitive thin film sensor for trace moisture measurement in gases. *Sens. Actuators B* **228**, 658–664.

Maneedaeng A. (2015) High uniformity of ZnO nanoparticles synthesized by surfactant-assisted solvothermal technique. *Adv. Mater. Res.* **1131**, 43–48.

Mao Y., Park T.-J., Zhang F., Zhou H., Wong S.S. (2007) Environmentally friendly methodologies of nanostructure synthesis. *Small* **3**(7), 1122–1139.

McCormick P.G., Tsuzuki T., Robinson J.S., Ding J. (2001) Nanopowders synthesized by mechanochemical processing. *Adv. Mater.* **13**(12–13), 1008–1010.

Meng L.-Y., Wang B., Ma M.-G., Lin K.-L. (2016) The progress of microwave-assisted hydrothermal method in the synthesis of functional nanomaterials. *Mater. Today Chem.* **1–2**, 63–83.

Minh N.Q., Takahashi T. (1995) *Science and Technology of Ceramic Fuel Cells.* Elsevier, Amsterdam, the Netherlands.

Mirzaei A., Neri G. (2016) Microwave-assisted synthesis of metal oxide nanostructures for gas sensing application: A review. *Sens. Actuators B* **237**, 749–775.

Mishra Y.K., Kaps S., Schuchardt A., Paulowicz I., Jin X., Gedamu D. et al. (2014) Versatile fabrication of complex shaped metal oxide nano-microstructure and their interconnected networks for multifunctional applications. *KONA Powder Particle J.* **31**, 92–110.

Modeshia D.R., Walton R.I. (2010) Solvothermal synthesis of perovskites and pyrochlores: Crystallization of functional oxides under mild conditions. *Chem. Soc. Rev.* **39**, 4303–4325.

Motshekga S.C., Pillai S.K., Ray S.S., Jalama K., Krause R.W.M. (2012) Recent trends in the microwave-assisted synthesis of metal oxide nanoparticles supported on carbon nanotubes and their applications. *J. Nanomater.* **12**, 1–12.

Nakaso K., Han B., Ahn K.H., Choi M., Okuyama K. (2003) Synthesis of non-agglomerated nanoparticles by an electrospray assisted chemical vapor deposition (ES-CVD) method. *J. Aerosol Sci.* **34**, 869–881.

Narendar Y., Messing G.L. (1997) Mechanisms of phase separation in gel-based synthesis of multicomponent metal oxides. *Catal. Today* **35**, 247–268.

Narimani K., Nayeri F.D., Kolahdouz M., Ebrahimi P. (2016) Fabrication, modeling and simulation of high sensitivity capacitive humidity sensors based on ZnO nanorods. *Sens. Actuators B* **224**, 338–343.

Nayral C., Viala E., Fau P., Senocq F., Jumas J.C. Maisonnat A., Chaudret B. (2000) Synthesis of tin and tin oxide nanoparticles of low size dispersity for application in gas sensing. *Chem. Eur. J.* **6**, 4082–4090.

Niederberger M. (2007) Nonaqueous sol-gel routes to metal oxide nanoparticles. *Acc. Chem. Res.* **40**, 793–800.

Niesen T.P., De Guire M.R. (2001) Review: Deposition of ceramic thin films at low temperatures from aqueous solutions. *J. Electroceram.* **6**, 169–207.

Olding T., Sayer M., Barrow D. (2001) Ceramic sol-gel composite coatings for electrical insulation. *Thin Solid Films* **398–399**, 581–586.

Panda A.B., Glaspell G., El-Shall M.S. (2006) Microwave synthesis of highly aligned ultra-narrow semiconductor rods and wires. *J. Am. Chem. Soc.* **128**, 2790–2791.

Parthibavarman M., Vallalperuman K., Sekar C., Rajarajan R., Logeswaran T. (2012) Microwave synthesis, characterization and humidity sensing properties of single crystalline $Zn_2SnO_4$ nanorods. *Vacuum* **86**, 1488–1493.

Piloto C., Shafiei M., Khan H., Gupta B., Tesfamichael T., Motta N. (2018) Sensing performance of reduced graphene oxide-Fe doped $WO_3$ hybrids to $NO_2$ and humidity at room temperature. *Appl. Surf. Sci.* **434**, 126–133.

Polshettiwar V., Baruwati B., Varma R.S. (2009) Self-assembly of metal oxides into three-dimensional nanostructures: Synthesis and application in catalysis. *ACS Nano.* **3**, 728–736.

Pope E.J.A., Sakka S., Klein L.C. (eds.) (1995) *Sol-Gel Science and Technology.* Ceramic Transactions, Vol. 55, Wiley, Chichester, UK.

Prado-Gonjal J., Schmidt R., Morán E. (2015) Microwave-assisted routes for the synthesis of complex functional oxides. *Inorganics* **3**, 101–117.

Ramkumar S., Rajarajan G. (2017) A comparative study of humidity sensing and photocatalytic applications of pure and nickel (Ni)-doped $WO_3$ thin films. *Appl. Phys. A* **123**, 401.

Reddy B.S. (ed.) (2011) *Advances in Nanocomposites: Synthesis, Characterization and Industrial Applications.* InTech, London, UK.

Ren Y., Liu Z., Pourpoint F., Armstrong A.R., Grey C.P., Bruce P.G. (2012) Nanoparticulate $TiO_2$(B): An anode for lithium-ion batteries. *Angew. Chem.* **124**, 2206–2209.

Rezlescu N., Rezlescu E., Doroftei C., Popa P.D. (2005) Study of some Mg-based ferrites as humidity sensors. *J. Phys.: Conf. Series* **15**, 296–299.

Roberts B.A., Strauss C.R. (2005) Toward rapid, "green," predictable microwave-assisted synthesis. *Acc. Chem. Res.* **38**, 653–661.

Sakka S. (ed.) (2003) *Sol-Gel Science and Technology.* Springer-Verlag, New York.

Sangaletti L., Depero L.E., Allieri B., Pioselli F., Angelucci R., Poggi A., Tagliani T., Nicoletti S. (1999) Microstructural development in pure and V-doped $SnO_2$ nanopowders. *J. Eur. Ceram. Soc.* **19**, 2073–2077.

Schanche J.S. (2003) Microwave synthesis solutions from personal chemistry. *Mol. Divers.* **7**, 293–300.

Šepelák V., Bégin-Colin S., Le Caër G. (2012) Transformations in oxides induced by high-energy ball-milling. *Dalton Trans.* **41**, 11927–11948.

Shafura A.K.S., Sin N.D., Mamat M.H., Mahmood M.R. (2013) Humidity sensor using CVD deposited $SnO_2$ thin film. *Adv. Mater. Res.* **667**, 415–420.

Skandan G., Chen Y.J., Glumac N., Kear B.H. (1999) Synthesis of oxide nanoparticles in low pressure flames. *NanoStruct. Mater.* **11**, 149–158.

Solans C., Izquierdo P., Nolla J., Azemar N., Garcia-Celma M.J. (2005) Nanoemulsions. *Curr. Opin. Colloid Interface Sci.* **10**, 102–110.

Somiya S., Roy R. (2000) Hydrothermal synthesis of fine oxide powders. *Bull. Mater. Sci.* **23**(6), 453–460.

Song K.C., Kim J.H. (2000) Synthesis of high surface area tin oxide powders via water-in- oil microemulsions. *Powder Technol.* **107**, 268–272.

Soultanidis N., Zhou W., Kiely C.J., Wong M.S. (2012) Solvothermal synthesis of ultrasmall tungsten oxide nanoparticles. *Langmuir.* **28**, 17771–17777.

Stankic S., Sterrer M., Hofmann P., Bernardi J., Diwald O., Knozinger E. (2005) Novel optical surface properties of $Ca^{2o+}_-$ doped MgO nanocrystals. *Nano Lett.* **5**, 1889–1893.

Stankic S., Suman S., Haque F., Vidic J. (2016) Pure and multi metal oxide nanoparticles: Synthesis, antibacterial and cytotoxic properties. *J. Nanobiotechnol.* **14**(1), 73.

Subramania A., Vijaya Kumar G., Sathiya Priya A.R., Vasudevan T. (2007) Polyolmediated thermolysis process for the synthesis of MgO nanoparticles and nanowires. *Nanotechnology* **18**, 225601.

Sui R., Charpentier P. (2012) Synthesis of metal oxide nanostructures by direct sol–gel chemistry in supercritical fluids. *Chem. Rev.* **112**, 3057–3058.

Suslick K.S. (1990) Sonochemistry. *Science* **247**, 1439–1445.

Tahar R.B.H., Ban T., Ohya Y., Takahashi Y. (1997) Optical, structural and electrical properties of indium oxide thin films prepared by the sol-gel method. *J. Appl. Phys.* **82**, 865–870.

Taurino A.M., Epifani M., Taccoli T., Iannotta S., Siciliano P. (2003) Innovative aspects in thin film technologies for nanostructured materials in gas sensor devices. *Thin Solid Films* **436**, 52–63.

Tian Y., Yu B., Li X., Li K. (2011) Facile solvothermal synthesis of monodisperse $Fe_3O_4$ nanocrystals with precise size control of one nanometre as potential MRI contrast agents. *J. Mater. Chem.* **21**, 2476–24781.

Titirici M.M., Antonietti M., Thomas A. (2006) A generalized synthesis of metal oxide hollow spheres using a hydrothermal approach. *Chem. Mater.* **18**, 3808–3812.

Tompsett G.A., Conner W.C., Yngvesson K.S. (2006) Microwave synthesis of nanoporous materials. *ChemPhysChem.* **7**, 296–319.

Troczynski T., Yang Q. (2001) Process for Making Chemically Bonded Sol-Gel Ceramics. U.S. Patent no. 6,284,682, May 2001.

Tschope A., Schaadt D., Birringer R., Yying J. (1997) Catalytic properties of nanostructured metal oxides. Synthesized by inert gas condensation. *NanoStruct. Mater.* **9**, 423–432.

Tsuzuki T., McCormick P.G. (2004) Mechanochemical synthesis of nanoparticles. *J. Mater. Sci.* **39**, 5143–5146.

Üzar N., Algün G., Akçay N., Akcan D., Arda L. (2017) Structural, optical, electrical and humidity sensing properties of (Y/Al) co-doped ZnO thin films. *Mater. Sci.: Mater. Electron.* **28**, 11861–11870.

Vahlas C., Caussat B., Serp P., Angelopoulos G.N. (2006) Principles and applications of CVD powder technology. *Mater. Sci. Eng. R* **53**, 1–72.

Verma S., Joy P., Khollam Y., Potdar H., Deshpande S. (2004) Synthesis of nanosized $MgFe_2O_4$ powders by microwave hydrothermal method. *Mater. Lett.* **58**(6), 1092–1095.

Vincenzini P. (ed.) (1995) *Advances in Inorganic Films and Coatings.* Techna SRL., Italy.

Viviani M., Buscaglia M.T., Buscaglia V., Leoni M., Nanni P. (2001) Barium perovskites as humidity sensing materials. *J. Eur. Cer. Soc.* **21**, 1981–1984.

Wang H., Ma Y., Yi G., Chen D. (2003) Synthesis of Mn-doped $Zn_2SiO_4$ rodlike nanoparticles through hydrothermal method. *Mater. Chem. Phys.* **82**(2), 414–418.

Wang W.W. (2008) Microwave-induced polyol-process synthesis of MIIFe$_2$O$_4$ (M = Mn, Co) nanoparticles and magnetic property. *Mater. Chem. Phys.* **108**, 227–231.

Wang W.W., Zhu Y.J., Yang L.X. (2007b) ZnO–SnO$_2$ hollow spheres and hierarchical nanosheets: Hydrothermal preparation, formation mechanism, and photocatalytic properties. *Adv. Func. Mater.* **17**(1), 59–64.

Wang Y., Xu G., Ren Z., Wei X., Weng W., Du P., et al. (2007a) Mineralizer-assisted hydrothermal synthesis and characterization of BiFeO$_3$ nanoparticles. *J. Am. Ceram. Soc.* **90**(8), 2615–2617.

Wang S.B., Min Y.L., Yu S.H. (2007c) Synthesis and magnetic properties of uniform hematite nanocubes. *J. Phys. Chem. C* **111**, 3551–3554.

Warren W.L., Lenahan P.M., Brinker C.J., Shaffen G.R., Ashley C.S., Reed S.T. (1990) Sol-gel thin film electronic properties. In: *Better Ceramics Through Chemistry IV.* MRS proceedings, Vol. **180**, pp. 413–419.

Wu W., He Q., Jiang C. (2008) Magnetic iron oxide nanoparticles: Synthesis and surface functionalization strategies. *Nanoscale Res. Lett.* **3**, 397–415.

Xie W., Liu B., Xiao S., Li H., Wang Y., Cai D., et al. (2015) High performance humidity sensors based on CeO$_2$ nanoparticles. *Sens. Actuators B* **215**, 125–132.

Yadav B.C., Chauhan K.S., Singh S., Sonker R.K., Sikarwar S., Kumar R. (2017) Growth and characterization of sol–gel processed rectangular shaped nanostructured ferric oxide thin film followed by humidity and gas sensing. *J. Mater. Sci.: Mater. Electron.* **28**, 5270–5280.

Yang H., McCormick P.G. (1998) Mechanically activated reduction of nickel oxide with graphite. *Metall. Mater. Trans. B* **29**, 449–455.

Yang Q., Lu Z., Liu J., Lei X., Chang Z., Luo L., Sun X. (2013) Metal oxide and hydroxide nanoarrays: Hydrothermal synthesis and applications as supercapacitors and nanocatalysts. *Progr. Nat. Sci.: Mater. Intern.* **23**(4), 351–366.

Yildirim O.A., Durucan C. (2010) Synthesis of zinc oxide nanoparticles elaborated by microemulsion method. *J. Alloys Compoun.* **506**, 944–949.

Yue Y., Gao Z. (2000) Synthesis of mesoporous TiO$_2$ with crystalline frame work. *Chem. Commun.* **18**, 1755–1756.

Yuk J., Troczynski T. (2003) Sol–gel BaTiO$_3$ thin film for humidity sensors. *Sens. Actuators B* **94**, 290–293.

Zakrzewska K. (2001) Mixed oxides as gas sensors. *Thin Solid Films* **391**, 229–238.

Zeigler J.M., Fearon F.W.G. (eds.) (1990) *Silicon Based Polymer Science: A Comprehensive Resource.* ACS Advances in Chemistry Series No. 224. American Chemical Society, Washington, DC.

Zhang G., Liu M. (2000) Effect of particle size and dopant on properties of SnO$_2$–based gas sensors. *Sens. Actuators B* **69**, 144–152.

Zhang M., Wei S., Ren W., Wu R. (2017b) Development of high sensitivity humidity sensor based on gray TiO$_2$/SrTiO$_3$ composite. *Sensors* **17**, 1310.

Zhang Y., He J., Yuan M., Jiang B., Li P., Tong Y., Zheng X. (2017a) Effect of annealing temperature on Bi$_{3.25}$La$_{0.75}$Ti$_3$O$_{12}$ powders for humidity sensing properties. *J. Electron. Mater.* **46**(1), 377–385.

# 21 Humidity Sensors Based on Individual Metal-Oxide 1D Structures

## Fabrication Features and Application Prospects

## 21.1 SENSORS BASED ON INDIVIDUAL METAL-OXIDE 1D STRUCTURES AND THEIR ADVANTAGES

There is a wide variety of nanostructures that can be used in the design of humidity sensors. These are 0D, 1D, 2D, and 3D structures, which include nanowires, nanotubes, nanobelts, nanorods, nanodots, and various hierarchical structures such as nanospheres, nanowhiskers, nanorings, and core-shell (Li et al. 2002; Jung et al. 2003; Varghese et al. 2003a, 2003b; Kam et al. 2004; Guha et al. 2004; Kuchibhatla et al. 2007; Barth et al. 2010; Soldano et al. 2012). SEM images of some nanostructures are shown in Figure 21.1.

However, in this chapter we consider sensors only on the basis of individual 1D structures. Humidity sensors based on 2D and hierarchical structures are discussed in Chapters 20 and 23. Compared to other low-dimensional systems, 1D structures have exceptional properties, not only due to their highly anisotropic shape, but also due to their large surface-to-volume ratio. In addition, only these sensors have a specific configuration and manufacturing technology that is radically different from the technology described earlier in Chapters 18 and 20. A typical example of a sensor based on individual 1D structures is shown in Figure 21.2. Sensors based on other nanostructures are manufactured according to the traditional technology and therefore, in their manufacture, it is possible to base on the regularities and recommendations proposed in Chapter 18. At the same time, this chapter will focus on inorganic, mainly metal-oxide 1D structures. As for other materials, that is, carbon-based 1D and 2D structures such as carbon nanotubes (CNTs) and graphene, they were discussed earlier in Chapter 4, 2D chalcogenide-based structures were considered in Chapter 5, and humidity sensors based on hierarchical structures were analyzed in Chapter 23. It is also important to note that the observed sensitivity of one-dimensional metal-oxide sensors is significantly higher than the reported sensitivity of carbon nanotubes (Kong et al. 2000). It is assumed that the difference is related to the nature of the metal-oxide surface, which can readily react with ambient species, compared to the inert sidewall of carbon nanotubes (Kong et al. 2000).

Interest in the use of individual 1D structures in gas sensors, including humidity sensors, was due to the following:

1. It was assumed that a well-defined geometry, single-crystallinity, small diameter of 1D structures, and a large area-to-volume ratio will ensure high sensitivity of the sensors (Law et al. 2004).

2. It was assumed that due to the lack of necks and grain boundaries (see Figure 21.3), such sensors would solve the problem of instability and temporal drift of the parameters of polycrystalline-based thin and thick-film sensors, caused by interaction and mass transfer on these intergrain interfaces (Hernandez-Ramirez et al. 2009; Sysoev et al. 2009). These processes may contribute to structural changes in the sensors. The high crystallinity of the nanowires and nanobelts structure should have contributed to the improvement of stability.

3. It was assumed that response dynamics should be faster compared to their polycrystalline counterpart since there is no need for gas diffusion in the gas-sensing matrix preliminary to the surface reaction (Hernandez-Ramirez et al. 2009). The absence of nooks and crannies in nanowire-based devices contributes to the direct adsorption/desorption of gas molecules.

4. It was assumed that through the use of individual 1D structures and self-heating effects it would be possible to significantly reduce the power consumed by the sensor, and thereby facilitate the integration of the sensors in portable systems (Strelcov et al. 2008; Prades et al. 2008; Kolmakov 2008; Hernandez-Ramirez et al. 2009). According to Meier et al. (2007a, 2007b) and Hernandez-Ramirez et al. (2007a, 2007b, 2007c), such approach allows to reduce the required power consumption from the milliwatt to the microwatt range. In addition, self-heating effects enable an experimental methodology to improve the selectivity of metal-oxide-based sensors based on the analysis of their fast response dynamics (Prades et al. 2011).

5. It was assumed that the control of the shape of 1D structures (metal-oxide 1D structures have a clearly defined cut depending on the synthesis conditions) would improve the selectivity of the sensors.

6. It was assumed that the measurement of individual nanowires (NWs) with well know properties may allow to understand the fundamentals of the gas-sensing

**FIGURE 21.1** SEM images of ZnO and Cu$_2$O nanostructures synthesized by different methods. (a) (From Lu, F. et al.: ZnO hierarchical micro/nanoarchitectures: Solvothermal synthesis and structurally enhanced photocatalytic performance. *Advanced Functional Materials.* 2008. 18. 1047–1056. Copyright Willey-VCH Verlag GmbH & Co. KGaA. Reproduced with permission); (b) (Reprinted from *J. Cryst. Grow.*, 310, Li, J. et al., Synthesis and luminescence properties of ZnO nanostructures produced by the sol–gel method, 599–603, Copyright 2008 with permission from Elsevier); (c, e, g) (Reprinted with permission from Xu, L. et al., *J. Phys. Chem. C*, 111, 11560–11565. Copyright 2007 American Chemical Society); (d) (Reprinted from *Mater. Chem. Phys.*, 115, Sepulveda-Guzman, S. et al., Synthesis of assembled ZnO structures by precipitation method in aqueous media, 172–178, Copyright 2009, with permission from Elsevier); (g) (Reprinted from *Thin Solid Films*, 515, Krunks, M. et al., Spray pyrolysis deposition of zinc oxide nanostructured layers, 1157–1160, Copyright 2006, with permission for Elsevier); (h) (Reprinted with permission from Orel, Z.C. et al., *Crystal Growth Design*, 7, 453–458, 2007. Copyright 2007 American Chemical Society); (i) (Reprinted with permission from Dev, A. et al., Optical and field emission properties of ZnO nanorod arrays synthesized on zinc foils by the solvothermal route, *Nanotechnol*, 17, 1533–1540, 2006. Copyright 2006, Institute of Physics.)

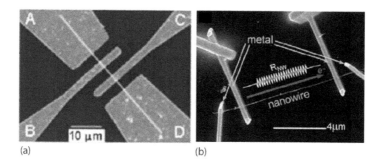

**FIGURE 21.2** SEM images of gas sensors in 4-probe DC configuration based on single-metal-oxide nanobelts and nanowires: (a) Gas sensor fabricated by deposition and patterning of contact materials on the SnO$_2$ nanobelt located on the surface of the dielectric substrate. (b) Metal stripes with well-defined shapes in the nanometer range and high electrical quality are fabricated using focused ion beam (FIB) deposition of metal. (a) (Reprinted with permission from Fields, L.L. et al., *Appl. Phys. Lett.*, 88, 263102, 2006. Copyright 2006 by the American Institute of Physics.) (b) (Hernandez-Ramirez, F. et al., *Phys. Chem. Chem. Phys.*, 11, 7105–7110. Reproduced by permission of The Royal Society of Chemistry.)

effect (Kolmakov and Moskovits 2004). The random aggregation of nanoparticles in poly(nano)crystalline metal oxides (MOXs), as well as the scatter in their size, make it difficult to study the gas-transduction phenomena accurately at the nanoscale. The single-crystallinity of the NWs, however, makes easier to interpret the experimental data (Figure 21.3). The geometry of individual single-crystal NWs constitutes an impediment to fine analysis of the gas–surface interactions, because there are no necks and boundaries.

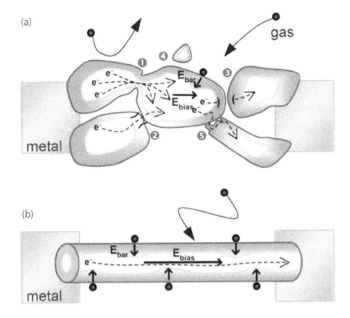

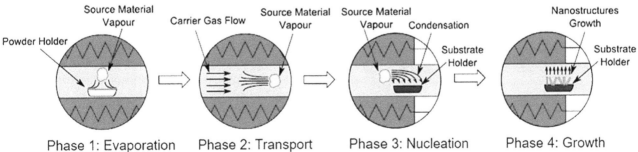

**FIGURE 21.3** Diagrams illustrating difference in gas-sensing effects in (a) polycrystalline material, and (b) individual nanowires: One can see that any intergrain necks or boundaries are absent in 1D-based sensors. Moreover, $E_{bar}$ and $E_{bias}$ fields are always orthogonal and independent. (Hernandez-Ramirez, F. et al., *Phys. Chem. Chem. Phys.*, 11, 7105–7110. Reproduced by permission of The Royal Society of Chemistry.)

That is why since 2000, the development of gas sensors based on 1D structures has become one of the most popular areas of research in the field of gas sensors (Kolmakov and Moskovits 2004; Zhang et al. 2005; Comini et al. 2009; Hernandez-Ramirez et al. 2009; Choi and Jang 2010; Li et al. 2010; Soldano et al. 2012).

## 21.2 SYNTHESIS OF METAL-OXIDE ONE-DIMENSIONAL NANOMATERIALS

There are many different techniques that can be used to prepare metal-oxide nanostructures, but essentially they can be grouped into two broad classes: "top-down" and "bottom-up" approaches (Comini et al. 2013). The basic idea of the "top-down" approaches is to use existing technology developed

in the semiconductor industry to microfabricate nanostructures. This class of techniques uses deposition, etching, and ion-beam milling on planar substrates in order to reduce the dimensions of the structures to a nanometer size. Electron beam, focused ion beam, X-ray lithography, nanoimprinting, and scanning probe microscopy techniques are used for the selective removal processes. Using these techniques, it is possible to obtain highly ordered nanostructures (Haghiri-Gosnet et al. 1999; Marrian and Tennant 2003; Candeloro et al. 2005), but they are very expensive both in terms of costs and preparation times.

The second approach is the "bottom-up," which consists of the assembly of molecular building blocks. Among many strategies for the synthesis of oxide NWs, bottom-up growth from the vapor phase has traditionally been considered the most cost-effective method of producing NWs in large quantities (Choi et al. 2008; Barth et al. 2010).

### 21.2.1 VAPOR-PHASE GROWTH

The growth of the vapor phase was one of the first methods developed for preparing micro- and nanostructures, studied in early 60s by Wagner and Ellis (1964). It consists in the evaporation of the source material in a tubular furnace, able to reach up to 1500°C. The evaporated source material is transported by a gas carrier toward the colder region, where it condensates and nucleates on the growth sites (see Figure 21.4). This idea was developed for silicon whiskers, and later adapted to other materials like metal oxides. It mainly consists in a tubular furnace, able to reach up to 1500°C. The furnace is connected to a rotary vacuum pump, in order to control the pressure inside the alumina furnace's tube. Two mass flow controllers inject the transport gases (argon or oxygen) inside the system.

Condensation on the target can occur by two different mechanisms: vapor–liquid–solid (VLS) or vapor–solid (VS). The VLS mechanism (Givargizov 1975; Kolasinski 2006; Roper et al. 2007) is named after the three different phases involved: the vapor-phase source material, the liquid catalyst droplet, and the solid crystalline-produced nanostructure. It is worth noting that the growth of NWs occurs at the solid–liquid interface. The ensuing growth is essentially along one particular crystallographic orientation that corresponds to the

**FIGURE 21.4** Schematic illustration of the mechanism of NWs growth. (Reprinted with permission from *Sens. Actuators B*, 179, Comini, E. et al., Metal oxide nanoscience and nanotechnology for chemical sensors, 3–20, Copyright 2013, with permission from Elsevier.)

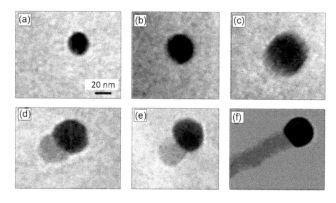

FIGURE 21.5 *In-situ* TEM images recorded during the process of nanowire growth: (a) Au nanoclusters in solid state at 500°C; (b) alloying initiates at 800°C. At this stage, Au is mostly solid-state; (c) liquid Au/Ge alloy; (d) nucleation of Ge nanocrystal on alloy surface; (e) Ge nanocrystal elongates with further Ge condensation, eventually forming a wire as shown in (f). (Reprinted with permission from Wu, Y.Y. and Yang, P.D., *J. Am. Chem. Soc.*, 123, 3165–3166, 2001. Copyright 2001 American Chemical Society.)

minimum atomic stacking energy (see Figure 21.5). Catalyst droplets may be deposited on the target substrates using many different techniques, that is, from colloidal solution or magnetron sputtering (Kolasinski 2002). Usually these catalysts are noble metals, like platinum, palladium, gold, and ruthenium. The dimensions of the catalyst clusters can determine the NW section either by direct matching of the size or by mechanisms involving the catalyst curvature, in which the strain and lattice matching are important.

Figure 21.6 illustrates three possible diffusion pathways in the development of nanowires (Wang et al. 2008). During the growth of NWs, the vapor molecules prefer to stay at the molten surface, and then are assembled into a solid wire. Figure 21.6a indicates a liquid droplet on the top of a NW that consists of vapor-phase material, leading to a supersaturation state. In the case of partially molten catalyst, as shown in Figure 21.6b, the growth of the NW is dominated by the surface diffusion of the source materials. The liquid

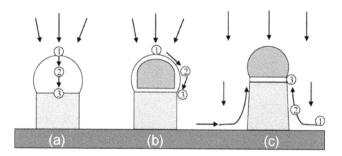

FIGURE 21.6 Diffusion models for molecules in vapor phase: (a) conventional VLS growth process; (b) the metal droplet is in a partially molten state. Its surface and interface are liquid, while the core of the droplet is solid; and (c) the metal catalyst is solid, but the interface is liquid. (Reprinted with permission from *Mater. Sci. Eng. R*, 60, Wang, N. et al., Growth of nanowires, 1–51, Copyright 2008, with permission from Elsevier.)

phase exists only at the interface of the catalyst and target materials, and diffusion occurs mainly at the NW surface, as shown in Figure 21.6c. There exist several atomic layers in semimolten state at the solid–liquid interface, which enable the atoms to move easily at the interface (Wang et al. 2008).

The VS growth takes place when the NW crystallization originates from the direct condensation from the vapor phase without the use of a catalyzer. Under high-temperature condition, the source materials are vaporized and then directly condensed on the target substrates, placed in the low-temperature region. Once the condensation process happens, the initially condensed molecules form seed crystals serving as nucleation sites (Wang 2004; Wang et al. 2008). Within the past few years, diverse nanostructures of various semiconductors, such as CuO, $Fe_3O_4$, ZnO, $In_2O_3$, GdO, and $SnO_2$, have been synthesized using the VS mechanism (Pan et al. 2001; Jiang et al. 2002; Lao et al. 2002; Kong and Wang, 2003; Zheng et al. 2005a; Bierman et al. 2008; Hwang et al. 2008). Without additional catalyst, the VS mechanism tends to have less complexity in the sample preparation; however, the size of the Q1D materials obtained is not easily controlled. The size distribution of VS-grown Q1D materials is broader and larger than that of VLS-grown samples.

Vapor-phase growth of 1D nanostructures can be realized using different experimental growth methods, including thermal chemical vapor deposition (CVD) (Wang et al. 2008), pulsed laser deposition (PLD) (Li et al. 2003a; Liu et al. 2003b), direct thermal evaporation (Dai et al. 2003), molecular beam epitaxy (MBE) (Wu et al. 2002b), and metal-organic chemical vapor deposition (MOCVD) (Park et al. 2002), and hot wire chemical deposition (Chakrapani 2016). The NWs growth using the vapor-phase technique is known to be very sensitive to process parameters like source type, substrate type and its orientation, catalyst and its size, reaction temperature and duration, temperature profile of the furnace, partial pressure of reactants and gas flow, and geometry of the tube furnace (Li et al. 2010; Chakrapani 2016). Figure 21.7 shows how much NWs can vary depending on the synthesis conditions. Most groups use an empirical approach to determine the suitable growth conditions for their respective tube furnaces, and the understanding of the growth mechanism from the kinetics or the thermodynamics point of view is still not sufficient and greatly missing. Importantly, nanostructures, grown in the vapor phase, usually exhibit higher characteristics in nanoscale electronic devices than those synthesized in solution, because of the better crystalline quality and the corresponding electron-transport properties.

These synthesis techniques, such as the VLS or VS growth, yield randomly oriented assemblies of NWs with the length of individual nanowires up to 10–500 μm (Li et al. 2002; Zhang et al. 2004a). One should note that most studies to date have been carried out on individual NWs selected from these assemblies, although it is difficult to integrate fabricated nanostructures on a planar substrate and have them well arranged and patterned (Hanrath and Korgel 2002).

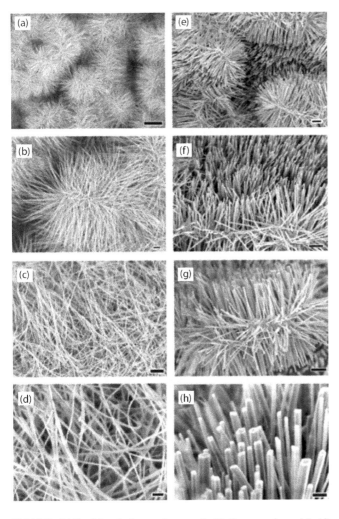

**FIGURE 21.7** Morphology of as-made ZnO nanowires: (a)–(d) show long (10–15 μm) and thin (30–60 nm) nanowires grown in the higher-temperature region of the furnace; (e)–(h) show short (1–2 μm) and thick (60–100 nm) nanowires grown in the lower-temperature region of the furnace. The scale bar for (a) is 10 μm, for (b), (c), (e), (f), and (g) 1 μm, and for (d) and (h) 200 nm. (Reprinted with permission from Banerjee, D. et al., Synthesis and photoluminescence studies on ZnO nanowires, *Nanotechnology* 15, 404–409, 2004. Copyright 2004, Institute of Physics.)

### 21.2.2 SOLUTION-PHASE GROWTH METHOD

In contrast to vapor–gas-phase growth, another main stream of 1D system synthesis is a solution-phase growth. Growth in solution phase is a simple and cost-effective way to produce large quantities of 1D nanostructures. There are many different approaches to produce 1D materials in liquid chemicals, including the polyol method (Jiang et al. 2004), the surfactant method (Miao et al. 2005), and the hydrothermal/solvothermal technique (Zou et al. 2006). Among these methods, the hydrothermal technique (read Chapter 20), usually carried out in an autoclave, is one of the most versatile and conventional methods, realized in industrial applications because of its high yield, low cost, and moderate-temperature processes. It requires simply precursors, mixed into solution and heated at proper temperature in an autoclave. The desired 1D

materials are then obtained after synthesizing chemical reactions. The drawback of this hydrothermal approach is its long synthesis time, which usually takes several hours. A variety of 1D nanostructures, such as Si, ZnO, $In_2O_3$, CuO, $Ga_2O_3$, $MnO_2$, $CeO_2$, $TiO_2$, $WO_3$, ZnS, and CdS, among others, have been explored (Li et al. 2010).

If we compare different NWs synthesized by different methods, then among the large numbers of MOX NWs synthesized up to now, $SnO_2$, $In_2O_3$, and especially ZnO NWs are usually considered to be the best candidates for developing gas and humidity sensors (Mathur et al. 2007; Huang and Choi 2007; Barth et al. 2010; Choi and Jang 2010). Interest in 1D ZnO structures is encouraged by the easy synthesis of high-quality and single-crystalline 1D ZnO nanostructures. The synthesis of 1D nanostructures based on other sensitive metal oxides such as $TiO_2$ and $WO_3$ has been reported to be difficult compared to other oxides.

### 21.3 FEATURES OF 1D STRUCTURE-BASED SENSOR FABRICATION

As it follows from previous discussions, most as-grown NW processes are incompatible with electrical characterization of individual wires. Additional processing is thus required to remove the NWs from their growth substrate and deposit them onto a platform that permits the study of single wires and use them as sensing element. Therefore, one of the most important issues during sensor fabrication is patterning the NWs with electrodes. An electrical contact to the sensors based on single NWs is often realized using a "pick and place" approach. NWs are first dispersed into solvents like methanol, ethanol, isopropanol, or water in very small concentrations (Figure 21.8a). Then, a drop of dilute suspension of wires is placed on a substrate and is dried. Substrate with NWs are then located under an optical microscope or scanning electron microscope (SEM), and the contacts are provided by depositing an electrode layer on the top of NW, using various methods.

Undoubtedly, NWs can be moved to the right place with the help of special manipulators (Figure 21.8b). For example, Li et al. (2007) offer to use for this electrostatic forces. Jimenez-Diaz et al. (2011) for these purposes used AC dielectrophoretic manipulation of NWs. A solution, containing NWs, is dropped on the top of the electrode array. The NWs aligned along the direction of electric field. Different species such as a NW or contaminant species will respond to different frequencies of AC bias. Therefore, applying an AC bias with a corrective frequency offers the advantage of selectively choosing NWs from other contaminant species in the solution (Nagahara et al. 2002). This approach to the formation of contacts to NWs has several advantages, since its implementation, which allows to properly position the NWs at a predetermined position of the sample and, more concretely, bridging two contacts, allows to significantly reduce the time of manufacture of the sensor. In this way, it would only be required to fix the NW to the contact, which is much less time-consuming process (Jimenez-Diaz et al. 2011). However, such an action

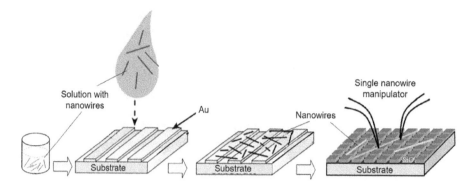

**FIGURE 21.8** Illustration of the process of transferring nanowires to the surface of the substrate: nanowires are removed into a suspension of solvent. A drop of the nanowire solution is dispersed on a template substrate and evaporated under a vacuum. Finally, nanowires with manipulator tips can be moved from a template substrate to the right place.

is very laborious and requires special equipment. In addition, in many cases the transfer is possible either for too small or for too large NWs. And this means that sensors based on such NWs will have low sensitivity. Therefore, as a rule, NWs in the manufacture of sensors are not subjected to any additional movements.

Nowadays, metal stripes with well-defined shapes in the nanometer range and high electrical quality, shown in Figure 21.8, are easily fabricated by various techniques such as focused ion beam (FIB) (Hernandez-Ramirez et al. 2006, 2007) (Figure 21.9), focused electron beam (FEB) (HernandezRamirez et al. 2006; Utkea et al. 2008), ultraviolet and shadow-mask lithography (Kalinin et al. 2005), and shadow-mask sputtering (Sysoev et al. 2006, 2007). Shadowmask sputtering enables fast engineering of advanced proof-of-concept devices. However, metallic contacts directly deposited over the NW bundles are unstable and do not guarantee the formation of a continuous metallic layer, thus inhibiting direct bonding of NWs and good electron transport. Electric contacts, fabricated by using, for instance, focused ion or electron beams, are much better.

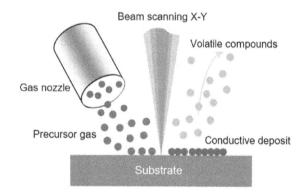

**FIGURE 21.9** Schematic representation of local deposition, assisted by FIB. The gas precursor introduced by the capillary is decomposed by the secondary electrons generated by the interaction of the primary ion beam with the target. (Reprinted with permission from Gierak, J., Focused ion beam technology and ultimate applications, *Semicond. Sci. Technol.*, 24, 043001, 2009. Copyright 2009, Institute of Physics.)

FIB is a powerful technique developed during the late 1970s and the early 1980s for the patterning, and later, for the deposition of materials, with resolution in the tens of nanometer range, and commonly used in circuit edit, mask repair, microsystem technology processes, and material characterization (Melngailis 1987; Tseng 2005; Gierak 2009). The basic principle of this technique is a focused ion beam of highly energetic particles that scans the sample's surface and sputters the material of the exposed area (FEI 2004). The scanning can be performed, similarly to an SEM, using electrostatic lenses and, thus, the milling occurs without the need of masks. Currently, gallium (Ga$^+$) ions accelerated to 30 kV are the most used particles in FIB technique. Gallium is a metallic element with a low melting temperature, which allows the fabrication of long life-time, high brightness and reliable metal ion source (LMIS) required in FIB technique. Moreover, the element gallium is positioned in the center of the periodic table, so its momentum-transfer capability is optimum for a wide variety of materials. On the contrary, lighter elements would be less efficient in milling heavier elements.

On the other hand, if a matalorganic compound is introduced in the beam path with the help of a so-called gas-delivery system by using a microneedle, decomposition occurs due to interaction of the compound with both secondary electrons and ions originated during the Ga$^+$ ion bombardment (Figure 21.9). Part of the compound can be deposited on the sample's surface (ion-assisted deposition) or can reactively assist the milling process (gas-assisted etching), while the rest is removed by the vacuum system. In this way, conductive and isolating materials can easily be deposited with FIB with nanometer precision (Gamo 1993; Gierak 2009). Many types of precursor gases have been used in FIB for making of a variety of metallic and ceramic structures. For example, the precursor gases WF$_6$ (or W(CO)$_6$), C$_7$H$_7$F$_6$O$_2$Au, (CH$_3$)$_3$NAlH$_3$, C$_9$H$_{16}$Pt, and TMOS+O$_2$ (TMOS=tetramethyloxysilane) have been used to produce W, Au, Al, Pt, and SiO$_2$, respectively (Melngailis 2001). Although the purity of the deposition is generally lower as compared to convention originated during the metalorganic decomposition (usually the deposited layer is contaminated by other elements such as oxygen or carbon from the background gas within the vacuum chamber, and by the elements forming the ion beam), the main advantage of this technique is its flexibility due to its direct writing capabilities and because masks are not required (FEI 2004; Gierak 2009). The experiment showed that materials such as W, Pt, Au, Al, Ta, C, and $SiO_2$ can be deposited in the FIB.

However, it is important to note that the ion bombardment necessary to decompose the metal–organic precursor for fabricating nanocontacts can produce the damage in the NWs (Ebbesen et al. 1996; Nam et al. 2005). Therefore, Hernandez-Ramirez et al. (2006) believe that the development of the so-called dual or cross-beam systems (conventional FIB with an SEM) has facilitated the use of FIB nanolithography enabling *in-situ* capture of electron images and simultaneous dissociation of metalorganic compounds with secondary electrons generated by the incident electron beam, giving the rise to electron-beam-induced deposition (Rotkina et al. 2003). Given the fact that the interaction between electrons and the sample is less destructive than using ions, performing electron-assisted deposition on the nanostructure for the electrical contacts and accomplishment of the rest of the contacts with the help of ions can avoid undesired surface damage and structure modification of the nanomaterial. Despite its promise, the electron and ion-beam combination are sparsely investigated and detailed studies are required to gain insight into the quality of the electrical contacts (Gopal et al. 2004) and to avoid false interpretations of the electrical properties of NWs, for instance, caused by contact resistance underestimations.

Alternatively, NWs are drop-casted onto substrates, containing predefined electrodes and are aligned using dielectrophoresis technique. NW samples are then subjected to annealing for better adherence. The stages of manufacturing such sensors are shown in Figure 21.10.

The previously discussed techniques are not scalable and do not fulfill the requirements for making the leap to commercialization. A typical view of the device based on an individual NW is shown in Figure 21.11. Sensors with such a configuration is really difficult to implement using methods developed for mass production. However, the approaches used to manufacture them are mostly suitable for research and development of device prototypes. But device prototypes are not intended for the market. For this reason, the scalability of NW-based devices and NW-based electronic circuits is currently under evaluation. Some examples of such approaches include self-assembly of NWs (Wang et al. 2007; Kumar 2010), electrospinning (Sawicka et al. 2005), and microcontact/ink-jet printing (Kim et al. 2006; Van Osch et al. 2008; Song et al. 2008). In the self-assembly approach, MOX NWs are deposited directly by a high-voltage-driven injection nozzle onto the electrodes. During printing, the metal electrodes are precisely placed on the top of the NWs. However, all these fabrication techniques are still in a preliminary stage of development, despite some promising results that have recently been reported (Song et al. 2008).

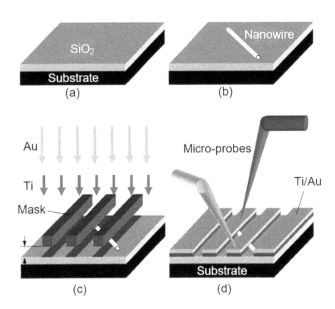

FIGURE 21.10 Protocol for resist-free fabrication of 1D MOX nanowire chemiresistors and chemFETS: (a) pristine $Si/SiO_2$ wafer; (b) nanostructures placed on wafer mechanically; (c) shadow masking to determine metal contacts; and (d) microprobes can be used to explore transport and sensing properties of the individual nanoresistors. (Idea from Kolmakov A., *Proc. SPIE*, 6370, 63700X, 2006. With permission.)

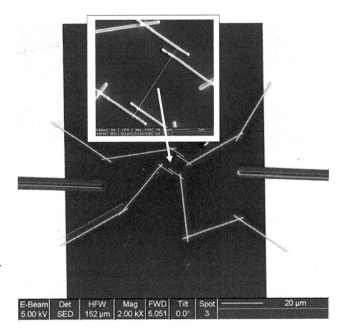

FIGURE 21.11 General view of the final device based on $SnO_2$ individual nanowire [Dimensions: L = 11 μm (length) and D = 55 ± 5 nm (diameter)] with Au/Ti/Ni microelectrodes. The position of the contacted NW is indicated by the arrow. (Reprinted with permission from *Sens. Actuators B*, 121, Hernandez-Ramirez, F. et al., High response and stability in CO and humidity measures using a single $SnO_2$ nanowire, 3–17, Copyright 2007a, with permission from Elsevier.)

## 21.4 LIMITATIONS OF TECHNOLOGY BASED ON INDIVIDUAL 1D NANOSTRUCTURES

As it was indicated before, at the first stage of the study of 1D nanostructures, most designers believed that the use of single crystalline individual NWs, as gas- and humidity-sensing material will significantly improve the parameters of gas and vapor sensors (Law et al. 2004; Hernandez-Ramirez et al. 2009; Sysoev et al. 2009). However, it turned out that the controlled separation, manipulation, and characterization of NWs are not straightforward processes due to the intrinsic problems of working at the nanoscale. These procedures are complex and require well-established methodologies, which are not yet fully developed. Several promising techniques have been reported to align or orientate NW assemblies with the help of microfluidics, electrostatic or magnetic fields, surface prepatterning, self-assembly, and templating. These architecture principles, which have made it possible to design prototype devices, have been comprehensively reviewed recently (Xia et al. 2003; Whang et al. 2003; Lieber 2003; Qi et al. 2003; Wang et al. 2007; Assad et al. 2012; Liu et al. 2012). The advantages and disadvantages of these approaches are

briefly summarized in Table 21.1. All these fabrication routes, however, are still in a preliminary stage of development. Thus, the use of NWs in real devices is still at a preliminary stage and needs a breakthrough in order to integrate them with low-cost industrial processes (Hernandez-Ramirez et al. 2009).

The study, however, showed the use of an array of 1D structure, placed in the form of monolayer mats (see Figure 21.12a), that allow the use of standard sensor technology for chemical-sensor fabrication, and it is already possible to realize a number of indisputable advantages of 1D nanostructures in gas sensors. This approach is currently the most common in the manufacture of sensors based on 1D structures (Sysoev et al. 2007). However, in this case, such sensors do not have fundamental differences from traditional polycrystalline sensors manufactured by thick-film technology. The sensor response, as well as in polycrystalline sensors, will be mainly controlled by the interwire potential barriers (Figure 21.12b). The only advantage of such sensors is a large porosity, that is, the better gas permeability of the sensing layer.

Another more progressive approach involves growing parallel NWs or nanotubes in a perpendicular direction on the substrates (Kolmakov and Moskovits 2004) (see Figure 21.13).

**TABLE 21.1**
**Summary of NW Assembly Technologies**

| NW Assembly Technologies | Advantages | Disadvantages |
|---|---|---|
| Flow-assisted alignment in microchannels | 1. Parallel and crossed NW arrays can be assembled; 2. Compatible with both rigid and flexible substrates. | 1. Area for NW assembly is limited by the size of fluidic microchannels; 2. Difficult to achieve very high density of NW arrays; 3. NW suspension needs to be prepared first. |
| Bubble-blown technique | 1. Area for NW assembly is large; 2. Compatible with both rigid and flexible substrates. | 1. It is difficult to achieve high-density NW arrays; 2. NW suspension needs to be prepared first. |
| Contact printing | 1. Area for NW assembly is large; 2. High-density NW arrays can be achieved; 3. Parallel and crossed NW arrays can be assembled; 4. Direct transfer NW from growth substrate to receiver substrate; 5. Compatible with both rigid and flexible substrates; 6. NW assembly process is fast. | 1. Growth substrate needs to be planar; 2. The process works the best for long NWs. |
| Differential roll printing | 1. Area for NW assembly is large; 2. High-density NW arrays can be achieved; 3. Direct transfer NW from growth substrate to receiver substrate; 4. Compatible with both rigid and flexible substrates; 5. NW assembly process is fast. | 1. Growth substrate needs to be cylindrical; 2. The process works the best for long NWs. |
| Langmuir–Blodgett technique | 1. Area for NW assembly is large; 2. High-density NW arrays can be achieved; 3. Parallel and crossed NW arrays can be assembled; 4. Compatible with both rigid and flexible substrates. | 1. NWs typically need to be functionalized with surfactant; 2. The assembly process is slow and has to be carefully controlled; 3. NW suspension needs to be prepared first. |
| Electric field-assisted orientation | 1. NWs can be placed at specific location; 2. Compatible with both rigid and flexible substrates; 3. NW assembly process is fast. | 1. Patterned electrode arrays are needed; 2. Area for NW assembly is limited by the electrode patterning; 3. NW density is limited; 4. It works the best for conductive NWs; 5. NW suspension needs to be prepared first. |

*Source:* Reprinted with permission from Liu, X. et al., *ACS Nano*, 6, 1888–1900, 2012. Copyright 2012 American Chemical Society.

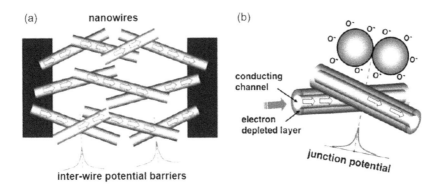

FIGURE 21.12 (a) Schematic diagrams for multinanowire-based chemical sensors; and (b) schematic illustration of the gas-sensing mechanism in a network of nanowires. (Reprinted with permission from Vomiero, A. et al., *Crystal Growth Design*, 7, 2500–2504, 2007. Copyright 2007 American Chemical Society.)

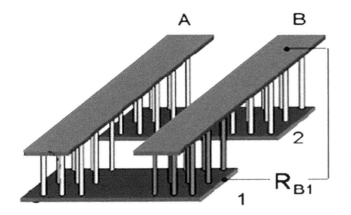

FIGURE 21.13 Schematic diagrams for multinanowire-based sensor with nanotubes oriented in a perpendicular direction. (Reprinted from Kolmakov, A. and Moskovits, M. *Annu. Rev. Mater. Res.*, 34, 151–180, 2004. Published by a non-profit scientific publisher as open access.)

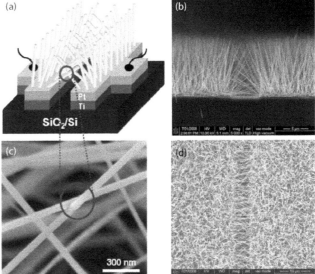

FIGURE 21.14 (a) The schematic illustration of ZnO-nanowire air bridges over the $SiO_2$/Si substrate; (b) side- and (d) top-view SEM images clearly show the selective growth of ZnO nanowires on Ti/Pt electrode; and (c) the junction between ZnO nanowires, grown on both electrodes. (Reprinted with permission from *Sens. Actuators B*, 138, Ahn, M.-W. et al., On-chip fabrication of ZnO-nanowire gas sensor with high gas sensitivity, 168–173, Copyright 2009, with permission from Elsevier.)

In this case, although the 1D structures array is used, such a configuration allows one to realize the advantages of the sensors characteristic of devices based on the individual 1D nanostructures. Varghese and Grimes (2003) and Varghese et al. (2009) applied such an approach to develop devices from $TiO_2$ nanotubes. This technique is similar to the manufacture of thin films, but promises higher accuracy in making the columnar nanoelements. Contacts can also be fabricated using methods of electrochemical deposition (Dimaggio and Pennelli 2016).

The approach proposed by Choi et al. (2008) and Ahn et al. (2008, 2009) is also interesting for gas- and humidity-sensor design (see Figure 21.14). Figure 21.14b and d show side- and top-view SEM images of ZnO NWs grown on patterned electrodes. ZnO NWs, grown only on the patterned electrodes, have many NW–NW junctions as seen in Figure 21.14c. These junctions act as electrical conducting path for electrons. The device structure in this work is very simple and efficient compared with those adopted by previous researchers, because the electrical contacts to NWs are self-assembled during the synthesis of NWs. Ahn et al. (2008, 2009) asserted that this method of on-chip fabrication of NW-based gas sensors is scalable and reproducible. However, this statement raises doubts.

The low sensitivity of the 1D-based sensors in comparison with conventional nanocrystallite-based sensors can also be attributed to the shortcomings of these devices. As in polycrystalline sensors, the sensitivity of 1D-based sensors depends on the diameter of NWs or nanobelts (see Figure 21.15); the smaller the diameter, the higher the sensitivity (Comini et al. 2002; Hernandez-Ramirez et al. 2007c; Dmitriev et al. 2007). Experiment and simulation have shown that to achieve the maximum response the Debye length ($L_D$) should be compared with the NW radius. For example, Hernandez-Ramirez et al. (2007c) have found that in real 1D $SnO_2$ structures, the Debye length is ~20–30 nm, and therefore the performance of NW sensors can be significantly improved only when the diameter is smaller than 25 nm. However, as a rule, NWs used

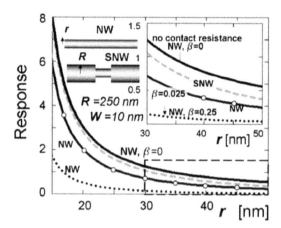

FIGURE 21.15  Response of straight and segmented nanowires as a function of their radius at various contact resistances ($b = 0$, 0.025, 0.25). For segmented nanowires (SNWs), the curve is drawn versus the radius of the smaller segment. The solid curve (top) corresponds to the nanowire with no contact resistance; the dashed curve corresponds to the SNW with thick segments of 500 nm diameter and $b = 0.025$; the solid curve marked with circles corresponds to the SNW with $b = 0.025$; the dotted curve corresponds to the SNW with $b = 0.25$. The depletion width is ~10 nm at all cases. (Reprinted with permission from Dmitriev, S. et al., Nanoengineered chemiresistors: the interplay between electron transport and chemisorption properties of morphologically encoded $SnO_2$ nanowires, *Nanotechnology*, 18, 055707, 2007. Copyright 2007, Institute of Physics.)

for the manufacture of sensors have a diameter in the range of 50–900 nm, while the grain size in polycrystalline-based sensors does not exceed 5–10 nm. Research carried out by Li et al. (2003b, 2003c) and Zhang et al. (2004b) confirmed this conclusion. They established that extremely low detection limit was achieved for $In_2O_3$ and ZnO NWs with 10 nm diameter only. Unfortunately, at present nobody wants to work with individual NWs with a size less than 20–30 nm (Comini et al. 2002; Ponzoni et al. 2006; Hernandez-Ramirez et al. 2007a, 2007c; Köck et al. 2009). Separation and manipulation of such small objects is too difficult (Hernandez-Ramirez et al. 2007a). Besides that, there are technological difficulties with the growth of such thin NWs ($d < 10$ nm) with bigger length. The controllable growth of very thin metal-oxide NWs still remains to be an experimental challenge (Wang et al. 2004). Low temperatures are necessary for the growth of such NWs. But in this case, the length of NWs will be insufficient for the creation of a bridge between two bonding pads on the measurement platform (Köck et al. 2009). Kolmakov and co-workers (Lilach et al. 2005) have found that one of the most promising ways to address this challenge is to fabricate a single-crystal quasi-1D chemiresistor with one (or a few) very fine segments, which adhere to $r \sim L_D$ condition. Since the narrow segment(s) will dominate the electron transport and sensing performance, such a device would have all advantages of ultra-thin single crystal NWs (see Figure 21.15). Studies carried out by Dmitriev et al. (2007) have shown that the narrow segments serve as ideal "necks," as observed between particles in conventional polycrystalline thin-film gas sensors, but provide

the significant advantages of greater morphological integrity and stability. Reports of the methodology of segmented oxide NW controllable growth via a programmable change in the local $Sn_xO_y$ ($x$, $y = 1$, 2, …) vapor-supersaturation ratio in the vicinity of the wires during their vapor solid growth have been published (Lilach et al. 2005). It is interesting, but more complicated approach to fabrication of 1D-based sensors.

The creation of a reliable low-resistance electric contact with such thin NWs is also an essential problem. Moreover, the reduced contact area between metal electrodes and NWs magnifies the contribution of the contact electrical properties and may hide the phenomena that takes place on the surface of NWs (Nam et al. 2005; Hernandez-Ramirez et al. 2006, 2007a, 2007b; Lin and Jian 2008). Therefore, Ebbesen et al. (1996) and Hernandez-Ramirez et al. (2006) believe that in order to reduce the influence of contacts on sensor readings, it is necessary to use four-probe contacts configuration of the sensors (see Figure 21.16a). It is also necessary to monitor the dissipated power in the process of testing, since even with powers exceeding 1 µW of heat dissipation, local heating is possible, which can originate many physical problems related to material diffusion, transformation, and fusion, that must be avoided to ensure nanosensor stability. Figure 21.16b shows the final electrical breakdown consequence of this phenomenon. Hernandez-Ramirez et al. (2007a) have also noted that before arrival to this failure state, the electrical characteristics drifted and showed strong degradation and non-repeatability. Comini et al. (2009) believe that the exactly technical difficulties in the implementation of reliable electrical contacts on one individual nanostructure in a controlled fabrication process at the nanoscale level limit the number of works on this topic.

The instability associated with the surface diffusion of the electrode material along the NW and the migration of surface clusters used for surface functionality can also be significantly increased; 1D structures do not contain edges and stages, which can serve as particle pinning centers that hinder the diffusion of noble metal clusters on the surface of metal-oxide supports.

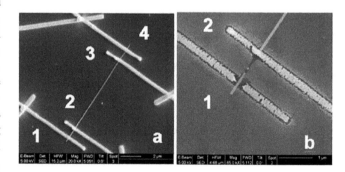

FIGURE 21.16  (a) A fabricated FIB four-probes contacted $SnO_2$ nanowire; and (b) a detail of the contact after electrical failure likely caused by excess of heat dissipation in the contacts. (Reprinted with permission from *Sens. Actuators B*, 121, Hernandez-Ramirez, F. et al., High response and stability in CO and humidity measures using a single $SnO_2$ nanowire, 3–17, Copyright 2007a, with permission from Elsevier.)

The reproducibility of the parameters of the 1D-based sensors should also be improved. Due to the sample to sample variation, it is very difficult to produce NWs-based sensors with identical parameters. A small variation in the size and the shape leads to a drastic variation in the electronic properties and hence one needs to average out the performance between different sensor devices.

## 21.5 HUMIDITY SENSORS BASED ON INDIVIDUAL 1D STRUCTURES

Humidity sensors based on 1D structures and their parameters are listed in Table 21.2. It is seen that among the large numbers of metal-oxide NWs, synthesized up to now, the $SnO_2$, $CeO_2$, CuO, and ZnO NWs are usually considered to be the best candidates for developing humidity sensors due to their relatively low cost of production and well-known properties (Mathur et al. 2007).

It is important to note that the development of humidity sensors uses the same approaches as in the development of gas sensors. This means that the humidity sensors based on individual 1D structures are characterized by the same limitations arising both during their manufacture and during their operation. So, sensors on the basis of individual 1D structures, in addition to technological difficulties, have a very low sensitivity (Table 21.2), which does not encourage their use in real devices. For capacitive sensors, low sensitivity is primarily associated with the features of the structure of the sensing element, in which capillary condensation of water vapors is impossible. For resistive humidity sensors, as well as for gas sensors, low sensitivity can be explained by the too-large

diameter of the NWs used. As a rule, NWs with an average diameter ranged from ~30 to 500 nm were used in such humidity sensors.

As can be seen from Table 21.2, most of the humidity sensors based on 1D structures were manufactured on the basis of NW arrays, formed by various methods. In most cases, thick-film technology was used, when 1D structures were separated from the substrate on which growth took place, and were later used instead of powders in the preparation of pastes and solutions intended for manufacturing humidity sensitive layers (Dai et al. 2007; Fu et al. 2007; Erol et al. 2011; Uzar et al. 2011). In a number of papers, the NW array was formed during growth on the corresponding substrate (Chang et al. 2010; Hsueh et al. 2012). For ZnO-based humidity sensors, this process is shown in Figure 21.17

Feng et al. (2012) have shown that with regard to humidity sensors, one can use the approach to the manufacture of devices proposed by Kolmakov and Moskovits 2004 (see Figure 21.13). The array of $Al_2O_3$ NWs was fabricated from anodic aluminum oxide (AAO) by a simple two-step anodization process. For this porous, the $Al_2O_3$ layer, formed using electrochemical anodization (read Chapter 3), was etched in 1.8 wt% phosphoric acid at 40°C. As a result of etching, NWs were formed. They are shown in Figure 21.18. Then coplanar interdigitated Au electrodes were formed on the surface of alumina nanowires (ANWs) by vacuum evaporation. The layout of the fabricated humidity sensor is shown in Figure 21.19.

Undoubtedly, the use of NW arrays instead of individual 1D structures significantly simplified the manufacturing process and, most importantly, contributed to an increase in sensitivity (Table 21.2). But even in this case, the sensitivity of

## TABLE 21.2
### Properties of Humidity Sensors Based on 1D Nanowires Classified by Sensing Material

| Material | Morphology | RH Range (%) | Sensor Signal | Hysteresis | Res./Rec. Time | References |
|---|---|---|---|---|---|---|
| $Al_2O_3$ | NW array | 10–95 | $S_C \sim 14$ | ~4% | ~40/70 s | Feng et al. (2012) |
| $CeO_2$ | NW array | 10–98 | $S_R \sim 2 \cdot 10^2$ | N/A | ~3/3 s | Fu et al. (2007) |
| | | 25–90 | $S_R \sim 1.2 \cdot 10^3$ | N/A | >120 s | Zhang et al. (2007) |
| CuO | NW-I | 20–80 | $S_R \sim 1.12$ | N/A | ~120 | Wang et al. (2012) |
| | NW array | 5–84 | $S_R \sim 2 \cdot 10^2$ | N/A | N/A | Yuan et al. (2010) |
| $Ga_2O_3$ | NW-I | 8–90 | $S_R \sim 1.2$ | N/A | >30 min | Domènech-Gil et al. (2017) |
| $SnO_2$ | NW-I | 30–80 | $S_R \sim 30$ | N/A | 120–170/20–60 s | Kuang et al. (2007) |
| | | 12–98 | $S_R \sim 1.3$ | N/A | 4/50 min | Hernandez-Ramirez et al. (2007a, 2007b) |
| $WO_3$ | NW array | 25–85 | $S_R \sim 1.2$ | N/A | N/A | Dai et al. (2007) |
| ZnO | NW array | 48–88 | $\Delta f \sim 90$ Hz | N/A | 100/125 s | Erol et al. (2011) |
| | | 25–90 | $S_R \sim 1.7$ | | N/A | Chang et al. (2010) |
| | | 20–90 | $S_R \sim 40$ | | 32/110 s | Hsueh et al. (2012) |
| ZnS | NW array | 33–100 | $S_R \sim 10^3$ | N/A | 14/18 s | Uzar et al. (2011) |
| | | | $\Delta f \sim 180$ Hz | | | |
| ZnSe | NW array | 11–98 | $S_R \sim 4$ | NW array | N/A | Leung et al. (2008) |
| $Bi_2S_3$ | NW array | 5–80 | $S_R < 1.4$ | N/A | ~50/60 s | Kunakova et al. (2015) |
| Si | NW array | 30–95 | $S_R \sim 1.03$ | N/A | >10 s | Hsueh et al. (2011) |
| Te | NW array | 11–97 | $S_R \sim 3$–3.5 | N/A | 54/65 | Erande and Late (2016) |

$S_R = R_{dry}/R_{wet}$ or $I_{dry}/I_{wet}$ and $S_C = C_{wet}/C_{dry}$); R—resistive; C—capacitance; NW—nanowire; I—individual; Frequency shift for QCM with $f_0 = 8MHz$.

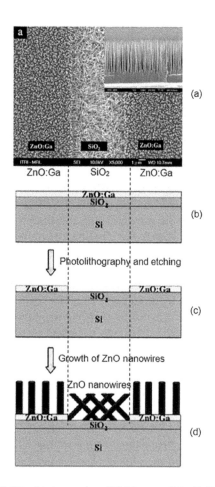

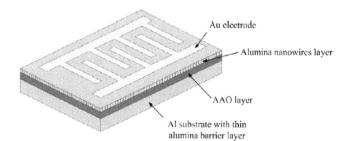

**FIGURE 21.19** Scheme of the $Al_2O_3$ NW-based humidity sensor. The real dimension is approximately $1 \times 1$ cm². (Reprinted with permission from Feng, Z.-S. et al., A novel humidity sensor based on alumina nanowire films, *J. Phys. D Appl. Phys.*, 45, 225305, 2012. Copyright 2012, Institute of Physics.)

the humidity sensors did not reach the level characteristic of sensors based on polycrystalline materials considered earlier. There is also no significant improvement in the kinetics of the sensor response, which suggests that the diffusion of water vapor in the humidity-sensitive layer does not always control the response and recovery time, especially for metal oxides and other solid-state materials. As for the other parameters of the humidity sensors based on 1D structures, such as stability, temperature dependence, and selectivity, they are controlled by the properties of the materials used and therefore do not differ from the parameters of conventional sensors discussed earlier.

## 21.6 SUMMARY

Summing up this review, it can be stated that despite of the declared prospects of 1D structures using in humidity sensors, these sensors are far from perfect. Of course, during last decade various fabrication and characterization strategies have been developed with purpose to accomplish electrical measurements of individual NWs free of parasitic effects, and to develop sensors competitive with conventional devices. For instance, low-current measurement protocols have been found to allow the devices to operate on a long-term scale without degradation of their performance (Hernandez-Ramirez et al. 2007a). So, it can be expected that the development of new NW-based technologies can lead to complete and well-controlled characterization of devices based on 1D structures, which were previously unattainable (Comini et al. 2009). However, it should be recognized that despite the progress achieved in the development of sensors based on individual 1D structures, it is not worthwhile to expect the appearance of 1D-based devices in the sensor market in the near future that could compete with conventional metal oxide and polymer humidity sensors.

**FIGURE 21.17** (a) A top-view SEM image of the ZnO nanowires grown on a ZnO:Ga/SiO₂/Si template and formed humidity sensitive element; and (b–d) schematic diagram of the growth and processing steps of the ZnO nanowire-based humidity sensor. (Reprinted with permission from *Superlattices Microstructures*, 47, Chang, S.-P. et al., A ZnO nanowire-based humidity sensor, 772–778, Copyright 2010, with permission from Elsevier.)

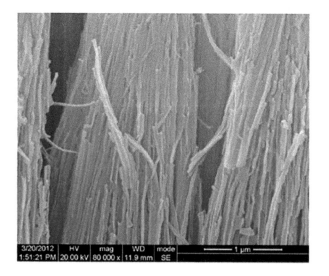

**FIGURE 21.18** The cross-sectional SEM image of the $Al_2O_3$ NWs. (Reprinted with permission from Feng, Z.-S. et al., A novel humidity sensor based on alumina nanowire films, *J. Phys. D: Appl. Phys.*, 45, 225305, 2012. Copyright 2012, Institute of Physics.)

### REFERENCES

Ahn M.-W., Park K.-S., Heo J.-H., Kim D.-W., Choi K.J., Park J.-G. (2009) On-chip fabrication of ZnO-nanowire gas sensor with high gas sensitivity. *Sens. Actuators B* **138**, 168–173.

Ahn M.-W., Park K.-S., Heo J.-H., Park J.-G., Kim D.-W., Choi K. J., Lee J.-H., Hong S.-H. (2008) Gas sensing properties of defect-controlled ZnO-nanowire gas sensor. *Appl. Phys. Lett.* **93**, 263103.

Assad O., Leshansky A.M., Wang B., Stelzner T., Christiansen S., Haick H. (2012) Spray coating route for highly aligned and large-scale arrays of nanowires. *ACS Nano* **6**, 4702–4712.

Banerjee D., Lao J.Y., Wang D.Z., Huang J.Y., Steeves D., Kimball B., Ren Z.F. (2004) Synthesis and photoluminescence studies on ZnO nanowires. *Nanotechnology* **15**, 404–409.

Barth S., Hernandez-Ramirez F., Holmes J.D., Romano-Rodriguez A. (2010) Synthesis and applications of one-dimensional semiconductors. *Prog. Mater. Sci.* **55**, 563–627.

Bierman M.J., Lau Y.K.A., Kvit A.V., Schmitt A.L., Jin S. (2008) Dislocation-driven nanowire growth and Eshelby twist. *Science* **320**, 1060–1063.

Candeloro P., Comini E., Baratto C., Faglia G., Sberveglieri G., R. Kumar R. (2005) SnO$_2$ lithographic processing for nanopatterned gas sensors. *J. Vac. Sci. Technol. B* **23**, 2784–2788.

Chakrapani V., Brier M., Puntambekar A., DiGiovanni T. (2016) Modulation of stoichiometry, morphology and composition of transition metal oxide nanostructures through hot wire chemical vapor deposition. *J. Mater. Res.* **31**(1), 17–27.

Chang S.-P., Chang S.-J., Lu C.-Y., Li M.-J., Hsu C.-L., Chiou Y.-Z., et al. (2010) A ZnO nanowire-based humidity sensor. *Superlattices Microstructures* **47**, 772–778.

Choi K.-J., Jang H.W. (2010) One-dimensional oxide nanostructures as gas-sensing materials: Review and issues. *Sensors* **10**, 4083–4099.

Choi Y.-J., Hwang I.-S., Park J.-G., Choi K.J., Park J.-H., Lee J.-H. (2008) Novel fabrication of an SnO$_2$ nanowire gas sensor with high sensitivity. *Nanotechnology* **19**, 095508.

Comini E., Baratto C., Concina I., Faglia G., Falasconi M., Ferroni M., et al. (2013) Metal oxide nanoscience and nanotechnology for chemical sensors. *Sens. Actuators B* **179**, 3–20.

Comini E., Faglia G., Sberveglieri G. (2009) Electrical-Based gas sensing. In: Comini E., Faglia G., Sberveglieri G. (eds.), *Solid State Gas Sensing*. Springer–Verlag, New York, NY, pp. 47–107.

Comini E., Faglia G., Sberveglieri G., Pan Z., Wang Z.L. (2002) Stable and highly sensitive gas sensors based on semiconducting oxide nanobelts. *Appl. Phys. Lett.* **81**(10), 1869–1871.

Dai C.-L., Ma Liu M.-C., Chen F.-S., Wu C.-C., Chang M.-W. (2007) A nanowire WO$_3$ humidity sensor integrated with micro-heater and inverting amplifier circuit on chip manufactured using CMOS-MEMS technique. *Sens. Actuators B* **123**, 896–901.

Dai Z.R., Pan Z.W., Wang Z.L. (2003) Novel nanostructures of functional oxides synthesized by thermal evaporation. *Adv. Funct. Mater.* **13**, 9–24.

Dev A., Kar S., Chakrabarti S., Chaudhuri S. (2006) Optical and field emission properties of ZnO nanorod arrays synthesized on zinc foils by the solvothermal route. *Nanotechnol.* **17**, 1533–1540.

Dimaggio E., Pennelli G. (2016) Reliable fabrication of metal contacts on silicon nanowire forests. *Nano Lett.* **16**, 4348–4354.

Dmitriev S., Lilach Y., Button B., Moskovits M., Kolmakov A. (2007) Nanoengineered chemiresistors: The interplay between electron transport and chemisorption properties of morphologically encoded SnO$_2$ nanowires. *Nanotechnol.* **18**, 055707.

Domènech-Gil G., Riera I.P., López-Aymerich E., Pellegrino P., Barth S., Gràcia I., et al. (2017) Individual gallium oxide nanowires for humidity sensing at low temperature. *Proceedings* **1**, 468.

Ebbesen T.W., Lezec H.J., Hiura H., Bennett J.W., Gaemi H.F., Thio T. (1996) Electrical conductivity of individual carbon nanotubes. *Nature* **382**, 54–56.

Erande M.B., Late D.J. (2016) Humidity and H$_2$O$_2$ sensing behavior of Te nanowires. *Adv. Dev. Mater.* **2**(1), 8–14.

Erol A., Okur S., Yagmurcukardes N., Arıkan M.C. (2011) Humidity-sensing properties of a ZnO nanowire film as measured with a QCM. *Sens. Actuators B* **152**, 115–120.

FEI (2004) Focused ion beam technology, capabilities and applications. FEI company. www.fei.com.

Feng Z.-S., Chen X.-J., Chen J.-J., Hu J. (2012) A novel humidity sensor based on alumina nanowire films. *J. Phys. D Appl. Phys.* **45**, 225305.

Fields L.L., Zheng J.P., Cheng Y., Xiong P. (2006) Room temperature low-power hydrogen sensor based on a single tin dioxide nanobelt. *Appl. Phys. Lett.* **88**, 263102.

Fu Q., Wang C., Yu H.C., Wang Y.G., Wang T.H. (2007) Fast humidity sensors based on CeO$_2$ nanowires. *Nanotechnol.* **18**, 145503.

Gamo K. (1993) Focused ion beam technology. *Semicond. Sci. Technol.* **8**, 1118–1123.

Gierak J. (2009) Focused ion beam technology and ultimate applications. *Semicond. Sci. Technol.* **24**, 043001.

Givargizov E. (1975) Fundamental aspects of VLS growth. *J. Crystal Growth* **31**, 20–30.

Gopal V., Radmilovic V.R., Daraio C., Jin S., Yang P., Stach E.A. (2004) Rapid prototyping of site-specific nanocontacts by electron and ion beam assisted direct-write nanolithography. *Nano Lett.* **4**, 2059–2063.

Guha P., Chakrabarti S., Chaudhuri S. (2004) Synthesis of b-Ga$_2$O$_3$ nanowire from elemental Ga metal and its photoluminescence study. *Physica E* **23**, 81–85.

Haghiri-Gosnet A.M., Vieu C., Simon G., Mejias M., Carcenac F., Launois H. (1999) Nanofabrication at a 10 nm length scale: limits of lift-off and electroplating transfer processes. *J. Phys. IV*, **9** (Pr. 2), 133–141.

Hanrath T., Korgel B.A. (2002) Nucleation and growth of germanium nanowires seeded by organic monolayer-coated gold nanocrystals. *J. Am. Chem. Soc. B* **124**, 1424–1429.

Hernandez-Ramirez F., Prades J.D., Jimenez-Diaz R., Fischer T., Romano-Rodriguez A., Mathur S., Morante J.R. (2009) On the role of individual metal oxide nanowires in the scaling down of chemical sensors. *Phys. Chem. Chem. Phys.* **11**, 7105–7110.

Hernandez-Ramirez F., Prades J.D., Tarancon A., Barth S., Casals O., Jimenez-Diaz R., et al. (2007c) Portable microsensors based on individual SnO$_2$ nanowires. *Nanotechnol.* **18**, 495–501.

Hernandez-Ramirez F., Tarancon A., Casals O., Arbiol J., Romano-Rodriguez A., Morante J.R. (2007a) High response and stability in CO and humidity measures using a single SnO$_2$ nanowire. *Sens. Actuators B* **121**, 3–17.

Hernandez-Ramirez F., Tarancon A., Casals O., Rodriguez J., Romano-Rodriguez A., Morante J.R., et al. (2006) Fabrication and electrical characterization of circuits based on individual tin oxide nanowires. *Nanotechnol.* **17**, 5577–5583.

Hsueh H.T., Hsueh T.J., Chang S.J., Hung F.Y., Hsu C.L., Dai B.T., et al. (2012) A flexible ZnO nanowire-based humidity sensor. *IEEE Trans. Nanotechnol.* **11**(3), 520–525.

Hsueh H.T., Hsueh T.J., Chang S.J., Hung F.Y., Weng W.Y., Hsu C.L., Dai B.T. (2011) Si nanowire-based humidity sensors prepared on glass substrate. *IEEE Sensors J.* **11**(11), 3036–3041.

Huang X.-J., Choi Y.-K. (2007) Chemical sensors based on nanostructured materials. *Sens. Actuators B* **122**, 659–671.

Hwang S.O., Kim C.H., Myung Y., Park S.-H., Park J., Kim J., Han C.-S., Kim J.-Y. (2008) Synthesis of vertically aligned manganese-doped Fe$_3$O$_4$ nanowire arrays and their excellent room-temperature gas sensing ability. *J. Phys. Chem. C* **112**, 13911–13916.

Jiang X.C., Herricks T., Xia Y.N. (2002) CuO nanowires can be synthesized by heating copper substrates in air. *Nano Lett.* **2**, 1333–1338.

Jiang X.C., Wang Y.L., Herricks T., Xia Y.N. (2004) Ethylene glycol-mediated synthesis of metal oxide nanowires. *J. Mater. Chem.* **14**, 695–703.

Jimenez-Diaz R., Prades J.D., Casals O., Andreu T., Moranter J.R., Mathur S., et al. (2011) From the fabrication strategy to the device integration of gas nanosensors based on individual nanowires. *Nanotechnol.* **2**, 204–207.

Jung S.W., Park W.I., Yi G.C., Kim M. (2003) Fabrication and controlled magnetic properties of Ni/ZnO nanorod heterostructures. *Adv. Mater.* **15**(15), 1358–1361.

Kalinin S.V., Shin J., Jesse S., Geohegan D., Baddorf A.P., Lilach Y., et al. (2005) Electronic transport imaging in a multiwire SnO$_2$ chemical field-effect transistor device. *J. Appl. Phys.* **98**, 044503.

Kam K.C., Deepak F.L., Cheetham A.K., Rao C.N.R. (2004) In$_2$O$_3$ nanowires, nanobouquets and nanotrees. *Chem. Phys. Lett.* **397**, 329–334.

Kim I.D., Rothschild A., Lee B.H., Kim D.Y., Jo S.M., Tuller H.L. (2006) Ultrasensitive chemiresistors based on electrospun TiO$_2$ nanofibers. *Nano Lett.* **6**, 2009–2013.

Köck A., Tischner A., Maier T., Kast M., Edtmaier C., Gspan C., Kothleitner G. (2009) Atmospheric pressure fabrication of SnO$_2$-nanowires for highly sensitive CO and CH$_4$ detection. *Sens. Actuators B* **138**, 160–167.

Kolasinski K.W. (2002) *Surface Science: Foundations of Catalysis and Nanoscience.* John Wiley & Sons, Chichester, UK.

Kolasinski K.W. (2006) Catalytic growth of nanowires: Vapor–liquid–solid, vapor–solid–solid, solution–liquid–solid and solid–liquid–solid growth. *Curr. Opinion Solid State Mater. Sci.* **10**, 182–191.

Kolmakov A. (2006) The effect of morphology and surface doping on sensitization of quasi-1D metal oxide nanowire gas sensors. *Proc. SPIE* **6370**, 63700X.1–63700X.8.

Kolmakov A. (2008) Some recent trends in fabrication, functionalisation and characterization of metal oxide nanowire gas sensors. *Int. J. Nanotechnol.* **5**, 450–474.

Kolmakov A., Moskovits M. (2004) Chemical sensing and catalysis by one-dimensional metal oxide nanostructures. *Annu. Rev. Mater. Res.* **34**, 151–180.

Kong J., Franklin N.R., Zhou C., Chapline M.G., Peng S., Cho K., Dai H. (2000) Nanotube molecular wires as chemical sensors. *Science* **287**, 622–625.

Kong X.Y., Wang Z.L. (2003) Spontaneous polarization-induced nanohelixes, nanosprings, and nanorings of piezoelectric nanobelts. *Nano Lett.* **3**, 1625–1631.

Krunks M., Dedova T., Açik I.O. (2006) Spray pyrolysis deposition of zinc oxide nanostructured layers. *Thin Solid Films* **515**, 1157–1160.

Kuang Q., Lao C., Wang Z.L., Xie Z., Zheng L. (2007) High-sensitivity humidity sensor based on a single SnO$_2$ nanowire. *J. Am. Chem. Soc.* **129**, 6070–6071.

Kumar P. (2010) Directed self-assembly: Expectations and achievements. *Nanoscale Res. Lett.* **5**(9), 1367–1376.

Kunakova G., Meija R., Bite I., Prikulis J., Kosmaca J., Varghese J., Holmes J.D., Erts D. (2015) Sensing properties of assembled Bi$_2$S$_3$ nanowire arrays. *Phys. Scripta* **90**, 094017.

Lao J.Y., Wen J.G., Ren Z.F. (2002) Hierarchical ZnO nanostructures. *Nano Lett.* **2**, 1287–1291.

Law M., Goldberger J., Yang P. (2004) Semiconductor nanowires and nanotubes. *Annu. Rev. Mater. Res.* **34**, 83–122.

Leung Y.P., Choy W.C.H., Yuk T.I. (2008) Linearly resistive humidity sensor based on quasi one-dimensional ZnSe nanostructures. *Chem. Phys. Lett.* **457**, 198–201.

Li C., Zhang D., Lei B., Han S., Liu X., Zhou C. (2003c) Surface treatment and doping dependence of In$_2$O$_3$ nanowires as ammonia sensors. *J. Phys. Chem. B* **107**, 12451–12455.

Li C., Zhang D., Liu X., Han S., Tang T., Han J., Zhou C. (2003b) In$_2$O$_3$ nanowires as chemical sensors. *Appl. Phys. Lett.* **82**, 1613–1615.

Li C., Zhang D.H., Han S., Liu X.L., Tang T., Zhou C.W. (2003a) Diameter-controlled growth of single crystalline In$_2$O$_3$ nanowires and their electronic properties. *Adv. Mater.* **15**, 143–146.

Li D., Chang P., Lu J.G. (2010) Quasi one-dimensional metal oxide structures: Synthesis, characterization and application as chemical sensors. In: Korotcenkov G. (ed.) *Chemical Sensors: Fundamental of Sensing Materials.* Vol. 2: *Nanostructured Materials.* Momentum Press, New York, pp. 29–86.

Li J., Srinivasan S., He G.N., Kang J.Y., Wu S.T., Ponce F.A. (2008) Synthesis and luminescence properties of ZnO nanostructures produced by the sol–gel method. *J. Cryst. Grow.* **310**, 599–603.

Li Q., Koo S.-M., Richter C.A., Edelstein M.D., Bonevich J.E., Kopanski J.J., et al. (2007) Precise alignment of single nanowires and fabrication of nanoelectromechanical switch and other test structures. *IEEE Trans. Nanotechnol.* **6**(2), 256–262.

Li Z.J., Li H.J., Chen X.L., Li L., Xu Y.P., Li K.Z. (2002) b-Ga$_2$O$_3$ nanowires on unpatterned and patterned MgO single crystal substrates. *J. Alloys Comp.* **345**, 275–279.

Liang J.H., Peng C., Wang X., Zheng X., Wang R.J., Qiu X.P.P., Nan C.W., Li Y.D. (2005) Chromate nanorods/nanobelts: General synthesis, characterization, and properties. *Inorg. Chem.* **44**, 9405–9415.

Lieber C.M. (2003) Nanoscale science and technology: Building a big future from small things. *MRS Bull.* **28**, 486–491.

Lilach Y., Zhang J.P., Moskovits M., Kolmakov A. (2005) Encoding morphology in oxide nanostructures during their growth. *Nano Lett.* **5**, 2019–2022.

Lin Y.-F., Jian W.-B. (2008) The impact of nanocontact on nanowire based nanoelectronics. *Nano Lett.* **8**, 3146–3150.

Liu X., Long Y.-Z., Liao L., Duan X., Fan Z. (2012) Large-scale integration of semiconductor nanowires for high-performance flexible electronics. *ACS Nano* **6**(3), 1888–1900.

Liu Z.Q., Zhang D.H., Han S., Li C., Tang T., Jin W., Liu X.L., Lei B., Zhou C.W. (2003b) Laser ablation synthesis and electron transport studies of tin oxide nanowires. *Adv. Mater.* **15**, 1754–1757.

Lu F., Cai W., Zhang Y. (2008) ZnO hierarchical micro/nanoarchitectures: Solvothermal synthesis and structurally enhanced photocatalytic performance. *Adv. Func. Mater.* **18**, 1047–1056.

Marrian C.R.K., Tennant D.M. (2003) Nanofabrication. *J. Vac. Sci. Technol. A* **21**, S207–S215.

Mathur S., Ganesan R., Grobelsek I., Shen H., Ruegamer T., Barth S. (2007) Plasma-assisted modulation of morphology and composition in tin oxide nanostructures for sensing applications. *Adv. Eng. Mater.* **9**, 658–663.

Meier D.C., Evju J.K., Boger Z., Raman B., Benkstein K.D., Martinez C.J., Montgomery C.B., Semancik S. (2007b) The potential for and challenges of detecting chemical hazards with temperature-programmed microsensors. *Sens. Actuators B* **121**, 282–294.

Meier D.C., Semancik S., Button B., Strelcov E., Kolmakov A. (2007a) Coupling nanowire chemiresistors with MEMS microhotplate gas sensing platforms. *Appl. Phys. Lett.* **91**, 063118.

Melngailis J. (1987) Focused ion beam technology and applications. *J. Vac. Sci. Technol. B* **5**, 469–495.

Melngailis J. (2001) Applications of ion microbeam lithography and direct processing. In: Helbert J.N. (ed.) *Handbook of VLSI Lithography*, 2nd ed., Noyes, Park Ridge, NJ, pp. 791–855.

Miao J.J., Wang H., Li Y.R., Zhu J.M., Zhu J.J. (2005) Ultrasonic-induced synthesis of CeO$_2$ nanotubes. *J. Cryst. Growth* **281**, 525–529.

Nagahara L.A., Amlani I., Lewenstein J., Tsui R.K. (2002) Directed placement of suspended carbon nanotubes for nanometer-scale assembly. *Appl. Phys. Lett.* **80**, 3826.

Nam C.Y., Tham D., Fischer J.E. (2005) Disorder effects in focused-ion-beam-deposited Pt contacts on GaN nanowires. *Nano Lett.* **5**, 2029–2033.

Orel Z.C., Anzlovar A., Drazic G., Zigon M. (2007) Cuprous oxide nanowires prepared by an additive-free polyol process. *Crystal Growth Design* **7**, 453–458.

Pan Z.W., Dai Z.R., Wang Z.L. (2001) Nanobelts of semiconducting oxides. *Science* **291**, 1947–1949.

Park W.I., Kim D.H., Jung S.W., Yi G.C. (2002) Metalorganic vapor-phase epitaxial growth of vertically well aligned ZnO nanorods. *Appl. Phys. Lett.* **80**, 4232–4234.

Ponzoni A., Comini E., Sberveglieri G., Zhou J., Deng S., Xu N., et al. (2006) Ultrasensitive and highly selective gas sensors using three-dimensional tungsten oxide nanowire networks. *Appl. Phys. Lett.* **88**, 203101.

Prades J.D., Hernández-Ramírez F., Fischer T., Hoffmann M., Müller R., López N., et al. (2011) Simultaneous CO and humidity quantification with self-heated nanowires in pulsed mode. *Procedia Eng.* **25**, 1485–1488.

Prades J.D., Jimenez-Diaz R., Hernandez-Ramírez F., Barth S., Cirera A., Romano-Rodrıguez A., Mathur S., Morante J.R. (2008) Ultralow power consumption gas sensors based on self-heated individual nanowires. *Appl. Phys. Lett.* **93**, 123110.

Qi P., Vermesh O., Grecu M., Javey A., Wang Q., Dai H. (2003) Toward large arrays of multiplex functionalized carbon nanotube sensors for highly sensitive and selective molecular detection. *Nano Lett.* **3**, 347–351.

Roper S.M., Davis S.H., Norris S.A., Golovin A.A., Voorhees P.W., Weiss M. (2007) Steady growth of nanowires via the vapor–liquid–solid method. *J. Appl. Phys.* **102**, 034304.

Rotkina L., Lin J.-F., Bird J.P. (2003) Nonlinear current-voltage characteristics of Pt nanowires and nanowire transistors fabricated by electron-beam deposition. *Appl. Phys. Let.* **83**, 4426

Sawicka K.M., Prasad A.K., Gouma P.I. (2005) Metal oxide nanowires for use in chemical sensing applications. *Sens. Lett.* **3**, 31–35.

Sepulveda-Guzman S., Reeja-Jayan B., de la Rosa E., Torres-Castro A., Gonzalez-Gonzalez V., Jose-Yacaman M. (2009) Synthesis of assembled ZnO structures by precipitation method in aqueous media. *Mater. Chem. Phys.* **115**, 172–178.

Soldano C., Comini E., Baratto C., Ferroni M., Faglia G., Sberveglieri G. (2012) Metal oxides mono-dimensional nanostructures for gas sensing and light emission. *J. Am. Ceram. Soc.* **95**(3), 831–850.

Song T., Choung J.W., Park J.G., Il Park W., Rogers J.A., Paik U. (2008) Surface polarity and shape-controlled synthesis of ZnO nanostructures on GaN thin films based on catalyst-free metalorganic vapor phase epitaxy. *Adv. Mater.* **20**, 4464–4469.

Strelcov E., Dmitriev S., Button B., Cothren J., Sysoev V., Kolmakov A. (2008) Evidence of the self-heating effect on surface reactivity and gas sensing of metal oxide nanowire chemiresistors. *Nanotechnol.* **19**, 355502.

Sysoev V.V., Button B.K., Wepsiec K., Dmitriev S., Kolmakov A. (2006) Toward the nanoscopic "electronic nose": Hydrogen vs. carbon monoxide discrimination with an array of individual metal oxide nano- and mesowire sensors. *Nano Lett.* **6**, 1584–1588.

Sysoev V.V., Goschnick J., Schneider T., Strelcov E., Kolmakov A. (2007) A gradient microarray electronic nose based on percolating SnO$_2$ nanowire sensing elements. *Nano Lett.* **7**, 3182–3188.

Tseng A.A. (2005) Recent developments in nanofabrication using focused ion beams. *Small* **1**(10), 924–939.

Utkea I., Hoffmann P., Melngailis J. (2008) Gas-assisted focused electron beam and ion beam processing and fabrication. *J. Vac. Sci. Technol B.* **26**(4), 1197–1276.

Uzar N., Okur S., Arikan M.C. (2011) Investigation of humidity sensing properties of ZnS nanowires synthesized by vapor liquid solid (VLS) technique. *Sens. Actuators A* **167**, 188–193.

Van Osch T.H.J., Perelaer J., De Laat A.W.M., Schubert U.S. (2008) Inkjet printing of narrow conductive tracks on untreated polymeric substrates. *Adv. Mater.* **20**, 343–345.

Varghese O.K., Gong D., Paulose M., Grimes C.A., Dickey E.C. (2003a) Crystallization and high-temperature structural stability of titanium oxide nanotube arrays. *J. Mater. Res.* **18**(1), 156–165.

Varghese O.K., Gong D., Paulose M., Ong K.G., Grimes C.A. (2003b) Hydrogen sensing using titania nanotubes. *Sens. Actuators B* **93**, 338–344.

Varghese O.K., Grimes C.A. (2003) Metal oxide nanoarchitectures for environmental sensing. *J. Nanosci. Nanotechnol.* **3**, 277–293.

Varghese O.K., Paulose M., Grimes C.A. (2009) Long vertically aligned titania nanotubes on transparent conducting oxide for highly efficient solar cells. *Nature Nanotechnol.* **4**, 592–597.

Vomiero A., Bianchi S., Comini E., Faglia G., Ferroni M., Sberveglieri G. (2007) Controlled growth and sensing properties of In$_2$O$_3$ nanowires. *Crystal Growth Design* **7**, 2500–2504.

Wagner R.S., Ellis W.C. (1964) Vapor–liquid–solid mechanism of single crystal growth. *Appl. Phys. Lett.* **4**, 89–90.

Wang D., Zhu R., Zhaoying Z., Ye X. (2007) Controlled assembly of zinc oxide nanowires using dielectrophoresis. *Appl. Phys. Lett.* **90**, 103110.

Wang N., Cai Y., Zhang R.Q. (2008) Growth of nanowires. *Mater. Sci. Eng. R* **60**, 1–51.

Wang S.-B., Hsiao C.-H., Chang S.-J., Lam K.-T., Wen K.-H., Young S.-J., et al. (2012) CuO nanowire-based humidity sensor. *IEEE Sensors J.* **12**(6), 1884–1888.

Wang X., Ding Y., Summers C.J., Wang Z.L. (2004) Large-scale synthesis of six nanometer-wide ZnO nanobelts. *J. Phys. Chem. B* **108**, 8773–8777.

Wang Z.L. (2004) Zinc oxide nanostructures: Growth, properties and applications. *J. Phys.—Condens. Matter* **16**, R829–R858.

Whang D., Jin S., Wu Y., Lieber C.M. (2003) Large-scale hierarchical organization of nanowire arrays for integrated nanosystems. *Nano Lett.* **3**, 1255–1259.

Wu Y.Y., Yang P.D. (2001) Direct observation of vapor-liquid-solid nanowire growth. *J. Am. Chem. Soc.* **123**, 3165–3166.

Wu Z.H., Mei X.Y., Kim D., Blumin M., Ruda H.E. (2002b) Growth of Au-catalyzed ordered GaAs nanowire arrays by molecular-beam epitaxy. *Appl. Phys. Lett.* **81**, 5177–5179.

Xia Y.N., Yang P. D., Sun Y.G., Wu Y.Y., Mayers B. (2003) One-dimensional nanostructures: Synthesis, characterization, and applications. *Adv. Mater.* **15**, 353–389.

Xu L., Chen Q., Xu D. (2007) Hierarchical ZnO nanostructures obtained by electrodeposition. *J. Phys. Chem. C* **111**, 11560–11565.

Yuan C., Xu Y., Deng Y., Jiang N., He N., Dai L. (2010) CuO based inorganic–organic hybrid nanowires: A new type of highly sensitive humidity sensor. *Nanotechnol.* **21**, 415501.

Zhang D.H., Liu Z.Q., Li C., Tang T., Liu X.L., Han S., Lei B., Zhou C.W. (2004b) Detection of NO$_2$ down to ppb levels using individual and multiple In$_2$O$_3$ nanowire devices. *Nano Lett.* **4**, 1919–1924.

Zhang Y., Ago H., Liu J., Yumura M., Uchida K., Ohshima S., Iijima S., Zhu J., Zhang X. (2004a) The synthesis of In, $In_2O_3$ nanowires and $In_2O_3$ nanoparticles with shape-controlled. *J. Cryst. Growth* **264**, 363–368.

Zhang Y., Kolmakov A., Libach Y., Moskovits M. (2005) Electronic control of chemistry and catalysis at the surface of an individual tin oxide nanowire. *J. Phys. Chem. B* **109**, 1923–1929.

Zhang Z., Hu C., Xiong Y., Yang R., Lin Wang Z.L. (2007) Synthesis of Ba-doped $CeO_2$ nanowires and their application as humidity sensors. *Nanotechnol.* **18**, 465504.

Zheng C.L., Wan J.G., Cheng Y., Cu D.H., Zhan Y.J. (2005a) Preparation of $SnO_2$ nanowires synthesized by vapor-solid mode and its growth mechanism. *Int. J. Mod. Phys. B* **19**, 2811–2816.

Zou G., Li H., Zhang Y., Xiong K., Qian Y. (2006) Solvothermal/hydrothermal route to semiconductor nanowires. *Nanotechnol.* **17**, S313–S320.

# 22 Nanofiber-Based Humidity Sensors and Features of Their Fabrication

Nanofiber-based humidity sensors are another promising approach to the development of highly sensitive and high-speed devices. Porous structures with their high surface areas are promising for applications in many different areas including humidity measurements. It was established that nanofibers are able to form such highly porous mesh with large surface-to-volume ratio. And this means that the use of nanofibers can provide improved sensor performance. It is important to note that a porous structure made out of nanofibers is a dynamic system where the pore size and shape can change, unlike conventional rigid porous structures. In addition, in comparison with other one-dimensional (1D) nanostructures such as nanorods and nanotubes, nanofibers are continuous, which possess them with high axial strength, combined with extreme flexibility. Therefore, the membranes assembled by electrospun nanofibers have excellent structural mechanical properties. Meanwhile, multiple extra functions can be incorporated into electrospun nanofibers to broaden their significances in applications (Ding et al. 2010; Choi et al. 2017; Patil et al. 2017; Thenmozhi et al. 2017). Nanofibers can also be linked to form a rigid structure if required. These nanofibers features open up additional possibilities for creating a humidity-sensitive matrix with optimal properties.

## 22.1 APPROACHES TO NANOFIBERS PREPARING

In the manufacture of nanofiber-based humidity sensors, polymer nanofibers are mainly used. For their preparation various methods can be used (see Table 22.1), including drawing, hard and soft-template synthesis, phase separation, self-assembly, and electrospinning (Jayaraman et al. 2004; Liu and Zhang 2009; Long et al. 2011; Patil et al. 2017). Among these methods, electrospinning seems to be the simplest and most versatile technique capable of generating quasi 1D nanostructures (nanofibers). Almost any soluble polymer with sufficiently high molecular weight can be electrospun.

### 22.1.1 ELECTROSPINNING

Electrospinning is the technique that uses a strong electric field to produce polymer nanofibers from polymer solution or polymer melt (see Figure 22.1). During the process of electrospraying, the liquid drop elongates with increasing electric field. When the repulsive force induced by the charge distribution on the surface of the drop is balanced with the surface tension of the liquid, the liquid drop distorts into a conical shape. Once the repulsive force exceeds the surface tension,

a charged jet of liquid is ejected from the cone tip and moves toward a grounded electrode. Unlike conventional spinning, the jet is only stable near the tip of the spinneret, after which the jet is subject to bending instability. Whether the jet will form a continuous fiber or disperse into droplets depends on polymer molecular weight, polymer-chain entanglement, and the solvent applied to the process (specifically, its evaporation rate). It is known from the literature that smooth fibers are produced when the product of intrinsic viscosity ($\eta$) and polymer concentration (c), known as the Berry's number, $Be = \eta c$, is greater than a certain critical value, which is characteristic of the polymer. Thus, small droplets are formed as a result of the varicose breakup of the jet in the case of low-viscosity and low-concentration liquids. Otherwise, a solid fiber is generated instead of breaking up into individual drops due to the electrostatic repulsions. As the charged jet accelerates toward regions of lower potential, the solvent evaporates while the entanglements of the polymer chains prevent the jet from breaking up. This results in fiber formation. Typical images of fibers are shown in Figures 22.1 and 22.2a. These fibers can be used for development different sensors including sensors based on a single nanofiber (Figure 22.2b). One should note that high-solution concentration for the electrospinning process is also not optimal. At high concentrations the formation of continuous fibers is prohibited because of the inability to maintain the flow of the solution at the tip of the needle resulting in the formation of larger fibers.

Generally, the electrospun fibers are deposited on a fixed collector in a three-dimensional (3D) nonwoven membrane structure with a wide range of fiber diameter distribution from several nanometers to a few micrometers. Currently, there are two standard electrospinning setups, horizontal and vertical (see Figure 22.1). With the expansion of this technology, several research groups have developed more sophisticated systems that can fabricate more complex nanofibrous structures in a more controlled and efficient manner. For example, aligned electrospun fibers were obtained by using a rotating or pre-patterned collector (Theron et al. 2001; Kameoka et al. 2003).

The ability to form porous fibers through electrospinning means that the surface area of the fiber mesh can be increased tremendously. Phase separation is proposed as the main mechanism behind the formation of porous fibers. When more volatile solvents are used, solvent-rich regions begin to form during electrospinning that transform into pores (Bognitzki et al. 2001a). Another method of producing porous nanofibers is the spinning of a blend of two different polymers. One of the polymers is removed after fiber formation by dissolution in a solvent in which the other polymer is insoluble

**TABLE 22.1**

**Synthesis Methods of Conducting Polymer Nanotubes and Nanofibers**

| Synthesis Methods | Advantages | Disadvantages | Examples |
|---|---|---|---|
| *Hard Physical Template Method:* Porous membranes | Aligned arrays of tubes and wires with controllable length and diameter | A post-synthesis process is needed to remove the template | PANI; PPY; P3MT; PEDOT; PPV nanotubes/wires; CdS-PPY; Au-PEDOT-Au; Ni/PPY; MnO₂/PEDOT nanowires; |
| *Soft Chemical Template Method:* Interfacial polymerization *Other methods:* Dilute polymerization; Template-free method; Rapidly mixed reaction; Reverse emulsion polymerization; Ultrasonic shake-up; Radiolytic synthesis | Simple self-assembly process without an external template | Relatively poor control of the uniformity of the morphology (shape, diameter); poorly or nonoriented 1D nanostructures | A variety of PANI and PPY micro/nanostructures such as tubes, wires/fibers, hollow microspheres, nanowire/tube junctions, and dendrites |
| *Electrospinning* | Low cost; large surface area; long continuous nanofibers can be produced | Jet instability; usually in form of nonwoven web; possible alignment | PANI/PEO; PPY/PEO; PANI; PPY; P3HT/PEO |
| *Nanoimprint Lithography or Embossing* | Rapid and low-cost | A micro-mold is needed | PEDOT nanowires; 2D nanodots of semiconducting polymer, and aligned CP arrays |
| *Directed Electrochemical Nanowire Assembly* | Electrode-wire-electrode or electrode-wire-target growth of CP micro/nanowires | Micro/nanowires with knobby structures | PPY; PANI; and PEDOT nanowires |
| *Other Methods of Nanofiber Fabrication:* Dip-pen nanolithography method; Molecular-combing method; Whisker method; Strong magnetic field-assisted chemical synthesis | | | |

*Source:* Reprinted from *Prog. Polymer Sci.*, 36, Long, Y.-Z. et al., Recent advances in synthesis, physical properties and applications of conducting polymer nanotubes and nanofibers, 1415–1442, Copyright 2011, with permission from Elsevier.

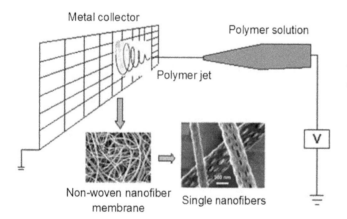

**FIGURE 22.1** Schematic diagram for the demonstration of polymer nanofibers by electrospinning. (Reprinted from *Compos. Sci. Technol.*, 63, Huang, Z.-M. et al., A review on polymer nanofibers by electrospinning and their applications in nanocomposites, 2223–2253, Copyright 2003a with permission from Elsevier.)

(Bognitzki et al. 2001b). Since stretching of the solution arises from repulsive charges, the electrospinning jet path is very chaotic and only nonwoven meshes are produced using a typical setup. Nevertheless, more ordered assemblies that allow the porosity of the mesh to be controlled have been produced

through clever manipulation of the setup and solution composition (Ramakrishna et al. 2005).

The experiment showed that for the successful implementation of electrospinning it is necessary to fulfill several conditions (Ramakrishna et al. 2005):

* Suitable solvent should be available for dissolving the polymer.
* The vapor pressure of the solvent should be suitable so that it evaporates quickly enough for the fiber to maintain its integrity when it reaches the target but not too quickly to allow the fiber to harden before it reaches the nanometer range.
* The viscosity and surface tension of the solvent must neither be too large to prevent the jet from forming nor be too small to allow the polymer solution to drain freely from the pipette.
* The power supply should be adequate to overcome the viscosity and surface tension of the polymer solution to form and sustain the jet from the pipette.
* The gap between the pipette and grounded surface should not be too small to create sparks between the electrodes but should be large enough for the solvent to evaporate in time for the fibers to form.

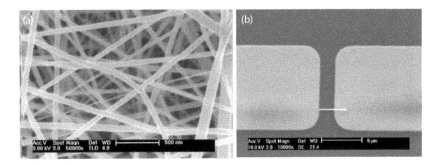

FIGURE 22.2 A SEM image of (a) typical electrospun fibers; and (b) single polypyrrole (PPY) nanofiber of 200-nm wide and 3.0-μm long. (a) (Reprinted from Ding, B. et al., *Sensors* 9, 1609–1624, 2009. Published by MDPI as open access, and (b) Reprinted with permission from Ramanathan, K. et al. *Nano Lett.*, 4, 1237–1239, 2004. Copyright 2004 American Chemical Society.)

## TABLE 22.2
### The Solvents Used in the Electrospinning Process to Form Polymer Nanofibers

| Polymer | Solvent |
|---|---|
| Polyimides (PI) | Phenol |
| Nylon 6 | Sulfuric acid or Formic acid |
| Polyamic acid (Kapton) | m-Cresol + Formic acid |
| Polyaramid | Sulfuric acid |
| Poly(p-phenylene terephthalamide) (PPTA) | Sulfuric acid |
| Polyetherimide (PEI) | Methylene chloride |
| Polyaniline (PANI) | Sulfuric acid; Hexafluoro isopropanol |
| Polyacrylonitrile (PAN) | Dimethylformamide (DMF) |
| Polyethylene-terephthalate (PET, PETE, or PETP) | Trifluoroacetic acid |
| Nylon | Dichloromethane |
| Poly(ethylene oxide) (PEO) | Water |
| Poly(methyl methacrylate) (PMMA) | Toluene |
| *Polyurethane* (PU) | Dimethylformamide (DMF) |
| Poly(acrylic acid) (PAA) | Ethanol + Formic acid |
| Poly(vinyl alcohol) (PVA) | Water |
| Cellulose acetate | Acetone/DMF/ Trifluoroethylene = 3:1:1 |

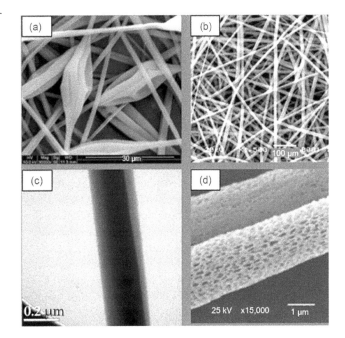

FIGURE 22.3 Different fiber morphologies: (a) beaded; (b) smooth; (c) core-shell; and (d) porous fibers. (Reprinted from *Mater. Today* 9, Ramakrishna, S. et al., Electrospun nanofibers: solving global issues, 40–50, Copyright 2006 with permission from Elsevier.)

As for solvents, which can be used in the process of electrospinning to form polymer nanofibers, some of them are listed in Table 22.2.

One of the most important advantages of the electrospinning technique is that it is relatively easy and not expensive to produce large numbers of different kinds of nanofibers (Ramakrishna et al. 2005; Lu et al. 2009; Abd Razak et al. 2015). For example, via using various deposition parameters, different fiber morphologies such as beaded, ribbon, porous, and core-shell fibers can be realized (see Figure 22.3). It was shown that electrospinning can be used to convert a large variety of polymers into nanofibers and may be the only process that has the potential for mass production. To date, it is believed that more than 100 different polymers have been successfully electrospun into nanofibers by this technique.

It is important to note that for nanofibers functionalization, all methods developed for polymeric materials can be used, including acid treatment, plasma treatment, decoration by noble metal clusters, and so on (Chen et al. 2017).

Other advantages of the electrospinning technique are the ability to control the fiber diameters, the high surface-to-volume ratio, high aspect ratio, and the pore size as nonwoven fabrics. Extensive research on electrospinning has shown that such parameters as the fiber collectability uniformity of fibers, average fiber diameter, fiber diameter distribution, and fiber porosity are strongly affected by the solution properties (Fong et al. 1999; Ding et al. 2002) and processing parameters (McCann et al. 2006; Ding et al. 2006b; Bhardwaj and Kundu 2010) such as polymer concentration, solution viscosity, solution conductivity, flow rate,

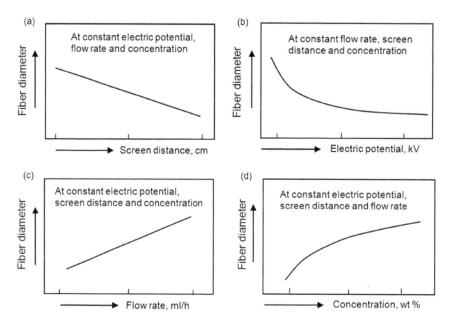

**FIGURE 22.4**  Effect of process parameters such as (a) screen distance, (b) electric potential, (c) flow rate, and (d) concentration of precursor on the diameter of the fiber, obtained by electrospinning.

applied voltage, the working distance between the collector and the needle tip, and air humidity (Zhang and Yu 2014). Correlations between diameter of the fiber and parameters of electrospinning process are shown in Figure 22.4 and discussed in Table 22.3. An ideal fiber diameter is below 20 nm for optimal performances in various applications; however, an electrospun fiber from the traditional process typically has a diameter in the range 100–500 nm. Therefore, many studies focus on the development of a robust method for manufacturing extremely small nanofibers in large quantities and with a uniform size.

It was established that nanofiber composites can also be made via the electrospinning technique (Lu et al. 2009; Choi et al. 2017). There is only one restriction. The second phase needs to be soluble or well dispersed in the initial solution. The advantage of the facile formation of 1D composite nanomaterials by electrospinning affords the materials multifunctional properties for various applications (Huang et al. 2003b; Ding et al. 2009; Abd Razak et al. 2015).

## 22.1.2  OTHER METHODS OF FORMING NANOFIBERS

As it was shown by Huang and Kaner (2004), polymer nanofibers can also by synthesized by a general chemical route using interfacial polymerization at an aqueous/organic interface. This approach, in particular, was realized for polyaniline (PANI) (Huang and Kaner 2004) and polypyrrole (PPy) (Huang et al. 2005). Nanofibers synthesized were 60–100 nm in diameter. In the template-free method reported, called simplified template-free method (STFM), the polymer nanotubes were obtained via a self-assembly process (Liu and Zhang 2009). This process schematically is shown in Figure 22.5. This figure illustrates

**TABLE 22.3**

**Electrospinning Parameters (Solution, Processing, and Ambient) and their Effects on Fiber Morphology**

| Parameters | | Effect on Fiber Morphology |
|---|---|---|
| Solution parameters | Viscosity | Low-beads generation, high-increase in fiber diameter, disappearance of beads. |
| | Polymer concentration | Increase in fiber diameter with increase of concentration. |
| | Molecular weight of polymer | Reduction in the number of beads and droplets with increase of molecular weight. |
| | Conductivity | Decrease in fiber diameter with increase in conductivity. |
| | Surface tension | No conclusive link with fiber morphology, high surface tension results in instability of jets. |
| Processing parameters | Applied voltage | Decrease in fiber diameter with increase in voltage. |
| | Distance between tip and collector | Generation of beads with too small and too large distance, minimum distance required for uniform fibers. |
| | Feed rate/Flow rate | Decrease in fiber diameter with decrease in flow rate, generation of beads with too high flow rate. |
| Ambient parameters | Humidity | High humidity results in circular pores on the fibers. |
| | Temperature | Increase in temperature results in decrease in fiber diameter. |

*Source:*  Reprinted from *Biotechnol. Adv.*, 28, Bhardwaj, N. and Kundu, S.C. Electrospinning: A fascinating fiber fabrication technique, 325–347, Copyright 2010 with permission from Elsevier.

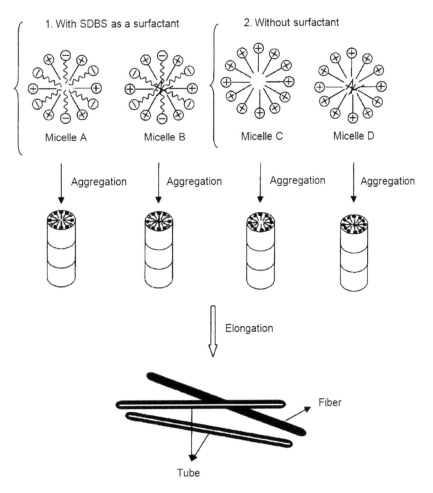

**FIGURE 22.5** The mechanism of formation of PANI nanotubes and nanofibers, synthesized by a self-assembly process. (Reprinted with permission from Zhang, Z.M. et al., *Macromolecules*, 35, 5937–5942, 2002 Copyright 2002: American Chemical Society.)

the formation mechanism of the PANI nanotubes and nanofibers with or without surfactant. In the presence of a surfactant, micelles formed by anilinium cations and surfactant anions were regarded as templates in the formation of the nanostructures. As in the absence of a surfactant, on the other hand, micelles formed by anilinium cations were considered as templates. However, the size of PANI nanostructures was slightly affected by the addition of the surfactant during the polymerization. Obviously, STFM is a facile and efficient approach to synthesize polymer, in particular PANI nanostructures, because it not only omits hard-template and post-treatment of the removal of the template, but also simplifies the reagents. However, the self-assembly mechanism of the conductive nanotubes of PANI by the STFM is not yet understood. It might be due to the formation of aniline dimer cation radicals, which could act as effective surfactants to shape the polyaniline morphology.

Regarding the template method, it can be said that the template synthesis can be organized using "soft templates," such as surfactants, organic dopants, or polyelectrolytes that assist in the self-assembly of polyaniline nanostructures,

and "hard templates" such as porous membranes or zeolites, where the templated polymerization occurs in the 1D nanochannels (Long et al. 2011). Nanoporous anodic aluminum oxide (AAO) membranes are the most extensively used templates for nanofibers synthesis (see Figure 22.6a). To synthesize nanofibers, materials have to be filled into the nanopores in some way. Electrochemistry is a powerful method for such applications and has been used to synthesize nanofibers consisting from various materials, including conducting polymers (Dan et al. 2007). This process schematically is shown in Figure 22.6b. However, this method cannot make one-by-one continuous nanofibers. In addition, the process requires many operations including the phase separation, which consists of dissolution, gelation, extraction using a different solvent, freezing, and drying resulting in a nanoscale porous foam. Thus, the template synthesis takes a relatively long period of time to transfer the solid polymer into the nanoporous foam. The self-assembly is a process in which individual, pre-existing components organize themselves into desired patterns and functions. However, as in phase separation, self-assembly is time-consuming when processing continuous polymer nanofibers.

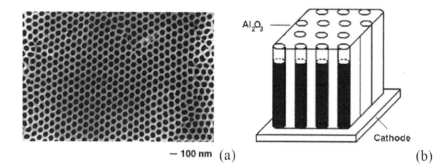

**FIGURE 22.6** (a) The SEM image of AAO membrane surface; and (b) AAO membrane used a template to make nanofibers by electroplating. (a) (Reprinted with permission from Jessensky, O. et al., *Appl. Phys. Lett.*, 72, 1173–1175, 1998 Copyright 1998 American Institute of Physics.)

## 22.2 POLYMER NANOFIBER-BASED HUMIDITY SENSORS

Of course, polymer nanofibers are not 1D structures in classical understanding as carbon nanotubes or metal-oxide nanowires. Studies, however, showed that the configuration features of polymer nanofibers and dimensional factor play a positive role in the development of polymer-based humidity sensors (Ding et al. 2009). Polymer nanofibers, which usually have a diameter in the range of 10–1000 nm and the length from several micrometers up to centimeters and meters, possess many unique properties, since these fibers have very large surface area per unit mass and small pore size. In addition, polymer nanofibers make it easier to add surface functionality compared to conventional polymer. For example, Huang et al. (2003a) and Virji et al. (2004) developed polyaniline nanofiber thin-film sensors and compared them to conventional PANI sensors. They found that the response of nanofiber-based sensors was higher and faster (Wang et al. 2004). Without any doubt, such a situation is conditioned by high surface area and high porosity of nanofiber-based sensing materials. The same situation is observed with humidity sensors. It should be noted that the decrease of response and recovery times is great advantage of nanofiber-based sensors, because slow response and recovery was one of the most important disadvantages of conventional polymer gas and humidity sensors.

In numerous studies it was shown that similarly to nanowire-based sensors discussed in previous chapters, ether fabric of nanofibers (or nanofiber mat) (see Figure 22.7) as well as a single nanofiber (Cheng et al. 2017) can be used for humidity-sensor design. The single nanofiber is more difficult to be utilized in the sensors (Liu et al. 2005), therefore this approach is practically not used in the manufacture of humidity sensors. One possible approach to solving this problem was proposed by Dong et al. (2005a, 2005b). They developed a new technology combined with nanoimprint lithography and lift-off process to fabricate a PPy nanofibers between microelectrodes. They also reported that the sensitivity and response time of single nanofibers were influenced by their diameter (Dong et al. 2005a). This means that the control of the diameter of the nanofibers, similarly to inorganic nanowire-based sensors,

is a required step in gas- and humidity-sensor fabrication to achieve acceptable reproducibility of sensor parameters.

Several examples of humidity sensors designed on the base of polymer nanofibers are listed in Table 22.4. As a rule, an electrospun is used for preparing polymer nanofibers. However, other methods such as sonocatalyzed-TEMPO-oxidation (Eyebe et al. 2017), dip coating (Cheng et al. 2017), chemical deposition (Zeng et al. 2010), plasma-stripping (Lei et al. 2015) and self-assembly methods (Mogera et al. 2014) can also be used for these purposes. Using indicated methods cellulose, PANI, polyimide (PI), and supramolecular (CS-DMV) nanofibers were formed. The possible structures of nanofibers for PANI are shown in Figure 22.7. As can be seen, the structure of nanofibers strongly depends on the method of preparation, and these differences should be considered when choosing the technology of manufacturing sensors and explaining obtained results.

Regarding the type of humidity sensors developed on the basis of polymer nanofibers, most of the sensors were resistive and mass sensitive-type sensors such as quartz crystal microbalance (QCM) and surface acoustic wave (SAW). There is also a message that on the base of polymer nanofibers it is possible to manufacture microwave humidity sensor with acceptable characteristics (Eyebe et al. 2017). At the same time, the experiment showed that due to the specific nature of the structure, the nanofiber mat is not the optimal structure for developing capacitive sensors. Due to too much porosity and a large distance between nanofibers, capillary condensation is possible only inside the nanofibers themselves and therefore the change in capacity with a change in humidity in nanofiber-mat-based structures is insignificant (Lei et al. 2015). According to Lei et al. (2015), the sensitivity of such sensors did not exceed ~0.01 pF/RH%.

As shown in Table 22.4, during the manufacture of the polymer nanofiber-based humidity sensors a sufficiently large number of polymers were tested. But the most used is still PANI and composites based on it. With appropriate doping, PANI-based humidity sensors have high sensitivity and good rate of response (Figure 22.8). In particular, Lin et al. (2012a, 2012b) have found that the blending of poly(ethylene oxide) (PEO) into PANI greatly modified the hydrophilicity

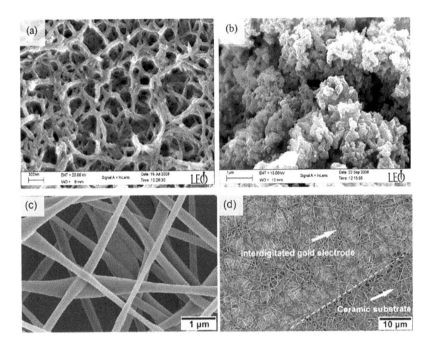

**FIGURE 22.7** SEM images of PANI prepared using different methods: (a) PANI films fabricated using conventional methods; (b) the PANI nanofibers prepared by chemical deposition method; (c) PANI nanofibers prepared using electrospun; and (d) humidity sensor fabricated using these nanofibers (heated at 100°C for 2 h). (a, b) (Reprinted from Sens. Actuators B, 143, Zeng, F.-W. et al., Humidity sensors based on polyaniline nanofibers, 530–534, Copyright 2010, with permission from Elsevier) (c, d) (Reprinted from Sens. Actuators B, 171–172, Lin, Q. et al., Investigations on the sensing mechanism of humidity sensors based on electrospun polymer nanofibers, 309–314, Copyright 2012, with permission from Elsevier.)

## TABLE 22.4
## Characteristics of Humidity Sensors Based on Polymer Nanofiber Mats Prepared Using Electrospun

| Polymer | Diameter (nm) | RH Range (%) | Max. Response | References |
|---|---|---|---|---|
| *Resistive and Impedance types* | | | | |
| PANI | ~50 | 0–98 | $S_R \sim 2.6$ | Zeng et al. (2010) |
| PANI | 100–250 | 11–98 | $S_R \sim 10^3$ | Lin et al. (2012a) |
| PANI:PSSA/PEO/PVB | 100–300 | 11–98 | $S_R \sim 10^3$ | Lin et al. (2012b) |
| PANI-PEO | 250–500 | 22–97 | $S_R \sim 3 \times 10^2$ | Li et al. (2009) |
| PANI-PHB | 100–600 | 10–70 | $S_R \sim 10$ | Macagnano et al. (2016) |
| Pedax:LiCl | 100–1000 | 11–95 | $S_R \sim 3 \times 10^3$ | Liang et al. (2015) |
| PVA-CNTs | 50–100 | 45–90 | $S_R \sim 5 \times 10^2$ | Ramos et al. (2016) |
| PEO:LiClO$_4$ | 400–1000 | 25–65 | $S_R \sim 10^2$ | Aussawasathien et al. (2005) |
| CS-DMV | 30–200 | 5–85 | $S_R \sim 4 \times 10^4$ | Mogera et al. (2014) |
| *Quartz Crystal Microbalance (QCM)* | | | | |
| PANI | 100–250 | 11–98 | $\Delta f = 180$ Hz | Lin et al. (2012a) |
| PAA-PVA | ~750 | 20–95 | $\Delta f = 450$ Hz | Wang et al. (2010) |
| PAA | ~250 | 20–95 | $\Delta f = 650$ Hz | Wang et al. (2010) |
| PA:CoCl$_2$ | 100 | 12–97 | $\Delta f = 7000$ Hz | You et al. (2017) |
| PA-PEI | ~150 | 2–95 | $\Delta f = 3200$ Hz | Wang et al. (2011b) |
| *Surface Acoustic Waves (SAW)* | | | | |
| PANI:PSSA-PVB | 100–200 | 20–90 | $\Delta f \sim 5$ MHz | Lin et al. (2012c) |
| PEO; PVP; PMMA; PVDF; PVB | 100–200 | 10–90 | $\Delta f \sim 8$ MHz (PVB) | Lin et al. (2012c) |
| Nafion-CNTs | N/A | 10–80 | $\Delta f \sim 28$ MHz | Lei et al. (2011) |

*Abbreviations:* CNT—carbon nanotube; CS-DMV—supramolecular system; CS—coronene tetracarboxylate; DMV—dodecyl methyl viologen derivative; PA—polyamide; PAA—polyacrylic acid; PANI—polyaniline; PEI—polyethyleneimine; PEO—polyethylene oxide; PHB—polyhydroxybutyrate; PMMA—polymethyl methacrylate; PSSA—poly(4-styrenesulfonic acid) (anionic polyelectrolyte); PVA—polyvinyl alcohol; PVB—polyvinylbutyral; PVDF—polyvinylidene difluoride; PVP—polyvinylpyrrolidone.

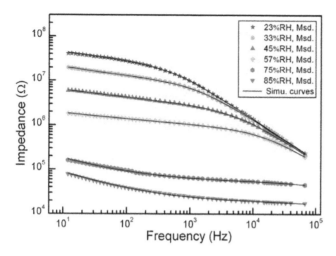

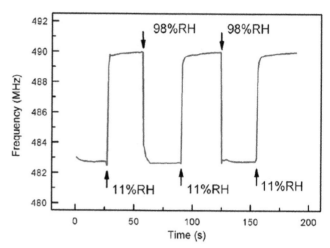

**FIGURE 22.8** Relationship between impedance of PANI nanofiber humidity sensor and frequency at different RH. Processable PANI was synthesized by solution polymerization of aniline with ammonium persulfate (APS) as an oxidant and PSSA as a soft template. Mass ratio of PSSA/aniline/APS was controlled at 4/1/0.7. (Reprinted from *Sens. Actuators B*, 171–172, Lin, Q. et al., Investigations on the sensing mechanism of humidity sensors based on electrospun polymer nanofibers, 309–314, Copyright 2012a, with permission from Elsevier.)

**FIGURE 22.9** Humidity-response transients of the SAW sensor based on electrospun PANI/PVB nanofibers (mass ratio of PANI:PVB = 2:3, measurement temperature: 30°C). (Reprinted from *Anal. Chim. Acta*, 748, Lin, Q. et al., Highly sensitive and ultrafast response surface acoustic wave humidity sensor based on electrospun polyaniline/poly(vinyl butyral) nanofibers, 73–80, Copyright 2012c, with permission from Elsevier.)

of the PANI-PEO nanofibers, and their humidity response. Moreover, it was established that the PANI-PEO nanofiber-based sensors revealed much higher sensitivity than their film counterparts with the impedance increasing by 3 orders of magnitude from 20% to 90% relative humidity (RH). In addition to revealed high sensitivity and fast response, humidity sensors based on PANI nanofibers with some beads and a small content of PEO had also small hysteresis.

Lin et al. (2012c) have shown that the SAW humidity sensors based on *polyaniline* (PANI) nanofibers can also be developed. For this purpose, they used electrospun polyaniline/poly(vinyl butyral) PANI/PVB blended nanofibers. It was established that the composite nanofiber SAW sensor revealed very high sensitivity of 75 kHz/% RH over a wide humidity range (20%–90%RH), good sensing linearity, and very fast response in both humidification and desiccation processes (see Figure 22.9). For comparison, the PVB-based humidity sensors had sensitivity ~5 kHz/%RH. According to Lin et al. (2012c), fabricated sensors could detect very dry atmosphere with good sensitivity (down to 0.5%RH). Moreover, the sensor exhibited ultrafast response, within 1 or 2 s. Lin et al. (2012c) believe that the high surface-to-volume ratio and the improved mass load, electroacoustic load, and viscoelastic load of the core-sheath-structured composite nanofibers of PANI blended with PVB were responsible for the attractive humidity-sensing properties of the SAW sensors. And most importantly, the use of PANI/PVB nanofibers significantly improved the stability of the sensor parameters. Doped PANI generally shows unsatisfying stability in the open air, leading to a short lifetime of PANI-based sensors. However, the PANI/PVB nanofiber sensor showed good stability. The investigation on the long-term stability of the SAW sensor based on PANI/

PVB nanofibers found that the sensor maintained its high sensitivity after storing for more than 7 months. Lin et al. (2012c) assumed that the chemically stable and relatively hydrophobic PVB effectively protects the PANI core.

One should note that the explanation of the gas-sensing effect in polymer nanofibers can be carried out within the framework of approaches developed for conventional polymer-based sensors. It should also be borne in mind that the effect of doping and the effect of the composition of composites on humidity-sensing characteristics of nanofiber-based sensors are also subject to regularities established for conventional polymer-based sensors.

## 22.3 INORGANIC NANOFIBERS AND HUMIDITY SENSORS

It is necessary to note that besides polymer and polymer-based composite nanofibers (Lu et al. 2009), different inorganic nanofibers can be fabricated using various methods (Shao et al. 2002; Ding et al. 2003, 2004b; Raible et al. 2005; Luoh and Hahn 2006; Yang et al. 2007a, 2007b; Mondal and Sharma 2016). In particular, using template methods, nanofibers consisting of metals (Favier et al. 2001; Murray et al. 2004; Im et al. 2006b), semiconductors (Routkevitch et al. 1996), and metal oxides (Miao et al. 2002) were synthesized. Electrospinning also can be used for semiconductor and metal-oxide nanofibers fabrication (Lim et al. 2010; Park et al. 2010). For example, there are reports related to nanofibers of $TiO_2$ (Kim et al. 2006; Li et al. 2008; Zhang et al. 2008; Landau et al. 2009), $SnO_2$ (Yang et al. 2007a; Wang et al. 2007; Zhang et al. 2008), $WO_3$ (Wang et al. 2006), ZnO (Yang et al. 2007b), $SrTi_{0.8}Fe_{0.2}O_{3-\delta}$ (Sahner et al. 2007),

**TABLE 22.5**

**Examples of Solvents and Precursors Used for the Fabrication of Metal-Oxide-Based Nanofibers**

| Metal Oxide | Precursor/Polymer | Solvent | Applied Voltage/ Diameter of Fibers |
|---|---|---|---|
| SnO$_2$ | Tin chloride/PVA | Water, 1-propanol and isopropanol | 5–10 kV/100 nm |
| TiO$_2$ | Titanium tetraisopropoxide/PVP | Acetic acid and ethanol | 10 kV/30–200 nm |
| ZrO$_2$ | Zirconium acetate /PVAc | DMF | 15 kV/200 nm |
| CeO$_2$ | Ceric ammonium nitrate/PVP | Deionized water and ethanol | 12 kV/>1000 nm |
| ZnO | Zinc acetate/ PVP | Ethanol | 20 kV/250 nm |
| NiO/SnO$_2$ | Nickel chloride hexahydrate and tin chloride dehydrate/PVP | DMF and ethanol | 10 kV/100–200 nm |
| ZnO/SnO$_2$ | Zinc nitrate and tin chloride dehydrate/PAN | DMF | 18 kV/150–315 nm |
| ZnO/SiO$_2$ | Zinc acetate and tetraethyl orthosilicate/PVP | DMF, DMSO, HCl, and ethanol | 15 kV/200 nm |
| NiTiO$_3$ | Nickel acetate, titanium isopropoxide/ PVP | Methanol and acetic acid | 12 kV/175 nm |
| MgTiO$_3$ | Magnesium ethoxide and titanium isopropoxide/PVAc | 2-methoxyethanol and DMF | 17 kV/200–400 nm |

BaTiO$_3$ (He et al. 2010) etc. All indicated nanofibers were used for gas-sensor design (Wang et al. 2002; Ding et al. 2004a, 2005; Aussawasathien et al. 2005; Dan et al. 2007; Chen et al. 2011; Choi et al. 2017), including humidity sensors. Usually for preparing metal-oxide nanofibers a hybrid solution, which is a mixture of the metal-oxide sol precursor, polymer, and solvent is used. In order to make the inorganic nanoparticles effectively disperse in polymer, sometimes a surfactant is needed. Table 22.5 gives several examples of the solvent, precursor, applied voltage, and diameter of inorganic fibers formed from electrospinning.

It should be borne in mind that to obtain metal-oxide fiber, the sintering at elevated temperatures usually is required. As a rule, such treatment is carried out at temperatures 500°C–900°C. This thermal treatment is necessary for both the transformation of hydroxides in oxides and decomposition and removing polymeric components used for electrospinning. Figure 22.10 shows WO$_3$ nanofibers prepared by deposition of W on SWCNT template with following annealing in an oxygen-containing atmosphere. The same images for In$_2$O$_3$-based nanofibers are shown in Figure 22.11. As it is seen, metal oxides in nanofibers are polycrystalline. The grain size in such nanofibers may vary in the range 10–50 nm. Extremely high porosity is the main advantage of these sensors, which show very good operating characteristics (great and fast response) in comparison with sensors based on conventional materials. This unique morphology facilitates effective penetration of the surrounding gas into the porous ceramic layer, which is believed to be the main reason for the exceptionally high gas sensitivity of metal-oxide gas sensors produced by this method (Kim et al. 2006). Unlike conventional screen-printing methods that produce mesoporous granular layers with densely packed nanoparticles that give rise to poor gas transport, sensors produced by electrospinning display a bimodal pore size distribution comprising both large and small pores that enhance gas transport and sensitivity in these layers (Kim et al. 2010).

Analyzing the content of Table 22.6, it can be concluded that, based on metal-oxide fibers, resistive- or impedance-type humidity sensors are mainly developed. The simplest method

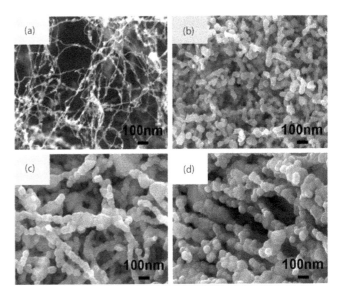

**FIGURE 22.10** (a) Morphology of the SWCNT template; (b–d) WO$_3$ nanowire morphology for W deposition times of 10 s, 20 s, and 60 s, respectively. The WO$_3$ structures were fabricated by oxidation at 700°C in air for 2 h. (Reprinted with permission from Vuong, N.M. et al., *J. Mater. Chem.*, 22, 6716–6725, 2012, Reproduced by permission of The Royal Society of Chemistry.)

of manufacturing such sensors is shown in Figure 22.12. However, interdigitated electrodes can also be used. Materials for the manufacture of electrodes commonly used are Au (Batool et al. 2013b; Imran et al. 2015), Ag/Pd (He et al. 2010), Al (Wang et al. 2011a), Ni (Batool et al. 2013b; Imran et al. 2015), Ag (Wang et al. 2011a), and Ti (Batool et al. 2013b). It would seem that the material of the electrodes should not affect the humidity-sensing characteristics. However, experiments carried out by Imran et al. (2015), Wang et al. (2011a), and Batool et al. (2013b), who compared Au and Ni electrodes, Ag and Al electrodes, and Ti, Au, and Ni electrodes, correspondingly, have shown that the electrode material can have a significant effect on sensor parameters. So, according to Imran et al. (2015) and Wang et al. (2011), sensors with Au

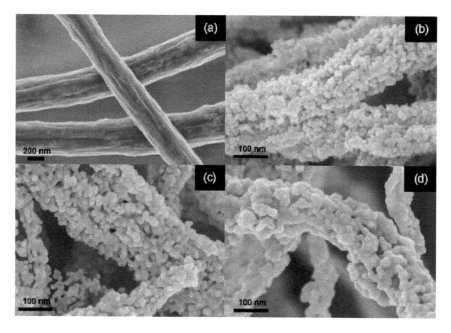

**FIGURE 22.11** SEM images of (a) as prepared PVA/indium acetate composite nanofibers; (b) after annealing at $T_{an}$ = 400°C; (c) $T_{an}$ = 500°C; and (d) $T_{an}$ = 600°C. (Reprinted from *Sens. Actuators B*, 149, Lim, S.K. et al., Preparation of mesoporous $In_2O_3$ nanofibers by electrospinning and their application as a CO gas sensor, 28–33, Copyright 2010, with permission from Elsevier.)

**TABLE 22.6**

**Characteristics of Humidity Sensors Based on Metal-Oxide Nanofiber Mats Prepared Using Electrospun**

| Metal Oxide | Diameter (nm) | RH Range (%) | Best Response | | References |
|---|---|---|---|---|---|
| | | | Max. Response | Res./Rec. time (s) | |
| *Resistive and Impedance types* | | | | | |
| $BaTiO_3$; $Ba_{0.8}Sr_{0.2}TiO_3$ | 50–200 nm | 11–95 | $S_R \sim 10^2-10^3$ | (5–30)/(4–9) | He et al. (2010); Wang et al. (2011a); Xia et al. (2012) |
| $(Na_{0.5}Bi_{0.5})_{0.94}TiO_3-Ba_{0.06}TiO_3$ | ~100 nm | 11–95 | $S_R \sim 4 \times 10^5$ | 2/50 | Li et al. (2015) |
| $CdTiO_3$; $La_2CuO_4$ | 50–200 nm | 40–98 | $S_R \sim 20-30$ | 4/6 | Imran et al. (2013); Hayat et al. (2016) |
| $KNbO_3$; $NaNbO_3$; $ZnFe_2O_4$ | 80–200 nm | 15–95 | $S_R \sim 10^4-10^5$ | (3–11)/(5–20) | Zhang et al. (2015); Zhuo et al. (2015); Ganeshkumar et al. (2017) |
| $NiO-SnO_2$ | 300–500 | 0–100 | $S_R \sim 10$ | 20/60 | Pascariu et al. (2016) |
| $SnO_2$:KCl | 100–200 | 11–95 | $S_R \sim 10^5$ | 5/6 | Song et al. (2009) |
| $SnO_2$:Li | 80–100 | 33–85 | $S_R \sim 50$ | ~1 | Yin et al. (2017) |
| $TiO_2-SnO_2$ | 100–180 | 11–95 | $S_R \sim 2 \times 10^3$ | 2/30 | Yang et al. (2015) |
| $TiO_2$:KCl; $TiO_2$:LiCl | 30–250 nm | 11–95 | $S_R \sim 2 \times 10^4$ | (1–3)/3 | Qi et al. (2008, 2009); Li et al. (2008) |
| $TiO_2$; $TiO_2$:Mg, Na | 100–200 | 11–95 | $S_R \sim (2-5) \times 10^2$ | (1–3)/(1–5) | Zhang et al. (2009); Jamil et al. (2012); Batool et al. (2013b) |
| ZnO; ZnO:Ce | 250–600 | 43–97 | $S_R \sim (5-7) \times 10^2$ | (15–17)/(13–15) | Liu et al. (2017) |
| ZnO | 100–150 | 40–90 | $S_R \sim 5$ | 3/5 | Imran et al. (2015) |
| ZnO:NaCl; ZnO:LiCl | 150–400 | 11–95 | $S_R \sim (2-3) \times 10^4$ | 3/6 | Wang et al. (2009); Zhang et al. (2010) |
| $ZrO_2$; $ZrO_2$:$TiO_2$, $Zr_{0.9}Mg_{0.1}O_{2-\delta}$ | 200–400 | 11–97 | $S_R \sim (3-4) \times 10^3$ | 5/(10–20) | Su et al. (2012a, 2012b) |
| Silica | 100–150 | 10–90 | $S_R \sim (1-2)10^2$ | (5–28)/(3–107) | Batool et al. (2013a); Xia et al. (2014) |
| *Quartz Crystal Microbalance (QCM)* | | | | | |
| ZnO | 80–120 nm | 10–90 | $\Delta f = 40$ Hz | 1/2 | Horzum et al. (2011) |
| *Surface Acoustic Waves (SAW)* ($f_0$ = 1.56GHz) | | | | | |
| $CeO_2$ | >1000 nm | 11–95 | $\Delta f = 2.5$ MHz | 16/16 | Liu et al. (2016) |
| *Piezoelectric* | | | | | |
| $NaNbO_3$ | 40–100 nm | 5–95 | 2 mV/%RH. | 12 | Gu et al. (2016) |

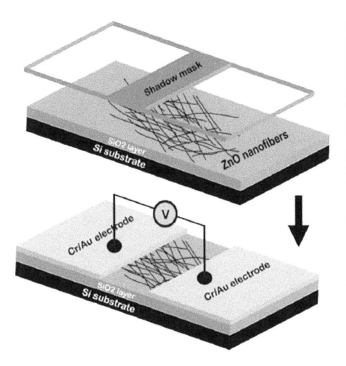

**FIGURE 22.12** Schematic illustration of the fabrication process of ZnO nanofiber-based humidity sensor with Au electrodes deposited on the nanofibers. (Reprinted from Imran, Z. et al., *AIP Adv.*, 5, 11721, 2015.)

and Ag electrodes exhibited improved sensing properties in comparison with sensors that used Ni and Al electrodes. At the same time, according to Batool et al. (2013b), sensors with Au electrodes showed the worst characteristics in comparison with devices using Ti and Ni electrodes. Batool et al. (2013b) believe that such situation takes place due to different porosity of electrode materials. According to Imran et al. (2015), the work function of electrode materials plays a more important role. However, there appear to be other factors that influence sensor parameters with different electrodes.

Table 22.6 testifies that there are also samples of QCM (Horzum et al. 2011), SAW (Liu et al. 2016), and capacitive humidity sensors (He et al. 2010; Li et al. 2015) based on metal-oxide-based fibers. But the number of such developments is very small. In addition, these sensors are inferior in their sensitivity to sensors developed on the basis of conventional technology. Gu et al. (2016) reported piezoresistive humidity sensors developed based on NaNbO₃ nanofibers. This is an interesting approach to the development of humidity sensors. The flexible active humidity sensors were fabricated by transferring the nanofibers from silicon to a soft polymer substrate. The sensors exhibited outstanding piezoelectric energy-harvesting performance with output voltage up to 2 V during the vibration process. The piezoelectric potential was generated due to bending motion. It was also found that the output voltage generated by the NaNbO₃ sensors had a negative correlation with the environmental humidity varying from 5% to 80%, where the peak-to-peak value of the output voltage generated by the sensors decreased from 0.40 to 0.07 V. This change corresponded to the sensitivity ~2 mV/%RH. The sensor also exhibited a short response time

(~12 s), good selectively against ethanol steam, and great temperature stability. According to Gu et al. (2016), the piezoelectric active humidity-sensing property could be attributed to the increased leakage current in the NaNbO₃ nanofibers, which was generated due to proton hopping among the H₃O⁺ groups in the absorbed H₂O layers under the driving force of the piezoelectric potential. Gu et al. (2016) believe that the active humidity sensor based on NaNbO₃ NFs provides an effective solution for self-powered sensor systems with high sensitivity, simple structure and fabrication processes, and low production costs. However, it seems that this is a difficult task, since it is difficult to find a stable source of mechanical energy in the natural environment. This means that the output signal will depend not only on humidity, but also on a lot of other difficult to control factors.

Typical humidity-sensing characteristics of TiO₂ fiber-based resistive sensors are shown in Figure 22.13. It is important to note that humidity-sensing characteristics of metal-oxide nanofiber-based sensors obey the same regularities as conventional metal-oxide humidity sensors discussed earlier in Chapters 10–14 (Volume 2). In particular, as seen in Figure 22.13, doping by alkali salts such as LiCl, KCl, and NaCl contributes to a significant increase in sensitivity. The formation of nanocomposites is also accompanied by an improvement in the parameters of the sensors, but not as strong as when doping with alkali salts. The mechanism of the effect of doping on humidity-sensing characteristics was discussed in some detail earlier in Chapter 11 (Volume 2). It was established that for the KCl-, LiCl-, or NaCl-doped metal-oxide nanofibers, besides the H₃O⁺, alkali salts can dissolve in the adsorbed water, and the dissociated ions (K⁺, Li⁺, Na⁺, and Cl⁻) from it can play the roles of conduction

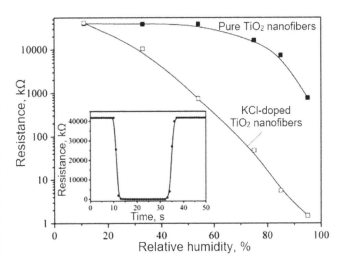

**FIGURE 22.13** The dependence of the resistance on the RH for pure and KCl-doped TiO₂ nanofibers. The inset is the response and recovery behavior of the KCl-doped TiO₂ nanofibers. The voltage and the frequency are 1 V and 100 Hz, respectively. KCl-doped TiO₂ nanofibers were prepared via electrospinning using a solution of tetrabutyl titanate, KCl, and PVP in ethanol and acetic acid for electrospinning, followed by calcination at $T = 500°C$ to obtain TiO₂/KCl composite nanofibers. (Reprinted with permission from Qi, Q. et al., *Appl. Phys. Lett.*, 93, 023105, 2008. Copyright 2008 American Institute of Physics.)

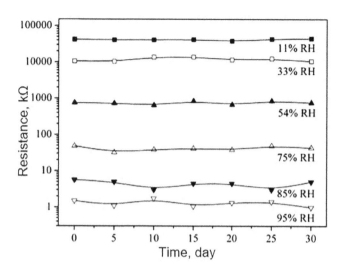

**FIGURE 22.14** The stability of the KCl-doped TiO₂ nanofiber-based humidity sensors. (Reprinted with permission from Qi, Q. et al., *Appl. Phys. Lett.*, 93, 023105, 2008. Copyright 2008 American Institute of Physics.)

carriers. In this case, an ion-controlled model should be used to explain the results obtained (Casalbore-Miceli et al. 2000; Qi et al. 2009). As well as conventional metal-oxide-based humidity sensors, metal-oxide nanofiber-based sensors have good stability of characteristics (see Figure 22.14). The only

difference is fast response and recovery after interaction with water vapor. In particular, for the analyzed TiO₂:KCl sensors developed by Qi et al (2008), the response and recovery times were about 3 s. There are messages that fiber-based humidity sensors had response times even smaller than 1 s (Horzum et al. 2011; Jamil et al. 2012; Yin et al. 2017). Hysteresis of humidity-sensing characteristics is largely determined by manufacturing technology and can vary from <2%RH (Wang et al. 2009) up to 5%–7% RH (He et al. 2010; Su et al. 2012a, 2012b; Imran et al. 2013).

The experiment showed that due to a special deposition method, the fiber mat has low adhesion to the substrate. To solve the problem of poor adhesion between fiber mats and the substrate, Kim et al. (2006) introduced an additional hot-pressing step after metal-oxide fiber deposition. Besides improving adhesion, this treatment was found to have an impact on the microstructure of the fibers, as shown in Figure 22.15 for TiO₂-based fibers. The as-spun metal oxide–polymer composite fibers exhibit a range of diameters from 200 to 500 nm (Figure 22.11a). When calcined without hot-pressing to remove the organic vehicle, a bundle structure composed of sheaths of 200–500 nm diameters was obtained. In some cases, the outer sheaths were broken, revealing cores filled with ~10-nm-thick fibrils as shown in Figure 22.15c. By introducing the hot-pressing step prior to calcination, an interconnected morphology of the TiO₂/polymer composite fibers was obtained, as illustrated in Figure 22.15b, due to the

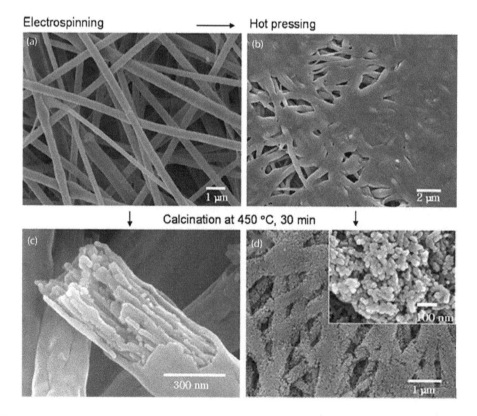

**FIGURE 22.15** Electrospinning and hot-pressing of metal-oxide materials: (a) SEM image of the as-spun TiO₂/PVAc composite fibers fabricated by electrospinning from a DMF solution; (b) SEM image of TiO₂ PVAc composite fibers after hot-pressing at 120°C for 10 min; (c) SEM image of unpressed TiO₂ nanofibers after calcination at 450°C; and (d) SEM image of hot-pressed TiO₂ nanofibers after calcinations at 450°C. (Reprinted with permission from Kim, I. et al., *Nano Lett.* 6, 2009–2013, 2006. Copyright 2006 American Chemical Society.)

partial melting of the polymer vehicle. Subsequent calcination resulted in the exceptionally high-surface-area structures, shown in Figure 22.15d. The mechanical pressure applied during the hot-pressing served to break the outer sheaths, thereby exposing the fibrils and leading to an exceptionally high surface-to-volume ratio. The combined hot-pressing/electrospinning technique was successfully transferred to $SnO_2$, $SnO_2$-$TiO_2$ composites, and other metal oxides (Kim et al. 2010; Sahner and Tuller 2010; Esfahani et al. 2017).

## 22.4 LIMITATIONS OF ELECTROSPINNING

There is no doubt that electrospinning is a powerful technique for preparing diverse nanostructured materials and highly sensitive sensors. Essential studies, nevertheless, are still required, and many challenges have to be faced. This is due to the fact that in addition to all advantages previously mentioned, the electrospinning process has some limitations. They are the following:

- First, for the production of polymer nanofibers by electrospinning, the selection of polymers is limited.
- Second, the direct electrospinning of metal and metal oxides is not possible, and a templating carrier polymer is always needed, which sometimes prevents this technique in the direct application of metal and metal-oxide nanofibers.
- Third, although electrospinning has already been introduced at the industrial level, in terms of the production rate of fibers, electrospinning is a much slower process compared to the other existing spinning techniques. In the case of practical applications, electrospinning is substandard to the usual methods owing to the high cost to produce thick fibers.
- Fourth, the variety of applications and performance of electrospun inorganic metal and metal-oxide nanofibers are restricted owing to their brittleness after calcination (Lee et al. 2009).
- Fifth, nanofibers have poor adhesion in the substrate.
- Sixth, electrospinning does not provide highly reproducible nanofiber parameters and sensors based on these nanofibers.
- Seventh, to date, it remains a challenge to produce nanofibers with diameters smaller than 10 nm by the existing electrospinning technique. Both the response time and the sensitivity were greatly improved with the decrease of diameter. Therefore, a smaller diameter leads to a faster response due to more rapid diffusion of gas molecules through the nanofiber.
- Eighth, and most importantly, electrospinning technology is not compatible with traditional mass-production processes.

Based on the above, we can conclude that this technology, despite some advantages, is unlikely to be used in the development of humidity sensors designed for the market.

## REFERENCES

Abd Razak S.I., Wahab I.F., Fadil F., Dahli F.N., Zahran Md Khudzari A.Z., Adeli H. (2015) A review of electrospun conductive polyaniline based nanofiber composites and blends: Processing features, applications, and future directions. *Adv. Mater. Sci. Eng.* **2015**, 1–19.

Aussawasathien D., Dong J.-H., Dai L. (2005) Electrospun polymer nanofiber sensors. *Synth. Met.* **154**, 37–40.

Batool S.S., Imran Z., Israr-Qadir M., Jamil-Rana S., Usman M., Jamil H. et al. (2013a) Silica nanofibers based impedance type humidity detector prepared on glass substrate. *Vacuum* **87**, 1–6.

Batool S.S., Imran Z., Qadir M.I., Usman M., Jamil H., Rafiq M.A. et al. (2013b) Comparative analysis of Ti, Ni, and Au electrodes on characteristics of $TiO_2$ nanofibers for humidity sensor application. *J. Mater. Sci. Technol.* **29**(5), 411–414.

Bhardwaj N., Kundu S.C. (2010) Electrospinning: A fascinating fiber fabrication technique. *Biotechnol. Adv.* **28**, 325–347.

Bognitzki M., Czado W., Frese T., Schaper A., Hellwig M., Steinhart M., Greiner A., Wendorff J.H. (2001a) Nanostructured fibers via electrospinning. *Adv. Mater.* **13**, 70–72.

Bognitzki M., Frese T., Steinhart M., Greiner A., Wendorff J.H., Schaper A., Hellwig M. (2001b) Preparation of fibers with nanoscaled morphologies: Electrospinning of polymer blends. *Polym. Eng. Sci.* **41**, 982–989.

Casalbore-Miceli G., Yang M.J., Camaioni N., Mari C.M., Li Y., Sun H., Ling M. (2000) Investigations on the ion transport mechanism in conducting polymer films. *Solid State Ionics* **131**, 311–321.

Chen C., Tang Y., Vlahovic B., Yan F. (2017) Electrospun polymer nanofibers decorated with noble metal nanoparticles for chemical sensing. *Nanoscale Res. Lett.* **12**, 451.

Chen D., Lei S., Chen Y. (2011) A single polyaniline nanofiber field effect transistor and its gas sensing mechanisms. *Sensors* **11**, 6509–6516.

Cheng Y.-T., Chen L.-C., Wang W.-C. (2017) Development of a fiber shape polymeric humidity sensor. *Proc. SPIE* **10167**, 101670K.

Choi S.-J., Persano L., Camposeo A., Jang J.-S., Koo W.-T., Kim S.-J., Cho H.-J., Kim I.-D., Pisignano D. (2017) Electrospun nanostructures for high performance chemiresistive and optical sensors. *Macromol. Mater. Eng.* **302**, 1600569.

Dan Y., Cao Y., Mallouk T.E., Johnson A.T., Evoy S. (2007) Dielectrophoretically assembled polymer nanowires for gas sensing. *Sens. Actuators B* **125**, 55–79.

Ding B., Kim C., Kim H., Seo M., Park S. (2004b) Titanium dioxide nanofibers prepared by using electrospinning method. *Fiber. Polym.* **5**, 105–109.

Ding B., Kim H., Kim C., Khil M., Park S. (2003) Morphology and crystalline phase study of electrospun $TiO_2$-$SiO_2$ nanofibers. *Nanotechnology* **14**, 532–537.

Ding B., Kim H., Lee S., Lee D., Choi K. (2002) Preparation and characterization of nanoscaled poly(vinyl alcohol) fibers via electrospinning. *Fiber. Polym.* **3**, 73–79.

Ding B., Kim J., Miyazaki Y., Shiratori S. (2004a) Electrospun nanofibrous membranes coated quartz crystal microbalance as gas sensor for $NH_3$ detection. *Sens. Actuators B* **101**, 373–380.

Ding B., Li C., Miyauchi Y., Kuwaki O., Shiratori S. (2006b) Formation of novel 2D polymer nanowebs via electrospinning. *Nanotechnology* **17**, 3685–3691.

Ding B., Wang M., Wang X., Yu J., Sun G. (2010) Electrospun nanomaterials for ultrasensitive sensors. *Mater. Today* **13**(11), 16–27.

Ding B., Wang M., Yu J., Sun G. (2009) Gas sensors based on electrospun nanofibers. *Sensors* **9**, 1609–1624.

Ding B., Yamazaki M., Shiratori S. (2005) Electrospun fibrous poly-acrylic acid membrane-based gas sensors. *Sens. Actuators B* **106**, 477–483.

Dong B., Krutschke M., Zhang X., Chi L.F., Fuchs H. (2005a) Fabrication of polypyrrole wires between microelectrodes. *Small* **1**, 520–524.

Dong B., Zhong D.Y., Chi L.F., Fuchs H. (2005b) Patterning of conducting polymers based on a random copolymer strategy: Toward the facile fabrication of nanosensors exclusively based on polymers. *Adv. Mater.* **17**, 2736–2741.

Esfahani H., Jose R., Ramakrishna S. (2017) Electrospun ceramic nanofiber mats today: Synthesis, properties, and applications. *Materials* **10**, 1238.

Eyebe G.A., Bideau B., Boubekeur N., Loranger E., Domingue F. (2017) Environmentally-friendly cellulose nanofibre sheets for humidity sensing in microwave frequencies. *Sens. Actuators B* **245**, 484–492.

Favier F., Walter E.C., Zach M.P., Benter T., Penner R.M. (2001) Hydrogen sensors and switches from electrodeposited palla-dium mesowire arrays. *Science* **293**, 2227–2231.

Fong H., Chun I., Reneker D. (1999) Beaded nanofibers formed dur-ing electrospinning. *Polymer* **40**, 4585–4592.

Ganeshkumar R., Cheah C.W., Zhao R. (2017) Annealing tempera-ture and bias voltage dependency of humidity nanosensors based on electrospun $KNbO_3$ nanofibers. *Surf. Interfaces* **8**, 60–64.

Gu L., Zhou D., Cheng Cao J.C. (2016) Piezoelectric active humidity sensors based on lead-free $NaNbO_3$ piezoelectric nanofibers. *Sensors* **16**, 833.

Hayat K., Niaz F., Ali S., Javid Iqbal M., Ajmal M., Ali M., Iqbal Y. (2016) Thermoelectric performance and humidity sensing characteristics of $La_2CuO_4$ nanofibers. *Sens. Actuators B* **231**, 102–109.

He Y., Zhang T., Zheng W., Wang R., Liu X., Xia Y., Zhao J. (2010) Humidity sensing properties of $BaTiO_3$ nanofiber prepared via electrospinning. *Sens. Actuators B* **146**, 98–102.

Horzum N., Tascioglu D., Okur S., Demir M.M. (2011) Humidity sensing properties of ZnO-based fibers by electrospinning. *Talanta* **85**, 1105–1111.

Huang J., Kaner R.B. (2004) Nanofiber formation in the chemi-cal polymerization of aniline: A mechanistic study. *Angew. Chem. Int. Ed.* **43**, 5817–5821.

Huang J.X., Virji S., Weiller B.H., Kaner R.B. (2003b) Polyaniline nanofibers: Facile synthesis and chemical sensors. *J. Am. Chem. Soc.* **125**, 314–315.

Huang K., Wan M.X., Long Y., Chen Z., d Wei Y. (2005) Multi-functional polypyrrole nanofibres via a functional dopant-introduced process. *Synth. Met.* **155**, 495–500.

Huang Z.-M., Zhang Y.-Z., Kotaki M., Ramakrishna S. (2003a) A review on polymer nanofibers by electrospinning and their applications in nanocomposites. *Compos. Sci. Technol.* **63**, 2223–2253.

Imran Z., Batool S.S., Jamil H., Usman M., Israr-Qadir M., Shah S.H. et al. (2013) Excellent humidity sensing properties of cadmium titanate nanofibers. *Ceram. Intern.* **39**, 457–462.

Imran Z., Rasool K., Batool S.S., Ahmad M., Rafiq M.A. (2015) Effect of different electrodes on the transport properties of ZnO nanofibers under humid environment. *AIP Adv.* **5**, 11721.

Jamil H., Batool S.S., Imran Z., Usman M., Rafiq M.A., Willander M., Hassan M.M. (2012) Electrospun titanium dioxide nano-fiber humidity sensors with high sensitivity. *Ceram. Intern.* **38**, 2437–2441.

Jayaraman K., Kotaki M., Zhang Y., Mo X., Ramakrishna S. (2004) Recent advances in polymer nanofibers. *J. Nanosci. Nanotechnol.* **4**(1–2), 52–65.

Jessensky O., Muller F., Gosele U. (1998) Self-organized formation of hexagonal pore arrays in anodic alumina. *Appl. Phys. Lett.* **72**, 1173–1175.

Kameoka J., Orth R., Yang Y., Czaplewski D., Mathers R., Coates G., Craighead H. (2003) A scanning tip electrospinning source for deposition of oriented nanofibers. *Nanotechnology* **14**, 1124–1129.

Kim I.-D., Jeon E.-K., Choi S.-H., Choi D.-K., Tuller H.L. (2010) Electrospun $SnO_2$ nano fiber mats with thermo-compression step for gas sensing applications. *J. Electroceram.* **25**, 159–167.

Kim I., Rothschild A., Lee B., Kim D., Jo S., Tuller H. (2006) Ultrasensitive chemiresistors based on electrospun $TiO_2$ nano-fibers. *Nano Lett.* **6**, 2009–2013.

Landau O., Rothschild A., Zussman E. (2009) Processing-microstructure-properties correlation of ultrasensitive gas sensors produced by electrospinning. *Chem. Mater.* **21**, 9–11.

Lee J.A., Krogman K.C., Ma M., Hill R.M., Hammond P.T., Rutledge G.C. (2009) Highly reactive multilayer-assembled $TiO_2$ coating on electrospun polymer nanofibers. *Adv. Mater.* **21**, 1252–1256.

Lei C., Tang L.C., Mao H.Y., Wang Y., Xiong J.J., Ou W. et al. (2015) Nanofiber forests as a humidity sensitive material. In: *Proceedings of the 28th IEEE Intern, Conference on Micro Electro Mechanical Systems, MEMS 2015*, 18–22 January, Estoril, Portugal, pp. 857–860.

Lei S., Chen D., Chen Y. (2011) A surface acoustic wave humidity sensor with high sensitivity based on electrospun MWCNT/Nafion nanofiber films. *Nanotechnology* **22**, 265504.

Li P., Li Y., Ying B., Yang M. (2009) Electrospun nanofibers of polymer composite as a promising humidity sensitive mate-rial. *Sens. Actuators B* **141**, 390–395.

Li P., Zheng X., Zhang Y., Yuan M., Jiang B., Deng S. (2015) Humidity sensor based on electrospun $(Na_{0.5}Bi_{0.5})_{0.94}TiO_3$–$Ba_{0.06}TiO_3$ nanofibers. *Ceram. Intern.* **41**, 14251–14257.

Li Z., Zhang H., Zheng W., Wang W., Huang H., Wang C., MacDiarmid A., Wei Y. (2008) Highly sensitive and stable humidity nanosensors based on LiCl doped $TiO_2$ electrospun nanofibers. *J. Am. Chem. Soc.* **130**, 5036–5037.

Liang S., He X., Wang F., Geng W., Fu X., Ren J., Jiang X. (2015) Highly sensitive humidity sensors based on LiCl–Pebax 2533composite nanofibers via electrospinning. *Sens. Actuators B* **208**, 363–368.

Lim S.K., Hwang S.H., Chang D., Kim S. (2010) Preparation of mesoporous $In_2O_3$ nanofibers by electrospinning and their application as a CO gas sensor. *Sens. Actuators B* **149**, 28–33.

Lin Q., Li Y., Yang M. (2012a) Investigations on the sensing mecha-nism of humidity sensors based on electrospun polymer nano-fibers. *Sens. Actuators B* **171–172**, 309–314.

Lin Q., Li Y., Yang M. (2012b) Polyaniline nanofiber humidity sensor prepared by electrospinning. *Sens. Actuators B* **161**, 967–972.

Lin Q., Li Y., Yang M. (2012c) Highly sensitive and ultrafast response surface acoustic wave humidity sensor based on electrospun polyaniline/poly(vinyl butyral) nanofibers. *Anal. Chim. Acta* **748**, 73–80.

Liu P., Zhang L. (2009) Hollow nanostructured polyaniline: Preparation, properties and applications. *Crit. Rev. Sol. St. Mater. Sci.* **34**, 75–87.

Liu X.L., Ly J. Han S., Zhang D.H., Requicha A., Thompson M.E., Zhou C.W. (2005) Synthesis and electronic properties of indi-vidual single-walled carbon nanotube/polypyrrole composite nanocables. *Adv. Mater.* **17**, 2727–2732.

Liu Y., Huang H., Wang L., Cai D., Liu B., Wang D. et al. (2016) Electrospun $CeO_2$ nanoparticles/PVP nanofibers based high-frequency surface acoustic wave humidity sensor. *Sens. Actuators B* **223**, 730–737.

Liu Y.-J., Zhang H.-D., Zhang J., Li S., Zhang J.-C., Zhu J.-W. et al. (2017) Effects of Ce doping and humidity on UV sensing properties of electrospun ZnO Nanofibers. *J. Appl. Phys.* **122**, 105102.

Long Y.-Z., Li M.-M., Gub C., Wan M., Duvail J.-L., Liue Z., Fan Z. (2011) Recent advances in synthesis, physical properties and applications of conducting polymer nanotubes and nanofibers. *Prog. Polymer Sci.* **36**, 1415–1442.

Lu X., Wang C., Wei Y. (2009) One-dimensional composite nanomaterials: Synthesis by electrospinning and their applications. *Small* **5**(21), 2349–2370.

Luoh R., Hahn H.T. (2006) Electrospun nanocomposite fiber mats as gas sensors. *Composites. Sci. Technol.* **66**, 2436–2441.

Macagnano A., Perri V., Zampetti E., Bearzotti A., De Cesare F. (2016) Humidity effects on a novel eco-friendly chemosensor based on electrospun PANi/PHB nanofibers. *Sens. Actuators B* **232**, 16–27.

McCann J., Marquez M., Xia Y. (2006) Highly porous fibers by electrospinning into a cryogenic liquid. *J. Am. Chem. Soc.* **128**, 1436–1437.

Miao Z., Xu D., Ouyang J., Guo G., Zhao X., Tang Y. (2002) Electrochemically induced sol-gel separation of single-crystalline TiO$_2$ nanowires. *Nano Lett.* **2**, 717–720.

Mogera U., Sagade A.A., George S.J., Kulkarni G.U. (2014) Ultrafast response humidity sensor using supramolecular nanofibre and its application in monitoring breath humidity and flow. *Sci. Rep.* **4**, 4103.

Mondal K., Sharma A. (2016) Recent advances in electrospun metal-oxide nanofiber-based interfaces for electrochemical biosensing. *RSC Adv.* **6**, 94595–94616.

Murray B.J., Walter E.C., Penner R.M. (2004) Amine vapor sensing with silver mesowires. *Nano Lett.* **4**, 665–670.

Park J.-A., Moon J., Lee S.-J., Kim S.H., Zyung T., Chu H.Y. (2010) Structure and CO gas sensing properties of electrospun TiO$_2$ nanofibers. *Mater. Lett.* **64**, 255–257.

Pascariu P., Airinei A., Olaru N., Petrila I., Nica V., Sacarescu L., Tudorache F. (2016) Microstructure, electrical and humidity sensor properties of electrospun NiO–SnO$_2$ nanofibers. *Sens. Actuators B* **222**, 1024–1031.

Patil J.V., Mali S.S., Kamble A.S., Hong C.K., Kim J.H., Patil P.S. (2017) Electrospinning: A versatile technique for making of 1D growth of nanostructured nanofibers and its applications: An experimental approach. *Appl. Surf. Sci.* **423**, 641–674.

Qi Q., Feng Y., Zhang T., Zheng X., Lu G. (2009) Influence of crystallographic structure on the humidity sensing properties of KCl-doped TiO$_2$ nanofibers. *Sens. Actuators B* **139**, 611–617.

Qi Q., Zhang T., Wang L. (2008) Improved and excellent humidity sensitivities based on KCl-doped electrospun nanofibers. *Appl. Phys. Lett.* **93**, 023105.

Raible I., Burghard M., Schlecht U., Yasuda A., Vossever T. (2005) V$_2$O$_5$ nanofibers: Novel gas sensors with extremely high sensitivity and selectivity to amines. *Sens. Actuators B* **106**, 730–735.

Ramakrishna S., Fujihara K., Teo W.-E., Lim T.-C., Ma Z. (eds.) (2005) *An Introduction to Electrospinning and Nanofibers.* World Scientific Publishing, Singapore.

Ramakrishna S., Fujihara K., Teo W.-E., Yong T., Ma Z., Ramaseshan R. (2006) Electrospun nanofibers: Solving global issues. *Mater. Today* **9**(3), 40–50.

Ramanathan K., Bangar M.A., Yun M., Chen W., Mulchandani A., Myung N. (2004) Individually addressable conducting polymer nanowires array. *Nano Lett.* **4**, 1237–1239.

Ramos P.G., Morales N.J., Goyanes S., Candal R.J., Rodríguez J. (2016) Moisture-sensitive properties of multi-walled carbon nanotubes/polyvinyl alcohol nanofibers prepared by electrospinning electrostatically modified method. *Mater. Lett.* **185**, 278–281.

Routkevitch D., Bigioni T., Moskovits M., Xu J. (1996) Electrochemical fabrication of CdS nanowire arrays in porous anodic aluminum oxide templates. *J. Phys. Chem.* **100**, 14307.

Sahner K., Gouma P., Moos R. (2007) Electrodeposited and sol-gel precipitated p-type SrTi$_{1-x}$Fe$_x$O$_{3-\delta}$ semiconductors for gas sensing. *Sensors* **7**, 1871–1886.

Sahner K., Tuller H.L. (2010) Novel deposition techniques for metal oxide: Prospects for gas sensing. *J. Electroceram.* **24**, 177–199.

Shao C., Kim H., Gong J., Lee D. (2002) A novel method for making silica nanofibers by using electrospun fibers of polyvinyl alcohol/silica composite as precursor. *Nanotechnology* **13**, 635–637.

Song X. F., Qi Q., Zhang T., Wang C. (2009) A humidity sensor based on KCl-doped SnO$_2$ nanofibers. *Sens. Actuators B* **138**, 368–373.

Su M.-Y., Wang J., Yao P.-J., Du H.-Y. (2012a) Performance humidity sensor based on electrospun Zr$_{0.9}$Mg$_{0.1}$O$_{2-\delta}$ nanofibers. *Chin. Phys. Lett.* **29**(11), 110701.

Su M., Wang J., Du H., Yao P., Zheng Y., Li X. (2012b) Characterization and humidity sensitivity of electrospun ZrO$_2$:TiO$_2$ hetero-nanofibers with double jets. *Sens. Actuators B* **161**, 1038–1045.

Thenmozhi S., Dharmaraj N., Kadirvelu K., Kim H.Y. (2017) Electrospun nanofibers: New generation materials for advanced applications. *Mater. Sci. Eng. B* **217**, 36–48.

Theron A., Zussman E., Yarin A.L. (2001) Electrostatic field-assisted alignment of electrospun nanofibers. *Nanotechnology.* **12**, 384–390.

Virji S., Huang J.X., Kaner R.B., Weiller B.H. (2004) Polyaniline nanofiber gas sensors: Examination of response mechanisms. *Nano Lett.* **4**, 491–496.

Vuong N.M., Jung H., Kim D., Kim H., Hong S.-K. (2012) Realization of an open space ensemble for nanowires: A strategy for the maximum response in resistive sensors. *J. Mater. Chem.* **22**, 6716–6725.

Wang G., Ji Y., Huang X., Yang X., Gouma P., Dudley M. (2006) Fabrication and characterization of polycrystalline WO$_3$ nanofibers and their application for ammonia sensing. *J. Phys. Chem. B* **110**, 23777–23782.

Wang J., Chan S., Carlson R.R., Luo Y., Ge G., Ries R.S., Heath J.R., Tseng G.R. (2004) Electrochemically fabricated polyaniline nanoframework electrode junctions that function as resistive sensors. *Nano Lett.* **4**, 1693–1697.

Wang L., He Y., Hu J., Qi Q., Zhang T. (2011a) DC humidity sensing properties of BaTiO$_3$ nanofiber sensors with different electrode materials. *Sens. Actuators B* **153**, 460–464.

Wang W., Li Z., Liu L., Zhang H., Zheng W., Wang Y. et al. (2009) Humidity sensor based on LiCl-doped ZnO electrospun nanofibers. *Sens. Actuators B* **141**, 404–409.

Wang X., Ding B., Yu J., Wang M. (2011b) Highly sensitive humidity sensors based on electro-spinning/netting a polyamide 6 nanofiber/net modified by polyethyleneimine. *J. Mater. Chem.* **21**, 16231–16238.

Wang X., Ding B., Yu J., Wang M., Pan F. (2010) A highly sensitive humidity sensor based on a nanofibrous membrane coated quartz crystal microbalance. *Nanotechnology* **21**, 055502.

Wang X., Drew C., Lee S.-H., Senecal K.J., Kumar J., Samuelson L.A. (2002) Electrospun nanofibrous membranes for highly sensitive optical sensors. *Nano. Lett.* **2**, 1273–1275.

Wang Y., Ramos I., Santiago-Aviles J. (2007) Detection of moisture and methanol gas using a single electrospun tin oxide nanofiber. *IEEE Sens. J.* **7**, 1347–1348.

Xia Y., Fei T., He Y., Wang R., Jiang F., Zhang T. (2012) Preparation and humidity sensing properties of Ba$_{0.8}$Sr$_{0.2}$TiO$_3$ nanofibers via electrospinning. *Mater. Lett.* **66**, 19–21.

Xia Y., Zhao H., Liu S., Zhang T. (2014) The humidity-sensitive property of MCM-48 self-assembly fiber prepared via electro-spinning. *RSC Adv.* **4**, 2807.

Yang A., Tao X., Wang R. (2007a) Room temperature gas sensing properties of SnO$_2$/multiwall-carbonnanotube composite nanofibers. *Appl. Phys. Lett.* **91**, 133110.

Yang M., Xie T., Peng L., Zhao Y., Wang D. (2007b) Fabrication and photoelectric oxygen sensing characteristics of electrospun Co doped ZnO nanofibers. *Appl. Phys. A-Mat. Sci. Process.* **89**, 427–430.

Yang Z., Zhang Z., Liu K., Yuan Q., Dong B. (2015) Controllable assembly of SnO$_2$ nanocubes onto TiO$_2$ electrospun nanofibers toward humidity sensing applications. *J. Mater. Chem. C* **3**, 6701.

Yin M., Yang F., Wang Z., Zhu M., Liu M., Xu X., Li Z. (2017) A fast humidity sensor based on Li$^+$-doped SnO$_2$ one-dimensional porous nanofibers. *Materials* **10**, 535.

You M.-H., Yan X., Zhang J., Wang X.-X., He X.-X., Yu M., Ning X., Long Y.-Z. (2017) Colorimetric humidity sensors based on electrospun polyamide/CoCl$_2$ nanofibrous membranes. *Nanoscale Res. Lett.* **12**, 360.

Zeng F.-W., Liu X.-X., Diamond D., Lau K.T. (2010) Humidity sensors based on polyaniline nanofibers. *Sens. Actuators B* **143**, 530–534.

Zhang C.-L., Yu S.-H. (2014) Nanoparticles meet electrospinning: Recent advances and future prospects. *Chem. Soc. Rev.* **43**, 4423–4448.

Zhang H. N., Li Z., Liu L., Wang C., Wei Y., MacDiarmid A.G. (2009) Mg$^{2+}$/Na$^+$-doped rutile TiO$_2$ nanofiber mats for high-speed and anti-fogged humidity sensors. *Talanta* **79**, 953–958.

Zhang H., Li Z., Wang W., Wang C. (2010) Na$^+$-doped zinc oxide nanofiber membrane for high speed humidity sensor. *J. Am. Ceram. Soc.* **93**(1), 142–146.

Zhang Y., He X., Li J., Miao Z., Huang F. (2008) Fabrication and ethanol-sensing properties of micro gas sensor based on electrospun SnO$_2$ nanofibers. *Sens. Actuators B* **132**, 67–73.

Zhang Y., Pan X., Wang Z., Hu Y., Zhou X., Hu Z., Gu H. (2015) Fast and highly sensitive humidity sensors based on NaNbO$_3$ nanofibers. *RSC Adv.* **5**, 20453.

Zhang Z.M., Wei Z.X., Wan M.X. (2002) Nanostructures of polyaniline doped with inorganic acids. *Macromolecules* **35**, 5937–5942.

Zhuo M., Yang T., Fu T., Li Q. (2015) High-performance humidity sensors based on electrospinning ZnFe$_2$O$_4$ nanotubes. *RSC Adv.* **5**, 68299.

# 23 Humidity Sensors Based on Metal-Oxide Mesoporous-Macroporous and Hierarchical Structures

## 23.1 MESOPOROUS-MACROPOROUS AND HIERARCHICAL STRUCTURES

### 23.1.1 GENERAL CONSIDERATION

Mesoporous-macroporous, and hierarchical metal-oxide structures are other modern directions in sensing materials design. It was established that the ability to create macroporous objects from nanoscaled components may create new resources for optimization of gas-sensor parameters. In particular, the presence of mesopores guarantees high sensitivity, and the presence of macropores facilitates fast response and recovery during the detection process.

The pores in 3D MOX structures are developed in the submicrometer or nanodimensional domain. Therefore, these structures are frequently called *mesoporous*. These rather new materials with extremely high surface area offer a high degree of versatility in terms of structure and texture. One should note that silica, zeolites, metal phosphates, metal-organic frameworks (MOFs), and supramolecular compounds described before in Chapters 7–12 are examples of mesoporous materials.

Similar to that for silica-based mesoporous materials, the most successful approaches to developing metal-oxide mesoporous structures are based on the synthesis of pore-containing particles through the soft-templating approach. However, due to the intrinsic properties of their elements, the precursors of metal oxides, such as metal chlorides and alkoxides, display much faster hydrolysis–condensation kinetics compared to silica precursors, which makes it difficult to assemble the metal oxides and the amphiphilic templates into ordered mesostructures. To date, many strategies have been developed to achieve a well-controlled assembly of template molecules and metal oxides without phase separation, including the use of complex molecules or acids as stabilizing agents, nonaqueous synthesis, assembly of presynthesized nanocrystals with template molecules, and so on. Ordered mesoporous materials with different compositions from pure inorganic or pure organic frameworks to organic–inorganic hybrid frameworks have been widely reported in the past two decades (Carreon and Guliants 2005). For example, mesoporous metal oxides and mixed oxides with semicrystalline frameworks, such as $TiO_2$, $ZnO$, $WO_3$, $SnO_2$, and $Al_2O_3$ are successfully prepared via a facile wet-chemical approach combining with an annealing process (Lin et al. 2012).

The hierarchical nanostructures also have extremely high surface areas and have little tendency to agglomerate, which allows one to employ them as high-performance gas-sensor

materials. A "hierarchical structure" means the higher dimension of a micro- or nanostructure composed of many, low-dimensional, nano-building blocks (Lee 2009). The various hierarchical structures can be classified according to the dimensions of nano-building blocks and the consequent hierarchical structures, referring to the dimensions, respectively, of the nano-building blocks and of the assembled hierarchical structures (see Figure 23.1). For example, "1–3 urchin" means that 1D nanowires/nanorods are assembled into a 3D urchin-like spherical shape, and "2–3 flower" indicates that a 3D flowerlike hierarchical structure is assembled from many 2D nanosheets. Under this framework, the hollow spheres can be regarded as the assembly of 1D nanoparticles into the 3D hollow spherical shape. Thus, strictly speaking, the 0–3 hollow spheres should be regarded as one type of hierarchical structure.

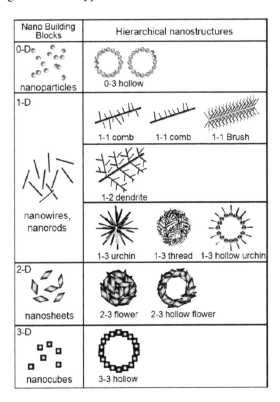

**FIGURE 23.1** Nomenclature of hierarchical structures according to the dimensions of the nano-building blocks (the former number) and of the consequent hierarchical structures (the latter number). (Reprinted with permission from *Sens. Actuators B*, 140, Lee, J.H., Gas sensors using hierarchical and hollow oxide nanostructures: Overview, 319–336, Copyright 2009, with permission from Elsevier.)

### 23.1.2 Technological Approaches to the Synthesis of Mesoporous-Macroporous and Hierarchical Structures

Various methods have been considered for synthesizing such hierarchical and hollow-particle structures, including spray-drying (Okuyama et al. 2006; Colombo et al. 2010), sol–gel (Hayashi et al. 2009), layer-by-layer (LbL) templating (Rothschild and Tuller 2006; Wang et al. 2008), electrodeposition (Wadea and Wegrowe 2005), vapor-phase impregnation (Yue and Zhou 2008), interface growth, and pulse-laser deposition (Sanchez et al. 2008). However, the most promising technologies apparently are methods based on sol–gel, aerosol spray, and LbL deposition (Lee 2009).

The conventional LbL process starts with electrostatic adsorption of a charged species onto the substrate, which is a priori charged with opposite sign. The species adsorption results in recharging of the substrate surface. Then the first layer is covered with the next one via deposition of an oppositely charged species and so on, until the required film thickness is achieved. The LbL approach is versatile for assembling various materials—polymers, lipids, proteins, dye molecules, etc.—on a number of substrates via not only electrostatic interactions, but also via hydrogen bonding, hydrophobic interactions, covalent bonding, and complementary base pairing (Ariga et al. 2007). The properties of LbL films, such as composition and thickness, can be readily adjusted by simply varying the species adsorbed, the number of layers deposited, and the conditions employed during the assembly process. Removal of the templating substrate following the LbL film formation can give rise to freestanding nanostructured materials with different morphologies and functions. Further details about this technique and its applications may be found elsewhere (Caruso et al. 1998; Ariga et al. 2007; Zhang et al. 2007; Wang et al. 2008). One route to fabricating mesoporous hierarchical structures using an aerosol spray method, and photos of the structures produced, are given in Figure 23.2.

Synthesis of mesoporous materials using liquid deposition techniques is shown in Figure 23.3. This overall complex transformation can be seen simply as direct polycondensation of the inorganic precursors around the organic micelles (or mesophase), which freezes the liquid-crystal mesostructure. Shimizu and coworkers (Hieda et al. 2008; Hayashi et al. 2009; Morio et al. 2009; Hyodo et al. 2010) have shown that by using a modified sol–gel technique, pyrolysis or a physical-vapor-deposition process employing a polymethylmethacrylate (PMMA) microsphere film as a macropore template, macroporous (*mp-*) films of various materials promising for gas sensor application can be fabricated. They established that different kinds of gas sensors fabricated with the *mp*–semiconductor films showed stronger gas responses as well as fast response and recovery speeds in comparison with those fabricated with a conventional film and powders without macropores (Hayashi et al. 2009; Yuan et al. 2011).

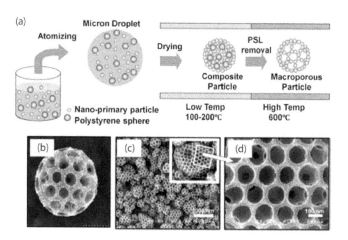

**FIGURE 23.2** (a) Experimental apparatus for developing particles containing ordered pores; and (b–d) photos of the particles. (Reprinted with permission from *Adv. Powder Tech.*, 20, Iskandar, F., Nanoparticle processing for optical applications—A review, 283–292, Copyright 2009, with permission from Elsevier.)

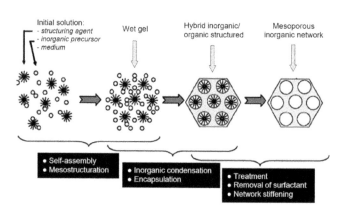

**FIGURE 23.3** Templating approach applied to make thin films via liquid deposition. (Reprinted with permission from Sanchez, C. et al., *Chem. Mater.*, 20, 682–737. Copyright 2008 American Chemical Society.)

Spherical colloids can also be employed as templates. The deposition of MOX films onto the outer surface of colloidal particles, which are then chemically or thermally removed, further gives the possibility to design hollow spherical structures like small capsules (Figure 23.4). This method permits good control over the properties of the hollow spheres (e.g., size, composition, thickness, permeability, function) by proper choice of the sacrificial colloids and the film components (Wang et al. 2008).

Colloidal crystals consisting of three-dimensional (3D) ordered arrays of monodispersed spheres can also be used as novel templates for the preparation of highly ordered macroporous inorganic solids that exhibit precisely controlled pore sizes and highly ordered 3D porous structures (Guliants et al. 2004). The macroscale templating approach typically consists of three steps. First, the interstitial voids of the monodisperse

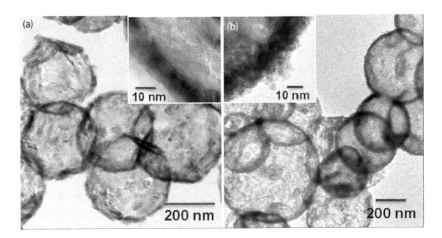

**FIGURE 23.4** (a) $TiO_2$ hollow spheres; and (b) $SnO_2$ hollow spheres prepared by the encapsulation of Ti and Sn precursors on Ni spheres and the removal of core metal templates by dilute HCl aqueous solution after heat treatment at 400°C. (Reprinted with permission from *Sens. Actuators B*, 140, Lee, J.H., Gas sensors using hierarchical and hollow oxide nanostructures: Overview, 319–336, Copyright 2009, with permission from Elsevier.)

sphere arrays are filled with precursors of various classes of materials, such as ceramics, semiconductors, metals, and monomers. In the second step, the precursors condense and form a solid framework around the spheres. Finally, the spheres are removed by either calcination or solvent extraction.

The success of forming macroporous-ordered structures is mainly determined by van der Waals interactions, wetting of the template surface, filling of the voids between the spheres, and volume shrinkage of the precursors during the solidification process. The colloidal crystal templating method may be used in combination with sol–gel, salt solution, nanocrystalline, and other precursors to produce the inorganic 3D macrostructures (Stein and Schroden 2001). The colloidal-crystal templates used to prepare 3D macroporous materials include monodisperse polystyrene (PS), PMMA, and silica spheres. A typical scanning electron microscopy (SEM) image of a colloidal array of polystyrene spheres used as a template in the synthesis of macroporous inorganic materials is shown in Figure 23.5. Prior to precursor infiltration, these monodisperse spheres are ordered into close-packed arrays by sedimentation, centrifugation, vertical deposition, or electrophoresis (Xia et al. 2000). The final inorganic macroporous structure after the removal of spheres contains the ordered interconnected pore structure shown in Figure 23.6.

As an alternative to the "soft-templating" route, the "nanocasting" (hard-templating) method (Wagner et al. 2013; Tomer and Duhan 2016) facilitates the synthesis of a much larger variety of mesoporous metal oxides. This method comprises a solid, mesoporous structure matrix, most frequently silica, which is used as the porogen (Figure 23.7). The desired product is created inside the pores of the matrix that is selectively removed later. Hence, the product remains as a negative replica of the matrix. For the synthesis of mesoporous metal oxides by this structure-replication procedure, a precursor compound, typically a metal salt, is filled into the pores of the silica matrix. The precursor compound is then converted

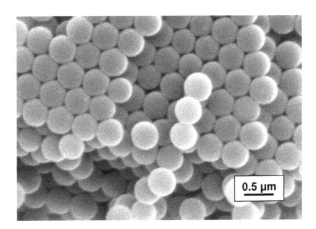

**FIGURE 23.5** SEM image of an ordered array of ~400 nm polystyrene spheres. (Reprinted with permission from Carreon, M.A. and Guliants V.V., *Chem. Mater.*, 14, 2670–2675, 2002. Copyright 2002 American Chemical Society.)

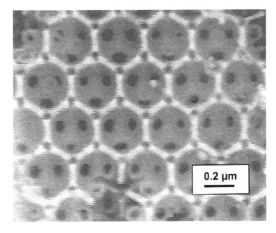

**FIGURE 23.6** SEM image of a macroporous inorganic material. The walls consist of ~20 nm $(VO)_2P_2O_7$ crystals. (Reprinted with permission from Carreon, M.A. and Guliants V.V., *Chem. Mater.*, 14, 2670–2675, 2002. Copyright 2002 American Chemical Society.)

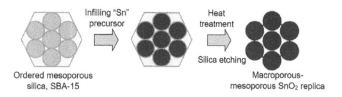

**FIGURE 23.7** Schematic representation for the synthesis of ordered mesoporous-macroporous SnO$_2$ via the hard-templating method. (Idea from Tomer, V.K. and Duhan, S., *Sens. Actuators B*, 223, 750–760, 2016.)

into the metal oxide by thermal decomposition, sometimes preceded by a pH-induced conversion into an intermediate phase (such as a hydroxide species). The silica matrix is then removed by chemical etching with either hydrofluoric acid (HF) or with a strong base, for example, concentrated sodium hydroxide solution. For example, Tomer and Duhan (2016) used for this 2M NaOH solution. Instead of silica, mesoporous carbon is sometimes used as the structure matrix, especially in those cases where the desired metal oxide is not stable against either HF or NaOH (as used for the etching of silica) (Roggenbuck et al. 2006).

For the formation of macroporous-mesoporous ordered structures, one can use more complex technological processes. For example, Chen et al. (2010) for the fabrication of ordered mesoporous TiO$_2$ sphere arrays suggested the two-step replication process shown in Figure 23.8. They noted that it is still challenging to fabricate 3D-ordered metal-oxide sphere arrays using traditional colloidal crystal self-assembly technique due to the difficulty in obtaining highly uniform and monodisperse mesoporous metal-oxide spheres. In particular, the mesoporous TiO$_2$ particles previously prepared are generally polydisperse and not spherical in some cases.

Planar substrates containing defined pore structures (e.g., macroporous membranes) have also been widely employed as templates to develop mesoporous MOXs (Wang et al. 2008).

Planar templates are typically fabricated using solid substrates such as quartz slides, Si wafers, and metal electrodes. The porous planar templates allow deposition of metals or MOXs with well-defined 3D morphologies. For example, membranes containing cylindrical pores provide the possibility to form (nano)tubes (hollow cylinders). The outer diameter, length, composition, and thickness of the tubes are controlled by the pore diameter, membrane thickness, the type of species deposited, and the number of layers assembled, respectively. The open ends and large surface area associated with the tubes make them useful for delivery applications. In particular, they can be readily loaded with large quantities of gas species.

Chemical methods can also be applied for hierarchical structures synthesis (Lee 2009). For example, the experiment showed that the hollow precursor or oxide particles can be prepared either by the chemically induced self-assembly of surfactants into micelle configuration (Zhao et al. 2006) (see Figure 23.9a) or by the polymerization of carbon spheres and subsequent encapsulation of metal hydroxide during the hydrothermal/solvothermal reaction (Yang et al. 2007). The core polymer parts are normally removed by heat treatment at elevated temperature (500°C–600°C). Thus, hollow oxide structures can be used stably as gas detection materials at the sensing temperature of 200°C–400°C without thermal degradation. Other approaches based on Ostwald ripening and the Kirkendall effect can also be used. These processes are illustrated in Figure 23.9b, c. The Kirkendall effect is realized during the oxidation of dense and crystalline metal particles, when the outward diffusion of metal cations through the oxide shell layers is very rapid compared to the inward diffusion of oxygen to the metal core (Liu and Zeng 2004; Fan et al. 2007). Solid evacuation is the common aspect of Ostwald ripening and the Kirkendall effect. However, in principle, the shell layers developed by the Kirkendall effect are denser and less permeable than those by Ostwald ripening.

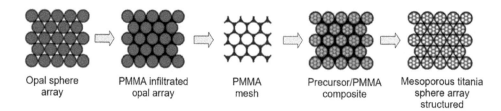

FIGURE 23.8 Schematic of two-step replication process for the fabrication of ordered arrays of mesoporous TiO$_2$ spheres. A piece of highly ordered single-crystal-like silica opal was immersed in methyl methacrylate (MMA) monomer with 1 wt% benzoyl peroxide (BPO) as an initiator. Polymerization was initially carried out at 40°C for 10 h and then 60°C for 12 h. The silica opal spheres were removed using a 10 wt% HF solution (24 h) to obtain a freestanding PMMA mesh (inverse opal). Then the PMMA mesh was used as the second-step template for the formation of ordered TiO$_2$ spheres, and amphiphilic triblock copolymer Pluronic P123 (EO$_{20}$PO$_{70}$EO$_{20}$) acted as a structure-directing agent to make the mesopores. The titania precursor solution was infiltrated into the spherical macropores of the PMMA mesh by immersing the mesh in the precursor solution for 4 h. After that, the precursor-filled PMMA mesh was removed from the solution, and then aged at room temperature for 3 days. Ordered mesoporous TiO$_2$ sphere arrays were obtained by heating the precursor/PMMA composite in open air to 400°C at 1°C/min, followed by a 4 h soak to remove the PMMA mesh template and the triblock copolymer surfactant. (Chen, J. et al., *Chem. Commun.* 46, 1872–1874, 2010. Reproduced by permission of The Royal Society of Chemistry.)

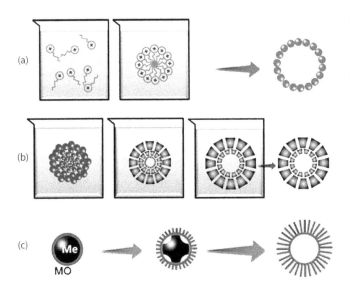

**FIGURE 23.9** Schematic diagrams for the preparation of hollow structures using the (a) self-assembled hydrothermal/solvothermal reaction; (b) Ostwald ripening of porous secondary particles; and (c) solid evacuation by the Kirkendall effect. (Reprinted with permission from *Sens. Actuators B*, 140, Lee, J.H., Gas sensors using hierarchical and hollow oxide nanostructures: Overview, 319–336, Copyright 2009, with permission from Elsevier.)

### 23.1.3 GAS SENSORS BASED ON MESOPOROUS AND HIERARCHICAL NANOSTRUCTURES

It should be noted that technologies for fabricating mesoporous and hierarchical nanostructures have been developed for all the basic MOXs (SnO₂, In₂O₃, TiO₂, WO₃, Fe₂O₃, etc.) utilized to develop various gas sensors (Lee 2009). The gas-sensing performance of sensors based on mesoporous and hollow nanostructures is well reviewed elsewhere (Tiemann 2007; Lee 2009), and it is agreed that such structures are

really attractive platforms for gas-sensing applications (Shimizu et al. 2004, 2005; Yue and Zhou 2008). Mesoporous and hollow structures have been reported to show very high gas-sensing response (Devi et al. 2002; Hyodo et al. 2003; Wagner et al. 2007) and fast response kinetics (Liu et al. 2007), which are attributed to their high surface area and well-defined porous architecture. A particularly large difference in the kinetics of the sensor response was observed compared with sensors fabricated using agglomerated powders. It was established that the hollow nanostructures follow the same basic trends as mentioned for the thin-film layers. When the shells are rather dense and thick, the gas-sensing reaction occurs only near the surface of the hollow spheres, and the inner parts of these spheres are inactive. However, if the shell is sufficiently thin, the primary particles in the entire hollow sphere are able to participate in gas-sensing reactions even when the shells are less permeable. In addition, the rate of sensor response of hollow spheres increases with the thinner shell configuration due to the faster gas diffusion. It has also been found that the sensor response and response kinetics of the mesoporous sensing materials, similarly to the conventional metal-oxide matrix, can be further improved by surface modification (Hyodo et al. 2003) and doping by catalytic materials (Rossinyol et al. 2007a, 2007b; He et al. 2009).

From our point of view, electrochemical etching of metal films with subsequent oxidation of fabricated porous structure is also a very promising approach to designing a mesoporous gas-sensing matrix. Such an approach was discussed with reference to TiO₂ and SnO₂ by Varghese et al. (2003b), Li et al. (2009), Rani et al. (2010), and Jeun and Hong (2010). Anodic oxidation is the most commonly employed method for the synthesis of self-ordered porous semiconductor structures, which is a relatively simple, low-cost, and high-yield process. The process used by Jeun and Hong (2010) is shown schematically in Figure 23.10. It was found that, after annealing

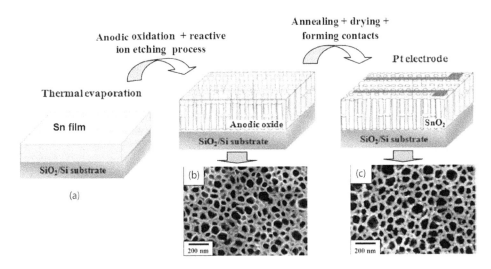

**FIGURE 23.10** (a) Schematic diagram of the experimental procedures of SnO₂-based sensor fabrication using Sn anodic etching with following oxidation and plain views of (b) as-etched and (c) as-annealed films. (Reprinted with permission from *Sens. Actuators B*, 151, Jeun J.-H. and Hong, S.-H., CuO-loaded nano-porous SnO₂ films fabricated by anodic oxidation and RIE process and their gas sensing properties, 1–7, Copyright 2010, with permission from Elsevier.)

at 700°C for 1 h, the pore walls of Sn films changed into a granular SnO$_2$ structure due to sintering, but the microscopic features of the nanosized pores and vertical nanochannels remained the same, indicating a high thermal stability of the anodized films (see Figure 23.10b, c).

It was established that the sensitivity of the previously mentioned mesoporous and hierarchical structures can be extremely high (Mor et al. 2004; Lin et al. 2012), and, similar to other gas-sensing materials, this sensitivity depends on the pore size, wall thickness, and tube length (Varghese et al. 2003b; Wagner et al. 2013). In particular, Varghese et al. (2003a, 2003b) have shown that the crystallized nanoscale walls and inter-tubular connecting points play critical roles in determining the remarkable hydrogen sensitivities of the TiO$_2$ nanotube arrays. The TiO$_2$ nanotube sample showing the highest sensitivity had a wall thickness of ~13 nm. However, we need to recognize that for commercialization of the results obtained, the continuation of research in this direction is required. Till now there have been technological problems, such as the reproducibility, the forming of low-resistance contacts, and structure stabilization during and following annealing, which need to be resolved. The anodization technique needs to be optimized as well (Li et al. 2009; Jeun and Hong 2010).

More detailed description of mesoporous-based metal-oxide gas sensors, including technology of their fabrication, is given in the reviews (Tiemann 2007; Lee 2009; Berenguer-Murcia 2013; Wagner et al. 2013; Hoa et al. 2015; Meng et al. 2016).

### 23.1.4 Structural Stability of Mesostructured Metal Oxides

It should be noted that the problem of structural instability exists for all types of the above structures, regardless of the material used. For example, Varghese et al. (2003a, 2003b) have found that the TiO$_2$ samples, prepared using anodization and consisting of well-defined nanotube arrays, were found to be stable up to the temperatures around 580°C (in oxygen ambient) even after crystallization of the tube walls. However, at higher temperatures, the crystallization of the titanium support disturbed the nanotube architecture, causing it to collapse and densify. At that it was found that a mesoporous metal oxide prepared by "soft templating" usually exhibit poor crystallinity and lower structural stability upon being exposed to higher temperatures than those prepared via the synthesis.

It was established that in general, the thermal stability of mesostructured metal-oxide phases depends on (Carreon and Guliants 2005): (1) the degree of charge-matching at the organic–inorganic interface, (2) the strength of interactions between inorganic species and surfactant head-groups, (3) the flexibility of the M–O–M bond angles in the constituent metal oxides, (4) the Tammann temperature of the metal oxide, and (5) the occurrence of redox reactions in the metal-oxide wall. In particular, the *charge-matching* at the organic–inorganic interface enables control over the wall composition and facilitates cross-linking of the inorganic species into a robust mesostructured network. The presence of strong *covalent bonds*

between metal-oxide species and the removal of surfactants can lead to the collapse of the mesostructure. Similarly, *rigid M–O–M bond angles* that are unable to accommodate the curvature of the inorganic–organic interface may result in the formation of only lamellar or dense metal-oxide phases. The mobility of metal ions or atoms in a crystalline metal oxide increases rapidly in the vicinity of its Tammann temperature, defined as 0.5–0.52 $T_m$, where $T_m$ is the melting point in Kelvin (Gray et al. 1957). Therefore, it is not surprising that the low Tamman temperature of some transition-metal oxides (Weast et al. 1986; Bell 2003) translates into a limited thermal stability of the corresponding mesostructures. Finally, the structural collapse of mesophases may be caused by *redox reactions* occurring in the metal-oxide wall during surfactant removal or catalytic reaction.

It is important to know that resolving the problem of instability does not have a universal approach. Unfortunately, every material used for mesoporous, macroporous, and hollow structure fabrication requires a specific approach for resolving it (Carreon and Guliants 2005). For example, the crystallinity and stability of porous metal oxides prepared by soft templating can be enhanced by filling of the pores with carbon and subsequent high-temperature treatment under oxygen-free conditions; the carbon serves as a "backbone" to prevent structural disintegration of the metal oxide and can later be removed (Roggenbuck and Tiemann 2005). Shimizu and coworkers (Hyodo et al. 2002) offered another approach. They established that the most important key to the drastic improvement of thermal stability of mesoporous (*m*-) SnO$_2$ powders is to immerse them in a phosphoric acid aqueous solution before calcination and consequently loading phosphorus components on the surface of *m*-SnO$_2$ crystallites. Such treatment enabled the preparation of the *m*-SnO$_2$ powders with small crystallite size (2–3 nm in diameter) and large specific surface area (>300 m$^2$/g) even after calcination at 600°C. Varghese et al. (2003a, b) have found that when subjected to rapid annealing in oxygen at up to 950°C, the structure of the TiO$_2$ nanotube sample prepared by electrochemical etching did not collapse completely, although a complete crystallization occurred; the tubes coalesced and formed a wormlike pattern.

Additional information related to different methods of mesoporous and macroporous materials preparation can be found in reviews (Xia et al. 2000; Stein and Schroden 2001; Guliants et al. 2004; Carreon and Guliants 2005; Tiemann 2008; Berenguer-Murcia 2013; Wei et al. 2017).

## 23.2 HUMIDITY SENSORS BASED ON MESOPOROUS AND HIERARCHICAL NANOSTRUCTURES

### 23.2.1 Mesoporous-Macroporous Metal-Oxide-Based Humidity Sensors

It should be noted that the manufacturing technology of the mesoporous-macroporous metal-oxide-based humidity sensors is no different from the technology used in the

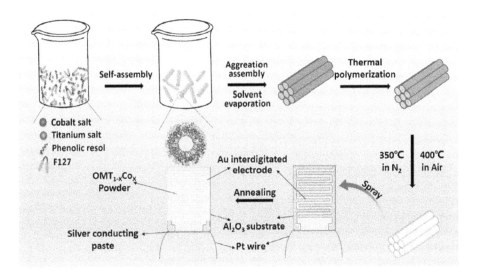

**FIGURE 23.11** The schematics showing the synthesis process of mesoporous Co-doped TiO$_2$-based sensors. (Reprinted with permission from *Appl. Surf. Sci.*, 412, Li, Z. et al., The effect of Co-doping on the humidity sensing properties of ordered mesoporous TiO$_2$, 638–647, Copyright 2017, with permission from Elsevier.)

manufacture of gas sensors. Schematically, this process is shown in Figure 23.11 for TiO$_2$:Co-based sensors developed by Li et al. (2017). Using F127 as the soft template for the self-assembly synthesis, 2 g F127 was dispersed into 12 ml of anhydrous ethanol with the condition of stirring at 40°C. Then, 2.5 g soluble resol was added to the solution and the stirring continued at the same temperature for 30 min. After that, a certain amount of cobalt acetate was added into the solution. This solution was named Solution-A. Next, 1.5 ml HCl and 3.4 g tetrabutyl titanate were added into 8 ml anhydrous ethanol, orderly. After stirring for 30 min, we obtained Solution-B. Then, it was slowly dropped into Solution-A with the condition

of stirring for 2 h at 40°C to obtain a homogeneous brick-red solution. Afterward, a dark red film was received by evaporating the solvents at room temperature and thermo-polymerizing at 100°C for 24 h. Then it was pyrolyzed in a tubular furnace at the temperature of 350°C for 4 h under the atmosphere of N$_2$ and heating rate of 1°C/min to convert tetrabutyl titanate to titanium oxide within the pores of the F127 template and obtained a puce polymer–TiO$_2$:Co composite. Finally, the polymer was removed at 400°C for 5 h in the air atmosphere and filtered the white pure ordered mesoporous TiO$_2$:Co.

Table 23.1 summarizes the results obtained during the development and testing of mesoporous-macroporous metal

**TABLE 23.1**

**Mesoporous-Macroporous Metal-Oxide-Based Resistive Humidity Sensors**

| Technology | Template | Metal Oxide | RH Range | Parameters | References |
|---|---|---|---|---|---|
| Multicomponent self-assembly procedure | Triblock copolymer Pluronic F127 | TiO$_2$:Co (3 mol%) | 9–90%RH | $S_R \sim 10^5$; $t_{res} \sim 24$ s | Li et al. (2017) |
| Nanocasting method using mesoporous silica as the hard template | SBA-15 | Bi$_{3.25}$La$_{0.75}$Ti$_3$O$_{12}$ | 11–95%RH | $S_R \sim 10^5$–$10^6$; $t_{res} \sim 32$ s | Zhang et al. (2016a, 2016b) |
| | | LaFeO$_3$ | 11–98 %RH | $S_R \sim 4 \cdot 10^5$; $t_{res} \sim 1$ s; $t_{rec} \sim 36$ s | Zhao et al. (2013) |
| | | SnO$_2$ | 11–98 %RH | $S_R \sim 10^5$; $t_{res} \sim 33$ s; $t_{rec} \sim 50$ s | Tomer, and Duhan (2015) |
| | | SnO$_2$:Ag | 11–98 %RH | $S_R \sim 6 \cdot 10^4$; $t_{res} \sim 4$ s; $t_{rec} \sim 7$ s | Tomer and Duhan (2016) |
| | KIT-6 | CeO$_2$:Gd | 20–80 %RH | $S_R \sim 10^4$; $t_{res} \sim 2$ s; | Almar et al. (2015) |
| | | In$_2$O$_3$ | 10–90 %RH | $S_R \sim 2 \cdot 10^3$; $t_{res} \sim 10$ s; $t_{rec} \sim 15$ s | Liu et al. (2010) |
| | | SnO$_2$:In/CN | 11–98 %RH | $S_R \sim 2 \cdot 10^5$; $t_{res} \sim 3$s; $t_{rec} \sim 2$ s | Malik et al. (2017) |

$S_R = R_{max}/R_{min}$.

**FIGURE 23.12** (a, b, c) TEM images of different samples of the mesoporous $In_2O_3$ particles used for humidity sensor development. An ordered mesoporous $In_2O_3$ material with crystalline walls has been synthesized through the nanocasting method. The pore and the pore wall sizes were estimated to be about 4 and 8 nm from the TEM images, respectively. (Reprinted with permission from *Sens. Actuators B*, 150, Liu, X. et al., Synthesis and characterization of mesoporous indium oxide for humidity-sensing applications, 442–448, Copyright 2010, with permission from Elsevier.)

oxide-based humidity sensors. It is seen that nanocasting method with using silica as the hard template is the most common method of synthesis such metal oxides. Typical transmission electron microscopy (TEM) images of such mesoporous metal oxides are shown in Figure 23.12. At that, $TiO_2$ and $SnO_2$ are the metal oxides most commonly found in articles devoted to such sensors. However, based on the above data, it is impossible to conclude about the advantages of one or another metal oxide, since even for the same material, the parameters of the sensors can vary within very wide limits (see Figure 23.13). The main thing is that all metal oxides specified in Table 23.1 are characterized by high stability of characteristics (Figure 23.14), what is most important for practical applications.

As follows from the data presented in Table 23.1, humidity sensors based on mesoporous-macroporous $SnO_2$, $In_2O_3$, $CeO_2$, $TiO_2$, and $Bi_{3.25}La_{0.75}Ti_3O_{12}$ metal oxides regardless of the material used have high sensitivity ($S_R = R_{max}/R_{min}$), reaching $10^4$–$10^6$ when changing humidity from 11% to 98%

relative humidity (RH). It should be noted that in many cases such sensitivity may be excessive. Even the sensitivity of $S_R \sim 10^2$–$10^3$ is quite sufficient for real applications.

As for response, there is also a large dispersion. Thus, the response time ($t_{res}$), depending on the technology used, can vary from 1 to 40 seconds, and the recovery time ($t_{rec}$) from 2 to 400 seconds. Such a large scatter is quite understandable, since the kinetics of response and recovery depends on the pore diameter, which can vary significantly in the structures studied, synthesized by various methods. For example, the kinetics should strongly depend not only on the diameter of pores in the hard template but also on the size of the particles of this hard template. It is the particle size that determines, on the one hand, the size of macropores, which contribute to the rapid penetration of water vapor into the humidity-sensitive matrix, and on the other hand, controls the surface area and the depth of mesopores, where the diffusion of water vapor is significantly slowed down and capillary condensation can

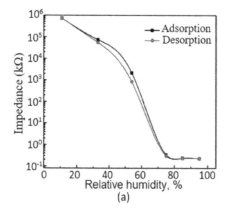
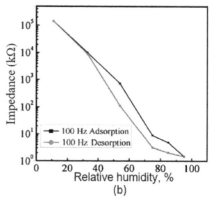

**FIGURE 23.13** Humidity hysteresis characteristic of the M-BLT humidity sensor at 100 Hz. Mesoporous structures are formed in the BLT nanorods via nanocasting method: (a) $Bi_{3.25}La_{0.75}Ti_3O_{12}$(BLT) nanorods are synthesized via electrospinning technique; and (b) $Bi_{3.25}La_{0.75}Ti_3O_{12}$(BLT) powder is synthesized via a metal-organic decomposition (MOD) method. (Reprinted with permission from *Sens. Actuators B*, 229, Zhang, Y. et al., Effect of mesoporous structure on $Bi_{3.25}La_{0.75}Ti_3O_{12}$ powder for humidity sensing properties, 453–460, Copyright 2016a, with permission from Elsevier; *Sens. Actuators B*, 237, Zhang, Y. et al., Humidity sensing and dielectric properties of mesoporous $Bi_{3.25}La_{0.75}Ti_3O_{12}$ nanorods, 41–48, Copyright 2016c, with permission from Elsevier.)

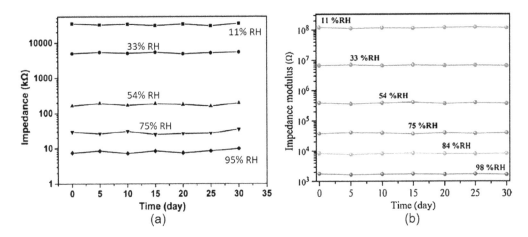

(a)    (b)

**FIGURE 23.14** Stability of the impedance response of (a) TiO$_2$; and (b) SnO$_2$:Ag mesoporous-macroporous samples to air humidity measured at 100 Hz: (a) (Reprinted with permission from Wang, Z. et al., The sol–gel template synthesis of porous TiO$_2$ for a high performance humidity sensor, *Nanotechnology*, 22, 275502, 2011a. Copyright 2011a, Institute of physics.) (b) (Reprinted with permission from *Sens. Actuators B*, 223, Tomer, V.K. and Duhan, S., A facile nanocasting synthesis of mesoporous Ag-doped SnO$_2$ nanostructures with enhanced humidity sensing performance, 750–760, Copyright 2016, with permission from Elsevier.)

take place. Therefore, to achieve acceptable results, these sizes must be optimized. Unfortunately, such studies are not analyzed in the articles. Unusually big recovery times for some mesoporous-macroporous structures (Li et al. 2017) may also be associated with imperfections in instrumentation.

Studies have shown that the sensitivity of the sensors, as expected, is controlled by active surface area, depending on the modes of synthesis and subsequent heat treatment (Wang et al. 2011a; Li et al. 2015; Tomer and Duhan 2016). Data obtained for TiO$_2$ samples synthesized by the sol–gel process

is shown in Figure 23.15. As a rule, the larger the surface area, the greater is the sensitivity. As well as for other humidity-sensing materials, the change in impedance under the influence of humidity decreases with increasing measurement frequency (Figure 23.16). Therefore, all measurements when testing humidity sensors are carried out at low frequencies, usually $f\sim 100$ Hz.

Doping can also be used to optimize the parameters of mesoporous-macroporous metal-oxide-based humidity sensors (Almar et al. 2015; Tomer and Duhan 2016; Li et al. 2017;

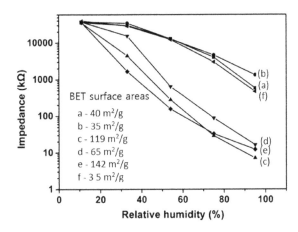

**FIGURE 23.15** Effect of humidity on the impedance of the TiO$_2$ samples synthesized by the sol–gel process ($f = 100$ Hz). Tetraethyl orthosilicate (TEOS) as a template was directly introduced into TiO$_2$ sol formed by the hydrolysis and condensation of titanium alkoxide; the following calcination led to the formation of TiO$_2$–SiO$_2$ composite, and the selective removal of SiO$_2$ by dilute HF solution led to the formation of porous structure in TiO$_2$. (Reprinted with permission from Wang, Z. et al., The sol–gel template synthesis of porous TiO$_2$ for a high-performance humidity sensor, *Nanotechnology*, 22, 275502, 2011a. Copyright 2011a, Institute of Physics.)

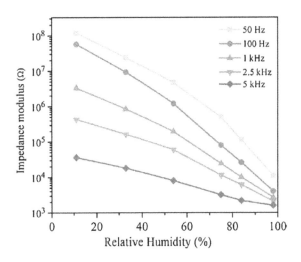

**FIGURE 23.16** Relationship of impedance of SnO$_2$:Ag-based mesoporous-macroporous samples and relative humidity at various frequencies. (Reprinted with permission from *Sens. Actuators B*, 223, Tomer, V.K. and Duhan, S., A facile nanocasting synthesis of mesoporous Ag-doped SnO$_2$ nanostructures with enhanced humidity sensing performance, 750–760, Copyright 2016, with permission from Elsevier.)

Malik et al. 2017). Apparently doping additives, as was indicated in Chapter 2, allow one to optimize the structure of humidity-sensitive metal oxides. It is important to note that to achieve the desired result, the concentration of these additives should not exceed a certain value. So in the case of $TiO_2$:Co, the Co concentration should not exceed 3% (Li et al. 2017). In the case of Ag-$SnO_2$/SBA-15 composite, the Ag concentration should not exceed 2 wt% (Tomer and Duhan 2016). Otherwise, the sensor parameters deteriorate.

Tomer and Duhan (2016), based on the data shown in Figure 23.17, concluded that the humidity sensors based on mesoporous-macroporous samples of $SnO_2$:Ag have higher selectivity toward humidity at room temperature. However, it is necessary to take into account that during these measurements the concentration of VOCs vapors varied in the range 0–1000 ppmV (0%–0.1%), and the concentration of water vapor varied in the range 3500–31600 ppmV (11–98 %RH). Comparing with the data obtained earlier, it can be argued that metal oxides with mesoporous-macroporous structure in their selectivity is no different from the previously considered metal-oxide-based humidity sensors. Their main advantage, as expected, is their high speed (see Table 23.1) and low hysteresis (Zhao et al. 2013). For the best samples, the hysteresis does not exceed 1% (Tomer and Duhan 2016; Malik et al. 2017).

## 23.2.2 HYBRID NANOCOMPOSITE-BASED HUMIDITY SENSORS

As it was shown in Tomer and Duhan (2015a) and Tomer et al. (2015, 2016), effective humidity sensors can be made without removing the template from the hybrid metal-oxide-template material. However, in this case, naturally, the amount of metal oxide in the composite should be substantially less than in the material used to make

mesoporous-macroporous metal oxides by removing the hard template. For example, Tomer and Duhan (2015b) studied $WO_3$/SBA-15 and found that when the $WO_3$ content was 1 wt.%, the impedance response was ~4 orders of magnitude changed toward the change of RH in the range 11%–98%; however, as the $WO_3$ content rose (10 wt. %), the response decreased and only 3 orders of change was observed due to the pore blockage of mesoporous channels with $WO_3$ nanoparticles and thus minimizing the possibility for the flow of charge carriers across the mesoporous channels of SBA-15. At the same time, the nanocomposite having 5 wt.% of $WO_3$ loaded in SBA-15 showed the highest change in the magnitude for impedance with best linearity.

Typical process of in-situ synthesis of metal-oxide-silica hybrid nanocomposite developed for highly sensitive $SnO_2$/SBA-15-based humidity sensors, can be described as follows (Tomer et al. 2016). 2 g P123 is dissolved in 70 ml distilled water at 40°C with vigorous stirring (1000 rpm), followed by addition of 10 ml HCl (2M). When a clear micellar solution was obtained after 3 h of stirring, 0.1 g of Tin chloride salt solution was added to it and stirred for another 2 h. Then, 4.5 g of TEOS was added and resultant solution was kept under stirring for 24 h at 45°C to ensure the homogeneous mixing of silica and tin species. The solid products were filtered, washed, and dried at 70°C and calcined at 550°C for 6 h in air to remove organic templates and thus pure mesoporous $SnO_2$/SBA-15 composite was obtained.

The experiment showed that humidity sensors based on hybrid nanocomposite are also highly sensitive with stable characteristics (Figure 23.18). With appropriate process optimization, these sensors also have a fast response (see

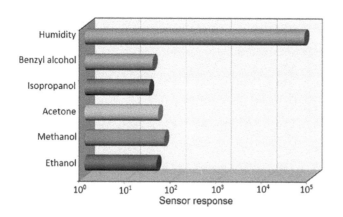

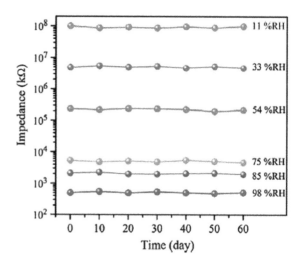

**FIGURE 23.17** Selectivity of the $SnO_2$:Ag-based sensor measured for a variety of VOCs and %RH. (Reprinted with permission from *Sens. Actuators B*, 223, Tomer, V.K. and Duhan, S., A facile nanocasting synthesis of mesoporous Ag-doped $SnO_2$ nanostructures with enhanced humidity sensing performance, 750–760, Copyright 2016, with permission from Elsevier.)

**FIGURE 23.18** Stability of the impedance response of $SnO_2$/SBA-15 samples to air humidity measured at 100 Hz. (Reprinted with permission from *Sens. Actuators B*, 223, Tomer, V.K. and Duhan, S., A facile nanocasting synthesis of mesoporous Ag-doped $SnO_2$ nanostructures with enhanced humidity sensing performance, 750–760, Copyright 2015, with permission from Elsevier.)

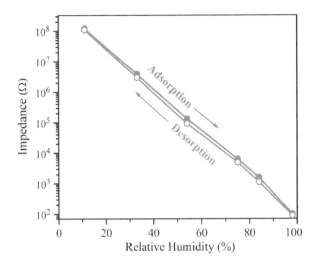

**FIGURE 23.19** Hysteresis curve, showing adsorption-desorption responses measured in the 11-98 RH% range for Ag-SnO₂/SBA-15 nanocomposite. (Reprinted with permission from *Sens. Actuators B*, 223, Tomer, V.K. and Duhan, S., A facile nanocasting synthesis of mesoporous Ag-doped SnO₂ nanostructures with enhanced humidity sensing performance, 750–760, Copyright 2015, with permission from Elsevier.)

Table 23.1) and low hysteresis (Figure 23.19). Therefore, based on the above data, it can be argued that hybrid nanocomposites have the same prospects for use in humidity sensors as mesoporous-macroporous metal oxides, discussed below (Tomer et al. 2016).

### 23.2.3 HUMIDITY SENSORS BASED ON METAL OXIDES WITH HIERARCHICAL STRUCTURE

Table 23.2 summarizes the results obtained in the study of humidity sensors based on various hierarchical structures. It is seen that a large number of different structures, synthesized mainly using the hydrothermal method, were tested. In this case, resistive sensors were the main type of sensors manufactured on the basis of such structures. Testing their humidity-sensitive characteristics showed that core-shell and flower-like nanospheres have the highest sensitivity, and the least sensitivity is characteristic of nanorod-based arrays. Although, Gu et al. (2011) showed that under certain conditions these structures also may have very high sensitivity (see Table 23.2).

## TABLE 23.2
## Examples of Humidity Sensors Based on Metal Oxides with Hierarchical Structure

| Hierarchical Structure | Metal Oxides | Technology | RH Range | Parameters | References |
|---|---|---|---|---|---|
| Core-shell nanospheres | Li₄Ti₅O₁₂/TiO₂ | Solvothermal synthesis and calcination | 11–95 %RH | $S_R$ ~10³; $t_{res}$ ~9 s; $t_{rec}$ ~17 s | Zhao et al. (2014) |
| Flower-like nanospheres | SnO₂:Fe (2%) | Surfactant-free hydrothermal synthesis | 11–95 %RH | $S_R$ ~6·10³; $t_{res}$ ~2 s; $t_{rec}$ ~5 s | Zhen et al. (2016) |
| Hollow core-shell nanospheres | WO₃/SnO₂ | Hydrothermal synthesis | 11–98 %RH | $S_R$ ~2·10²; $t_{res}$ ~29 s; $t_{rec}$ ~8 s | Li et al. (2014) |
| Urchin-like | CuO/rGO | Microwave-assisted hydrothermal synthesis | 11–98 %RH | $S_R$ ~2·10⁴; $t_{res}$ ~2 s; $t_{rec}$ ~17 s | Wang et al. (2014) |
| 3D dodecahedral nanocrystals | SnO₂ | Hydrothermal synthesis | 11–95 %RH | $S_R$ ~2·10⁴; $t_{res}$ <1 s; $t_{rec}$ ~13 s | Feng et al. (2017) |
| Twig-like | SnO₂ | Hydrothermal synthesis | 33–98 %RH | $S_R$ >2·10² | Qu et al. (2017) |
| Flower-like | WO₃ | Hydrothermal synthesis | 30–90 %RH | Δf = 25 kHz (SAW) | Pang et al. (2013) |
| Nanosheets | ZnO | Hydrothermal synthesis | 12–96 %RH | $S_R$ ~2·10²; $t_{res}$ ~600 s; $t_{rec}$ ~3 s | Tsai and Wang (2014) |
| Nanorod arrays | ZnO/MWCNTs | Hydrothermal synthesis and electrostatic layer-by-layer self-assembly | 11–97 %RH | $S_R$ <2; $t_{res}$ ~4 s; $t_{rec}$ ~30–70 s | Zhang et al. (2016c) |
| Nanorod arrays | ZnO:Sn | Sonicated sol–gel immersion method | 40–90 %RH | $S_R$ ~3; $t_{res}$ ~230 s; $t_{rec}$ ~30 s | Ismail et al. (2016) |
| Nanorod arrays on the fibers | TiO₂/ZnO | Hydrothermal synthesis | 10–98 %RH | $S_R$ <3 | Araújo et al. (2017) |
| Nanorod arrays | ZnO/VO₂ | Sol–gel | 11–95 %RH | $S_R$ ~10³; $t_{res}$ ~7 s; $t_{rec}$ ~35 s | Li et al. (2016) |
| Core/shell nanorod arrays | TiO₂/ZnO | Hydrothermal synthesis | 11–95 %RH | $S_C$ ~10⁵; $t_{res}$ ~770 s; $t_{rec}$ ~20 s | Gu et al. (2011) |

$S_R = R_{max}/R_{min}$; $S_C = C_{max}/C_{min}$.

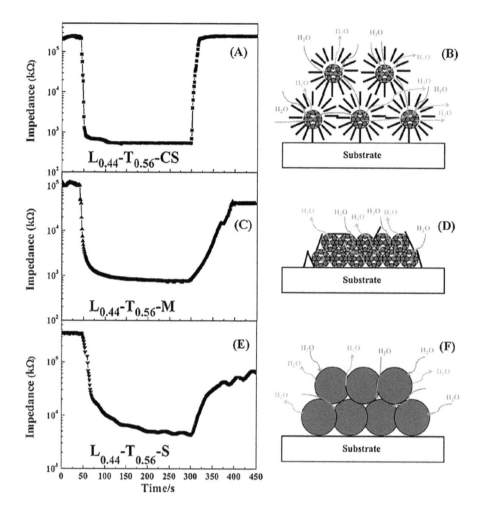

**FIGURE 23.20** Response/recovery characteristics of $Li_4Ti_5O_{12}$-based humidity sensors developed on the material with different structure: (a and b) $Li_{0.44}$-$T_{0.56}$-Core Shell; (c and d) $Li_{0.44}$-$T_{0.56}$-M (mesoporous); and (e and f) $Li_{0.44}$-$T_{0.56}$-S (densified), respectively. (Reprinted with permission from *Sens. Actuators B*, 203, Zhao, Y. et al., Highly sensitive and quickly responsive core-shell hierarchical porous $Li_4Ti_5O_{12}$-$TiO_2$ humidity sensors, 122–129, Copyright 2014, with permission from Elsevier.)

These studies also showed that the main advantage of core-shell and flower-like nanosphere-based sensors, similarly to mesoporous-macroporous-based sensors, is fast response and recovery. This is understandable, since such structures are highly porous and have good gas permeability. It is seen that time of sensor response and recovery for many humidity sensors does not exceed 1–2 and 3–8 sec, correspondingly (Table 23.2). Illustration of the influence of the structure of humidity film on the kinetics of sensor response in relation to $L_{0.44}$-$T_{0.56}$-based sensors is shown in Figure 23.20. It is seen that ultrafast response is characterized for core-shell structures. According to Zhao et al. (2014), constructing the shell of $Li_4Ti_5O_{12}$-CS from many nanosheets all pointing toward the center of the sphere, creates channels for facile water transmission (Figure 23.20b). The many macropores accelerate the transmission of water molecules. Compared with mesoporous $Li_4Ti_5O_{12}$-M (Figure 23.20d), when adjacent spheres contact, the nanosheets of $Li_4Ti_5O_{12}$-CS cross each other. This avoids close contact and prevents the clogging of channels. Thus,

water molecules can easily pass through the channels, contributing to a quicker response/recovery behavior. $Li_4Ti_5O_{12}$-S with its solid spherical structure has the longest response and recovery behavior, because it lacks the porous water transmission channels and the capacity for absorption and dissociation (Figure 23.20f).

How important it is to find the optimal morphology and surface architecture of humidity-sensitive material can also be judged by the results reported by Feng et al. (2017). They compared humidity-sensing characteristics of $SnO_2$-based 3D hierarchical dodecahedral nanocrystals (DNCs), 3D hierarchical nanorods (NRs), and nanoparticles (NPs) (Figure 23.21) and found that only the 3D hierarchical $SnO_2$ DNCs showed excellent humidity-sensing performance, including fast response and recovery times (see Figure 23.22), narrow hysteresis loop, high sensitivity, great linearity response, and good stability (Figure 23.23). They believe that such parameters were achieved due to the unique 3D open structure and high chemical activity of the exposed {101} facets.

**FIGURE 23.21** SEM images of as-synthesized (a) 3D hierarchical SnO$_2$NRs; (b) SnO$_2$NPs; and (c, d) 3D hierarchical SnO$_2$DNCs. Indicated structures were synthesized by a hydrothermal method. (Reprinted with permission from *Sens. Actuators B*, 243, Feng, H. et al., Three-dimensional hierarchical SnO$_2$ dodecahedral nanocrystals with enhanced humidity sensing properties, 704–714, Copyright 2017, with permission from Elsevier.)

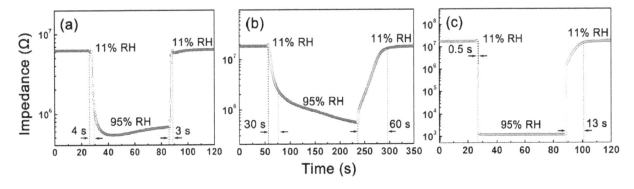

**FIGURE 23.22** (a) Humidity response and recovery curves for 3D hierarchical SnO$_2$NRs; (b) SnO$_2$NPs; and (c) 3D hierarchical SnO$_2$DNCs. (Reprinted with permission from *Sens. Actuators B*, 243, Feng, H. et al., Three-dimensional hierarchical SnO$_2$ dodecahedral nanocrystals with enhanced humidity sensing properties, 704–714, Copyright 2017, with permission from Elsevier.)

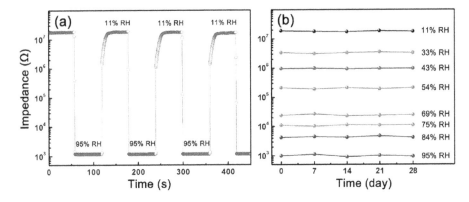

**FIGURE 23.23** (a) Repeated response and recovery characteristics of 3D hierarchical SnO$_2$DNCs; (b) responses of 3D hierarchical SnO$_2$DNCs monitored at different humidity conditions for 28 days. (Reprinted with permission from *Sens. Actuators B*, 243, Feng, H. et al., Three-dimensional hierarchical SnO$_2$ dodecahedral nanocrystals with enhanced humidity sensing properties, 704–714, Copyright 2017, with permission from Elsevier.)

## 23.3  SUMMARY

The analysis shows that metal oxides with mesoporous-mac-roporous and hierarchical structures are very attractive materials for the development of humidity sensors with improved sensitivity and kinetics of response. However, based on the above results, it is impossible to make an unequivocal conclusion about the advantage of a particular structure for humidity sensors, since in many cases it is not the type of structure that determines the humidity sensitive characteristics, but its dimensions and parameters of the crystallites that form it. How much these parameters can vary depending on the technology used can be seen from the data given in Tables 23.1 and 23.2. In addition, the data presented in Table 23.3 indicate that, after appropriate optimization of the technology and composition, the humidity sensors may have similar parameters, regardless of the structure

of the humidity-sensitive layer used. This suggests that the improvement of synthesis technology in order to produce metal oxides with the required structural parameters should be the main direction in the development of humidity sensors. At the same time, this improvement should take into account all factors influencing the structure of metal oxides, since even a small change in the synthesis conditions can be accompanied by a fundamental change in the structure of the material (Pang et al. 2013). For example, changing the ratio between the capping agent of ammonium benzoate (AB) and sodium tungstate (TS) in the process of synthesis of WO$_3$, the crystal microstructures could be changed from flower-shape to star-shape (see Figure 23.24). At that, the crystal phases are changed from orthorhombic WO$_3$·0.33H$_2$O to hexagonal WO$_3$ with the increase in the concentration of AB (Pang et al. 2013).

**TABLE 23.3**

**Information about the Structure of SnO$_2$ Flower-Like Materials Reported in Published Articles**

| Materials | Synthesis Method | Surfactant or Template | Diameter of Sphere (μm) | Width/Length of Nanorod (nm) | References |
|---|---|---|---|---|---|
| SnO$_2$ flower-like nanospheres | Hydrothermal method | PEG 400 | 0.5–1.0 | 200–300/ 200–300 | Zhang et al. (2013) |
| SnO$_2$/Ag flower-like heterostructures | Microwave hydrothermal treatment | Sodium citrate | 0.8–1.0 | <50/300–400 | Wang et al. (2013) |
| SnO$_2$ sphere-like structures | Hydrothermal method | No | ~5 | ~80/2500 | Qin et al. (2008) |
| SnO$_2$ flower-like nanostructure | One step hydrothermal method | No | 0.15 | <5 | Wang et al. (2011b) |
| SnO$_2$:Fe flower-like nanospheres | Hydrothermal method | NO | ~0.2 | 40/150 | Zhen et al. (2016) |

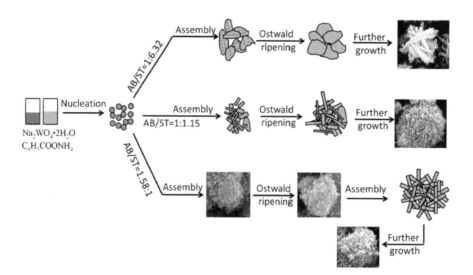

**FIGURE 23.24**  Illustration of the self-assembly growth process of the tungsten-oxide (hydrate) nanocrystal at different mole ratios of AB to ST. (With kind permission from Springer Science+Business Media: *Appl. Phys. A*, Hierarchical structured tungsten oxide nanocrystals via hydrothermal route: Microstructure, formation mechanism and humidity sensing, 112, 1033–1042, Pang, H.-F. et al. (2013).)

# REFERENCES

Almar L., Tarancón A., Andreu T., Torrell M., Hu Y., Dezanneau G., Morata A. (2015) Mesoporous ceramic oxides as humidity sensors: A case study for gadolinium-doped ceria. *Sens. Actuators B* **216**, 41–48.

Araújo E.S., Libardi J., Faia P.M., de Oliveira H.P. (2017) Humidity-sensing properties of hierarchical TiO$_2$: ZnO composite grown on electrospun fibers. *J. Mater. Sci. Mater. Electron.* **28**, 16575–16583.

Ariga K., Hill J.P., Ji Q. (2007) Layer-by-layer assembly as a versatile bottom-up nanofabrication technique for exploratory research and realistic application. *Phys. Chem. Chem. Phys.* **9**, 2319–2340.

Bell A.T. (2003) The impact of nanoscience on heterogeneous catalysis. *Science* **299**, 1688–1691.

Berenguer-Murcia A. (2013) Ordered porous nanomaterials: The merit of small. *Nanotechnology* **2013**, 257047.

Carreon M.A., Guliants V.V. (2002) Macroporous vanadium phosphorus oxide phases displaying three-dimensional arrays of spherical voids. *Chem. Mater.* **14**(6), 2670–2675.

Carreon M.A., Guliants V.V. (2005) Ordered meso- and macroporous binary and mixed metal oxides. *Eur. J. Inorg. Chem.* **2005**, 27–43.

Caruso F., Caruso R.A., Möhwald H. (1998) Nanoengineering of inorganic and hybrid hollow spheres by colloidal templating. *Science* **282**, 1111–1114.

Chen J., Hua Z., Yan Y., Zakhidov A.A., Baughman R.H., Xu L. (2010) Template synthesis of ordered arrays of mesoporous titania spheres. *Chem. Commun.* **46**, 1872–1874.

Colombo P., Vakifahmetoglu C., Costacurta S. (2010) Fabrication of ceramic components with hierarchical porosity. *J. Mater. Sci.* **45**, 5425–5455.

Devi G.S., Hyodo T., Shimizu Y., Egashira M. (2002) Synthesis of mesoporous TiO$_2$-based powders and their gas-sensing properties. *Sens. Actuators B* **87**, 122–129.

Fan H.J., Gosele Y., Zacharias M. (2007) Formation of nanotubes and hollow nanoparticles based on Kirkendall and diffusion processes: A review. *Small* **3**, 1660–1671.

Feng H., Li C., Li T., Diao F., Xin T., Liu B., Wang Y. (2017) Three-dimensional hierarchical SnO$_2$ dodecahedral nanocrystals with enhanced humidity sensing properties. *Sens. Actuators B* **243**, 704–714.

Gray T.J., Detwiler D.P., Rase D.E., Lawrence W.G., West R.R., Jennings T.J. (1957) *The Defect Solid State*, Interscience, New York, p. 96.

Gu L., Zheng K., Zhou Y., Li J., Mo X., Patzke G.R., Chen G. (2011) Humidity sensors based on ZnO/TiO$_2$ core/shell nanorod arrays with enhanced sensitivity. *Sens. Actuators B* **159**, 1–7.

Guliants V.V., Carreon M.A., Lin Y.S. (2004) Ordered mesoporous and macroporous inorganic films and membranes. *J. Membrane Sci.* **235**, 53–72.

Hayashi M., Hyodo T., Shimizu Y., Egashira M. (2009) Effects of microstructure of mesoporous SnO$_2$ powders on their H$_2$ sensing properties. *Sens. Actuators B* **141**, 465–470.

He L., Jia Y., Meng F., Li M., Liu J. (2009) Development of sensors based on CuO-doped SnO$_2$ hollow spheres for ppb level H$_2$S gas sensing. *J. Mater. Sci.* **44**, 4326–4333.

Hieda K., Hyodo T., Shimizu Y., Egashira M. (2008) Preparation of porous tin dioxide powder by ultrasonic spray pyrolysis and their application to sensor materials. *Sens. Actuators B* **133**, 144–150.

Hoa N.D., Duy N.V., El-Safty S.A., Hieu N.V. (2015) Meso-nanoporous semiconducting metal oxides for gas sensor applications. *J. Nanomaterials* **2015**, 972025.

Hyodo T., Inoue H., Motomura H., Matsuo K., Hashishin T., Tamaki J., Shimizu Y., Egashira M. (2010) NO$_2$ sensing properties of macroporous In$_2$O$_3$-based powders fabricated by utilizing ultrasonic spray pyrolysis employing polymethylmethacrylate microspheres as a template. *Sens. Actuators B* **151**, 265–273.

Hyodo T., Shimizu Y., Egashira M. (2003) Gas-sensing properties of ordered mesoporous SnO$_2$ and effects of coating thereof. *Sens. Actuators B* **93**, 590–600.

Iskandar F. (2009) Nanoparticle processing for optical applications—A review. *Adv. Powder Tech.* **20**, 283–292.

Ismail A.S., Mamat M.H., Sin N.D.M., Malek M.F., Zoolfakar A.S., Suriani A.B., et al. (2016) Fabrication of hierarchical Sn-doped ZnO nanorod arrays through sonicated sol-gel immersion for room temperature, resistive-type humidity sensor applications. *Ceram. Intern.* **l42**, 9785–9795.

Jeun J.-H., Hong S.-H. (2010) CuO-loaded nano-porous SnO$_2$ films fabricated by anodic oxidation and RIE process and their gas sensing properties. *Sens. Actuators B* **151**, 1–7.

Lee J.H. (2009) Gas sensors using hierarchical and hollow oxide nanostructures: Overview. *Sens. Actuators B* **140**, 319–336.

Li D., Zhang J., Shen L., Dong W., Feng C., Liu C., Ruan S. (2015) Humidity sensing properties of SrTiO$_3$ nanospheres with high sensitivity and rapid response. *RSC Adv.* **5**, 22879–22883.

Li H., Liu B., Cai D., Wang Y., Liu Y., Mei L., et al. (2014) High-temperature humidity sensors based on WO$_3$–SnO$_2$ composite hollow nanospheres. *J. Mater. Chem. A* **2**, 6854–6862.

Li W., Ma Y., Ji S., Sun G., Jin P. (2016) Synthesis and humidity sensing properties of the VO$_2$(B)@ZnO heterostructured nanorods. *Ceram. Intern.* **42**, 9234–9240.

Li Y., Yu X., Yang Q. (2009) Fabrication of TiO$_2$ nanotube thin films and their gas sensing properties. *J. Sensors* **2009**, 402174.

Li Z., Haidry A.A., Gao B., Wang T., Yao Z.J. (2017) The effect of Co-doping on the humidity sensing properties of ordered mesoporous TiO$_2$. *Appl. Surf. Sci.* **412**, 638–647.

Lin Z., Song W., Yang H. (2012) Highly sensitive gas sensor based on coral-like SnO$_2$ prepared with hydrothermal treatment. *Sens. Actuators B* **173**, 22–27.

Liu B., Zeng H.C. (2004) Fabrication of ZnO "dandelions" via a modified Kirkendall process. *J. Am. Ceram. Soc.* **126**, 16744–16746.

Liu Q., Zhang W.-M., Cui Z.-M., Zhang B., Wan L.-J., Song W.-G. (2007) Aqueous route for mesoporous metal oxides using inorganic metal source and their applications. *Micropor. Mesopor. Mater.* **100**, 233–240.

Liu X., Wang R., Zhang T., He Y., Tu J., Li X. (2010) Synthesis and characterization of mesoporous indium oxide for humidity-sensing applications. *Sens. Actuators B* **150**, 442–448.

Malik R., Tomer V.K., Chaudhary V., Dahiya M.S., Sharma A., Nehra S.P., et al. (2017) An excellent humidity sensor based on In–SnO$_2$ loaded mesoporous graphitic carbon nitride. *J. Mater. Chem. A* **5**, 14134–14143.

Meng F.L., Jin Z., Li M.Q., Liu J.H. (2016) Hierarchically-structured metal oxide-based gas sensors for volatile organic compound detection. *Sci. Foundation China* **24**(1), 63–80.

Mor G.K., Carvalho M.A., Varghese O.K., Pishko M.V., Grimes C.A. (2004) A room-temperature TiO$_2$-nanotube hydrogen sensor able to self-clean photoactively from environmental contamination. *J. Mater. Res.* **19**, 628–634.

Morio M., Hyodo T., Shimizu Y., Egashira M. (2009) Effect of macrostructural control of an auxiliary layer on the CO$_2$ sensing properties of NASICON-based gas sensors. *Sens. Actuators B* **139**, 563–569.

Okuyama K., Abdullan M., Llenggoro I.W., Iskandar F. (2006) Preparation of functional nanostructured particles by spray drying. *Adv. Powder. Technol.* **17**, 587–611.

Pang H.-F., Li Z.-J., Xiang X., Fu Y.-Q., Placido F., Zu X.-T. (2013) Hierarchical structured tungsten oxide nanocrystals via hydrothermal route: Microstructure, formation mechanism and humidity sensing. *Appl. Phys. A* **112**, 1033–1042.

Qin L., Xu J., Dong X., Pan Q., Cheng Z., Xiang Q., F. Li F. (2008) The template-free synthesis of square-shaped $SnO_2$ nanowires: The temperature effect and acetone gas sensors. *Nanotechnology* **19**, 185705.

Qu X., Wang M.-H., Chen Y., Sun W.-J., Yang R., Zhang H.-P. (2017) Facile synthesis of hierarchical $SnO_2$ twig-like microstructures and their applications in humidity sensors. *Mater. Letters* **186**, 182–185.

Rani S., Roy S.C., Paulose M., Varghese O.K., Mor G.K., et al. (2010) Synthesis and applications of electrochemically self-assembled titania nanotube arrays. *Phys. Chem. Chem. Phys.* **12**, 2780–2800.

Roggenbuck J., Koch G., Tiemann M. (2006) Synthesis of mesoporous magnesium oxide by CMK-3 carbon structure replication. *Chem. Mater.* **18**(17), 4151–4156.

Roggenbuck J., Tiemann M. (2005) Ordered mesoporous magnesium oxide with high thermal stability synthesized by exotemplating using CMK-3 carbon. *J. Am. Chem. Soc.* **127**(4), 1096–1097.

Rossinyol E., Prim A., Pellicer E., Arbiol J., Hernandez-Ramirez F., Peiry F., et al. (2007b) Synthesis and characterization of chromium-doped mesoporous tungsten oxide for gas sensing applications. *Adv. Funct. Mater.* **17**, 1801–1806.

Rossinyol E., Prim A., Pellicer E., Rodriguez J., Peiry F., Cornet A., et al. (2007a) Mesostructured pure and copper-catalyzed tungsten oxide for $NO_2$ detection. *Sens Actuators B* **126**, 18–23.

Rothschild A., Tuller H.L. (2006) Gas sensors: new materials and processing approaches. *J. Electroceram.* **17**, 1005–1012.

Sanchez C., Boissière C., Grosso D., Laberty C., Nicole L. (2008) Design, synthesis, and properties of inorganic and hybrid thin films having periodically organized nanoporosity. *Chem. Mater.* **20**, 682–737.

Shimizu Y., Hyodo T., Egashira M. (2004) Mesoporous semiconducting oxides for gas sensor application. *J. Eur. Ceram. Soc.* **24**, 1389–1398.

Shimizu Y., Jono A., Hyodo T., Egashira M. (2005) Preparation of large mesoporous $SnO_2$ powders for gas sensor application. *Sens. Actuators B* **108**, 56–61.

Stein A., Schroden R.C. (2001) Colloidal crystal templating of three dimensionally ordered macroporous solids: Materials for photonics and beyond. *Curr. Opin. Solid State Mater. Sci.* **5**, 553–564.

Tiemann M. (2007) Porous metal oxides as gas sensors. *Chem. A Eur. J.* **13**(30), 8376–8388.

Tiemann M. (2008) Repeated templating. *Chem. Mater.* **20**(3), 961–971.

Tomer V.K., Devi S., Malik R., Nehra S.P., Duhan S. (2016) Fast response with high performance humidity sensing of Ag-$SnO_2$/SBA-15 nanohybrid sensors. *Microporous Mesoporous Mater.* **219**, 240–248.

Tomer V.K., Duhan S. (2015a) In-situ synthesis of $SnO_2$/SBA-15 hybrid nanocomposite as highly efficient humidity sensor. *Sens. Actuators B* **212**, 517–525.

Tomer V.K., Duhan S. (2015b) Highly sensitive and stable relative humidity sensors based on $WO_3$ modified mesoporous silica. *Appl. Phys. Lett.* **106**, 063105.

Tomer V.K., Duhan S. (2016) A facile nanocasting synthesis of mesoporous Ag-doped $SnO_2$ nanostructures with enhanced humidity sensing performance. *Sens. Actuators B* **223**, 750–760.

Tomer V.K., Duhan S., Malik R., Nehra S.P., Devi S. (2015) A novel highly sensitive humidity sensor based on ZnO/SBA-15 hybrid nanocomposite. *J. Am. Ceram. Soc.* **98**, 3719–3725.

Tsai F.-S., Wang S.-J. (2014) Enhanced sensing performance of relative humidity sensors using laterally grown ZnO nanosheets. *Sens. Actuators B* **193**, 280–287.

Varghese O.K., Gong D., Paulose M., Ong K.G., Grimes C.A. (2003b) Hydrogen sensing using titania nanotubes. *Sens. Actuators B* **93**, 338–344.

Wadea T.L., Wegrowe J.-E. (2005) Template synthesis of nanomaterials. *Eur. Phys. J. Appl. Phys.* **29**, 3–22.

Wagner T., Haffer S., Weinberger C., Klaus D., Tiemann M. (2013) Mesoporous materials as gas sensors. *Chem. Soc. Rev.* **42**, 4036–4053.

Wagner T., Waitz T., Roggenbuck J., Froeba M., Kohl C.-D., Tiemann M. (2007) Ordered mesoporous ZnO for gas sensing. *Thin Solid Films* **515**, 8360–8363.

Wang H., Liang Q., Wang W., An Y., Li J., Guo L. (2011b) Preparation of flower-like $SnO_2$ nanostructures and their applications in gas-sensing and Lithium storage. *Cryst. Growth Des.* **11**, 2942–2947.

Wang X., Fan H., Ren P. (2013) Self-assemble flower-like $SnO_2$/Ag heterostructures: Correlation among composition, structure and photocatalytic activity. *Colloids Surf. A* **419**, 140–146.

Wang Y., Angelatos A.S., Caruso F. (2008) Template synthesis of nanostructured materials via layer-by-layer assembly. *Chem. Mater.* **20**, 848–858.

Wang Z., Shi L., Wu F., Yuan S., Zhao Y., Zhang M. (2011a) The sol–gel template synthesis of porous $TiO_2$ for a high-performance humidity sensor. *Nanotechnology* **22**, 275502.

Wang Z., Xiao Y., Cui X., Cheng P., Wang B., Gao Y., et al. (2014) Humidity-sensing properties of urchinlike CuO nanostructures modified by reduced graphene oxide. *ACS Appl. Mater. Interfaces* **6**, 3888–3895.

Weast C., Astle M.J., Beyer W.H. (eds.). (1986) *CRC Handbook of Chemistry and Physics: A Ready-Reference Book of Chemical and Physical Data*, 67th edn. CRC Press, Boca Raton, FL.

Wei J., Sun Z., Luo W., Li Y., Elzatahry A.A., Al-Enizi A.M., et al. (2017) New insight into the synthesis of large-pore ordered mesoporous materials. *J. Am. Chem. Soc.* **139**, 1706–1713.

Xia Y., Gates B., Yin Y., Lu Y. (2000) Monodispersed colloidal spheres: Old materials with new applications. *Adv. Mater.* **12** (10), 693–713.

Yang H.X., Qian J.F., Chen Z.X., Ai X.P., Cao Y.L. (2007) Multilayered nanocrystalline $SnO_2$ hollow microspheres synthesized by chemically induced self-assembly in the hydrothermal environment. *J. Phys. Chem.* **111**, 14067–14071.

Yuan L., Hyodo T., Shimizu Y., Egashira M. (2011) Preparation of mesoporous and/or macroporous $SnO_2$-based powders and their gas-sensing properties as thick film sensors. *Sensors* **11**(2), 1261–1276.

Yue W., Zhou W. (2008) Crystalline mesoporous metal oxide. *Progr. Nat. Sci.* **18**, 1329–1338.

Zhang D., Yin N., Xia B., Sun Y., Liao Y., He Z., Hao S. (2016c) Humidity-sensing properties of hierarchical ZnO/MWCNTs/ZnO nanocomposite film sensor based on electrostatic layer-by-layer self-assembly. *Mater. Sci.: Mater. Electron.* **27**, 2481–2487.

Zhang X., Chen H., Zhang H.Y. (2007) Layer-by-layer assembly: From conventional to unconventional methods. *Chem. Commun.* **2007**, 1395–1405.

Zhang X.H., Huang M.X., Qiao Y.J. (2013) Synthesis of $SnO_2$ single-layered hollow microspheres and flowerlike nanospheres through a facile template-free hydrothermal method. *Mater. Lett.* **95**, 67–69.

Zhang Y., Jiang B., Yuan M., Li P., Zheng X. (2016b) Humidity sensing and dielectric properties of mesoporous $Bi_{3.25}La_{0.75}Ti_3O_{12}$ nanorods. *Sens. Actuators B* **237**, 41–48.

Zhang Y., Yuan M., Jiang B., Li P., Zheng X. (2016a) Effect of mesoporous structure on $Bi_{3.25}La_{0.75}Ti_3O_{12}$ powder for humidity sensing properties. *Sens. Actuators B* **229**, 453–460.

Zhao J., Liu Y., Li X., Lu G., You L., Liang X., Liu F., Zhang T., Du Y. (2013) Highly sensitive humidity sensor based on high surface area mesoporous $LaFeO_3$ prepared by a nanocasting route. *Sens. Actuators B* **181**, 802–809.

Zhao Q., Gao Y., Bai X., Wu C., Xie Y. (2006) Facile synthesis of $SnO_2$ hollow nanospheres and applications in gas sensors and electrocatalysts. *Eur. J. Inorg. Chem.* **8**, 1643–1648.

Zhao Y., Ding Y., Chen X., Yang W. (2014) Highly sensitive and quickly responsive core-shell hierarchical porous $Li_4Ti_5O_{12}$-$TiO_2$ humidity sensors. *Sens. Actuators B* **203**, 122–129.

Zhen Y., Sun F.-H., Zhang M., Jia K., Li L., Xue Q. (2016) Ultrafast breathing humidity sensing properties of low-dimensional Fe-doped $SnO_2$ flower-like spheres. *RSC Adv.* **6**, 27008–27015.

# 24 Packaging, Air Cleaning, and Storage of Humidity Sensors

## 24.1 HUMIDITY-SENSOR PACKAGING: GENERAL APPROACH

Sensor packaging is a very important issue in the sensor development, since no sensor can be designed without considering the final package and its cost (Mastrangelo and Tang 1994; Hsu 2004; Dean et al. 2005; Frank 2013). The nature of any sensor requires the sensing element to be exposed to the sensing environment. This requirement differs from that of common microelectronic components, which are often completely encapsulated or sealed. Consequently, the material and the dimensions of the encapsulation are not as critical as those for a sensor. We need to note that the packaging of humidity sensors has two functions. First, the packaging must allow the water vapor to enter the sensor and interact with the sensing element. When packaging a sensor, any area, which should not be exposed, must be covered properly. Second, the packaging should protect the sensor from harmful environments and destruction (Mastrangelo and Tang 1994). Third, the package must prevent dirt and debris from clogging. In the case of MEMS humidity sensors, the task of packaging is even more complicated as the MEMS sensor may contain fragile structures that must be protected from mechanical contact that could otherwise damage the device. Undoubtedly, the packaging of each type of sensor is different and has different requirements. Furthermore, the packaging varies considerably from manufacturer to manufacturer, with no standard form factor. However, we believe that when choosing the packaging and ways to mount the humidity sensors, it is necessary to take into account the following issues:

1. The housing should not prevent the access of the water vapor to the sensor.
2. The packaging process needs to avoid the presence of gaseous components, which may poison the sensor active layer.
3. The materials utilized to package the sensor cannot emit gaseous components during sensor operation. As a rule, materials that outgas or emit odor have the potential to affect the operation of the sensor.
4. The materials utilized to package the sensors have to be inert in relation to the atmosphere in which the sensor will operate.
5. In the development of humidity sensors, different materials are commonly used. Thus, the choice of a proper packaging platform, such as silicon, silicon dioxide, alumina, glass-based dielectric materials, or other, for example, epoxy, should consider

the adhesive properties of these materials as well as the thermal match and mismatch of these materials in order to ensure the overall integrity of the sensor (Liu 1996).

When using one or another case for packaging the humidity sensor, it must be borne in mind that the packaging used should not affect the measurement accuracy and sensor response time. For example, a lid with many large holes means that the sensor is in direct contact with the atmosphere, but also that dust can readily fill the internal volume and impact measurements. At the same time, restricting the air flow to reduce the introduction of external particles may dramatically increase the time of response, especially if packaging materials that adsorb moisture are used. In particular, water-absorbing materials such as epoxies must be used to protect wire bonds from corrosion caused by liquid condensation. Packaging is essential to establishing the reliability of the sensor as well. Therefore, the reliability requirements must be taken into account in the design of the sensor package, especially for custom packages in specific applications. Testing a combination of sensors and circuits also requires combining the testing capabilities of both technologies.

Experience showed that for humidity-sensor packaging, a wide variety of packages can be used. Their appearance is determined by the scope of application of a specific sensor and the capabilities of the sensor manufacturer. For example, Figure 24.1 shows the packages that are used for the packaging of HMX2000 RH sensor, developed by Hygrometrix Inc. (2000) (www.hydrometrics.com). To manufacture the humidity-sensor packages, metals or polymers such as polypropylene and polybutylene terephthalate are commonly used.

One should note that the use of standard packages of commercial microelectronics (see Figure 24.2) and related packaging technologies is the most effective way for relative humidity (RH) sensor packaging, since in this case they are easily handled with existing printed circuit-board-assembly equipment (Hsu 2004). For example, cylindrical TO-5 and TO-8 metal-can semiconductor packages as well as ceramic and plastic surface-mount packages such as MO8A have been adapted for using as sensor packages. The caps of the packages have holes for contact with surrounding atmosphere. All sensor packages have a small size. In particular, the eight-pin MO8A surface-mount package measures <1 × 1 cm. Metal, ceramic, and plastic packages are fully compatible with high-volume assembly equipment for die attach and die wire bonding. In both packages, the sensor is glued to the socket, and the

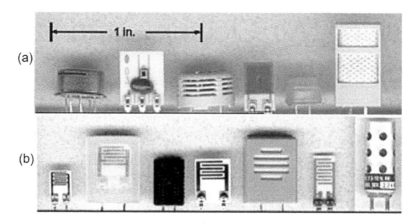

**FIGURE 24.1**   RH sensors can be produced in a wide range of specifications, sizes, and shapes including integrated monolithic electronics: (a) Capacitance humidity sensors; (b) Resistive humidity sensors. The sensors shown here are from various manufacturers. (Modified from https://www.fierceelectronics.com/components/choosing-a-humidity-sensor-a-review-three-technologies and https://www.ohmicinstruments.com/article-choosing-a-humidity-sensor.)

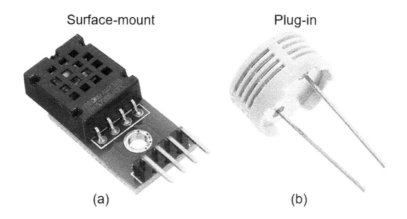

**FIGURE 24.2**   (a) AM2320 module digital temperature and humidity sensor in standard SOIC package with a plastic top with through-holes. The holes that allow the sensor to interact with the external environment are clearly visible in the lid of the device (Modified from https://www.adafruit.com/product/3721); and (b) Humidity sensor HS1101 fabricated using standard TO-like package for plug-in socket or board-soldered applications (Modified from http://shopsimplytronics.com/humidity-sensor-hs1101/.)

electrical contacts are bonded to the pins. Taking into account aggressive environments, the material used for this connection should be stable in the relevant atmosphere. Therefore, depending on the operating temperature and carrier gas, Pt, Au, or Al wire is commonly used for these purposes.

It is believed that the metal transistor outline (TO) header and the can are the most reliable packages for humidity sensors, while surface-mount packages are less expensive than the TO-like ones. The TO-like package allows the circuit designer to install a socket on the circuit board to permit plug-and-play sensor replacement. The leads extend through the base of the package through the glass mounting seals. The use of gold-plating material and hard-die attachment to the glass substrate of the sensor provides one of the most media-compatible sensors. However, in the case of using surface-mount ceramic packages, the process of encapsulating humidity sensors is much easier to automate (Kranz et al. 2001). In this case, it is possible to abandon the use of thermocompression bonding, which is typical technological operation during packaging in the TO-like package. The experience of the semiconductor industry has shown

that packaging methods that use molded plastic packages can also be adapted for mass production. At present, the lead-frame and molding techniques used in plastic packages provide the lowest cost of semiconductor packages. A single mold forms both the body and the back of the chip carrier. The patented unibody package provides lower cost, fewer process steps, and greater media compatibility compared to earlier versions made of separate body and metal back plate (Ristic 1994). The lead-frame assembly technology makes it possible to easily process several devices at the same time and automate such assembly operations, such as die-bond, wire-bond, and gel-filling operations. Automation allows one to implement rigorous process controls and provides high performance.

## 24.2   SEMICONDUCTOR PACKAGING APPLIED TO HUMIDITY SENSORS

Sensor packages fabricated using modern technologies usually have basic requirements that are similar to those of semiconductor devices. The variety of harsh sensor applications makes packaging more difficult than packaging for a semi-conductor device. However, the basic package operations take place in a similar manner (Frank 2013).

A completed sensor wafer has a final processing step that prepares it for packaging (Van Zant 1990). That step could include thinning the wafer and attaching a backside metal such as a gold-silicon eutectic. Sensors tested at the wafer level that do not meet minimum specifications are identified as rejected units by an ink dot. Sensors are then separated into individual dice from the wafer by sawing or scribe-and-break techniques. Good sensor dices are placed in carriers that allow one to automatically select and place machines for transfer-ring the dice from the carrier to the final package, where a die bond attaches the sensor firmly to the package. Wire bonds connect the electrical contacts on the die surface to the leads of the package, which allows the sensor to interface to exter-nal components. Then individual parts of the sensor structure, if necessary, can be sealed or covered with protective coat-ings. This is followed by lead plating and trim, followed by marking and final testing.

For sensor packaging, two approaches can be used: die-level and wafer-level packaging. The difference in these approaches is clear from the diagrams given in Figure 24.3. For humidity sensors, die-level packaging is commonly used. However, wafer-level packaging is also used, but mainly for enclosing MEMS sensors. In this case, much smaller pack-ages are possible.

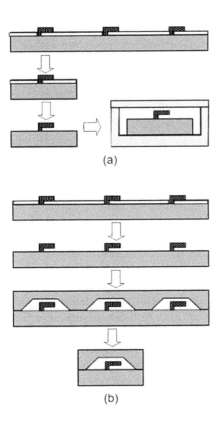

**FIGURE 24.3** Diagram illustration (a) die-level; and (b) wafer-level packaging.

### 24.2.1 Increased Pin Count

Additional circuitry, whether simple signal conversion or more complex approaches that include microcontroller unit (MCU) capabilities, affect sensor packaging (Frank and Staller 1990). Extended functionality through additional cir-cuitry, either on the same chip or on a separate chip included in the final package or sensor module, affects the pin count, normally increasing the number of package pins. Therefore, one of the more difficult problems that must be solved when additional electronic circuitry is integrated with or interfaced to the humidity sensor is the requirement for additional pin outs. However, one should note that integrated circuits have industry-accepted standard packages that allow for a large number of pin outs.

### 24.2.2 Ceramic Packaging and Ceramic Substrates

Ceramic packages use a lead-frame that is attached to the ceramic base through a glass layer. After die and wire bonding, a ceramic top is glass-sealed to the base. The same technique is used for other form-factor ceramic packages, including the ceramic flat pack. Ceramic packages are usually used for high-reliability applications and are much more expensive than other semiconductor packaging techniques. They are very use-ful in the development phase of a sensor because the silicon die does not have to be encapsulated. That allows various test points on the die to be easily probed and measured in pack-aged form (Ristic 1994). Hybrid packages, such as ceramic multichip packages, are also used to manufacture sensors.

The ceramic substrate, usually an aluminum oxide, pro-vides a firm mounting platform for the sensor die. Figure 24.4 shows potential packaging techniques that could possibly be used for sensors (Frank and Staller 1990). These approaches include (a) conventional chip-and-wire, (b) flip-chip, and (c) conventional tape-automated bonding (TAB). Chip-and-wire is the standard die-on-substrate packaging technique. Flip-chip packaging technology, commonly used in integrated circuits (ICs), is starting to receive attention in sensor design (Markus et al. 1994). In particular, this approach was used in the manu-facture of the HMX2000 RH sensor (see Figure 24.5). Flip-chip technology allows the electronics to be fabricated on one chip, which can be attached to the MEMS chip through either a flux-less process or flux-assisted solder reflow process. This is an approach to combining different silicon-processing tech-nologies instead of increasing masking layers and die size to achieve a required functionality of the sensor. TAB packaging

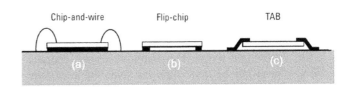

**FIGURE 24.4** Bare-die-mounting techniques that can be used for multichip-module fabrication: (a) chip-and-wire; (b) flip-chip; and (c) TAB.

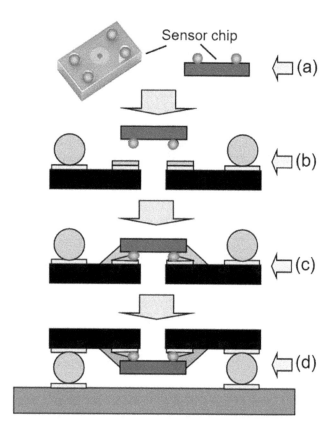

**FIGURE 24.5** Standard RH sensor-packaging process: (a) Using a gold ball bonder, place gold stud bumps on sensor die; (b) Flip and align die to the Sn/Ag coated carrier. Bonding is performed in a thermocompression bonder, but the bond is developed through the formation of a eutectic between gold ball and solder; (c) Apply underfill around sensor die and cure; (d) Align packaged device with site on printed circuit board and reflow along with other SMT microelectronics.

eliminates wiring bonding from the die to a lead by directly attaching a lead to the top of the die. Any of these methods is a potential candidate for packaging sensor chips. Stress isolation can be obtained by utilizing a compliant silicone for the die attachment. The ceramic substrate allows laser trimming of thick-film resistors deposited on the ceramic surface to provide the calibration for a signal-conditioned sensor.

As shown in Figure 24.5, one of the basic operations of sensor-chip mounting is the reflow. According to the requirements of many humidity-sensor developers, the reflow should follow JEDEC standards for lead-free solder and reflow with a peak temperature of less than 260°C. JEDEC was formerly known as the Joint Electron Device Engineering Council. According to JEDEC standards, "no clean" solder paste should be used. Further, it is recommended that the humidity sensors only be soldered with standard reflow (no hand soldering or hot air tools). The humidity-sensor opening should also generally be covered during soldering to prevent flux from getting on the sensor surface. The hydrophobic filter is compatible with standard reflow soldering. It provides lifetime protection against dust and liquids and should not be removed after soldering. If the hydrophobic filter cover is

not used, Kapton tape will serve the same purpose, although it has to be removed after soldering. Soldering iron touchup is possible if liquid flux is not required, and care should be taken to avoid excessive heating. The use of an ultrasonic bath with alcohol for cleaning after soldering is specifically not recommended. The humidity sensor opening must be kept clean and free of particulates during assembly; the preinstalled white filter cover will protect the opening from particulates. The sensor opening should not be into the contact with conformal coatings. The humidity sensors should not be exposed to volatile organic compounds or solvents. The use of water-soluble flux and water rinse after soldering is permissible if done with care. The use of deionized water is recommended. If the hydrophobic cover is used, a spray pressure of less than 40 PSI will prevent water entry into the cover. Without the cover, care should be taken to avoid particles from the water or from blow drying to contaminate the sensor area.

## 24.3 PACKAGING OF HIGH-FREQUENCY HUMIDITY SENSORS

Undoubtedly, ultra-high-frequency humidity sensors, such as surface acoustic wave (SAW) or film bulk acoustic resonator (FBAR) humidity sensors require their specific approaches to the packaging process (Giangu et al. 2014a, 2014b). SAW and FBAR sensor structures must be packaged both for physical protection and convenient assembly into electronic systems as regular super high frequency (SHF) components. The experiment showed that low-temperature co-fired ceramic (LTCC) technology offers an opportunity for packaging SAW and FBAR devices, providing excellent electrical performance, good reliability, and direct access to the surface-assembling techniques (Jantunen 2001; Fournier 2010). LTCC has become an attractive material system due to its moderate production costs, good heat conductivity (Muller et al. 2011), high-quality-factor passive devices, and low-loss transmission lines for microwave circuits (Wang et al. 2009). Moreover, LTCC packages use materials with reasonable dielectric constants, offering wider microwave transmission lines, thus leading to lower conductor loss than the circuits on Si or other materials (Golonoka 2006). The main advantage of LTCC ceramic tapes is their mechanical-processing flexibility, enabling the configuration of complex 3D structures with channels, membranes, cavities, holes, and vertical interconnections between layers using laser or mechanical drilling.

For example, Giangu et al. (2014a, 2014b) developed six-layer packages that allowed for a unified technological process for SAW and FBAR sensors, despite the fact that their configurations are quite different (Figure 24.6a,b). Each package consisted of an open cavity used for chip assembly as well as for direct sensing of environmental conditions. Vertical metal-filled via holes connect the inner signal and ground pads to their corresponding external pads.

The SAW LTCC package developed by Giangu et al. (2014a, 2014b) had inside the cavity two wide pads used to connect locally the SAW structure with gold wires; further, connections to the bottom pads were obtained through three

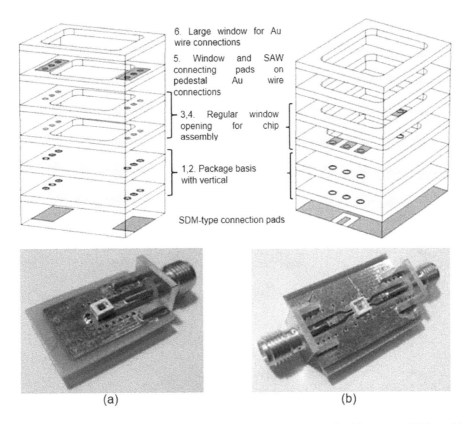

**FIGURE 24.6** LTCC packages and test circuits for (a) single-port SAW; and (b) two-port FBAR sensors. SDM—solder mask defined pads. (Reprinted from Giangu, I. et al., *Romanian J. Inform. Sci. Technol.*, 17, 320–339, 2014b. Published by Romanian Academy as open access.)

metal-plated via holes for each pad, in order to decrease their equivalent inductance. The bottom part of the LTCC package for FBAR structure had a coplanar waveguide (CPW) configuration, taking into account the general CPW design rules, regarding line widths and corresponding gaps. The LTCC packages were manufactured from DuPont 9 K7 ceramic material (7.1 relative permittivity and 0.001 loss tangent at 10 GHz). Horizontal and vertical shrinkage was specified at 9.1% and 11.8%, respectively. The dimensions of both LTCC packages were decided after 3D electromagnetic analysis and optimization process performed with CST Microwave Studio, in order to reduce the effect of parasitic elements of inductive and capacitive type (Buiculescu et al. 2014; Giangu et al. 2014b). The dimension of the SAW chips meant to be packaged are 920 × 850 × 500 μm (L × W × H); hence, the external size of the corresponding LTCC package is 3.1 × 2.6 × 1.5 mm (L × W × H). For the FBAR chip, the dimensions are 2000 × 1800 × 100 μm (L × W × H); thus, the external LTCC package size is 3.4 × 3.3 × 1.5 mm (L × W × H). The packaged structure of the FBAR sensor is shown in Figure 24.6b.

## 24.4 FEATURES OF THE PACKAGING OF HUMIDITY SENSORS FABRICATED ON FLEXIBLE SUBSTRATES

De Rooij and co-workers (Courbat et al. 2010; Molina-Lopez et al. 2012, 2013) have shown that humidity sensors made on flexible substrates also require their own specific approach to packaging. Due to the nature of the materials involved, standard techniques used for packaging such as eutectic (Wolffenbuttel and Wise 1994), anodic (Wei et al. 2003), or fusion bonding (Barth 1990) have to be discarded, since they are based on nonflexible substrates and require high temperatures. Low-temperature bonding processes such as solder (Yang et al. 2005), ultrasonic (Kim et al. 2009), or adhesive (Niklaus et al. 2006) bonding were also developed. However, the two previous methods usually require a metallic layer to properly solder the two parts involved, while for the latter, the polymeric materials are often spin coated to reach a uniform thickness (Oberhammer and Stemme 2005). These techniques might not be compatible with preexisting structures, such as functionalized surfaces used in chemical sensors.

De Rooij and co-workers (Raible et al. 2006; Courbat et al. 2010; Molina-Lopez et al. 2013) developed two approaches to packaging of the humidity sensors fabricated on flexible substrates. In Courbat et al. (2010), they proposed a packaging method based on prepatterned structures, made from dry photoresist (PerMX3000 from DuPont) films and their lamination at low temperature without the need of an intermediate binding layer. This process is shown in Figure 24.7. The dry film PerMX3000 from DuPont is a negative-tone photosensitive epoxy film. Parameters of PerMX3000 are listed in Table 24.1. Figure 24.7A shows the process of rims forming, and Figure 24.7B illustrates the process of the encapsulation of the sensors. Once the rims were patterned and laid on their polyethylene (PE) foil, they were

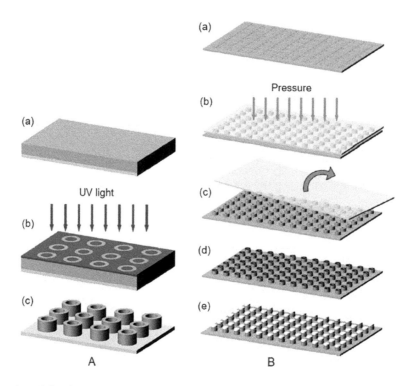

**FIGURE 24.7** (A) Processing of the rims made of a dry photoresist: (a) dry photoresist on its handling PE film; (b) UV exposure of the film through a Cr mask by standard photolithography; (c) rims on the handling film after development. (B) Foil-level packaging of the MOX sensors: (a) polyimide foil with gas-sensing structures; (b) alignment and lamination of the pre-patterned rims; (c) removal of the handling film; (d) glue dispensing on the rims; and (e) gas-permeable membrane deposition. (Reprinted with permission from Courbat, J. et al., Foil level packaging of a chemical gas sensor. *J. Micromech. Microeng.* 20, 055026, 2010. Copyright 2010, Institute of Physics.)

### TABLE 24.1
### Properties of PerMX3000

| Properties | PerMX3000 |
| --- | --- |
| Aspect ratio | 4:1 |
| Glass transition temperature (°C) | 220 |
| 5% weight loss temperature (°C) | 346 |
| Coefficient of thermal expansion (ppm/°C) | 72 |
| Young's modulus (GPa) | 2.0 |
| Tensile strength (MPa) | 50 |
| Elongation (%) | 11 |
| Dielectric constant (1 GHz, 50% RH) | 3.2 |
| Dielectric strength (V/μm) | 220 |
| Volume resistivity (Ω cm) | $3-7 \times 10^{16}$ |
| Shelf life (years) | 1 |

*Source:* Data extracted from Courbat, J. et al. *J. Micromech. Microeng.*, 20, 055026, 2010.

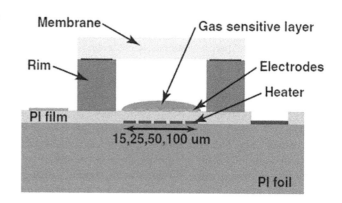

**FIGURE 24.8** Cross-sectional schematic of the packaged gas sensor. (Reprinted with permission from Courbat, J. et al., Foil level packaging of a chemical gas sensor. *J. Micromech. Microeng.* 20, 055026, 2010. Copyright 2010, Institute of Physics.)

assembled with the polyimide sheet embedding the MOX sensors [Figure 24.7B(a)]. They were then aligned and laminated at 75°C in a press for 3 min [Figure 24.7B(b)]. Prior to the lamination, an optional activation of the surface of the polyimide layer can be performed with an $O_2$ plasma to enhance the adhesion. The final step was the dispensing of an epoxy glue (Epotek H70E) on the rims [Figure 24.7B(d)] before the deposition of a Teflon-based gas-permeable membrane [Figure 24.7B(e)], which was manually drop coated with a needle. A cross-sectional view of the encapsulated sensor is schematized in Figure 24.8. By being compatible with roll-to-roll processing, this technique preserves the advantage of the cost reduction conferred by the fabrication of devices at the foil level. These dry films present a combination of several advantages over conventional liquid photoresists: excellent adhesion to a wide range of substrates (silicon, glass, PET, polyimide, etc.), unlimited shapes, good conformability, less solvent involved, uniform thickness, no edge bead, low price, and short processing time.

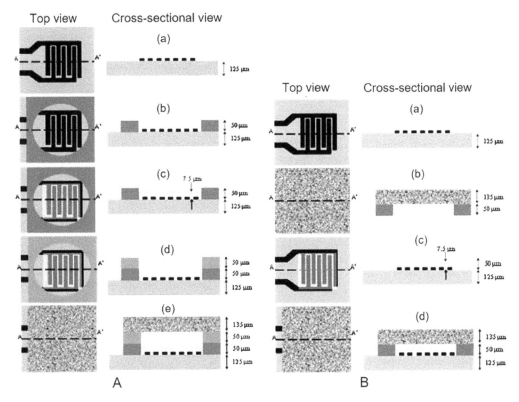

**FIGURE 24.9** (A) Top and cross-sectional view of the process flow at foil level for the first-generation encapsulation: (a) inkjet printing of sensor; (b) PerMX3050 lamination; (c) CAB printing; (d) ARClear bonding; and (e) alignment and bonding of the Versapor film. (B) Top and cross-sectional view of the process flow at foil level for the second optimized generation: (a) inkjet printing of the sensor; (b) PerMX3050 lamination onto the Versapor film; (c) CAB printing; and (d) alignment and bonding of the Versapor film onto the PET-sensor substrate. ARClear—dry adhesive 50-μm-thick film; CAB—cellulose acetate butyrate; PerMX3050 is used as a protection and adhesive layer; PET—polyethylene terephthalate; Versapor film—gas-permeable porous 135-μm-thick acrylic copolymer film. (Reprinted with permission from Molina-Lopez, F. et al., Large-area compatible fabrication and encapsulation of inkjet-printed humidity sensors on flexible foils with integrated thermal compensation. *J. Micromech. Microeng.*, 23, 025012, 2013. Copyright 2013, Institute of Physics.)

In Molina-Lopez et al. (2013), the previously described process was improved. An optimized encapsulation method was based on lamination of plasma-treated PET foils at low temperature (85°C), and in this process an addition adhesive layer was not required. The process flow of the encapsulations of first- and second generations are depicted in Figure 24.9. It is important that developed processes of packaging can also be applied to sensors fabricated on solid substrates such as silicon (Raible et al. 2006).

## 24.5 PROTECTIVE COVER AND FILTERS

As follows from the previous section, external effects and contaminants, such as dust, corrosive gases, oil vapors, and liquids, can affect sensor performance. Therefore, humidity sensors require protection over the entire life of the product. As a rule, such functions are performed by some type of covers or filters. The experiment showed that careful sensor cover design can lessen this issue and even solve the problem of protecting the sensor during manufacture. To perform protective functions, the pore size in such covers and filters should be sufficiently small to trap harmful aerosol particles (in a maritime environment sea-salt particles may be present in significant quantity

down to a diameter of 0.1 μm), and the porosity should be sufficient to allow an adequate diffusion rate.

Usually to protect against external influences and contaminants, the top of the package or the holes in the lid are made of wire mesh for air flow to the sensor. In addition to the wire mesh, a water-vapor-permeable filter is usually used to present contamination of the sensor. Although the diffusion of water vapor through some materials, such as some cellulose products, is theoretically more rapid than for still air, porous-hydrophobic membranes achieve better diffusion rates in practice. It was established that a material like Gore-Tex™ can be used as such a membrane. Gore-Tex is a waterproof, breathable fabric membrane and registered trademark of W. L. Gore and Associates. Invented in 1969, Gore-Tex is able to repel liquid water while allowing water vapor to pass through and is designed to be a lightweight, waterproof fabric for all-weather use. It is composed of stretched polytetrafluoroethylene (PTFE), which is more commonly known as the generic trademark Teflon. Thus, the use of PTFE layer allows the water vapor through to the sensor, while preventing degradation performance due to appearance of particulates or oil (Dean et al. 2005). In particular, such an approach was used by Silicon Labs (www.silabs.com/humidity-sensor) for protection

(a)                         (b)                         (c)

**FIGURE 24.10** (a) Optional cover offers lifetime protection for the Si7007 relative humidity sensor (Modified from https://www.silabs.com/documents/public/data-sheets/Si7007-A20.pdf); (b) SHT3x humidity sensor with protective cover. The membrane is an PTFE film that protects the sensor opening from water and dust (Modified from https://www.sensirion.com/en/environmental-sensors/humidity-sensors/digital-humidity-sensors-for-various-applications/); (c) SHT1x humidity sensors developed by Sensirion Com. with cover. The cavity inside the filter cap is designed so that the volume between membrane and sensor is negligible; hence the impact on response time for humidity measurements is reduced to a minimum. The cap is available in white and is made of a single piece of polypropylene (PP) and a filter membrane welded to the single piece. The filter cap is designed to be mounted after soldering by clipping the two pins on to the PCB and welding them from the back. Together with the sensor, it provides a compact entity that can serve as an adaptor to the device housing. With the addition of adhesive for sealing to the PCB and use of an o-ring, it provides a waterproof mounting solution (Modified from https://www.sensirion.com/en/environmental-sensors/humidity-sensors/filterkappen-sf1-1/).

of the Si7005 Relative Humidity Sensor (see Figure 24.10a), by Hygrometrix Inc. (2000) (www.hygrometrix.net/) for protection of the HMX2000 RH sensor (Figure 24.10b), and by Sensirion (www.sensirion.com) for protection of humidity sensors in the SHTx series (Figure 24.10c). In last case, the PTFE-based membrane cover was attached to the sensor package by a double-faced adhesive tape. The developers claim that the cover has a minimal effect on the response time. The cavity inside the filter cap is designed so that the volume between membrane and sensor is negligible; hence, the impact on response time for humidity measurements is reduced to a minimum. The sensor can also be protected by sealing it in an envelope of a thin (less than 0.5 mm) silicone membrane. The envelope should not touch the sensor surface. As it was shown in the previous section, Versapor® film can also be used as a protective cover for humidity sensors (Courbat et al. 2010; Molina-Lopez et al. 2015). The Versapor film is an acrylic copolymer membrane. It is Repel™ treated for superior oleophobicity and hydrophobicity. Membranes have excellent air permeability, broad chemical compatibility, and good compatible with a variety of sealing methods.

The hydrophobic filter also prevents liquid water from penetrating, so condensation on the outside of the part will generally not result in condensation on the active layer in humidity sensor. However, if condensation forms on the top of the filter, readings will be high until it evaporates. Therefore, integrating the sensor with a heater can significantly reduce the chance of condensation and will also contribute to the rapid evaporation of condensation.

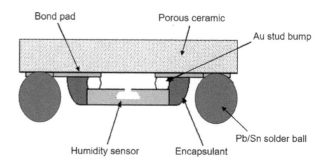

**FIGURE 24.11** Illustration of the porous ceramic substrate packaging concept. (Modified from Dean, R.N. et al., *IEEE Trans. Comp. Packag. Manufact. Technol.*, 1, 428–435, 2011.)

Of course, the presence of several filter layers increases the volume of the sensor and complicates the packaging technology. This approach requires the filter structure to be separately fabricated and then attached to the sensor die or package. Additionally, the utilized filter materials must be compatible with the subsequent lid-attachment processes. The size of the package tends to grow because room must be left for a sealing ring for the lid. Overall, the added processing leads to higher packaging costs and increased package size. One way to reduce the package volume and simplify the technology is to make the lid out of a porous material such as porous ceramic (Dean et al. 2011). An illustration of the packaging concept utilized for the porous ceramic package concept is presented in Figure 24.11. One should note that porous ceramic, used in humidity sensors, differs from traditional thick-film ceramic substrates in that it is fabricated to leave voids or pores in the substrate material through which gas can travel. Porous ceramic materials with open porosity have been successfully generated using a fairly wide variety of fabrication techniques, including casting, hot-press sintering, direct incorporation of porous components, coagulation from colloidal suspensions, and sol-gel processes (Galassi 2006). In this case, the lid, fabricated from porous ceramics, provides both mechanical protection and vapor filtration. It allows moist air to flow through the porous substrate to the moisture-sensitive portions of the sensor die while protecting the sensor from both mechanical contact (which could damage the fragile micromachined structures) and dust and other particulate contaminants that could settle onto the sensor microstructures and degrade performance. Dean et al. (2011) suggests using for this purpose CoorsTek ADS-96R. CoorsTek ADS-96R has a reasonably good balance between pore size, porosity, and strength. This material has a pore size ranging from 0.25 to 0.70 μm. That preclude the surfaces of the humidity sensor from contaminating by dust particles larger than 1 μm. The Coors material also has a rated porosity of 24.2%. This permitted air flow through the ceramic without severely degrading the response time of the sensor. One problem with this approach is that packages are usually marketed after packaging on the lid, and it is difficult to inexpensively print information on a porous lid (Hsu 2004). Information such as the batch number or individual calibration parameters are of particular interest for sensors.

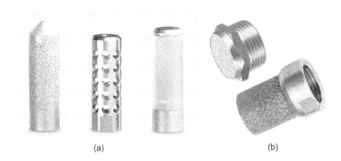

(a)                                               (b)

**FIGURE 24.12** (a) Protective filters for Vaisala's humidity instruments (Modified from https://www.vaisala.com/sites/default/files/documents/CEN-TIA-G-Param-How-to-Choose-Application-note-B211203EN-A_0.pdf); (b) stainless-steel replacement humidity filters designed by BAPI for duct and outside air humidity sensors (Modified from https://www.bapihvac.com/product/replacement-humidity-filters-for-duct-and-outside-air-humidity-sensors/).

Sintered metal filters may be also used (Figure 24.12), but they should be heated to avoid problems with condensation within the material. This is not normally appropriate for a relative humidity sensor but is quite acceptable for a dewpoint sensor. Sintered-metal filters are robust and well suited for aspirated applications, which allows using a filter, having a large surface area and, consequently, an acceptably small pressure differential.

However, if the sensor does not have artificial aspiration, the use of a filter reduces the response of the sensor by preventing the bulk movement of air and by relying upon molecular diffusion through the filter material (WMO 2008; Zhao et al. 2011). For example, semi-permeable Gore-Tex or silicone membranes can lengthen the response time of the thin-film humidity sensors from about 30–60 s to about 10–20 minutes. The size of the filter also affects the overall diffusion rate. Diffusion will be enhanced by aspiration, but it must be remembered that this technique relies upon maintaining low air pressure on the sensing side of the filter, and that this can have a significant effect on the measurement. If diffusion is not enhanced by artificial aspiration, the relation of the surface area of the filter to the volume of the air being sampled by the sensor must be considered as well. In the case of a typical sorption sensor composed of a flat substrate, a flat membrane positioned close to the sensor surface will provide the optimum configuration. In the case of a cylindrical-sensing surface, a cylindrical filter is appropriate.

It should also be borne in mind that the PTFE-based cover and porous ceramic, which block liquids and particles, and thus allow sensor use under harsh environmental conditions, where spray water and high exposure to dust is challenging for accurate sensor performance, do not in general prevent contamination from volatile chemicals. They may be effective if the fume concentration is not high. There have been numerous attempts to develop filters that eliminate the effect of organic solvent vapors on sensor readings. But most of these attempts were fruitless. Although, Liu et al. (2014) claim that they managed to solve this problem by using the graphene oxide (GO)

membrane. They established that the introduction of the GO filter membrane into the sensor structure greatly reduced the cross-sensitivity of the humidity sensor to other gases such as ethanol and acetone. The measuring errors of carbon nanotube (CNT)-based quartz crystal microbalance humidity sensors in the presence of the interference gas (1000 ppm ethanol or 1000 ppm acetone) at the humidity point of 54% RH were only about 3% and 1.7% RH, respectively. Liu et al. (2014) believe that this effect takes place since the GO membrane is selectively permeable to water molecules.

## 24.6 STORAGE OF HUMIDITY SENSORS

It is of great importance to understand that a humidity sensor is not a standard electronic component and needs to be handled with care. Chemical vapors at high concentration in combination with long exposure times may offset the sensor reading. Therefore, in manufacturing, transport and storage all sensors should be prevented of high concentration of chemical solvents and long exposure times. Bleach, hydrogen peroxide, ammonia, and other chemicals in the surrounding atmosphere can also affect or damage the sensor. Out-gassing of glues, adhesive tapes, and stickers or out-gassing packaging material such as bubble foils and foams must be avoided. The manufacturing area should be well ventilated.

One should note that every manufacturer sets its own requirements for the storage of sensors based on the materials used and the principles of measurement. For example, according to Silicon Labs (www.silabs.com), the Si70xx humidity sensors should be shipped in sealed anti-static bags. The sensors may be stored in a humidity- and temperature-controlled (RH: 20%–60%, Temperature: 10°C–35°C) environment for up to 1 year after being removed from the bag prior to assembly. Humidity sensors should not be stored in polyethylene bags (typically blue, yellow, or pink) because these emit gases that can affect the sensor. Metallic, anti-static, sealable, moisture-barrier bags are recommended for storage. Sealants or tapes to seal inside the packaging should not be used as well.

## REFERENCES

Barth P.W. (1990) Silicon fusion bonding for fabrication of sensors, actuators and microstructures. *Sens. Actuators* A **21–23**, 919–926.

Buiculescu V., Bechtold F., Giangu I., Müller A. (2014) LTCC packages optimized for use with SAW and FBAR sensors in environmental parameters monitoring. In: *Proceedings of Smart System Integration—SSI*, March 26–27, Vienna, Austria, pp. 191–197.

Courbat J., Briand D., de Rooij N.F. (2010) Foil level packaging of a chemical gas sensor. *J. Micromech. Microeng.* **20**, 055026.

Dean R., Pack J., Sanders N., Reiner P. (2005) Micromachined LCP for packaging MEMS sensors. In: *Proceedings of Industrial Electronics Conference, IECON*, 6–10 November, Raleigh, NC, pp. 2363–2367.

Dean R.N., Surgnier S., Pack J., Sanders N., Reiner P., Long C.W., et al. (2011) Porous ceramic packaging for a MEMS humidity sensor requiring environmental access. *IEEE Trans. Comp. Packag. Manufact. Technol.* **1**(3), 428–435.

Fournier Y. (2010) 3D Structuration Techniques of LTCC for Microsystems Applications, PhD Thesis, École Polytechnique Fédérale De Lausanne.

Frank R. (2013) *Understanding Smart Sensors.* Artech House, Boston, MA.

Frank R., Staller J. (1990) The Merging of micromachining and microelectronics. In: *Proceedings of 3rd International Forum on ASIC and Transducer Technology,* May 20–23, Banff, Alberta, Canada, pp. 53–60.

Galassi C. (2006) Processing of porous ceramics: Piezoelectric materials. *J. Eur. Ceram. Soc.* **26**(14), 2951–2958.

Giangu I., Buiculescu V., Konstantinidis G., Szaciowski K., Stefanescu A., Bechtold F. et al. (2014a) Acoustic wave sensing devices and their LTCC packaging. In: *Proceedings of the International Semiconductor Conference, CAS,* Sinae, Romania, pp. 147–150.

Giangu I., Buiculescu V., Bechtold F., Konstantinidis G., Szaciowski K., Stefanescu A. et al. (2014b) Development and experimental results on LTCC packages for acoustic wave sensing devices. *Romanian J. Inform. Sci. Technol.* **17**(4), 320–339.

Golonoka L.J. (2006) Technology and applications of low temperature Co-fired ceramic (LTCC) based sensors and microsystems. *Bull. Polish Acad. Sci.: Techn. Sci.* **54**(2), 221–231.

Hsu T.-R. (ed.) (2004) *MEMS Packaging.* The Institution of Electrical Engineers, London, UK, pp. 92–97.

Hygrometrix Inc. (2000) *DSHMX2000, HMX2000 Relative Humidity/Moisture Sensor,* Product Data Sheet, Hygrometrix Inc.

Jantunen H. (2001) A novel low temperature co-firing ceramic (LTCC) material for telecommunication devices, PhD Thesis, University of Oulu, Department of Electrical Engineering and Infotech Oulu.

Kim J., Jeong B., Chiao M., Lin L. (2009) Ultrasonic bonding for MEMS sealing and packaging. *IEEE Trans. Adv. Packag.* **32**, 461–467.

Kranz M., Legowik R., Bowers W., Dean R., Garrison H., Shultz N. (2001) Micro-packaging of COTS MEMS for remote monitoring systems. In: *Proceedings of 3rd Advanced Technology Workshop on Packaging of MEMS and Related Micro Integrated Nano Systems,* IMAPS, Scotts Valley, CA, November 8–10.

Liu B., Chen X.-D., Li N., Li X.-Y., Yao Y. (2014) Cross-sensitivity reduction of QCM humidity sensor using graphene oxide membrane as filter layer. *Electron. Lett.* **50**(20), 1447–1449.

Liu C.-C. (1996) Electrochemical sensors: microfabrication techniques. In: Taylor R.F., Schultz J.S. (eds.) *Handbook of Chemical and Biological Sensors.* IOP, Bristol, UK, pp. 419–434.

Markus K.W., Dhuler V., Cowen A. (1994) Smart MEMS: Flip chip integration of MEMS and electronics. In: *Proceedings of Sensors Expo,* September 20–22, Cleveland, Ohio, pp. 559–564.

Mastrangelo C., Tang W. (1994) Semiconductor sensor technologies. In: Sze S. (ed.) *Semiconductor Sensors.* Wiley, New York, pp. 17–95.

Molina-Lopez F., Briand D., de Rooij N.F. (2012) All additive inkjet-printed humidity sensors on plastic substrate. *Sens. Actuators B* **166–167**, 212–222.

Molina-Lopez F., Vasquez Quintero A., Mattana G., Briand D., de Rooij N.F. (2013) Large-area compatible fabrication and encapsulation of inkjet-printed humidity sensors on flexible foils with integrated thermal compensation. *J. Micromech. Microeng.* **23**, 025012.

Muller R., Wollenschlager F., Schulza A., Elkhouly M., Trautwein U., Hein M.A., et al. (2011) 60 GHz ultrawideband front-ends with gain control, phase shifter, and wave guide transition in LTCC technology. In: *Proceedings of IEEE European Conference on Antennas and Propagation (EUCAP),* April 11–15, Rome, Italy, pp. 3255–3259.

Niklaus F., Stemme G., Lu J.-Q., Gutmann R.J. (2006) Adhesive wafer bonding. *J. Appl. Phys.* **99**, 031101-1–28.

Oberhammer J., Stemme G. (2005) BCB contact printing for patterned adhesive full-wafer bonded 0-level packages. *J. Microelectromech. Syst.* **14**, 419–425.

Raible S., Briand D., Kappler J., de Rooij N.F. (2006) Wafer level packaging of micromachined gas sensors. *IEEE Sensors J.* **6**, 1232–1235.

Ristic L.J. (1994) *Sensor Technology and Devices.* Artech House, Norwood, MA.

Van Zant P. (1990) *Microchip Fabrication.* McGraw-Hill, New York.

Wang Z., Li P., Xu R., Lin W. (2009) A compact X-band receiver front-end module based on low temperature co-fired ceramic technology. *Prog. Electromagn. Res.* **92**, 167–180.

Wei J., Xie H., Nai M.L., Wong C.K., Lee L.C. (2003) Low temperature wafer anodic bonding. *J. Micromech. Microeng.* **13**, 217–222.

WMO-No.8 (2008) *Guide to Meteorological Instruments and Methods of Observation,* World Meteorological Organization, Geneva, Switzerland.

Wolffenbuttel R.F., Wise K.D. (1994) Low-temperature silicon wafer-to-wafer bonding using gold at eutectic temperature. *Sens. Actuators A* **43**, 223–229.

Yang H.-A., Wu M., Fang W. (2005) Localized induction heating solder bonding for wafer level MEMS packaging. *J. Micromech. Microeng.* **15**, 394–399.

Zhao C.L., Qin M., Huang Q.A. (2011) A fully packaged CMOS interdigital capacitive humidity sensor with polysilicon heaters. *IEEE Sensors J.* **11**(11), 2986–2992.

# Section III

*Calibration and Market of Humidity Sensors*

# 25 Humidity-Sensor Selection and Operation Guide

## 25.1 WHAT IS AN IDEAL HUMIDITY SENSOR?

The main purpose of any humidity sensor is to provide a reliable real-time information about the humidity of the surrounding environment (Wexler 1965; WMO 2011). As has been shown in this book, various devices based on different materials and operating on diverse principles can be used for this task (see Table 25.1). Ideally, a device designed for humidity sensing should operate continuously and reversibly, without perturbing a sample. An ideal humidity sensor should be small, low-cost, portable, and fool-proof, working with perfect and instantaneous selectivity. Humidity sensors should be able to detect and recognize a water vapor in a mixture of a large number of different gases. The device should produce a measurable signal output at any required concentration of water vapor in the air and should have low energy consumption. Simple fabrication, fast response, attainment of minimal measuring uncertainty, relative temperature insensitivity, high resistance to contamination and poisoning, and low noise are also useful requirements to the humidity-sensor designed for the gas-sensor market. In addition, minimal specialized training should be required for personnel who must manage the operation and maintenance of the sensor. Humidity sensors should provide repetitive measurements for a long time in multiple or remote locations, supporting environmental monitoring. The ideal sensor should function continuously and reliably without requiring recalibration for a long time. Such a sensor should be able to function in an industrial environment, and it should be replaceable or renewable at a reasonable cost. Also, the sensor should be easy to place into a multipoint system, and either a controller or a controlled distribution system must be able to manage its functioning simply. The needed degree of quantitative reliability, such as precision and accuracy, should be provided as well.

However, no such ideal humidity sensor can be found in the sensor market today, in spite of great advances achieved over the past decades. One will never find a humidity sensor that satisfies all these requirements; there will always be some disadvantages. This situation is a result of the complexity of humidity-sensor operation and application. There are a great number of requirements due to the necessity to control and characterize different processes and environments. For example, humidity measurements at the Earth's surface, required for meteorological analysis and forecasting, for climate studies, and for many special applications in hydrology, agriculture, aeronautical services, and environmental studies require sensors with parameters listed in Table 25.1.

## 25.2 SOME PRACTICAL ADVICE ON CHOOSING AND USING A HUMIDITY SENSOR

As was shown in this series, the detection of air humidity has always been a complex subject, making the choice of an appropriate gas-monitoring instrument a difficult task. Therefore, we have compiled this guide to provide information about several aspects of sensor selection and using. In preparing this guide, we used the information provided in numerous references (Wexler 1965; Taylor 1987; Woodfin 1994; Scott et al. 1996; BS EN 1999; COGDEM, 1999; Chou 2000; White 2000; Brett 2001; Ho et al. 2001; Walsh et al. 2001; L138 2003; Brown 2006; WMO 2011; Bell 2012). This should help in deciding which type of sensor is best suited for a particular application and how this sensor can be used with maximum efficiency. Some criteria to consider when choosing a sensor are shown in Figure 25.1.

### 25.2.1 SENSOR SELECTION

As was shown in this book, we do not have humidity sensors that are 100% specific for water vapor. Each sensor has certain capabilities and limitations, and therefore the suitability of this sensor depends largely on the application in which it should be used. This means that technical characteristics and the chosen sensor's performance should correspond to exploitation conditions. For example, some sensors may have physical constraints that limit their location or mounting. In particular, gas sensors for detecting air pollutants must be able to operate stably under deleterious conditions, including chemical and/or thermal attack. Some humidity sensors have been designed for intrinsic safety, that is, for use in potentially explosive atmospheres. Therefore, it is necessary to be sure that a particular device meets the relevant intrinsic safety requirements.

The area of monitoring is also important for sensor selection. The devices operate over different monitoring distances. Conductometric, capacitive, quartz crystal microbalance, surface acoustic wave, or electrochemical sensors are single-point devices. These detectors operate best at a monitoring radius of 0.5–2 m. As a result, they are installed near the place where the air humidity should be monitored. Open-path IR-based humidity detectors, as opposed to point detectors, protect larger areas. Typically, such devices have sampling paths from 10 to 100 m and even more.

The composition of the surrounding atmosphere is also a very important factor for choosing the optimal humidity

**TABLE 25.1**

**Summary of Performance Requirements for Surface Humidity**

| Requirement | Wet-Bulb Temperature | Relative Humidity | Dew-Point Temperature |
|---|---|---|---|
| Range | −10°C–35°C | 0 5%–100% | At least 50 K in the range −60°C–35°C |
| Target accuracy[a] (uncertainty) | 0.1 K high RH | 12% high RH | 0.1 K high RH |
|  | 0.2 K mid RH | 5% mid RH | 0.5 K mid RH |
| Achievable observing uncertainty[b] | 0.2 K | 3%c–5%c | 0.5 K[c] |
| Reporting code resolution | 0.1 K | 1% | 0.1 K |
| Sensor time-constant[d] | 20 s | 40 s | 20 s |
| Output averaging time[e] | 60 s | 60 s | 60 s |

*Source:* Data extracted from WMO, *Technical Regulations.* Vol. 1—General Meteorological Standards and Recommended Practices, (WMO-No. 49), Geneva, Switzerland, 2011.

[a] Accuracy is the given uncertainty stated as two standard deviations.

[b] At mid-range, relative humidity for well-designed and well-operated instruments; difficult to achieve in practice.

[c] If measured directly.

[d] For climatological use, a time-constant of 60 s is required (for 63% of a step change).

[e] For climatological use, an averaging time of 3 min is required.

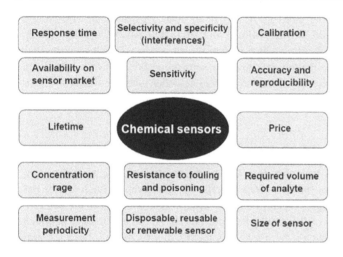

**FIGURE 25.1** Important aspects in choosing chemical sensors. (Reprinted with permission from Korotcenkov, G. and Cho, B.K., Chemical sensors selection and operation guide. In: *Chemical Sensors: Comprehensive Sensor Technologies*, Vol. 6, *Chemical Sensors Applications*, Korotcenkov G. (ed.), Momentum Press, New York, 281–348, 2011.)

sensor. Sensor selectivity (as opposed to cross-sensitivity) relates to the ability of the sensor to differentiate among various gases. The main cause of sensor failure is the presence of background gases, which the device manufacturer did not take into account. Sensor manufacturers can provide data for each sensor, indicating the concentration of interfering gas that is required to create a sensor response equivalent to the response obtained from a given concentration of water vapor. This is an important criterion for a sensor that is cross-sensitive to other gases that may be found in the same area; depending on the sensor type chosen, one may encounter varying degrees of interference from other gases that are present in the environment. The presence of background gases, which cannot be determined, should toughen our requirements to sensor selectivity.

Obviously, the sensor should have sensitivity sufficient for detecting water vapors at the required concentration. Response time is another important sensor parameter. The required response time depends on the location, the purpose of the system, and the speed of development of the expected problem. However, in the case of humidity sensors, we do not have too-strict requirements to the rate of response, since humidity measurements are usually not related to safety systems. Water vapor is not toxic. The lowest detection limit is also an important parameter of humidity sensors that must be taken into account during sensor selection. Several manufacturers provide a published range of detection, but they do not provide this minimum detection limit, due to their inability to detect low concentration levels. But one should recognize that the ability to detect low-level concentrations is required for humidity sensors only in specific applications. Lifetime, the time between calibrations, and watt consumption are other parameters that one needs to know before purchasing humidity sensors. These parameters will determine the costs associated with the operation of these devices.

Conditions of exploitation are also important for sensor selection. Some sensors are susceptible to interference from radio-frequency and magnetic fields. Sensors and cables, connecting sensors to controllers or other monitoring equipment, handle signals of very low magnitude. Especially at the low end of the sensor's sensitivity scale, the signal-to-noise ratio is very low and electromagnetic interference or radio-frequency interference can cause false alarms if sensors, controllers, and cables are not shielded. The total cost of ownership is an important criterion in selecting a sensor system or technology as well. In addition to first purchase cost, it is necessary to consider carefully the cost of expendable parts, maintenance replacements, integration with other systems, expandability, and downtime due to maintenance and false alarms.

Of course, it is difficult to mention every possible consideration, but a carefully evaluated and studied application can save both time and money. Therefore, when choosing a sensor, always follow the criteria below, which will provide you with a guideline for choosing the right sensors for their specific applications:

1. *What do you want to measure?* Is it relative humidity, dew point, or something else? Many instruments display the results converted into several alternative units, but the measurement will be intrinsically just one of these.

2. *Configuration*: A hygrometer may be hand held or mounted in a duct or on a wall, used on a bench top or under some other arrangement. It may need to be mounted in a particular orientation. The sensor may be housed in a remote probe (which may be specially shaped to suit a particular purpose), or it may be located in the main body of the instrument. There may be a limitation on the length of the cable joining the probe to the main body of the instrument. The size of the hygrometer relative to the sampling space may also be relevant. So, what kind of instrument do you need?

3. *Environmental condition*: Is the sensor suited for the environmental condition it will inhabit? What are the unique environmental conditions? Are any chemicals or contaminants present? What is the operating temperature range?

   Each sensor has strengths and weaknesses with certain chemicals, so it is recommended that you identify a list of all potential chemicals where the sensor will be installed, as well as the concentration of those chemicals, and compare them with the sensor manufacturer's compatibility charts. Doing this will help you to select the right sensor, determine the expected life of the sensor, and choose the best mounting location.

   It is also important to know—is the unit being installed in a room where temperature ranges are typically narrow, such as an operating room, or will it be installed in a duct or outdoor application, where it may see wider ranges in temperature? Consider the extremes prior to making your selection.

4. *Range*: What relative humidity (RH) range is needed to measure? What is the measurement limit of the sensor? Will the target be within the range? Will the sensor see high humidity or condensation? Will it see low humidity? How low?

5. *Resolution*: What accuracy is required? What are the minimum humidity changes that need to be controlled?

   This really depends on the application. Accuracy may be critical or it may not be all that important, so you need to understand the implications of accuracy. For example, it is extremely important to have a higher degree of accuracy in highly controlled environments such as labs and hospitals. If high accuracy is not a primary concern, selecting a less accurate RH sensor may be an acceptable option.

6. *Stability*: What drift rate is acceptable? Does the unit have a temperature effect? Is it temperature compensated? Are there any contaminants that affect the accuracy? What happens to the sensor specification as the device ages?

   Drift is inherent to all RH sensors, and it occurs over time due to a shift in the sensor characteristics. This happens regardless of the sensor manufacturer, so it is recommended to always select a sensor with a low drift value of less than 1% per year. In addition, manufacturers must compensate the circuitry of all RH sensors because they are all affected by temperature. This is usually called "temperature effect," which is expressed as %°F or %°C. Most sensors are factory calibrated at 77°F (22°C), so it is recommended to select a product with a low temperature-effect value. This helps to ensure RH accuracy at high- and low-probe temperatures.

7. *Power required and source*: What kind of power supply is required for sensor operation—battery or main electricity? Can the sensor operate unpowered?

8. *Repeatability*: Is constant monitoring necessary or can periodic measurements be made? How often should measurements be taken?

   The main purpose of any humidity sensors is to provide a reliable real-time information about the humidity of the surrounding environment. Therefore, ideally, a device designed for humidity sensing should operate continuously and reversibly, without perturbing a sample. However, as a rule, humidity changes slowly, and therefore humidity measurements can be performed periodically, with averaging time of several minutes. For the same reason, high speed is not required for humidity sensors. For example, for climatological use, a time constant of 60 seconds is required for humidity sensors.

9. *Form factor*: How much physical space is available for the sensor and what form best fits the application?

10. *Calibration*: How does the manufacturer calibrate the sensor? Can the sensor be recalibrated in the field?

    A calibration report does not guarantee satisfactory performance under actual operating conditions, but it is an important starting point, and it tells a lot about the quality of a sensor. It is necessary to be prepared for the fact that in real-time applications of humidity measurement instruments, specifications of a manufacturer and calibration data of a standards laboratory lose some, and often much, of their significance (Wiederhold 1997). This is because each manufacturer has its own way of measuring and specifying accuracy, and the user is often left confused. Although most manufacturers are not deliberately misstating accuracy claims, competition forces them to specify accuracy of their instruments in the most favorable way, sometimes only over a narrow range, at a fixed temperature, or under ideal laboratory conditions. For example, many perform manual calibrations at only 3 RH values (low-30%, medium-50% & high-80%). However, the best RH sensors are digitally calibrated at multiple RH values (8 or more). This ensures better accuracy over the entire measurement range.

11. *The field of application*: As mentioned earlier, we do not have an ideal humidity sensor. All sensors will always have some disadvantages. Therefore, while designing humidity sensors, first of all, the designers should consider the requirements for the devices used in the planned applications.

For example, according to Potyrailo and Mirsky (2008), high reliability, adequate long-term stability, a low false-positive rate, and good sensitivity form the priority list for industrial sensor users, whereas often the size and the cost of sensors are the least important factors. Environmental monitoring should be available for a long time and everywhere under conditions of constantly changing temperature. Therefore, stability, reliability, low cost, opportunity for wide-ranging manufacturing, and good matching with measuring chains are the main requirements to humidity sensors to be used for environmental monitoring. At the same time, humidity sensors intended for these uses are not required to have extremely high sensitivity. Therefore, humidity sensors that differ in sensitivity, selectivity, or other characteristics can be used for environmental monitoring. Wireless sensing is the field of application with different requirements. Wireless sensors are devices in which sensing electronic transducers are spatially and galvanically separated from their associated readout/display components, that is, sensors have limited or no on-board electrical power for sensor operation. This means that such systems require sensors with extremely low power consumable, because the communication distance depends first of all on the power stored on-board the sensor or delivered to the sensor and the power needed for the sensor to operate at a predetermined signal-to-noise ratio.

12. *Control interface*: What type of controller interface and switching logic is required? What data output format is required—display, analogue voltage or current, digitally read output, or data logged and stored will be downloaded later?

13. *Hazardous environments require more than the sensor*: Understand that sensors are only one segment of the entire control system; connectors and cables must also be considered.

14. *Price and support:* Some methods are more expensive than others, and the best method doesn't have to be the most expensive. What are we willing to pay for the required performance? What should we expect from the manufacturer for support after the sale? Due to drift, quality, and installation environment, RH sensors may need to be recalibrated or

replaced. These costs are often overlooked but can easily surpass the original purchase price if not considered at the beginning of a project.

15. *Ease of use*: Some hygrometers are straightforward to use. Others require some skill.

Thus, considering the wide variety of environmental variables that influence water vapor and affect measurement, it's important to know precisely which type of instrument and technology will enable you to measure most accurately in any given application. It is also important to know the various factors of the application that will affect different types of measurement technology. A few basic examples of how instruments might be used in particular applications are shown in Table 25.2.

## 25.2.2  FEATURES OF HUMIDITY MEASUREMENT

It is believed that the measurement of humidity is one of the most difficult problems in basic metrology (Makkonen and Laakso 2005). Two thermometers can be compared by immersing them both in an insulated vessel of water (or alcohol, for temperatures below the freezing point of water) and stirring vigorously to minimize temperature variations. A high-quality liquid-in-glass thermometer if handled with care should remain stable for some years. Hygrometers must be calibrated in air, which is a much less effective heat-transfer medium than is water, and many types are subject to drift, so regular calibration is necessary. A further difficulty is that most hygrometers sense RH rather than the absolute amount of water present, but RH is a function of both the temperature and absolute moisture content, so small fluctuations in air temperature in the test chamber will lead to changes in RH. In addition, in a cold and humid environment, sublimation of ice may occur on the sensor head, whether it is a hair, dew cell, mirror, capacitance-sensing element, or dry-bulb thermometer of an aspiration psychrometer. The ice on the probe matches the reading to the saturation humidity with respect to ice at that temperature, that is, the frost point. However, a conventional hygrometer is unable to measure properly above the frost point, and the only way to go around this fundamental problem is to use a heated humidity probe.

This means that there are no secondary factors when measuring air humidity. To obtain reliable information, it is necessary to take into account all factors, including the location of the sensor, sampling during humidity measurement, etc.

### 25.2.2.1  Sensor Location and Installation

There are no established rules as to where humidity sensors should be installed, but one should always follow the opinion of trained personnel and common sense. When installing and using the sensor, consider the following:

- Always choose a measuring point that is representative of the environment being measured, avoiding any hot or cold spots. A transmitter mounted under direct sunlight, near a door, humidifier, heat source

**TABLE 25.2**

**Examples of Humidity Sensors Using in Specific Applications**

| Application | Requirements | Consider | Possible Solution |
|---|---|---|---|
| Room humidity monitoring | Long-term record of temperature and humidity in a moderate range | Unsupervised operation? Would data at intervals be enough, for example, hourly? Real-time information needed, or record to view later? | Automated "data logger" whose readings can be downloaded easily to a computer |
| An industrial oven | Humidity measured at high temperatures to show progress of heat treatment such as drying or baking | Will there be dust, or other vapors, apart from water? | Relative humidity probes can be used (but only those specified for high temperatures). It may be possible to extract a sample flow of gas to a cooler temperature (but avoiding condensation). Check sensitivity to any chemicals given off during heating |
| A compressed air supply line | Monitor air supply inside a pipeline to confirm level of dryness. May have some level of dust or oil in the gas stream | Unsupervised use? Reading remotely? Any alarms needed? | Typically a dew-point capacitive probe would be used. The sensor usually has a sintered filter (and upstream filtering of the gas flow may help to protect the sensor). Dew-point probes vary in their speed of response and long-term performance |
| A climatic test chamber | A range of temperatures and humidities. May need to track rapid changes. There may be condensation at times (depending on the test performed) | Response time. Robustness of instrument at hot and wet extremes. Check the actual planned range of test conditions | A few relative humidity sensors (probes) are designed to cope with a wide humidity and temperature range. Wet- and dry-bulb hygrometers (psychrometers) can be suitable, although they need careful setting up and maintenance |

*Source:* Data extracted from Bell, S., *A Beginner's Guide to Humidity Measurement*, National Physical Laboratory, Teddington, Middlesex, UK, TW11 OLW, 2012.

such as radiators, or air conditioning inlet will be subject to rapid humidity changes and may appear unstable. Install the transmitter in a location where it will be exposed to an unrestricted air circulation that is representative of the average humidity and/or temperature of the controlled environment.

- Locations of sensors should provide easy access for maintenance and calibration. Sensors that are properly installed can save hours of maintenance and provide trouble-free operation.
- Avoid installing sensors too close to walls or floors. These surfaces can absorb and emit gases with changes in temperature, affecting the reading of the sensor.
- The sensors or sample points should not be positioned where they may be susceptible to excess vibration or heat, contamination, mechanical damage, or water damage. This will reduce the operational life of the sensor.
- Take into account that solid-state humidity sensors operate best at a monitoring radius of 0.5–2 m. Therefore, sensors should be located at or near points where the air humidity should be monitored.
- In larger rooms, be sure to take RH readings in different areas of the room. The RH value of a large room may be different from one end of the room to the other.
- In case of outdoor use, the mounting location should be in a sheltered area. Ideally, the transmitter should

be located on the north-side of the building (under an eave) to prevent sun-heated air from rising up the building wall and affecting the RH of the sensor.

- Reliable application of humidity sensors depends on a system that pairs the detectors and sampling techniques to the target gas and ambient conditions of the monitored area. Do not install sensors near sources of steam; steam will damage the sensors. Do not install sensors next to sources that are constantly leaking gas.
- Limit moisture condensation on the sensor. In applications with high humidity (>90%RH), where condensation can occur, a warmed-sensor head probe should be used.
- As RH is strongly temperature dependent, it is very important that the humidity sensor is at the same temperature as the measured air or gas.
- The temperature range, in which the sensor is to be installed, should be within the sensor specifications.
- Only products with appropriate certification can be used in potentially explosive areas.
- Continuous exposure to extreme temperatures may affect sensor and probe materials over time. It is therefore very important to select a suitable product for demanding environments. In temperatures above 60°C, the transmitter electronics should be mounted outside the process and only a suitable high-temperature probe should be inserted into the high-temperature environment. Moreover, built-in temperature compensation is required to minimize

the errors caused by large temperature swings or operation at temperature extremes.

- In areas of new construction, sensors should be installed after operations such as sandblasting, painting, or welding are completed. During such operations, the sensors present should be protected.
- Wherever possible, the probe should be mounted in the actual process to achieve the most accurate measurements and a rapid response time. However, direct installations are not always feasible. In such situations, sample cells installed in-line provide an entry point for a suitable measurement probe (Figure 25.2). In many applications, it is advisable to isolate the probe from the process with a ball valve to enable the removal of the probe for maintenance without shutting down the process (see Figure 25.2).

### 25.2.2.2 Sampling

Note that external sampling systems should not be used to measure RH because the change in temperature will affect the measurement. Sampling systems can instead be used with dew-point probes. When measuring dew point, sampling systems are typically used to lower the temperature of the process gas to protect the probe against particulate contamination, or to enable easy connection and disconnection of the instrument without ramping down the process. However, in the case of the development of humidity sensors, their calibration and testing of sampling systems are necessary.

Sampling format of a hygrometer can mean its configuration (a probe may sample free air, or a flow of gas is sampled through a tube). More generally, correct sampling is about making sure the measurement is representative of the condition you want to measure. For example, Bell (2012) gives the following advice: Avoid water being spuriously added to the measured gas (e.g., don't measure near pooled water). Avoid unintended removal of water from sample gas (e.g., don't have accidental condensation in tubing upstream of the hygrometer). Wherever there is a risk of condensation, localized heating (such as electrical "trace heating" of sample tubing) protects against this by keeping the gas above condensation temperature (above the dew point or frost point).

The accuracy of a measurement does not just depend on the sensor accuracy itself but also on the setup of the sensing system. Thus, it is important that the local conditions at the sensor correspond to the conditions under test. Therefore, the sensor should be connected as well as possible to the environmental air. In particular, the configuration of the sensor system shown in Figure 25.3 does not contribute to this. The system case has too small hole for connection with external surrounding atmosphere and too big dead volume (see Figure 25.3). Housing designs with a large dead volume and/or small aperture may act as a separation of the sensor and environment. As a result, sensor conditions may differ from external environment conditions, and the sensor response may be characterized by highly increased response times.

Therefore, in order to monitor outside humidity by using the sensor mounted in the device, the airflow design around the sensor is favorable in terms of response times. In Figure 25.4, there are some possible designs that optimize the air flow to measure the outside humidity. In order to achieve a fast response time, it is necessary to follow the following rules:

- Place the sensor as close to the environment as possible.
- A design that allows an airflow over the sensor is preferred to a design with a single aperture.
- The dead volume should be as small as possible.
- The aperture(s) should be as large as possible.
- There should be no material that can absorb humidity inside of the dead volume.

For measuring very dry gases (frost points around −40°C, −50°C, or below; water vapor fraction around 100 parts per million, or less), sample tubing and all materials in the

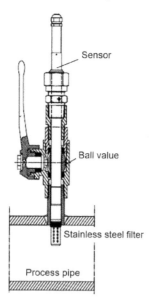

**FIGURE 25.2** A ball-valve installation in a process pipeline. (Modified from https://www.vaisala.com/sites/default/files/documents/CEN-TIA-G-Param-How-to-Choose-Application-note-B211203EN-A_0.pdf.)

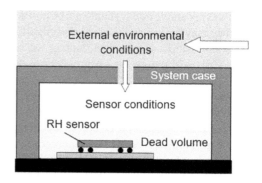

**FIGURE 25.3** Non-optimal configuration of the sensor system. (Modified from Granozio, D. *Humidity Sensor: Application Report. Texas Instrument.* SNAA216–July, 2014. http://www.ti.com/lit/an/snaa216/snaa216.pdf.)

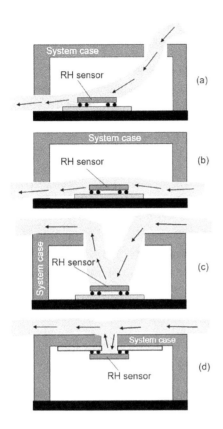

**FIGURE 25.4** Same sensor conditions and external environmental conditions guarantee good RH measurements: (a) multi windows (openings not aligned); (b) multi windows (openings aligned); (c) single wide window; and (d) single window (flipped device) (Modified from Granozio D. *Humidity Sensor: Application Report. Texas Instrument.* SNAA216–July 2014. http://www.ti.com/lit/an/snaa216/snaa216.pdf.)

gas-flow path are critical. In this range, even tiny amounts of stray water released from surfaces can significantly add to the gas-moisture content and can give badly misleading results. To avoid this, use clean, moisture-neutral materials; the smallest possible volume of pipework; and long flushing times. Sometimes heating is used to drive off traces of surface water. If using a pump—suck, don't blow. Put the pump at the far end (outlet) of a system to avoid contamination.

For an RH measurement, the temperature must be representative too. Measuring at a "cold spot" in a room can give an overestimate of the typical RH, even though water vapor might be evenly spread across the space.

If conditions are varying (such as often happens under cycling control of an air-conditioning system), take repeated readings over a period of time to get an idea of typical range. From repeated readings, you can calculate an average or mean. This can also reduce the uncertainty due to short-term instability of the instrument.

### 25.2.2.3 Uncertainty of Measurement

Every measurement is subject to some uncertainty (Bell 2012). Uncertainty in humidity measurement can come from various causes. It depends partly on the hygrometer, which might suffer from drift, contamination, short-term "noise," limited resolution of the thermometer, etc. Therefore, calibration uncertainty needs to be taken into account. If the condition being measured is unstable, this too contributes to uncertainty in the result—for example, temperature changes can cause uncertainty in RH values.

There is no "typical uncertainty" for a humidity measurement. But in general, in ideal conditions, RH measurements might achieve uncertainty within ± 2%RH to ± 3%RH. Under difficult conditions, this could be ± 5%RH to ± 10%RH, or even worse. The best dew-point measurements could be as good as ± 0.05°C to ± 0.1°C, in a lab, using the best condensation hygrometers. In worse conditions, and using less sophisticated dew-point probes, uncertainty of several degrees is not unusual, especially at very dry levels (all uncertainties at 95% confidence).

### 25.2.2.4 Contaminations

Unfortunately, every humidity measurement technology is subject to degradation due to contamination. All humidity sensors must be in contact with the gas they measure. Therefore, anything in the gas that "disagrees" with the sensor can alter the sensor's performance. For example, small oil droplets in an aerosol cleaning agent can coat the sensor, forming a barrier that limits water-vapor permeability. Dust can accumulate on the sensor with a similar effect. This process will result in a reading drift over time. There is no doubt that degradation of the sensor to varying degrees depends on the nature of the environment. For example, in most industrial processes there is a high potential for contamination either by direct, process gas-borne particulates or by soluble contaminants contained within the very moisture content that it is necessary to measure. Unfortunately, contaminated sensors cannot inform us about this and therefore are considered as sensors representing reliable information about the state of the environment. Therefore, with the majority of humidity sensors, it is essential that periodic maintenance should include checks of response and accuracy. This means that to mitigate these risks, analyzing and understanding the environment that you measure are absolutely essential.

The most difficult contaminants are chemicals that interfere with, or change the nature of, the humidity-sensitive material. Unfortunately, this is the nature of all solid-state humidity sensors. These contaminants can be sneaky; when sensors are exposed to them, they create measurement error. In some cases, when the sensor is removed from service for calibration, the contaminants may "outgas" and the measurement error disappears. Other contaminants, especially at high levels of pollutants, may cause permanent damage to the sensing material, including even dielectric material. It is necessary to know that some materials known to be harmful to the sensor include plastics, such as delrin, acetate, PVC, polystyrene, adhesive tapes, and adhesives other than silicone or epoxy.

Therefore, the sensor should not get in close contact with volatile chemicals, such as solvents or other organic compounds. Especially high concentration and long exposure must be avoided. Ketenes, acetone, ethanol, isopropyl alcohol, toluene, etc. are known to cause drift of the humidity reading—irreversibly in most of the cases. Please note that such chemicals are integral part of epoxies, glues, adhesives, etc. and outgas during baking and curing. These chemicals are also added as plasticizers into plastics, used for packaging materials, and do outgas for some period. Additives may also be added to materials that are listed for recommended use. For high safety, device housing and shipment packaging must be qualified. If a material emits a strong odor, you should not use it. Acids and bases may affect the sensor irreversibly and should be avoided: $HCl$, $H_2SO_4$, $HNO_3$, $NH_3$ etc. Also ozone in high concentration or $H_2O_2$ have the same effect and therefore must be avoided. Please note that the preceding examples do not represent a complete list of harmful substances. The sensor should not get in contact with cleaning agents (e.g., PCB board wash after soldering) or strong air blasts from an air pistol (not oil-free air). Applying cleaning agents to the sensor may lead to drift of the reading or complete breakdown of the sensor.

One should take into account that sensors from different manufacturers may react differently to environments because of differences in the sensing material used or sensor design (http://www.vaisala.com; https://able.co.uk), but sorting them out is nearly impossible without direct testing.

Two approaches are adopted to try to accommodate contamination affects: one approach is to devise a sensor where the detrimental effect of contamination is reduced, thereby prolonging the active life of the sensor. This may be inherent in the sensor design itself (this is the concept behind the bulk polymer, resistive-RH sensor) or may be affected by introducing some form or filter or sheath into the system. The more physical barriers you put between the sensor and the environment, however, the more problems you encounter in trying to make a viable and accurate measurement. Once contaminated and blocked, a filter may have the effect of creating an unrepresentative microenvironment between itself and the sensor. The measurement is therefore limited in terms of accuracy and response time, and the filter will only intercept particulate contamination.

Alternatively, the second approach is to recognize that contamination will occur, and therefore develop a way to monitor it and, if possible, compensate. One measurement technique that falls into the latter category is the optical dew-point hygrometer, which can incorporate a self-checking feature that may be operated either manually or automatically (in the case of the most sophisticated designs) within the electronic control unit (https://able.co.uk). Such devices were considered in Chapter 5 (Volume 2). The optical monitoring system constantly monitors the surface of the mirror and responds to contaminants that deposit on the mirror. As a result, optical dew-point hygrometers can generally operate continuously and unattended for longer periods of time than most other humidity-measurement systems and provide what is probably the most accurate, repeatable, and reliable humidity measurement available for process-industry humidity monitoring, particularly in heavily contaminated atmospheres. It is only the relatively higher initial cost of this type of hygrometer that prevents it from being used more widely, perhaps, to solve industrial humidity-measurement problems. This means that if the initial cost is not a determining factor in any particular application for measuring humidity, then the chilled mirror optical dew-point hygrometer seems to provide the most versatile method for measuring humidity, having built-in functions that allow you to control the degree of contamination occurring and adjust its performance to compensate.

### 25.2.2.5 Some Recommendations in Humidity Measurement

Bell (2012), for the avoidance of errors in measuring humidity, offers the following:

*Do...*

- Use manufacturer's guidelines and recommendations.
- Have the instrument calibrated, and take account of any calibration corrections required.
- Let the instrument reach the temperature of the location.
- Allow time for the humidity reading to stabilize (especially for dry gases).
- Record measurement results with suitable care.
- Be clear when expressing humidity differences. A change of 10% of reading is not the same as a change of 10%RH.
- Check any hygrometer that has been exposed to extreme conditions or contaminants.

*Don't ...*

- Don't handle humidity sensors roughly. Don't abuse well-mounted sensors by treating them as coat hooks!
- Don't use sensors outside the temperature or humidity range specified by the manufacturer—this could cause a shift in calibration.
- Don't expose sensors to condensation—unless you know they are can definitely tolerate this
- Don't "mix and match" humidity probes with different electronics units—unless they are definitely designed for this. Instrument suppliers can advise if a probe is a "self-contained" interchangeable unit with its own signal processing inside.

*Watch out for ...*

- Droplets or stray water in any form: if present, humidity measurements may be misleading.
- Dust—most instrument types.
- Pressure differences (dew point) or temperature differences (relative humidity).

*Take into account....*

- When measuring gas at one temperature or pressure and using it with others, the measured values may need to be converted or interpreted.
- Humidity measurements can be made below atmospheric pressure but need some thought to set up and interpret.
- Humidity is measured in a wide variety of gases, such as fuel gas ("natural gas") and various process gases. Take care to choose a measurement method suitable for the particular gas. Beware of humidity readings displayed in units that are gas dependent, and be ready to interpret (convert) for the gas used.

## REFERENCES

Bell S. (2012) *A Beginner's Guide to Humidity Measurement*, National Physical Laboratory, Teddington, Middlesex, UK, TW11 OLW (http://www.npl.co.uk; https://www.rotronic.com/).

Brett C.M.A. (2001) Electrochemical sensors for environmental monitoring. Strategy and examples. *Pure Appl. Chem.* **73**(12), 1969–1977.

Brown V.R. (2006) Sensor selection for hand-held portable gas detection. In: *The Grey House Safety and Security Directory*. Grey House Publishing, Millerton, NY, 291–293 (www.enmet.com).

BS EN 50241 (1999) *Specification for Open Path Apparatus for the Detection of Combustible or Toxic Gases and Vapours. Part 1—General Requirements and Test Methods. Part 2—Performance Requirements for Apparatus for the Detection of Combustible Gases*. British Standards Institution (www.bsi-global.com).

Chou J. (2000) *Hazardous Gas Monitors: A Practical Guide to Selection, Operation and Application*. McGraw-Hill, New York.

COGDEM (1999) *Gas Detection and Calibration Guide*. The Council of Gas Detection and Environmental Monitoring, Hitchin, UK (www.cogdem.org.uk).

Granozio D. (2014) Humidity Sensor: Application Report. Texas Instrument. SNAA216–July 2014 (http://www.ti.com).

Ho C.K., Itamura M.T., Kelley M., Hughes R.C. (2001) Review of chemical sensors for in-situ monitoring of volatile contaminants. Sandia Report SAND2001-0643, Unlimited release, Sandia National Laboratories, Albuquerque, NM (www.sandia.gov/sensor).

Korotcenkov G., Cho B.K. (2011) Chemical sensors selection and operation guide. In: Korotcenkov G. (ed.), *Chemical Sensors: Comprehensive Sensor Technologies*. Vol. 6. *Chemical Sensors Applications*, Momentum Press, New York, pp. 281–348.

L138 (2003) Dangerous substances and explosive atmospheres. In: *Dangerous, Substances and Explosive Atmospheres Regulations*. HSE Books, ISBN 0 7176 2203 7 (www.hse.gov.uk).

Makkonen, L., Laakso, T (2005) Humidity measurements in cold and humid environments. *Bound.-Layer Meteorol.* **116**, 131–147.

Potyrailo R.A., Mirsky V.M. (2008) Combinatorial and high-throughput development of sensing materials: The first 10 years. *Chem. Rev. B* **108**, 770–813.

Scott M., Bell S., Bannister M., Cleraver K., Cridland D., Dadachanju F., et al. (1996) *A Guide to the Measurement of Humidity*. Institute of Measurement and Control, London, UK.

Taylor J.K. (ed.) (1987) *Sampling and Calibration for Atmospheric Measurements*, ASTM STP 957. American Society for Testing and Materials, Philadelphia, PA.

Walsh P. T., Hedley D., Pritchard D.K. (2001) Framework for HSE guidance on gas detectors. (On-line checking of flammability monitoring equipment–Final report). HSL Internal Report EC/01/10, FM/01/01.

Wexler A. (ed.) (1965) *Humidity and Moisture*. Vols. 1 and 3, Reinhold, New York.

White L.T. (2000) *Hazardous Gas Monitoring: A Guide for Semiconductor and Other Hazardous Occupancies*. Noyes Publications/William Andrew Publishing, Norwich, NY.

Wiederhold P.R. (1997) *Water Vapor Measurement, Methods and Instrumentation*. Marcel Dekker, New York.

WMO (2011) World Meteorological Organization, *Technical Regulations*. Vol. 1—General Meteorological Standards and Recommended Practices (WMO-No. 49), Geneva, Switzerland.

Woodfin W.J. (1994) Portable electrochemical sensor methods. In: Eller P.M., Cossinelli M.E. (eds.), *NIOSH Manual Analytical Methods*, 4th ed. U.S. Department of Health and Human Services, Cincinnati, OH, pp. 70–73.

# 26 Humidity-Sensor Testing and Calibration

## 26.1 HUMIDITY-SENSOR TESTING: RELIABILITY IMPLICATIONS

Humidity sensors require tests similar to those performed on integrated circuits (ICs) and unique qualification tests to verify that they will have acceptable performance of both the sensor chip and the packaging for their intended applications. For example, several tests have been developed for humidity sensors based on the need to detect potential failures due to the environment in which the devices will operate (Frank 2013). Key tests include but are not limited to the following:

- Operational life, such as pulse, humidity cycling;
- High humidity, high temperature;
- High- and low-temperature storage life;
- Temperature cycling;
- Mechanical shock;
- Variable-frequency vibration; and
- Solderability.

Those tests use accelerated life and mechanical-integrity testing to determine the lifetime reliability statistics for sensors. Potential failure mechanisms are determined by the materials, processes, and process variability that can occur in the manufacturing of a particular sensor. Identifying minimum expectations and critical application factors that limit the life of a device is part of a methodology known as a physical failure approach for reliability testing. The physics of failure approach involves analyzing the potential failure mechanisms and modes. For example, Figure 26.1 shows 10 product-related and 8 process-related areas with 73 different items that can affect reliability (Maudie and Tucker 1991; Maudie and Wertz 1997; Frank 2013).

Once this is understood, application-specific packaging development for a particular application can continue, appropriate test conditions can be established, and critical parameters can be measured and verified (Maudie et al. 1997; Frank 2013). In particular, one key failure mechanism for humidity sensors that operate in harsh media applications is corrosion, specifically, galvanic corrosion caused by dissimilar metals in electrical contact in an aqueous solution. Using a physics of failure approach, one can start to ask the appropriate questions. Which environmental factors contribute to the failure mechanism? What accelerates it? What should be done to the sensor package to prevent corrosion or minimize its impact over the sensor's lifetime? What is the expected sensor lifetime? Answers to these questions allow one to simulate and analyze costs before long and potentially costly testing begins. The testing will ultimately prove the acceptability of a particular proposal. Unless the proper mechanisms are known, overdesigning can add unnecessary cost to the packaging,

such as a passivation layer, a hard die attachment, or any non-standard process. In addition, the design trade-offs that follow may not prove beneficial.

At present, establishing media compatibility for sensors usually uses in-situ monitoring under accelerated aging (Frank 2013). A typical three-step approach for verifying media compatibility is to (1) characterize, (2) expose to environment, and (3) retest for conformity to specification. Accelerated testing procedures for organic compounds used in the sensor packaging, in the sensor element, and in the sensor-integrated circuits are also widely used without removing the sensor from the test chamber. New sensor materials and modified packaging designs have been shown to survive in a wide variety of harsh environments typical of humidity-sensor applications. Previously, a comparable confidence level would have been possible only after achieving millions of unit hours in the actual application or experiencing a high rate of warranty returns. End-of-life techniques require even longer testing to ensure that a product life (e.g., 20 years) is achievable. In addition, normal test techniques require already packaged units.

One should note that for humidity sensors manufactured using semiconductor technology, it is possible to control the wafer-level sensor reliability (Dellin et al. 1993). Test structures on the wafer or monitor wafers allow evaluation of critical process steps as they occur and provide rapid feedback to detect and correct process errors. The cost of testing at the wafer level is considerably less than testing at the completed assembly level. Wafer-level reliability testing is one of the ways that semiconductor humidity sensors can continue to achieve lower product cost and continuous improvement of sensor performance. This feature is especially important for smarter sensors containing microcontrollers.

## 26.2 HUMIDITY-SENSOR CALIBRATION: INTRODUCTION

Like all measuring instruments (Chou 2000), humidity sensors should be calibrated (Bell 1996, 2012; Wiederhold 1997; Heinonen 2006; Majewski 2016). One should also take into account that an essential part of humidity and moisture measurements is the calibration against a standard. With regard to humidity measurements, it's commonly accepted that a standard is a system or device that either can produce a gas stream of known humidity by reference to fundamental base units (e.g., temperature, mass, and pressure) or is an instrument that can measure humidity in a gas in a fundamental way, using similar base units. The most fundamental standard that is used by national calibration laboratories is the so-called gravimetric hygrometer (read Chapter 2, Volume 2). Using this method, a certain amount of dry gas is weighed and compared with the weight of the test gas in the same volume. From this,

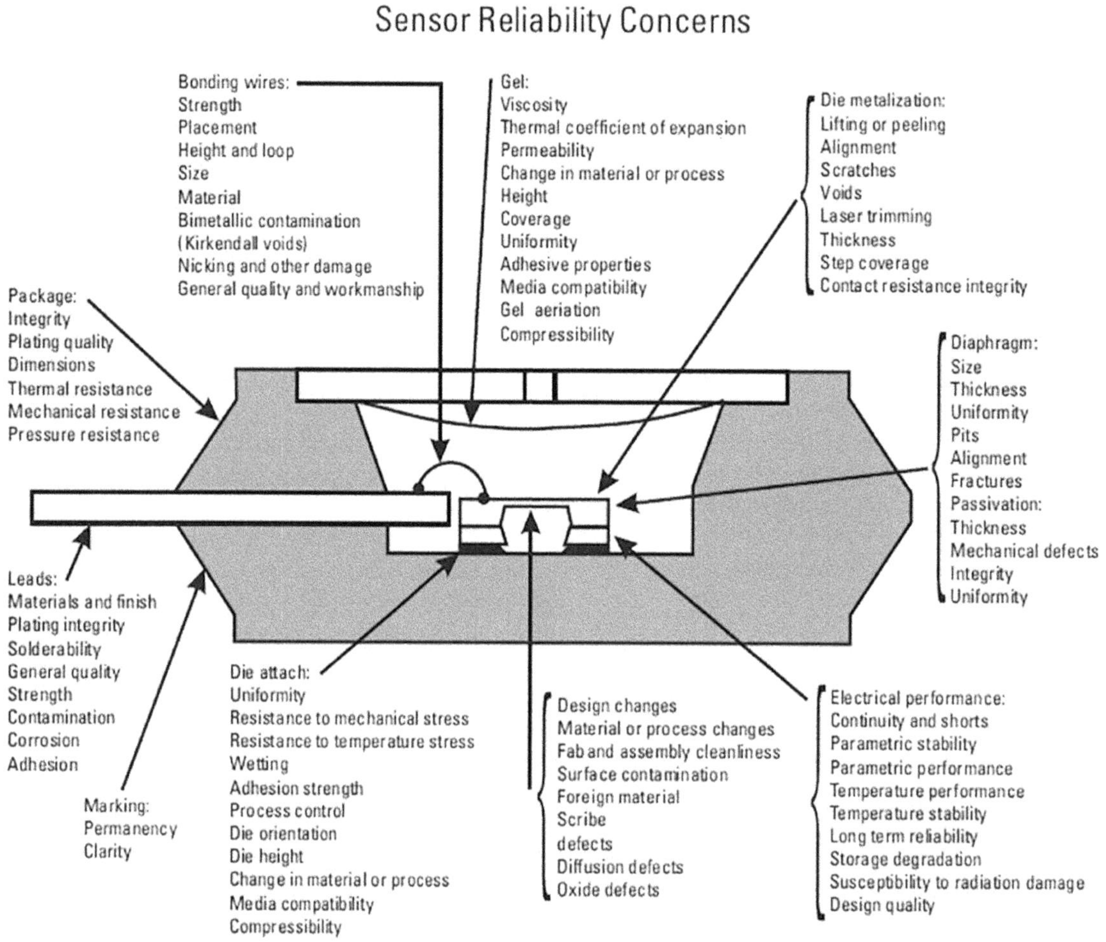

## Sensor Reliability Concerns

**Bonding wires:**
Strength
Placement
Height and loop
Size
Material
Bimetallic contamination
(Kirkendall voids)
Nicking and other damage
General quality and workmanship

**Gel:**
Viscosity
Thermal coefficient of expansion
Permeability
Change in material or process
Height
Coverage
Uniformity
Adhesive properties
Media compatibility
Gel aeration
Compressibility

**Die metalization:**
Lifting or peeling
Alignment
Scratches
Voids
Laser trimming
Thickness
Step coverage
Contact resistance integrity

**Package:**
Integrity
Plating quality
Dimensions
Thermal resistance
Mechanical resistance
Pressure resistance

**Diaphragm:**
Size
Thickness
Uniformity
Pits
Alignment
Fractures
Passivation:
Thickness
Mechanical defects
Integrity
Uniformity

**Leads:**
Materials and finish
Plating integrity
Solderability
General quality
Strength
Contamination
Corrosion
Adhesion

**Marking:**
Permanency
Clarity

**Die attach:**
Uniformity
Resistance to mechanical stress
Resistance to temperature stress
Wetting
Adhesion strength
Process control
Die orientation
Die height
Change in material or process
Media compatibility
Compressibility

Design changes
Material or process changes
Fab and assembly cleanliness
Surface contamination
Foreign material
Scribe
defects
Diffusion defects
Oxide defects

**Electrical performance:**
Continuity and shorts
Parametric stability
Parametric performance
Temperature performance
Temperature stability
Long term reliability
Storage degradation
Susceptibility to radiation damage
Design quality

**FIGURE 26.1** Reliability areas in plastic-packaged sensors. (Data extracted from Maudie, T. and Tucker, B., Reliability issues for silicon pressure sensors, in *Proceedings of Sensors Expo*, October 1–3, Chicago, IL, pp. 101–1 to 101–8, 1991, and Frank, R., *Understanding Smart Sensors*, Artech House, Boston, MA, 2013.)

the amount of water is determined, and the vapor pressure is calculated. The method can provide the most accurate measurements possible, but the system is cumbersome, expensive, and time consuming to use. For example, at low humidity levels, the device can require many hours of operation to obtain a large enough sample. Some national laboratories, such as the National Institute for Standards Testing (NIST) in the US, the National Physics Laboratory (NPL) in the UK, and the National Research Laboratory of Metrology (NRLM) in Japan, have access to gravimetric hygrometers (Wiederhold 1997). However, these laboratories use the system only to calibrate other, slightly less accurate, standards that are easier and faster to use for day-to-day calibrations, such as the two-pressure humidity generator (Section 26.2.2.1), a precision-chilled mirror hygrometer (Chapter 5, Volume 2), or a carefully designed psychrometer (Chapter 4, Volume 2). Some of these national standards and the approximate ranges that they can cover are shown in Table 26.1. Among these instruments, the chilled-mirror hygrometer is probably the most widely used transfer standard.

Other methods of humidity measurements such as impedance hygrometers, polymer film capacitance relative humidity

**TABLE 26.1**
**Instrument Classifications**

| Type | Class | Range | Typical Accuracy |
|---|---|---|---|
| Gravimetric | Primary | 100°C/−50°C | 0.1°C dew point |
| Chilled mirror hygrometer | Fundamental (transfer) | 90°C/−90°C dew point | 0.2°C dew point |
| Electrolytic hygrometer | Fundamental | 1 to 2000 ppmv | 5% of reading ppmv |
| Psychrometer | Fundamental | 5%–95% RH; 0°C/100°C | 2% RH |

(RH) sensors, aluminum-oxide-based RH sensors, and so on, are secondary devices. These devices are nonfundamental and must be calibrated against a transfer standard or other fundamental system. To obtain accurate data from them, these devices must be recalibrated frequently. Secondary systems are rarely used for laboratory calibration but have many applications. Properly manufactured, calibrated, and operated, these devices can provide good online service for many years. However, calibration is required on a regular basis, and adjustments are often needed. In most applications, the accuracy of these devices is modest and much lower than that provided by transfer-standard-type instruments.

Calibrations against reference hygrometers, using humidity generators, can be done in laboratories. A calibration lab may have accreditation giving assurance that calibrations are traceable and competently performed. For some hygrometers, there are also methods of "field calibration"—for example using salt solutions to generate values of RH—which can also be made traceable. Most hygrometers are calibrated at a fixed ambient temperature. This may vary from manufacturer to manufacturer, but the temperature is usually around 25°C ± 1°C.

One should note that calibrations need to be repeated, because hygrometers can drift—due to conditions of use, or just due to sensor aging. Therefore, periodic calibration is essential for accurate humidity measurement. Calibration intervals may vary from sensor to sensor (Chou 2000). Choosing how often to calibrate depends on usage, risk of drift, and the importance of measurement. Some experts suggest every year, some every half year, and others, every quarter. RH sensors might be calibrated every 6 to 12 months. Less stable types like aluminum-oxide sensors must be calibrated every 6 months, or sooner if desired (Bell 1996). A reference hygrometer, such as condensation dew-point meters and psychrometers in a lab, might be calibrated every 1–2 years. Some sensor technology requires frequent calibrations, because the calibration is an important factor in the accuracy of an analyzer reading. For example, some companies perform a span calibration every other week. Therefore, they are able to achieve high accuracies of the measurements. Time between recalibrations or maintenance also relates to sensor life. Generally, the manufacturer of the sensor recommends a time interval between calibrations. Therefore, you should do recalibration in accordance with the recommendations of the sensor manufacturer. However, it is good general practice to check the sensor more closely during the first 30 days after installation. During this period, it is possible to observe how well the sensor is adapting to its new environment (Chou 2000). Experience shows that a sensor surviving 30 days after the initial installation will have a good chance of performing its function for the expected duration.

### 26.2.1 REQUIREMENTS OF A TEST SYSTEM

A frequent method of calibrating an RH instrument is to place the humidity sensor in a closed container, the humidity chamber, with both temperature and humidity control, where an instrument is compared with a calibrated standard instrument like a mirror dew-point meter or a psychrometer. Of course, the construction of a test system is important for obtaining reliable information during humidity-sensor calibration and testing. However, at present there are no strong general requirements for a test chamber aimed at calibration or study of humidity sensors. Every chamber is designed for a specific task, and therefore configuration of the chamber

first of all has to simulate or reproduce the conditions of exploitation and provide control of specific functional sensor parameters in the proper way. As a result, depending on the task, the outward appearance of the chamber for humidity-sensor calibration and testing can change considerably. Every company and research team designs a test system according to their own notion of the optimal construction, and therefore at present in the literature one can find a lot of different setups used for sensor calibration and testing (see, e.g., Harvey et al. 1989; Demarne et al. 1990; Mousa-Bahia et al. 1993; Nieuwenhuizen and Harteveld 1994; Endres et al. 1995; Magliulo 1996; Tang et al. 1998; COGDEM 1999; Pilling et al. 2003; Sawaguchi et al. 2005; Brahim-Belhouari et al. 2005; Afgan et al. 2006; Heinonen 2006; Korposh et al. 2006; Berry and Brunet 2008; Maziarz and Pisarkiewicz 2008). Several of them are shown in Figure 26.2. It can be seen that the chambers can differ in configuration, size, and component parts. At that, the industry can use for testing and calibrating devices an environmental chamber that can have impressive dimensions needed to simultaneously calibrate a large number of instrument (see Figure 26.3).

Taking into account the existing situation, we cannot offer any universal solution that is acceptable for all applications. We offer only one requirement: to ensure qualified performance of humidity-sensor measurements, the test system must adjust all the conditions that are essential for the behavior of the sensor in various applications. This is the only way to get the correct information about the properties of the sensor and its behavior under the influence of gases, which is necessary for understanding, developing, and optimizing the sensor. A detailed description of test systems and approaches used for their design was given by Endres et al. (1995). According to Endres et al. (1995), to obtain gas mixtures similar to application conditions, the test system must have a gas-mixing system, which should mix many gases in a wide range of combinations and concentrations that cover all desired concentration ranges. Further, the accuracy of the

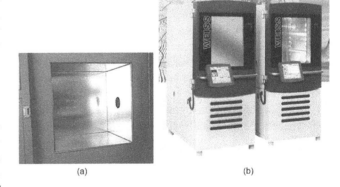

(a)                                                        (b)

**FIGURE 26.2** (a) Test chamber in *2500 Benchtop humidity generator.* (Modified from www.thunderscientific.com/humidity_equipment/model_2500.html); (b) the climatic camera of laboratory WKL realizes essays to scale of laboratory, with a rank of temperatures of −75°C to 180°C and control of humidity in the rank of 10 to 98%RH. (Modified from https://www.weiss-technik.com/en/product-configurator.)

**FIGURE 26.3** Sensor calibration stand used by Michell Instruments. (Modified from http://www.michell.com/uk/documents/RH_Catalogue_Version_3.pdf.)

gas combinations has to be maximized. In addition, for better reproducibility, the gas mixture in the test chamber first of all has to be as homogeneous as possible. This effect is mainly influenced by the geometry of the measurement chamber. This means that the sensor chamber should be optimally designed for the gas flow. For example, it is known that solid-state gas sensors, including humidity sensors, respond to the local sample concentration on the surface of the sensitive layer rather than to the average concentration inside the chamber. This means that data collected during characterization of these sensors will be affected by sensor position and concentration transients. Rapid establishment of a uniform concentration is therefore desirable to avoid poor measurement reproducibility due to random fluctuations, as happens when recirculating regions are present. Di Francesco et al. (2005) note correctly that quite often the exposure of the sensor to the sample is not considered a particularly critical phase of the measurement process; hence, resources devoted to the design of these devices are limited. As a result, in many cases the chamber design and sensor arrangement do not guarantee identical exposure conditions.

It is obvious that all thermodynamic processes occurring at the sensor are influenced by the physical conditions in the measurement chamber. Therefore, all important parameters, for example, temperature or pressure, have to be controlled or rather actively directed by the test system to attain stable conditions in all measurements. Undoubtedly, all these requirements for a test-gas system necessitate automatic control of all system functions and automatic processing of the sensor data. In addition, without a doubt, only computer-based measuring units can guarantee precise control of environmental conditions and allow receiving and storing a large amount of measurement data necessary for the development and optimization

of gas sensors. Thus, the optimal construction of the test system should include three main parts: the gas-mixing system, the measurement chamber (cell), and an electronic system based on a personal computer aimed at controlling all parts of the system, including the gas-mixing system and the experimental setup used for measuring, controlling, and analyzing the gas-sensor parameters (see Figure 26.4). Of course, the test system may also contain additional instruments for monitoring the humidity of the test gas, and for analyzing the composition of the exhaust gases.

## 26.2.2 SOURCES OF WATER VAPOR IN MEASUREMENT CHAMBER

### 26.2.2.1 Humidity Generators

There are various methods to generate a fixed concentration of water vapor in a chamber with a limited volume (Barratt 1981; Vitenberg et al. 1984; Lin et al. 1996; Heinonen 1996, 1999; Nakamura et al. 1999; ISO 2001a, 2001b, 2003a, 2003b, 2005, 2009a, 2009b, 2009c). For example, a small ampoule containing the necessary volume of the water is placed into the chamber. After the chamber is evacuated, the ampoule is broken. The water is evaporated and the partial pressure $p_i$ is created in the chamber. However, for the purposes previously mentioned, methods such as two-flow, the two-temperature, and the two-pressure are commonly used. Briefly, in the first method, a test chamber is fed by two streams of air, one being dry, the second one saturated with water at a known temperature. Table 26.2 shows the approximate mass of water (in grams) contained in a cubic meter ($m^3$) of saturated air at a total pressure of 101 325 Pa (1013.25 mbar). Unsaturated air will be contained comparatively less. Thus, it is possible to

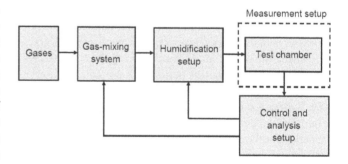

**FIGURE 26.4** Principal scheme of gas-sensor measurement setup.

**TABLE 26.2**

**Mass of Water Vapor per Cubic Meter of Saturated Air**

| Temperature, °C | 0 | 5 | 10 | 15 | 20 | 25 | 30 | 35 |
|---|---|---|---|---|---|---|---|---|
| Water vapor, g/m³ | 4.9 | 6.8 | 9.4 | 12.9 | 17.4 | 23.1 | 30.5 | 39.8 |

*Source:* Bell, S., *A beginner's guide to humidity measurement*, National Physical Laboratory, Teddington, Middlesex, UK, TW11 0LW (http://www.npl.co.uk; https://www.rotronic.com/).

dilute the obtained saturated vapor of the concentration $X_S$ using dry, clean air to the desired concentration $X_i$. The resulting vapor concentration (also known as volumetric humidity or absolute humidity) $(X_i)$ can be calculated from the two flow rates. For this purpose, one can use Eq. (26.1)

$$X_i = X_s \frac{v_s \cdot t_s}{\left(v_s \cdot t_s + v_a \cdot t_a\right)}, \qquad (26.1)$$

where $X_S$ is the concentration of water vapor in saturated air (see Table 26.2); $v_S$ and $v_a$ are the flow rates of dry and saturated air; and $t_S$ and $t_a$ are the working time of the corresponding channels.

The two-temperature method uses air that has been saturated with water vapor at a well-known temperature, after which the air is heated isobarically to a higher temperature for use (Figure 26.5). A description of such a humidity generator can be found in Heinonen (1996). Measurements of the temperature and pressure of the cool, saturated-gas stream, and in the warmer test chamber, are all that is required to determine the resulting humidity content of the gas stream. This method is very accurate, but typically is slow; one measurement point per day (http://www.vaisala.com). According to Heinonen (1996), the uncertainty of 0.06°C dew point can be reached in the middle of the measurement range by planning and preparing the measurements carefully.

In a two-pressure humidity generator, air is saturated with water vapor at elevated pressure to determine the dew point. Then the saturated high-pressure air flows from the saturator, through a pressure reducing valve, where the air is isothermally reduced to test pressure, usually normally atmospheric pressure, at the test temperature (see Figure 26.6). Both temperature and pressure of the saturator and the test chamber are measured accurately (Eq. 26.2). Since we know the pressure change, we can determine the change in the dewpoint and calculate RH. The uncertainty of the system depends on the accurate measurement of temperature and pressure and the stability of these measurements. This method is fast, but expensive (http://www.vaisala.com).

$$RH = \frac{p_c \dot{e}(t_s) \cdot f(p_s, t_s)}{p_s \dot{e}(t_c) . f(p_c, t_c)} \cdot 100, \qquad (26.2)$$

where $p_S$ and $p_C$ are the pressure in saturator and chamber, respectively; $e(t_S)$ and $e(t_C)$ are saturation vapor pressure at temperature $t_S$ and $t_C$ in the saturator and chamber, respectively; and $f(p_S, t_S)$ and $f(p_C, t_C)$ are enhancement factors at corresponding temperature and pressure in saturator, respectively.

In general, precision-humidity generators are not transportable (see Figure 26.7), so intercomparisons have to be

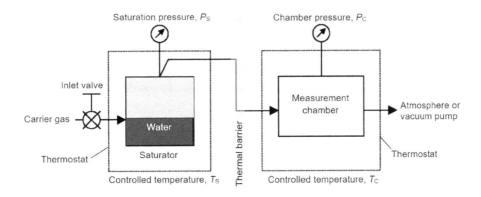

**FIGURE 26.5** Schematic diagram of two-temperature humidity generator.

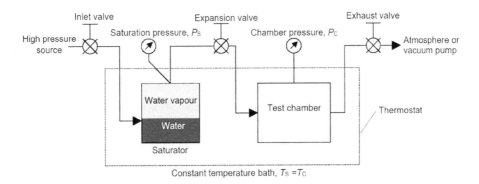

**FIGURE 26.6** Schematic diagram of two-pressure humidity generator.

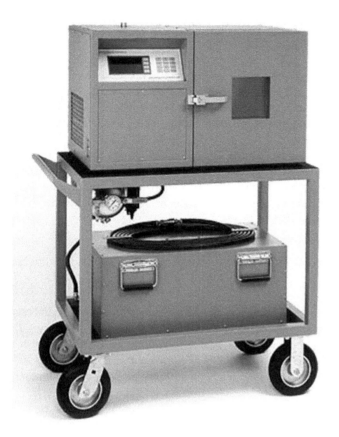

**FIGURE 26.7** The view of "two-pressure" humidity generator. Model 2500 Benchtop/Mobile. This system is capable of continuously supplying accurately known humidity values with 0.5%RH uncertainty for instrument calibration, evaluation, and verification, as well as for environmental testing: Relative humidity—10–95%RH; Chamber temperature range—0°C–70°C; Chamber temperature uniformity—±0.1°C; Chamber temperature uncertainty—0.06°C. (Modified from http://www.thunderscientific.com/humidity_equipment/model_2500.html.)

made with a transfer standard of high accuracy. A good, if not the only, choice is a standard mirror dew-point meter.

### 26.2.2.2 Saturated Salt Solutions

Saturated (or diluted) salt solutions can also be used to create a constant RH in a confined space used for humidity-sensor calibration and testing (Lu and Chen 2007). Calibration of instruments for measuring the RH using saturated salt solutions is a simpler and cheaper method in comparison with methods previously described. It was established that in confined space the water-vapor concentration, and therefore the RH over a salt solution is less than that over pure water. In addition, various salts generate different RH (Greenspan 1977). Moreover, at any temperature, the concentration of a saturated solution is fixed and does not have to be determined. By providing excess solute, the solution will remain saturated even in the presence of modest moisture sources and sinks. Therefore, using various saturated salt solutions it is possible to form an atmosphere in the chamber with different predetermined humidity.

Such situation takes place because water is present in both the gas and the liquid phase, whereas the scarcely volatile salt molecules are only present in the liquid. They dilute the water and hinder escape of water molecules into the air. The rate of return of water molecules to the liquid surface is proportional to their concentration in the gas, where there are no salt ions to interfere. The system therefore adjusts to an equilibrium where there are fewer water molecules in the air than there would be over a pure water surface. The RH is therefore lower than 100%.

RH above various saturated salt solutions are listed in Tables 26.3 and 26.4. It is seen that in using different salt solutions, it is possible to calibrate sensors in practically the entire range of possible changes in air humidity. Good choices for these purposes are common salts (at 20°C) such as LiCl

## TABLE 26.3
## Relative Humidity Above Saturated Salt Solutions Used for Humidity-Sensor Calibration

| Temperature (°C) | Relative Humidity (%) | | | | | | |
|---|---|---|---|---|---|---|---|
| | LiCl | CH$_3$CO$_2$K | MgCl$_2$ | K$_2$CO$_3$ | Mg(NO$_3$)$_2$ | NaCl | K$_2$SO$_4$ |
| 0 | * | * | 33.7 ± 0.3 | 43.1 ± 0.7 | 60.4 ± 0.6 | 75.5 ± 0.3 | 98.8 ± 1.1 |
| 5 | * | * | 33.6 ± 0.3 | 43.1 ± 0.5 | 58.9 ± 0.4 | 75.7 ± 0.3 | 98.5 ± 0.9 |
| 10 | * | 23.4 ± 0.5 | 33.5 ± 0.2 | 43.1 ± 0.4 | 57.4 ± 0.3 | 75.7 ± 0.2 | 98.2 ± 0.8 |
| 15 | * | 23.4 ± 0.3 | 33.3 ± 0.2 | 43.1 ± 0.3 | 55.9 ± 0.3 | 75.6 ± 0.2 | 97.9 ± 0.6 |
| 20 | 11.3 ± 0.3 | 23.1 ± 0.25 | 33.1 ± 0.2 | 43.2 ± 0.3 | 54.4 ± 0.2 | 75.5 ± 0.1 | 97.6 ± 0.5 |
| 25 | 11.3 ± 0.3 | 22.5 ± 0.3 | 32.8 ± 0.2 | 43.2 ± 0.4 | 52.9 ± 0.2 | 75.3 ± 0.1 | 97.3 ± 0.5 |
| 30 | 11.3 ± 0.2 | 21.6 ± 0.5 | 32.4 ± 0.1 | 43.2 ± 0.5 | 51.4 ± 0.2 | 75.1 ± 0.1 | 97.0 ± 0.4 |
| 35 | 11.3 ± 0.2 | | 32.1 ± 0.1 | | 49.9 ± 0.3 | 74.9 ± 0.1 | 96.7 ± 0.4 |
| 40 | 11.2 ± 0.2 | | 31.6 ± 0.1 | | 48.4 ± 0.4 | 74.7 ± 0.1 | 96.2 ± 0.4 |
| 45 | 11.2 ± 0.2 | | 31.1 ± 0.1 | | 46.9 ± 0.5 | 74.5 ± 0.2 | 96.1 ± 0.4 |
| 50 | 11.1 ± 0.2 | | 30.5 ± 0.1 | | 45.4 ± 0.6 | 74.4 ± 0.2 | 95.8 ± 0.5 |

*Source:* Patissier, B. and Walters, D., Basics of relative humidity calibration for Humirel HS1100/HS1101 sensors. Humirel, Toulouse Cedex, France, 1999, https://www.omega.com.

## TABLE 26.4
### Relative Humidity Above Saturated Water Salt Solutions in a Closed Space at the Temperature of 25°C

| Saturated Water Salt Solution | Relative Humidity (%) |
|---|---|
| Lead chloride | 99 |
| Potassium nitrate | 95 |
| Tin-lead solder chloride | 96 |
| Potassium chloride | 84 |
| Tin chloride | 77 |
| Sodium chloride | 75 |
| Potassium iodide | 69 |
| Copper chloride | 62 |
| Nickel chloride | 59 |
| Iron chloride | 52 |
| Aluminum chloride | 48 |
| Zinc chloride | 21 |
| Lithium chloride | 11 |
| Potassium hydroxide | 8 |

*Source:* Greenspan, L., Humidity fixed points of binary saturated aqueous solutions, *J. Res. Nat. Bur. Stand. A*, 81A, 89–96, 1977; https://www.omega.com.

(lithium chloride, 11.3%RH), $MgCl_2$ (magnesium chloride, 33.1%RH), NaCl (sodium chloride, 75.5%RH), and $K_2SO_4$ (potassium sulfate, 97.6%RH).

One should note that at present in the market you can find various devices for calibrating humidity sensors, developed on the basis of the method discussed. One such device, developed by Driesen + Kern GmbH, is shown in Figure 26.8. The MHT Series Humidity Checks allow one to test and calibrate a variety of humidity-measuring instruments, such as probes, hand-held instruments, or transducers. The Humidity Check contains a saturated salt solution that maintains an equilibrium humidity within the cartridges for every salt solution. Using different salts it is possible to manufacture reference cartridges for the entire measurement range of 0...100%RH. The MHT Humidity Checks were designed to be used in the field as well as in laboratories. They are small, handy, and can operate in any position. This allows one to calibrate humidity probes without removing them from their respective facilities. Special adapters for probes with different diameters hermetically seal the Humidity Checks during calibration.

As for the disadvantages of this method, they are as follows (https://www.omega.com). Saturated salt solutions are difficult to wash out of the objects after accidents. The salt tends to crystallize at the edges of the container. The crystals form a labyrinth that allows solution to rise by capillarity and crystallize further up the wall. Eventually the salt gets everywhere. A less well-known problem is that the salt solution is bad at dehumidifying. During this process, the saturated solution begins to absorb water from the air. This dilutes the solution at the surface. The less dense surface solution is gravitationally stable and does not mix with the bulk of saturated solution, so the RH drifts upward from the theoretical value for the saturated solution toward 100%. If the room temperature is unstable, the salt solution behaves quite unpredictably, causing considerable errors when such solutions are used in experimental work to measure absorption isotherms or to calibrate instruments.

In addition, one should note that many of the salts previously mentioned are hazardous to one's health. Therefore, the properties of such agents should always be brought to the attention of the personnel working with them. All chemicals should be stored in safe and clearly labeled containers

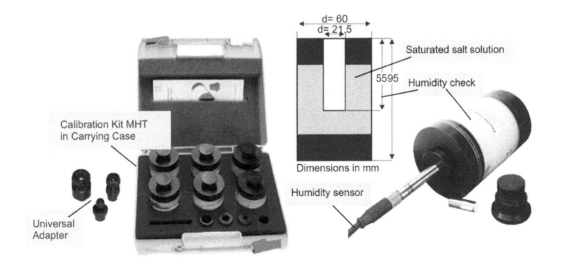

**FIGURE 26.8** Humidity Calibration Kit MHT. The Basic Kit contains 3 Humidity Checks whereas the Professional Kit comes with 6 Humidity Checks: MHT0 Humidity Check 0.8%RH; MHT11 Humidity Check 11.3%RH; MHT33 Humidity Check 33.1%RH; MHT54 Humidity Check 54.0%RH; MHT75 Humidity Check 75.5%RH; and MHT97 Humidity Check 97.5%RH. (Modified from https://www.driesen-kern.com/downloads/dk-product-line-of-humidity-and-temperature.pdf)

and in an appropriate environment. The notes that follow give some guidance for the safe use of some commonly used salts:

a. Barium chloride ($BaCl_2$): Colorless crystals; very soluble in water; stable, but may emit toxic fumes in a fire; no hazardous reaction with water, acids, bases, oxidizers, or with combustible materials; ingestion causes nausea, vomiting, stomach pains, and diarrhea; harmful if inhaled as dust and if it comes into contact with the skin; irritating to eyes; treat with copious amounts of water and obtain medical attention if ingested.

b. Calcium chloride ($CaCl_2$): Colorless crystals; deliquescent; very soluble in water, dissolves with increase in heat; will initiate exothermic polymerization of methyl vinyl ether; can react with zinc to liberate hydrogen; no hazardous reactions with acids, bases, oxidizers, or combustibles; irritating to the skin, eyes, and respiratory system; ingestion causes gastric irritation; ingestion of large amounts can lead to hypocalcaemia, dehydration, and renal damage; treat with copious amounts of water and obtain medical attention.

c. Lithium chloride (LiCl): Colorless crystals; stable if kept dry; very soluble in water; may emit toxic fumes in a fire; ingestion may affect ionic balance of blood leading to anorexia, diarrhea, vomiting, dizziness, and central nervous system disturbances; kidney damage may result if sodium intake is low (provide plenty of drinking water and obtain medical attention); no hazardous reactions with water, acids, bases, oxidizers, or combustibles.

d. Magnesium nitrate [$Mg(NO_3)_2$]: Colorless crystals; deliquescent; very soluble in water; may ignite combustible material; can react vigorously with deoxidizers; can decompose spontaneously in dimethylformamide; may emit toxic fumes in a fire (fight the fire with a water spray); ingestion of large quantities can have fatal effects (provide plenty of drinking water and obtain medical attention); may irritate the skin and eyes (wash with water).

e. Potassium nitrate ($KNO_3$): White crystals or crystalline powder; very soluble in water; stable but may emit toxic fumes in a fire (fight the fire with a water spray); ingestion of large quantities causes vomiting, but it is rapidly excreted in urine (provide plenty of drinking water); may irritate eyes (wash with water); no hazardous reaction with water, acids, bases, oxidizers, or combustibles.

f. Sodium chloride (NaCl): Colorless crystals or white powder; very soluble in water; stable; no hazardous reaction with water, acids, bases, oxidizers, or combustibles; ingestion of large amounts may cause diarrhea, nausea, vomiting, deep and rapid breathing, and convulsions (in severe cases obtain medical attention).

### 26.2.2.3  Other Secondary Methods

Water-sulfuric acid mixtures (Wilson 1921) also produce an atmosphere of RH, which depends on the composition and temperature. Two techniques may be employed. The liquid can be exposed in a suitable tray in a sealed chamber to provide an equilibrium vapor pressure of the mixture, or air may be bubbled or otherwise brought into intimate contact with the liquid. Wilson's data (1921) are reproduced in Table 26.5.

The technique employed water-glycerine mixtures also works equally well with water-sulfuric acid mixtures. It was established that water-glycerine mixtures (Wexler and Brombacher 1972) also similarly produce atmospheres of known RH. The RH obtainable at 25°C from various water-glycerine solutions is given in Table 26.6.

**TABLE 26.5**

**Relative Humidity Obtained from Water–Sulfuric Acid Solutions**

| Relative Humidity | Percentage of $H_2SO_4$ (by weight) at Temperature | | | |
|---|---|---|---|---|
| % | 0°C | 25°C | 50°C | 75°C |
| 10 | 63.1 | 64.8 | 66.6 | 68.3 |
| 25 | 54.3 | 55. 9 | 57.5 | 59.0 |
| 35 | 49.4 | 50.9 | 52.5 | 54.0 |
| 50 | 42.1 | 43.4 | 44.8 | 46.2 |
| 6.5 | 34.8 | 36.0 | 37.1 | 38.3 |
| 75 | 29.4 | 30.4 | 31.4 | 32.4 |
| 90 | 17.8 | 18.5 | 19.2 | 20.0 |

*Source:*  Wilson, R.E., *J. Ind. Eng. Chem.*, 13, 326–331, 1921.

**TABLE 26.6**

**Relative Humidity Obtained from Water–Glycerine Mixtures at 26°C**

| Relative Humidity (%) | Glycerine (by weight) (%) | Specific Gravity |
|---|---|---|
| 10 | 95 | 1.245 |
| 20 | 92 | 1.237 |
| 30 | 89 | 1.229 |
| 40 | 84 | 1.216 |
| 50 | 79 | 1.203 |
| 60 | 72 | 1.184 |
| 70 | 64 | 1.162 |
| 80 | 51 | 1.127 |
| 90 | 33 | 1.079 |

*Source:*  Wexler, A. and Brombacher, W.G., Methods of measuring humidity and testing hygrometers: A review and bibliography, in: Bloss R.L., Orloski M.J. (eds.) *Precision Measurement and Calibration: Mechanics*, National Bureau of Standard, Washington, DC, pp. 261–280, https://www.nist.gov/.

## 26.3 DYNAMIC METHOD OF HUMIDITY-SENSOR CALIBRATION AND TESTING

### 26.3.1 GAS-MIXING SYSTEMS

All of the above methods used to create the necessary atmosphere for calibration and testing of humidity sensors are static. The main disadvantage of using static methods for preparing humidity standards is the absorptive losses on surfaces such as bags and gas cylinder walls (Jardine et al. 2010). This loss is difficult to explain, especially in the parts-per-billion range and is particularly significant for water vapor with low vapor pressures. Endres et al. (1995) believe that an optimal gas-mixing system can be designed only on the basis of a dynamic mixing flow method and provide a constant gas flow through the measurement chamber. The main benefits of this method are that a constant gas flow avoids any problem with gas consumption by the sensor itself, and also minimizes problems with adsorption and desorption at the walls of the test system. In addition, a constant gas flow prevents any distortion of the measurement due to changes in gas-exchange time constants, forced cooling of heated sensors, and other flow-dependent effects. Thus, pure dynamic methods minimize wall effects, allow rapid variations in concentration, and enable the generation of standards when needed.

A variety of systems have been designed for injection of known, controlled amounts of the water vapor into the flowing diluents stream (Takahashi and Kitano 1996; Su and Wu 2004; Wang et al. 2006). At present, such flow-humidity generator systems are used in many industries, calibration laboratories, and testing laboratories to generate a constant humidity atmosphere for calibration of hygrometers and humidity sensors. One of the possible systems used for calibration and testing of humidity sensors is shown in Figure 26.9. Usually a dynamic gas-dilution system has sources of water vapor and diluent gases (air), and several channels with gas-flow controllers and mixing chambers. In particular, the system shown in Figure 26.9 has two additional channels intended for cross-calibration. At constant pressure $p$ and constant temperature $T$, the flows $q_i$ of gases are mixed continually. The concentration of water vapor in the output stream of dilution system depends on the initial concentration of water vapor in calibration flow and the ratio of gas flows in the channels. If the values of flows $q_i$ are known, the relative concentration $X_i$ can be calculated from Eq. (26.3).

$$X_i = \frac{q_i}{q_i + q_n} = M, \qquad (26.3)$$

where $M$ is the dilution constant determined by the ratio of the flows $q_i$ and $q_n$.

As a rule, in devices for testing humidity sensors, calibration mixtures are not used, but mixing air saturated with water vapor with dry air is used, as is done in the two-flow method described in Section 26.2. In this case, the most common method of saturating air with water vapor is the bubbler

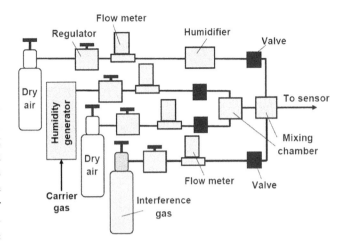

**FIGURE 26.9** Schematic diagram of gas system used for humidity-sensor calibration. (Adapted with permission from Korotcenkov, G. and Cho, B.K., Chemical sensors selection and operation guide, in: *Chemical Sensors: Comprehensive Sensor Technologies*. Vol. 6. *Chemical Sensors Applications*, Korotcenkov G. (ed.), Momentum Press, New York, pp. 281–348, 2011. Copyright 2011: Momentum Press.)

method, which is distinguished by its simplicity. As is known, in a closed volume containing water, a liquid-gas equilibrium is established, which is characterized by temperature-dependent values, such as saturated vapor pressure and density of saturated water vapor. At thermodynamic equilibrium, the concentration in the vapor phase corresponds to that of the saturation state (Boublik et al. 1984). The vapor pressure of water may be approximated by the following relation (Eq. 26.4):

$$P = \exp\left(20.386 - \frac{5132}{T}\right), \qquad (26.4)$$

where $P$ is the vapor pressure (mm Hg), and $T$ is the temperature in Kelvin. Data related to water saturated vapor pressure and density of saturated water vapor are presented in Table 26.7.

An important point in such bubblers is to achieve a steady, controllable flux of a water vapor (Takahashi and Kitano 1996; Su and Wu 2004). Common restriction is the formation of aerosols (Kim et al. 2007). However, it is possible to minimize this problem through an optimized bubbler design (Smith et al. 2004) and correcting via a system calibration (Mackay-Sim et al. 1993). To achieve a state of equilibrium, it may be necessary to consistently implement several stages of concentration enrichment in the vapor phase. A conventional realization of a calibration humidity generator, working according to a three-step saturation principle, is shown in Figure 26.10. Purified compressed air passes sequentially through three bottles filled with water, the temperature of which is controlled. Of course, one- or two-step saturation can be used as well (Kim et al. 2007).

It is worth noting that systems with modern gas flow controllers can have dilution constants exceeding 100. If greater

**TABLE 26.7**

**Water Saturated Vapor Pressure and Saturated Vapor Density**

| Temperature (°C) | Saturated Vapor Pressure | | Saturated Vapor Density |
|---|---|---|---|
| | (10³ Pa) | (mm Hg) | (g/m³) |
| 0 | 0.6105 | 4.58 | 4.85 |
| 5 | 0.8722 | 6.54 | 6.8 |
| 10 | 1.228 | 9.21 | 9.4 |
| 20 | 2.338 | 17.54 | 17.3 |
| 30 | 4.243 | 31.8 | 30.4 |
| 40 | 7.376 | 55.3 | 51.1 |
| 50 | 12.33 | 92.5 | 83.1 |
| 60 | 19.92 | 149.4 | 130.5 |
| 70 | 31.16 | 233.7 | 198.6 |
| 80 | 47.34 | 355.1 | 293.8 |
| 90 | 70.10 | 525.8 | 424.1 |
| 100 | 101.3 | 760.0 | 598.0 |

*Source:* Data extracted from www.vivoscuola.it/US and www.engineering Toolbox.com.

**TABLE 26.8**

**Transmission Rates of Water Vapor in Some Basic Plastic Materials**

| Plastic Material | Water Vapor Transmission Rate (38°C, 90%RH), g·mm/m²·day |
|---|---|
| Polyvinylidene chloride (PVDC) | 0.01–0.08 |
| Polychlorotrifluoroethylene (PCTFE) | 0.015 |
| Polypropylene (oriented) | 0.16 |
| Polypropylene | 0.26 |
| Polyethylene terephthalate (oriented) (PET) | 0.8 |
| Polyvinylchloride (PVC), rigid | 0.88 |
| Nylon MXD6 (oriented) | 1.1 |
| Nylon 66 | 1.5 |
| High nitrile resins | 1.6 |
| Ethylene vinyl alcohol based materials (EVOH) | 1.3–3.4 |
| Nylon 6 | 4.3 |
| Cellophane | 137 |
| Polyvinyl alcohol | 750 |

*Source:* Sidmell, J.A., Food contact polymeric materials, RARPA review report. RARPA Technology Ltd, Shrewsbury, Shropshire, UK, 1992.

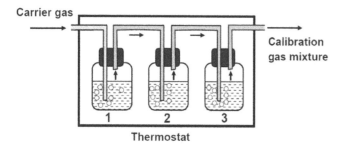

**FIGURE 26.10** Setup of a three-step calibration humidity generator according to the saturation method.

dilutions are desired, it is possible to connect the dilution units to a cascade. However, multistage dilution techniques that require more than two dilution steps become considerably more difficult (Mitchell 1987). As more dilution steps are added, the system pressure builds up, and the required amount of gas of interest becomes the main problem. In such arrangements, as much as 99% of the test gas could be bled off and not used in the output stream. In addition, when multistage systems are used, at the low end of the concentration range additional problems should be taken into account that may arise from the following (Mitchell 1987):

- *Insufficient flow-rate control.*
- *Back diffusion.* Diffusion of the sample outward or contaminant or even a major constituent inward— probably one of the most frequent sources of error in gas analysis. The permeabilities of the calibration system to both the gases of interest and unwanted contaminants have to be determined if reliable results are to be obtained.
- *Surface absorption.* This problem is most critical when considering trace levels of reactive gases. Losses in calibration mixtures may occur either by primary absorption on container walls or connecting gas pipes, or by secondary absorption that occurs through chemical reaction of component of interest with previously absorbed species.

The mechanical features of the sampling system can also introduce various errors if they go undetected. The sources of these errors are leaks, faulty seals, corrosion of vital parts, etc. It should also be borne in mind that many materials, such as polymers, are permeable to water vapor (see Table 26.8), and therefore the choice of materials used for the manufacture of gas systems is important (Hienonen and Lahtinen 2007). Even good quality plastics always let a certain amount of water through. Moreover, it only takes a few hours or weeks for water to penetrate a layer of plastic of 1.0 mm thick. It follows that the use of metal for these purposes is preferred. Therefore, for reasons of desired accuracy, it is not recommended to mix the flows with a ratio greater than $q_i{:}q_n \leq 1{:}100$.

Examples of market-acceptable humidity generators that provide a constant gas flow with controlled humidity are shown in Figure 26.11. Diagram illustrating configuration of one of such humidity generators is shown in Figure 26.12.

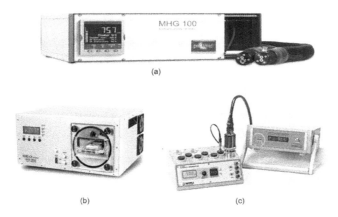

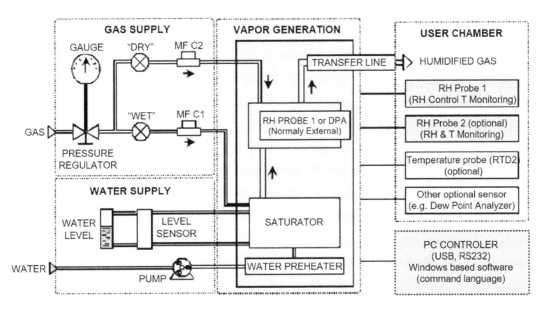

**FIGURE 26.11** (a) Humidity generator MHG100 (Modified from https://proumid.com/en/humidity-generator/); (b) portable humidity generator MODEL 2000 (Modified from https://www.geocalibration.com/humidity-generators/); and (c) Mitchell S-503 Portable Relative Humidity Generator (Modified from https://www.instrumart.com/products/40213/michell-instruments-s503-relative-humidity-generator). As a rule, such generators provide accuracy ±2% RH or better in the range of 5–95% RH.

**FIGURE 26.12** Operational schematic of humidity generator HumiSys HF. (Modified from http://www.instruquest.com/HumiSys.html.)

### 26.3.2 DIFFUSION AND PERMEATION SYSTEMS

One should note that other than dynamic dilution, when testing humidity sensors in dynamic mode, other methods can be used to prepare the gas-calibration mixtures with desired air humidity. Such methods include diffusion, permeation, and chemical reactions (Mitchell 1987). A typical example of a method based on chemical reactions is the catalytic reaction of hydrogen oxidation on a platinum catalyst (Mermoud et al. 1991). Another example is the catalytic oxidation of hydrogen on the surface of copper oxide in $N_2$ atmosphere. $H_2$ is oxidized stoichiometrically to $H_2O$ at medium temperature (350°C–450°C) effecting reduction of the metal-oxide couple. GR2L company (www.GR2L.co.uk) developed a moisture generator based on this method. It is argued that the accuracy of $H_2O$ depends only on the accuracy of the original $H_2$ mixture, and typically this accuracy is better than 1%. The metal-oxide couple can be regenerated with air.

Diffusion systems have been used for a long time in analytical chemistry for preparing calibration gas mixtures (Nakamoto et al. 1991; ISO 2005; Zhou and Gai 2006; Pratzler et al. 2010). A schematic diagram of a diffusion cell is shown in Figure 26.13. In our case, water is filled into a diffusion vessel,

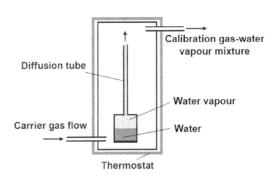

**FIGURE 26.13** Schematic diagram of a diffusion cell.

which is provided with a diffusion tube and is surrounded by the carrier gas flow. Water molecules from the liquid phase evaporate into the gaseous phase until the saturated water-vapor pressure is reached. Then, molecules in the gas phase rise into the diffusion tube and, after diffusion through the diffusion tube, mix with the carrier gas in one step. The quantity of water molecules that diffuses into the carrier gas depends on the temperature, the geometry of the diffusion tube, and the water-vapor pressure. The diffusion rate can be determined by weighing the diffusion vessel and is thus directly traceable to the mass. Zhou and Gai (2006) have shown that the relative uncertainty of the composition of the calibration gas mixture prepared using diffusion tubes is about 5%.

However, at present, the methods based on chemical reactions and diffusion are used only rarely for humidity-sensor calibration. Permeation systems found wider application for these purposes. But this application is also limited.

### 26.3.2.1 Permeation Systems

The operation of permeation systems is very similar to that of diffusion systems (Staude 1992). However, their application is much more convenient. A *permeation device* is a sealed small container filled with a pure chemical compound, water in our case, which is in a two-phase equilibrium between its gas phase and liquid phase (Scace and Miller 2008). The water molecules are permeated either through a permeable container wall or through the end cap and mixed with a carrier gas to obtain a known gas-water vapor mixture used as reference in testing equipment. This phenomenon leads to a very versatile method of producing ultralow concentrations of calibration gas-water vapor mixtures. A permeation device usually presents a polymeric container, typically fabricated from polytetrafluoroethylene (Teflon). Construction and some typical permeation tubes are shown in Figure 26.14. The rate at which the water molecules permeate depends on the permeability of the material and the temperature. Moreover, the rate of permeation is constant over long periods of time if the temperature, concentration gradient, and tube geometry remain constants. The rate of flow of moisture emitted from the tube is measured by holding the tube at constant temperature in a stream of dry gas for an extended period of time and periodically weighing the tube to measure the rate of weight loss. The emission rate of any tube can be varied by changing the operating temperature of the tube.

Permeation devices have several advantages (Scace and Miller 2008). They are small, inexpensive, and convenient to use. They are ideal for lab environments. Usually, permeation devices are the size of a pencil (see Figure 26.14). With these devices, there is a low degree of complexity, a wide range of permeation rates can be achieved, along with accurate, stable concentrations ranging from parts per billion to high parts per million. Permeation devices are ideal for use in the field, and high precision and excellent accuracy may be obtained for many gases (Mitchell 1987). And, after they are depleted, they can be disposed of easily. However, permeation tubes continuously emit chemicals at a constant rate, thus creating a storage and safety problem (Chou 2000). Also, the rate of permeation for a given gas of interest can be too high or too

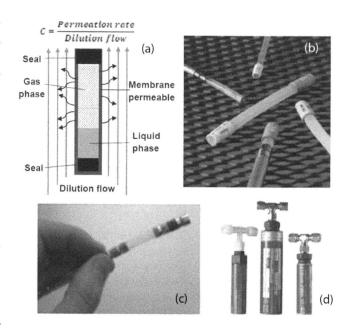

**FIGURE 26.14** (a) Diagram illustrating construction and operation of permeation tube; and (b–d) examples of permeation tubes fabricated by various companies. (Modified from https://www.vici.com/calib/perm_dyna.php and https://www.owlstoneinc.com/products/permeation-tubes-and-diffusion-tubes/.)

low for a given application. In our case, the rate of permeation is low, and therefore, this method is used only to create trace concentration moisture standards (ws680.nist.gov; low ppm and ppb range) for calibration hygrometers in electronic industry (Huang and Kacker 2003; Scace and Miller 2008).

### 26.3.3 FEATURES OF TESTING THE SENSOR IN THE MEASURING CELL OF FLOW TYPE

Theoretically, the flow rate should not have a direct effect on the measured level of humidity in the gas system, but in practice, the flow rate of a gas can affect the system's accuracy. Therefore, care should be taken to ensure that the sampling system can accommodate the required flow rate for the measurement instrument (Wiederhold 1997). It must be borne in mind that an excessive flow rates in piping systems can introduce pressure gradients and influence the temperature of the sensor. However, too-low flow rate can be accompanied by such side effects as:

- Back diffusion: If an open-ended sampling system is used, ambient air can flow back into the system.
- Ineffective purging of the sampling system: In a complex system, inadequate flow can allow pockets of undisturbed wet gas to remain in the sampling system or sensor, which will gradually be released into the sample flow. In worst-case conditions, liquid water may exist for long periods in moisture traps in the sample handling system. This can happen after hydraulic testing or after a high-moisture fault condition.

- Increased influence of adsorption and desorption effects in the volume of gas passing through the sampling system.

As is known, sensitivity and response and recovery times are important parameters of any gas sensors, including humidity sensors. However, it is necessary to admit that most researchers direct their efforts toward the measuring magnitude of sensor response only, and therefore they usually use test chambers with large volumes. In the literature, one can find descriptions of measurement chambers with volumes up to $(1–5) \times 10^3$ cm$^3$ and higher. In industry, the volume of chamber can achieve more than $10^6$ cm$^3$. However, as shown by modeling and experiment, carried out by Lezzi et al. (2001), for test chambers with volumes of 1000 and 233 cm$^3$ and for flow from 100 to 500 cm$^3$/min, using a large-volume chamber is a faulty approach even for experiments in steady-state conditions. Numerical results confirmed that geometric parameters such as the measurement chamber volume, the chamber shape, and the inlet and outlet positions play important roles for the fluid flow and the concentration fields inside the chamber as well. With a poorly designed test chamber, characterized by large volumes of almost stagnant fluid and/or by a peripheral position of the sensor with respect to the gas stream, the concentration of calibration gas near the sensor surface does not correspond to the concentration of the calibration gas after preparation, and the calibration gas concentration on the sensor surface may continue to increase for quite a long time. It has been established that in many cases the time taken for replacement of the gas atmosphere in a gas chamber from base air to sample gases is much longer than the time required for the sensor to reach steady state. In this case, the device response time is more a measure of the change of tested gas concentration inside the chamber than a characteristic feature of the device. This means that these chambers cannot be used for measuring operating characteristics of gas sensors in transient mode and determining their response and recovery times. Even chambers with volumes of 100 cm$^3$ are too large to quickly introduce gases into the chamber within a sufficiently short time.

There are two ways to exchange the ambient quickly. Either the chamber has a normal volume and the gas has extremely high flow, or the gas flow is normal and the chamber is very small. As is known, the time constant of gas exchange in a measurement cell depends on its volume, the construction of the gas tubes, and the flow rate. Approximately, the purging time ($t_{purge}$) of the measurement chamber can be calculated according to Eq. (26.5) (Endres et al. 1995)

$$t_{purge} = 3 \frac{V_{chamber}}{v_{tot}} \quad (26.5)$$

where $V_{chamber}$ is the volume of the chamber, $v_{tot}$ is the gas flow, and 3 is a security factor. The first principle was realized by Gerblinger and Meixner (1991), who mounted nozzles of 10 injection valves into a wall of a sensor chamber. However, this approach is not optimal because a gas stream, moving at high speed, strongly affects the temperature of the heated

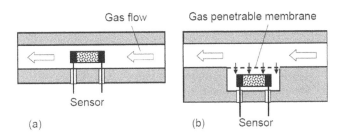

(a)                                      (b)

**FIGURE 26.15** Variants of the gas-sensor position relative to the gas stream: (a) the sensor is located directly in the gas stream, (b) the sensor is located outside the gas stream and it is separated by a gas permeable membrane. (Reprinted with permission from Korotcenkov, G. and Cho, B.K., Chemical sensors selection and operation guide, in: *Chemical Sensors: Comprehensive Sensor Technologies.* Vol. 6. *Chemical Sensors Applications*, Korotcenkov G. (ed.), Momentum Press, New York, pp. 281–348, 2011. Copyright 2011: Momentum Press.)

sensor. The experiment showed that in this case the decrease of measurement cell is a more acceptable approach.

The location of the sensor relative to the gas flow can also affect the test results. There are two options for the position of the humidity sensor relative to the gas flow, which contribute to a significant reduction in the influence of the gas flow on the temperature parameters of the sensor. The most commonly utilized variant is shown in Figure 26.15b. However, due to the additional volume in the measurement cell and the presence of an additional membrane between the gas flow and the gas sensor, the time to establish a steady state in such a system after exchanging the gas atmosphere is longer than in the flow system shown in Figure 26.15a. It is necessary to note that test chambers developed to study fast processes in the humidity-sensor response, as a rule, are designed to use unpackaged sensors, which do not have additional filters and membranes that prevent the test gas from penetrating into the active element of the humidity sensor.

## 26.4 SOME GENERAL RULES FOR HUMIDITY-SENSOR CALIBRATION AND TESTING

Cleaver (2001) analyzed the sources of uncertainty in analytical procedures and found that they can appear at all stages of the measurement process, including sampling, sample preparation, calibration of the instrument, analysis and data acquisition, data processing, and interpretation of results. Factors such as insufficient gas-mixture homogeneity, limited number of replicate samples taken, variations in temperature and pressure during sampling, and the uncertainty of the certified reference material used in the measuring system may cause errors when calibrating and testing chemical sensors. According to Chou (2000), the calibration procedure should be simple, straightforward, and easily executed by regular personnel. Therefore, to avoid this situation, the process of calibration and testing should be carried out in accordance with all rules established for these processes. It is recommended to follow the following rules (Scott et al. 1996; Chou 2000; Rotronic 2005; Heinonen 2006; Bell 1996, 2012):

- Calibrate in terms of the quantity to be measured (such as RH, or dew point, or other …). Calibrate for the range and conditions of use.
- Wherever possible, calibrations should be performed under the intended conditions of use, that is, at similar values of humidity and temperature, and preferably in similar conditions of pressure, air flow, etc.
- If it is possible, synthetic air ($N_2 + O_2$) should not be used for calibration. Besides, nitrogen and oxygen in normal air contains traces of other gases such as hydrocarbons, carbon monoxide, carbon dioxide, and possibly other gases that can interference with water vapors. Therefore, it is much more realistic and practical to use the air surrounding the sensor for sensor calibration.
- Obtaining equilibrium conditions is one of the most critical requirements of the method. This means that there should be no difference of temperature between the humidity sensor, the solution and the head space above the solution. Unstable temperatures during calibration will not permit this. A temperature stability of 0.02°C/min or better is required during the calibration process for the method to be accurate. Once the sensor is inserted, wait for at least 30 minutes at stable temperature so that the instrument and the salt solutions reach thermal equilibrium with the environment. The measurement chamber must be closed; otherwise, the equilibrium cannot be reached.
- In an environmental chamber, the errors of measurement can result of a poor choice in the mounting location of the thermometers. This is the case when the thermometers are installed too closely to a source of moisture or to the walls of the chamber. It is important to mount the thermometers at a location where conditions are fairly representative of the average conditions inside the chamber.
- The RH values generated by the different solutions used to the purpose of calibration are affected by temperature. Therefore, a correction must be made for the temperature of calibration. A saturated solution results of the dynamic equilibrium of two processes: formation of a solution and crystallization. If a saturated salt solution is used for calibration, attention must be paid to this equilibrium. Because solubility depends on temperature, a variation in the temperature of a saturated solution can disturb the dynamic equilibrium process in the solution.
- Where a hygrometer consists of separate parts (e.g., probe and electronics), the pair should be calibrated together "as one item" and used together.
- Time must be allowed for equilibrium when the temperature of a saturated salt solution is changed.

This time period is usually significantly longer than the time required by the temperature of the solution to stabilize at a new value. Nonsaturated solutions adapt faster to a change in temperature because they do not require a dynamic equilibrium between a crystal and a liquid. This is convenient for use in field calibrations where temperature is not precisely known ahead of time. Most solutions release or absorb heat during their preparation. For that reason, solutions should be prepared in advance. Do not add more water than is needed to make the salt look damp.

- Especially for dew-point probes used for more than one pressure or gas type, check whether calibration in air at atmospheric pressure is applicable.
- Measurements in the low ppm and ppb range require special attention. There are several special problems associated with calibrating trace-moisture monitors. Of particular significance are (a) assuring low background concentration of water vapor in carrier gas, (b) excluding all external moisture, (c) obtaining adequate dry-down of the system, (d) transmitting concentration changes to the analyzer, (e) preventing transient shifts, and (f) interfacing to the analyzer.
- To obtain good dry-down characteristics of the measurement system, the choice of components and materials is of extreme importance. Where low water content has to be measured, special attention should be given to the material and cleanliness of the pipes used. The lower the moisture content, the more significant the effects are (Figure 26.16). It was found that the materials used in all components must be of high quality, surface area should be minimized, and all components must be thoroughly cleaned of any possible impurities. Tubing used in the system should

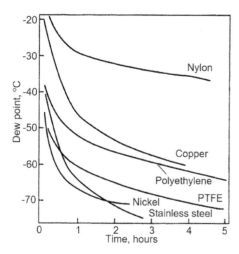

**FIGURE 26.16** Illustration of the moisture given off by different tubing materials when flushed with very dry gas after being at ambient humidity. (Data extracted from Guide, *A Guide to the Measurement of Humidity*, The Institute of Measurement and Control, London, UK, 1996; https://www.rotronic.com.)

be internally electropolished. Tubing can be somewhat passivated to moisture adsorption by heating in the presence of high-purity nitrogen. Once a system has reached dry-down, it is also important to test the system to be sure that it will transmit changes in moisture concentration. Moisture should be sequentially added to and removed from the system to insure that the overall system will respond rapidly to changes in moisture concentration. High-quality stainless-steel components, if they are clean and polished at the internal surface, will transmit adequately. If a sluggish response is observed, it should be assumed that there is some area in the system that has not been adequately cleaned or that some undesirable material has been inadvertently included in the system.

- A check is not a calibration. But checks against another instrument (if available) are highly useful. A check can help in assessing whether a hygrometer is functioning as expected, and whether repair or recalibration is needed.

# REFERENCES

Afgan N.H., Carvalho M.G., Pilavachi P.A., Tourlidakis A., Olkhonski G.G., Martins N. (2006) An expert system concept for diagnosis and monitoring of gas turbine combustion chambers. *Appl. Thermal Eng.* **26**, 766–771.

Barratt R.S. (1981) The preparation of standard gas mixtures: A review. *Analyst* **106**, 817–849.

Bell S. (2012) A beginner's guide to humidity measurement, National Physical Laboratory, Teddington, Middlesex, UKTW11 OLW (http://www.npl.co.uk; https://www.rotronic.com/).

Bell S.A. (1996) *A Guide to the Measurement of Humidity*, The Institute of Measurement and Control, London, UK.

Berry L., Brunet J. (2008) Oxygen influence on the interaction mechanisms of ozone on $SnO_2$ sensors. *Sens. Actuators B* **129**, 450–458.

Boublik T., Fried V., Hala E. (1984) *The Vapour Pressures of Pure Substances*. Elsevier, Amsterdam, the Netherlands.

Brahim-Belhouari S., Bermak A., Shi M., Chan P.C.H. (2005) Fast and robust gas identification system using an integrated gas sensor technology and Gaussian mixture models. *IEEE Sensors J.* **5**(6), 1433–1444.

Chou J. (2000) *Hazardous Gas Monitors: A Practical Guide to Selection, Operation and Application.* McGraw-Hill, New York.

Cleaver K.D. (2001) The analysis of process gases: A review. *Accred. Qual. Assur.* **6**, 8–15.

COGDEM (1999) *Gas Detection and Calibration Guide.* The Council of Gas Detection and Environmental Monitoring, Hitchin, UK (www.cogdem.org.uk).

Dellin T.A., Miller W.M., Pierce D.G., Snyder E.S. (1993) Wafer level reliability. In: *Proc. SPIE* Vol. 1802, *Microelectronics Manufacturing and Reliability*, Society of Photo-Optical Instrumentation Engineers, Bellingham, Washington, US, pp. 144–154.

Demarne V., Grisel A., Sanjines R., Levy F. (1990) Integrated semiconductor gas sensors evaluation with an automatic test system. *Sens. Actuators B* **1**, 87–92.

Di Francesco F., Falcitelli M., Marano L., Pioggia G. (2005) A radially symmetric measurement chamber for electronic noses. *Sens. Actuators B* **105**, 295–303.

Endres H.E., Jander H.D., Gottler W. (1995) A test system for gas sensors. *Sens. Actuators B* **23**, 163–172.

Frank R. (2013) *Understanding Smart Sensors.* Artech House, Boston, MA.

Gerblinger J., Meixner H. (1991) Fast oxygen sensor on sputtered strontium titanate. *Sens. Actuators B* **4**, 99–102.

Greenspan L. (1977) Humidity fixed points of binary saturated aqueous solutions. *J. Res. Nat. Bur. Stand. A* **81A**(1), 89–96.

Guide (1996) *A Guide to the Measurement of Humidity*, The Institute of Measurement and Control, London, UK.

Harvey I., Coles G., Watson J. (1989) The development of an environmental chamber for the characterization of gas sensors. *Sens. Actuators* **16**(4), 393–405.

Heinonen M. (1996) The CMA humidity standard. *Measurement* **17** (3), 183–188.

Heinonen M. (1999) A humidity generator with a test chamber system. *Measurement* **25**(4), 307–313.

Heinonen M. (2006) Uncertainty in humidity measurements, Publication of the EUROMET Workshop P758, Mittatekniikan keskus, Espoo 2006, Publication J4/2006.

Hienonen R., Lahtinen R. (2007) *Corrosion and Climatic Effects in Electronics.* VTT Technical Research Centre of Finland, Espoo, Finland.

Huang P.H., Kacker R. (2003) Repeatability and reproducibility standard deviations in the measurement of trace moisture generated using permeation tubes. *J. Res. NIST* **108**(3), 235–240.

ISO 6142 (2001a) *Gas Analysis—Preparation of Calibration Gas Mixtures—Gravimetric Method.* www.iso.org.

ISO 6144 (2003a) *Gas Analysis—Preparation of Calibration Gas Mixtures—Static Volumetric Method.* www.iso.org.

ISO 6145-2 (2001b) *Gas Analysis—Preparation of Calibration Gas Mixtures Using Dynamic Volumetric Methods: Part 2. Volumetric Pumps.* www.iso.org.

ISO 6145-5 (2009a) *Gas Analysis—Preparation of Calibration Gas Mixtures Using Dynamic Volumetric Methods: Part 5. Capillary Calibration Devices.* www.iso.org.

ISO 6145-6 (2003b) *Gas Analysis—Preparation of Calibration Gas Mixtures Using Dynamic Volumetric Methods: Part 6. Critical Orifices.* www.iso.org.

ISO 6145-7 (2009b) *Gas Analysis—Preparation of Calibration Gas Mixtures Using Dynamic Volumetric Methods: Part 7. Thermal Mass-Flow Controllers.* www.iso.org.

ISO 6145-8 (2005) *Gas Analysis—Preparation of Calibration Gas Mixtures Using Dynamic Volumetric Methods: Part 8. Diffusion Method.* www.iso.org.

ISO 6145-9 (2009c) *Gas Analysis—Preparation of Calibration Gas Mixtures Using Dynamic Volumetric Methods: Part 9. Saturation Method.* www.iso.org.

Jardine K.J., Henderson W.M., Huxman T.E., Abrell L., Shartsis T. (2010) Dynamic solution injection: A new method for preparing pptv-ppbv standard atmospheres of volatile organic compounds. *Atmos. Meas. Tech. Discuss.* **3**, 3047–3066.

Kim Y.S., Ha S.-C., Yang H., Kim Y.T. (2007) Gas sensor measurement system capable of sampling volatile organic compounds (VOCs) in wide concentration range. *Sens. Actuators B* **122**, 211–218.

Korotcenkov G., Cho B.K. (2011) Chemical sensors selection and operation guide. In: Korotcenkov G. (ed.), *Chemical Sensors: Comprehensive Sensor Technologies.* Vol. 6. *Chemical Sensors Applications*, Momentum Press, New York, pp. 281–348.

Korposh S.O., Takahara N., Ramsden J.J., Lee S.-W., Kunitake T. (2006) Nano-assembled thin film gas sensors. I. Ammonia detection by a porphyrin-based multilayer film. *J. Bio. Phys. Chem.* **6**, 125–132.

Lezzi A.M., Beretta G.P., Comini E., Faglia G., Galli G., Sberveglieri G. (2001) Influence of gaseous species transport on the response solid state gas sensors within enclosures. *Sens. Actuators B* **78**, 144–150.

Lin Y.F., Yeh T.I., Chan K.H., Chen T.S. (1996) The automatic calibration system of humidity fixed points at CMS. *Measurement* **19**(2), 65–71.

Lu T., Chen C. (2007) Uncertainty evaluation of humidity sensors calibrated by saturated salt solutions. *Measurement* **40**, 591–599.

Mackay-Sim A., Kennedy T.R., Bushell G.R., Thiel D.V. (1993) Sources of variability arising in piezoelectric odorant sensors. *Analyst* **118**, 1393–1398.

Magliulo V. (1996) Automatic scanning of multiple leaf chambers for gas exchange measurements. *Comput. Electron. Agric.* **15**, 149–160.

Majewski J. (2016) Polymer-based sensors for measurement of low humidity in air and industrial gases. *Przeglad Electrotechniczny* **92**(8), 74–77.

Maudie T., Monk D.J., Frank R. (1997) Packaging considerations for predictable lifetime sensors. In: *Proceedings of Sensors Expo*, May 13–15, Boston, MA, pp. 167–172.

Maudie T., Tucker B. (1991) Reliability issues for silicon pressure sensors. In: *Proceedings of Sensors Expo*, October 1–3, Chicago, IL, pp. 101–1 to 101–8.

Maudie T., Wertz J. (1997) Pressure sensor performance and reliability. *IEEE Industry Applications Magazine* **3**(3), 37–43.

Maziarz W., Pisarkiewicz T. (2008) Gas sensors in a dynamic operation mode. *Meas. Sci. Technol.* **19**, 055205.

Mermoud F., Brandt M.D., McAndrew J. (1991) Low-level moisture generation. *Anal. Chem.* **63**, 198–202.

Mitchell G.D. (1987) Trace gas calibration systems using permeation devices. In: Taylor J.K. (ed.), *Sampling and Calibration for Atmospheric Measurements*, ASTM STP 957. American Society for Testing and Materials, Philadelphia, PA, pp. 110–120.

Mousa-Bahia A.A., Coles G.S.V., Watson J. (1993) A gas injector for an automatic environmental test chamber for the characterization of gas sensors. *Sens. Actuators B* **2**, 141–145.

Nakamoto T., Fukuda T., Moriizumi T. (1991) Gas identification system using plural sensors with characteristics of plasticity. *Sens. Actuators B* **3**, 1–6.

Nakamura K., Nakamoto T., Moriizumi T. (1999) Prediction of quartz crystal microbalance gas sensor responses using a computational chemistry method. *Sens. Actuators B* **61**, 6–11.

Nieuwenhuizen M.S., Harteveld J.L.N. (1994) An automated SAW gas sensor testing system. *Sens. Actuators A* **44**, 219–229.

Patissier B., Walters D. (1999) Basics of relative humidity calibration for Humirel HS1100/HS1101sensors. Humirel, Toulouse Cedex, France.

Pilling R.S., Bernhardt G., Kim C.S., Duncan J., Crothers C.B.H., Kleinschmidt D., et al. (2003) Quantifying gas sensor and delivery system response time using GC/MS. *Sens. Actuators B* **96**, 200–214.

Pratzler S., Knopf D., Ulbig P., and Scholl S. (2010) Preparation of calibration gas mixtures for the measurement of breath alcohol concentration. *J. Breath Res.* **4**, 036004.

Rotronic (2005) *The Rotronic Humidity Handbook*, Rotronic Instrument Corporation. www.rotronic-usa.com

Sawaguchi N., Shin W., Izu N., Matsubara I., Murayama N. (2005) Effect of humidity on the sensing property of thermoelectric hydrogen sensor. *Sens. Actuators B* **108**, 461–466.

Scace G.E., Miller W.W. (2008) Reducing the uncertainty of industrial trace humidity generators through NIST permeation-tube calibration. *Int. J. Thermophys.* **29**, 1544–1554.

Scott M., Bell S., Bannister M., Cleraver K., Cridland D., Dadachanju F., et al. (1996) *A Guide to the Measurement of Humidity.* Institute of Measurement and Control. London, UK.

Sidmell J.A. (1992) Food contact polymeric materials, RARPA review report. RARPA Technology Ltd., Shrewsbury, Shropshire, UK.

Smith L.M., Odedra R., Kingsley A.J., Coward K.M., Rushworth S.A., Williams G., et al. (2004) Trimethylindium transport studies: The effect of different bubbler designs. *J. Crystal Growth* **272**, 37–41.

Staude E. (1992) *Membranen und Membranprozesse.* Verlag Chemie, Weinheim, Germany.

Su P.-G., Wu R.-J. (2004) Uncertainty of humidity sensors testing by means of divided-flow generator. *Measurement* **36**, 21–27.

Takahashi C., Kitano H. (1996) Calibration method of flow meters for a divided flow humidity generator. *Sens. Actuators B* **35–36**, 522–527.

Tang Z., Fung S.K.H., Wong D.T.W., Chan P.C.H., Sin J.K.O., Cheung P.W. (1998) An integrated gas sensor based on tin oxide thin-film and improved micro-hotplate. *Sens. Actuators B* **46**, 174–179.

Vitenberg A.G., Kostkina M.I., Ioffe B.V. (1984) Preparation of standard vapor-gas mixtures for gas chromatography: Continuous gas extraction. *Anal. Chem.* **56**, 2496–2500.

Wang Y., Besant R.W., Simonson C.J., Shang W. (2006) Application of humidity sensors and an interactive device. *Sens. Actuators B* **115**, 93–101.

Wexler A., Brombacher W.G. (1972) Methods of measuring humidity and testing hygrometers. A review and bibliography. In: Bloss R.L., Orloski M.J. (eds.) *Precision Measurement and Calibration: Mechanics*, National Bureau of Standard, Washington, DC, pp. 261–280.

Wiederhold P.R. (1997) *Water Vapor Measurement: Methods and Instrumentation.* Marcel Dekker, New York.

Wilson R.E. (1921) Humidity control by means of sulphuric acid solutions with critical compilation of vapor pressure data. *J. Ind. Eng. Chem.* **13**(4), 326–331.

Zhou Z., Gai L. (2006) Preparation of calibration gas mixtures of water and nitrogen by using diffusion tubes. *Accred. Qual. Assur.* **11**, 205–207.

# 27 Comparative Analysis of Humidity Sensors and Their Advantages and Shortcomings

## 27.1 WHICH HUMIDITY SENSOR IS BETTER?

One of the most frequently asked questions regarding sensors is, "Which humidity sensor is better?" There is no simple answer to this question. In terms of the working characteristics of humidity sensors, every type of sensor has specific advantages, for specific areas of application, and one type of humidity sensor may compete with another type. Comparative characterization of different methods acceptable for humidity measurement is presented in Tables 27.1 and 27.2. It is seen that we do not have ideal sensors, and no single sensor can satisfy every application. For example, possible areas of humidity-sensor application covers 6 orders of magnitude of change in humidity, when considered in terms of parts per million (ppm) by volume of water vapor, equivalent to an overall range of −85 to +100°C dew point. It is very unlikely that one measurement technique can cover the entire range.

Comparative characterization of popular humidity sensors is presented in Table 27.2. It is seen that all humidity-sensing technology is characterized by its own set of advantages and disadvantages. Resistive sensors have hive sensitivity and are cost effective. Capacitive sensors provide wide relative humidity (RH) range and condensation tolerance. Thermal conductivity sensors perform well in corrosive environments and at high temperatures. This means that when choosing a sensor, one should always look for a compromise, depending on the purpose of use, the range of possible changes in humidity and the operating conditions of the sensors. Each application has different set of requirements for humidity instruments, such as required measurement range, tolerance to extreme temperature and pressure conditions, ability to recover from condensation, ability to operate in hazardous environments, and options for installation and calibration. Some applications require continuous and long-term control over a wide range of humidity with good resolution and low long-term drift. For other applications, periodic measurements are sufficient. In some cases, humidity measurements can be carried out by fixed installations. While field measurements require portable instruments that consume low power, are compact in size, and incur low electronics overhead (Wilson et al. 2001). Operation temperature is also important. Temperature and environmental aspects of the individual application have significant impact to sensor lifetime requirements. Often, humidity sensors show their weaknesses if it comes to high or low temperatures, to temperature cycling, or to elevated chemical sustainability. It is also necessary to take into account the speed of humidity change, the possibility of the appearance of any unusual gases, and compatibility of humidity sensors with these gases. Some requirements to humidity sensors, depending on the field of application, are listed in Table 27.3. This means that each field requires a specific humidity sensor. In this case, of course, it is necessary to take into account the costs associated with sensor replacement, field and in-house calibrations, ease of use or level of skill required for user, and the complexity and reliability of the signal conditioning and data acquisition circuitry.

## 27.2 COMPARATIVE CHARACTERIZATION OF HUMIDITY SENSORS

The main advantage of capacitive readout is the capability of high enough sensitivity combined with the simplicity of the relevant electronics that enables miniaturization of sensor arrays, linear output response, smaller temperature dependence, low-power consumption, high stability, and integration of the readout setup in the sensor (Britton et al. 2000; Kang and Wise 2000; Lim et al. 2007). Capacitance-type devices are passive elements and therefore do not require any additional power other than what is required by their electronic readout. They are ideal in applications where the devices are operated from a battery source, that is, portable devices, or where power is harvested from the environment, for example, using a vibrational power generator or solar cells. This is often the case when long-term, autonomous, and unattended operation is required, as in the monitoring of stored goods. In addition, sensors of this type have low cost, because capacitance-type chemical sensors are relatively simple devices, usually requiring only a few steps for their fabrication. Moreover, the use of polymers as chemically sensitive materials allows for easy and low-cost adaptation of the same basic structure to many different applications (Chatzandroulis et al. 2011). The main disadvantage of capacitive humidity sensors is the high sensitivity to the presence of vapors of solvents, such as acetone, alcohol, benzene, and ethyl alcohol, capable of condensing in a porous matrix even at low concentrations. We also need to take into account that the sensitivity of capacitance sensors is significantly lower in comparison with condutometric humidity sensors.

Regarding microcantilever-based capacitive humidity sensors, one can say that the main disadvantages of these devices

## TABLE 27.1
### Characteristics of Some Humidity Sensors

| Type | Class | Measurement Range | Typical Measurement Accuracy |
|---|---|---|---|
| Psychrometer | Fundamental | 5% RH–95% RH 0°C–100°C ambient | 1%–5% RH |
| Electrolytic hygrometer ($P_2O_5$) | Fundamental | 1 to 2000 PPMv | 5% of reading PPMv |
| Chilled-mirror hygrometer | Fundamental | −90°C–90°C dew point | 0.2°C dew point |
| Gravimetric hygrometer | Primary | −50°C–100°C | 0.1°C dew point |
| Mechanical hygrometer | Secondary | 10% RH–100% RH | 2%–10% RH |
| Resistance hygrometer | Secondary | 5% RH–100% RH | 1%–5% RH |
| Polymer RH sensor | Secondary | 5% RH–95% RH | 2%–5% RH |

*Source:* WMO reports (WMO 2008 and WMO 2011).

are associated mainly with the possibility of sticking when the capacitor plates are brought too close together (Chatzandroulis et al. 2002). The need to use as sensitive-film polymers, which are not stable in hazard environment, can also be considered as disadvantage of microcantilever-based capacitance sensors (Chatzandroulis et al. 2004).

A number of developers believe that resistive sensors, developed on the basis of the hygrometric effect (Chapter 13, Volume 2) using a piezoelectric transformation, also have great prospects for application (Gerlach and Sager 1994). In such sensors, piezoresistors are configured in a Wheatstone-bridge arrangement within the silicon diaphragm to transduce the diaphragm deflection, caused by the swelling effect in polyimide film, into an output voltage measurement of strain. An advantage of this sensor design over conventional capacitive and resistive sensors is that the polymeric film can be electrically isolated from the sensing electrodes using special passivation layer between the sensing film and underlying silicon diaphragm. Yamazoe (1986) reported that humidity-dependent changes of a polymer-sensing film could strongly affect the accuracy and long-term stability of capacitive sensors when the film is in electrical contact with the electrodes.

It is known that control and measurement electronics overhead can play a major role in increasing the size and power consumption of the overall system beyond what is practical; in many sensing systems, electronics overhead can consume 40% or more of the total space and often contribute at least this much to the overall power consumption as well. The resistive humidity sensors due to its simplicity of operation and using perhaps are the most attractive humidity sensor type for portable applications. However, it also presents significant obstacles in terms of noise, drift, aging, and sensitivity to environmental parameters.

The advantages of capacitive and resistive-humidity sensors can also include the possibility of their manufacture on

## TABLE 27.2
### Comparative Characteristics of Methods Used for Humidity Measurement

| Sensor Type | Advantages | Disadvantages |
|---|---|---|
| *Resistive* | Small and cheap; mass production possible; high sensitivity; low cost; often easy to use | Nonlinear characteristics; limited range (typically 15% to 95% RH); there are difficulties in measuring low values (<5% RH); High-temperature dependence; poor stability, significant drift; sensitive to contamination, condensation and interference; the change in impedance is too high, and hence it is difficult to control the dynamics and temperature effects significantly |
| *Capacitive* | Small in size; low cost; mass-production possible; require very little maintenance; usually tolerate condensation: though calibration may shift; wide measurement range 0%–100% RH; wide temperature range (up to 200°C); low-temperature dependence; low hysteresis (below 1%); low drift; good linearity; relatively fast response; highly resistant to contaminants; auto-calibration is possible | Can be limited by distance from electronics to sensor; loss of relative accuracy at low end (<5%); requires electronics to convert capacitance to relative humidity; significant drift; may suffer calibration shifts and hysteresis if used at high temperatures and or high humidities; degradation under influence of electrical shocks |
| *Resistive; capacitive* (ceramics) | Resistant to chemical attacks; safe in explosive installations; thermal and physical stability | Need for periodic regeneration due to aging effects; hysteresis; contamination |
| *Resistive; capacitive* (polymer) | Low cost; miniaturization; high sensitivity | Degradation upon exposure to some solvents, aggressive chemicals, reactive gases |
| *Gravimetric* dew point (QCM, SAW) | Wide operation range; low sensitivity to contamination; measurements of very low frost points; good accuracy; low drift | Nonlinear characteristics; high-temperature dependence; require complicated systems for signal processing of sensor outputs; high cost; difficulties with integration |

*Source:* Scott, M., et al., *A Guide to the Measurement of Humidity*, Institute of Measurement and Control, London, UK, 1996.

**TABLE 27.3**

**Humidity-Sensor Markets per Requirement**

| Parameter | Remark | Automotive | Energy | Medical | Building Automation | Paper and Wood | Building Materials | Food Processing | Storage and Refrigeration | Cloth Drivers | IT and Computing | Agriculture | Heating Ventilation Air | Test and Measurement | Snowmaking | Weather Forecasting |
|---|---|---|---|---|---|---|---|---|---|---|---|---|---|---|---|---|
| **Humidity Level** | | | | | | | | | | | | | | | | |
| Very dry | Humidity warning | | X | | | X | X | | | | | | | X | | X |
| Dry | Drying | X | | | X | X | X | X | | X | | | | X | | X |
| Medium | Humidity control | X | | X | X | X | X | | X | X | X | X | X | X | | X |
| Humid | Humidifying | | | X | | X | | | | X | | X | | X | X | X |
| Very humid | Humidity guarantee | | | | | X | | | | | | | X | X | | X |
| **Response Time** | | | | | | | | | | | | | | | | |
| <1 s | Fast air stream | | | X | | | | | | | | | | | | X |
| 1–5 s | Fast fan | X | | X | | | | | | | | | X | | X | |
| 5–10 s | Ventilated | X | | | X | X | X | X | | X | | X | X | X | X | |
| 10–30 s | Convection | | X | | X | X | X | X | X | X | X | X | | X | | X |
| >30 s | Air diffusion | | X | | X | X | | X | X | X | X | | | X | | X |
| **Temperature Range** | | | | | | | | | | | | | | | | |
| –80°C to –50°C | Upper atmosphere | | | | | | | | | | | | | | | X |
| –50°C to –25°C | Deep freeze | X | X | | | | | | X | | | | | | | X |
| –25°C–0°C | Winter | X | X | | X | | X | X | | | | X | X | X | X | X |
| 0°C–60°C | Normal | X | X | X | X | X | X | X | | | X | X | X | X | | X |
| 60°C–85°C | Elevated | X | X | X | X | X | X | X | | X | | | | X | X | |
| 85°C–100°C | Boiling | X | X | X | | X | X | X | | X | | | | | | |
| 100°C–125°C | Electronics limit | X | X | | | X | X | | | X | | | | | | |
| 120°C–150°C | Extreme dissipation | X | X | | | | X | | | X | | | | | | |
| 150°C–190°C | Active heated | | X | | | | X | | | | | | | | | |

*Source:* Innovative Sensor Technology (www.ist-ad.com).

flexible (plastic) substrates (Oprea et al. 2009; Briand et al. 2011) and what opens up new opportunities for the application of humidity sensors, for example, applications in mobile autonomous systems, like smart RFID tags. The example of capacitive-based humidity sensors fabricated on plastic substrate (polyimide foils) is shown in Figure 27.1. The characteristics of plastic platforms are of high interest for the realization of ultra-low-power devices that could be processed at low-cost using printing processes (Briand et al. 2011).

It should be noted that the surface acoustic wave (SAW) devices are also of great interest. Humidity sensors based on SAW are in demand in various application fields because they do not require a power source and have wireless function. In addition to the fact that SAW devices are highly sensitive humidity sensors, they can be used to develop absolute humidity sensors. If the atmosphere of a SAW sensor is ambient and temperature controlled, water will condense on it at the dew-point temperature, thereby making it an effective dew-point sensor for applications in medical diagnostics, pharmaceuticals, and meteorology. Dew-point hygrometers based on direct mass measurements of condensation have the potential to provide more accurate measurement of dew condensation and temperature with high resolution. Such SAW devices can be cooled using a Peltier element. When the water-vapor condensation appears on the Rayleigh-wave propagation path, it induces a substantial attenuation of the wave amplitude and a shift in the associate oscillator's frequency (mass loading). At that, the amplitude depends minimally on the temperature while it decreases rapidly when sufficient water vapor condenses on the quartz plate. For humidity detection, a thin film of hygroscopic-selective coating such as polyvinyl alcohol and polyvinyl pyrrolidone must be deposited on the sensors.

Although polymer-based humidity sensors have been developed, the response/recovery time for many polymer coatings were shown to be rather long, frequently on the order of several seconds or even minutes. As a result, despite numerous studies in the field of humidity-sensor design, electronic humidity sensors continue to suffer from slow response, low accuracy (%RH), and considerable long-term drift, especially in areas of high humidity levels. Therefore, these sensors

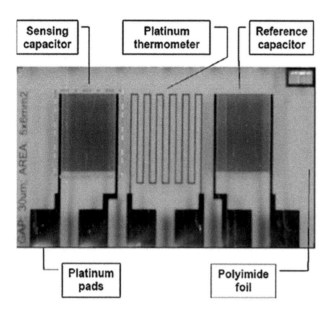

**FIGURE 27.1** Sensor platform with Pt thermometer, electrodes, and connection pads. The interdigital electrode structures realize two plane capacitive transducers, a sensing one (left) and a reference one (right). The area reserved for the polymer-sensing layer is surrounded by a dot-line frame. (Reprinted with permission from *Sens. Actuators B*, 140, Oprea, A. et al., Temperature, humidity and gas sensors integrated on plastic foil for low power applications, 227–232, Copyright 2009, with permission from Elsevier.)

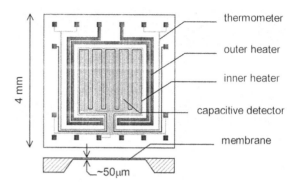

**FIGURE 27.2** Silicon dew-point hygrometer with integrated heater and temperature sensors. (Reprinted with permission from *Sens. Actuators B*, 85, Jachowicz, R.S. and Weremczuk, J., Sub-cooled detection in silicon dew point hygrometer, 75–83, Copyright 2000, with permission from Elsevier.)

require frequent calibration. A clear understanding of underlying physicochemical phenomena together with accurate modeling and simulation of humidity sensor characteristics, as well as searching humidity-sensitive materials with long-term stability are, therefore, extremely important now for fabrication humidity sensors acceptable for market.

In addition, it should be recognized that humidity sensors typically do not respond well to high humidity conditions due to water condensation, and sometimes it takes 24 hours or more to "dry down" after brief exposure to almost saturated conditions. To speed up the recovery of characteristics after condensation in a number of works, it is proposed to use heat (Pascal-Delannoy et al. 1998) for which to build in a sensor an additional heater and use cyclic operation (Jachowicz and Weremczuk 2000). To reduce the thermal inertia of devices Jachowicz and Weremczuk (2000) suggest using the sensor on a membrane fabricated using micromachining technology (Figure 27.2), thus, reducing the thermal capacitance and improving the response time of the sensor. This device is a typical example of the achievements of microsystem technology because it constitutes a successful attempt to beat the difficulties in the hygrometer by integrating different components on a single chip.

Thus, we have stated, there are no ideal sensors, not a single type of sensor demonstrates optimal performance in general, and therefore, none can satisfy all the requirements for workable devices. It was found that every analyzed type of sensor has its own specific technical limitations in achieving the required parameters. This means that which technology

is most suited to a given application depends on the operating requirements for that application (Wiederhold 1997). For example, polymer-based capacitance RH sensors, fabricated by Rotronic Instrument Corp. (www.rotronic-usa.com), are sensitive to organic solvents and especially to the presence in the air of ethylene oxide ($C_2H_4O$), $Cl_2$, $NO_x$, $SO_2$, and especially ozone. At 500 ppm ozone in atmosphere the sensors have a lifetime of approximately 1 month (Rotronic 2005). Ceramic-based humidity sensors do not have this disadvantage, and that is why their use is preferable if the previously mentioned additives are present in the atmosphere. Thus, the unique combinations of advantages and disadvantages related to individual sensor types largely determines the range of capabilities and potential applications that each sensor type provides for the analysis of air and gas humidity in specific operating situations. The correct choice of humidity sensors should be based on a number of inputs: ambient conditions, required sensitivity, response time, price, size, maintenance cycle, and method of operation, among others (read Chapter 26).

As was shown in Volume 1 of present series, the optical and fiber-optic technologies represent great opportunities for the development of a variety of devices capable of controlling the humidity (Yeo et al. 2006, 2008; Alwis et al. 2013; Kolpakov et al. 2014). These devices can have a sufficiently high sensitivity and good operation speed necessary for in-situ monitoring (Muto et al. 2003), and the sensing element may have a very small size of probe used for measurements (Table 27.4). But these efforts are insufficient, as most of the studies carried out in this area are still aimed at demonstrating the feasibility of the humidity measurement. This means that the individual samples were made, which were not intended for sensor market. The fabrication of such sensors does not come to agreement with mass production. In addition, special sources and radiation detectors, and PCs with special programs for signal processing, are required for their functioning. This means that the detection systems and measuring systems may be complex, and their operation requires precise installation procedures and qualified

**TABLE 27.4**

**Advantages and Disadvantages of Optical and Fiber-Optic Humidity Sensors**

| Type | Advantages | Disadvantages |
|---|---|---|
| Spectroscopic (optical) | Suitable for use with almost any gas, including corrosive and reactive ones; can be used to measure concentration of other substance at the same time; water vapors at high temperatures can be measured; noncontact measurement; fast response; high sensitivity; no hysteresis | Sophisticated technology; expensive instrumentation; difficult to calibrate; high cost |
| Spectroscopic: IR, transmittance | Can be designed for use in hostile environments; unaffected by contaminants; can measure extremely low concentration of water vapor; unaffected by contaminants; broad measuring range; in-situ use; good selectivity | Long path length required for sensitivity at low moisture levels; deterioration of the optical windows over time; carbon dioxide may interfere with the humidity measurement if present in quantity |
| Spectroscopic: UV, transmittance | Fast response; accurate measurements at high altitude; in-situ use; cloud measurements; good selectivity | Sensitivity to oxygen at low humidity; require magnesium fluoride windows which degrade with use; high cost |
| Fiber-optic | Low cost; high sensitivity; small size; robustness; flexibility; ability for remote monitoring as well as multiplexing; used even in the presence of unfavorable environmental conditions such as noise, strong electromagnetic fields, high voltages, nuclear radiation, in explosive or chemically corrosive media, at high temperatures; low hysteresis | Inherent losses; dispersion; nonlinearity; birefringence; reproducibility |
| Humidity indicators (color change) | No battery or electrical power needed; can be observed remotely (if visible) without disturbing environment; cheap and simple | Only gives a very coarse estimate of humidity (uncertainty may be in the region of $\pm(10-20)\%RH$); difficult to calibrate |

*Source:* Scott, M., et al., *A Guide to the Measurement of Humidity*, Institute of Measurement and Control, London, UK, 1996; Moreno-Bondi, M.C. et al., *Optical Sensors: Industrial Environmental and Diagnostic Applications*, Fiber-optic sensors for humidity monitoring, Springer, Berlin, Germany, 2004.

professionals. This naturally substantially restricts the use of such facilities, since such sensors and devices, using them, can be expensive. Therefore, it is difficult to expect that the optical and fiber-optic sensors displace from the market of electronic and electrical humidity sensors, which, as well as optical and fiber-optic sensors, do not possess the required selectivity, but are considerably cheaper in production and in operation. Currently, there are only a few commercially available fiber-optic humidity sensors. Therefore, there is a constant need to develop a compact fiber-optic humidity sensing system with high reliability, high sensitivity, long lifetime, increased reproducibility of the parameters, and low cost.

Optical, in particular, integrated optical (IO) sensors, have other features. The strong points of IO humidity sensors arise from optical sensing, and the use of waveguides and integrated structures. By sensing optically, electromagnetic interference and the danger of explosion are (almost) excluded. As well as this, optical methods often offer a high sensitivity and high selectivity. By using waveguides, the light path can be controlled simply, without the necessity of using bulky components like mirrors or beam splitters. Also, propagation of light through waveguides offers excellent possibilities of influencing its optical characteristics both for sensing and processing purposes. The use of integrated optic systems instead of fiber-optic ones adds another set of strong points. There is a large flexibility in the choice of construction materials and dimensions.

The systems are small and rigid, and by using silicon wafers as substrates, integration of optical and electrical functions can be achieved, offering the prospect of monolithic smart multisensor systems. It also has economic advantages, which are common for most microtechnological devices: a batchwise mass production, in which all basic functions required for the sensing system can be realized simultaneously. However, the rather complicated smart multisensory systems are still in an initial state, and a lot of research will have to be done before they will be ready for production. Moreover, the technology of IO sensors is complicated and expensive. This means that the IO humidity sensor, in general, will not be very cheap. Of course, they will be much more expensive than, for example, capacitance, field effect transistor (FET), and resistive humidity sensors. Therefore, their use may be restricted to applications where sensors previously mentioned cannot meet the specifications.

As for conventional devices discussed in the Volume 2 of this series, then their advantages and disadvantages are reflected in Tables 27.5 and 27.6. It should be noted that many of these devices are also manufactured in large quantities (read Chapter 28), indicating their widespread use. Moreover, some of them such as gravimetric and chilled-mirror hygrometers are instruments required for use in calibrating electronic and electrical humidity sensors (read Chapter 26). Their parameters are listed in Table 27.6. As is seen in Figure 27.3, these devices have the best absolute accuracy in humidity measurements.

**TABLE 27.5**

**Comparative Characteristics of Conventional Methods Used for Humidity Measurement**

| Sensor Type | Advantages | Disadvantages |
|---|---|---|
| Mechanical hygrometric (hair, synthetic fibers, etc.) | May not need electrical power or batteries; chart types provide a permanent record; may be cheap to buy; simple use | Slow response to changes in humidity; may suffer from hysteresis; temperature dependence; easily upset by vibration or transportation; dust, oil, pollutants, ammonia, exhaust gas, rain and fog contamination decrease the sensitivity and accuracy of the devices; hysteresis; drifts over time; different expansion coefficient for each hair; increase in the response time with increasing T and RH; after the hair hygrometer is exposed in low temperature and low humidity for a long time, reading error increase due to the increasing of delay |
| Psychrometer wet- and dry-bulb | Relatively simple, cheap, reliable and robust instrument; can have good stability; wide range of humidity; tolerates high temperatures and condensation | Some skill is usually required to use and to maintain the instrument; the results can be calculated from temperature readings (although some are automatic); a large air sample is required for the measurement; the sample will be humidified by wet-sock evaporation; airborne particles or water impurity may contaminate the wick; measurements can be complicated below about 10°C, Whirling types are prone to serious errors |
| Saturated lithium chloride (LiCl) | Simple; can be cheap; low sensitivity to contaminants | Slow response; limited measuring range; will not operate below 10%RH; moderate accuracy; needs frequent maintenance; does not tolerate condensation; affected by contamination from hygroscopic materials or solvents; some skill required, especially for maintenance |
| Electrolytic (phosphorous pentoxide) | Compatible with some corrosive gases; can have good sensitivity | Measuring cells have a limited life, after which they must be recoated (and recalibrated); damaged by exposure to normal ambient humidities, and by periods of extremes of dryness (below 1 ppmv); slow response at lowest humidities; hydrocarbons, butadiene, ammonia and some other contaminants prevent proper operation; adds traces of hydrogen and oxygen to the gas sample, which can recombine, leading to errors |
| Dew-point mirror hygrometer | Broad measuring range; high precision and accuracy; no hysteresis; good long-term performance; long-term calibration stability; fundamental measurement | High cost; not for in-situ use; can be slow in response; contamination (last variants have automatic calibration); dew points below 0°C require careful interpretation; usually required some skill to operate |

*Source:* Scott, M., et al., *A Guide to the Measurement of Humidity*, Institute of Measurement and Control, London, UK, 1996; Moreno-Bondi, M.C. et al., *Optical Sensors: Industrial Environmental and Diagnostic Applications*, Fiber-optic sensors for humidity monitoring, Springer, Berlin, Germany, 2004.

**TABLE 27.6**

**Comparison of Psychrometers and Chilled-Mirror Hygrometer Parameters**

| Device Parameter | Psychrometer | Chilled-Mirror Hygrometer |
|---|---|---|
| Quantity measured | Wet- and dry-bulb temperatures | Dew-point temperature |
| Range humidity (%RH) | 20–100 | −15 to +93 |
| Cost, including signal conditioning | High | Very high |
| Accuracy (%RH) | 3 to 4 | High |
| Temperature range (°C) | 0 to < +100 | −15 to +93 |
| Temperature effect (%RH/°C) | <0.5 | Very low |
| Long-term stability (%RH/yr) | 0.01 | Very good |
| Response time (sec) | 120–300 | Medium |
| Hysteresis (%RH at 25°C) | Poor | Low |
| Linearity (%RH) | Poor | Very good |
| Interchangeability (%RH) | Excellent | Good |
| Lead effect | Medium | Medium |
| Resistance to contamination | Fair | Fair |
| Resistance to condensation | Very good | Excellent |
| Cleanability | Good | Fair |
| Calibration ease | Excellent | Very good |
| Size | Medium to large | Large |
| Packaging | Custom probe | Custom probe |

*Source:* Data extracted from WMO reports. WMO, World Meteorological Organization, *Guide to Meteorological Instruments and Methods of Observation*, Appendix 4B, WMO–No. 8 (CIMO Guide), Geneva, Switzerland, 2008; WMO, World Meteorological Organization, *Technical Regulations*. Vol. 1—General Meteorological Standards and Recommended Practices, (WMO-No. 49), Geneva, Switzerland, 2011.

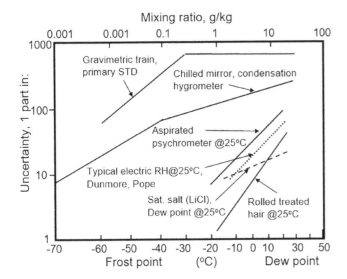

**FIGURE 27.3** The inaccuracy of moisture measurement by different methods. This graph was originally developed by the Mechanical/Humidity Section at the National Bureau of Standards in the 1960s and shows typical uncertainties for several popular humidity-sensor technologies. (Modified from www.nist.gov.)

## REFERENCES

Alwis L., Sun T., Grattan K.T.V. (2013) Optical fibre-based sensor technology for humidity and moisture measurement: Review of recent progress. *Measurement* **46**, 4052–4074.

Briand D., Oprea A., Courbat J., Bârsan N. (2011) Making environmental sensors on plastic foil. *Mater. Today* **14**(9), 416–423.

Britton C.L. Jr., Jones R.L., Oden P.I., Hu Z., Warmack R.J., Smith S.F., Bryan W.L., Rochelle J.M. (2000) Multiple-input microcantilever sensors. *Ultramicroscopy* **82**, 17–21.

Chatzandroulis S., Tserepi A., Goustouridis D., Normand P., Tsoukalas D. (2002) Fabrication of single crystal Si cantilevers using a dry release process and application in a capacitive-type humidity sensor. *Microelectron. Eng.* **61–62**, 955–961.

Chatzandroulis S., Tegou E., Goustouridis D., Polymenakos S., Tsoukalas D. (2004) Capacitive-type chemical sensors using thin silicon-polymer bimorph membranes. *Sens. Actuators B* **103**, 392–396.

Chatzandroulis S., Tsouti V., Raptis I., Goustouridis D. (2011) Capacitance-type chemical sensors. In: Korotcenkov G. (ed.) *Chemical Sensors: Comprehensive Sensor Technologies*, Vol. 4, *Solid State Sensors*. Momentum Press, New York, pp. 229–260.

Gerlach G., Sager K. (1994) A piezoresistive humidity sensor. *Sens. Actuators A* **43**, 181–184.

Jachowicz R.S., Weremczuk J. (2000) Sub-cooled detection in silicon dew point hygrometer. *Sens. Actuators B* **85**, 75–83.

Kang U., Wise K.D. (2000) A high-speed capacitive humidity sensor with on-chip thermal reset. *IEEE Trans. Electron. Dev.* **47** (4), 702–710.

Kolpakov S.A., Gordon N.T., Mou C., Zhou K. (2014) Toward a new generation of photonic humidity sensors. *Sensors* **14**, 3986–4013.

Lim S.H., Jaworski J., Satyanarayana S., Wang F., Raorane D., Lee S.-W., Majumdar A. (2007) Nanomechanical chemical sensor platform. In: *Proceedings of the 2nd IEEE International Conference on Nano/Micro Engineered and Molecular Systems*, January 16–19, Bangkok, Thailand, pp. 886–889.

Moreno-Bondi M.C., Orellana G., Bedoya M. (2004) Fiber-optic sensors for humidity monitoring. In: Narayanaswamy R., Wolfbeis O.S. (eds.) *Optical Sensors: Industrial Environmental and Diagnostic Applications*. Springer, Berlin, Germany, pp. 251–280.

Muto S., Suzuki O., Amano T., Morisawa M. (2003) A plastic optical fiber sensor for real-time humidity monitoring. *Meas. Sci. Technol.* **14**, 746–750.

Oprea A., Courbat J., Bârsan N., Briand D., de Rooij N.F., Weimar U. (2009) Temperature, humidity and gas sensors integrated on plastic foil for low power applications. *Sens. Actuators B* **140**, 227–232.

Pascal-Delannoy R., Sacka A., Giani A., Foucaran A., Boyer A. (1998) Fast humidity sensor using opto-electronic detection on pulsed Peltier device. *Sens. Actuators A* **65**, 165–170.

Rotronic. (2005) *The Rotronic Humidity Handbook*, Rotronic Instrument Corporation. (www.rotronic-usa.com).

Scott M., Bell S., Bannister M., Cleraver K., Cridland D., Dadachanju F., et al. (1996) *A Guide to the Measurement of Humidity*. Institute of Measurement and Control. London, UK.

Wiederhold P.R. (1997) *Water Vapor Measurement: Methods and Instrumentation*. Marcel Dekker, New York.

Wilson D.M., Hoyt S., Janata J., Booksh K., Obando L. (2001) Chemical sensors for portable, handheld field instruments. *IEEE Sensors J.* **1**, 256–274.

WMO (2008) World Meteorological Organization, Guide to Meteorological Instruments and Methods of Observation, Appendix 4B, WMO–No. 8 (CIMO Guide), Geneva, Switzerland.

WMO (2011) World Meteorological Organization, *Technical Regulations*. Vol. 1—General Meteorological Standards and Recommended Practices, (WMO-No. 49), Geneva, Switzerland.

Yamazoe N. (1986) Humidity sensors: Principles and applications. *Sens. Actuators* **10**, 379–398.

Yeo T.L., Eckstein D., McKinley B., Boswell L.F., Sun T., Grattan K.T.V. (2006) Demonstration of a fiber-optic sensing technique for the measurement of moisture absorption in concrete. *Smart Mater. Struct.* **15**, N40–N45.

Yeo T.L., Sun T., Grattan K.T.V. (2008) Fiber-optic sensor technologies for humidity and moisture measurement. *Sens. Actuators A* **144**, 280–295.

# 28 Market of Electronic Humidity Sensors

## 28.1 ELECTRONIC HUMIDITY SENSORS AND THEIR SPECIFICATIONS

At present, relative humidity (RH) sensors of capacitive and resistive types are produced in a wide range of specifications, sizes, and shapes, including integrated monolithic electronics. Examples of such sensors from various manufacturers are shown in Figure 28.1. Important parameters of different types of electronic humidity sensors presented in sensor market are listed in Table 28.1.

It should be noted that despite the rather developed market of humidity sensors, specifications of RH sensors quite often are unsatisfying: There is no common agreed standard for specifying humidity sensors. Hence, it is difficult to navigate in specifications, statements may be misleading, testing sensors against specification needs clarification, and specifications of different sensor types are almost impossible to compare. Therefore, in the present section, we will give some details on how to understand the terms in the specifications of humidity sensors. In this description, recommendations were presented on the site SENSIRION (2010).

- **Capacitance:** The nominal capacitance is the capacity of the sensor at a certain relative humidity, at temperature of 20°C or 30°C and operating frequency of 20 kHz. The humidity capacitive sensors can operate within the specified frequency limits. For the best results, an operating frequency of 20 kHz is recommended.
- **Accuracy:** For a simple but thorough understanding, the accuracy of RH sensors may be divided into three different, rather independent, terms: calibration accuracy, hysteresis, and long-term drift.

Calibration accuracy is the main component of an accuracy specification. It provides information on deviation of the individual sensor readings in an equilibrium state against a high-precision reference at the time of calibration. As a rule, calibration accuracy is measured against a dew-point mirror—a high-precision reference (read Chapter 26).

The hysteresis value is the difference of measured values of the same sensor at a certain log point accruing from a dry environment on one hand, and humid environment on the other hand—giving enough dwell time (Figure 28.2). In other words, humidity sensors carry some memory of conditions experienced in recent past: sensors with a dry history carry some negative offset, while sensors with a humid history carry some positive offset. Hysteresis appears due to composition and design of the sensor element. Because the hysteresis does not depend on quality of calibration, but it depends on the exposure range in the application, this value is considered to be additional to calibration accuracy.

Calibration accuracy and hysteresis values usually are determined by running the sensor in a full humidity loop 10% → 30% → 50% → 70% → 90% → 90% → 70% → 50% → 30% → 10% with dwell times of half an hour at each log point. For the determination of calibration accuracy at a certain humidity value, the mean value is calculated from measured values of ascending and descending path. The difference of these measured values and the mean value determine the value for hysteresis.

Long-term drift is the result of the aging of the sensor, which may lead to a drift in the measured value compared to the reference. Such a long-term drift is random—it can move to the upper or lower side or change direction over time. The long-term drift value is a maximal limit for such drift per year. As a rule, long-term drift is determined by exposing a sample of sensors to the so-called high-temperature operating lifetime. For example, Sensirion for these purposes uses operation at 125°C during 408 h. The experiment showed that the exposure at 125°C corresponds to aging at 25°C during a much longer time period. In particular, it was found that the storage of the humidity-sensor element at 125°C during 408 h corresponds to the operation of the sensor at 25°C during of 27–71 years depending on the type of sensor. Long-term drift can also be used to characterize the stability of sensor parameters.

- **Repeatability:** Short-term stability may be characterized by repeatability—repeated measurements with same sensor at constant conditions. The measure for such a term is the standard deviation for the sample of repeatability measurements.
- **Non-linearity:** This term stands for systematic deviations outside the calibrated log points. Such deviations of the sensor output as well as temperature compensation may be corrected by a linearization formula. Hence, non-linearity may be made very small—by giving the right formula. Non-linearity values are included by the accuracy tolerance and shall not be considered as an additional term.
- **Response time:** Response time usually specifies as a so-called 63%τ (also 1/eτ) time or 90%τ. For a sensor exposed to an abruptly changing environment (step function of measured physical value), the sensor reading approaches the final value typically on an exponential function over time. The 63%τ and 90%τ time extend from the moment of the environmental change at the sensor until the sensor readings complete 63% or 90% of the step height, respectively (see Figure 28.3).

To test the response time, it is important that all other parameters, besides the one that is to be tested, remain constant. As it was indicated in Chapter 26, for measuring the response time of the RH sensor, the volume of the climate chamber, where the sensor is placed in, must be kept very small.

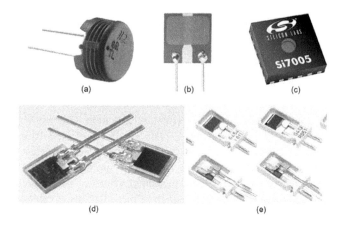

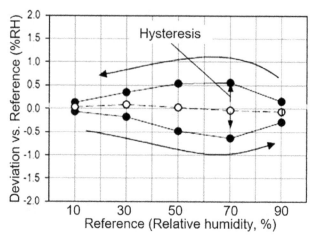

**FIGURE 28.1** Examples of capacitive humidity sensors fabricated by different manufactures: (a) "Measurement Specialties" (Modified from https://www.te.com/usa-en/product-CAT-HSC0006.html?q=humidity%2Bsensor&source=header); (b) "Michell Instruments" (Modified from http://www.michell.com/uk/products/h8000.htm); (c) "Silicon Labs" (Modified from https://www.silabs.com/documents/public/data-sheets/Si7007-A20.pdf); and (d) Smartec (Modified from https://smartec-sensors.com/cms/media/Datasheets/Humidity/HUMIDITY_datasheet.pdf); (e) Vaisala (Modified from https://store.vaisala.com/eu/vaisala-intercap-humidity-sensor/15778HM/dp).

The step function of the changing environment can be created by a rotary valve. In addition, one should be sure there is no dead time between step function initiated by the system and the sensor experiencing the step function.

- **Supply voltage:** The Supply Voltage ($V_{DD}$) range is defined with an upper and a lower limit plus a typical value. Any supply voltage in that range may be used for continuous operation. Absolute maximum

**FIGURE 28.2** The example for hysteresis measurement. A path from dry to humid and one from humid to dry is measured—full dots on graph. Dwell time at each log point is 30 min. Open dots determine mean values—representing calibration accuracy. (Modified from https://www.mouser.com/datasheet/2/682/Sensirion_Humidity_Sensors_Specification_Statement-1139772.pdf)

voltages, which may be applied during limited time, for some sensors, are specified in the User's Guide of the respective Datasheet. The typical value defines the supply voltage at which the sensors are calibrated and at which outgoing quality control is performed.

- **Supply current:** In operation, the sensor pulls a certain Supply Current, $I_{DD}$. This current is different for sleep mode and measuring/communicating. Furthermore, in the sample of sensors, there is a certain variation of current consumption—the average is specified as typical value while with minimum and maximum values the upper and lower limit is defined.

## TABLE 28.1
### Typical Parameters of Electronic-Related Humidity Sensors

| Parameter | Polymer (thermoset) | Polymer (thermoplastic) | Polymer (thermoplastic) | Ceramic ($Al_2O_3$) | Ceramic ($Al_2O_3$) | LiCl Film |
|---|---|---|---|---|---|---|
| Substrate | Si, ceramic, glass | Si, ceramic, polyester, etc. | Ceramic, $SiO_2$ | Al | Ceramic, $SiO_2$ | Ceramic, glass |
| Sensed parameter | Capacitance | Capacitance | Resistance | Capacitance | Resistance | Conductivity |
| RH change | 0%–100% | 0%–100% | 20%–100% (non-condensing) | 0.001 ppm to 0.2% | 2%–90% | 11%–<100% |
| RH accuracy | ±1% to ±5% | ±3% to ±5% | ±3% to ±10% | | ±1% to ±5% | ±5% |
| Hysteresis | <1% to 3% RH | 2% to 5% RH | 3% to 6% RH | | <2% RH | Very poor |
| Linearity | ±1% RH | ±1% RH | poor | | poor | Very poor |
| Response time | 15 s–60 s | 15 s–90 s | 20–60 s | | 3–5 min | 3–5 min |
| Temperature range | −40°C to 200°C | −30°C to 200°C | 10°C to 60°C | | −10°C to 75°C | – |
| Long term stability | ±1%RH/5 year | ±1%RH/1 year | ±1%RH/1 year | | ±3%RH/1 year | >1%RH/1°C |
| Conversion and signal processing | Complex | Complex | Simple | Complex | Simple | Simple |
| Measurement | Continuous | Continuous | Continuous | Continuous | Continuous | Intermittently |

*Source:* Modified from https://www.engineersgarage.com/article_page/humidity-sensor/.

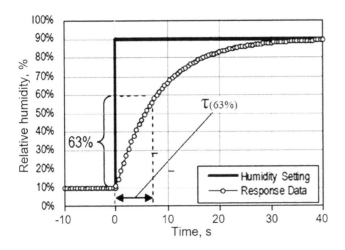

**FIGURE 28.3** Measurement profile of response time testing for relative humidity. (Modified from https://www.mouser.com/datasheet/2/682/Sensirion_Humidity_Sensors_Specification_Statement-1139772.pdf.)

- **Current and energy consumption:** Power Dissipation (P) is typically calculated from Supply Current values and Supply Voltage, taking into account the time necessary for measuring and communication. One should also take into account that additional power is consumed during start-up phase.

As an example, Table 28.2 gives the parameters of capacitive humidity sensors SHTW2 (RH&T sensor in chip-scale package) and HIH-4602-A/C (TO-5 can housing), developed

by Sensirion AG (www.sensirion.com) and Honeywell (www.honeywell.com/sensing), correspondingly.

## 28.2 COMPANIES OPERATING IN THE MARKET FOR HUMIDITY SENSORS

The list of some companies operating in the of electronic humidity sensors is listed in Table 28.3. Most popular are Panametrics, Ametek, General Eastern, Endress & Hauser,

**TABLE 28.3**

**Companies Developing Electronic Humidity Sensors and Devices Based on Them**

| Company | Website |
|---|---|
| Alpha Moisture Systems | www.amsystems.co.uk/ |
| Ambient Weather | www.ambientweather.com/ |
| Ametek Process Instruments | www.ametekpi.com; www.thermox.com |
| Amphenol: Advanced Sensors | http://amphenol-sensors.com/en/products/humidity |
| Chino Works America Inc. | www.chinoamerica.com |
| Davis Instruments | www.weathershop.com/davis_accessories.htm |
| Decagon | www.decagon.com/en/soils/ |
| Delta-T Devices | www.delta-t.co.uk/ |
| E+E Electronik | www.epluse.com |
| Elmwood Electronics | https://elmwoodelectronics.ca |
| Endress &Hauser | www.us.endress.com |
| General Eastern | www.geinet.com |
| GE Measurement | www.gemeasurement.com |
| GHI Electronics | www.ghielectronics.com/catalog/ |
| Hygrometrix | www.hygrometrix.com |
| Hydronix | www.hydronix.com/products/ |
| Innovative Sensor Technology | www.ist-usadivision.com/sensors/humidity/ |
| Kahn Instruments, Inc., | www.kahn.com/ |
| Measurement Specialties | www.meas-spec.com/ |
| MEMS Vision com. | www.mems-vision.com/products-services/ |
| Michell Instruments | www.michell.com/uk/ |
| Onset Computer Corporation | www.onsetcomp.com/products/ |
| PASCO | www.pasco.com/prodCatalog/ |
| Panametrics Inc. | www.panametrics.com |
| Rotronic Instrument Corp. | www.rotronic-usa.com |
| Shinyei Kaisha | www.shinyei.com.hk |
| Siemens Ltd. | www.usa.siemens.com |
| Silicon Labs. | www.silabs.com/ |
| Sensirion AG | www.sensirion.com/ |
| Spectrum Technologies | www.specmeters.com/weather-monitoring/ |
| Stevens Water Monitoring Systems Inc. | www.stevenswater.com/products/sensors/ |
| Systech Instrumentation Inc. | www.systechinst.com/ |
| TES Electrical Electronic Corp. | www.tes.com.tw/en/about.asp |
| Testo | www.testo.com |
| Tews-elektronik | www.tews-elektronik.com/en/ |
| Vaisala | www.Vaisala.com |
| Vegetronix | www.vegetronix.com/Products/ |
| Vernier | www.vernier.com/products/sensors/ |

**TABLE 28.2**

**Parameters of Electronic Modules of Capacitive Humidity Sensors. Electrical**

| Parameter | SHTW2 | HIH-4602-°C |
|---|---|---|
| | **Value** | |
| Size | $1.3 \times 0.7 \times 0.5$ mm³ | $9 \times 9 \times 7$ mm³ |
| Operating range (% RH) | 0–100 | 0–100 |
| Typical accuracy (% RH) | ±3 | ±3.5 |
| Hysteresis (% RH) | ±1 | ±3 |
| Typical long-term drift (% RH/yr) | <0.25 | 0.25 |
| Response time τ63% (s) | 8 | 50 |
| Operating range (°C) | −30 to 100 | −40 to 85 |
| Typical accuracy (°C) | ±0.4 | ±1 |
| Typical long-term drift (°C/yr) | <0.02 | |
| Response time τ63% (s) | 5–30 | 50 |
| Interface | I²C (digital) | analog |
| Supply voltage range (V) | 1.62–1.98 | 4–5.8 |
| Measurement duration (ms) (high/low) | 10.8 (high) 0.7 (low) | 70 |
| Average current consumption (μA) (high/low) | 4.8 (high) 0.9 (low) | 500 200 |
| Power consumption (μW) | 8.6 (average) | |
| Idle current (μA) | 0.77 | |
| Price, $ | 1–3.4 (500–1.55) | 1–43 (500–36) |

Rotronic, and Vaisala. The key players dominating this market include also Aptina Imaging Corporation, Autoliv Inc., BEI Sensors, Bourns Inc., Continental AG, Corrsys-Datron Sensorsystem GmbH, CTS Corporation, Custom Sensors & Technologies Inc., Delphi Corporation, Denso Corporation, Freescale Semiconductor Inc., GE Sensing & Inspection Technologies, Hamamatsu Photonics KK, Hamlin Electronics LP, Hella KGaA Hueck & Co., Hitachi Automotive Systems, Honeywell Sensing & Control, Infineon Technologies North America Corp., Kavlico Corporation, Melexis Microelectronic Integrated Systems N.V, Melexis Inc., Omron Corporation, OSRAM Opto Semiconductors GmbH, Robert Bosch GmbH, etc. It should be noted that according to a new market research report "Digital Temperature and Humidity Sensor Market to 2020" prepared by "MarketsandMarkets," the market size of humidity sensors will grow substantially. For example, it is expected that the market of digital temperature and humidity sensors for automotive applications until 2020 will grow from $1,084.38 million in 2013 to $2,129.68 million in 2020, what is ~11% per year. It is believed that digital sensors offer many advantages over conventional sensors, such as high performance and accuracy coupled with low cost, easy implementation, no requirement for complex calibration, and low-power consumption, among others. The acoustic-wave humidity-sensor market was valued at $15.0 million in 2013 and is expected to reach a market value of $38.0 million in 2018, growing at a CAGR of 21.0% from 2013 to 2018.

As a rule, these companies, along with sensors (see Figure 28.4), develop electronic modules that provide preliminary processing of information coming from the sensor, and humidity/temperature transmitter, designed directly

for use. It should be noted that these electronic modules can differ significantly from each other, since each type of humidity sensor has unique needs that dictate incorporating circuitry for signal conversion, linearization, and buffering to make it a practical device having an easily measured output signal. A separate bias supply is needed for power. Functional features of capacitive humidity sensor module are shown in Figure 28.5.

Examples of electronic modules and electronic devices developed on the basis of humidity sensors by various companies are given in Figures 28.6 through 28.8. Manufactures claim that their devices are the guarantee for: (1) best accuracy over the whole working range; (2) display and output of RH, temperature, dew-point, and frost-point temperature; (3) small hysteresis; (4) excellent long-term stability, and (5) highest resistance to pollutants. A comparison of the parameters of

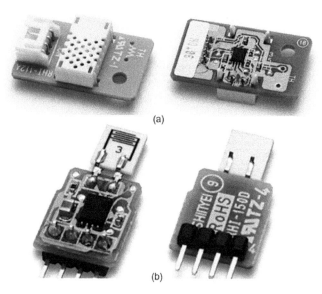

(a)

(b)

**FIGURE 28.6** Humidity and temperature-sensing module: (a) RHI-112A; and (b) RHI-150D designed by Shinyei Kaisha: operating temperatures—0–60°C; operating humidity—10%RH–90%RH (<95%RH); accuracy—±3%RH at 25°C; price—(a) $6, and (b) $36. (Modified from https://www.shinyei.com.hk/eng/biz_humidity_RHI-150D.html.)

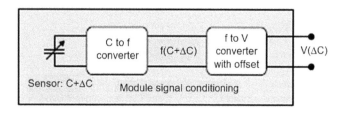

**FIGURE 28.4**   Capacitive humidity sensor module.

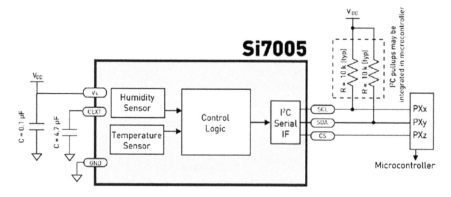

**FIGURE 28.5**   Si7005 humidity sensor provides a single-chip solution for humidity measurement. (Modified from Silicon Labs, https://eu.mouser.com/new/silicon-labs/silabs-si7005/.)

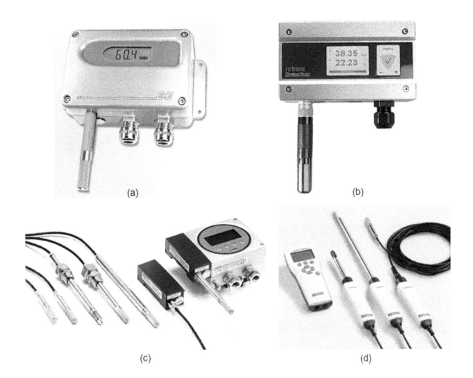

FIGURE 28.7 Humidity/temperature transmitters for industrial applications fabricated by different companies: (a) EE23 (EE Electronik, Modified from https://www.epluse.com/en/products/humidity-instruments/industrial-humidity-transmitters/ee23/); (b) HygroFlex5-HF5 (Rotronic, Modified from https://www.rotronic.com/en/hygroflex5-hf5.html); (c) HUMICAP HMT360 (Vaisala, Modified from https://www.vaisala.com/sites/default/files/documents/HMT360-User-Guide-in-English-M010056EN.pdf); and (d) HM70 hand-held meter (Vaisala, Modified from https://www.vaisala.com/en/products/instruments-sensors-and-other-measurement-devices/instruments-industrial-measurements/hm70).

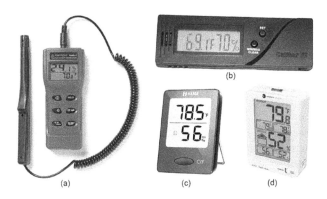

FIGURE 28.8 Humidity sensors of wide application: (a) Amprobe THWD-5 Temperature and Relative Humidity Meter with Wet Bulb and Dew Point (Modified from https://www.tequipment.net/AmprobeTHWD-5.html); (b) Cigar Oasis Caliber IV Digital Hygrometer (Modified from https://www.cigaroasis.com/images/products/manuals/Caliber-IV_v6.18.pdf); (c) Habor Digital Hygrometer Indoor Thermometer (Modified from http://www.ihabor.com/category/humidity-monitor/); (d) Ambient Weather WS-2063-W Indoor Temperature & Humidity Monitor with Backlight (Modified from https://www.ambientweather.com/amws2063w.html).

these devices, presented in Table 28.4, shows that, as a rule, these devices have very similar parameters and differ only slightly. However, different equipment configurations and the availability of additional probes that ensure the possibility of using devices in various conditions can significantly raise the price of these devices.

## TABLE 28.4
## Selected Parameters of Humidity/Temperature Transmitters Shown in Figure 28.7

| | Humidity/Temperature Transmitter | | |
| --- | --- | --- | --- |
| Parameter | EE23 (EE Electronik) | HMT360 (Vaisala) | HygroFlex5-HF5 (Rotronic) |
| Working range, RH | 0%–100%RH | 0%–100%RH | 0%–100%RH |
| Working range, T (for probe) | −40°C to +120°C | −70°C to +180°C | −100°C to +200°C |
| Accuracy, RH | −15°C…40°C: ±1.3% RH (≤90% RH) and ±2.3% RH (>90%RH) | +15°C…25°C: ±1%RH (<90%RH) and ±1.7%RH (>90%RH) | 23°C: ±0.8%RH |
| Accuracy, T | 0.2°C | 0.2°C | 0.1°C |
| Response time at 20°C, $t_{90}$ | with metal grid filter <15 s | 8 s with grid filter, 20 s with grid + steel netting, and 40 s with sintered filter | <5 s |
| Price, $ | 1,700 | 3,100 | 560 |

*Source:* Data extracted from http://downloads.epluse.com/fileadmin/data/product/ee23/datasheet_EE23.pdf, https://www.rotronic.com/en/hygroflex5-hf5.html and https://www.vaisala.com/sites/default/files/documents/HMT360-Datasheet-B210956EN.pdf.

## 28.3 ELECTRONIC HUMIDITY SENSORS AND THEIR CHARACTERIZATION

Analyzing the market for humidity sensors, we have to state that despite the large variety of humidity sensors offered by various companies, these sensors are mainly capacitive- and resistive-type sensors made from polymers or porous alumina ($Al_2O_3$) (Roveti 2001). It should be noted that an alpha alumina is the first material, the use of which allowed to develop the world's first drift-free humidity sensor. This material is now used in many commercial humidity sensors, because, on the one hand, the etching technology of Al porosification is well established (Masuda et al. 1993; Nahar and Khanna 1998), and on the other hand, $Al_2O_3$ has proven to be stable at elevated temperature and at high-humidity level in comparison with many other humidity-sensitive materials (Nahar 2000).

### 28.3.1 Al₂O₃-Based Humidity Sensors

The most famous companies currently producing $Al_2O_3$-based humidity sensors are Systech Illinois (www.systechillinois.com), COSA+Hentaur (http://cosaxentaur.com), Michel Instruments (www.michell.com/uk), GE Panametrics Inc. (www.gemeasurement.com), Alpha Moisture Systems (http://www.amsystems.co.uk/), and Teledyne Analytical Instruments (http://www.teledyne-ai.com). Typically, these sensors share the same basic operating principle: the capacitance measured between the sensor's aluminum core and a gold film deposited on the top of the oxide layer varies with the water vapor content in the pores of the oxide layer. The appearance of some of the sensors manufactured by these companies is shown in Figure 28.9. Typical parameters of these sensors are given in Table 28.5.

These sensors are also characterized by high stability, low hysteresis and temperature coefficients, high selectivity for moisture, and the possibility to operate over a wide range of temperature. Aluminum-oxide sensors have the capability to be installed at high pressure and their footprint is quite compact. The sensors however are seldom installed directly in the pipeline. Instead, an extraction-type sampling system is utilized. Testing has shown that in general, $Al_2O_3$-based sensors provide excellent response to moisture changes in the dry-to-wet direction. They however have significant response times in the wet-to-dry direction. As it was indicated in Chapter 3, aluminum-oxide sensors are subject to drift over time. The typical drift is around 2°C per year, and this drift should be managed by a regime of recalibration. Since aluminum-oxide sensors are economical, very often the users maintain additional sensors that are rotated in and out of service, thus always maintaining the in-service sensors within their recommended recalibration interval (typically one year).

In addition to the traditional manufacturing technology of $Al_2O_3$-based humidity sensors, many companies such as COSA+Hentaur и GE Panametrics also use thin-film technology, the so-called hyper-thin-film (HTF) aluminum-oxide technology, which by reducing the thickness of the porous layer of $Al_2O_3$ contributes to a significant increase in

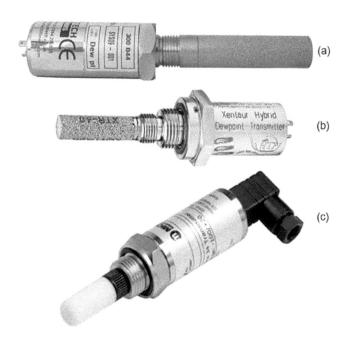

**FIGURE 28.9** (a) Moisture Transducer MM300 fabricated by Systech Illinois (Modified from https://www.systechillinois.com/en/products/mm300-moisture-transducer); (b) Xentaur's Dew Point Meter fabricated by COSA+Hentaur (Modified from https://cosaxentaur.com/product/517/XENTAUR%20LPDT); and (c) Easidew 34 Dew-Point Transmitter (Modified from http://www.michell.com/us/products/easidew-34-transmitter.htm).

**TABLE 28.5**
**Typical Parameters of Al₂O₃-Based Humidity Sensors (parameters are given for sensor MM300)**

| Parameter | Value |
|---|---|
| Operating humidity | 0–100% RH |
| Measurement range | −100°C to +20°C dew point; 0.1 ppm–1000 ppm |
| Accuracy | ±2°C dew point |
| Response time | 90% of reading ($t_{95}$) less than 5 min (dry to wet) |
| Operating temperature | −40°C to +60°C |
| Flow rate | 0.5–7 NL/min |
| Power requirements | 12–28 VDC |
| Signal outputs | 4–20 mA (2 wire) |

*Source:* Modified from https://www.systechillinois.com/fr/products/mm300-moisture-transducer.

sensitivity and decrease in response time (see Figures 28.10 and 28.11a). The developers claim that the HTF-based sensors have a capacitance change, several orders of magnitude larger than that of conventional aluminum-oxide sensors due to the hyper-thin film, a sharp transition layer and a special pore geometry (Figure 28.11b). The larger sensitivity of HTF™ sensors makes Xentaur HTF™ more stable with better repeatability and almost completely immune to other influences, such as temperature, electrical noise, and even long-term drift. Additionally, this change is quasi-linear. Also, the measurement system is less prone to noise and drift, and signal conditioning is kept to a minimum.

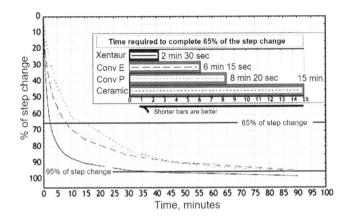

**FIGURE 28.10** Response times of Al$_2$O$_3$ sensors to a step change (DOWN) from −39.4°C(dp) to −62.2°C (dp). (Modified from https://cosaxentaur.com/resources/files/634/htf_sensor.pdf.)

Based on these sensors, a large series of moisture analyzers has been developed, such as Moisture Analyzers MM400-500 (www.systechillinois.com), X-STREAM Process Gas Analyzers (http://www2.emersonprocess.com/), Xentaur's Dew Point Meters (http://cosaxentaur.com), and Easidew online dewpoint hygrometers (www.michell.com/uk) (see Figure 28.12). Developers recommend using data of moisture analyzers in air dryers, chemical manufacturing, processes using compressed air, corrosive gases, natural gas and industrial specialty gases, glove boxes, clean-room environments, heat-treating furnaces, medical gas analysis, metallurgy, plastics manufacturing, welding reactive, and refractory metals.

### 28.3.2 CAPACITIVE AND RESISTIVE HUMIDITY SENSORS

Capacitive humidity sensors designed for the sensor market consist of a substrate on which a thin film of polymer or metal oxide is deposited or formed between two conductive electrodes (Figure 28.13). Porous ceramic films are formed on substrates using thick-film screen printing, vapor

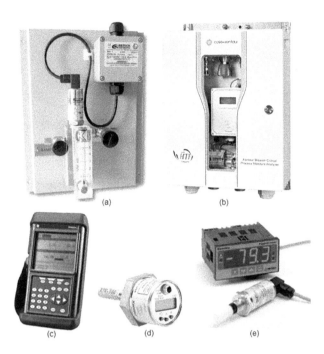

**FIGURE 28.12** (a) Moisture analyzer MM300 (Modified from https://www.systechillinois.com/en/products/mm300-moisture-transducer); (b) Mission Critical Process Moisture Analyser, designed to handle rigorous demands of moisture measurement in natural gas applications, provides fully automatic self-calibration dew point system, and provides cost-effective solution for demanding and critical moisture monitoring applications. The system has an integrated, fully automatic self-calibration procedure, in which the sensor is periodically exposed to a NIST-certified calibration gas and recalibrated. Thus, there can be high NIST traceable confidence in the measurement, and the sensors do not need to be returned to the factory for calibration (Modified from https://cosaxentaur.com/product/586/XENTAUR%20Mission%20Critical); (c) PM880 Portable Hygrometer/Moisture Analyzer (Modified from http://veronics.com/products/Dewpoint_portables/pm880.pdf); (d) Xentaur's Dew Point Meter, Model LPDT, is the world's smallest loop powered (two-wire) dew-point transmitter with a display (Modified from https://cosaxentaur.com/product/517/XENTAUR%20LPDT); and (e) Easidew online dew-point hygrometer (Modified from http://www.michell.com/us/products/easidew-online.htm).

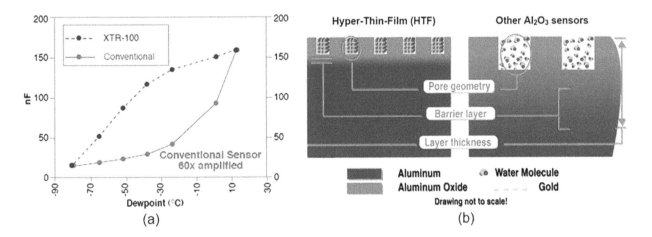

**FIGURE 28.11** (a–b) The change of capacitance with moisture of Xentaur HTF™ sensors. (Modified from https://cosaxentaur.com/resources/files/634/htf_sensor.pdf.)

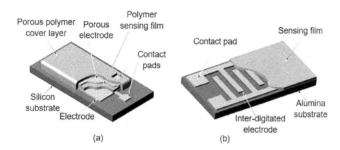

**FIGURE 28.13** (a, b) Standard configuration of commercial capacitive and resistive humidity sensors. Parallel plate design for capacitive RH microsensors can include a porous polymer cover layer to exclude particulates. (Modified from Hydrometrix Applications Note 2004-2: A comparison of relative humidity sensing technologies, www.hygrometrix.com.)

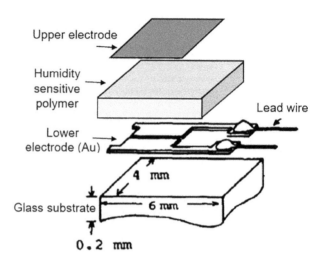

**FIGURE 28.14** Configuration of "Hunucape" humidity sensor Vaisala. (Modified from https://www.vaisala.com/sites/default/files/documents/HUMICAP-Technology-description-B210781EN-C.pdf.)

deposition, or direct anodization of an aluminum or silicon substrate. Thick films are usually printed onto an alumina substrate as a paste or conductive ink with film thickness greater than 10 μm. Dopants can be added to the mixture as reaction catalysts to promote the dissociation of adsorbed water into hydrogen and hydroxyl ions. Alumina and porous-silicon films can also be formed by directly modifying the top layers of the substrate through anodization (for alumina) or electrochemical etching (for silicon). Polymers used in capacitive sensors are essentially electrical nonconductors, with bulk resistances 18 orders of magnitude greater than metals. The sensing surface is coated with a porous-metal electrode to protect it from contamination and exposure to condensation. The substrate is typically glass, ceramic, or silicon. Most of the latest models of capacitive sensors also have an additional polymer layer deposited on a metal electrode, usually porous platinum. The top polymer layer is a porous film and acts as mechanical filter against sensor contamination from dust, dirt, and oils.

It should be noted that in addition to the standard configuration of capacitive sensors, there is a configuration of capacitive sensors proposed by Vaisala in Finland. The sensor configuration is illustrated in Figure 28.14. Lower twin electrodes are attached on a glass substrate by evaporation. A humidity-sensitive material, such as cellulose acetate, dissolving ethylene dichloride, is then applied in a thin layer (about 1.0-μm thick). On the top of this gold is evaporated as the upper electrode, which is thin (10–20 nm) and porous enough to permit quick lateral moisture transport. The upper electrode acts as a common counterelectrode to the lower twin electrodes, and the composite capacitance of two bulk capacitors, which are connected in series with the upper electrode, is measured between the lower twin electrodes. This configuration eliminates difficulties in contacting a lead wire to the thin upper electrode. For a long time, such sensors were widely used in radio-zonde and many other humidity-measuring instruments.

Testing and analysis of the data presented in Tables 28.1 and 28.3 show that capacitive type sensors are very linear, and hence can measure RH from 0% to 100%, but require complex

circuit and also need regular calibration. At that, it is seen that thermoset polymer-based capacitive sensors, as opposed to thermoplastic-based capacitive sensors, allow higher operating temperatures. In addition, such sensors provide better resistivity against chemical liquids and vapors such as isopropyl, benzene, toluene, formaldehydes, oils, and common cleaning agents. For standard capacitance-type sensors, the change in capacitance is typically 0.2–0.5 pF for a 1% RH change, while the bulk capacitance is between 100 and 500 pF at 50% RH at 25°C. As a rule, capacitive sensors on the sensor market are characterized by low temperature coefficient, ability to function at high temperatures (up to 200°C), full recovery from condensation, and reasonable resistance to chemical vapors. The typical uncertainty of capacitive sensors is ±2% RH from 5% to 95% RH with two-point calibration. The response time ranges from 15 to 60 s for a 63% RH step change. State-of-the-art techniques for producing capacitive sensors take advantage of many of the principles used in semiconductor manufacturing to yield sensors with minimal long-term drift and hysteresis. As a rule, the manufacturer indicates that these sensors are designed to work in the range of 0%–100% RH. However, in practice, acceptable measurement accuracy is achieved in the range from 5% to 95% RH. In this range, RH with two-point calibration, the typical uncertainty of capacitive sensors, is ±2% RH. However, when not calibrated, inaccuracy becomes a major factor. It should be noted that when measuring very dry gases (frost points around −40°C, −50°C, or below; water-vapor fraction around 100 ppm or less) besides the additional calibration of sensors, it is necessary to carefully control the conditions in which measurements are made. In this range, even tiny amounts of stray water released from surfaces can significantly add to the gas-moisture content and can give badly misleading results. To avoid this, it is necessary to use clean, moisture-neutral materials, the minimum possible volume of the pipeline and long flushing times. Sometimes heating is required to remove traces of surface water. If using a pump, use the suction instead of blowing and put the pump at the far end (outlet) of a system, to avoid contamination.

Approximately the same accuracy, according to the information provided by the manufacturer, humidity measurements provide resistive-type sensors (Roveti 2001). Resistive sensors are based on an interdigitated or bifilar winding. Resistive sensors usually consist of noble metal electrodes either deposited on a substrate by photoresist techniques or wire-wound electrodes on a plastic or glass cylinder. The substrate is coated with a salt (LiBr or LiCl) or conductive polymer. A leading sensor manufacturer currently offers an AC-resistive sensor based on the using a proprietary polyelectrolyte (PE) film. PEs are a special class of modified polymers in which one type of an ionic chemical radical group is fixed to the repeat units of the polymer backbone to form a single-ion conducting material. The introduction of water vapor to a PE film under a voltage bias will hydrolyze the ionic groups, resulting in a flow of ions. Film conductivity can be measured as ionic impedance (i.e., AC resistance) and will vary in proportion to the water-vapor concentration present. The measurement scheme is AC-based to avoid polarizing the film over time. Polyelectrolytes that contain strong acidic or basic radical groups in their structure, tend to be very hydrophilic polymers that dissolve readily in water. Cross-linking reagents may be added to convert them into a water-insoluble compound to ensure long life as a sensing film. The specified operating range is 0–100% RH with a ±1%RH inherent accuracy. The latest development in resistive humidity sensors uses a ceramic coating to overcome limitations in environments, where condensation occurs. The substrate surface is coated with a conductive polymer/ceramic binder mixture, and the sensor is installed in a protective plastic housing with a dust filter. A distinct advantage of resistive RH sensors is their interchangeability, usually within ±2% RH, which allows the electronic signal conditioning circuitry to be calibrated by a resistor at a fixed RH point. This eliminates the need for humidity calibration standards, so resistive humidity sensors are generally field replaceable.

The response time for most resistive sensors ranges from 5 to 30 s for a 63% step change. After full condensation, the typical recovery time of a ceramic-coated resistive sensor to its initial state is a few minutes from 5 to 15 min, depending on air velocity. The impedance range of typical resistive elements varies from 1 k$\Omega$ to 100 M$\Omega$. Most resistive sensors use symmetrical AC excitation voltage ($f$ in the range of 30 Hz–10 kHz) with no DC bias to prevent polarization of the sensor. The resulting current flow is converted and rectified to a DC voltage signal for additional scaling, amplification, linearization, or A/DR conversion.

The working temperature range depends on the used humidity-sensitive material. For example, nominal operating temperature of resistive sensors ranges from −40°C to 100°C. Resistive humidity sensors have significant temperature dependencies when installed in an environment with large (>3°C–5°C) temperature fluctuations. Therefore, to compensate the temperature effect and increase the accuracy of measurements, most of the sensors on the market include

a temperature sensor. Also, like capacitive sensors, resistive-type humidity sensors can have small size, low cost, and long-term stability. In residential and commercial environments, the life expectancy of these sensors is >>5 years, but exposure to chemical vapors and other contaminants such as oil mist may lead to premature failure.

Unlike capacitance-type sensors, resistive-type sensors do not require a complex circuit for signal processing. However, they find difficulty in measuring low values (below 5% RH), the change in impedance is too high, and hence it is difficult to control the dynamics, and the temperature affects the properties significantly. In addition, they suffer from significant hysteresis between high and low humidity (memory effects) and accompanying long-term drift. Some believe that advances in electronics can mitigate the problems of temperature effects and high impedance change. However, despite this, the amount of humidity sensors of the resistive type on the sensor market is significantly smaller than the amount of capacitive sensors. Consequently, capacitive RH sensors dominate both in atmospheric and technological measurements. Because of their low-temperature effect, they are often used over wide temperature ranges without active temperature compensation. At the same time, resistive-type sensors are used only in the applications where information is needed, like humidity (yes/no), and accuracy is not required. The inherent drift makes their use difficult for any measurement application.

The most advanced humidity sensors on the market, such as Sensirion humidity sensors (see Figure 28.15), manufactured using technology combined with complementary metal-oxide semiconductor (CMOS) chip technology (Christian 2002). One approach to using CMOS process technology for the manufacture of CMOS-MEMS-based humidity sensors is described in Dennis et al. (2015). Comparative characteristics of CMOS technology with traditional technologies used in the manufacture of microminiature humidity sensors are given in Table 28.6.

The result of this combination is a highly integrated and extremely small humidity sensor on a single silicon chip. The temperature sensor and the humidity sensor together form a single unit. This also enables an accurate and point-precise determination of the dew point, without incurring errors due to temperature gradients between the humidity and the

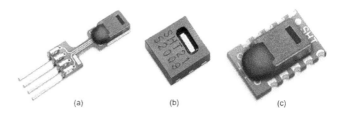

**FIGURE 28.15** Digital humidity sensors of capacitance type designed by Sensirion AG. Sensors have the size of (a) 19.5 × 5.08 × 3.1 mm; (b) 3 × 3 × 1.1 mm; and (c) 7.5 × 4.9 × 2.6 mm. (Modified from https://www.sensirion.com/en/environmental-sensors/humidity-sensors/.)

**TABLE 28.6**

**Comparison of Microsensor Manufacturing Technologies**

| Manufacturing Technology | Film Thickness (μm) | Critical Dimension (μm) | Aspect Ratio | Topology (μm) | Device Size (μm) |
|---|---|---|---|---|---|
| CMOS | <1 | 0.17 | 2:1 | <1 | 1 |
| MEMS | 2–6 | 1.00 | 6:1 | 4–10 | 100 |
| Thick film | >5 | >500 | >50:1 | >200 | >1,000 |

*Source:* Data extracted from Fenner R. and Zdankiewicz, E., *IEEE Sensor J.*, 1, 309–317, 2001.

temperature sensors. Manufacturers claim that a significant improvement in parameters is achieved through the unique linkage of these sensor elements with the signal amplifier unit, the analog-to-digital converter, the calibration data memory, as well as the digital, bus-ready interface, all on a surface area of a few square millimeters (www.silabs.com; www.sensirion.com). According to Sensirion AG (www.sensirion.com) and Silicon Labs (www.silabs.com), the calibration data loaded on the sensor chip guarantee that their humidity sensors have identical specifications and thus can be replaced 100%. In addition, there is no need for customers to calibrate the sensor, which makes the system highly cost efficient. Beyond this, the little mass and area of the microsensor element leads to a very short response time (~8 s), high precision (±2% to ±5% according to configuration), as well as very low power consumption (<3–80 μW standby). Finally, the sensor can be connected directly, without additional external components to any microprocessor system available in many applications, by means of the digital two-wire interface. For example, thin-film capacitive sensors may include monolithic signal conditioning circuitry integrated onto the substrate. The most widely used signal conditioner incorporates a CMOS timer to pulse the sensor and to produce a near-linear voltage output. The exponential response of the resistive sensor can also be linearized by a signal conditioner for direct meter reading or process control. The analog-to-digital conversion, which is also performed "in place," makes the humidity signal extremely insensitive to noise and decouples it from the supply voltage. Table 28.7 summarizes the benefits of using a monolithic solution versus a legacy-discrete approach.

It is important to note that sensors of the same type on the market, but manufactured by different companies, may differ significantly in their parameters, including even temperature limits (see Figure 28.16). Most humidity sensors have a defined temperature range where their function is guaranteed within limits of accuracy. If a humidity sensor is used beyond its temperature limits, it may display erroneous values or get permanently damaged. The temperature limits of a sensor device may determine the application areas of the entire gauge consisting of sensor, probe, and instrument. A wide temperature range is advantageous also for accuracy within that range. All sensors tested had the highest accuracy in the center of their temperature range. They were considerably less accurate at their lower and upper temperature limits. Analysis of the data given in Figure 28.16 testifies that Hygromer C-94 fabricated by Rotronic has the widest temperature range of application.

But this does not mean that the Hygromer C-94 sensors outperform other sensors in all parameters. For example, comparison shows that the sensors Humicap 180 fabricated by VAISALA had the fastest response. The sensor VAISALA Humicap 180 took 40 s to go from 50%RH to 70%RH. The response time for 90%RH was 55 s. Thus, VAISALA Humicap 180 was the fastest of all evaluated sensors both on the wet and on the dry side. The response time of a humidity sensor is the time it takes to respond to a changing set of humidity conditions, deviating from a standard 50% RH. A short response time is advantageous in process control, where it leads to higher process linearity. In applications where large numbers of humidity measurements must be run in sequence, short response times help save time and cost.

**TABLE 28.7**

**Benefits of Monolithic Sensors Compared to Legacy Discrete Designs**

| Key Monolithic Sensor IC Feature | Benefits |
|---|---|
| Higher integration than discrete, hybrid or multi-chip modules (MCMs) | Considerably lower component count reduces design complexity; Smaller overall footprint and minimal height and weight; Faster time to market |
| Standard CMOS fabrication and DFN package | Industry-standard CMOS provides higher yield and reliability; Low-profile DFN enables automated pick-and-place high-volume assembly and manufacturing |
| IC factory calibration with individual coefficients stored in on-chip memory | Shorter customer production/test time and lower solution cost; Production line and field replaceable/interchangeable |
| Optional hydrophobic protective cover | Protects the exposed sensor before, during and after assembly; No materials or labor cost to install the protective cover post-reflow or need to apply/remove protective tape. |

*Source:* Data extracted from Silicon Labs (www.silabs.com)

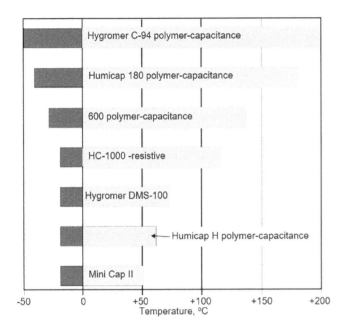

**FIGURE 28.16** Temperature limits of humidity sensors. (Data from the *Rotronic Humidity Handbook*, 2005, https://www.rotronic.com/.)

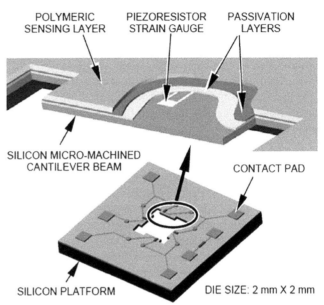

**FIGURE 28.17** The Hygrometrix RH sensor is an advanced resistive device with four semiconductor strain gauges and full Wheatstone Bridge all on one chip. The eight-pin sensor chip has a surface area of only 2 × 2 mm. (Modified from Hydrometrix Applications Note 2004-2: A comparison of relative humidity sensing technologies, www.hygrometrix.com.)

At the same time, tests have shown that the sensors Hygrolyt DMS-100 had the shortest reset rate. The reset rate determines the time (in hours) a humidity sensor takes to reset to 50%RH after it was exposed to 95%RH for 24 h. A drawback of most polymeric hygroscopic materials used as dielectrics in capacitive humidity sensors is that they retain humidity longer than desirable. Thus, polymeric sensors hesitate to reset, and consequently indicate higher values than they should, for quite long periods of time. Hygrolyt DMS-100 took a half hour to reset to 50%...51%RH, a factor of 20 faster than any other sensor system. A quick reset rate is a determining factor in rough climate, when humidity is high and monitoring of fog, dew point, etc. must be done in short intervals. In non-meteo applications such as food processing or water adsorption the reset rate is a key factor for process control.

### 28.3.3 ADVANCED RESISTIVE (PIEZORESISTIVE) HUMIDITY SENSORS

It should be noted that in addition to the previously indicated capacitive and resistive sensors, so-called advanced-resistive (piezoresistive) humidity sensors are also present on the market; for example, the HMX2000-HT humidity-temperature sensor, which exploits the volumetric changes in polymeric films due to water vapor (Figure 28.17). These sensors have been developed by Hygrometrix (www.hygrometrix.com), and in their manufacture the principles of micromachining technology were used. Typical CMOS steps performed on the wafer included chemical vapor deposition, oxidation, doping, diffusion, and metallization. A hybrid CMOS process was used to deposit, pattern, and activate the polymeric sensing element. The final hybrid process step was a plasma etch that released the microcantilever beams from the surrounding silicon.

The vapor-sensing element is constructed from a thin polymer film, deposited and bonded to the top surface of four cantilever beams that are bulk micromachined from the surrounding substrate of a silicon chip. It can be assumed that these sensors use as humidity-sensitive material a hygroscopic material such as polyimide. Each microbeam contains an electrically isolated and diffused-in semiconductor strain gauge that measures the beam stress using a piezoresistive effect. Semiconductor strain gauges are a well-demonstrated, very mature, and stable technology with a proven track record. They have been used for over 20 years in piezoresistive pressure sensors with over 200 million such devices deployed in the automotive industry alone. Adsorption and desorption of water vapor causes the polymer film to expand and contract, inducing a stress in the underlying silicon microbeam through the surface shear stress coupling at the polymer–silicon interface. Behavior of this structure with respect to water vapor is similar to a classical bimorph structure, responding to a temperature gradient at which differences in thermal expansion of the material cause a deflection. The concentration of water vapor is transmitted and linearly measured as mechanical strain. The process is fully regenerative and reversible, depending solely upon van der Waal adsorption of water vapor.

It is known that humidity-dependent changes of a polymer-sensing film could strongly affect the accuracy and long-term stability of conventional capacitive and resistive sensors, when the sensing film is in direct electrical contact with the electrodes. An advantage of the Hygrometrix sensor design over conventional capacitive and resistive sensors is that the polymeric film is electrically isolated from the sensing electrodes. A nitride and oxide passivation layer separates the polymer film from the strain gauges and associated metallization to ensure long-term stability and accuracy. The Hygrotron humidity and temperature sensor from Hygrometrix Inc., Alpine, Calif. (hygrometrix.com), is smaller than a dime but can determine RH between 0 and 100% with 1.5% accuracy after being frozen, totally immersed in water, or exposed to high levels of air pollution and diesel fumes.

Sensor response shows high linearity and low hysteresis (±1%). The sensor can measure full RH levels including 100% RH with water condensation at temperatures from −40°C to 85°C, and can be stored for long periods at temperatures from −55°C to 125°C without damage. Signal conditioning for the Hygrometrix RH sensor is very simple, because the four embedded strain gauges are electrically connected together into a full Wheatstone Bridge circuit directly on the sensor chip. Bridge excitation is provided by a 1.2 to 2.5 V DC voltage source that is external to the device. Sensor output is measured as bridge output voltage that is linearly proportional to the excitation voltage, and ranges from 0 to 72 mV Full Scale (FS) for RH values from 0% to 100%, respectively. With a drive current requirement of only 0.285 mA/V, the Hygrotron sensor consumes only 0.5 mW of power at 1.25 V excitation. It operates in pulsed or DC modes with an average response time of less than 5 s. The sensor is available in SOIC and TO5 style packages (see Figure 28.18). Sensor samples start at

$75 in single units, and the evaluation kit is $400. Additional details on the sensor and its operation may be found elsewhere (Hygrometrix Inc. 2004).

### 28.3.4  THERMAL-CONDUCTIVITY-BASED HUMIDITY SENSORS

Humidity sensors ABS-300 of thermal conductivity type developed by Ohmic Instruments Inc. (www.ohmicinstruments.com) are also present in the sensor market. These sensors (see Figure 28.19) measure the absolute humidity by quantifying the difference between the thermal conductivity of dry air and that of air containing water vapor (read Chapter 14, Volume 2). Thermal-conductivity humidity sensors (or absolute humidity sensors) consist of two matched negative temperature coefficient thermistor elements in a bridge circuit; one is hermetically encapsulated in dry nitrogen and the other is exposed to the environment (see Figure 28.19). As current passes through the thermistors, resistive heating increases their temperatures. The sealed sensor dissipates more heat than the exposed sensor. Since the heat dissipated yields different operating temperatures, the difference in resistance of the thermistors is proportional to the absolute humidity (see Figure 28.20). If temperature and pressure are known, the absolute humidity easily converts to relative humidity.

Absolute humidity sensors are very durable, operate at temperatures up to 300°C, and are resistant to chemical vapors by virtue of the inert materials used for their construction, that is, glass, semiconductor material for the thermistors, high-temperature plastics, or aluminum. In general, absolute humidity sensors provide greater resolution at temperatures >100°C than do capacitive and resistive sensors, and may be used in applications where these sensors would not survive. The typical accuracy of an absolute humidity sensor is ±3 g/m³; this converts to about ±5% RH at 40°C and ±0.5% RH at 100°C. An interesting feature of thermal conductivity sensors is that they respond to any gas that has thermal properties different from those of dry nitrogen; this will affect the measurements.

In addition to sensors, Ohmic Instruments Inc. offers Thermal Conductivity Absolute Humidity Transmitters,

## Pin connection (Top view)

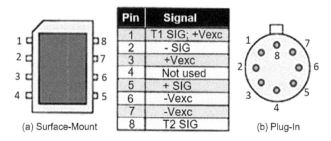

| Pin | Signal |
|-----|--------|
| 1 | T1 SIG; +Vexc |
| 2 | - SIG |
| 3 | +Vexc |
| 4 | Not used |
| 5 | + SIG |
| 6 | -Vexc |
| 7 | -Vexc |
| 8 | T2 SIG |

(a) Surface-Mount

(b) Plug-In

**FIGURE 28.18** Standard packaging options are available for surface-mount (a), and plug-in socket or board-soldered (b) applications.

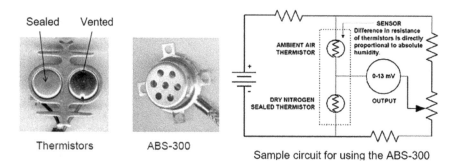

Sealed   Vented

Thermistors        ABS-300

Sample circuit for using the ABS-300

**FIGURE 28.19** Photo of thermal conductivity humidity sensor. Thermistor chambers are left, and absolute humidity sensors are in the center. In thermal conductivity sensors, two matched thermistors are used in a DC bridge circuit. One sensor is sealed in dry nitrogen, and the other is exposed to ambient. The bridge output voltage is directly proportional to absolute humidity. (Modified from https://www.ohmicinstruments.com/humidity-sensors and https://www.fierceelectronics.com/components/choosing-a-humidity-sensor-a-review-three-technologies.)

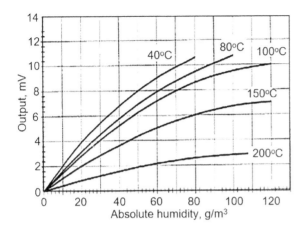

FIGURE 28.20 The output signal of the thermal conductivity sensor is affected by the operating temperature. Maximum output is at 600°C; output at 200°C drops by 70%. (Modified from https://www.ohmicinstruments.com/humidity-sensors and https://www.fierceelectronics.com/components/choosing-a-humidity-sensor-a-review-three-technologies.)

ANT-200, developed on the basis of ABS-300 sensors (see Figure 28.21): Range from 0 to 130 g/m³ (0%–100% RH at 60°C); operation from 25°C to 200°C; recovers from condensation. The developer claims that AHT-200 Series transmitters overcome many of the limitations of humidity sensors that cannot be used or fail prematurely in corrosive industrial applications. Absolute humidity can be converted to other psychrometric units such as %RH, PPMv, PPMw, and dew point.

Absolute humidity sensors of thermal conductivity type are commonly used in appliances such as clothes dryers and both microwave and steam-injected ovens. There has been an increase in the number of industrial and process applications for absolute humidity detectors due to their ability to operate at high temperatures, recover from condensation, and their excellent immunity to many chemical and physical contaminants. Absolute humidity measurement, using thermal-conductivity sensors, is an economical way of monitoring and controlling many industrial processes such as drying wood, textiles, paper and chemical solids, catalytic converters, pharmaceutical production, material curing, catalyst production, cooking, sterilization, food dehydration, and desiccant heat recovery. Since one of the by-products of combustion and fuel cell operation is water vapor, particular interest has been shown in using absolute humidity sensors to monitor the efficiency of those reactions.

### 28.3.5 QUARTZ CRYSTAL MICROBALANCE-BASED HUMIDITY SENSORS

As for the quartz crystal microbalance (QCM)-based humidity sensors on the market they are represented by such sensors as Quartz Crystal Microbalance moisture analyzers QMA401 and QMA601, developed by Michell Instruments (www.michell.com/uk/) (see Figure 28.22). This model was designed to provide reliable, fast, and accurate measurement of trace moisture content in a wide variety of process applications, where keeping moisture levels as low as possible is of critical importance. This device provides fast and reliable measurement from 0.1–2000 ppmv and accuracy of ±0.1 ppmv at <1 ppmv and 10% of reading from 1 to 2000 ppmv, and maintenance-free for 3 years.

Users of quartz microbalance sensors (www.gemeasurement.com) mark that QCMs have a certain degree of hysteresis and must be "re-zeroed" periodically. The measurement system therefore requires a "zero gas." While no gas supply can have an absolute value of zero, the zero gas may be defined as a gas that is closer to zero than any significant amount of

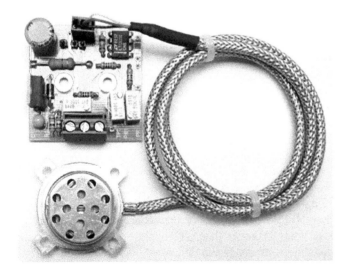

FIGURE 28.21 AHT-200 series transmitters. (Modified from https://www.ohmicinstruments.com/humidity-instrumentation.)

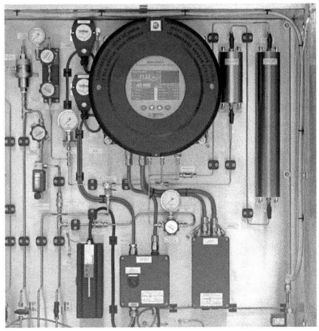

FIGURE 28.22 Advanced quartz crystal microbalance analyzer from Michell Instruments. (Modified from http://www.michell.com/us/products/qma601.htm.)

water. Some measurement modes employ a non-equilibrium technique, where the sensor alternates from being exposed to the zero gas and the process gas. The offline time spent on the zero gas should be factored into response time requirements. The sensing surface is also susceptible to contamination and must remain clean. Therefore, a suitable sampling system must be employed.

According to developers, the new analyzer utilizes a new generation of precision crystal oscillators, guaranteeing a highly accurate measurement, which is completely insensitive to changes in the composition of the background gas. Its configuration is shown in Figure 28.23. The basis of the sensor is a hygroscopic-coated quartz crystal with specific sensitivity to water vapor. For maximum stability, all critical components of the QMA601—the moisture generator, sensor and flow control devices—are precisely temperature controlled. This ensures that fluctuations in the sample gas or environmental temperature have no influence on the measurement. The analyzer utilizes a mass flow controller to ensure a precise control of the sample and reference gas flows to ±0.1 mL/min. Coupled with a pressure transducer, this system ensures continued accuracy of measured and calculated parameters, even during fluctuations in the sample pressure. The analyzer incorporates automatic (or manual) verification, using an integrated moisture generator reference source.

Currently, QMA601 moisture analyzers are used in the processes related to glycol dehydration of natural gas, molecular sieve dehydration of natural gas, natural gas transmission and storage, refinery catalytic reforming, recycle gas monitoring, ethylene and propylene production, and LNG production/revaporization.

AMETEK (www.ametekpi.com) also works in the same area, manufacturing on the basis of QCM-based sensors Moisture Analyzers 3050-AP, 5910, and 5920. The heart of the indicated analyzers is a QCM sensor and sampling system developed by AMETEK specifically for highly accurate moisture measurements. The sensor consists of a quartz crystal disc coated with a hygroscopic polymer. Model 3050-AP is shown in Figure 28.24. The accuracy of the 3050-AP is 0.1 ppmv or 10% of reading from 1 to 2500 ppmv with standard calibration. Normally, the 3050-AP provides a fast and

**FIGURE 28.24** Model 3050-AP Moisture Analyzer developed by AMETEK. (Modified from https://pdf.directindustry.com/pdf/ametek-process-instruments/model-3050-ap/14271-689695.html.)

**FIGURE 28.25** Model 5910 UHP Moisture Analyzer. The heart of the 5910 analyzer is a QCM sensor and sampling system developed by AMETEK specifically for highly accurate moisture measurements. (Modified from https://www.ametekpi.com/products/brands/process-instruments/5910-uhp-moisture-analyzer.)

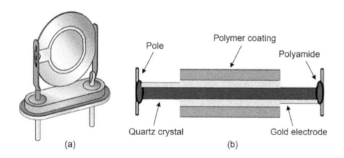

**FIGURE 28.23** (a–b) Configuration of QCM sensor used in QMA601 analyzer. (Modified from http://www.michell.com/uk/documents/QMA601-97490-UK-Datasheet-v4.pdf.)

accurate analysis of the sample gas's moisture concentration, while quickly cycling between the sample gas and the dry reference gas created from the sample gas passing through the supplied dryer. If contaminants are a concern in your application, the 3050-AP can be operated in its sensor saver mode and equipped with a contaminant trap prior to the dryer. This configuration allows the sensor to see the contaminant free reference gas for an extended period of time.

Model 5910 is shown in Figure 28.25. The Model 5910 is sensitive to changes in moisture concentration of less than 500 pptv. It was established that the nominal noise level is below 50 pptv. The result of these phenomenal abilities is an accuracy of better than ±100 pptv or ±10% of reading. These moisture analyzers are ideal for critical moisture-measurement applications, including high-purity gas production and semiconductor gases.

### 28.3.6 SAW-Based Dew-Point Hygrometers

There are not many proposals in this area compared to hygrometers developed on the basis of capacitive and resistive humidity sensors. The main developers of surface acoustic wave (SAW)-based hygrometers are Jet Propulsion Laboratory (JPL) (https://ghrc.nsstc.nasa.gov), which in 1998 developed for its projects CAMEX-3 JPL Surface Acoustic Wave (SAW) Hygrometer, Midwest Micro-Tek (www.midwestmicro-tek.com), which offers the Ultra DP5 and DP5RH SAW hygrometer (Figure 28.26a), and Vaisala (www.vaisala.com), which has developed and presents on the market DM500 Precision SAW Hygrometer (Figure 28.26b).

All indicated hygrometers use a SAW device (read Chapter 12, Volume 2) as a sensitive detector of condensation in lieu of the more traditional chilled mirror and light-beam approach. Since the oscillation in a SAW device is quite sensitive to surface effects, condensation produces a measurable shift in the resonance frequency of the SAW device. A two- or three-stage thermoelectric converter heats or cools the SAW device, while a platinum resistor or thermistor of another type is used to monitor the temperature. The SAW hygrometer measures dew point by establishing equilibrium between evaporation and condensation on the surface of the SAW device. Using a SAW device as a fast, high-sensitivity moisture sensor, a feedback controller measures the condensation on the sensor surface and responds by heating or cooling the sensor to maintain equilibrium. The equilibrium temperature under feedback control is a measure of dew point or frost point, depending on the phase of the condensed moisture. Using special cooling techniques, CAMEX-3 JPL measured the dew point/frost point as low as −65°C. The Ultra DP5 can continuously monitor dew points between −20°C and 60°C, while DM501, depending on the modification, can measure dew-point temperature in the range from −40°C to +60°C (DM500S) to −75°C to +60°C (DM500X).

As the main advantages of SAW-based dew-point hygrometers compared to chilled-mirror hygrometers, developers highlight the speed of response (https://ghrc.nsstc.nasa.gov), lower price (www.midwestmicro-tek.com), high accuracy, and a reliable dew-point measurement with excellent repeatability and fast response time (www.vaisala.com). In addition, Vaisala claims that their technology can withstand particulate contamination and distinguish between dew and frost directly.

### 28.3.7 Wireless Humidity Sensors

It should be noted that, in addition to the traditional approach to the development and use of humidity sensors described earlier, the wireless humidity sensors market is currently developing intensively. In generally, wireless sensors are not sensors per se, but rather are autonomous data acquisition nodes to which traditional sensors can be attached (Figure 28.27). Wireless sensors are best viewed as a platform in which mobile computing and wireless communication elements converge with the sensing transducer (Lynch and Loh 2006; Ruiz-Garcia et al. 2009; Aqeel-ur-Rehman et al. 2014). As shown in Figure 28.27, all wireless sensors are measurement tools equipped with sensor module and transmitters to convert signals from sensor into a radio transmission. The radio signal is interpreted by a receiver, which then converts the wireless signal to a specific, desired output, such as an analog current or data analysis via computer software. External memory is an additional module that could be needed in case of data storage requirement for local decision making. Thus, with computational power coupled with the sensor, wireless sensors are capable of autonomous operation. Without a physical link, existing between individual wireless sensors and the remainder of the wireless sensor network, wireless sensors must know when to act autonomously or collaboratively. Software embedded in the wireless sensor's computational core is responsible for its autonomous operation, including the

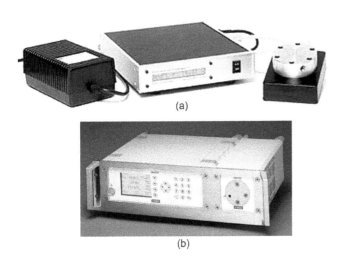

(a)

(b)

FIGURE 28.26 (a) DP5RH SAW hygrometer (Modified from http://www.midwestmicro-tek.com/microtek/products/instrumentation/ultradp5/ultradp5.html); and (b) DM500 Precision SAW Hygrometer (Modified from https://www.vaisala.com/sites/default/files/documents/VN160_Vaisala_DM500_Precision_SAW_Hygrometer.pdf).

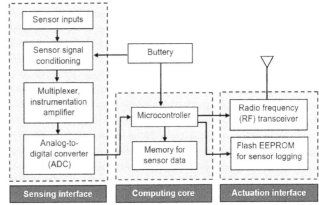

FIGURE 28.27 Functional elements of a wireless sensors. (Idea from Lynch, J.P., and Loh, K.J., *Shock vib. Digest*, 38, 91–128, 2006.)

collection and storage of data, interrogation of measurement data, and deciding when and what to communicate to other wireless sensors in the wireless sensor network.

It is necessary to recognize that wired sensor networks are very reliable and stable communication systems for instruments and controls. However, wireless technology promises lower installation costs than wired devices, since required cabling engineering is very costly. Installation of wireless sensors does not require extension wire, conduit, and other costly accessories. Wireless sensors are flexible and can be easily reconfigured. In particular, they are useful for obtaining data in hard to access locations or in locations that are difficult to access due to extreme conditions. Wireless sensors can be also used to form a web/network that would allow an engineer to monitor a number of different locations from one station. This provides a centralized control of a required area. Additionally, a number of wireless sensors have the ability to create a unique web page making up-to-the-minute data, accessible anywhere in the world. This is the most important advantage of using such technology for monitoring. Wireless sensor technology also allows micro-electro-mechanical systems (MEMS) sensors to be integrated with signal conditioning and radio units to form "motes"—all for a low cost, a small size, and with low-power requirements.

One should note that the design of wireless sensor requires many considerations like energy conservation, scalability, size, housing, etc. Therefore, the selection of sensor type for such devices is carried out considering the major requirements of application domain, the power consumption, and the available energy sources. Wireless communication technologies, like ZigBee, Bluetooth, Wibree, and WiFi are a part of several sensor-network-based research works. These technologies have different capabilities and properties on which they are complemented. A brief comparison is given in Table 28.8.

ZigBee wireless communication technology (IEEE 802.15.4) is preferred over other technologies for the development of wireless sensor network due to its low cost and low power consumption property. It was introduced in May 2003 and operates on Industrial, Scientific and Medical (ISM) band, that is, 2.4 GHz globally. There are 16 ZigBee channels of 5 MHz bandwidth each in 2.4 GHz band.

Analysis of the available information shows that a number of commercial wireless sensor platforms have emerged in recent years that are well-suited for humidity measurement. There are many companies in this market, including E+E Elektronik, MONNIT, Omega, APAP, Strike, and many others. Examples of wireless humidity sensors developed in these companies are shown in Figure 28.28. They differ in functionality, size, and price. This information is presented on the sites of these companies.

Usually, wireless transmitters (EE244, EE245, EE242, UWRH-2, UWRH-2-NEMA, etc.) are based on the IEEE 802.15.4 protocol with a transmission frequency of 2.4 GHz. Of course, other protocols and frequencies can also be used. For example, wireless sensors developed by Monnit (www.monnit.com) operate at the frequency ~900 MHz. As for humidity sensing parameters of wireless sensors, they are determined by the humidity sensors used in these devices. According to their characteristics, they are no different from the parameters of traditional humidity sensors, discussed earlier in this chapter. Depending on the power of the device, wireless sensors can provide radio transmission to a host receiver in the range from 40 to 200 m (open space)—depending on the local radio propagation conditions: the type and the thickness of walls, doors, etc. In some cases, the radio range can be increased up to 400 m and more.

The power sources of such sensors are batteries (usually 1.5 V AA standard or 3.6 V lithium batteries) or external

## TABLE 28.8
### Comparison of Wireless Communication Technologies

| / | Communication Technology | | | |
| --- | --- | --- | --- | --- |
| | **ZigBee** | **Bluetooth** | **Wibree** | **WiFi** |
| Frequency band | 2.4 GHz | 2.4 GHz | 2.4 GHz | 2.4 GHz |
| Range | 30 m–1.6 km | 9–90 m | Up to 3 m | 30–45 m |
| Data rate | 250 kbps | 1 Mbps | 1 Mbps | 11–54 Mbps |
| Power consumption | Low | Medium | Low | High |
| Cost | Low | Low | Low | High |
| Modulation/protocol | DSSS, CSMA/CA | FHSS | FHSS | DSSS/CCK, OFDM |
| Security | 128 bit | 64 or 128 bit | 128 bit | 128 bit |

*Source:* Data extracted from Aqeel-ur-Rehman, A.Z. et al., Building a smart university using RFID technology, in: *Proceedings of International Conference on Computer Science and Software Engineering* (CSSE 2008), Wuhan, China, 641–644, 2008.

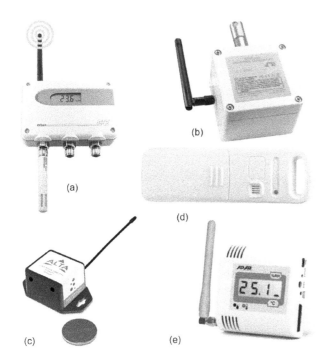

FIGURE 28.29 Wireless Sensor Tag (13-bit Temperature and Humidity) developed by Strike. (Modified from https://store.wirelesstag.net/products/wireless-tag-13-bit-temperature-and-humidity.)

FIGURE 28.28 Wireless humidity sensors: (a) wireless humidity and temperature transmitter EE244 (Modified from https://www.epluse.com/en/products/humidity-instruments/wireless-sensors-1/ee244/); (b) wireless relative humidity/temperature transmitter UWRH-2-NEMA (Modified from https://www.omega.com/en-us/control-and-monitoring-devices/wireless-monitoring-devices/wireless-transmitters/p/UWRH-2-NEMA-Series); (c) ALTA wireless humidity sensor—coin cell powered MNS2-9-W1-HU-RH (Modified from https://www.monnit.com/Product/MNS2-9-W1-HU-RH); (d) AcuRite 06002M wireless temperature and humidity sensor (Modified from https://www.acurite.com/temperature-and-humidity-sensor-06002rm-592txr.html); and (e) wireless temperature and humidity sensor AR432 (Modified from https://www.apar.pl/en/wireless-temperature-sensor-ar432.html).

power supply. The developers claim that wireless sensors can operate for a long time without changing batteries. For example, according to APAR (www.apar.pl), wireless sensors AR431-AR436 with a new battery can operate at room temperature, with measurement period >30 min and uninterrupted radio transmission up to 4 years.

Recognizing power consumption to be a major limitation of wireless sensors, operating on batteries, recently some research teams are exploring the development of power-free wireless sensors, known as radio-frequency identification (RFID) sensors. RFID sensors are a passive radio technology, which capture radio energy, emanated from a remote reader, so that it can communicate its measurement back. It is important to note that research aimed at developing RFID sensors for measuring humidity is also being conducted (read Chapter 20, Volume 2). Moreover, currently such sensors developed by Strike (www.strike.com.au) are already on the market (see Figure 28.29).

## 28.4 MARKET OF CONVENTIONAL HYGROMETERS

It should be noted that the conventional humidity-sensing devices (hygrometers) described earlier in Chapters 2 through 9, Volume 2 are also manufactured and widely used. Table 28.9 provides a selective listing of the methods and manufacturers, which propose devices for humidity measurement. It is seen that there are many companies working in this field, including AMETEK, Kahn Instruments, Inc., Fisher Scientific, Avagadro's Lab Supply Inc., Yankee Environmental Systems, Inc., and so on. However, some of these devices are not widely used because of their cost and complexity.

For example, Figure 28.30 shows examples of chilled-mirror hygrometers fabricated by different companies. There are also prices of some devices. It is seen that chilled-mirror hygrometers are really expensive. Typical prices for other conventional hygrometers are listed in Table 28.10.

The appearance of the hygrometer using the coulometric method of measuring humidity is shown in Figure 28.31. The Model 303B Moisture Monitor is based on the using $P_2O_5$ coulometric humidity sensor (read Chapter 7, Volume 2). This is a high-precision portable device for measuring trace humidity in the range from 0 to 1000 ppmv in gas streams at 100 mL/min sample flow through the cell at 25°C and 101.4 kPa (0–2000 ppmv range is possible with reduced sample flow). Sensitivity: 0.1 ppm.

Typical hygrometers for using in a library, archive, or museum are shown in Figure 28.32. Model H-311 uses a human hair bundle to measure RH and a bimetallic device to measure temperature. Hygrothermographs are available

**TABLE 28.9**

**Devices for Measuring Humidity and Moisture on the Market**

| Method | Medium | Range | Manufacturer |
|---|---|---|---|
| Mechanical (hair) | g | 0%–100%RH | Lambrecht, Thies, Haenni, Jumo, Sato, Casella, Pekly et Richard |
| Condensation dew point | g | −80°C to +100°C dew point | Ametek, General Eastern, Michell Instr., EG&G, E+H, MBW, Protimeter, Panametrics |
| Dry and wet bulb | g | 10%–100% RH | Lambrecht, Thies, Haenni, ASI, Jenway, Casella, Ultrakust, IMAG-DLO |
| Lithium chloride | g | −45°C to +95°C dew point | Honeywell, Jumo, Lee Engineering, Siemens, Philips, Weiss |
| Thermal conductivity | g | 0–130 g/m³ | Shibaura Electronics Co. Ltd |
| $Al_2O_3$/silicon | g, l | −80°C to +20°C dp | E+H, General Eastern, Panametrics, Michell Instr., MCM, Shaw |
| Phosphorous pentoxide | g | 0.5–10,000 PPM | Ametek, Anacon, Beckman, DuPont |
| Infrared absorbance | g, l | 0–50 PPM up to 65°C dew point | Siemens, H&B, ADC, Anacon, Kent, Horiba, Sieger, Beckman, Li-Cor |
| Infrared reflectance | s | 0.02%–100% | Anacon, Infrared Engin., Moisture Systems Corp., Pier Electronic, Zeltex, Bran & Luebbe, Buhler |
| Nuclear magnetic resonance (NMR) | l, s | 0.05%–100% | Oxford Anal. Instr., Bruker |
| Neutron moderation | s | >0.5% | Kay Ray Inc., Berthold, Nuclear Ent. |
| Microwave attention | s | 0%–85% | Mutec, Scanpro, Kay Ray Inc., BFMRA |
| Time-domain reflectometry (TDR) | s | 0%–100% | IMKO GmbH, Campbell Sci. Inc., Soil Moisture, Tektronix |
| Frequency domain (FD) | s | 0%–100% | ΔT Devices, IMAG-DLO, VITEL, Troxler |

*Source:* Data extracted from Brodgesell, A. and Liptak, B.G. Moisture in air: Humidity and Dew point. in: Liptak B.G. (ed.) *Analytycal Instrumentation*, Chilton Book Company, Radnor, PA, 215–226, 1993 and Visscher, G.J.W., Humidity and moisture measurement. in: Webster J.G. and Eren H. (eds.) *The Measurement, Instrumentation and Sensors Handbook*, 2nd edn.. CRC Press, Boca Raton, FL, 2014.

*Abbreviations:* E+H—Endress+Hauser; MBW—MBW Electronik AG; ASI—Automatic Systems Laboratories Ltd; PCRC—Physical and Chemical Research Corporation; MCM—Moisture Control and Measurement Ltd; H&B—Elsag Bailey Harmann & Braun; ADC—Analytical Development Company; IMAG-DLO—Institute for Agricultural and Environmental Engineering; dp—dew point; g, l, s—gas, liquid, solid;

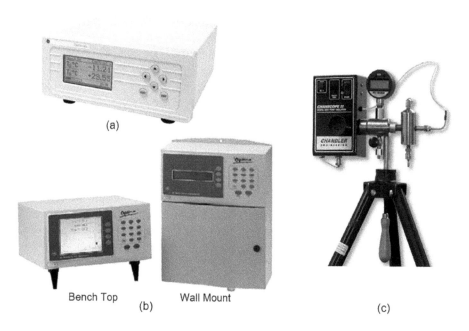

**FIGURE 28.30** Chilled-mirror hygrometers: (a) GE General Eastern OptiSonde chilled mirror hygrometer. Compact, rugged, chilled-mirror hygrometer for metrology labs and industrial applications ($4,500) (Modified from https://www.instrumart.com/products/23780/panametrics-optisonde-chilled-mirror-hygrometer); (b) GE General Eastern Optica series chilled mirror hygrometer ($8,500) (Modified from http://www.ssco.com.tw/General_Eastern/ChilledMirror_Optica/Optica.pdf); and (c) manually operated dew-point chilled-mirror devices developed by AMETEK (Modified from https://www.ametekpi.com/products/brands/chandler-engineering/chanscope-ii-dew-point-tester).

**TABLE 28.10**

**Devices (Humidity Transmitters) for Measuring Air Humidity Offered on the Market**

| Method | Manufacturer | Approx. price, $ |
|---|---|---|
| Mechanical (hair) | Lambrecht, Thies, Haenni, Jumo, Sato, Casella, Pekly et Richard | 0.1–1 k |
| Condensation dew point | General Eastern, Michell Instr., EG&G, E+H, MBW, Protimeter, Panametrics | 2–20 k |
| Dry and wet bulb | Lambrecht, Thies, Haenni, ASI, Jenway, Casella, Ultrakust, IMAG-DLO | 0.1–3 k |
| Lithium chloride | Honeywell, Jumo, Lee Engineering, Siemens, Philips, Weiss | 0.7–2 k |
| Pope cell | General Eastern | 0.5–0.7 k |
| Thermal conductivity | Shibaura Electronics Co. Ltd | 0.1–1 k |
| $Al_2O_3$/silicon | E+H, General Eastern, Panametrics, Michell Instr., MCM, Shaw | 1.5–3 k |
| Phosphorous pentoxide | Anacon, Beckman, DuPont | 2–6 k |
| Karl Fisher titrator | Mettler Toledo, Metroohm | 10–20 k |

*Source:* Data extracted from Visscher, G.J.W., Humidity and moisture measurement. in: Webster J.G. and Eren H. (eds.) *The Measurement, Instrumentation and Sensors Handbook*, 2nd edn. CRC Press, Boca Raton, FL, 2014.

**FIGURE 28.31** Moisture Monitor Model 303B developed by AMETEK. (Modified from https://www.ametekpi.com/products/brands/process-instruments/303b-moisture-analyzer.)

with circular charts, but linear charts (also called drum charts) are preferred since they are easier to string together and read for annual reviews. Like humidity sensors and other hygrometers, hygrothermographs also require periodic calibration, preferably quarterly. Variants of psychrometers fabricated by different companies are shown in Figure 28.33.

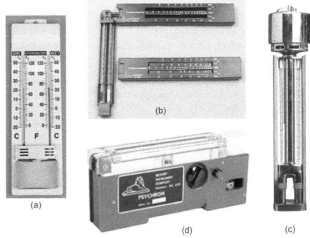

(a)

(b)

(d)

(c)

**FIGURE 28.33** (a) Psychrometers for home (Modified from http://www.psscientific.com/shop/humidity/moisture-meters/reedmod-elr6013woodmoisturemeter.aspx); (b) portable psychrometer fabricated by Avagadro's Lab Supply Inc. (Modified from https://www.avogadro-lab-supply.com/item/Compact_Sling_Psychrometer_Wet_Dry_Bulb_Hygrometer/1003/c124); (c) Aspiration Psychrometer by Assmann (Modified from http://www.th-friedrichs.de/en/products/humidity/relative-humidity-psychrometer/aspiration-psychrometer-by-assmann/); and (d) Portable battery-powered and motor-aspirated psychrometer, Psychron, designed by Belfort Instrument (Modified from http://www.belfort-inst.com/Model_566.htm).

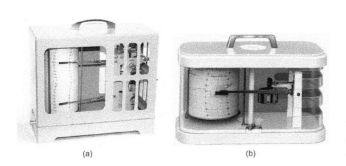

(a)

(b)

**FIGURE 28.32** Recording (a) hygrothermograph and (b) hygrograph: (a) Model 225-5020-A Hi-Q (Modified from https://novalynx.com/store/pc/225-5020-A-Hi-Q-Hygrothermograph-p642.htm); (b) Model 3210.0000 (Theodor Friedrichs & Co, Modified from http://www.th-friedrichs.de/en/products/humidity/hygrograph-en-GB/hygrograph-en-GB/.)

## 28.5 SUMMARY

Our analysis has shown that the market for electronic and electrical humidity sensors and conventional-based hygrometers is developing dynamically, and the area of their use is constantly expanding. Besides conventional applications such as the building sector, weather stations, irrigation control, and industrial applications, new applications are arising in consumer electronics, household appliances as well as in the environmental and medical sector. Rapid advancements in semiconductor technology have made possible to fabricate highly accurate, reliable, and stable solid-state humidity sensors with resistance to chemicals and physical contaminants at economical prices. Therefore, the manufacturers of today's household and industry appliances incorporate in their designs a new generation of humidity sensors, since these devices improve their energy efficiency, safety, and performance. It is important, that miniaturization of electronic device is one of the major trends in the global humidity sensor market. Mechanical sensors and many conventional humidity sensors are being replaced by solid-state devices that form an integral part of smaller, less expensive, and more reliable microsystems. These in turn can be readily integrated with standard microcontrollers to make home and industry appliances more functional and user friendly. For example, low-cost and small RH sensors can be deployed throughout a home refrigerator to provide microclimate control based on the type of food being stored. Many of today's refrigerators keep food fresher longer by providing sealed storage drawers for select food groups. Optimal temperature is maintained in each drawer by circulating the cold air through baffles around the drawer instead of blowing it directly onto the food. Water-vapor sensors can transform standard refrigerator temperature controls into environmental or microclimate controls that provide added protection against food spoilage. The same principle can be applied to other appliances as well. For example, an RH sensor in the return air duct of a microwave oven downstream of the air filter could monitor the water vapor being driven off the cooking food item, and thus, could improve the quality of food cooking (see Figure 28.34).

Atmospheric conditions are constantly changing; therefore, measurements and calculations should be carried out frequently. Continuous measurement is one reason why digital, all-in-one instruments based on electronic humidity sensors are quickly becoming popular. Digital sensors offer many advantages over conventional sensors such as high performance and accuracy coupled with low cost, easy implementation, no requirement for complex calibration, and low-power consumption among others. It is important to note that the possibility of automatic calibration in the measurement process is an important task facing developers of humidity sensors and devices using them. (www.ist-ag.com). Currently this feature, allowing to calibrate out the inherent measurement errors due to aging, drift, and surface-adhesion problems, is being attempted to be realized on the basis of highly integrated sensor modules. However, this approach requires very sophisticated algorithms and at present achievements on this path are observed only with the use of capacitive sensors. In particular, Vaisala uses an auto-calibration procedure to accurately measure the dew point. Autocalibration includes several steps shown in Figure 28.35: (1) Auto-calibration start: sensor warmed RH value decreases when temperature increases; (2) Sensor starts to cool down after short warming. RH reading starts to increase; (3) RH and T values logged during cooling phase; (4) When drawing a straight line, the cooling phase through the collected RH and T points the offset correction value is received as a result; (5) Auto-calibration ends: Sensor is cooled and normal measurement mode gets active again using the correction value; and (6) After set interval (once/hour), auto-calibration repeated. Thus, autocalibration warms up the sensor automatically, and based on the data obtained while simultaneously measuring the humidity and temperature during the cooling of the sensor, makes the offset (= zero point) correction. The developers claim that this calibration guarantees the very good accuracy in low dew points and offers excellent long term stability.

The use of digital sensors also greatly simplifies the process of measuring and calculating critical environmental parameters. In addition, when using these digital instruments, it is possible not only to display these values, but also these values can be stored in the gage's memory along with the date and time. Another advantage that digital instruments provide is that they take much of the guesswork out of measuring. Many models have alarms that automatically alert the user, when the surface temperature is too close to the dew-point temperature; this feature signals the high risk of moisture formation.

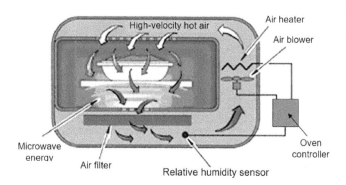

**FIGURE 28.34** Microwave oven with installed RH sensor. (Based on Babyak, R.J. (2000) New generation appliances combine speed with quality, Racing recipes. In: *Appliance Manufacturer*, Cahners Publishing Company, Des Plaines, IL, 233–239, 2000)

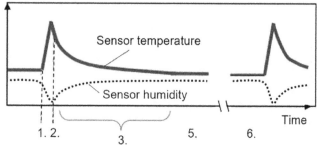

**FIGURE 28.35** Temperature profile used for autocalibration of humidity sensors.

**TABLE 28.11**

**Typical and Recommended Fields of Applications for Humidity Sensors and Hygrometers**

| Devices | Recommended Fields of Application |
|---|---|
| Wet bulb/dry bulb psychrometer | Climatic/environmental chambers; humidity calibration standards; |
| Displacement sensors (hair hygrometers) | Climate control in libraries, museums, apartments, and storages |
| Polymer resistive and capacitive sensors | Heating, ventilating, and air conditioning (HVAC) energy management; computer room/clean room control; handheld devices; environmental and meteorological monitoring |
| Metal-oxide resistive and capacitive sensors | HVAC energy management; high temperature, portable devices; microwave ovens; climate control; control of industry processes |
| QCM and SAW sensors | High-purity natural and semiconductor gases production (molecular sieve dehydration, gas transmission and storage, etc.); chemical industry (refinery catalytic reforming—recycle gas monitoring, etc.) |
| Thermal conductivity | Drying wood, clothes, textiles, paper, chemical solids, etc.; microwave and steam-injected ovens; high temperatures; catalyst production; cooking, sterilization; food dehydration; desiccant heat recovery, etc. |
| Saturated salt lithium-chloride sensor | Refrigeration controls; dryers; dehumidifiers; airline monitoring; pill coaters |
| Aluminium-oxide dew-point sensors | Aggressive gases; metallurgy, plastics manufacturing; moisture measurements in liquids (hydrocarbons); intrinsically safe and explosion-proof installations; petrochemical and power industries where low dew points are to be monitored in line and where the reduced accuracies and other limitations are acceptable |
| Chilled-mirror (optical condensation) hygrometer | Medical air lines; liquid-cooled electronics; cooled computers; heat-treating furnaces; smelting furnaces; clean-room controls; dryers; humidity calibration standards; engine test beds |
| Electrolytic hygrometer (phosphorous pentoxide, $P_2O_5$) | Ozone generators; dry air lines; nitrogen transfer systems; inert gas welding |
| Optical and microwave hygrometers, LIDAR | Atmosphere monitoring; meteorology; remote control |
| Fiber-optic sensors | Chemical industry, medicine and security systems, where electronic sensors cannot be applied (radiation, high electromagnetic fields, etc.); remote control |

However, it cannot be argued that electronic humidity sensors will force all other types of instruments designed for humidity monitoring out of the market. The variety of applications with specific requirements for monitoring devices leaves room for all types of humidity-sensitive devices. To confirm this statement, you can look at Table 28.11, where recommended fields of application are given for each type of such devices.

## REFERENCES

Aqeel-ur-Rehman, Abbasi A.Z., Islam N., Shaikh Z.A. (2014) A review of wireless sensors and networks' applications in agriculture. *Comput Stand Inter.* **36**, 263–270.

Aqeel-ur-Rehman A.Z., Shaikh Z.A. (2008) Building a smart university using RFID technology. In: *Proceedings of International Conference on Computer Science and Software Engineering* (CSSE 2008), Wuhan, China, pp. 641–644.

Babyak R.J. (2000) New generation appliances combine speed with quality, Racing recipes. In: *Appliance Manufacturer*, Cahners Publishing Company, Des Plaines, IL, pp. 233–239.

Brodgesell A., Liptak B.G. (1993) Moisture in air: Humidity and Dew point. In: Liptak B.G. (ed.) *Analytycal Instrumentation*, Chilton Book Company, Radnor, PA, pp. 215–226.

Christian S. (2002) New generation of humidity sensors. *Sensor Rev.* **22**(4), 300–302.

Dennis J.-O., Ahmed A.-Y., Khir M.-H. (2015) Fabrication and characterization of a CMOS-MEMS humidity sensor. *Sensors* **15**, 16674–16687.

Fenner R., Zdankiewicz E. (2001) Micromachined water vapor sensors: A review of sensing technologies. *IEEE Sensor J.* **1**(4), 309–317.

Hygrometrix Inc. (2004) Application Note No. 2004-1: Implementing the Hygrotron™ Humidity Sensor Into Your Design.

Lynch J.P., Loh K.J. (2006) A summary review of wireless sensors and sensor networks for structural health monitoring. *Shock Vibration Digest* **38**(2), 91–128.

Masuda H., Nishio K., Baba N. (1993) Fabrication of a one-dimensional microhole array by anodic oxidation of aluminium. *Appl. Phys. Lett.* **63**, 3155–3157.

Nahar R.K. (2000) Study of the performance degradation of thin film aluminium oxide sensor at high humidity. *Sens. Actuators B* **63**, 49–54.

Nahar R.K., Khanna V.K. (1998) Ionic doping and inversion of the characteristic thin film porous $Al_2O_3$ humidity sensor. *Sens. Actuators B* **46**, 35–41.

Rotronic (2005) The Rotronic Humidity Handbook (www.rotronic-usa.com).

Roveti D.K. (2001) Humidity/moisture: Choosing a humidity sensor: A review of three technologies. *Sensors* (Peterborough) **18**(7), 54–58.

Ruiz-Garcia L., Lunadei L., Barreiro P., Ignacio Robla J.I. (2009) A review of wireless sensor technologies and applications in agriculture and food industry: State of the art and current trends. *Sensors* **9**, 4728–4750.

SENSIRION (2010) Sensor Specification Statement: How to Understand Specification of Relative Humidity Sensors. *Sensirion* (www.sensirion.com).

Visscher G.J.W. (2014) Humidity and moisture measurement. In: Webster J.G., Eren H. (eds.) *The Measurement, Instrumentation and Sensors Handbook*, 2nd edn. CRC Press, Boca Raton, FL.

# Index

Note: Page numbers in italic and bold refer to figures and tables, respectively.